STUDY GUIDE AND SOLUTIONS MANUAL
TO ACCOMPANY

ORGANIC CHEMISTRY

TENTH EDITION

STUDY GUIDE AND SOLUTIONS MANUAL TO ACCOMPANY

ORGANIC CHEMISTRY

TENTH EDITION

T. W. GRAHAM SOLOMONS

University of South Florida

CRAIG B. FRYHLE

Pacific Lutheran University

ROBERT G. JOHNSON

Xavier University

JOHN WILEY & SONS, INC.

This book was set in 10/12 Times Roman by Aptara Delhi and printed and bound by Bind-Rite. The cover was printed by Bind-Rite.

Library of Congress Cataloging-in-Publication Data

Main Text
Solomons, T. W. Graham
Organic Chemistry/T. W. Graham Solomons.—10th ed./Craig B. Fryhle.
p. cm
Includes index.
ISBN 978-0-470-40141-5 (cloth)
Binder-ready version ISBN 978-0-470-55659-7

1. Chemistry, Organic—Textbooks. I. Fryhle, Craig B. II. Title.

QD253.2.S65 2011
547—dc22 2009032800

Study Guide and Solutions Manual
ISBN 978-0-470-47839-4

Printed in the United States of America

10 9 8 7 6 5 4 3 2 1

To the Student

Contrary to what you may have heard, organic chemisty does not have to be a difficult course. It will be a rigorous course, and it will offer a challenge. But you will learn more in it than in almost any course you will take—and what you learn will have a special relevance to life and the world around you. However, because organic chemistry can be approached in a logical and systematic way, you will find that with the right study habits, mastering organic chemistry can be a deeply satisfying experience. Here, then, are some suggestions about how to study:

1. **Keep up with your work from day to day-never let yourself get behind.** Organic chemistry is a course in which one idea almost always builds on another that has gone before. It is essential, therefore, that you keep up with, or better yet, be a little ahead of your instructor. Ideally, you should try to stay one day ahead of your instructor's lectures in your own class preparations. The lecture, then, will be much more helpful because you will already have some understanding of the assigned material. Your time in class will clarify and expand ideas that are already familiar ones.

2. **Study material in small units, and be sure that you understand each new section before you go on to the next.** Again, because of the cumulative nature of organic chemistry, your studying will be much more effective if you take each new idea as it comes and try to understand it completely before you move on to the next concept.

3. **Work all of the in-chapter and assigned problems.** One way to check your progress is to work each of the in-chapter problems when you come to it. These problems have been written just for this purpose and are designed to help you decide whether or not you understand the material that has just been explained. You should also carefully study the Solved Problems. If you understand a Solved Problem and can work the related in-chapter problem, then you should go on; if you cannot, then you should go back and study the preceding material again. Work all of the problems assigned by your instructor from the end of the chapter, as well. Do all of your problems in a notebook and bring this book with you when you go to see your instructor for extra help.

4. **Write when you study.** Write the reactions, mechanisms, structures, and so on, over and over again. Organic chemistry is best assimilated through the fingertips by writing, and not through the eyes by simply looking, or by highlighting material in the text, or by referring to flash cards. There is a good reason for this. Organic structures, mechanisms, and reactions are complex. If you simply examine them, you may think you understand them thoroughly, but that will be a misperception. The reaction mechanism may make sense to you in a certain way, but you need a deeper understanding than this. You need to know the material so thoroughly that you can explain it to someone else. This level of understanding comes to most of us (those of us without photographic memories) through writing. Only by writing the reaction mechanisms do we pay sufficient attention to their details, such as which atoms are connected to which atoms, which bonds break in a reaction and which bonds form, and the three-dimensional aspects of the structures. When we write reactions and mechanisms, connections are made in our brains that provide the long-term memory needed for success in organic chemistry. We virtually guarantee that your grade in the course will be directly proportional to the number of pages of paper that you fill with your own writing in studying during the term.

5. **Learn by teaching and explaining.** Study with your student peers and practice explaining concepts and mechanisms to each other. Use the *Learning Group Problems* and other exercises your instructor may assign as vehicles for teaching and learning interactively with your peers.

6. **Use the answers to the problems in the** ***Study Guide*** **in the proper way.** Refer to the answers only in two circumstances: (1) When you have finished a problem, use the Study Guide to check your answer. (2) When, after making a real effort to solve the problem, you find that you are completely stuck, then look at the answer for a clue and go back to work out the problem on your own. The value of a problem is in solving it. If you simply read the problem and look up the answer, you will deprive yourself of an important way to learn.

7. **Use molecular models when you study.** Because of the three-dimensional nature of most organic molecules, molecular models can be an invaluable aid to your understanding of them. When you need to see the three-dimensional aspect of a particular topic, use the Molecular Visions™ model set that may have been packaged with your textbook, or buy a set of models separately. An appendix to the *Study Guide* that accompanies this text provides a set of highly useful molecular model exercises.

8. **Make use of the rich online teaching resources in** ***WileyPLUS*** (www.wileyplus.com) and do any online exercises that may be assigned by your instructor.

INTRODUCTION

"Solving the Puzzle"

or

"Structure Is Everything (Almost)"

As you begin your study of organic chemistry it may seem like a puzzling subject. In fact, in many ways organic chemistry is like a puzzle—a jigsaw puzzle. But it is a jigsaw puzzle with useful pieces, and a puzzle with fewer pieces than perhaps you first thought. In order to put a jigsaw puzzle together you must consider the shape of the pieces and how one piece fits together with another. In other words, solving a jigsaw puzzle is about **structure.** In organic chemistry, molecules are the pieces of the puzzle. Much of organic chemistry, indeed life itself, depends upon the fit of one molecular puzzle piece with another. For example, when an antibody of our immune system acts upon a foreign substance, it is the puzzle-piece-like fit of the antibody with the invading molecule that allows "capture" of the foreign substance. When we smell the sweet scent of a rose, some of the neural impulses are initiated by the fit of a molecule called geraniol in an olfactory receptor site in our nose. When an adhesive binds two surfaces together, it does so by billions of interactions between the molecules of the two materials. Chemistry is truly a captivating subject.

As you make the transition from your study of general to organic chemistry, it is important that you solidify those concepts that will help you understand the structure of organic molecules. A number of concepts are discussed below using several examples. It is suggested that you consider the examples and the explanations given, and refer to information from your general chemistry studies when you need more elaborate information. There are also occasional references below to sections in your text, Solomons and Fryhle's *Organic Chemistry,* because some of what follows foreshadows what you will learn in the course.

SOME FUNDAMENTAL PRINCIPLES WE NEED TO CONSIDER

What do we need to know to understand the structure of organic molecules? First, we need to know where electrons are located around a given atom. To understand this we need to recall from general chemistry the ideas of **electron configuration** and **valence shell electron orbitals,** especially in the case of atoms such as carbon, hydrogen, oxygen, and nitrogen. We also need to use **Lewis valence shell electron structures.** These concepts are useful because the shape of a molecule is defined by its constituent atoms, and the placement of the atoms follows from the location of the electrons that bond the atoms. Once we have a Lewis structure for a molecule, we can consider **orbital hybridization** and **valence shell electron pair repulsion (VSEPR) theory** in order to generate a three-dimensional image of the molecule.

Secondly, in order to understand why specific organic molecular puzzle pieces fit together we need to consider the attractive and repulsive forces between them. To understand this we need to know how electronic charge is distributed in a molecule. We must use tools such as **formal charge** and **electronegativity.** That is, we need to know which parts of a molecule are relatively positive and which are relatively negative—in other words, their **polarity.** Associations between molecules strongly depend on both shape and the complementarity of their electrostatic charges (polarity).

When it comes to organic chemistry it will be much easier for you to understand why organic molecules have certain properties and react the way they do if you have an appreciation for the structure of the molecules involved. Structure is, in fact, almost everything, in that whenever we want to know why or how something works we look ever more deeply into its structure. This is true whether we are considering a toaster, jet engine, or an organic reaction. If you can visualize the shape of the puzzle pieces in organic chemistry (molecules), you will see more easily how they fit together (react).

SOME EXAMPLES

In order to review some of the concepts that will help us understand the structure of organic molecules, let's consider three very important molecules—water, methane, and methanol (methyl alcohol). These three are small and relatively simple molecules that have certain similarities among them, yet distinct differences that can be understood on the basis of their structures. Water is a liquid with a moderately high boiling point that does not dissolve organic compounds well. Methanol is also a liquid, with a lower boiling point than water, but one that dissolves many organic compounds easily. Methane is a gas, having a boiling point well below room temperature. Water and methanol will dissolve in each other, that is, they are miscible. We shall study the structures of water, methanol, and methane because the principles we learn with these compounds can be extended to much larger molecules.

Water

HOH

Let's consider the structure of water, beginning with the central oxygen atom. Recall that the atomic number (the number of protons) for oxygen is eight. Therefore, an oxygen atom also has eight electrons. (An ion may have more or less electrons than the atomic number for the element, depending on the charge of the ion.) Only the valence (outermost) shell electrons are involved in bonding. Oxygen has six valence electrons—that is, six electrons in the second principal shell. (Recall that the number of valence electrons is apparent from the group number of the element in the periodic table, and the row number for the element is the principal shell number for its valence electrons.) Now, let's consider the electron configuration for oxygen. The sequence of atomic orbitals for the first three shells of any atom is shown below. Oxygen uses only the first two shells in its lowest energy state.

$1s, 2s, 2p_x, 2p_y, 2p_z, 3s, 3p_x, 3p_y, 3p_z$

The p orbitals of any given principal shell (second, third, etc.) are of equal energy. Recall also that each orbital can hold a maximum of two electrons and that each equal energy orbital must accept one electron before a second can reside there (Hund's rule). So, for oxygen we place two electrons in the $1s$ orbital, two in the $2s$ orbital, and one in each of the $2p$ orbitals, for a subtotal of seven electrons. The final eighth electron is paired with another in one of the $2p$ orbitals. The configuration for the eight electrons of oxygen is, therefore

$1s^2\ 2s^2\ 2p_x{}^2\ 2p_y{}^1\ 2p_z{}^1$

where the superscript numbers indicate how many electrons are in each orbital. In terms of relative energy of these orbitals, the following diagram can be drawn. Note that the three $2p$ orbitals are depicted at the same relative energy level.

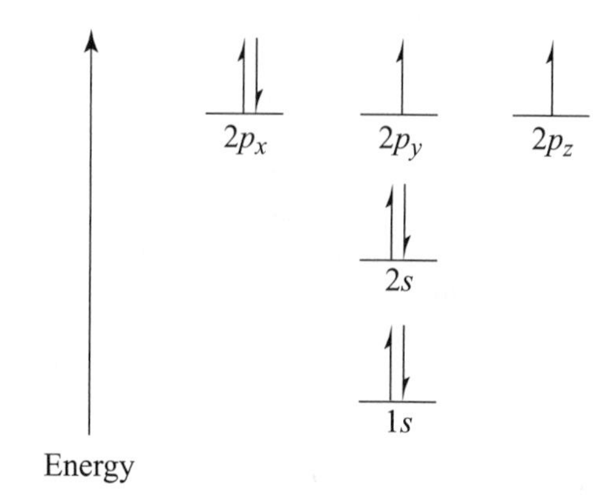

Now, let's consider the shape of these orbitals. The shape of an ***s* orbital** is that of a sphere with the nucleus at the center. The shape of each ***p* orbital** is approximately that of a dumbbell or lobe-shaped object, with the nucleus directly between the two lobes. There is one pair of lobes for each of the three p orbitals (p_x, p_y, p_z) and they are aligned along the x, y, and z coordinate axes, with the nucleus at the origin. Note that this implies that the three p orbitals are at 90° angles to each other.

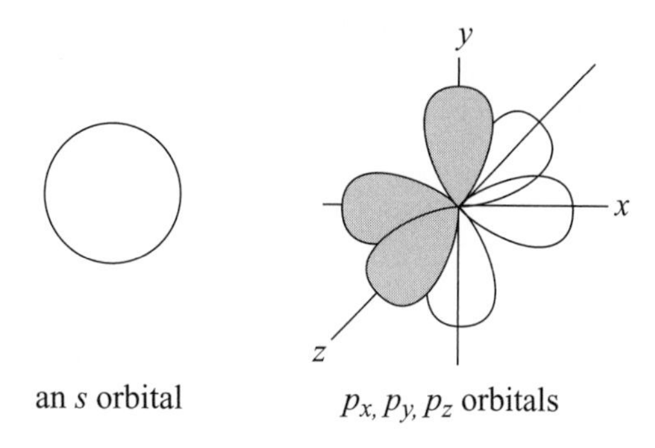

an s orbital $\quad$ p_x, p_y, p_z orbitals

Now, when oxygen is bonded to two hydrogens, bonding is accomplished by the sharing of an electron from each of the hydrogens with an unpaired electron from the oxygen. This type of bond, involving the sharing of electrons between atoms, is called a **covalent bond.** The formation of covalent bonds between the oxygen atom and the two hydrogen atoms is advantageous because each atom achieves a full valence shell by the sharing of these electrons. For the oxygen in a water molecule, this amounts to satisfying the octet rule.

A **Lewis structure** for the water molecule (which shows only the valence shell electrons) is depicted in the following structure. There are two nonbonding pairs of electrons around the oxygen as well as two bonding pairs.

O O
x x
H H H H

In the left-hand structure the six valence electrons contributed by the oxygen are shown as dots, while those from the hydrogens are shown as x's. This is done strictly for bookkeeping purposes. All electrons are, of course, identical. The right-hand structure uses the convention that a bonding pair of electrons can be shown by a single line between the bonded atoms.

This structural model for water is only a *first approximation,* however. While it is a proper Lewis structure for water, it is not an entirely correct three-dimensional structure. It might appear that the angle between the hydrogen atoms (or between any two pairs of electrons in a water molecule) would be 90°, but this is not what the true angles are in a water molecule. The angle between the two hydrogens is in fact about 105°, and the nonbonding electron pairs are in a different plane than the hydrogen atoms. The reason for this arrangement is that groups of bonding and nonbonding electrons tend to repel each other due to the negative charge of the electrons. Thus, the ideal angles between bonding and nonbonding groups of electrons are those angles that allow maximum separation in three-dimensional space. This principle and the theory built around it are called the **valence shell electron pair repulsion (VSEPR) theory.**

VSEPR theory predicts that the ideal separation between four groups of electrons around an atom is 109.5°, the so-called **tetrahedral angle.** At an angle of 109.5° all four electron groups are separated equally from each other, being oriented toward the corners of a regular tetrahedron. The exact tetrahedral angle of 109.5° is found in structures where the four groups of electrons and bonded groups are identical.

In water, there are two different types of electron groups—pairs bonding the hydrogens with the oxygen and nonbonding pairs. Nonbonding electron pairs repel each other with greater force than bonding pairs, so the separation between them is greater. Consequently, the angle between the pairs bonding the hydrogens to the oxygen in a water molecule is compressed slightly from 109.5°, being actually about 105°. As we shall see shortly, the angle between the four groups of bonding electrons in methane (CH_4) *is* the ideal tetrahedral angle of 109.5°. This is because the four groups of electrons and bound atoms are identical in a methane molecule.

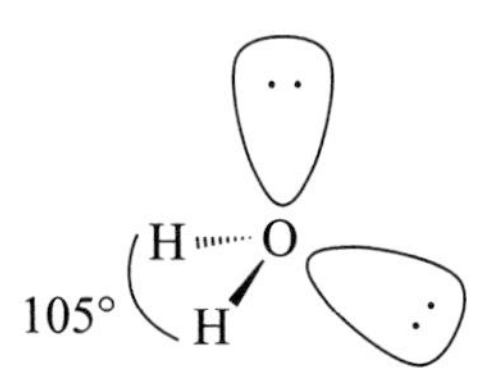

Orbital hybridization is the reason that 109.5° is the ideal tetrahedral angle. As noted earlier, an *s* orbital is spherical, and each *p* orbital is shaped like two symmetrical lobes aligned along the *x, y,* and *z* coordinate axes. Orbital hybridization involves taking a weighted average of the valence electron orbitals of the atom, resulting in the same number of new hybridized orbitals. With four groups of valence electrons, as in the structure of water, one *s* orbital and three *p* orbitals from the second principal shell in oxygen are hybridized (the 2*s* and $2p_x$, $2p_y$, and $2p_z$ orbitals). The result is four new hybrid orbitals of equal energy designated as sp^3 orbitals (instead of the original three *p* orbitals and one *s* orbital). Each of the four sp^3 orbitals has roughly 25% *s* character and 75% *p* character. The geometric result is that the major lobes of the four sp^3 orbitals are oriented toward the corners of a tetrahedron with an angle of 109.5° between them.

sp^3 hybrid orbitals
(109.5° angle between lobes)

In the case of the oxygen in a water molecule, where two of the four sp^3 orbitals are occupied by nonbonding pairs, the angle of separation between them is larger than 109.5° due to additional electrostatic repulsion of the nonbonding pairs. Consequently, the angle between the bonding electrons is slightly smaller, about 105°.

More detail about orbital hybridization than provided above is given in Sections 1.9–1.15 of *Organic Chemistry.* With that greater detail it will be apparent from consideration of orbital hybridization that for three groups of valence electrons the ideal separation is 120° (trigonal planar), and for two groups of valence electrons the ideal separation is 180° (linear). VSEPR theory allows us to come to essentially the same conclusion as by the mathematical hybridization of orbitals, and it will serve us for the moment in predicting the three-dimensional shape of molecules.

Methane

CH_4

Now let's consider the structure of methane (CH_4). In methane there is a central carbon atom bearing four bonded hydrogens. Carbon has six electrons in total, with four of them being valence electrons. (Carbon is in Group IVA in the periodic table.) In methane each valence electron is shared with an electron from a hydrogen atom to form four covalent bonds. This information allows us to draw a Lewis structure for methane (see below). With four groups of valence electrons the VSEPR theory allows us to predict that the three-dimensional shape of a methane molecule should be tetrahedral, with an angle of 109.5° between each of the bonded hydrogens. This is indeed the case. Orbital hybridization arguments can also be used to show that there are four equivalent sp^3 hybrid orbitals around the carbon atom, separated by an angle of 109.5°.

```
      H               H               H
     · x              |               |
  H ˣ. C .ₓH      H—C—H          H''''''C
     x ·              |              ╱   ╲
      H               H             H     H
```

All H-C-H angles are 109.5°

The structure at the far right above uses the dash-wedge notation to indicate three dimensions. A solid wedge indicates that a bond projects out of the paper toward the reader. A dashed bond indicates that it projects behind the paper away from the viewer. Ordinary lines represent bonds in the plane of the paper. The dash-wedge notation is an important and widely used tool for depicting the three-dimensional structure of molecules.

Methanol

CH_3OH

Now let's consider a molecule that incorporates structural aspects of both water and methane. Methanol (CH_3OH), or methyl alcohol, is such a molecule. In methanol, a central carbon atom has three hydrogens and an O–H group bonded to it. Three of the four valence electrons of the carbon atom are shared with a valence electron from the hydrogen atoms, forming three C—H bonds. The fourth valence electron of the carbon is shared with a valence electron from the oxygen atom, forming a C–O bond. The carbon atom now has an octet of valence electrons through the formation of four covalent bonds. The angles between these four covalent bonds is very near the ideal tetrahedral angle of 109.5°, allowing maximum separation between them. (The valence orbitals of the carbon are sp^3 hybridized.) At the oxygen atom, the situation is very similar to that in water. The oxygen uses its two unpaired valence electrons to form covalent bonds. One valence electron is used in the bond with the carbon atom, and the other is paired with an electron from the hydrogen to form the O–H bond. The remaining valence electrons of the oxygen are present as two nonbonding pairs, just as in water. The angles separating the four groups of electrons around the oxygen are thus near the ideal angle of 109.5°, but reduced slightly in the C–O–H angle due to repulsion by the two nonbonding pairs on the oxygen. (The valence orbitals of the oxygen are also sp^3 hybridized since there are four groups of valence electrons.) A Lewis structure for methanol is shown below, along with a three-dimensional perspective drawing.

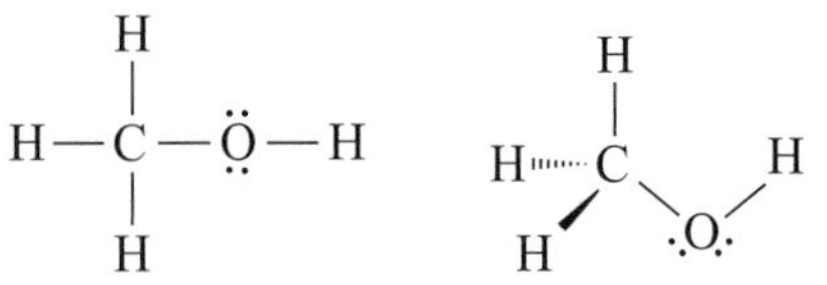

THE "CHARACTER" OF THE PUZZLE PIECES

With a mental image of the three-dimensional structures of water, methane, and methanol, we can ask how the structure of each, as a "puzzle piece," influences the interaction of each molecule with identical and different molecules. In order to answer this question we have to move one step beyond the three-dimensional shape of these molecules. We need to consider not only the location of the electron groups (bonding and nonbonding) but also the distribution of electronic charge in the molecules.

First, we note that nonbonding electrons represent a locus of negative charge, more so than electrons involved in bonding. Thus, water would be expected to have some partial negative charge localized in the region of the nonbonding electron pairs of the oxygen. The same would be true for a methanol molecule. The lower case Greek δ (delta) means "partial."

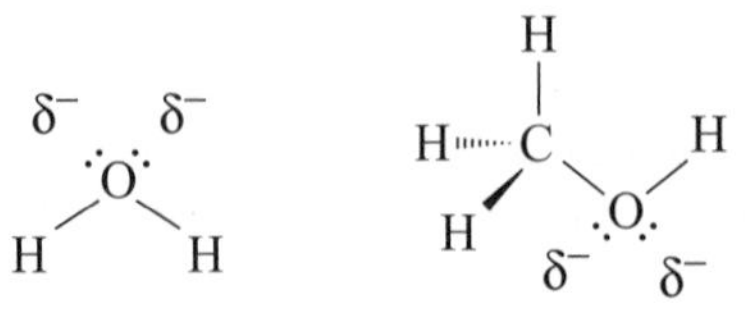

Secondly, the phenomenon of electronegativity influences the distribution of electrons, and hence the charge in a molecule, especially with respect to electrons in covalent bonds. **Electronegativity** is the propensity of an element to draw electrons toward it in a covalent bond. The trend among elements is that of increasing electronegativity toward the upper right corner of the periodic table. (Fluorine is the most electronegative element.) By observing the relative locations of carbon, oxygen, and hydrogen in the periodic table, we can see that oxygen is the most electronegative of these three elements. Carbon is more electronegative than hydrogen, although only slightly. Oxygen is significantly more electronegative than hydrogen. Thus, there is substantial separation of charge in a water molecule, due not only to the nonbonding electron pairs on the oxygen but also to the greater electronegativity of the oxygen with respect to the hydrogens. The oxygen tends to draw electron density toward itself in the bonds with the hydrogens, leaving the hydrogens partially positive. The resulting separation of charge is called **polarity.** The oxygen–hydrogen bonds are called **polar covalent bonds** due to this separation of charge. If one considers the net effect of the two nonbonding electron pairs in a water molecule as being a region of negative charge, and the hydrogens as being a region of relative positive charge, it is clear that a water molecule has substantial separation of charge, or polarity.

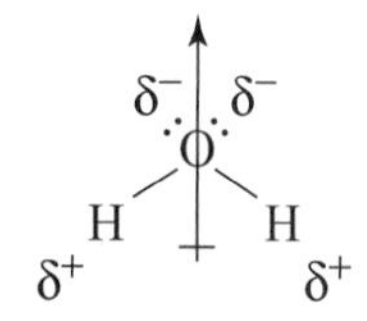

An analysis of polarity for a methanol molecule would proceed similarly to that for water. Methanol, however, is less polar than water because only one O–H bond is present. Nevertheless, the region of the molecule around the two nonbonding electron pairs of the oxygen is relatively negative, and the region near the hydrogen is relatively positive. The electronegativity difference between the oxygen and the carbon is not as large as that between oxygen and hydrogen, however, so there is less polarity associated with the C–O bond. Since there is even less difference in electronegativity between hydrogen and carbon in the three C–H bonds, these bonds contribute essentially no polarity to the molecule. The net effect for methanol is to make it a polar molecule, but less so than water due to the nonpolar character of the CH_3 region of the molecule.

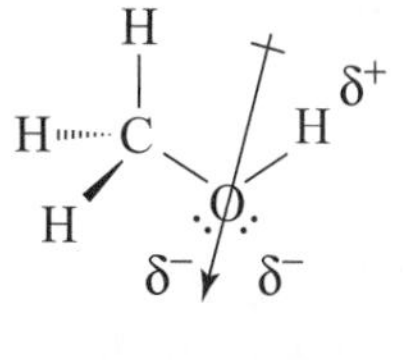

Now let's consider methane. Methane is a nonpolar molecule. This is evident first because there are no nonbonding electron pairs, and secondly because there is relatively little electronegativity difference between the hydrogens and the central carbon. Furthermore, what little electronegativity difference there is between the hydrogens and the central carbon atom is negated by the symmetrical distribution of the C–H bonds in the tetrahedral shape of methane. The slight polarity of each C–H bond is canceled by the symmetrical orientation of the four C–H bonds. If considered as vectors, the vector sum of the four slightly polar covalent bonds oriented at 109.5° to each other would be zero.

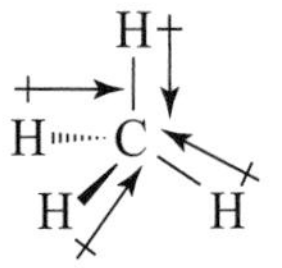

Net dipole is zero.

The same analysis would hold true for a molecule with identical bonded groups, but groups having electronegativity significantly different from carbon, so long as there were symmetrical distribution of the bonded groups. Tetrachloromethane (carbon tetrachloride) is such a molecule. It has no net polarity.

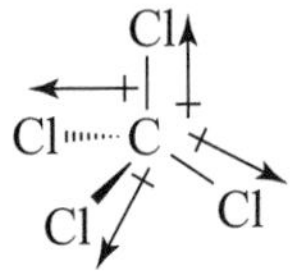

Net dipole is zero.

INTERACTIONS OF THE PUZZLE PIECES

Now that you have an appreciation for the polarity and shape of these molecules it is possible to see how molecules might interact with each other. The presence of polarity in a molecule bestows upon it attractive or repulsive forces in relation to other molecules. The negative part of one molecule is attracted to the positive region of another. Conversely, if there is little polarity in a molecule, the attractive forces it can exert are very small [though not completely nonexistent, due to van der Waals forces (Section *2.13B* in *Organic Chemistry*)]. Such effects are called **intermolecular forces** (forces between molecules), and strongly depend on the polarity of a molecule or certain bonds within it (especially O—H, N—H, and other bonds between hydrogen and more electronegative atoms with nonbonding pairs). Intermolecular forces have profound effects on physical properties such as **boiling point, solubility,** and **reactivity.** An important manifestation of these properties is that the ability to isolate a pure compound after a reaction often depends on differences in boiling point, solubility, and sometimes reactivity among the compounds present.

Boiling Point

An intuitive understanding of boiling points will serve you well when working in the laboratory. The polarity of water molecules leads to relatively strong intermolecular attraction between water molecules. One result is the moderately high boiling point of water (100°C, as compared to 65°C for methanol and –162°C for methane, which we will discuss shortly). Water has the highest boiling point of these three example molecules because it will strongly associate with itself by attraction of the partially positive hydrogens of one molecule (from the electronegativity difference between the O and H) to the negatively charged region in another water molecule (where the nonbonding pairs are located).

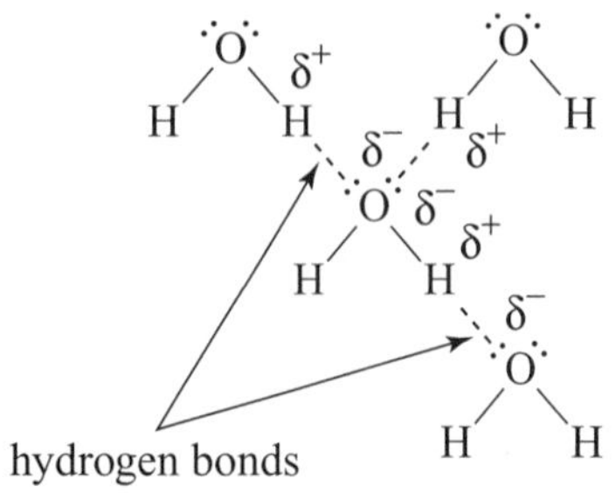

hydrogen bonds

The specific attraction between a partially positive hydrogen atom attached to a heteroatom (an atom with both nonbonding and bonding valence electrons, e.g., oxygen or nitrogen) and the nonbonding electrons of another heteroatom is called **hydrogen bonding.** It is a form of **dipole-dipole attraction** due to the polar nature of the hydrogen–heteroatom bond. A given water molecule can associate by hydrogen bonding with several other water molecules, as shown above. Each water molecule has two hydrogens that can associate with the non-bonding pairs of other water molecules, and two nonbonding pairs that can associate with the hydrogens of other water molecules. Thus, several hydrogen bonds are possible for each water molecule. It takes a significant amount of energy (provided by heat, for example) to give the molecules enough kinetic energy (motion) for them to overcome the polarity-induced attractive forces between them and escape into the vapor phase (evaporation or boiling).

Methanol, on the other hand, has a lower boiling point (65 °C) than water, in large part due to the decreased hydrogen bonding ability of methanol in comparison with water. Each methanol molecule has only one hydrogen atom that can participate in a hydrogen bond with the nonbonding electron pairs of another methanol molecule (as compared with two for each water molecule). The result is reduced intermolecular attraction between methanol molecules and a lower boiling point since less energy is required to overcome the lesser intermolecular attractive forces.

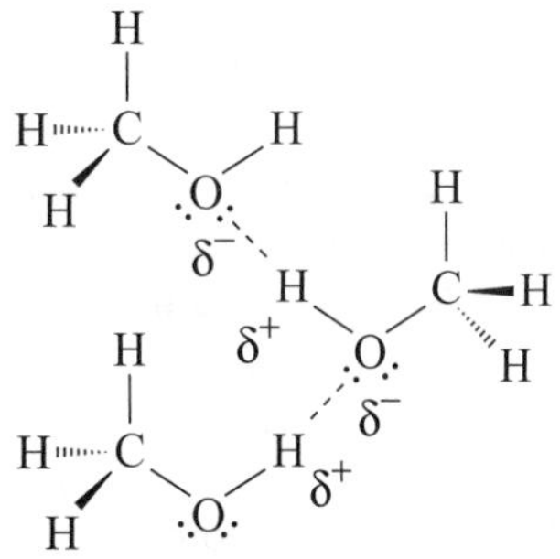

The CH_3 group of methanol does not participate in dipole–dipole attractions between molecules because there is not sufficient polarity in any of its bonds to lead to significant partial positive or negative charges. This is due to the small electronegativity difference between the carbon and hydrogen in each of the C–H bonds.

Now, on to methane. Methane has no hydrogens that are eligible for hydrogen bonding, since none is attached to a heteroatom such as oxygen. Due to the small difference in electronegativity between carbon and hydrogen there are no bonds with any significant polarity. Furthermore, what slight polarity there is in each C–H bond is canceled due to the tetrahedral symmetry of the molecule. [The minute attraction that is present between methane molecules is due to van der Waals forces, but these are negligible in comparison to dipole–dipole interactions that exist when significant differences in electronegativity are present in molecules such as water and methanol.] Thus, because there is only a very weak attractive force between methane molecules, the boiling point of methane is very low (−162 °C) and it is a gas at ambient temperature and pressure.

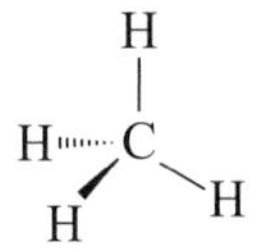

Solubility

An appreciation for trends in solubility is very useful in gaining a general understanding of many practical aspects of chemistry. The ability of molecules to dissolve other molecules or solutes is strongly affected by polarity. The polarity of water is frequently exploited during the isolation of an organic reaction product because water will not dissolve most organic compounds but will dissolve salts, many inorganic materials, and other polar byproducts that may be present in a reaction mixture.

As to our example molecules, water and methanol are miscible with each other because each is polar and can interact with the other by dipole–dipole hydrogen bonding interactions. Since methane is a gas under ordinary conditions, for the purposes of this discussion let's consider a close relative of methane–hexane. Hexane (C_6H_{14}) is a liquid having only carbon—carbon and carbon—hydrogen bonds. It belongs to the same chemical family as methane. Hexane is *not* soluble in water due to the essential absence of polarity in its bonds. Hexane is slightly soluble in methanol due to the compatibility of the nonpolar CH_3 region of methanol with hexane. The old saying "like dissolves like" definitely holds true. This can be extended to solutes, as well. Very polar substances, such as ionic compounds, are usually freely soluble in water. The high polarity of salts generally prevents most of them from being soluble in methanol, however. And, of course, there is absolutely no solubility of ionic substances in hexane. On the other hand, very nonpolar substances, such as oils, would be soluble in hexane.

Thus, the structure of each of these molecules we've used for examples (water, methanol, and methane) has a profound effect on their respective physical properties. The presence of nonbonding electron pairs and polar covalent bonds in water and methanol versus the complete absence of these features in the structure of methane imparts markedly different physical properties to these three compounds. Water, a small molecule with strong intermolecular forces, is a moderately high boiling liquid. Methane, a small molecule with only very weak intermolecular forces, is a gas. Methanol, a molecule combining structural aspects of both water and methane, is a relatively low boiling liquid, having sufficient intermolecular forces to keep the molecules associated as a liquid, but not so strong that mild heat can't disrupt their association.

Reactivity

While the practical importance of the physical properties of organic compounds may only be starting to become apparent, one strong influence of polarity is on the **reactivity** of molecules. It is often possible to understand the basis for a given reaction in organic chemistry by considering the relative polarity of molecules and the propensity, or lack thereof, for them to interact with each other.

Let us consider one example of reactivity that can be understood at the initial level by considering structure and polarity. When chloromethane (CH_3Cl) is exposed to hydroxide ions (OH^-) in water a reaction occurs that produces methanol. This reaction is shown below.

$$CH_3Cl \quad + \quad HO^- \text{(as NaOH dissolved in water)} \quad \rightarrow \quad HOCH_3 \quad + \quad Cl^-$$

This reaction is called a substitution reaction, and it is of a general type that you will spend considerable time studying in organic chemistry. The reason this reaction occurs readily can be understood by considering the principles of structure and polarity that we have been discussing. The hydroxide ion has a negative charge associated with it, and thus should be attracted to a species that has positive charge. Now recall our discussion of electronegativity and polar covalent bonds, and apply these ideas to the structure of chloromethane. The chlorine atom is significantly more electronegative than carbon (note its position in the periodic table). Thus, the covalent bond between the carbon and the chlorine is polarized such that there is partial negative charge on the chlorine and partial positive charge on the carbon. This provides the positive site that attracts the hydroxide anion!

$$H\ddot{O}:^- \quad + \quad \delta^+\ \mathrm{CH_3}-\ddot{Cl}:\ \delta^- \longrightarrow HO-\mathrm{CH_3} \quad + \quad :\ddot{Cl}:^-$$

The intimate details of this reaction will be studied in Chapter 6 of your text. Suffice it to say for the moment that the hydroxide ion attacks the carbon atom using one of its nonbonding electron pairs to form a bond with the carbon. At the same time, the chlorine atom is pushed away from the carbon and takes with it the pair of electrons that used to bond it to the carbon. The result is substitution of OH for Cl at the carbon atom and the synthesis of methanol. By calculating **formal charges** (Section 1.7 in the text) one can show that the oxygen of the hydroxide anion goes from having a formal negative charge in hydroxide to zero formal charge in the methanol molecule. Similarly, the chlorine atom goes from having zero formal charge in chloromethane to a formal negative charge as a chloride ion after the reaction. *The fact that the reaction takes place at all rests largely upon the complementary polarity of the interacting species. This is a pervasive theme in organic chemistry.*

Acid-base reactions are also very important in organic chemistry. Many organic reactions involve at least one step in the overall process that is fundamentally an acid-base reaction. Both Brønsted-Lowry acid-base reactions (those involving proton donors and acceptors) and Lewis acid-base reactions (those involving electron pair acceptors and donors, respectively) are important. In fact, the reaction above can be classified as a Lewis acid-base reaction in that the hydroxide ion acts as a Lewis base to attack the partially positive carbon as a Lewis acid. It is strongly recommended that you review concepts you have learned previously regarding acid-base reactions. Chapter 3 in *Organic Chemistry* will help in this regard, but it is advisable that you begin some early review about acids and bases based on your previous studies. Acid–base chemistry is widely applicable to understanding organic reactions.

JOINING THE PIECES

Finally, while what we have said above has largely been in reference to three specific compounds, water, methanol, and methane, the principles involved find exceptionally broad application in understanding the structure, and hence reactivity, of organic molecules in general. You will find it constantly useful in your study of organic chemistry to consider the electronic structure of the molecules with which you are presented, the shape caused by the distribution of electrons in a molecule, the ensuing polarity, and the resulting potential for that molecule's reactivity. What we have said about these very small molecules of water, methanol, and methane can be extended to consideration of molecules with 10 to 100 times as many atoms. You would simply apply these principles to small sections of the larger molecule one part at a time. The following structure of Streptogramin A provides an example.

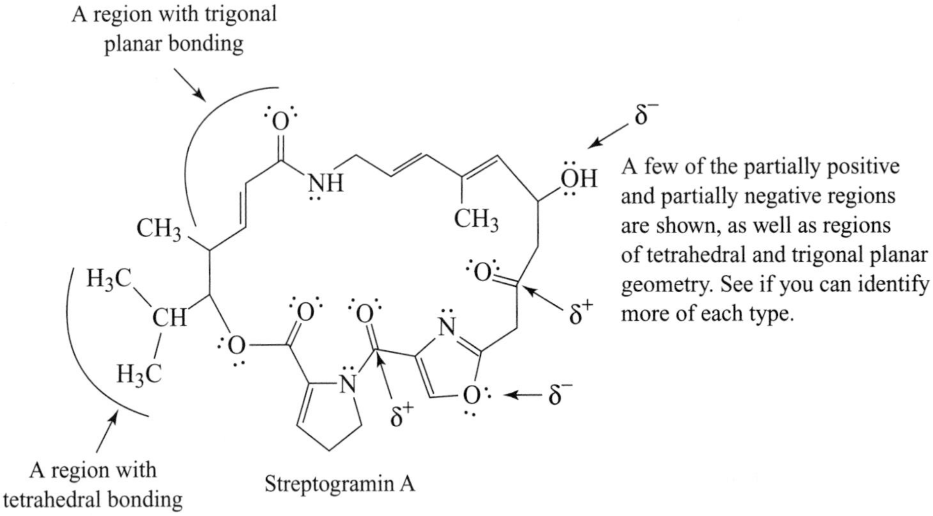

A region with trigonal planar bonding

A few of the partially positive and partially negative regions are shown, as well as regions of tetrahedral and trigonal planar geometry. See if you can identify more of each type.

A region with tetrahedral bonding

Streptogramin A

A natural antibacterial compound that blocks protein synthesis at the 70S ribosomes of Gram-positive bacteria.

We have not said much about how overall shape influences the ability of one molecule to interact with another, in the sense that a key fits in a lock or a hand fits in a glove. This type of consideration is also extremely important, and will follow with relative ease if you have worked hard to understand the general principles of structure outlined here and expanded upon in the early chapters of *Organic Chemistry.* An example would be the following. Streptogramin A, shown above, interacts in a hand-in-glove fashion with the 70S ribosome in bacteria to inhibit binding of transfer RNA at the ribosome. The result of this interaction is the prevention of protein synthesis in the bacterium, which thus accounts for the antibacterial effect of Streptogramin A. Other examples of hand-in-glove interactions include the olfactory response to geraniol mentioned earlier, and the action of enzymes to speed up the rate of reactions in biochemical systems.

FINISHING THE PUZZLE

In conclusion, if you pay attention to learning aspects of structure during this initial period of "getting your feet wet" in organic chemistry, much of the three-dimensional aspects of molecules will become second nature to you. You will immediately recognize when a molecule is tetrahedral, trigonal planar, or linear in one region or another. You will see the potential for interaction between a given section of a molecule and that of another molecule based on their shape and polarity, and you will understand why many reactions take place. Ultimately, you will find that there is much less to memorize in organic chemistry than you first thought. You will learn how to put the pieces of the organic puzzle together, and see that structure is indeed almost everything, just applied in different situations!

ACKNOWLEDGMENTS

We are grateful to those people who have made many helpful suggestions for various editions of this study guide. These individuals include: George R. Jurch, George R. Wenzinger, and J. E. Fernandez at the University of South Florida; Darell Berlin, Oklahoma State University; John Mangravite, West Chester State College; J. G. Traynham, Louisiana State University; and Desmond M. S. Wheeler, University of Nebraska.

We are especially grateful to Chris Callam (The Ohio State University) for providing many fine new problems and their solutions for the 10th edition, and for providing suggestions about existing problems. We also appreciate the assistance and problem reviews provided by Sean Hickey (University of New Orleans), Neal Tonks (College of Charleston), and Justin Wyatt (College of Charleston).

R. G. Johnson also thanks his wife Ginny for her patience and understanding.

T. W. Graham Solomons

C. B. Fryhle

R. G. Johnson

CONTENTS

1 THE BASICS: BONDING AND MOLECULAR STRUCTURE

SOLUTIONS TO PROBLEMS

Another Approach to Writing Lewis Structures

When we write Lewis structures using this method, we assemble the molecule or ion from the constituent atoms showing only the valence electrons (i.e., the electrons of the outermost shell). By having the atoms share electrons, we try to give each atom the electronic structure of a noble gas. For example, we give hydrogen atoms two electrons because this gives them the structure of helium. We give carbon, nitrogen, oxygen, and fluorine atoms eight electrons because this gives them the electronic structure of neon. The number of valence electrons of an atom can be obtained from the periodic table because it is equal to the group number of the atom. Carbon, for example, is in Group IVA and has four valence electrons; fluorine, in Group VIIA, has seven; hydrogen, in Group 1A, has one. As an illustration, let us write the Lewis structure for CH_3F. In the example below, we will at first show a hydrogen's electron as x, carbon's electrons as o's, and fluorine's electrons as dots.

Example A

3 H˟, ∘C∘ (with four o electrons), and ·F: (with seven dot electrons) are assembled as

```
   H            H
 H:C:F:  or  H:C:F:
   H            H
```

If the structure is an ion, we add or subtract electrons to give it the proper charge. As an example, consider the chlorate ion, ClO_3^-.

Example B

:Cl·, and :O: and an extra electron × are assembled as

```
[    :O:    ]-        [    :O:    ]-
[ :O:Cl×O:  ]    or   [ :O:Cl:O:  ]
```

1.1 ^{14}N, 7 protons and 7 neutrons; ^{15}N, 7 protons and 8 neutrons

1.2 (a) one (b) seven (c) four (d) three (e) eight (f) five

1.3 (a) covalent (b) ionic (c) covalent (d) covalent

1.4

:O:
||
$^{-}$:O—P—O:$^{-}$
|
:O:$^{-}$

1.5

(a) H—F:

(b) :F—F:

(c)
H
|
H—C—F:
|
H

(d) H—O—N=O:

(e)
:O:
||
H—O—S—O—H

(f)
[H
|
H—B—H
|
H]$^{-}$

(g)
:O:
||
H—O—P—O—H
|
:O—H

(h)
:O:
||
C
H—O / \ O—H

1.6

(a)
H
|
H—C—O:$^{-}$
|
H

(b)
H—N:$^{-}$
|
H

(c) $^{-}$:C≡N:

(d)
:O:
//
H—C
\
:O:$^{-}$

(e)
:O:
||
C
H—O / \ :O:$^{-}$

(f) H—C≡C:$^{-}$

1.7 (a) (b) (c) (d) (e) (f) (g) (h)

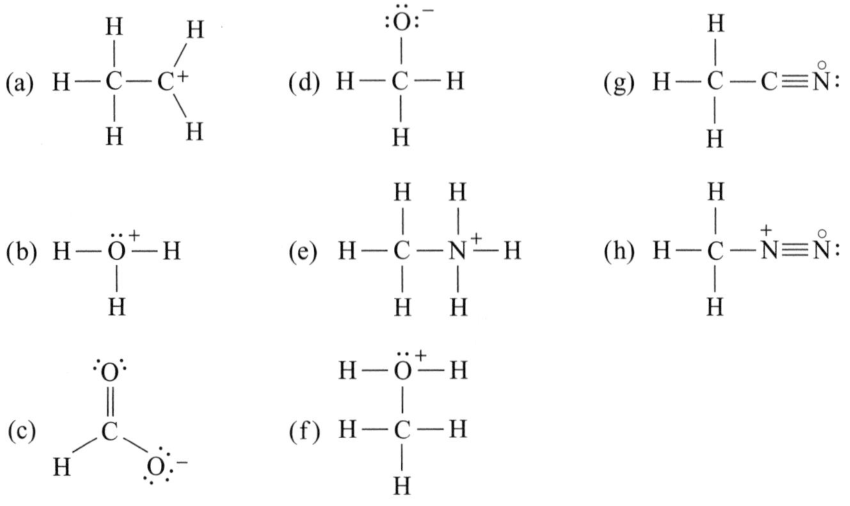

1.8 (a)

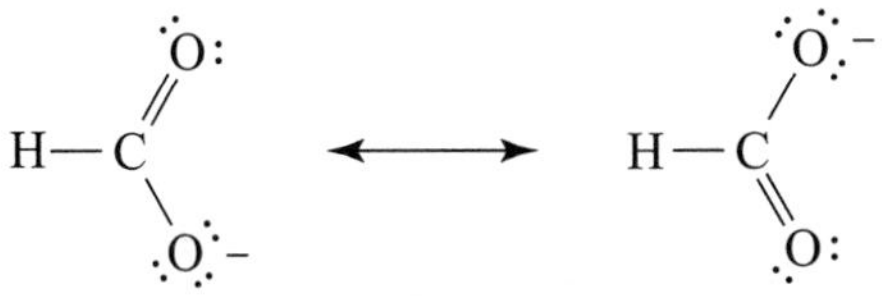

(b) and (c). Since the two resonance structures are equivalent, each should make an equal contribution to the overall hybrid. The C—O bonds should therefore be of equal length (they should be of bond order 1.5), and each oxygen atom should bear a 0.5 negative charge.

1.9 (a) (b) (c) (d)

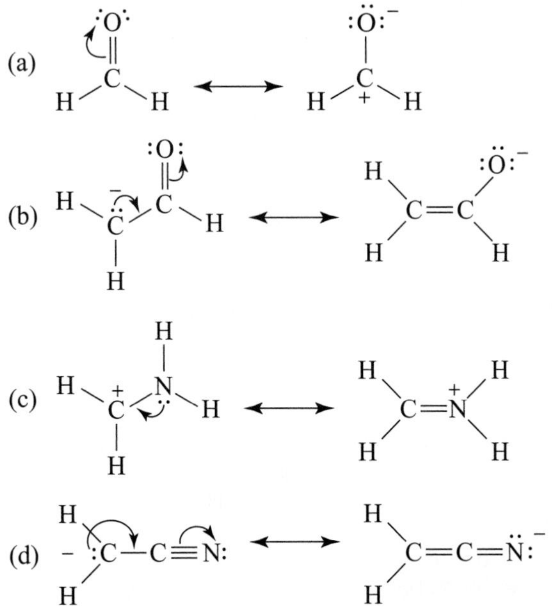

1.10 (a) $CH_3CH{=}CH{-}CH{=}\overset{+}{O}H \longleftrightarrow CH_3CH{=}CH{-}\overset{+}{C}H{-}\ddot{O}H \longleftrightarrow$

$CH_3\overset{+}{C}H{-}CH{=}CH{-}\ddot{O}H$

$\overset{\delta+}{CH_3CH}\text{---}CH\text{---}\overset{\delta+}{CH}\text{---}\overset{\delta+}{OH}$

(b) $CH_2{=}CH{-}\underset{+}{C}H{-}CH{=}CH_2 \longleftrightarrow {}^{+}CH_2{-}CH{=}CH{-}CH{=}CH_2 \longleftrightarrow$

$CH_2{=}CH{-}CH{=}CH{-}\overset{+}{C}H_2$

$\overset{\delta+}{CH_2}\text{---}CH\text{---}\overset{\delta+}{CH}\text{---}CH\text{---}\overset{\delta+}{CH_2}$

(c) 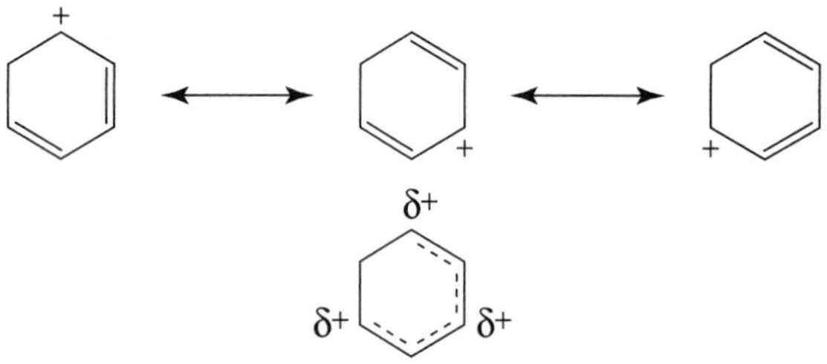 $\longleftrightarrow$ $\longleftrightarrow$

δ+ δ+ δ+

(d) $CH_2{=}CH{-}\ddot{Br}: \longleftrightarrow {}^{-}:CH_2{-}CH{=}\ddot{Br}:^{+}$

$\overset{\delta-}{CH_2}\text{---}CH\text{---}\overset{\delta+}{Br}$

(e) 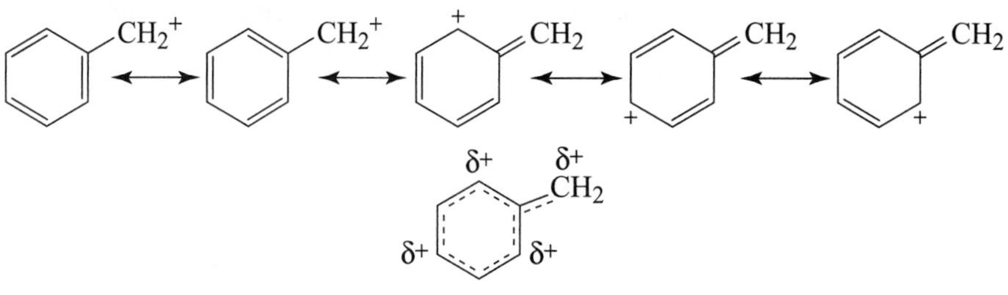 CH_2^{+} $\longleftrightarrow$ CH_2^{+} $\longleftrightarrow$ CH_2 $\longleftrightarrow$ CH_2 $\longleftrightarrow$ CH_2

δ+ δ+ CH_2 δ+ δ+

(f) 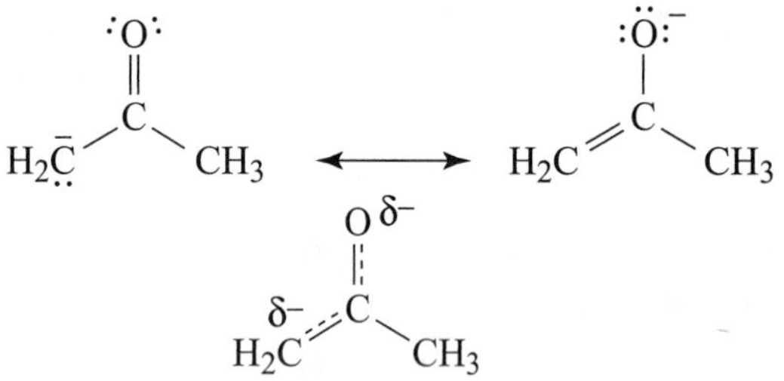 $H_2\bar{C}{-}C({=}O){-}CH_3 \longleftrightarrow H_2C{=}C(O^{-}){-}CH_3$

$H_2\overset{\delta-}{C}\text{---}C(\text{---}\overset{\delta-}{O}){-}CH_3$

(g) $CH_3{-}\ddot{S}{-}CH_2^{+} \longleftrightarrow CH_3{-}\overset{+}{S}{=}CH_2$

$CH_3{-}\overset{\delta+}{S}\text{---}\overset{\delta+}{CH_2}$

(h) $CH_3-\overset{+}{N}(=\ddot{O}:)(-\ddot{O}:^-)$ ⟷ $CH_3-\overset{+}{N}(-\ddot{O}:^-)(=\ddot{O}:)$ ⟷ $CH_3-\overset{2+}{N}(-\ddot{O}:^-)(-\ddot{O}:^-)$ (minor)

$CH_3-\overset{+}{N}(\cdots O^{\delta-})(\cdots O^{\delta-})$

1.11 (a) $CH_2=\overset{+}{N}(CH_3)_2$ because all atoms have a complete octet (rule 3), and there are more covalent bonds (rule 1).

(b) $CH_3-C(=\ddot{O})-\ddot{O}-H$ because it has no charge separation (rule 2).

(c) $:NH_2-C\equiv N:$ because it has no charge separation (rule 2).

1.12 (a) In its ground state, the valence electrons of carbon might be disposed as shown in the following figure.

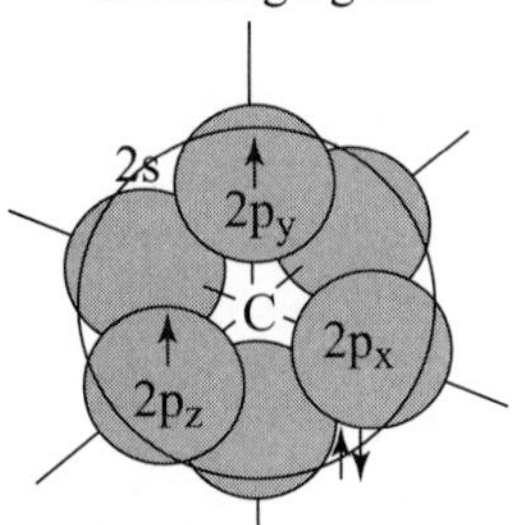

The electronic configuration of a ground state carbon atom: The p orbitals are designated $2p_x$, $2p_y$, and $2p_z$ to indicate their respective orientations along the x, y, and z axes. The assignment of the unpaired electrons to the $2p_y$ and $2p_x$ orbitals is arbitrary. They could also have been placed in the $2p_x$ and $2p_z$ or $2p_y$ and $2p_z$ orbitals. (To have placed them both in the same orbital would not have been correct, however, for this would have violated Hund's rule.) (Section 1.10A)

The formation of the covalent bonds of methane *from individual atoms* requires that the carbon atom overlap its orbitals containing *single electrons* with 1s orbitals of hydrogen atoms (which also contain a single electron). If a ground state carbon atom were to combine with hydrogen atoms in this way, the result would be that depicted below. *Only two carbon-hydrogen bonds would be formed, and these would be at right angles to each other.*

The hypothetical formation of CH_2 from a carbon atom in its ground state:

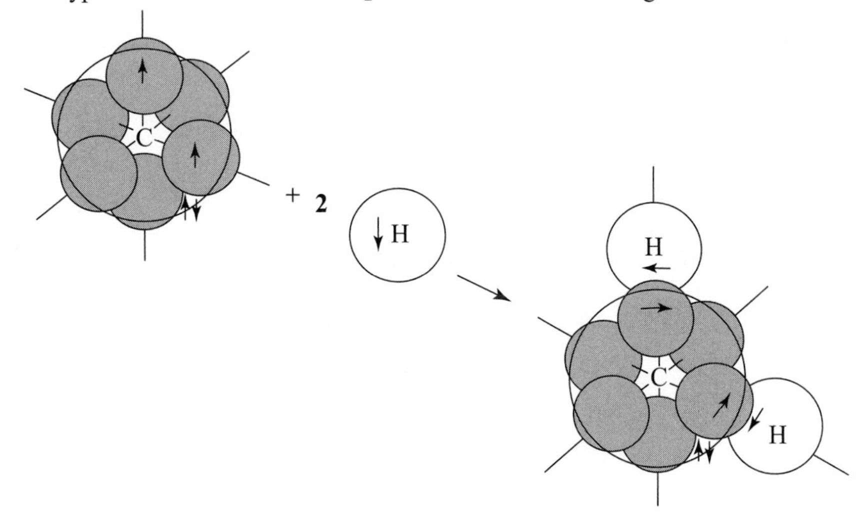

(b) An excited-state carbon atom might combine with four hydrogen atoms as shown in the figure above.

The promotion of an electron from the $2s$ orbital to the $2p_z$ orbital requires energy. The amount of energy required has been determined and is equal to 400 kJ mol^{-1}. This expenditure of energy can be rationalized by arguing that the energy released when two additional covalent bonds form would more than compensate for that required to excite the electron. No doubt this is true, but it solves only one problem. The problems that cannot be solved by using an excited-state carbon as a basis for a model of methane are the problems of the carbon-hydrogen bond angles and the apparent equivalence of all four carbon-hydrogen bonds. Three of the hydrogens—those overlapping their $1s$ orbitals with the three p orbitals—would, in this model, be at angles of 90° with respect to each other; the fourth hydrogen, the one overlapping its $1s$ orbital with the 2s orbital of carbon, would be at some other angle, probably as far from the other bonds as the confines of the molecule would allow. Basing our model of methane on this excited state of carbon gives us a carbon that is tetravalent *but one that is not tetrahedral*, and it predicts a structure for methane in which one carbon-hydrogen bond differs from the other three.

The hypothetical formation of CH_4 from an excited-state carbon atom:

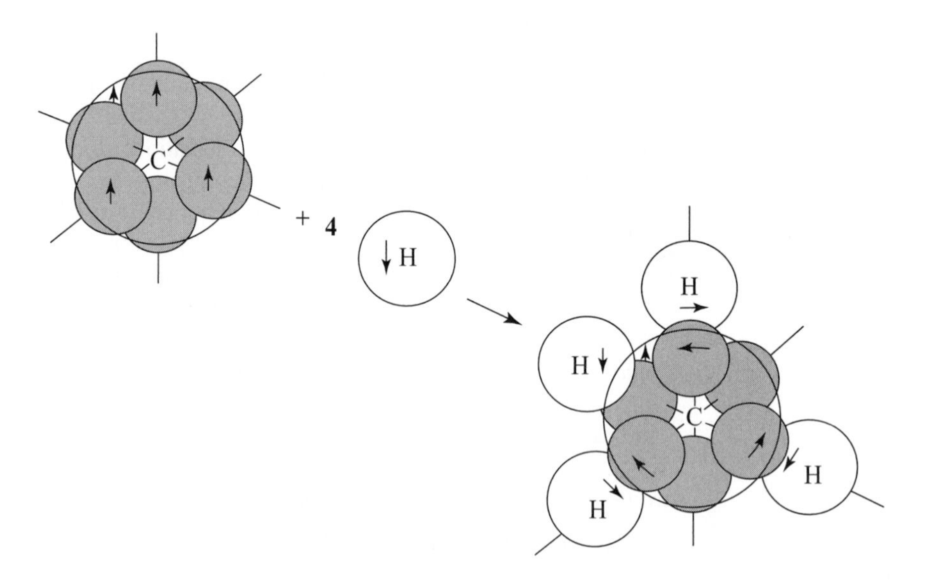

1.13 (a) Cis-trans isomers are not possible.

(b) $CH_3(H)C{=}C(H)CH_3$ (cis: CH_3 and CH_3 on the same side) and $CH_3(H)C{=}C(H)CH_3$ (trans: CH_3 and CH_3 on opposite sides)

CH3 CH3 / C=C / H H and CH3 H / C=C / H CH3

(c) Cis-trans isomers are not possible.

(d) CH3CH2 Cl / C=C / H H and CH3CH2 H / C=C / H Cl

1.14 *sp*^{3}

1.15 *sp*^{3}

1.16 *sp*^{2}

1.17 *sp*

1.18 (a)

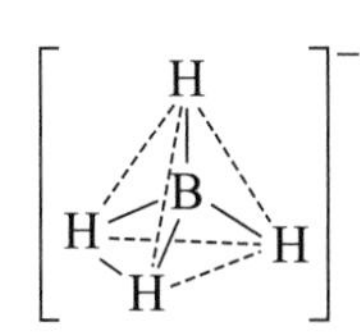

There are four bonding pairs.
The geometry is tetrahedral.

(b) :F̤̈—Be—F̤̈:

There are two bonding pairs about the central atom. The geometry is linear.

(c) $[NH_4]^+$

There are four bonding pairs.
The geometry is tetrahedral.

(d) H_2S (two nonbonding pairs on S)

There are two bonding pairs and two nonbonding pairs. The geometry is angular.

(e) BH_3

There are three bonding pairs. The geometry is trigonal planar.

(f) CF_4

There are four bonding pairs around the central atom. The geometry is tetrahedral.

(g)

There are four bonding pairs around the central atom.
The geometry is tetrahedral.

(h)

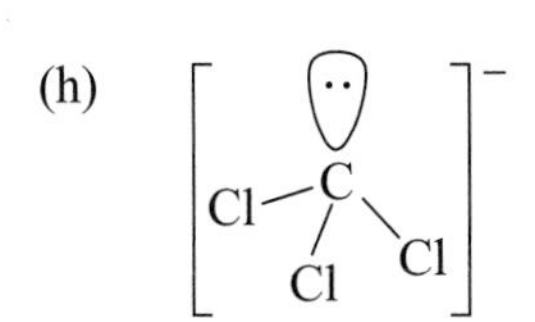

There are three bonding pairs and one nonbonding pair around the central atom. The geometry is trigonal pyramidal.

1.19 (a) $F_2C{=}CF_2$ (120°, 120°) trigonal planar at each carbon atom

(b) $CH_3{-}C{\equiv}C{-}CH_3$ (180°) linear (c) $H{-}C{\equiv}N{:}$ (180°) linear

1.20 $H{-}CH_2{-}O{-}CH_2{-}CH_2{-}H$ (H–C(H)(H)–O–C(H)(H)–C(H)(H)–H)

1.21 $CH_3CH(CH_3)CH(CH_3)CHCH_3$ with CH_3 or $(CH_3)_2CHCH(CH_3)CH(CH_3)_2$

1.22 (a) $CH_3CH(CH_3)CH_2CH_3$ =

(b) $CH_3CH(CH_3)CH_2CH_2OH$ = OH

(c) $(CH_3)_2C{=}CHCH_2CH_3$ =

(d) $CH_3CH_2CH_2CH_2CH_3$ =

(e) $CH_3CH_2CH(OH)CH_2CH_3$ = OH

(f) $CH_2{=}C(CH_2{-}CH_3)_2$ =

(g) $CH_3C(=O)CH_2CH_2CH_2CH_3$ = O

(h) $CH_3CH(Cl)CH_2CH(CH_3)CH_3$ = Cl

1.23 (a) and (d) are constitutional isomers with the molecular formula C_5H_{12}.

(b) and (e) are constitutional isomers with the molecular formula $C_5H_{12}O$.

(c) and (f) are constitutional isomers with the molecular formula C_6H_{12}.

1.24 (a)

```
     H  Cl  H  H  O  H
     |  |   |  |  ‖  |
 H — C — C — C — C — C — C — H
     |  |   |  |     |
     H  H   H  H     H
```

(c)

```
                  H         H
              H   |    H     \     H
               \  C H   \     C — H
    H          //   \    C = C
     \        C      C       \   H
      C            |          H
      |          C — H
  H — C ——————— C
      |          \
      H           H
```

(b)

```
        H    O — H
         \  /
   H      C      H
    \   /   \  /
 H — C       C — H
 H   |       |   H
   \ C       C /
  H /  \   /  \ H
     H — C      H
           \
            C — H
     H    /  \    H
      \  C    C /
     H /  \  / \ H
          H H
```

1.25 (a)

```
      Cl
      |
      C ····· H
    /  \
   H    H
```

(Note that the **Cl** atom and the three **H** atoms may be written at any of the four positions.)

(b)

```
      Cl                    Cl
      |                     |
      C ····· H     or      C ····· H     and so on
    /  \                  /  \
  Cl    H                H    Cl
```

(c)

```
      Cl
      |
      C ····· H     and others
    /  \
  Br    H
```

(d)

```
        H
        |
  H ····· C         Cl
       /   \      /
      H      C           and others
           /  \
          H    H
```

Problems

Electron Configuration

1.26 (a) Na^+ has the electronic configuration, $1s^22s^22p^6$, of Ne.

(b) Cl^- has the electronic configuration, $1s^22s^22p^63s^23p^6$, of Ar.

(c) F^+ and (h) Br^+ do not have the electronic configuration of a noble gas.

(d) H^- has the electronic configuration, $1s^2$, of He.

(e) Ca^{2+} has the electronic configuration, $1s^22s^22p^63s^23p^6$, of Ar.

(f) S^{2-} has the electronic configuration, $1s^22s^22p^63s^23p^6$, of Ar.

(g) O^{2-} has the electronic configuration, $1s^22s^22p^6$, of Ne.

Lewis Structures

1.27 (a) $:\ddot{Cl}-\ddot{S}(=\ddot{O})-\ddot{Cl}:$ (b) $:\ddot{Cl}-P(=\ddot{O})(-\ddot{Cl}:)-\ddot{Cl}:$ (c) PCl_5 (P bonded to five $:\ddot{Cl}:$) (d) $H-\ddot{O}-\overset{+}{N}(=\ddot{O}:)-\ddot{O}:^-$

1.28 (a) $CH_3-\ddot{O}-S(=\ddot{O})_2-\ddot{O}:^-$ (c) $^-:\ddot{O}-S(=\ddot{O})_2-\ddot{O}:^-$

(b) $CH_3-\overset{+}{\ddot{S}}(-\ddot{O}:^-)-CH_3$ (d) $CH_3-S(=\ddot{O})_2-\ddot{O}:^-$

Structural Formulas and Isomerism

1.29 (a) $(CH_3)_2CHCH_2OH$ (c) H_2C-CH_2 / $HC=CH$ (four-membered ring)

(b) $(CH_3)_2CHC(=O)CH(CH_3)_2$ (d) $(CH_3)_2CHCH_2CH_2OH$

1.30 (a) $C_4H_{10}O$ (c) C_4H_6

(b) $C_7H_{14}O$ (d) $C_5H_{12}O$

1.31 (a) Different compounds, not isomeric (d) Same compound

(b) Constitutional isomers (e) Same compound

(c) Same compound (f) Constitutional isomers

(g) Different compounds, not isomeric

(h) Same compound

(i) Different compounds, not isomeric

(j) Same compound

(k) Constitutional isomers

(l) Different compounds, not isomeric

(m) Same compound

(n) Same compound

(o) Same compound

(p) Constitutional isomers

1.32 (a) 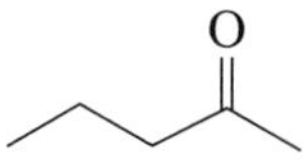 O

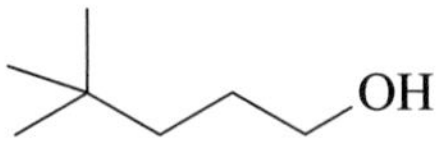

(d) 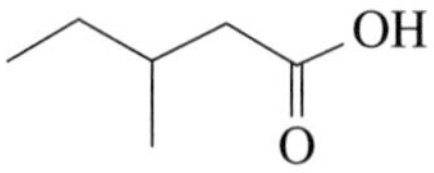 OH O

(b)

(e)  or

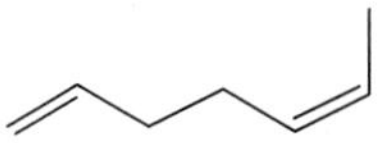

(c) OH

(f) 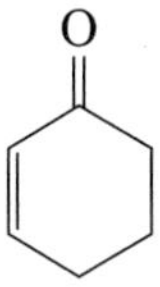 O

1.33

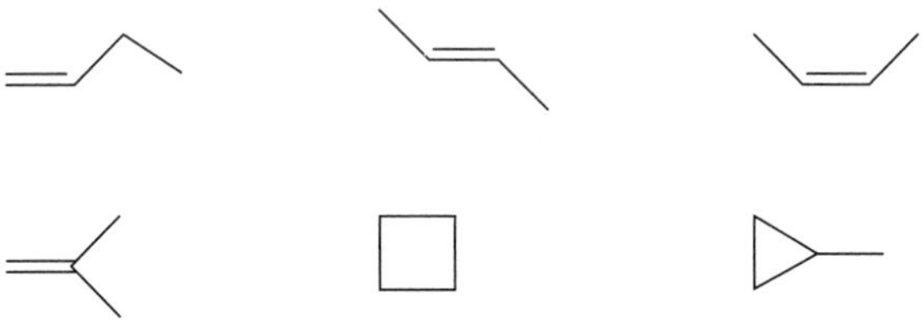

1.34

H—C(H)(H)—N⁺(=O:)—O:⁻ ⟷ H—C(H)(H)—N⁺(—O:⁻)=O:

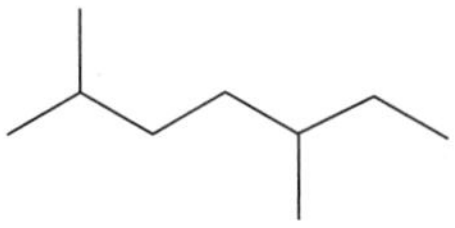

H—O—C(=O)—N(—H)—H

H—C(H)(H)—O—N=O:

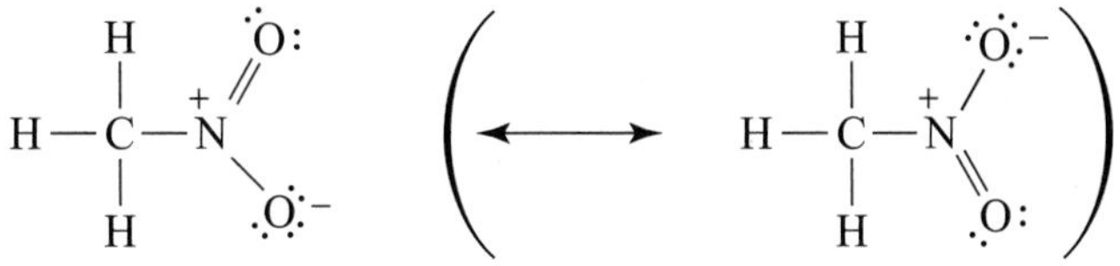

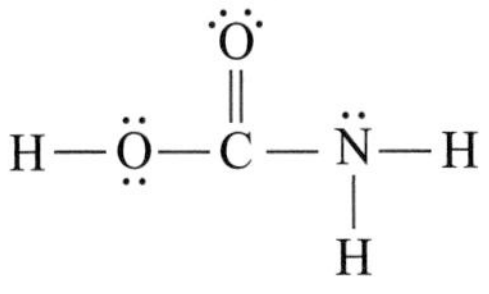

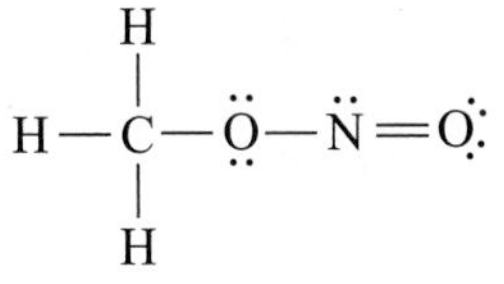

(Other structures are possible.)

Resonance Structures

1.35 (a)  (b) 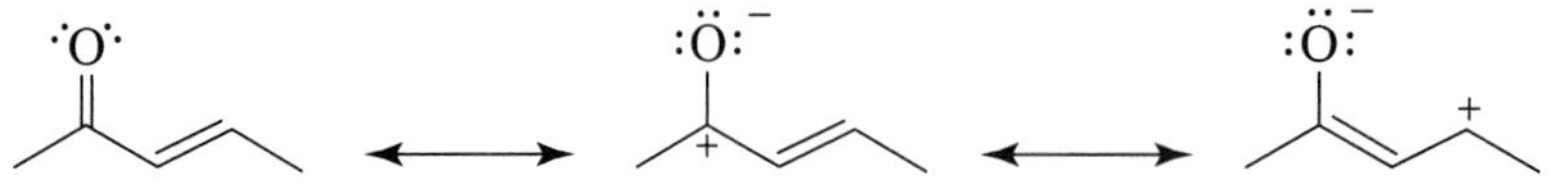 (c) 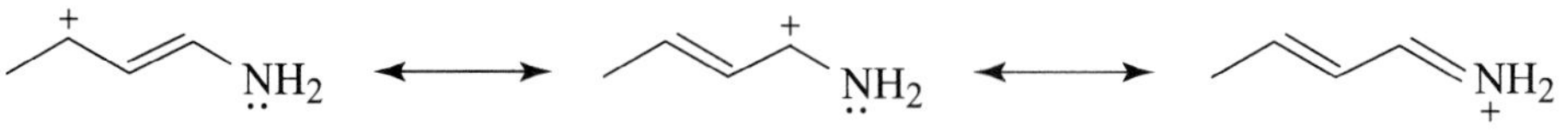 (d) 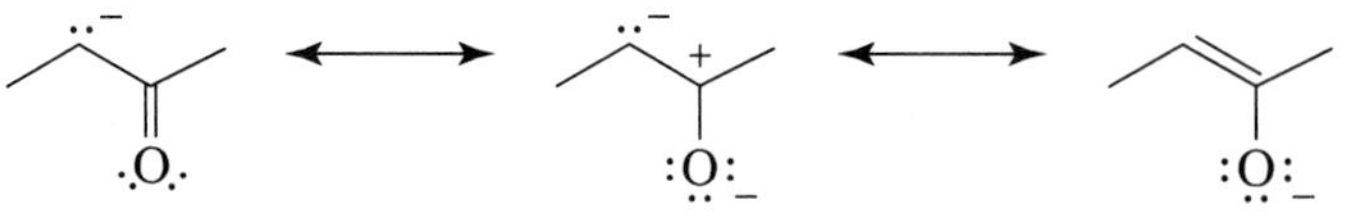 (e) 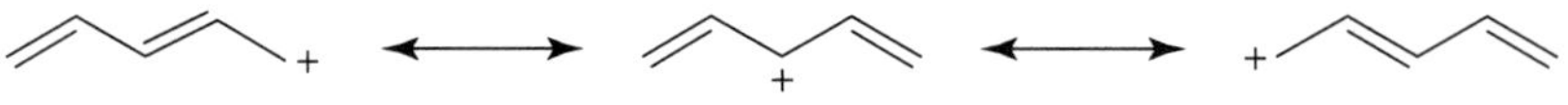 (f) 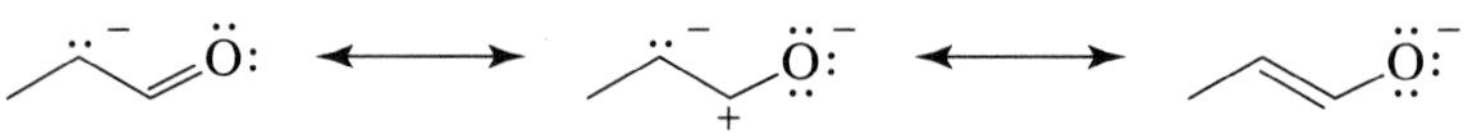 (g) 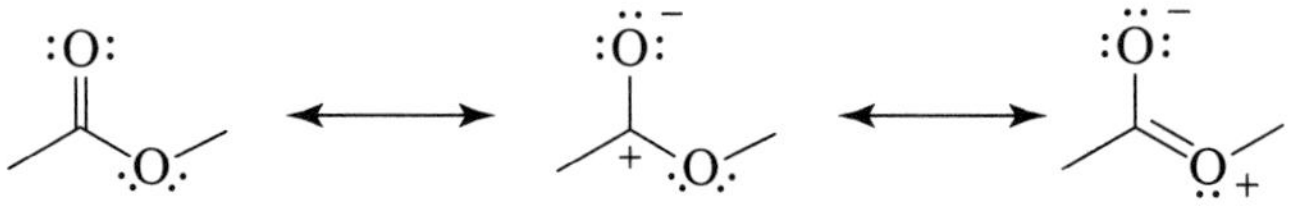 (h)

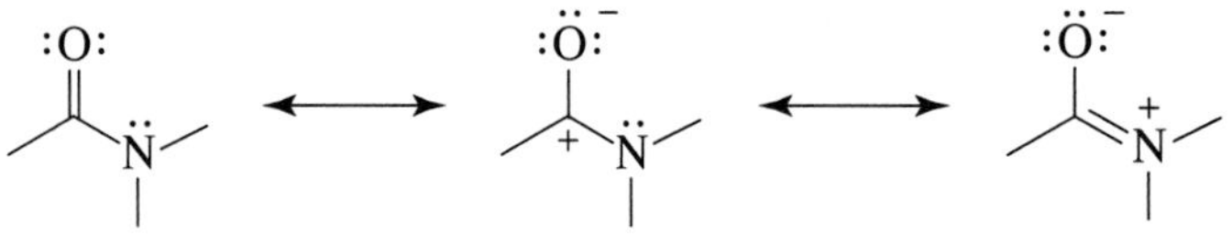

1.36

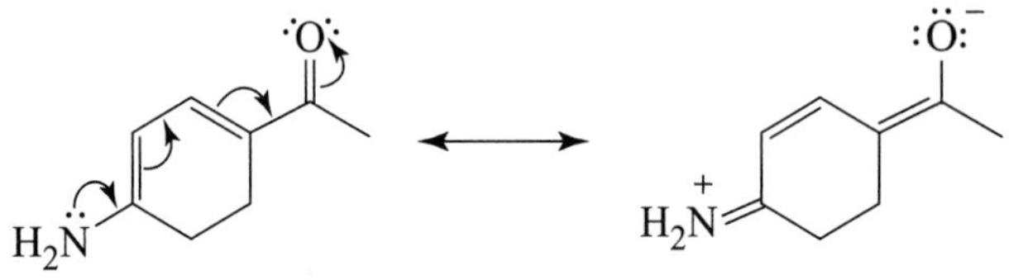

1.37

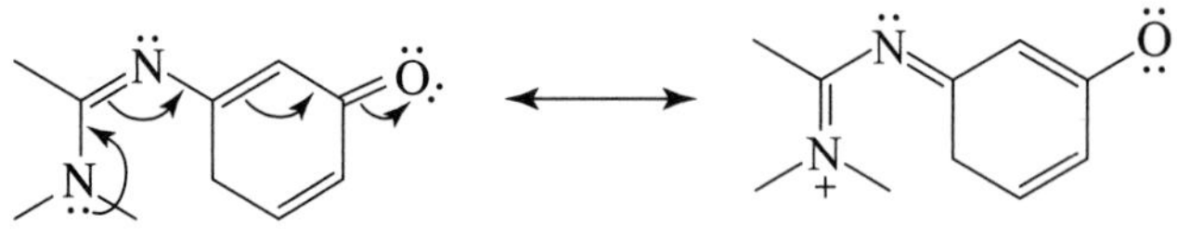

1.38 (a) While the structures differ in the position of their electrons, they also differ in the positions of their nuclei and thus *they are not resonance structures*. (In cyanic acid the hydrogen nucleus is bonded to oxygen; in isocyanic acid it is bonded to nitrogen.)

(b) The anion obtained from either acid is a resonance hybrid of the following structures: $^{-}\!:\!\ddot{O}\!-\!C\!\equiv\!N\!:\;\longleftrightarrow\;\ddot{O}\!=\!C\!=\!\ddot{N}\!:^{-}$

1.39

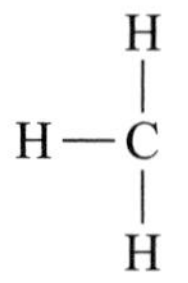

```
    H
    |
H — C
    |
    H
```

(a) A +1 charge. ($F = 4 - {}^{6}/_{2} - 2 = +1$)

(b) A +1 charge. (It is called a methyl cation.)

(c) Trigonal planar, that is,

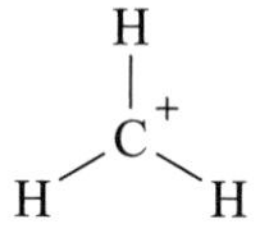

```
   H
   | +
   C
  / \
 H   H
```

(d) sp^2

1.40

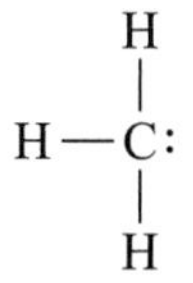

```
    H
    |
H — C:
    |
    H
```

(a) A −1 charge. ($F = 4 - 6/2 - 2 = -1$)

(b) A −1 charge. (It is called a methyl anion.)

(c) Trigonal pyramidal, that is

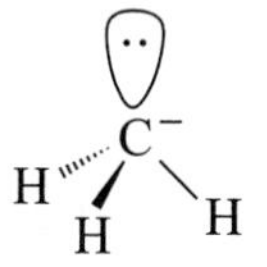

```
    (··)
     C⁻
 H'''/ \
    H   H
```

(d) sp^3

1.41

```
    H
    |
H — C·
    |
    H
```

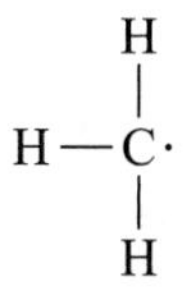

(a) No formal charge. ($F = 4 - {}^{6}/_{2} - 1 = 0$)

(b) No charge.

(c) sp^2, that is,

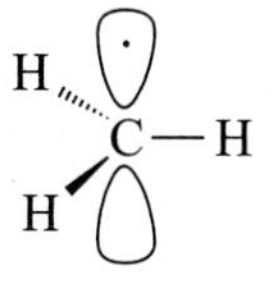

```
 H  (·)
   ''C — H
 H   ( )
```

1.42 (a) and (b)

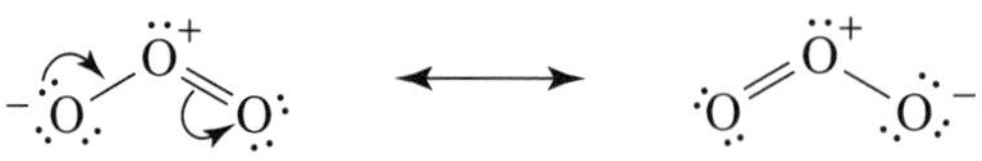

(c) Because the two resonance structures are equivalent, they should make equal contributions to the hybrid and, therefore, the bonds should be the same length.

(d) Yes. We consider the central atom to have two groups or units of bonding electrons and one unshared pair.

1.43 $:N{\equiv}\overset{+}{N}{-}\ddot{\underset{..}{N}}:^{2-} \longleftrightarrow {}^{-}:\ddot{N}{=}\overset{+}{N}{=}\ddot{N}:^{-} \longleftrightarrow {}^{2-}:\ddot{\underset{..}{N}}{-}\overset{+}{N}{\equiv}N:$

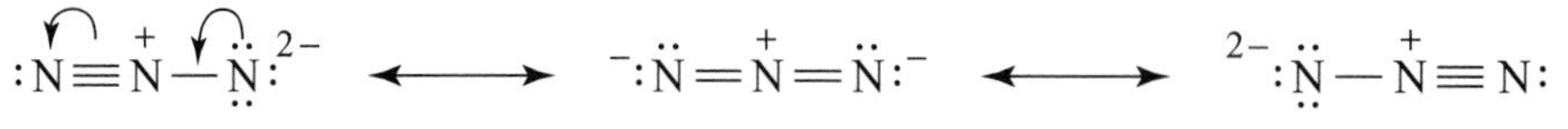

A **B** **C**

Structures **A** and **C** are equivalent and, therefore, make equal contributions to the hybrid. The bonds of the hybrid, therefore, have the same length.

1.44 (a) OH OH O O

OH OH O

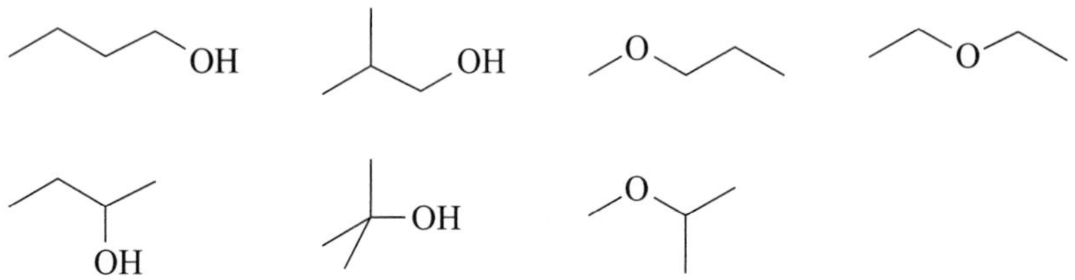

(b) $(CH_3)_2NH$ $CH_3CH_2NH_2$

(c) $(CH_3)_3N$ $CH_3CH_2NHCH_3$ $CH_3CH_2CH_2NH_2$ CH_3CHCH_3 (with NH_2 on the central carbon)

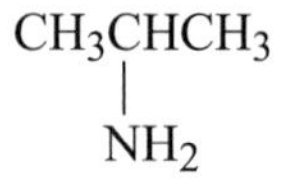

(d)

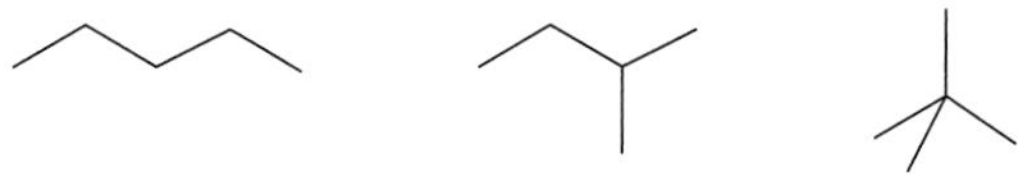

1.45 (a) constitutional isomers (b) the same

(c) resonance forms (d) constitutional isomers

(e) resonance forms (f) the same

<u>**Challenge Problems**</u>

1.46 (a) $\ddot{O}{=}\overset{+}{N}{=}\ddot{O}$

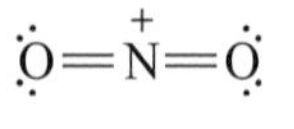

(b) Linear

(c) Carbon dioxide

1.47 Set A: :Br: Br Br :Br: Br Br Br Br

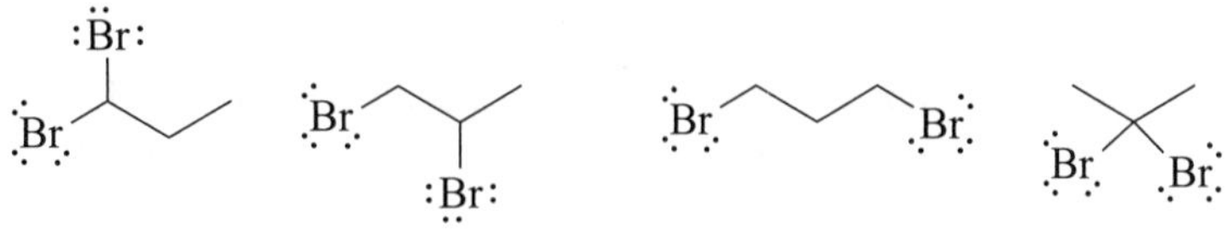

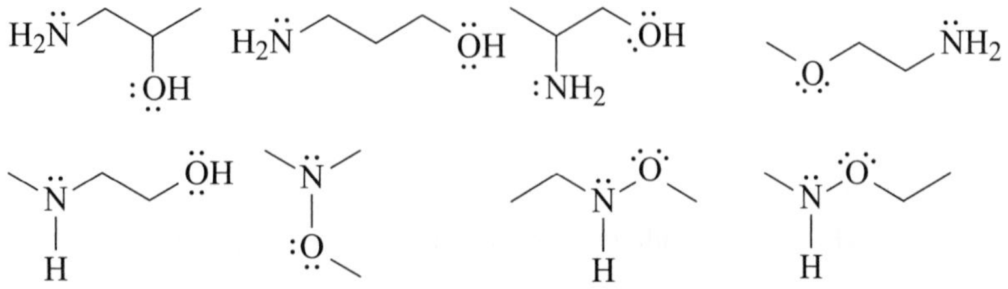

Set B: $H_2\ddot{N}$:OH $H_2\ddot{N}$ OH OH :NH_2 O NH_2

N OH N :O N O N O

H H H

Set C: H O O O: OH H H OH H

a b C H O c O

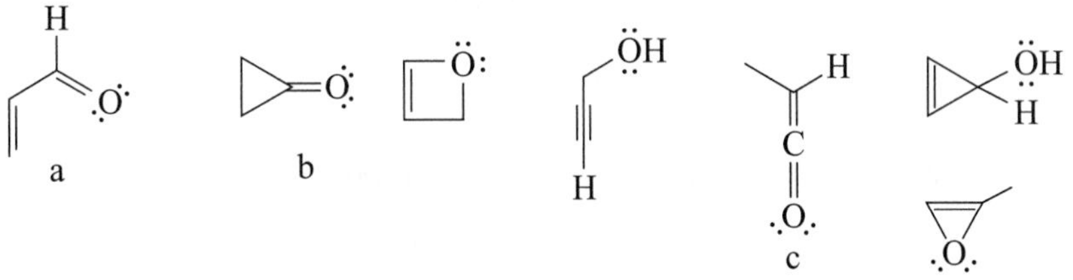

[and unstable enol forms of a, b, and c]

Set D: $\overset{+}{N}H_3$ N + H H

Set E: :− (i.e., $CH_3CH_2\ddot{C}H_2$ and $CH_3\ddot{C}HCH_3$)

1.48 (a) Dimethyl ether Dimethylacetylene *cis*-1,2-Dichloro-1,2-difluoroethene

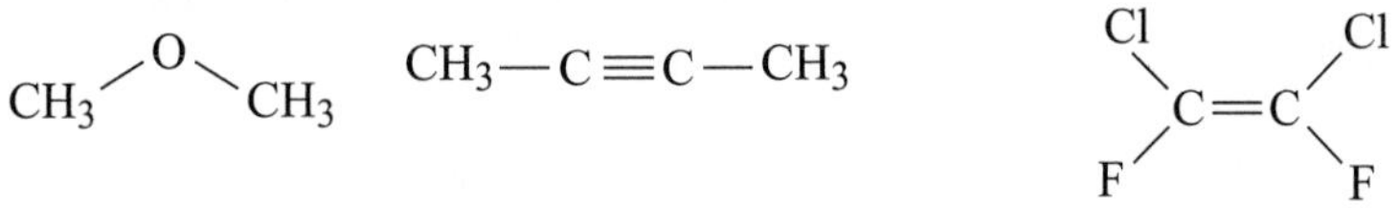

CH_3-O-CH_3 $CH_3-C\equiv C-CH_3$ Cl Cl C=C F F

(b) O Cl Cl F F

(c) H C O C H H H H H H H H—C—C≡C—C—H H H Cl Cl C=C F F or Cl Cl F C=C F

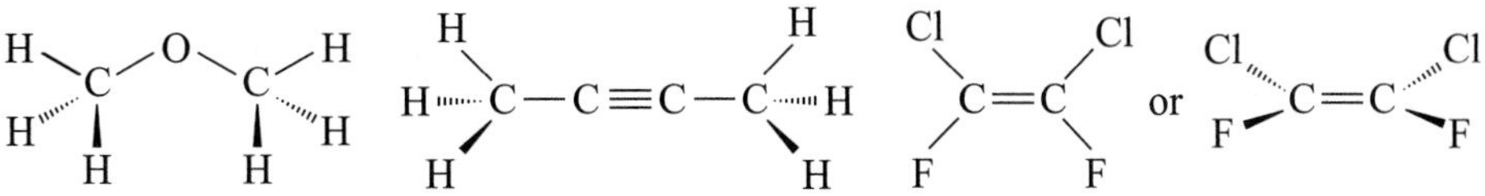

1.49 The large lobes centered above and below the boron atom represent the $2p$ orbital that was not involved in hybridization to form the three $2sp^2$ hybrid orbitals needed for the three boron-fluorine covalent bonds. This orbital is not a pure $2p$ atomic orbital, since it is not an isolated atomic p orbital but rather part of a molecular orbital. Some of the other lobes in this molecular orbital can be seen near each fluorine atom.

1.50 The two resonance forms for this anion are $^{-}\!:CH_2—CH=\ddot{O}:$ and $CH_2=CH—\ddot{O}:^{-}$. The MEP indicates that the resonance contributor where the negative charge on the anion is on the oxygen is more important, which is what we would predict based on the fact that oxygen is more electronegative than carbon.

Resonance hybrid, $\overset{\delta-}{C}H_2 \cdots CH \cdots \overset{\delta-}{O}$

QUIZ

1.1 Which of the following is a valid Lewis dot formula for the nitrite ion (NO_2^-)?

(a) $^{-}:\ddot{O}—\ddot{N}=\ddot{O}:$ (b) $:\ddot{O}=\ddot{N}—\ddot{O}:^{-}$ (c) $:\ddot{O}—\ddot{N}\equiv\ddot{O}:$ (d) Two of these

(e) None of the above

1.2 What is the hybridization state of the boron atom in BF_3?

(a) *s* (b) *p* (c) *sp* (d) sp^2 (e) sp^3

1.3 BF_3 reacts with NH_3 to produce a compound, $F_3B—NH_3$ (F—B(F)(F)—N(H)(H)—H). The hybridization state of B is

(a) *s* (b) *p* (c) *sp* (d) sp^2 (e) sp^3

1.4 The formal charge on N in the compound given in Problem 1.3 is

(a) −2 (b) −1 (c) 0 (d) +1 (e) +2

1.5 The correct bond-line formula of the compound whose condensed formula is $CH_3CHClCH_2CH(CH_3)CH(CH_3)_2$ is

(a) 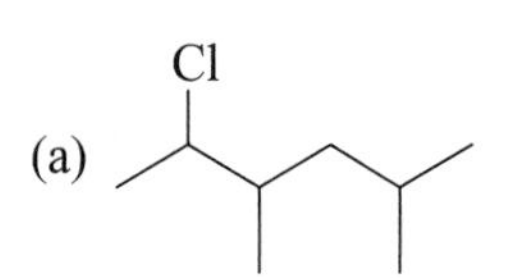

(b) 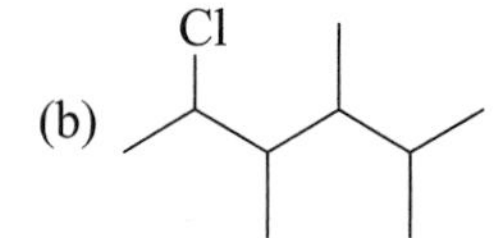

(c) 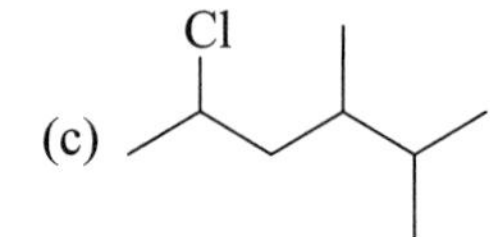

(d) 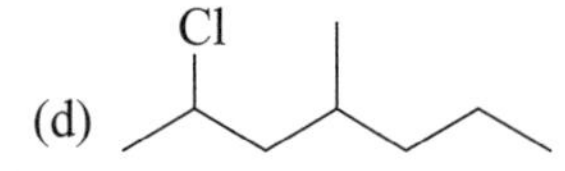

(e) 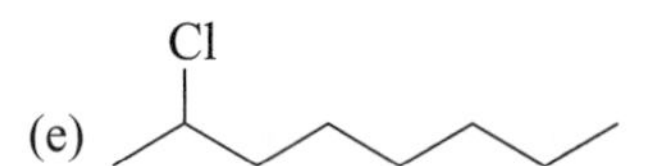

1.6 Write another resonance structure for the acetate ion.

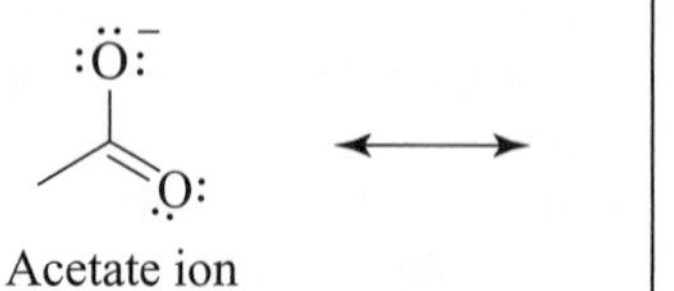

Acetate ion

1.7 In the boxes below write condensed structural formulas for constitutional isomers of $CH_3(CH_2)_3CH_3$.

1.8 Write a three-dimensional formula for a constitutional isomer of compound **A** given below. Complete the partial structure shown.

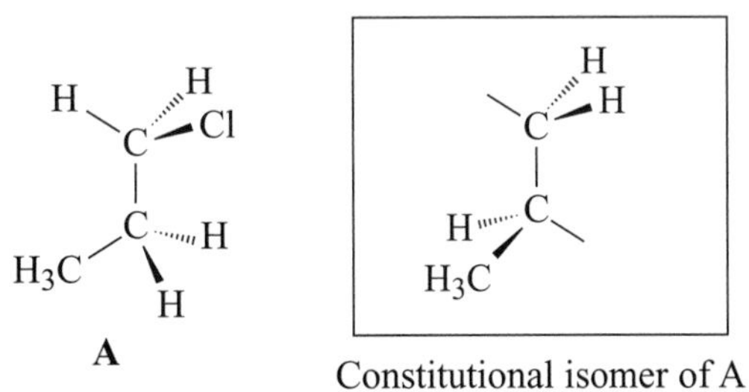

Constitutional isomer of A

1.9 Consider the molecule $(CH_3)_3B$ and give the following:

(a) Hybridization state of boron

(b) Hybridization state of carbon atoms

(c) Formal charge on boron

(d) Orientation of groups around boron

(e) Dipole moment of $(CH_3)_3B$

1.10 Give the formal charge on oxygen in each compound.

(a) $CH_3-\ddot{O}-CH_3$ with CH_3 attached below O

(c) $\ddot{\underset{..}{O}}:^-$ (on an ethyl chain)

(b) O (ketone, acetone skeletal structure)

1.11 Write another resonance structure in which all of the atoms have a formal charge of zero.

$:\ddot{O}:^-$, H, $=\overset{+}{N}$, H, H $\longleftrightarrow$

1.12 Indicate the direction of the net dipole moment of the following molecule.

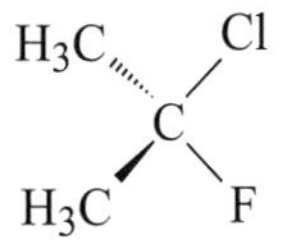

H_3C Cl
C
H_3C F

1.13 Write bond-line formulas for all compounds with the formula C_3H_6O.

2 FAMILIES OF CARBON COMPOUNDS: FUNCTIONAL GROUPS, INTERMOLECULAR FORCES, AND INFRARED (IR) SPECTROSCOPY

SOLUTIONS TO PROBLEMS

2.1 The four carbon atoms occupy different positions in the two representations (cf. rule 2, Sec. 1.8A).

2.2 (a) H—F (with dipole arrow $\longmapsto$) or $\overset{\delta+}{H}—\overset{\delta-}{F}$

(c) Br—Br
$\mu = 0$ D

(b) I—Br (with dipole arrow $\longmapsto$) or $\overset{\delta+}{I}—\overset{\delta-}{Br}$

(d) F—F
$\mu = 0$ D

2.3 VSEPR theory predicts a planar structure for BF_3.

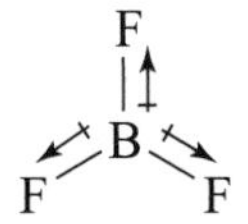

$\mu = 0$ D

The vector sum of the bond moments of a trigonal planar structure would be zero, resulting in a prediction of $\mu = 0$ D for BF_3. This correlates with the experimental observation and confirms the prediction of VSEPR theory.

2.4 The shape of $CCl_2{=}CCl_2$ (below) is such that the vector sum of all of the C—Cl bond moments is zero.

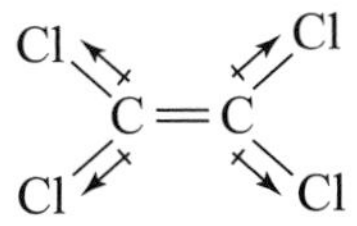

2.5 The fact that SO_2 has a dipole moment indicates that the molecule is angular, not linear.

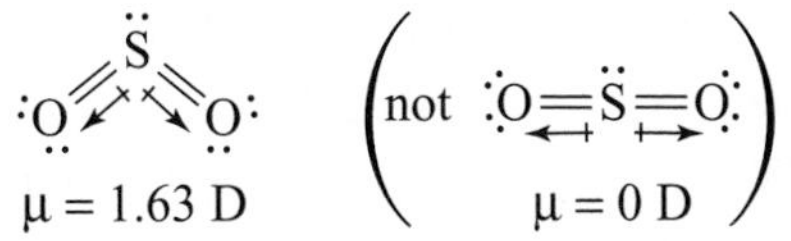

$\mu = 1.63$ D (not O=S=O, $\mu = 0$ D)

An angular shape is what we would expect from VSEPR theory, too.

2.6 Again, this is what VSEPR theory predicts.

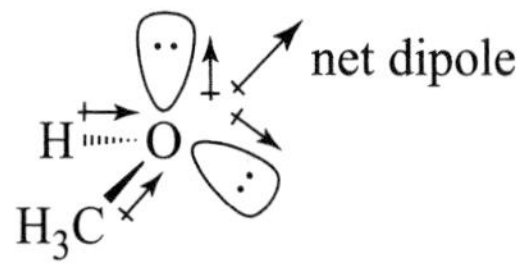

net dipole

H—O—H_3C

2.7 In $CFCl_3$ the large C—F bond moment opposes the C—Cl moments, leading to a net dipole moment in the direction of the fluorine. Because hydrogen is much less electronegative than fluorine, no such opposing effect occurs in $CHCl_3$; therefore, it has a net dipole moment that is larger and in the direction of the chlorine atoms.

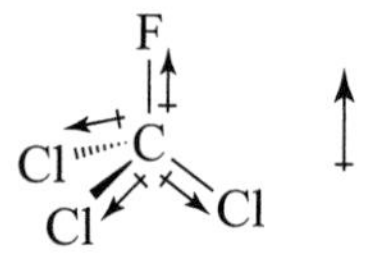

Smaller net dipole moment

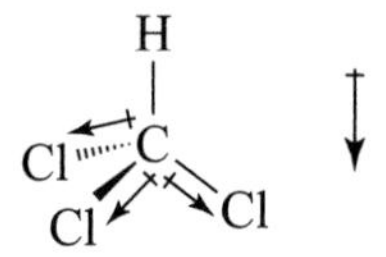

Larger net dipole moment

2.8 (a) net dipole moment

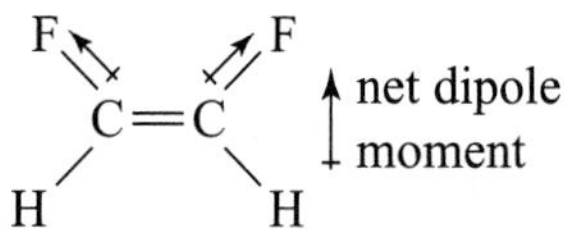

(b) $\mu = 0$ D

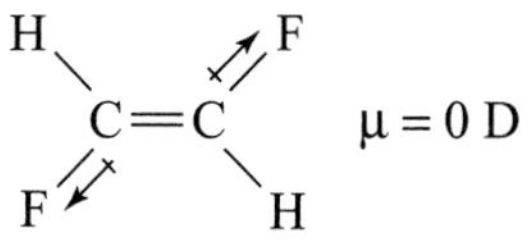

(c) net dipole moment

(d) $\mu = 0$ D

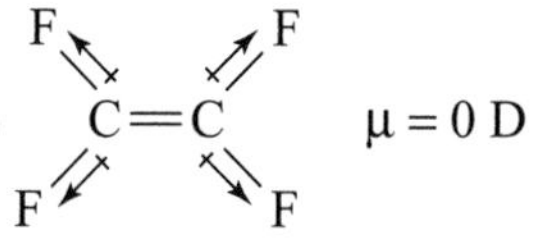

2.9 (a) net dipole moment

cis — net dipole moment

trans — $\mu = 0$ D

Cis-trans isomers

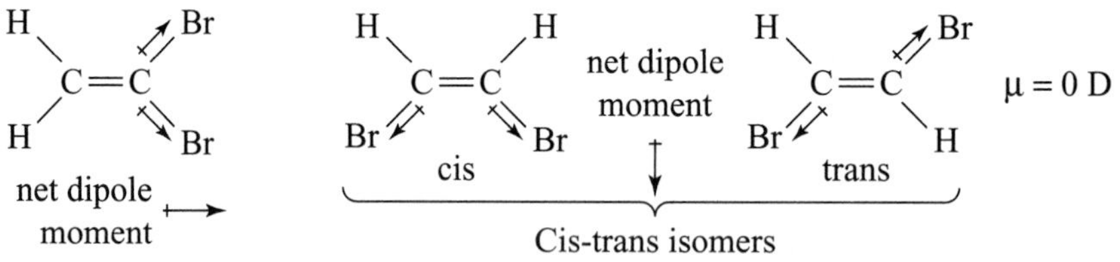

(b) net dipole moment

cis — net dipole moment

trans — $\mu = 0$ D

Cis-trans isomers

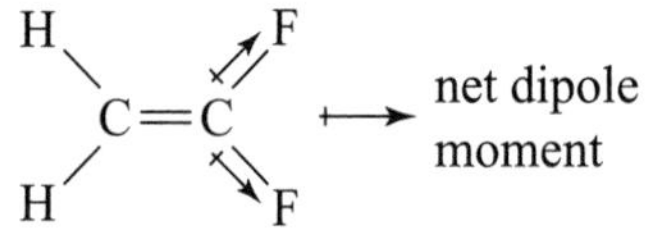

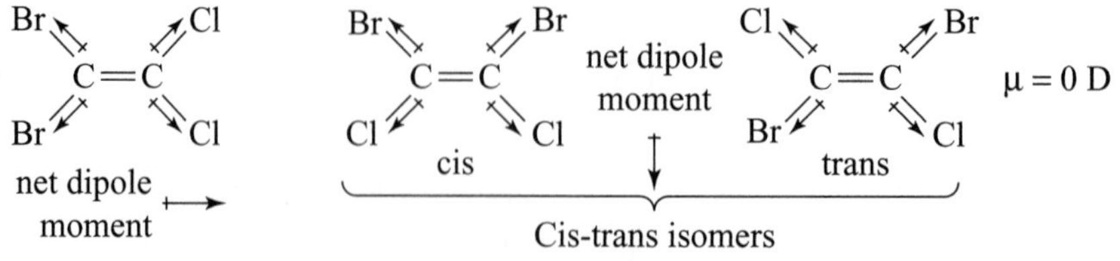

2.10 (a) 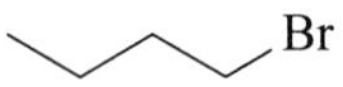 Br and Br

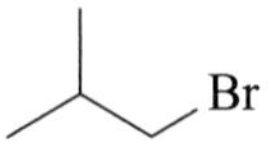

(b) 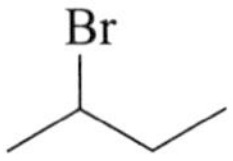 Br (c) 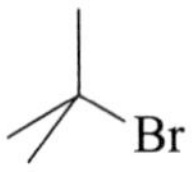 Br

2.11 (a)  F (b) 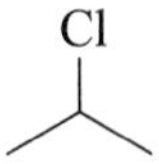 Cl

(c) Propyl bromide (d) Isopropyl fluoride

(e) Phenyl iodide

2.12 (a) 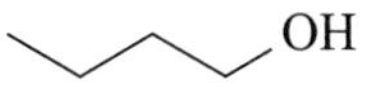 OH and OH

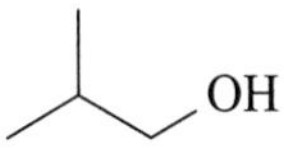

(b) 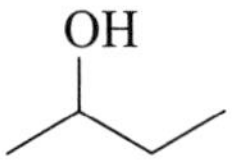 OH (c) 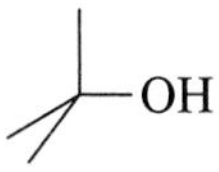 OH

2.13 (a) 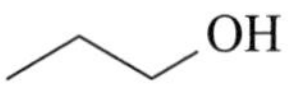 OH (b) 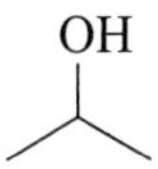 OH

2.14 (a)  O (b) 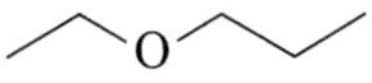 O

(c) 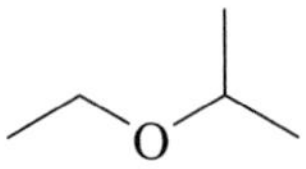 O (d) Methyl propyl ether

(e) Diisopropyl ether (f) Methyl phenyl ether

2.15 (a) Isopropylpropylamine (b) Tripropylamine

(c) Methylphenylamine (d) Dimethylphenylamine

(e)  NH_2 (f) $CH_3-N(CH_3)-CH_3$ or $(CH_3)_3N$

(g) 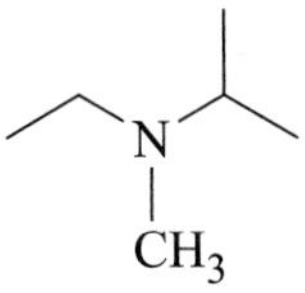 N, CH_3

2.16 (a) (e) only (b) (a, c) (c) (b, d, f, g)

2.17 (a) $(CH_3)_3N: + H-\ddot{Cl}: \longrightarrow (CH_3)_3\overset{+}{N}-H + :\ddot{Cl}:^-$

(b) sp^3

2.18

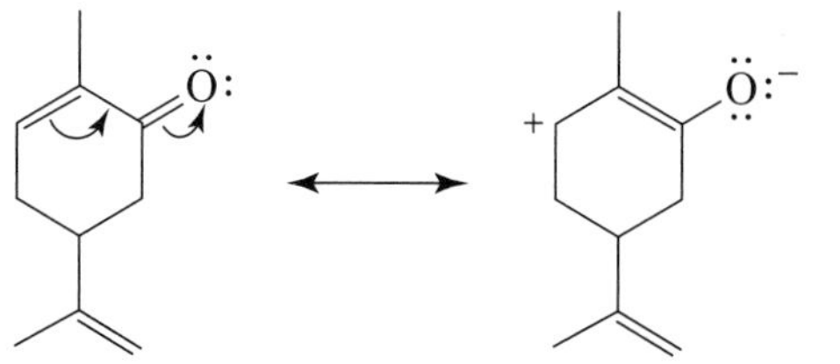

2.19 (a)

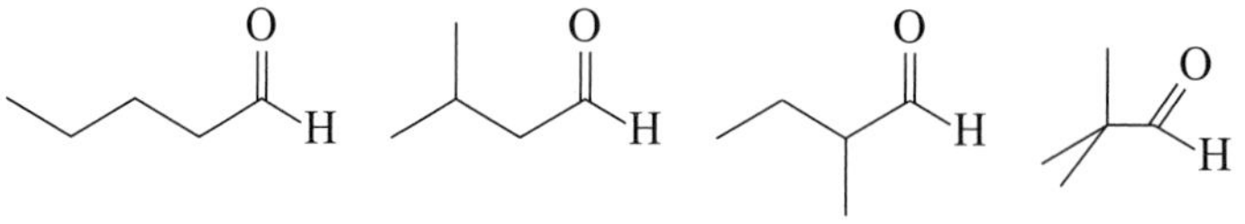

(b)

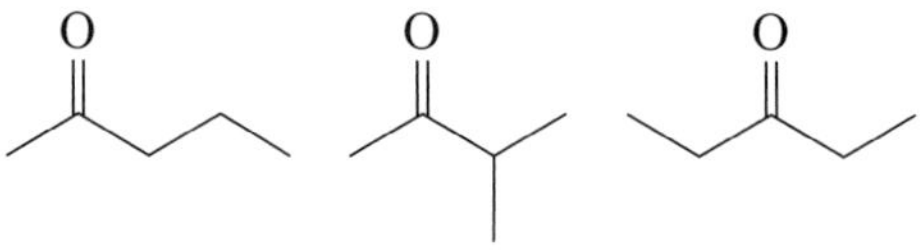

2.20 H—C(=O:)—O—H + :B⁻ ⟶ H—C(=O:)—O:⁻ + HB

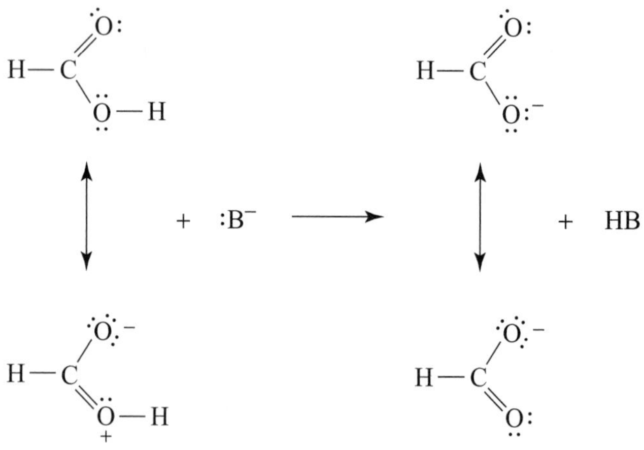

The formate ion is more stabilized by resonance because its two resonance structures are equivalent (Rule 4, Sec. 1.8A).

2.21

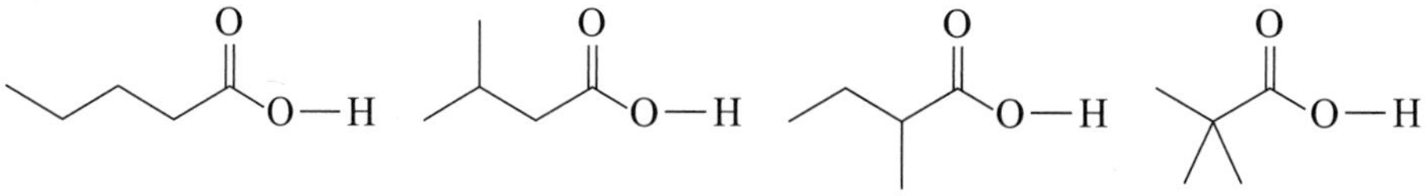

2.22 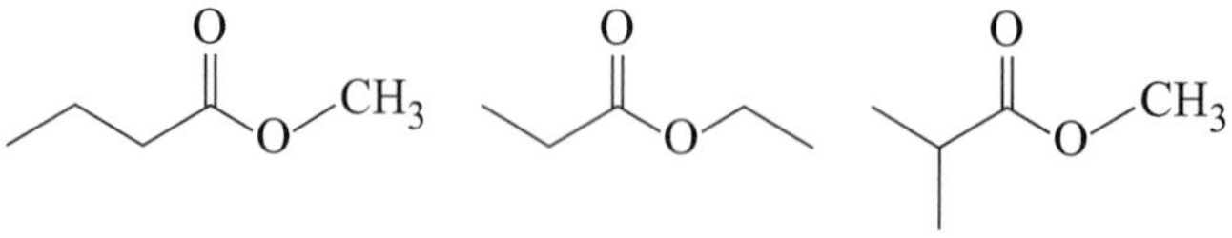 + others

2.23 CH_3—C(=O:)—O—CH_2CH_3 ⟷ CH_3—C(—O:⁻)=O⁺—CH_2CH_3

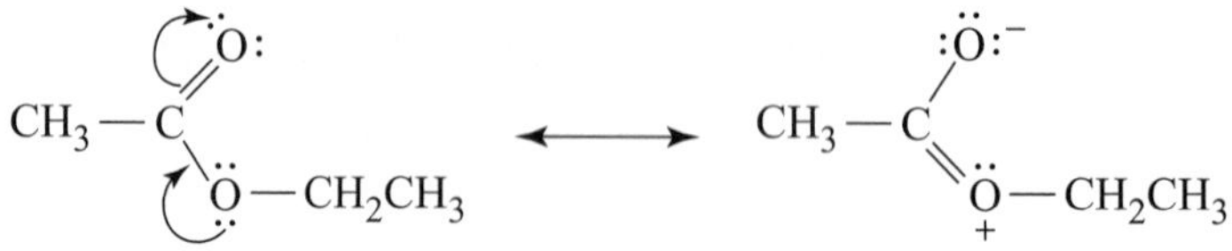

2.24 $CH_3-C(=\ddot{O}:)-\ddot{N}H_2 \longleftrightarrow CH_3-C(-\ddot{O}:^-)=\overset{+}{N}H_2$

2.25 (a) $CH_3CH_2CH_2CH_2OH$ would boil higher because its molecules can form hydrogen bonds to each other through the $-\ddot{O}-H$ group.

(b) $CH_3CH_2CH_2N(H)CH_3$ would boil higher because its molecules can form hydrogen bonds to each other through the $-\ddot{N}-H$ group.

(c) $HOCH_2CH_2CH_2CH_2OH$ because by having two $-\ddot{O}-H$ groups, it can form more hydrogen bonds.

2.26 Cyclopropane would have the higher melting point because its cyclic structure gives it a rigid compact shape that would permit stronger crystal lattice forces.

2.27 $d < a < b < c$

(c) has the highest boiling point due to hydrogen bonding involving its O—H group.

(b) is a polar molecule due to its C=O group, hence higher boiling than the essentially non-polar (a) and (d).

(a) has a higher boiling point than (d) because its unbranched structure permits more van der Waals attractions.

2.28 If we consider the range for carbon-oxygen double bond stretching in an aldehyde or ketone to be typical of an unsubstituted carbonyl group, we find that carbonyl groups with an oxygen or other strongly electronegative atom bonded to the carbonyl group, as in carboxylic acids and esters, absorb at somewhat higher frequencies. On the other hand, if a nitrogen atom is bonded to the carbonyl group, as in an amide, then the carbonyl stretching frequency is lower than that of a comparable aldehyde or ketone. The reason for this trend is that strongly electronegative atoms increase the double bond character of the carbonyl, while the unshared electron pair of an amide nitrogen atom contributes to the carbonyl resonance hybrid to give it less double bond character.

Functional Groups and Structural Formulas

2.29 (a) Ketone (b) Alkyne (c) Alcohol (d) Aldehyde
(e) Alcohol (f) Alkene

2.30 (a) Three carbon-carbon double bonds (alkene) and a 2° alcohol

(b) Phenyl, carboxylic acid, amide, ester, and a 1° amine

(c) Phenyl and a 1° amine

(d) Carbon-carbon double bond and a 2° alcohol

(e) Phenyl, ester, and a 3° amine

(f) Carbon-carbon double bond and an aldehyde

(g) Carbon-carbon double bond and 2 ester groups

2.31

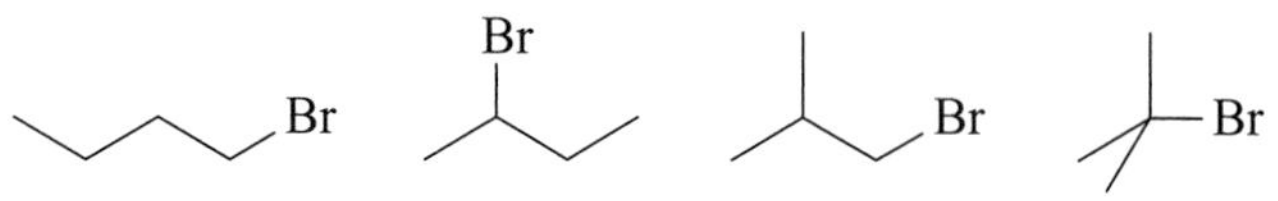

Br Br Br Br

1° Alkyl bromide 2° Alkyl bromide 1° Alkyl bromide 3° Alkyl bromide

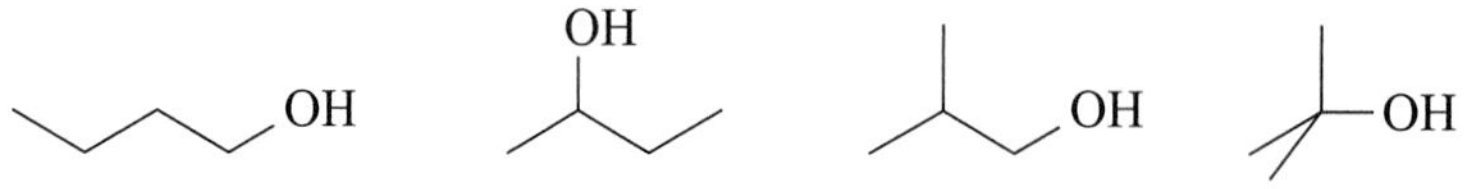

2.32

OH OH OH OH

1° Alcohol 2° Alcohol 1° Alcohol 3° Alcohol

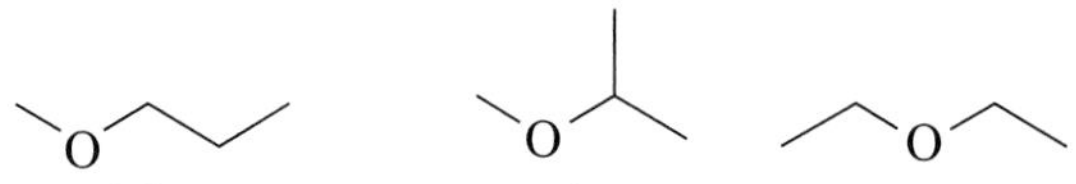

O O O

Ether Ether Ether

2.33 (a) 1° (b) 2° (c) 3° (d) 3° (e) 2°

2.34 (a) 2° (b) 1° (c) 3° (d) 2° (e) 2° (f) 3°

2.35 (a) 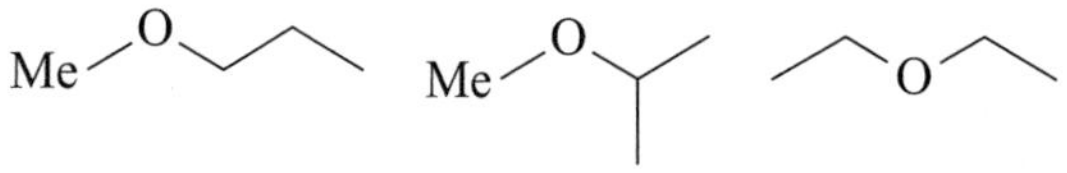 Me O Me O O

(b) 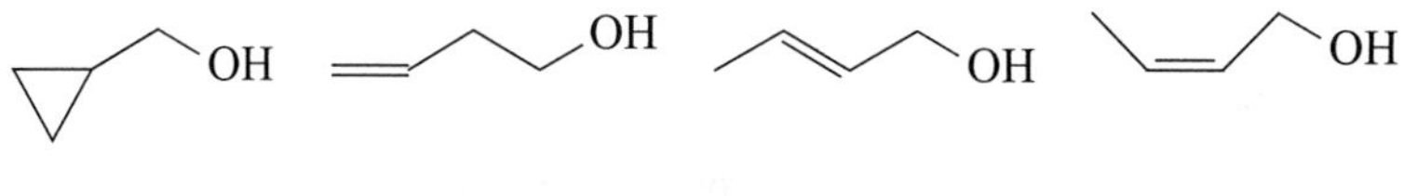 OH OH OH OH

(c) 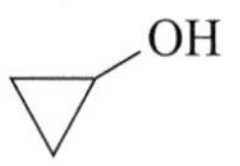 OH

(d) 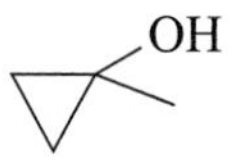 OH

(e)  O H O O O Me

(f) Br Br Br Br

(g) Br Br Br

(h) Br

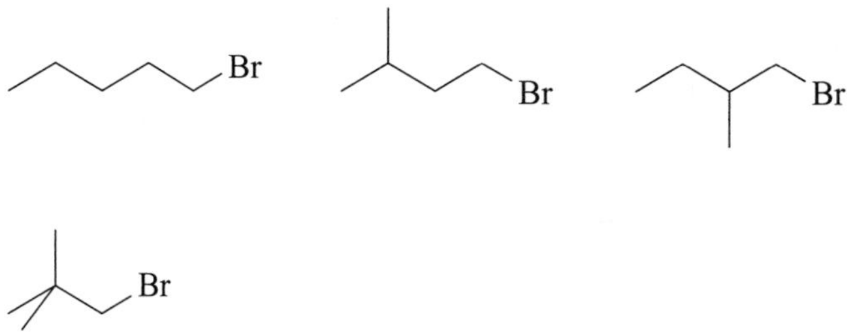

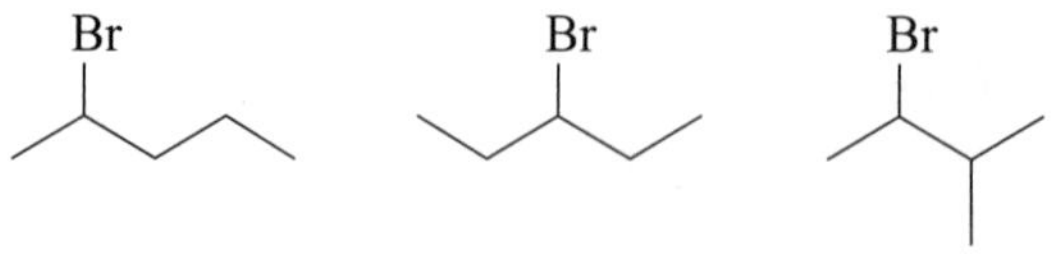

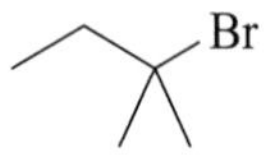

(i)

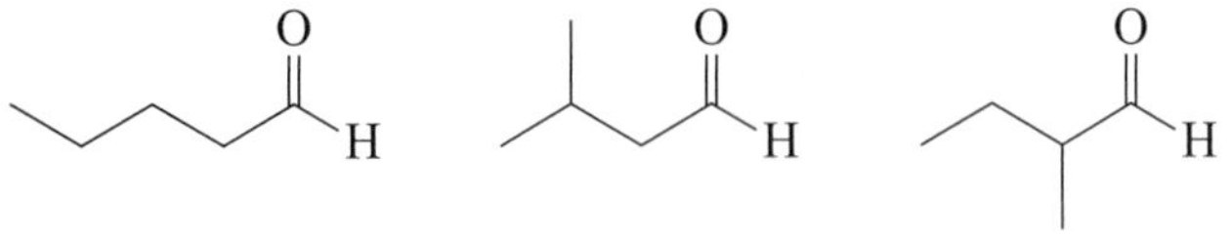

(j)

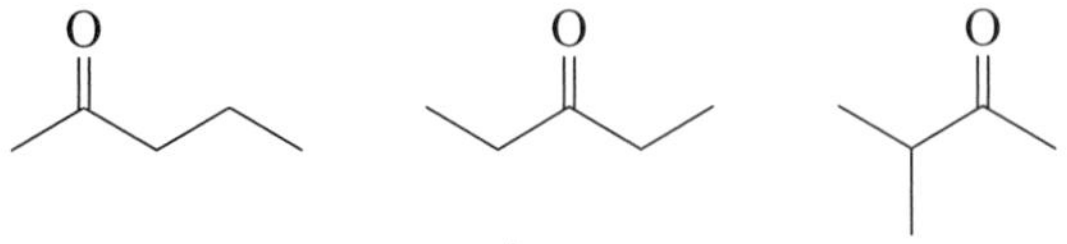

(k) 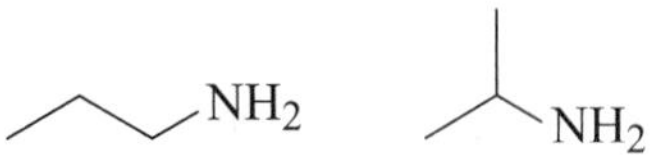 NH_2 NH_2

(l) 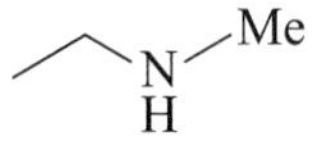 Me N H

(m) Me_3N

(n) 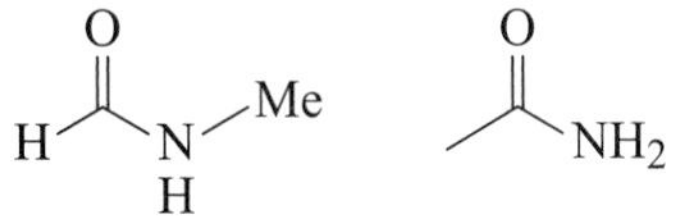 H N Me H NH_2

Physical Properties

2.36 (a) The O—H group of Vitamin A is the hydrophilic portion of the molecule, but the remainder of the molecule is not only hydrophobic but much larger. The multiple van der Waals attractions outweigh the effect of hydrogen bonding to water through a single hydroxyl group. Hence, Vitamin A is not expected to be water soluble.

(b) For Vitamin B_3, there are multiple hydrophilic sites. The carbonyl oxygen and the O—H of the acid function as well as the ring nitrogen can all hydrogen bond to water. Since the hydrophobic portion (the ring) of the molecule is modest in size, the molecule is expected to be water soluble.

2.37 The attractive forces between hydrogen fluoride molecules are the very strong dipole-dipole attractions that we call *hydrogen bonds*. (The partial positive charge of a hydrogen fluoride molecule is relatively exposed because it resides on the hydrogen nucleus. By contrast, the positive charge of an ethyl fluoride molecule is buried in the ethyl group and is shielded by the surrounding electrons. Thus the positive end of one hydrogen fluoride molecule can approach the negative end of another hydrogen fluoride molecule much more closely, with the result that the attractive force between them is much stronger.)

2.38 The cis isomer is polar while the trans isomer is nonpolar ($\mu = 0$ D). The intermolecular attractive forces are therefore greater in the case of the cis isomer, and thus its boiling point should be the higher of the two.

2.39 Because of its ionic character—it is a true salt—the compound is water-soluble. The organic cation and the bromide ion are well-solvated by water molecules in a fashion similar to sodium bromide. The compound also is soluble in solvents of low polarity such as diethyl ether (though less so than in water). The hydrophobic alkyl groups can now be regarded as lipophilic—groups that seek a nonpolar environment. Attractive forces between the alkyl groups of different cations can be replaced, in part, by attractive forces (van der Waals forces) between these alkyl groups and ether molecules.

2.40 (a) and (b) are polar and hence are able to dissolve ionic compounds. (c) and (d) are non-polar and will not dissolve ionic compounds.

2.41

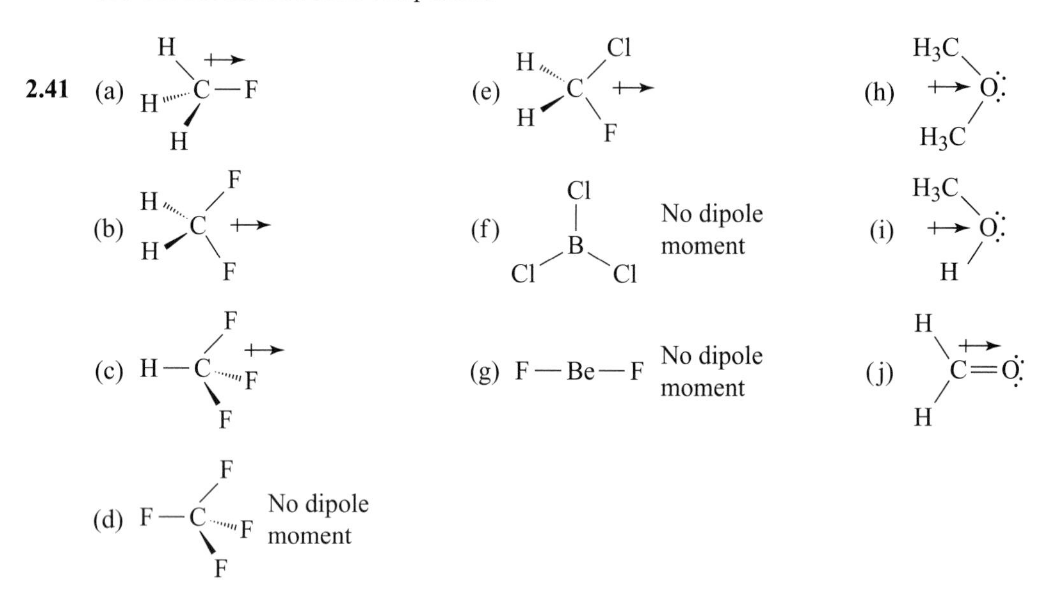

2.42 (a) Dimethyl ether: There are four electron pairs around the central oxygen: two bonding pairs and two nonbonding pairs. We would expect sp^3 hybridization of the oxygen with a bond angle of approximately 109.5° between the methyl groups.

H_3C—O—H_3C $\mu > 0$ D

(b) Trimethylamine: There are four electron pairs around the central nitrogen: three bonding pairs and one nonbonding pair. We would expect sp^3 hybridization of the nitrogen with a bond angle of approximately 109.5° between the methyl groups.

H_3C—N(CH_3)—H_3C $\mu > 0$ D

(c) Trimethylboron: There are only three bonding electron pairs around the central boron. We would expect sp^2 hybridization of the boron with a bond angle of 120° between the methyl groups.

H_3C—B(CH_3)—CH_3 $\mu = 0$ D

(d) Dimethylberyllium: There are only two bonding electron pairs around the central beryllium atom. We would expect sp hybridization of the beryllium atom with a bond angle of 180° between the methyl groups.

H_3C—Be—CH_3 $\mu = 0$ D

2.43 This is a special case of a Lewis acid (Ag^+) — Lewis base (alkene) reaction. The product can be represented by $\left[\begin{array}{c} \diagdown C \diagup \\ \Vert \!\!\rightarrow Ag \\ \diagup C \diagdown \end{array} \right]^+$. The filled π orbital of the alkene overlaps with the empty 5s orbital of Ag^+ to form a σ bond. A π bond results from the overlap of the filled 4d orbital of Ag^+ with an empty antibonding π^* orbital of the alkene.

2.44 Without one (or more) polar bonds, a molecule cannot possess a dipole moment and, therefore, it cannot be polar. If the bonds are directed so that the bond moments cancel, however, the molecule will not be polar even though it has polar bonds.

2.45 Crixivan has the following functional groups:

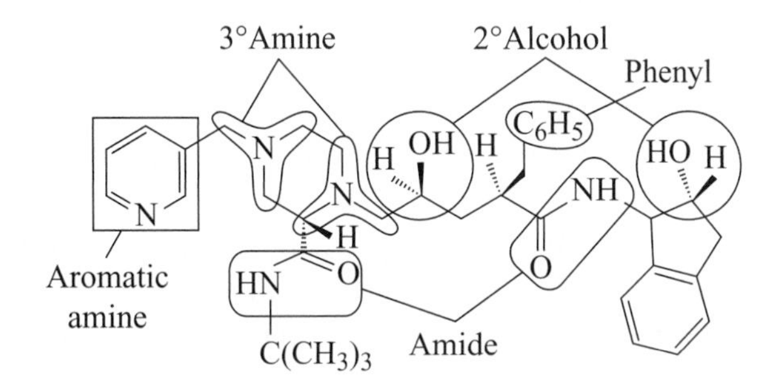

2.46 (a) $\diagup\diagdown\diagup$OH because its molecules can form hydrogen bonds to each other through its $-\ddot{\underset{\cdot\cdot}{O}}-H$ group.

(b) HO$\diagdown\diagup\diagdown$OH because with two $-\ddot{\underset{\cdot\cdot}{O}}-H$ groups, its molecules can form more hydrogen bonds with each other.

(c) $\diagup\diagdown\diagup\diagdown\diagup$OH because its molecules can form hydrogen bonds to each other.

(d) $\triangleright$—OH [same reason as (c)].

(e) (piperidine ring with NH) because its molecules can form hydrogen bonds to each other through the $-\ddot{N}-H$ group.

(f) F, F (cis) because its molecules will have a larger dipole moment. (The trans compound will have $\mu = 0$ D.)

(g) $\diagup\diagdown\overset{O}{\overset{\Vert}{C}}\diagdown$OH [same reason as (c)].

(h) Nonane, because of its larger molecular weight and larger size, will have larger van der Waals attractions.

(i) 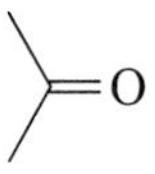 $\rangle\!\!=$O because its carbonyl group is far more polar than the double bond of $\rangle\!\!=$.

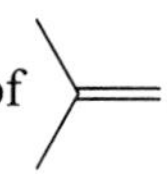

IR Spectroscopy

2.47 (a) The alcohol would have a broad absorption from the O—H group in the 3200 to 3500 cm^{-1} region of its IR spectrum. The ether would have no such absorption.

(c) The ketone would have a strong absorption from its carbonyl group near 1700 cm^{-1} in its IR spectrum. The alcohol would have a broad absorption due to its hydroxyl group in the 3200 to 3500 cm^{-1} region of its IR spectrum.

(d) Same rationale as for (a).

(e) The secondary amine would have an absorption near 3300 to 3500 cm^{-1} arising from N—H stretching. The tertiary amine would have no such absorption in this region since there is no N—H group present.

(g) Both compounds would exhibit absorptions near 1710 to 1780 cm^{-1} due to carbonyl stretching vibrations. The carboxylic acid would also have a broad absorption somewhere between 2500 and 3500 cm^{-1} due to its hydroxyl group. The ester would not have a hydroxyl absorption.

(i) The ketone would have a strong absorption from its carbonyl group near 1700 cm^{-1} in its IR spectrum. The alkene would have no such absorption but would have an absorption between 1620 and 1680 cm^{-1} due to C═C stretching.

2.48 For absorption in the infrared to occur there must be a change in the molecular dipole moment during the stretching process. The 3-hexyne molecule is symmetrical about the triple bond and so there is no change in the dipole moment accompanying the stretching. Hence, there is no IR absorption.

2.49

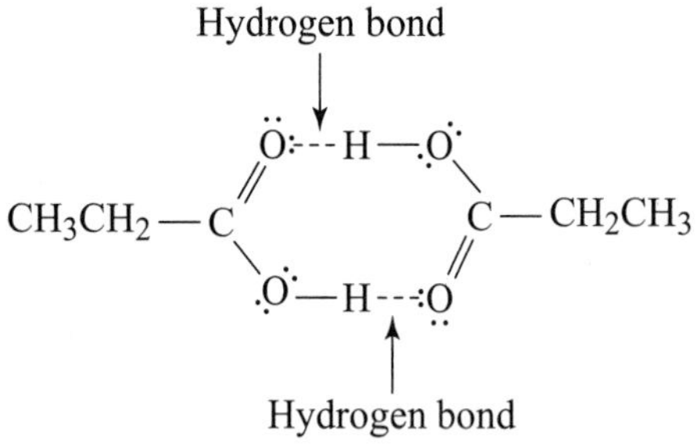

Hydrogen bond

CH_3CH_2—C(═O⋯H—O)(—O—H⋯O═)C—CH_2CH_3

Hydrogen bond

2.50 There are two peaks as a result of the asymmetric and symmetric stretches of the carbonyl groups.

Multiconcept Problems

2.51 Any four of the following:

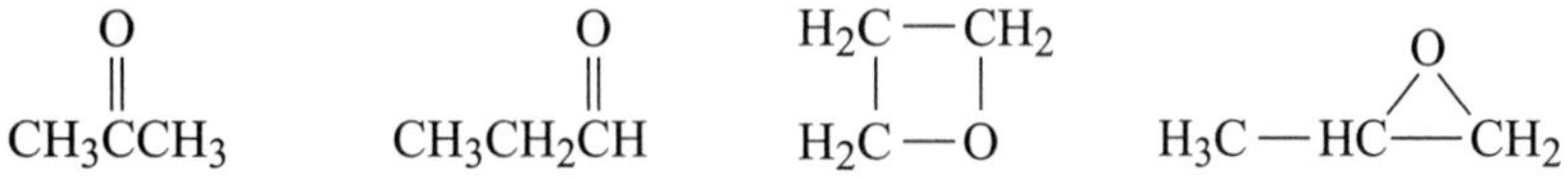

CH_3COCH_3 (Ketone); CH_3CH_2CHO (Aldehyde); $H_2C—CH_2 / H_2C—O$ (four-membered ring) (Ether); $H_3C—HC—CH_2$ with O bridging (epoxide) (Ether)

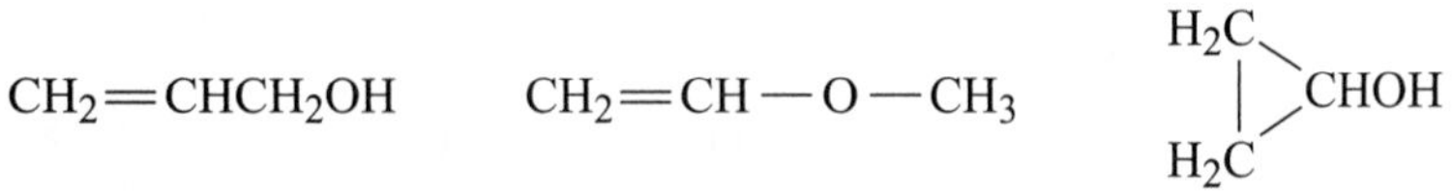

$CH_2{=}CHCH_2OH$ (Alkene, alcohol); $CH_2{=}CH—O—CH_3$ (Alkene, ether); cyclopropanol, H_2C/H_2C ring with CHOH (Alcohol)

The ketone carbonyl absorption is in the 1680–1750 cm^{-1} range; that for the aldehyde is found in the 1690–1740 cm^{-1} region. The C—O absorption for the ethers is observed at about 1125 cm^{-1}. The C=C absorption occurs at approximately 1650 cm^{-1}. Absorption for the (hydrogen-bonded) O—H group takes the form of a broad band in the 3200–3550 cm^{-1} region.

2.52 (a) **A**: $CH_3CH_2C(=O)NH_2$ **B**: $CH_3C(=O)N(H)CH_3$ **C**: $HC(=O)N(H)CH_2CH_3$ **D**: $HC(=O)N(CH_3)CH_3$

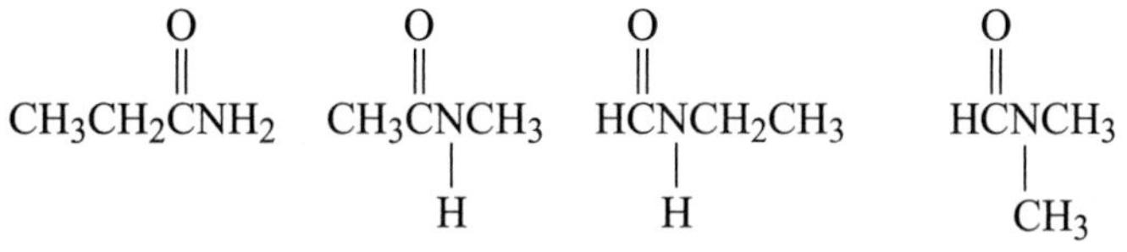

(b) **D**, because it does not have a hydrogen that is covalently bonded to nitrogen and, therefore, its molecules cannot form hydrogen bonds to each other. The other molecules all have a hydrogen covalently bonded to nitrogen and, therefore, hydrogen-bond formation is possible. With the first molecule, for example, hydrogen bonds could form in the following way:

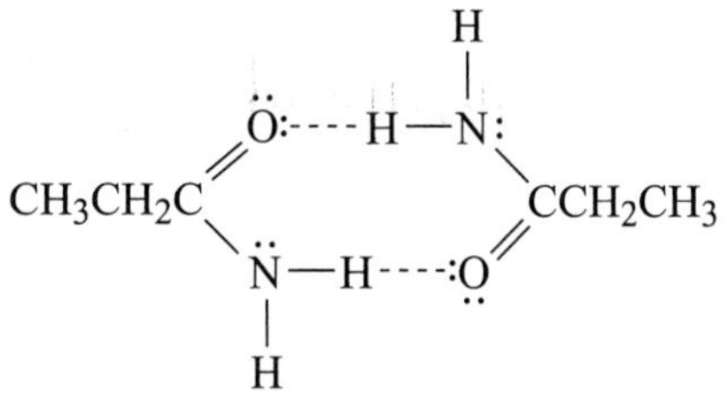

$CH_3CH_2C(=O:)—N(H)—H \cdots :O=C(CH_2CH_3)—N(H)—H \cdots$ (cyclic dimer: each C=O oxygen hydrogen-bonded to an N—H of the other molecule)

(c) All four compounds have carbonyl group absorption at about 1650 cm^{-1}, but the IR spectrum for each has a unique feature.

A shows two N—H bands (due to symmetrical and asymmetrical stretching) in the 3100–3400 cm^{-1} region.

B has a single stretching absorption band in that same region since it has only a single N—H bond.

C has two absorption bands, due to the H—C bond of the aldehyde group, at about 2820 cm^{-1} and 2920 cm^{-1}, as well as one for the N—H bond.

D does not absorb in the 3100–3500 cm^{-1} region, as the other compounds do, since it does not possess a N—H bond.

2.53 The molecular formula requires unsaturation and/or one or more rings. The IR data exclude the functional groups: —OH, C=O and C=C. Oxygen (O) must be present in an ether linkage and there can be either a triple bond or two rings present to account for the low hydrogen-to-carbon ratio. These are the possible structures:

$HC{\equiv}CCH_2OCH_3$ $HC{\equiv}COCH_2CH_3$ $CH_3C{\equiv}COCH_3$

O O O O CH3 O CH3

2.54 —C—C(=O)—O—C— $(CH_2)_n$ (a cyclic ester)

Challenge Problems

2.55

A B B′

O =O O H

The 1780 cm^{-1} band is in the general range for C=O stretching so structure B′ is considered one of the possible answers, but only B would have its C=O stretch at this high frequency (B′ would be at about 1730 cm^{-1}).

2.56 (a) HO OH H H *cis* HO H H OH *trans*

(b) The cis isomer will have the 3572 cm^{-1} band because only in it are the two hydroxyl groups close enough to permit intramolecular hydrogen-bonding. (Intermolecular hydrogen-bonding is not possible at high dilution in a non-polar solvent like CCl_4.)

2.57 (a)

C

(b)

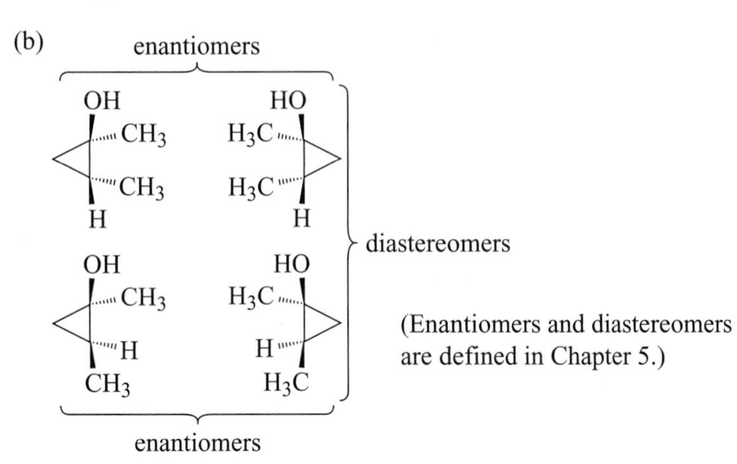

diastereomers

(Enantiomers and diastereomers are defined in Chapter 5.)

2.58 The helical structure results from hydrogen bonds formed between amide groups—specifically between the carbonyl group of one amide and the N—H group of another.

QUIZ

2.1 Which of the following pairs of compounds is *not* a pair of constitutional isomers?

(a) 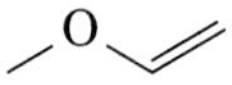 [skeletal structure] and [skeletal structure]

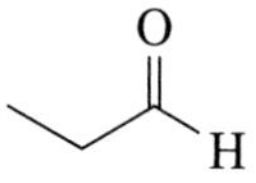

(b) 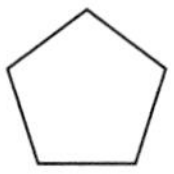 [skeletal structure] and [skeletal structure]

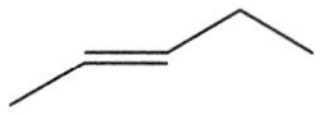

(c) 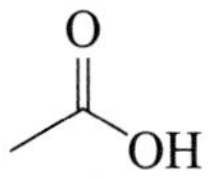 [skeletal structure] and [skeletal structure]

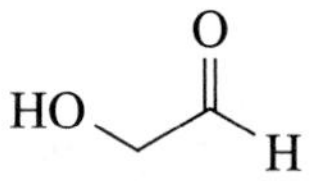

(d) $CH_3CH_2C{\equiv}CH$ and $CH_3CH{=}C{=}CH_2$

(e) $CH_3CH(CH_3)CH(CH_3)_2$ and $(CH_3)_2CHCH(CH_3)_2$

2.2 Which of the answers to Problem 2.1 contains an ether group?

2.3 Which of the following pairs of structures represents a pair of isomers?

(a) 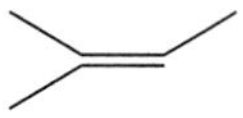 [skeletal structure] and [skeletal structure]

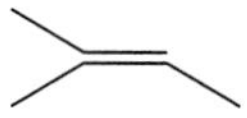

(b)  [skeletal structure] and [skeletal structure]

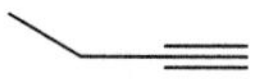

(c) $CH_3CH_2CH(CH_3)CH_2CH_3$ and $CH_3CH_2CH(CH_2CH_3)CH_3$

(d) More than one of these pairs are isomers.

2.4 Give a bond-line formula for each of the following:

(a) A tertiary alcohol with the formula $C_5H_{12}O$

(b) An *N*,*N*-disubstituted amide with the formula C_4H_9NO

(c) The alkene isomer of $C_2H_2Cl_2$ that has no dipole moment

(d) An ester with the formula $C_2H_4O_2$

(e) The isomer of $C_2H_2Cl_2$ that cannot show cis-trans isomerism

(f) The isomer of C_3H_8O that would have the lowest boiling point

(g) The isomer of $C_4H_{11}N$ that would have the lowest boiling point

2.5 Write the bond-line formula for a constitutional isomer of the compound shown below that does not contain a double bond.

$CH_3CH_2CH{=}CH_2$

2.6 Circle the compound in each pair that would have the higher boiling point.

(a)
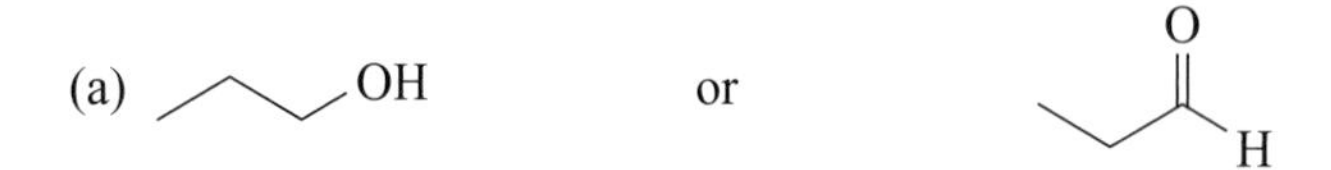

(b) N—H or N—CH_3

(c) O O—CH_3 or O OH

(d) CH_3—O—OH or CH_3—O—O—CH_3

(e) O N—CH_3 H or O N—CH_3 CH_3

2.7 Give an acceptable name for each of the following:

(a) C_6H_5—O

(b) N CH_3 C_6H_5

(c) —NH_2

3 AN INTRODUCTION TO ORGANIC REACTIONS AND THEIR MECHANISMS: ACIDS AND BASES

SOLUTIONS TO PROBLEMS

3.1 (a) 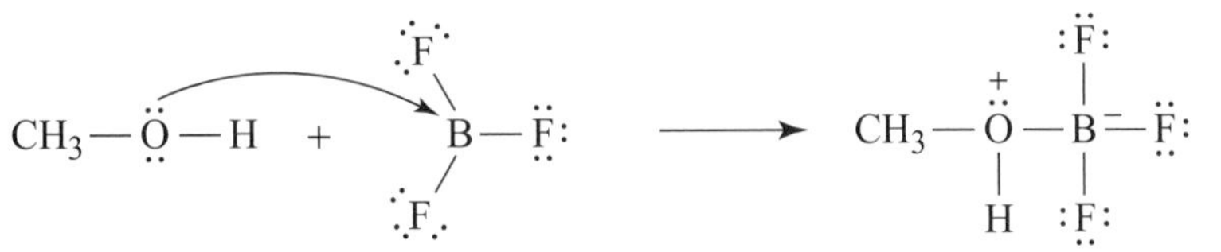 $CH_3—\ddot{O}—H + BF_3 \longrightarrow CH_3—\overset{+}{O}(H)—\overset{-}{B}F_3$

(b) $CH_3—\ddot{Cl}: + AlCl_3 \longrightarrow CH_3—\overset{+}{Cl}—\overset{-}{Al}Cl_3$

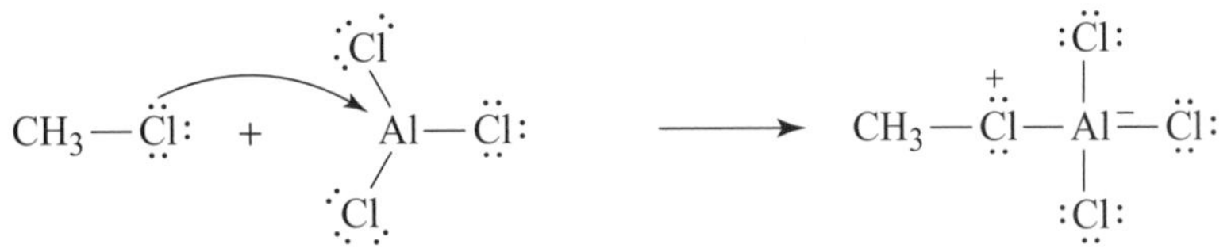

(c) 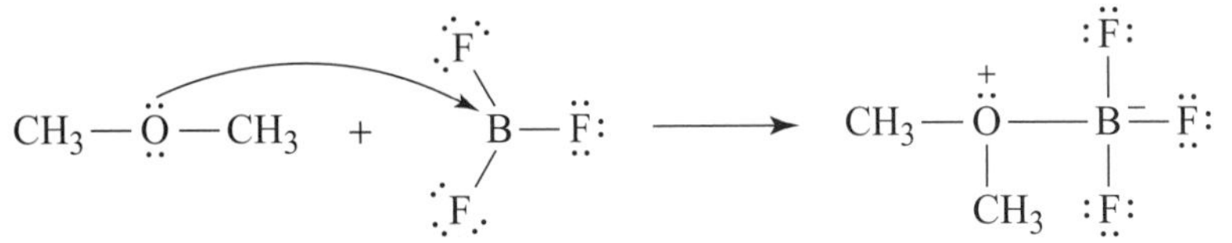 $CH_3—\ddot{O}—CH_3 + BF_3 \longrightarrow CH_3—\overset{+}{O}(CH_3)—\overset{-}{B}F_3$

3.2 (a) Lewis base

(b) Lewis acid

(c) Lewis base

(d) Lewis base

(e) Lewis acid

(f) Lewis base

3.3 $CH_3—\ddot{N}H—CH_3 + BF_3 \longrightarrow CH_3—\overset{+}{N}H(CH_3)—\overset{-}{B}F_3$

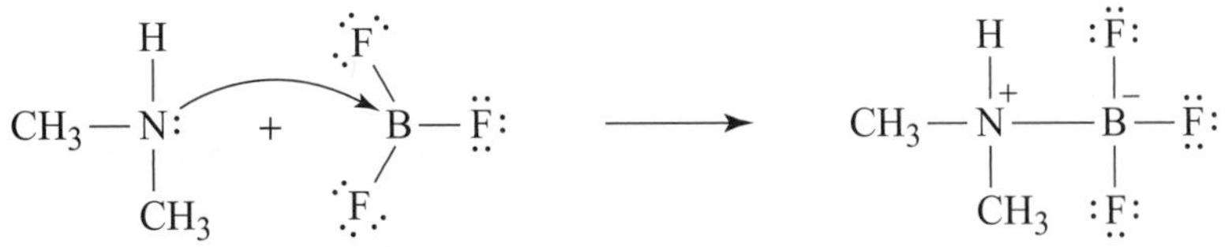

Lewis base   Lewis acid

3.4 (a) $K_a = \dfrac{[H_3O^+][HCO_2^{\;-}]}{[HCO_2H]} = 1.77 \times 10^{-4}$

Let $x = [H_3O^+] = [HCO_2^{\;-}]$ at equilibrium
then, $0.1 - x = [HCO_2H]$ at equilibrium
but, since the K_a is very small, x will be very small and $0.1 - x \simeq 0.1$

Therefore,

$$\frac{(x)(x)}{0.1} = 1.77 \times 10^{-4}$$
$$x^2 = 1.77 \times 10^{-5}$$
$$x = 0.0042 = [H_3O^+] = [HCO_2^-]$$

(b) $\% \text{ Ionized} = \frac{[H_3O^+]}{0.1} \times 100 \quad \text{or} \quad \frac{[HCO_2^-]}{0.1} \times 100$

$$= \frac{.0042}{0.1} \times 100 = 4.2\%$$

3.5 (a) $pK_a = -\log 10^{-7} = -(-7) = 7$

(b) $pK_a = -\log 5.0 = -0.669$

(c) Since the acid with a $K_a = 5$ has a larger K_a, it is the stronger acid.

3.6 When H_3O^+ acts as an acid in aqueous solution, the equation is

$$H_3O^+ + H_2O \rightleftharpoons H_2O + H_3O^+$$

and K_a is

$$K_a = \frac{[H_2O][H_3O^+]}{[H_3O^+]} = [H_2O]$$

The molar concentration of H_2O in pure H_2O, that is $[H_2O] = 55.5$; therefore, $K_a = 55.5$

The pK_a is

$$pK_a = -\log 55.5 = -1.74$$

3.7 (a) 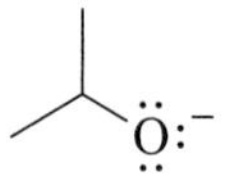 (c) 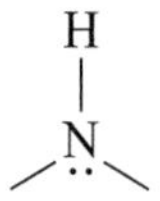 H N

(b) 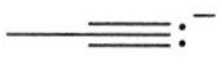 (d) 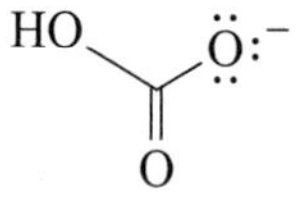 HO Ö:⁻ O

3.8 The pK_a of the methylaminium ion is equal to 10.6 (Section 3.6C). Since the pK_a of the anilinium ion is equal to 4.6, the anilinium ion is a stronger acid than the methylaminium ion, and aniline ($C_6H_5NH_2$) is a weaker base than methylamine (CH_3NH_2).

3.9 + $:\ddot{N}H_2^-$ ⟶ $:^-$ + $:NH_3$

3.10 R—C(=O)—O—H + Na^+ $^-$:O—C(=O)—OH ⟶ R—C(=O)—O:$^-$ Na^+ + HO—C(=O)—OH

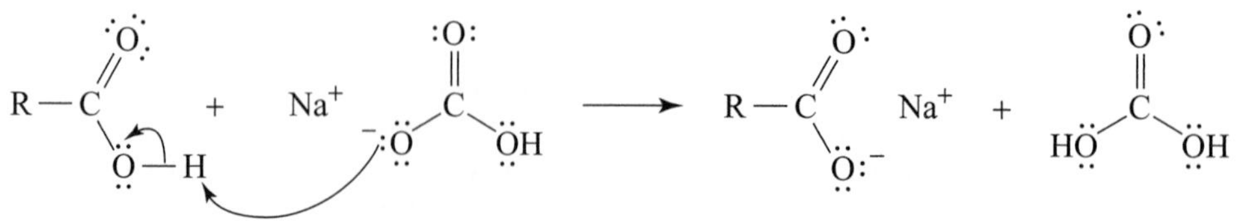

3.11 (a) Negative. Because the atoms are constrained to one molecule in the product, they have to become more ordered.

(b) Approximately zero.

(c) Positive. Because the atoms are in two separate product molecules, they become more disordered.

3.12 (a) If $K_{eq} = 1$
then,

$$\log K_{eq} = 0 = \frac{-\Delta G^\circ}{2.303RT}$$

$$\Delta G^\circ = 0$$

(b) If $K_{eq} = 10$
then,

$$\log K_{eq} = 1 = \frac{-\Delta G^\circ}{2.303RT}$$

$$\Delta G^\circ = -(2.303)(0.008314 \text{ kJ mol}^{-1} \text{ K}^{-1})(298 \text{ K}) = -5.71 \text{ kJ mol}^{-1}$$

(c) $\Delta G^\circ = \Delta H^\circ - T\Delta S^\circ$

$\Delta G^\circ = \Delta H^\circ = -5.71 \text{ kJ mol}^{-1}$ if $\Delta S^\circ = 0$

3.13 Structures **A** and **B** make equal contributions to the overall hybrid. This means that the carbon-oxygen bonds should be the same length and that the oxygens should bear equal negative charges.

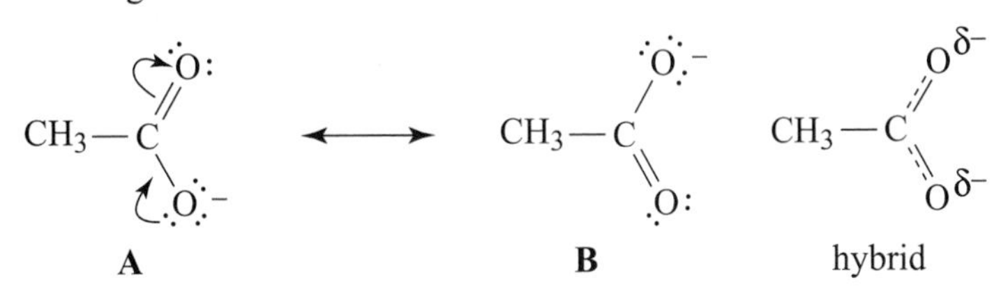

3.14 (a) $CHCl_2CO_2H$ would be the stronger acid because the electron-withdrawing inductive effect of *two chlorine atoms* would make its hydroxyl proton more positive. The electron-withdrawing effect of the two chlorine atoms would also stabilize the dichloroacetate ion more effectively by dispersing its negative charge more extensively.

(b) CCl_3CO_2H would be the stronger acid for reasons similar to those given in (a), except here there are three versus two electron-withdrawing chlorine atoms involved.

(c) CH_2FCO_2H would be the stronger acid because the electron-withdrawing effect of a fluorine atom is greater than that of a bromine atom (fluorine is more electronegative).

(d) CH_2FCO_2H is the stronger acid because the fluorine atom is nearer the carboxyl group and is, therefore, better able to exert its electron-withdrawing inductive effect. (*Remember:* Inductive effects weaken steadily as the distance between the substituent and the acidic group increases.)

3.15 All compounds containing oxygen and most compounds containing nitrogen will have an unshared electron pair on their oxygen or nitrogen atom. These compounds can, therefore, act as bases and accept a proton from concentrated sulfuric acid. When they accept a proton, these compounds become either oxonium ions or ammonium ions, and having become ionic, they are soluble in the polar medium of sulfuric acid. The only nitrogen compounds that do not have an electron pair on their nitrogen atom are quaternary ammonium compounds, and these, already being ionic, also dissolve in the polar medium of concentrated sulfuric acid.

3.16 (a) $CH_3\ddot{O}{-}H + :H^- \xrightarrow{\text{methanol}} CH_3\ddot{O}:^- + H_2$

Stronger acid ($pK_a = 16$); Stronger base (from NaH); Weaker base; Weaker acid ($pK_a = 35$)

(b) $CH_3CH_2\ddot{O}{-}H + :\ddot{N}H_2^- \xrightarrow{\text{ethanol}} CH_3CH_2\ddot{O}:^- + :NH_3$

Stronger acid ($pK_a = 16$); Stronger base (from $NaNH_2$); Weaker base; Weaker acid ($pK_a = 38$)

(c) $H{-}\ddot{N}(H){-}H + {}^-:CH_2CH_3 \xrightarrow{\text{hexane}} :\ddot{N}H_2^- + CH_3CH_3$

Stronger acid ($pK_a = 38$); Stronger base (from CH_3CH_2Li); Weaker base; Weaker acid ($pK_a = 50$)

(d) $H{-}\overset{+}{N}H_2{-}H + :\ddot{N}H_2^- \xrightarrow{\text{liq. } NH_3} :NH_3 + :NH_3$

Stronger acid ($pK_a = 9.2$) (from NH_4Cl); Stronger base (from $NaNH_2$); Weaker base; Weaker acid ($pK_a = 38$)

(e) $H{-}\ddot{O}{-}H + {}^-:\ddot{O}C(CH_3)_3 \xrightarrow{H_2O} H{-}\ddot{O}:^- + HOC(CH_3)_3$

Stronger acid ($pK_a = 15.7$); Stronger base [from $(CH_3)_3CONa$]; Weaker base; Weaker acid ($pK_a = 18$)

(f) No appreciable acid-base reaction would occur because HO^- is not a strong enough base to remove a proton from $(CH_3)_3COH$.

3.17 (a) $HC{\equiv}CH + NaH \xrightarrow{\text{hexane}} HC{\equiv}CNa + H_2$

(b) $HC{\equiv}CNa + D_2O \longrightarrow HC{\equiv}CD + NaOD$

(c) $CH_3CH_2Li + D_2O \xrightarrow{\text{hexane}} CH_3CH_2D + LiOD$

(d) $CH_3CH_2OH + NaH \xrightarrow{\text{hexane}} CH_3CH_2ONa + H_2$

(e) $CH_3CH_2ONa + T_2O \longrightarrow CH_3CH_2OT + NaOT$

(f) $CH_3CH_2CH_2Li + D_2O \xrightarrow{\text{hexane}} CH_3CH_2CH_2D + LiOD$

Problems

Brønsted-Lowry Acids and Bases

3.18 (a) $:\ddot{N}H_2^-$ (the amide ion)

(b) $H-\ddot{\underset{..}{O}}:^-$ (the hydroxide ion)

(c) $:H^-$ (the hydride ion)

(d) $H-C\equiv C:^-$ (the ethynide ion)

(e) $CH_3\ddot{\underset{..}{O}}:^-$ (the methoxide ion)

(f) $H_2\ddot{\underset{..}{O}}$ (water)

3.19 $:\ddot{N}H_2^- > :H^- > H-C\equiv C:^- > CH_3\ddot{\underset{..}{O}}:^- \approx H-\ddot{\underset{..}{O}}:^- > H_2O$

3.20 (a) H_2SO_4

(b) H_3O^+

(c) $CH_3NH_3^+$

(d) NH_3

(e) CH_3CH_3

(f) CH_3CO_2H

3.21 $H_2SO_4 > H_3O^+ > CH_3CO_2H > CH_3NH_3^+ > NH_3 > CH_3CH_3$

Lewis Acids and Bases

3.22 (a) $CH_3CH_2-\ddot{\underset{..}{Cl}}: + AlCl_3 \longrightarrow CH_3CH_2-\overset{+}{\ddot{\underset{..}{Cl}}}-\overset{-}{Al}(:\ddot{\underset{..}{Cl}}:)_3$

Lewis base (CH₃CH₂Cl); Lewis acid (AlCl₃)

(b) $CH_3-\ddot{\underset{..}{O}}H + BF_3 \longrightarrow CH_3-\overset{+}{\ddot{O}}(H)-\overset{-}{B}(:\ddot{\underset{..}{F}}:)_3$

Lewis base (CH_3OH); Lewis acid (BF_3)

(c) $(CH_3)_3C^+ + H_2\ddot{O}: \longrightarrow (CH_3)_3C-\ddot{O}H_2^+$

Lewis acid ($(CH_3)_3C^+$); Lewis base (H_2O)

Curved-Arrow Notation

3.23 (a) $CH_3—\ddot{O}H + H—\ddot{I}: \longrightarrow CH_3—\ddot{O}^+(H)—H + :\ddot{I}:^-$

(b) $CH_3—\ddot{N}H_2 + H—\ddot{Cl}: \longrightarrow CH_3—N^+H_3 + :\ddot{Cl}:^-$

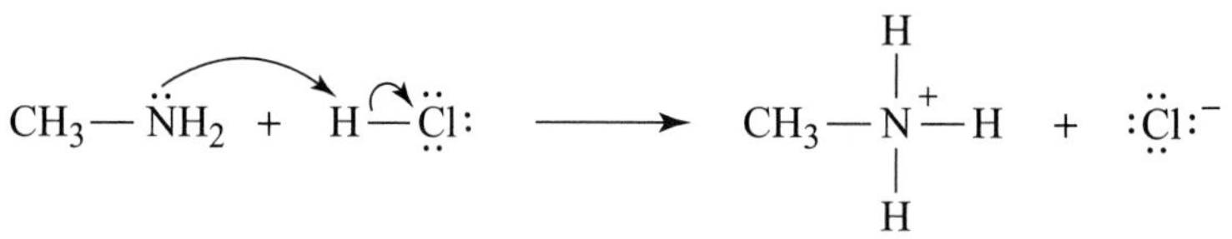

(c) $H_2C{=}CH_2 + H—\ddot{F}: \longrightarrow H_2C^+—CH_3 + :\ddot{F}:^-$

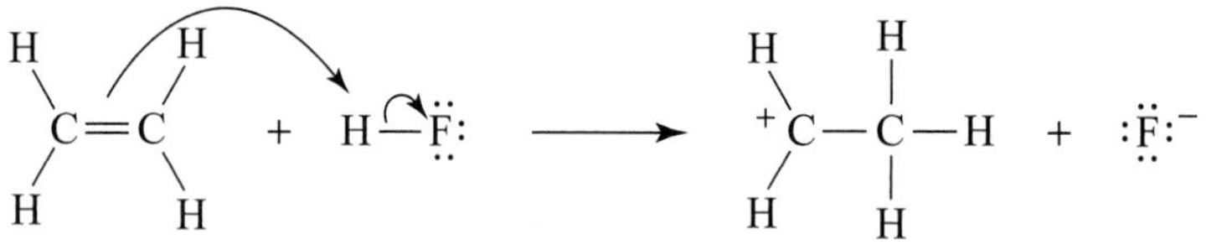

3.24 (a) $(CH_3)_2C{=}\ddot{O}: + BF_3 \longrightarrow (CH_3)_2C{=}\ddot{O}^+:\bar{B}F_3$

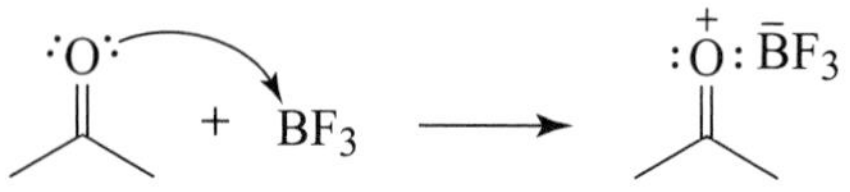

(b) $CH_3\ddot{O}CH_3 + BF_3 \longrightarrow CH_3\dot{O}^+(\bar{B}F_3)CH_3$

(c) $CH_3C(=\ddot{O}:)OH + H—Cl \longrightarrow CH_3C(=\ddot{O}^+H)OH\ \ Cl^-$

(d) $CH_3CH_2\ddot{O}—H + CH_3CH_2CH_2CH_2—Li \longrightarrow CH_3CH_2\ddot{O}:^-\ Li^+ + CH_3CH_2CH_2CH_3$

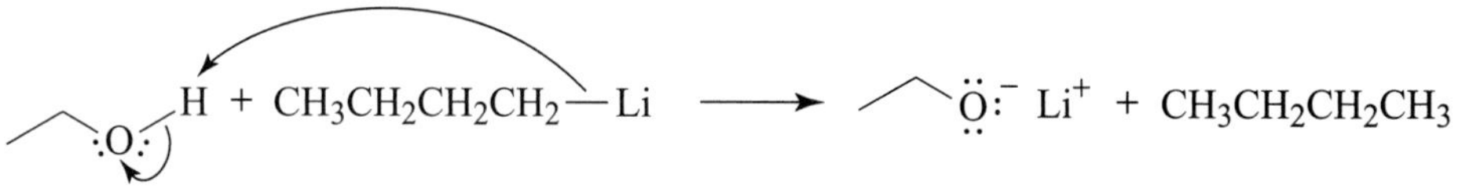

3.25 (a) $CH_3CH_2—C(=O)—\ddot{O}—H + {}^-:\ddot{O}—H \longrightarrow CH_3CH_2—C(=O)—\ddot{O}:^- + H—\ddot{O}—H$

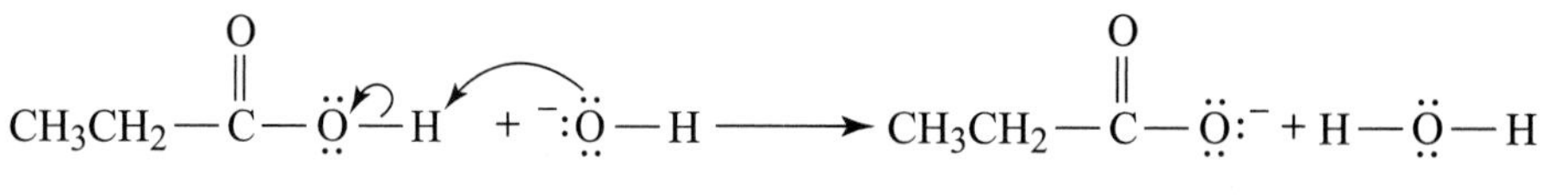

(b) $C_6H_5—S(=O)_2—\ddot{O}—H + {}^-:\ddot{O}—H \longrightarrow C_6H_5—S(=O)_2—\ddot{O}:^- + H—\ddot{O}—H$

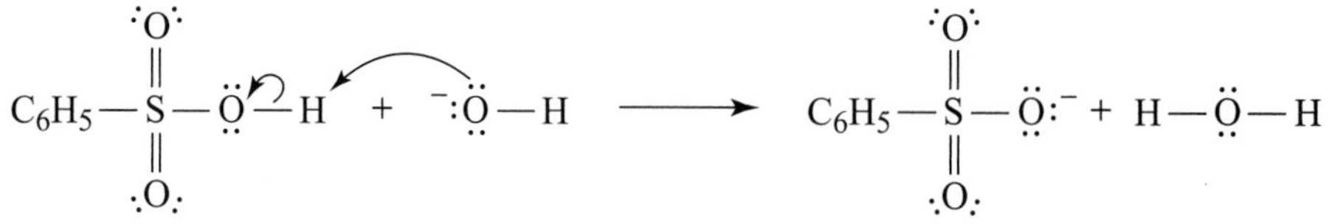

(c) No appreciable acid-base reaction takes place because CH_3CH_2ONa is too weak a base to remove a proton from ethyne.

(d) $H—C{\equiv}C—H + {}^-:CH_2CH_3$ (from $LiCH_2CH_3$) $\xrightarrow{\text{hexane}} H—C{\equiv}C:^- + CH_3CH_3$

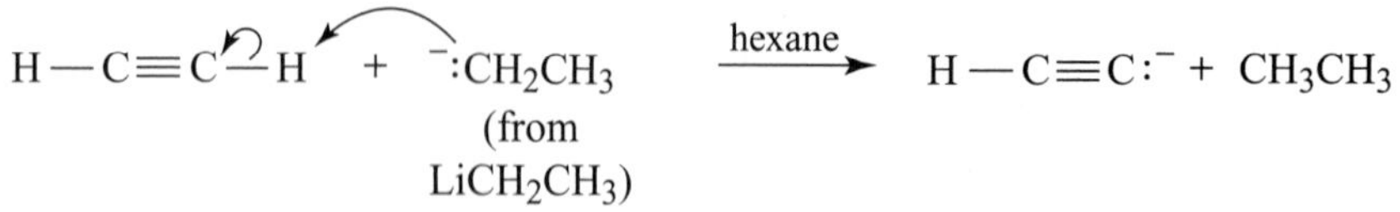

(e) $CH_3—CH_2—\ddot{O}—H + {}^-:CH_2CH_3$ (from $LiCH_2CH_3$) $\xrightarrow{\text{hexane}} CH_3—CH_2—\ddot{O}:^- + CH_3CH_3$

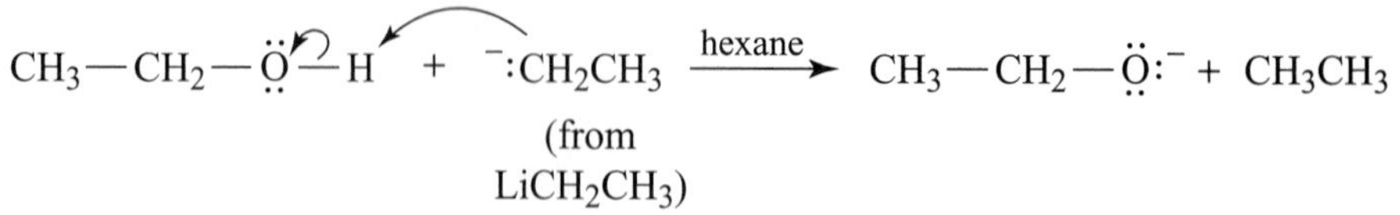

Acid-Base Strength and Equilibrium

3.26 Because the proton attached to the highly electronegative oxygen atom of CH_3OH is much more acidic than the protons attached to the much less electronegative carbon atom.

3.27 $$CH_3CH_2{-}\ddot{\underset{\cdot\cdot}{O}}{-}H + {}^{-}{:}C{\equiv}C{-}H \xrightarrow{\text{liq. } NH_3} CH_3CH_2{-}\ddot{\underset{\cdot\cdot}{O}}{:}^{-} + H{-}C{\equiv}C{-}H$$

3.28 (a) $pK_a = -\log 1.77 \times 10^{-4} = 4 - 0.248 = 3.75$

(b) $K_a = 10^{-13}$

3.29 (a) HB is the stronger acid because it has the smaller pK_a.

(b) Yes. Since A^- is the stronger base and HB is the stronger acid, the following acid-base reaction will take place.

$$\underset{\substack{\text{Stronger}\\\text{base}}}{A{:}^{-}} + \underset{\substack{\text{Stronger}\\\text{acid}\\pK_a = 10}}{H{-}B} \longrightarrow \underset{\substack{\text{Weaker}\\\text{acid}\\pK_a = 20}}{A{-}H} + \underset{\substack{\text{Weaker}\\\text{base}}}{B{:}^{-}}$$

3.30 (a) $C_6H_5{-}C{\equiv}C{-}H + NaNH_2 \xrightarrow{\text{ether}} C_6H_5{-}C{\equiv}C{:}^{-}\ Na^{+} + NH_3$

then

$C_6H_5{-}C{\equiv}C{:}^{-}\ Na^{+} + T_2O \longrightarrow C_6H_5{-}C{\equiv}C{-}T + NaOT$

(b) $CH_3{-}\underset{\substack{|\\CH_3}}{CH}{-}O{-}H + NaH \longrightarrow CH_3{-}\underset{\substack{|\\CH_3}}{CH}{-}O^{-}Na^{+} + H_2$

then

$CH_3{-}\underset{\substack{|\\CH_3}}{CH}{-}O^{-}Na^{+} + D_2O \longrightarrow CH_3{-}\underset{\substack{|\\CH_3}}{CH}{-}O{-}D + NaOD$

(c) $CH_3CH_2CH_2OH + NaH \longrightarrow CH_3CH_2CH_2O^{-}Na^{+} + H_2$

then

$CH_3CH_2CH_2O^{-}Na^{+} + D_2O \longrightarrow CH_3CH_2CH_2OD + NaOD$

3.31 (a) $CH_3CH_2OH > CH_3CH_2NH_2 > CH_3CH_2CH_3$

Oxygen is more electronegative than nitrogen, which is more electronegative than carbon. The O-H bond is most polarized, the N-H bond is next, and the C-H bond is least polarized.

(b) $CH_3CH_2O^{-} < CH_3CH_2NH^{-} < CH_3CH_2CH_2^{-}$

The weaker the acid, the stronger the conjugate base.

3.32 (a) $CH_3C{\equiv}CH > CH_3CH{=}CH_2 > CH_3CH_2CH_3$

(b) $CH_3CHClCO_2H > CH_3CH_2CO_2H > CH_3CH_2CH_2OH$

(c) $CH_3CH_2OH_2^+ > CH_3CH_2OH > CH_3OCH_3$

3.33 (a) $CH_3NH_3^+ < CH_3NH_2 < CH_3NH^-$

(b) $CH_3O^- < CH_3NH^- < CH_3CH_2^-$

(c) $CH_3C{\equiv}C^- < CH_3CH{=}CH^- < CH_3CH_2CH_2^-$

General Problems

3.34 The acidic hydrogens must be attached to oxygen atoms. In H_3PO_3, one hydrogen is bonded to a phosphorus atom:

$$H{-}\ddot{O}{-}P(=O)(OH){-}\ddot{O}{-}H \qquad H{-}\ddot{O}{-}P(=O)(OH){-}H$$

3.35 (a) $HC(=O)OH + {}^-{:}\ddot{O}{-}H \longrightarrow HC(=O)O^- + H{-}\ddot{O}{-}H$

(b) $HC(=O)OCH_3 + {}^-{:}\ddot{O}{-}H \longrightarrow H{-}C(O^-)(OCH_3){-}\ddot{O}{-}H$

(c) $H{-}C(O^-)(OCH_3){-}\ddot{O}{-}H \longrightarrow HC(=O){-}\ddot{O}{-}H + {}^-{:}\ddot{O}{-}CH_3$

(d) $H{-}\ddot{O}{:}^- + CH_3{-}\ddot{I}{:} \longrightarrow H{-}\ddot{O}{-}CH_3 + {:}\ddot{I}{:}^-$

(e) $H{-}\ddot{O}{:}^- + H{-}CH_2{-}C(CH_3)_2{-}\ddot{Cl}{:} \longrightarrow CH_2{=}C(CH_3)_2 + {:}\ddot{Cl}{:}^- + H{-}\ddot{O}{-}H$

3.36 (a) Assume that the acidic and basic groups of glycine in its two forms have acidities and basicities similar to those of acetic acid and methylamine. Then consider the equilibrium between the two forms:

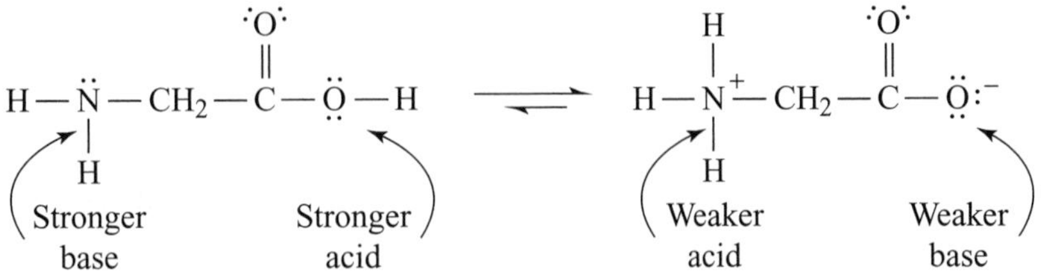

$$H-\ddot{N}H-CH_2-C(=O)-\ddot{O}-H \rightleftharpoons H-N^{+}H_2-CH_2-C(=O)-\ddot{O}:^{-}$$

Stronger base, Stronger acid ⇌ Weaker acid, Weaker base

We see that the ionic form contains the groups that are the weaker acid and weaker base. The equilibrium, therefore, will favor this form.

(b) The high melting point shows that the ionic structure better represents glycine.

3.37 (a) The second carboxyl group of malonic acid acts as an electron-withdrawing group and stabilizes the conjugate base formed (i.e., $HO_2CCH_2CO_2^-$) when malonic acid loses a proton. [Any factor that stabilizes the conjugate base of an acid always increases the strength of the acid (Section 3.11C).] An important factor here may be an entropy effect as explained in Section 3.10.

(b) When $^-O_2CCH_2CO_2H$ loses a proton, it forms a dianion, $^-O_2CCH_2CO_2^-$. This dianion is destabilized by having two negative charges in close proximity.

3.38 HB is the stronger acid.

3.39

$$\begin{aligned}
\Delta G^\circ &= \Delta H^\circ - T\Delta S^\circ \\
&= 6.3 \text{ kJ mol}^{-1} - (298 \text{ K})(0.0084 \text{ kJ mol}^{-1}\text{K}^{-1}) \\
&= 3.8 \text{ kJ mol}^{-1} \\
\log K_{eq} &= \log K_a = -\text{p}K_a = -\frac{\Delta G^\circ}{2.303RT} \\
\text{p}K_a &= \frac{\Delta G^\circ}{2.303RT} \\
\text{p}K_a &= \frac{3.8 \text{ kJ mol}^{-1}}{(2.303)(0.008314 \text{ kJ mol}^{-1}\text{K}^{-1})(298 \text{ K})} \\
\text{p}K_a &= 0.66
\end{aligned}$$

3.40 The dianion is a hybrid of the following resonance structures:

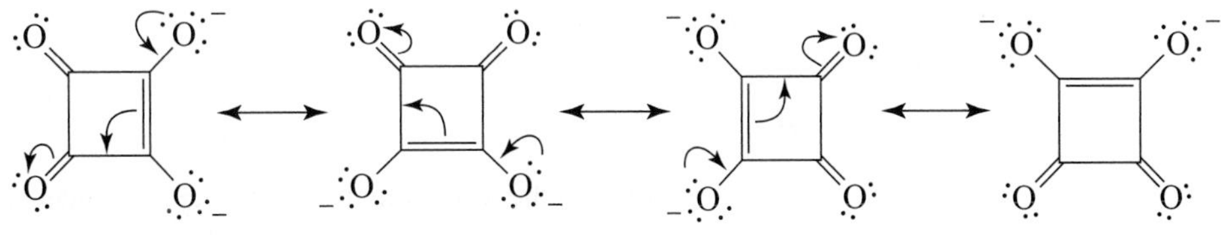

If we mentally fashion a hybrid of these structures, we see that each carbon-carbon bond is a single bond in three structures and a double bond in one. Each carbon-oxygen bond is a double bond in two structures and a single bond in two structures. Therefore, we would expect all of the carbon-carbon bonds to be equivalent and of the same length, and exactly the same can be said for the carbon-oxygen bonds.

Challenge Problems

3.41 (a) *A* is $CH_3CH_2S^-$ *B* is CH_3OH

C is $CH_3CH_2SCH_2CH_2O^-$ *D* is $CH_3CH_2SCH_2CH_2OH$

E is OH^-

(b) $CH_3CH_2{-}\ddot{S}{-}H + CH_3{-}\ddot{O}{:}^- \longrightarrow CH_3CH_2{-}\ddot{S}{:}^- + CH_3{-}\ddot{O}{-}H$

$CH_3CH_2{-}\ddot{S}{:}^- + CH_2{-}CH_2 \text{ (bridged by } {:}\ddot{O}{:}) \longrightarrow CH_3CH_2{-}\ddot{S}{-}CH_2CH_2{-}\ddot{O}{:}^-$

$CH_3CH_2{-}\ddot{S}{-}CH_2CH_2{-}\ddot{O}{:}^- + H{-}\ddot{O}{-}H \longrightarrow$

$CH_3CH_2{-}\ddot{S}{-}CH_2CH_2{-}\ddot{O}{-}H + H{-}\ddot{O}{:}^-$

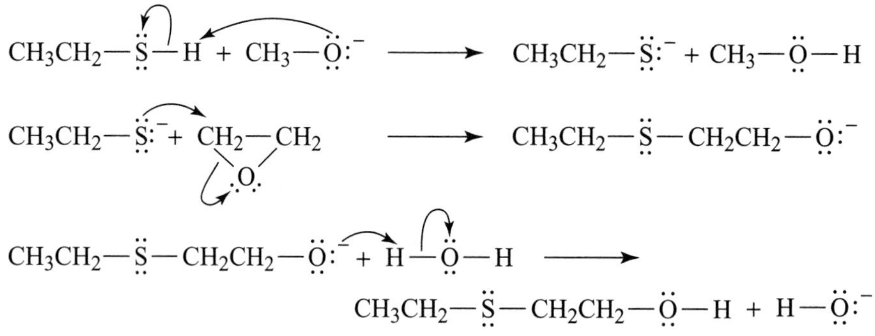

3.42 (a) $CH_3(CH_2)_8OD + CH_3(CH_2)_8Li \longrightarrow CH_3(CH_2)_8O^- \, Li^+ + CH_3(CH_2)_8D$

Hexane could be used as solvent. Liquid ammonia and ethanol could not because they would compete with $CH_3(CH_2)_8OD$ and generate mostly non-deuterio-labelled $CH_3(CH_2)_7CH_3$.

(b) $NH_2^- + CH_3C{\equiv}CH \longrightarrow NH_3 + CH_3C{\equiv}C{:}^-$

Hexane or liquid ammonia could be used; ethanol is too acidic and would lead to $CH_3CH_2O^-$ (ethoxide ion) instead of the desired alkynide ion.

(c) $HCl + C_6H_5{-}\ddot{N}H_2 \longrightarrow C_6H_5{-}\overset{+}{N}H_3 + Cl^-$

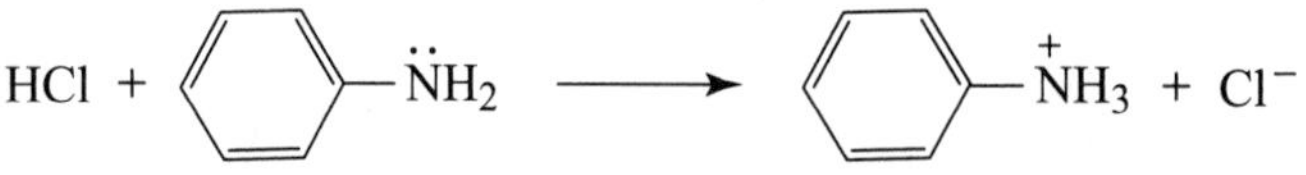

Hexane or ethanol could be used; liquid ammonia is too strong a base and would lead to NH_4^+ instead of the desired anilinium ion.

3.43 (a,b) $H{-}C({=}\ddot{O}{:}){-}\ddot{N}(CH_3)_2 \longleftrightarrow H{-}C({-}{:}\ddot{O}{:}^-){=}\overset{+}{N}(CH_3)_2$

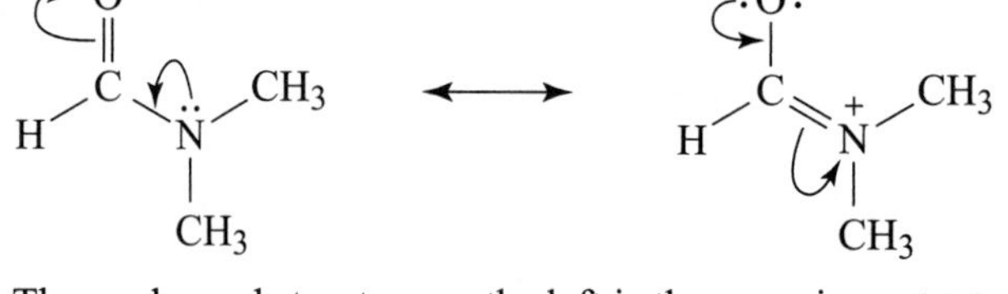

The uncharged structure on the left is the more important resonance form.

(c) Since DMF does not bind with (solvate) anions, their electron density remains high and their size small, both of which make nucleophiles more reactive.

3.44 (a) 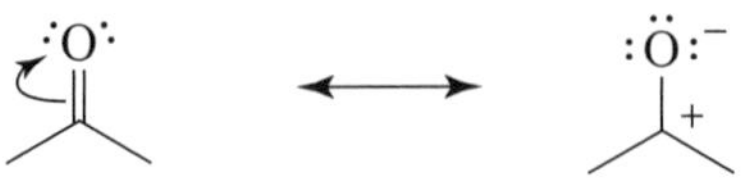 $\ddot{O}$ ⟷ $:\ddot{O}:^-$ $+$

(b)  $\ddot{O}$ $:^-$ ⟷ $:\ddot{O}:^-$

(c) 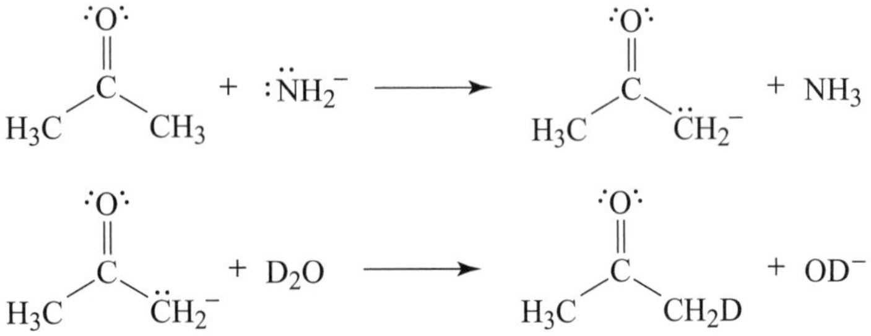 $$H_3C{-}C({=}\ddot{O}){-}CH_3 + :\ddot{N}H_2^- \longrightarrow H_3C{-}C({=}\ddot{O}){-}\ddot{C}H_2^- + NH_3$$

$$H_3C{-}C({=}\ddot{O}){-}\ddot{C}H_2^- + D_2O \longrightarrow H_3C{-}C({=}\ddot{O}){-}CH_2D + OD^-$$

3.45 The most acidic hydrogen atoms in formamide are bonded to the nitrogen atom. They are acidic due to the electron-withdrawing effect of the carbonyl group and the fact that the resulting conjugate base can be stabilized by resonance delocalization of the negative charge into the carbonyl group. The electrostatic potential map shows deep blue color near the hydrogen atoms bonded to the nitrogen atom, consistent with their relative acidity.

QUIZ

3.1 Which of the following is the strongest acid?

(a) $CH_3CH_2CO_2H$ (b) CH_3CH_3 (c) CH_3CH_2OH (d) $CH_2{=}CH_2$

3.2 Which of the following is the strongest base?

(a) CH_3ONa (b) $NaNH_2$ (c) CH_3CH_2Li (d) $NaOH$ (e) CH_3CO_2Na

3.3 Dissolving $NaNH_2$ in water will give:

(a) A solution containing solvated Na^+ and NH_2^- ions.

(b) A solution containing solvated Na^+ ions, OH^- ions, and NH_3.

(c) NH_3 and metallic Na.

(d) Solvated Na^+ ions and hydrogen gas.

(e) None of the above.

3.4 Which base is strong enough to convert $(CH_3)_3COH$ into $(CH_3)_3CONa$ in a reaction that goes to completion?

(a) $NaNH_2$ (b) CH_3CH_2Na (c) NaOH (d) CH_3CO_2Na

(e) More than one of the above.

3.5 Which would be the strongest acid?

(a) $CH_3CH_2CH_2CO_2H$ (b) $CH_3CH_2CHFCO_2H$ (c) $CH_3CHFCH_2CO_2H$

(d) $CH_2FCH_2CH_2CO_2H$ (e) $CH_3CH_2CH_2CH_2OH$

3.6 Which would be the weakest base?

(a) CH_3CO_2Na (b) CF_3CO_2Na (c) CHF_2CO_2Na (d) CH_2FCO_2Na

3.7 What acid-base reaction (if any) would occur when NaF is dissolved in H_2SO_4?

3.8 The pK_a of $CH_3NH_3^+$ equals 10.6; the pK_a of $(CH_3)_2NH_2^+$ equals 10.7. Which is the stronger base, CH_3NH_2 or $(CH_3)_2NH$?

3.9 Supply the missing reagents.

$CH_3CH_2C{\equiv}CH$ + (a) [ ] $\xrightarrow{\text{hexane}}$ $CH_3CH_2C{\equiv}C{:}^- \; Li^+$ + CH_3CH_3

$\downarrow$ (b) [ ]

$CH_3CH_2C{\equiv}CD$ + LiOD

3.10 Supply the missing intermediates and reagents.

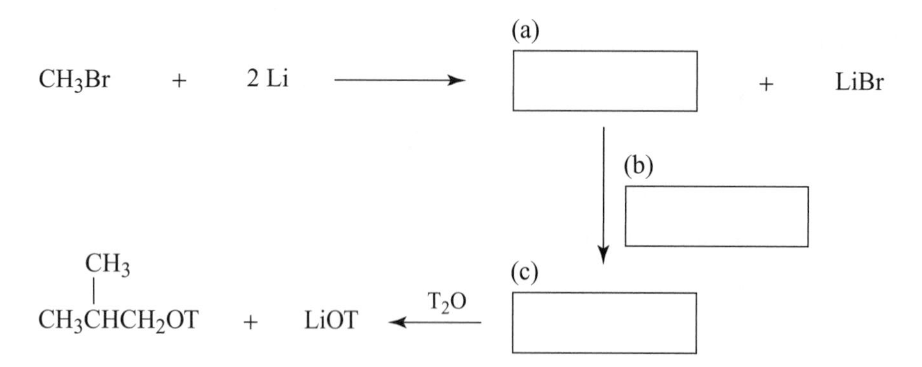

4 NOMENCLATURE AND CONFORMATIONS OF ALKANES AND CYCLOALKANES

SOLUTIONS TO PROBLEMS

4.1

or

$CH_3(CH_2)_5CH_3$
Heptane

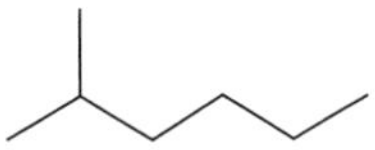

or

$(CH_3)_2CHCH_2CH_2CH_2CH_3$
2-Methylhexane

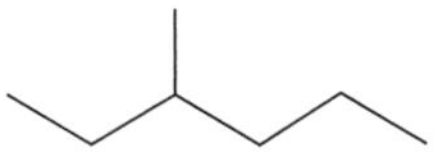

or

$CH_3CH_2CH(CH_3)CH_2CH_2CH_3$
3-Methylhexane

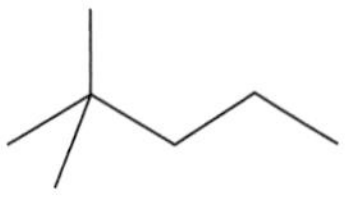

or

$(CH_3)_3CCH_2CH_2CH_3$
2,2-Dimethylpentane

or

$(CH_3CH_2)_2C(CH_3)_2$
3,3-Dimethylpentane

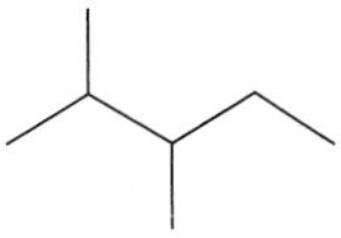

or

$(CH_3)_2CHCH(CH_3)CH_2CH_3$
2,3-Dimethylpentane

or

$(CH_3)_2CHCH_2CH(CH_3)_2$
2,4-Dimethylpentane

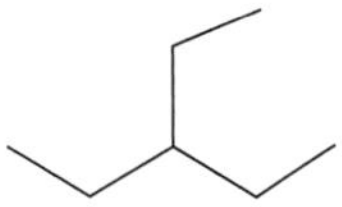

or

$(CH_3CH_2)_3CH$
3-Ethylpentane

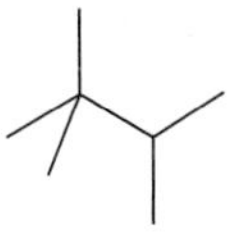

or

$(CH_3)_3CCH(CH_3)_2$
2,2,3-Trimethylbutane

4.2 (d); it represents 3-methylpentane

4.3 $CH_3CH_2CH_2CH_2CH_2CH_3$ hexane

4.4 (a,b)

2-Methylheptane 3-Methylheptane 4-Methylheptane

2,2-Dimethylhexane 2,3-Dimethylhexane 2,4-Dimethylhexane

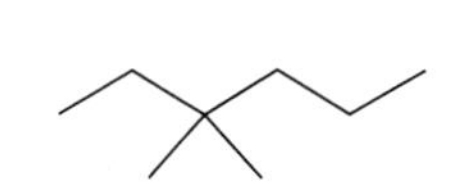

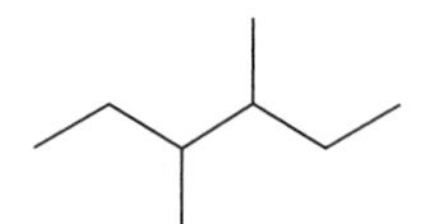

2,5-Dimethylhexane 3,3-Dimethylhexane 3,4-Dimethylhexane

3-Ethylhexane

2,2,3-Trimethylpentane 2,2,4-Trimethylpentane

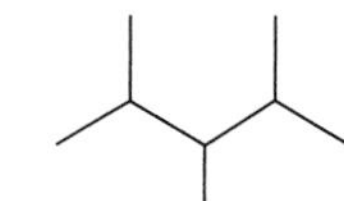

2,3,3-Trimethylpentane 2,3,4-Trimethylpentane

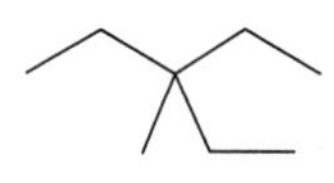

3-Ethyl-2-methylpentane 3-Ethyl-3-methylpentane

2,2,3,3-Tetramethylbutane

4.5 (a)

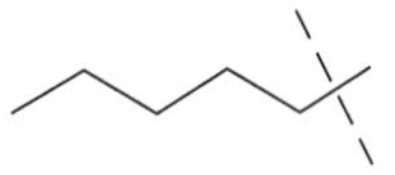

Pentyl

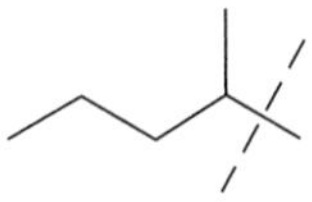

1-Methylbutyl

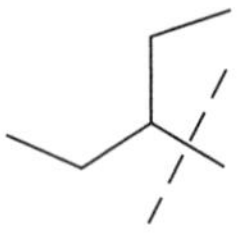

1-Ethylpropyl

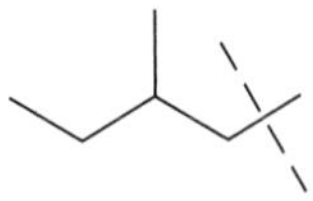

2-Methylbutyl

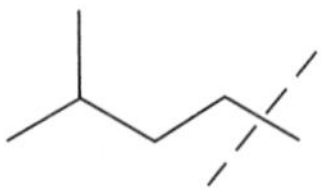

3-Methylbutyl

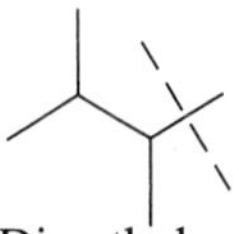

1,2-Dimethylpropyl

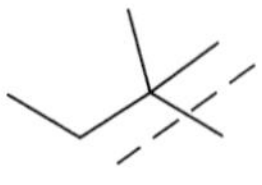

1,1-Dimethylpropyl

(b) See the answer to Review Problem 4.1 for the formulas and names of C_7H_{16} isomers.

4.6 (a)

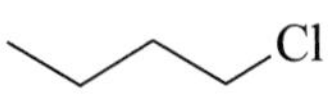

Cl

1-Chlorobutane

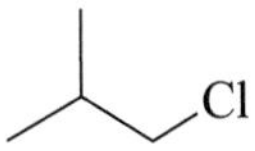

Cl

1-Chloro-2-methylpropane

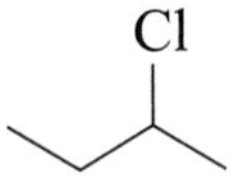

Cl

2-Chlorobutane

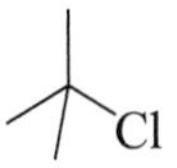

Cl

2-Chloro-2-methylpropane

(b)

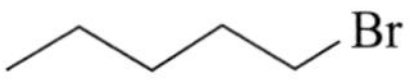

Br

1-Bromopentane

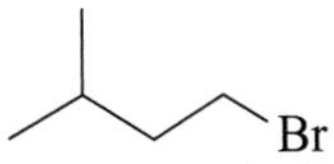

Br

1-Bromo-3-methylbutane

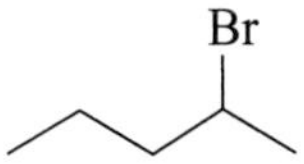

Br

2-Bromopentane

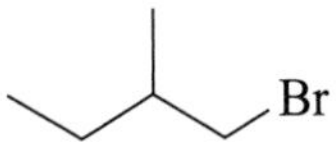

Br

1-Bromo-2-methylbutane

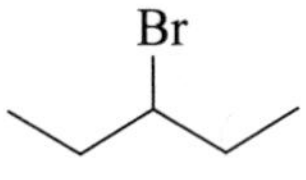

Br

3-Bromopentane

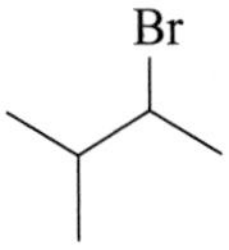

Br

2-Bromo-3-methylbutane

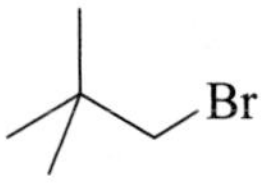

Br

1-Bromo-2,2-dimethylpropane

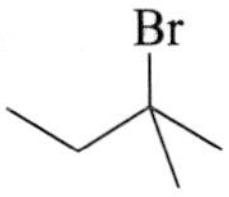

Br

2-Bromo-2-methylbutane

4.7 (a)

1-Butanol

2-Methyl-1-propanol

2-Butanol

2-Methyl-2-propanol

(b)

1-Pentanol

3-Methyl-1-butanol

2-Pentanol

2-Methyl-1-butanol

3-Pentanol

3-Methyl-2-butanol

2,2-Dimethyl-1-propanol

2-Methyl-2-butanol

4.8 (a) 1,1-Dimethylethylcyclopentane or *tert*-butylcyclopentane

(b) 1-Methyl-2-(2-methylpropyl)cyclohexane or 1-isobutyl-2-methylcyclohexane

(c) Butylcyclohexane

(d) 1-Chloro-2,4-dimethylcyclohexane

(e) 2-Chlorocyclopentanol

(f) 3-(1,1-Dimethylethyl)cyclohexanol or 3-*tert*-butylcyclohexanol

4.9 (a) 2-Chlorobicyclo[1.1.0]butane

(b) Bicyclo[3.2.1]octane

(c) Bicyclo[2.1.1]hexane

(d) 9-Chlorobicyclo[3.3.1]nonane

(e) 2-Methylbicyclo[2.2.2]octane

(f) Bicyclo[3.1.0]hexane or

bicyclo[2.1.1]hexane

4.10 (a) *trans*-3-Heptene
(b) 2,5-Dimethyl-2-octene
(c) 4-Ethyl-2-methyl-l-hexene
(d) 3,5-Dimethylcyclohexene
(e) 4-Methyl-4-penten-2-ol
(f) 2-Chloro-3-methylcyclohex-3-en-1-ol

4.11 (a) (b) (c)

(d) Cl (e) Br Br (f) Br

(g) (h) (i) Cl Cl

(j) Cl Cl

4.12

1-Hexyne 2-Hexyne 3-Hexyne 3,3-Dimethyl-1-butyne

3-Methyl-1-pentyne 4-Methyl-1-pentyne 4-Methyl-2-pentyne

4.13

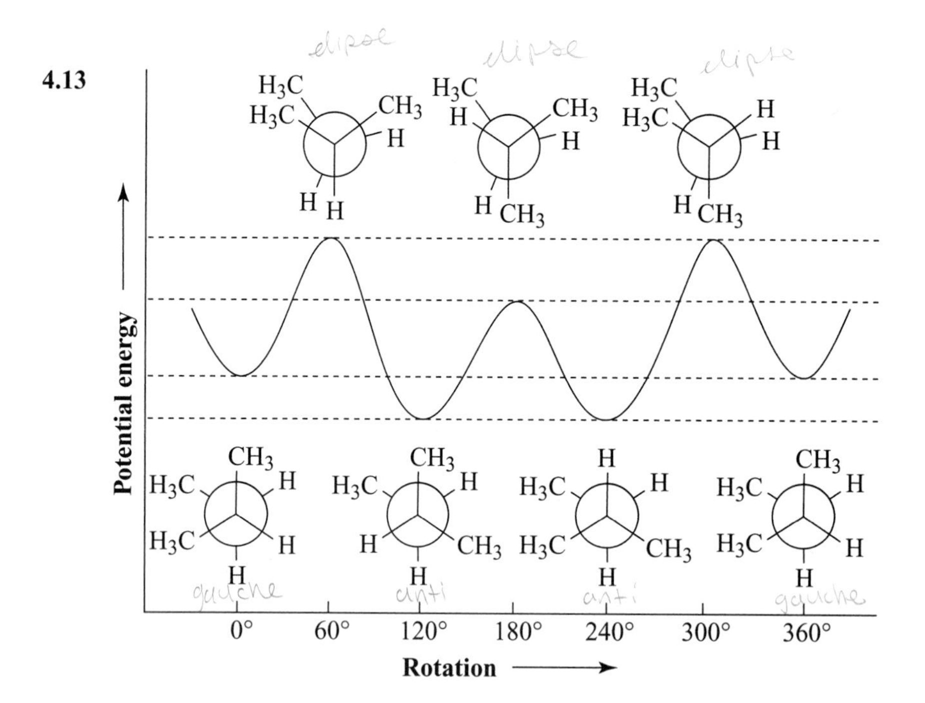

4.14 $\log K_{eq} = \dfrac{\Delta G^\circ}{-2.303RT} = \dfrac{-7600 \text{ J}}{(-2.303)(8.314 \text{ J K}^{-1})(298 \text{ K})} = 1.32$

$K_{eq} = 21.38$

Let e = amount of equatorial form

and a = amount of axial form

then, $K_{eq} = \dfrac{e}{a} = 21.38$

$e = 21.38a$

$\% e = \dfrac{21.38a}{a + 21.38a} \times 100 = 95.5\%$

4.15 (a)

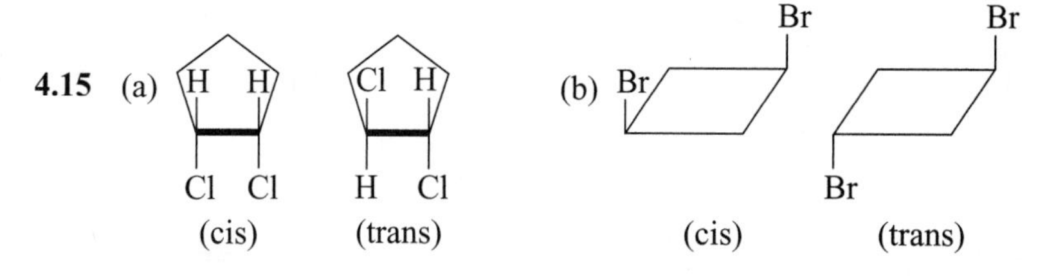

(c) No

4.16 (a-d)

More stable because larger group is equatorial and so preferred at equilibrium

Less stable because larger group is axial

4.17

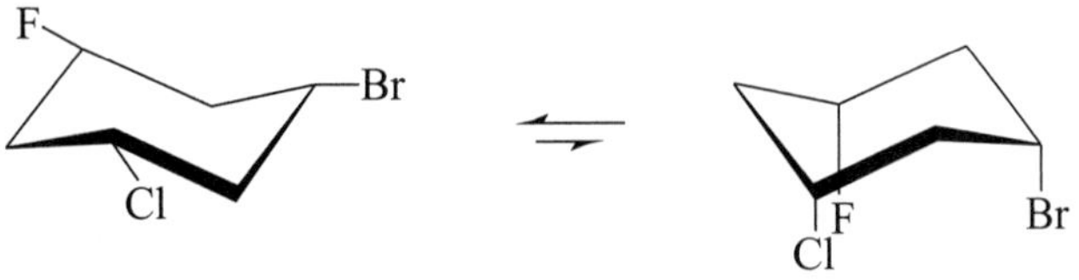

F, Br, Cl ⇌ Cl, F, Br

all equatorial
(more stable)

all axial
(less stable)

4.18 (a,b)

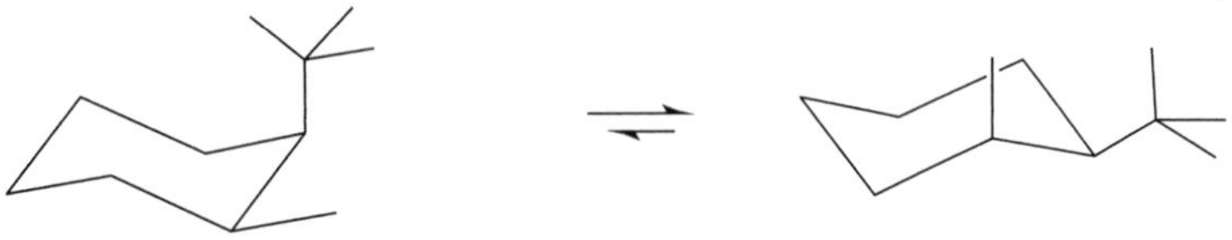

Less stable because the large *tert*-butyl group is axial (more potential energy)

More stable because the large *tert*-butyl group is equatorial (less potential energy)

4.19

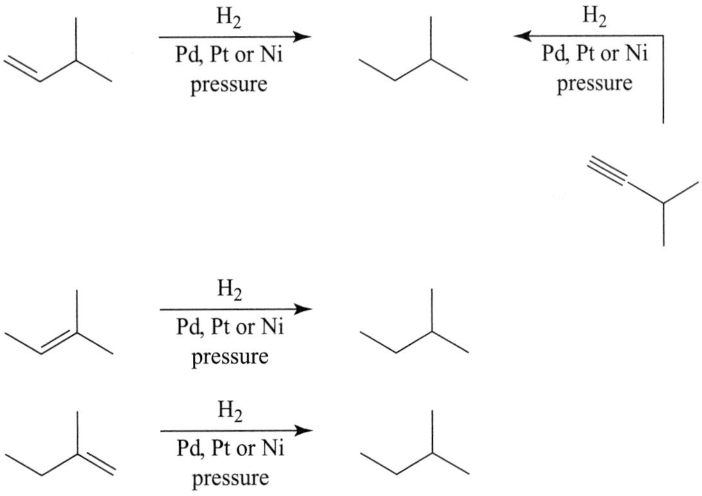

H_2 / Pd, Pt or Ni / pressure

H_2 / Pd, Pt or Ni / pressure

H_2 / Pd, Pt or Ni / pressure

H_2 / Pd, Pt or Ni / pressure

4.20 (a) C_6H_{14} = formula of alkane
$\underline{C_6H_{12}}$ = formula of 2-hexene
H_2 = difference = 1 pair of hydrogen atoms
Index of hydrogen deficiency = 1

(b) C_6H_{14} = formula of alkane
$\underline{C_6H_{12}}$ = formula of methylcyclopentane
H_2 = difference = 1 pair of hydrogen atoms
Index of hydrogen deficiency = 1

(c) No, all isomers of C_6H_{12}, for example, have the same index of hydrogen deficiency.

(d) No

(e) C_6H_{14} = formula of alkane
$\underline{C_6H_{10}}$ = formula of 2-hexyne
H_4 = difference = 2 pairs of hydrogen atoms

Index of hydrogen deficiency = 2

(f) $C_{10}H_{22}$ (alkane)
$\underline{C_{10}H_{16}}$ (compound)
H_6 = difference = 3 pairs of hydrogen atoms

Index of hydrogen deficiency = 3

The structural possibilities are thus

3 double bonds
1 double bond and 1 triple bond
2 double bonds and 1 ring
1 double bond and 2 rings
3 rings
1 triple bond and 1 ring

4.21 (a) $C_{15}H_{32}$ = formula of alkane
$\underline{C_{15}H_{24}}$ = formula of zingiberene
H_8 = difference = 4 pairs of hydrogen atoms

Index of hydrogen deficiency = 4

(b) Since 1 mol of zingiberene absorbs 3 mol of hydrogen, one molecule of zingiberene must contain three double bonds. (We are told that molecules of zingiberene do not contain any triple bonds.)

(c) If a molecule of zingiberene has three double bonds and an index of hydrogen deficiency equal to 4, it must have one ring. (The structural formula for zingiberene can be found in Review Problem 23.2.)

4.22 CH_3O … O   molecular formula $C_{10}H_{16}O_2$

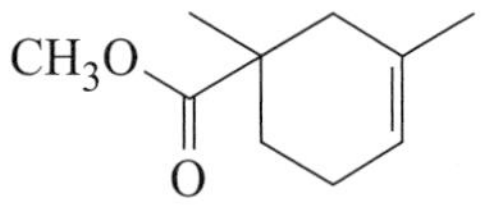

$C_{10}H_{22}$ = formula for alkane
$\underline{C_{10}H_{16}}$ = formula for unsaturated ester (ignoring oxygens)
H_6 = difference = 3 pairs of hydrogen atoms
IHD = 6

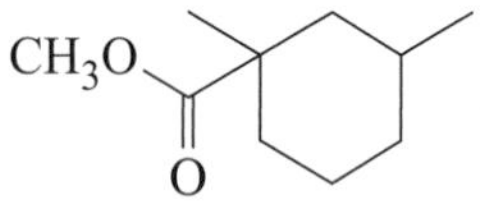

CH_3O … O   molecular formula $C_{10}H_{18}O_2$

$C_{10}H_{22}$ = formula for alkane
$\underline{C_{10}H_{18}}$ = formula for hydrogenation product (ignoring oxygens)
H_4 = difference = 2 pairs of hydrogen atoms
IHD = 4

Nomenclature and Isomerism

4.23 (a) Cl Cl

(b) Br

(c)

(d)

(e)

(f) Cl Cl

(g) H H CH_3 CH_3

(h) H CH_3 CH_3 H

(i) OH

(j)

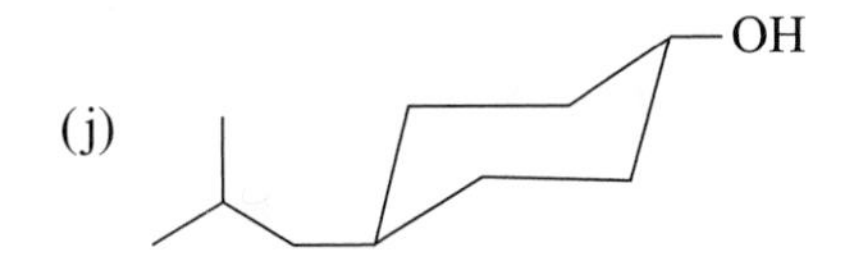

(k)

(l)

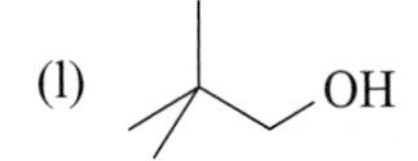

(m) or

(n)

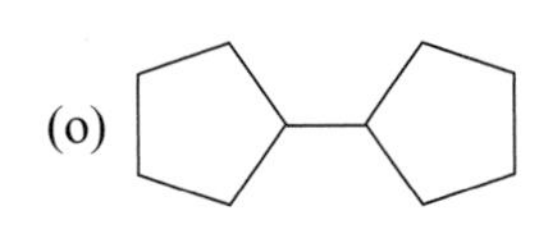

(o)

4.24 (a) 3,3,4-Trimethylhexane

(b) 2,2-Dimethyl-1-butanol

(c) 3,5,7-Trimethylnonane

(d) 3-Methyl-4-heptanol

(e) 2-Bromobicyclo[3.3.1]nonane

(f) 2,5-Dibromo-4-ethyloctane

(g) Cyclobutylcyclopentane

(h) 7-Chlorobicyclo[2.2.1]heptane

4.25 The two secondary carbon atoms in *sec*-butyl alcohol are equivalent; however, there are three five-carbon alcohols (pentyl alcohols) that contain a secondary carbon atom.

4.26 (a) 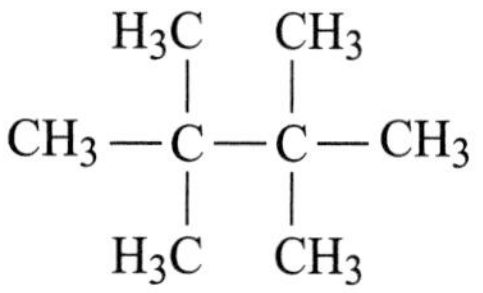 $CH_3-C(CH_3)_2-C(CH_3)_2-CH_3$

2,2,3,3-Tetramethylbutane

(b) 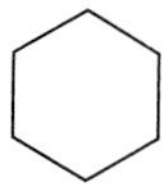 Cyclohexane

(c)  1,1-Dimethylcyclobutane

(d) 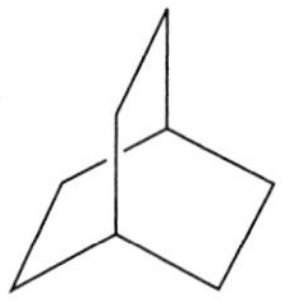 Bicyclo[2.2.2]octane

4.27 $CH_3-C(CH_3)_2-CH_2CH(CH_3)CH_3$ or 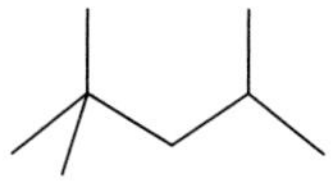 2,2,4-trimethylpentane

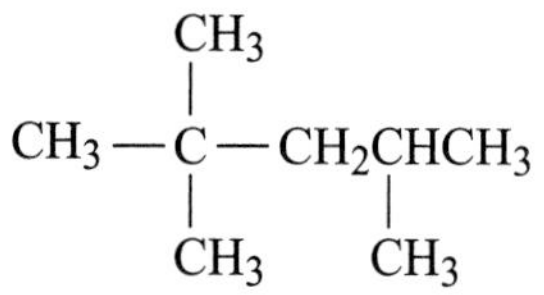

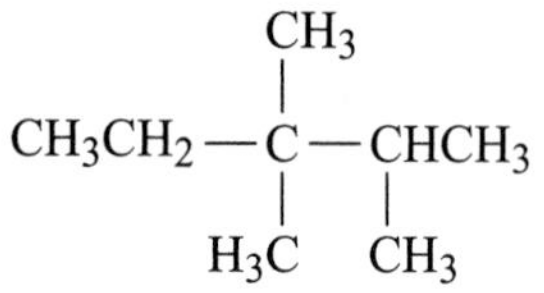

$CH_3CH_2-C(CH_3)_2-CH(CH_3)CH_3$ or 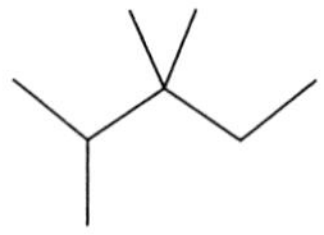 2,3,3-trimethylpentane

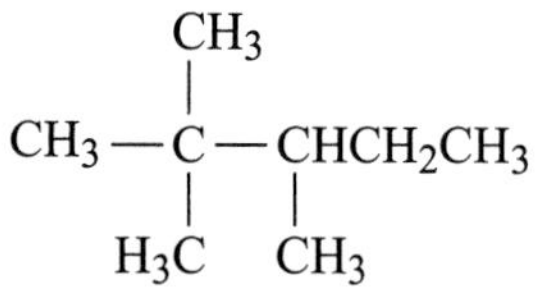

$CH_3-C(CH_3)_2-CH(CH_3)CH_2CH_3$ or 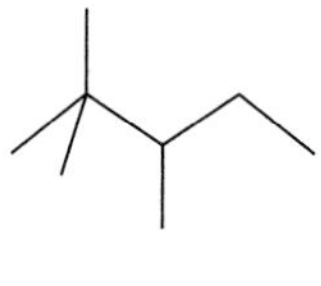 2,2,3-trimethylpentane

4.28

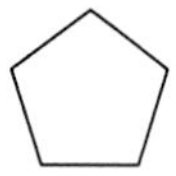

Cyclopentane

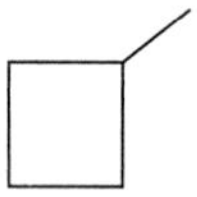

Methylcyclobutane

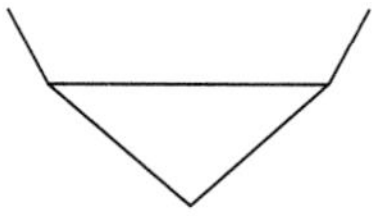

cis-1,2-Dimethylcyclopropane

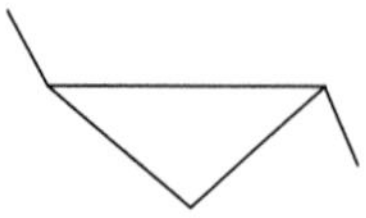

trans-1,2-Dimethylcyclopropane

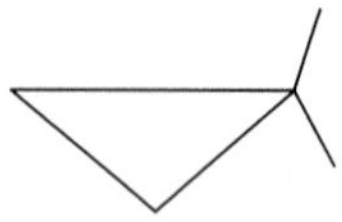

1,1-Dimethylcyclopropane

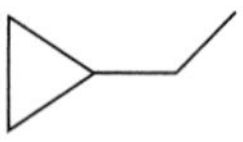

Ethylcyclopropane

4.29 (a) (b) (c) (d)

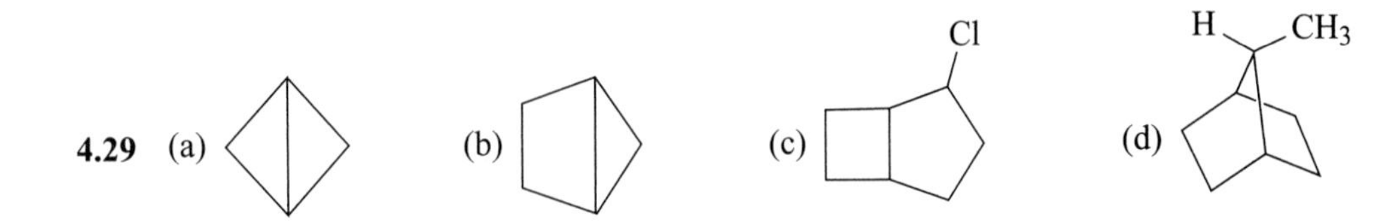

4.30 S − A + 1 = N
For cubane, S = 12 and A = 8. Thus 12 − 8 + 1 = N; N = 5 rings in cubane.

4.31 (a)

(b)

4.32 A homologous series is one in which each member of the series differs from the one preceding it by a constant amount, usually a CH_2 group. A homologous series of alkyl halides would be the following:

CH_3X
CH_3CH_2X
$CH_3(CH_2)_2X$
$CH_3(CH_2)_3X$
$CH_3(CH_2)_4X$
etc.

Hydrogenation

4.33

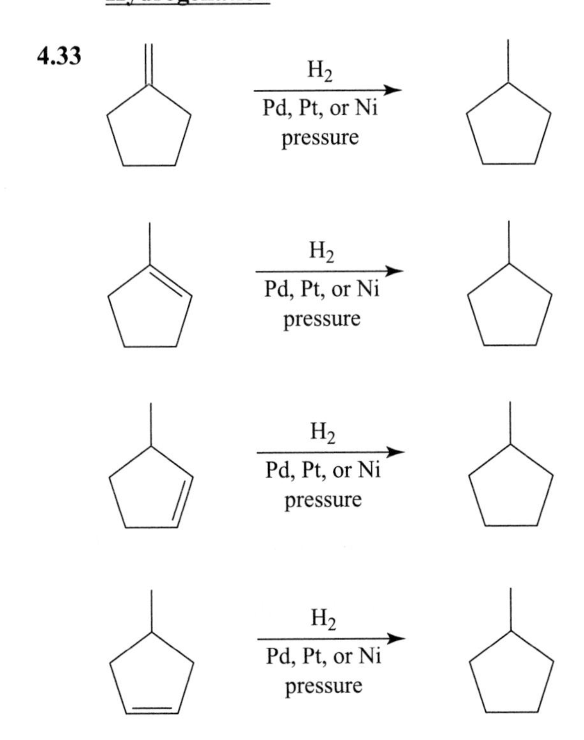

4.34 (a) Each of the desired alkenes must have the same carbon skeleton as 2-methylbutane,

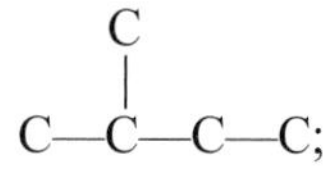

C—C(—C)—C—C; they are therefore

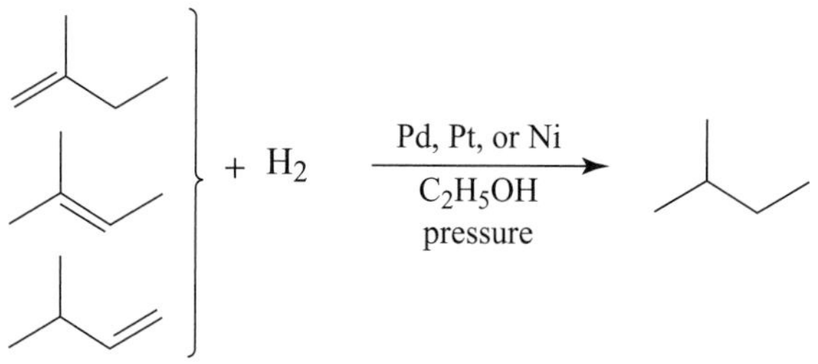

[three alkenes] + H_2 $\xrightarrow[\text{C}_2\text{H}_5\text{OH, pressure}]{\text{Pd, Pt, or Ni}}$ [2-methylbutane]

(b)

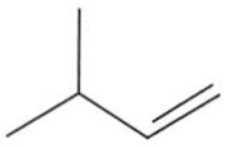

4.35

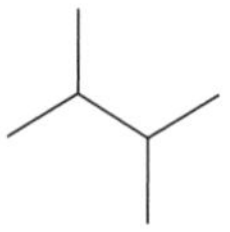

2,3-Dimethylbutane

From two alkenes

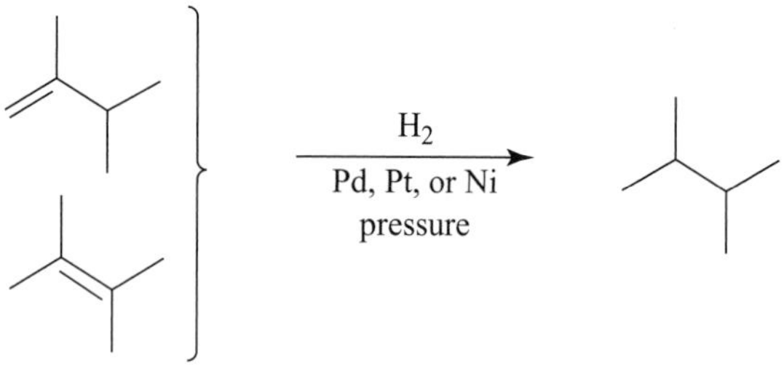

$\xrightarrow[\text{Pd, Pt, or Ni, pressure}]{\text{H}_2}$

Conformations and Stability

4.36

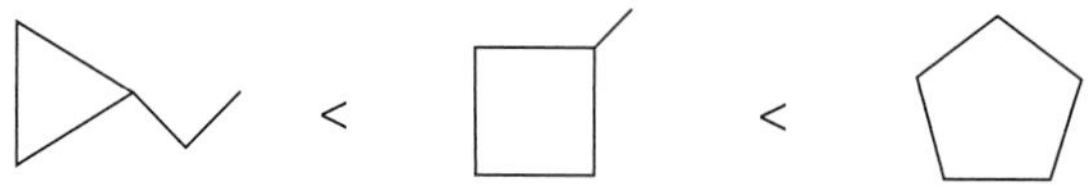

< <

Least stable　　　　Most stable

4.37

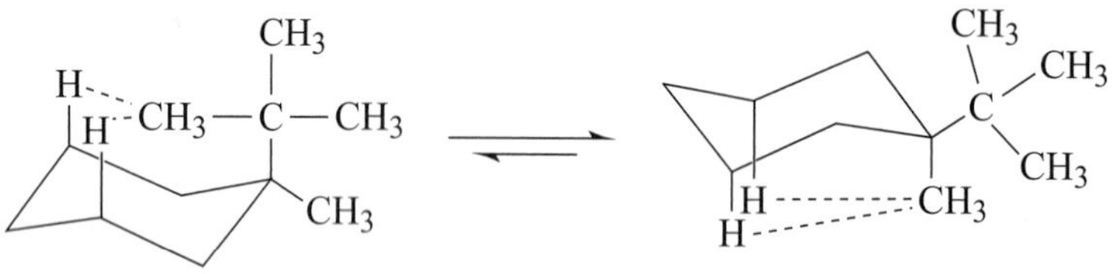

H
H CH$_3$—C(CH$_3$)—CH$_3$
CH$_3$

CH$_3$
CH$_3$
C
CH$_3$
H CH$_3$
H

This conformation is *less stable* because 1,3-diaxial interactions with the large *tert*-butyl group cause considerable repulsion.

This conformation is *more stable* because 1,3-diaxial interactions with the smaller methyl group are less repulsive.

4.38

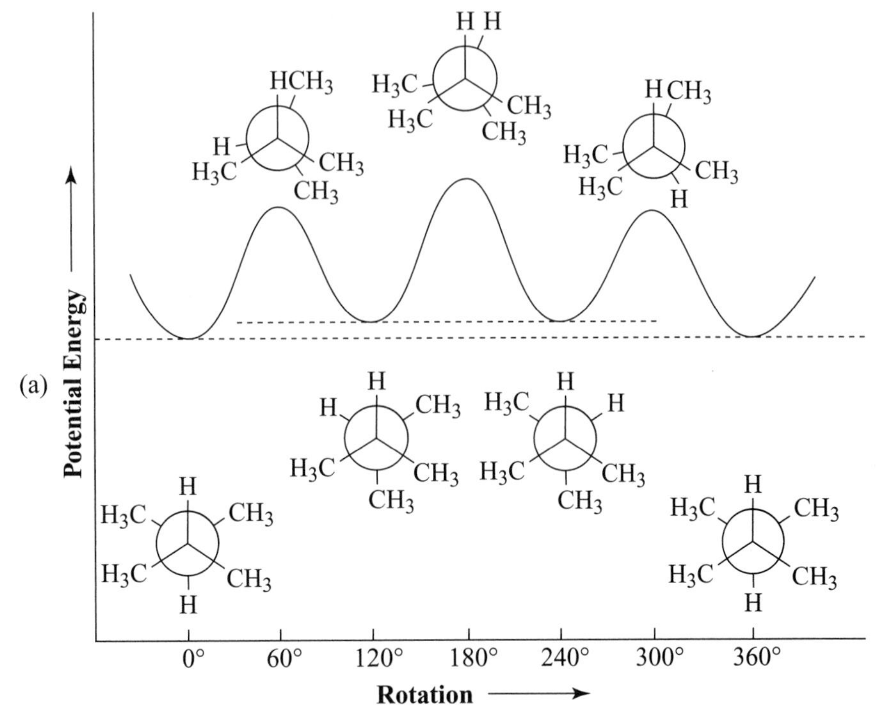

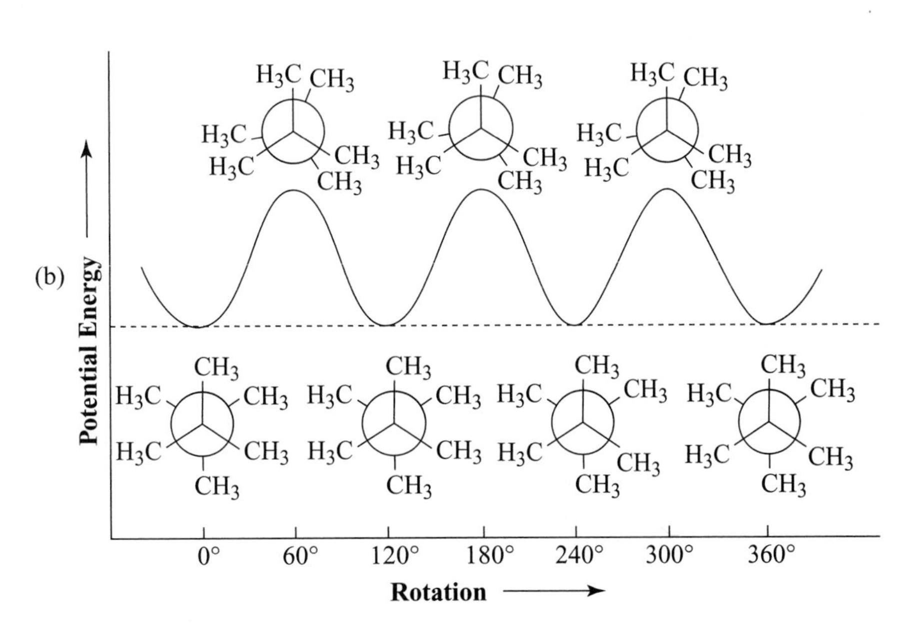

4.39 (a) Pentane would boil higher because its chain is unbranched. Chain-branching lowers the boiling point.

(b) Heptane would boil higher because it has the larger molecular weight and would, because of its larger surface area, have larger van der Waals attractions.

(c) 2-Chloropropane because it is more polar and because it has a larger molecular weight.

(d) 1-Propanol would boil higher because its molecules would be associated with each other through hydrogen-bond formation.

(e) Propanone (CH_3COCH_3) would boil higher because its molecules are more polar.

4.40 C_4H_6

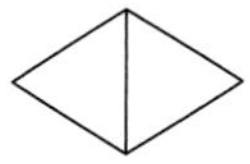

Bicyclo[1.1.0]butane  1-Butyne

The IR stretch at ~ 2250 cm^{-1} for the alkyne C≡C bond distinguishes these two compounds.

4.41 *trans*-1,2-Dimethylcyclopropane would be more stable because there is less crowding between its methyl groups.

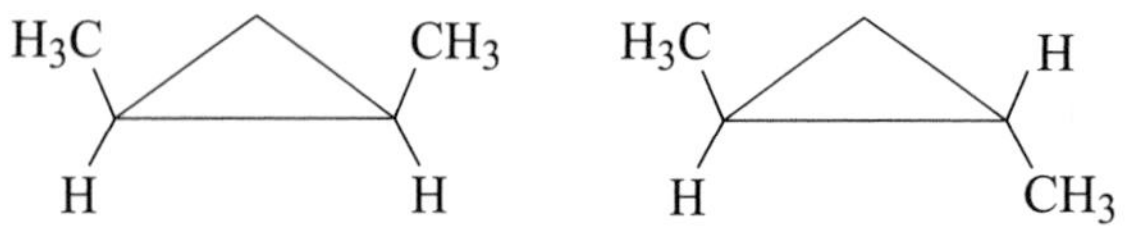

Less stable More stable

4.42 For 1,2-disubstituted cyclobutanes, the trans isomer is *e*,*e*; the cis isomer is *a*,*e*, and so the less stable of the two. For 1,3-disubstituted cyclobutanes, the cis isomer is *e*,*e* and more stable than the *a*,*e* trans isomer.

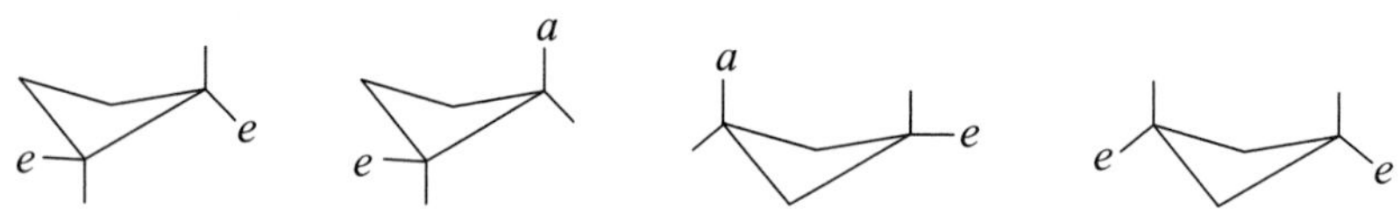

Trans Cis Trans Cis

4.43 (a)

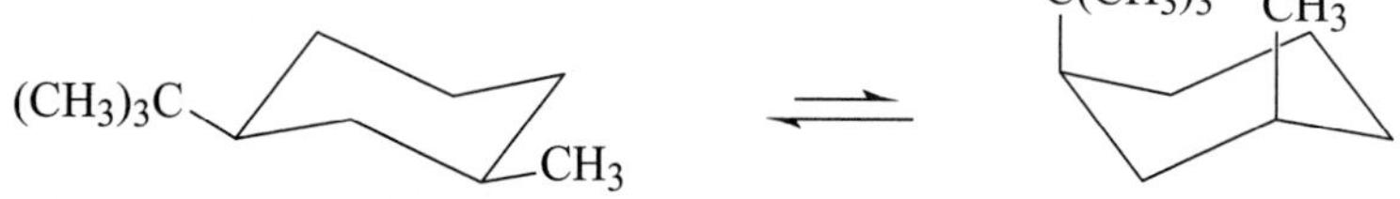

More stable conformation because both alkyl groups are equatorial

(b)

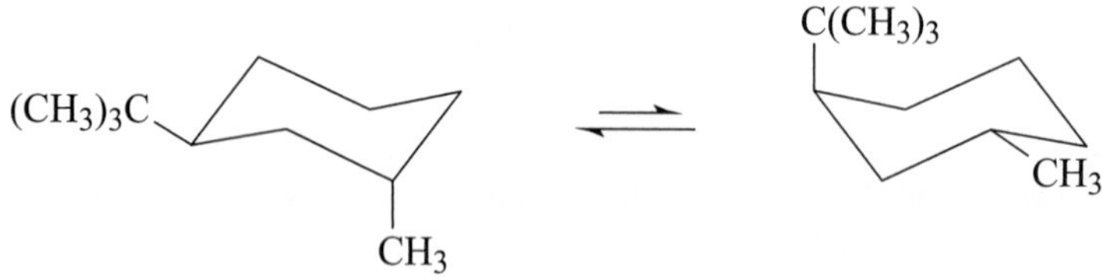

More stable because larger group is equatorial

(c)

$(CH_3)_3C$ … CH_3 ⇌ $C(CH_3)_3$ … CH_3

More stable conformation because both alkyl groups are equatorial

(d)

$(CH_3)_3C$ … CH_3 ⇌ $C(CH_3)_3$ … CH_3

More stable because larger group is equatorial

4.44 Certainly it is expected that the alkyl groups would prefer the equatorial disposition in a case such as this, and indeed this is true in the case of *all-trans*-1,2,3,4,5,6-hexaethylcyclohexane, which does have all ethyl groups equatorial.

Apparently, the torsional and steric effects resulting from equatorial isopropyl groups destabilize the all-equatorial conformation to a greater degree than axial isopropyl groups destabilize the all-axial conformation. The fully axial structure assigned on the basis of X-ray studies is also supported by calculations.

4.45 If the cyclobutane ring were planar, the C—Br bond moments would exactly cancel in the trans isomer. The fact that *trans*-1,3-dibromocyclobutane has a dipole moment shows the ring is not planar.

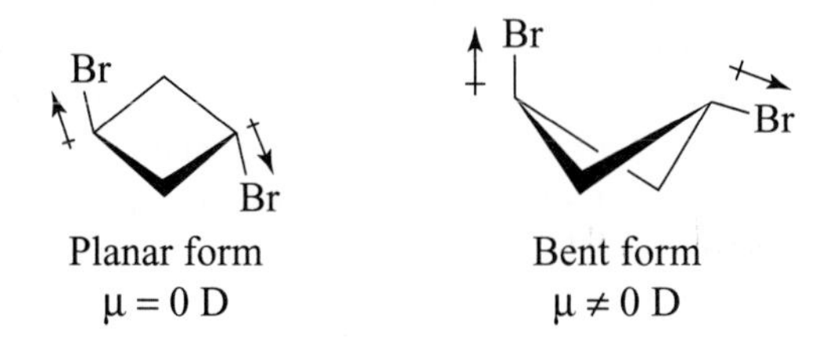

Planar form $\mu = 0$ D

Bent form $\mu \neq 0$ D

Synthesis

4.46 (a) H_2; Pd, Pt, or Ni catalyst, pressure

(b) H_2; Pd, Pt, or Ni catalyst, pressure

(c) Predicted IR absorption frequencies for reactants in parts (a) and (b) are the following:

(1) *trans*-5-Methyl-2-hexene has an absorption at approximately 964 cm^{-1}, characteristic of C—H bending in a trans-substituted alkene.

(2) The alkene double bond of the reactant is predicted to have a stretching absorption between 1580 and 1680 cm^{-1}.

Challenge Problems

4.47 If *trans*-1,3-di-*tert*-butylcyclohexane were to adopt a chair conformation, one *tert*-butyl group would have to be axial. It is, therefore, more energetically favorable for the molecule to adopt a twist boat conformation.

4.48 (a) More rules are needed (see Chapter 7) to indicate relative stereochemistry for these 1-bromo-2-chloro-1-fluoroethenes.

(b) Bicyclo[4.4.0]decane (or decalin)

(c) Bicyclo[4.4.0]dec-2-ene (or Δ^1-octalin)

(d) Bicyclo[4.4.0]dec-1-ene (or $\Delta^{1(8a)}$-octalin)
NOTE: The common name decalin comes from decahydronaphthalene, the derivative of naphthalene

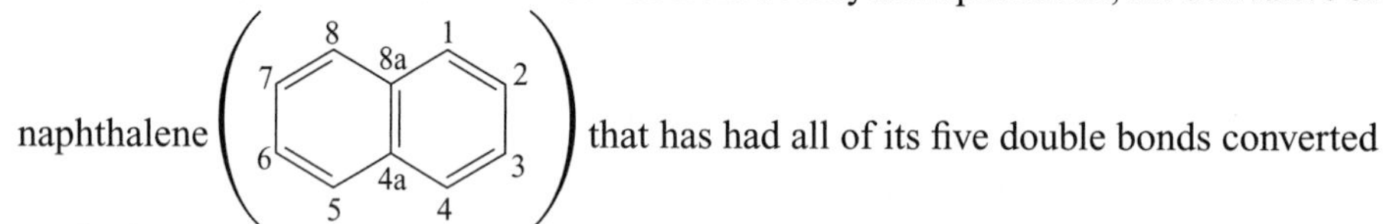

that has had all of its five double bonds converted to single bonds by addition of 10 atoms of hydrogen. Octalin similarly comes from octahydronaphthalene and contains one surviving double bond. When using these common names derived from naphthalene, their skeletons are usually numbered like that of naphthalene. When, as in case (d), a double bond does not lie between the indicated carbon and the next higher numbered carbon, its location is specified as shown.

Also, the symbol Δ is one that has been used with common names to indicate the presence of a double bond at the position specified by the accompanying superscript number(s).

4.49 The cyclohexane ring in *trans*-1-*tert*-butyl-3-methylcyclohexane is in a chair conformation. The ring in *trans*-1,3-di-*tert*-butylcyclohexane is in a twist-boat conformation. In *trans*-1-*tert*-butyl-3-methylcyclohexane the *tert*-butyl group can be equatorial if the methyl group is axial. The energy cost of having the methyl group axial is apparently less than that for adopting the twist-boat ring conformation. In *trans*-1,3-di-*tert*-butylcyclohexane the potential energy cost of having one *tert*-butyl group equatorial and the other axial is apparently greater than having the ring adopt a twist-boat conformation so that both can be approximately equatorial.

4.50 All of the nitrogen-containing five-membered rings of Vitamin B_{12} contain at least one sp^2-hybridized atom (in some cases there are more). These atoms, because of the trigonal planar geometry associated with them, impose some degree of planarity on the nitrogen-containing five-membered rings in B_{12}. Furthermore, 13 atoms in sequence around the cobalt are sp^2-hybridized, a fact that adds substantial resonance stabilization to this part of the ring system. The four five-membered nitrogen-containing rings that surround the cobalt in roughly a plane and whose nitrogens lend their unshared pairs to the cobalt comprise what is called a corrin ring. The corrin ring may look familiar to you, and for good reason, because it is related to the porphyrin ring of heme. (Additional question: What geometry is associated with the cobalt atom and its six bound ligands? Answer: octahedral, or square bipyramidal.)

4.51 The form of a benzene ring occurs 16 times in buckminsterfullerene. The other eight facets in the 24 faces of a "buckyball" are five-membered rings. Every carbon of buckminsterfullerene is approximately sp^2-hybridized.

QUIZ

4.1 Consider the properties of the following compounds:

NAME	FORMULA	BOILING POINT (°C)	MOLECULAR WEIGHT
Ethane	CH_3CH_3	−88.2	30
Fluoromethane	CH_3F	−78.6	34
Methanol	CH_3OH	+64.7	32

Select the answer that explains why methanol boils so much higher than ethane or fluoromethane, even though they all have nearly equal molecular weights.

(a) Ion-ion forces between molecules.

(b) Weak dipole-dipole forces between molecules.

(c) Hydrogen bonding between molecules.

(d) van der Waals forces between molecules.

(e) Covalent bonding between molecules.

4.2 Select the correct name of the compound whose structure is

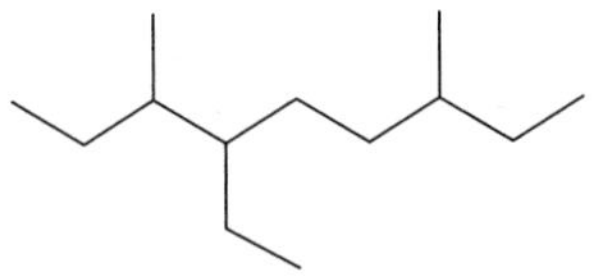

(a) 2,5-Diethyl-6-methyloctane

(b) 4,7-Diethyl-3-methyloctane

(c) 4-Ethyl-3,7-dimethylnonane

(d) 6-Ethyl-3,7-dimethylnonane

(e) More than one of the above

4.3 Select the correct name of the compound whose structure is $CH_3\overset{\overset{\displaystyle CH_3}{|}}{C}HCH_2Cl$.

(a) Butyl chloride

(b) Isobutyl chloride

(c) *sec*-Butyl chloride

(d) *tert*-Butyl chloride

(e) More than one of the above

4.4 The structure shown in Problem 4.2 has:

(a) 1°, 2°, and 3° carbon atoms

(b) 1° and 2° carbon atoms only

(c) 1° and 3° carbon atoms only

(d) 2° and 3° carbon atoms only

(e) 1°, 3°, and 4° carbon atoms only

4.5 How many isomers are possible for C_3H_7Br?

(a) 1 (b) 2 (c) 3 (d) 4 (e) 5

4.6 Which isomer of 1,3-dimethylcyclohexane is more stable?

(a) cis (b) trans (c) Both are equally stable (d) Impossible to tell

4.7 Which is the lowest energy conformation of *trans*-1,4-dimethylcyclohexane?

(a) 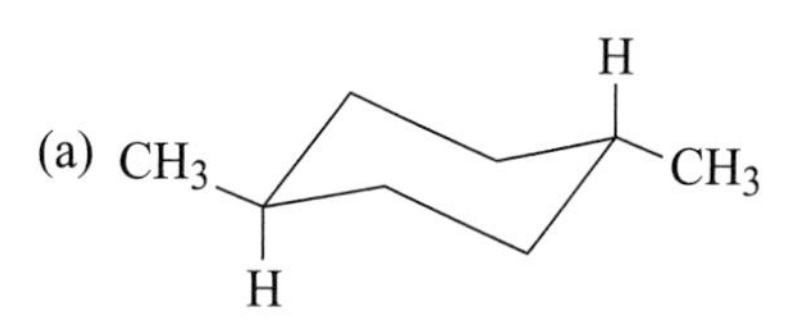

(b)

(c)

(d)

(e) More than one of the above

4.8 Supply the missing structures

(a) 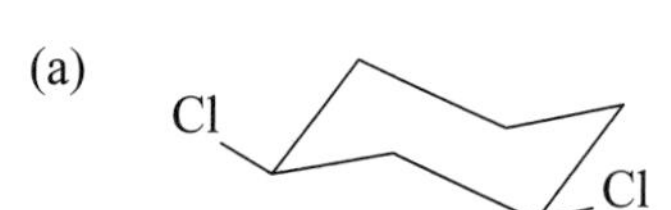

ring flip

(b) 

2-Bromobicyclo[2.2.1]heptane

(c) Newman projection for a gauche form of 1,2-dibromoethane

4.9 Supply the missing reagents in the box:

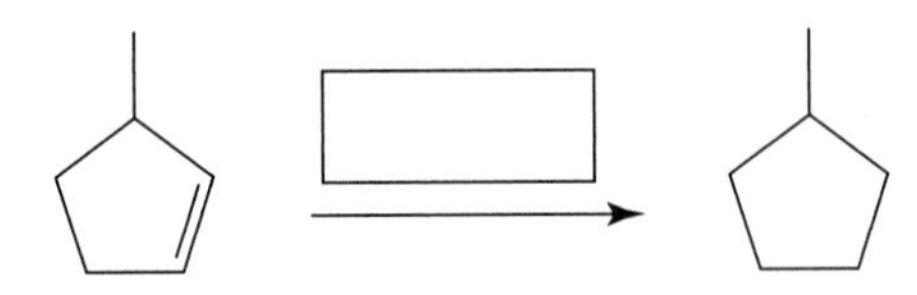

4.10 The most stable conformation of *trans*-1-isopropyl-3-methylcyclohexane:

5 STEREOCHEMISTRY: CHIRAL MOLECULES

SOLUTIONS TO PROBLEMS

5.1 (a) Achiral (b) Achiral (c) Chiral (d) Chiral (e) Chiral (f) Chiral (g) Chiral (h) Achiral

5.2 (a) Yes (b) No (c) No

5.3 (a) They are the same molecule. (b) They are enantiomers.

5.4 (a), (b), (e), (g), and (i) do not have chirality centers.

(c)

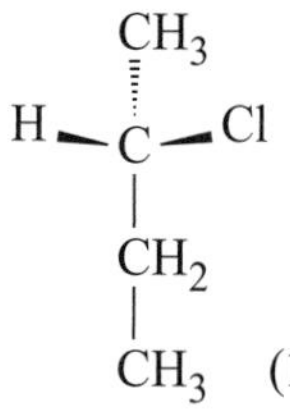

CH_3
H—C—Cl
CH_2
CH_3 **(1)**

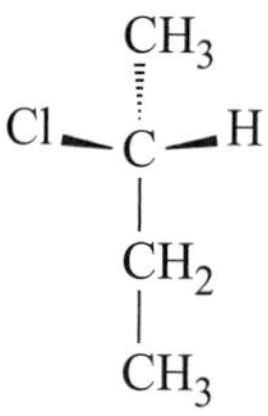

CH_3
Cl—C—H
CH_2
CH_3 **(2)**

(d)

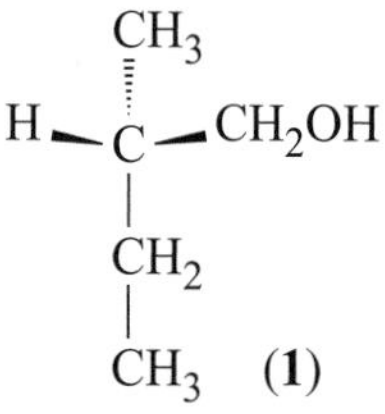

CH_3
H—C—CH_2OH
CH_2
CH_3 **(1)**

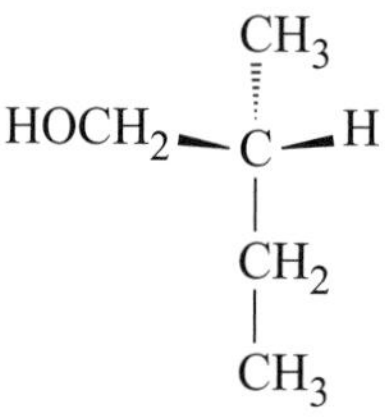

CH_3
$HOCH_2$—C—H
CH_2
CH_3 **(2)**

(f)

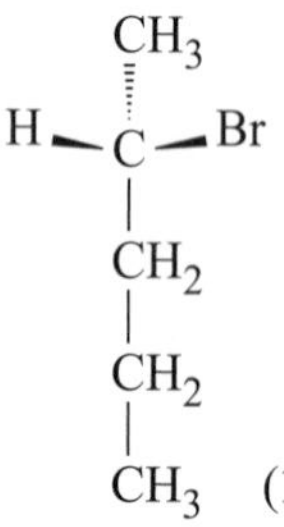

CH_3
H—C—Br
CH_2
CH_2
CH_3 **(1)**

CH_3
Br—C—H
CH_2
CH_2
CH_3 **(2)**

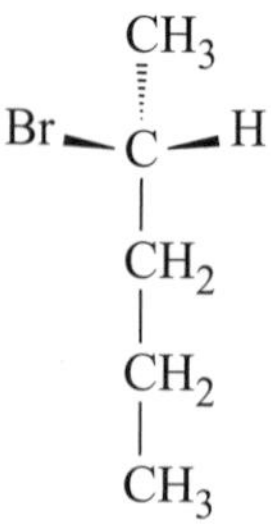

(h) 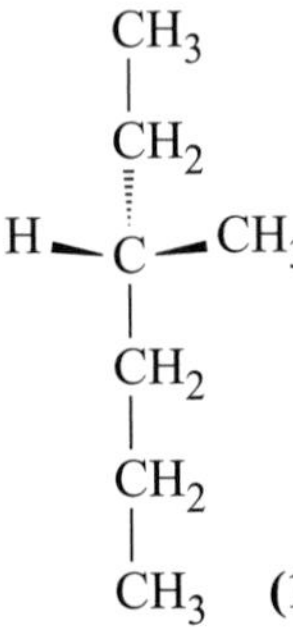 (1) 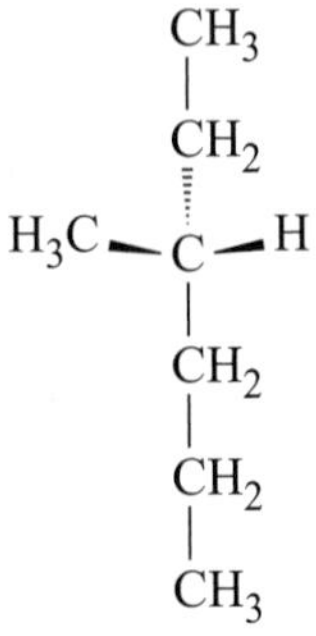 $H-C(CH_2CH_3)(CH_3)-CH_2CH_2CH_3$ (2) $H_3C-C(CH_2CH_3)(H)-CH_2CH_2CH_3$

(j) 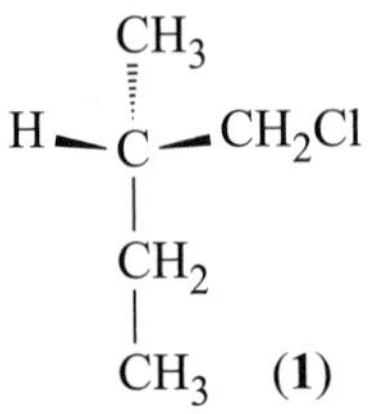 (1) 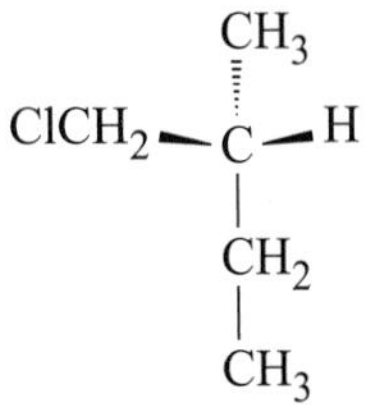 $H-C(CH_3)(CH_2Cl)-CH_2CH_3$ (2) $ClCH_2-C(CH_3)(H)-CH_2CH_3$

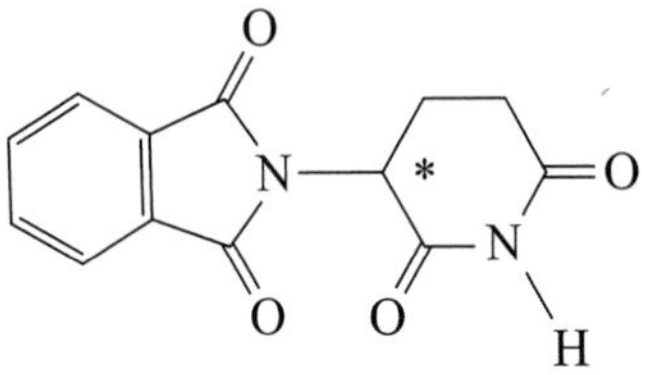

5.5 (a) 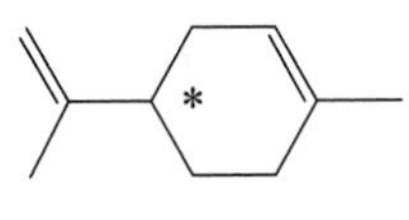 Limonene (R) (S)

(b) Thalidomide (S) (R)

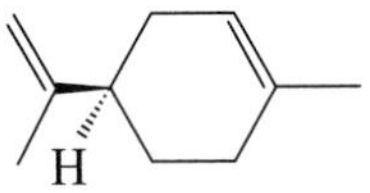

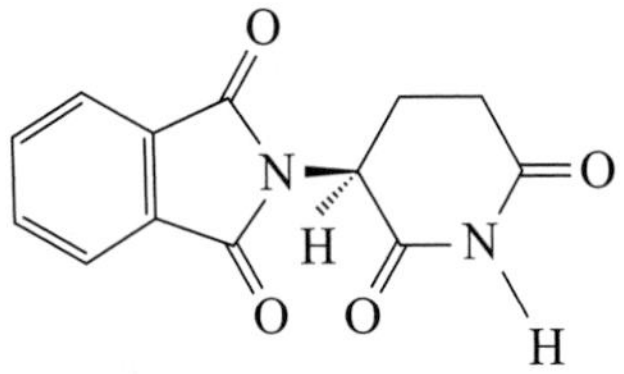

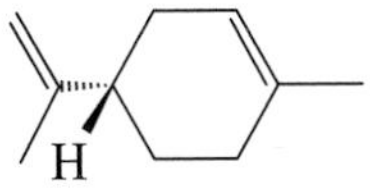

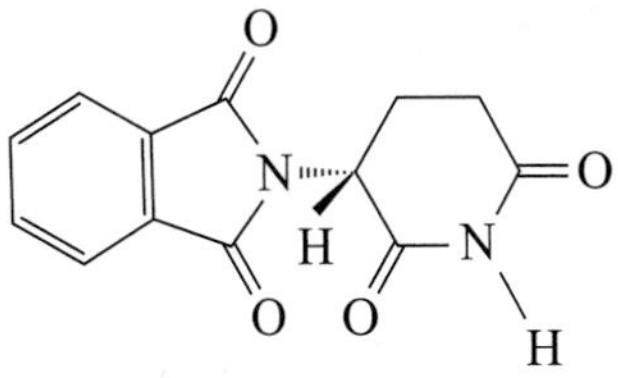

5.6 (a) 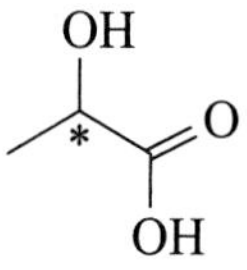 (b) 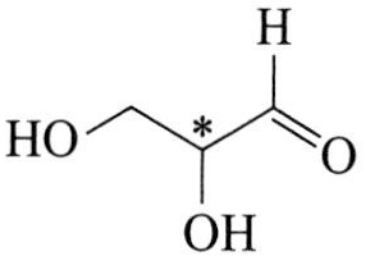 (c) 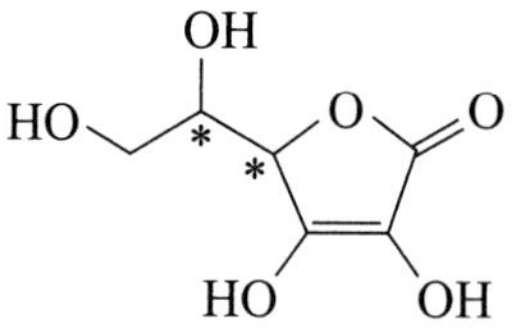 (d)

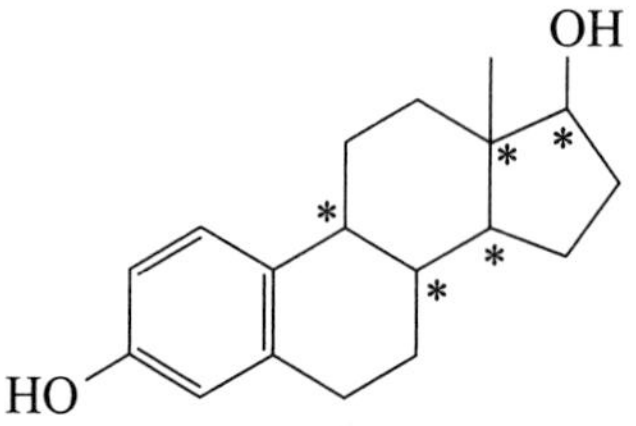

5.7 The following items possess a plane of symmetry, and are, therefore, achiral.

(a) A screwdriver

(b) A baseball bat (ignoring any writing on it)

(h) A hammer

5.8 In each instance below, the plane defined by the page is a plane of symmetry.

(a) 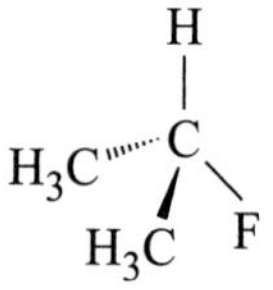 (b) 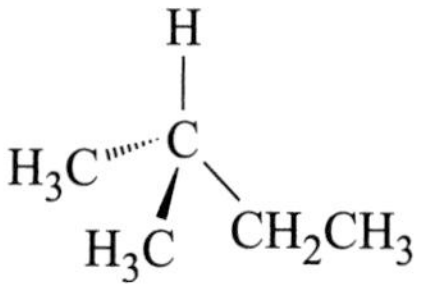 (e) 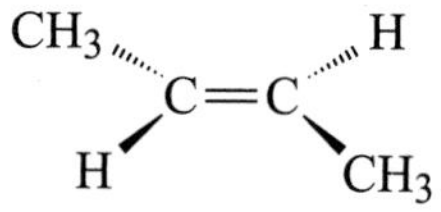 (g) 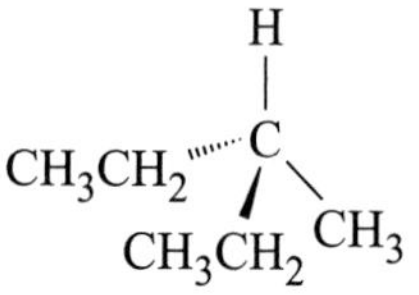 (i)

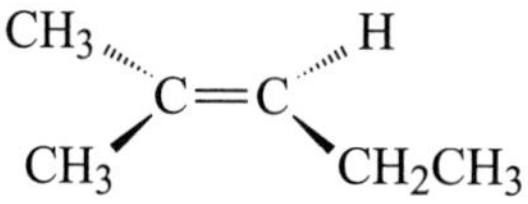

5.9

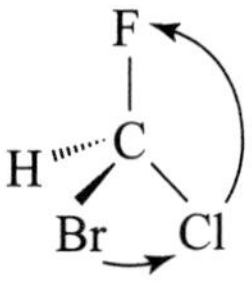

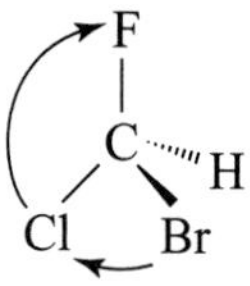

(*S*) (*R*)

5.10 (c) (1) is (*S*)
(2) is (*R*)

(d) (1) is (*S*)
(2) is (*R*)

(f) (1) is (*S*)
(2) is (*R*)

(h) (1) is (*S*)
(2) is (*R*)

(j) (1) is (*S*)
(2) is (*R*)

5.11 (a) $-Cl > -SH > -OH > -H$

(b) $-CH_2Br > -CH_2Cl > -CH_2OH > -CH_3$

(c) $-OH > -CHO > -CH_3 > -H$

(d) $-C(CH_3)_3 > -CH{=}CH_2 > -CH(CH_3)_2 > -H$

(e) $-OCH_3 > -N(CH_3)_2 > -CH_3 > -H$

(f) $-OPO_3H_2 > -OH > -CHO > -H$

5.12 (a) (*S*) (b) (*R*) (c) (*S*) (d) (*R*)

5.13 (a) Enantiomers

(b) Two molecules of the same compound

(c) Enantiomers

5.14

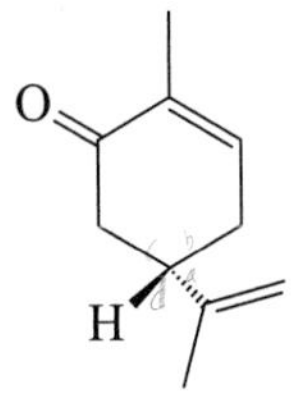

(*S*)-(+)-Carvone

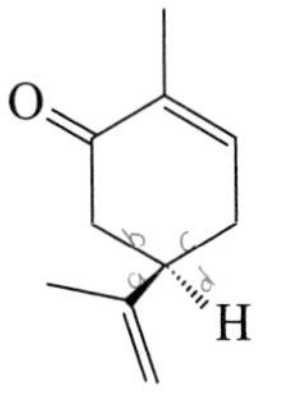

(*R*)-(–)-Carvone

5.15 (a) $$\text{Enant. Excess} = \frac{\text{observed specific rotation}}{\text{specific rotation of pure enantiomer}} \times 100$$

$$= \frac{+1.151}{+5.756} \times 100$$

$$= 20.00\%$$

(b) Since the (*R*) enantiomer (see Section 5.8C) is +, the (*R*) enantiomer is present in excess.

5.16 (a)

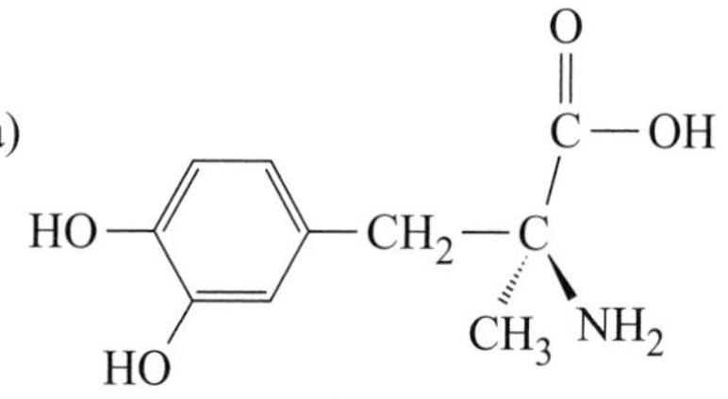

O
‖
C—OH
HO—C₆H₃—CH_2—C
HO
CH_3 NH_2

(*S*)-Methyldopa

(b)

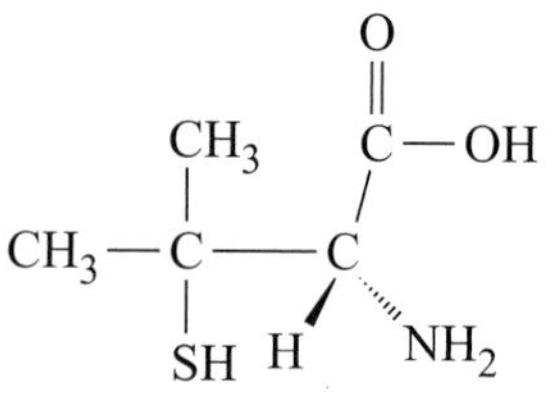

CH_3 O
‖
C—OH
CH_3—C——C
SH H NH_2

(*S*)-Penicillamine

(c)

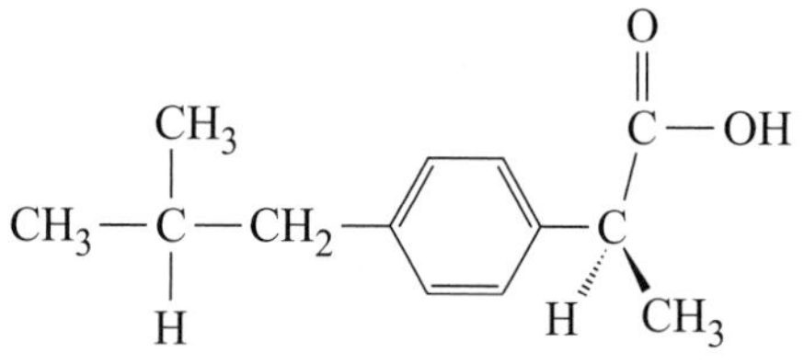

CH_3 O
‖
C—OH
CH_3—C—CH_2—C₆H₄—C
H H CH_3

(*S*)-Ibuprofen

5.17

HO H

(*R*) configuration

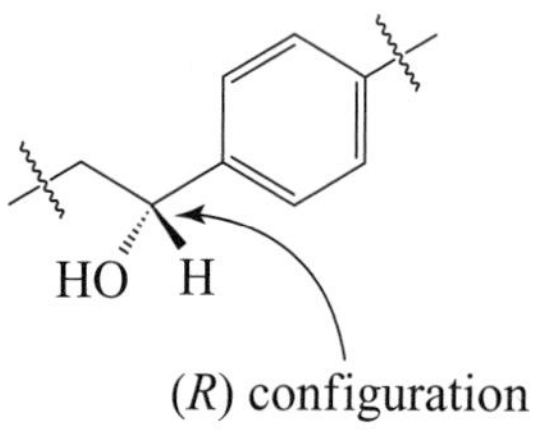

5.18 The configuration is *R* at both chirality centers.

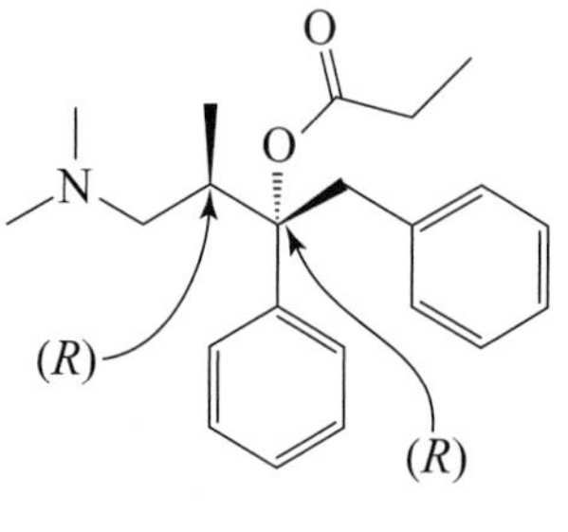

O
O
N
(*R*)
(*R*)

5.19 (a) Diastereomers.

(b) Diastereomers in each instance.

(c) No, diastereomers have different melting points.

(d) No, diastereomers have different boiling points.

(e) No, diastereomers have different vapor pressures.

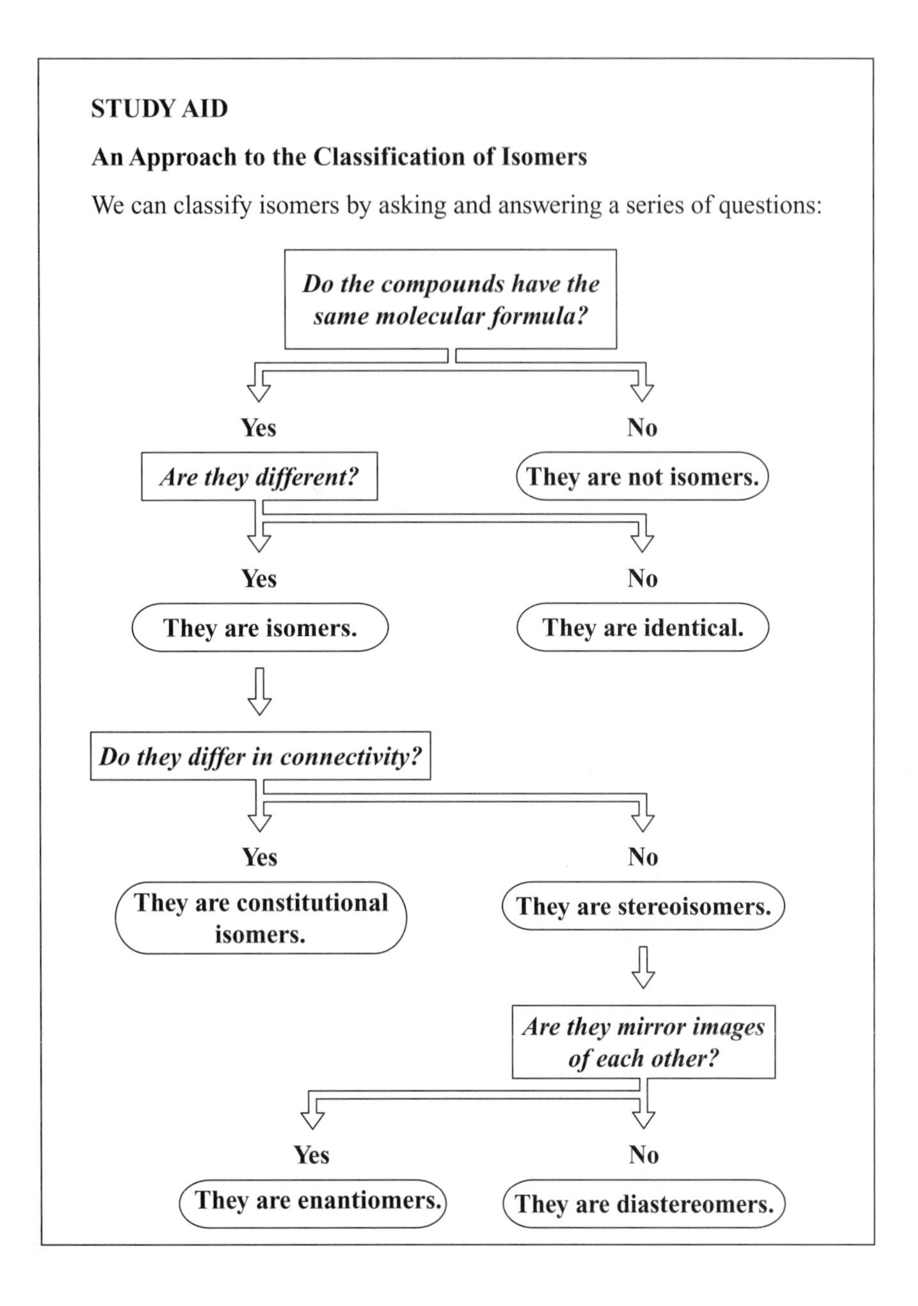

5.20 (a) It would be optically active.

(b) It would be optically active.

(c) No, because it is a meso compound.

(d) No, because it would be a racemic form.

5.21 (a) Represents **A**

(b) Represents **C**

(c) Represents **B**

5.22

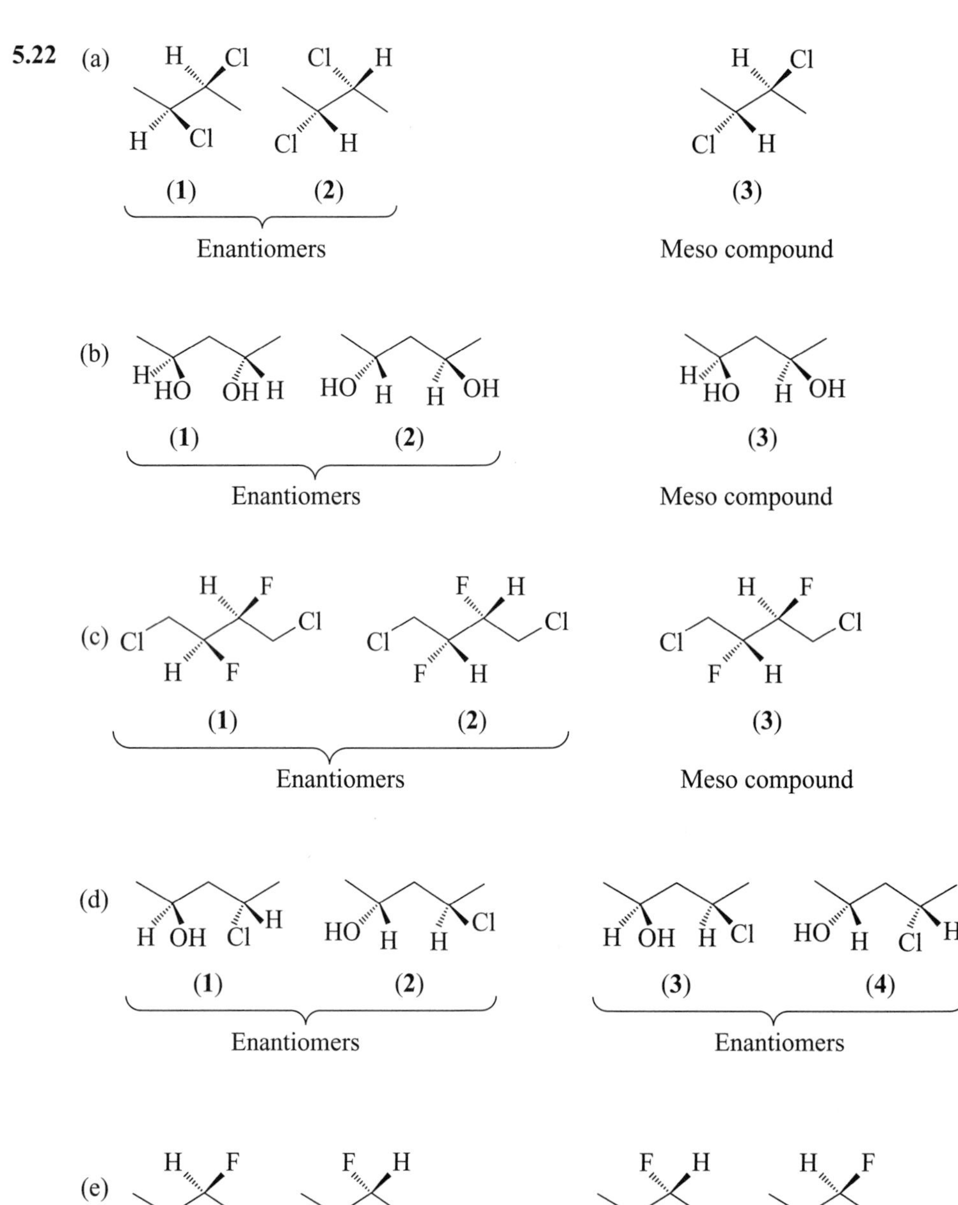
(a)
(1)
(2)
Enantiomers
(3)
Meso compound
(b)
(1)
(2)
Enantiomers
(3)
Meso compound
(c)
(1)
(2)
Enantiomers
(3)
Meso compound
(d)
(1)
(2)
Enantiomers
(3)
(4)
Enantiomers

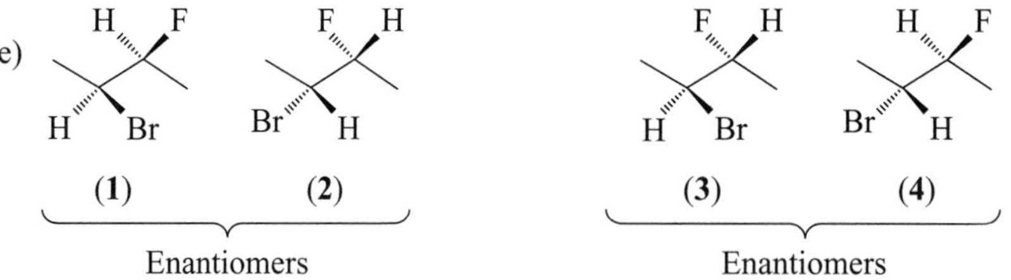
(e)
(1)
(2)
Enantiomers
(3)
(4)
Enantiomers

(f)

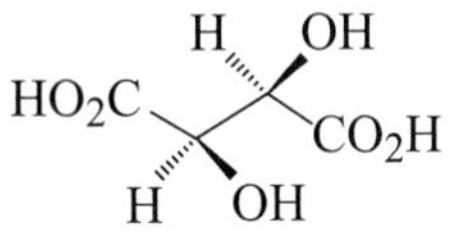

(1) 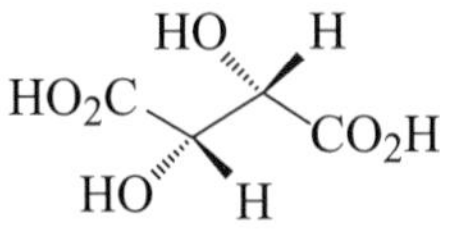 **(2)**

Enantiomers

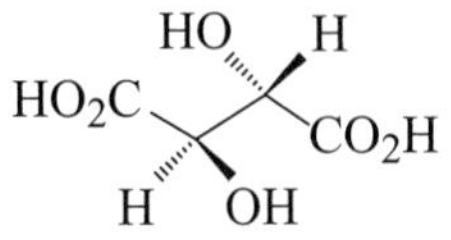

(3)
Meso compound

5.23 **B** is (2*S*,3*S*)-2,3-Dibromobutane
C is (2*R*,3*S*)-2,3-Dibromobutane

5.24 (a) **(1)** is (2*S*,3*S*)-2,3-Dichlorobutane
(2) is (2*R*,3*R*)-2,3-Dichlorobutane
(3) is (2*R*,3*S*)-2,3-Dichlorobutane

(b) **(1)** is (2*S*,4*S*)-2,4-Pentanediol
(2) is (2*R*,4*R*)-2,4-Pentanediol
(3) is (2*R*,4*S*)-2,4-Pentanediol

(c) **(1)** is (2*R*,3*R*)-1,4-Dichloro-2,3-difluorobutane
(2) is (2*S*,3*S*)-1,4-Dichloro-2,3-difluorobutane
(3) is (2*R*,3*S*)-1,4-Dichloro-2,3-difluorobutane

(d) **(1)** is (2*S*,4*S*)-4-Chloro-2-pentanol
(2) is (2*R*,4*R*)-4-Chloro-2-pentanol
(3) is (2*S*,4*R*)-4-Chloro-2-pentanol
(4) is (2*R*,4*S*)-4-Chloro-2-pentanol

(e) **(1)** is (2*S*,3*S*)-2-Bromo-3-fluorobutane
(2) is (2*R*,3*R*)-2-Bromo-3-fluorobutane
(3) is (2*S*,3*R*)-2-Bromo-3-fluorobutane
(4) is (2*R*,3*S*)-2-Bromo-3-fluorobutane

(f) **(1)** is (2*R*,3*R*)-Butanedioic acid
(2) is (2*S*,3*S*)-Butanedioic acid
(3) is (2*R*,3*S*)-Butanedioic acid

5.25

Chloramphenicol

5.26 (a) **A** C1, *R*; C2, *R*
B C1, *S*; C2, *S*

(b,c)

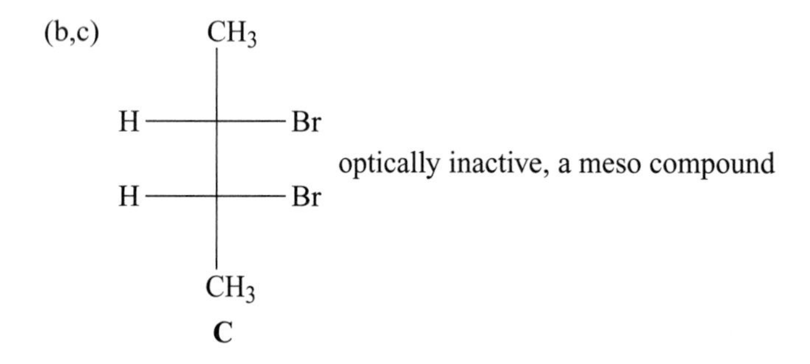

5.27 (a) No (b) Yes (c) No (d) No (e) Diastereomers
(f) Diastereomers

5.28

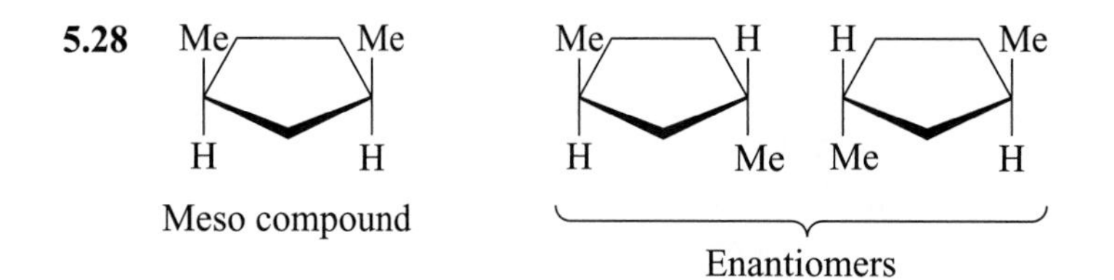

5.29 (a)

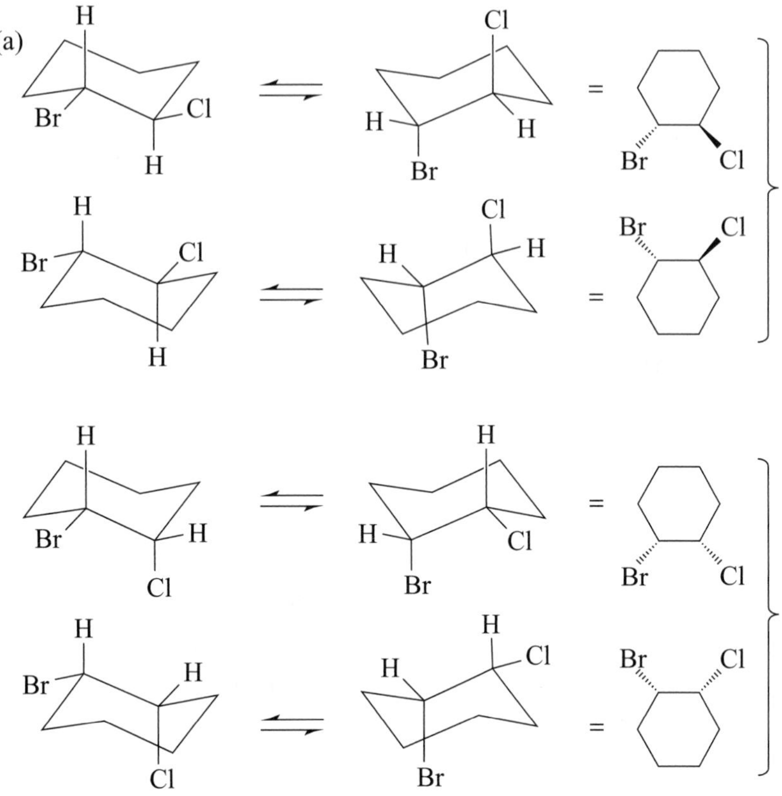

(1*R*,2*R*)

Enantiomers (both trans)

(1*S*,2*S*)

(1*R*,2*S*)

Enantiomers (both cis)

(1*S*,2*R*)

(b)

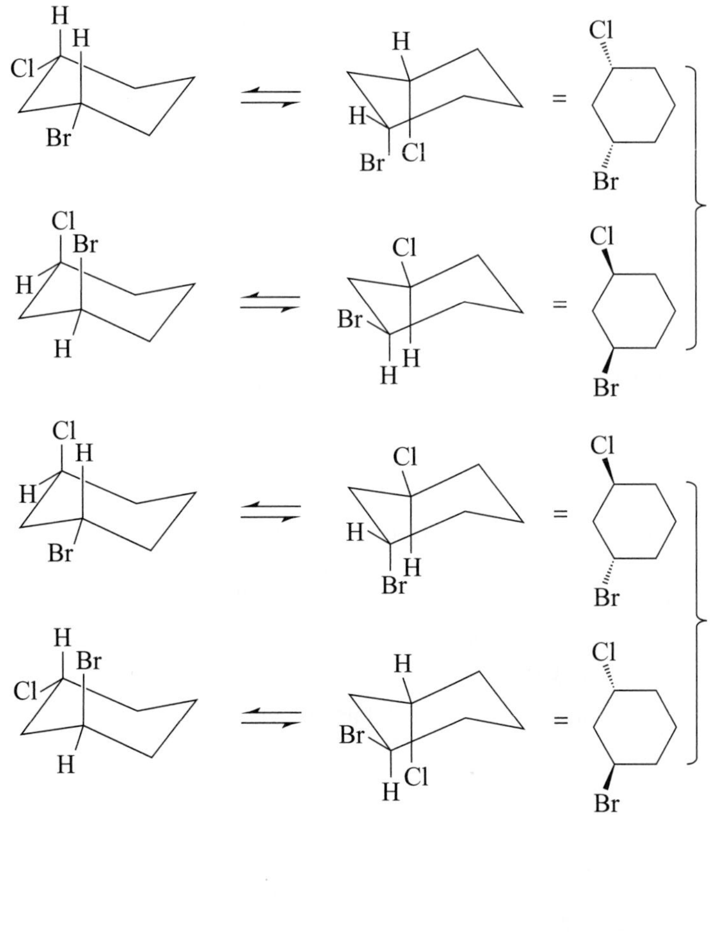

(1*S*,3*R*)

Enantiomers (both cis)

(1*R*,3*S*)

(1*S*,3*S*)

Enantiomers (both trans)

(1*R*,3*R*)

(c)

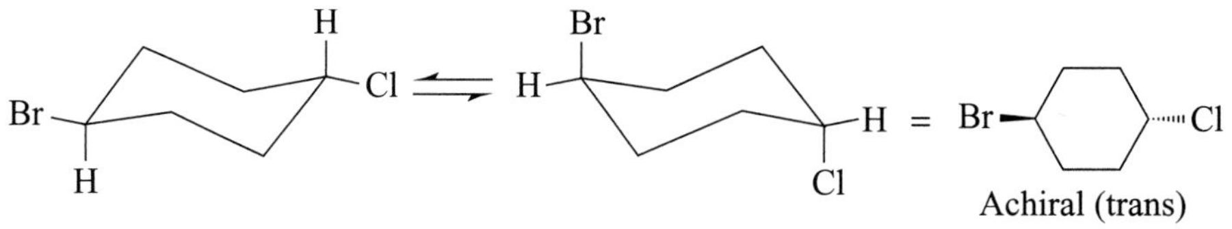

Achiral (trans)

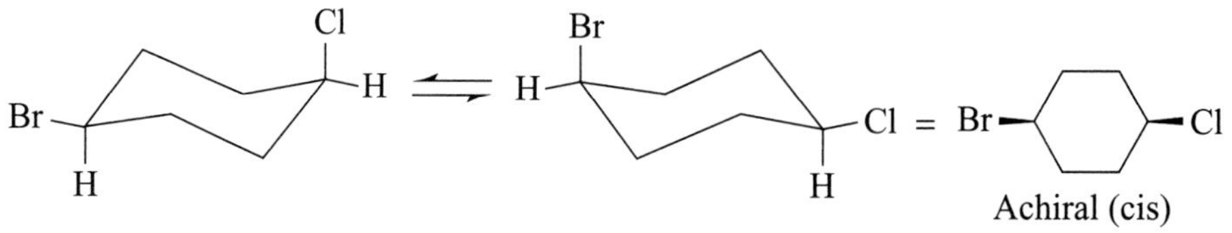

Achiral (cis)

5.30 See Problem 5.29. The molecules in (c) are achiral, so they have no (*R*-*S*) designation.

5.31

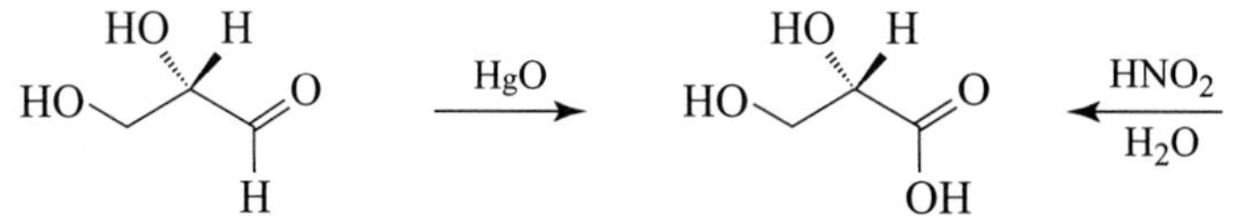

(*S*)-(−)-Glyceraldehyde (*S*)-(+)-Glyceric acid

(*S*)-(−)-Isoserine (see the following reaction also)

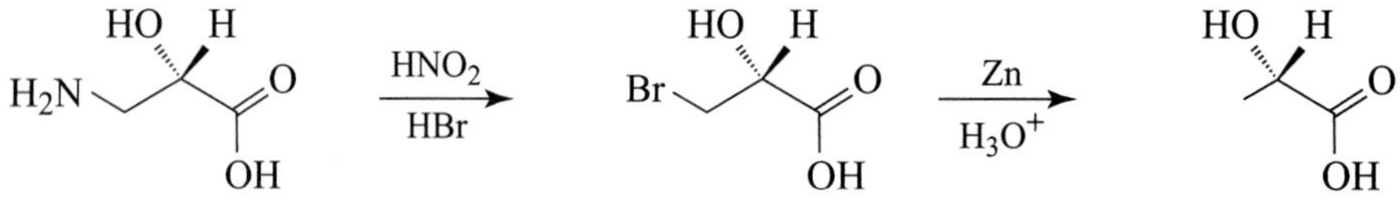

(*S*)-(−)-Isoserine (*R*)-(+)-3-Bromo-2-hydroxypropanoic acid (*S*)-(+)-Lactic acid

5.32 (a)

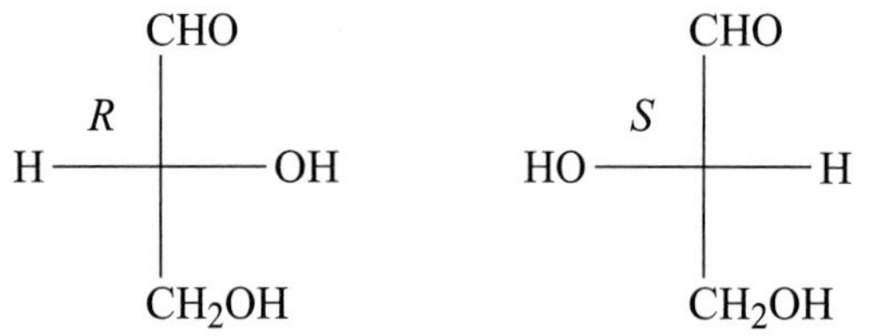

(*R*)-Glyceraldehyde (*S*)-Glyceraldehyde

(b)

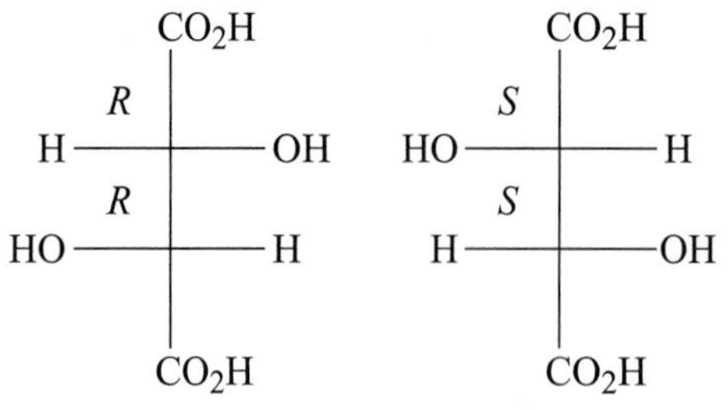

(+)-Tartaric acid (−)-Tartaric acid

(c)

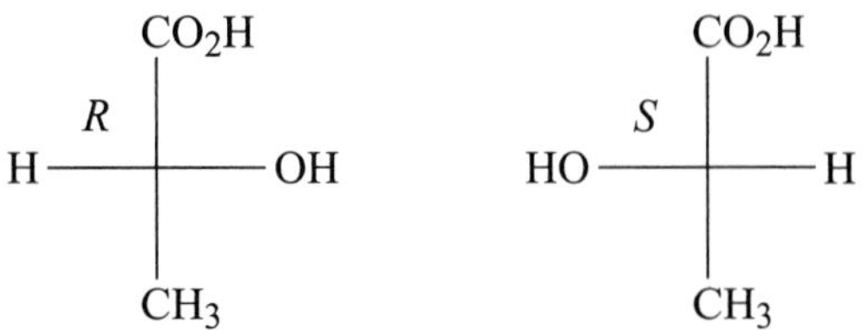

(*R*)-Lactic acid (*S*)-Lactic acid

Problems

Chirality and Stereoisomerism

5.33 (a), (b), (f), and (g) only

5.34 (a) Seven.

(b) (*R*)- and (*S*)-3-Methylhexane and (*R*)- and (*S*)-2,3-dimethylpentane.

5.35

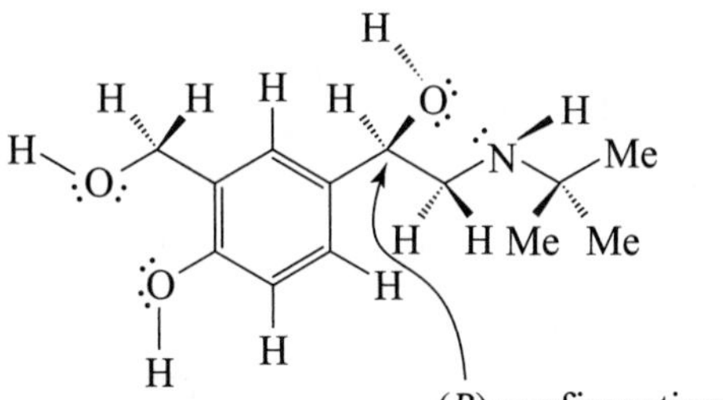

(*R*) configuration

5.36 (a)

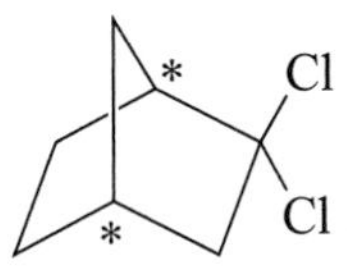

(b) Two, indicated by asterisks in (a)

(c) Four

(d) Because a trans arrangement of the one carbon bridge is structurally impossible. Such a molecule would have too much strain.

5.37 (a) **A** is (2*R*,3*S*)-2,3-dichlorobutane; **B** is (2*S*,3*S*)-2,3-dichlorobutane; **C** is (2*R*,3*R*)-2,3-dichlorobutane.

(b) **A**

5.38 (a) 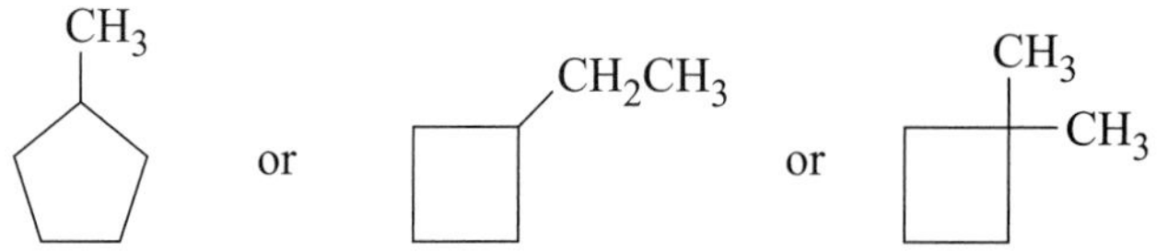 CH_3 or CH_2CH_3 or CH_3 CH_3 etc.

(b) 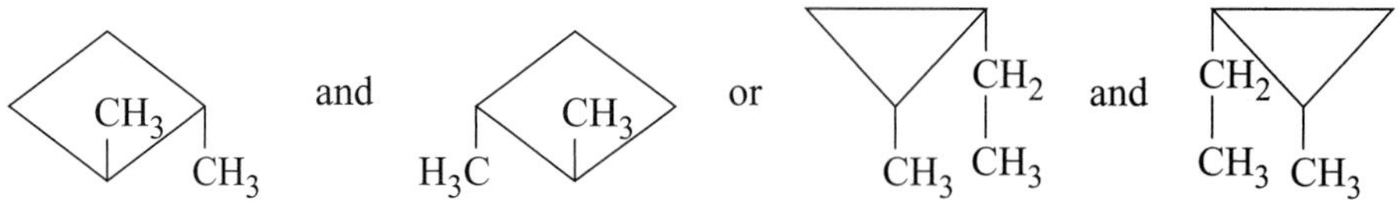 CH_3 CH_3 and H_3C CH_3 or CH_2 CH_3 CH_3 and CH_2 CH_3 CH_3

(other answers are possible)

(c) 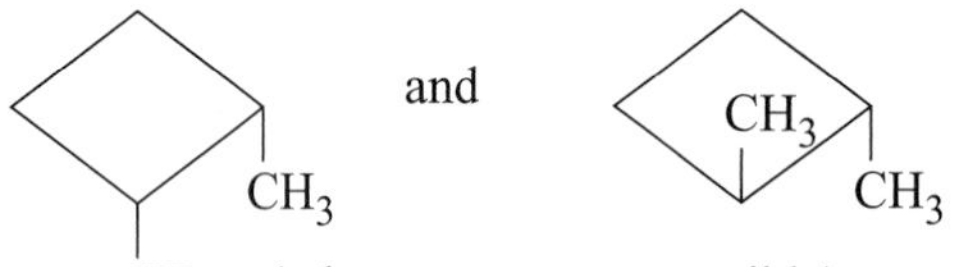 CH_3 CH_3 and CH_3 CH_3 (other answers are possible)

(d) 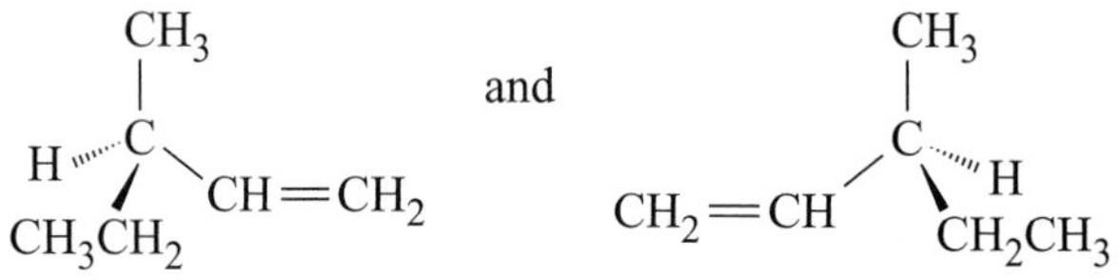 CH_3 H C $CH=CH_2$ CH_3CH_2 and CH_3 C H $CH_2=CH$ CH_2CH_3

(e)  H_3C C=C $CH_2CH_2CH_3$ H H and H_3C C=C H H $CH_2CH_2CH_3$

(other answers are possible)

5.39 (a) Same: (*S*)

(b) Enantiomers: left (*S*); right (*R*)

(c) Diastereomers: left (1*S*, 2*S*); right (1*R*, 2*S*)

(d) Same: (1*S*, 2*S*)

(e) Diastereomers: left (1*S*, 2*S*); right (1*S*, 2*R*)

(f) Constitutional isomers: both achiral

(g) Diastereomers: left, cis (4*S*); right, trans (4*R*)

(h) Enantiomers: left (1*S*, 3*S*); right (1*R*, 3*R*)

(i) Same: no chirality centers

(j) Different conformers of the same molecule (interconvertable by a ring flip): (1*R*, 2*S*)

(k) Diastereomers: left (1*R*, 2*S*); right (1*R*, 2*R*)

(l) Same: (1*R*, 2*S*)

(m) Diastereomers: no chirality center in either

(n) Constitutional isomers: no chirality center in either

(o) Diastereomers: no chirality center in either

(p) Same: no chirality center

(q) Same: no chirality center

5.40 All of these molecules are expected to be planar. Their stereochemistry is identical to that of the corresponding chloroethenes. (a) can exist as cis and trans isomers. Only one compound exists in the case of (b) and (c).

5.41 (a) diastereomers (c) enantiomers

(b) enantiomers (d) same compound

5.42

$CH_2{=}C(H)$ CH_2CH_3 C CH_3 H $CH_2{=}C(H)$ CH_2CH_3 C H CH_3

D (racemic)

CH_3CH_2 CH_2CH_3 C CH_3 H

E (achiral)

5.43 CH_3 C=C=C CH_3 H H $\xrightarrow[\text{Pd, Pt, or Ni, pressure}]{H_2}$ $CH_3CH_2CH_2CH_2CH_3$

(or enantiomer) **F**   (achiral) **G**

5.44 CH_3 $\xrightarrow[\text{Pd, Pt, or Ni, pressure}]{H_2}$ CH_3

(or enantiomer) **H**   (achiral) **I**

or

CH_2CH_3 $\xrightarrow[\text{Pd, Pt, or Ni, pressure}]{H_2}$ CH_2CH_3

(or enantiomer)   (achiral)

5.45

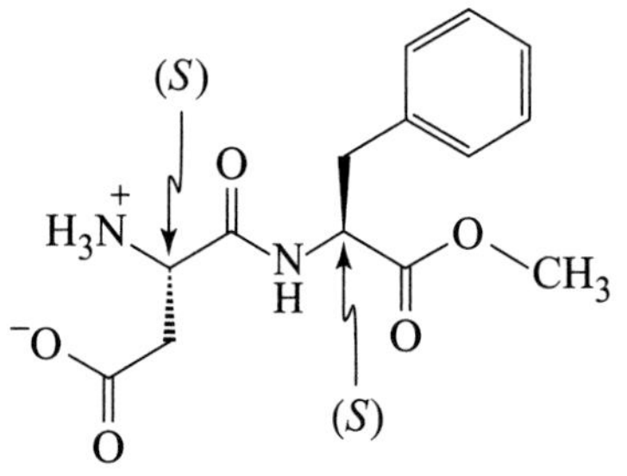

(*S*) (*S*)

Aspartame

5.46 (a)

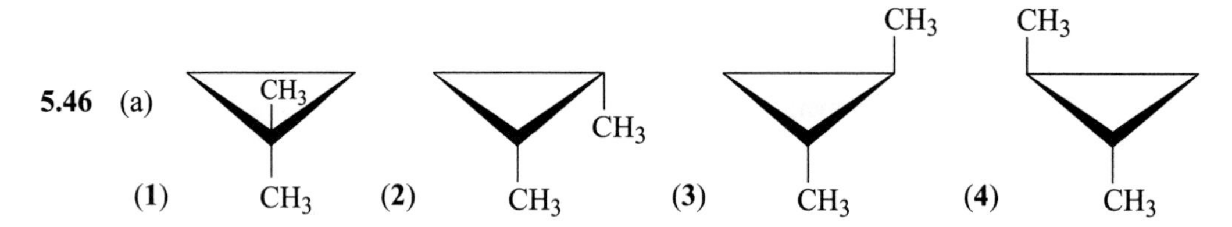

(b) (**3**) and (**4**) are chiral and are enantiomers of each other.

(c) Three fractions: a fraction containing (**1**), a fraction containing (**2**), and a fraction containing (**3**) and (**4**) [because, being enantiomers, (**3**) and (**4**) would have the same vapor pressure].

(d) None

5.47 (a)

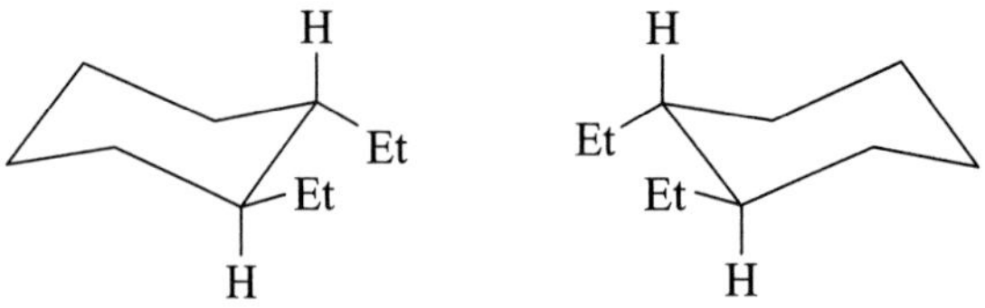

(b) No, they are not superposable.

(c) No, and they are, therefore, enantiomers of each other.

(d)

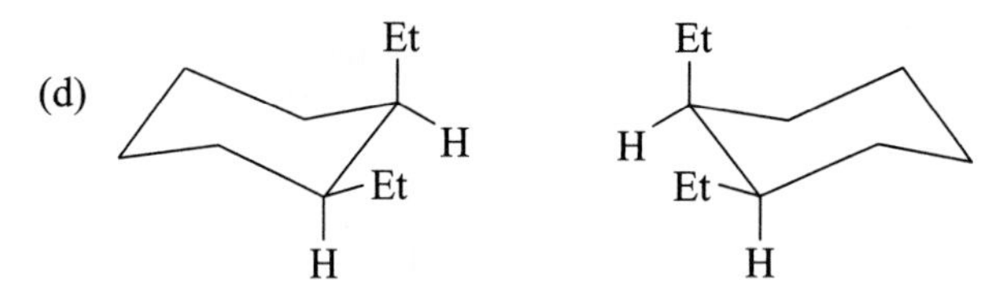

(e) No, they are not superposable.

(f) Yes, and they are, therefore, just different conformations of the same molecule.

5.48 (a)

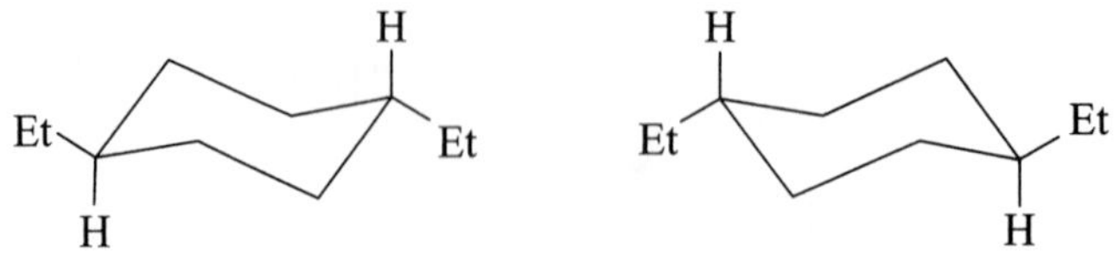

(b) Yes, and, therefore, *trans*-1.4-diethylcyclohexane is achiral.

(c) No, they are different orientations of the same molecule.

(d) Yes, *cis*-1,4-diethylcyclohexane is a stereoisomer (a diastereomer) of *trans*-1.4-diethylcyclohexane.

cis-1,4-Diethylcyclohexane

(e) No, it, too, is superposable on its mirror image. (Notice, too, that the plane of the page constitutes a plane of symmetry for both *cis*-1,4-diethylcyclohexane and for *trans*-1,4-diethylcyclohexane as we have drawn them.)

5.49 *trans*-1,3-Diethylcyclohexane can exist in the following enantiomeric forms.

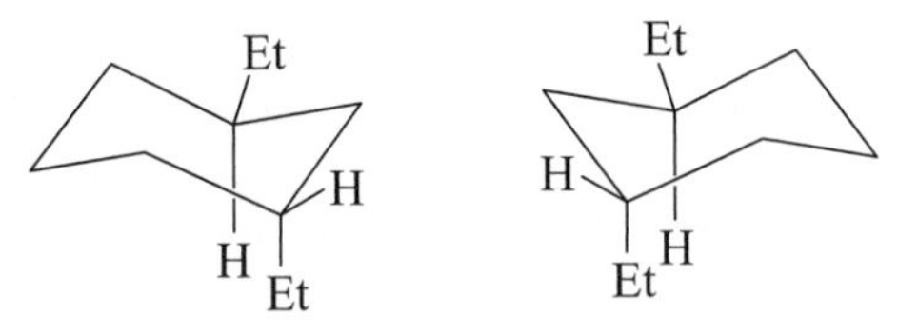

trans-1,3-Diethylcyclohexane enantiomers

cis-1,3-Diethylcyclohexane consists of achiral molecules because they have a plane of symmetry. [The plane of the page (below) is a plane of symmetry.]

cis-1,3-Diethylcyclohexane (meso)

Challenge Problems

5.50 (a) Since it is optically active and not resolvable, it must be the meso form:

(meso) (b) (*R*, *R*) (*S*, *S*)

(c) No (d) A racemic mixture

5.51 (a) $[\alpha]_D = \dfrac{-30}{(0.10\,\text{g/mL})(1.0\,\text{dm})} = -300$

(b) $[\alpha]_D = \dfrac{+165}{(0.05\,\text{g/mL})(1.0\,\text{dm})} = +3300$

The two rotation values can be explained by recognizing that this is a powerfully optically active substance and that the first reading, assumed to be −30, was really +330. Making this change the $[\alpha]_D$ becomes +3300 in both cases.

(c) No, the apparent 0 rotation could actually be + or −360 (or an integral multiple of these values).

5.52 Yes, it could be a *meso* form or an enantiomer whose chirality centers, by rare coincidence, happen to cancel each other's activities.

5.53 A compound $C_3H_6O_2$ has an index of hydrogen deficiency of 1. Thus, it could possess a carbon-carbon double bond, a carbon-oxygen double bond, or a ring.

The IR spectral data rule out a carbonyl group but indicate the presence of an —OH group.

No stable structure having molecular formula $C_3H_6O_2$ with a C=C bond can exist in stereoisomeric forms but 1,2-cyclopropanediol can exist in three stereoisomeric forms.

Only ethylene oxide (oxirane) derivatives are possible for Y.

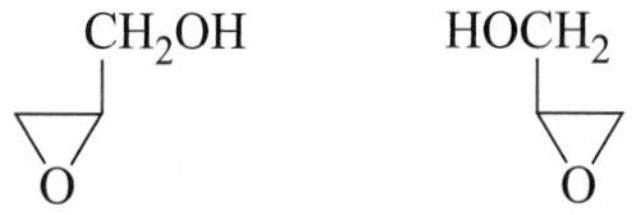

CH_2OH $HOCH_2$

O   O

QUIZ

5.1 Describe the relationship between the two structures shown.

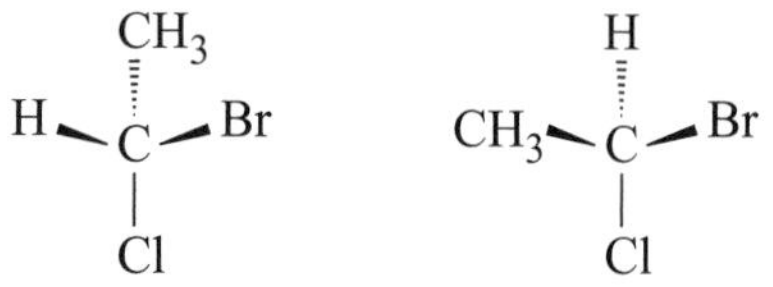

CH_3 H

H—C—Br CH_3—C—Br

Cl   Cl

(a) Enantiomers (b) Diastereomers (c) Constitutional isomers

(d) Conformations (e) Two molecules of the same compound

5.2 Which of the following molecule(s) possess(es) a plane of symmetry?

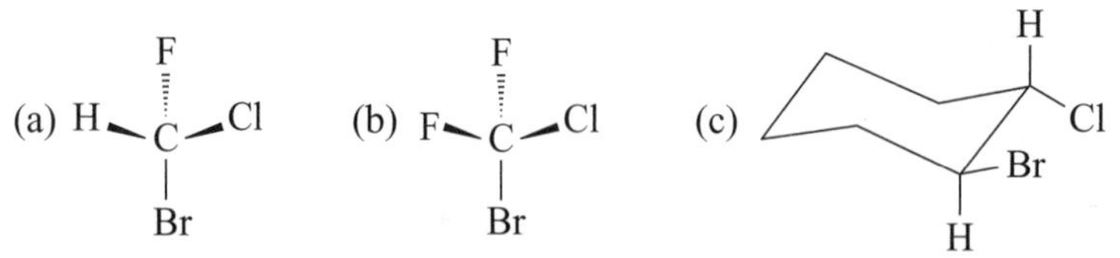

(a) F, H—C—Cl, Br   (b) F, F—C—Cl, Br   (c) H, Cl, Br, H

(d) More than one of these (e) None of these

5.3 Give the (*R*-*S*) designation of the structure shown:

O CH$_3$
HO—C—C—H
Cl

(a) (*R*) (b) (*S*) (c) Neither, because this molecule has no chirality center.

(d) Impossible to tell

5.4 Select the words that best describe the following structure:

CH$_3$
H—C—Cl
H—C—Cl
CH$_3$

(a) Chiral (b) Meso form (c) Achiral (d) Has a plane of symmetry

(e) More than one of these

5.5 Select the words that best describe what happens to the optical rotation of the alkene shown when it is hydrogenated to the alkane according to the following equation:

H H
CH$_3$CH$_2$—C—CH$_3$ + H$_2$ $\xrightarrow[\text{pressure}]{\text{Ni}}$ CH$_3$CH$_2$—C—CH$_3$
(*R*) CH=CH$_2$ CH$_2$CH$_3$

(a) Increases (b) Changes to zero (c) Changes sign

(d) Stays the same (e) Impossible to predict

5.6 There are two compounds with the formula C_7H_{16} that are capable of existing as enantiomers. Write three-dimensional formulas for the (*S*) isomer of each.

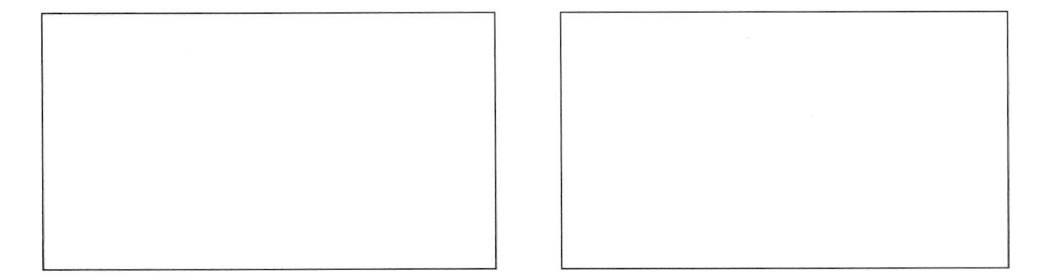

5.7 Compound A is optically active and is the (*S*) isomer.

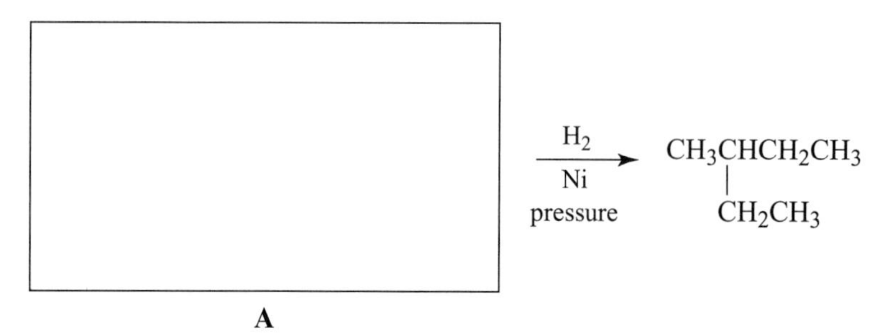

A

5.8 Compound B is a hydrocarbon with the minimum number of carbon atoms necessary for it to possess a chirality center and, as well, alternative stereochemistries about a double bond.

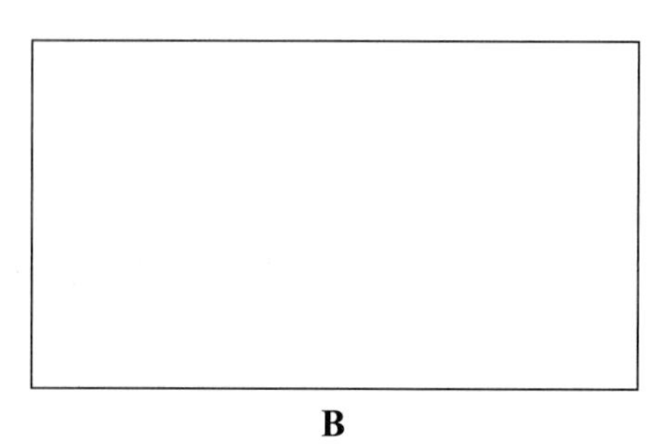

B

5.9 Which is untrue about the following structure?

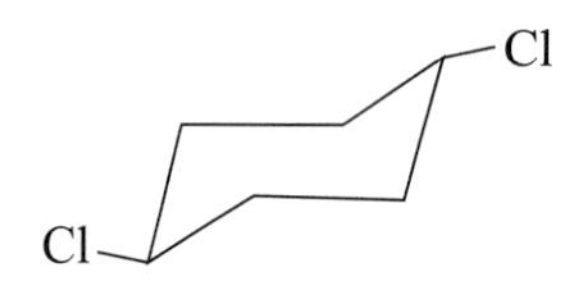

(a) It is the most stable of the possible conformations.

(b) $\mu = 0$ D

(c) It is identical to its mirror image.

(d) It is optically active.

(e) (*R*,*S*) designations cannot be applied.

5.10

$$
\begin{array}{c}
\mathrm{CH_2OH} \\
| \\
\mathrm{H-C-OH} \\
| \\
\mathrm{HO-C-H} \\
| \\
\mathrm{H-C-OH} \\
| \\
\mathrm{CH_3}
\end{array}
$$

is a Fischer projection of one of ________ stereoisomers.

(a) 2 (b) 3 (c) 4 (d) 7 (e) 8

6 IONIC REACTIONS — NUCLEOPHILIC SUBSTITUTION AND ELIMINATION REACTIONS OF ALKYL HALIDES

SOLUTIONS TO PROBLEMS

6.1 (a) *cis*-1-Bromo-2-methylcyclohexane

(b) *cis*-1-Bromo-3-methylcyclohexane

(c) 2,3,4-Trimethylheptane

6.2 (a) 3° (b) vinylic (c) 2° (d) aryl (e) 1°

6.3 (a) $CH_3{-}\ddot{\underset{..}{I}}{:} + CH_3CH_2{-}\ddot{\underset{..}{O}}{:}^- \longrightarrow CH_3{-}\ddot{\underset{..}{O}}{-}CH_2CH_3 + {:}\ddot{\underset{..}{I}}{:}^-$

Substrate; Nucleophile; Leaving group

(b) ${:}\ddot{\underset{..}{I}}{:}^- + CH_3CH_2{-}\ddot{\underset{..}{Br}}{:} \longrightarrow CH_3CH_2{-}\ddot{\underset{..}{I}}{:} + {:}\ddot{\underset{..}{Br}}{:}^-$

Nucleophile; Substrate; Leaving group

(c) $2\ CH_3\ddot{\underset{..}{O}}H + (CH_3)_3C{-}\ddot{\underset{..}{Cl}}{:} \longrightarrow (CH_3)_3C{-}\ddot{\underset{..}{O}}{-}CH_3 + {:}\ddot{\underset{..}{Cl}}{:}^- + CH_3OH_2^+$

Nucleophile; Substrate; Leaving group

(d) (propyl bromide, Br:) $+\ {}^-{:}C{\equiv}N{:} \longrightarrow$ (propyl–C≡N:) $+\ {:}\ddot{\underset{..}{Br}}{:}^-$

Substrate; Nucleophile; Leaving group

(e) (benzyl bromide, Br:) $+\ 2\ \ddot{N}H_3 \longrightarrow$ (benzyl–$\ddot{N}H_2$) $+\ {:}\ddot{\underset{..}{Br}}{:}^- + NH_4^+$

Substrate; Nucleophile; Leaving group

6.4

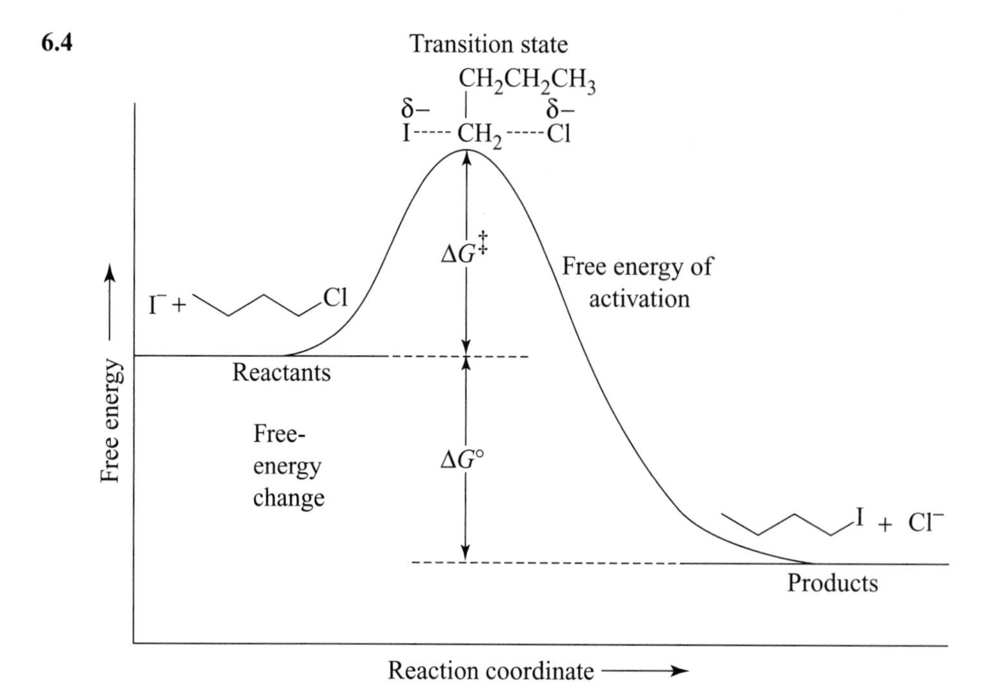

6.5

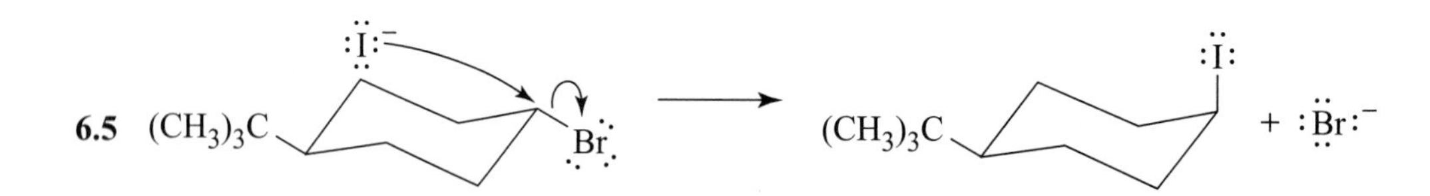

6.6 (a) We know that when a secondary alkyl halide reacts with hydroxide ion by substitution, the reaction occurs with *inversion of configuration* because the reaction is S_N2. If we know that the configuration of (−)-2-butanol (from Section 5.8C) is that shown here, then we can conclude that (+)-2-chlorobutane has the opposite configuration.

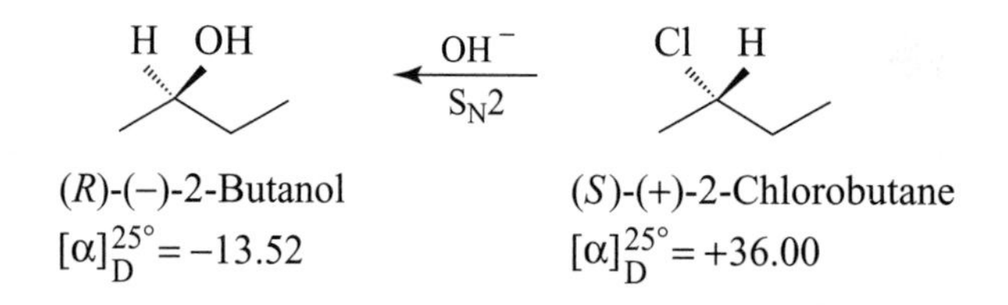

(b) Again the reaction is S_N2. Because we now know the configuration of (+)-2-chlorobutane to be (*S*) [cf., part (a)], we can conclude that the configuration of (−)-2-iodobutane is (*R*).

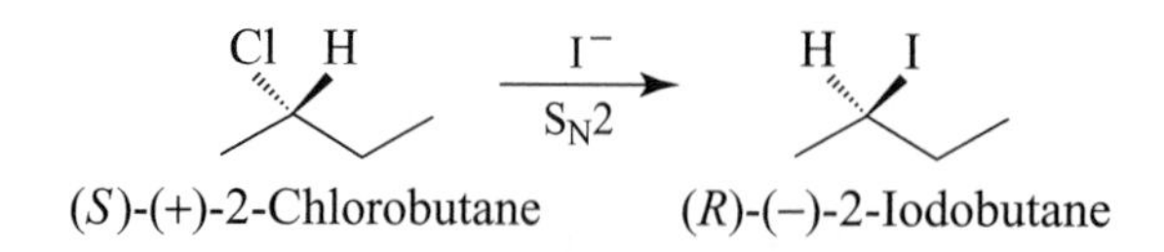

(+)-2-Iodobutane has the (*S*) configuration.

6.7 (b) < (c) < (a) in order of increasing stability

6.8 (a, b)

$(CH_3)_3C$–[cyclohexane]–I, CH_3 $\xrightarrow[S_N1]{H_2O}$ $(CH_3)_3C$–[cyclohexyl cation]$^+$–CH_3 + $:\ddot{O}H_2$ (paths (a) and (b))

$+ :A^-$ ↓ $- HA$

$(CH_3)_3C$–[cyclohexane]–CH_3, OH + $(CH_3)_3C$–[cyclohexane]–OH, CH_3

By path (a) By path (b)

6.9 $(CH_3)_3C$–[cyclohexane]–CH_3, OCH_3 and $(CH_3)_3C$–[cyclohexane]–OCH_3, CH_3

6.10 (c) is most likely to react by an S_N1 mechanism because it is a tertiary alkyl halide, whereas (a) is primary and (b) is secondary.

6.11 (a) Being primary halides, the reactions are most likely to be S_N2, with the nucleophile in each instance being a molecule of the solvent (i.e., a molecule of ethanol).

(b) Steric hindrance is provided by the substituent or substituents on the carbon β to the carbon bearing the leaving group. With each addition of a methyl group at the β carbon (below), the number of pathways open to the attacking nucleophile becomes fewer.

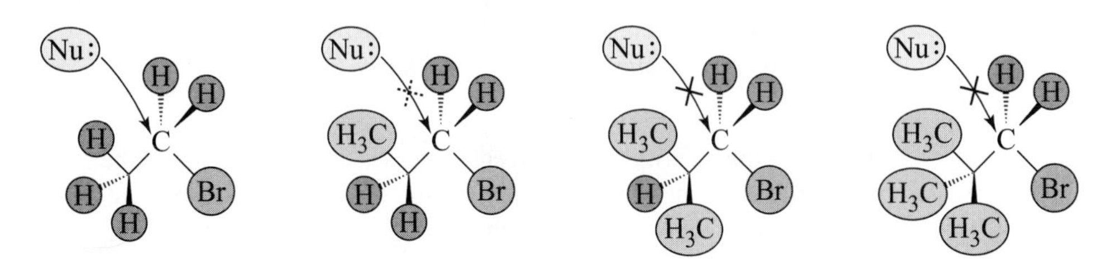

6.12 $CN^- > CH_3O^- > CH_3CO_2^- > CH_3CO_2H > CH_3OH$
Order of decreasing nucleophilicity in methanol

6.13 $CH_3S^- > CH_3O^- > CH_3CO_2^- > CH_3SH > CH_3OH$
Order of decreasing nucleophilicity in methanol

6.14 **Protic solvents** are those that have an H bonded to an oxygen or nitrogen (or to another strongly electronegative atom). Therefore, the protic solvents are formic acid, $HC(=O)OH$; formamide, $HC(=O)NH_2$; ammonia, NH_3; and ethylene glycol, $HOCH_2CH_2OH$.
Aprotic solvents lack an H bonded to a strongly electronegative element. Aprotic solvents in this list are acetone, $CH_3C(=O)CH_3$; acetonitrile, $CH_3C{\equiv}N$; sulfur dioxide, SO_2; and trimethylamine, $N(CH_3)_3$.

6.15 The reaction is an S_N2 reaction. In the polar aprotic solvent (DMF), the nucleophile (CN^-) will be relatively unencumbered by solvent molecules, and, therefore, it will be more reactive than in ethanol. As a result, the reaction will occur faster in *N,N*-dimethylformamide.

6.16 (a) CH_3O^-

(b) H_2S

(c) $(CH_3)_3P$

6.17 (a) Increasing the percentage of water in the mixture increases the polarity of the solvent. (Water is more polar than methanol.) Increasing the polarity of the solvent increases the rate of the solvolysis because separated charges develop in the transition state. The more polar the solvent, the more the transition state is stabilized (Section 6.13D).

(b) In an S_N2 reaction of this type, the charge becomes dispersed in the transition state:

$$I^- + CH_3CH_2{-}Cl \longrightarrow \left[\overset{\delta-}{I}\cdots\cdots C(CH_3)(H)(H)\cdots\cdots\overset{\delta-}{Cl}\right]^{\ddagger} \longrightarrow ICH_2CH_3 + Cl^-$$

Reactants
Charge is concentrated

Transition state
Charge is dispersed

Increasing the polarity of the solvent increases the stabilization of the reactant I^- more than the stabilization of the transition state, and thereby increases the free energy of activation, thus decreasing the rate of reaction.

6.18 $CH_3OSO_2CF_3 > CH_3I > CH_3Br > CH_3Cl > CH_3F > {}^{14}CH_3OH$

(Most reactive) (Least reactive)

6.19 (a)

$$CH_3CH_2O^- Na^+ + (S)\text{-}CH_3(CH_3CH_2)(H)C\text{—}Br \xrightarrow[S_N2]{\text{inversion}} (R)\text{-}CH_3CH_2O\text{—}C(CH_3)(H)(CH_2CH_3) + Na^+ Br^-$$

(b)

$$CH_3\overset{O}{\overset{\|}{C}}O^- Na^+ + (S)\text{-}CH_3(CH_3CH_2)(H)C\text{—}Br \xrightarrow[S_N2]{\text{inversion}} (R)\text{-}CH_3\overset{O}{\overset{\|}{C}}O\text{—}C(CH_3)(H)(CH_2CH_3) + Na^+ Br^-$$

(c)

$$HS^- Na^+ + (S)\text{-}CH_3(CH_3CH_2)(H)C\text{—}Br \xrightarrow[S_N2]{\text{inversion}} (R)\text{-}HS\text{—}C(CH_3)(H)(CH_2CH_3) + Na^+ Br^-$$

(d)

$$CH_3S^- Na^+ + (S)\text{-}CH_3(CH_3CH_2)(H)C\text{—}Br \xrightarrow[S_N2]{\text{inversion}} (R)\text{-}CH_3S\text{—}C(CH_3)(H)(CH_2CH_3) + Na^+ Br^-$$

Relative Rates of Nucleophilic Substitution

6.20 (a) 1-Bromopropane would react more rapidly because, being a primary halide, it is less hindered.

(b) 1-Iodobutane, because iodide ion is a better leaving group than chloride ion.

(c) 1-Chlorobutane, because the carbon bearing the leaving group is less hindered than in 1-chloro-2-methylpropane.

(d) 1-Chloro-3-methylbutane, because the carbon bearing the leaving group is less hindered than in 1-chloro-2-methylbutane.

(e) 1-Chlorohexane because it is a primary halide. Phenyl halides are unreactive in S_N2 reactions.

6.21 (a) Reaction (1) because ethoxide ion is a stronger nucleophile than ethanol.

(b) Reaction (2) because the ethyl sulfide ion is a stronger nucleophile than the ethoxide ion in a protic solvent. (Because sulfur is larger than oxygen, the ethyl sulfide ion is less solvated and it is more polarizable.)

(c) Reaction (2) because triphenylphosphine [$(C_6H_5)_3P$] is a stronger nucleophile than triphenylamine. (Phosphorus atoms are larger than nitrogen atoms.)

(d) Reaction (2) because in an S_N2 reaction the rate depends on the concentration of the substrate and the nucleophile. In reaction (2) the concentration of the nucleophile is twice that of the reaction (1).

6.22 (a) Reaction (2) because bromide ion is a better leaving group than chloride ion.

(b) Reaction (1) because water is a more polar solvent than methanol, and S_N1 reactions take place faster in more polar solvents.

(c) Reaction (2) because the concentration of the substrate is twice that of reaction (1). The major reaction would be E2. (However, the problem asks us to consider that small portion of the overall reaction that proceeds by an S_N1 pathway.)

(d) Considering only S_N1 reactions, as the problem specifies, both reactions would take place at the same rate because S_N1 reactions are independent of the concentration of the nucleophile. The predominant process in this pair of reactions would be E2, however.

(e) Reaction (1) because the substrate is a tertiary halide. Phenyl halides are unreactive in S_N1 reactions.

Synthesis

6.23 (a) Br + NaOH ⟶ OH + NaBr

(b) Br + NaI ⟶ I + NaBr

(c) Br + ONa ⟶ O + NaBr

(d) Br + CH_3SNa ⟶ S CH_3 + NaBr

(e) Br + O ONa ⟶ O O + NaBr

(f) Br + NaN_3 ⟶ N_3 + NaBr

(g) Br + $:N(CH_3)_3$ ⟶ $\overset{+}{N}(CH_3)_3\ Br^-$

(h) Br + NaCN ⟶ N + NaBr

(i) Br + NaSH ⟶ SH + NaBr

6.24 Possible methods are given here.

(a) $CH_3Cl \xrightarrow[CH_3OH,\ S_N2]{I^-} CH_3I$

(b) Cl $\xrightarrow[CH_3OH,\ S_N2]{I^-}$ I

(c) $CH_3Cl \xrightarrow[CH_3OH/H_2O \; S_N2]{OH^-} CH_3OH$

(d) ethyl chloride (Cl) $\xrightarrow[CH_3OH/H_2O \; S_N2]{OH^-}$ ethanol (OH)

(e) $CH_3Cl \xrightarrow[CH_3OH \; S_N2]{SH^-} CH_3SH$

(f) ethyl chloride (Cl) $\xrightarrow[CH_3OH \; S_N2]{SH^-}$ ethanethiol (SH)

(g) $CH_3I \xrightarrow[DMF]{CN^-} CH_3CN$

(h) ethyl bromide (Br) $\xrightarrow[DMF]{CN^-}$ propanenitrile (CN)

(i) $CH_3OH \xrightarrow[(-H_2)]{NaH} CH_3ONa \xrightarrow[CH_3OH]{CH_3I} CH_3OCH_3$

(j) ethanol (OH) $\xrightarrow[(-H_2)]{NaH}$ sodium ethoxide (ONa) $\xrightarrow{CH_3I}$ ethyl methyl ether (OMe)

(k) chlorocyclopentane (Cl) $\xrightarrow[\text{ethanol (OH)}]{\text{sodium ethoxide (ONa)}}$ cyclopentene

6.25 (a) The reaction will not take place because the leaving group would have to be a methyl anion, a very powerful base, and a very poor leaving group.

(b) The reaction will not take place because the leaving group would have to be a hydride ion, a very powerful base, and a very poor leaving group.

(c) The reaction will not take place because the leaving group would have to be a carbanion, a very powerful base, and a very poor leaving group.

$HO^- +$ cyclobutane $\not\rightarrow$ $^-$CH₂CH₂CH₂CH₂OH → butanol (OH)

(d) The reaction will not take place by an S_N2 mechanism because the substrate is a tertiary halide, and is, therefore, not susceptible to S_N2 attack because of the steric hindrance. (A very small amount of S_N1 reaction may take place, but the main reaction will be E2 to produce an alkene.)

(e) The reaction will not take place because the leaving group would have to be a CH_3O^- ion, a strong base, and a very poor leaving group.

$$:NH_3 + CH_3{-}OCH_3 \not\longrightarrow CH_3NH_3^+ + CH_3O^-$$

$$\downarrow$$

$$CH_3NH_2 + CH_3OH$$

(f) The reaction will not take place because the first reaction that would take place would be an acid-base reaction that would convert the ammonia to an ammonium ion. An ammonium ion, because it lacks an electron pair, is not nucleophilic.

$$NH_3 + CH_3OH_2^+ \rightleftharpoons NH_4^+ + CH_3OH$$

6.26 The better yield will be obtained by using the secondary halide, 1-bromo-1-phenylethane, because the desired reaction is E2. Using the primary halide will result in substantial S_N2 reaction as well, producing the alcohol as well as the desired alkene.

6.27 Reaction (2) would give the better yield because the desired reaction is an S_N2 reaction, and the substrate is a methyl halide. Use of reaction (1) would, because the substrate is a secondary halide, result in considerable elimination by an E2 pathway.

6.28 (a)

OH → NaH, Et_2O $(-H_2)$ → O^-Na^+

Br → (−NaBr) → O

(b) SH → NaH, Et_2O $(-H_2)$ → S^-Na^+ → Br, (−NaBr) → S

(c) OH —NaH, ethyl ether ($-H_2$)→ $O^- Na^+$ —CH_3I ($-NaI$)→ $O-CH_3$

(d) OH —NaH, ethyl ether ($-H_2$)→ $O^- Na^+$ —CH_3I ($-NaI$)→ $O-CH_3$

(e) Br + Na^+ $^-$:CN: —ethanol (OH) ($-NaBr$)→ CN:

(f) O, Ö:$^-$ Na^+ + Br —CH_3CO_2H ($-NaBr$)→ O, O

(g) Na^+ OH^- + H Br —acetone ($-NaBr$)→ HO H

(R)-2-Bromopentane (S)-2-Pentanol

(h) Na^+ I^- Cl H —acetone ($-NaCl$)→ I H

(S)-2-Chloro-4-methylpentane (R)-2-Iodo-4-methylpentane

(i) Br —$EtO^- Na^+$, EtOH ($-NaBr$)→ + OH

(j) Br —Na^+OH^-, H_2O/CH_3OH ($-NaBr$)→ OH

(k) Na^+ $^-$:C≡N: + Br H —ethanol (OH) ($-NaBr$)→ H CN

(S)-2-Bromobutane

(l) Cl + Na^+ :Ï:$^-$ —acetone ($-NaCl$)→ I

General S_N1, S_N2, and Elimination

6.29 (a) The major product would be [structure] (by an S_N2 mechanism) because the substrate is primary and the nucleophile-base is not hindered. Some [structure] would be produced by an E2 mechanism.

(b) The major product would be [structure] (by an E2 mechanism), even though the substrate is primary, because the base is a hindered strong base. Some [structure] would be produced by an S_N2 mechanism.

(c) For all practical purposes, [structure] (by an E2 mechanism) would be the only product because the substrate is tertiary and the base is strong.

(d) Same answer as (c) above.

(e) [structure: t-Bu, I] (formed by an S_N2 mechanism) would, for all practical purposes, be the only product. Iodide ion is a very weak base and a good nucleophile.

(f) Because the substrate is tertiary and the base weak, an S_N1 reaction (solvolysis) will occur, accompanied by elimination (E1). At 25°C, the S_N1 reaction would predominate.

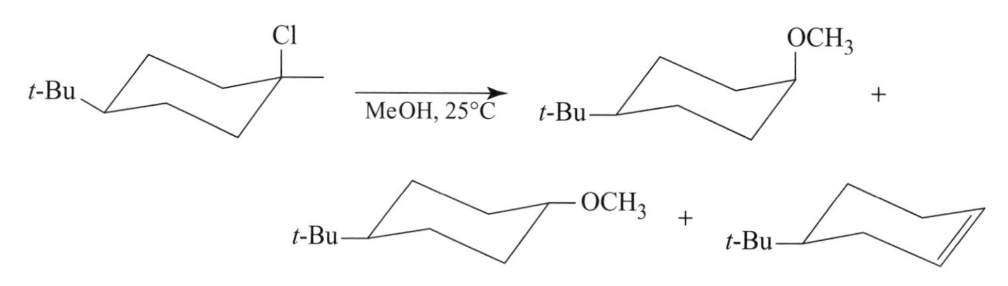

(g) [structure] [also (*Z*)] (by an E2 mechanism) would be the major product because the substrate is secondary and the base/nucleophile is a strong base. Some of the ether [structure: OCH_3] would be formed by an S_N2 pathway.

(h) The major product would be [structure: O, O] (by an S_N2 mechanism) because the acetate ion is a weak base. Some [structure] and [structure] might be formed by an E2 pathway.

(i) [structure] [also (*Z*)] and [structure] (by E2) would be major products, and [structure: HO H] [(*S*) isomer] (by S_N2) would be the minor products.

(j) OCH_3 (by S_N1) would be the major product. [also (Z)],

[also (Z)], and (by E1) would be minor products.

(k) H I (by S_N2) would be the only product.

6.30 (a), (b), and (c) are all S_N2 reactions and, therefore, proceed with inversion of configuration. The products are

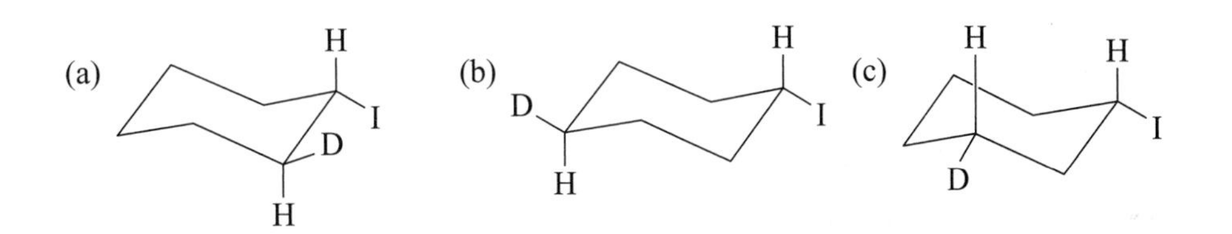

(d) is an S_N1 reaction. The carbocation that forms can react with either nucleophile (H_2O or CH_3OH) from either the top or bottom side of the molecule. Four substitution products (below) would be obtained. (Considerable elimination by an E1 path would also occur.)

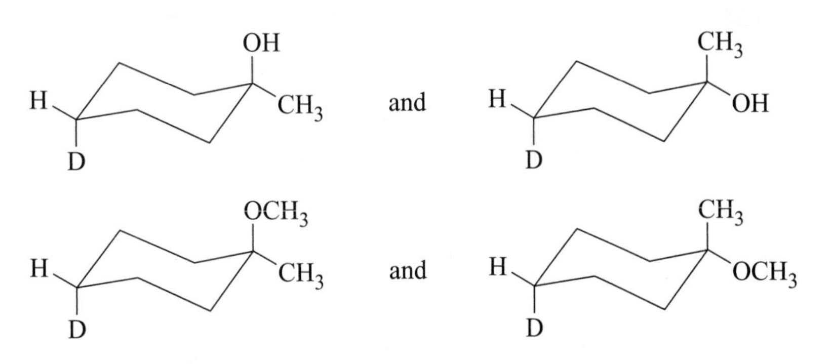

6.31 Isobutyl bromide is more sterically hindered than ethyl bromide because of the methyl groups on the β carbon atom.

CH_3, β, α, H—C—CH_2—Br, CH_3

Isobutyl bromide

H, H—C—CH_2—Br, H

Ethyl bromide

This steric hindrance causes isobutyl bromide to react more slowly in S_N2 reactions and to give relatively more elimination (by an E2 path) when a strong base is used.

6.32 (a) S_N2 because the substrate is a 1° halide.

(b) $\text{Rate} = k\,[CH_3CH_2Cl][I^-]$

$$= 5 \times 10^{-5}\ \text{L mol}^{-1}\text{s}^{-1} \times 0.1\ \text{mol L}^{-1} \times 0.1\ \text{mol L}^{-1}$$

$\text{Rate} = 5 \times 10^{-7}\ \text{mol L}^{-1}\text{s}^{-1}$

(c) $1 \times 10^{-6}\ \text{mol L}^{-1}\text{s}^{-1}$

(d) $1 \times 10^{-6}\ \text{mol L}^{-1}\text{s}^{-1}$

(e) $2 \times 10^{-6}\ \text{mol L}^{-1}\text{s}^{-1}$

6.33 (a) CH_3NH^- because it is the stronger base.

(b) CH_3O^- because it is the stronger base.

(c) CH_3SH because sulfur atoms are larger and more polarizable than oxygen atoms.

(d) $(C_6H_5)_3P$ because phosphorus atoms are larger and more polarizable than nitrogen atoms.

(e) H_2O because it is the stronger base.

(f) NH_3 because it is the stronger base.

(g) HS^- because it is the stronger base.

(h) OH^- because it is the stronger base.

6.34 (a) $HOCH_2CH_2Br + OH^- \rightleftharpoons {}^-OCH_2CH_2Br \longrightarrow$ (epoxide) $+ Br^-$

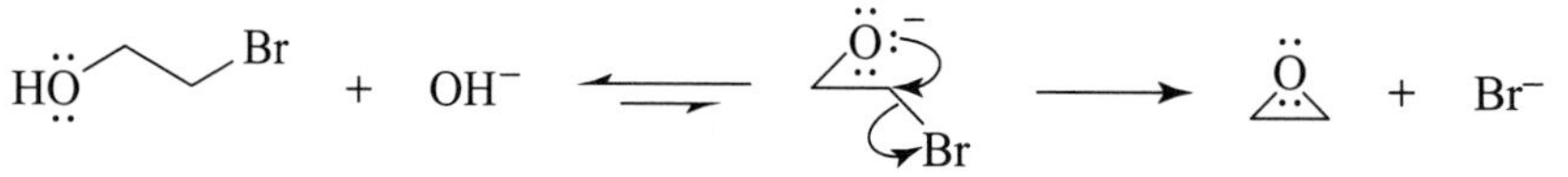

(b) (4-bromobutylamine, $\ddot{N}H_2$) $\longrightarrow$ (pyrrolidinium, $\overset{+}{N}H_2$) Br^- $\xrightarrow{OH^-}$ (pyrrolidine, N–H) $+ H_2O + Br^-$

6.35 $:N{\equiv}C:^- + CH_3CH_2\ddot{Br}: \longrightarrow \left[\, :N{\equiv}\overset{\delta-}{C} \cdots \overset{CH_3}{C}(H)(H) \cdots \overset{\delta-}{\ddot{Br}}: \,\right]^{\ddagger} \longrightarrow :N{\equiv}CCH_2CH_3 + :\ddot{Br}:^-$

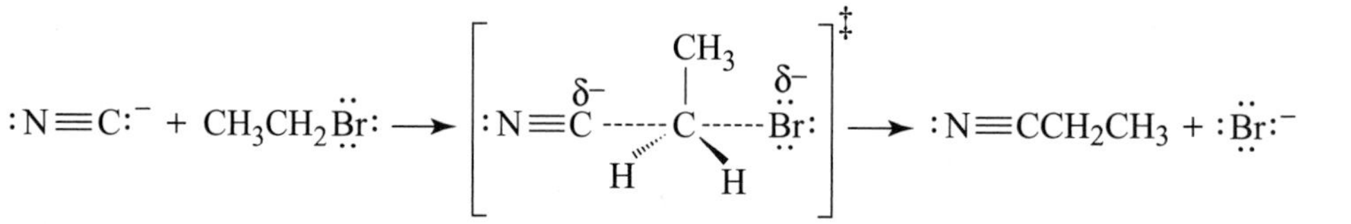

6.36 Iodide ion is a good nucleophile and a good leaving group; it can rapidly convert an alkyl chloride or alkyl bromide into an alkyl iodide, and the alkyl iodide can then react rapidly with another nucleophile. With methyl bromide in water, for example, the following reaction can take place:

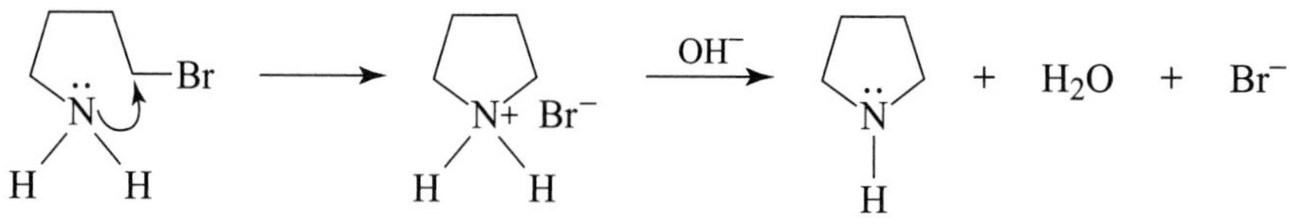

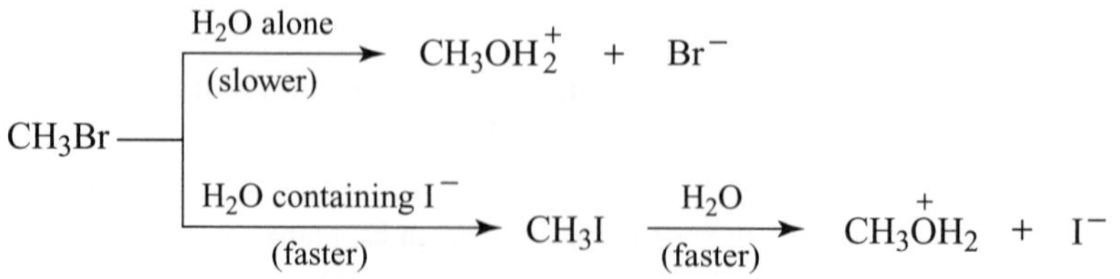

$$CH_3Br \xrightarrow[\text{(slower)}]{H_2O\ \text{alone}} CH_3\overset{+}{O}H_2 + Br^-$$

$$CH_3Br \xrightarrow[\text{(faster)}]{H_2O\ \text{containing}\ I^-} CH_3I \xrightarrow[\text{(faster)}]{H_2O} CH_3\overset{+}{O}H_2 + I^-$$

6.37 *tert*-Butyl alcohol and *tert*-butyl methyl ether are formed via an S_N1 mechanism. The rate of the reaction is independent of the concentration of methoxide ion (from sodium methoxide). This, however, is only one reaction that causes *tert*-butyl bromide to disappear. A competing reaction that also causes *tert*-butyl bromide to disappear is an E2 reaction in which methoxide ion reacts with *tert*-butyl bromide. This reaction is dependent on the concentration of methoxide ion; therefore, increasing the methoxide ion concentration causes an increase in the rate of disappearance of *tert*-butyl bromide.

6.38 (a) You should use a strong base, such as RO^-, at a higher temperature to bring about an E2 reaction.

(b) Here we want an S_N1 reaction. We use ethanol as the solvent *and as the nucleophile*, and we carry out the reaction at a low temperature so that elimination will be minimized.

6.39 1-Bromobicyclo[2.2.1]heptane is unreactive in an S_N2 reaction because it is a tertiary halide and its ring structure makes the backside of the carbon bearing the leaving group completely inaccessible to attack by a nucleophile.

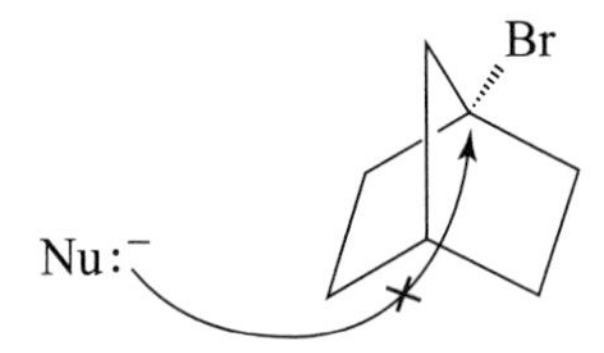

1-Bromobicyclo[2.2.1]heptane is unreactive in an S_N1 reaction because the ring structure makes it impossible for the carbocation that must be formed to assume the required trigonal planar geometry around the positively charged carbon. Any carbocation formed from 1-bromobicyclo[2.2.1]heptane would have a trigonal pyramidal arrangement of the $-CH_2-$ groups attached to the positively charged carbon (make a model). Such a structure does not allow stabilization of the carbocation by overlap of sp^3 orbitals from the alkyl groups (see Fig. 6.7).

6.40 The cyanide ion has two nucleophilic atoms; it is what is called an ambident nucleophile.

$$^{-}{:}C{\equiv}N{:}$$

It can react with a substrate using either atom, although the carbon atom is more nucleophilic.

$$Br{-}CH_2CH_3 + {}^{-}{:}C{\equiv}N{:} \longrightarrow CH_3CH_2{-}C{\equiv}N{:}$$

$$^{-}{:}C{\equiv}N{:} + CH_3CH_2{-}Br \longrightarrow CH_3CH_2{-}\ddot{N}{=}C{:}$$

6.41 (a) 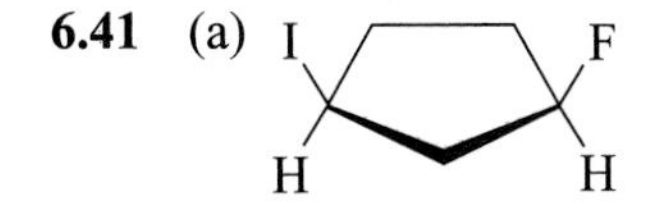

(Formation of this product depends on the fact that bromide ion is a much better leaving group than fluoride ion.)

(b) 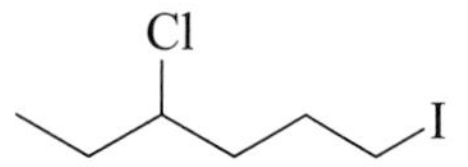 [structure: 1-iodo-4-chlorohexane skeleton with Cl and I]

(Formation of this product depends on the greater reactivity of 1° substrates in S_N2 reactions.)

(c)  [structure: 1,4-dithiane ring with two S]

(Here two S_N2 reactions produce a cyclic molecule.)

(d) 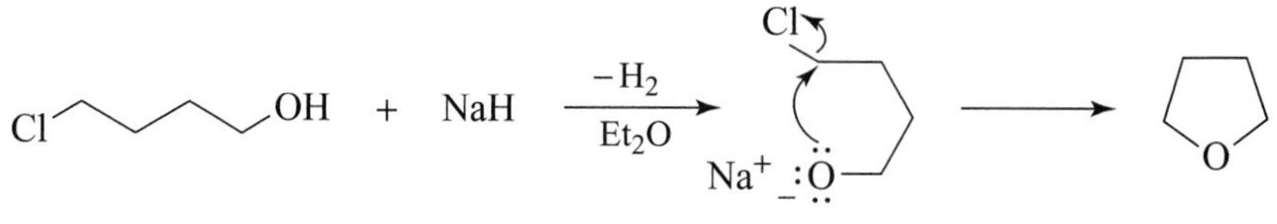 Cl∕∕∕∕OH + NaH $\xrightarrow[Et_2O]{-H_2}$ Na$^+$ $^-$:Ö– (ring with Cl) ⟶ O (tetrahydrofuran)

(e) 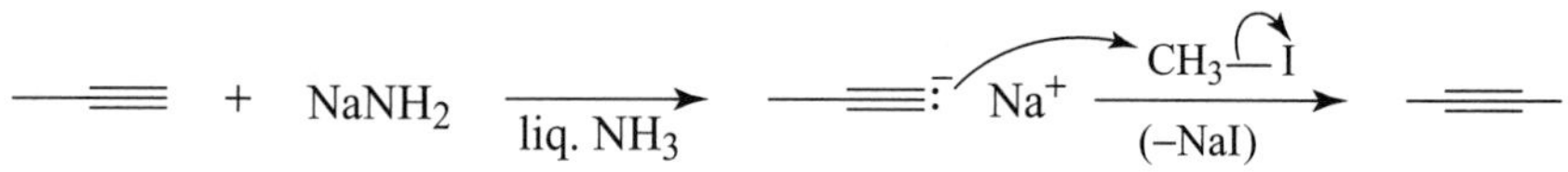 —≡— + $NaNH_2$ $\xrightarrow[\text{liq. } NH_3]{}$ —≡:$^-$ Na$^+$ $\xrightarrow[(-NaI)]{CH_3-I}$ —≡—

6.42 The rate-determining step in the S_N1 reaction of *tert*-butyl bromide is the following:

$$(CH_3)_3C-Br \xrightarrow{\text{slow}} (CH_3)_3C^+ + Br^-$$

$$(CH_3)_3C^+ \xrightarrow{H_2O} (CH_3)_3COH_2^+$$

$(CH_3)_3C^+$ is so unstable that it reacts almost immediately with one of the surrounding water molecules, and, for all practical purposes, no reverse reaction with Br^- takes place. Adding a common ion (Br^- from NaBr), therefore, has no effect on the rate.

Because the $(C_6H_5)_2CH^+$ cation is more stable, a reversible first step occurs and adding a common ion (Br^-) slows the overall reaction by increasing the rate at which $(C_6H_5)_2CH^+$ is converted back to $(C_6H_5)_2CHBr$.

$$(C_6H_5)_2CHBr \rightleftharpoons (C_6H_5)_2CH^+ + Br^-$$

$$(C_6H_5)_2CH^+ \xrightarrow{H_2O} (C_6H_5)_2CHOH_2^+$$

6.43 Two different mechanisms are involved. $(CH_3)_3CBr$ reacts by an S_N1 mechanism, and apparently this reaction takes place faster. The other three alkyl halides react by an S_N2 mechanism, and their reactions are slower because the nucleophile (H_2O) is weak. The reaction rates of CH_3Br, CH_3CH_2Br, and $(CH_3)_2CHBr$ are affected by the steric hindrance, and thus their order of reactivity is $CH_3Br > CH_3CH_2Br > (CH_3)_2CHBr$.

6.44 The nitrite ion is an *ambident nucleophile*; that is, it is an ion with two nucleophilic sites. The equivalent oxygen atoms and the nitrogen atom are nucleophilic.

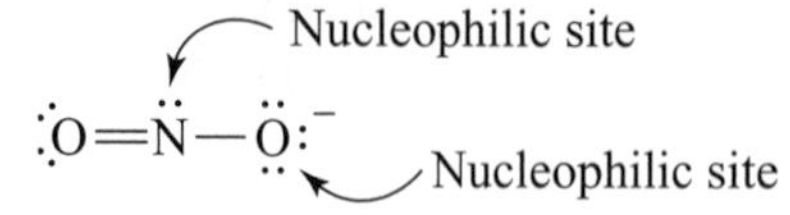

$$:\ddot{O}=\ddot{N}-\ddot{O}:^-$$

Nucleophilic site (N)

Nucleophilic site (O)

6.45 (a) The transition state has the form:

$$\overset{\delta^+}{\text{Nu}}\text{----R-----}\overset{\delta^-}{\text{L}}$$

in which charges are developing. The more polar the solvent, the better it can solvate the transition state, thus lowering the free energy of activation and increasing the reaction rate.

(b) The transition state has the form:

$$\overset{\delta^+}{\text{R}}\text{---------}\overset{\delta^+}{\text{L}}$$

in which the charge is becoming dispersed. A polar solvent is less able to solvate this transition state than it is to solvate the reactant. The free energy of activation, therefore, will become somewhat larger as the solvent polarity increases, and the rate will be slower.

6.46 (a) Cl I

(b) HO Cl + some alkene

6.47 (a) In an S_N1 reaction the carbocation intermediate reacts rapidly with any nucleophile it encounters in a Lewis acid-Lewis base reaction. In the case of the S_N2 reaction, the leaving group departs only when "pushed out" by the attacking nucleophile and some nucleophiles are better than others.

(b) CN^- is a much better nucleophile than ethanol and hence the nitrile is formed in the S_N2 reaction of Cl. In the case of Cl, the *tert*-butyl cation reacts chiefly with the nucleophile present in higher concentration, here the ethanol solvent.

Challenge Problems

6.48 (a) The entropy term is slightly favorable. (The enthalpy term is highly unfavorable.)

(b)

$$\begin{aligned}\Delta G^\circ &= \Delta H^\circ - T\Delta S^\circ \\ &= 26.6 \text{ kJ mol}^{-1} - (298)(0.00481 \text{ kJ mol}^{-1}) \\ &= 25.2 \text{ kJ mol}^{-1}\end{aligned}$$

The hydrolysis process will not occur to any significant extent.

(c)

$$\begin{aligned}\log K_{eq} &= \frac{-\Delta G^\circ}{2.303RT} \\ &= \frac{-25.2 \text{ kJ mol}^{-1}}{(2.303)(0.008314 \text{ kJ mol}^{-1}\text{K}^{-1})(298 \text{ K})} \\ &= -4.4165 \\ K_{eq} &= 10^{-4.4165} = 3.85 \times 10^{-5}\end{aligned}$$

(d) The equilibrium is very much more favorable in aqueous solution because solvation of the products (ethanol, hydronium ions, and chloride ions) takes place and thereby stabilizes them.

6.49 The mechanism for the reaction, in which a low concentration of OH^- is used in the presence of Ag_2O, involves the participation of the carboxylate group. In step 1 (see following reaction) an oxygen of the carboxylate group attacks the chirality center from the back side and displaces bromide ion. (Silver ion aids in this process in much the same way that protonation assists the ionization of an alcohol.) The configuration of the chirality center inverts in step 1, and a cyclic ester called an α-lactone forms.

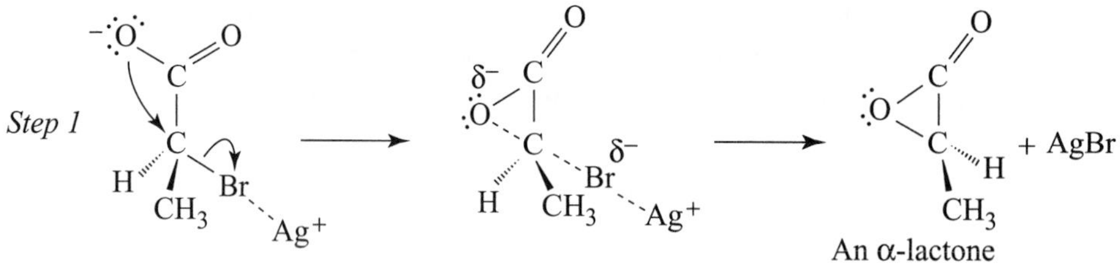

Step 1

+ AgBr

An α-lactone

The highly strained three-membered ring of the α-lactone opens when it is attacked by a water molecule in step 2. *This step also takes place with an inversion of configuration.*

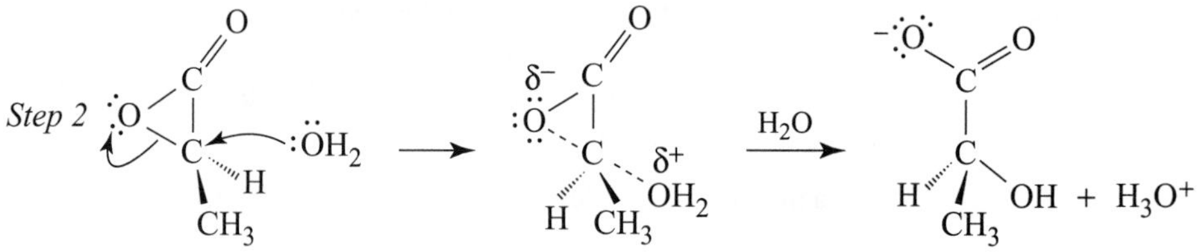

Step 2

H_2O

$OH + H_3O^+$

The net result of two inversions (in steps 1 and 2) is an overall *retention of configuration.*

6.50 (a) and (b)

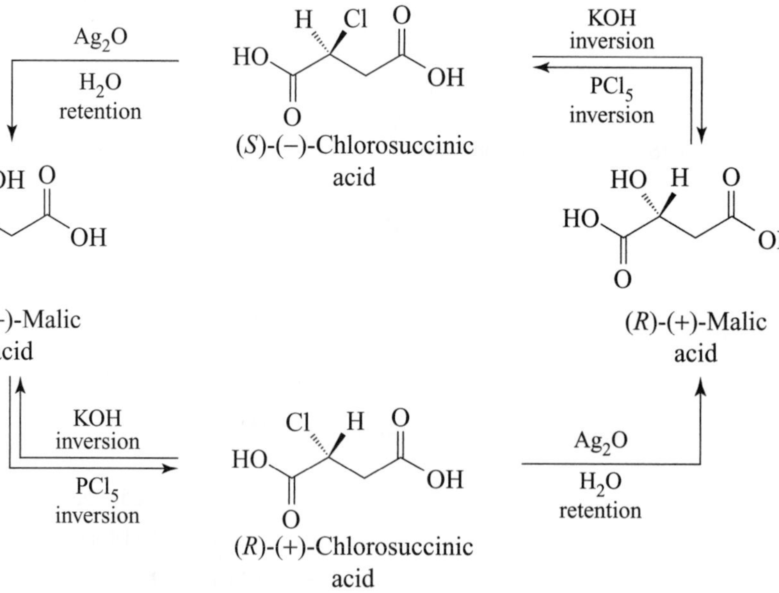

Ag_2O / H_2O retention

(*S*)-(−)-Chlorosuccinic acid

KOH inversion / PCl_5 inversion

(*S*)-(−)-Malic acid

(*R*)-(+)-Malic acid

KOH inversion / PCl_5 inversion

Ag_2O / H_2O retention

(*R*)-(+)-Chlorosuccinic acid

(c) The reaction takes place with retention of configuration.

(d)

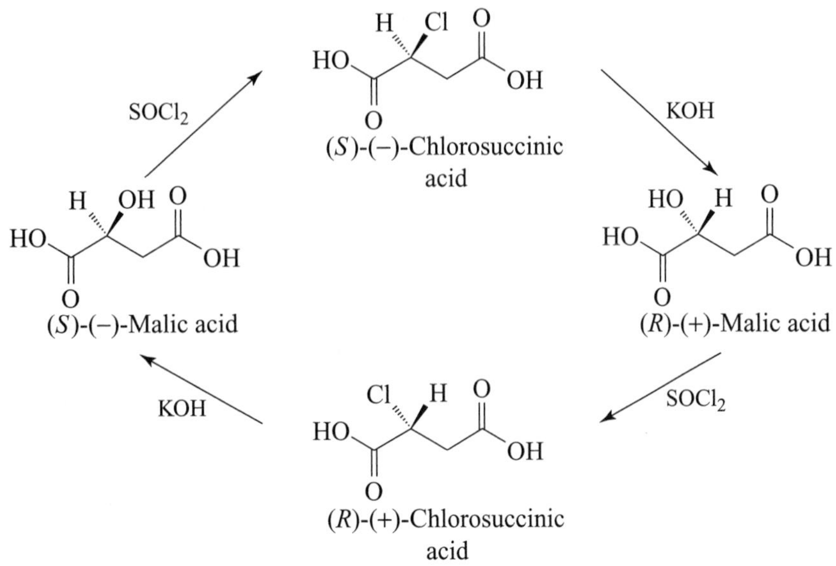

H Cl O
HO OH
O
(S)-(−)-Chlorosuccinic acid

$SOCl_2$

KOH

H OH O
HO OH
O
(S)-(−)-Malic acid

HO H O
HO OH
O
(R)-(+)-Malic acid

KOH

$SOCl_2$

Cl H O
HO OH
O
(R)-(+)-Chlorosuccinic acid

6.51 (a) H CH_3 CH_3O Cl **A** $\xrightarrow{N_3^-}$ H CH_3 CH_3O N_3 **B**

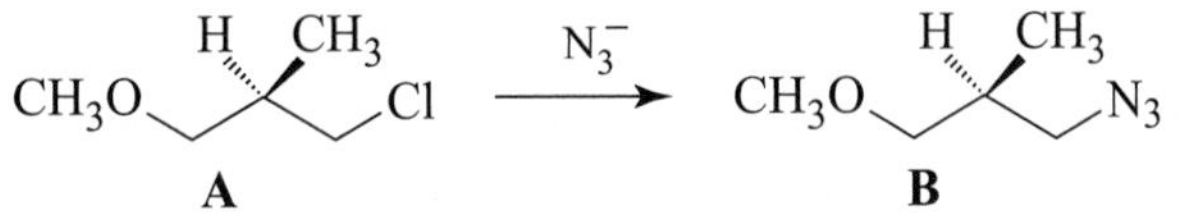

(b) No change of configuration occurs, just a change in the relative priority of a group at the chirality center.

6.52 Comparison of the molecular formulas of starting material and product indicates a loss of HC1. The absence of IR bands in the 1620–1680 cm^{-1} region rules out the presence of the alkene function.

A nucleophilic substitution agrees with the evidence:

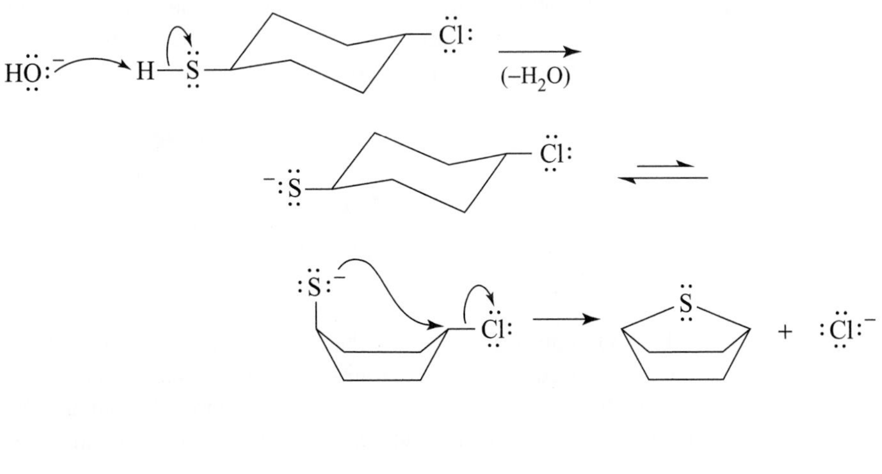

$HO:^-$ H—S Cl: $\xrightarrow{(-H_2O)}$

$^-:S$ Cl: $\rightleftharpoons$

$:S:^-$ Cl: $\longrightarrow$ S + $:Cl:^-$

6.53 The IR evidence indicates that C possesses both an alkene function and a hydroxyl group. An E2 reaction on this substrate produces enantiomeric unsaturated alcohols.

tert-BuÖ:⁻ (a) (b)

C (racemic)

(a) (*R*) + (b) (*S*)

6.54 Regarding the S_N2 reaction, there is extreme steric hindrance for attack by the nucleophile from the back side with respect to the leaving group due to atoms on the other side of the rigid ring structure, as the following model shows.

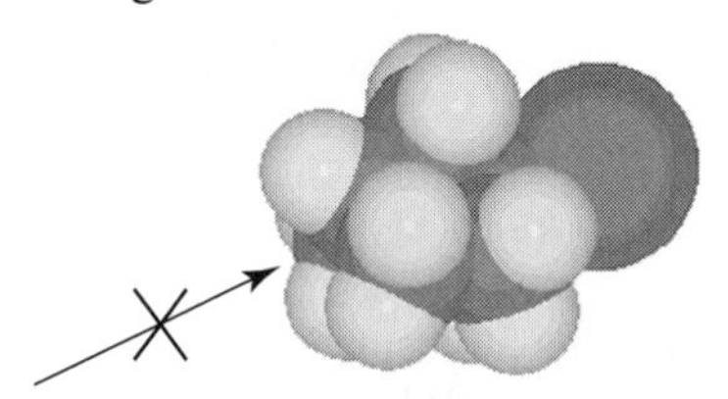

For the S_N1 reaction, formation of a carbocation would require that the bridgehead carbon approach trigonal planar geometry, which would lead to a carbocation of extremely high energy due to the geometric constraints of the bicyclic ring.

6.55 The lobe of the LUMO that would accept electron density from the nucleophile is buried within the bicyclic ring structure of 1-bromobicyclo[2.2.1]heptane (the large blue lobe), effectively making it inaccessible for approach by the nucleophile.

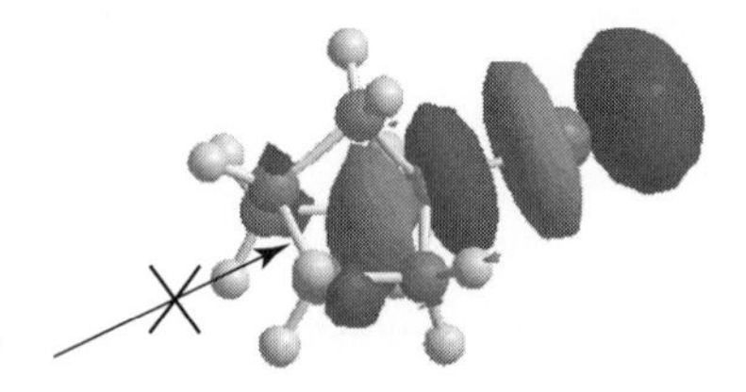

6.56 (a) The LUMO in an S_N1 reaction is the orbital that includes the vacant *p* orbital in our simplified molecular orbital diagrams of carbocations. (b) The large lobes above and below the trigonal planar carbon atom of the isopropyl group are the ones that would interact with a nucleophile. These are the lobes associated with stylized *p* orbitals we draw in simplified diagrams of carbocations. (c) The HOMO for this carbocation shows the contribution of electron density from a nonbonding electron pair of the ether oxygen that is adjacent to the carbocation, This is evident by the lobes that extend over these two atoms and encompass the bond between them. In effect, this orbital model represents the resonance hybrid we can draw where a nonbonding electron pair from oxygen is shifted to the bonding region between the carbon and oxygen.

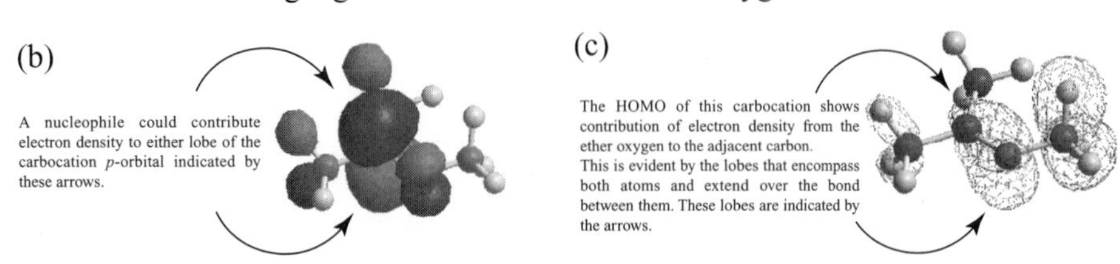

(b)
A nucleophile could contribute electron density to either lobe of the carbocation p-orbital indicated by these arrows.
(c)
The HOMO of this carbocation shows contribution of electron density from the ether oxygen to the adjacent carbon.
This is evident by the lobes that encompass both atoms and extend over the bond between them. These lobes are indicated by the arrows.

QUIZ

6.1 Which set of conditions would you use to obtain the best yield in the reaction shown?

Br $\xrightarrow{?}$

(a) H_2O, heat

(b) ONa, OH, heat

(c) Heat alone

(d) H_2SO_4

(e) None of the above

6.2 Which of the following reactions would give the best yield?

(a) CH_3ONa + Br ⟶ CH_3 O

(b) ONa + CH_3Br ⟶ CH_3 O

(c) CH_3OH + Br $\xrightarrow{\text{heat}}$ CH_3 O

6.3 A kinetic study yielded the following reaction rate data:

Experiment Number	Initial Concentrations $[OH^-]$	$[R{-}Br]$	Initial Rate of Disappearance of R—Br and Formation of R—OH
1	0.50	0.50	1.00
2	0.50	0.25	0.50
3	0.25	0.25	0.25

Which of the following statements best describe this reaction?

(a) The reaction is second order.

(b) The reaction is first order.

(c) The reaction is S_N1.

(d) Increasing the concentration of OH^- has no effect on the rate.

(e) More than one of the above.

6.4 There are four compounds with the formula C_4H_9Br. List them in order of decreasing reactivity in an S_N2 reaction.

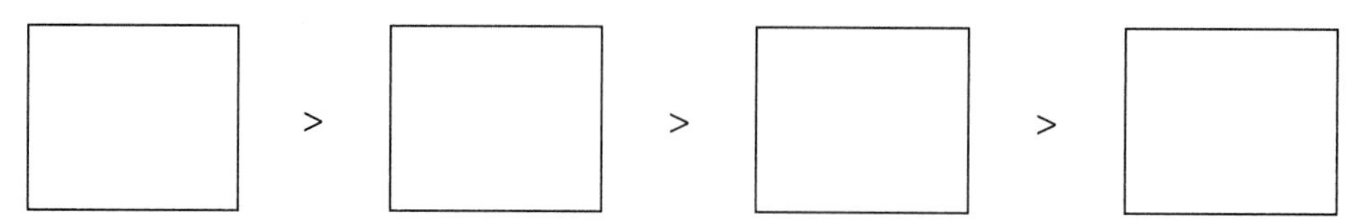

6.5 Supply the missing reactants, reagents, intermediates, or products.

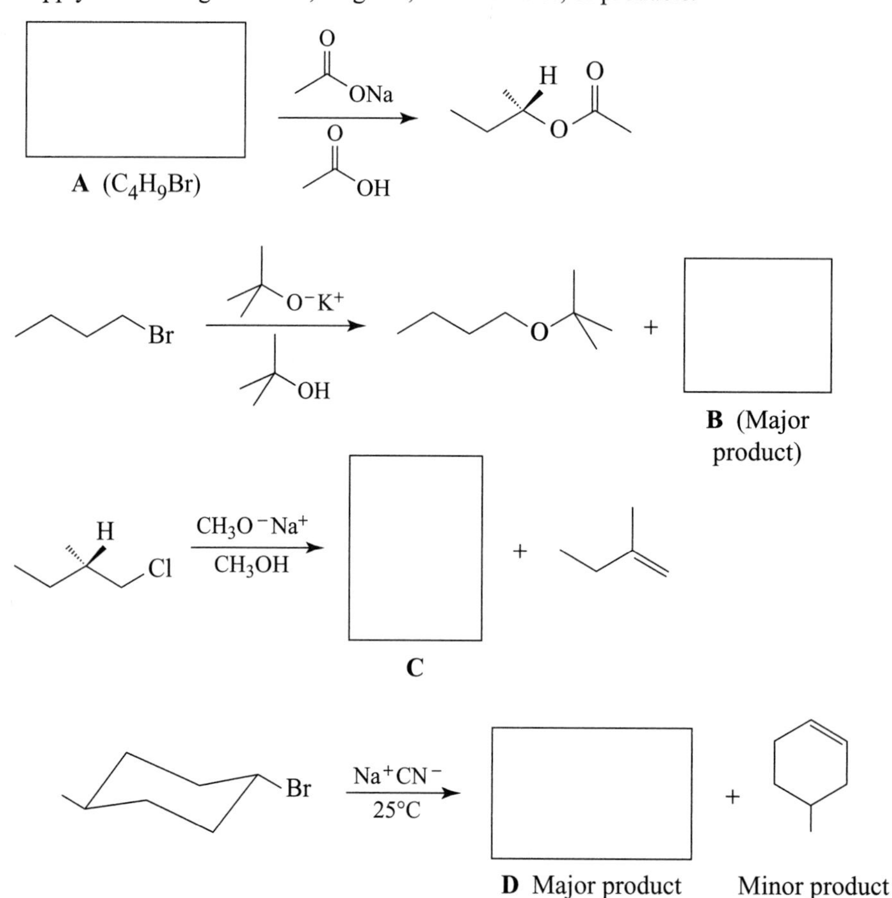

6.6 Which S_N2 reaction will occur most rapidly. (Assume the concentrations and temperatures are all the same.)

(a) CH_3O^- + ethyl fluoride (F) ⟶ $CH_3CH_2OCH_3$ + F^-

(b) CH_3O^- + ethyl iodide (I) ⟶ $CH_3CH_2OCH_3$ + I^-

(c) CH_3O^- + ethyl chloride (Cl) ⟶ $CH_3CH_2OCH_3$ + Cl^-

(d) CH_3O^- + ethyl bromide (Br) ⟶ $CH_3CH_2OCH_3$ + Br^-

6.7 Provide three-dimensional structures for the missing boxed structures and formulas for missing reagents.

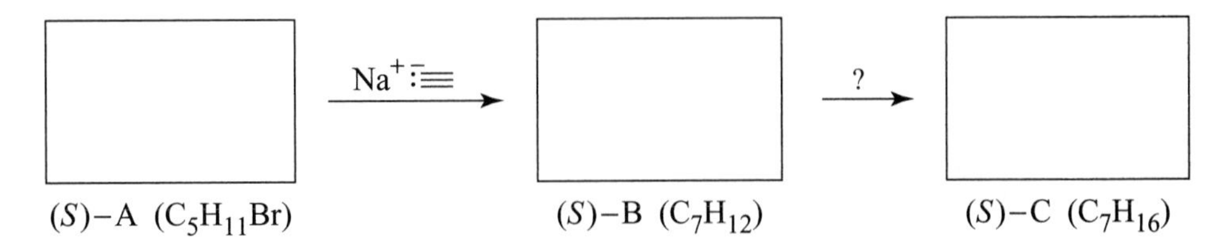

7 ALKENES AND ALKYNES I: PROPERTIES AND SYNTHESIS. ELIMINATION REACTIONS OF ALKYL HALIDES

SOLUTIONS TO PROBLEMS

7.1 (a) (*E*)-1-Bromo-1-chloro-1-pentene or (*E*)-1-Bromo-1-chloropent-1-ene

(b) (*E*)-2-Bromo-1-chloro-1-iodo-1-butene or (*E*)-2-Bromo-1-chloro-1-iodobut-1-ene

(c) (*Z*)-3,5-Dimethyl-2-hexene or (*Z*)-3,5-Dimethylhex-2-ene

(d) (*Z*)-1-Chloro-1-iodo-2-methyl-1-butene or (*Z*)-1-Chloro-1-iodo-2-methylbut-1-ene

(e) (*Z*,4*S*)-3,4-Dimethyl-2-hexene or (*Z*,4*S*)-3,4-Dimethylhex-2-ene

(f) (*Z*,3*S*)-1-Bromo-2-chloro-3-methyl-1-hexene or (*Z*,3*S*)-1-Bromo-2-chloro-3-methylhex-1-ene

7.2 < < Order of increasing stability

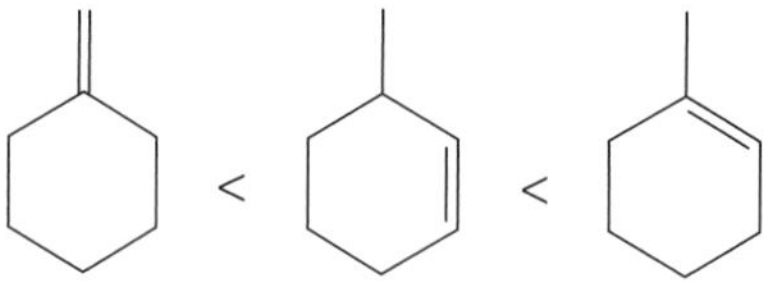

7.3 (a), (b)

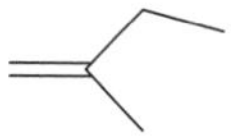

2-Methyl-1-butene (disubstituted) $\xrightarrow[\text{Pt, pressure}]{H_2}$ $\Delta H° = -119$ kJ mol^{-1}

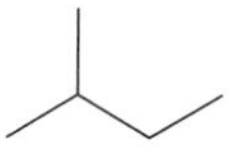

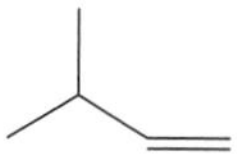

3-Methyl-1-butene (monosubstituted) $\xrightarrow[\text{Pt, pressure}]{H_2}$ $\Delta H° = -127$ kJ mol^{-1}

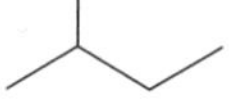

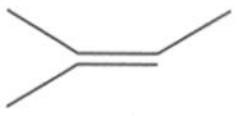

2-Methyl-2-butene (trisubstituted) $\xrightarrow[\text{Pt, pressure}]{H_2}$ $\Delta H° = -113$ kJ mol^{-1}

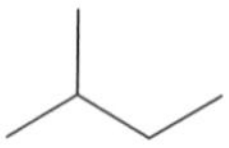

(c) Yes, because hydrogenation converts each alkene into the same product.

(d) 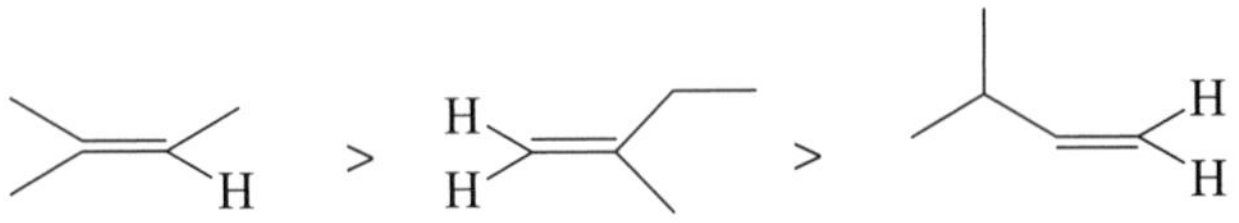 > >

(trisubstituted) (disubstituted) (monosubstituted)

Notice that this predicted order of stability is confirmed by the heats of hydrogenation. 2-Methyl-2-butene evolves the least heat; therefore, it is the most stable. 3-Methyl-1-butene evolves the most heat; therefore, it is the least stable.

(e)

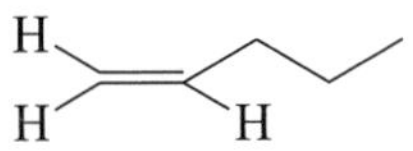

1-Pentene 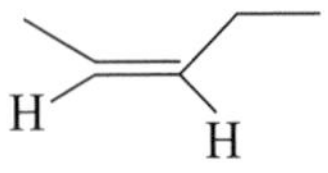 *cis*-2-Pentene 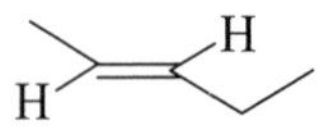 *trans*-2-Pentene

(f) Order of stability: *trans*-2-pentene > *cis*-2-pentene > 1-pentene

7.4 (a) 2,3-Dimethyl-2-butene would be the more stable because the double bond is tetrasubstituted. 2-Methyl-2-pentene has a trisubstituted double bond.

(b) *trans*-3-Hexene would be the more stable because alkenes with trans double bonds are more stable than those with cis double bonds.

(c) *cis*-3-Hexene would be more stable because its double bond is disubstituted. The double bond of 1-hexene is monosubstituted.

(d) 2-Methyl-2-pentene would be the more stable because its double bond is trisubstituted. The double bond of *trans*-2-hexene is disubstituted.

7.5 The location of IR absorptions between 600 cm^{-1} and 1000 cm^{-1} due to out-of-plane bending of alkene C—H bonds can be the basis of differentiation.

(a) 2-Methyl-2-pentene, ~800 cm^{-1}
2,3-Dimethyl-2-butene, no alkene C—H bonds

(b) *cis*-3-Hexene, 650–750 cm^{-1}
trans-3-Hexene, ~960 cm^{-1}

(c) 1-Hexene, ~900 cm^{-1} and ~1000 cm^{-1}
cis-3-Hexene, 650–750 cm^{-1}

(d) *trans*-2-Hexene, ~960 cm^{-1}
2-Methyl-2-pentene, ~800 cm^{-1}

7.6 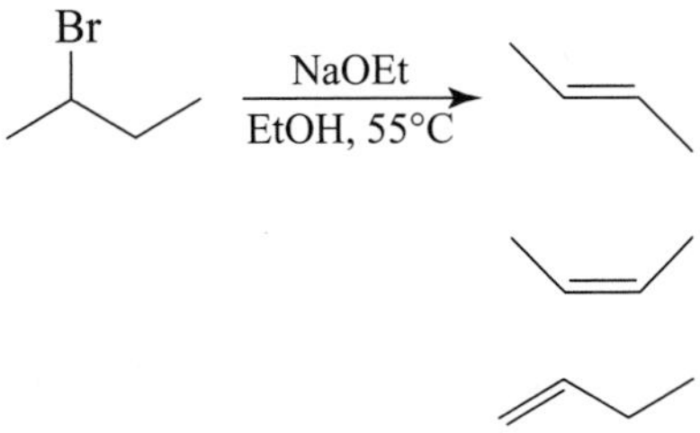 Br —NaOEt / EtOH, 55°C→ most

less

least

7.7 (a)

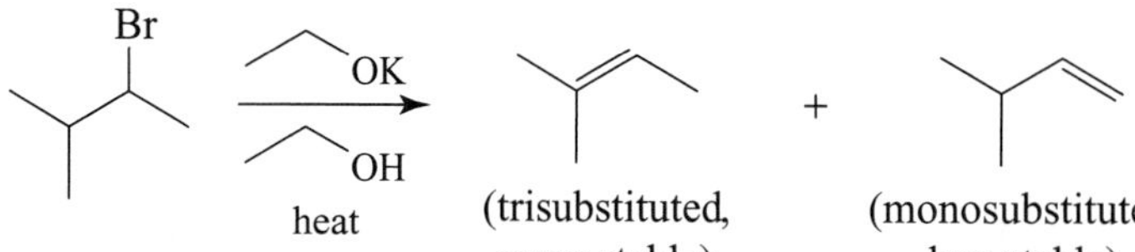

Br; OK / OH, heat

(trisubstituted, more stable) Major product + (monosubstituted, less stable) Minor product

(b)

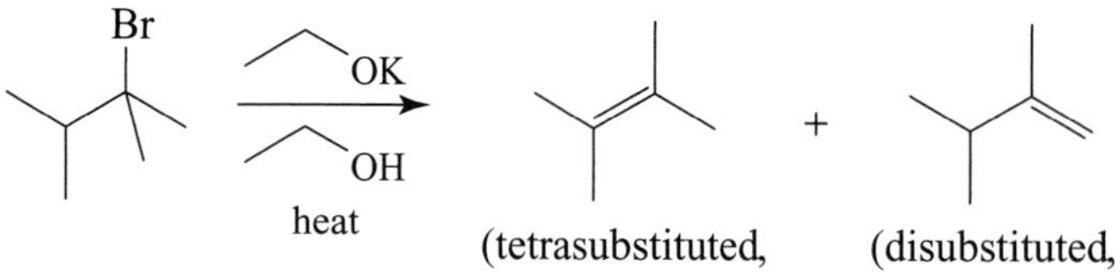

Br; OK / OH, heat

(tetrasubstituted, more stable) Major product + (disubstituted, less stable) Minor product

7.8 *t*-BuOK in *t*-BuOH

7.9 An anti coplanar transition state allows the molecule to assume the more stable staggered conformation,

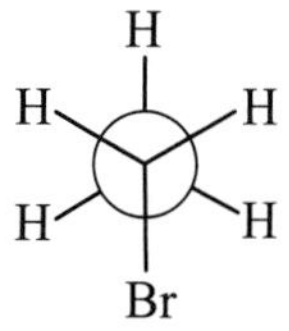

whereas a syn coplanar transition state requires the molecule to assume the less stable eclipsed conformation.

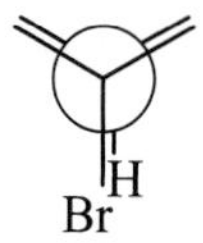

7.10 *cis*-1-Bromo-4-*tert*-butylcyclohexane can assume an anti coplanar transition state in which the bulky *tert*-butyl group is equatorial.

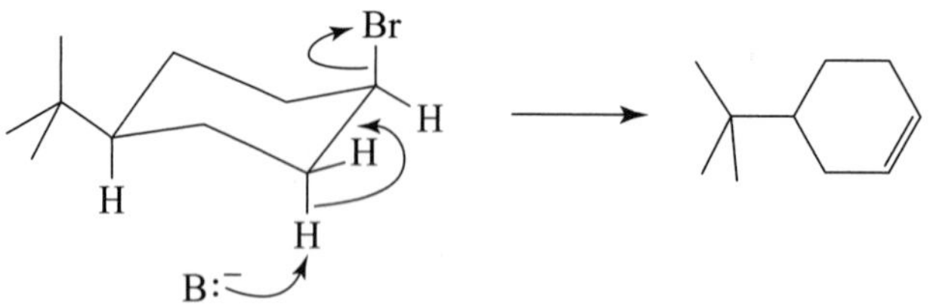

The conformation (above), because it is relatively stable, is assumed by most of the molecules present, and, therefore, the reaction is rapid.

On the other hand, for *trans*-1-bromo-4-*tert*-butylcyclohexane to assume an anti coplanar transition state, the molecule must assume a conformation in which the large *tert*-butyl group is axial:

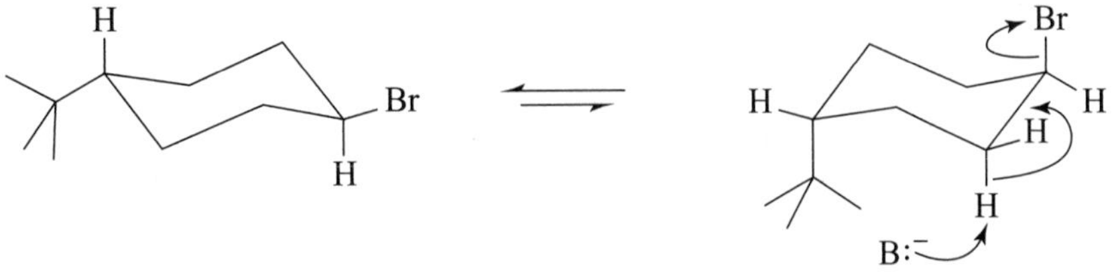

Such a conformation is of high energy; therefore, very few molecules assume this conformation. The reaction, consequently, is very slow.

7.11 (a) Anti coplanar elimination can occur in two ways with the cis isomer.

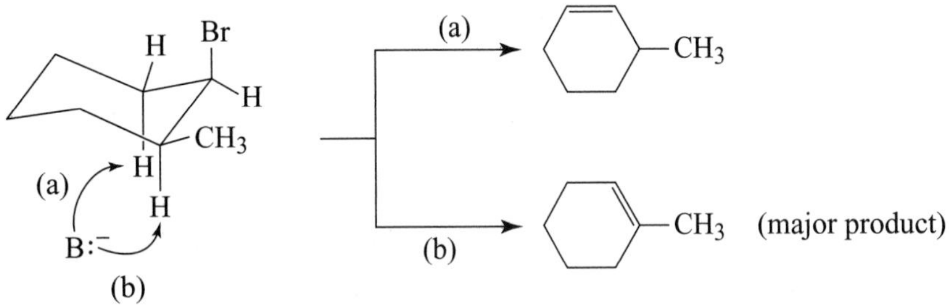

(a) → 3-methylcyclohexene ($-CH_3$)

(b) → 1-methylcyclohexene ($-CH_3$) (major product)

cis-1-Bromo-2-methylcyclohexane

(b) Anti coplanar elimination can occur in only one way with the trans isomer.

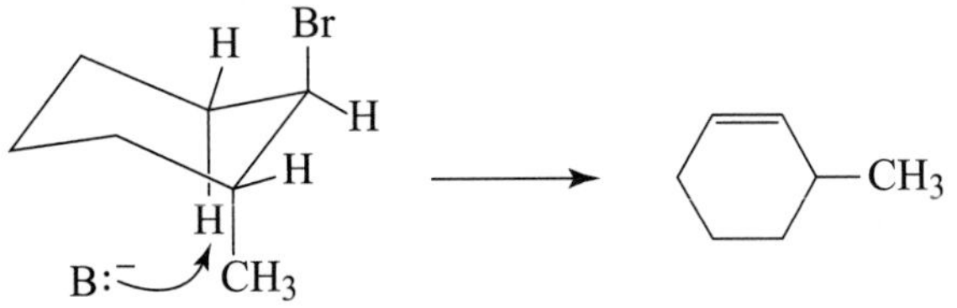

→ 3-methylcyclohexene ($-CH_3$)

trans-1-Bromo-2-methylcyclohexane

7.12 (a) (1) $CH_3-CH(CH_3)-\ddot{O}H + H-\ddot{O}-SO_2-\ddot{O}-H \rightleftharpoons CH_3-CH(CH_3)-\overset{+}{O}H_2 + {}^{-}:\ddot{O}-SO_2-\ddot{O}-H$

(2) $CH_3-CH(CH_3)-\overset{+}{O}H_2 \rightleftharpoons CH_3-\overset{+}{C}H-CH_3 + H_2O$

(3) $CH_3-\overset{+}{C}H-CH_2-H + {}^{-}:\ddot{O}SO_3H \rightleftharpoons CH_3-CH=CH_2 + HOSO_3H$

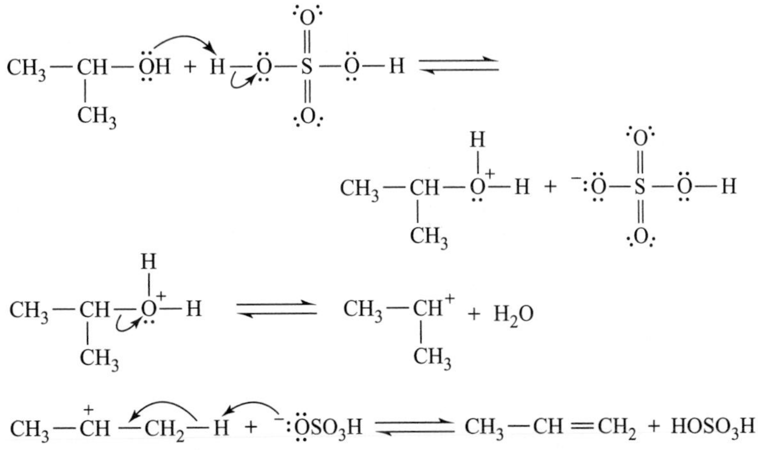

(b) By donating a proton to the —OH group of the alcohol in step (1), the acid allows the loss of a relatively stable, weakly basic, leaving group (H_2O) in step (2). In the absence of an acid, the leaving group would have to be the strongly basic OH^- ion, and such steps almost never occur.

7.13 (cyclohexylmethanol, OH) < (2-methylcyclohexanol, OH) < (1-methylcyclohexanol, OH)

1° 2° 3°

Order of increasing case of dehydration

7.14 (1) $CH_3C(CH_3)_2CH_2\ddot{O}H + H{-}A \rightleftharpoons CH_3C(CH_3)_2CH_2\overset{+}{O}H_2 + :A^-$

(2) $CH_3C(CH_3)_2CH_2{-}\overset{+}{O}H_2 \rightleftharpoons CH_3C(CH_3)_2\overset{+}{C}H_2 + H_2O$

1° Carbocation

(3) $CH_3C(CH_3)_2\overset{+}{C}H_2 \longrightarrow \left[CH_3C(CH_3){-}CH_2\right]^{\ddagger} \longrightarrow CH_3\overset{+}{C}(CH_3){-}CH_2{-}CH_3$

1° Carbocation — Transition state — 3° Carbocation

[Steps (2) and (3), ionization and rearrangement, may occur simultaneously.]

(4) $CH_3{-}\overset{+}{C}(CH_3){-}CH(H){-}CH_3 \; :A^- \longrightarrow (CH_3)_2C{=}C(H)CH_3 + HA$

2-Methyl-2-butene

7.15

$$CH_3CH_2CH(CH_3)CH_2-\ddot{O}H + H-\overset{+}{\ddot{O}}(H)-H \underset{(+H_2O)}{\overset{(-H_2O)}{\rightleftharpoons}} CH_3CH_2CH(CH_3)CH_2-\overset{+}{\ddot{O}}H_2 \underset{(+H_2O)}{\overset{(-H_2O)}{\rightleftharpoons}}$$

2-Methyl-1-butanol

$$CH_3CH_2-C(H)(CH_3)-\overset{+}{C}H_2 \xrightarrow[\text{shift*}]{\text{1,2-hydride}} CH_3-CH(H)-\overset{+}{C}(CH_3)-CH_3 \overset{:\ddot{O}H_2}{\rightleftharpoons}$$

1° Cation

3° Cation

$$CH_3CH{=}C(CH_3)-CH_3 + H_3O^+$$

2-Methyl-2-butene

$$CH_3CH(CH_3)CH_2CH_2-\ddot{O}H + H-\overset{+}{\ddot{O}}(H)-H \underset{(+H_2O)}{\overset{(-H_2O)}{\rightleftharpoons}} CH_3CH(CH_3)CH_2CH_2-\overset{+}{O}H_2 \underset{(+H_2O)}{\overset{(-H_2O)}{\rightleftharpoons}}$$

3-Methyl-1-butanol

$$CH_3CH(CH_3)-CH(H)-\overset{+}{C}H_2 \xrightarrow[\text{shift*}]{\text{1,2-hydride}} CH_3-C(H)(CH_3)-\overset{+}{C}H-CH_3 \overset{:\ddot{O}H_2}{\rightleftharpoons}$$

$$CH_3C(CH_3){=}CH-CH_3 + H_3O^+$$

2-Methyl-2-butene

* The hydride shift may occur simultaneously with the preceding step.

7.16 HO (Isoborneol) = CH_3 CH_3 CH_3 HO $\xrightarrow{H_3O^+}$ $\xrightarrow{(-H_2O)}$ CH_3 CH_3 CH_3 + →

Isoborneol

CH_3 CH_3 CH_3 + = H $:\ddot{O}H_2$ $\overset{+}{C}H_2$ CH_2 CH_3 CH_3 → CH_2 CH_2 CH_3 CH_3 $+ H_3O^+$ ≡ CH_2

Camphene

7.17 (a) $CH_3CH{=}CH_2 + NaNH_2 \longrightarrow$ No reaction

(b) $CH_3C{\equiv}C{-}H + Na^+ \ {}^-{:}\ddot{N}H_2 \longrightarrow CH_3C{\equiv}C{:}^-Na^+ + {:}NH_3$

Stronger acid — Stronger base — Weaker base — Weaker acid

(c) $CH_3CH_2CH_3 + NaNH_2 \longrightarrow$ No reaction

(d) $CH_3C{\equiv}C{:}^- + H{-}\ddot{O}CH_2CH_3 \longrightarrow CH_3C{\equiv}CH + {}^-{:}\ddot{O}CH_2CH_3$

Stronger base — Stronger acid — Weaker acid — Weaker base

(e) $CH_3C{\equiv}C{:}^- + H{-}\overset{+}{N}H_3 \longrightarrow CH_3C{\equiv}CH + {:}NH_3$

Stronger base — Stronger acid — Weaker acid — Weaker base

7.18 Acetophenone $\xrightarrow[0°C]{PCl_5}$ $C_6H_5CCl_2CH_3$ $\xrightarrow[\text{mineral oil, heat; (2) HA}]{\text{(1) 3 equiv. } NaNH_2}$ $C_6H_5C{\equiv}CH$

7.19 (a) $CH_3\overset{O}{\overset{\|}{C}}CH_3 \xrightarrow[0°C]{PCl_5} CH_3CCl_2CH_3 \xrightarrow[\text{mineral oil, heat; (2) } NH_4^+]{\text{(1) 3 } NaNH_2} CH_3C{\equiv}CH$

(b) $CH_3CH_2CHBr_2 \xrightarrow[\text{mineral oil, heat; (2) } NH_4^+]{\text{(1) 3 } NaNH_2} CH_3C{\equiv}CH$

(c) $CH_3CHBrCH_2Br \xrightarrow{\text{[same as (b)]}} CH_3C{\equiv}CH$

(d) $CH_3CH{=}CH_2 \xrightarrow[CCl_4]{Br_2} CH_3CHBrCH_2Br \xrightarrow{\text{[same as (b)]}} CH_3C{\equiv}CH$

7.20 $(CH_3)_3C{-}C{\equiv}C{-}H + Na^+ \ {}^-{:}\ddot{N}H_2 \xrightarrow{(-NH_3)} (CH_3)_3C{-}C{\equiv}C{:}^- Na^+ \xrightarrow{CH_3{-}I} (CH_3)_3C{-}C{\equiv}C{-}CH_3$

(Starting the synthesis with 1-propyne and attempting to alkylate with a *tert*-butyl substrate would not work because elimination would occur instead of substitution.)

7.21

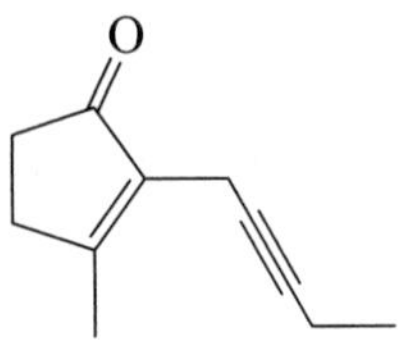

Compound A

7.22

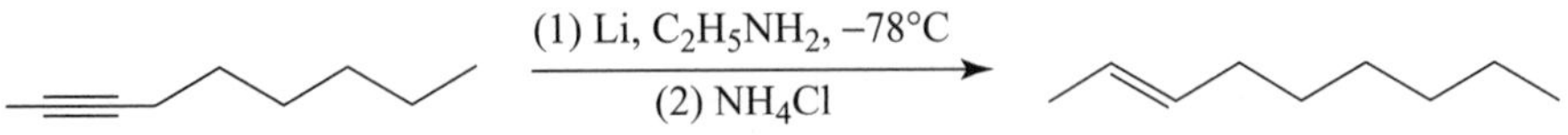

2-Nonyne $\xrightarrow[\text{(2) } NH_4Cl]{\text{(1) Li, } C_2H_5NH_2\text{, }-78°C}$ (E)-2-Nonene

7.23 Route 1

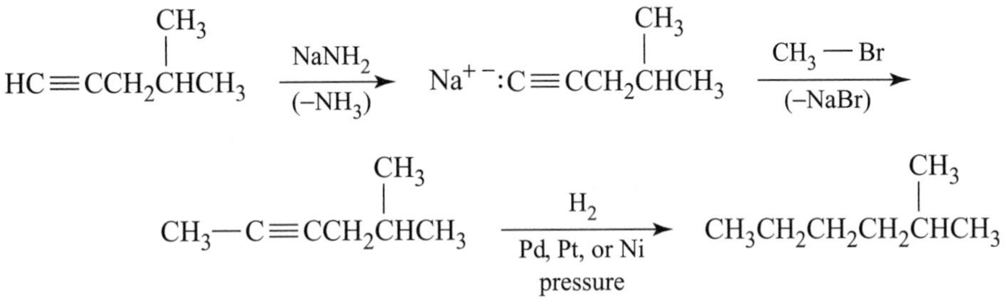

$$HC{\equiv}CCH_2CH(CH_3)CH_3 \xrightarrow[(-NH_3)]{NaNH_2} Na^{+\,-}{:}C{\equiv}CCH_2CH(CH_3)CH_3 \xrightarrow[(-NaBr)]{CH_3-Br}$$

$$CH_3-C{\equiv}CCH_2CH(CH_3)CH_3 \xrightarrow[\text{Pd, Pt, or Ni, pressure}]{H_2} CH_3CH_2CH_2CH_2CH(CH_3)CH_3$$

Route 2

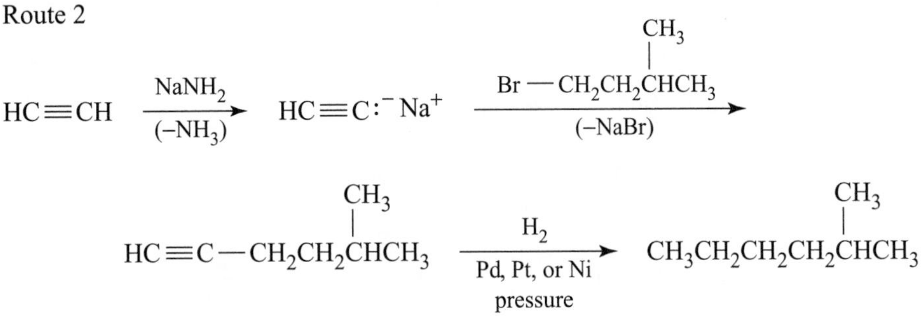

$$HC{\equiv}CH \xrightarrow[(-NH_3)]{NaNH_2} HC{\equiv}C{:}^{-}\,Na^{+} \xrightarrow[(-NaBr)]{Br-CH_2CH_2CH(CH_3)CH_3}$$

$$HC{\equiv}C-CH_2CH_2CH(CH_3)CH_3 \xrightarrow[\text{Pd, Pt, or Ni, pressure}]{H_2} CH_3CH_2CH_2CH_2CH(CH_3)CH_3$$

Route 3

$$HC{\equiv}CCH(CH_3)CH_3 \xrightarrow[(-NH_3)]{NaNH_2} Na^{+\,-}{:}C{\equiv}CCH(CH_3)CH_3 \xrightarrow[(-NaBr)]{CH_3CH_2Br}$$

$$CH_3CH_2-C{\equiv}CCH(CH_3)CH_3 \xrightarrow[\text{Pd, Pt, or Ni}]{H_2} CH_3CH_2CH_2CH_2CH(CH_3)CH_3$$

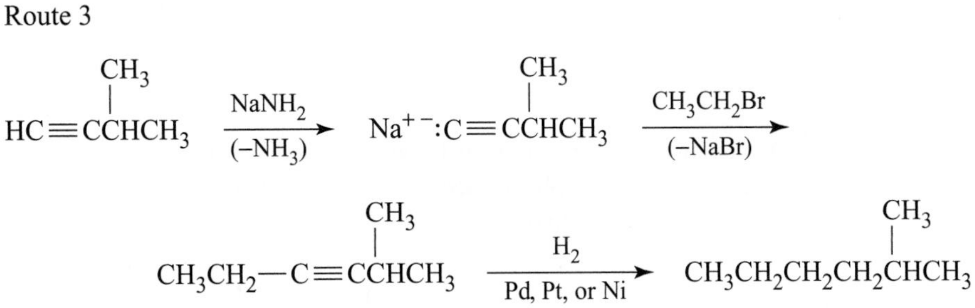

7.24 (a) Undecane

Etc. (using the other alkyne and alkyl halide homologue pairs)
HC≡C:⁻ + X(CH₂)₈CH₃
CH₃C≡C:⁻ + X(CH₂)₇CH₃
CH₃CH₂C≡C:⁻ + X(CH₂)₆CH₃
CH₃CH₂CH₂C≡C:⁻ + X(CH₂)₅CH₃

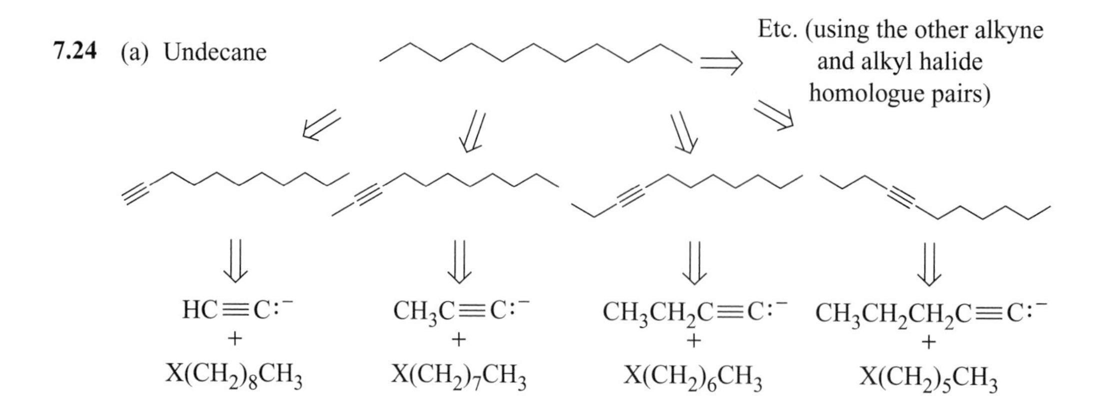

2-Methylheptadecane

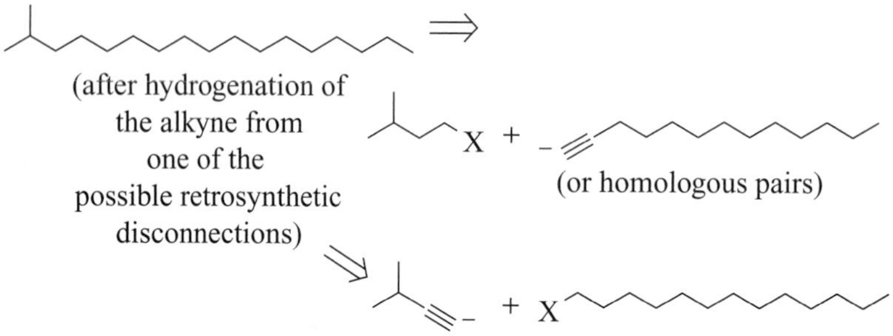

(after hydrogenation of the alkyne from one of the possible retrosynthetic disconnections)
X +
(or homologous pairs)
+ X

(Note that is not a good choice because the alkyl halide is branched at the carbon adjacent to the one which bears the halogen. Neither would work because the alkyl halide is secondary. Both of these routes would lead to elimination instead of substitution.)

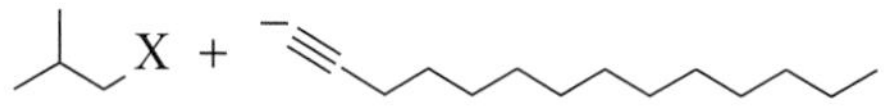

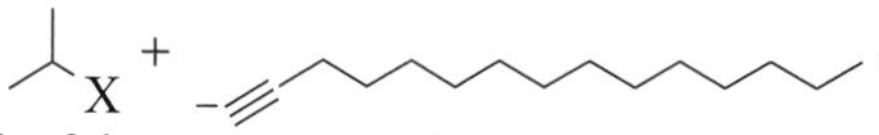

(b) For any pair of reactants above that is a feasible retrosynthetic disconnection, the steps for the synthesis would be

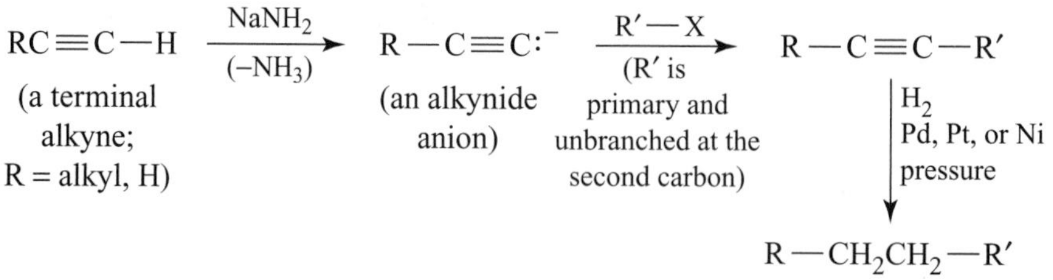

RC≡C—H
(a terminal alkyne; R = alkyl, H)
NaNH₂
(–NH₃)
R—C≡C:⁻
(an alkynide anion)
R′—X
(R′ is primary and unbranched at the second carbon)
R—C≡C—R′
H₂
Pd, Pt, or Ni
pressure
R—CH₂CH₂—R′

7.25 (a) We designate the position of the double bond by using the *lower* number of the two numbers of the doubly bonded carbon atoms, and the chain is numbered from the end nearer the double bond. The correct name is *trans*-2-pentene.

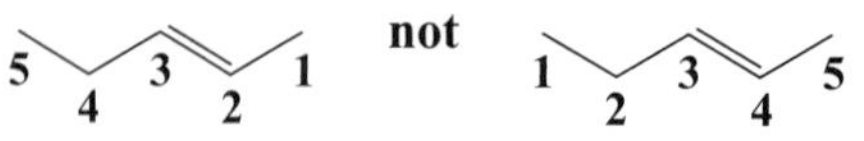

5 4 3 2 1
not
1 2 3 4 5

(b) We must choose the longest chain for the base name. The correct name is 2-methylpropene.

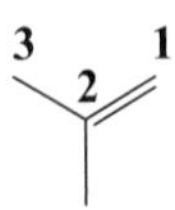

(c) We use the lower number of the two doubly bonded carbon atoms to designate the position of the double bond. The correct name is 1-methylcyclohexene.

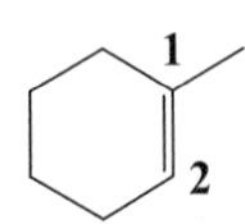

(d) We must number the ring starting with the double bond in the direction that gives the substituent the lower number. The correct name is 3-methylcyclobutene.

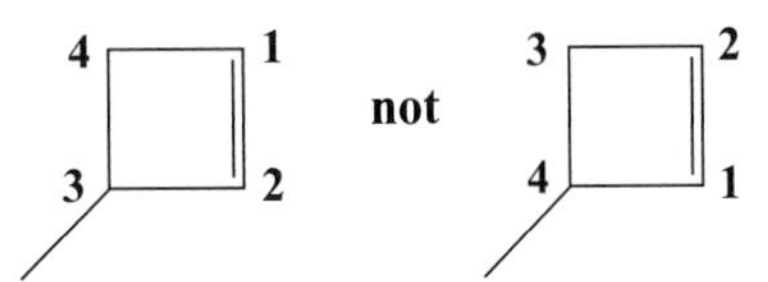

(e) We number in the way that gives the double bond *and the substituent* the lower number. The correct name is (Z)-2-chloro-2-butene or (Z)-2-chlorobut-2-ene.

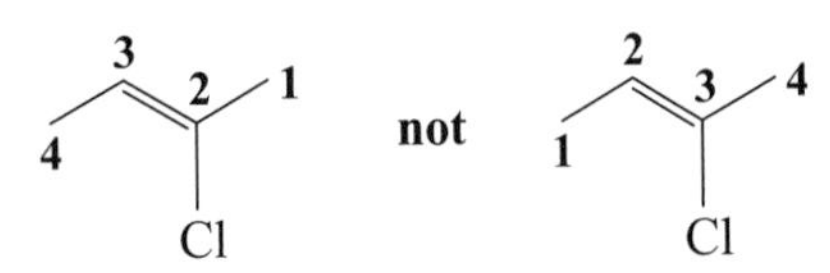

(f) We number the ring starting with the double bond so as to give the substituents the lower numbers. The correct name is 3,4-dichlorocyclohexene.

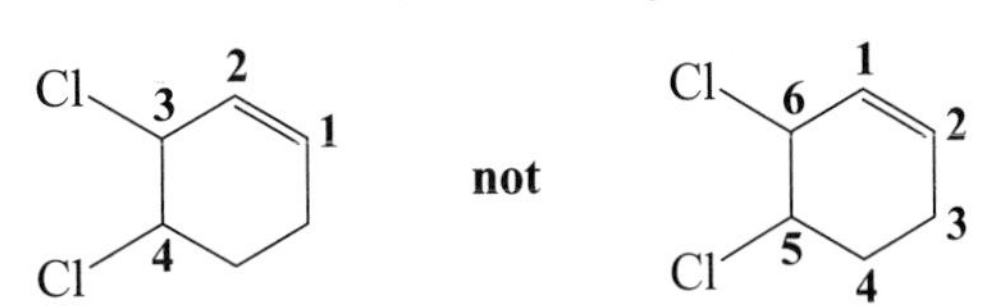

7.26 (a) (b) (c)

(d) (e) (f) Br Br Br

(g) H (h) Cl H (i)

(j) (k) H Cl (l)

7.27 (a)

H Br

(2*Z*,4*R*)-4-Bromo-2-hexene
or
(2*Z*,4*R*)-4-Bromohex-2-ene

H Br

(2*Z*,4*S*)-4-Bromo-2-hexene
or
(2*Z*,4*S*)-4-Bromohex-2-ene

H Br

(2*E*,4*R*)-4-Bromo-2-hexene
or
(2*E*,4*R*)-4-Bromohex-2-ene

H Br

(2*E*,4*S*)-4-Bromo-2-hexene
or
(2*E*,4*S*)-4-Bromohex-2-ene

(b)

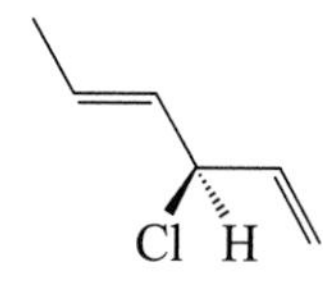

(3*R*,4*Z*)-3-Chloro-1,4-hexadiene
or
(3*R*,4*Z*)-3-Chlorohexa-1,4-diene

H Cl

(3*S*,4*Z*)-3-Chloro-1,4-hexadiene
or
(3*S*,4*Z*)-3-Chlorohexa-1,4-diene

Cl H

(3*R*,4*E*)-3-Chloro-1,4-hexadiene
or
(3*R*,4*E*)-3-Chlorohexa-1,4-diene

H Cl

(3*S*,4*E*)-3-Chloro-1,4-hexadiene
or
(3*S*,4*E*)-3-Chlorohexa-1,4-diene

(c) Cl

Cl H

(2*E*,4*R*)-2,4-Dichloro-2-pentene
or
(2*E*,4*R*)-2,4-Dichloropent-2-ene

Cl

Cl H

(2*Z*,4*R*)-2,4-Dichloro-2-pentene
or
(2*Z*,4*R*)-2,4-Dichloropent-2-ene

Cl

H Cl

(2*E*,4*S*)-2,4-Dichloro-2-pentene
or
(2*E*,4*S*)-2,4-Dichloropent-2-ene

Cl

H Cl

(2*Z*,4*S*)-2,4-Dichloro-2-pentene
or
(2*Z*,4*S*)-2,4-Dichloropent-2-ene

(d)

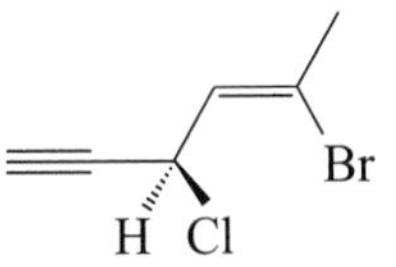

Br
H Cl

(3*R*,4*Z*)-5-Bromo-3-chloro-4-hexen-1-yne
or
(3*R*,4*Z*)-5-Bromo-3-chlorohex-4-en-1-yne

Br
Cl H

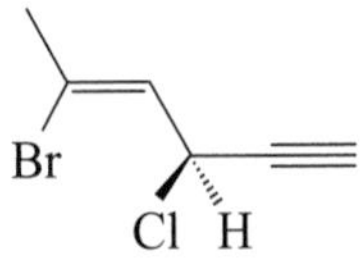

(3*S*,4*Z*)-5-Bromo-3-chloro-4-hexen-1-yne
or
(3*S*,4*Z*)-5-Bromo-3-chlorohex-4-en-1-yne

Br
H Cl

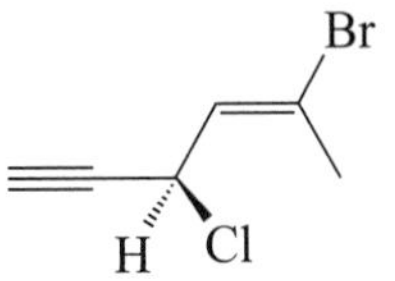

(3*R*,4*E*)-5-Bromo-3-chloro-4-hexen-1-yne
or
(3*R*,4*E*)-5-Bromo-3-chlorohex-4-en-1-yne

Br
Cl H

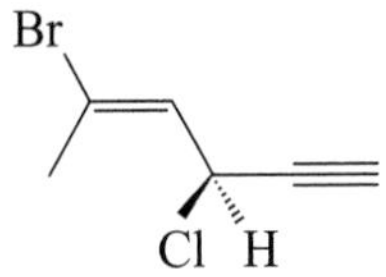

(3*S*,4*E*)-5-Bromo-3-chloro-4-hexen-1-yne
or
(3*S*,4*E*)-5-Bromo-3-chlorohex-4-en-1-yne

An IUPAC rule covers those cases in which a double bond and a triple bond occur in the same molecule:

> Numbers as low as possible are given to double and triple bonds as a set, even though this may at times give "-yne" a lower number than "-ene." If a choice remains, preference for low locants is given to the double bonds.*

*International Union of Pure and Applied Chemistry, http://www.acdlabs.com/iupac/nomenclature/93/r93_280.htm (accessed March 2003).

7.28 (a) (*E*)-3,5-Dimethyl-2-hexene or (*E*)-3,5-dimethylhex-2-ene

(b) 4-Chloro-3-methylcyclopentene

(c) 6-Methyl-3-heptyne or 6-methylhept-3-yne

(d) 1-*sec*-Butyl-2-methylcyclohexene or 1-methyl-2-(1-methylpropyl)cyclohexene

(e) (4*Z*,3*R*)-3-Chloro-4-hepten-1-yne or (4*Z*,3*R*)-3-chlorohept-4-en-1-yne

(f) 2-Pentyl-1-heptene or 2-pentylhept-1-ene

7.29 1-Pentanol > 1-pentyne > 1-pentene > pentane
(See Section 3.8 for the explanation.)

Synthesis

7.30 (a) 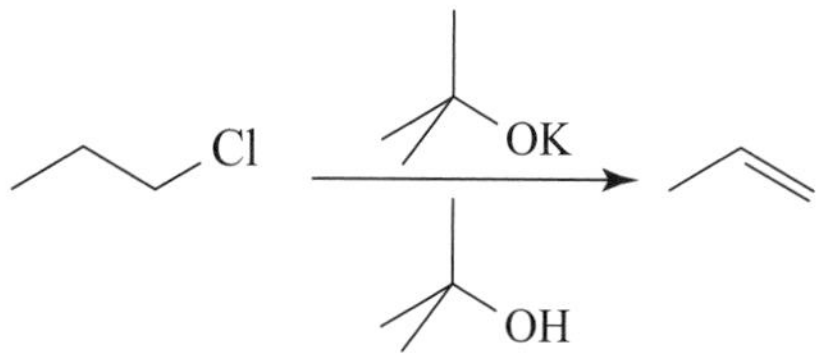 Cl $\xrightarrow[\text{OH}]{\text{OK}}$

(b) 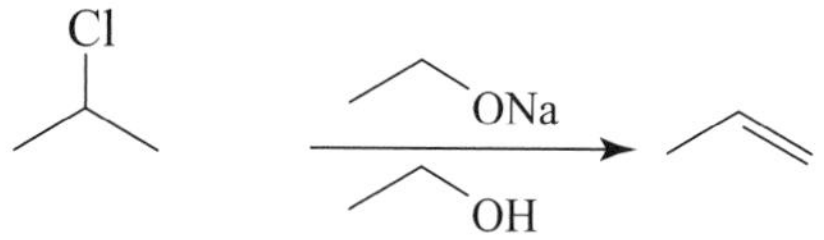 Cl $\xrightarrow[\text{OH}]{\text{ONa}}$

(c)  OH $\xrightarrow{\text{HA, heat}}$

(d) 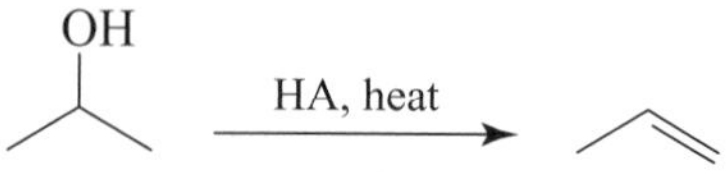 OH $\xrightarrow{\text{HA, heat}}$

(e) 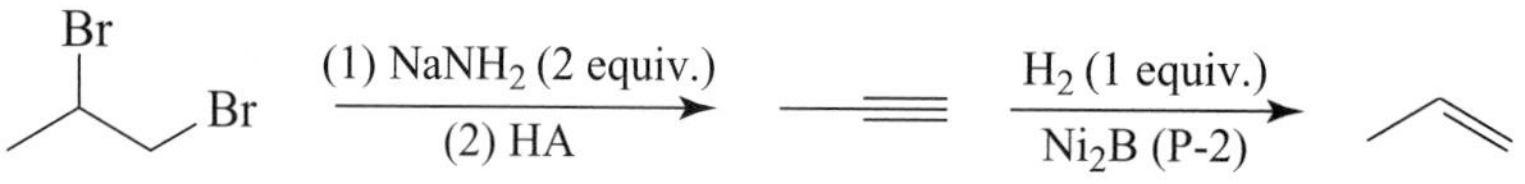 Br, Br $\xrightarrow[\text{(2) HA}]{\text{(1) NaNH}_2\text{ (2 equiv.)}}$ $\xrightarrow[\text{Ni}_2\text{B (P-2)}]{\text{H}_2\text{ (1 equiv.)}}$

(f) 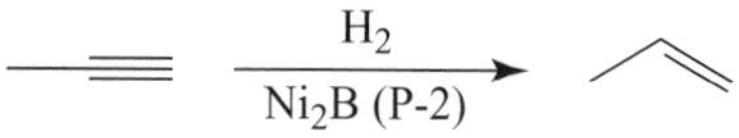 $\xrightarrow[\text{Ni}_2\text{B (P-2)}]{\text{H}_2}$

7.31 (a) 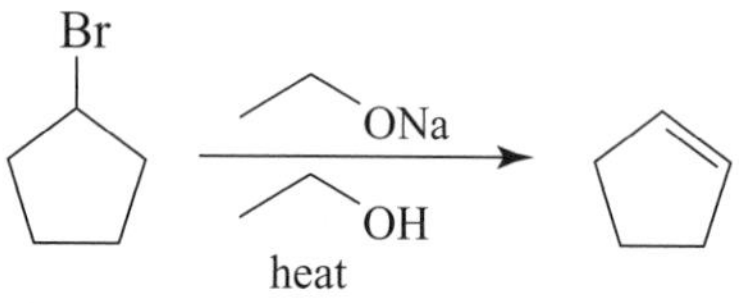 Br $\xrightarrow[\text{OH, heat}]{\text{ONa}}$

(b) 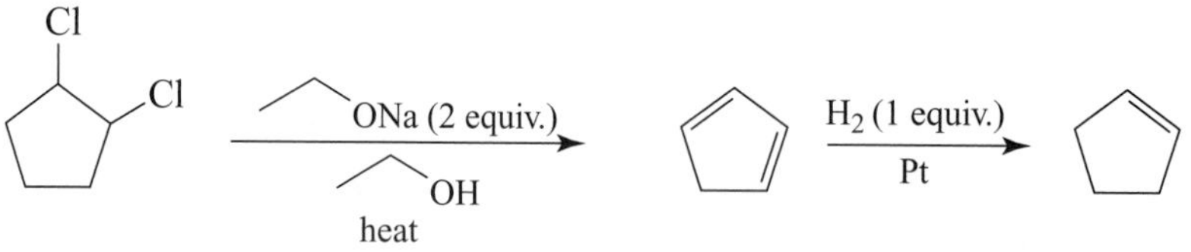 Cl, Cl $\xrightarrow[\text{OH, heat}]{\text{ONa (2 equiv.)}}$ $\xrightarrow[\text{Pt}]{\text{H}_2\text{ (1 equiv.)}}$

(c) 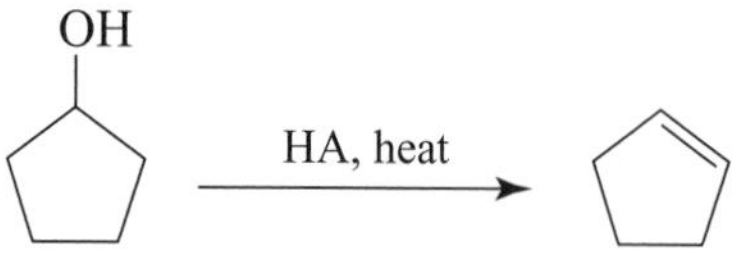 OH $\xrightarrow{\text{HA, heat}}$

7.32 (a) 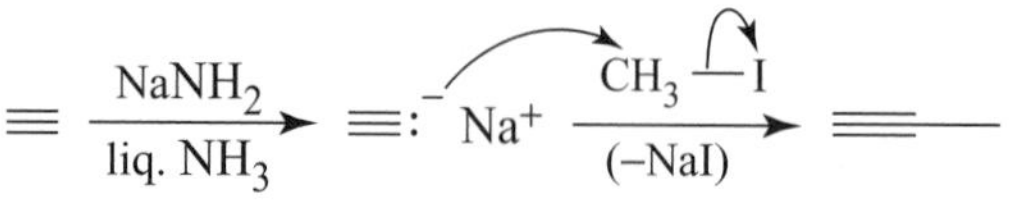 $\equiv$ $\xrightarrow[\text{liq. NH}_3]{\text{NaNH}_2}$ $\equiv:^- \text{Na}^+$ $\xrightarrow[\text{(–NaI)}]{\text{CH}_3\text{—I}}$ $\equiv$—

(b) $\xrightarrow[\text{liq. } NH_3]{NaNH_2}$:$^-$ Na$^+$ Br $\xrightarrow{(-NaBr)}$

(c) $\xrightarrow[\text{liq. } NH_3]{NaNH_2}$:$^-$ Na$^+$ CH_3—I $\xrightarrow{(-NaI)}$
[from (a)]

(d) $\xrightarrow[Ni_2B\ (P\text{-}2)]{H_2}$
[from (c)]

(e) $\xrightarrow[(2)\ NH_4Cl]{(1)\ Li,\ NH_2}$
[from (c)]

(f) :$^-$ Na$^+$ Br $\xrightarrow{(-NaBr)}$
[from (a)]

(g) $\xrightarrow[\text{liq. } NH_3]{NaNH_2}$:$^-$ Na$^+$ CH_3—I $\xrightarrow{(-NaI)}$
[from (f)]

(h) $\xrightarrow[Ni_2B\ (P\text{-}2)]{H_2}$
[from (g)]

(i) $\xrightarrow[(2)\ NH_4Cl]{(1)\ Li,\ NH_2}$
[from (g)]

(j) :$^-$ Na$^+$ Br $\xrightarrow{(-NaBr)}$ $\xrightarrow[\text{liq. } NH_3]{NaNH_2}$:$^-$ Na$^+$ $\xrightarrow{Br}$
[from (a)]

(k) :$^-$ Na$^+$ $\xrightarrow{D_2O}$ D
[from (j)]

(l) $\xrightarrow[Ni_2B\ (P\text{-}2)]{D_2}$ D D
[from (c)]

7.33 We notice that the deuterium atoms are cis to each other, and we conclude, therefore, that we need to choose a method that will cause a syn addition of deuterium. One way would be to use D_2 and a metal catalyst (Section 7.14)

CH_3 $\xrightarrow[Pt]{D_2}$ D CH_3 D H

7.34

(a) Br Br → $3NaNH_2$ / mineral oil, heat → $:^-$ Na^+ → NH_4Cl →

(b) Br Br → $3NaNH_2$ / mineral oil, heat → $:^-$ Na^+ → NH_4Cl →

(c) → Br_2 → Br Br → $3NaNH_2$ / mineral oil, heat → $:^-$ Na^+ → NH_4Cl →

(d) O → PCl_5 → Cl Cl → $3NaNH_2$ / mineral oil, heat → $:^-$ Na^+ → NH_4Cl →

Dehydrohalogenation and Dehydration

7.35 $Et\ddot{O}:^-$ H, H, CH_3, C–C, CH_3, CH_3, Br → [$EtO^{\delta-}$ H, H, C=C, CH_3, CH_3, CH_3, $Br^{\delta-}$]‡ → H, CH_3, C=C, CH_3, CH_3

7.36 Dehydration of *trans*-2-methylcyclohexanol proceeds through the formation of a carbocation (through an E1 reaction of the protonated alcohol) and leads preferentially to the more stable alkene. 1-Methylcyclohexene (below) is more stable than 3-methylcyclohexene (the minor product of the dehydration) because its double bond is more highly substituted.

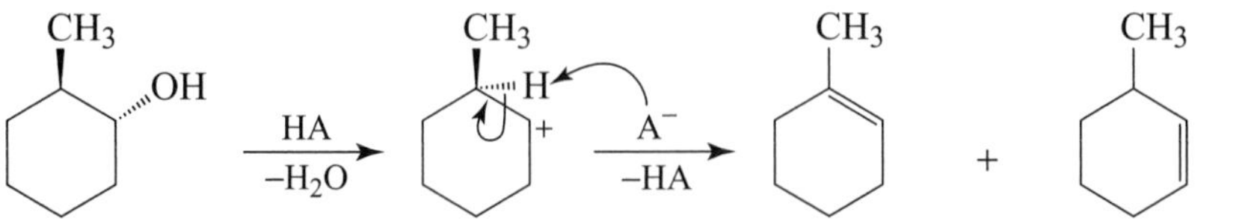

CH_3 OH — HA, $-H_2O$ → CH_3 H + — A^-, $-HA$ → CH_3 (major) Trisubstituted double bond + CH_3 (minor) Disubstituted double bond

Dehydrohalogenation of *trans*-1-bromo-2-methylcyclohexane is an E2 reaction and must proceed through an anti coplanar transition state. Such a transition state is possible only for the elimination leading to 3-methylcyclohexene (cf. Review Problem 7.11).

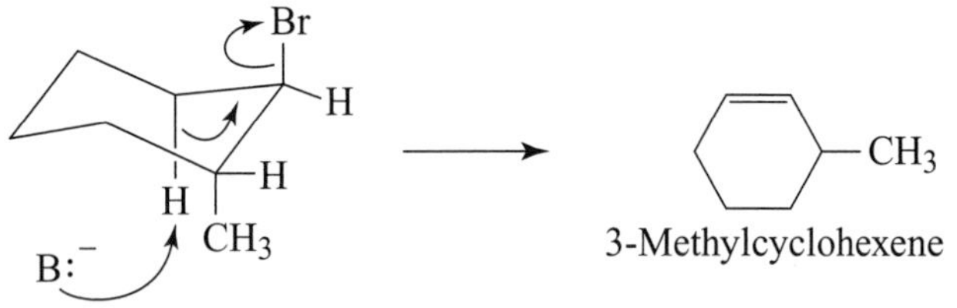

Br, H, H, H, CH_3, $B:^-$ → CH_3 3-Methylcyclohexene

7.37 (a) 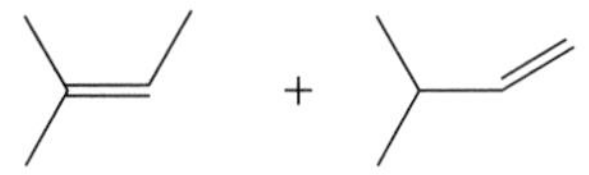 major + minor

(b) only product

(c) major (+ stereoisomer) + minor

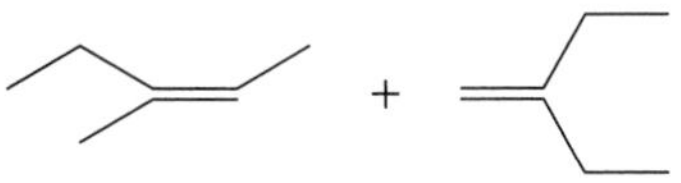

(d) major + minor

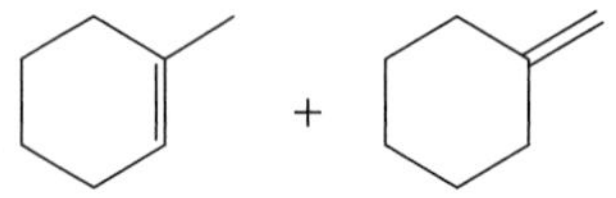

(e) major + minor

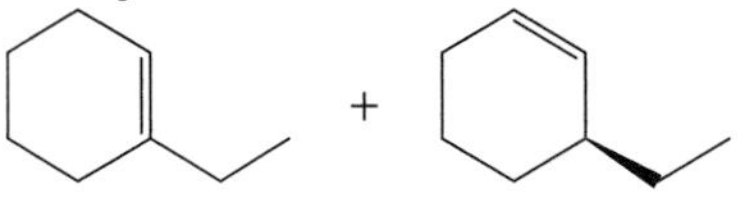

(f) only product

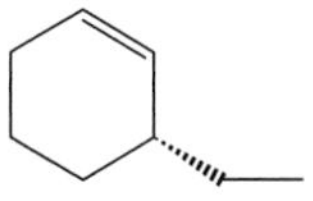

7.38 (a) major + minor

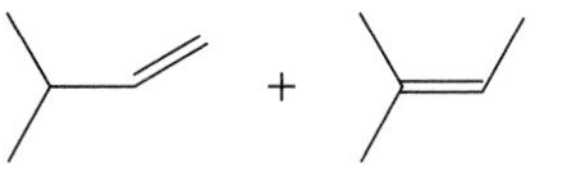

(b) only product

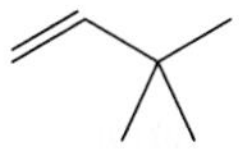

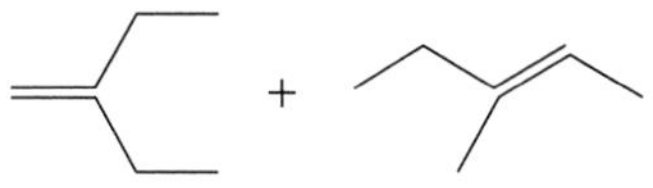

(c) major + minor (+ stereoisomer)

(d) major + minor (+ stereoisomer)

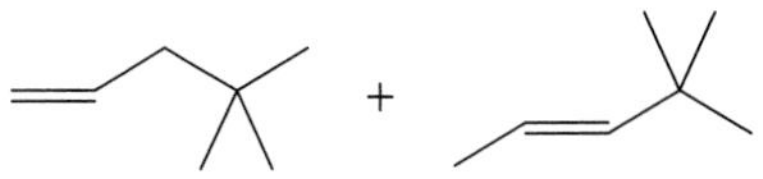

(e) 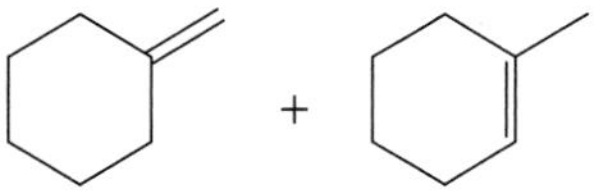 +

major minor

7.39 (a) 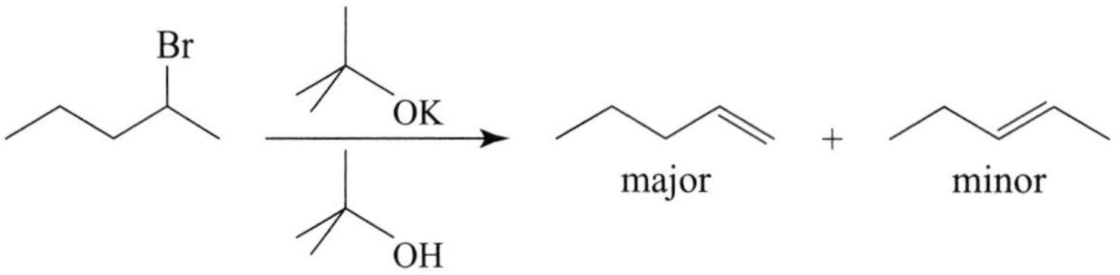 Br OK → major + minor [+(*Z*)]

OH

(b) 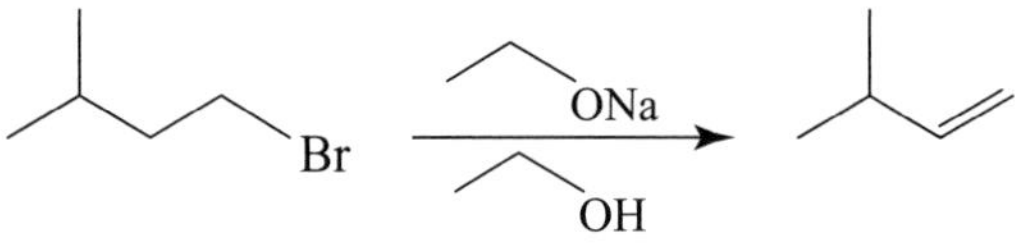 Br ONa → only product

OH

(c) 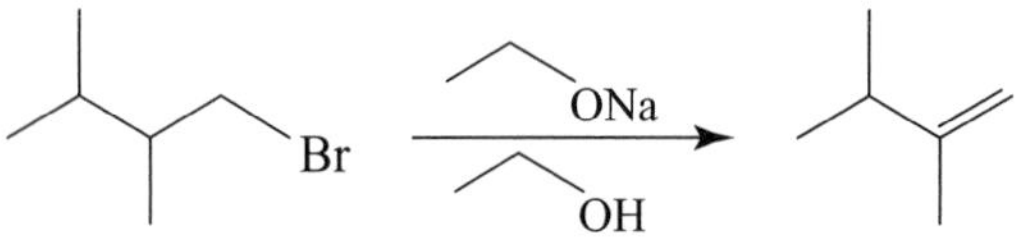 Br ONa → only product

OH

(d) 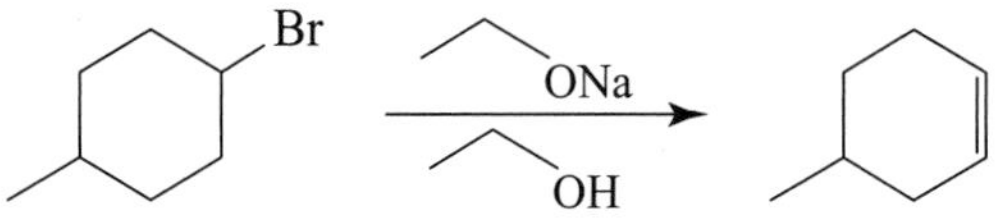 Br ONa → only product

OH

(e) 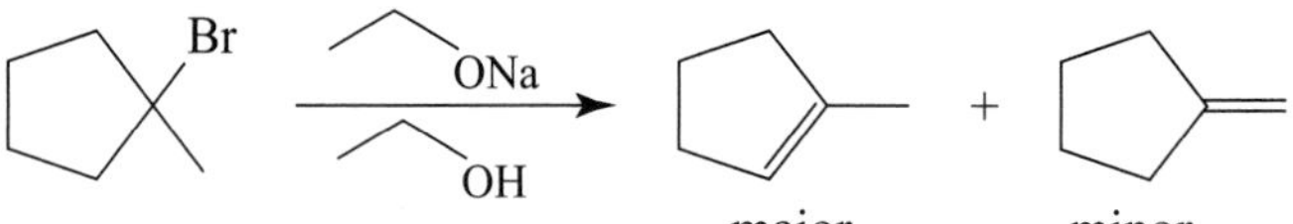 Br ONa → +

OH major minor

7.40 $CH_3C(CH_3)(OH)CH_2CH_3$ 3° > $CH_3CH(OH)CH(CH_3)CH_3$ 2° > $CH_3CH_2CH_2CH_2CH_2OH$ 1°

7.41 (a) 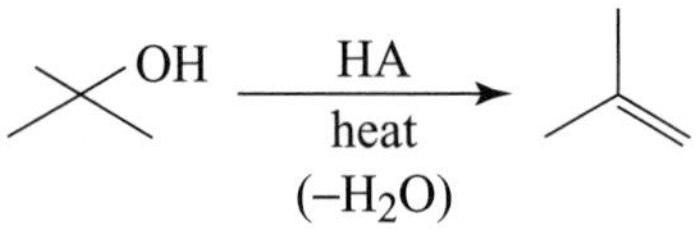 OH HA heat $(-H_2O)$ →

(b) 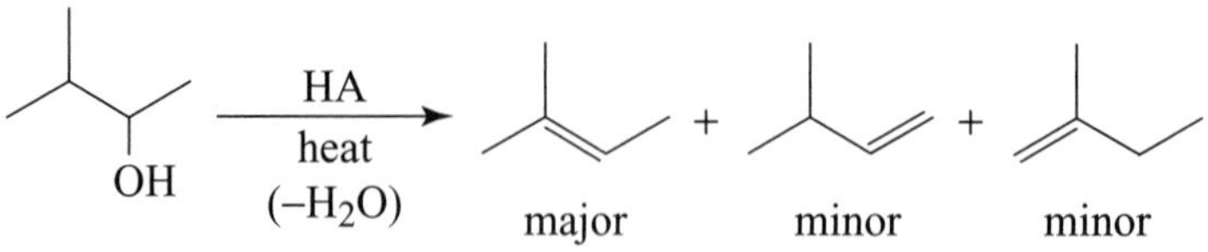 OH HA heat $(-H_2O)$ → + +

major minor minor

(c) 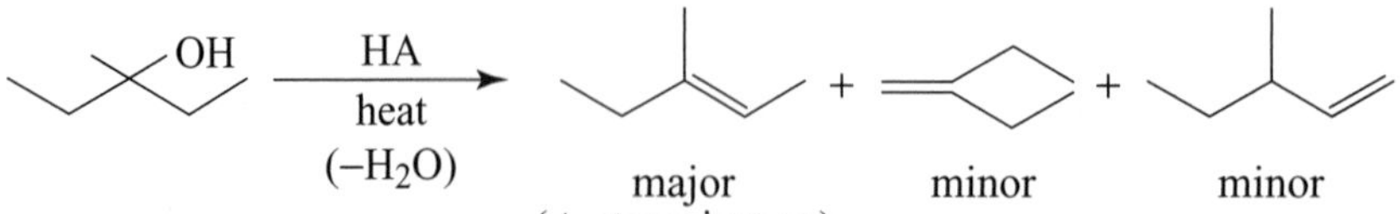 OH HA heat $(-H_2O)$ → + +

major (+ stereoisomer) minor minor

(d) 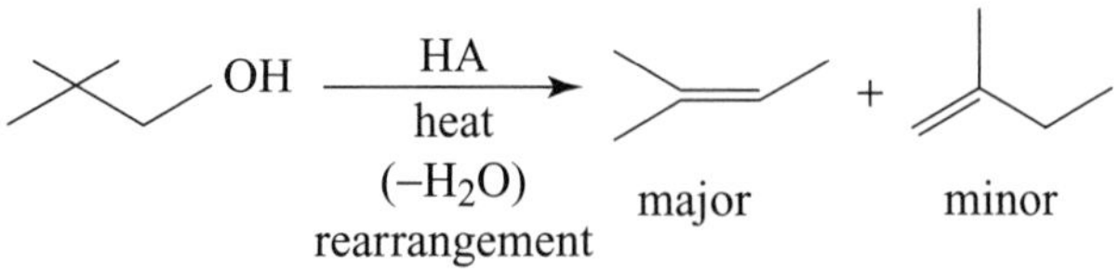 OH $\xrightarrow[\text{(–}H_2O\text{) rearrangement}]{\text{HA, heat}}$ major + minor

(e) 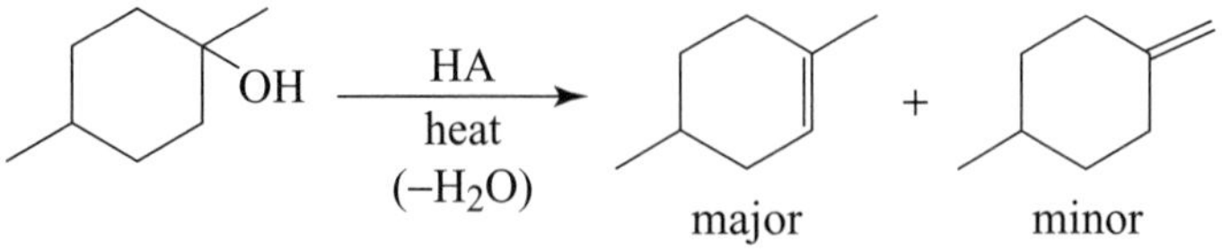 OH $\xrightarrow[\text{(–}H_2O\text{)}]{\text{HA, heat}}$ major + minor

7.42 The alkene cannot be formed because the double bond in the product is too highly strained. Recall that the atoms at each carbon of a double bond prefer to be in the same plane.

7.43 Only the deuterium atom can assume the anti coplanar orientation necessary for an E2 reaction to occur.

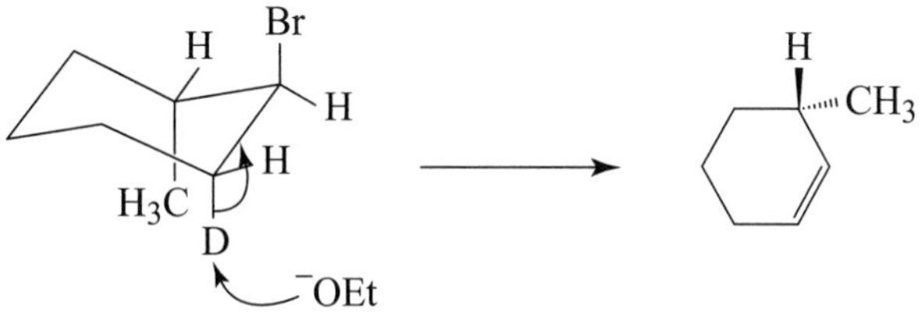

H, Br, H, H, H_3C, D, ^{-}OEt ⟶ H, CH_3

7.44 (a) A hydride shift occurs.

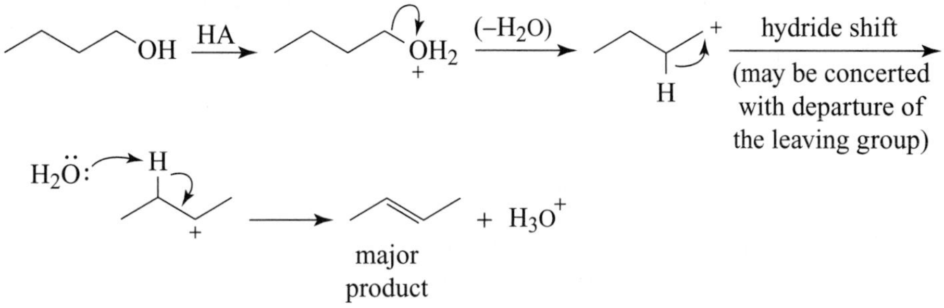

OH $\xrightarrow{\text{HA}}$ $\overset{+}{O}H_2$ $\xrightarrow{(-H_2O)}$ H, + $\xrightarrow{\text{hydride shift}}$ (may be concerted with departure of the leaving group)

$H_2\ddot{O}$: H, + ⟶ major product + H_3O^+

(b) A methanide shift occurs.

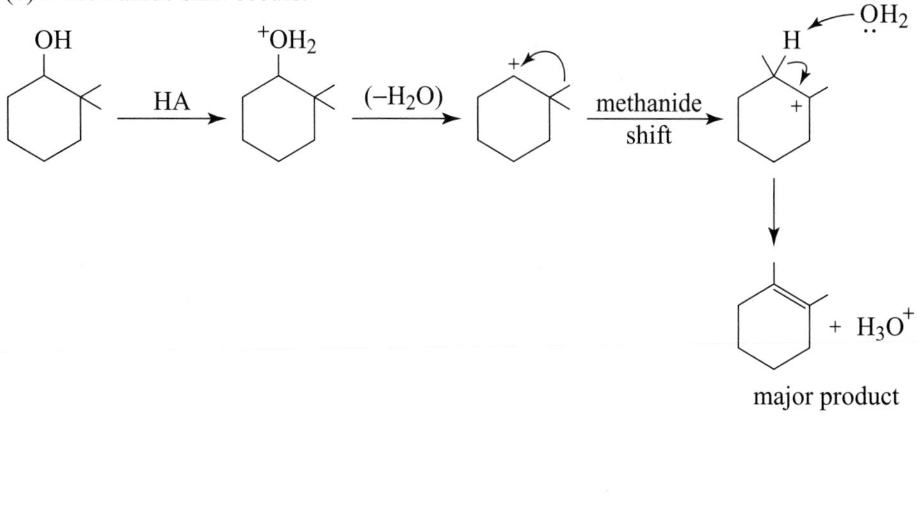

OH $\xrightarrow{\text{HA}}$ $^{+}OH_2$ $\xrightarrow{(-H_2O)}$ + $\xrightarrow{\text{methanide shift}}$ H, +, $\ddot{O}H_2$ ⟶ + H_3O^+

major product

(c) A methanide shift occurs.

I, $AgNO_3$, I Ag^+, $(-AgI)$, +, methanide shift, +, H, HÖEt, $(-Et\overset{+}{O}H_2)$

major product

(d) The required anti coplanar transition state leads only to (E) alkene:

Ph, H, :Ö⁻, Ph, Br, H, ONa, OH, heat, Ph, Ph, + Br^- + Na^+

(E) only

Index of Hydrogen Deficiency

7.45 (a) Caryophyllene has the same molecular formula as zingiberene (Review Problem 4.21); thus it, too, has an index of hydrogen deficiency equal to 4. That 1 mol of caryophyllene absorbs 2 mol of hydrogen on catalytic hydrogenation indicates the presence of two double bonds per molecule.

(b) Caryophyllene molecules must also have two rings. (See Review Problem 23.2 for the structure of caryophyllene.)

7.46 (a) $C_{30}H_{62}$ = formula of alkane
$C_{30}H_{50}$ = formula of squalene
H_{12} = difference = 6 pairs of hydrogen atoms
Index of hydrogen deficiency = 6

(b) Molecules of squalene contain six double bonds.

(c) Squalene molecules contain no rings. (See Review Problem 23.2 for the structural formula of squalene.)

Structure Elucidation

7.47 That **I** and **J** rotate plane-polarized light in the same direction tells us that **I** and **J** are not enantiomers of each other. Thus, the following are possible structures for **I, J,** and **K.** (The enantiomers of **I, J,** and **K** would form another set of structures, and other answers are possible as well.)

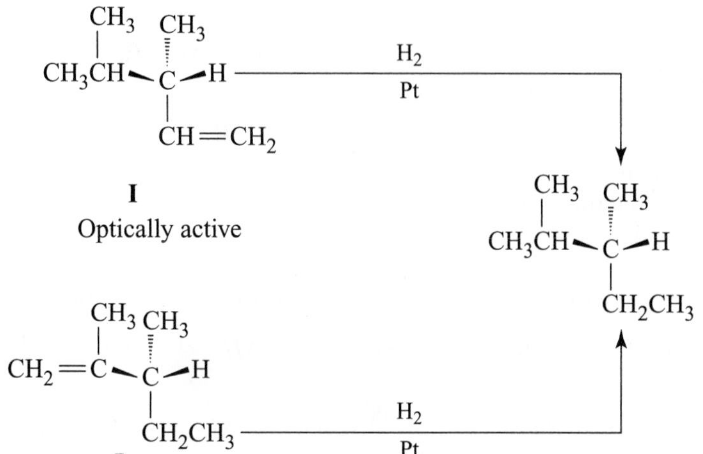

CH_3 CH_3
CH_3CH—C—H
$CH{=}CH_2$

$\xrightarrow[Pt]{H_2}$

I
Optically active

CH_3 CH_3
CH_3CH—C—H
CH_2CH_3

K
Optically active

CH_3 CH_3
$CH_2{=}C$—C—H
CH_2CH_3

$\xrightarrow[Pt]{H_2}$

J
Optically active

7.48 The following are possible structures:

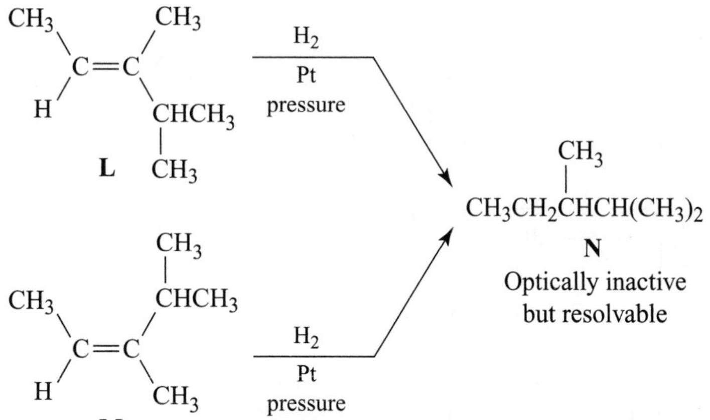

CH_3 CH_3
C=C
H $CHCH_3$
L CH_3

$\xrightarrow[Pt, pressure]{H_2}$

CH_3
$CH_3CH_2CHCH(CH_3)_2$
N
Optically inactive but resolvable

CH_3
CH_3 $CHCH_3$
C=C
H CH_3
M

$\xrightarrow[Pt, pressure]{H_2}$

(other answers are possible as well)

Challenge Problems

7.49

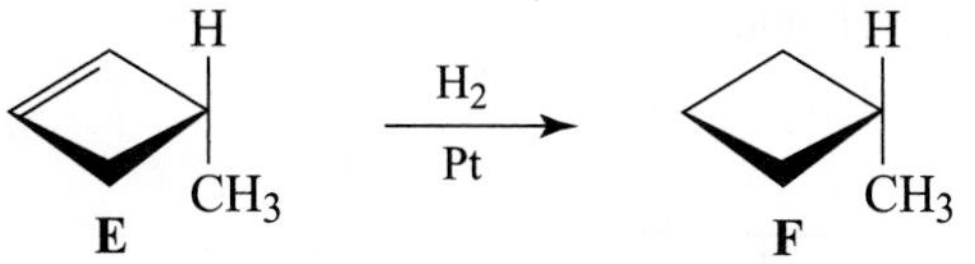

H
CH_3
E

$\xrightarrow[Pt]{H_2}$

H
CH_3
F

Optically active (the enantiomeric form is an equally valid answer)

Optically inactive and nonresolvable

$$\underset{\textbf{G}}{\text{H(CH}_3\text{CH}_2\text{)C=C=C(CH}_3\text{)H}} \xrightarrow[\text{Pt, pressure}]{H_2} \underset{\textbf{H}}{CH_3CH_2CH_2CH_2CH_2CH_3}$$

G: Optically active (the enantiomeric form is an equally valid answer)

H: Optically inactive and nonresolvable

7.50 (a) We are given (Section 7.3A) the following heats of hydrogenation:

$$cis\text{-2-Butene} + H_2 \xrightarrow{Pt} \text{butane} \quad \Delta H^\circ = -120 \text{ kJ mol}^{-1}$$

$$trans\text{-2-Butene} + H_2 \xrightarrow{Pt} \text{butane} \quad \Delta H^\circ = -115 \text{ kJ mol}^{-1}$$

Thus, for

$$cis\text{-2-Butene} \longrightarrow trans\text{-2-butene} \quad \Delta H^\circ = -5.0 \text{ kJ mol}^{-1}$$

(b) Converting *cis*-2-butene into *trans*-2-butene involves breaking the π bond. Therefore, we would expect the energy of activation to be at least as large as the π-bond strength, that is, at least 264 kJ mol^{-1}.

(c)

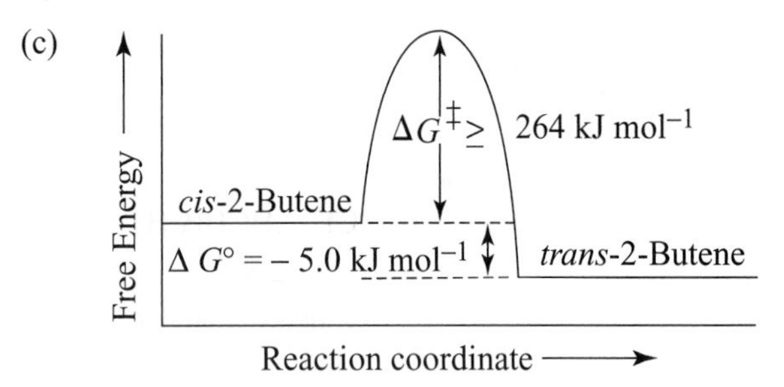

7.51 (a) With either the (1*R*,2*R*)- or the (1*S*,2*S*)-1,2-dibromo-1,2-diphenylethane, only one conformation will allow an anti coplanar arrangement of the H- and Br-. In either case, the elimination leads only to (*Z*)-1-bromo-1,2-diphenylethene:

(1*R*,2*R*)-1,2-Dibromo-1,2-diphenylethane (anti coplanar orientation of H- and -Br) $\longrightarrow$ (*Z*)-1-Bromo-1,2-diphenylethene

(1*S*,2*S*)-1,2-Dibromo-1,2-diphenylethane (anti coplanar orientation of H- and -Br) $\longrightarrow$ (*Z*)-1-Bromo-1,2-diphenylethene

(b) With (1*R*,2*S*)-1,2-dibromo-1,2-diphenylethane, only one conformation will allow an anti coplanar arrangement of the H- and Br-. In either case, the elimination leads only to (*E*)-1-bromo-1,2-diphenylethene:

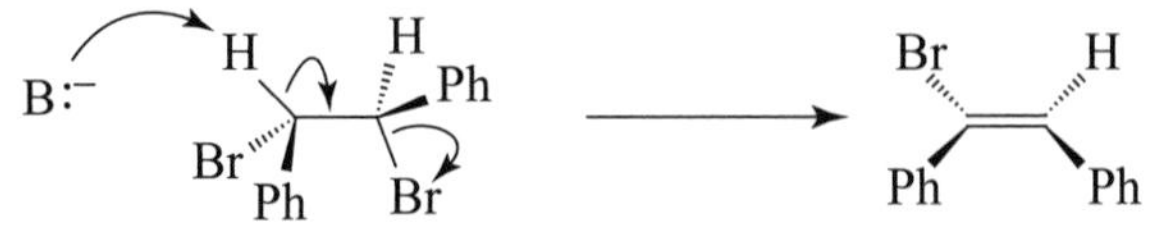

B:⁻ H Br Ph H Ph Br → Br H Ph Ph

(1*R*,2*S*)-1,2-Dibromo-1,2-diphenylethane (anti coplanar orientation of H and Br) (*E*)-1-Bromo-1,2-diphenylethene

(c) With (1*R*,2*S*)-1,2-dibromo-1,2-diphenylethane, only one conformation will allow an anti coplanar arrangement of both bromine atoms. In this case, the elimination leads only to (*E*)-1,2-diphenylethene:

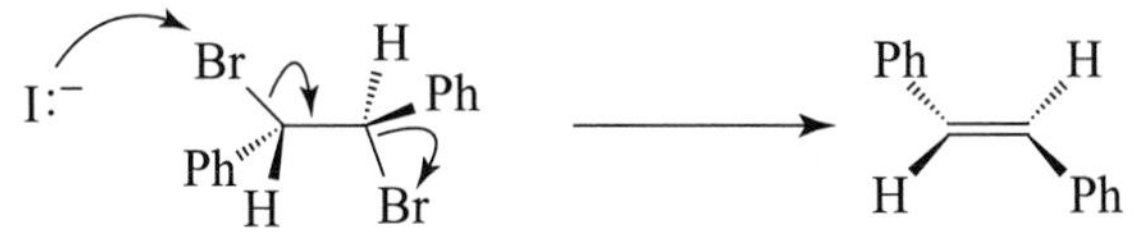

I:⁻ Br Ph H H Ph Br → Ph H H Ph

(1*R*,2*S*)-1,2-Dibromo-1,2-diphenylethane (anti coplanar orientation of both -Br atoms) (*E*)-1,2-Diphenylethene

7.52 (a) H_2, Ni_2B(P-2) or Na/NH_3 → HA → – HA

:A^- H +

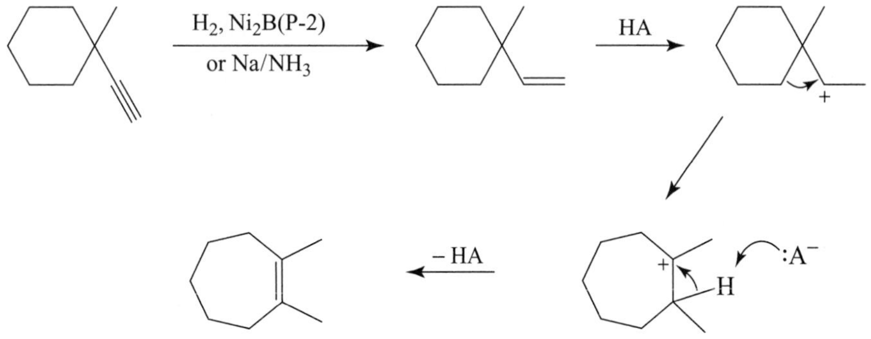

(b) No, tetrasubstituted double bonds usually show no C=C stretching absorption in their infrared spectra.

7.53 OH → OH HA →

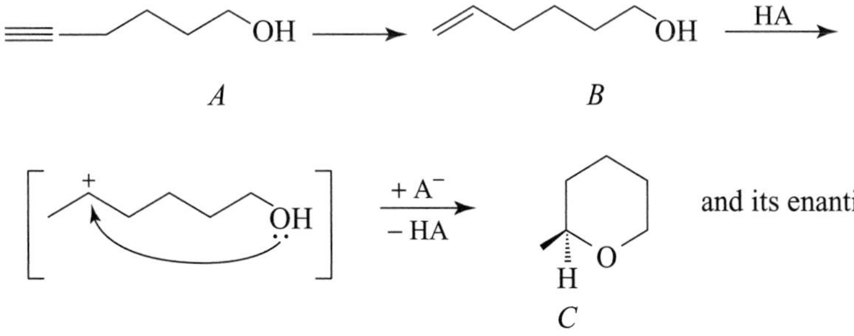

A *B*

OH + A^- – HA → O H *C* and its enantiomer

7.54 (a) Three

(b) Six

QUIZ

7.1 Which conditions/reagents would you employ to obtain the best yields in the following reaction?

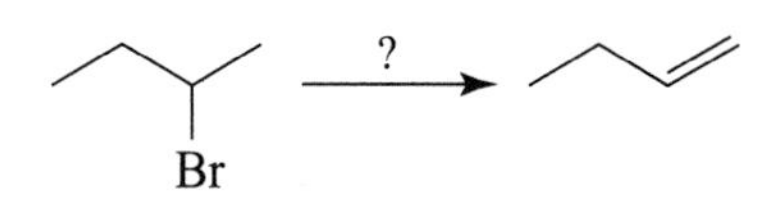

(a) H_2O/heat

(b) ONa / OH

(c) OK / OH, heat

(d) Reaction cannot occur as shown

7.2 Which of the following names is incorrect?

(a) 1-Butene (b) *trans*-2-Butene (c) (*Z*)-2-Chloro-2-pentene

(d) 1,1-Dimethylcyclopentene (e) Cyclohexene

7.3 Select the major product of the reaction

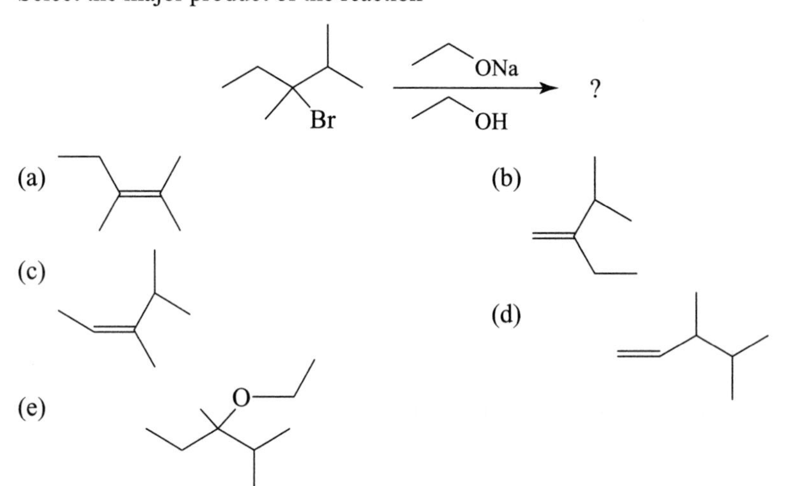

7.4 Supply the missing reagents.

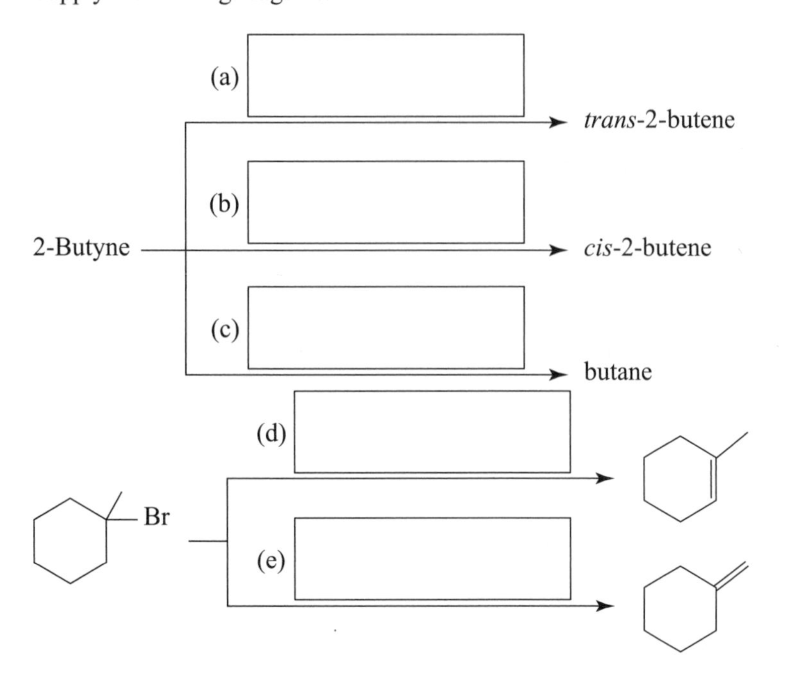

7.5 Arrange the following alkenes in order of decreasing stability. 1-Pentene, *cis*-2-pentene, *trans*-2-pentene, 2-methyl-2-butene

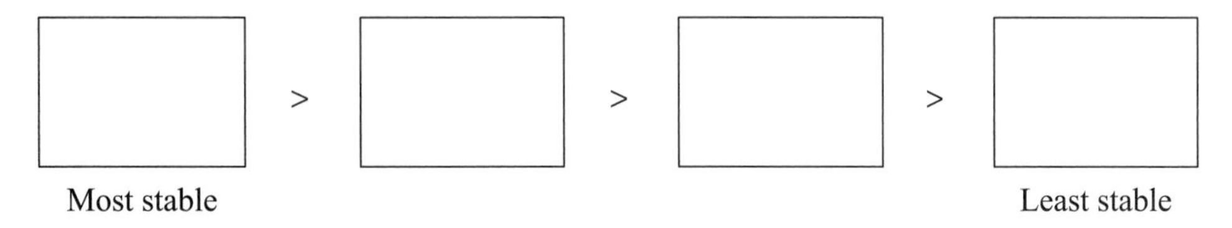

7.6 Complete the following synthesis.

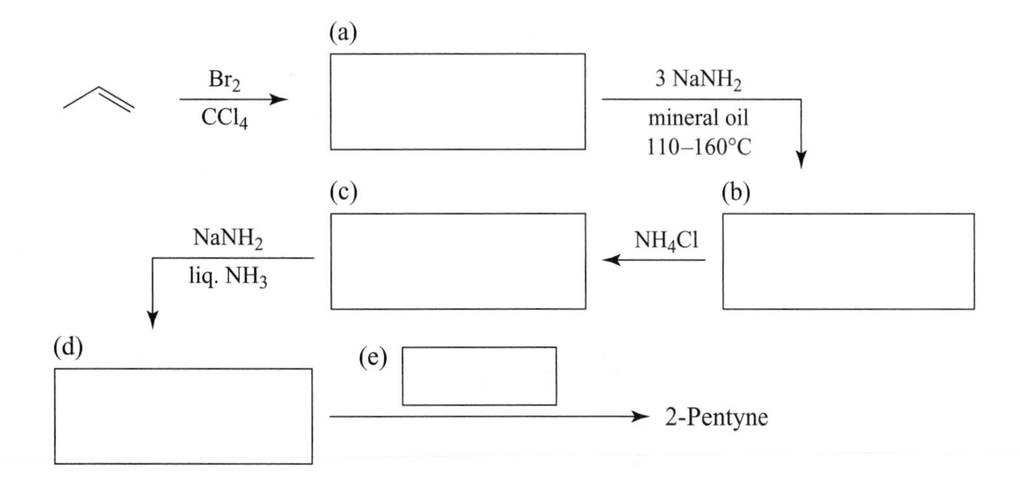

8 ALKENES AND ALKYNES II: ADDITION REACTIONS

SOLUTIONS TO PROBLEMS

8.1 CH_3CHCH_2I with Br on C-2

2-Bromo-1-iodopropane

8.2 (a) + $\overset{\delta+}{H}—\overset{\delta-}{Br}$ ⟶ (carbocation) $\xrightarrow{Br^-}$ (product with Br)

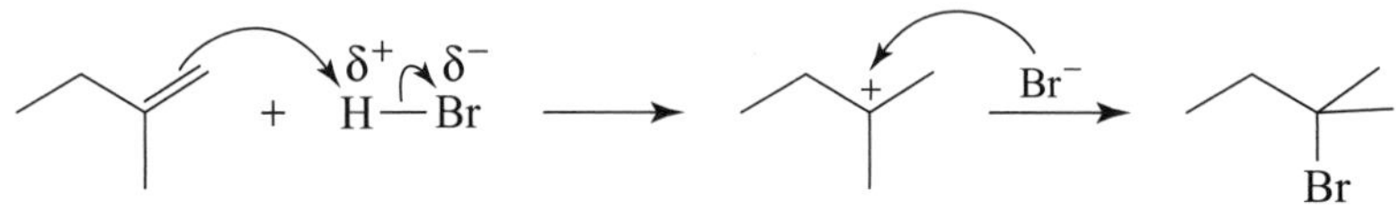

(b) + $\overset{\delta+}{I}—\overset{\delta-}{Cl}$ ⟶ (carbocation with I) $\xrightarrow{Cl^-}$ (product with Cl and I)

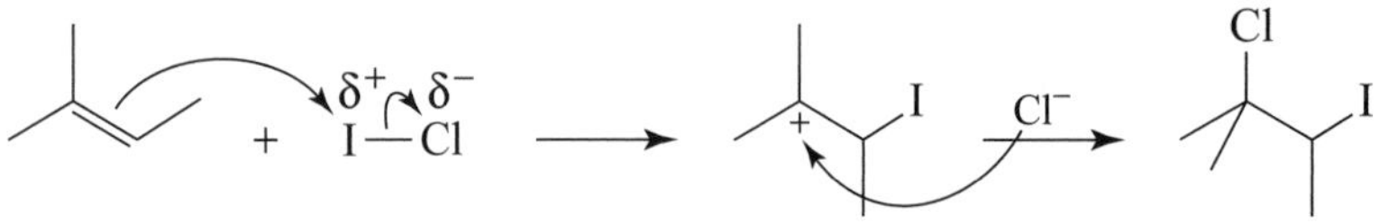

(c) ⟶ $\overset{\delta+}{H}—\overset{\delta-}{I}$ ⟶ (carbocation) $\xrightarrow{I^-}$ (product with I)

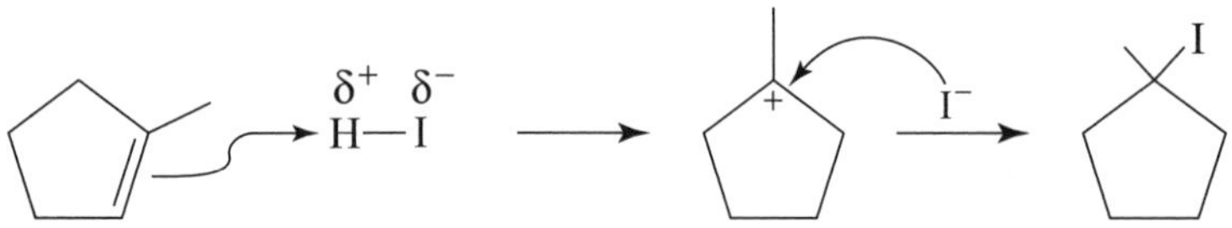

8.3 (a) + $\overset{\delta+}{H}—\overset{\delta-}{Cl}$ ⟶ 2° Carbocation $\xrightarrow[\text{shift}]{\text{1, 2-hydride}}$ 3° Carbocation

Cl^-

2-Chloro-2-methylbutane
(from rearranged carbocation)

Unrearranged
2° Carbocation $\xrightarrow{Cl^-}$ 2-Chloro-3-methylbutane
(from unrearranged carbocation)

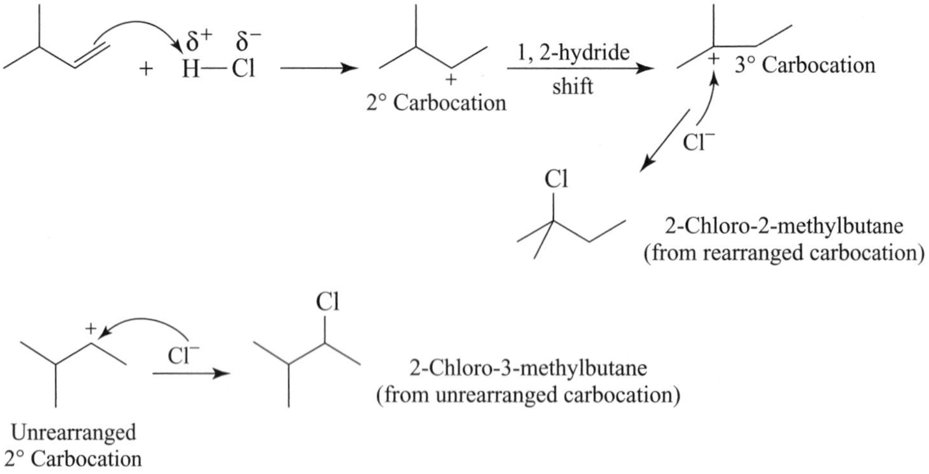

(b) δ^+ δ^- + H—Cl ⟶ (carbocation) $\xrightarrow{Cl^-}$

1,2-methanide shift

3-Chloro-2,2-dimethylbutane (from unrearranged carbocation)

$\xleftarrow{Cl^-}$

2-Chloro-2,3-dimethylbutane (from rearranged carbocation)

8.4 $CH_2{=}CH_2 + H_2SO_4 \longrightarrow CH_3CH_2OSO_3H \xrightarrow[\text{heat}]{H_2O} CH_3CH_2OH + H_2SO_4$

8.5 (a) cyclohexene + $H{-}\ddot{O}^+{-}H$ (H) (from dilute H_2SO_4) ⇌ cyclohexyl cation + $:\ddot{O}{-}H$ (H) ⇌

protonated cyclohexanol + $:\ddot{O}{-}H$ (H) ⇌ cyclohexanol + H_3O^+

(b) Use a high concentration of water because we want the carbocation produced to react with water. And use a strong acid whose conjugate base is a very weak nucleophile. (For this reason we would not use HI, HBr, or HCl.) An excellent method, therefore, is to use dilute sulfuric acid.

(c) Use a low concentration of water (i.e., use concentrated H_2SO_4) and use a higher temperature to encourage elimination. Distill cyclohexene from reaction mixture as it is formed, so as to draw the equilibrium toward product.

(d) 1-Methylcyclohexanol would be the product because a 3° carbocation would be formed as the intermediate.

1-methylcyclohexene + $H{-}\ddot{O}^+{-}H$ (H) ⇌ 1-methylcyclohexyl cation + $:\ddot{O}{-}H$ (H) ⇌

protonated 1-methylcyclohexanol + $:\ddot{O}{-}H$ (H) ⇌ 1-methylcyclohexanol (OH) + H_3O^+

8.6

$$H-\overset{+}{\underset{..}{O}}(H)-H \quad \text{methanide migration} \quad + \ H_2O \qquad + \ H_3O^+$$

8.7 The order reflects the relative ease with which these alkenes accept a proton and form a carbocation. $(CH_3)_2C{=}CH_2$ reacts faster because it leads to a tertiary cation,

$$(CH_3)_2C{=}CH_2 \xrightarrow{H_3O^+} CH_3-\overset{+}{C}(CH_3)(CH_3) \qquad 3^\circ \text{ Carbocation}$$

$CH_3CH{=}CH_2$ leads to a secondary cation,

$$CH_3CH{=}CH_2 \xrightarrow{H_3O^+} CH_3-\overset{+}{C}(H)(CH_3) \qquad 2^\circ \text{ Carbocation}$$

and $CH_2{=}CH_2$ reacts most slowly because it leads to a primary carbocation.

$$CH_2{=}CH_2 \xrightarrow{H_3O^+} CH_3-\overset{+}{C}(H)(H) \qquad 1^\circ \text{ Carbocation}$$

Recall that formation of the carbocation is the rate-determining step in acid-catalyzed hydration and that the order of stabilities of carbocations is the following:

$$3^\circ > 2^\circ > 1^\circ > {}^{+}CH_3$$

8.8

$$H-OSO_3H \qquad Me-\ddot{O}-H \qquad Me \qquad H \quad :\ddot{O}SO_3H \qquad (-H_2SO_4) \qquad Me$$

8.9 (a) $\xrightarrow[\text{(2) NaBH}_4\text{, OH}^-]{\text{(1) Hg(OAc)}_2\text{/THF–H}_2\text{O}}$ OH

(b) $\xrightarrow[\text{(2) NaBH}_4\text{, OH}^-]{\text{(1) Hg(OAc)}_2\text{/THF–H}_2\text{O}}$ OH

(c) $\xrightarrow[\text{(2) NaBH}_4\text{, OH}^-]{\text{(1) Hg(OAc)}_2\text{/THF–H}_2\text{O}}$ OH

(can also be used.)

8.10 (a) $\text{C}{=}\text{C} + {}^{+}\text{HgOCCF}_3$ (O) ⟶ δ^+ –C–C– HgOCCF_3 (O) δ^+ $\xrightarrow{\text{R–}\ddot{\text{O}}\text{–H}}$ R, H, $\ddot{\text{O}}^{+}$, :A$^-$; –C–C– HgOCCF_3 (O)

$\xrightarrow{\text{–HA}}$ RO –C–C– HgOCCF_3 (O)

(b) $CH_3{-}C(CH_3){=}CH_2 \xrightarrow[\text{THF-CH}_3\text{OH}]{\text{Hg(OCCF}_3)_2\ (\text{O})}$ solvomercuration $CH_3{-}C(CH_3)(OCH_3){-}CH_2HgOCCF_3$ (O) $\xrightarrow[\text{demercuration}]{\text{NaBH}_4\text{/OH}^-}$

$CH_3C(CH_3)(OCH_3)CH_3 + Hg + CF_3COO^-$

(c) The electron-withdrawing fluorine atoms in mercuric trifluoroacetate enhance the electrophilicity of the cation. Experiments have demonstrated that for the preparation of tertiary alcohols in satisfactory yields, the trifluoroacetate must be used rather than the acetate.

8.11 (a) 3 $\xrightarrow{\text{BH}_3\text{:THF}}$ B

(b) 3 [isobutylene] $\xrightarrow{BH_3:THF}$ [triisobutylborane]

(c) 3 [2-butene] (or cis isomer) $\xrightarrow{BH_3:THF}$ [tri-sec-butylborane]

(d) 3 [1-methylcyclohexene] $\xrightarrow[\text{syn addition, anti Markovnikov}]{BH_3:THF}$ [trialkylborane] + enantiomer

8.12 $2\ CH_3\overset{CH_3}{\overset{|}{C}}{=}CHCH_3 \xrightarrow{BH_3:THF} \left(CH_3\overset{CH_3}{\overset{|}{C}H}{-}\underset{CH_3}{\underset{|}{C}H}{-}\right)_2 BH$

Disiamylborane

8.13 (a) [1-pentene] $\xrightarrow[(2)\ H_2O_2,\ OH^-]{(1)\ BH_3:THF}$ [1-pentanol] OH

(b) [2-methyl-1-pentene] $\xrightarrow[(2)\ H_2O_2,\ OH^-]{(1)\ BH_3:THF}$ [2-methyl-1-pentanol] OH

(c) [3-methyl-2-pentene] [or (*Z*) isomer] $\xrightarrow[(2)\ H_2O_2,\ OH^-]{(1)\ BH_3:THF}$ [3-methyl-2-pentanol] OH

(d) [2-methyl-2-pentene] $\xrightarrow[(2)\ H_2O_2,\ OH^-]{(1)\ BH_3:THF}$ [2-methyl-3-pentanol] OH

(e) [1-methylcyclobutene] CH_3 $\xrightarrow[(2)\ H_2O_2,\ OH^-]{(1)\ BH_3:THF}$ [trans-2-methylcyclobutanol] CH_3, OH + enantiomer

(f) 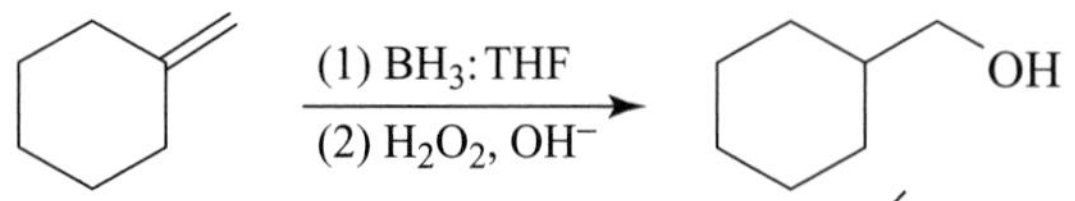 methylenecyclohexane $\xrightarrow[\text{(2) } H_2O_2,\ OH^-]{\text{(1) } BH_3:THF}$ cyclohexylmethanol (OH)

8.14 (a) $3\ CH_3CH(CH_3)CH{=}CH_2 \xrightarrow{BH_3:THF} (CH_3CH(CH_3)CH_2CH_2)_3B \xrightarrow{CH_3CO_2D}$

$3\ CH_3CH(CH_3)CH_2CH_2D$

(b) $3\ CH_3C(CH_3){=}CHCH_3 \xrightarrow{BH_3:THF} CH_3CH(CH_3){-}CH(B){-}CH_3 \xrightarrow{CH_3CO_2D}$

$3\ CH_3CH(CH_3)CHDCH_3$

(c) 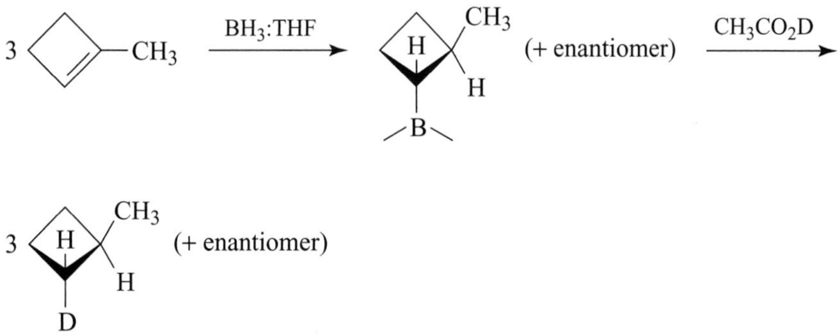 3 1-methylcyclobutene ($-CH_3$) $\xrightarrow{BH_3:THF}$ (H, CH_3, H, B) (+ enantiomer) $\xrightarrow{CH_3CO_2D}$

3 (H, CH_3, H, D) 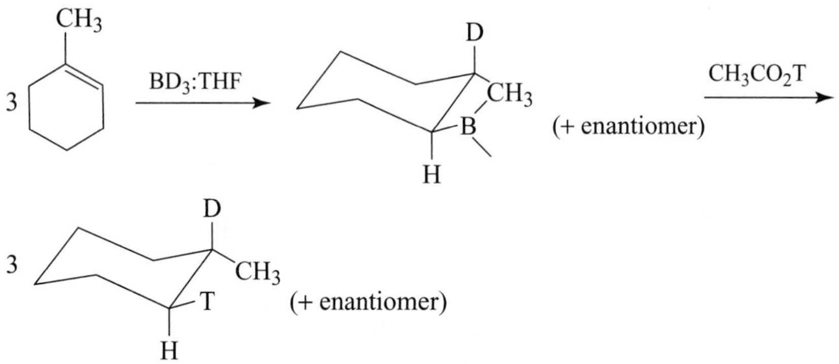 (+ enantiomer)

(d) 3 1-methylcyclohexene (CH_3) $\xrightarrow{BD_3:THF}$ (D, CH_3, B, H) (+ enantiomer) $\xrightarrow{CH_3CO_2T}$

3 (D, CH_3, T, H) (+ enantiomer)

8.15 + $\ddot{:Br}^{\delta+}$ — $\ddot{Br}:^{\delta-}$ ⇌ [(a) (b) $\ddot{Br}^+$ (a) $:\ddot{O}H_2$ (b)] →

Bromonium ion

$\overset{+}{O}H_2$, Br + Br, $\overset{+}{O}H_2$ —H_2O→ OH, Br + Br, OH

from (a)

from (b)

Racemic mixture

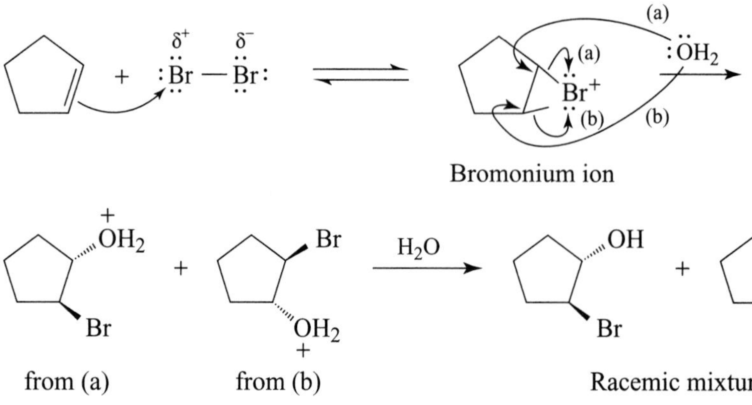

Because paths (a) and (b) occur at equal rates, these enantiomers are formed at equal rates and in equal amounts.

8.16 ═ + $:\ddot{Br}^{\delta+}$—$\ddot{Br}:^{\delta-}$ ⟶ $.\ddot{Br}.^+$ + $:\ddot{Br}:^-$

Nu:

$.\ddot{Br}.^+$

Nu = H_2O → Br—CH₂CH₂—$\overset{+}{O}H_2$ —(:A, —HA)→ Br—CH₂CH₂—OH

or Br^- → Br—CH₂CH₂—Br

or Cl^- → Br—CH₂CH₂—Cl

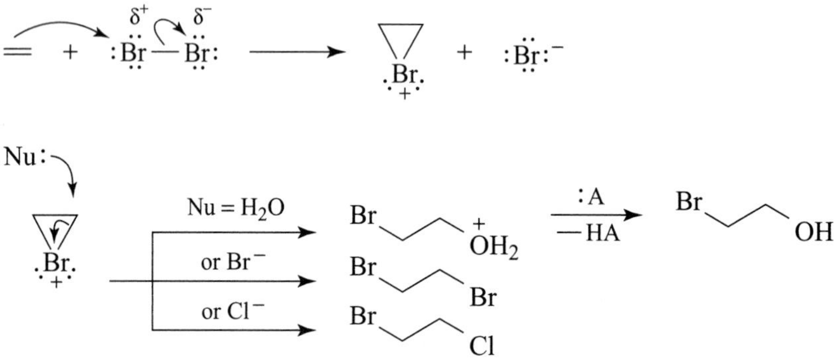

8.17 (a) —(*t*-BuOK, $CHCl_3$)→ Cl, Cl, H, H + enantiomer

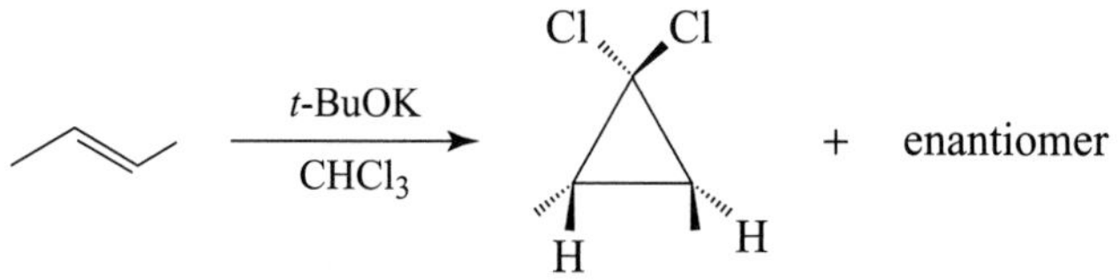

(b) —(*t*-BuOK, $CHBr_3$)→ Br, Br

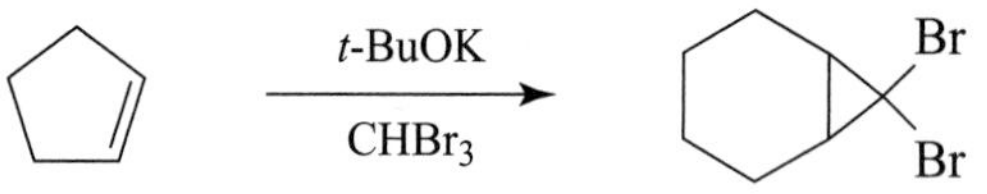

(c) —($CH_2I_2/Zn(Cu)$, Et_2O)→ I, I

8.18 —($CHBr_3$, *t*-BuOK)→ Br, Br

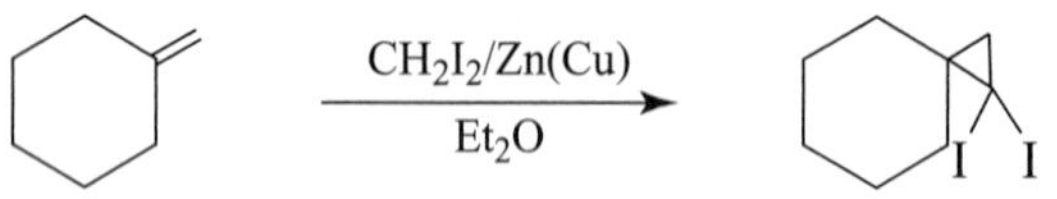

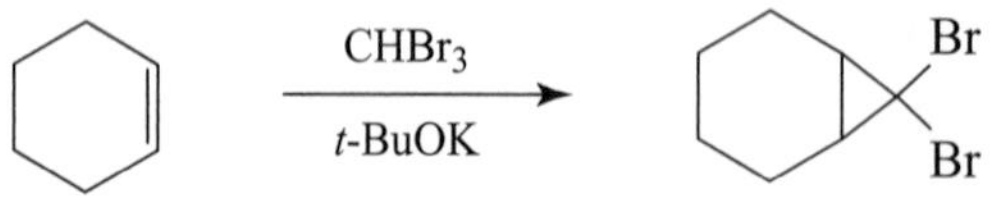

8.19 $\xrightarrow[\text{Zn(Cu)}]{CH_3CHI_2}$ +

H H

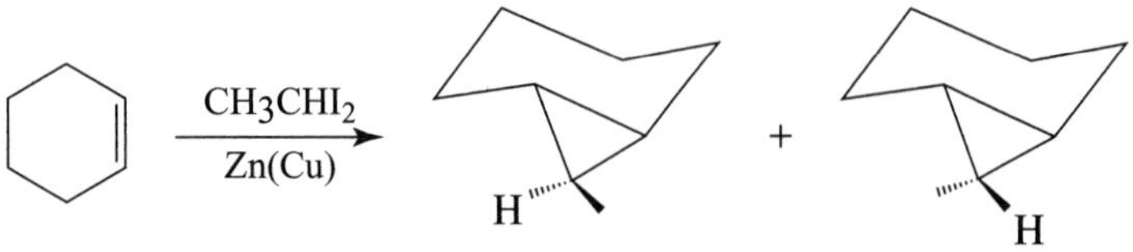

8.20 (a) $\xrightarrow[\text{(2) NaHSO}_3]{\text{(1) OsO}_4\text{, pyridine, 25°C}}$ OH OH

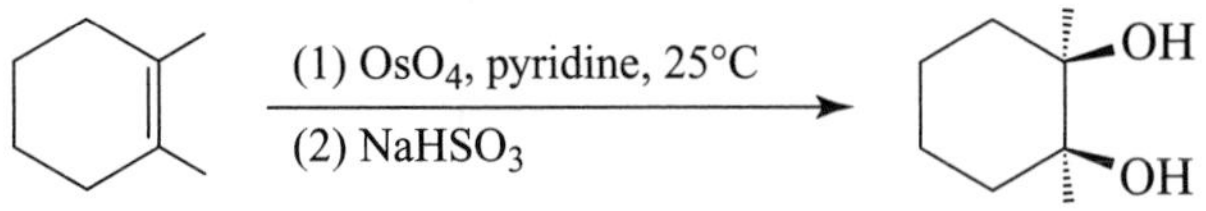

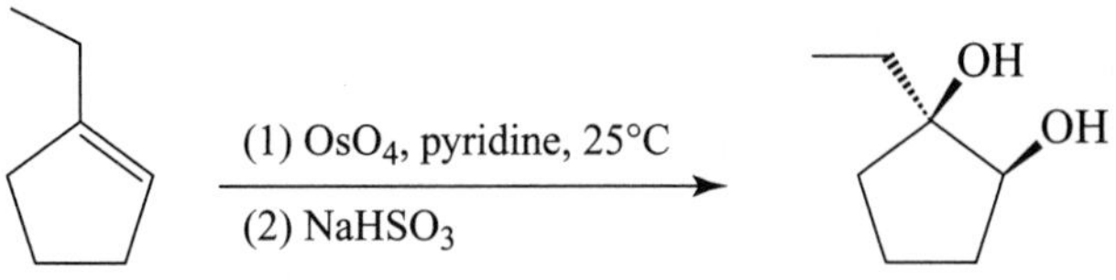

(b) $\xrightarrow[\text{(2) NaHSO}_3]{\text{(1) OsO}_4\text{, pyridine, 25°C}}$ OH OH (racemic)

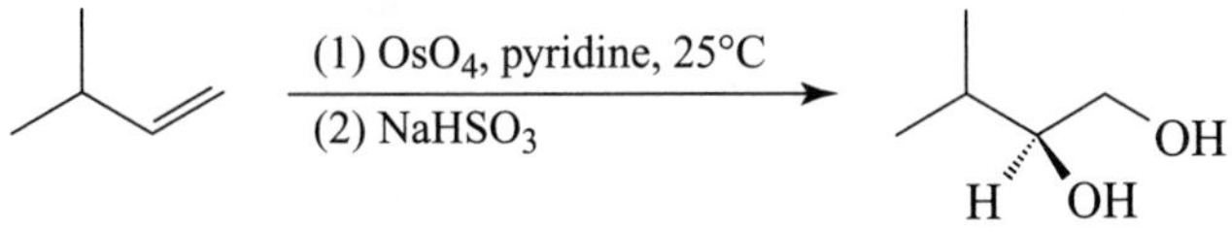

(c) $\xrightarrow[\text{(2) NaHSO}_3]{\text{(1) OsO}_4\text{, pyridine, 25°C}}$ OH H OH (racemic)

8.21 (a) $\xrightarrow[\text{(2) Me}_2\text{S}]{\text{(1) O}_3}$ O + O H

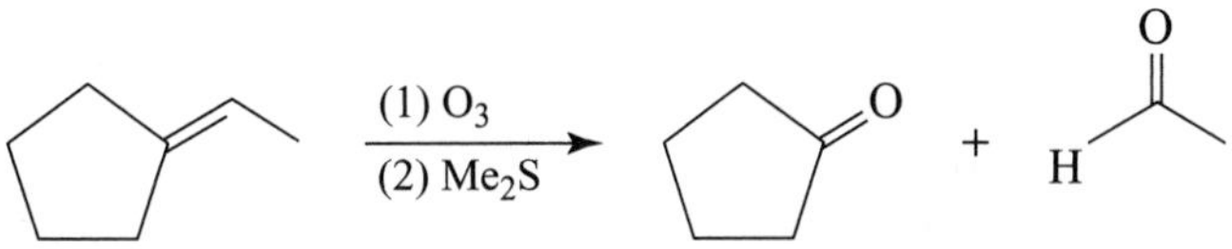

(b) $\xrightarrow[\text{(2) Me}_2\text{S}]{\text{(1) O}_3}$ O O

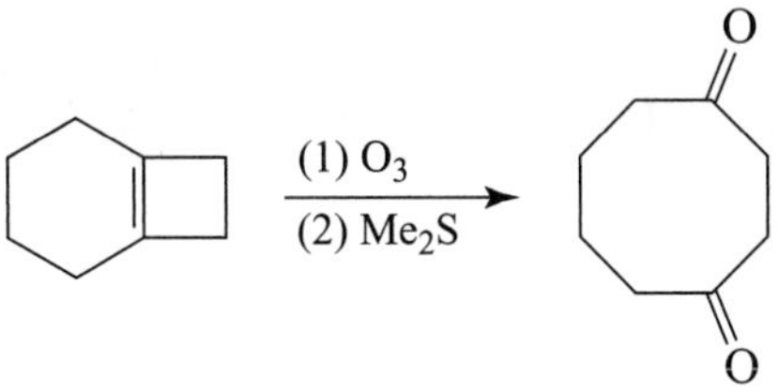

(c) $\xrightarrow[\text{(2) Me}_2\text{S}]{\text{(1) O}_3}$ O H H O O + O

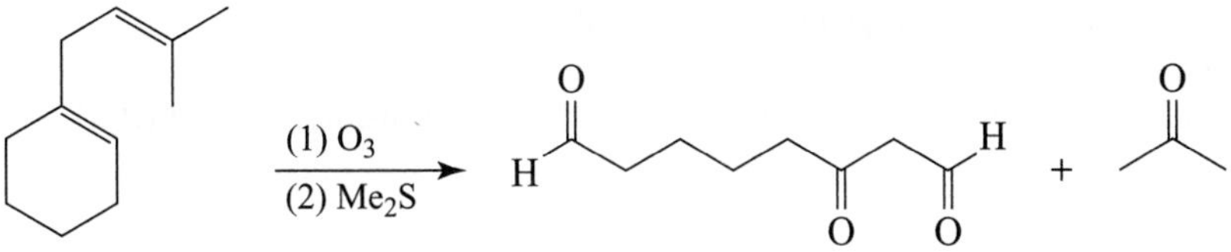

8.22 (a) $\xrightarrow[\text{(2) Me}_2\text{S}]{\text{(1) O}_3}$ O + O H

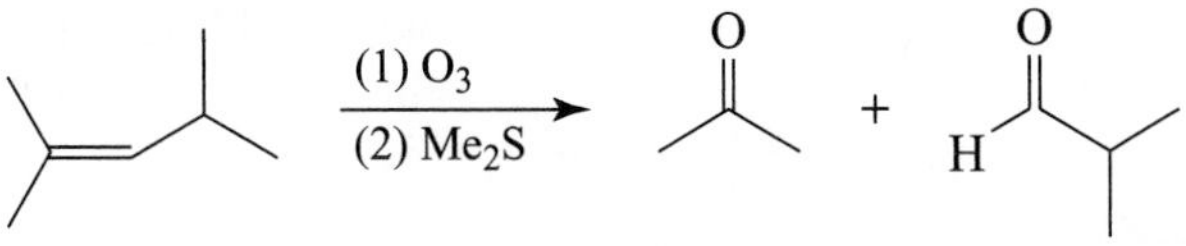

(b) $\xrightarrow[\text{(2) Me}_2\text{S}]{\text{(1) O}_3}$ 2 O H

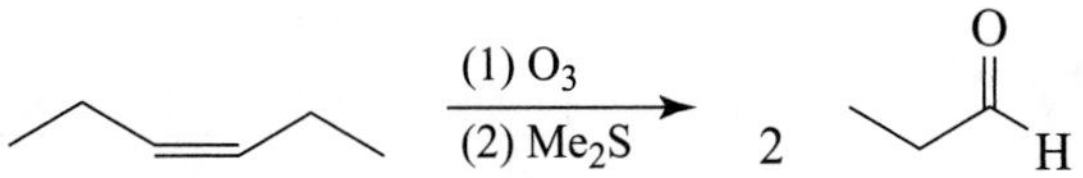

[or (*E*) isomer]

(c) [methylenecyclopentane] $\xrightarrow[(2)\ Me_2S]{(1)\ O_3}$ [cyclopentanone] + $HCHO$

8.23 Ordinary alkenes *are* more reactive toward electrophilic reagents. But the alkenes obtained from the addition of an electrophilic reagent to an alkyne have at least one electronegative atom (Cl, Br, etc.) attached to a carbon atom of the double bond.

$$-C{\equiv}C- \xrightarrow{HX} \text{(H)C}{=}\text{C(X)}$$

or

$$-C{\equiv}C- \xrightarrow{X_2} \text{(X)C}{=}\text{C(X)}$$

These alkenes are less reactive than alkynes toward electrophilic addition because the electronegative group makes the double bond "electron poor."

8.24 The molecular formula and the formation of octane from **A** and **B** indicate that both compounds are unbranched octynes. Since **A** yields only [butanoic acid] on ozonolysis, **A** must be the symmetrical octyne [4-octyne]. The IR absorption for **B** shows the presence of a terminal triple bond. Hence **B** is [1-octyne].

Since **C** (C_8H_{12}) gives HO(C=O)(CH$_2$)$_6$(C=O)OH on ozonolysis, **C** must be cyclooctyne. This is supported by the molecular formula of **C** and the fact that it is converted to **D**, C_8H_{16} (cyclooctane), on catalytic hydrogenation.

8.25 By converting the 3-hexyne to *cis*-3-hexene using H_2/Ni_2B (P-2).

[3-hexyne] $\xrightarrow[Ni_2B(P\text{–}2)]{H_2}$ [*cis*-3-hexene]

Then, addition of bromine to *cis*-3-hexene will yield (3*R*,4*R*), and (3*S*,4*S*)-3,4-dibromohexane as a racemic form.

[*cis*-3-hexene] $\xrightarrow[\text{anti addition}]{Br_2}$ (3*R*, 4*R*) + (3*S*, 4*S*)

Racemic 3,4-dibromohexane

Alkenes and Alkynes Reaction Tool Kit

8.26 (a) 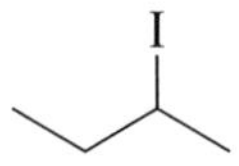 I (b) 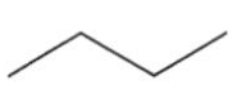 (c) 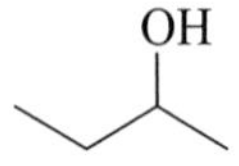 OH (d) 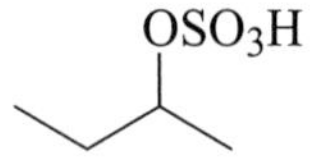 OSO_3H

(e) 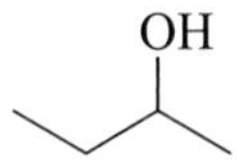 OH (f) 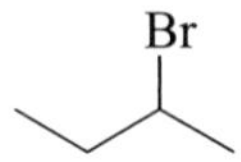 Br (g) 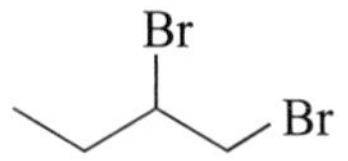 Br Br (h) 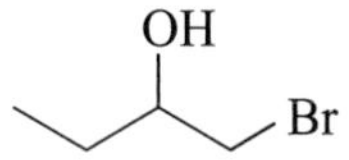 OH Br

(i) 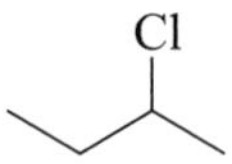 Cl (j) 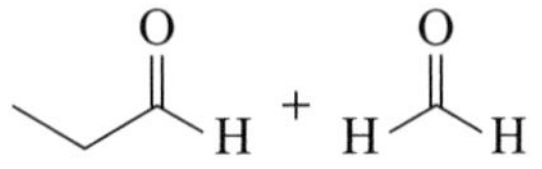 O H + O H H (k) 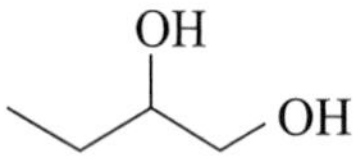 OH OH

(l) 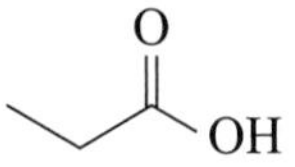 O OH + CO_2 (m) 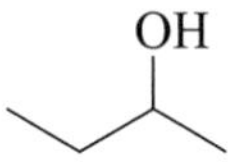 OH (n) 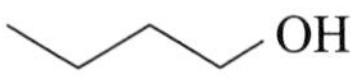 OH

8.27 (a) 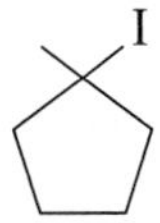 I (b) 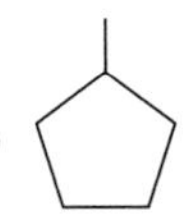 (c) 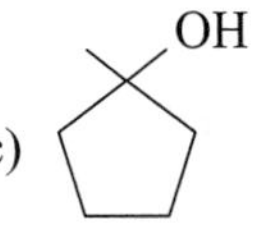 OH (d) 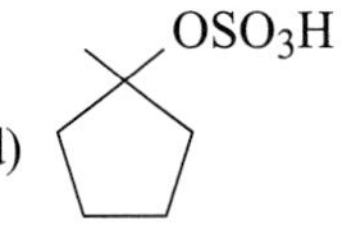 OSO_3H

(e) 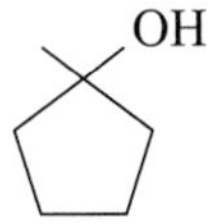 OH (f) 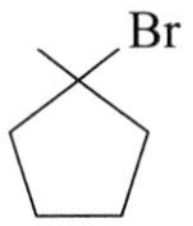 Br (g) 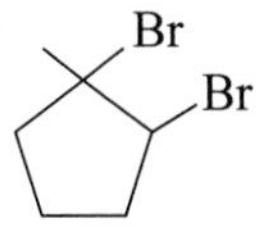 Br Br (h) 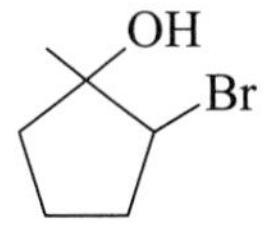 OH Br

(i) 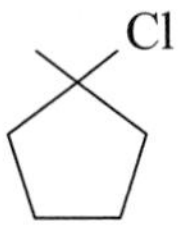 Cl (j) 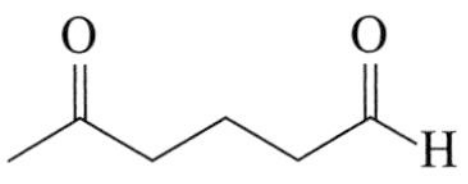 O O H (k) 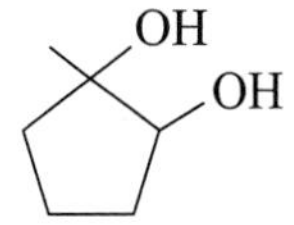 OH OH

(l) 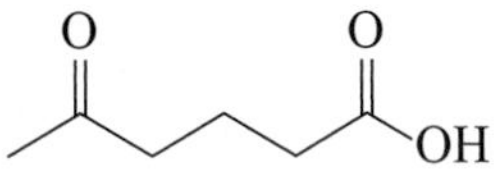 O O OH (m) 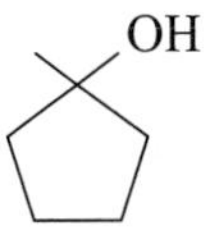 OH (n) 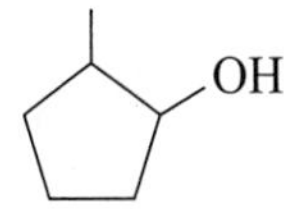 OH

8.28 (a) 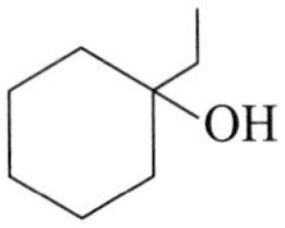 OH (b) 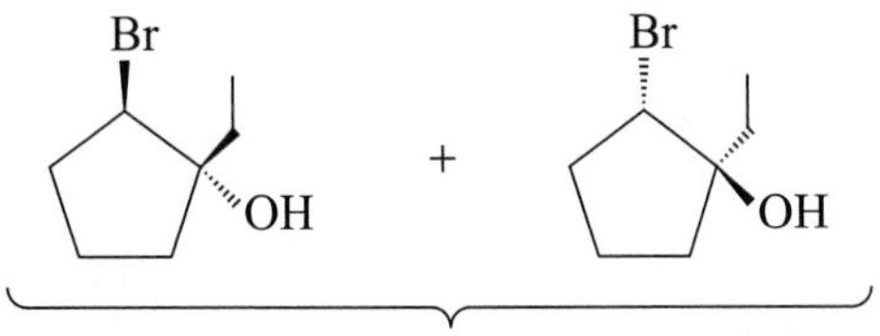 Br OH + Br OH

enantiomers

(c) 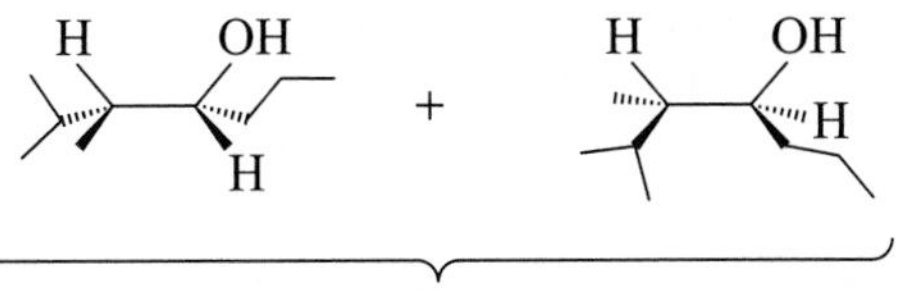 H OH H + H OH H

enantiomers

(d) 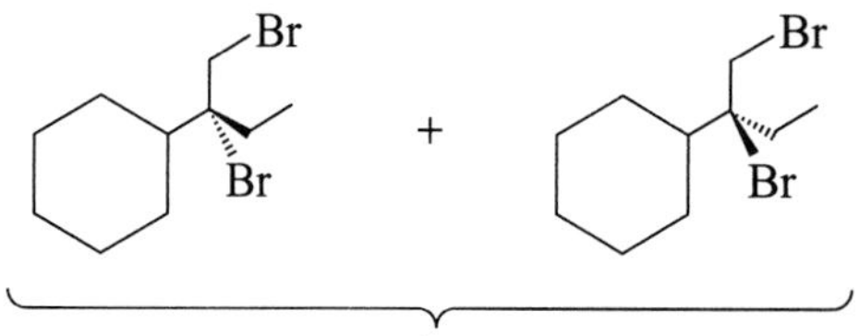 Br, Br + Br, Br

enantiomers

(e) 2 O H

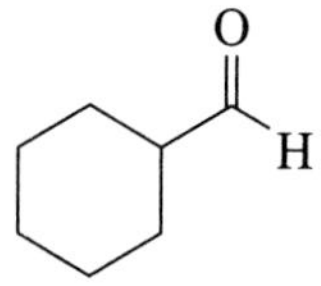

8.29 (a) 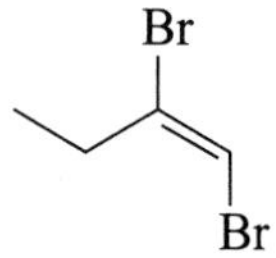 Br Br (b) 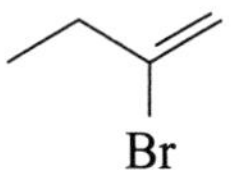 Br (c) 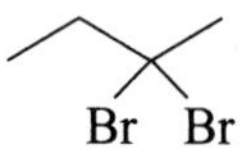 Br Br

(d)  (e) 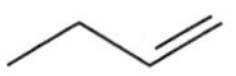 (f)

(g) 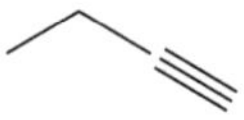 and  [An E2 reaction would take place

when :$^{-}$Na^{+} is treated with Br]

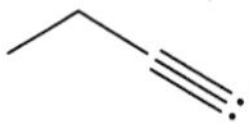

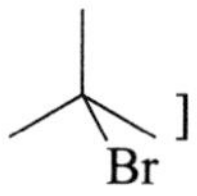

8.30 (a) 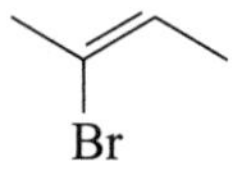 Br (b) 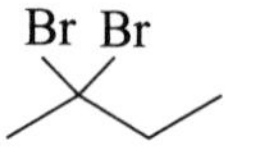 Br Br (c) 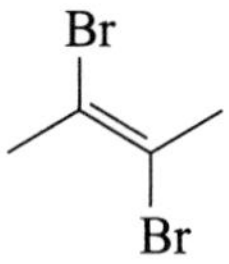 Br Br

(d) 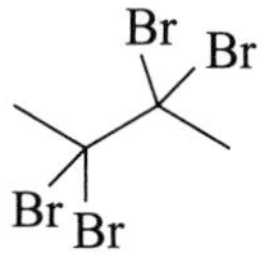 Br Br Br Br (e) 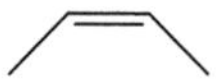 (f) 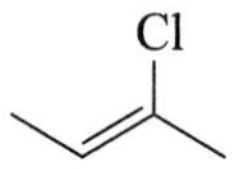 Cl

(g) 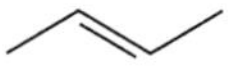 (h)  (i)

(j) 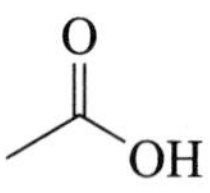 O OH (2 molar equivalents) (k) 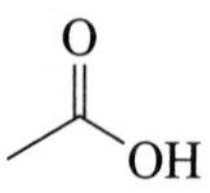 O OH (2 molar equivalents) (l) No reaction

8.31 (a) Cl Cl (b) Br Br

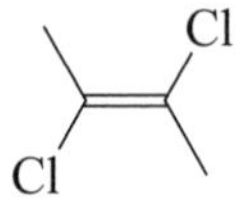

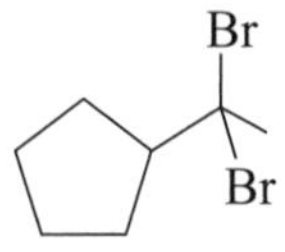

(c) 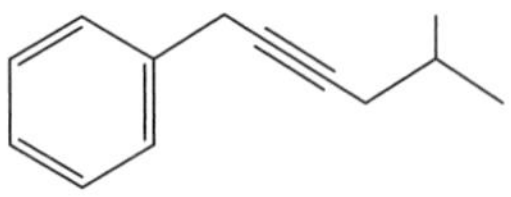 (d) OH O + HO O

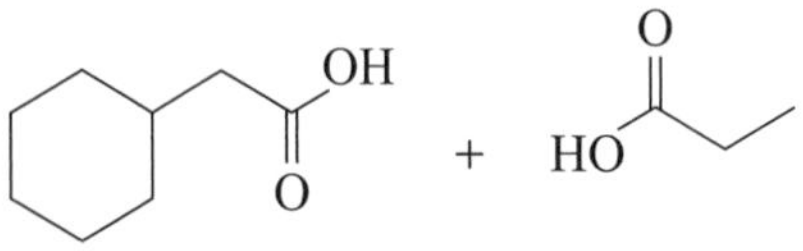

8.32 (a) $\xrightarrow[CCl_4]{Br_2}$ Br Br $\xrightarrow[\text{mineral oil, heat}]{3\ NaNH_2}$

$:^{-}\ Na^{+}$ $\xrightarrow{NH_4Cl}$

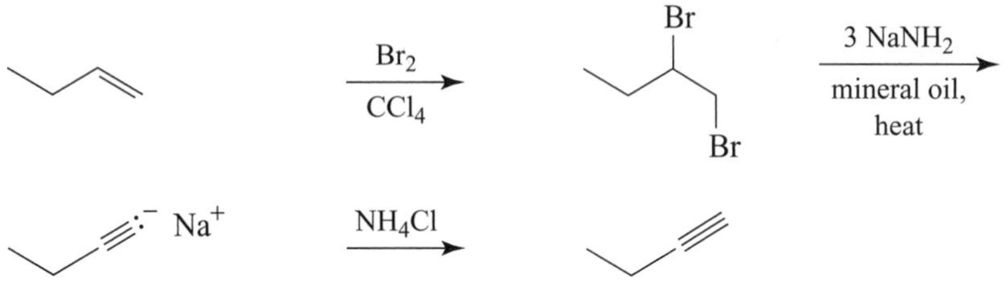

(b) Cl $\xrightarrow[t\text{-BuOH, heat}]{t\text{-BuOK}}$ Then as in (a)

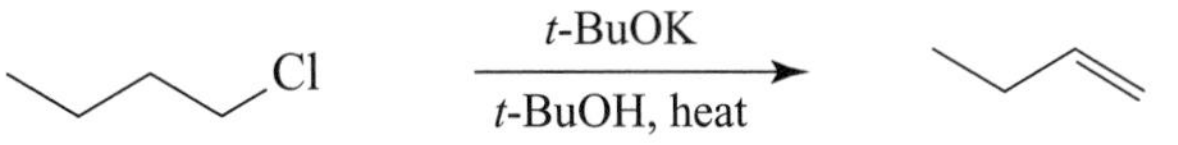

(c) Cl $\xrightarrow[\text{mineral oil, heat}]{2\ NaNH_2}$ $:^{-}\ Na^{+}$ $\xrightarrow{NH_4Cl}$

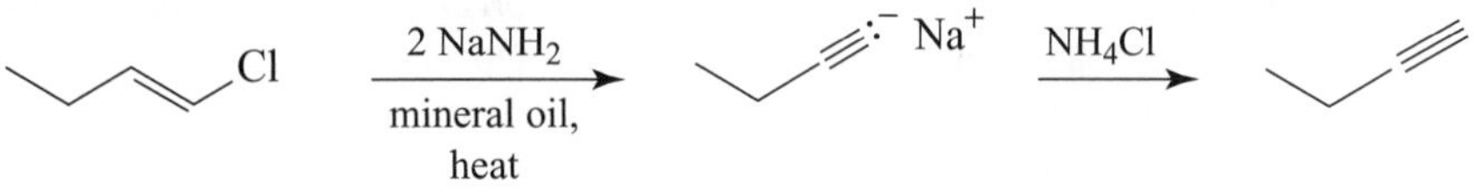

(d) Cl Cl $\xrightarrow[\text{mineral oil, heat}]{3\ NaNH_2}$ $:^{-}\ Na^{+}$ $\xrightarrow{NH_4Cl}$

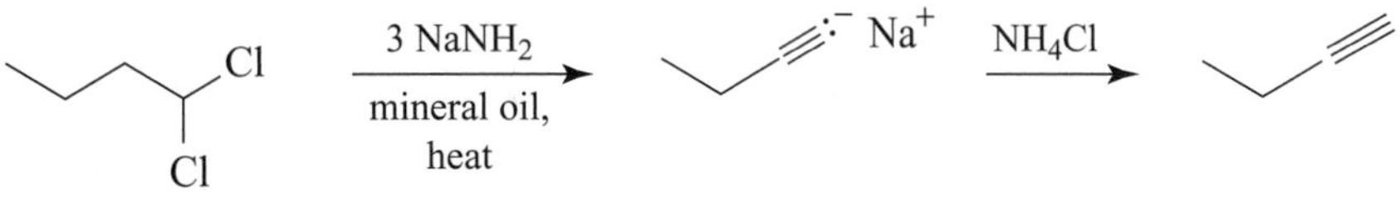

(e) $\xrightarrow[\text{liq. } NH_3]{NaNH_2}$ $\equiv:^{-}\ Na^{+}$ $\xrightarrow{\text{Br}}$

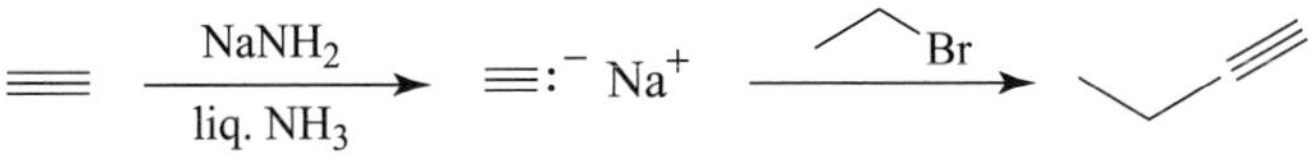

8.33 (a) $\xrightarrow{H_3O^+,\ H_2O}$ OH

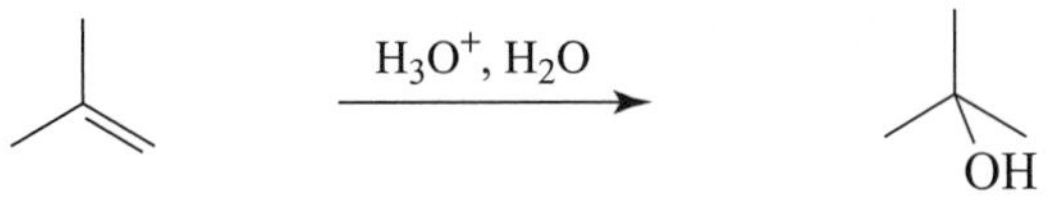

(b) $\xrightarrow[\text{(no peroxides)}]{HBr}$ Br

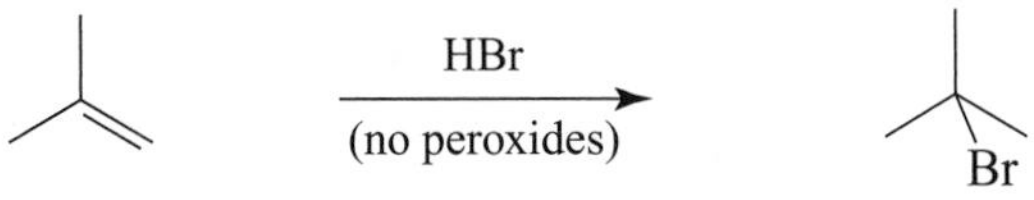

(c) $\xrightarrow{Cl_2,\ H_2O}$ Cl OH

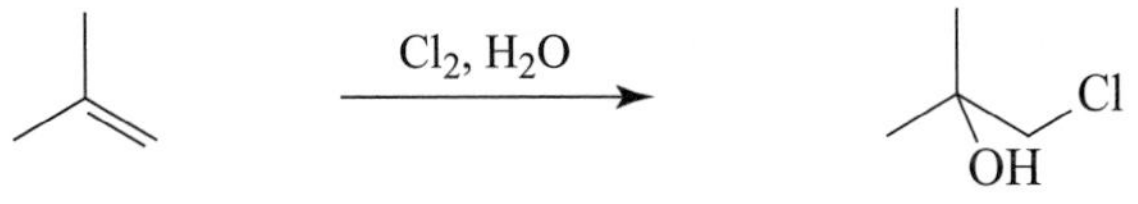

(d) $\xrightarrow[(2)\ H_2O_2,\ OH^-]{(1)\ BH_3:THF}$ OH

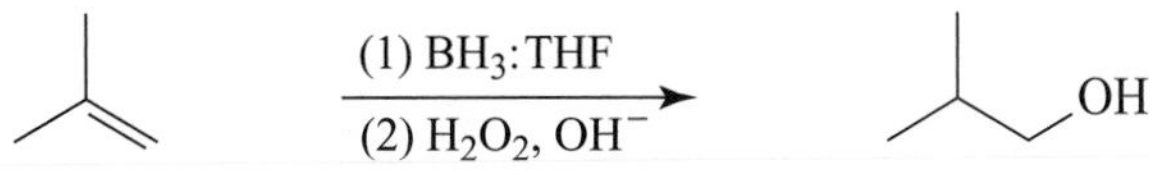

8.34 (a) 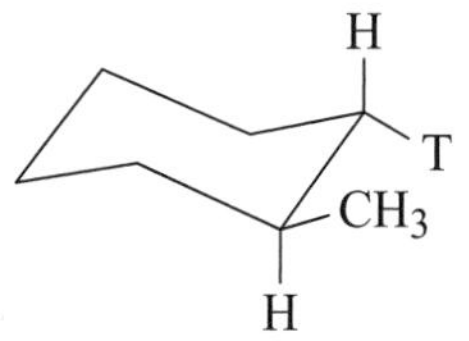 H, T, CH_3, H (b) 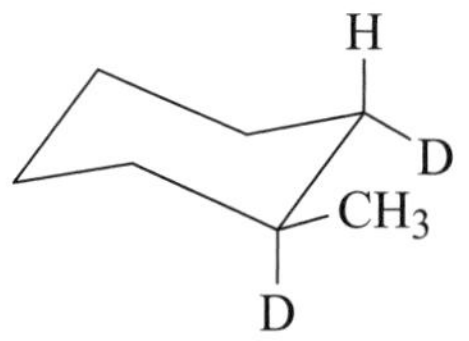 H, D, CH_3, D (c) 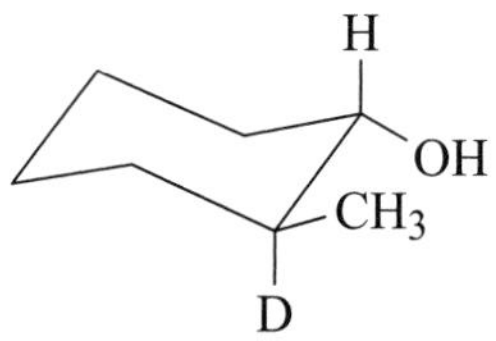 H, OH, CH_3, D

8.35 + H—Cl → + ; $H\ddot{O}—CH_2CH_3$; Cl^- ; Cl ; :Ö ; –HCl ; H ; Cl^- ; Ö+

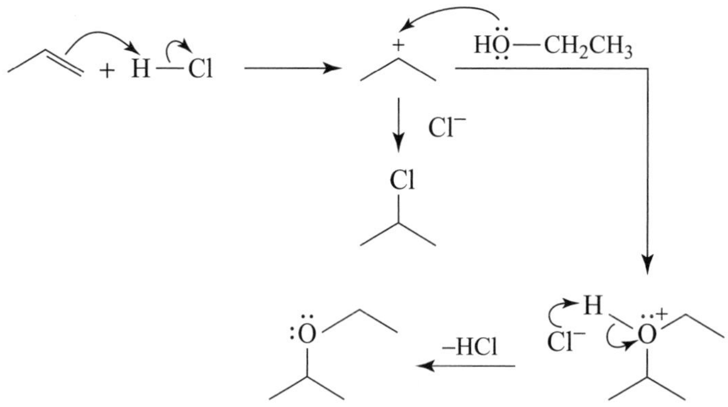

8.36 The rate-determining step in each reaction is the formation of a carbocation when the alkene accepts a proton from HI. When 2-methylpropene reacts, it forms a 3° carbocation (the most stable); therefore, it reacts fastest. When ethene reacts, it forms a 1° carbocation (the least stable); therefore, it reacts the slowest.

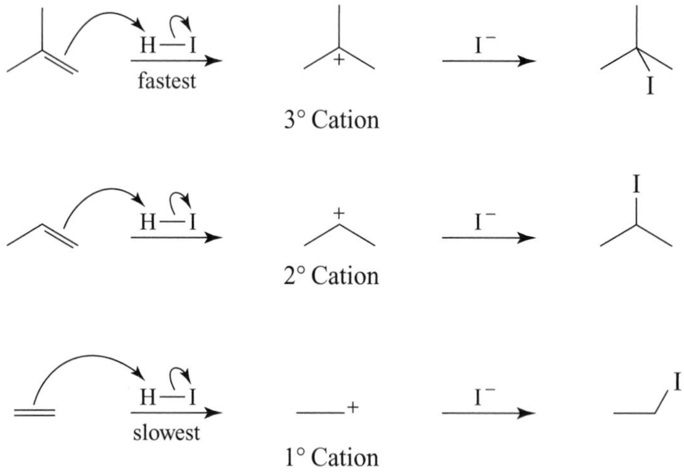

H—I (fastest) → 3° Cation → I^- → I

H—I → 2° Cation → I^- → I

H—I (slowest) → 1° Cation → I^- → I

8.37

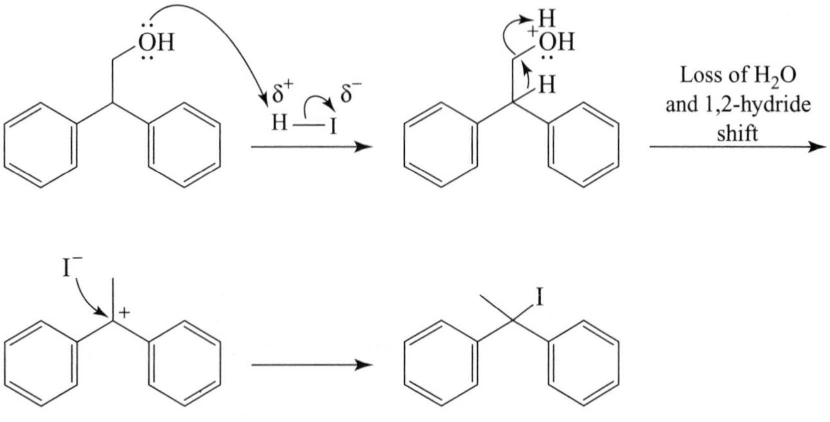

$\ddot{O}H$ δ^+ δ^- H—I → $\overset{+}{O}H$, H, H

Loss of H_2O and 1,2-hydride shift

I^- + → I

8.38

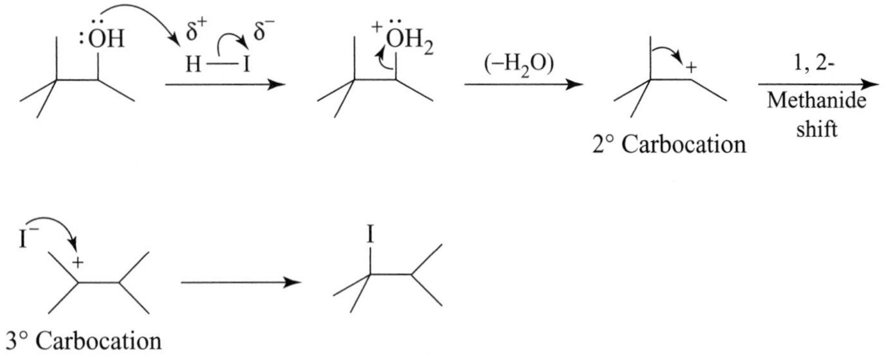

:$\ddot{O}H$ δ^+ δ^- H—I → $^+\ddot{O}H_2$

(–H_2O) → +

2° Carbocation

1, 2- Methanide shift

I^- + → I

3° Carbocation

8.39 (a) (1) OsO_4 (2) $NaHSO_3$, H_2O → OH OH, H H + enantiomer

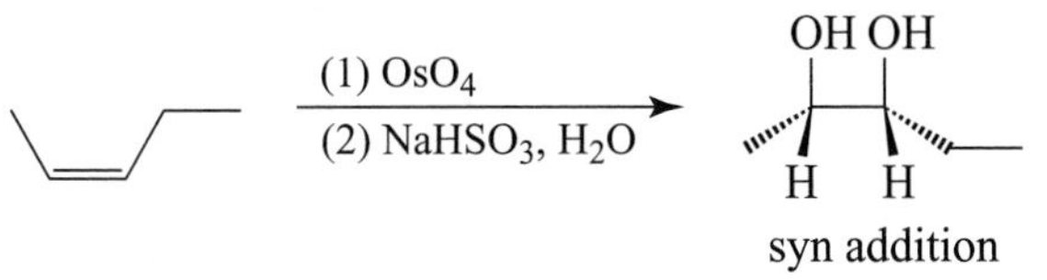

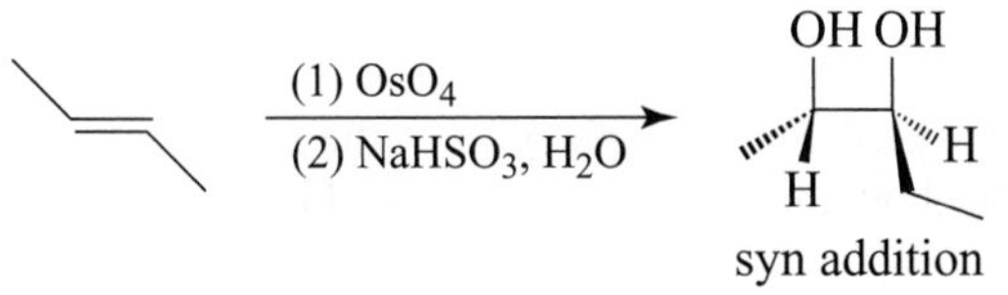

syn addition

(b) (1) OsO_4 (2) $NaHSO_3$, H_2O → OH OH, H, H + enantiomer

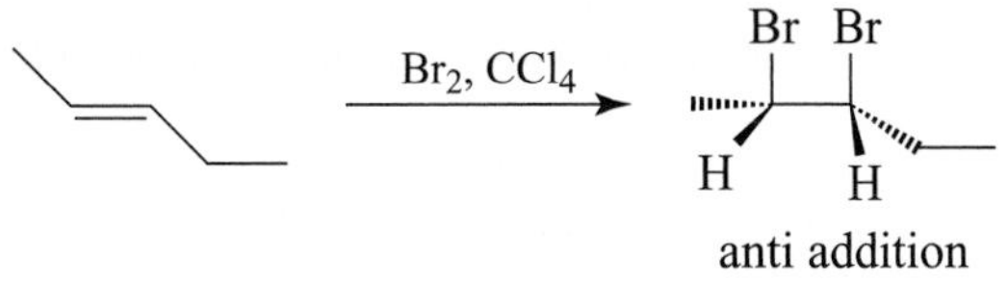

syn addition

(c) Br_2, CCl_4 → Br Br, H H + enantiomer

anti addition

(d) Br_2, CCl_4 → Br Br, H, H + enantiomer

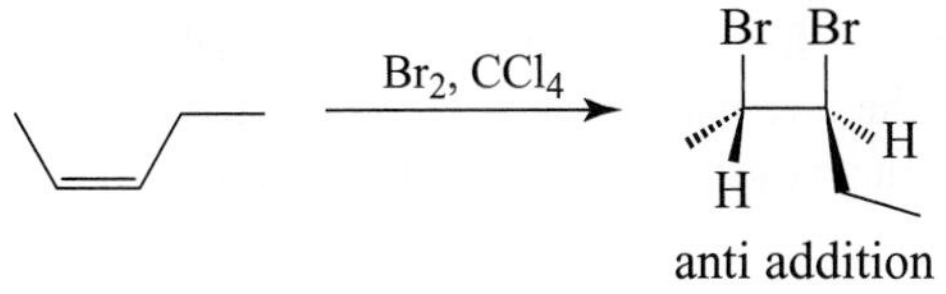

anti addition

8.40 (a) (2*S*, 3*R*)- [the enantiomer is (2*R*, 3*S*)-]

(b) (2*S*, 3*S*)- [the enantiomer is (2*R*, 3*R*)-]

(c) (2*S*, 3*R*)- [the enantiomer is (2*R*, 3*S*)-]

(d) (2*S*, 3*S*)- [the enantiomer is (2*R*, 3*R*)-]

8.41 Because of the electron-withdrawing nature of chlorine, the electron density at the double bond is greatly reduced and attack by the electrophilic bromine does not occur.

8.42 The bicyclic compound is a *trans*-decalin derivative. The fused nonhalogenated ring prevents the ring flip of the bromine-substituted ring necessary to give equatorial bromines.

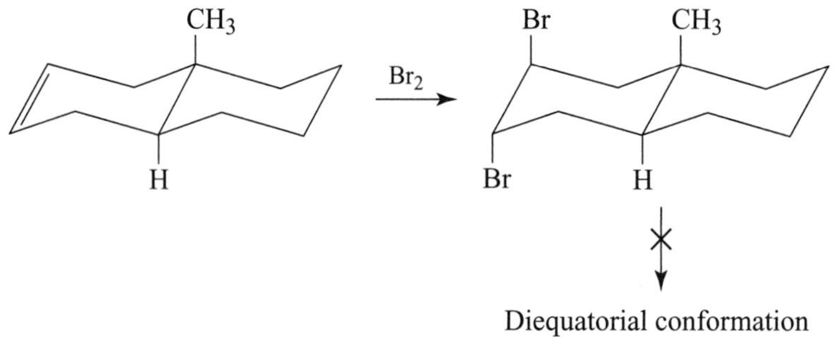

CH_3 ... H $\xrightarrow{Br_2}$ Br, CH_3, Br, H

Diequatorial conformation

8.43 $\xrightarrow[\text{cat.}]{H_2SO_4}$ + $\longrightarrow$ + $\xrightarrow{HSO_4^-}$

I (major) + II (minor)

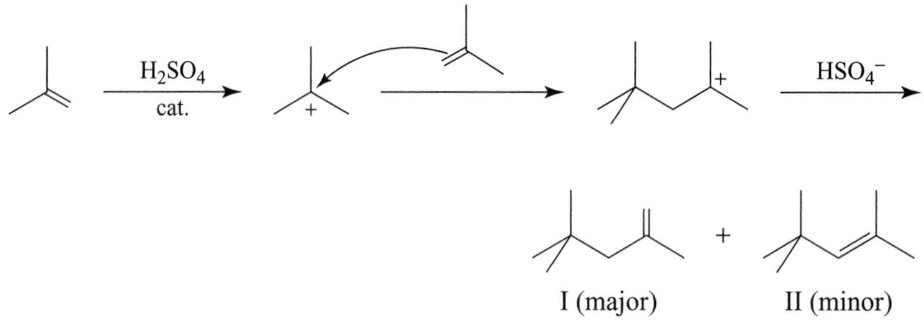

Though II is the product predicted by application of the Zaitsev rule, it actually is less stable than I due to crowding about the double bond. Hence I is the major product (by about a 4:1 ratio).

8.44 The terminal alkyne component of the equilibrium established in base is converted to a salt by $NaNH_2$, effectively shifting the equilibrium completely to the right.

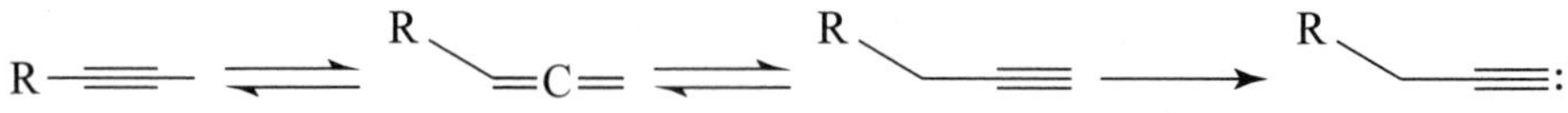

R—≡— ⇌ R ═C═ ⇌ R ≡ ⟶ R ≡:⁻

NaOH is too weak a base to form a salt with the terminal alkyne. Of the equilibrium components with NaOH, the internal alkyne is favored since it is the most stable of these structures. Very small amounts of the allene and the terminal alkyne are formed.

8.45 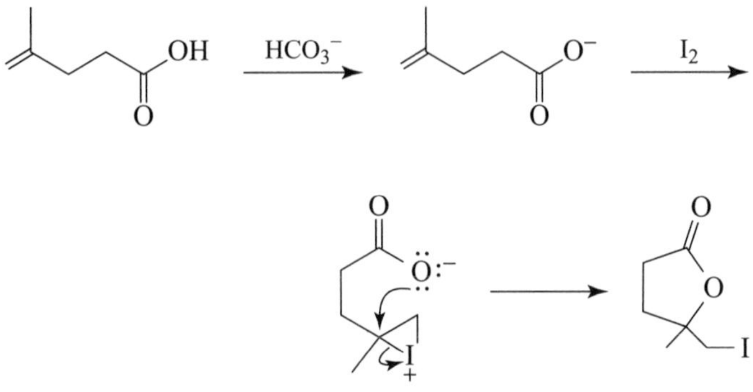 OH $\xrightarrow{HCO_3^-}$ O^- $\xrightarrow{I_2}$

O O

O I

8.46 $\xrightarrow{H_2SO_4}$ $\longrightarrow$ $\xrightarrow{H_2\ddot{O}:}$

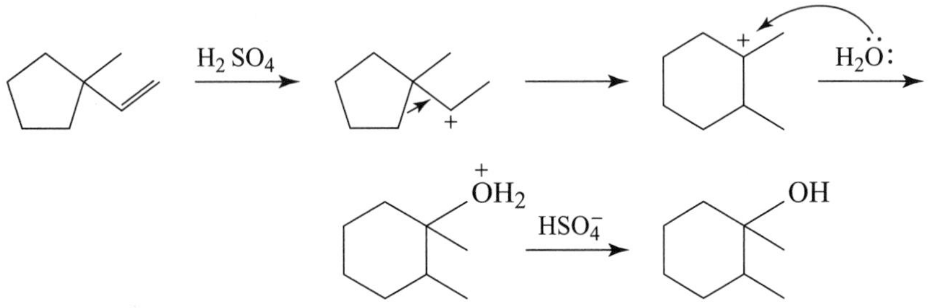

$\overset{+}{O}H_2$ $\xrightarrow{HSO_4^-}$ OH

8.47 $\overset{\delta+}{Br}$—$\overset{\delta-}{Br}$ $\longrightarrow$ $\overset{+}{Br}$ $\xrightarrow{Br^-}$ Br, Br

$\xrightarrow{H_2O}$ Br, $\overset{+}{O}H_2$ $\xrightarrow{H_2O}$ Br, OH

$\xrightarrow{Cl^-}$ Br, Cl

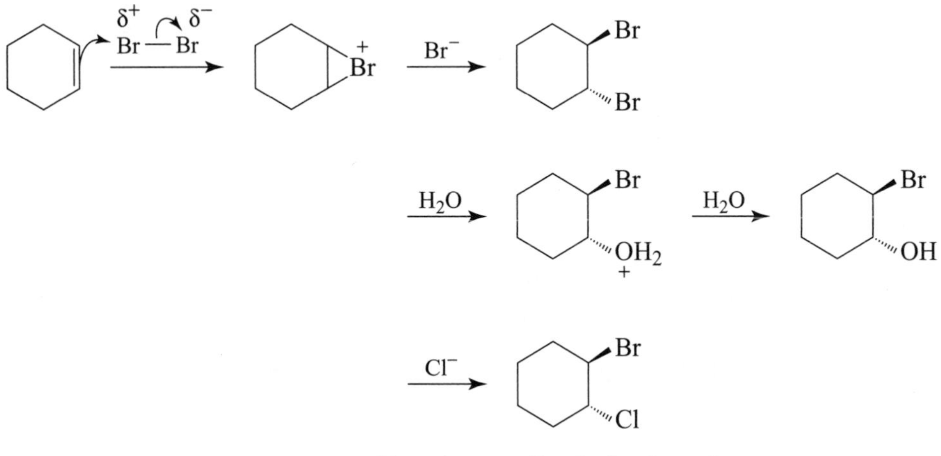

Enantiomers of each also formed.

8.48 (a) $C_{10}H_{22}$ (saturated alkane)
$C_{10}H_{16}$ (formula of myrcene)
H_6 = 3 pairs of hydrogen atoms
Index of hydrogen deficiency (IHD) = 3

(b) Myrcene contains no rings because complete hydrogenation gives $C_{10}H_{22}$, which corresponds to an alkane.

(c) That myrcene absorbs three molar equivalents of H_2 on hydrogenation indicates that it contains three double bonds.

(d) 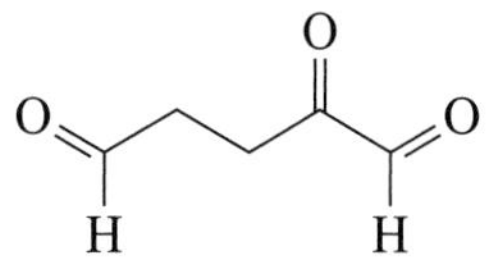 O O O H H

(e) Three structures are possible; however, only one gives 2,6-dimethyloctane on complete hydrogenation. Myrcene is therefore

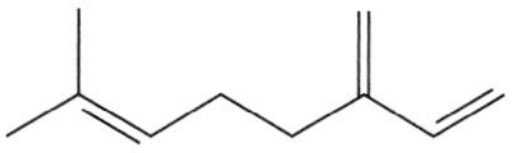

8.49 (a) 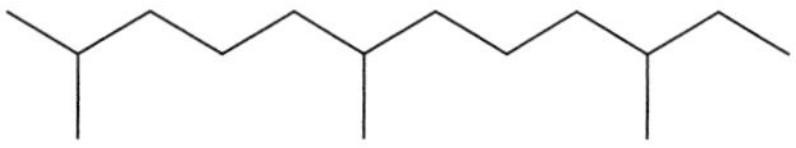 2,6,10-Trimethyldodecane (b) Four

8.50 H O (1) O_3 (2) Me_2S → O + O O H + O H O H

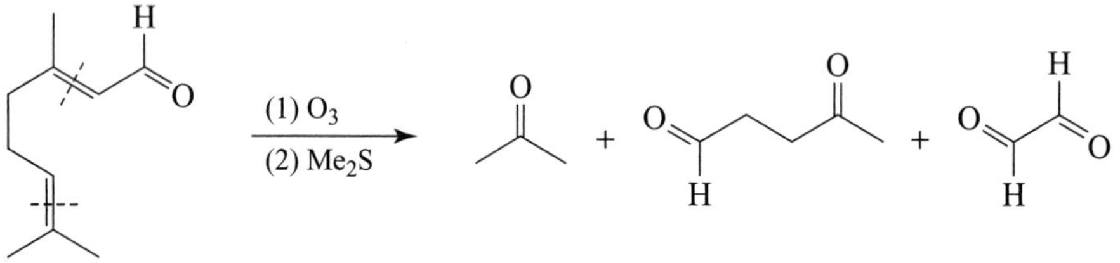

8.51 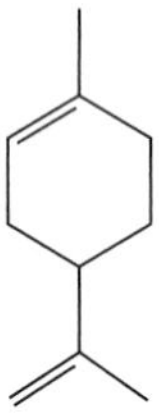 Limonene

8.52 The hydrogenation experiment discloses the carbon skeleton of the pheromone.

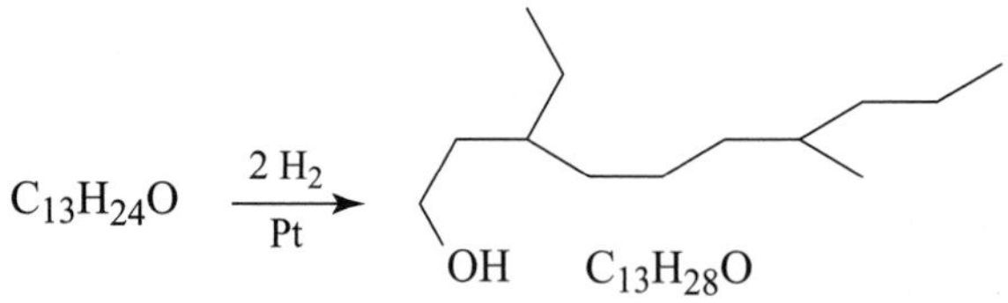

$C_{13}H_{24}O$ $\xrightarrow[\text{Pt}]{2\ H_2}$ OH $C_{13}H_{28}O$

Codling moth pheromone

3-Ethyl-7-methyl-1-decanol

The ozonolysis experiment allows us to locate the position of the double bonds.

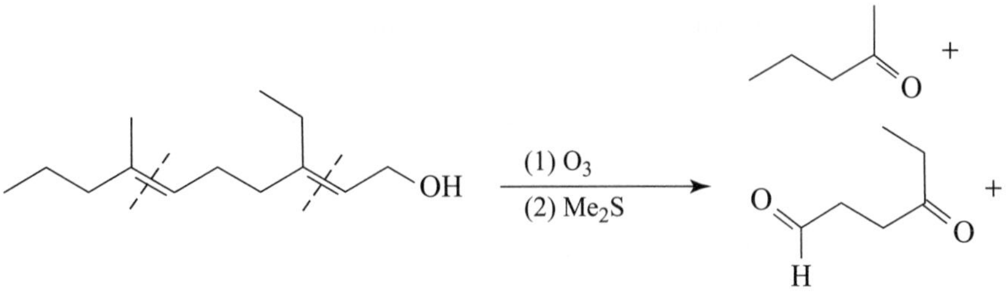

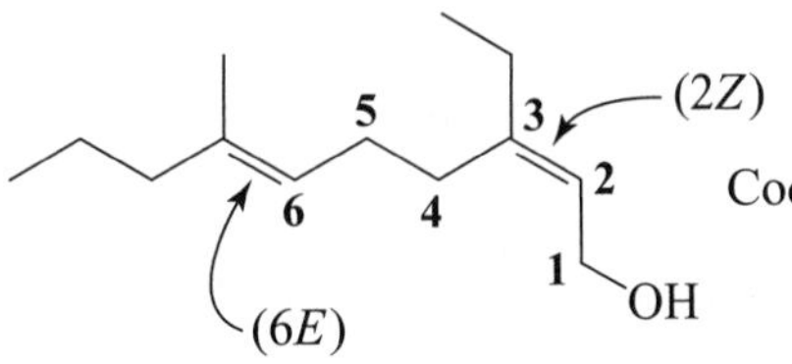

(1) O_3
(2) Me_2S

OH

Codling moth pheromone

(2Z) (6E)

General Problems

8.53 ***Retrosynthetic analysis***

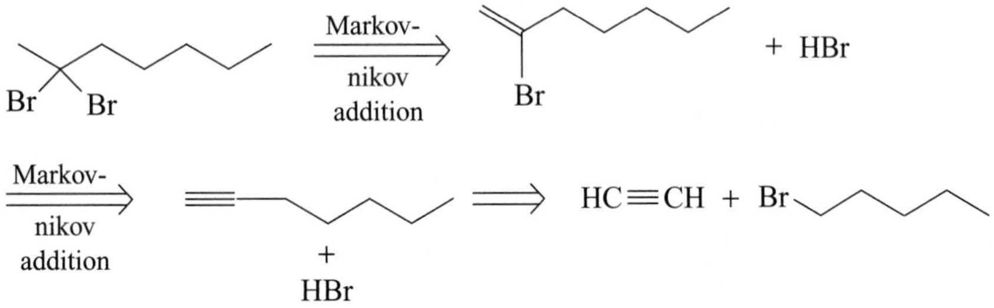

Markov-nikov addition

+ HBr

$HC{\equiv}CH$ + Br

Synthesis

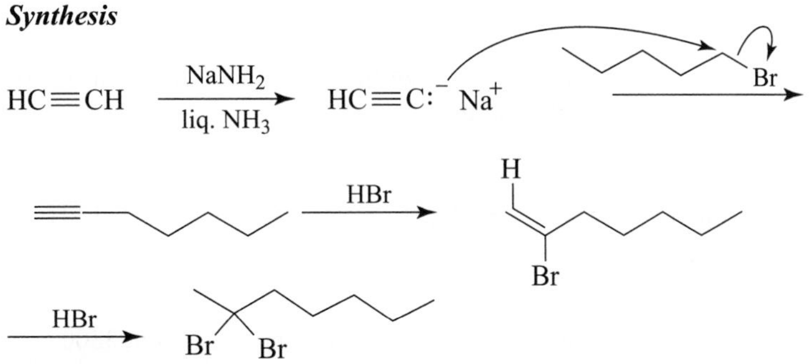

$HC{\equiv}CH \xrightarrow[\text{liq. } NH_3]{NaNH_2} HC{\equiv}C{:}^- \; Na^+$

HBr

HBr

8.54 Syn hydrogenation of the triple bond is required. So use H_2 and Ni_2B(P-2) or H_2 and Lindlar's catalyst.

8.55 (a) 1-Pentyne has IR absorption at about 3300 cm^{-1} due to its terminal triple bond. Pentane does not absorb in that region.

(b) 1-Pentene absorbs in the 1620–1680 cm^{-1} region due to the alkene function. Pentane does not exhibit absorption in that region.

(c) See parts (a) and (b).

(d) 1-Bromopentane shows C-Br absorption in the 515–690 cm^{-1} region while pentane does not.

(e) For 1-pentyne, see (a). The interior triple bond of 2-pentyne gives relatively weak absorption in the 2100–2260 cm^{-1} region.

(f) For 1-pentene, see (b). 1-Pentanol has a broad absorption band in the 3200–3550 cm^{-1} region.

(g) See (a) and (f).

(h) 1-Bromo-2-pentene has double bond absorption in the 1620–1680 cm^{-1} region which 1-bromopentane lacks.

(i) 2-Penten-1-ol has double bond absorption in the 1620–1680 cm^{-1} region not found in 1-pentanol.

8.56 The index of hydrogen deficiency of **A, B,** and **C** is two.

$$\begin{array}{r} C_6H_{14} \\ \underline{C_6H_{10}} \\ H_4 \end{array} \quad = \quad \text{2 pairs of hydrogen atoms}$$

This result suggests the presence of a triple bond, two double bonds, a double bond and a ring, or two rings.

$$\textbf{A, B, C} \xrightarrow[CCl_4]{Br_2} \text{all decolorize}$$

$$\textbf{A, B, C} \xrightarrow{\text{cold concd } H_2SO_4} \text{all soluble}$$

The fact that **A, B,** and **C** all decolorize Br_2/CCl_4 and dissolve in concd. H_2SO_4 suggests they all have a carbon-carbon multiple bond.

$$\underset{(C_6H_{10})}{\textbf{A, B}} \xrightarrow[Pt]{xs\ H_2} \underset{(C_6H_{14})}{\text{hexane}}\ ; \quad \underset{(C_6H_{10})}{\textbf{C}} \xrightarrow[Pt]{xs\ H_2} C_6H_{12}$$

A must be a terminal alkyne, because of IR absorption at about 3300 cm^{-1}.

Since **A** gives hexane on catalytic hydrogenation, **A** must be 1-hexyne.

$$\text{hexane} \xleftarrow[Pt]{2H_2} \underset{\textbf{A}}{\text{1-hexyne}} \xrightarrow[(2)\ H_3O^+]{(1)\ KMnO_4,\ OH^-,\ heat} \text{pentanoic acid} + CO_2\uparrow$$

This is confirmed by the oxidation experiment

B $\xrightarrow[(2)\ H_3O^+]{(1)\ KMnO_4,\ OH^-,\ heat}$ 2 (propanoic acid)

Hydrogenation of **B** to hexane shows that its chain is unbranched, and the oxidation experiment shows that **B** is 3-hexyne.

Hydrogenation of **C** indicates a ring is present.

Oxidation of **C** shows that it is cyclohexene.

C (cyclohexene) $\xrightarrow[(2)\ H_3O^+]{(1)\ KMnO_4,\ OH^-,\ heat}$ HO(O=)C–(CH$_2$)$_4$–C(=O)OH

8.57 (a) Four

(b) $CH_3(CH_2)_5$–CH(OH)–CH_2–CH=CH–$(CH_2)_7CO_2H$ (cis) + enantiomer

$CH_3(CH_2)_5$–CH(OH)–CH_2–CH=CH–$(CH_2)_7CO_2H$ (trans) + enantiomer

8.58 Hydroxylations by OsO_4 are syn hydroxylations (cf. Section 8.16). Thus, maleic acid must be the cis-dicarboxylic acid:

Maleic acid $\xrightarrow[\text{syn hydroxylation}]{(1)\ OsO_4,\ pyr;\ (2)\ NaHSO_3/H_2O}$ *meso*-Tartaric acid

Fumaric acid must be the trans-dicarboxylic acid:

Fumaric acid $\xrightarrow[\text{syn hydroxylation}]{(1)\ OsO_4,\ pyr;\ (2)\ NaHSO_3/H_2O}$ (±)-Tartaric acid

8.59 (a) The addition of bromine is an anti addition. Thus, fumaric acid yields a meso compound.

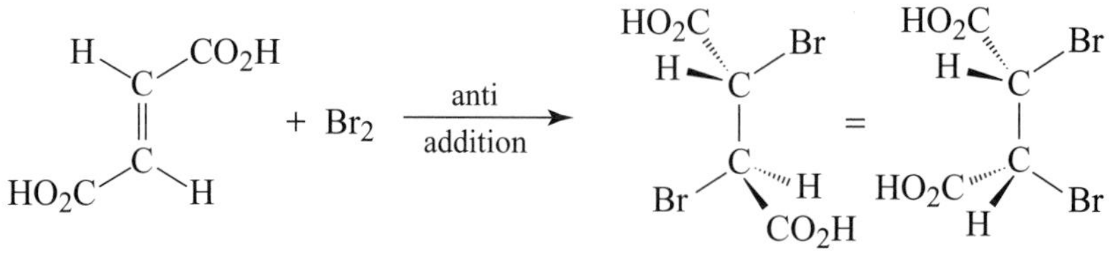

A meso compound

(b) Maleic acid adds bromine to yield a racemic form.

8.60

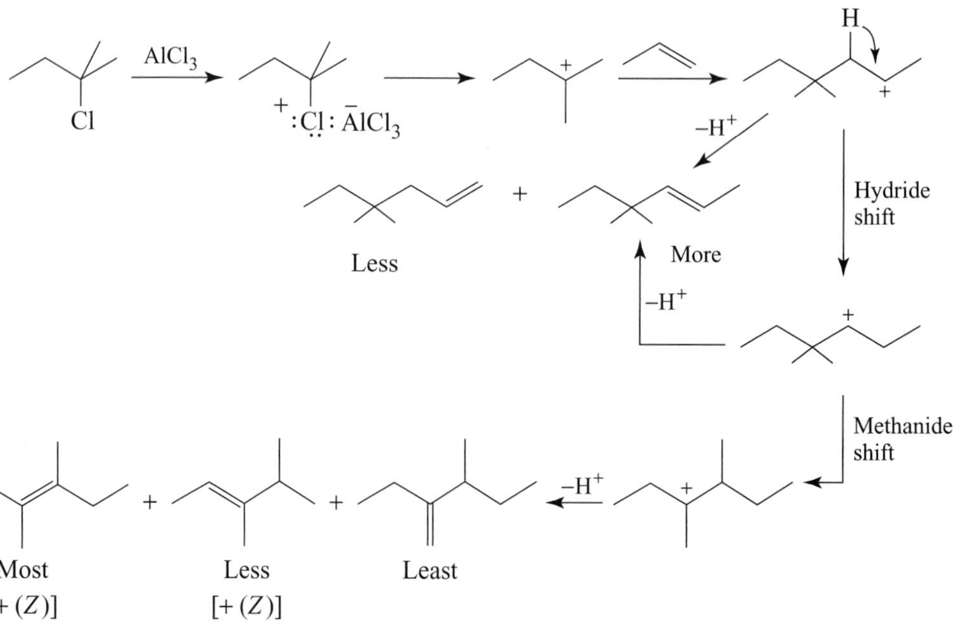

Each carbocation can combine with chloride ion to form a racemic alkyl chloride.

8.61 The catalytic hydrogenation involves syn addition of hydrogen to the predominantly less hindered face of the cyclic system (the face lacking 1,3-diaxial interactions when the molecule is adsorbed on the catalyst surface). This leads to I, even though II is the more stable of the two isomers since both methyl groups are equatorial in II.

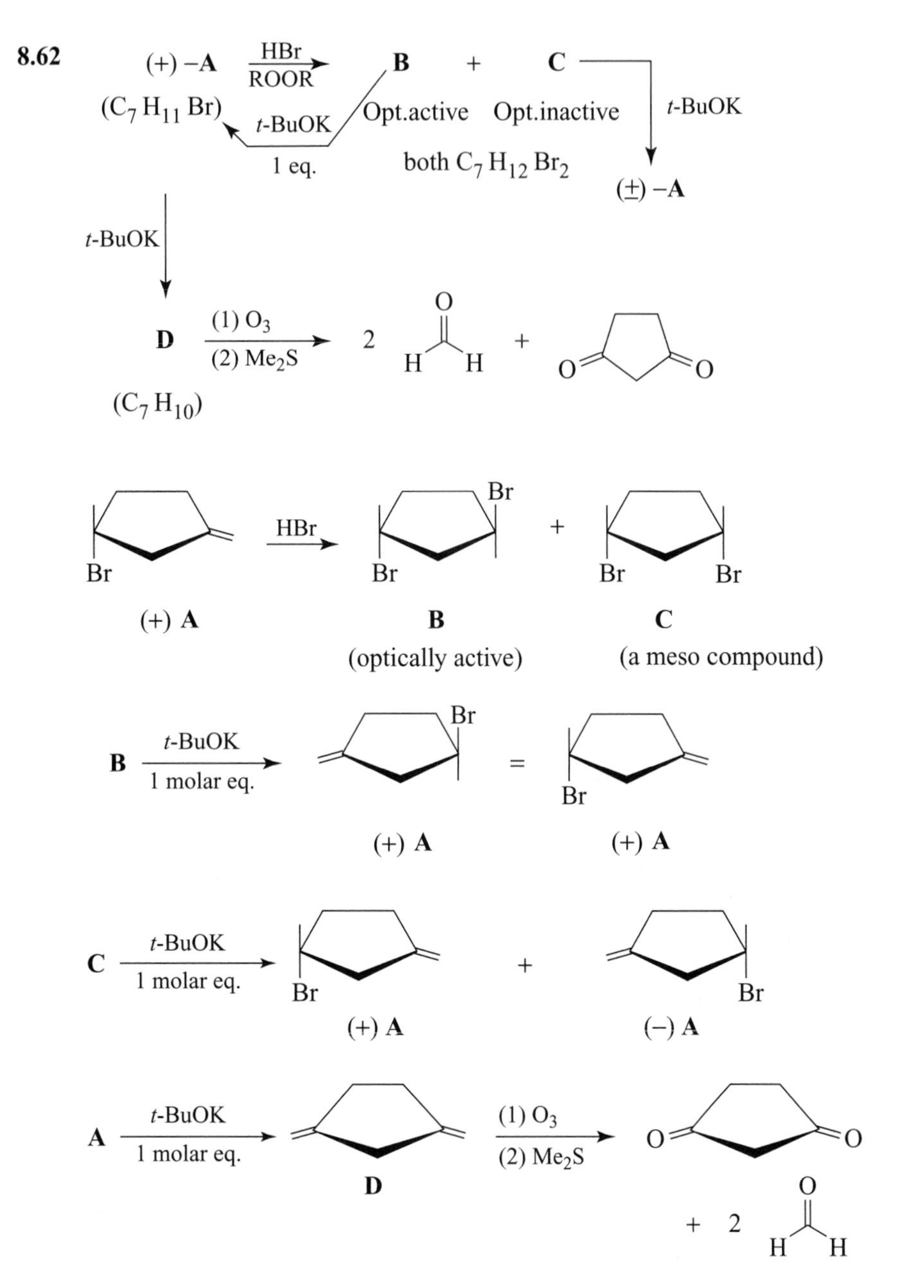

8.63 $HC{\equiv}C{-}C{\equiv}C(H)C{=}C{=}C(H)(CH{=}CH)_2CH_2CO_2H$

$HC{\equiv}C{-}C{\equiv}C(H)C{=}C{=}C(H)(CH{=}CH)_2CH_2CO_2H$

8.64

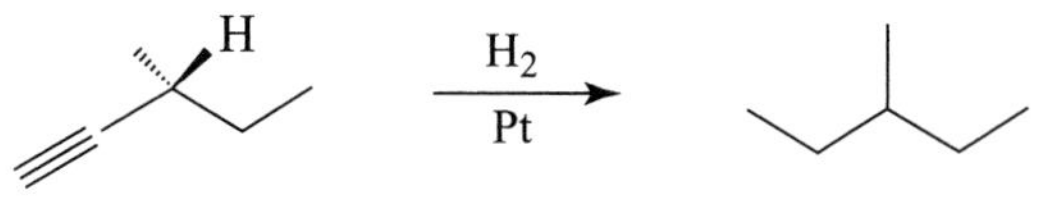

H

$\xrightarrow[\text{Pt}]{H_2}$

D **E**

Optically active
(the other enantiomer is
an equally valid answer)

Optically inactive
(nonresolvable)

8.65 (a) 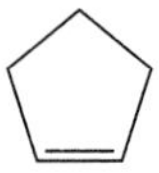 $\xrightarrow[H_2O]{Br_2}$ **A** $\xrightarrow[\text{1 eq.}]{NaOH/H_2O}$ **B** $\xrightarrow[H^+ \text{ cat.}]{MeOH}$ **C**

C_5H_8O $C_6H_{12}O_2$

(no 3590–3650 cm^{-1} IR absorption) (3590–3650 cm^{-1} IR absorption)

IR data indicate **B** does not possess an —OH group, but **C** does.

(b)

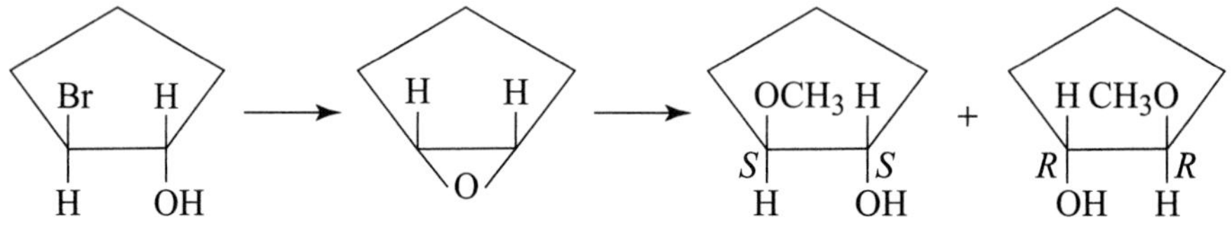

Br H / H OH → H H / O → OCH_3 H / *S* *S* / H OH + H CH_3O / *R* *R* / OH H

A (and enantiomer) **B** **racemate C**

(c) **C**, in contrast to its cis isomer, would exhibit no intramolecular hydrogen-bonding. This would be proven by the absence of infrared absorption in the 3500- to 3600-cm^{-1} region when studied as a very dilute solution in CCl_4. **C** would only show free O—H stretch at about 3625 cm^{-1}

Challenge Problems

8.66

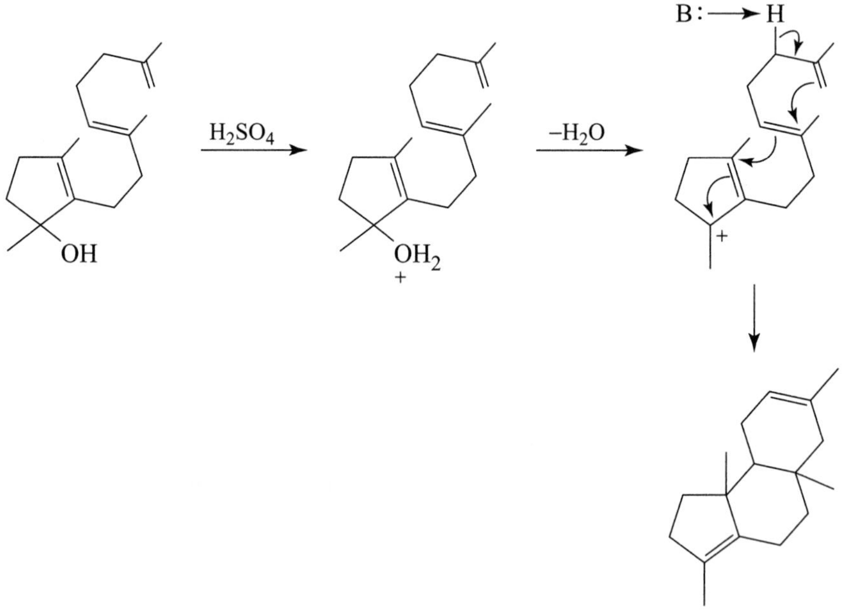

OH $\xrightarrow{H_2SO_4}$ $\overset{+}{O}H_2$ $\xrightarrow{-H_2O}$ B:→H +

↓

8.67

D $\xrightarrow{-C_2H_4}$ E $\xrightarrow{-Cl^-}$ $\xrightarrow{H_2O}$ $\xrightarrow{-H^+}$ $\xrightarrow{-Cl^-}$ $\xrightarrow{-H^+}$ F

E: $(CH_3CH_2)_2N{-}CHCl_2$

F: $(CH_3CH_2)_2N{-}CHO$

8.68 The HOMO and LUMO orbitals of ethene and BH_3 are shown here. In our discussion of the mechanism of hydroboration we mentioned the importance of the "vacant *p* orbital" of the boron. This is the LUMO of BH_3 in actuality, and it is this orbital that interacts with the pi bond of ethene as the HOMO.

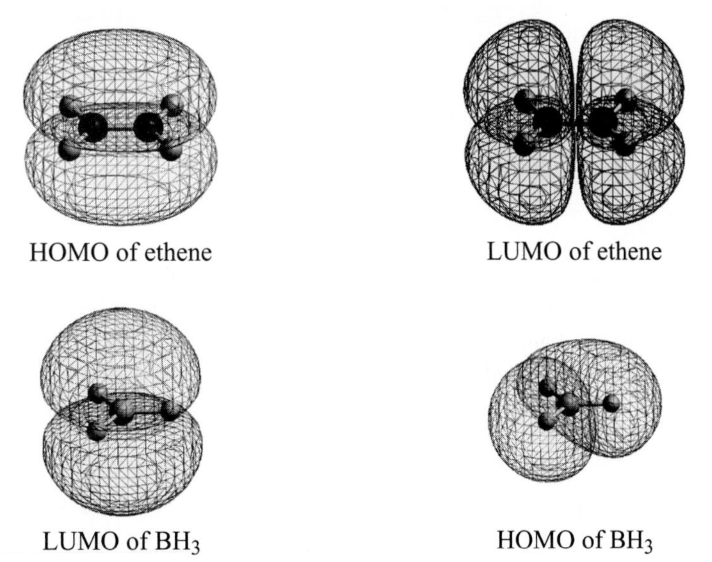

8.69 The LUMO of the BH_3:THF complex has a lobe (in red in the Chem3D version) which extends outward from the boron, much like the *p* orbital that is the LUMO of a hypothetical BH_3 monomer. This orbital is in position to accept the donation of electrons from a Lewis base such as an alkene pi bond.

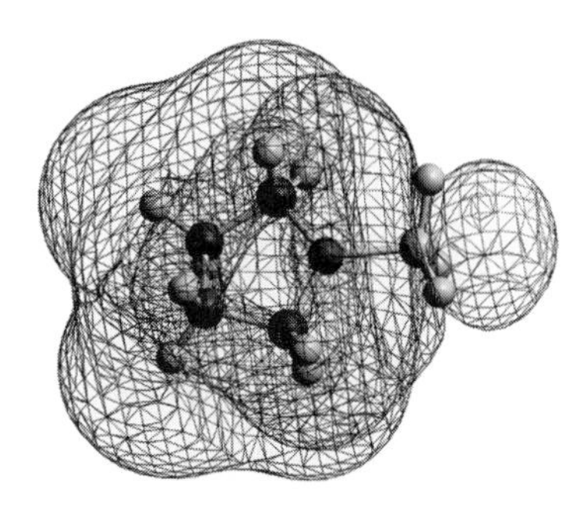

8.70 It is the LUMO of diborane that interacts with the HOMO of an alkene (the bonding pi molecular orbital) or other Lewis base. Inspection of the LUMO for diborane shows that its lobes include part of the region between the partially bonded bridging hydrogen atoms and each boron atom, but that the major portion of each lobe is oriented away from these hydrogens and the boron such that electron density from a Lewis base could interact with the LUMO without hindrance. In a sense, the LUMO appears almost as if it is a "squashed" *p* orbital similar to what would be available with BH_3 as a monomer. The HOMO of diborane, on the other hand, has lobes localized about four of the boron—hydrogen bonds. As occupied orbitals, they are not available for bonding with the occupied orbital of a Lewis base.

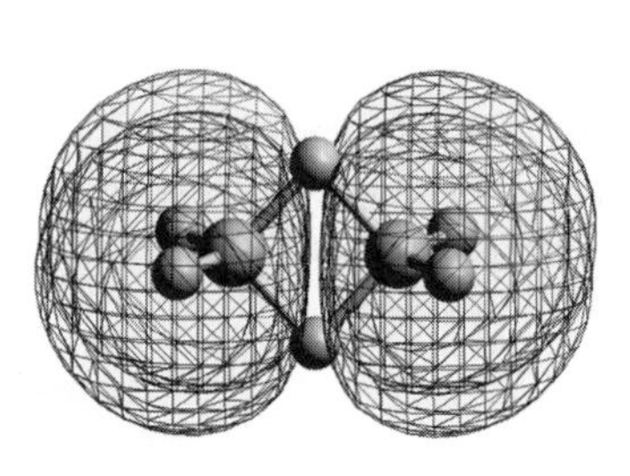

HOMO of B_2H_6

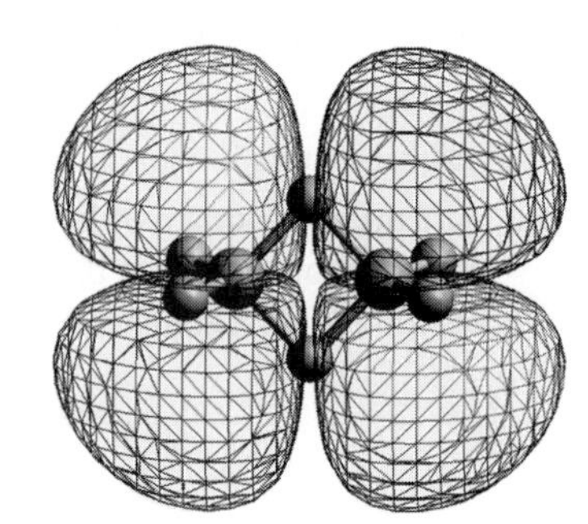

LUMO of B_2H_6

QUIZ

8.1 A hydrocarbon whose molecular formula is C_7H_{12}, on catalytic hydrogenation (excess H_2/Pt), yields C_7H_{16}. The original hydrocarbon adds bromine and also exhibits an IR absorption band at 3300 cm^{-1}. Which of the following is a plausible choice of structure for the original hydrocarbon?

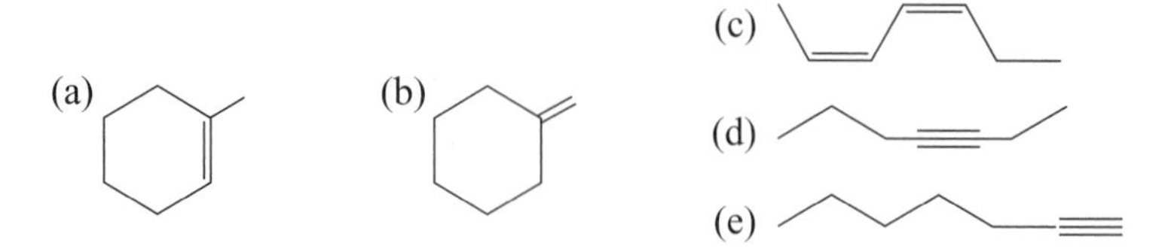

8.2 Select the major product of the dehydration of the alcohol, 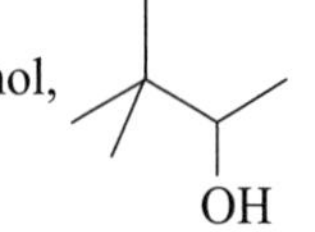

(a) 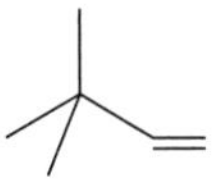 (b) 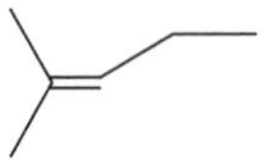 (c)

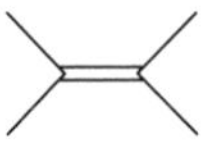

(d) 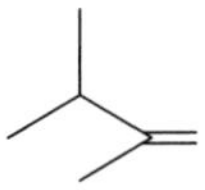 (e)

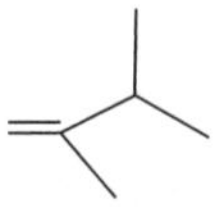

8.3 Give the major product of the reaction of *cis*-2-pentene with bromine.

(a) CH_3; H—C—Br; H—C—Br; CH_2CH_3

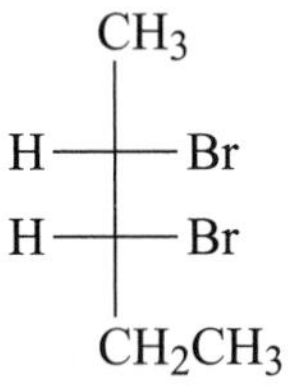

(b) CH_3; Br—C—H; Br—C—H; CH_2CH_3

(c) CH_3; H—C—Br; Br—C—H; CH_2CH_3

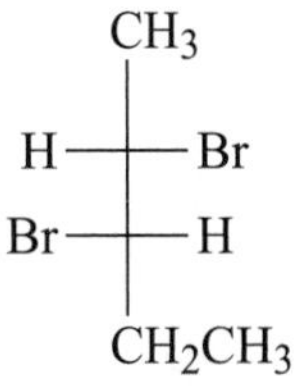

(d) CH_3; Br—C—H; H—C—Br; CH_2CH_3

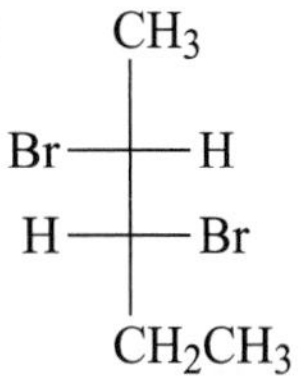

(e) A racemic mixture of (c) and (d)

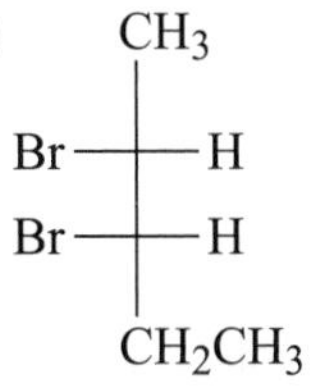

8.4 The compound shown here is best prepared by which sequence of reactions?

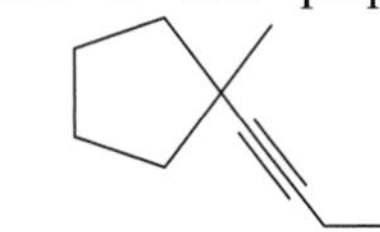

(a) + $NaNH_2$ ⟶ then Br ⟶ product

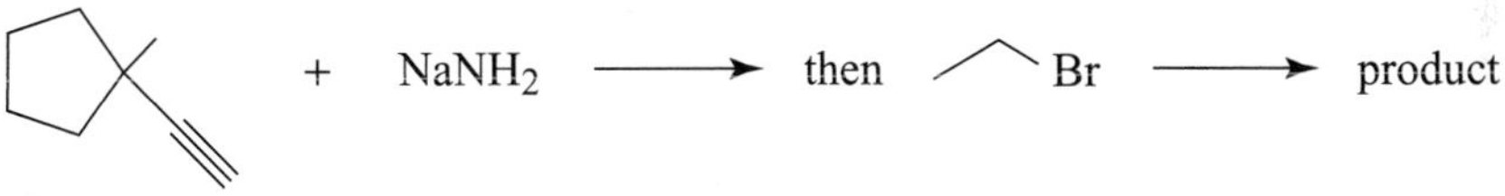

(b) + $NaNH_2$ ⟶ then Br ⟶ product

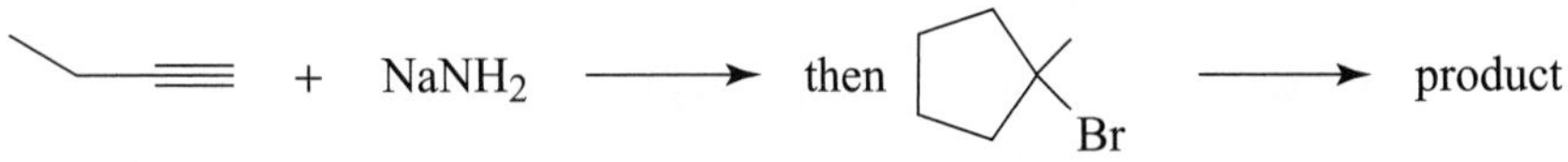

(c) + H_2 $\xrightarrow{Pt}$ product

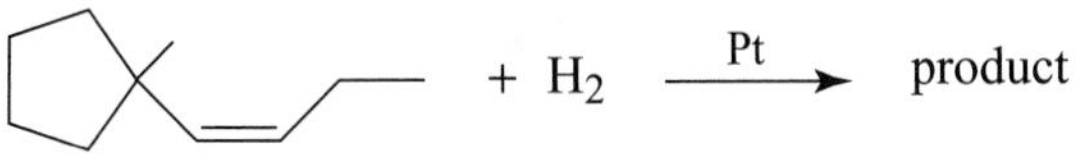

(d) Br $\xrightarrow[C_2H_5OH]{NaOC_2H_5}$ product

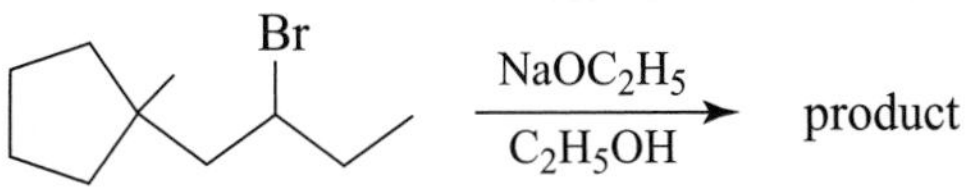

8.5 A compound whose formula is C_6H_{10} (Compound **A**) reacts with H_2/Pt in excess to give a product C_6H_{12}, which does not decolorize Br_2/CCl_4. Compound **A** does not show IR absorption in the 3200–3400 cm^{-1} region.

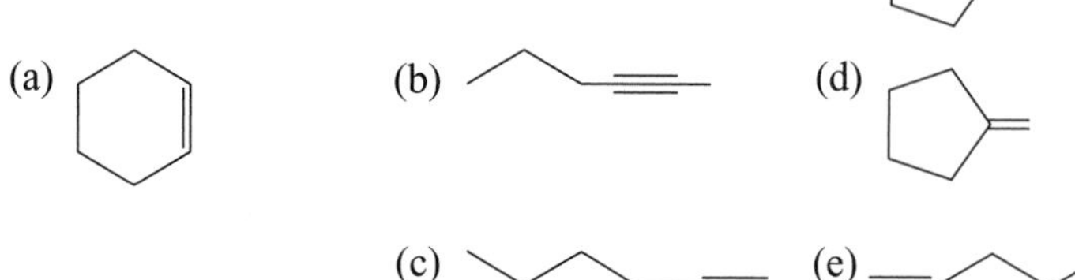

Ozonolysis of A gives 1 mol of
O
H
H and 1 mol of
O. Give the structure of A.

(a) (b) (d)

(c) (e)

8.6 Compound **B** (C_5H_{10}) does not dissolve in cold, concentrated H_2SO_4. What is **B**?

(a) (b)

(c) (d)

8.7 Which reaction sequence converts cyclohexene to *cis*-1,2-cyclohexanediol? That is,

? H OH H OH

(a) H_2O_2 (b) (1) O_3 (2) Me_2S

(c) (1) OsO_4 (2) $NaHSO_3/H_2O$ (d) (1) R–C(=O)–O–O–H (2) H_3O^+/H_2O

(e) More than one of these

8.8 Which of the following sequences leads to the best synthesis of the compound ? (Assume that the quantities of reagents are sufficient to carry out the desired reaction.)

(a) $\xrightarrow{Br_2}$ $\xrightarrow[H_2O]{NaOH}$

(b) $\xrightarrow{Br_2}$ $\xrightarrow{NaNH_2}$

(c) Br Br $\xrightarrow{H_2SO_4}$

(d) $\xrightarrow[light]{Br_2}$ $\xrightarrow{NaNH_2}$

(e) $\xrightarrow{O_3}$ $\xrightarrow{Me_2S}$

9 NUCLEAR MAGNETIC RESONANCE AND MASS SPECTROMETRY: TOOLS FOR STRUCTURE DETERMINATION

SOLUTIONS TO PROBLEMS

9.1 $CH_3C(=O)OCH_2CH_3$ — CH_3 (*b*), OCH_2 (*c*), CH_3 (*a*)

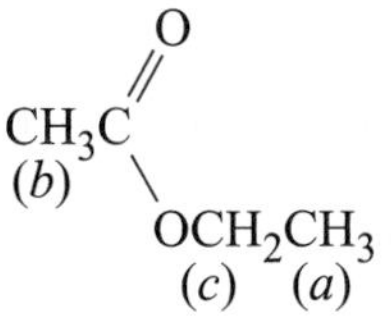

(a) δ 0.8–1.10
(b) δ 2.1–2.6
(c) δ 3.3–3.9

9.2 The presence of two signals in the ^{1}H NMR spectrum signifies two unique proton environments in the molecule. The chemical shift of the downfield signal, (a), is consistent with protons and a chlorine on the same carbon. Its triplet nature indicates two hydrogens on the adjacent carbon.

The upfield quintet (b) indicates four adjacent and equivalent hydrogens, which requires two hydrogens on each of two carbons.

Integration data indicate a ratio of (approximately) 2:1, which is actually 4 : 2 in this case.

Of the isomeric structures with the formula $C_3H_6Cl_2$, only Cl–CH₂(*a*)–CH₂(*b*)–CH₂(*a*)–Cl fits the evidence.

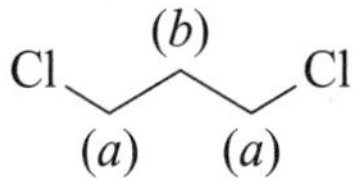

9.3 (a) We see below that if we replace a methyl hydrogen by Cl, then rotation of the methyl group or turning the whole molecule end-for-end gives structures that represent the same compound. This means that all of the methyl hydrogens are equivalent.

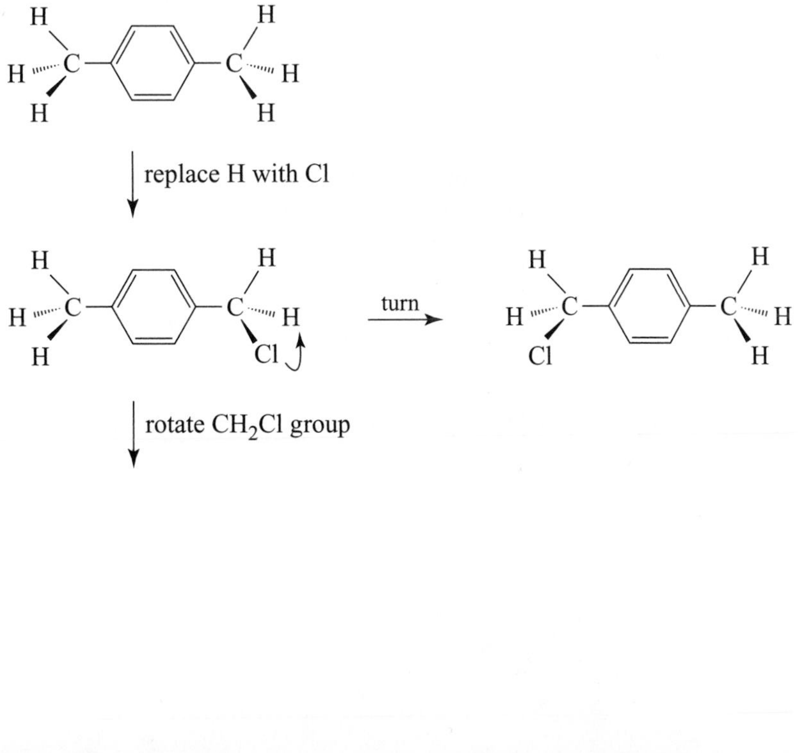

replace H with Cl

turn

rotate CH_2Cl group

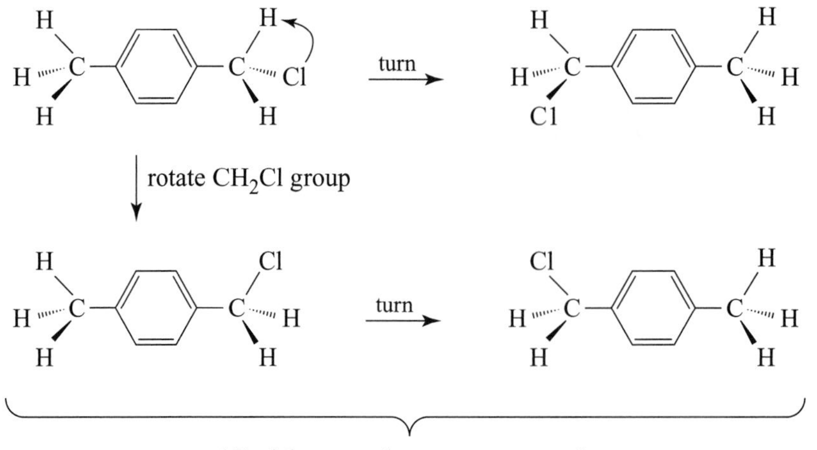

turn

rotate CH_2Cl group

turn

All of these are the same compound.

Replacing a ring hydrogen by Cl gives another compound, but replacing each ring hydrogen in turn gives the same compound. This shows that all four ring hydrogens are equivalent and that 1,4-dimethylbenzene has only two sets of chemical shift equivalent protons.

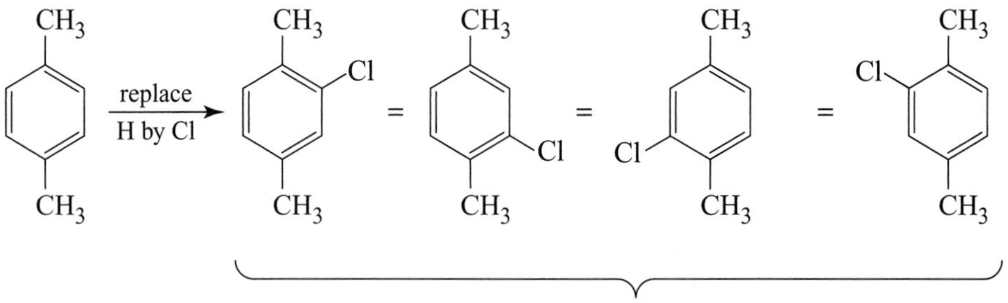

replace H by Cl

All of these are the same compound.

(b) As in (a) we replace a methyl hydrogen by Cl yielding the compound shown. Essentially free rotation of the $-CH_2Cl$ group means that all orientations of that group give the same compound.

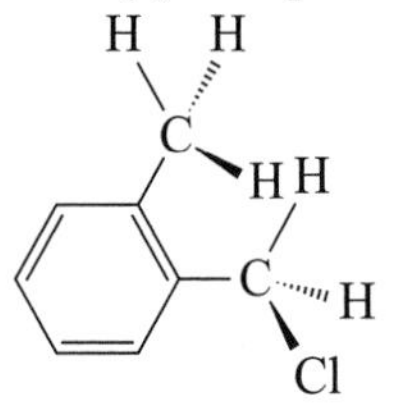

Substitution into the other methyl group produces the same compound, as seen when we flip and rotate to form that structure.

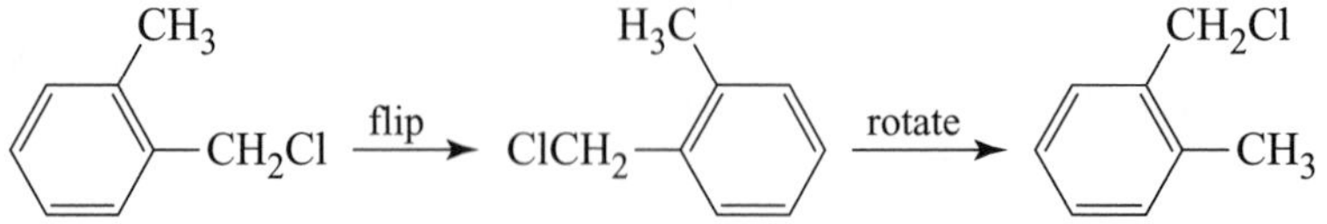

CH_3, CH_2Cl —flip→ H_3C, $ClCH_2$ —rotate→ CH_2Cl, CH_3

Thus the methyl hydrogens are chemical shift equivalent.

Ring substitution produces two different compounds.

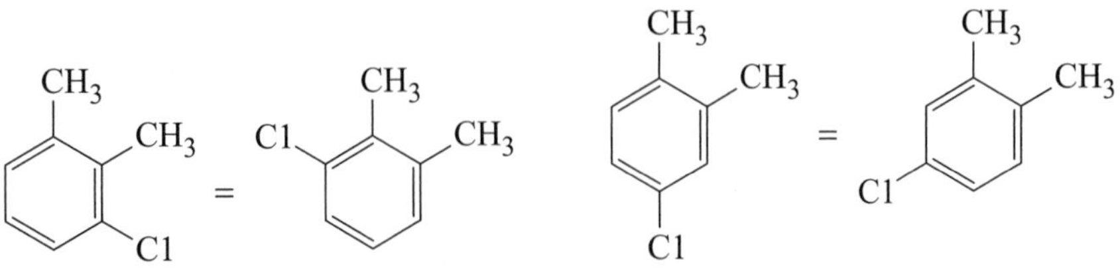

CH_3 CH_3 Cl = CH_3 CH_3 Cl CH_3 CH_3 Cl = CH_3 CH_3 Cl

Again, flipping and rotating members of each pair demonstrates their equivalence. There are then, three sets of chemical shift equivalent protons—one for the methyl hydrogens and one each for the two different ring hydrogens.

(c) 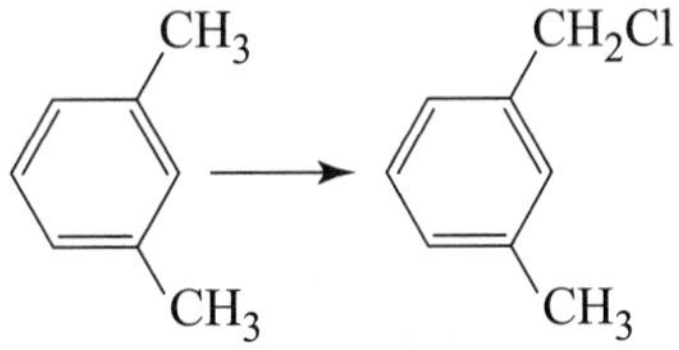 CH_3 CH_3 → CH_2Cl CH_3 on replacement of a methyl hydrogen by Cl.

Again, different orientations (conformations) do not result in different compounds. Substitution in the other methyl group yields the same compound.

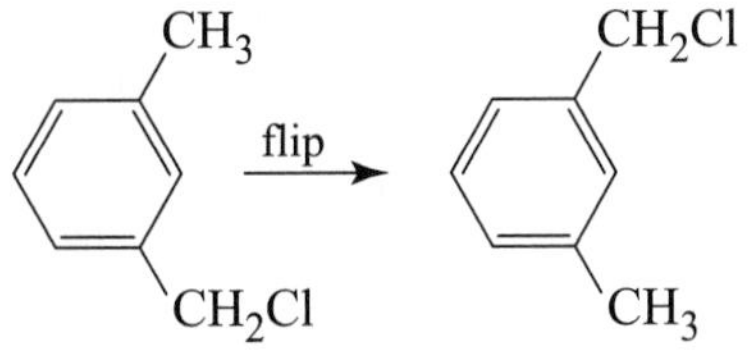

CH_3 CH_2Cl —flip→ CH_2Cl CH_3

Replacement of ring hydrogens produces three compounds:

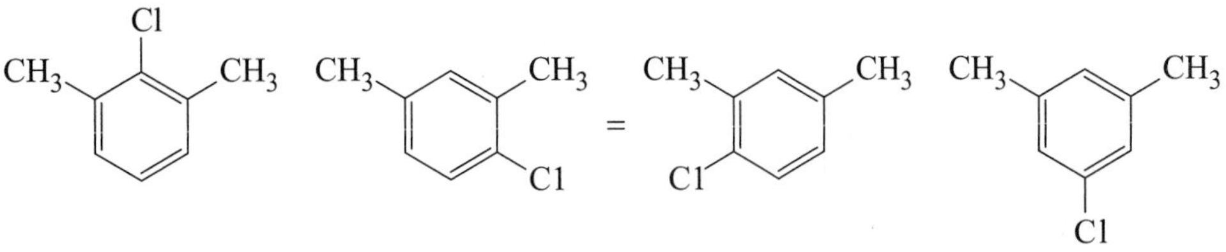

Cl CH_3 CH_3 CH_3 CH_3 Cl = CH_3 CH_3 Cl CH_3 CH_3 Cl

Four sets of chemical shift equivalent hydrogens are present in this isomer: one for the methyl hydrogens and three for the chemically non-equivalent ring hydrogens.

9.4 (a) One
(b) Two
(c) Two
(d) One
(e) Two
(f) Two

9.5 (a)

CH_3 HO—C—H H—C—H CH_3 —replacement by Q→ CH_3 HO—C—H H—C—Q CH_3 CH_3 HO—C—H Q—C—H CH_3

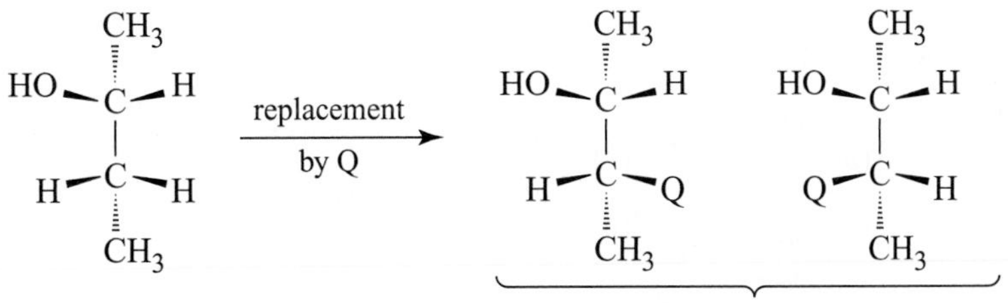

Diastereomers

(b) Six

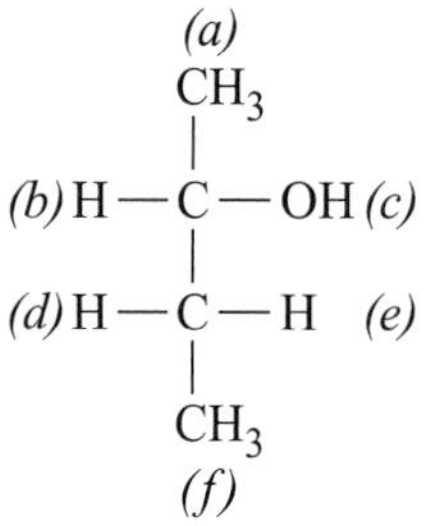

(a) CH_3
(b) H—C—OH (c)
(d) H—C—H (e)
CH_3 (f)

(c) Six signals

9.6 (a) Two (b) Three 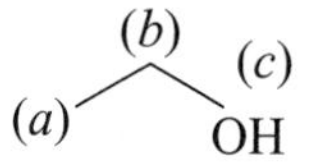 (c) Four

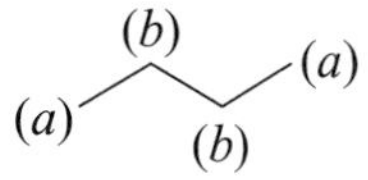

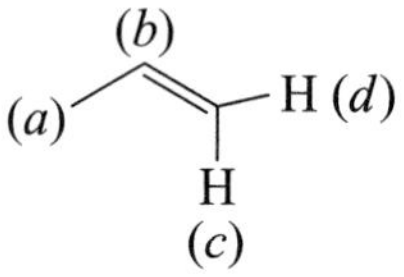

(d) Two 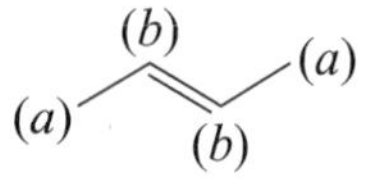 (e) Four 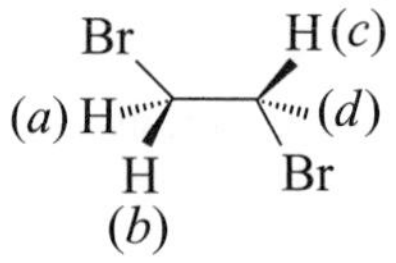 (f) Two

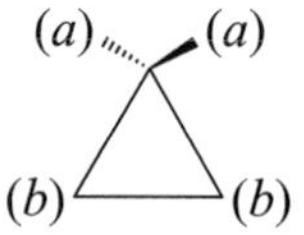

(g) Three 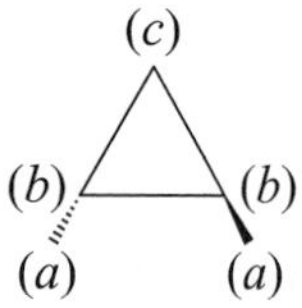 (h) Four 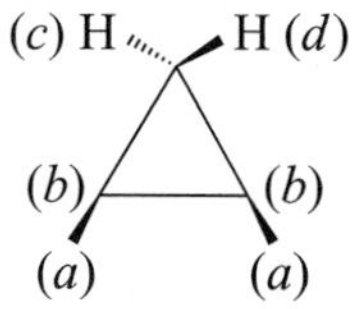 (i) Six

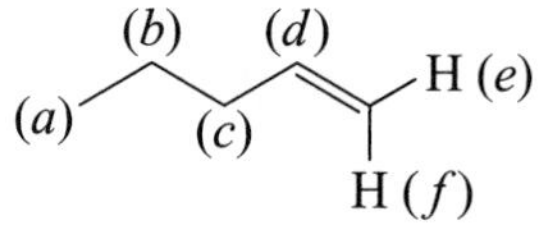

(j) Five 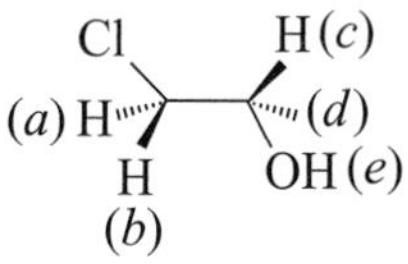 (k) Three 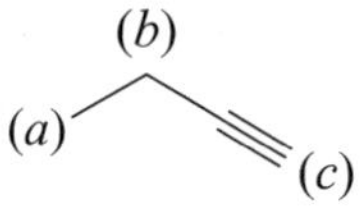 (l) Six

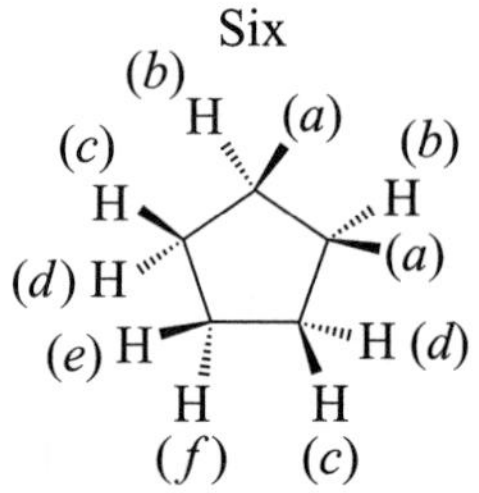

(m) Five 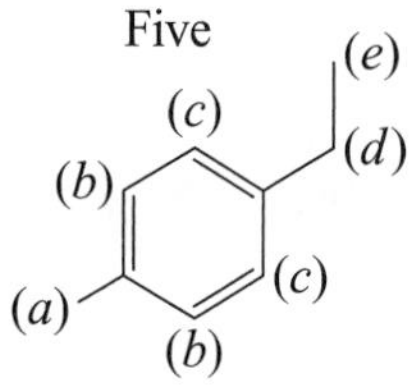 (n) Three 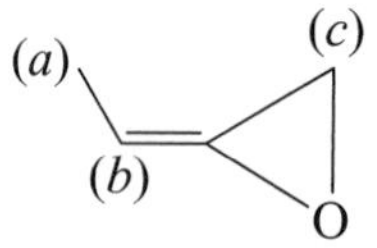 (o) Three

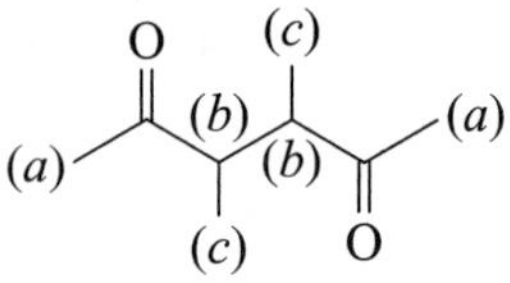

(p) Four

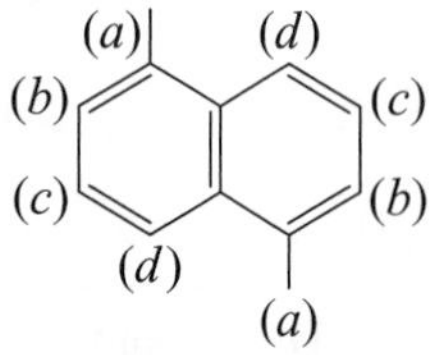

9.7 The determining factors here are the number of chlorine atoms attached to the carbon atoms bearing protons and the deshielding that results from chlorine's electronegativity. In 1,1, 2-trichloroethane the proton that gives rise to the triplet is on a carbon atom that bears two chlorines, and the signal from this proton is downfield. In 1,1,2,3,3-pentachloropropane, the proton that gives rise to the triplet is on a carbon atom that bears only one chlorine; the signal from this proton is upfield.

9.8

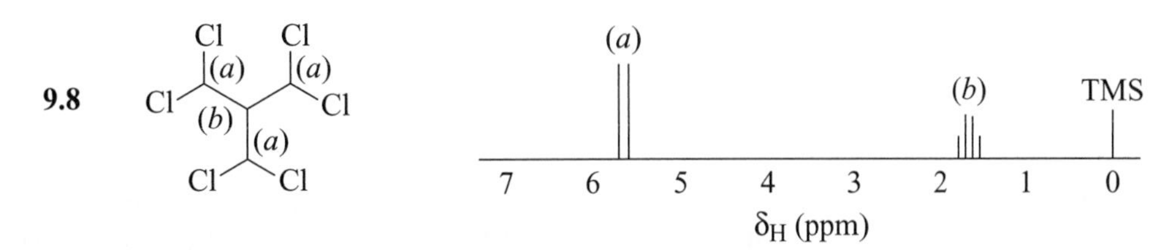

The signal from the three equivalent protons designated (*a*) should be split into a doublet by the proton *(b)*. This doublet, because of the electronegativity of the two attached chlorines, should occur downfield ($\delta \sim 5\text{–}6$). It should have a relative area (integration value) three times that of (b).

The signal for the proton designated (*b*) should be split into a quartet by the three equivalent protons (*a*). The quartet should occur upfield ($\delta \sim 1\text{–}2$).

9.9 **A** C_3H_7I is (*a*) (*b*) (*a*), with I on (*b*)

(a) doublet δ 1.9

(b) septet δ 4.35

B $C_2H_4Cl_2$ is (*a*) (*b*) Cl, with Cl on (*b*)

(a) doublet δ 2.08

(b) quartet δ 5.9

C $C_3H_6Cl_2$ is Cl (*a*) (*b*) (*a*) Cl

(a) triplet δ 3.7

(b) quintet δ 2.2

9.10 (a)

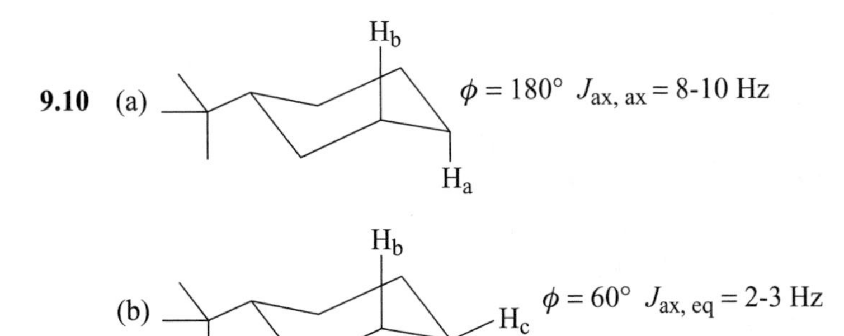

(b)

9.11 A chair conformation with both the bromine and the chlorine in equatorial positions is consistent with $J_{1,2} = 7.8$ Hz, since the hydrogens would be axial.

H_{ax}, Br, Cl, H_{ax}

$J_{ax,ax} = 7.8$ Hz

trans-1-Bromo-2-chlorocyclohexane

9.12 NMR coupling constants would be distinctive for the protons indicated. Compound **A** would have an equatorial-axial proton NMR coupling constant of 2-3 Hz between the indicated protons, whereas **B** would have an axial-axial coupling constant of 8-10 Hz.

A

$J_{a,b} = 2\text{-}3$ Hz

B

$J_{a,b} = 8\text{-}10$ Hz

9.13 (a) $J_{ab} = 2J_{bc}$

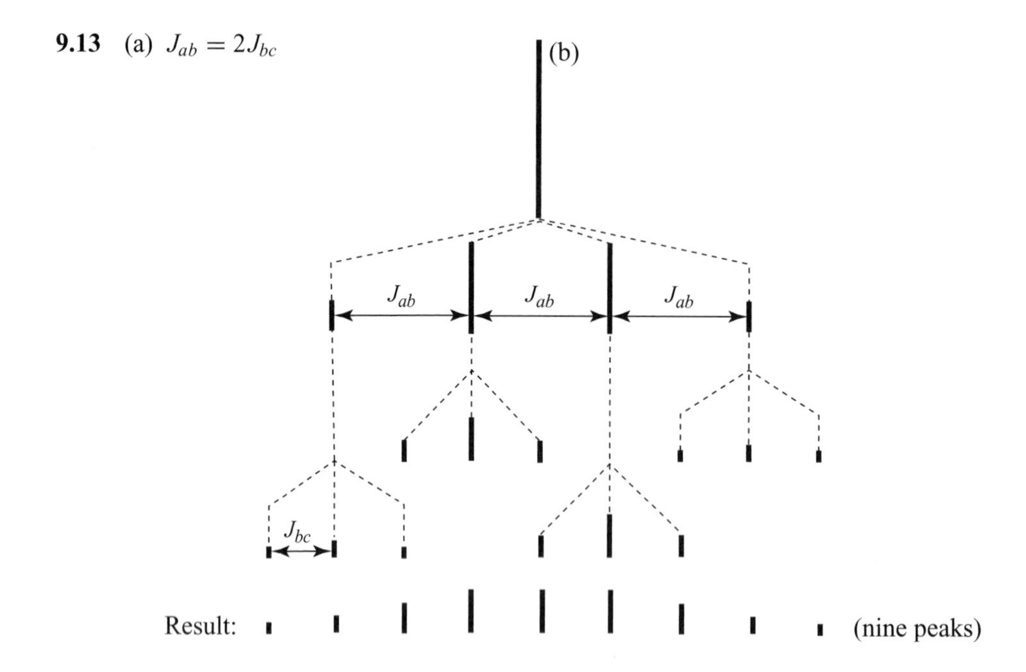

(b) $J_{ab} = J_{bc}$

(b)

J_{ab} J_{ab} J_{ab}

J_{bc} J_{bc}

J_{bc} J_{bc} J_{bc} J_{bc}

J_{bc} J_{bc}

Result: (six peaks)

9.14 A single unsplit signal, because the environment of the proton rapidly changes, reversibly, from axial to equatorial.

9.15 **A** is 1-bromo-3-methylbutane. The following are the signal assignments:

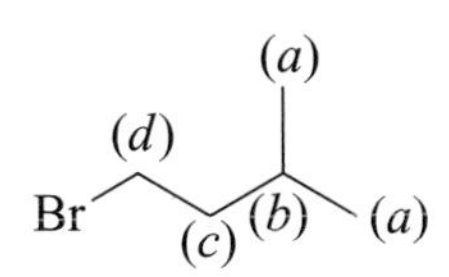

(a) δ 23
(b) δ 27
(c) δ 32
(d) δ 42

B is 2-bromo-2-methylbutane. The following are the signal assignments:

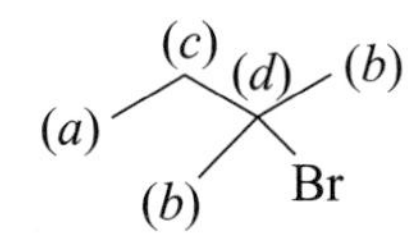

(a) δ 11
(b) δ 33
(c) δ 40
(d) δ 68

C is 1-bromopentane. The following are the signal assignments.

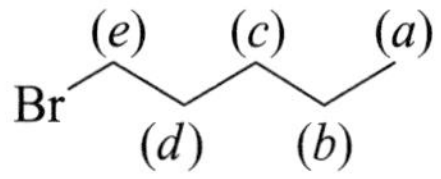

(e) (c) (a)
Br
(d) (b)

(a) δ 14
(b) δ 23
(c) δ 30
(d) δ 33
(e) δ 34

9.16 A peak at $M^{\ddagger}-15$ involves the loss of a methyl radical and the formation of a 1° or 2° carbocation.

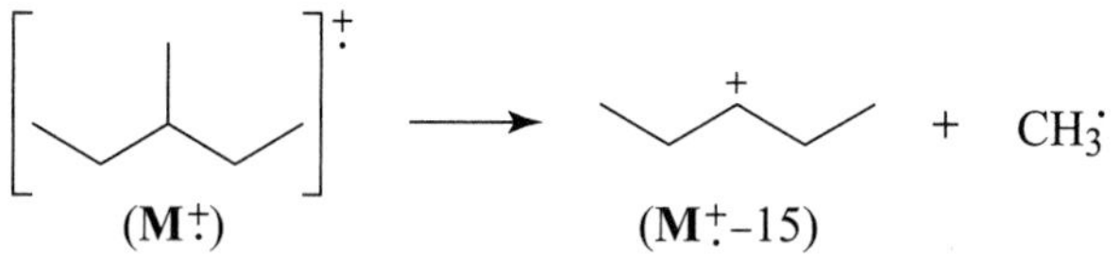

(**M**‡) → (**M**‡–15) + $CH_3^{\cdot}$

or

(**M**‡) → (**M**‡–15) + $CH_3^{\cdot}$

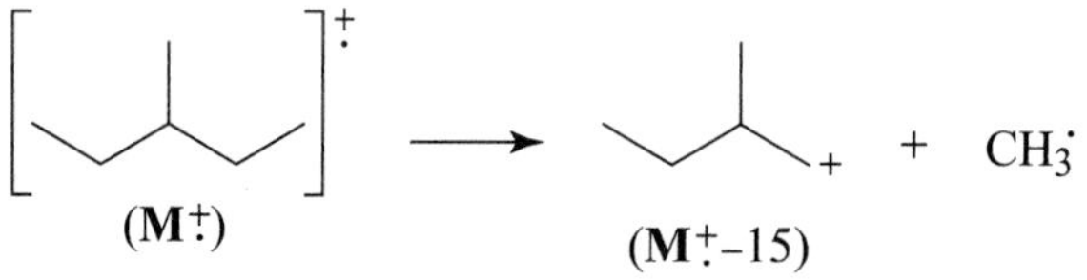

A peak at $M^{\ddagger}-29$ arises from the loss of an ethyl radical and the formation of a 2° carbocation.

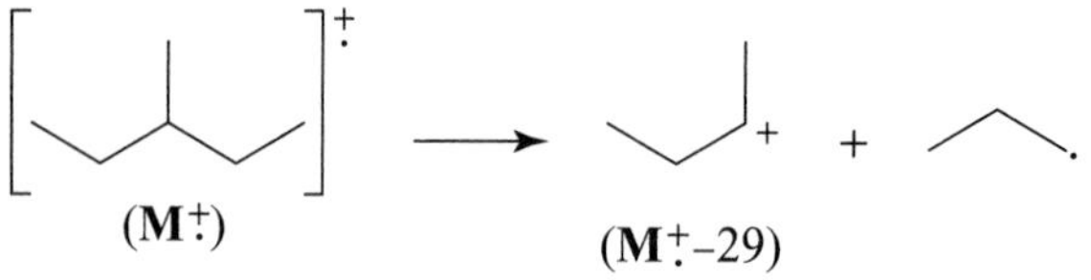

(**M**‡) → (**M**‡–29) +

Since a secondary carbocation is more stable than a methyl carbocation, and since there are two cleavages that form secondary carbocations by loss of an ethyl radical, the peak at $M^{\ddagger}-29$ is more intense.

9.17 After loss of a pi bonding electron through electron impact ionization, both peaks arise from allylic fragmentations:

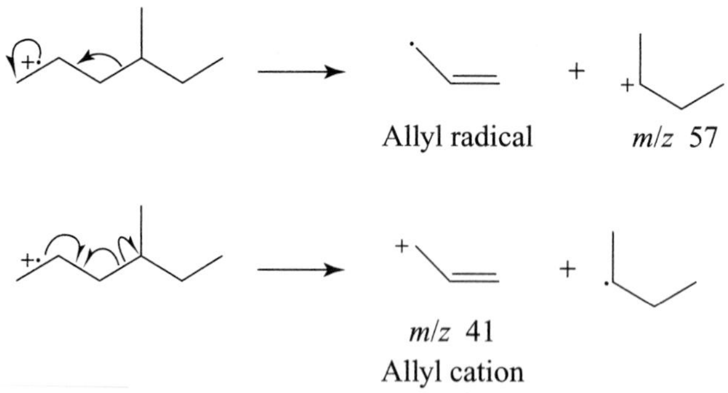

→ +

Allyl radical   *m*/*z* 57

→ +

m/*z* 41
Allyl cation

9.18 The spectrum given in Fig. 9.40 is that of butyl isopropyl ether. The main clues are the peaks at *m*/*z* 101 and *m*/*z* 73 due to the following fragmentations.

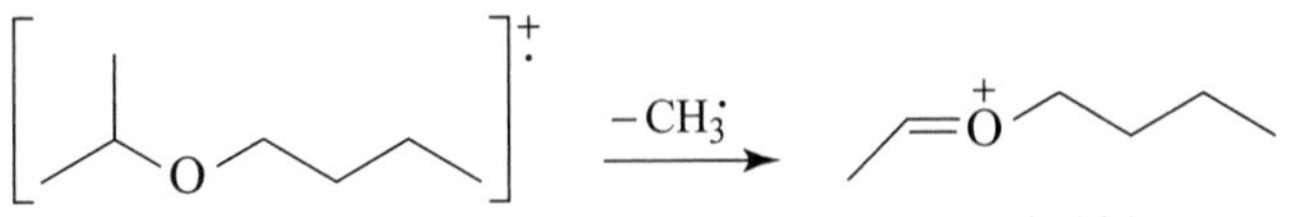

$-CH_3^{\cdot}$

m/z 101

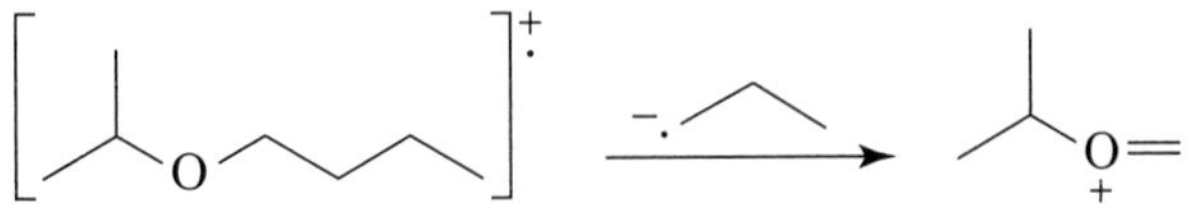

m/z 73

Butyl propyl ether (Fig. 9.41) has no peak at m/z 101 but has a peak at m/z 87 instead.

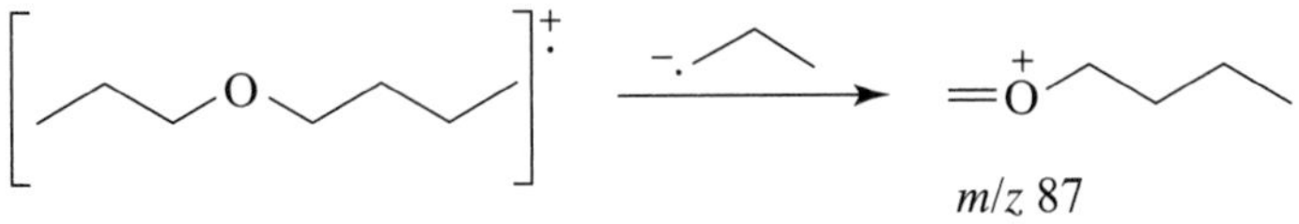

m/z 87

9.19 First, we recalculate the intensities of the peaks so as to base them on the $M^{\ddagger}$ peak:

m/z		INTENSITY % of $M^{\ddagger}$
86 $M^{\ddagger}$	$10.0/10.0 \times 100 =$	100
87	$0.56/10.0 \times 100 =$	5.6
88	$0.04/10.0 \times 100 =$	0.4

1. Since $M^{\ddagger}$ is even, the compound must contain an even number of nitrogen atoms (i.e., 0, 2, 4, etc.).
2. The value of the $M^{\ddagger}+1$ peak gives the number of carbon atoms.

$$\text{Number of carbon atoms} = 5.6/1.1 \simeq 5$$

The compound must contain no nitrogen atoms because $C_5N_2 = (5 \times 12) + (2 \times 14) = 88$, and the molecular weight of the compound (from the $M^{\ddagger}$ peak) is only 86.

3. The very low value of the $M^{\ddagger}+2$ peak (0.4%) tells us that the compound does not contain S, Cl, or Br.
4. If the compound were composed only of C and H, it would have to be C_5H_{26}:

$$H = 86 - (5 \times 12) = 26$$

But C_5H_{26} is impossible.
However, a formula with one oxygen gives a reasonable number of hydrogen atoms,

$$H = 86 - (5 \times 12) - 16 = 10$$

and thus our compound has the formula $C_5H_{10}O$.

9.20 Recalculating the intensities to base on $M^{\ddagger}$

PEAK	m/z	% of BASE PEAK	% of $M^{\ddagger}$
$M^{\ddagger}$	73	86.1	100
$M^{\ddagger}+1$	74	3.2	3.72
$M^{\ddagger}+2$	75	0.2	0.23

These data best fit the formula C_3H_7NO.

9.21 (a) First calculate the expected masses of the compound shown for $C_6H_4 + {}^{79}Br + {}^{79}Br$ (M) = 234 m/z, $C_6H_4 + {}^{81}Br + {}^{79}Br$ (M+2) = 236 m/z, and $C_6H_4 + {}^{81}Br + {}^{81}Br$ (M+4) = 238 m/z.

The relative ratio of ^{79}Br and ^{81}Br is 1:1. Therefore if you have two bromines in the molecule we must simplify the expression of (1:1)(1:1) = 1:2:1.

The ratio of the peaks will be M (234), M+2 (236), M+4 (238) = (1:2:1)

(b) First calculate the expected masses of the compound shown for $C_6H_4 + {}^{35}Cl + {}^{35}Cl$ (M) =146 m/z, $C_6H_4 + {}^{35}Cl + {}^{37}Cl$ (M+2) = 148 m/z, and $C_6H_4 + {}^{37}Cl + {}^{37}Cl$ (M+4) = 150 m/z.

The relative ratio of ^{35}Cl and ^{37}Cl is 3:1. Therefore if you have two chlorines in the molecule we must simplify the expression of (3:1)(3:1) = 9:6:1.

The ratio of the peaks will be M (146), M+2 (148), M+4 (150) = (9:6:1).

9.22 (a) First recalculating the intensities so as to base them on the $M^{+\cdot}$ peak:

m/z		INTENSITY % of $M^{+\cdot}$
78 $M^{+\cdot}$	$24/24 \times 100 =$	100
79	$0.8/24 \times 100 =$	3.3
80	$8/24 \times 100 =$	33

1. Since $M^{+\cdot}$ is even, the compound contains an even number of nitrogen atoms.
2. Number of carbon atoms = $(M^{+\cdot} + 1)/1.1 = 3.3/1.1 = 3$.
3. The intensity of the $M^{+\cdot}$ +2 peak (33%) tells us that the compound contains one chlorine atom.
4. We use the molecular weight (from the $M^{+\cdot}$ peak) to calculate the number of hydrogen atoms

$$H = 78 - (3 \times 12) - 35 = 7$$

Thus, the formula for the compound is C_3H_7Cl.

(b) Cl

NMR Spectroscopy

9.23 (a) 2 H 3 1 H 4

(b)

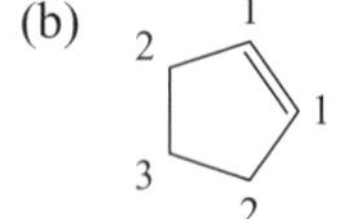

(c)

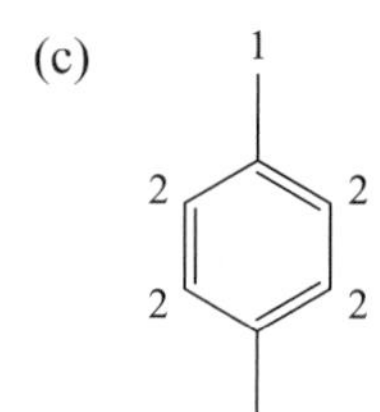

(d)

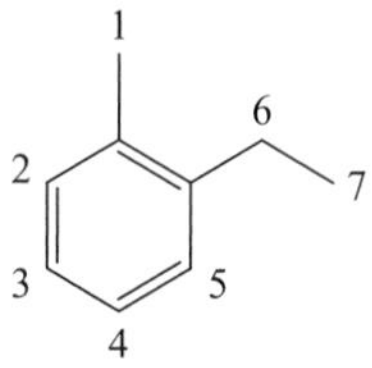

(e)

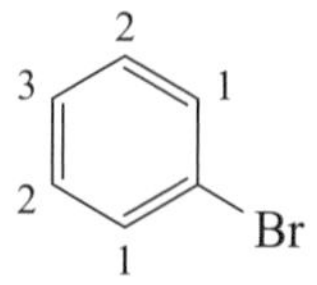

(f)

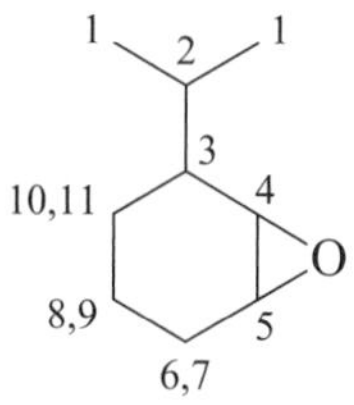

(g)

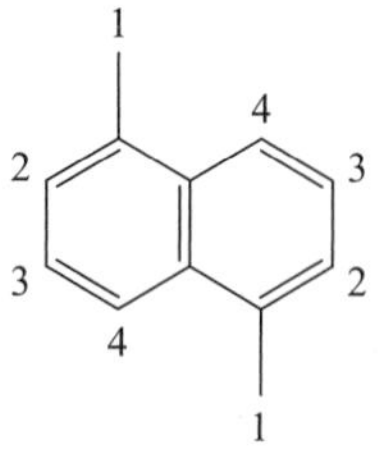

(h)

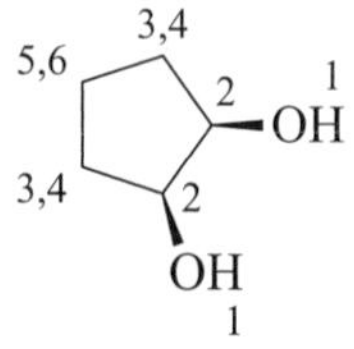

(i)

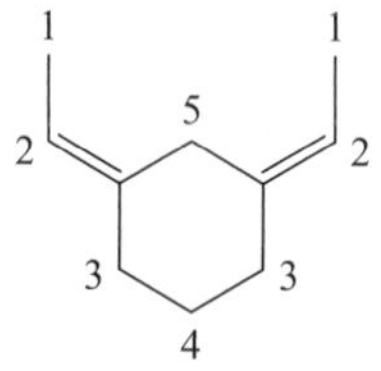

* Two numbers at the same carbon indicates diastereotopic protons.

9.24 (a)

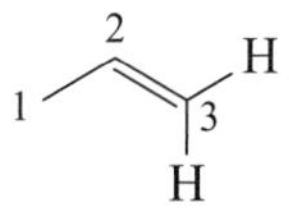

(b)

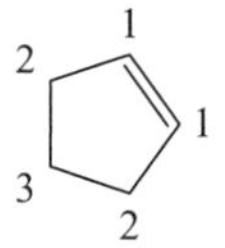

(c)

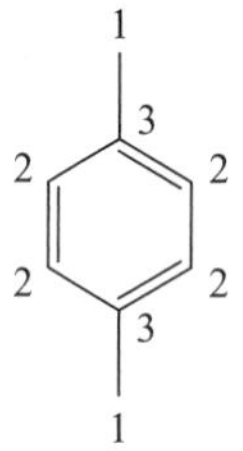

(d)

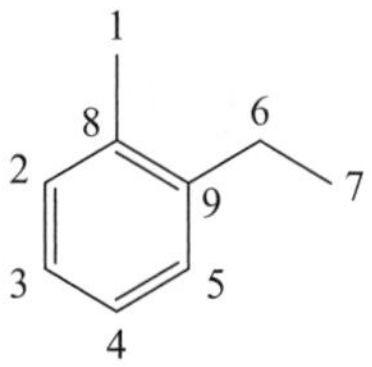

(e)

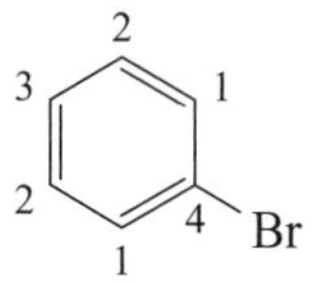

(f)

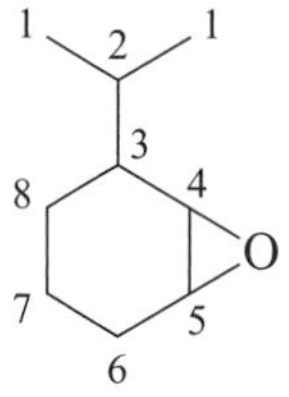

(g)

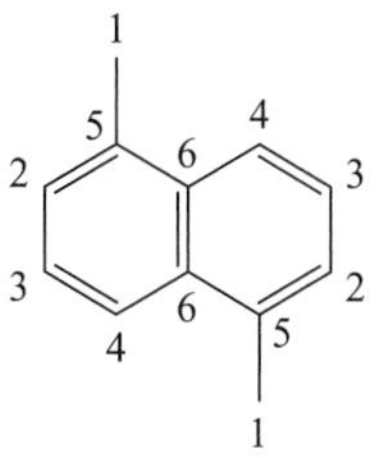

(h)

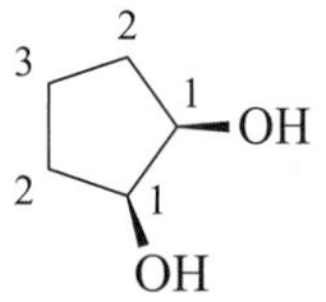

(i)

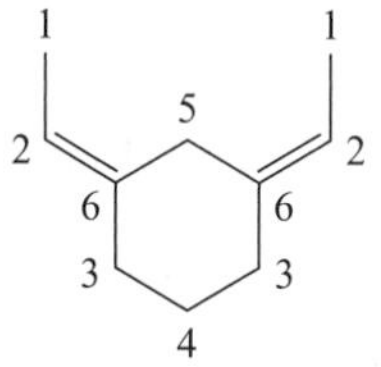

9.25

	Chemical Shift	Splitting	Integration	Structure
(a)	0.9	t	3 H	CH_3 adjacent to CH_2
(b)	1.2	s	6 H	Two equivalent CH_3 groups adjacent to no hydrogens
(c)	1.3	q	2H	CH_2 adjacent only to CH_3
(d)	2.0	s	1H	OH

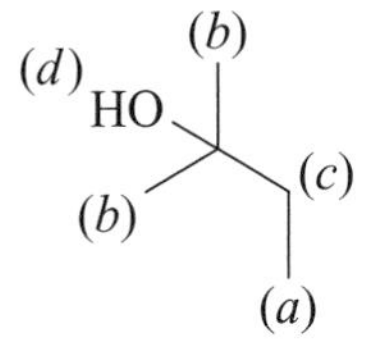

(*d*) HO (*b*) (*c*) (*b*) (*a*)

9.26 Compound **G** is 2-bromobutane. Assignments are as follows:

Br (*a*) (*b*) (*d*) (*c*)

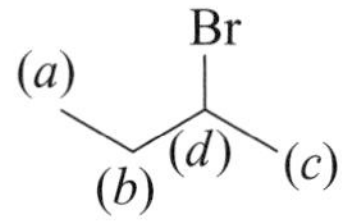

(*a*) triplet δ 1.05
(*b*) multiplet δ 1.82
(*c*) doublet δ 1.7
(*d*) multiplet δ 4.1

Compound **H** is 2,3-dibromopropene. Assignments are as follows:

(*b*) H (*a*) Br H Br (*c*)

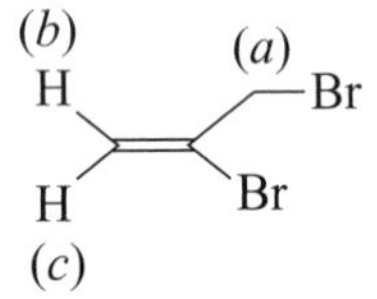

(*a*) δ 4.2
(*b*) δ 6.05
(*c*) δ 5.64

Without an evaluation of the alkene coupling constants, it would be impossible to specify whether the alkene has (*E*), (*Z*), or geminal substitution. We do know from the integral values and number of signals that the alkene is disubstituted. For the analysis of the alkene substitution pattern these typical alkene coupling constants are useful: H—C=C—H (trans), $J = 12–18$ Hz; H—C=C—H (cis), $J = 6–12$ Hz; C=CH_2 (geminal), $J = 0–3$ Hz. Since the alkene signal for compound **H** shows little or no coupling, the data support the designation of **H** as 2, 3-dibromopropene.

9.27 Run the spectrum with the spectrometer operating at a different magnetic field strength (i.e., at 30 or at 100 MHz). If the peaks are two singlets the distance between them—*when measured in hertz*—will change because chemical shifts *expressed in hertz* are proportional to the strength of the applied field (Section 9.7A). If, however, the two peaks represent a doublet, then the distance that separates them, expressed in hertz, will not change because this distance represents the magnitude of the coupling constant and coupling constants are independent of the applied magnetic field.

9.28 Compound **O** is 1,4-cyclohexadiene and **P** is cyclohexane.

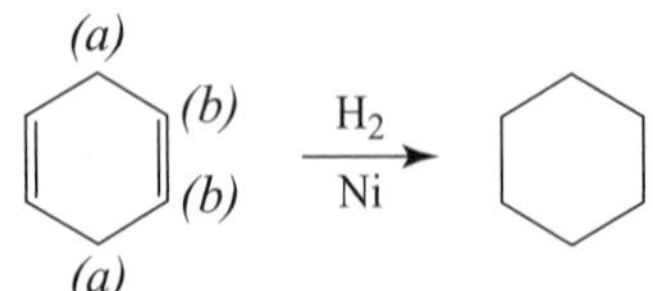

(*a*) (*b*) (*b*) (*a*) **O** $\xrightarrow[\text{Ni}]{H_2}$ **P**

(*a*) δ 26.0 CH_2 (DEPT)
(*b*) δ 124.5 CH (DEPT)

9.29 The molecular formula of **Q** (C_7H_8) indicates an index of hydrogen deficiency (Section 4.17) of four. The hydrogenation experiment suggests that **Q** contains two double bonds (or one triple bond). Compound **Q,** therefore, must contain two rings.

Bicyclo[2.2.1]hepta-2,5-diene. The following reasoning shows one way to arrive at this conclusion: There is only one signal (δ 144) in the region for a doubly bonded carbon. This fact indicates that the doubly bonded carbon atoms are all equivalent. That the signal at δ 144 is a CH group in the DEPT spectra indicates that each of the doubly bonded carbon atoms bears one hydrogen atom. Information from the DEPT spectra tells us that the signal at δ 85 is a —CH_2— group and the signal at δ 50 is a $\geq$C—H group.

The molecular formula tells us that the compound must contain two $\geq$C—H groups, and since only one signal occurs in the ^{13}C spectrum, these $\geq$C—H groups must be equivalent. Putting this all together, we get the following:

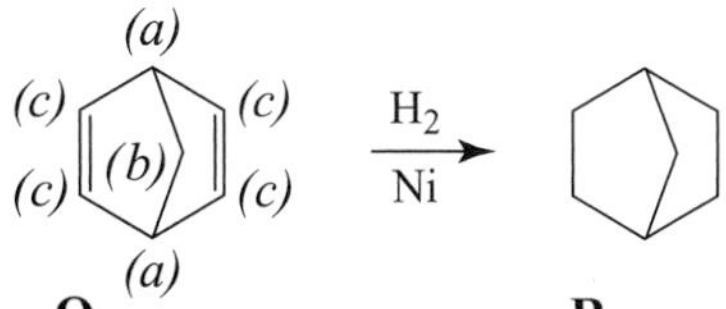

(a) (c) (c) (b) (c) (c) (a) **Q** $\xrightarrow[\text{Ni}]{H_2}$ **R**

(a) δ 50	CH (DEPT)
(b) δ 85	CH_2 (DEPT)
(c) δ 144	CH (DEPT)

9.30 (a) ^{1}H NMR

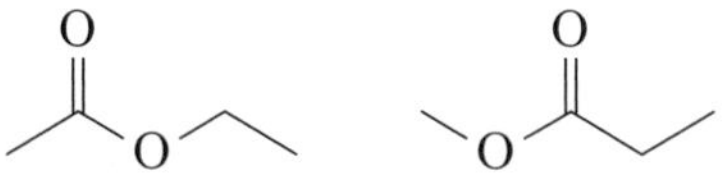

A B

Compound A = δ 0.9 (t, 3 H), δ 2.0 (s, 3 H), δ 4.1 (q, 2 H)
Compound B = δ 0.9 (t, 3 H), δ 2.0 (q, 2 H), δ 4.1 (s, 3 H)

(b)

^{1}H NMR

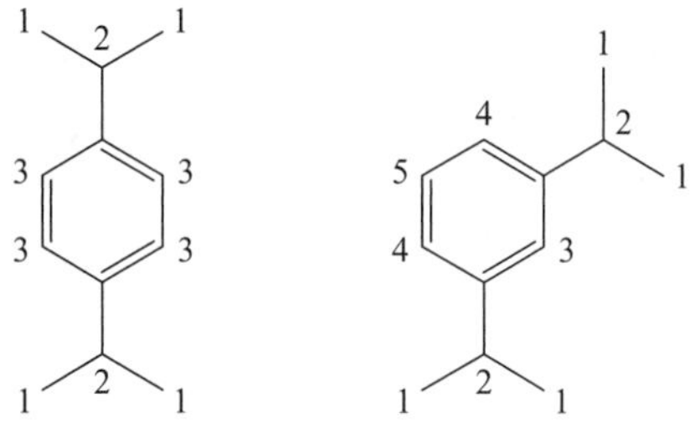

^{13}C NMR

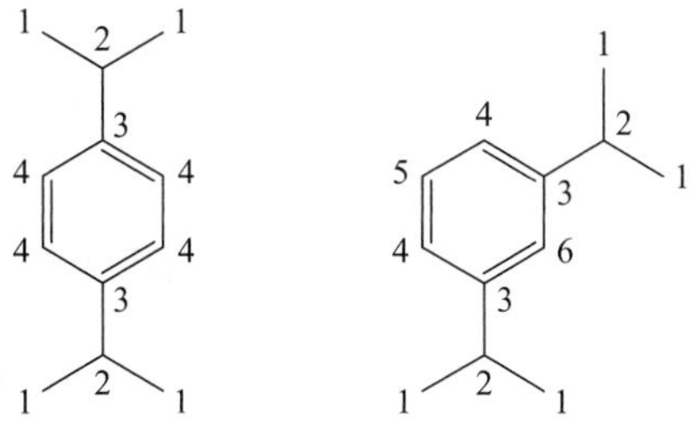

(c) ^{13}C NMR

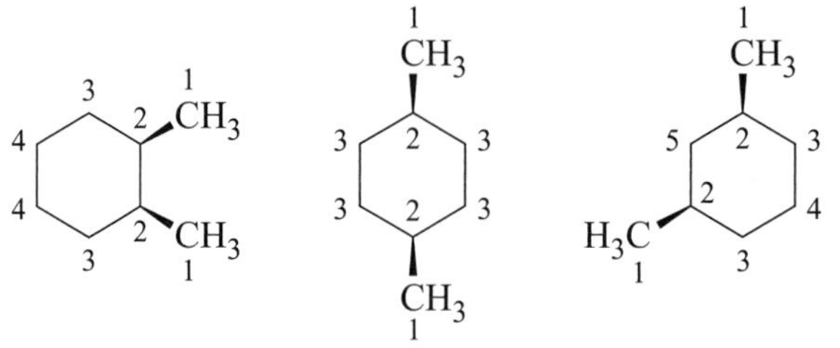

9.31 That **S** decolorizes bromine indicates that it is unsaturated. The molecular formula of **S** allows us to calculate an index of hydrogen deficiency equal to 1. Therefore, we can conclude that **S** has one double bond.

The ^{13}C spectrum shows the doubly bonded carbon atoms at δ 130 and δ 135. In the DEPT spectra, one of these signals (δ 130) is a carbon that bears no hydrogen atoms; the other (δ 135) is a carbon that bears one hydrogen atom. We can now arrive at the following partial structure.

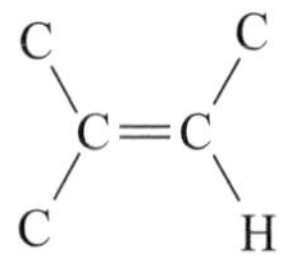

The three most upfield signals (δ 19, δ 28, and δ 31) all arise from methyl groups. The signal at δ 32 is a carbon atom with no hydrogen atoms. Putting these facts together allows us to arrive at the following structure.

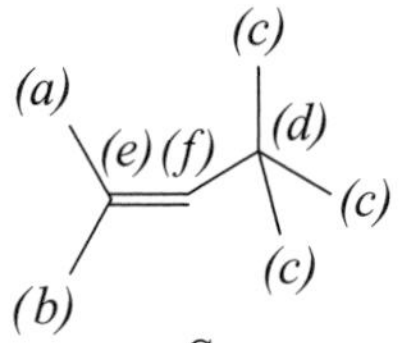

S

(a) δ 19	*(d)* δ 32
(b) δ 28	*(e)* δ 130
(c) δ 31	*(f)* δ 135

Although the structure just given is the actual compound, other reasonable structures that one might be led to are

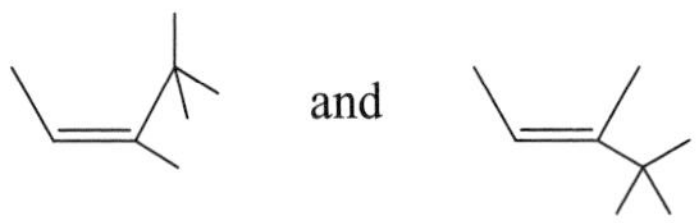

and

Mass Spectrometry

9.32 The compound is butanal. The peak at m/z 44 arises from a McLafferty rearrangement.

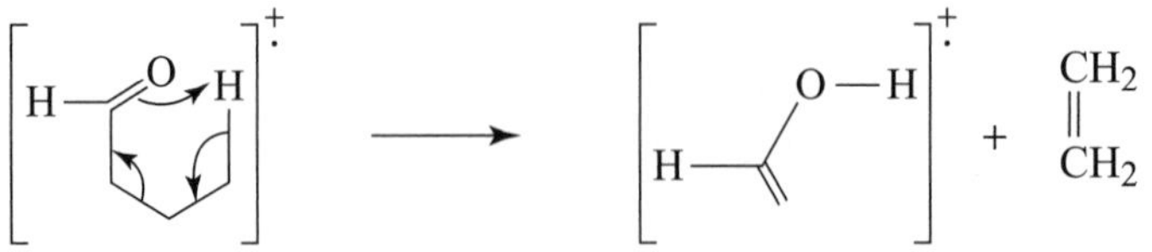

m/z 72 (**M**$^{+\cdot}$)

m/z 44 (**M**$^{+\cdot}$–28)

The peak at *m/z* 29 arises from a fragmentation producing an acylium ion.

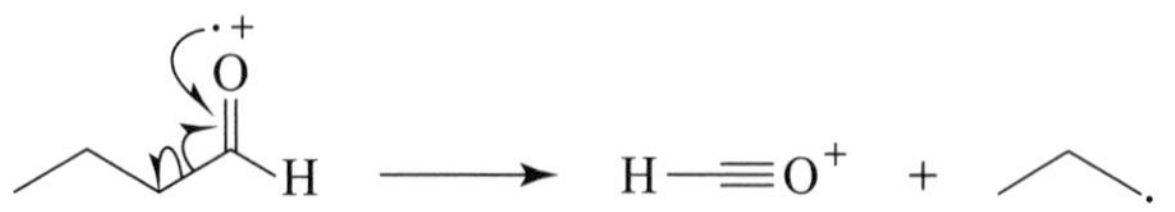

$H{-}{\equiv}O^+$ + ·

m/z 29

9.33

+ H_2O

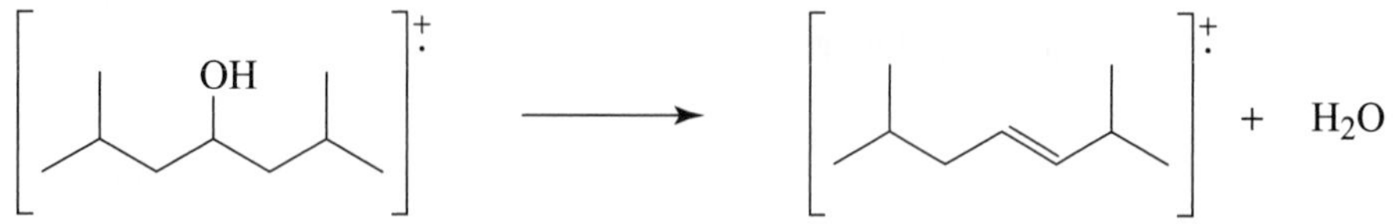

m/z: 144 (100.0%), 145 (10.0%)

m/z: 126 (100.0%), 127 (9.9%)

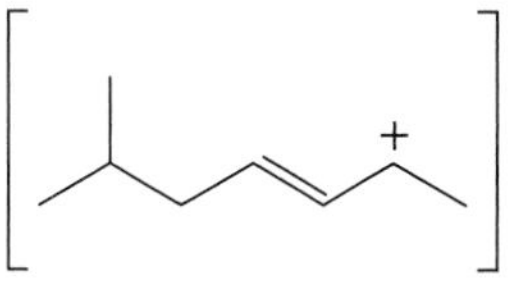

m/z: 111 (100.0%), 112 (8.8%)

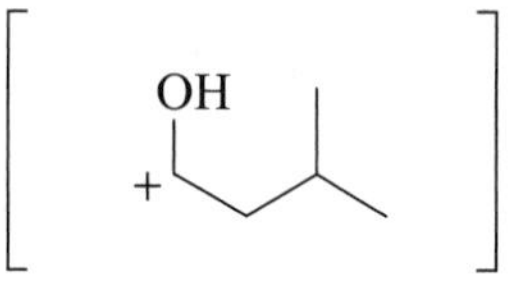

OH

m/z: 87 (100.0%), 88 (5.6%)

9.34

O

OH

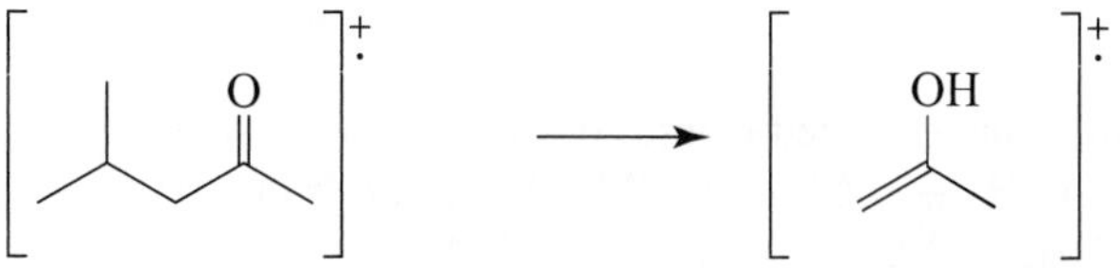

McLafferty rearrangement

m/z: 100 (100.0%), 101 (6.7%)

m/z: 58 (100.0%), 59 (3.4%)

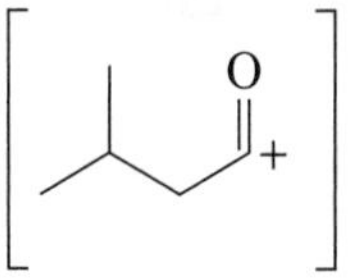

O

m/z: 85 (100.0%), 86 (5.5%)

O

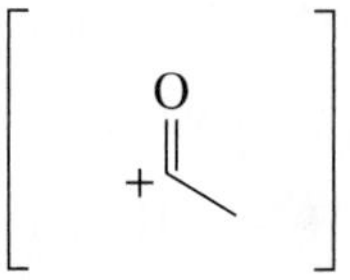

m/z: 43 (100.0%), 44 (2.2%)

9.35 (a)

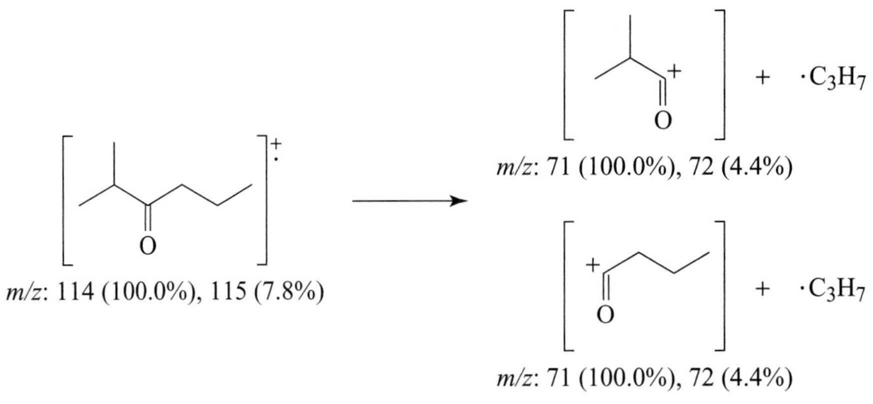

+ $\cdot C_3H_7$

m/z: 71 (100.0%), 72 (4.4%)

m/z: 114 (100.0%), 115 (7.8%)

+ $\cdot C_3H_7$

m/z: 71 (100.0%), 72 (4.4%)

(b)

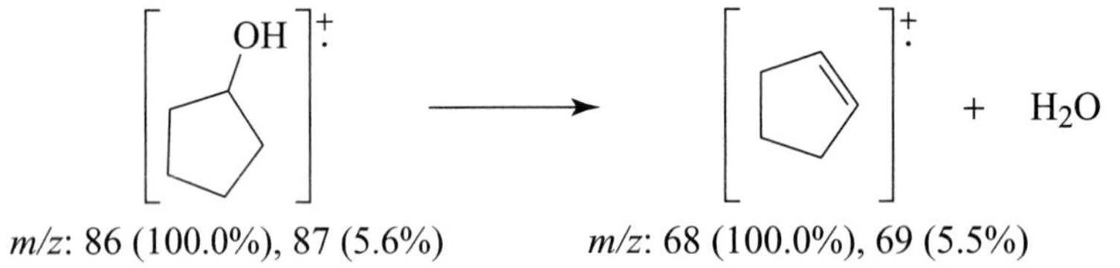

+ H_2O

m/z: 86 (100.0%), 87 (5.6%) m/z: 68 (100.0%), 69 (5.5%)

(c)

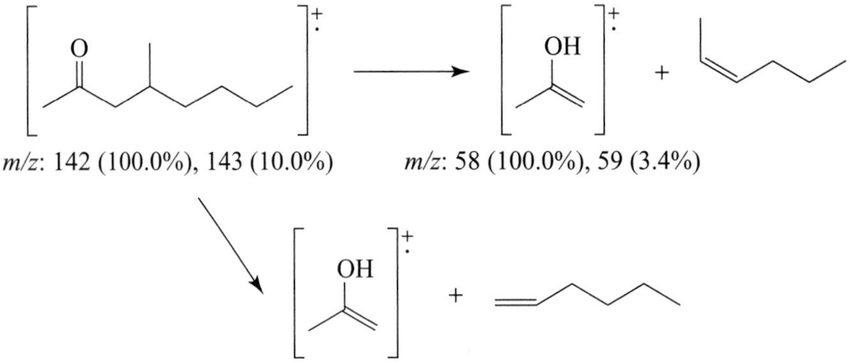

m/z: 142 (100.0%), 143 (10.0%) m/z: 58 (100.0%), 59 (3.4%)

9.36 First calculate the expected masses of the compound shown for C_6H_4 + $^{79}Br + ^{35}Cl$ (M) = 190 m/z, C_6H_4 + $^{37}Cl + ^{79}Br$ (M+2) = 192 m/z or C_6H_4 + $^{35}Cl + ^{81}Br$ (M+2) = 192 m/z, and C_6H_4 + $^{37}Cl + ^{81}Br$ (M+4) = 194 m/z.

The relative ratio of ^{79}Br and ^{81}Br is 1:1 and ^{35}Cl and ^{37}Cl is 3:1. Therefore if you have 1 bromine and 1 chlorine in the molecule we must simplify the expression of (3:1)(1:1) = 3:4:1.

The ratio of the peaks will be M (190), M+2 (192), M+4 (194) = (3:4:1) or (100%: 77%: 24%).

9.37 Ethyl bromide will have a significant M+2 for the ^{81}Br isotope at 110. The ratio of the peak at m/z 108 and m/z 110 will be approximately 1:1. The spectra will also show peaks at M+1 and M+3 for the presence of a ^{13}C isotope (2.2 %)

Methoxybenzene will have a small peak, M+1, for the ^{13}C isotope (7.8 %)

9.38 The ion, $CH_2=\overset{+}{N}H_2$, produced by the following fragmentation.

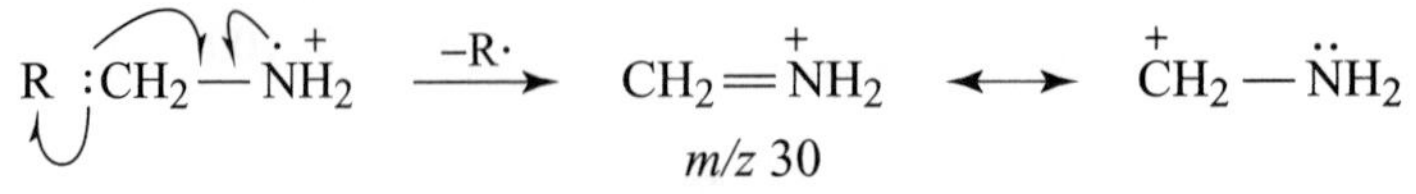

$$R\ {:}CH_2-\overset{\cdot +}{N}H_2 \xrightarrow{-R\cdot} CH_2=\overset{+}{N}H_2 \longleftrightarrow \overset{+}{C}H_2-\ddot{N}H_2$$

m/z 30

Integrated Structure Elucidation

9.39 (a)

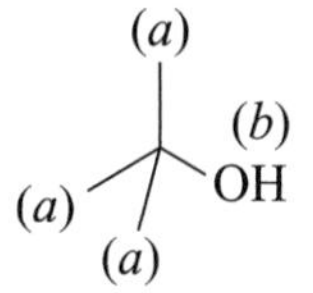

(a) Singlet, δ 1.28 (9H)
(b) Singlet, δ 1.35 (1H)

(b)

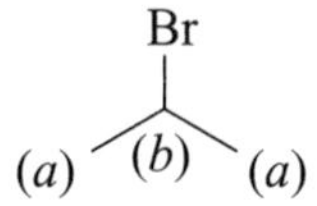

(a) Doublet, δ 1.71 (6H)
(b) Septet, δ 4.32 (1H)

(c)

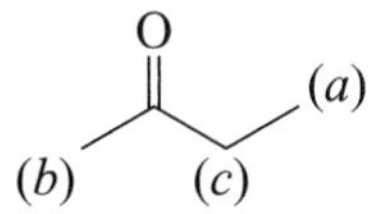

(a) Triplet, δ 1.05 (3H) C=O, near 1720 cm^{-1} (s)
(b) Singlet, δ 2.13 (3H)
(c) Quartet, δ 2.47 (2H)

(d)

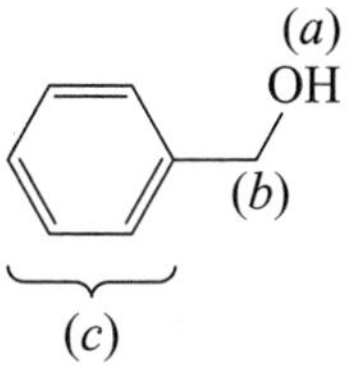

(a) Singlet, δ 2.43 (1H) O—H, 3200–3550 cm^{-1} (br)
(b) Singlet, δ 4.58 (2H)
(c) Multiplet, δ 7.28 (5H)

(e)

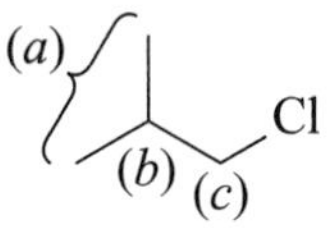

(a) Doublet, δ 1.04 (6H)
(b) Multiplet, δ 1.95 (1H)
(c) Doublet, δ 3.35 (2H)

(f)

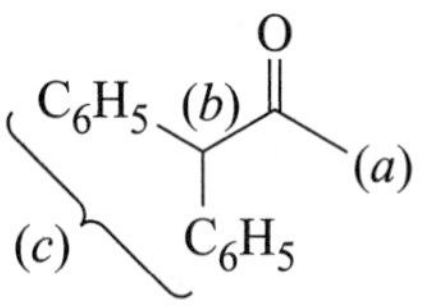

(a) Singlet, δ 2.20 (3H) C=O, near 1720 cm^{-1} (s)
(b) Singlet, δ 5.08 (1H)
(c) Multiplet, δ 7.25 (10H)

(g)

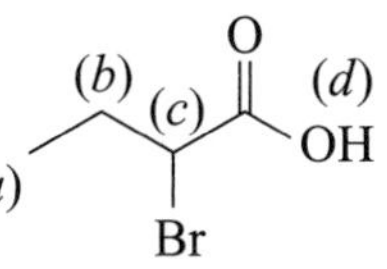

(a) Triplet, δ 1.08 (3H) C=O (acid), 1715 cm^{-1} (s)
(b) Multiplet, δ 2.07 (2H) O—H, 2500–3500 cm^{-1} (br)
(c) Triplet, δ 4.23 (1H)
(d) Singlet, δ 10.97 (1H)

(h)

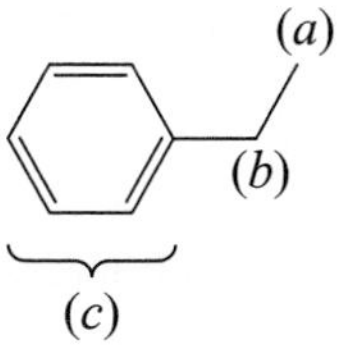

(a) Triplet, δ 1.25 (3H)
(b) Quartet, δ 2.68 (2H)
(c) Multiplet, δ 7.23 (5H)

(i)

(a) Triplet, δ 1.27 (3H)
(b) Quartet, δ 3.66 (2H)
(c) Singlet, δ 4.13 (2H)
(d) Singlet, δ 10.95 (1H)

C=O (acid), 1715 cm^{-1} (s)
O—H, 2500–3550 cm^{-1} (br)

(j)

(a) Doublet, δ 1.55 (6H)
(b) Septet, δ 4.67 (1H)

(k)

(a) Singlet, δ 3.25 (6H)
(b) Singlet, δ 3.45 (4H)

(l)

(a) Doublet, δ 1.10 (6H)
(b) Singlet, δ 2.10 (3H)
(c) Septet, δ 2.50 (1H)

C=O, near 1720 cm^{-1} (s)

(m)

(a) Doublet, δ 2.00 (3H)
(b) Quartet, δ 5.15 (1H)
(c) Multiplet, δ 7.35 (5H)

9.40 Compound **E** is phenylacetylene, $C_6H_5C{\equiv}CH$. We can make the following assignments in the IR spectrum:

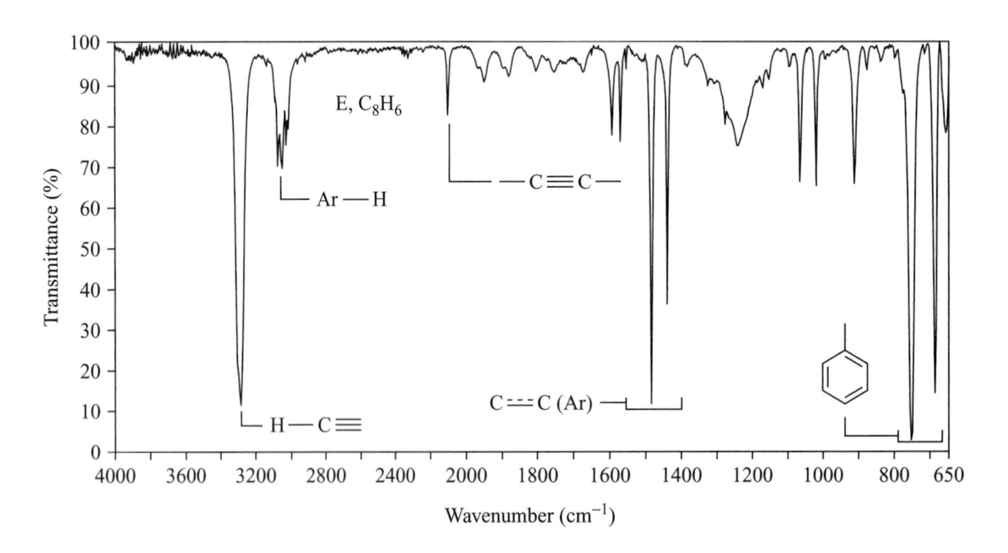

The IR spectrum of compound **E** (Problem 9.40). Compound **F** is $C_6H_5CBr_2CHBr_2$.

9.41 Compound **J** is *cis*-1,2-dichloroethene.

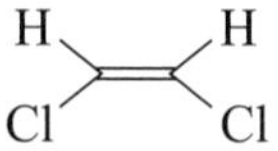

We can make the following IR assignments:

3125 cm^{-1}, alkene C—H stretching
1625 cm^{-1}, C=C stretching
695 cm^{-1}, out-of-plane bending of cis C—H bonds

The ^{1}H NMR spectrum indicates the hydrogens are equivalent.

9.42 (a) Compound **K** is,

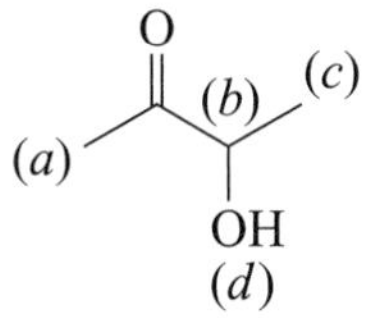

(a) Singlet, δ 2.15 C=O, near 1720 cm^{-1} (s)
(b) Quartet, δ 4.25
(c) Doublet, δ 1.35
(d) Singlet, δ 3.75

(b) When the compound is dissolved in D_2O, the —OH proton (*d*) is replaced by a deuteron, and thus the ^{1}H NMR absorption peak disappears.

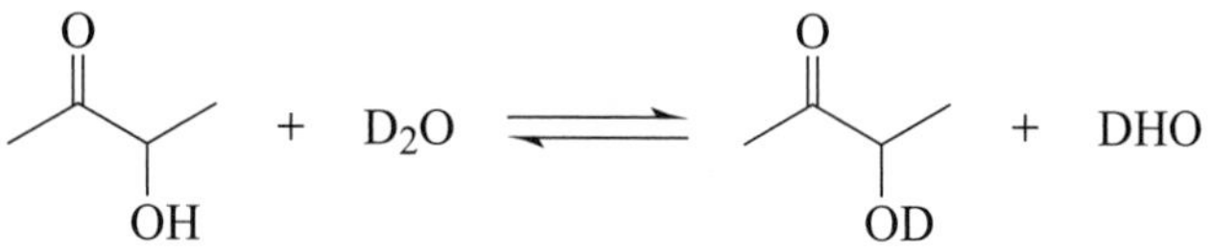

+ D_2O ⇌ + DHO

9.43 The IR absorption band at 1745 cm^{-1} indicates the presence of a C=O group in a five-membered ring, and the signal at δ 220 can be assigned to the carbon of the carbonyl group.

There are only two other signals in the ^{13}C spectrum; the DEPT spectra suggest two equivalent sets of two —CH_2— groups each. Putting these facts together, we arrive at cyclopentanone as the structure for **T.**

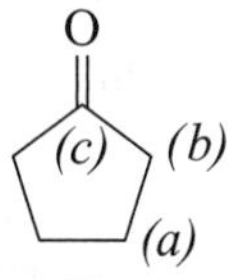

T

(a) δ 23
(b) δ 38
(c) δ 220

9.44 (1) Molecular ion = 96 *m*/*z*

Potential molecular formula = C_7H_{14} or C_6H_8O

(2) The presence of a strong C=O absorption at 1685 cm^{-1} in the IR and the integrals of the ^{1}H NMR spectra totaling 8, and the appearance of 6 unique carbons in the ^{13}C spectra lead to C_6H_8O as the correct molecular formula.

(3) Degree of unsaturation for $C_6H_8O = 3$

(4) ^{1}H NMR

Letter	ppm	Splitting	Integration	Conclusion
(a)	1.9	s	3H	CH_3 (allylic)
(b)	2.2	q	2H	CH_2 (allylic)
(c)	3.0	t	2H	CH_2 (adjacent to ketone and CH_2)
(d)	6.5	t	1H	CH (adjacent to CH_2)

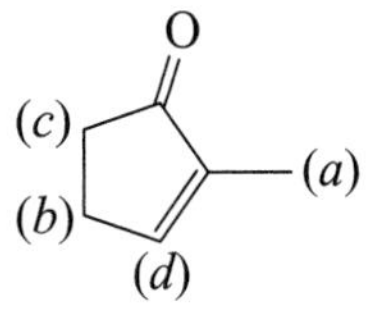

(5) ^{13}C NMR

1. δ207 – (ketone)
2. δ145 – (CH – alkene)
3. δ139 – (C – alkene)
4. δ37 – (CH_2 alpha to ketone)
5. δ25 – (CH_2 allylic)
6. δ16 – (CH_3 – allylic)

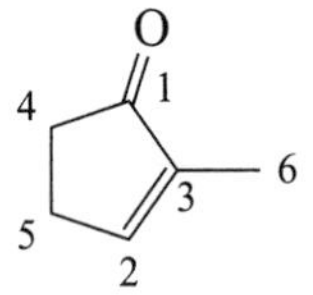

9.45 (1) Molecular ion = 148 m/z

Potential molecular formula = $C_{10}H_{12}O$

(2) IR – 3065 cm^{-1} (C-H sp^2 aromatic), 2960 cm^{-1} (C-H sp^3), 2760 cm^{-1} (C-H aldehyde), 1700 cm^{-1} (C=O conjugate aldehyde), 1600 cm^{-1} (C=C aromatic)

(3) Degree of unsaturation for $C_{10}H_{12}O = 5$

(4) ^{1}H NMR

Letter	ppm	Splitting	Integration	Conclusion
(a)	1.3	d	6H	Two CH_3 adjacent to C-H
(b)	3.1	septet	1H	C-H adjacent to two CH_3
(c)	7.2	d	2H	Aromatic ring disub - para
(d)	7.8	d	2H	Aromatic ring disub - para
(e)	9.8	s	1H	Aldehyde

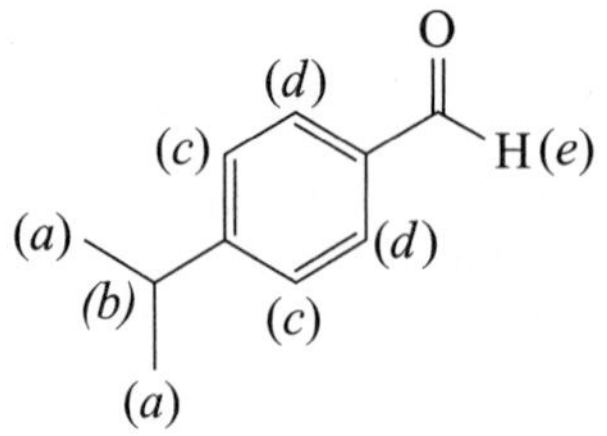

(5) ^{13}C NMR

1. δ191 – CH-aldehyde
2. δ154 – C aromatic
3. δ134 – C aromatic
4. δ130 – CH aromatic
5. δ127 – CH aromatic
6. δ36 – CH aliphatic
7. δ23 – CH_3 aliphatic

9.46 (1) Molecular ion = 204 m/z

Potential molecular formula = $C_{15}H_{24}$

(2) IR-3065 cm^{-1} (C—H sp^2 aromatic), 2960 cm^{-1} (C—H sp^3) 1600 cm^{-1} (C=C aromatic)

(3) Degree of unsaturation for $C_{15}H_{24}$ = 4

(4) ^{1}H NMR

Letter	ppm	Splitting	Integration	Conclusion
(a)	1.3	d	6 H	Two CH_3 next to CH
(b)	2.8	septet	1 H	CH next to two CH_3
(c)	6.9	s	1 H	CH aromatic

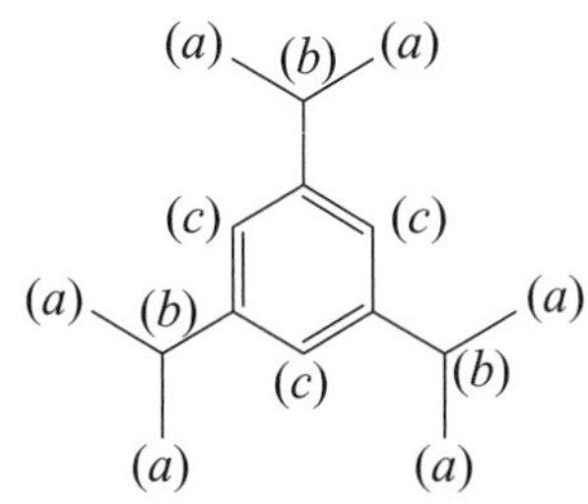

(5) ^{13}C NMR

1. δ148 – (C-aromatic)
2. δ122 – (CH-aromatic)
3. δ37 – (CH-aliphatic)
4. δ23 – (CH_3 – aliphatic)

9.47 (1) Molecular Formula = $C_{10}H_{12}O_3$

(2) IR – 3065 cm^{-1} (C—H sp^2 aromatic), 2960 cm^{-1} (C—H sp^3), 1740 cm^{-1} (C=O ester), 1600 cm^{-1} (C=C aromatic)

(3) Degree of unsaturation for $C_{10}H_{12}O_3$ = 5

(4) ^{1}H NMR

Letter	ppm	Splitting	Integration	Conclusion
(a)	3.5	s	2H	CH_2
(b)	3.7	s	3H	CH_3
(c)	3.8	s	3H	CH_3
(d)	6.7	d	2H	CH (aromatic para-sub)
(e)	6.9	d	2H	CH (aromatic para -sub)

The low ppm of the aromatic protons indicates electron donating groups are attached to the ring.

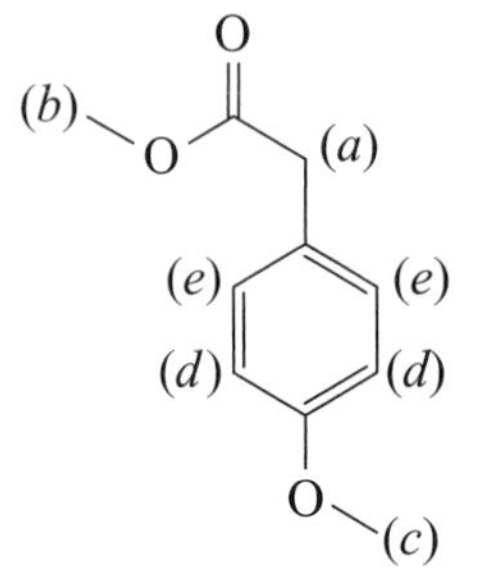

(4) ^{13}C NMR

1. δ171 – (C=O ester)
2. δ160 – (C- aromatic)
3. δ130 – (CH aromatic)
4. δ127 – (C-aromatic)
5. δ115 – (CH aromatic)
6. δ56 – (CH_3 - aliphatic)
7. δ52 – (CH_3 ester)
8. δ46 – (CH_2 - benzylic)

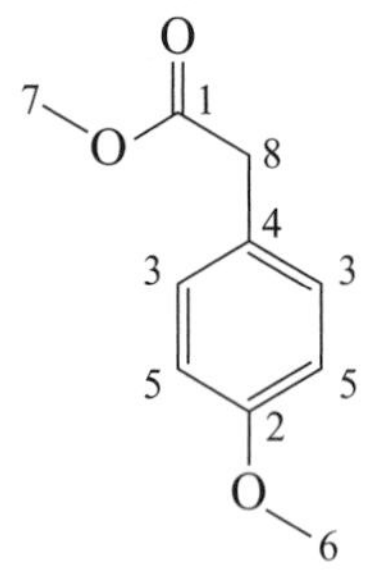

Challenge Problems

9.48 The product of the protonation is a relatively stable allylic cation. The six methyl, four methylene, and single methine hydrogens account for the spectral features.

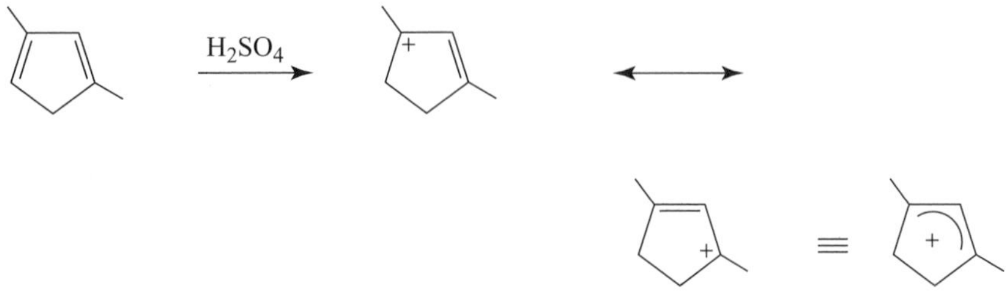

H_2SO_4

9.49 A McLafferty rearrangement accounts for this outcome, as shown below for butanoic acid. In the case of longer chain carboxylic acids the additional carbons are eliminated in the alkene formed.

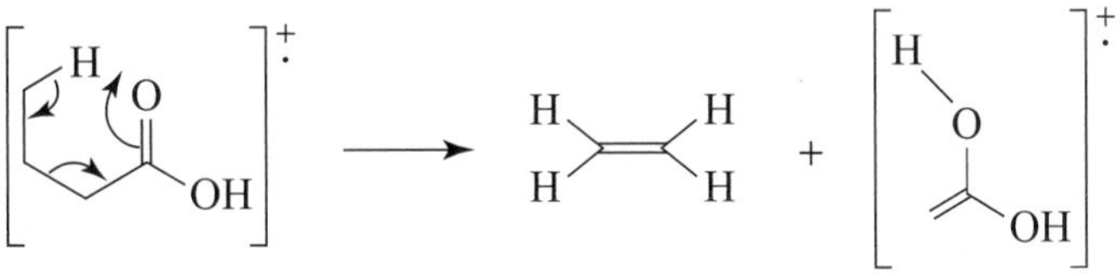

m/*z* 60

9.50 As in the case of IR spectroscopy of alcohols (Section 2.16B), the degree of intermolecular association of the hydroxyl groups is the principal determining factor. If the alcohol is examined in the vapor phase or at very low concentration in a nonpolar solvent such as carbon tetrachloride, the —OH hydrogen absorption occurs at about δ 0.5.

Increasing concentrations of the alcohol up to the neat state (solvent is absent) leads to a progressive shift of the absorption toward the value of δ 5.4.

Hydrogen bonding, R—O—H$^{\delta+}$---O$^{\delta-}$(H)(R), results in a decrease of electron density (deshielding) at the hydroxyl proton. Thus, the proton peak is shifted downfield. The magnitude of the shift, at some particular temperature, is a function of the alcohol concentration.

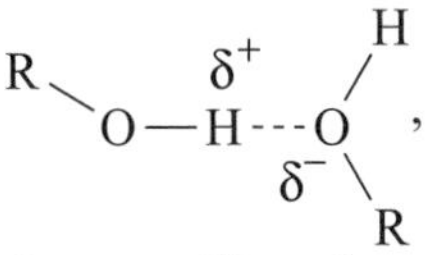

9.51 At the lower temperature a DMF molecule is effectively restricted to one conformation due to the resonance contribution of II, which increases the bond order of the C—N bond. At ~ 300 K there is insufficient energy to bring about significant rotation about that bond.

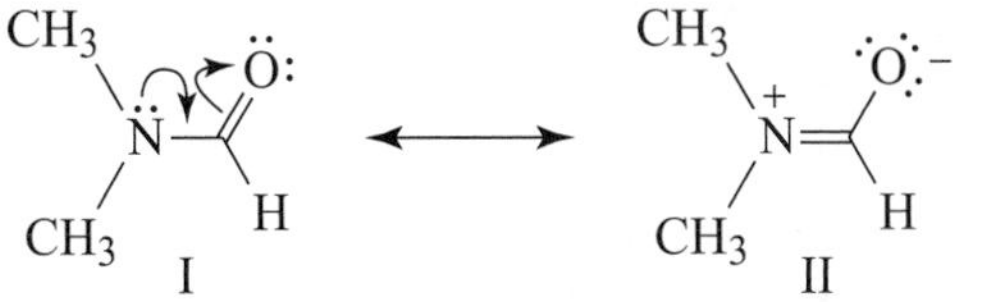

I ⟷ II

The methyl groups clearly are nonequivalent since each has a unique relationship to the single hydrogen atom. Hence each set of methyl group hydrogens is represented by its own signal (though they are quite close).

At a sufficiently high temperature (> 130°C) there is enough energy available to overcome the barrier to rotation due to the quasi-double bond, C⋯N, and substantially free rotation occurs. The differences between the two sets of methyl group hydrogens are no longer discernable and all six methyl hydrogens are represented by a single signal.

9.52 Formed by fragmentation of the molecular ion, $C_4H_3^+$ possesses the requisite mass. Possible structures are $[CH_2{=}CH{-}C{=}C]^+$ and $[HC{=}CH{-}C{\equiv}CH]^+$, each of which is resonance-stabilized.

9.53

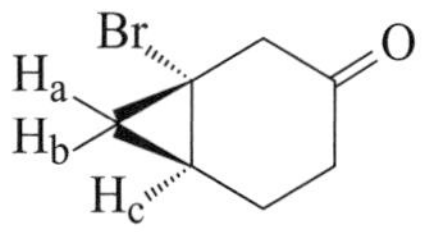

$J_{ab} = 5.3$ Hz
$J_{ac} = 8.2$ Hz
$J_{bc} = 10.7$ Hz

(a) 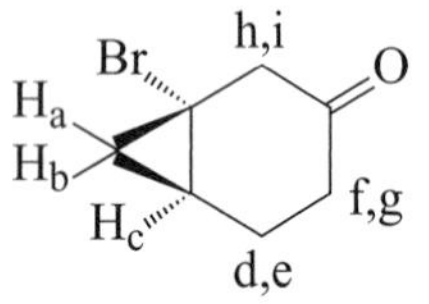 9 signals

(b) 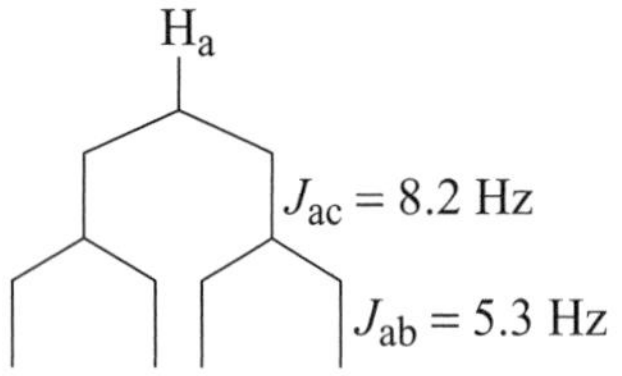 $J_{ac} = 8.2$ Hz
$J_{ab} = 5.3$ Hz

QUIZ

9.1 Propose a structure that is consistent with each set of the following data.

(a) C_4H_9Br — ^1H NMR spectrum
Singlet δ 1.7

(b) $C_4H_7Br_3$ — ^1H NMR spectrum
Singlet δ 1.95 (3H)
Singlet δ 3.9 (4H)

(c) C_8H_{16} — ^1H NMR spectrum
Singlet δ 1.0 (9H)
Singlet δ 1.75 (3H)
Singlet δ 1.9 (2H)
Singlet δ 4.6 (1H)
Singlet δ 4.8 (1H)

IR spectrum
3040, 2950, 1640 cm^{-1}
and other peaks.

(d) $C_9H_{10}O$	^{1}H NMR spectrum Singlet δ 2.0 (3H) Singlet δ 3.75 (2H) Singlet δ 7.2 (5H)	IR spectrum 3100, 3000, 1720, 740, 700 cm^{-1} and other peaks.
(e) $C_5H_7NO_2$	^{1}H NMR spectrum Triplet δ 1.2 (3H) Singlet δ 3.5 (2H) Quartet δ 4.2 (2H)	IR spectrum 2980, 2260, 1750 cm^{-1} and other peaks. This compound has a nitro group.

9.2 How many ^{1}H NMR signals would the following compound give?

(a) One (b) Two (c) Three (d) Four (e) Five

9.3 How many ^{1}H NMR signals would 1,1-dichlorocyclopropane give?

(a) One (b) Two (c) Three (d) Four (e) Five

9.4 Which of these C_6H_{14} isomers has the greatest number of ^{13}C NMR signals?

(a) Hexane
(b) 2-Methylpentane
(c) 3-Methylpentane
(d) 2,2-Dimethylbutane
(e) 2,3-Dimethylbutane

9.5 How many ^{13}C NMR signals would be given by the following compound?

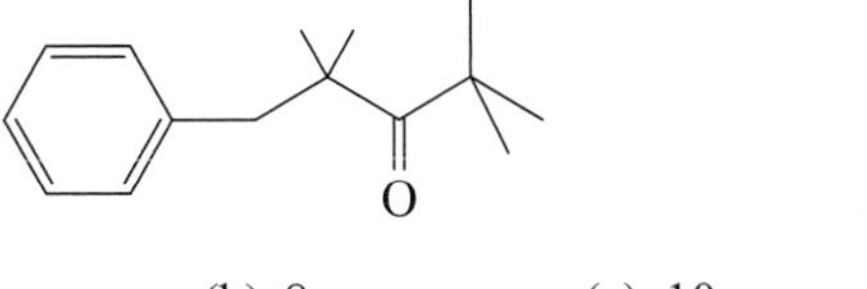

(a) 7 (b) 8 (c) 10 (d) 11 (e) 13

9.6 Which of these is a true statement concerning mass spectrometry? (There may be more than one.)

(a) The M^+ peak is always the most prominent (largest m/z).
(b) Only liquids can be analyzed by MS.
(c) Unlike IR and NMR, MS is a destructive method of analysis.
(d) The molecular ion is assigned a value of 100 on the vertical scale.
(e) The initial event in the determination of a mass spectrum in the EI mode is the formation of a radical cation.

9.7 What is the structure of a compound C_5H_{12} which exhibits a prominent MS peak at m/z 57?

10 RADICAL REACTIONS

SOLUTIONS TO PROBLEMS

10.1 (a)

$$\underset{(DH^\circ = 436)}{H—H} + \underset{(DH^\circ = 159)}{F—F} \longrightarrow \underset{2(DH^\circ = 570)}{2\,H—F}$$

$+595$ kJ mol^{-1} is required for bond cleavage

-1140 kJ mol^{-1} is evolved in formation of the bonds in 2 mol of HF

$$\Delta H^\circ = +595 - 1140 = -545 \text{ kJ mol}^{-1}$$

(b)

$$\underset{(DH^\circ = 440)}{CH_3—H} + \underset{(DH^\circ = 159)}{F—F} \longrightarrow \underset{(DH^\circ = 461)}{CH_3—F} + \underset{(DH^\circ = 570)}{H—F}$$

$+599$ kJ mol^{-1} is required for bond cleavage

-1031 kJ mol^{-1} is evolved in bond formation

$$\Delta H^\circ = +599 - 1031 = -432 \text{ kJ mol}^{-1}$$

(c)

$$\underset{(DH^\circ = 440)}{CH_3—H} + \underset{(DH^\circ = 243)}{Cl—Cl} \longrightarrow \underset{(DH^\circ = 352)}{CH_3—Cl} + \underset{(DH^\circ = 432)}{H—Cl}$$

$+683$ kJ mol^{-1} is required for bond cleavage

-784 kJ mol^{-1} is evolved in bond formation

$$\Delta H^\circ = +683 - 784 = -101 \text{ kJ mol}^{-1}$$

(d)

$$\underset{(DH^\circ = 440)}{CH_3—H} + \underset{(DH^\circ = 193)}{Br—Br} \longrightarrow \underset{(DH^\circ = 293)}{CH_3—Br} + \underset{(DH^\circ = 366)}{H—Br}$$

$+633$ kJ mol^{-1} is required for bond cleavage

-659 kJ mol^{-1} is evolved in bond formation

$$\Delta H^\circ = +633 - 659 = -26 \text{ kJ mol}^{-1}$$

(e) $CH_3{-}H \quad + \quad I{-}I \longrightarrow CH_3{-}I \quad + \quad H{-}I$

$(DH^\circ = 440) \quad (DH^\circ = 151) \qquad\qquad (DH^\circ = 240) \quad (DH^\circ = 298)$

$+591\ \text{kJ mol}^{-1}$ is required for bond cleavage; $-538\ \text{kJ mol}^{-1}$ is evolved in bond formation

$$\Delta H^\circ = +591 - 538 = +53\ \text{kJ mol}^{-1}$$

(f) $CH_3CH_2{-}H \quad + \quad Cl{-}Cl \longrightarrow CH_3CH_2{-}Cl \quad + \quad H{-}Cl$

$(DH^\circ = 421) \quad (DH^\circ = 243) \qquad\qquad (DH^\circ = 353) \quad (DH^\circ = 432)$

$+664\ \text{kJ mol}^{-1}$ is required for bond cleavage; $-785\ \text{kJ mol}^{-1}$ is evolved in bond formation

$$\Delta H^\circ = +664 - 785 = -121\ \text{kJ mol}^{-1}$$

(g) $(CH_3)_2CH{-}H \quad + \quad Cl{-}Cl \longrightarrow (CH_3)_2CH{-}Cl \quad + \quad H{-}Cl$

$(DH^\circ = 413) \quad (DH^\circ = 243) \qquad\qquad (DH^\circ = 355) \quad (DH^\circ = 432)$

$+656\ \text{kJ mol}^{-1}$ is required for bond cleavage; $-787\ \text{kJ mol}^{-1}$ is evolved in bond formation

$$\Delta H^\circ = +656 - 787 = -131\ \text{kJ mol}^{-1}$$

(h) $(CH_3)_3C{-}H \quad + \quad Cl{-}Cl \longrightarrow (CH_3)_3C{-}Cl \quad + \quad H{-}Cl$

$(DH^\circ = 400) \quad (DH^\circ = 243) \qquad\qquad (DH^\circ = 349) \quad (DH^\circ = 432)$

$+643\ \text{kJ mol}^{-1}$ is required for bond cleavage; $-781\ \text{kJ mol}^{-1}$ is evolved in bond formation

$$\Delta H^\circ = +643 - 781 = -138\ \text{kJ mol}^{-1}$$

10.2 $(CH_3)_2\dot{C}CH_2CH_3 \; > \; CH_3CH_2\dot{C}HCH_3 \; > \; (CH_3)_2CH\dot{C}H_2 \; > \; CH_3\cdot$

3° > **2°** > **1°** > **Methyl**

10.3 The compounds all have different boiling points. They could, therefore, be separated by careful fractional distillation. Or, because the compounds have different vapor pressures, they could easily be separated by gas chromatography. GC/MS (gas chromatography/mass spectrometry) could be used to separate the compounds as well as provide structural information from their mass spectra.

10.4 Their mass spectra would show contributions from the naturally occurring ^{35}Cl and ^{37}Cl isotopes. The natural abundance of ^{35}Cl is approximately 75% and that of ^{37}Cl is approximately 25%. Thus, for CH_3Cl, containing only one chlorine atom, there will be an $M^{+\cdot}$ peak and an $M^{+\cdot}+2$ peak in roughly a 3 : 1 (0.75 : 0.25) ratio of intensities. For CH_2Cl_2 there will be $M^{+\cdot}$, $M^{+\cdot}+2$, and $M^{+\cdot}+4$ peaks in roughly a 9 : 6 : 1 ratio, respectively. [The probability of a molecular ion $M^{+\cdot}$ with both chlorine atoms as ^{35}Cl is (.75)(.75) = .56, the probability of an $M^{+\cdot}+2$ ion from one ^{35}Cl and one ^{37}Cl is 2(.75)(.25) = .38, and the probability of an $M^{+\cdot}+4$ ion peak from both chlorine atoms as ^{37}Cl is (.25)(.25) = 0.06; thus, their ratio is 9 : 6 : 1.] For $CHCl_3$ there will be $M^{+\cdot}$, $M^{+\cdot}+2$, and $M^{+\cdot}+4$, and $M^{+\cdot}+6$ peaks in approximately a 27 : 27 : 9 : 1 ratio, respectively (based on a calculation by the same method). (This calculation does not take into account the contribution of ^{13}C, ^{2}H, and other isotopes, but these are much less abundant.)

10.5 The use of a large excess of chlorine allows all of the chlorinated methanes (CH_3Cl, CH_2Cl_2, and $CHCl_3$) to react further with chlorine.

10.6 ***Chain Initiation***

Step 1 $$\underset{(DH^\circ = 159)}{F{-}F} \longrightarrow 2\,F\cdot \qquad \Delta H^\circ = +159 \text{ kJ mol}^{-1}$$

Chain Propagation

Step 2 $$\underset{(DH^\circ = 440)}{CH_3{-}H} + F\cdot \longrightarrow CH_3\cdot + \underset{(DH^\circ = 570)}{H{-}F} \qquad \Delta H^\circ = -130 \text{ kJ mol}^{-1}$$

Step 3 $$CH_3\cdot + \underset{(DH^\circ = 159)}{F{-}F} \longrightarrow \underset{(DH^\circ = 461)}{CH_3{-}F} + F\cdot \qquad \Delta H^\circ = -302 \text{ kJ mol}^{-1}$$

Chain Termination

$$CH_3\cdot + F\cdot \longrightarrow \underset{(DH^\circ = 461)}{CH_3{-}F} \qquad \Delta H^\circ = -461 \text{ kJ mol}^{-1}$$

$$CH_3\cdot + CH_3\cdot \longrightarrow \underset{(DH^\circ = 378)}{CH_3{-}CH_3} \qquad \Delta H^\circ = -378 \text{ kJ mol}^{-1}$$

$$F\cdot + F\cdot \longrightarrow \underset{(DH^\circ = 159)}{F{-}F} \qquad \Delta H^\circ = -159 \text{ kJ mol}^{-1}$$

10.7

$$CH_3{-}H + \cancel{F\cdot} \longrightarrow \cancel{CH_3\cdot} + H{-}F \qquad \Delta H^\circ = -130 \text{ kJ mol}^{-1}$$

$$\cancel{CH_3\cdot} + F{-}F \longrightarrow CH_3{-}F + \cancel{F\cdot} \qquad \Delta H^\circ = -302 \text{ kJ mol}^{-1}$$

$$CH_3{-}H + F{-}F \longrightarrow CH_3{-}F + H{-}F \qquad \Delta H^\circ = -432 \text{ kJ mol}^{-1}$$

10.8 (a) Reactions (3), (5), and (6) should have $E_{act} = 0$ because these are gas-phase reactions in which small radicals combine to form molecules.

(b) Reactions (1), (2), and (4) should have $E_{act} > 0$ because in them covalent bonds are broken.

(c) Reactions (1) and (2) should have $E_{act} = \Delta H^\circ$ because in them bonds are broken homolytically but no bonds are formed.

10.9 (a)

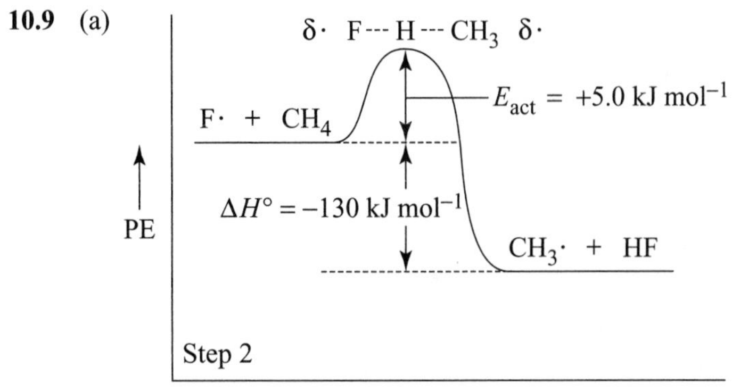

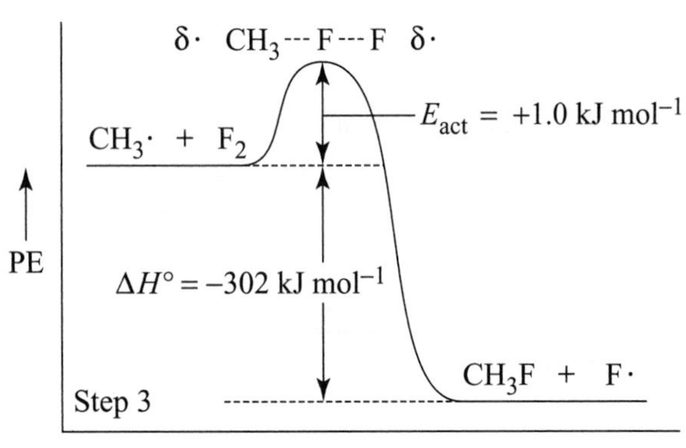

(b)

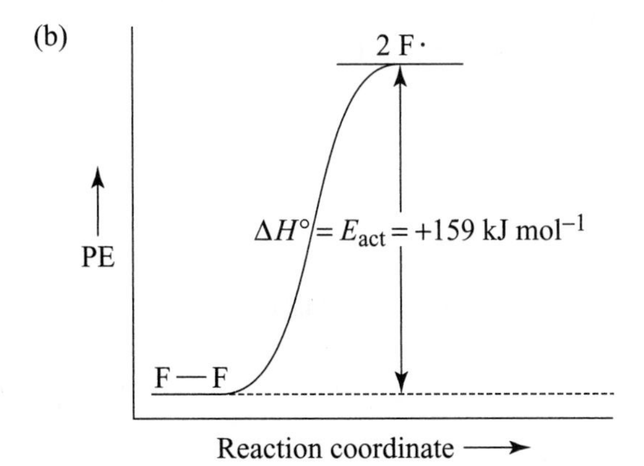

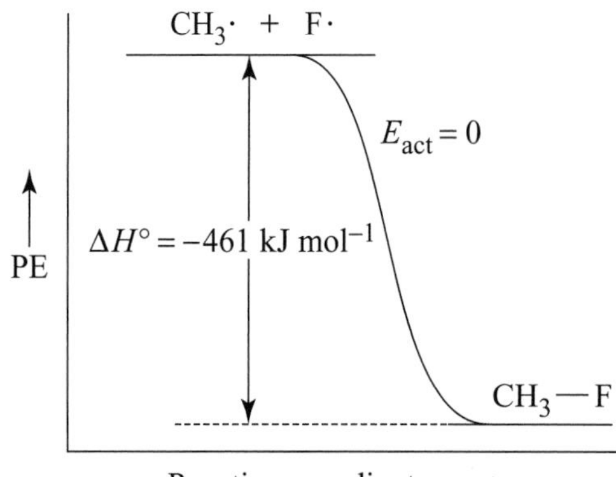

(c) Notice that this is the reverse of Step (2) in part (a)

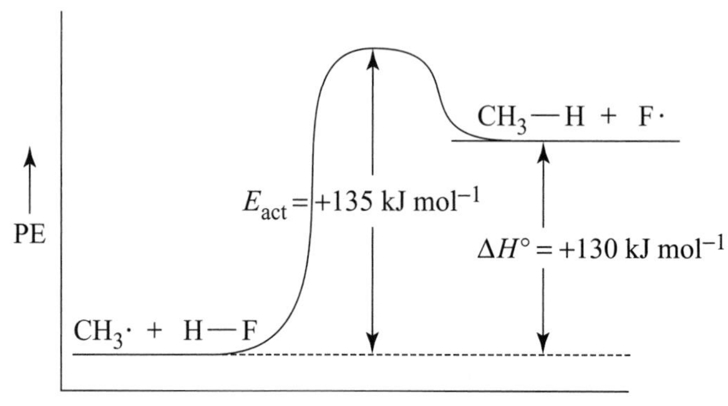

10.10 (a) $CH_3CH_2{-}H \; + \; Cl\cdot \longrightarrow CH_3CH_2\cdot \; + \; H{-}Cl$

$(DH^\circ = 421)$ $\qquad\qquad$ $(DH^\circ = 432)$

$$\Delta H^\circ = -432 + 421 = -11 \text{ kJ mol}^{-1}$$

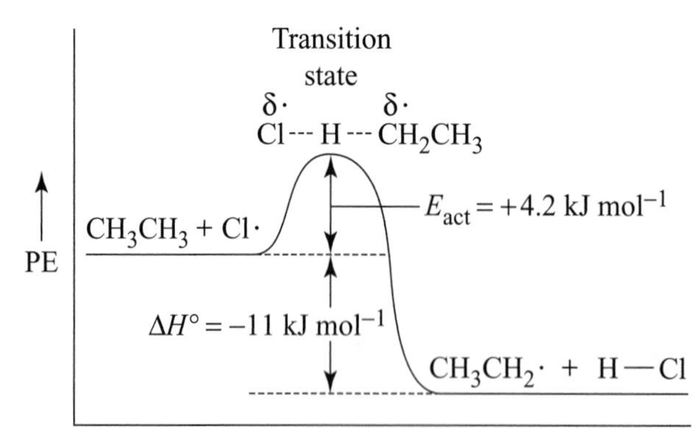

(b) The hydrogen abstraction step for ethane,

$$CH_3CH_2{-}H + Cl\cdot \longrightarrow CH_3CH_2\cdot + HCl \quad (E_{act} = 4.2\ kJ\ mol^{-1})$$

has a much lower energy of activation than the corresponding step for methane:

$$CH_3{-}H + Cl\cdot \longrightarrow CH_3\cdot + HCl \quad (E_{act} = 16\ kJ\ mol^{-1})$$

Therefore, ethyl radicals form much more rapidly in the mixture than methyl radicals, and this leads to the more rapid formation of ethyl chloride.

10.11

$$Cl_2 \xrightarrow[\text{or heat}]{h\nu} 2\ Cl\cdot$$

Step 2a $Cl\cdot$ + H (on CH of CH_3CHCl chain, Cl) ⟶ H—Cl + Cl (radical at CH)

Step 3a (radical at CH, Cl) + Cl—Cl ⟶ (Cl, Cl on same carbon) + $Cl\cdot$

1,1-dichloroethane

Step 2b $Cl\cdot$ + H (on CH_2, chain to Cl) ⟶ H—Cl + $\cdot$ (chain to Cl)

Step 3b $\cdot$ (chain to Cl) + Cl—Cl ⟶ Cl (chain) Cl + $Cl\cdot$

1,2-dichloroethane

10.12 (a) There is a total of eight hydrogen atoms in propane. There are six equivalent 1° hydrogen atoms, replacement of any one of which leads to propyl chloride, and there are two equivalent 2° hydrogen atoms, replacement of any one of which leads to isopropyl chloride.

(propane) + Cl_2 ⟶ (propyl chloride) Cl + (isopropyl chloride) Cl

If all the hydrogen atoms were equally reactive, we would expect to obtain 75% propyl chloride and 25% isopropyl chloride:

$$\%\ \text{Propyl chloride} = {}^{6}\!/_{8} \times 100 = 75\%$$

$$\%\ \text{Isopropyl chloride} = {}^{2}\!/_{8} \times 100 = 25\%$$

(b) Reasoning in the same way as in part (a), we would expect 90% isobutyl chloride and 10% *tert*-butyl chloride, if the hydrogen atoms were equally reactive.

(isobutane) + Cl_2 ⟶ (isobutyl chloride) Cl + (tert-butyl chloride) Cl

$$\%\ \text{Isobutyl chloride} = {}^{9}\!/_{10} \times 100 = 90\%$$

$$\%\ \textit{tert}\text{-Butyl chloride} = {}^{1}\!/_{10} \times 100 = 10\%$$

(c) In the case of propane (see Section 10.6), we actually get more than twice as much isopropyl chloride (55%) than we would expect if the 1° and 2° hydrogen atoms were equally reactive (25%). Clearly, then, 2° hydrogen atoms are more than twice as reactive as 1° hydrogen atoms.

In the case of isobutane, we get almost four times as much *tert*-butyl chloride (37%) as we would get (10%) if the 1° and 3° hydrogen atoms were equally reactive. The order of reactivity of the hydrogens then must be

$$3^\circ > 2^\circ > 1^\circ$$

10.13 The hydrogen atoms of these molecules are all equivalent. Replacing any one of them yields the same product.

cyclopropane + Cl_2 $\xrightarrow{hv}$ chlorocyclopropane (+ more highly chlorinated products)

cyclobutane + Cl_2 $\xrightarrow{hv}$ chlorocyclobutane (+ more highly chlorinated products)

We can minimize the amounts of more highly chlorinated products formed by using a large excess of the cyclopropane or cyclobutane. (And we can recover the unreacted cyclopropane or cyclobutane after the reaction is over.)

10.14 (a) cyclopentane (b) 2,2,3,3-tetramethylbutane

10.15 (a) (*S*)-2-Chloropentane $\xrightarrow[\text{light}]{Cl_2}$ (2*S*,4*S*)-2,4-Dichloropentane + (2*R*,4*S*)-2,4-Dichloropentane

(b) They are diastereomers. (They are stereoisomers, but they are not mirror images of each other.)

(c) No, (2*R*,4*S*)-2,4-dichloropentane is achiral because it is a meso compound. (It has a plane of symmetry passing through C3.)

(d) No, the achiral meso compound would not be optically active.

(e) Yes, by fractional distillation or by gas chromatography. (Diastereomers have different physical properties. Therefore, the two isomers would have different vapor pressures.)

(f and g) In addition to the (2*S*,4*S*)-2,4-dichloropentane and (2*R*,4*S*)-2,4-dichloropentane isomers described previously, we would also get (2*S*,3*S*)-2,3-dichloropentane, (2*S*,3*R*)-2,3-dichloropentane and the following:

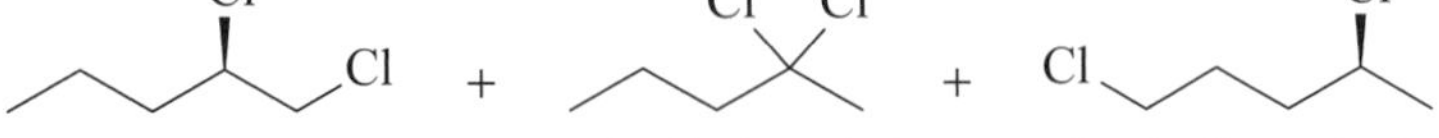

Cl, Cl + Cl Cl + Cl, Cl

(optically active) (optically inactive) (optically active)

10.16 (a) The only fractions that would contain chiral molecules (as enantiomers) would be those containing 1-chloro-2-methylbutane and the two diastereomers of 2-chloro-3-methylbutane. These fractions would not show optical activity, however, because they would contain racemic forms of the enantiomers.

(b) Yes, the fractions containing 1-chloro-2-methylbutane and the two containing the 2-chloro-3-methylbutane diastereomers.

(c) Yes, each fraction from the distillation could be identified on the basis of ^{1}H NMR spectroscopy. The signals related to the carbons where the chlorine atom is bonded would be sufficient to distinguish them. The protons at C1 of 1-chloro-2-methylbutane would be a doublet due to splitting from the single hydrogen at C2. There would be no proton signal for C2 of 2-chloro-2-methylbutane since there are no hydrogens bonded at C2 in this compound; however there would be a strong singlet for the six hydrogens of the geminal methyl groups. The proton signal at C2 of 2-chloro-3-methylbutane would approximately be a quintet, due to combined splitting from the three hydrogens at C1 and the single hydrogen at C3. The protons at C1 of 1-chloro-3-methylbutane would be a triplet due to splitting by the two hydrogens at C2.

10.17 Head-to-tail polymerization leads to a more stable radical on the growing polymer chain. In head-to-tail coupling, the radical is 2° (actually 2° benzylic, and as we shall see in Section 15.12A this makes it even more stable). In head-to-head coupling, the radical is 1°.

10.18 (a)

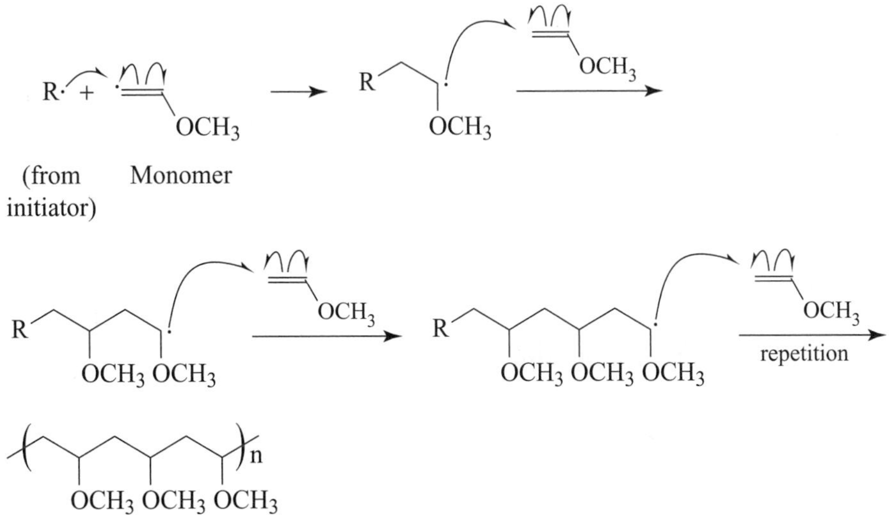

R· + OCH_3 ⟶ R OCH_3 $\xrightarrow{OCH_3}$

(from initiator) Monomer

R OCH_3 OCH_3 $\xrightarrow{OCH_3}$ R OCH_3 OCH_3 OCH_3 $\xrightarrow[\text{repetition}]{OCH_3}$

OCH_3 OCH_3 OCH_3 n

(b)

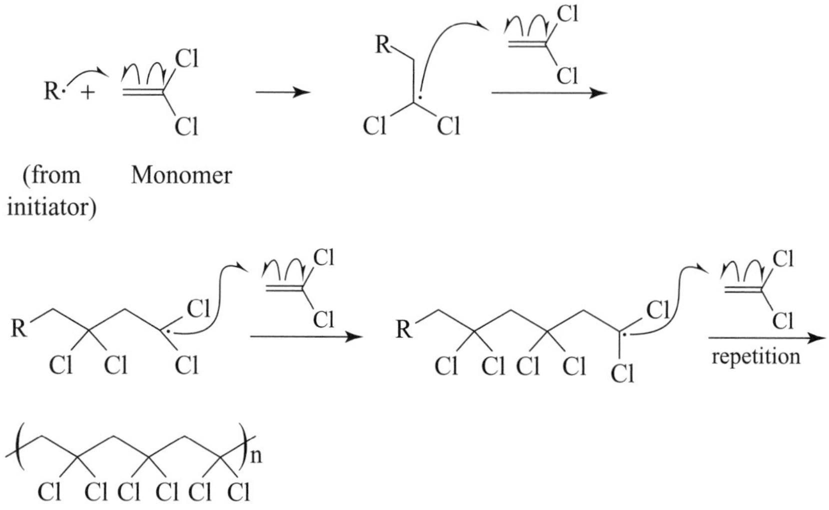

R· + (from initiator) + Monomer (Cl₂C=CH₂) → ... → repetition

10.19 In the cationic polymerization of isobutylene (see text), the growing polymer chain has a stable 3° carbocation at the end. In the cationic polymerization of ethene, for example, the intermediates would be much less stable 1° cations.

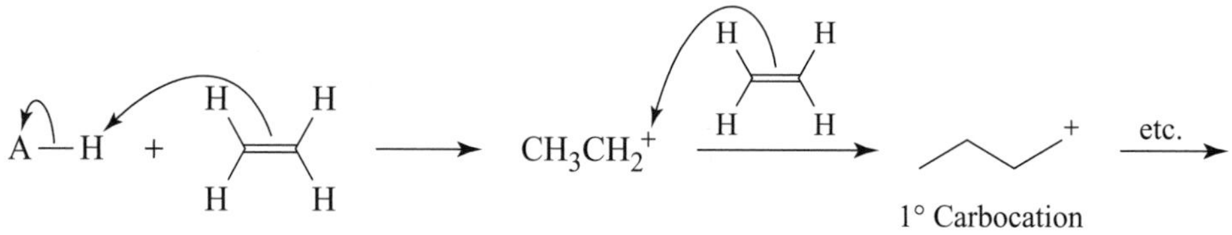

A—H + $H_2C=CH_2$ → $CH_3CH_2^+$ → 1° Carbocation → etc.

With vinyl chloride and acrylonitrile, the cations at the end of the growing chain would be destabilized by electron-withdrawing groups.

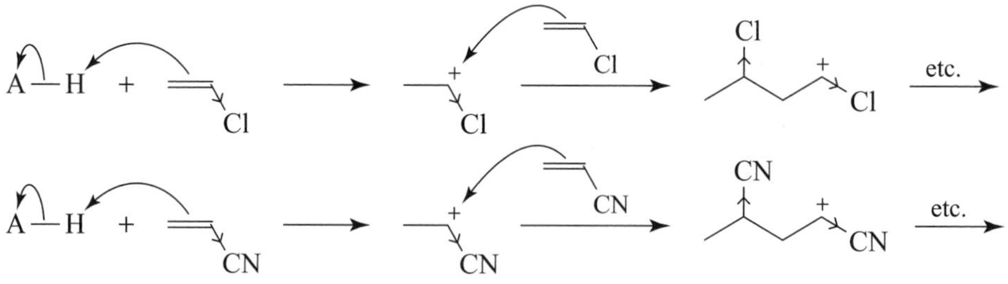

Radical Mechanisms and Properties

10.20 (1) $Br_2 \xrightarrow{h\nu} 2\ Br^{\cdot}$

(2) $Br^{\cdot}$ + toluene → benzyl radical + HBr

(3) [benzyl radical] + Br_2 ⟶ [benzyl bromide] + $Br^{\cdot}$

then 2,3 2,3 etc.

10.21 For formation of [2-bromo-2,3-dimethylbutane], the rate-determining step is:

$Br^{\cdot}$ + [2,3-dimethylbutane, H] ⟶ $[\ \delta^{\cdot} \cdots H \cdots \delta^{\cdot} Br\ ]^{\ddagger}$ ⟶ [3° radical]

$A^{\ddagger}$ A

For formation of [1-bromo-2,3-dimethylbutane], the rate-determining step is:

$Br^{\cdot}$ + [2,3-dimethylbutane, H] ⟶ $[\ \delta^{\cdot} \cdots H \cdots \delta^{\cdot} Br\ ]^{\ddagger}$ ⟶ [1° radical]

$B^{\ddagger}$ B

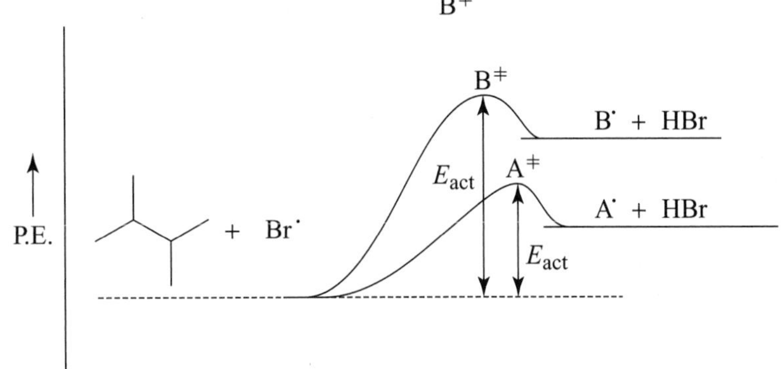

Reaction coordinate

$A^{\bullet}$ is a 3° alkyl radical and more stable than $B^{\bullet}$, a 1° alkyl radical, a difference anticipated by the relative energies of the transition states.

As indicated by the potential energy diagram, the activation energy for the formation of $A^{\ddagger}$ is less than that for the formation of $B^{\ddagger}$. The lower energy of $A^{\ddagger}$ means that a greater fraction of bromine atom-alkane collisions will lead to $A^{\bullet}$ rather than to $B^{\bullet}$.

Note there is a statistical advantage to the formation of $B^{\bullet}$ (6:1) but this is outweighed by the inherently greater reactivity of a 3° hydrogen.

10.22 [chlorocyclopentane] and [iodoethane] are the only monosubstitution products which can be formed from cyclopentane and ethane, respectively. (However, direct iodination of alkane is not a feasible reaction, practically speaking.)

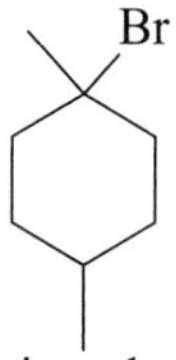

Br and Br would be formed in amounts much larger than the isomeric alternatives due to the highly selective nature of bromine.

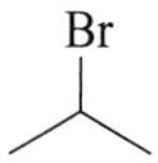

Formation of Br and Cl would be acompanied by considerable amounts of isomeric byproducts in each case.

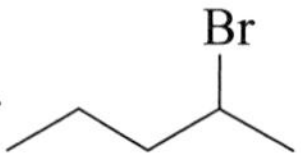

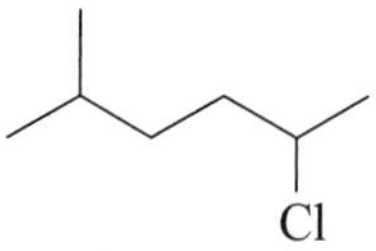

10.23 ***Chain-Initiating Step***

$$Cl_2 \xrightarrow[\text{light}]{\text{heat, } h\nu} 2\ Cl\cdot$$

Chain-Propagating Steps

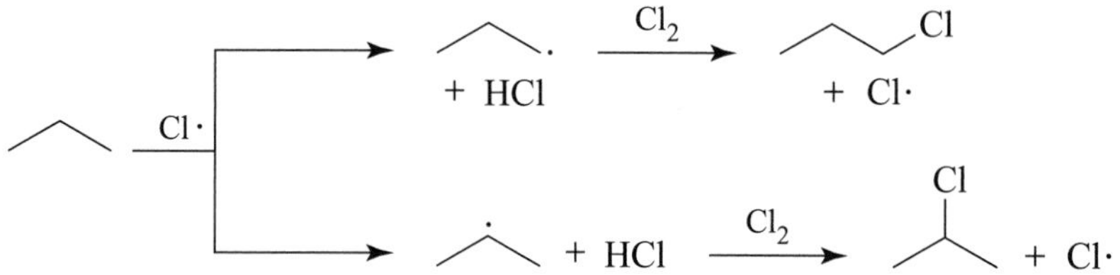

Cl· → + HCl $\xrightarrow{Cl_2}$ Cl + Cl·

→ + HCl $\xrightarrow{Cl_2}$ Cl + Cl·

10.24 (a) Three

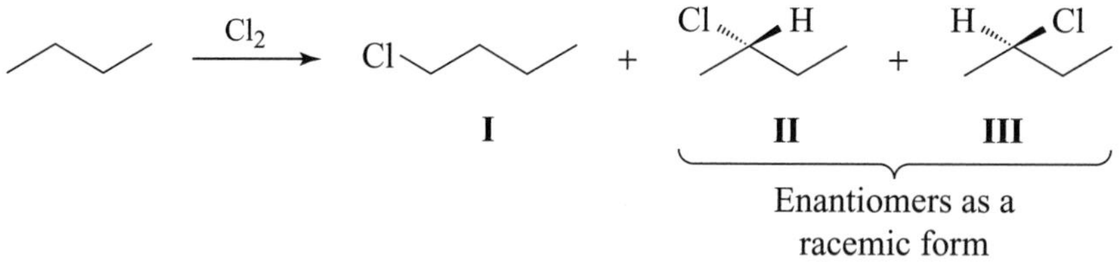

$\xrightarrow{Cl_2}$ Cl **I** + Cl H **II** + H Cl **III**

Enantiomers as a racemic form

(b) Only two: one fraction containing I, and another fraction containing the enantiomers II and III as a racemic form. (The enantiomers, having the same boiling points, would distill in the same fraction.)

(c) Both of them.

(d) The fraction containing the enantiomers.

(e) In the ^{1}H spectrum for 1-chlorobutane the signal furthest downfield would be that for $—CH_2Cl$; it would be a triplet. The corresponding signal for —CH(Cl)— in either enantiomer of 2-chlorobutane (also furthest downfield) would be an approximate sextet.

The DEPT spectra for 1-chlorobutane would identify one CH_3 group and three CH_2 groups; for 2-chlorobutane, two (non equivalent) CH_3 groups, one CH_2 group, and one CH group would be specified.

(f) Molecular ions from both 1-chlorobutane and the 2-chlorobutane enantiomers would be present (but probably of different intensities). $M^{\ddagger}$ +2 peaks would also be present. Both compounds would likely undergo C—Cl bond cleavage to yield $C_4H_9^+$ cations. The mass spectrum of 1-chlorobutane would probably show loss of a propyl radical by C—C bond cleavage adjacent to the chlorine, resulting in an *m/z* 49 peak for CH_2Cl^+ (and *m/z* 51 from ^{37}Cl). Similar fragmentation in 2-chlorobutane would produce an *m/z* 63 peak for CH_3CHCl^+ (and *m/z* 65).

10.25 (a) Five

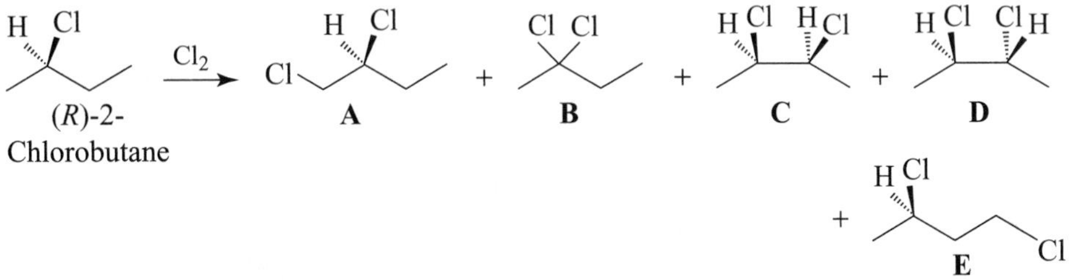

(*R*)-2-Chlorobutane $\xrightarrow{Cl_2}$ **A** + **B** + **C** + **D** + **E**

(b) Five. None of the fractions would be a racemic form.

(c) The fractions containing **A**, **D**, and **E**. The fractions containing **B** and **C** would be optically inactive. (**B** contains no chirality center and **C** is a meso compound.)

10.26 (a) Oxygen-oxygen bonds are especially weak, that is,

HO—OH $DH^\circ = 214\ \text{kJ mol}^{-1}$

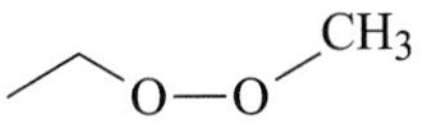

$CH_3CH_2O—OCH_3$ $DH^\circ = 184\ \text{kJ mol}^{-1}$

This means that a peroxide will dissociate into radicals at a relatively low temperature.

$$RO—OR \xrightarrow{100\text{-}200\,^\circ C} 2\ RO\cdot$$

Oxygen-hydrogen single bonds, on the other hand, are very strong. (For HO—H, $DH^\circ = 499\ \text{kJ mol}^{-1}$.) This means that reactions like the following will be highly exothermic.

$$RO\cdot + R—H \longrightarrow RO—H + R\cdot$$

(b) *Step 1* $(CH_3)_3CO—OC(CH_3)_3 \xrightarrow{heat} 2\ (CH_3)_3CO\cdot$

Step 2 $(CH_3)_3CO\cdot + R—H \longrightarrow (CH_3)_3COH + R\cdot$

Chain Initiation

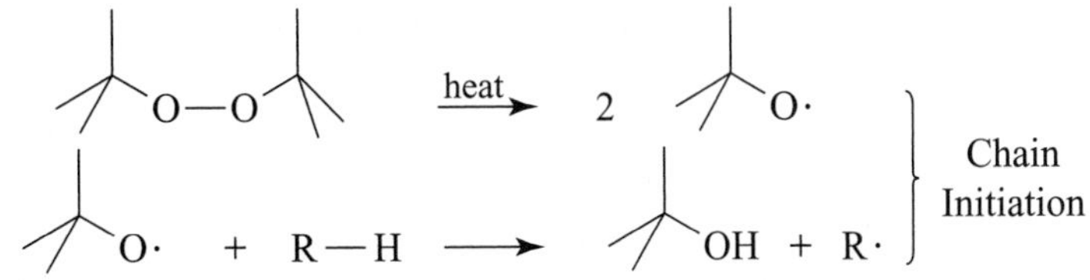

Step 3 $R\cdot + Cl—Cl \longrightarrow R—Cl + Cl\cdot$

Step 4 $Cl\cdot + R—H \longrightarrow H—Cl + R\cdot$

Chain Propagation

10.27  (3°) > (2°) > (1°) ~ (1°)

10.28 *R—CH_3 bond dissociation to form a primary radical:*

$$BDE[CH_3CH_2CH_2CH_2{-}CH_3] = 80.9 + 147 - (-146.8) = 375\ kJ/mol$$

R—CH_3 bond dissociation to form a secondary radical:

$$BDE[CH_3CH_2CH(CH_3){-}CH_3] = 69 + 147 - (-153.7) = 370\ kJ/mol$$

R—CH_3 bond dissociation to form a tertiary radical:

$$BDE[(CH_3)_3C{-}CH_3] = 48 + 147 - (-167.9) = 363\ kJ/mol$$

These calculations are consistent with the relative stability of radicals presented in Section 10.2B based on C—H bond dissociation energy comparisons.

10.29 Single-barbed arrows show conversion of the enediyne system to a 1,4-benzenoid diradical via the Bergman cycloaromatization. Each alkyne contributes one electron from a pi bond to the new sigma bond. The remaining electrons in each of the pi bonds become the unpaired electrons of the 1,4-benzenoid diradical. The diradical is a highly reactive intermediate that reacts further to abstract hydrogen atoms from the DNA sugar-phosphate backbone. The new radicals formed on the DNA lead to bond fragmentation along the backbone and to double-stranded cleavage of the DNA.

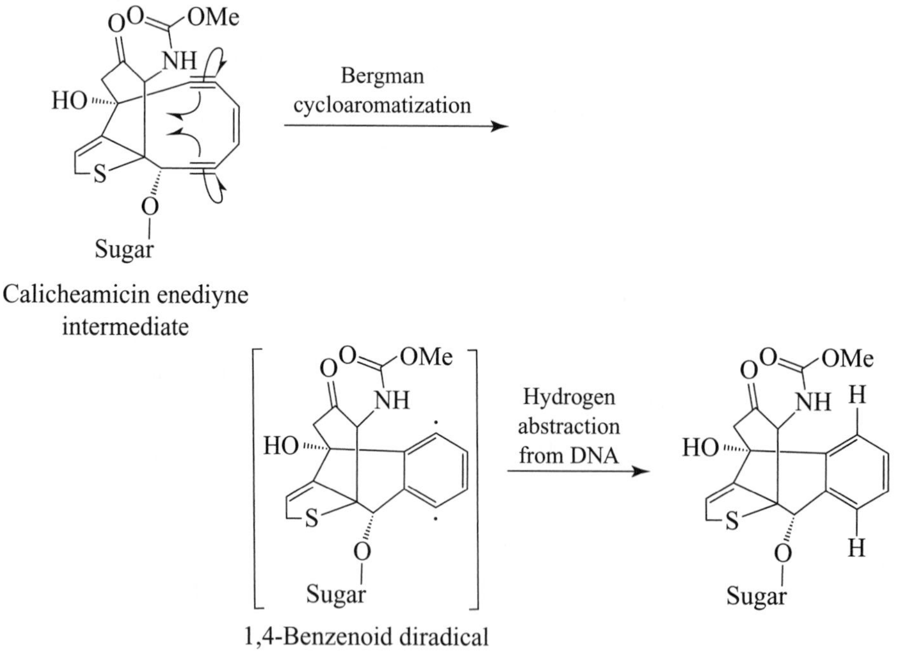

Bergman cycloaromatization

Calicheamicin enediyne intermediate

Hydrogen abstraction from DNA

1,4-Benzenoid diradical

10.30

	$DH°$ (kJ mol^{-1})
$CH_2=CH—H$	465
$(CH_3)_2CH—H$	413
$CH_2=CHCH_2—H$	369

(a) It is relatively difficult to effect radical halogenation at H_a because of the large $DH°$ for dissociation of a bond to a vinylic hydrogen.

(b) Substitution of H_b occurs more readily than H_c because $DH°$ for the generation of an allylic radical is significantly smaller than that for the formation of a simple 2° radical.

10.31 (1) O, O, Br, heat, O, O$^\cdot$ + Br$^\cdot$

(2) O, O$^\cdot$ → $\cdot$ + CO_2

(3) O, O, Br + → Br + O, O$^\cdot$

then 2,3 2,3 etc.

Synthesis

10.32 (a) CH_3CH_3 $\xrightarrow[\text{heat, light}]{Br_2}$ Br $\xrightarrow[(S_N2)]{NaI}$ I

(b) Br [from part (a)] $\xrightarrow[(S_N2)]{NaOH}$ OH $\xrightarrow[(-H_2)]{NaH}$ O^-Na^+ → (Br) → O

(c) $\xrightarrow[\text{heat, light}]{Br_2}$ Br $\xrightarrow[\text{OH, heat (E2)}]{\text{OK}}$

(d) $\xrightarrow[\text{heat, light}]{Br_2}$ Br $\xrightarrow[\text{OH, heat (E2)}]{\text{ONa}}$

HBr, ROOR, heat, light

Br

(by anti-Markovnikov addition)

(e) CH_4 $\xrightarrow[\text{heat, light}]{Br_2}$ CH_3Br

H—≡—H $\xrightarrow[\text{liq. } NH_3 \text{ 2) } CH_3Br]{\text{1) } NaNH_2}$ H—≡— $\xrightarrow[\text{liq. } NH_3 \text{ 2) } CH_3Br]{\text{1) } NaNH_2}$ —≡—

(f) H—≡—H $\xrightarrow[\text{liq. } NH_3 \text{ 2) } \text{Br [from part (a)]}]{\text{1) } NaNH_2}$ —≡—H $\xrightarrow[\text{or } H_2\text{, Lindlar's catalyst}]{H_2\text{, } Ni_2B}$

HA, H_2O

OH

(g) Br [from part (a)] + $Na^+ \, ^-N{=}\overset{+}{N}{=}N^-$ $\xrightarrow[S_N2]{}$ $—N{=}\overset{+}{N}{=}N^-$

10.33 (a) $\xrightarrow[h\nu]{Br_2}$ Br $\xrightarrow[\text{OH}]{\text{ONa}}$ $\xrightarrow[H_2SO_4]{H_2O}$ OH

(b) [from (a)] $\xrightarrow[\text{ROOR}]{\text{HBr}}$ Br

(c) [from (a)] $\xrightarrow[\text{(2) } H_2O_2\text{, } OH^-]{\text{(1) } BH_3 : THF}$ OH

(d) 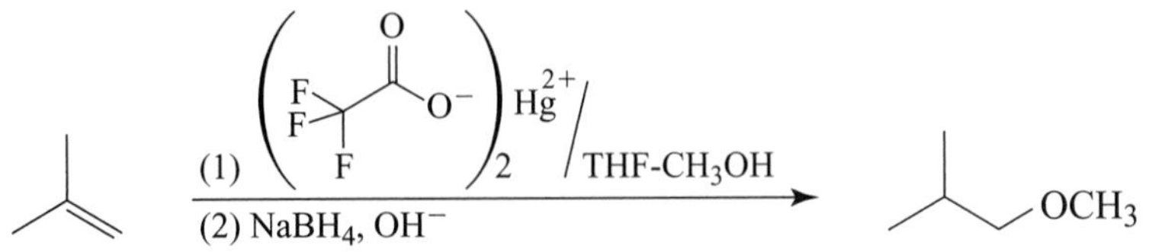 [from (a)] (1) $(CF_3CO_2^-)_2Hg^{2+}$ / THF-CH_3OH; (2) $NaBH_4$, OH^- → OCH_3

(e) 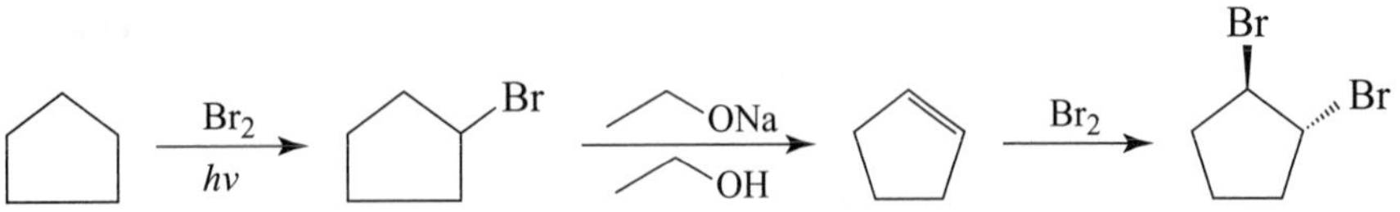 $\xrightarrow[hv]{Br_2}$ Br $\xrightarrow[\text{EtOH}]{\text{EtONa}}$ $\xrightarrow{Br_2}$ Br, Br

+ enantiomer

(f) 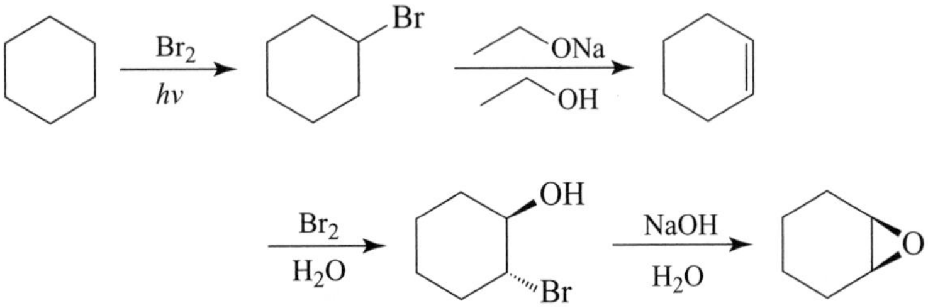 $\xrightarrow[hv]{Br_2}$ Br $\xrightarrow[\text{EtOH}]{\text{EtONa}}$

$\xrightarrow[H_2O]{Br_2}$ OH, Br $\xrightarrow[H_2O]{NaOH}$ O

+ enantiomer

10.34 (1) 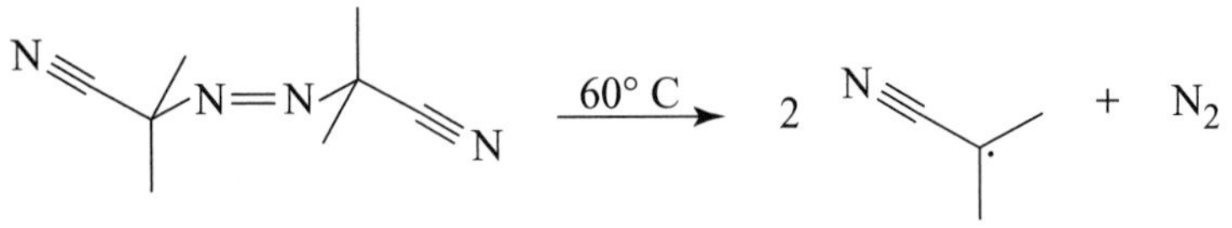 N≡C–C(CH₃)₂–N=N–C(CH₃)₂–C≡N $\xrightarrow{60°\ C}$ 2 N≡C–Ċ(CH₃)₂ + N_2

AIBN

(2) 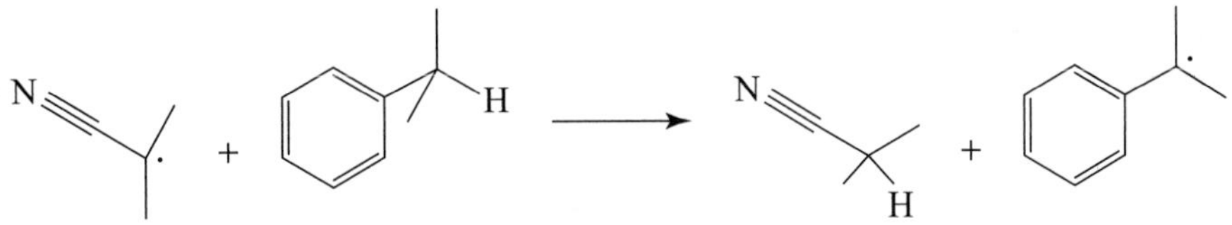 N≡C–Ċ(CH₃)₂ + PhC(CH₃)₂–H → N≡C–C(CH₃)₂–H + PhĊ(CH₃)₂

(3) 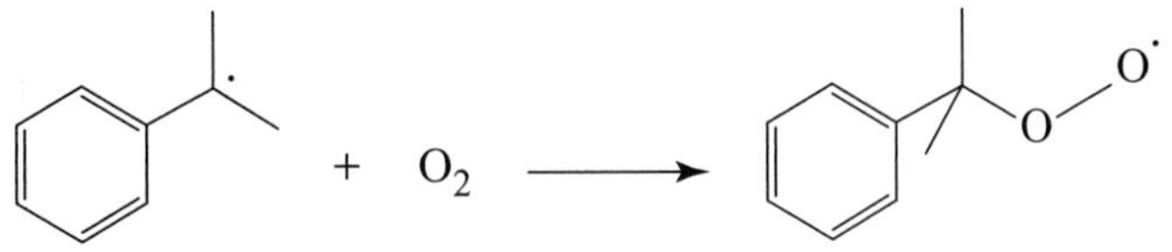 PhĊ(CH₃)₂ + O_2 → PhC(CH₃)₂–O–O·

(4) 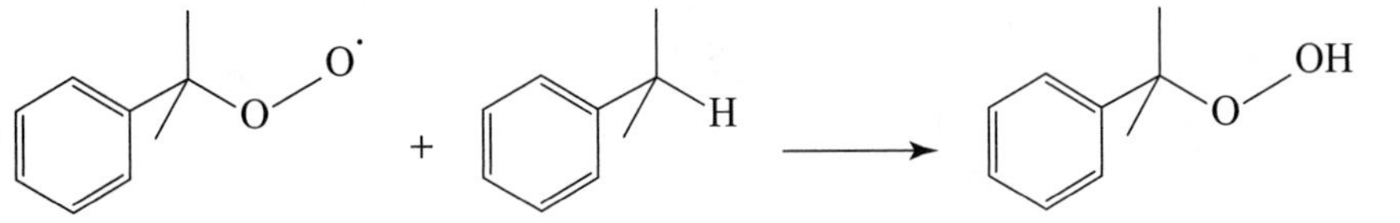 PhC(CH₃)₂–O–O· + PhC(CH₃)₂–H → PhC(CH₃)₂–O–OH + PhĊ(CH₃)₂

then 3,4 3,4 etc.

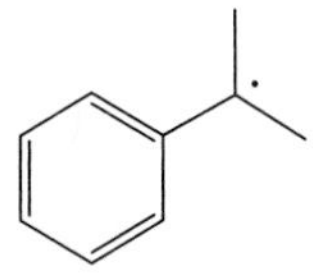

10.35 Besides direct H· abstraction from C5 there would be many H· abstractions from the three methyl groups, leading to:

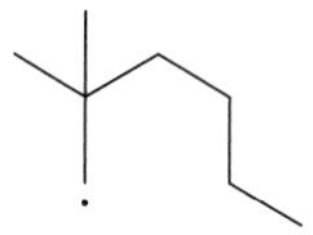

Any of these radicals could then, besides directly attacking chlorine, intramolecularly abstract H· from C5 (analogous to the "back biting" that explains branching during alkene radical polymerization).

10.36 (1) Ph–C(=O)–O–O–C(=O)–Ph $\xrightarrow{\text{heat}}$ 2 Ph–C(=O)–O·

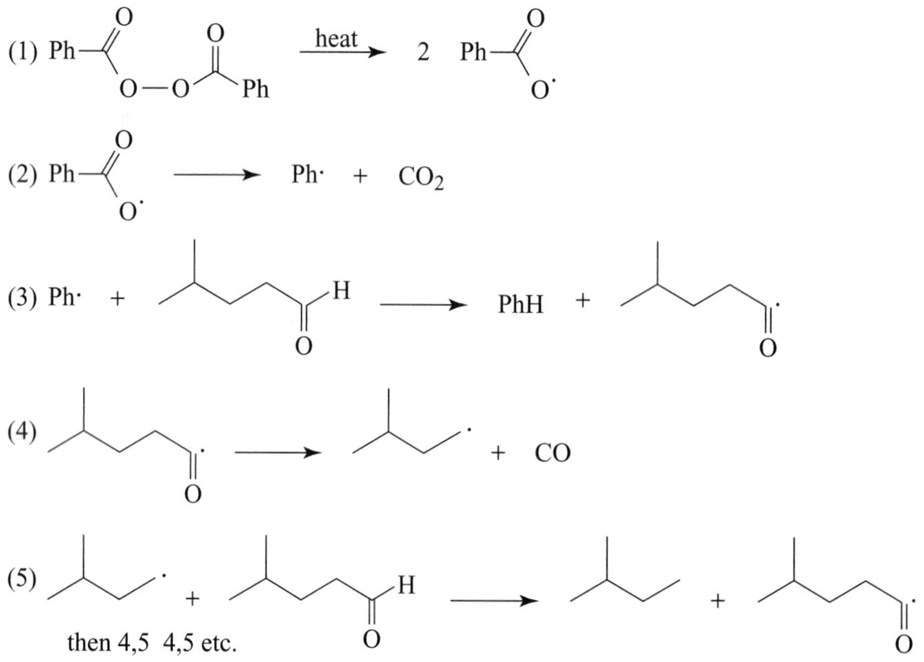

(2) Ph–C(=O)–O· ⟶ Ph· + CO_2

(3) Ph· + [aldehyde, C(=O)H] ⟶ PhH + [acyl radical, C(=O)·]

(4) [acyl radical, C(=O)·] ⟶ [alkyl radical] + CO

(5) [alkyl radical] + [aldehyde, C(=O)H] ⟶ [alkane] + [acyl radical, C(=O)·]

then 4,5 4,5 etc.

10.37 HO–C(CH$_3$)$_3$ + HO· ⟶ HO–C(CH$_3$)$_2$–CH$_2$· + HOH

dimerization ⟶ HO–C(CH$_3$)$_2$–CH$_2$CH$_2$–C(CH$_3$)$_2$–OH

X

10.38 (1) N≡C–C·(CH$_3$)$_2$ (from AIBN) + Bu_3SnH ⟶ N≡C–CH(CH$_3$)$_2$ + Bu_3Sn·

(2) Bu_3Sn· + PhCH$_2$I ⟶ Bu_3SnI + PhCH$_2$·

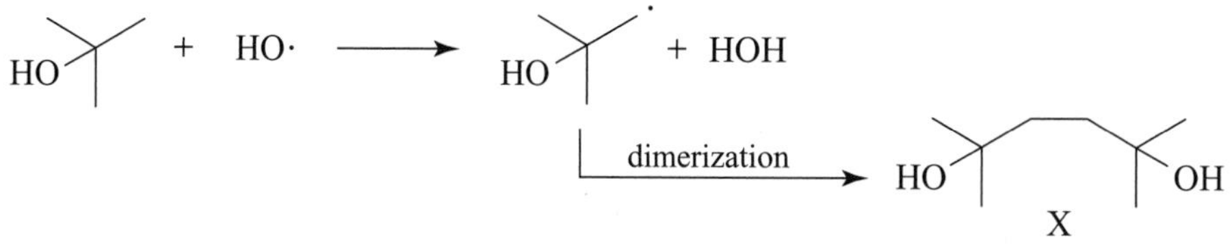

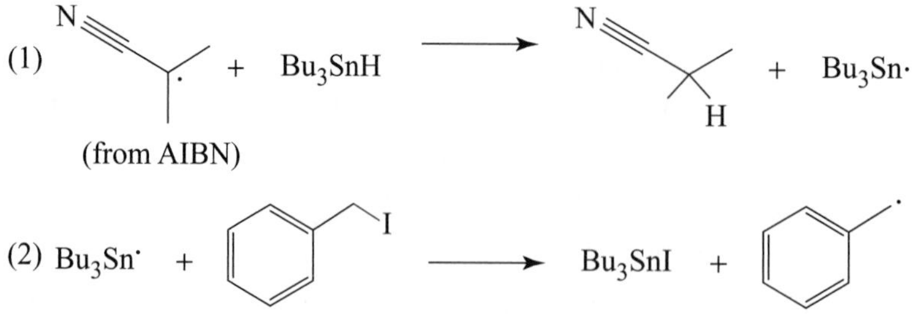

(3) + Bu_3SnH ⟶ + $Bu_3Sn^{\cdot}$

then 2,3 2,3 etc.

10.39 $PhC(=O)O{-}OC(=O)Ph$ ⟶ 2 $PhC(=O)O^{\cdot}$

$PhC(=O)O^{\cdot}$ ⟶ $Ph^{\cdot}$ + CO_2

$Ph^{\cdot}$ + Bu_3SnH ⟶ PhH + $Bu_3Sn^{\cdot}$

$Bu_3Sn^{\cdot}$ + Br ⟶ Bu_3SnBr +

⟶

+ Bu_3SnH ⟶ + $Bu_3Sn^{\cdot}$

+ Bu_3SnH ⟶ + $Bu_3Sn^{\cdot}$

10.40 Unpaired electron density in the methyl radical is localized solely above and below the carbon atom, in the region corresponding to the *p* orbital of the approximately sp^2-hybridized carbon atom. The ethyl radical shows some unpaired electron density at the adjacent hydrogen atoms, especially the hydrogen atom that in the conformation shown has its H—C sigma bond aligned parallel to the unpaired electron density of the *p* orbital of the radical. The larger size of the spin density lobe of the hydrogen with its H—C bond parallel to the *p* orbital of the radical indicates hyperconjugation with the radical. This effect is even more pronounced in the *tert*-butyl radical, where three hydrogen atoms with H—C sigma bonds parallel to the radical *p* orbital (two hydrogens above the carbon plane and one below in the conformation shown) have larger unpaired electron density volumes than the other hydrogen atoms.

10.41 The sequence of molecular orbitals in O_2 is $\sigma 1s$ (HOMO-7), $\sigma^* 1s$ (HOMO-6), $\sigma 2s$ (HOMO-5), $\sigma^* 2s$ (HOMO-4), $\pi 2p_y$ (HOMO-3), $\pi 2p_z$ (HOMO-2), $\sigma 2p_x$ (HOMO-1), $\pi^* 2p_y$ (HOMO), $\pi^* 2p_z$ (LUMO). Therefore (a) HOMO-3 and HOMO-2 represent bonding pi molecular orbitals, (b) HOMO-1 is a bonding sigma molecular orbital comprised of overlap of the p_x orbitals on each oxygen, and (c) the HOMO and LUMO represent the antibonding pi molecular orbital counterparts to the bonding pi molecular orbitals represented by HOMO-3 and HOMO-2. Note that the orbitals in s and p orbitals in O_2 are not hybridized. A diagram of the orbitals and their respective energy levels is shown below.

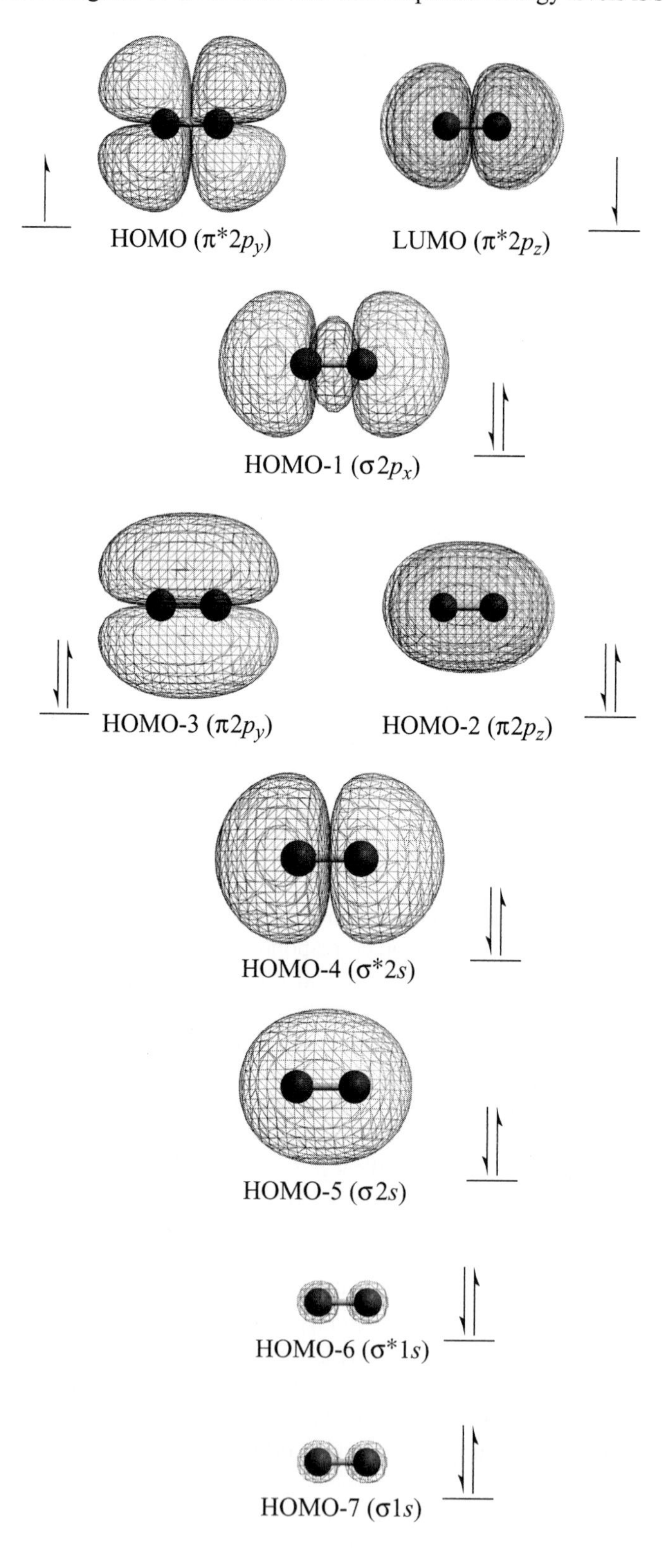

QUIZ

Use the single-bond dissociation energies of Table 10.1 (page 202):

10.1 On the basis of Table 10.1, what is the order of decreasing stability of the radicals,

$HC{\equiv}C\cdot \quad CH_2{=}CH\cdot \quad CH_2{=}CHCH_2\cdot$?

(a) $HC{\equiv}C\cdot > CH_2{=}CH\cdot > CH_2{=}CHCH_2\cdot$

(b) $CH_2{=}CH\cdot > HC{\equiv}C\cdot > CH_2{=}CHCH_2\cdot$

(c) $CH_2{=}CHCH_2\cdot > HC{\equiv}C\cdot > CH_2{=}CH\cdot$

(d) $CH_2{=}CHCH_2\cdot > CH_2{=}CH\cdot > HC{\equiv}C\cdot$

(e) $CH_2{=}CH\cdot > CH_2{=}CHCH_2\cdot > HC{\equiv}C\cdot$

10.2 In the radical chlorination of methane, one propagation step is shown as

$$Cl\cdot + CH_4 \longrightarrow HCl + \cdot CH_3$$

Why do we eliminate the possibility that this step goes as shown below?

$$Cl\cdot + CH_4 \longrightarrow CH_3Cl + H\cdot$$

(a) Because in the next propagation step, H· would have to react with Cl_2 to form Cl· and HCl; this reaction is not feasible.

(b) Because this alternative step has a more endothermic ΔH° than the first.

(c) Because free hydrogen atoms cannot exist.

(d) Because this alternative step is not consistent with the high photochemical efficiency of this reaction.

10.3 Pure (*S*)-$CH_3CH_2CHBrCH_3$ is subjected to monobromination to form several isomers of $C_4H_8Br_2$. Which of the following is *not* produced?

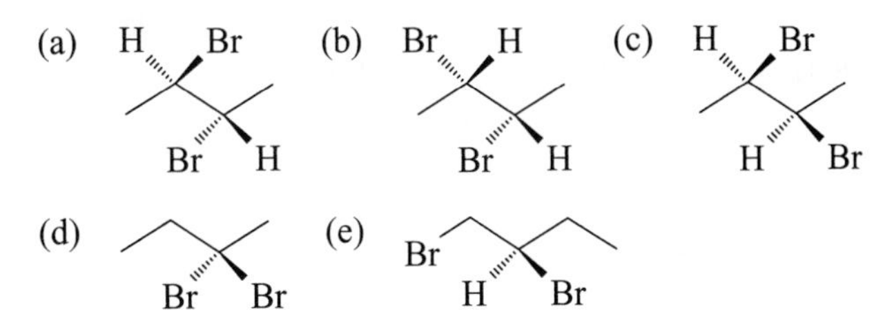

10.4 Using the data of Table 10.1, calculate the heat of reaction, ΔH°, of the reaction,

$$CH_3CH_3 + Br_2 \longrightarrow \text{/\\Br} + HBr$$

(a) 47 kJ mol^{-1} (b) −47 kJ mol^{-1} (c) 1275 kJ mol^{-1}

(d) −1275 kJ mol^{-1} (e) −157 kJ mol^{-1}

Table 10.1 Single-bond homolytic dissociation energies DH° at 25°C

$$A:B \longrightarrow A\cdot + B\cdot$$

Compound	kJ mol^{-1}	Compound	kJ mol^{-1}
H—H	436	$(CH_3)_2CH$—Br	298
D—D	443	$(CH_3)_2CH$—I	222
F—F	159	$(CH_3)_2CH$—OH	402
Cl—Cl	243	$(CH_3)_2CH$—OCH_3	359
Br—Br	193	$(CH_3)_2CHCH_2$—H	422
I—I	151	$(CH_3)_3C$—H	400
H—F	570	$(CH_3)_3C$—Cl	349
H—Cl	432	$(CH_3)_3C$—Br	292
H—Br	366	$(CH_3)_3C$—I	227
H—I	298	$(CH_3)_3C$—OH	400
CH_3—H	440	$(CH_3)_3C$—OCH_3	348
CH_3—F	461	$C_6H_5CH_2$—H	375
CH_3—Cl	352	$CH_2{=}CHCH_2$—H	369
CH_3—Br	293	$CH_2{=}CH$—H	465
CH_3—I	240	C_6H_5—H	474
CH_3—OH	387	$HC{\equiv}C$—H	547
CH_3—OCH_3	348	CH_3—CH_3	378
CH_3CH_2—H	421	CH_3CH_2—CH_3	371
CH_3CH_2—F	444	$CH_3CH_2CH_2$—CH_3	374
CH_3CH_2—Cl	353	CH_3CH_2—CH_2CH_3	343
CH_3CH_2—Br	295	$(CH_3)_2CH$—CH_3	371
CH_3CH_2—I	233	$(CH_3)_3C$—CH_3	363
CH_3CH_2—OH	393	HO—H	499
CH_3CH_2—OCH_3	352	HOO—H	356
$CH_3CH_2CH_2$—H	423	HO—OH	214
$CH_3CH_2CH_2$—F	444	$(CH_3)_3CO$—$OC(CH_3)_3$	157
$CH_3CH_2CH_2$—Cl	354		
$CH_3CH_2CH_2$—Br	294		
$CH_3CH_2CH_2$—I	239	$C_6H_5C(=O)O$—$OC(=O)C_6H_5$	139
$CH_3CH_2CH_2$—OH	395	CH_3CH_2O—OCH_3	184
$CH_3CH_2CH_2$—OCH_3	355	CH_3CH_2O—H	431
$(CH_3)_2CH$—H	413		
$(CH_3)_2CH$—F	439		
$(CH_3)_2CH$—Cl	355	$CH_3C(=O)$—H	364

10.5 Which gas-phase reaction would have $E_{act} = 0$?

(a) $CH_3\cdot$ + ⟶ CH_4 +

(b) $CH_3\cdot$ + CH_3CH_3 ⟶ CH_4 + $CH_3CH_2\cdot$

(c) $CH_3CH_2\cdot$ + $CH_3CH_2\cdot$ ⟶

(d) $Br\cdot$ + H—Cl ⟶ H—Br + $Cl\cdot$

(e) $Br\cdot$ + H—I ⟶ H—Br + $I\cdot$

10.6 What is the most stable radical that would be formed in the following reaction?

10.7 The reaction of 2-methylbutane with chlorine would yield a total of _________ different monochloro products (including stereoisomers).

10.8 For which reaction would the transition state most resemble the products?

(a) CH_4 + $F\cdot$ ⟶ $CH_3\cdot$ + HF

(b) CH_4 + $Cl\cdot$ ⟶ $CH_3\cdot$ + HCl

(c) CH_4 + $Br\cdot$ ⟶ $CH_3\cdot$ + HBr

(d) CH_4 + $I\cdot$ ⟶ $CH_3\cdot$ + HI

11 ALCOHOLS AND ETHERS

SOLUTIONS TO PROBLEMS

Note: A mixture of bond-line and condensed structural formulas is used for solutions in this chapter so as to aid your facility in using both types.

11.1 These names mix two systems of nomenclature (functional class and substitutive; see Section 4.3F). The proper names are: isopropyl alcohol (functional class) or 2-propanol (substitutive), and *tert*-butyl alcohol (functional class) or 2-methyl-2-propanol (substitutive). Names with mixed systems of nomenclature should not be used.

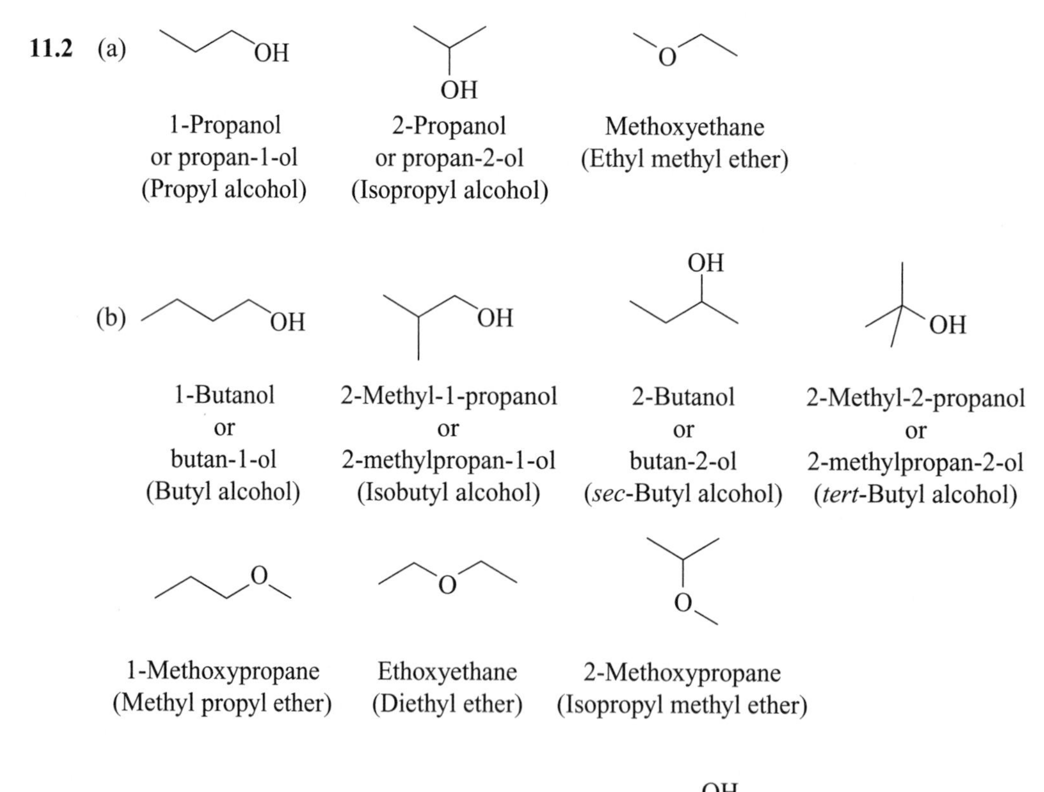

11.3 (a) OH (b) OH (c) OH

11.4 A rearrangement takes place.

(a) 1,2-methanide shift

$:\ddot{O}H_2$ $:\ddot{O}H_2$

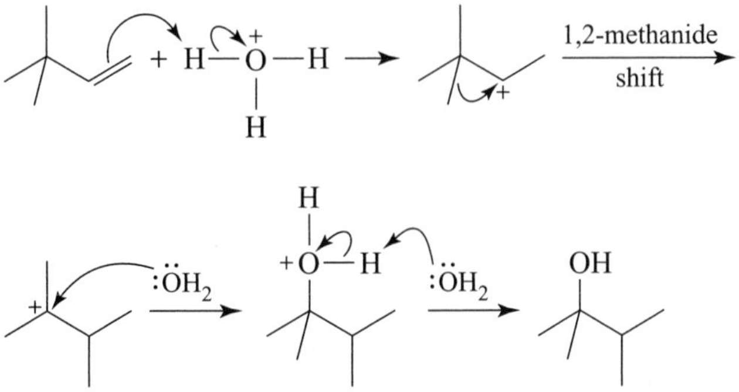

2,3-Dimethyl-2-butanol (major product)

(b) (1) $Hg(OAc)_2/THF\text{-}H_2O$; (2) $NaBH_4$, OH^- (oxymercuration-demercuration)

11.5 (a) $CH_3CH_2OH + NaNH_2 \longrightarrow CH_3CH_2O^- Na^+ + NH_3$

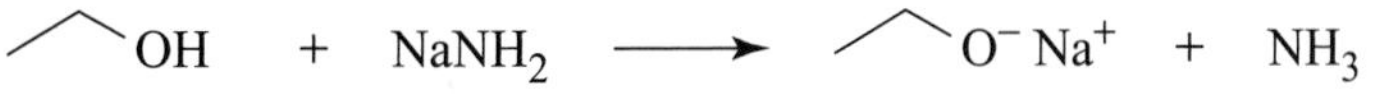

Stronger acid	Stronger base	Weaker base	Weaker acid

(b) $CH_3CH_2OH + H{-}C{\equiv}C{:}^- Na^+ \longrightarrow CH_3CH_2O^- Na^+ + H{-}C{\equiv}C{-}H$

Stronger acid	Stronger base	Weaker base	Weaker acid

(c) $CH_3CH_2OH + CH_3CO_2^- Na^+ \not\rightleftarrows CH_3CH_2O^- Na^+ + CH_3CO_2H$

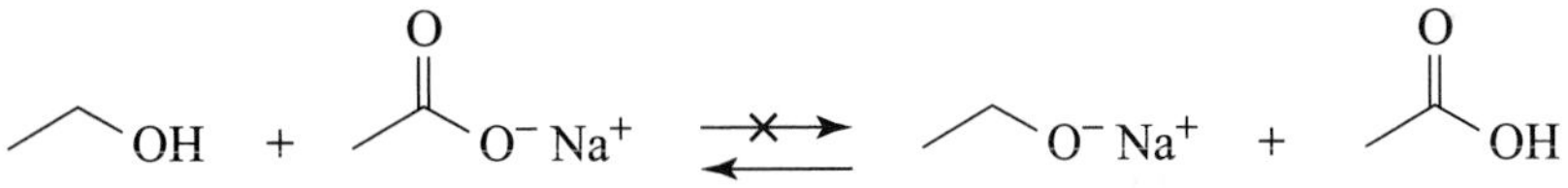

Weaker acid	Weaker base	Stronger base	Stronger acid

(d) $CH_3CH_2OH + NaOH \not\rightleftarrows CH_3CH_2O^- Na^+ + H_2O$

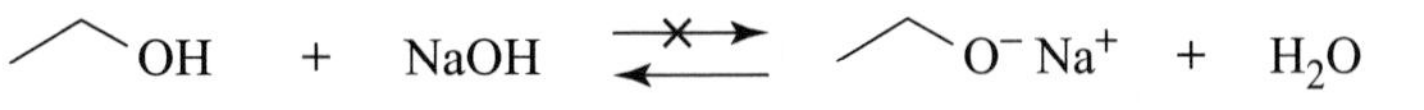

Weaker acid	Weaker base	Stronger base	Stronger acid

11.6 H—Br $-H_2\ddot{O}:$ $:\ddot{Br}:^-$ Br

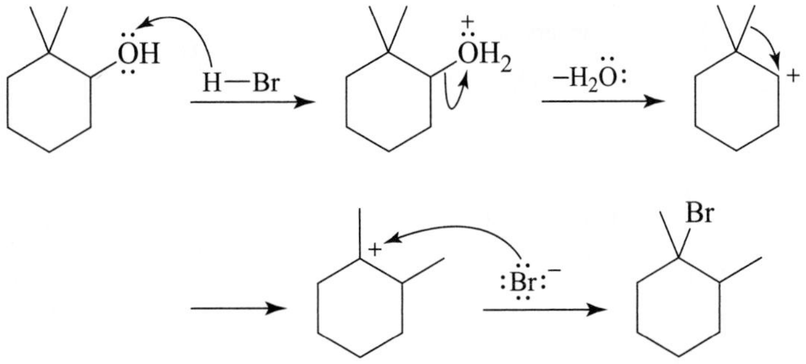

11.7 (a) Tertiary alcohols react faster than secondary alcohols because they form more stable carbocations; that is, 3° rather than 2°:

$$(CH_3)_3C{-}\overset{+}{O}H_2 \rightleftharpoons (CH_3)_3C^+ + H_2O$$

$$(CH_3)_3C^+ \xrightarrow{X^-} (CH_3)_3C{-}X$$

(b) CH_3OH reacts faster than 1° alcohols because it offers less hindrance to S_N2 attack. (Recall that CH_3OH and 1° alcohols must react through an S_N2 mechanism.)

11.8 (a) $H_3C{-}C_6H_4{-}SO_2OH \xrightarrow[(-POCl_3,\ -HCl)]{PCl_5} H_3C{-}C_6H_4{-}SO_2Cl$

$\xrightarrow[\text{base }(-HCl)]{CH_3OH} H_3C{-}C_6H_4{-}SO_2OCH_3$

(b) $CH_3SO_2OH \xrightarrow[(-POCl_3,\ -HCl)]{PCl_5} CH_3SO_2Cl \xrightarrow[\text{base }(-HCl)]{(CH_3)_2CHCH_2OH}$

$(CH_3)_2CHCH_2OSO_2Me$

(c) $CH_3SO_2Cl \xrightarrow[\text{base }(-HCl)]{(CH_3)_3COH} (CH_3)_3COSO_2Me$

11.9 (a) 2-butanol (OH) + TsCl (pyr.) $\xrightarrow[(-HCl)]{\text{retention}}$ 2-butyl tosylate (OTs) **X**

$HO:^-$ + **X** (OTs) $\xrightarrow[S_N2]{\text{inversion}}$ 2-butanol (OH, inverted) **Y** + $:\ddot{O}Ts^-$

(b) *cis*-2-Methylcyclohexanol (OH) $\xrightarrow[\text{(pyr.)}]{\text{TsCl, retention}}$ **A** (OTs) $\xrightarrow[\text{inversion}]{:\ddot{Cl}^-}$ **B** (Cl) + $:\ddot{O}Ts^-$

11.10 Use an alcohol containing labeled oxygen. If all of the label appears in the sulfonate ester, then one can conclude that the alcohol C—O bond does not break during the reaction:

$$R{-}^{18}O{-}H \; + \; R'{-}SO_2Cl \xrightarrow[(-HCl)]{\text{base}} R{-}^{18}O{-}SO_2{-}R'$$

11.11 HO–C(CH_3)$_3$ $\xrightarrow{HA}$ $H_2\overset{+}{O}$–C(CH_3)$_3$ $\xrightarrow{-H_2O}$ $(CH_3)_3C^+$ $\xrightarrow[\text{(1° only)}]{R\text{–}CH_2\text{–}OH}$ R–CH_2–$\overset{+}{O}$(H)–C(CH_3)$_3$ $\xrightarrow[-HA]{A^-}$ R–CH_2–O–C(CH_3)$_3$

This reaction succeeds because a 3° carbocation is much more stable than a 1° carbocation. Consequently, mixing the 1° alcohol and H_2SO_4 does not lead to formation of appreciable amounts of a 1° carbocation. However, when the 3° alcohol is added, it is rapidly converted to a 3° carbocation, which then reacts with the 1° alcohol that is present in the mixture.

11.12 (a)

(1) $(CH_3)_2CH\ddot{O}{:}^-$ + CH_3–L ⟶ $(CH_3)_2CH$–O–CH_3 + L^-

(L = X, OSO_2R, or OSO_2OR)

(2) $CH_3\ddot{O}{:}^-$ + $(CH_3)_2CH$–L ⟶ $(CH_3)_2CH$–O–CH_3 + L^-

(L = X, OSO_2R, or OSO_2OR)

(b) Both methods involve S_N2 reactions. Therefore, method (1) is better because substitution takes place at an unhindered methyl carbon atom. In method (2) where substitution must take place at a relatively hindered secondary carbon atom, the reaction would be accompanied by considerable elimination.

$CH_3\ddot{O}{:}^-$ + H–CH_2–CH(L)–CH_3 ⟶ CH_2=CH–CH_3 + CH_3OH + L^-

11.13 (a) HO^- + $H\ddot{O}$–CH_2CH_2–Cl ⇌ H_2O + $^-{:}\ddot{O}$–CH_2CH_2–Cl ⟶ (ethylene oxide) + Cl^-

(b) The —$\ddot{\underset{..}{O}}$:$^-$ group must displace the Cl$^-$ from the backside,

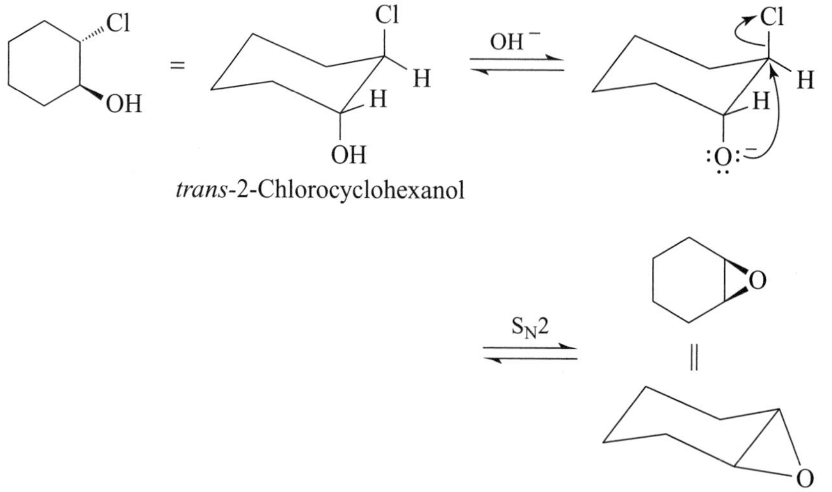

Cl OH = Cl H H OH

trans-2-Chlorocyclohexanol

$\xrightarrow{OH^-}$ Cl H H :O:$^-$

$\xrightarrow{S_N2}$ O

(c) Backside attack is not possible with the cis isomer (below); therefore, it does not form an epoxide.

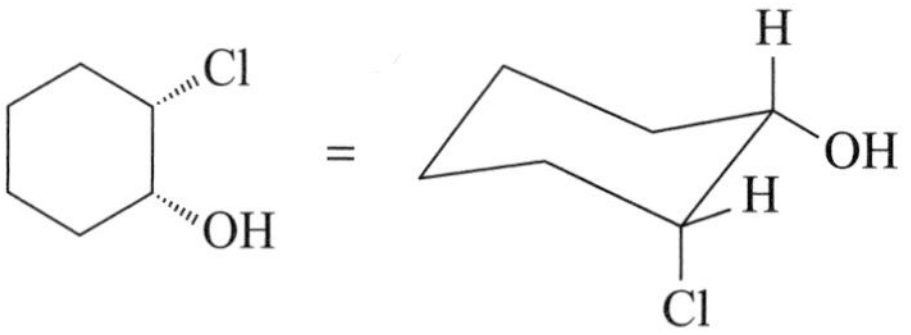

Cl OH = H OH H Cl

cis-2-Chlorocyclohexanol

11.14 OH $\xrightarrow[(-H_2)]{K°}$ O$^-$ K$^+$ **A** $\xrightarrow{\text{EtBr}}$ O **B**

TsCl, pyr.

OTs **C** $\xrightarrow[K_2CO_3]{\text{EtOH}}$ O **D**

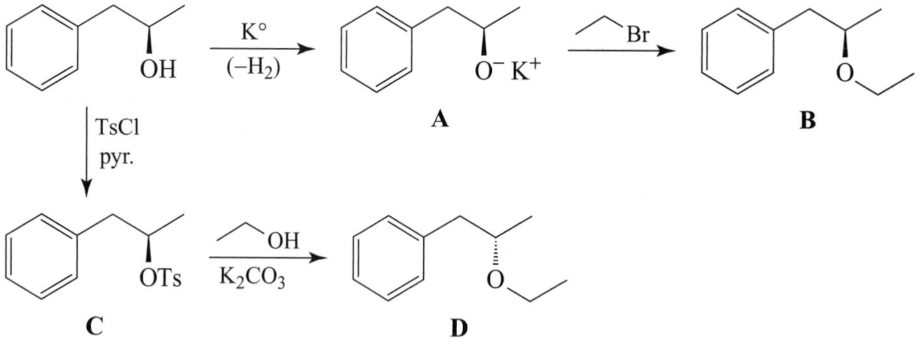

11.15 (a)

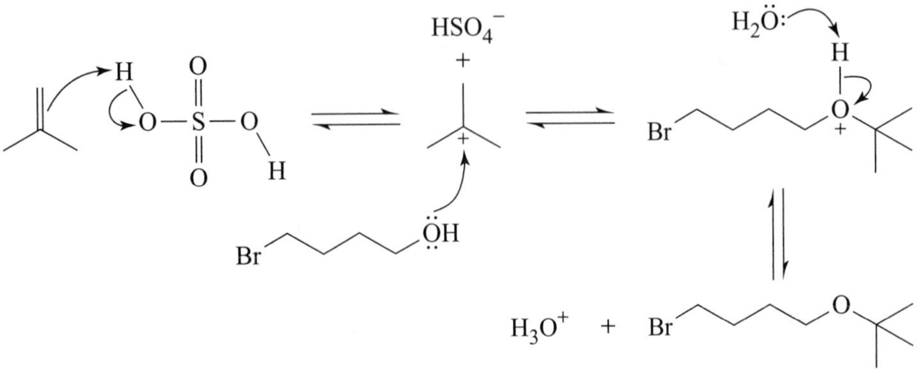

H O O—S—O H O ⇌ HSO_4^- + ⇌ Br O+ H $H_2\ddot{O}$:

Br ÖH

⇌ H_3O^+ + Br O

(b)

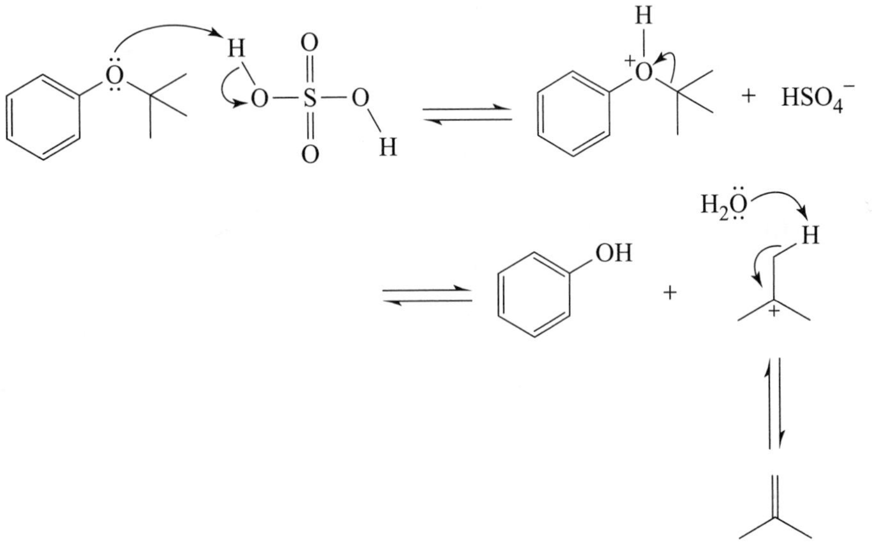

Ö H O O—S—O H O ⇌ H +O + HSO_4^-

⇌ OH + $H_2\ddot{O}$ H ⇌

11.16 (a) Me O + HI → $:\ddot{I}:^-$ + H Me—O+ →

MeI + OH

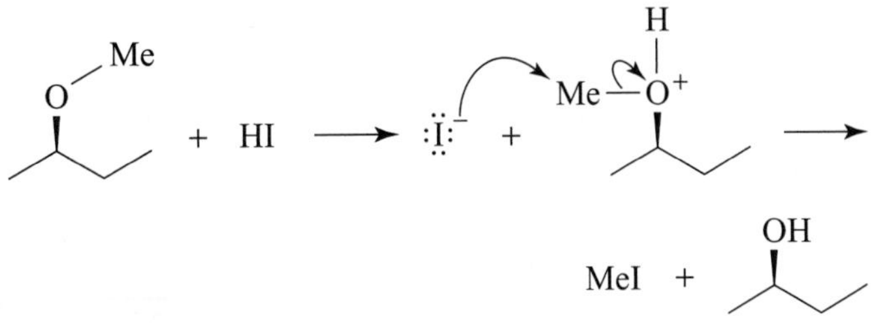

S_N2 attack by I^- occurs at the methyl carbon atom because it is less hindered; therefore, the bond between the *sec*-butyl group and the oxygen is not broken.

(b)

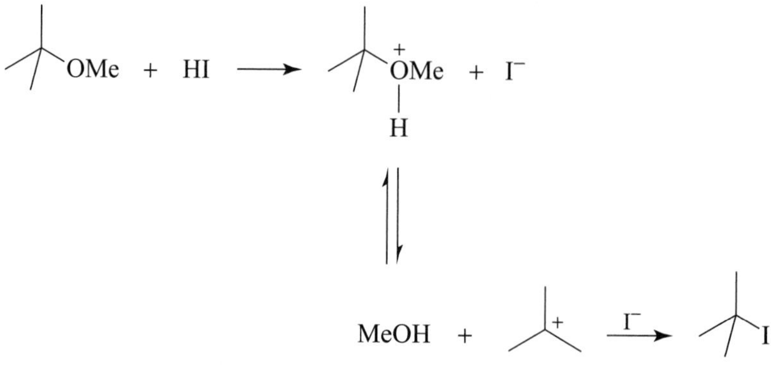

OMe + HI → $\overset{+}{\text{O}}$Me(H) + I^- ⇌ MeOH + (CH₃)₃C⁺ —I^-→ (CH₃)₃C–I

In this reaction the much more stable *tert*-butyl cation is produced. It then combines with I^- to form *tert*-butyl iodide.

11.17

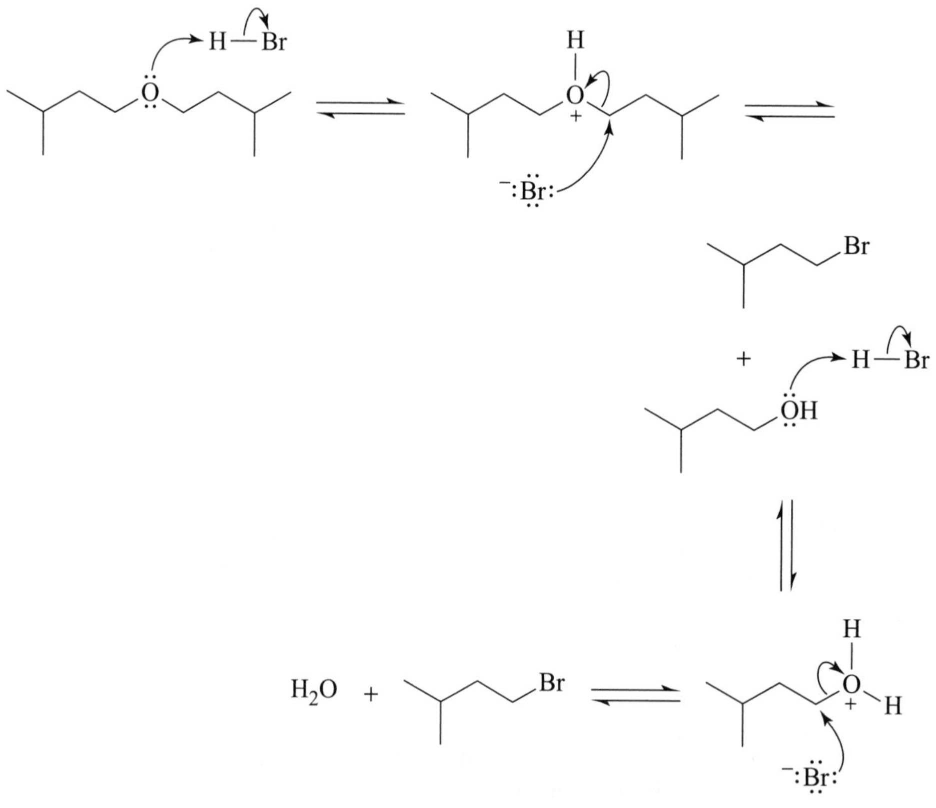

H—Br, :Ö: ⇌ H, O+, ⁻:Br: ⇌ Br + H—Br, ÖH ⇌ H, O+, H, ⁻:Br: ⇌ H_2O + Br

11.18

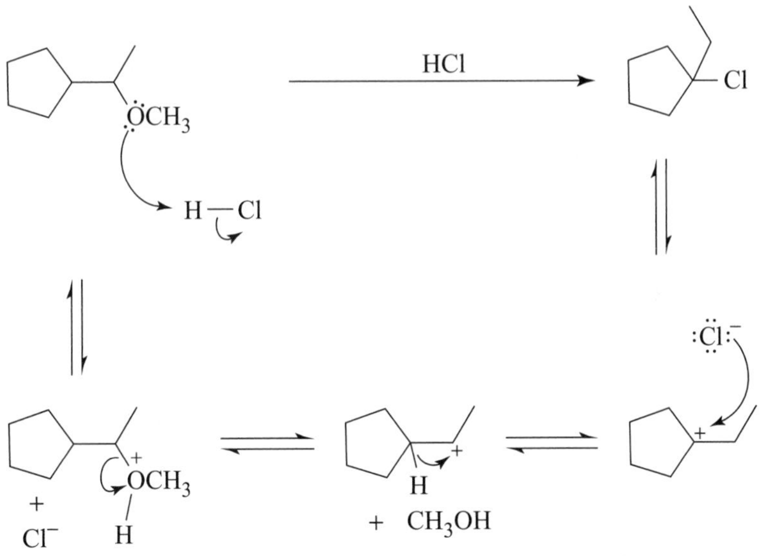

$\ddot{O}CH_3$ → HCl → Cl

H—Cl

$:\ddot{Cl}:^-$

+ Cl^- $\overset{+}{O}CH_3$ H ⇌ H + CH_3OH ⇌ +

11.19 (a) O → HA → H, ^+O → $H\ddot{O}$—Me → HO, $\overset{+}{O}$—Me, H

→ A^- → HO, OMe

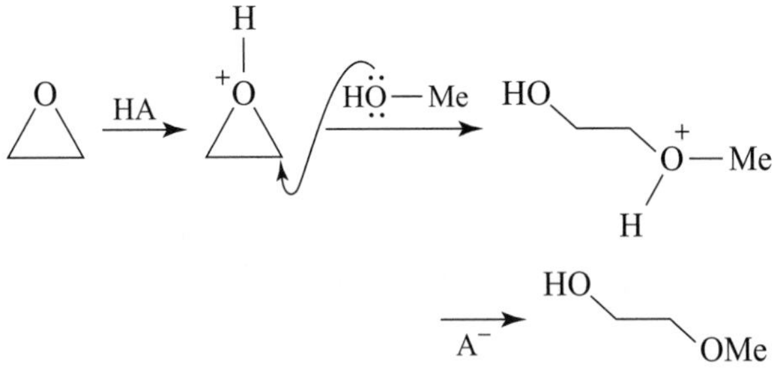

Methyl Cellosolve

(b) An analogous reaction yields ethyl cellosolve, HO O .

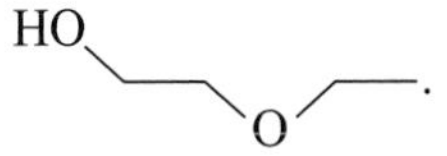

(c) O → $:\ddot{I}:^-$ → ^-O I → H_2O → HO I + OH^-

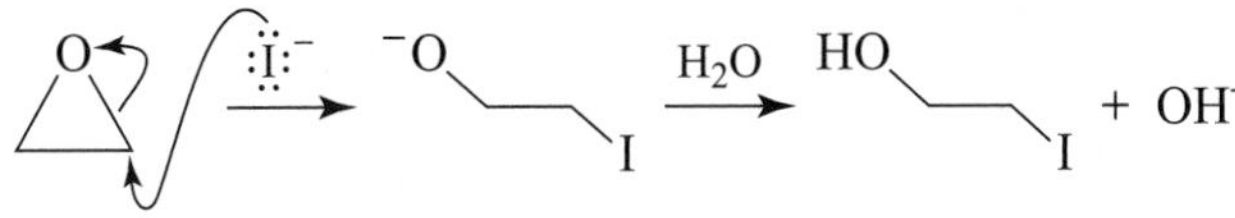

(d) O → $:NH_3$ → ^-O $_+NH_3$ → HO NH_2

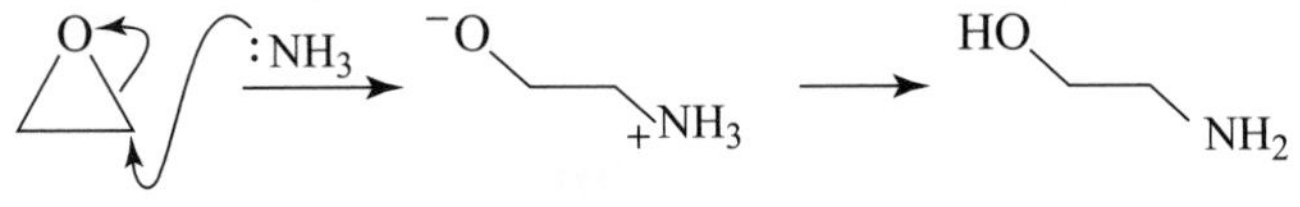

(e) O → $^-:\ddot{O}Me$ → ^-O OMe → MeOH → HO OMe + CH_3O^-

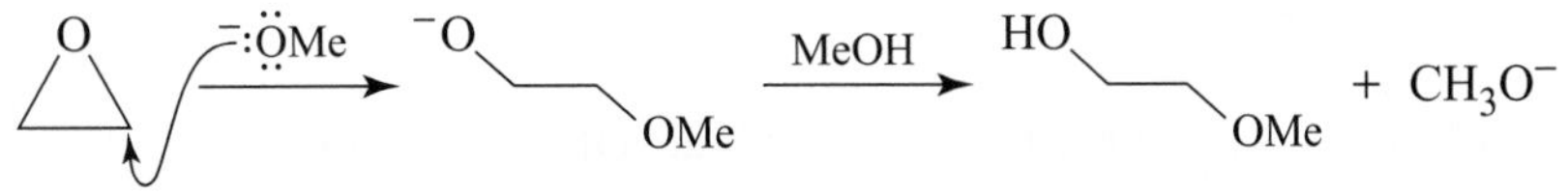

11.20 The reaction is an S_N2 reaction, and thus nucleophilic attack takes place much more rapidly at the primary carbon atom than at the more hindered tertiary carbon atom.

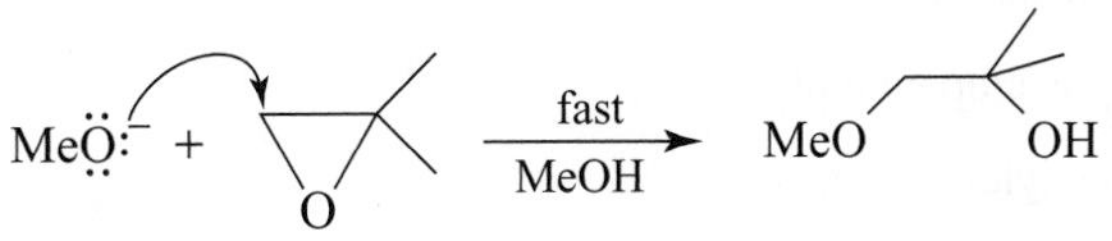

$Me\ddot{O}:^-$ + O → fast, MeOH → MeO OH Major product

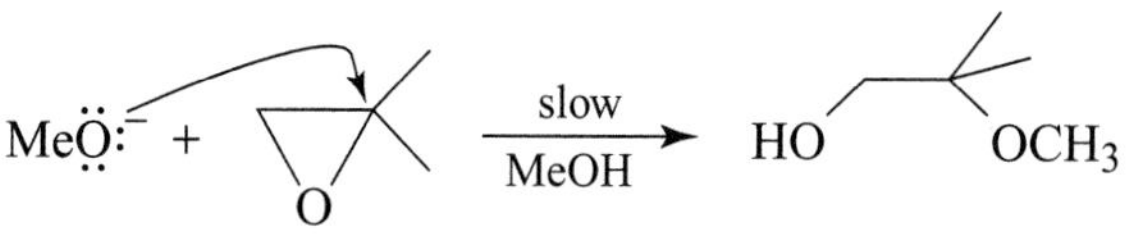

$MeÖ:^{-}$ + (2,2-dimethyloxirane) $\xrightarrow[\text{MeOH}]{\text{slow}}$ HO–CH_2–$C(CH_3)_2$–OCH_3 Minor product

11.21 Ethoxide ion attacks the epoxide ring at the primary carbon because it is less hindered, and the following reactions take place.

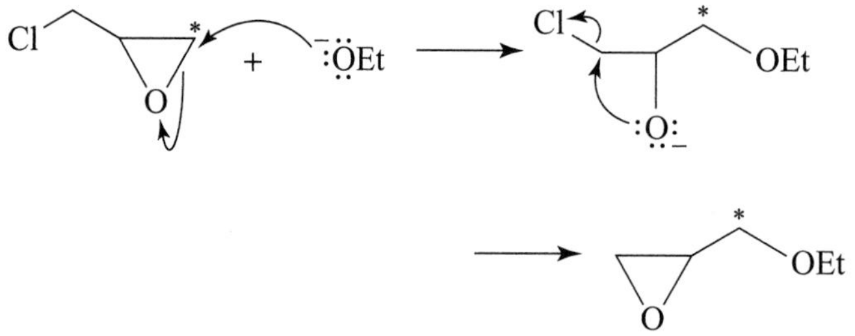

Cl–CH_2–(epoxide)* + $:\ddot{O}Et^{-}$ ⟶ Cl–CH_2–CH($:\ddot{O}:^{-}$)–CH_2*–OEt

⟶ (epoxide)–CH_2*–OEt

11.22 (cyclopentene oxide) + H_3O^+ ⇌ (protonated epoxide, OH^+) + $H_2\ddot{O}:$ ⟶ (trans-2-hydroxycyclopentyl, H_2O^+)

$\xrightarrow[-H_3O^+]{H_2O}$ (trans-1,2-cyclopentanediol) (plus enantiomer, by attack at the other epoxide carbon)

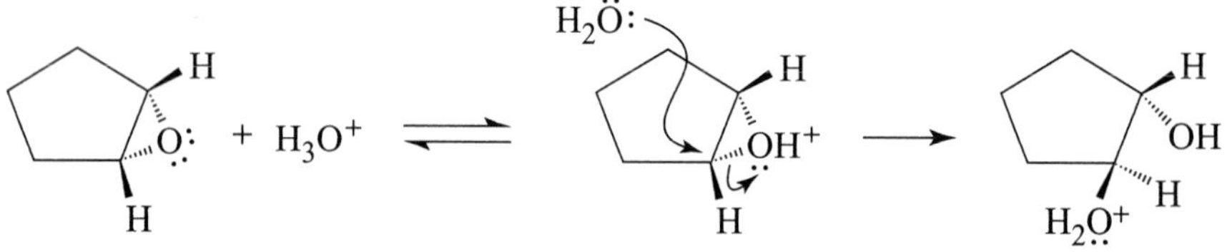

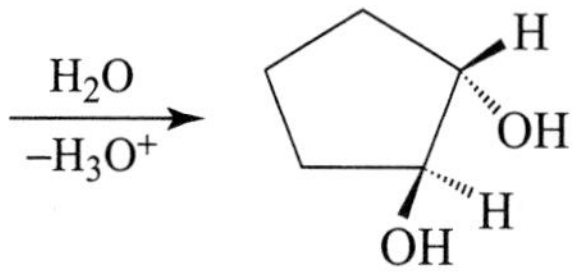

11.23 A: 2-Butyne

B: H_2, Ni_2B (P – 2)

C: MCPBA

D: (cis-2,3-dimethyloxirane)

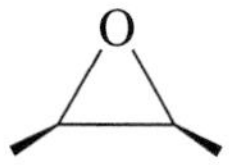

E: MeOH, cat. acid

11.24 (a) 15-Crown-5 (b) 12-Crown-4

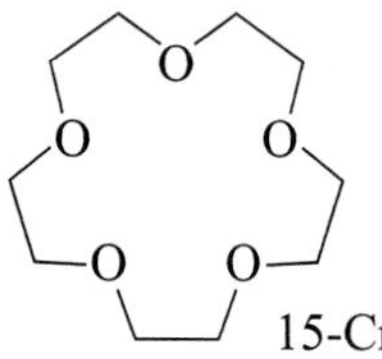

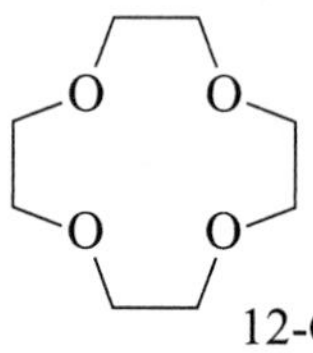

Problems

Nomenclature

11.25 (a) 3,3-Dimethyl-1-butanol or 3,3-dimethylbutan-1-ol

(b) 4-Penten-2-ol or pent-4-en-2-ol

(c) 2-Methyl-1,4-butanediol or 2-methylbutan-1,4-diol

(d) 2-Phenylethanol

(e) 1-Methylcyclopentanol

(f) *cis*-3-Methylcyclohexanol

11.26

(a) 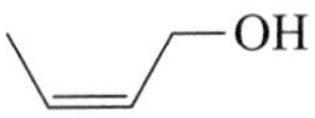 OH (b) 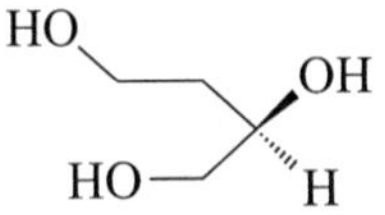 HO, OH, HO, H

(c) 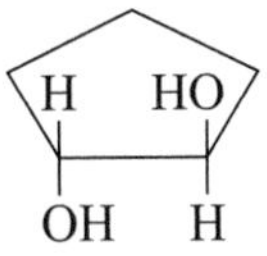 H HO, OH H (d) 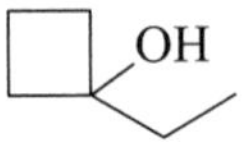 OH

(e) 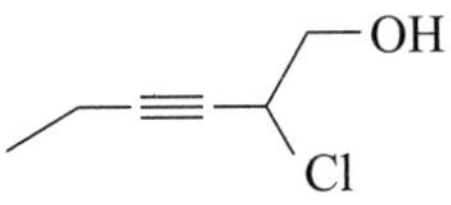 OH, Cl (f) 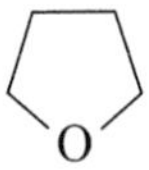 O

(g) 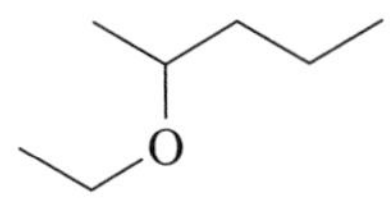 O (h) O

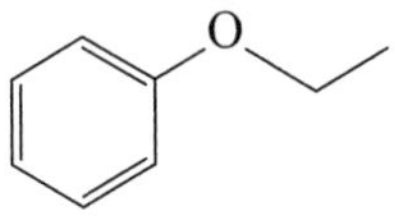

(i) 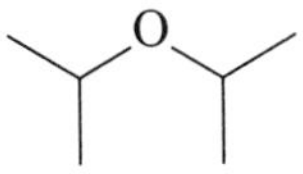 O (j) 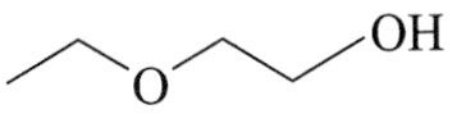 O, OH

Reactions and Synthesis

11.27 (a)  (c) 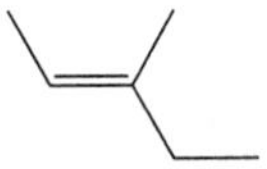 or

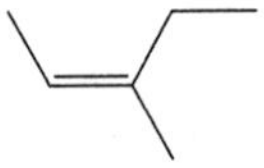

(b) 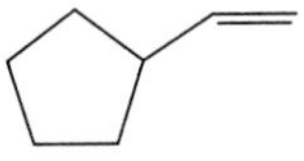 (d)

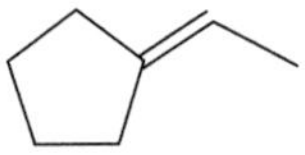

11.28 (a) 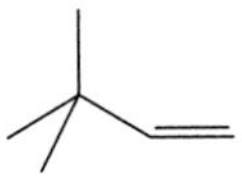 (c)

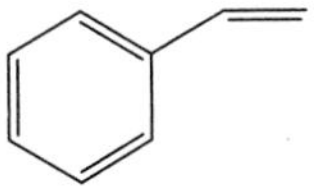

(b)  (d)

11.29 (a) 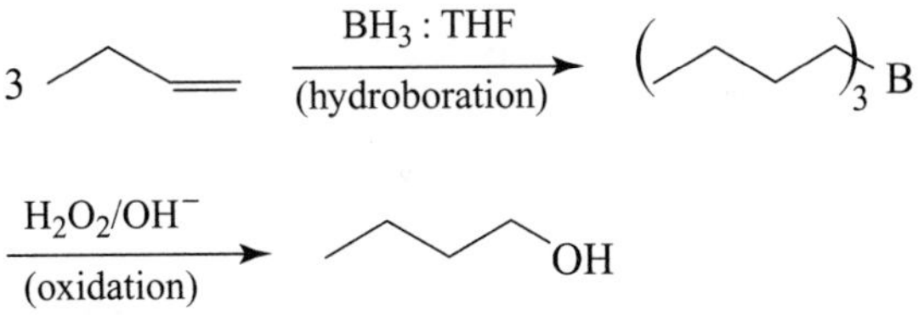 3 $\xrightarrow[\text{(hydroboration)}]{BH_3 : THF}$ ( )$_3$B

$\xrightarrow[\text{(oxidation)}]{H_2O_2/OH^-}$ OH

(b) Cl $\xrightarrow{OH^-}$ OH

(c) 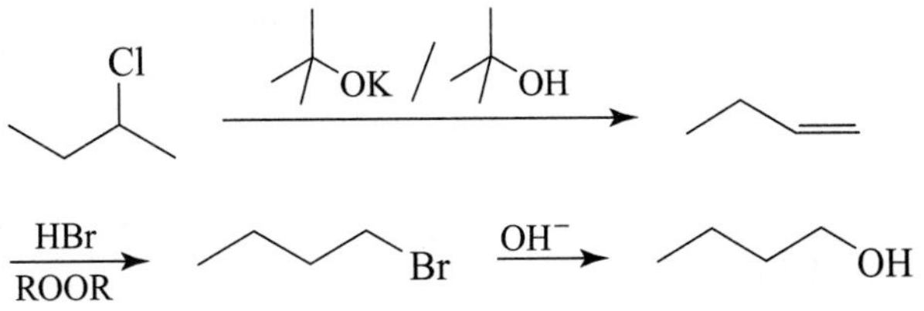 Cl $\xrightarrow{\text{OK} / \text{OH}}$

$\xrightarrow[\text{ROOR}]{HBr}$ Br $\xrightarrow{OH^-}$ OH

(d) 1-butyne $\xrightarrow[\text{Ni}_2\text{B (P-2)}]{H_2}$ 1-butene

$\xrightarrow{\text{[as in (a)]}}$ 1-butanol (OH)

11.30 (a) 3 2-butanol (OH) + PBr_3 ⟶ 3 2-bromobutane (Br) + H_3PO_3

(b) 1-butanol (OH) $\xrightarrow{PBr_3}$ 1-bromobutane (Br) $\xrightarrow[\text{tert-butyl alcohol (OH)}]{\text{potassium tert-butoxide (OK)}}$

1-butene $\xrightarrow[\text{(no peroxides)}]{\text{HBr}}$ 2-bromobutane (Br)

(c) See (b) above.

(d) 1-butyne $\xrightarrow[\text{Ni}_2\text{B (P-2)}]{H_2}$ 1-butene $\xrightarrow[\text{(no peroxides)}]{\text{HBr}}$ 2-bromobutane (Br)

11.31 (a) 2-methylpropene $\xrightarrow[\text{(2) } H_2O_2\text{, } OH^-]{\text{(1) } BH_3 : THF}$ 2-methyl-1-propanol (OH)

(b) 2-methylpropene $\xrightarrow[\text{(2) } CH_3C(O)OT]{\text{(1) } BH_3 : THF}$ (T-labeled isobutane, T)

(c) 2-methylpropene $\xrightarrow{BD_3 : THF}$ (D-labeled alkyl–BR_2; R, B, R, D) $\xrightarrow{CH_3C(O)OT}$ (D, T-labeled isobutane; D, T)

(d) 2-methyl-1-propanol (OH) [from (a)] $\xrightarrow{\text{Na}}$ (ONa) $\xrightarrow{\text{ethyl bromide (Br)}}$ ethyl isobutyl ether (O)

11.32 (a) cyclohexanol (OH) + $SOCl_2$ ⟶ chlorocyclohexane (Cl) + SO_2 + HCl

(b) cyclohexene + HCl ⟶ chlorocyclohexane (Cl)

(c) 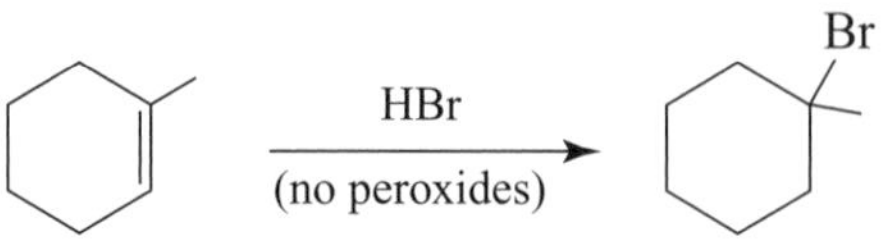 HBr (no peroxides)

Br

(d) 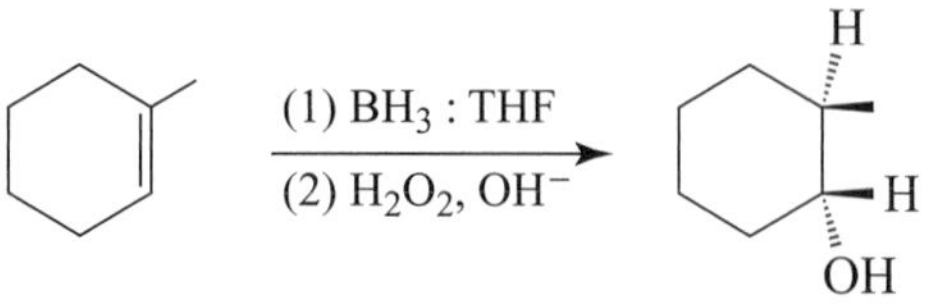 (1) BH_3 : THF (2) H_2O_2, OH^-

H H OH + enantiomer

(e) 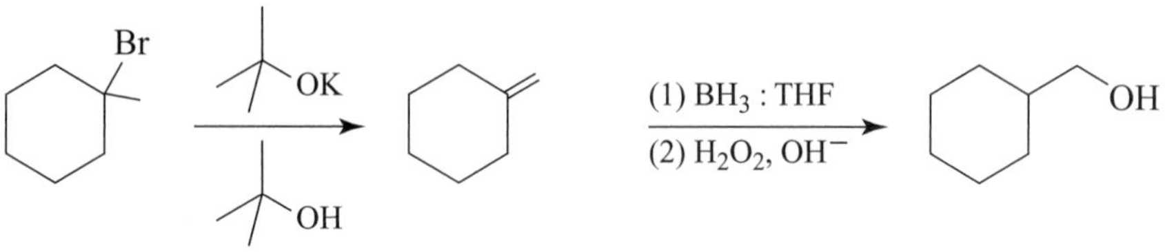 Br OK OH (1) BH_3 : THF (2) H_2O_2, OH^- OH

11.33 (a) CH_3Br + Br (c) 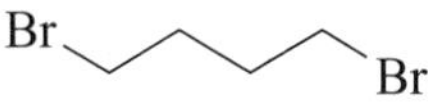 Br Br

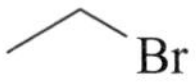

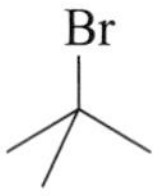

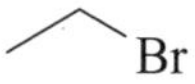

(b) Br + Br (d) 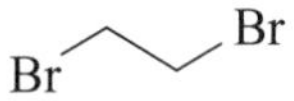 Br Br (2 molar equivalents)

11.34 **A:** O^- Na^+

B: O

C: O SO_2CH_3

D: O CH_3 + $CH_3SO_3^-$ Na^+

E: O SO_2

F: I + SO_3^- Na^+

G: O

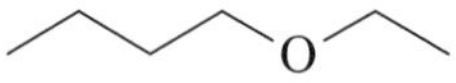

H: O Si

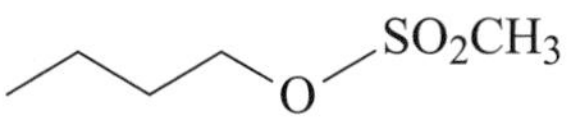

I: OH + F Si

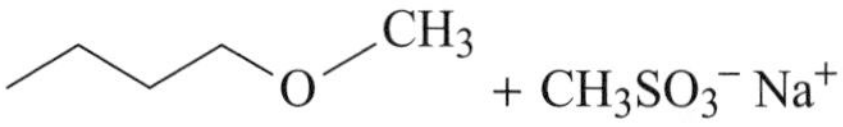

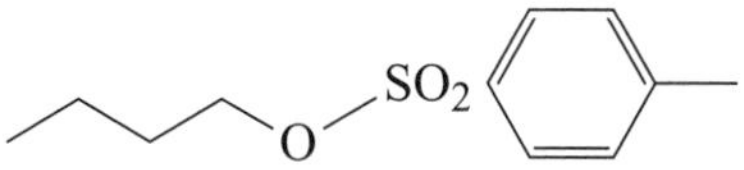

J: Br

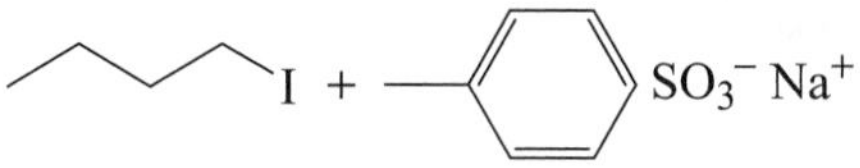

K: Cl

L: Br

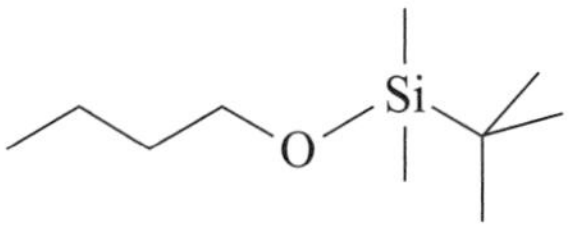

11.35 **A:** O^- Na^+

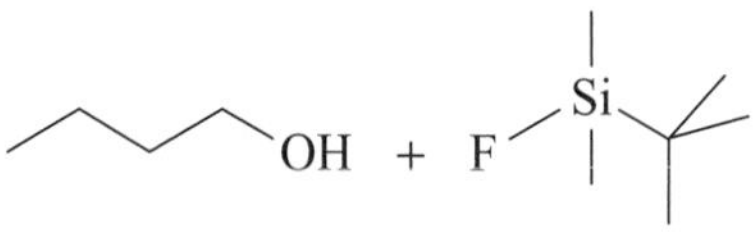

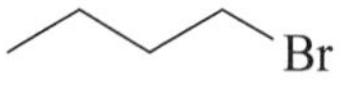

B: O

C: O SO_2CH_3

D: O CH_3 + $CH_3SO_3^-Na^+$

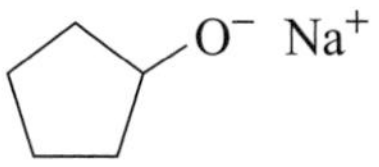

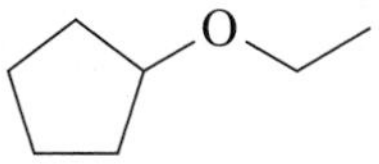

E: O SO_2

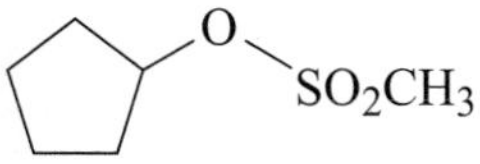

F: I + $SO_3^-Na^+$

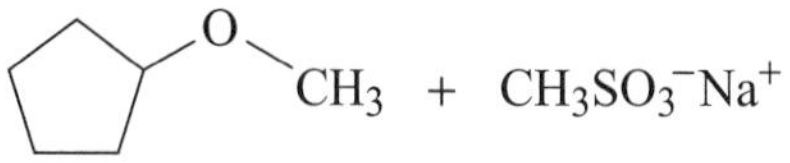

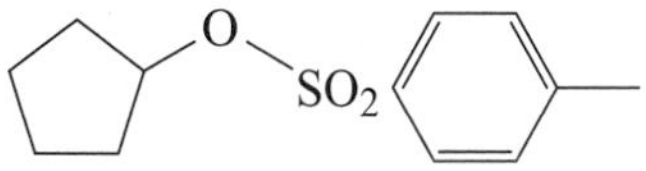

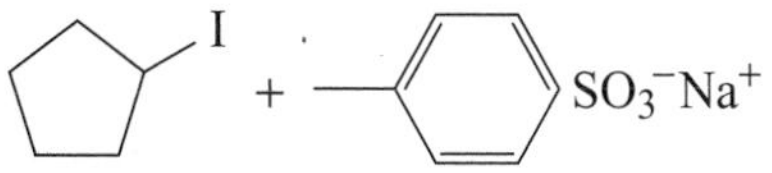

G: O +

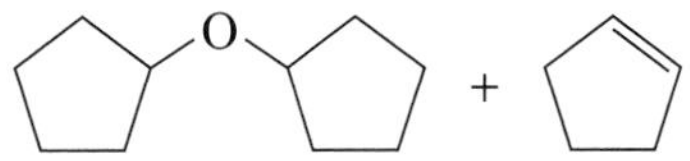

H: O Si

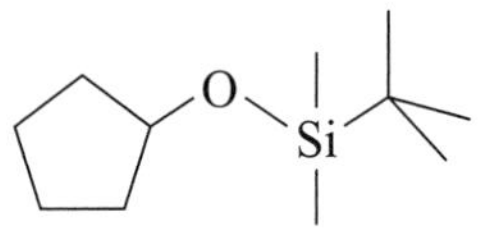

I: OH + F Si

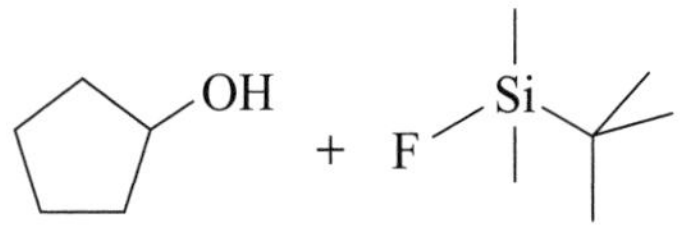

J: Br

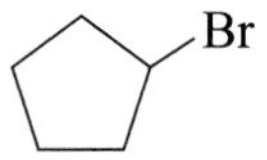

K: Cl

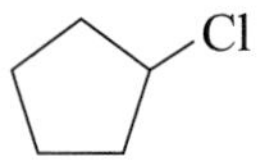

L: Br

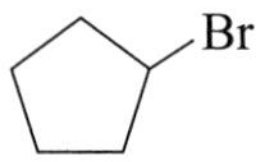

11.36 (a) $\xrightarrow[\text{heat, } h\nu]{Br_2}$ Br

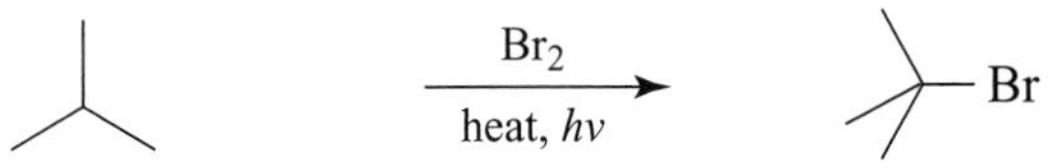

(b) Br $\xrightarrow[\text{OH}]{\text{ONa}}$

(c) $\xrightarrow[\text{peroxides}]{HBr}$ Br

(d) Br $\xrightarrow[\text{acetone}]{KI}$ I

(e) Br $\xrightarrow[H_2O]{OH^-}$ OH

or

$\xrightarrow[(2)\ H_2O_2,\ OH^-]{(1)\ (BH_3)_2}$ OH

(f) $\xrightarrow[\text{(no peroxides)}]{HBr}$ Br

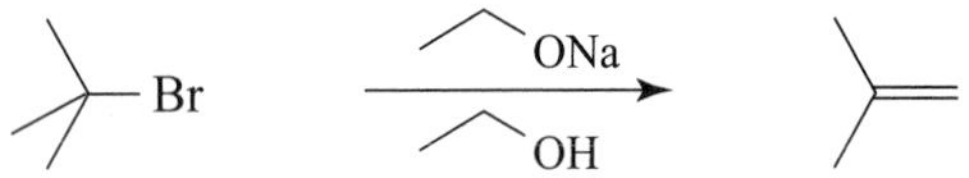

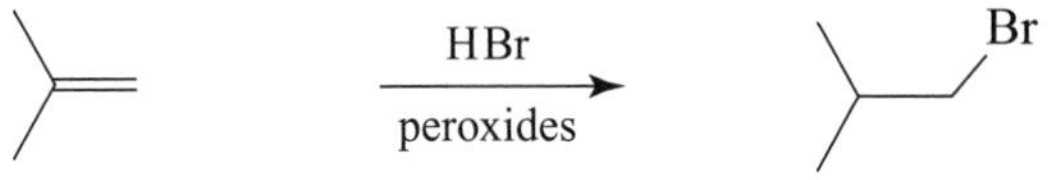

(g) Br $\xrightarrow[CH_3OH]{CH_3ONa}$ O CH_3

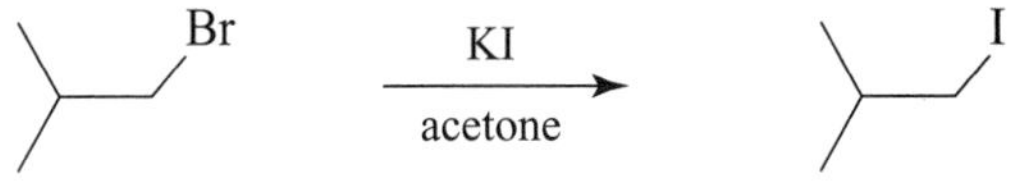

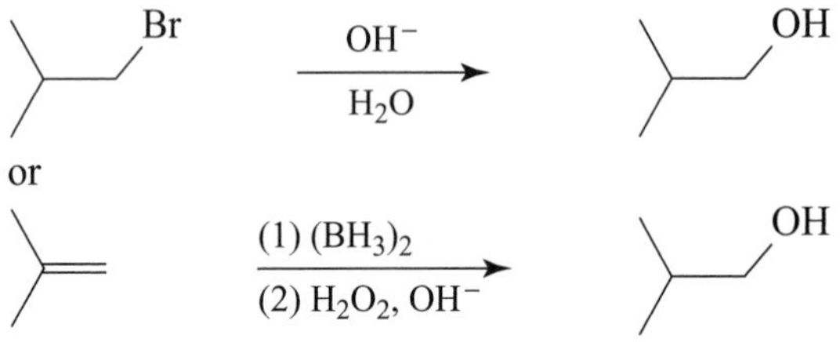

(h) Br $\xrightarrow[\text{O, OH}]{\text{O, ONa}}$ O O

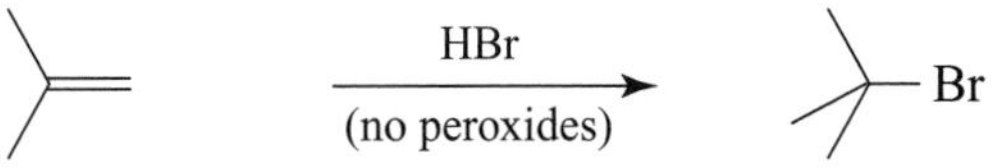

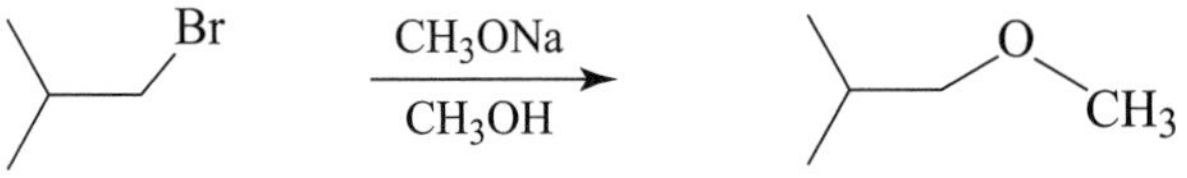

(i) Br $\xrightarrow{NaCN}$ CN

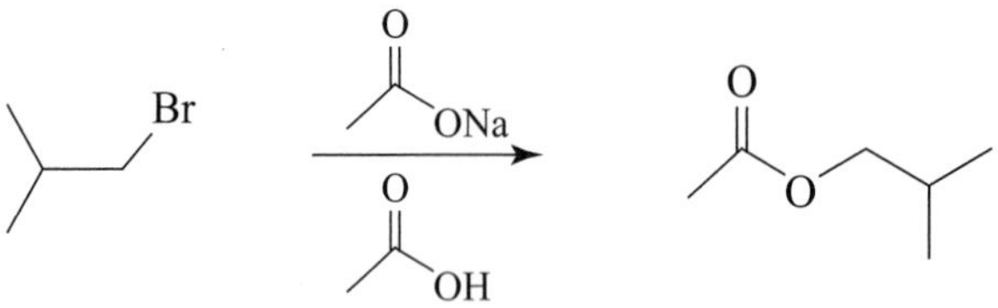

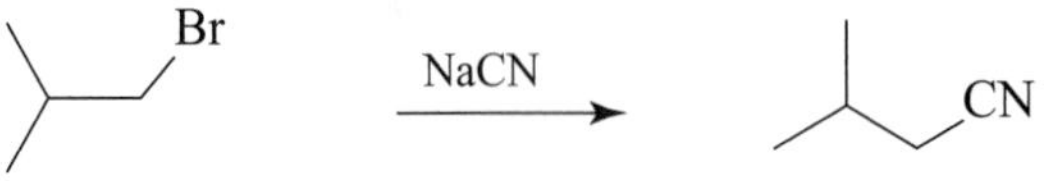

(j) Br $\xrightarrow{CH_3SNa}$ SCH_3

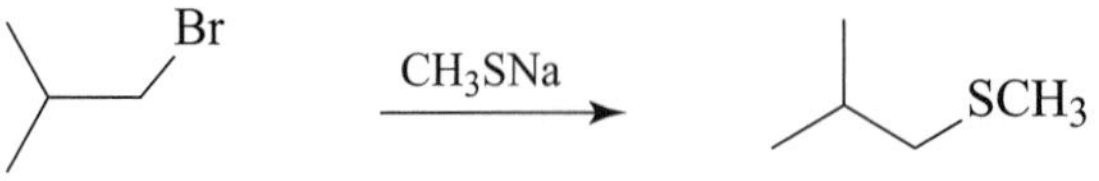

(k) 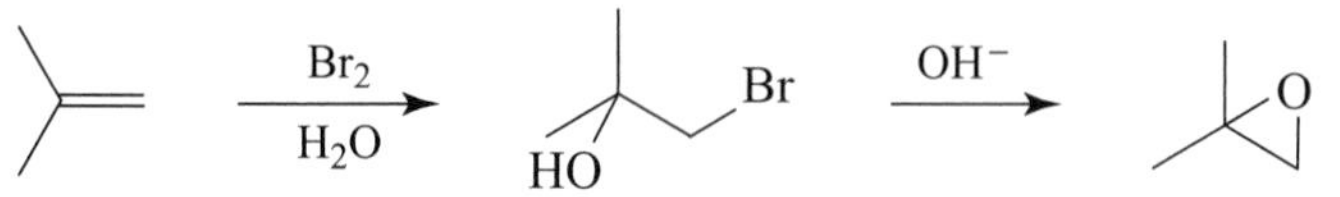 $\xrightarrow[H_2O]{Br_2}$ HO, Br $\xrightarrow{OH^-}$ O

(l) 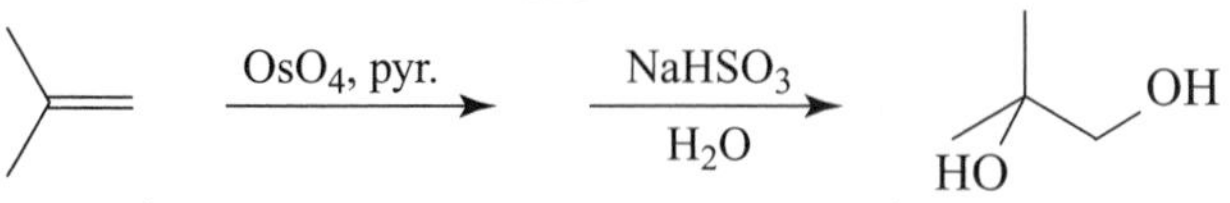 $\xrightarrow{OsO_4,\ pyr.}$ $\xrightarrow[H_2O]{NaHSO_3}$ HO, OH

(m) 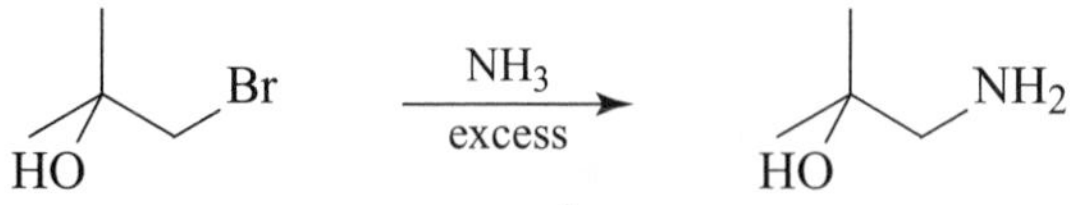 HO, Br $\xrightarrow[excess]{NH_3}$ HO, NH_2

(n) 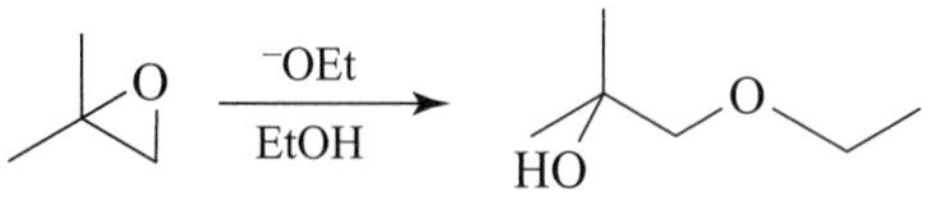 O $\xrightarrow[EtOH]{^-OEt}$ HO, O

11.37 H—≡—H $\xrightarrow[liq.\ NH_3]{NaNH_2}$ H—≡—Na $\xrightarrow{Br}$ H

A (C_9H_{16})

$\xrightarrow[liq.\ NH_3]{NaNH_2}$ Na **B** ($C_9H_{15}Na$) $\xrightarrow{(\ )_4\ Br}$

C ($C_{19}H_{36}$) 4

$\xrightarrow[Ni_2B\ (P\text{-}2)]{H_2}$ H H **D** ($C_{19}H_{38}$) 4 $\xrightarrow{MCPBA}$

H O H Disparlure ($C_{19}H_{38}O$) 4

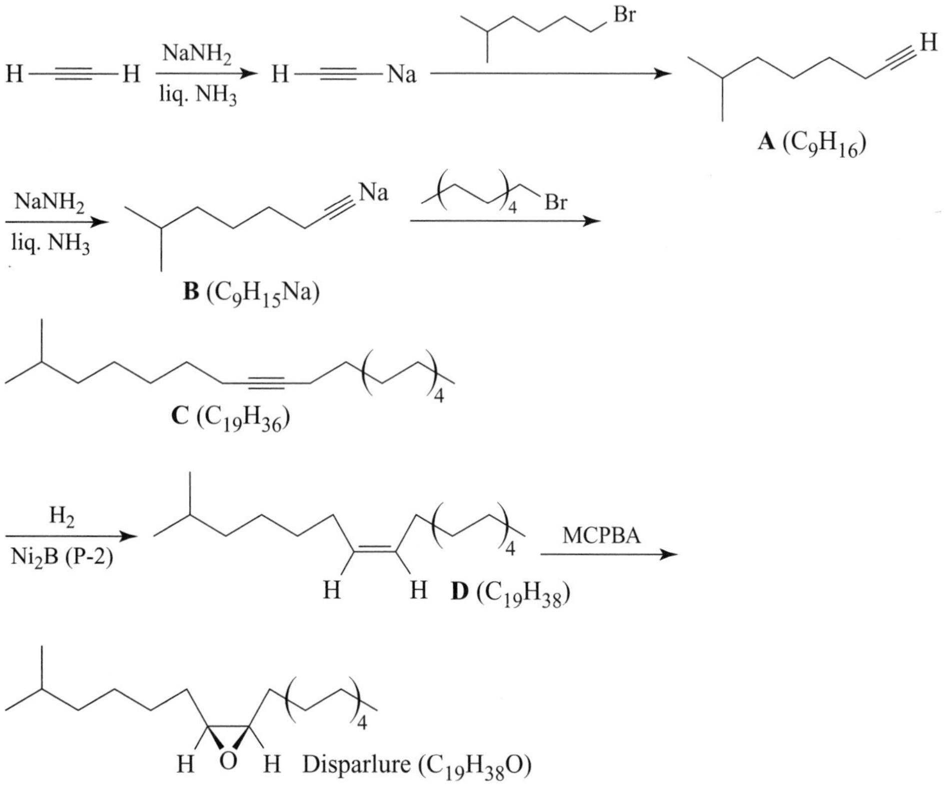

11.38 (a) 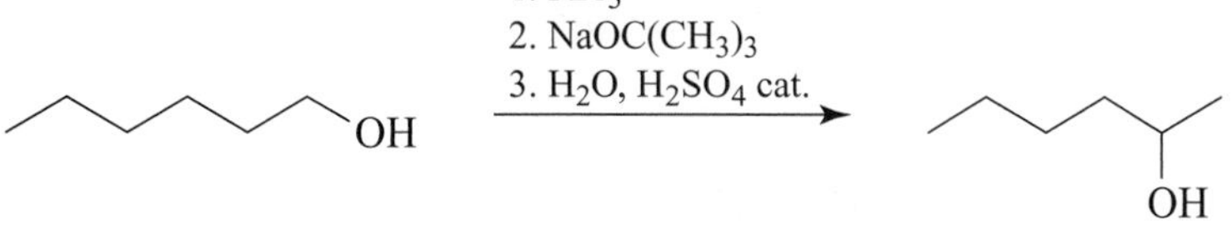 OH $\xrightarrow{1.\ PBr_3\quad 2.\ NaOC(CH_3)_3\quad 3.\ H_2O,\ H_2SO_4\ cat.}$ OH

(b) 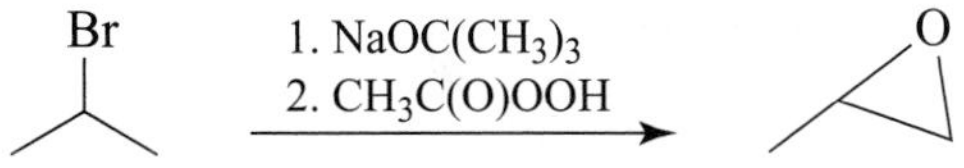 Br $\xrightarrow{1.\ NaOC(CH_3)_3\quad 2.\ CH_3C(O)OOH}$ O

(c) 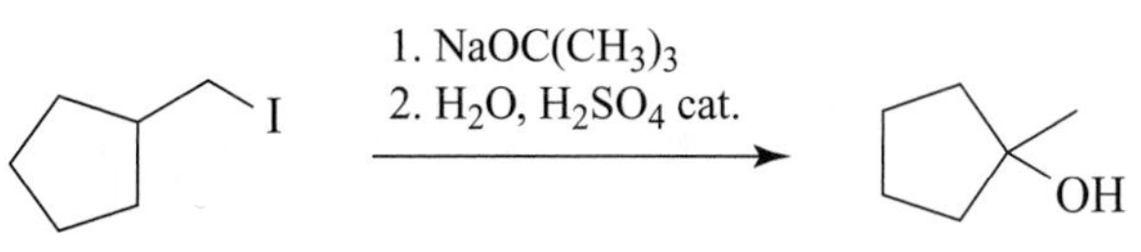 I $\xrightarrow{1.\ NaOC(CH_3)_3\quad 2.\ H_2O,\ H_2SO_4\ cat.}$ OH

(d) 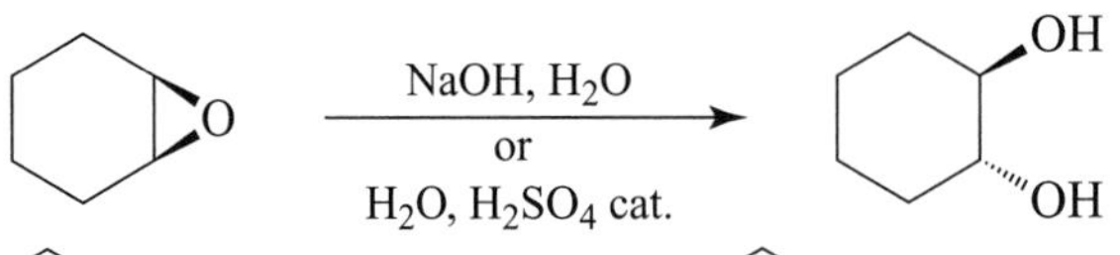 NaOH, H_2O or H_2O, H_2SO_4 cat. → OH, OH

(e)  MCPBA → O

(f) 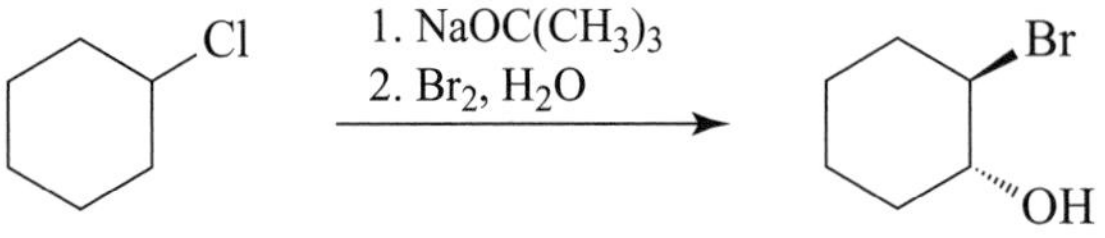 Cl 1. $NaOC(CH_3)_3$ 2. Br_2, H_2O → Br, OH

11.39 (a) 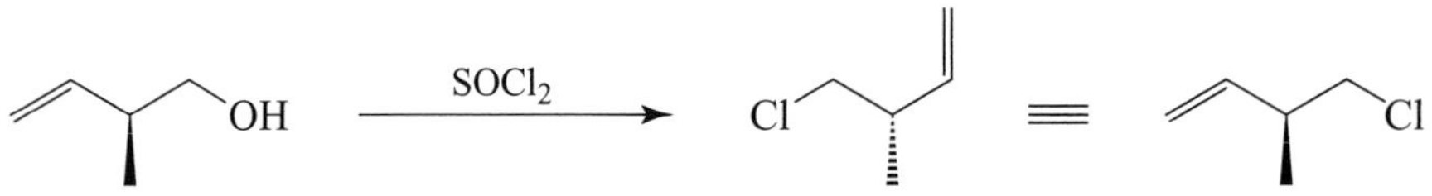 OH $SOCl_2$ → Cl ≡ Cl

(b) 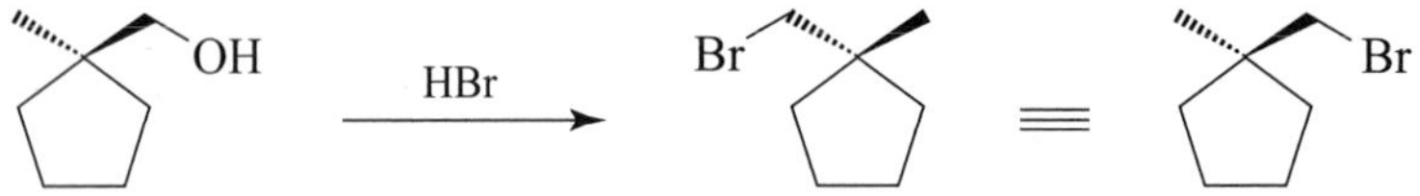 OH HBr → Br ≡ Br

(c)  OH $NaNH_2$ → O^-Na^+ + NH_3

(d) 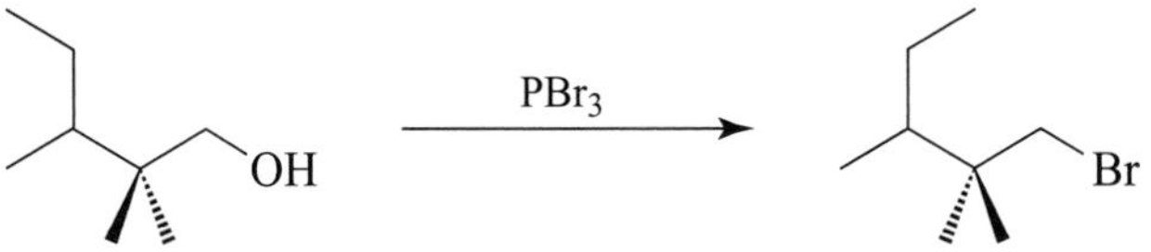 OH PBr_3 → Br

(e) 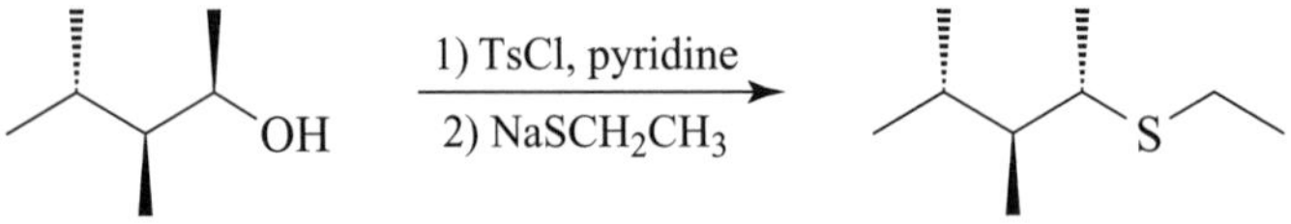 OH 1) TsCl, pyridine 2) $NaSCH_2CH_3$ → S

(f) 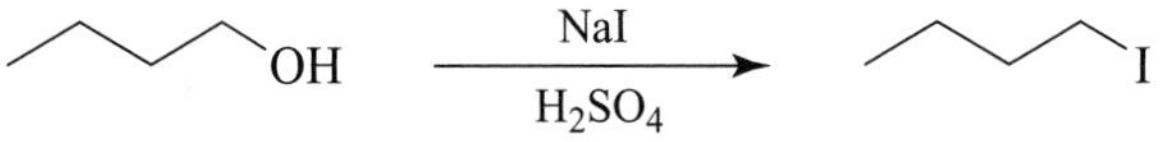 OH NaI H_2SO_4 → I

11.40 (a) 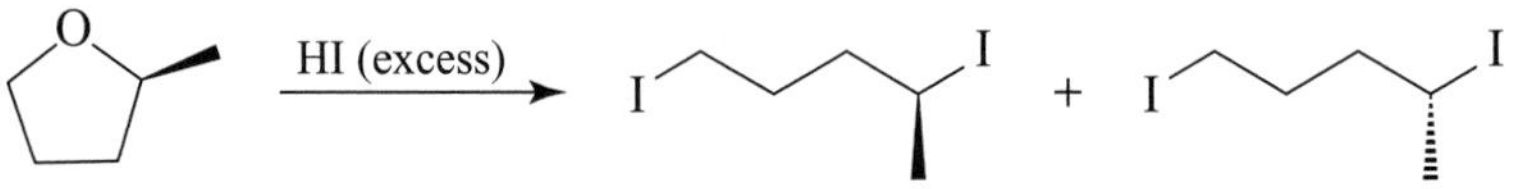 O HI (excess) → I I + I I

(b) 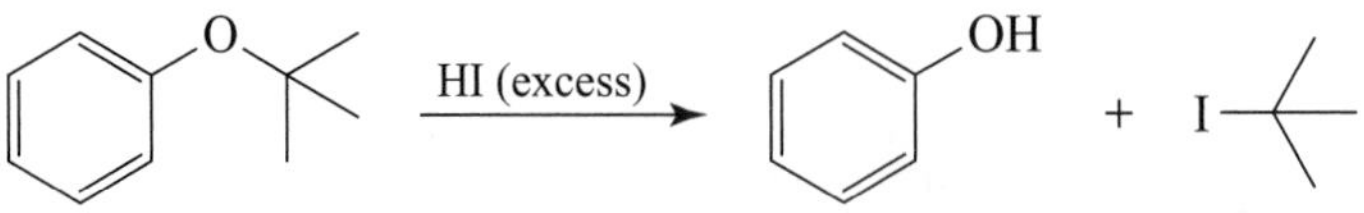 O HI (excess) → OH + I

(c) 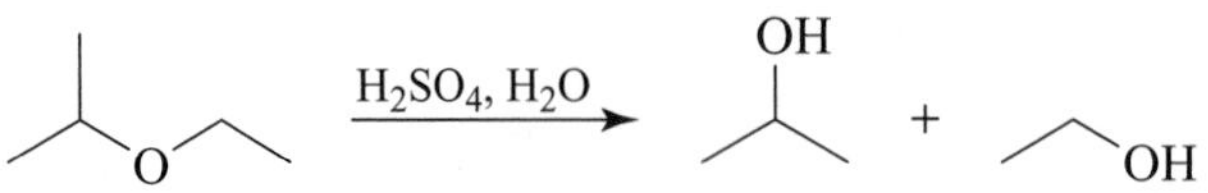 O H_2SO_4, H_2O → OH + OH

(d) 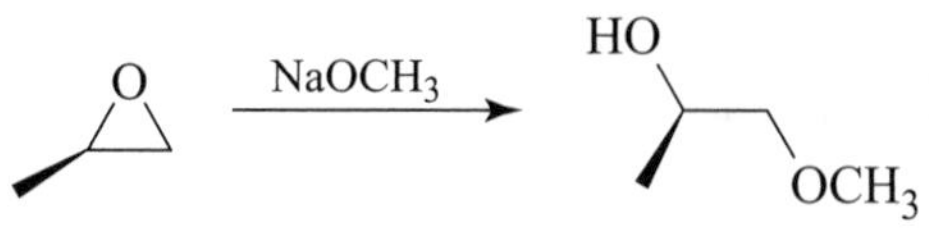 NaOCH$_3$ → HO, OCH$_3$

(e) 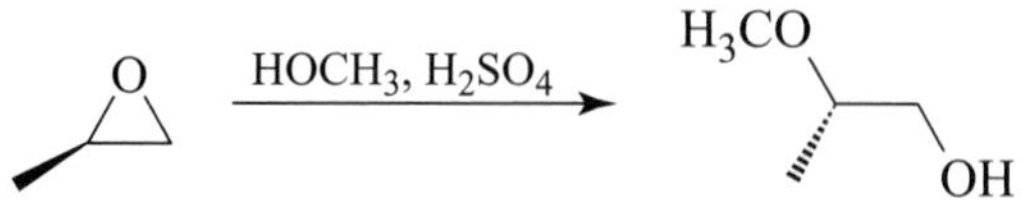 HOCH$_3$, H$_2$SO$_4$ → H$_3$CO, OH

(f) 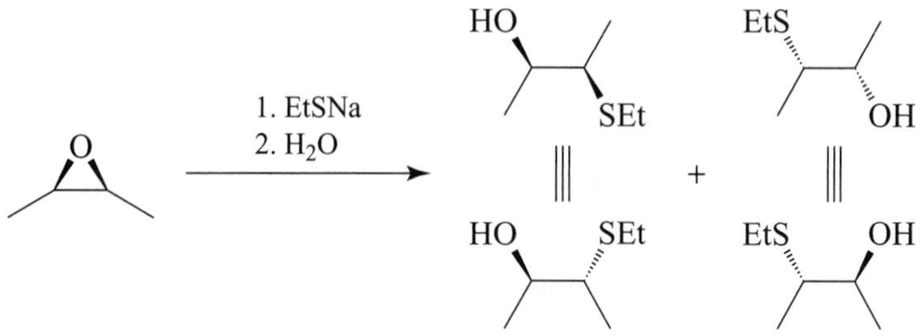 1. EtSNa 2. H$_2$O → HO, SEt ≡ HO, SEt + EtS, OH ≡ EtS, OH

(g) 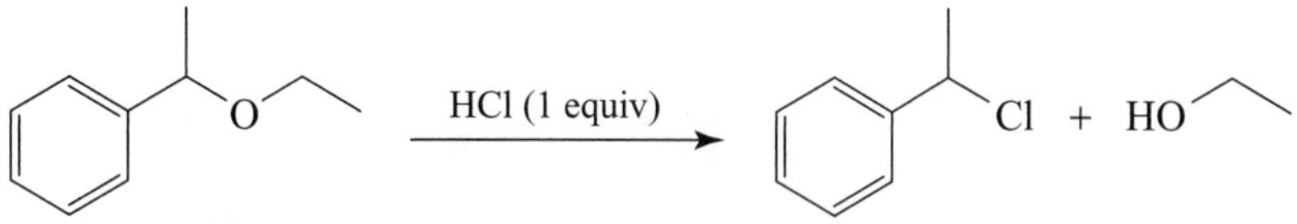 HCl (1 equiv) → Cl + HO

(h) MeONa → no rxn

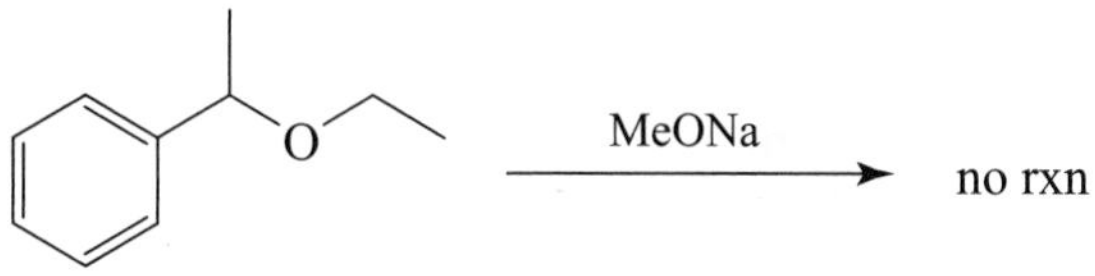

(i) 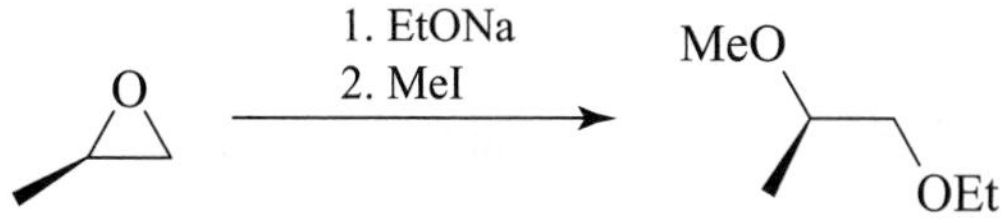 1. EtONa 2. MeI → MeO, OEt

(j) 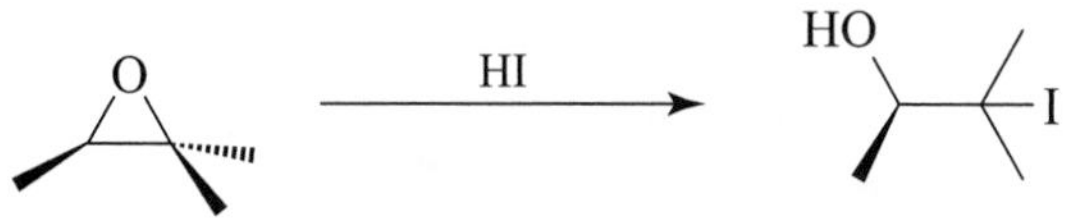 HI → HO, I

11.41 (a) 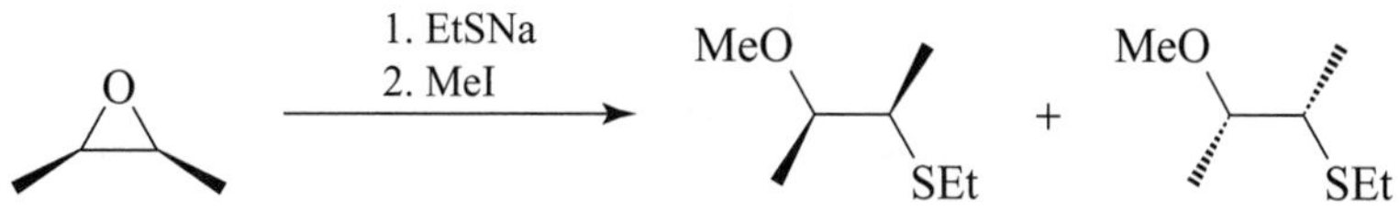 1. EtSNa 2. MeI → MeO, SEt + MeO, SEt

(b) 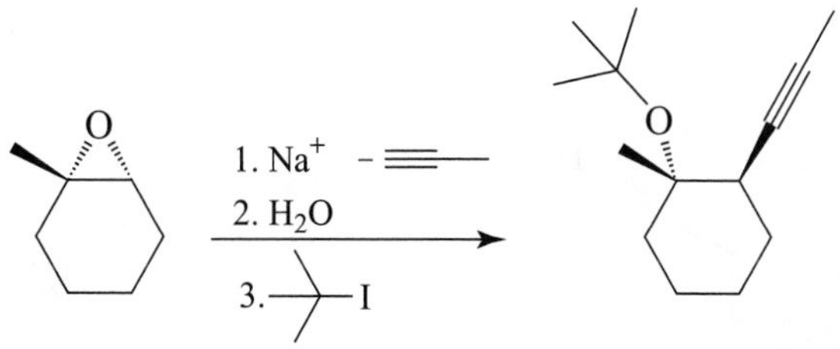 1. Na$^+$ $^-$≡— 2. H$_2$O 3. —I → O

(c) 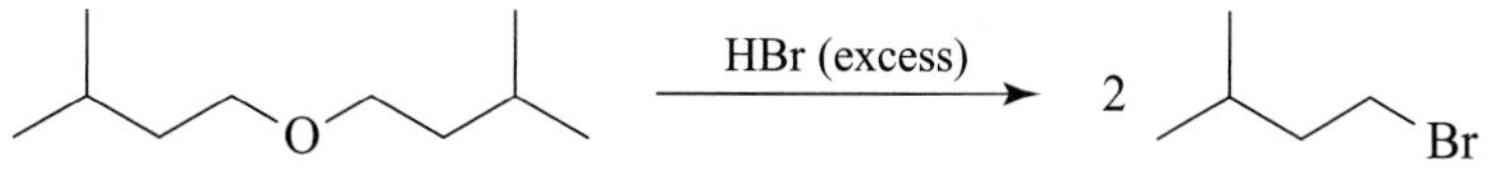 O, HBr (excess) → 2 Br

(d) 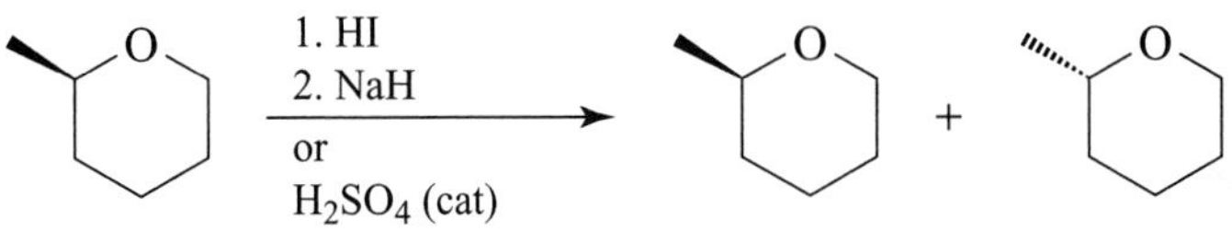 1. HI
2. NaH
or
H_2SO_4 (cat)
+

11.42 Cl_2 400°C Cl Cl_2 H_2O OH Cl Cl

OH^-, 1 eq **(b)**

OH HO OH OH^- xs **(a)** Cl O

Glycerol Epichlorohydrin

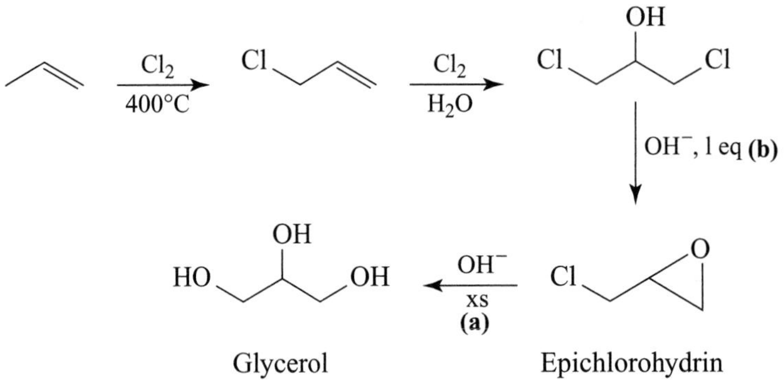

11.43 (a) **A** = 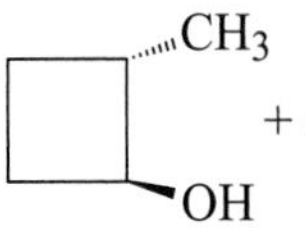 CH_3 OH + enantiomer **B** = 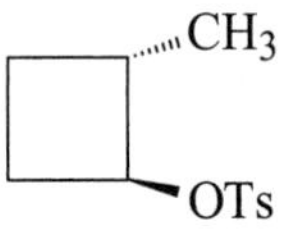 CH_3 OTs + enantiomer

C = 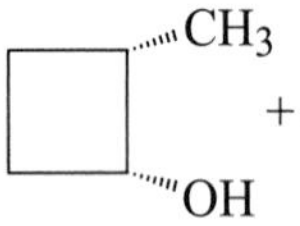 CH_3 OH + enantiomer

D = 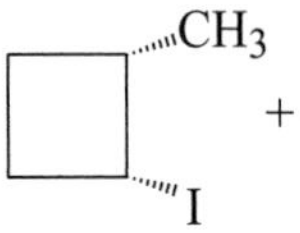 CH_3 I + enantiomer **A** and **C** are diastereomers.

(b) **E** = 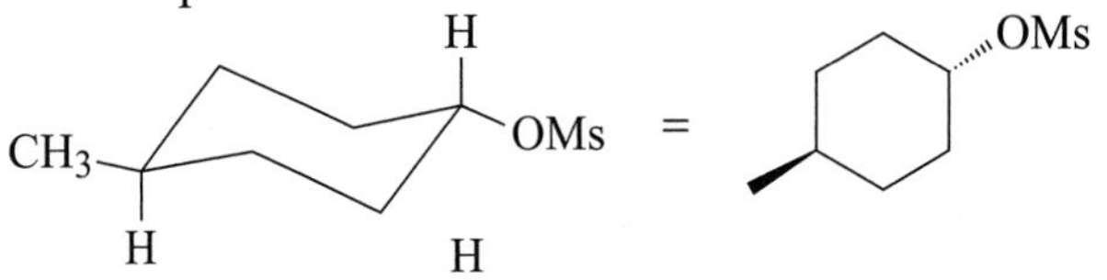 CH_3 H H OMs = OMs

F = CH_3 H H C C H H ⇌ CH_3 H C≡CH H

‖

H

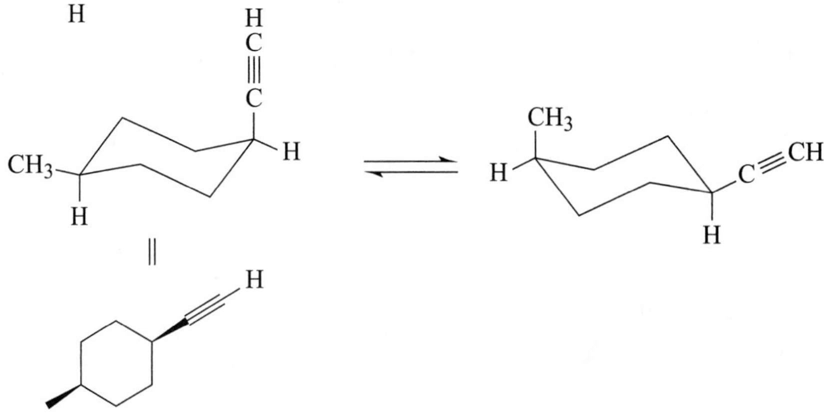

(c) **G** = $O^- Na^+$ **H** = 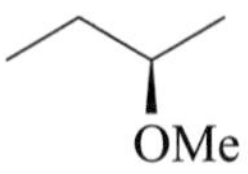 OMe **I** = 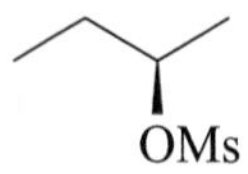 OMs

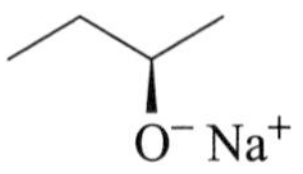

J = OMe

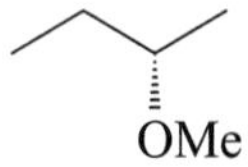

H and **J** are enantiomers.

Mechanisms

11.44 OH HA $\overset{+}{O}H_2$ $-H_2O$ H + H_2O H

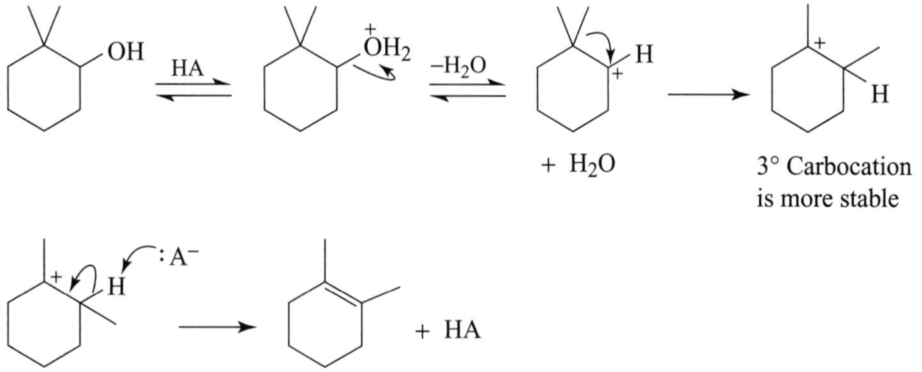

3° Carbocation is more stable

H $:A^-$ + HA

11.45 OH H O O—S—O O H $\ddot{O}H$

+O H O $^-:\ddot{O}$—S—O O H O

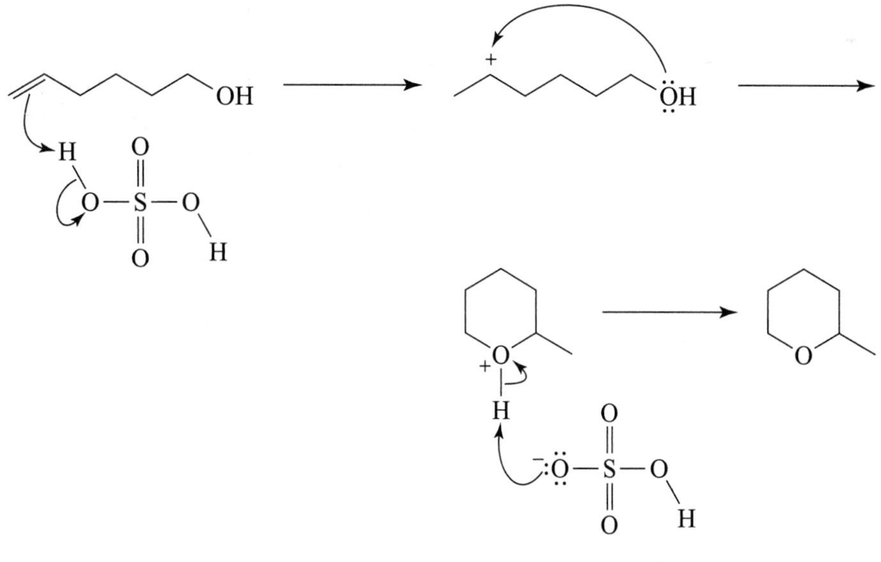

11.46 Br—Br

OH

$^+$Br

:ÖH

:Br:$^-$

H

$^+$O

Br

O

Br

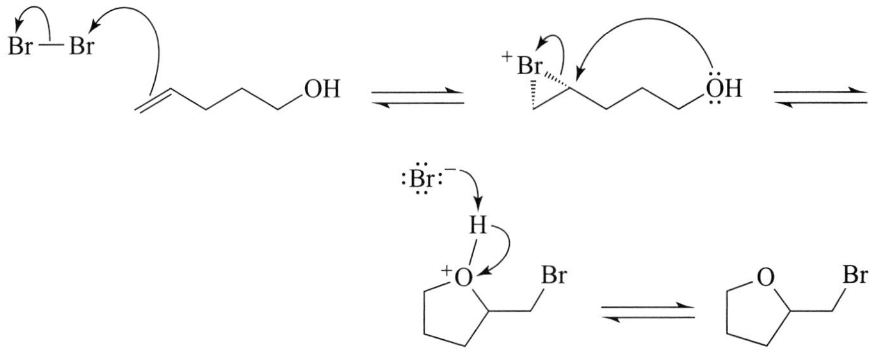

11.47

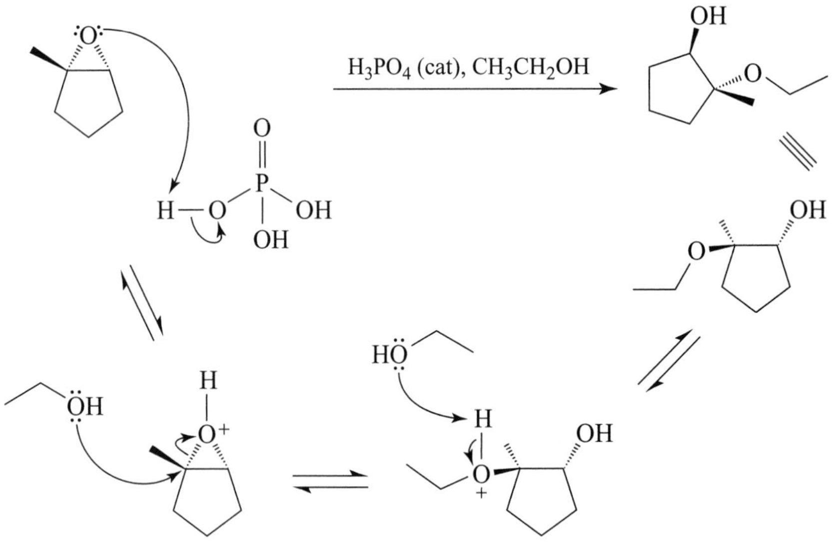

:O:

H_3PO_4 (cat), CH_3CH_2OH

OH

O

O

H—O P OH

OH

OH

O

OH

HÖ

ÖH

H

O$^+$

H

OH

O

+

11.48 (a)

+ H O—OH ⟶ O $\xrightarrow{HCl}$

Cl

+ enantiomer

OH

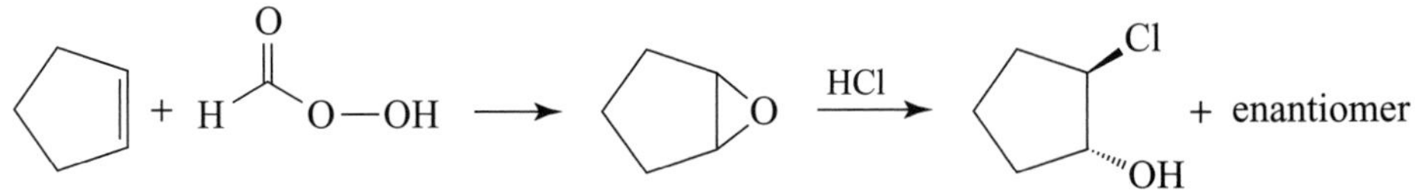

(b) The trans product because the Cl^- attacks anti to the epoxide and an inversion of configuration occurs.

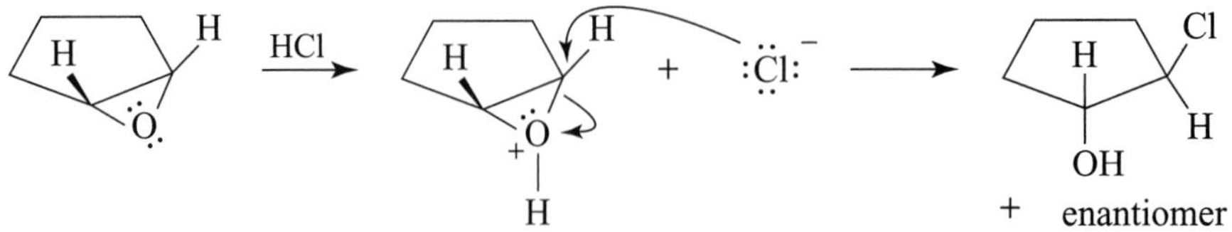

H H $\xrightarrow{HCl}$ H H + :Cl:$^-$ ⟶ Cl H H OH

O

O$^+$

H

+ enantiomer

11.49

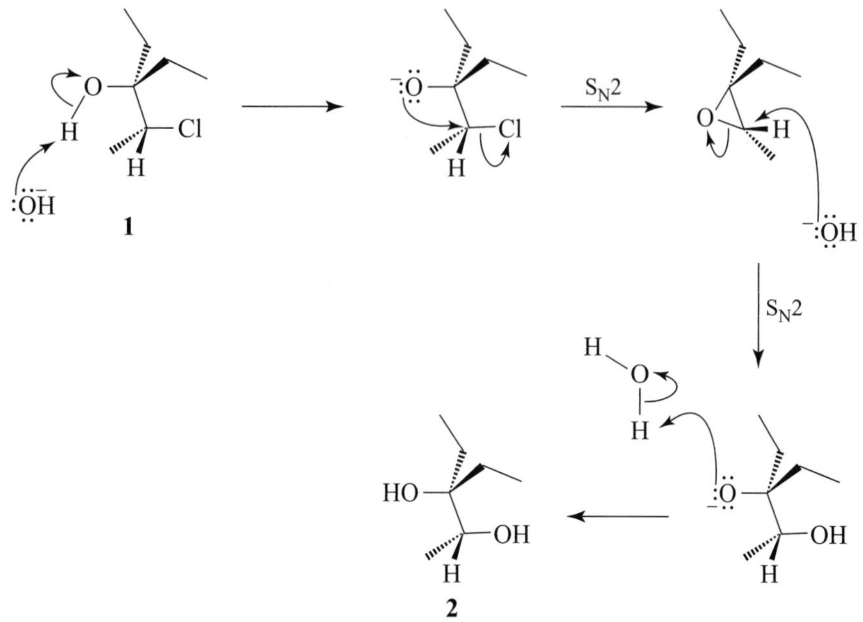

Two S_N2 reactions yield retention of configuration.

11.50 Collapse of the α-haloalcohol by loss of a proton and expulsion of a halide ion leads to the thermodynamically-favored carbonyl double bond. Practically speaking, the position of the following equilibrium is completely to the right.

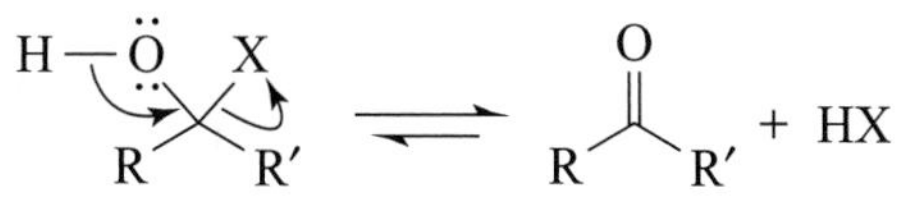

11.51

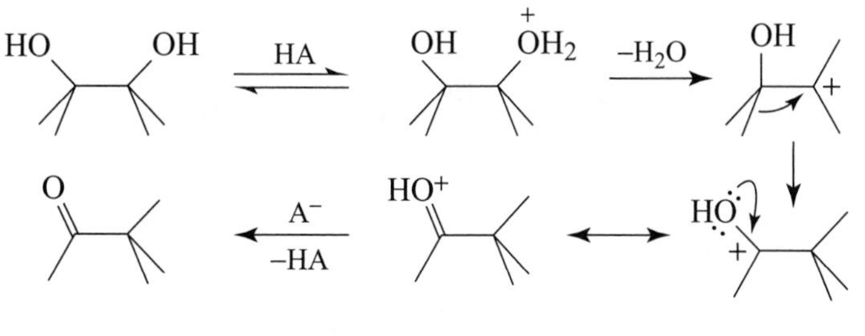

The reaction, known as the pinacol rearrangement, involves a 1,2-methanide shift to the positive center produced from the loss of the protonated —OH group.

11.52 The angular methyl group impedes attack by the peroxy acid on the front of the molecule (as drawn in the problem). II results from epoxidation from the back of the molecule—the less hindered side.

11.53 For ethylene glycol, hydrogen bonding provides stabilization of the gauche conformer. This cannot occur in the case of gauche butane.

H H Ö H δ+ H O δ− H H H

H H H H H only van der Waals repulsive forces

Challenge Problems

11.54 The reactions proceed through the formation of bromonium ions identical to those formed in the bromination of *trans*- and *cis*-2-butene (see Section 8.12A).

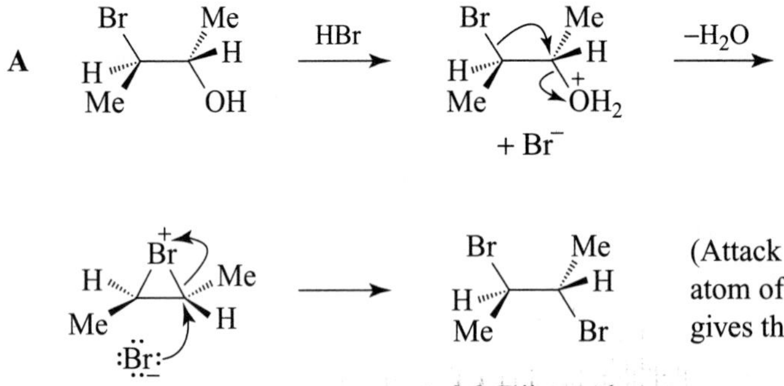

A

Br Me H H Me OH —HBr→ Br Me H H Me $\overset{+}{O}H_2$ + Br^- —$-H_2O$→

$\overset{+}{Br}$ H Me Me H :Br:⁻ ⟶ Br Me H H Me Br

meso-2,3-Dibromobutane

(Attack at the other carbon atom of the bromonium ion gives the same product.)

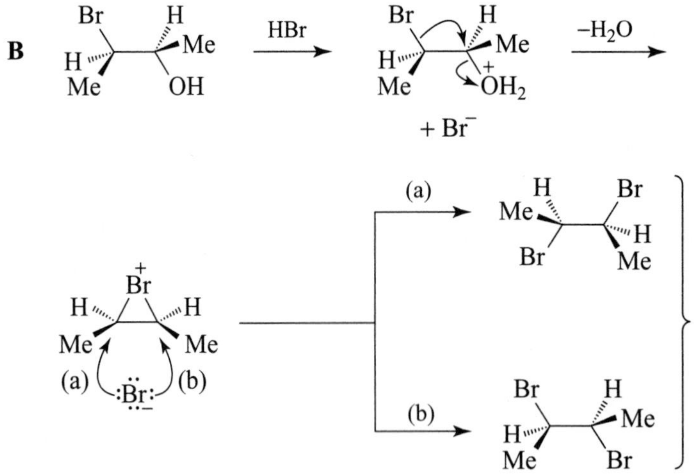

B

Br H H Me Me OH —HBr→ Br H H Me Me $\overset{+}{O}H_2$ + Br^- —$-H_2O$→

$\overset{+}{Br}$ H H Me Me (a) :Br:⁻ (b)

(a) → H Br Me H Br Me

(b) → Br H H Me Me Br

2,3-Dibromobutane (racemic)

11.55

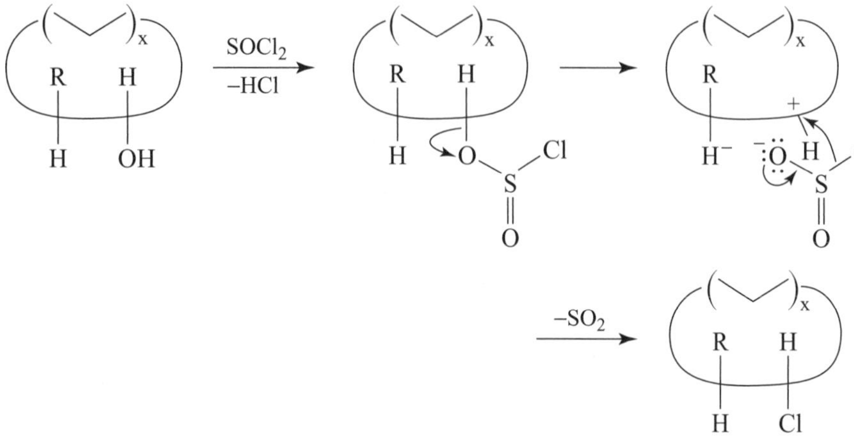

SOCl$_2$ / –HCl → –SO$_2$

11.56

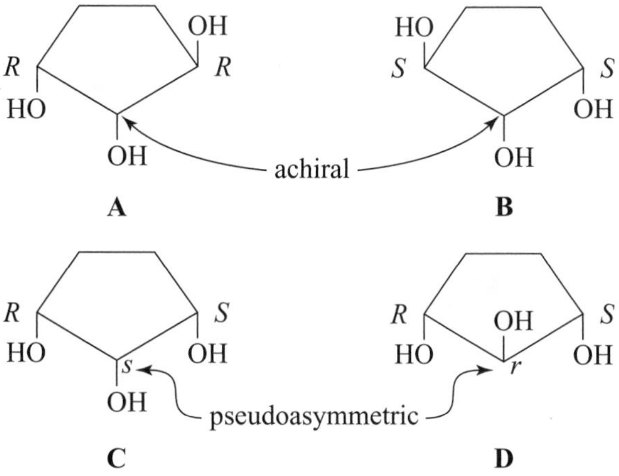

achiral

A B

pseudoasymmetric

C D

A and **B** are enantiomers
A, C, and **D** are all diastereomers
B, C, and **D** are all diastereomers
C is meso
D is meso

11.57

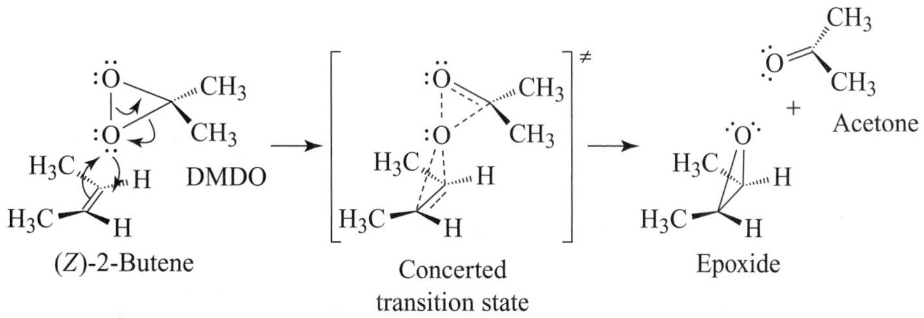

(*Z*)-2-Butene DMDO Concerted transition state Epoxide Acetone

11.58 The interaction of DMDO with (*Z*)-2-butene could take place with "syn" geometry, as shown below. In this approach, the methyl groups of DMDO lie over the methyl groups of (*Z*)-2-butene. This approach would be expected to have higher energy than that shown in the solution to Problem 11.57, an "anti" approach geometry. Computations have been done that indicate these relative energies. (Jenson, C.; Liu, J.; Houk, K.; Jorgenson, W. *J. Am. Chem. Soc.* **1997**, *199*, 12982–12983.)

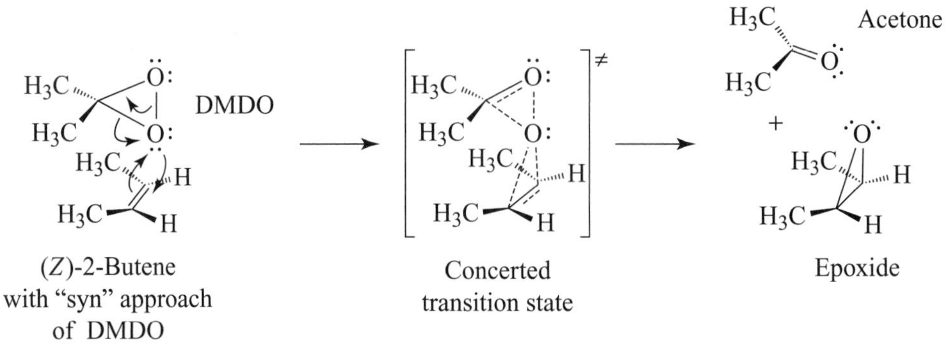

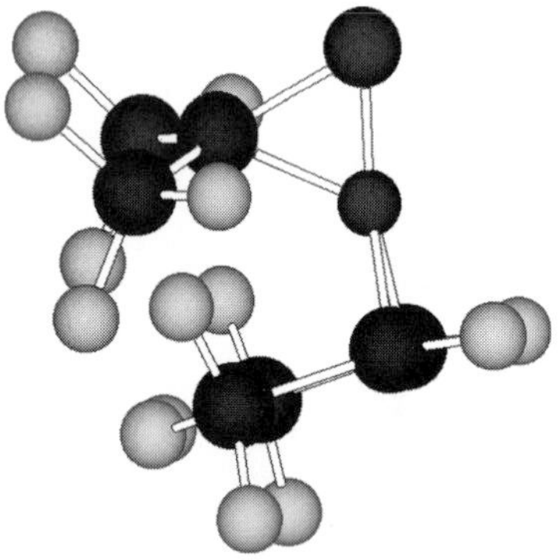

DMDO

(*Z*)-2-Butene with "syn" approach of DMDO

Concerted transition state

Acetone

Epoxide

Syn transition state

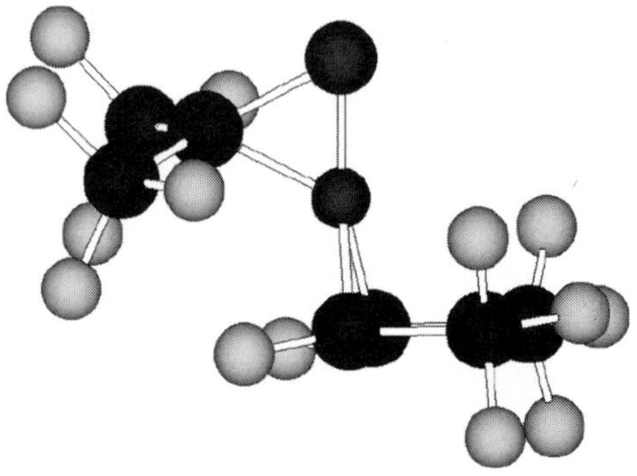

Anti transition state

QUIZ

11.1 Which set of reagents would effect the conversion,

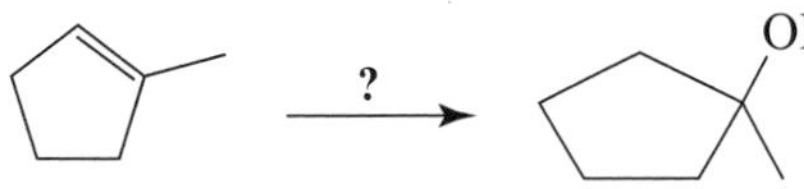

(a) BH_3:THF, then H_2O_2/OH^- (b) $Hg(OAc)_2$, THF-H_2O, then $NaBH_4/OH^-$

(c) H_3O^+, H_2O, heat (d) More than one of these (e) None of these

11.2 Which of the reagents in item 11.1 would effect the conversion,

+ enantiomer

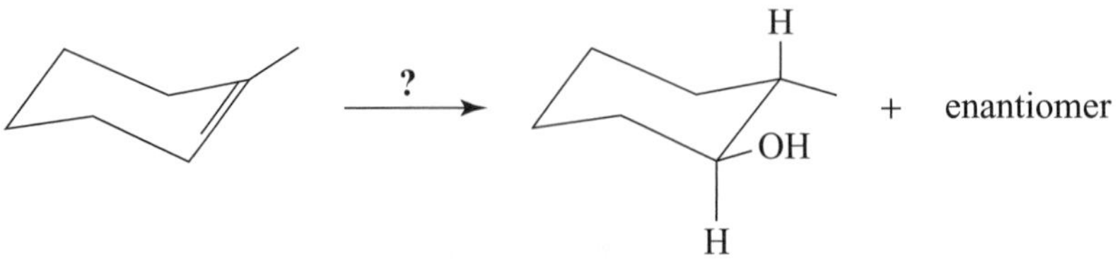

11.3 The following compounds have identical molecular weights. Which would have the lowest boiling point?

(a) 1-Butanol

(b) 2-Butanol

(c) 2-Methyl-1-propanol

(d) 1,1-Dimethylethanol

(e) 1-Methoxypropane

11.4 Complete the following synthesis:

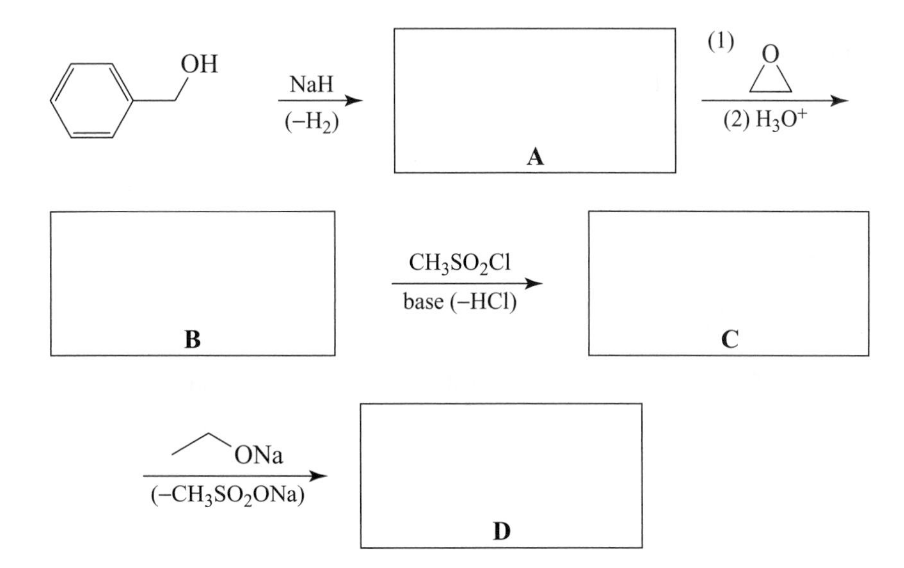

12 ALCOHOLS FROM CARBONYL COMPOUNDS: OXIDATION-REDUCTION AND ORGANOMETALLIC COMPOUNDS

SOLUTIONS TO PROBLEMS

12.1 (a)

Ethanol	Acetaldehyde	Acetic acid
Bonds to C_1	Bonds to C_1	Bonds to C_1
2 to H $= -2$	1 to H $= -1$	3 to O $= +3$
1 to O $= +1$	2 to O $= +2$	$=$ Oxid. state of C_1
Total $= -1$	Total $= +1$	Bonds to C_2
$=$ Oxid. state of C_1	$=$ Oxid. state of C_1	3 to H $= -3$
Bonds to C_2	Bonds to C_2	$=$ Oxid. state of C_2
3 to H $= -3$	3 to H $= -3$	
$=$ Oxid. state of C_2	$=$ Oxid. state of C_2	

(b) Only the carbon of ethanol bonded to the $-OH$ group undergoes a change in oxidation state, from -1 to $+1$ to $+3$. The oxidation state of the carbon of the CH_3- group remains unchanged.

12.2 (a)

Propene	Propane
Bonds to C_1	Bonds to C_1
2 to H $= -2$	3 to H $= -3$
$=$ Oxid. state of C_1	$=$ Oxid. state of C_1
Bonds to C_2	Bonds to C_2
1 to H $= -1$	2 to H $= -2$
$=$ Oxid. state of C_2	$=$ Oxid. state of C_2
Bonds to C_3	
3 to H $= -3$	
$=$ Oxid. state of C_3	

The oxidation states of both C_1 and C_2 decrease as a result of the addition of hydrogen to the double bond. Thus, the reaction can be considered as both an addition reaction and as a reduction reaction.

(b) The hydrogenation of acetaldehyde is not only an addition reaction, but it is also a *reduction* because the carbon atom of the C=O group goes from a $+1$ to a -1 oxidation state. The reverse reaction (the *dehydrogenation* of ethanol) is not only an *elimination* reaction, but also an *oxidation*.

Ion-Electron Half-Reaction Method for Balancing Organic Oxidation-Reduction Equations

Only two simple rules are needed:

Rule 1 Electrons (e^-) together with protons (H^+) are arbitrarily considered the reducing agents in the half-reaction for the reduction of the oxidizing agent. Ion charges are balanced by *adding electrons to the left-hand side.* (If the reaction is run in neutral or basic solution, add an equal number of OH^- ions to both sides of the balanced half-reaction to neutralize the H^+, and show the resulting $H^+ + OH^-$ as H_2O.)

Rule 2 Water (H_2O) is arbitrarily taken as the formal source of oxygen for the oxidation of the organic compound, producing *product, protons*, and *electrons* on the right-hand side. (Again, use OH^- to neutralize H^+ in the *balanced* half-reaction in neutral or basic media.)

EXAMPLE 1

Write a balanced equation for the oxidation of RCH_2OH to RCO_2H by $Cr_2O_7^{2-}$ in acid solution.

Reduction half-reaction:

$$Cr_2O_7^{2-} + H^+ + e^- = 2Cr^{3+} + 7H_2O$$

Balancing atoms and charges:

$$Cr_2O_7^{2-} + 14H^+ + 6e^- = 2Cr^{3+} + 7H_2O$$

Oxidation half-reaction:

$$RCH_2OH + H_2O = RCO_2H + 4H^+ + 4e^-$$

The least common multiple of a 6-electron uptake in the reduction step and a 4-electron loss in the oxidation step is 12, so we multiply the first half-reaction by 2 and the second by 3, and add:

$$3RCH_2OH + 3H_2O + 2Cr_2O_7^{2-} + 28H^+ = 3RCO_2H + 12H^+ + 4Cr^{3+} + 14H_2O$$

Canceling common terms, we get:

$$3RCH_2OH + 2Cr_2O_7^{2-} + 16H^+ \longrightarrow 3RCO_2H + 4Cr^{3+} + 11H_2O$$

This shows that the oxidation of 3 mol of a primary alcohol to a carboxylic acid requires 2 mol of dichromate.

EXAMPLE 2

Write a balanced equation for the oxidation of styrene to benzoate ion and carbonate ion by MnO_4^- in alkaline solution.

Reduction half-reaction:

$$MnO_4^- + 4H^+ + 3e^- = MnO_2 + 2H_2O \text{ (in acid)}$$

Since this reaction is carried out in basic solution, we must add 4 OH^- to neutralize the $4H^+$ on the left side, and, of course, 4 OH^- to the right side to maintain a balanced equation.

$$MnO_4^- + 4H^+ + 4\,OH^- + 3e^- = MnO_2 + 2H_2O + 4\,OH^-$$

or, $\quad MnO_4^- + 2H_2O + 3e^- = MnO_2 + 4\,OH^-$

Oxidation half-reaction:

$$ArCH{=}CH_2 + 5H_2O = ArCO_2^- + CO_3^{2-} + 13H^+ + 10e^-$$

We add 13 OH^- to each side to neutralize the H^+ on the right side,

$$ArCH{=}CH_2 + 5H_2O + 13\,OH^- = ArCO_2^- + CO_3^{2-} + 13H_2O + 10e^-$$

The least common multiple is 30, so we multiply the reduction half-reaction by 10 and the oxidation half-reaction by 3 and add:

$$3ArCH{=}CH_2 + 39\,OH^- + 10MnO_4^- + 20H_2O = 3ArCO_2^- + 3CO_3^{2-} +$$
$$24H_2O + 10MnO_2 + 40\,OH^-$$

Canceling:

$$3ArCH{=}CH_2 + 10MnO_4^- \longrightarrow 3ArCO_2^- + 3CO_3^{2-} + 4H_2O + 10MnO_2 + OH^-$$

SAMPLE PROBLEMS

Using the ion-electron half-reaction method, write balanced equations for the following oxidation reactions.

(a) Cyclohexene + $MnO_4^- + H^+ \xrightarrow{\text{(hot)}} HO_2C(CH_2)_4CO_2H + Mn^{2+} + H_2O$

(b) Cyclopentene + $MnO_4^- + H_2O \xrightarrow{\text{(cold)}}$ *cis*-1,2-cyclopentanediol + $MnO_2 + OH^-$

(c) Cyclopentanol + $HNO_3 \xrightarrow{\text{(hot)}} HO_2C(CH_2)_3CO_2H + NO_2 + H_2O$

(d) 1,2,3-Cyclohexanetriol + $HIO_4 \xrightarrow{\text{(cold)}} OCH(CH_2)_3CHO + HCO_2H + HIO_3$

SOLUTIONS TO SAMPLE PROBLEMS

(a) Reduction:

$$MnO_4^- + 8\,H^+ + 5e^- = Mn^{2+} + 4\,H_2O$$

Oxidation:

$$\text{cyclohexene (H, H)} + 4\,H_2O = \text{HO}_2\text{C(CH}_2)_4\text{CO}_2\text{H} + 8\,H^+ + 8\,e^-$$

The least common multiple is 40:

$$8\,MnO_4^- + 64\,H^+ + 40\,e^- = 8Mn^{2+} + 32\,H_2O$$

$$5\ \text{cyclohexene (H, H)} + 20\,H_2O = 5\ \text{HO}_2\text{C(CH}_2)_4\text{CO}_2\text{H} + 40\,H^+ + 40\,e^-$$

Adding and canceling:

$$5\ \text{cyclohexene (H, H)} + 8\,MnO_4^- + 24\,H^+ \longrightarrow 5\ \text{HO}_2\text{C(CH}_2)_4\text{CO}_2\text{H} + 8\,Mn^{2+} + 12\,H_2O$$

(b) Reduction:

$$MnO_4^- + 2\,H_2O + 3e^- = MnO_2 + 4\,OH^-$$

Oxidation:

$$\text{cyclopentene} + 2\,OH^- = \text{cyclopentane-1,2-diol (OH, OH)} + 2\,e^-$$

The least common multiple is 6:

$$2MnO_4^- + 4\,H_2O + 6e^- = 2\,MnO_2 + 8\,OH^-$$

$$3\ \text{cyclopentene} + 6\,OH^- = 3\ \text{cyclopentane-1,2-diol (OH, OH)} + 6\,e^-$$

Adding and canceling:

$$3\ \text{cyclopentene} + 2\,MnO_4^- + 4\,H_2O \longrightarrow 3\ \text{cyclopentane-1,2-diol (OH, OH)} + 2\,MnO_2 + 2\,OH^-$$

(c) Reduction:

$HNO_3 + H^+ + e^- = NO_2 + H_2O$

Oxidation:

$$\text{cyclopentanol (OH)} + 3\,H_2O = \text{(CH}_2)_3(CO_2H)_2 + 8\,H^+ + 8\,e^-$$

The least common multiple is 8:

$8\,HNO_3 + 8\,H^+ + 8\,e^- = 8\,NO_2 + 8\,H_2O$

$$\text{cyclopentanol (OH)} + 3\,H_2O = \text{(CH}_2)_3(CO_2H)_2 + 8\,H^+ + 8\,e^-$$

Adding and canceling:

$$\text{cyclopentanol (OH)} + 8\,HNO_3 \longrightarrow \text{(CH}_2)_3(CO_2H)_2 + 8\,NO_2 + 5\,H_2O$$

(d) Reduction:

$HIO_4 + 2\,H^+ + 2e^- = HIO_3 + H_2O$

Oxidation:

$$\text{cyclohexane-1,2,3-triol (HO, H; OH, H; OH, H)} + H_2O = \text{(CH}_2)_3(CHO)_2 + HC(=O){-}OH + 4\,H^+ + 4\,e^-$$

The least common multiple is 4:

$2\,HIO_4 + 4\,H^+ + 4\,e^- = 2\,HIO_3 + 2H_2O$

$$\text{cyclohexane-1,2,3-triol (HO, H; OH, H; OH, H)} + H_2O = \text{(CH}_2)_3(CHO)_2 + HC(=O){-}OH + 4\,H^+ + 4\,e^-$$

Adding and canceling:

$$\text{cyclohexane-1,2,3-triol (HO, H; OH, H; OH, H)} + 2\,HIO_4 = \text{(CH}_2)_3(CHO)_2 + HC(=O){-}OH + 2\,HIO_3 + H_2O$$

12.3 (a) $LiAlH_4$

(b) $LiAlH_4$

(c) $NaBH_4$

12.4 (a) 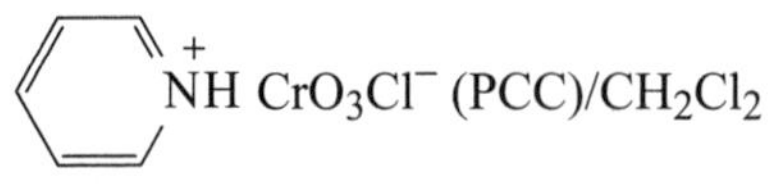 $\overset{+}{N}H$ CrO_3Cl^- (PCC)/CH_2Cl_2

(b) $KMnO_4$, OH^-, H_2O, heat; then H_3O^+ [or conditions as in (c) below]

(c) H_2CrO_4/acetone [or conditions as in (b) above]

(d) (1) O_3 (2) Me_2S

12.5 (a) 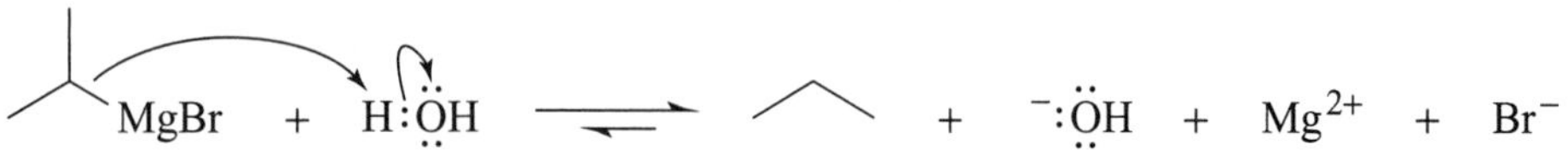 MgBr + H:ÖH ⇌ + $^-$:ÖH + Mg^{2+} + Br^-

pK_a 15.7 pK_a ~ 50

Stronger base Stronger acid Weaker acid Weaker base

(b) 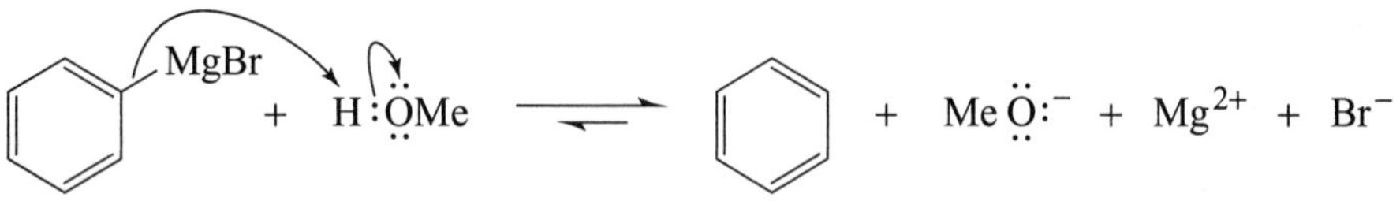 MgBr + H:ÖMe ⇌ + Me Ö:$^-$ + Mg^{2+} + Br^-

pK_a 15.5 pK_a ~ 50

Stronger base Stronger acid Weaker acid Weaker base

(c) 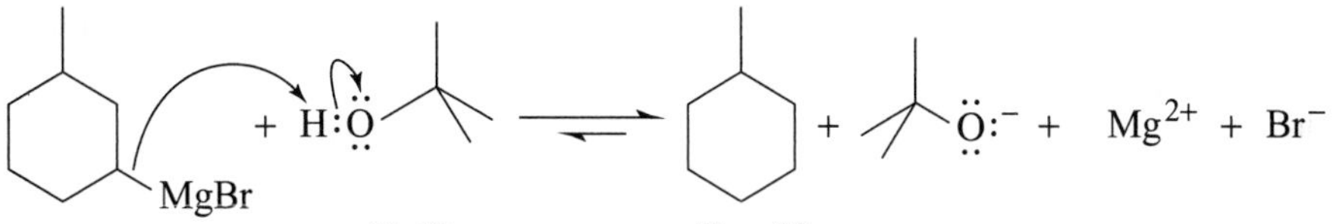 MgBr + H:Ö ⇌ + Ö:$^-$ + Mg^{2+} + Br^-

pK_a 18 pK_a ~ 50

Stronger base Stronger acid Weaker acid Weaker base

(d)  Li + H:Ö ⇌ + Ö:$^-$ + Li^+

pK_a 18 pK_a ~ 50

Stronger base Stronger acid Weaker acid Weaker base

12.6 (a) 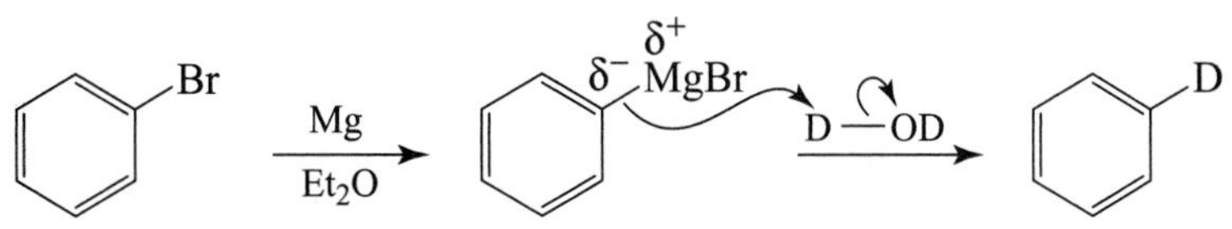 Br $\xrightarrow[Et_2O]{Mg}$ δ^- $\overset{\delta^+}{MgBr}$ $\xrightarrow{D—OD}$ D

(b) 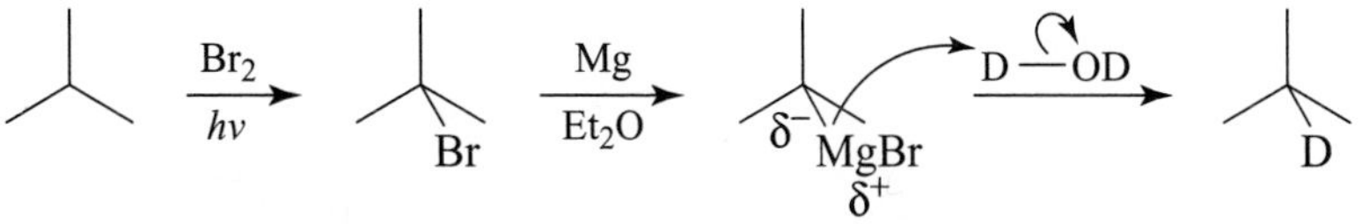 $\xrightarrow[hv]{Br_2}$ Br $\xrightarrow[Et_2O]{Mg}$ δ^- $\underset{\delta^+}{MgBr}$ $\xrightarrow{D—OD}$ D

12.7 Benzoyl chloride + BrMg–Ph → [tetrahedral intermediate: Cl, O^- $\overset{+}{Mg}Br$] $\xrightarrow{-MgBrCl}$ [benzophenone] $\xrightarrow{BrMgPh}$ $Ph_3C–O^-$ $\overset{+}{Mg}Br$ $\xrightarrow{NH_4Cl/H_2O}$ $Ph_3C–OH$

12.8 (a) (1) 2-methyl-2-butanol (OH) ⟹ 2-butanone (O) + CH_3MgI

2-butanone (O) + CH_3MgI $\xrightarrow[(2)\ NH_4^+/H_2O]{(1)\ ether}$ 2-methyl-2-butanol (OH)

(2) 2-methyl-2-butanol (OH) ⟹ acetone (O) + ethyl MgBr

acetone (O) + ethyl MgBr $\xrightarrow[(2)\ NH_4^+/H_2O]{(1)\ ether}$ 2-methyl-2-butanol (OH)

(3) 2-methyl-2-butanol (OH) ⟹ ethyl propanoate (O, O) + 2 CH_3MgI

ethyl propanoate (O, O) + 2 CH_3MgI $\xrightarrow[(2)\ NH_4^+/\ H_2O]{(1)\ ether}$ 2-methyl-2-butanol (OH)

(b) (1) 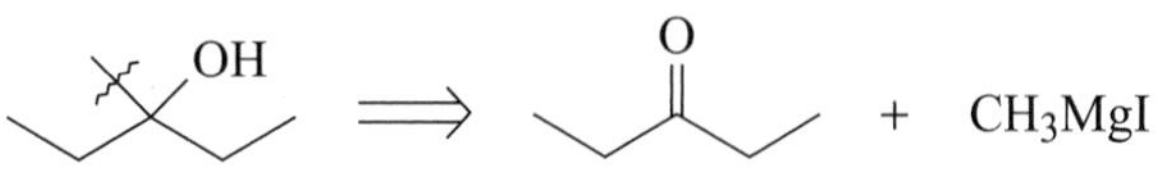 OH ⟹ O + CH_3MgI

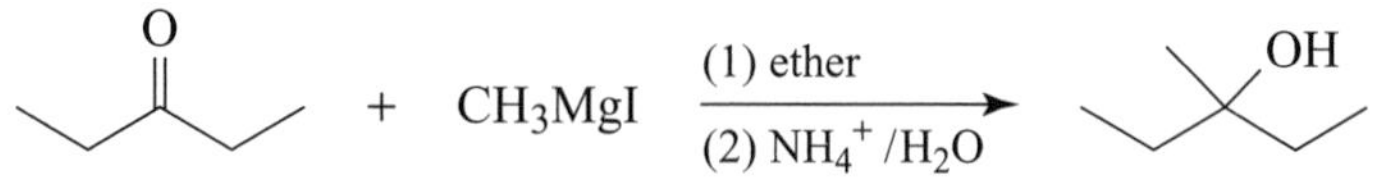

O + CH_3MgI (1) ether (2) 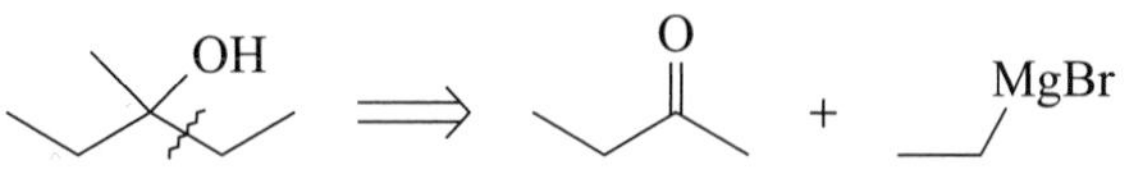 NH_4^+/H_2O → OH

(2) OH ⟹ O + MgBr

O + MgBr (1) ether (2) NH_4^+/H_2O → OH

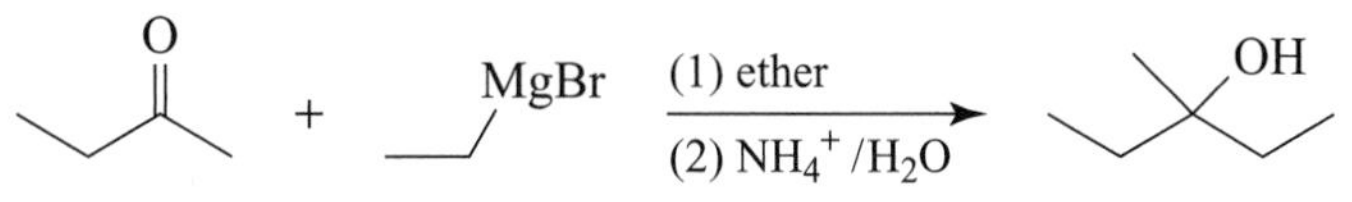

(3) 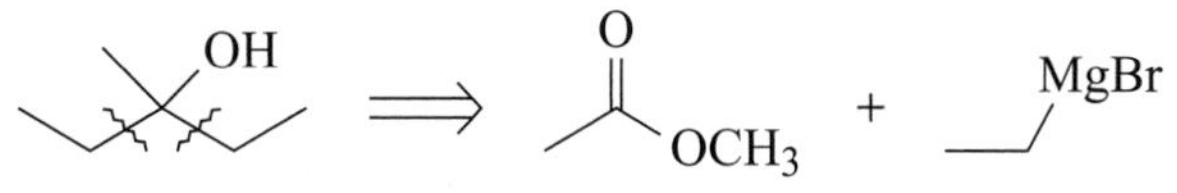 OH ⟹ O OCH_3 + MgBr

O OCH_3 + 2 MgBr (1) ether (2) NH_4^+/H_2O → OH

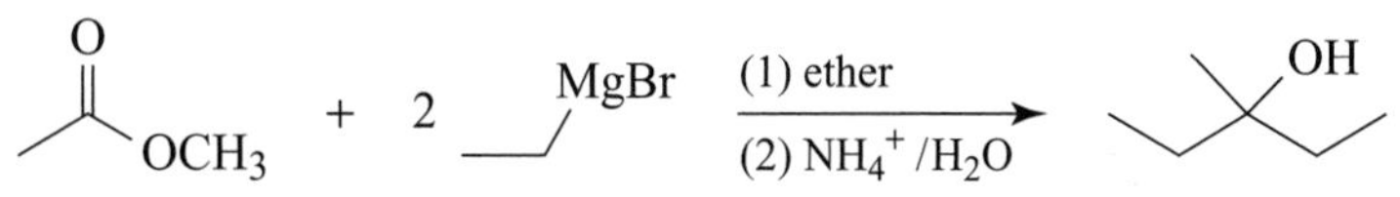

(c) (1) 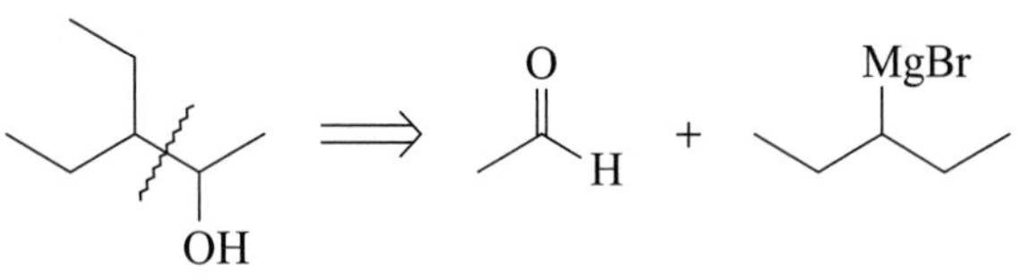 OH ⟹ O H + MgBr

O H + MgBr (1) ether (2) 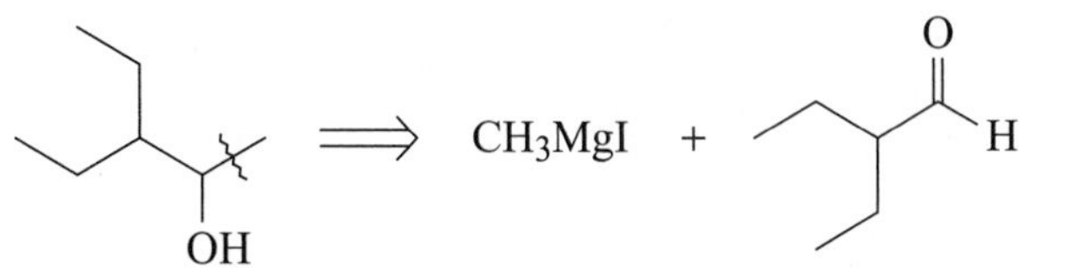 H_3O^+ → OH

(2) OH ⟹ CH_3MgI + O H

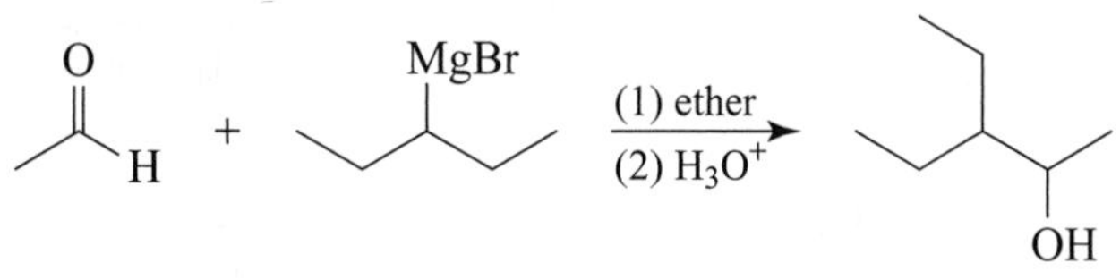

CH_3MgI + O H (1) ether (2) H_3O^+ → OH

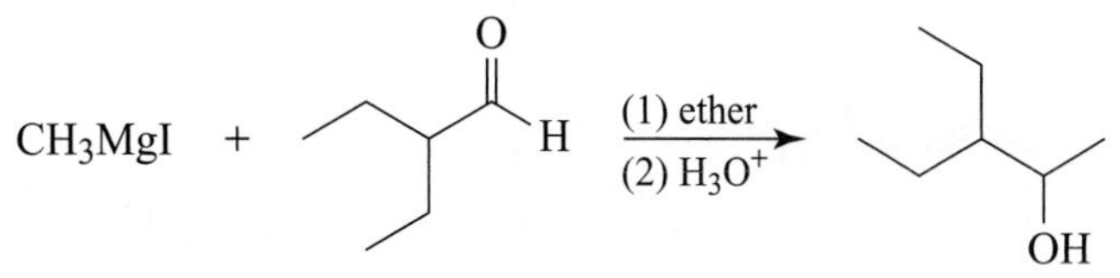

(d) (1) ⟹ + MgBr

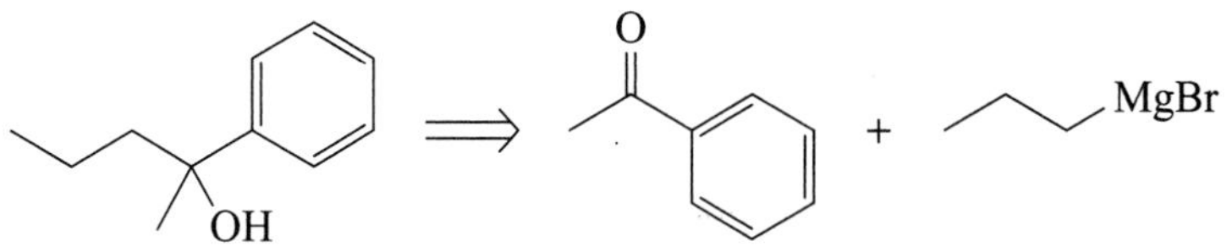

+ MgBr $\xrightarrow[\text{(2) } NH_4^+/H_2O]{\text{(1) ether}}$ OH

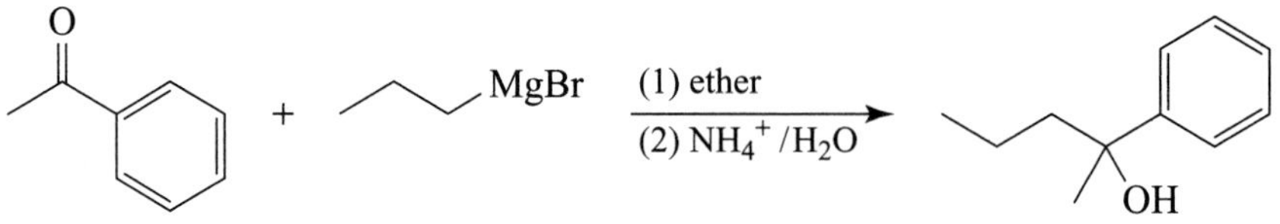

(2) ⟹ + CH_3MgI

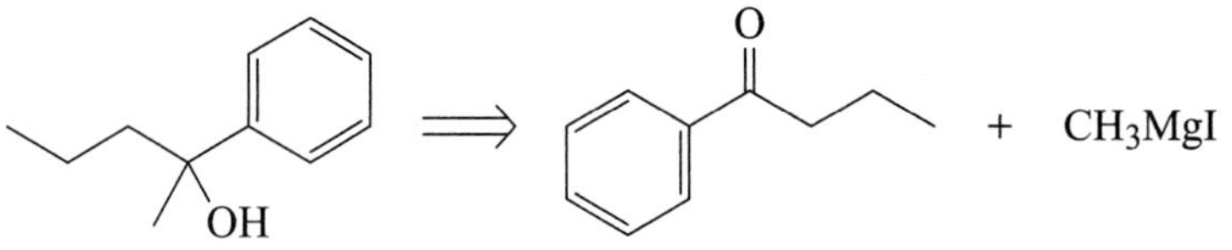

+ CH_3MgI $\xrightarrow[\text{(2) } NH_4^+/H_2O]{\text{(1) ether}}$ OH

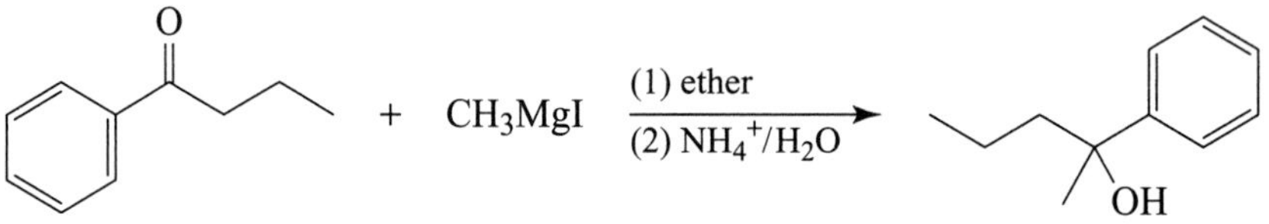

(3) ⟹ + MgBr

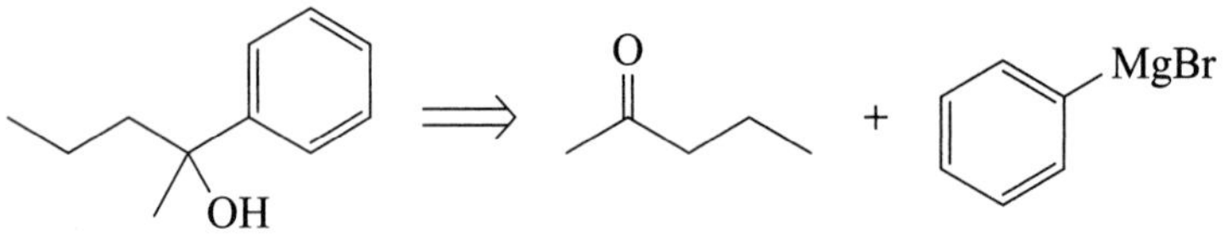

+ MgBr $\xrightarrow[\text{(2) } NH_4^+/H_2O]{\text{(1) ether}}$ OH

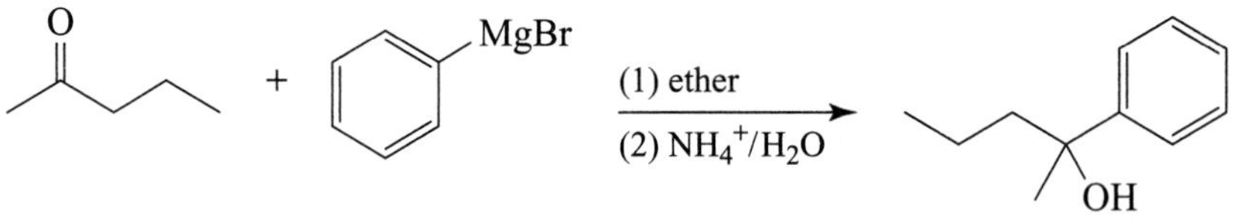

(e) (1) OH ⟹ + 2 MgBr

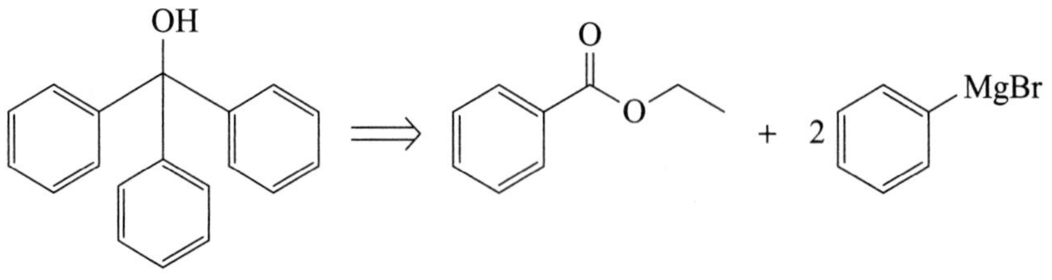

+ 2 MgBr $\xrightarrow[\text{(2) } NH_4^+]{\text{(1) ether}}$ OH

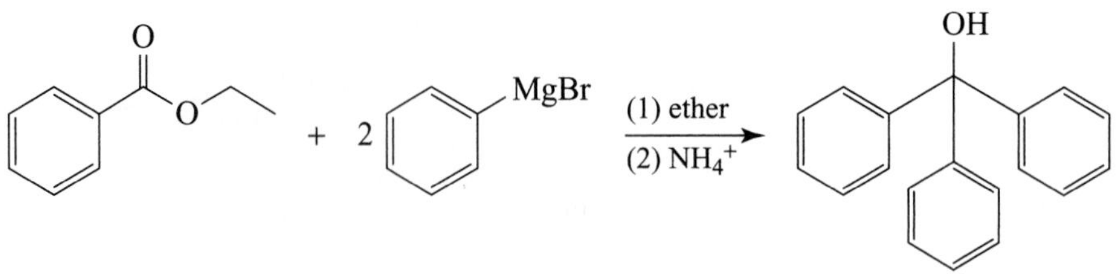

(2) OH ⟹ O + MgBr

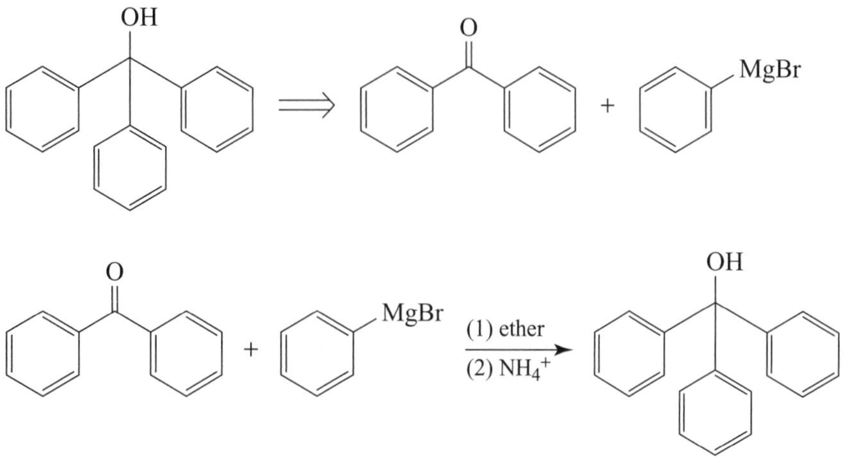

O + MgBr $\xrightarrow[(2)\ NH_4^+]{(1)\ ether}$ OH

(f) (1) OH ⟹ MgBr + O, H H ⟹

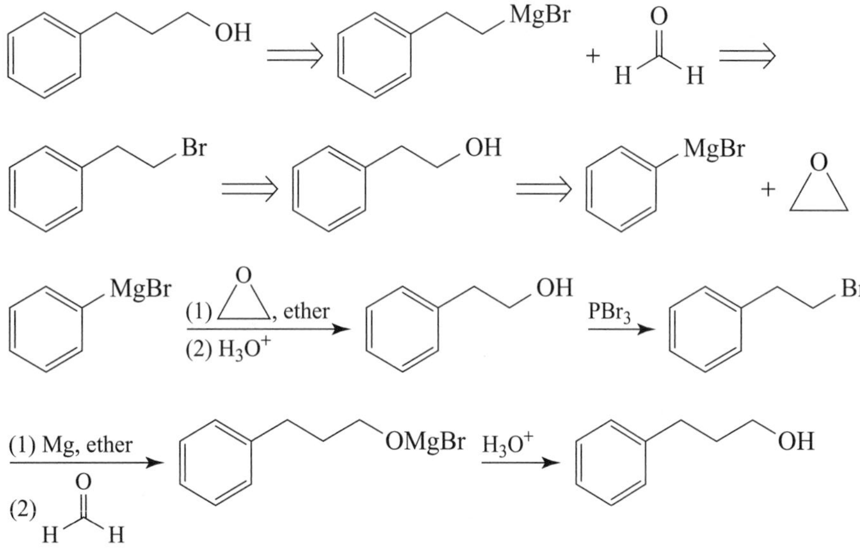

Br ⟹ OH ⟹ MgBr + O

MgBr $\xrightarrow[(2)\ H_3O^+]{(1)\ \text{O}, ether}$ OH $\xrightarrow{PBr_3}$ Br

$\xrightarrow[(2)\ \text{O, H H}]{(1)\ Mg,\ ether}$ OMgBr $\xrightarrow{H_3O^+}$ OH

(2) OH ⟹ MgBr + O ⟹ Br

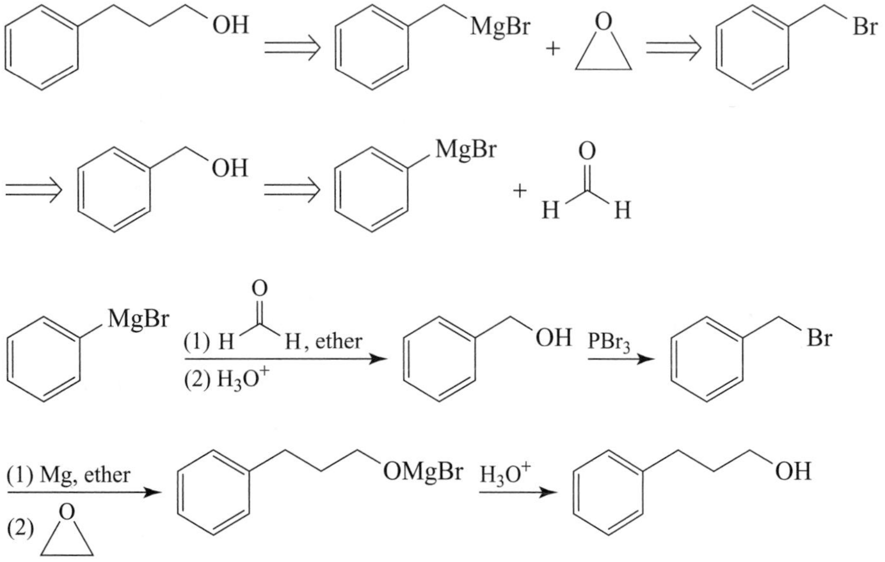

⟹ OH ⟹ MgBr + O, H H

MgBr $\xrightarrow[(2)\ H_3O^+]{(1)\ \text{H O H}, ether}$ OH $\xrightarrow{PBr_3}$ Br

$\xrightarrow[(2)\ \text{O}]{(1)\ Mg,\ ether}$ OMgBr $\xrightarrow{H_3O^+}$ OH

12.9 (a)

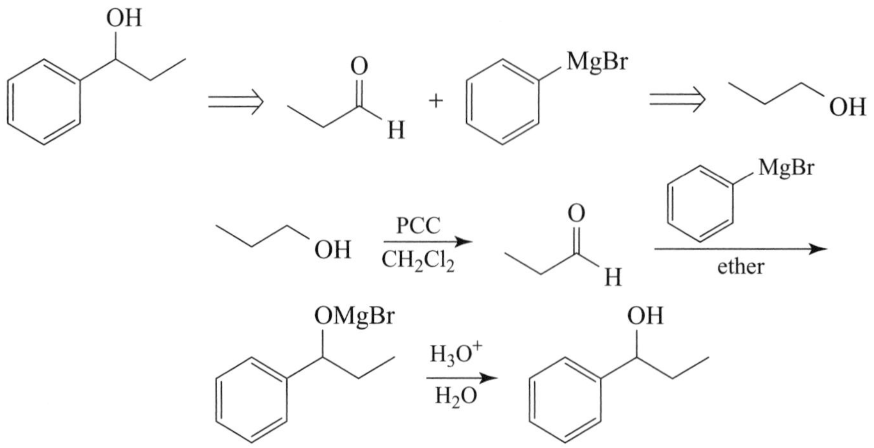

OH; H; MgBr; OH

OH; PCC; CH_2Cl_2; O; H; MgBr; ether

OMgBr; H_3O^+; H_2O; OH

(b)

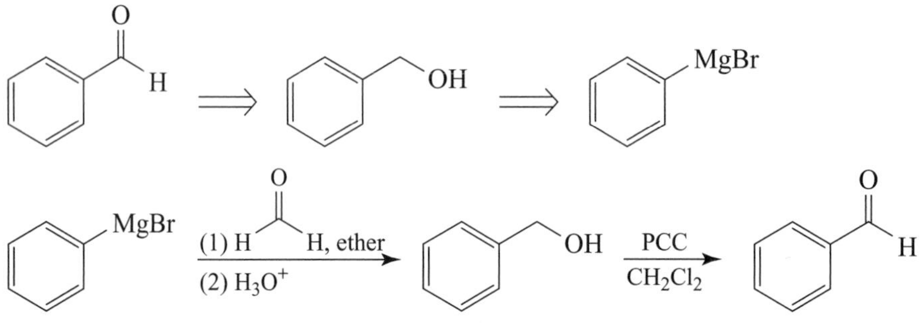

O; H; OH; MgBr

MgBr; (1) H, O, H, ether; (2) H_3O^+; OH; PCC; CH_2Cl_2; O; H

(c)

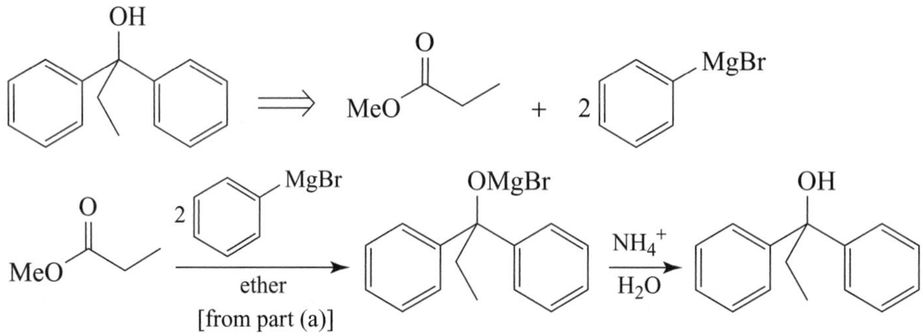

OH; O; MeO; + 2; MgBr

O; MeO; 2; MgBr; ether; [from part (a)]; OMgBr; NH_4^+; H_2O; OH

(d)

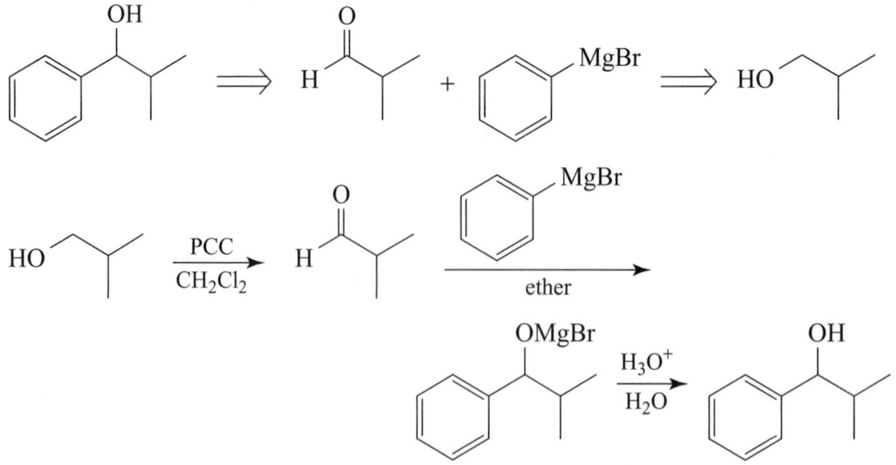

OH; O; H; +; MgBr; HO

HO; PCC; CH_2Cl_2; O; H; MgBr; ether

OMgBr; H_3O^+; H_2O; OH

Problems

Reagents and Reactions

12.10 (a) CH_3CH_3 (b) D (c) OH Ph

(d) OH Ph Ph (e) OH Ph (f) OH Ph

(g) CH_3CH_3 + OH

12.11 (a)

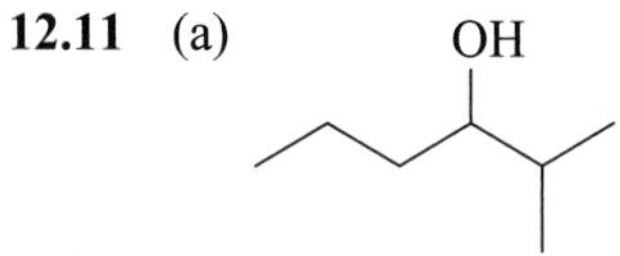

(b) OH

(c) + OH (d) + OLi

(e) D + O OLi

12.12 (a) OH + (b) N (c)

(d) OCH_3 + (e) OH (f) OH

(g) OH (h) OH (i) OH

(j) + CH_3OLi (k) + H—≡—Li

12.13 (a) $LiAlH_4$ (d) (1) $KMnO_4$, OH^-, heat, (2) H_3O^+
(b) $NaBH_4$ (or H_2CrO_4/acetone)
(c) $LiAlH_4$ (e) PCC/CH_2Cl_2

12.14 (a) EtMgBr + EtO(C=O)OEt ⟶ [EtO–C(OMgX)(Et)–OEt] $\xrightarrow{-EtOMgX}$ [Et(C=O)OEt] $\xrightarrow{EtMgX}$ [Et–C(OMgX)(Et)–OEt] $\xrightarrow{-EtOMgX}$ [Et(C=O)Et] $\xrightarrow{EtMgX}$ Et–C(OMgX)(Et)–Et $\xrightarrow[H_2O]{NH_4^+}$ Et_3C–OH

(b) MeMgBr + H(C=O)OEt ⟶ [H–C(OMgX)(Me)–OEt] $\xrightarrow{-EtOMgX}$ [H(C=O)Me] $\xrightarrow{MeMgX}$ H–C(OMgX)(Me)–Me $\xrightarrow{H_3O^+}$ H–C(Me)(Me)–OH

12.15 (a) OH H ... O $\xrightarrow[2.\ H_2O]{1.\ NaBH_4,\ EtOH}$ OH ... OH

(b) O O $\xrightarrow[2.\ H_2O/H_2SO_4]{1.\ LiAlH_4,\ THF}$ OH OH

(c) 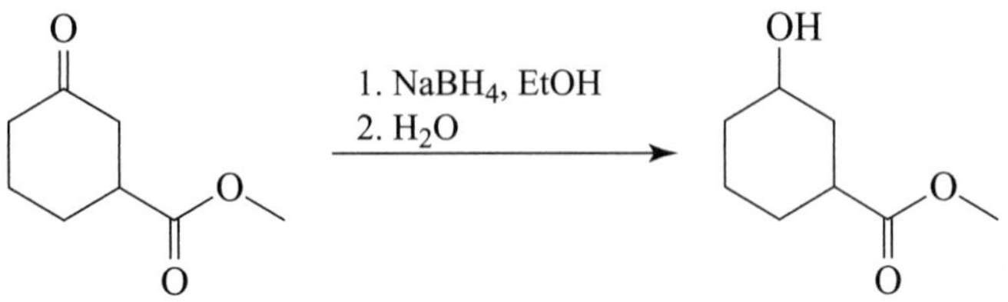 O ... O, O → OH ... O, O

1. $NaBH_4$, EtOH
2. H_2O

12.16 (a) 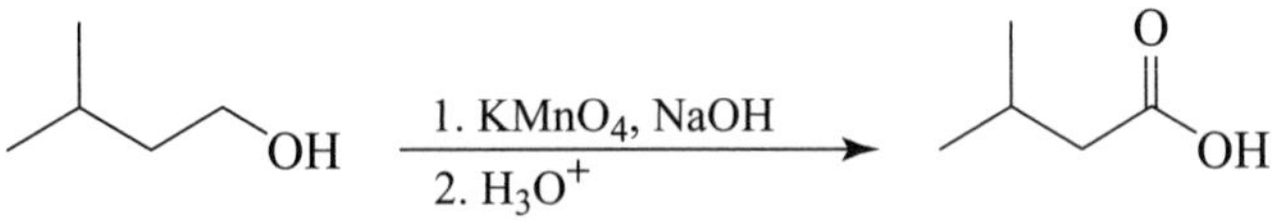 OH → O, OH

1. $KMnO_4$, NaOH
2. H_3O^+

(b) 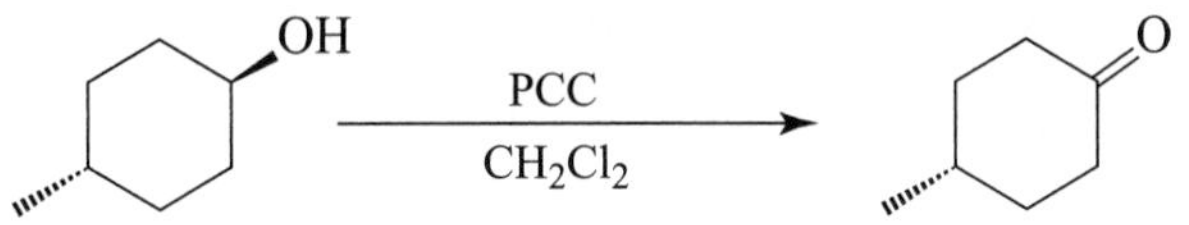 OH → O

PCC
CH_2Cl_2

(c) 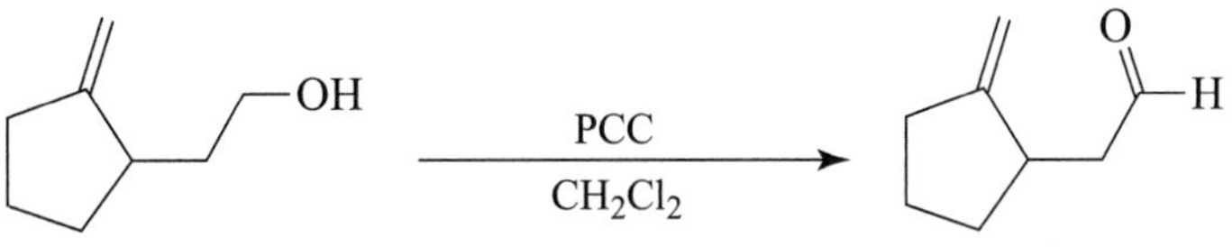 OH → O, H

PCC
CH_2Cl_2

(d) 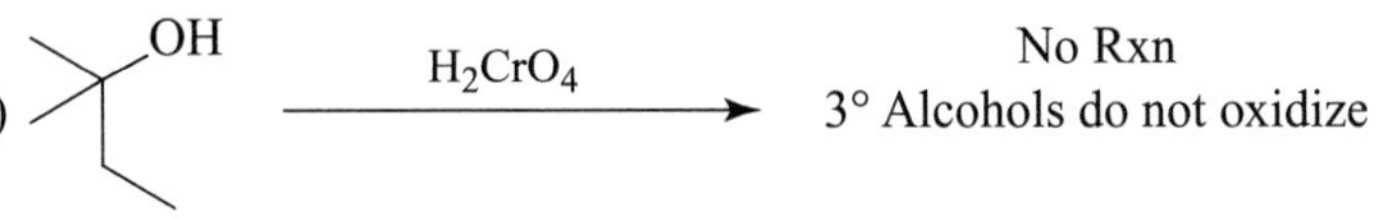 OH

H_2CrO_4

No Rxn
3° Alcohols do not oxidize

(e) 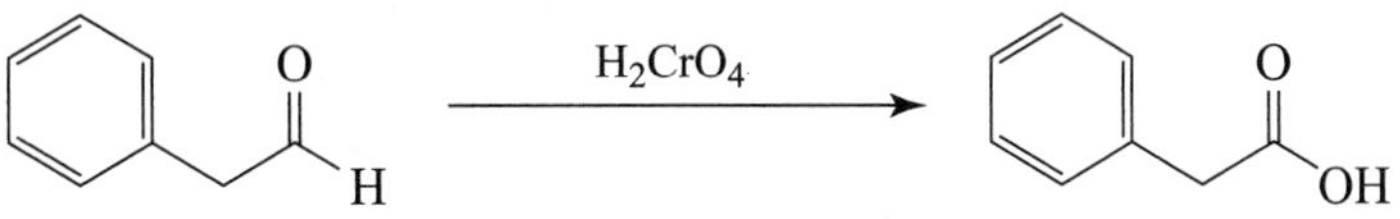 O, H → O, OH

H_2CrO_4

12.17 (a) 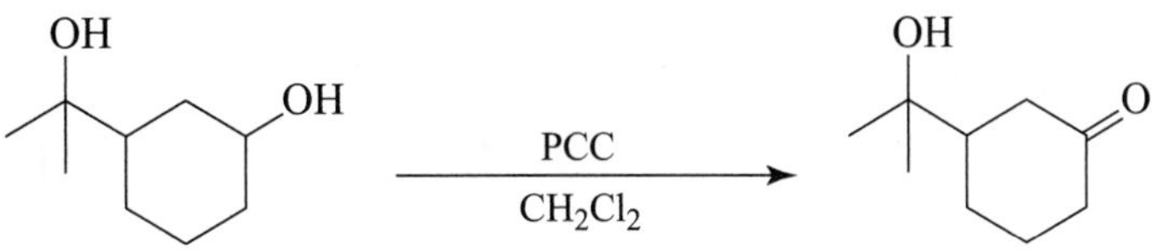 OH, OH → OH, O

PCC
CH_2Cl_2

(b) 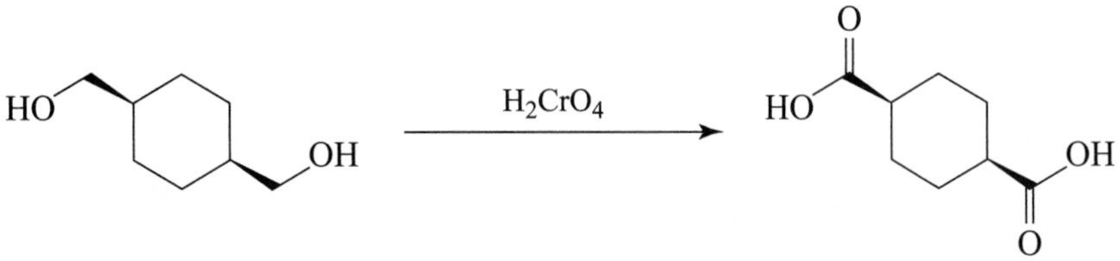 HO, OH → O, HO, OH, O

H_2CrO_4

(c) 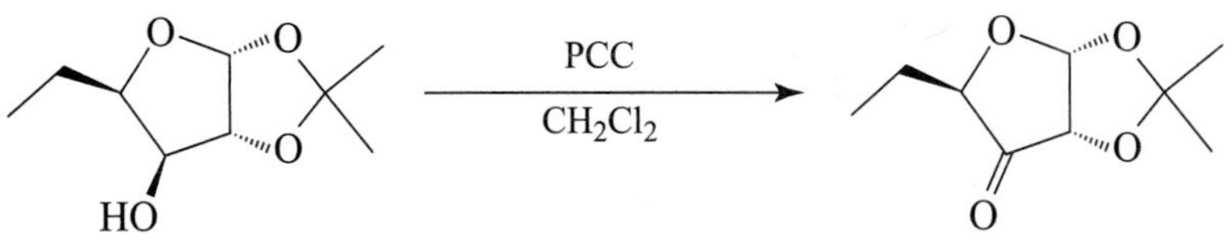 O, O, O, HO → O, O, O, O

PCC
CH_2Cl_2

(d) 1. LAH
2. aq. H_2SO_4
OH O OH

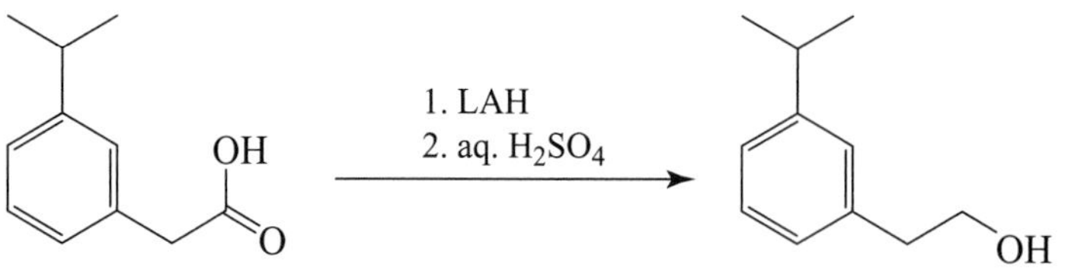

(e) O NaBH$_4$ OH

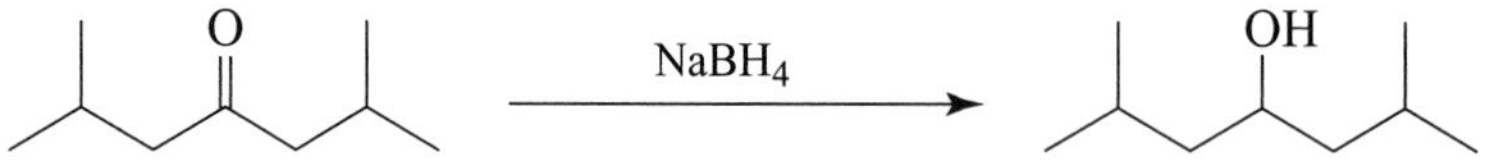

12.18 (a) (1) CH_3MgBr
(2) H_3O^+
H O HO

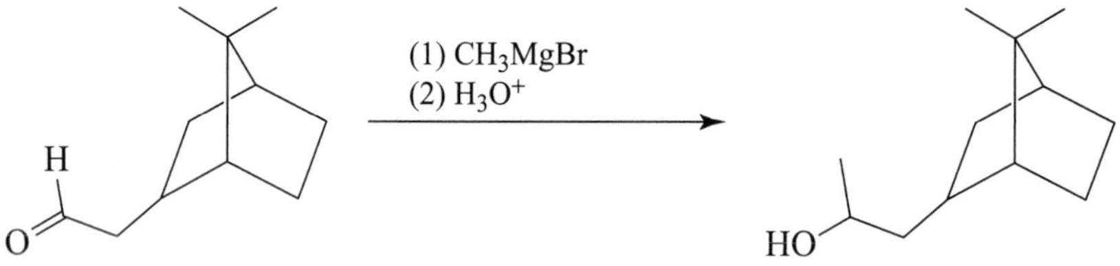

(b) O (1) MgBr
(2) NH_4Cl, H_2O
HO

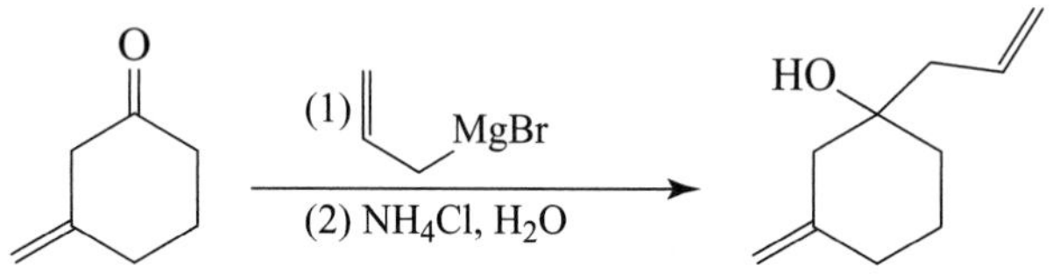

(c) O OH (1) MgBr O (2) H_3O^+ O OH O + O

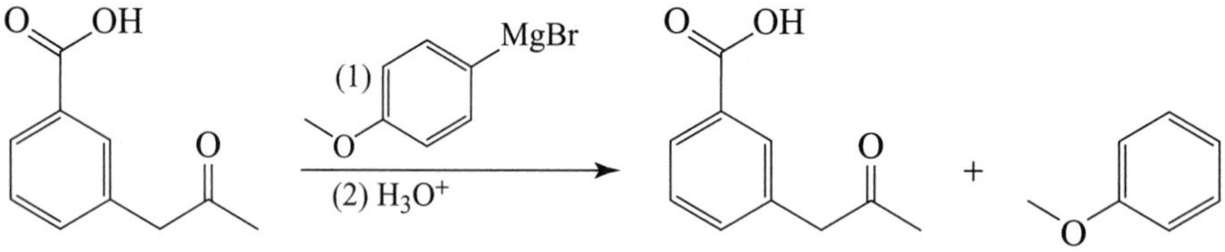

(d) O O (1) CH_3CH_2Li (excess)
(2) NH_4Cl, H_2O
OH HO

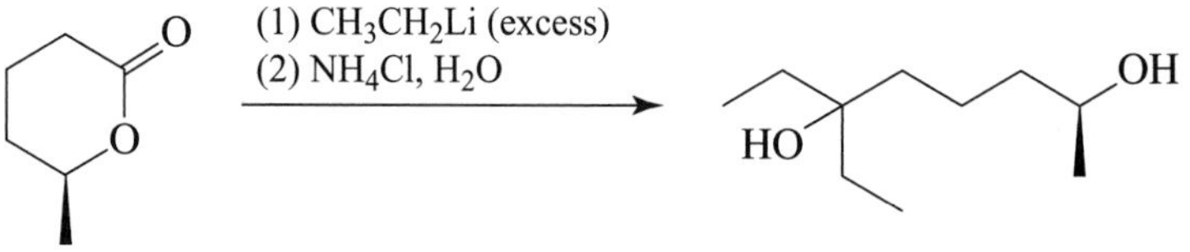

12.19 (a) O O (1) MeMgBr (excess)
(2) NH_4Cl, H_2O
OH + MeOH

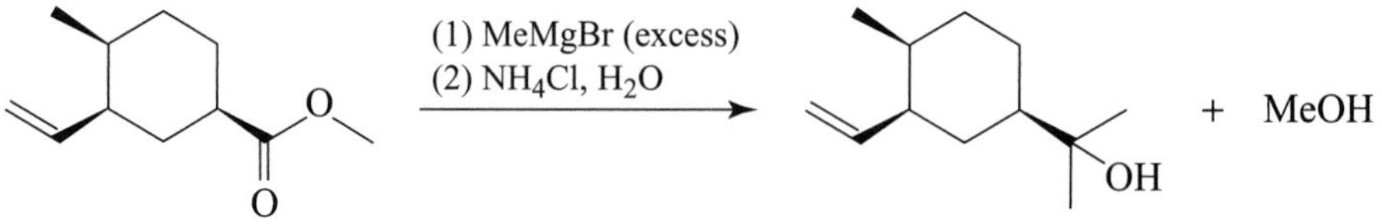

(b) Br (1) Mg
O
(2) H H
(3) H_3O^+
OH

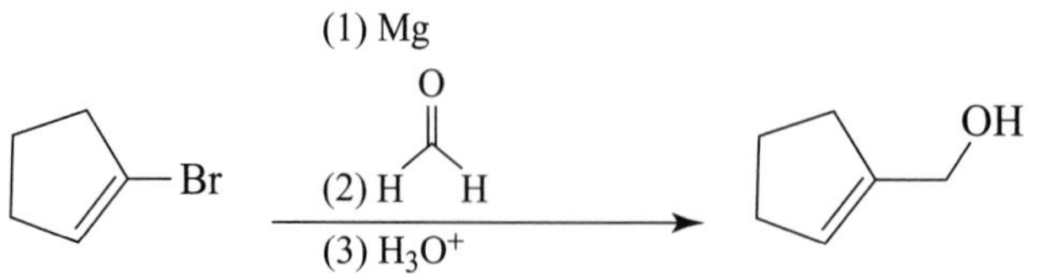

(c) OH (1) PBr_3 (2) Mg (3) H_3O^+

(d) OH (1) PCC (2) MgBr (3) H_3O^+ OH

(e) H O (1) EtMgBr (2) H_3O^+ (3) NaH (4) CH_3Br O

(f) O H O (1) O O MgBr (excess) (2) NH_4Cl, H_2O (3) PCC O O O O O

12.20 O O 1. BrMg MgBr (1 equiv) 2. H_3O^+ OH + MeOH

<u>**Mechanisms**</u>

12.21 O 1. $NaBD_4$ 2. H_3O^+ HO D

O 1. $NaBH_4$ 2. D_3O^+ DO H

O 1. $NaBD_4$ 2. D_3O^+ DO D

12.22

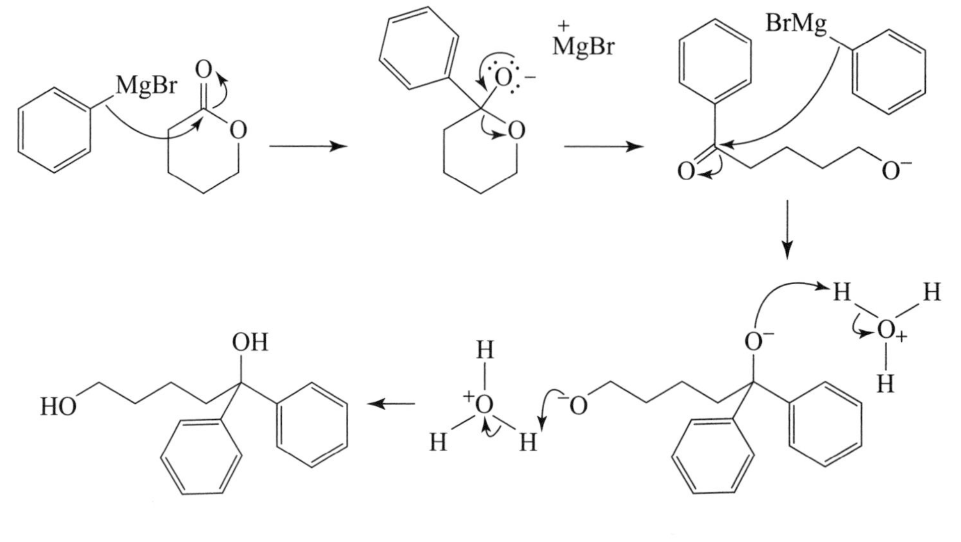

MgBr O O → + MgBr O⁻ O → BrMg O O⁻

↓

HO OH ← H +O H H O⁻ O⁻ H H O+ H

12.23 H:⁻ O O → H O⁻ O → H:⁻ H O O⁻

↓

HO OH ← H +O H H ⁻O O⁻ H H O+ H

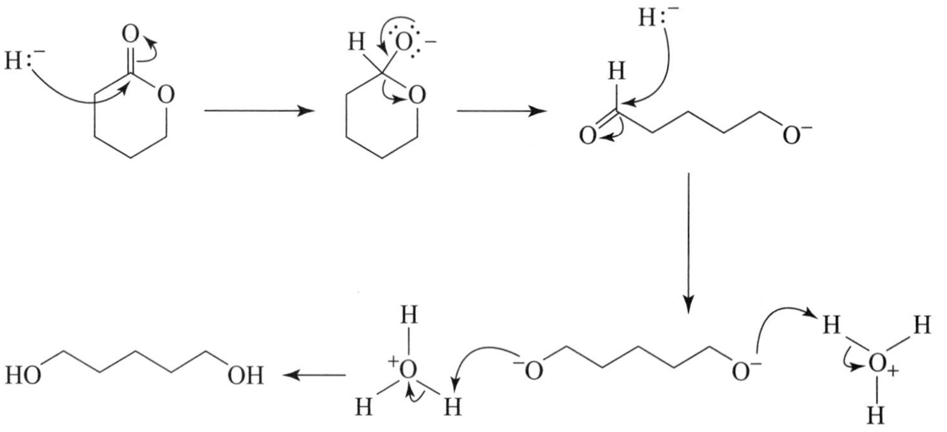

12.24 The three- and four-membered rings are strained, and so they open on reaction with RMgX or RLi. THF possesses an essentially unstrained ring and hence is far more resistant to attack by an organometallic compound.

12.25 R_1 C O Mg R R_2 R X Mg X → R_1 C O Mg R R_2 R + Mg X X

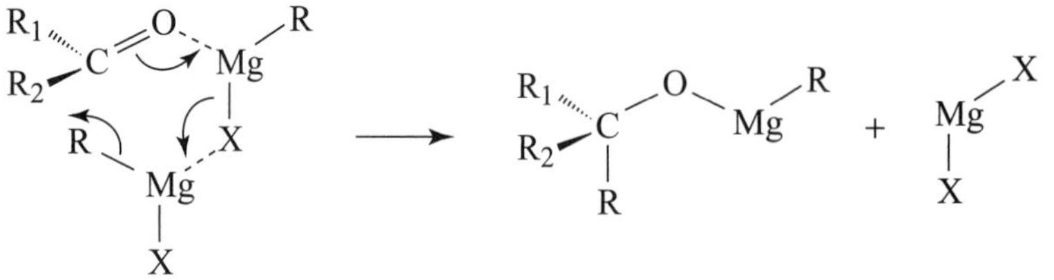

The alkylmagnesium group of the alkoxide can go on to react as a nucleophile with another carbonyl group.

Synthesis

12.26 (A) CH_4 + Li≡

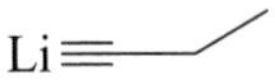

(B) 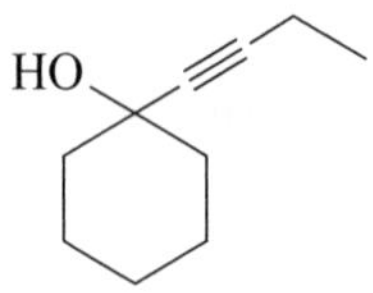 HO

(C) 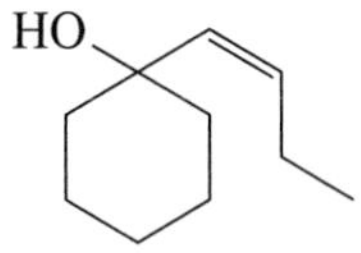 HO

(D) 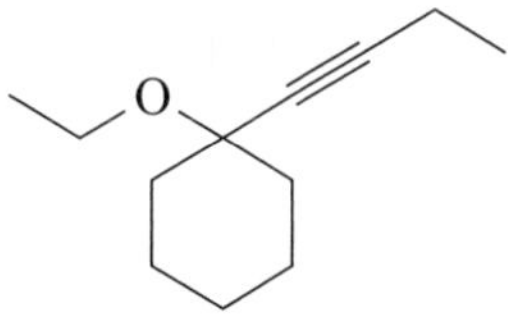 O + $Na^{+-}OSO_2Me$

(E) 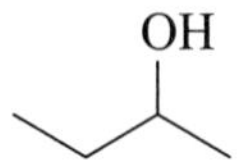 OH

(F) 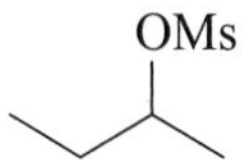 OMs

(G) 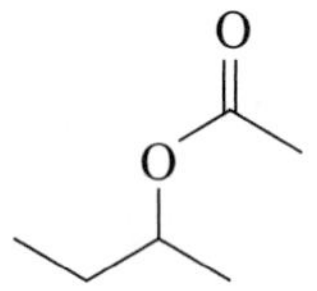 O O + $Na^{+-}OSO_2Me$

(H) 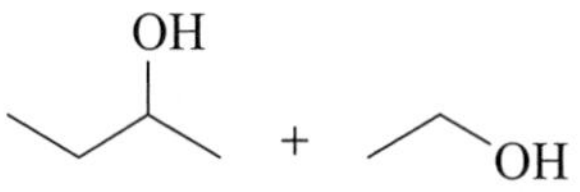 OH + OH

12.27 (a) 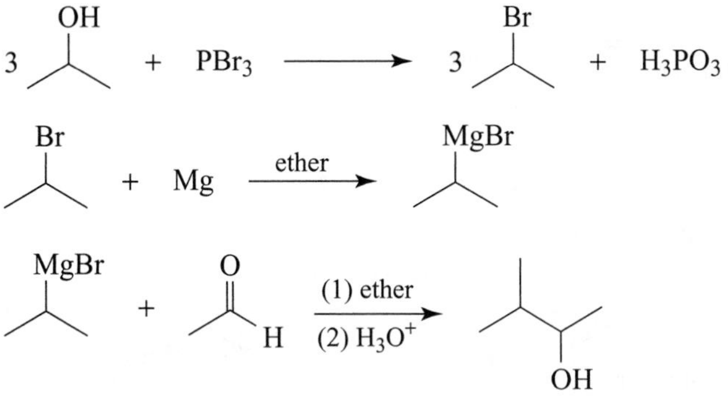 3 OH + PBr_3 ⟶ 3 Br + H_3PO_3

Br + Mg --ether--> MgBr

MgBr + O H --(1) ether, (2) H_3O^+--> OH

(b) 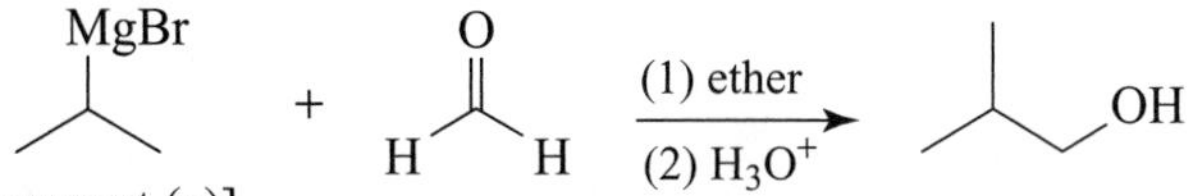 MgBr [from part (a)] + O H H --(1) ether, (2) H_3O^+--> OH

(c) 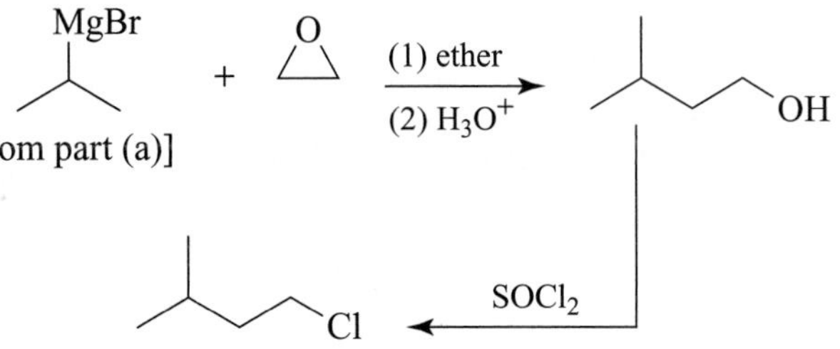 MgBr [from part (a)] + O --(1) ether, (2) H_3O^+--> OH

Cl ⟵ $SOCl_2$

(d) MgBr [from part (a)] + H–C(=O)–CH(CH$_3$)$_2$ $\xrightarrow{(1)\ \text{ether};\ (2)\ H_3O^+}$ OH

(e) MgBr [from part (a)] + D_2O ⟶ D

12.28 (a) OH $\xrightarrow{PBr_3}$ Br

(b) Br [from (a)] $\xrightarrow{\text{OK} / \text{OH}}$

(c) [from (b)] $\xrightarrow{(1)\ Hg(OAc)_2/\text{THF-}H_2O;\ (2)\ NaBH_4/OH^-}$ OH

(d) [from (b)] $\xrightarrow{H_2/\text{Pt, Pd, or Ni, pressure}}$

(e) OH [from (c)] $\xrightarrow{PBr_3}$ Br

(f) Br [from (a)] $\xrightarrow{Mg,\ Et_2O}$ MgBr $\xrightarrow{(1)\ H–C(=O)–H;\ (2)\ H_3O^+}$ OH

(g) MgBr [from (f)] $\xrightarrow{(1)\ \text{ethylene oxide};\ (2)\ H_3O^+}$ OH

(h) OH $\xrightarrow{PCC,\ CH_2Cl_2}$ C(=O)H

(i) 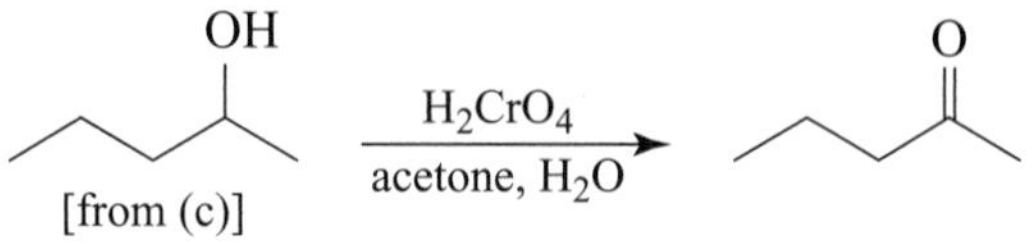 [from (c)] $\xrightarrow[\text{acetone, } H_2O]{H_2CrO_4}$

(j) 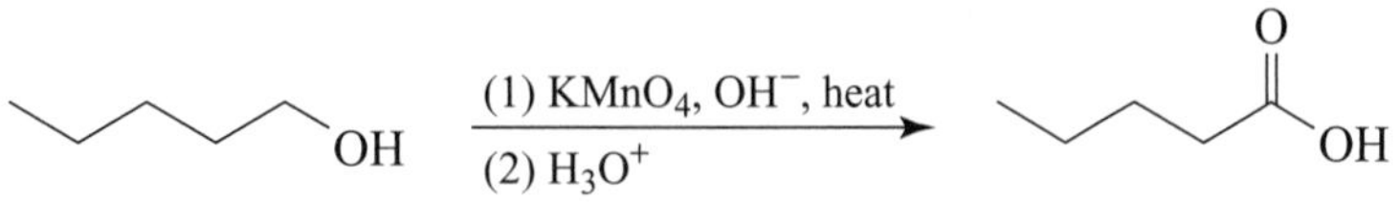 $\xrightarrow[(2)\ H_3O^+]{(1)\ KMnO_4,\ OH^-,\ \text{heat}}$

(k) (1) 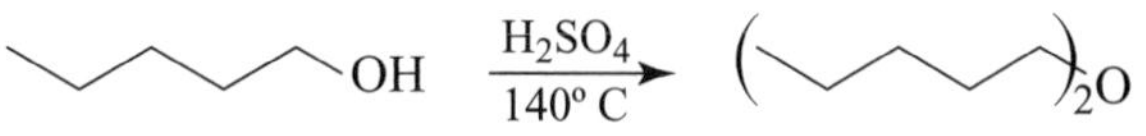 $\xrightarrow[140^\circ C]{H_2SO_4}$

(2) 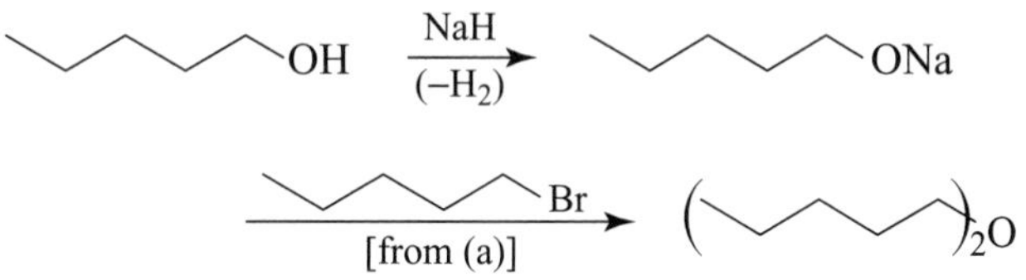 $\xrightarrow[(-H_2)]{NaH}$ ONa

Br [from (a)]

(l) 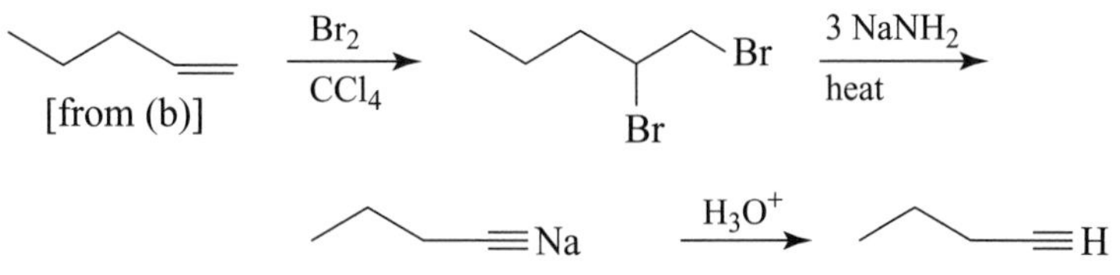 [from (b)] $\xrightarrow[CCl_4]{Br_2}$ Br Br $\xrightarrow[\text{heat}]{3\ NaNH_2}$

≡Na $\xrightarrow{H_3O^+}$ ≡H

(m) 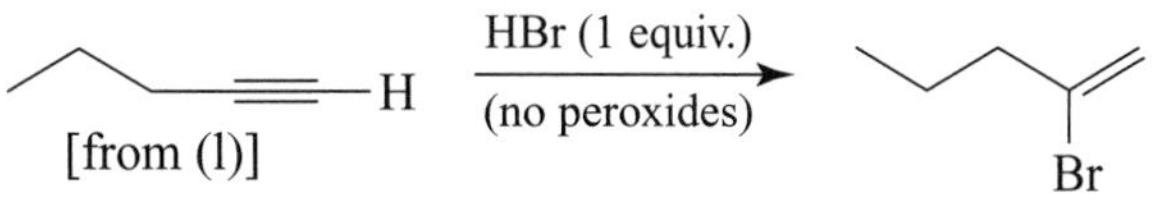 [from (l)] $\xrightarrow[\text{(no peroxides)}]{\text{HBr (1 equiv.)}}$ Br

(n) 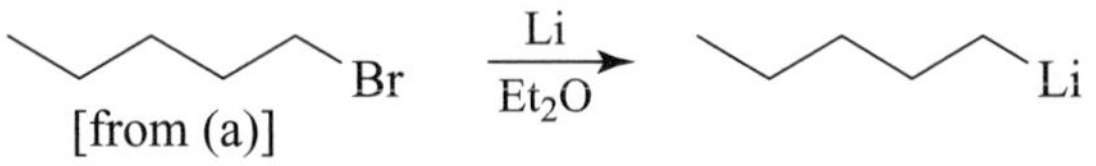 Br [from (a)] $\xrightarrow[Et_2O]{Li}$ Li

(o) 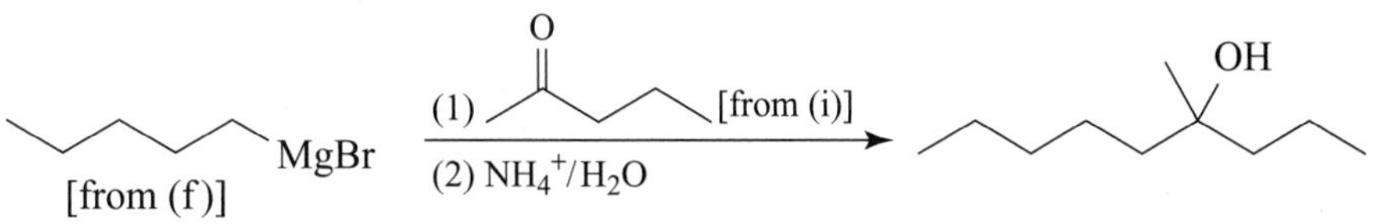 MgBr [from (f)] (1) [from (i)] (2) NH_4^+/H_2O OH

12.29 (a) $NaBH_4$, ^-OH

(b) 85% H_3PO_4, heat

(c) H_2/Ni_2B - (P−2)

(d) (1) BH_3:THF
(2) H_2O_2, ^-OH

(e) (1) NaH (2) $(CH_3)_2SO_4$

(f) (1) NaH (2)  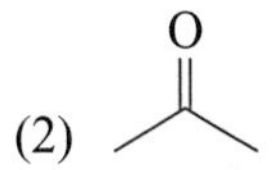 (3) HA

(g) (1) CH_3MgBr/Et_2O

(2)

(3) NH_4Cl/H_2O

12.30 (a) ***Retrosynthetic Analysis***

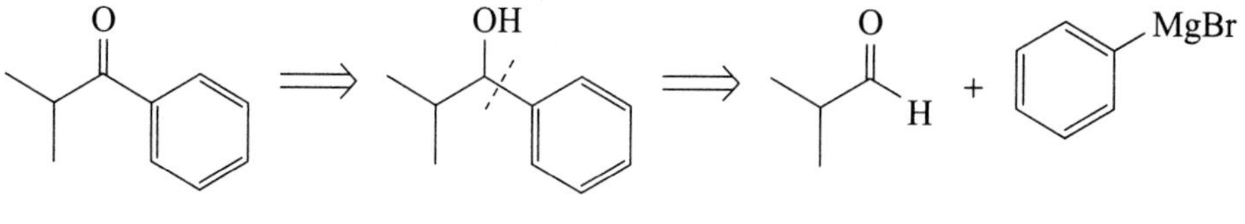

O OH O MgBr

⟹ ⟹ H +

Synthesis

O OH

OH $\xrightarrow[CH_2Cl_2]{PCC}$ H MgBr $\xrightarrow{H_3O^+}$

Br Mg / Et_2O

H_2CrO_4 acetone

O

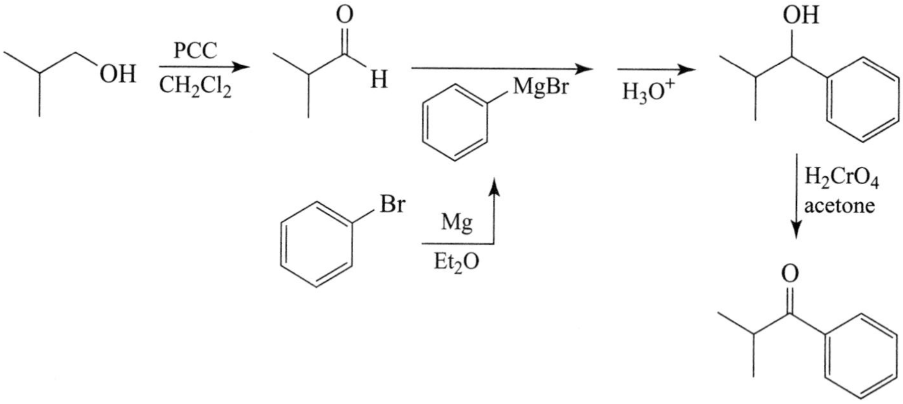

(b) ***Retrosynthetic Analysis***

HO

⟹ 2 MgBr + O OCH$_3$

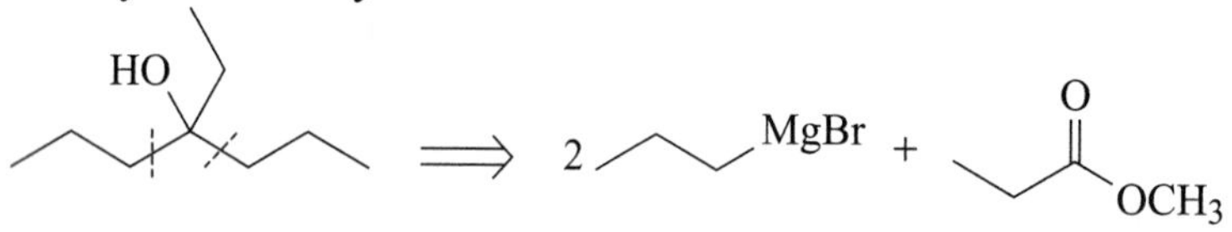

Synthesis

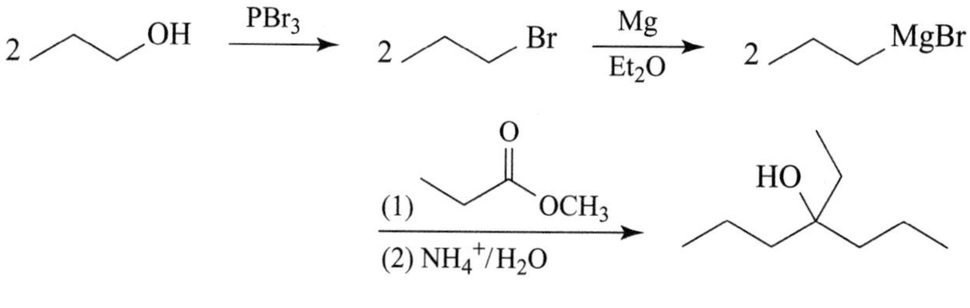

2 OH $\xrightarrow{PBr_3}$ 2 Br $\xrightarrow[Et_2O]{Mg}$ 2 MgBr

(1) O OCH$_3$

(2) NH_4^+/H_2O → HO

(c) ***Retrosynthetic Analysis***

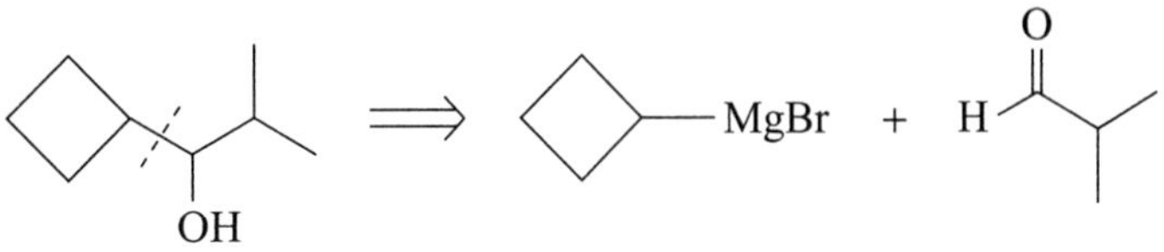

cyclobutyl–CH(OH)–CH(CH_3)$_2$ ⟹ cyclobutyl–MgBr + H–C(=O)–CH(CH_3)$_2$

Synthesis

cyclobutanol $\xrightarrow{PBr_3}$ bromocyclobutane $\xrightarrow[Et_2O]{Mg}$ cyclobutylmagnesium bromide

$\xrightarrow[\text{[from (a)]}]{(CH_3)_2CHCHO}$ cyclobutyl–CH(OMgBr)–CH(CH_3)$_2$ $\xrightarrow{H_3O^+}$ cyclobutyl–CH(OH)–CH(CH_3)$_2$

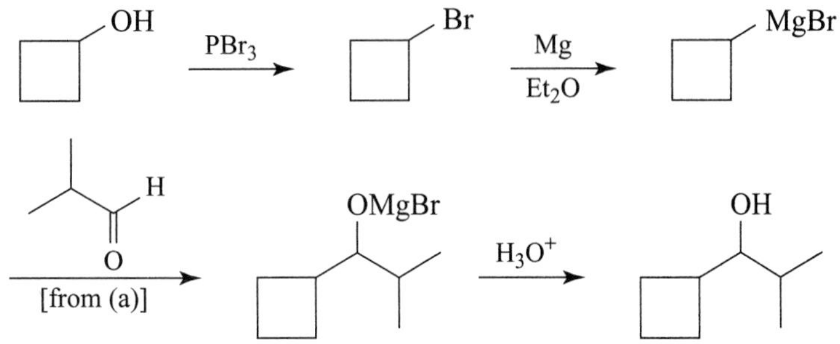

(d) ***Retrosynthetic Analysis***

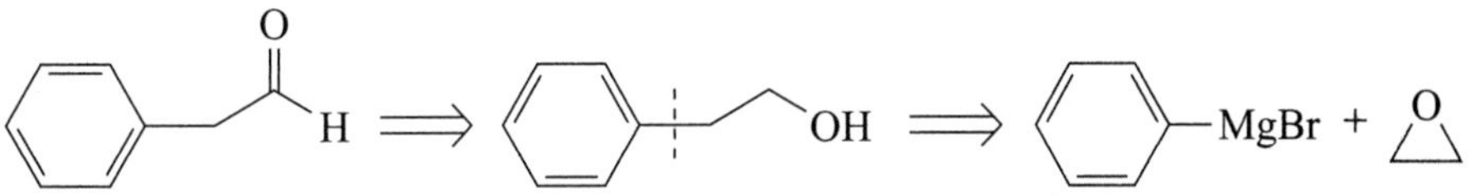

$C_6H_5CH_2CHO$ ⟹ $C_6H_5CH_2CH_2OH$ ⟹ C_6H_5MgBr + ethylene oxide

Synthesis

C_6H_5Br $\xrightarrow[Et_2O]{Mg}$ C_6H_5MgBr $\xrightarrow[(2)\ H_3O^+]{(1)\ \text{ethylene oxide}}$

$C_6H_5CH_2CH_2OH$ $\xrightarrow[CH_2Cl_2]{PCC}$ $C_6H_5CH_2CHO$

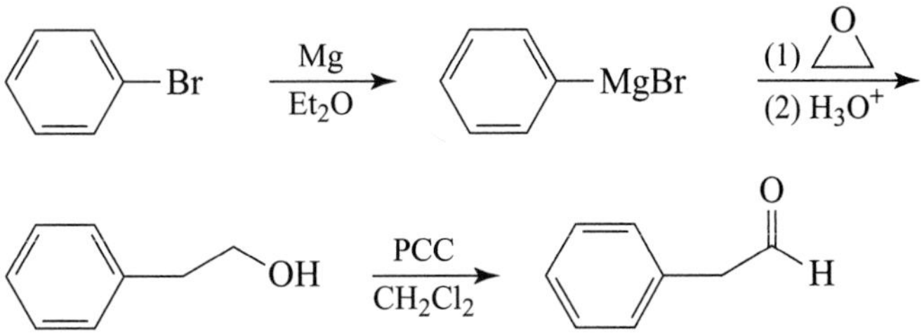

(e) ***Retrosynthetic Analysis***

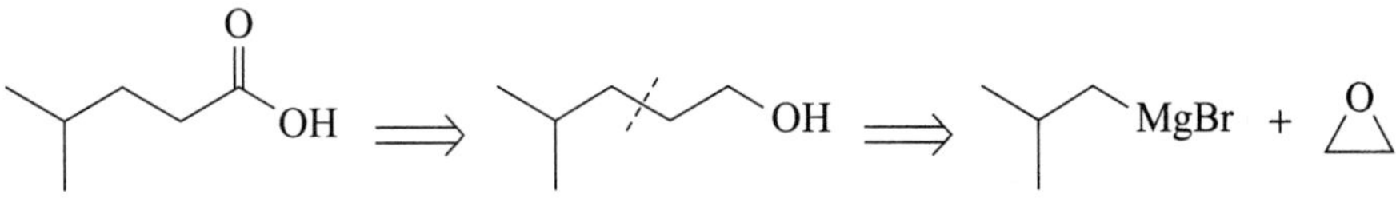

$(CH_3)_2CHCH_2CH_2COOH$ ⟹ $(CH_3)_2CHCH_2CH_2CH_2OH$ ⟹ $(CH_3)_2CHCH_2MgBr$ + ethylene oxide

Synthesis

$(CH_3)_2CHCH_2OH$ $\xrightarrow{PBr_3}$ $(CH_3)_2CHCH_2Br$ $\xrightarrow[Et_2O]{Mg}$ $(CH_3)_2CHCH_2MgBr$

$\xrightarrow[(2)\ H_3O^+]{(1)\ \text{ethylene oxide}}$ $(CH_3)_2CHCH_2CH_2CH_2OH$ $\xrightarrow[(2)\ H_3O^+]{(1)\ KMnO_4,\ OH^-,\ heat}$ $(CH_3)_2CHCH_2CH_2COOH$

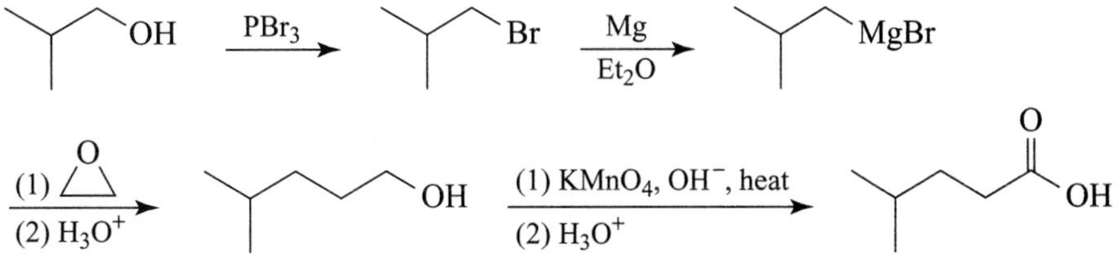

(f) ***Retrosynthetic Analysis***

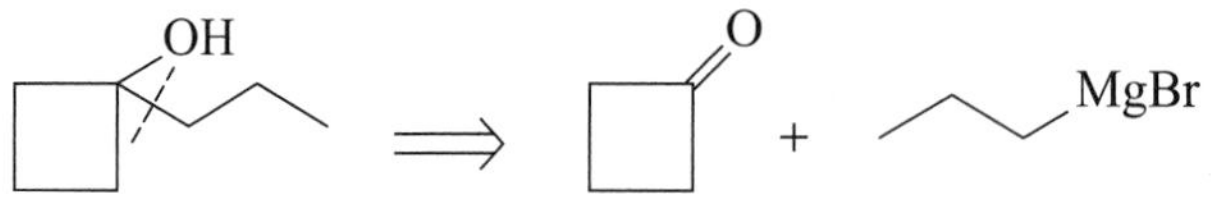

OH

⟹ O + MgBr

Synthesis

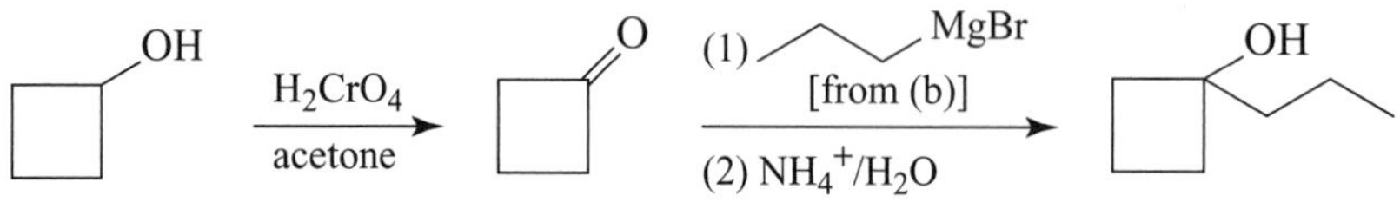

OH $\xrightarrow[\text{acetone}]{H_2CrO_4}$ O $\xrightarrow[(2)\ NH_4^+/H_2O]{(1)\ \text{MgBr [from (b)]}}$ OH

(g) ***Retrosynthetic Analysis***

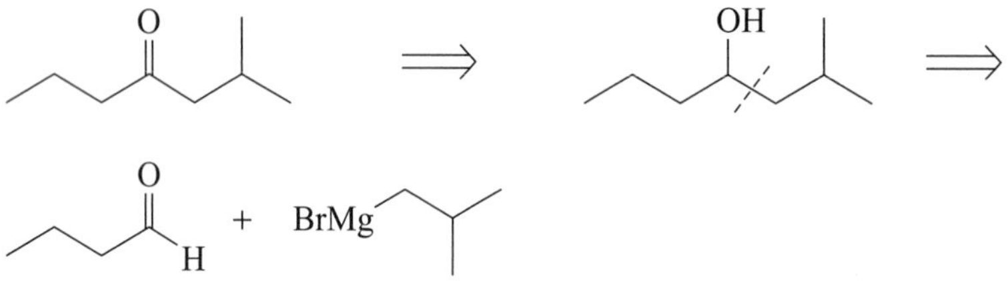

O ⟹ OH ⟹

O H + BrMg

Synthesis

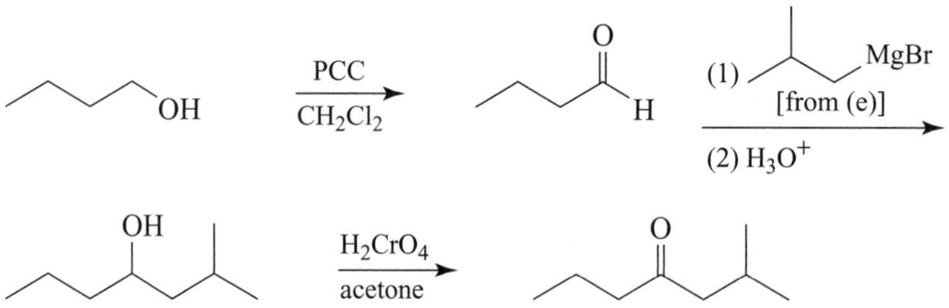

OH $\xrightarrow[CH_2Cl_2]{PCC}$ O H $\xrightarrow[(2)\ H_3O^+]{(1)\ \text{MgBr [from (e)]}}$

OH $\xrightarrow[\text{acetone}]{H_2CrO_4}$ O

(h) ***Retrosynthetic Analysis***

Br ⟹ OH ⟹

MgBr + O ⟹ OH ⟹

H O + MgBr

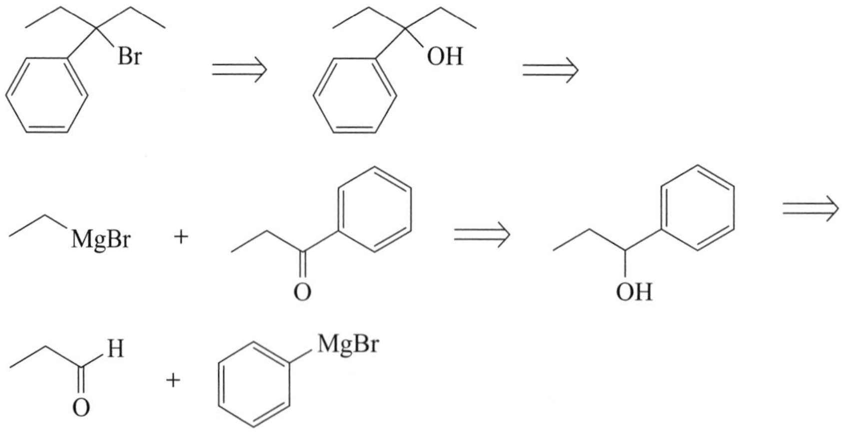

Synthesis

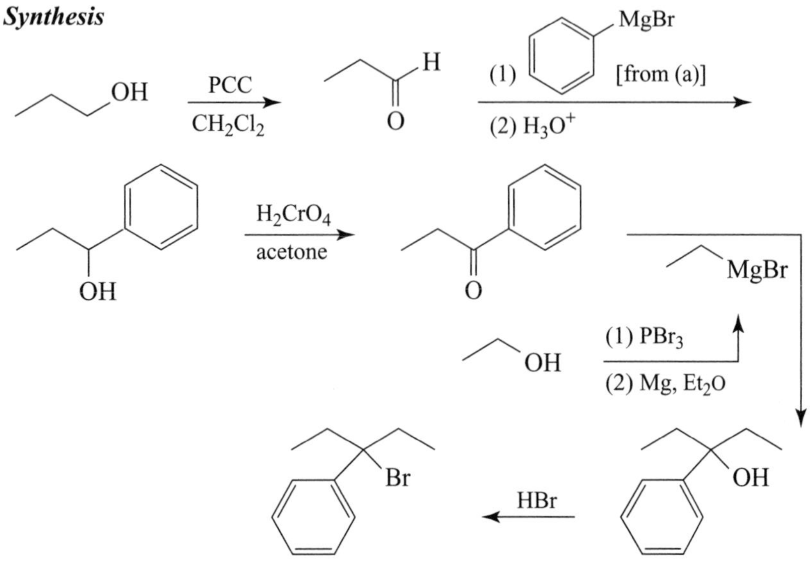

OH —PCC, CH_2Cl_2→ H, O —(1) MgBr [from (a)] (2) H_3O^+→

OH —H_2CrO_4, acetone→ O —MgBr→

OH —(1) PBr_3 (2) Mg, Et_2O→ MgBr

Br ←HBr— OH

12.31 (a) OH ⟹ H O + MgBr or O H + MgBr

H O + MgBr —(1) Et_2O (2) H_3O^+→ OH or O H + MgBr —(1) Et_2O (2) H_3O^+→ OH

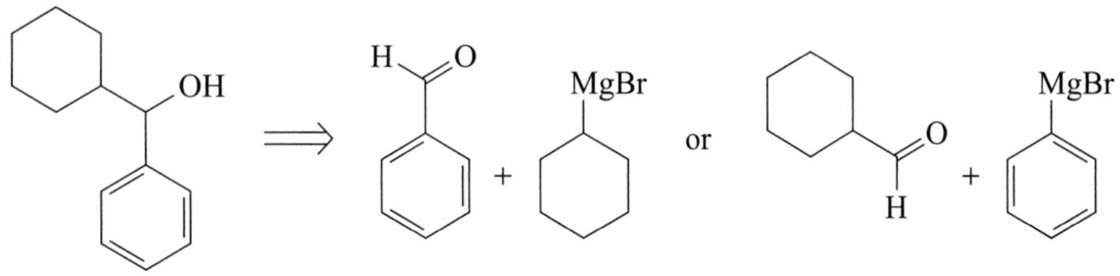

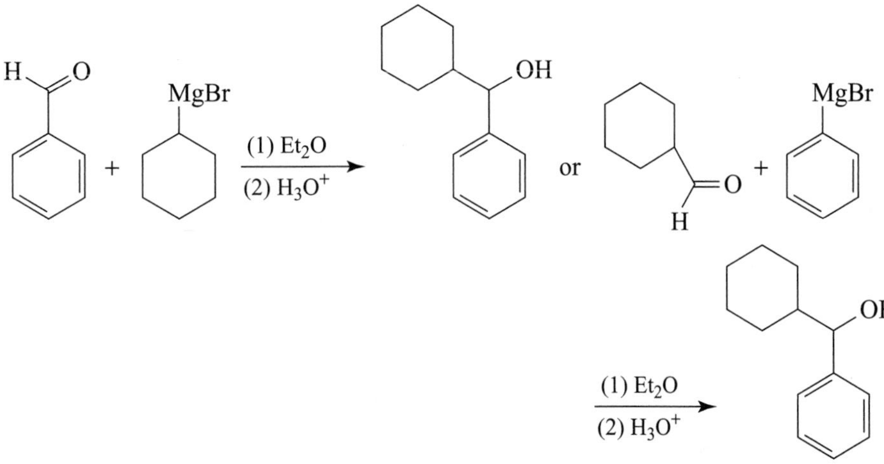

(b) OH ⟹ O + BrMg or O + BrMg

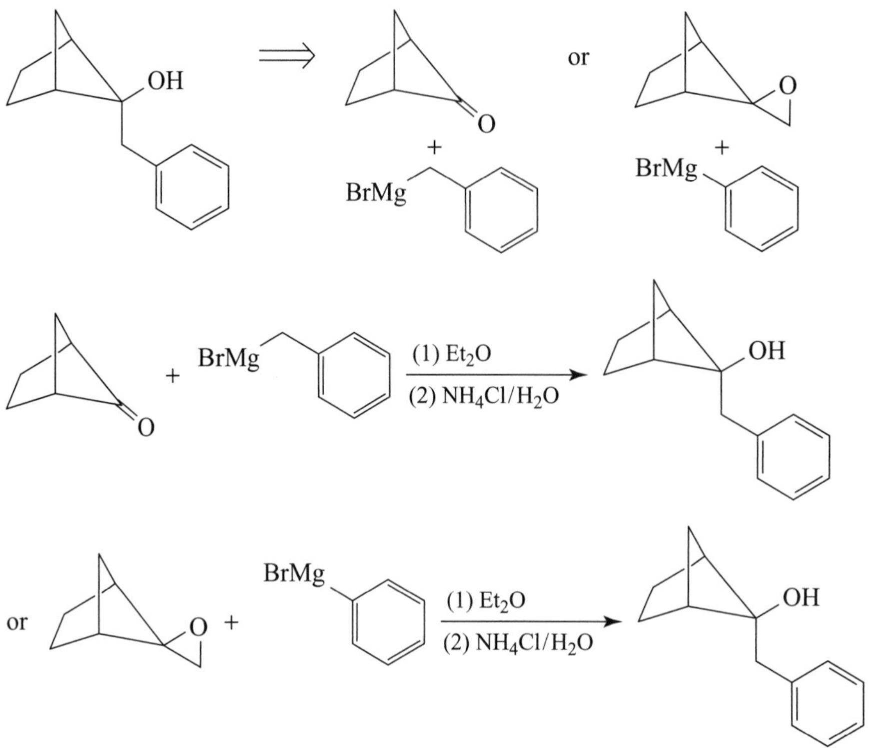

+ BrMg (1) Et_2O (2) NH_4Cl/H_2O ⟶ OH

or O + BrMg (1) Et_2O (2) NH_4Cl/H_2O ⟶ OH

(c) OH OH ⟹ O O + CH_3CH_2MgBr (excess)

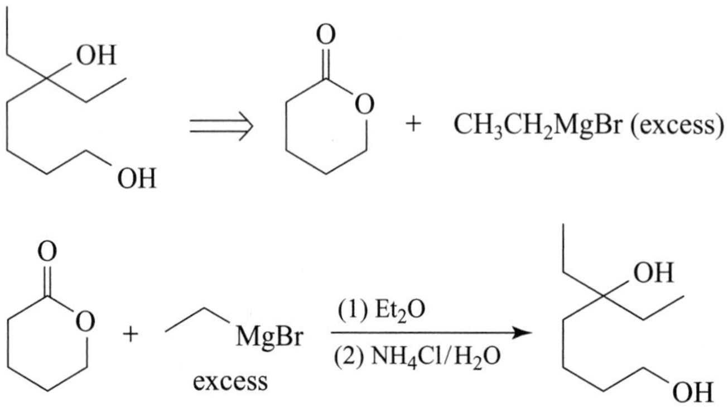

O O + MgBr excess (1) Et_2O (2) NH_4Cl/H_2O ⟶ OH OH

12.32 There are multiple strategies to approach this these problems.

(a)

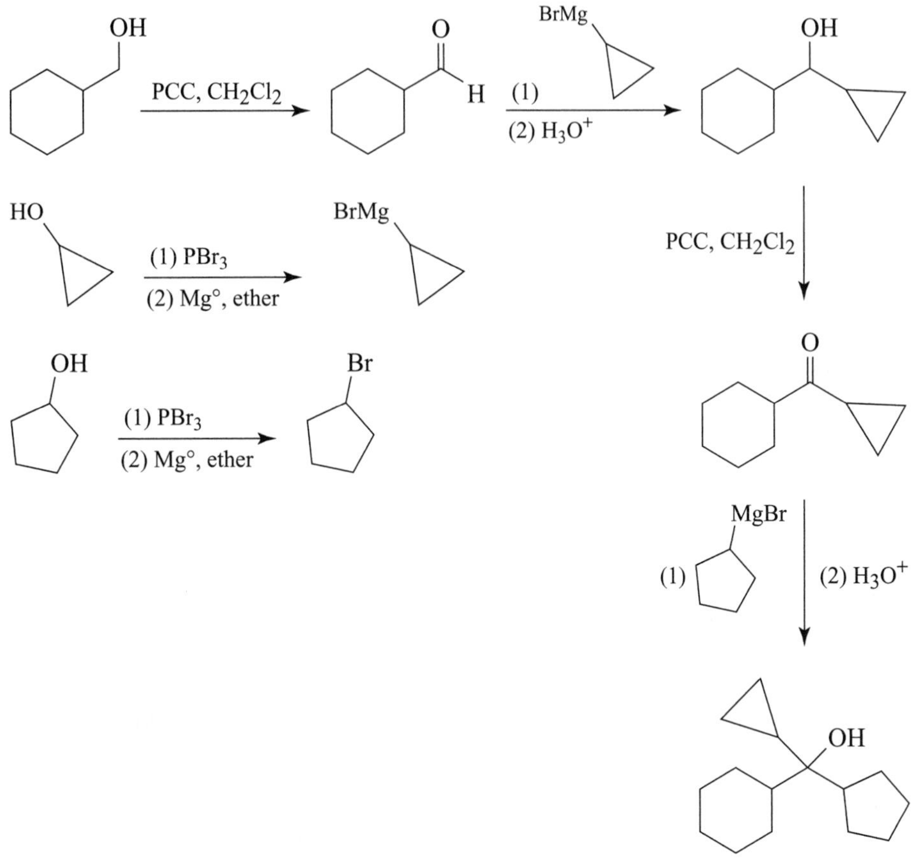

PCC, CH_2Cl_2

(1) BrMg–cyclopropyl (2) H_3O^+

(1) PBr_3 (2) Mg°, ether

(1) PBr_3 (2) Mg°, ether

PCC, CH_2Cl_2

(1) cyclopentyl-MgBr (2) H_3O^+

(b)

PCC, CH_2Cl_2

(1) CH_3MgBr (2) H_3O^+

CH_3OH $\xrightarrow[(2)\ Mg°,\ ether]{(1)\ PBr_3}$ CH_3MgBr

PCC, CH_2Cl_2

(1) CH_3MgBr (2) H_3O^+

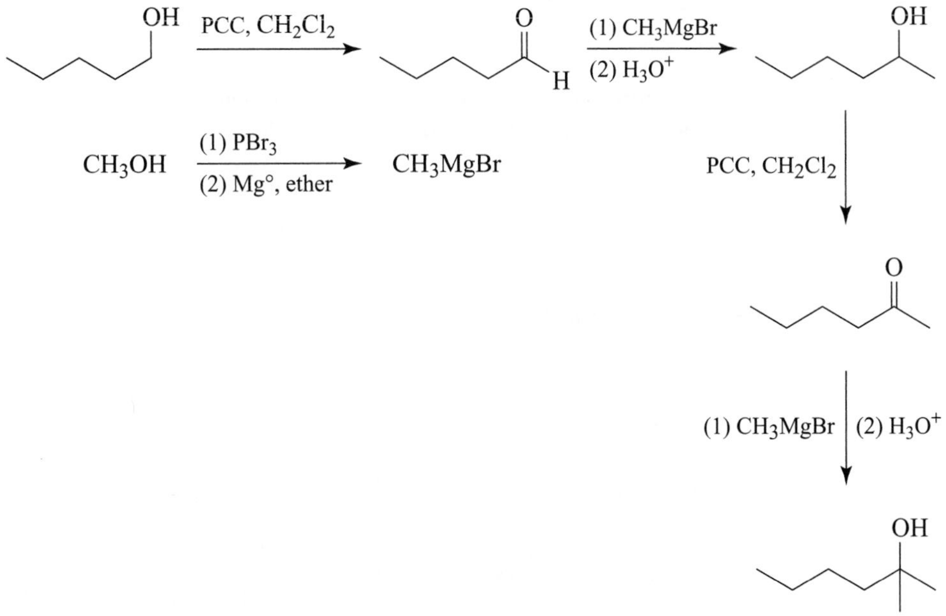

(c) OH — PCC, CH_2Cl_2 → O, H — (1) MgBr, (2) H_3O^+ → OH

Br — (1) Mg, ether; (2) O, H H; (3) H_3O^+ → OH — (1) PBr_3; (2) Mg, ether → MgBr

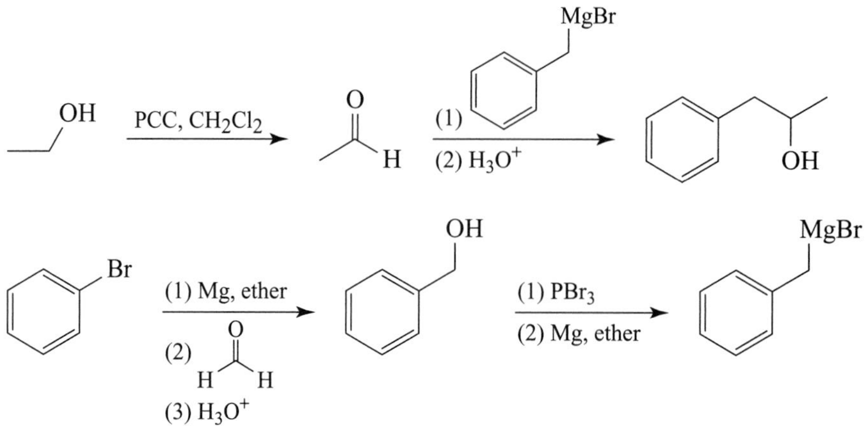

12.33 ***Retrosynthetic Analysis***

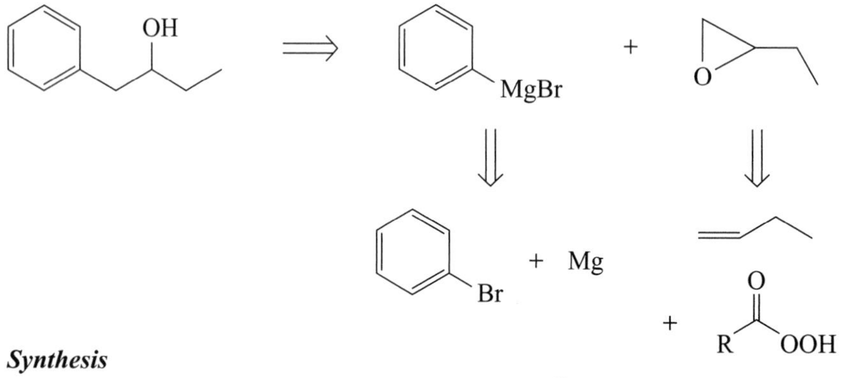

OH ⟹ MgBr + O

Br + Mg

\+ O, R OOH

Synthesis

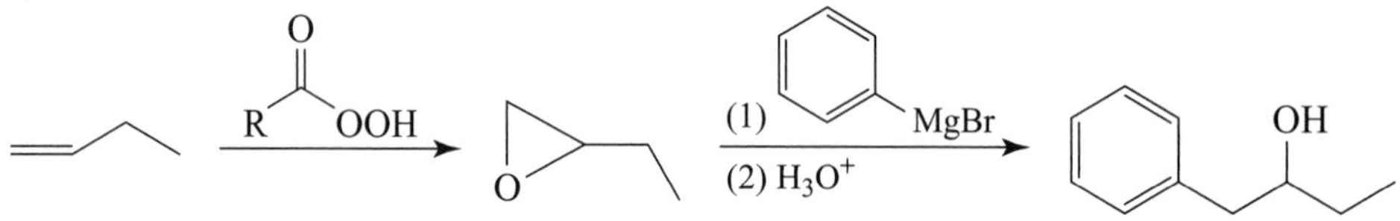

O, R OOH → O — (1) MgBr; (2) H_3O^+ → OH

12.34 The starting compound is a cyclic ester. Addition of two molar equivalents of CH_3MgI will (after acidification) furnish the desired product.

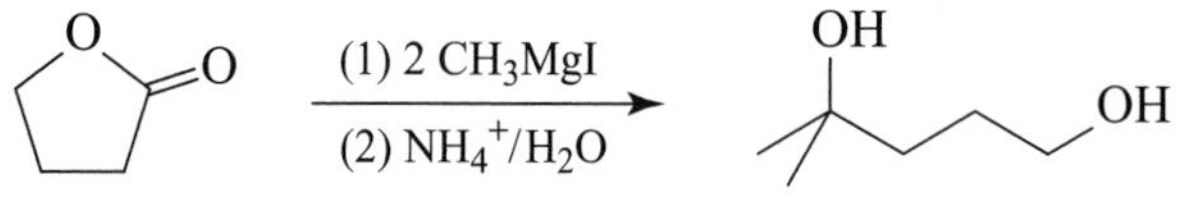

O, O — (1) 2 CH_3MgI; (2) NH_4^+/H_2O → OH, OH

12.35 ***Retrosynthetic Analysis***

OH ⟹ O + Na^+ $^-$:≡

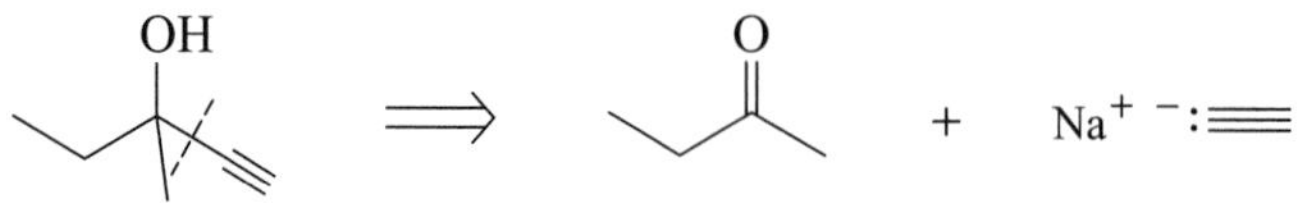

Synthesis

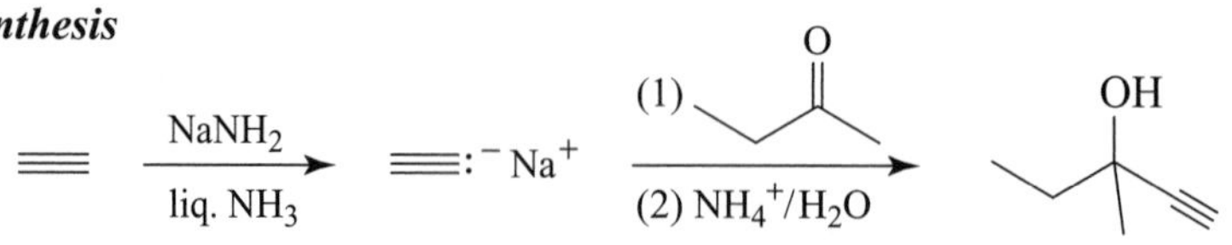

≡ — $NaNH_2$, liq. NH_3 → ≡:$^-$ Na^+ — (1) O; (2) NH_4^+/H_2O → OH

12.36 ***Retrosynthetic Analysis***

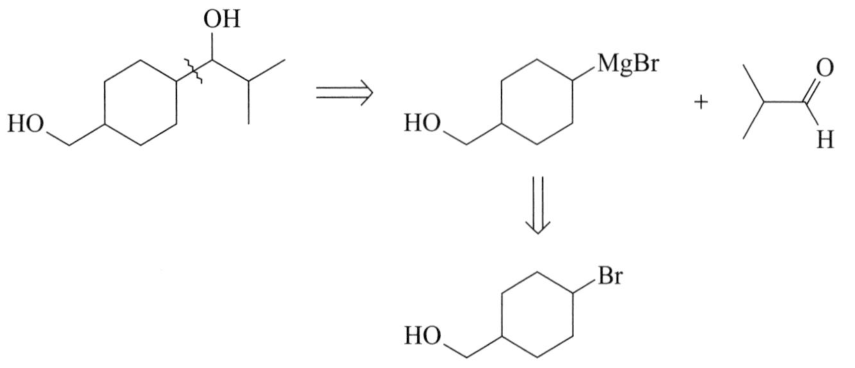

OH

HO ⟹ HO MgBr + O H

⇓

HO Br

Synthesis

HO Br

Si—Cl

imidazole

SiO Br

(1) Mg, Et_2O

(2) O H

(3) H_3O^+

SiO OH

$Bu_4\overset{+}{N}\overset{-}{F}$

HO OH

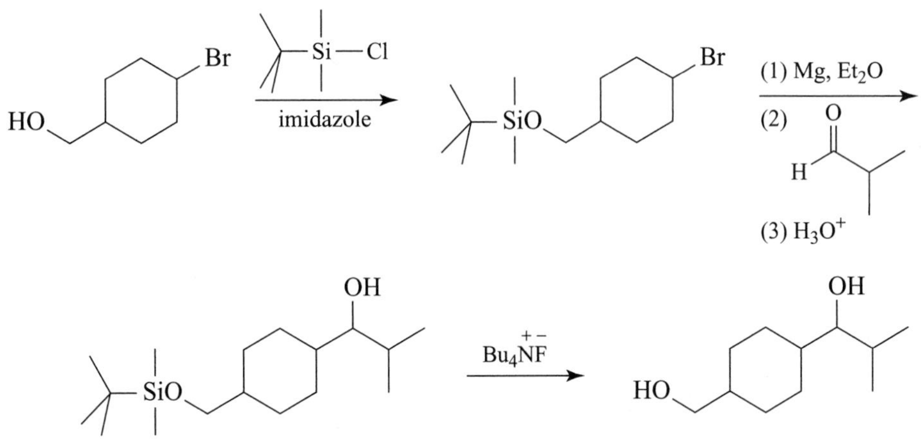

Before converting the reactant to a Grignard reagent it is first necessary to mask the alcohol, such as by converting it to a *tert*-butyldimethylsilyl ether. After the Grignard reaction is over, the protecting group is removed.

12.37

Br_2, *hv*; Br; H—≡—H, $NaNH_2$; H; (1) BH_3 (2) NaOH, H_2O_2; O, H

Br; Mg, Et_2O; MgBr; (1) O, H; (2) NH_4Cl, H_2O; HO; NaH, EtI; O

Challenge Problems

12.38 2-Phenylethanol, 1,2-diphenylethanol, and 1,1-diphenylethanol are distinct from 2,2-diphenylethanoic acid and benzyl 2-phenylethanoate in that they do not have carbonyl groups. IR spectroscopy can be used to segregate these five compounds into two groups according to those that do or do not exhibit carbonyl absorptions.

^{1}H NMR can differentiate among all of the compounds. In the case of the alcohols, in the ^{1}H NMR spectrum of 2-phenylethanol there will be two triplets of equal integral value, whereas for 1,2-diphenylethanol there will be a doublet and a triplet in a 2:1 area ratio. The triplet will be downfield of the doublet. 1,1-Diphenylethanol will exhibit a singlet for the unsplit methyl hydrogens.

The broadband proton-decoupled ^{13}C NMR spectrum of 2-phenylethanol should show 6 signals (assuming no overlap), four of which are in the chemical shift region for aromatic carbons. 1,2-Diphenylethanol should exhibit 10 signals (assuming no overlap), 8 of which are in the aromatic region. 1,1-Diphenylethanol should show 6 signals (assuming no overlap), four of which would be in the aromatic region. The DEPT ^{13}C NMR spectra would give direct evidence as to the number of attached hydrogens on each carbon.

Regarding the carbonyl compounds, both 2,2-diphenylethanoic acid and benzyl 2-phenylethanoate will show carbonyl absorptions in the IR, but only the former will also have a hydroxyl absorption. The ^{1}H NMR spectrum of 2,2-diphenylethanoic acid should show a broad absorption for the carboxylic acid hydrogen and a sharp singlet for the unsplit hydrogen at C2. Their integral values should be the same, and approximately one-tenth the integral value of the signals in the aromatic region. Benzyl 2-phenylethanoate will exhibit two singlets, one for each of the unsplit CH_2 groups. These signals will have an area ratio of 2 : 5 with respect to the signal for the 10 aromatic hydrogens. The broadband ^1H decoupled ^{13}C NMR spectrum for 2,2-diphenylethanoic acid should show four aromatic carbon signals, whereas that for benzyl 2-phenylethanoate (assuming no overlapping signals) would show 8 signals in the aromatic carbon region. Aside from the carbonyl and aromatic carbon signals, benzyl 2-phenylethanoate would show two additional signals, whereas 2,2-diphenylethanoic acid would show only one. DEPT ^{13}C NMR spectra for these two compounds would also distinguish them directly.

12.39 It makes it impossible to distinguish between aldehyde and ketone type sugars (aldoses and ketoses) that had been components of the saccharide. Also, because the R groups of these sugars contain chirality centers, reduction of the ketone carbonyl will be stereoselective. This will complicate the determination of the ratio of sugars differing in configuration at C2.

12.40 The IR indicates the presence of OH and absence of C=C and C=O. The MS indicates a molecular weight of 116 amu and confirms the presence of hydroxyl. The reaction data indicate X contains 2 protons per molecule that are acidic enough to react with a Grignard reagent, meaning two hydroxyl groups per molecule. (This analytical procedure, the Zerewitinoff determination, was routinely done before the advent of NMR.)

Thus X has a partial structure such as:

$C_6H_{10}(OH)_2$ with one ring, or
$C_5H_6O(OH)_2$ with two rings, or (less likely)
$C_4H_2O_2(OH)_2$ with three rings.

QUIZ

12.1 Which of the following could be employed to transform ethanol into OH?

(a) Ethanol + HBr, then Mg/diethyl ether, then H_3O^+

(b) Ethanol + HBr, then Mg/diethyl ether, then H(C=O)H, then H_3O^+

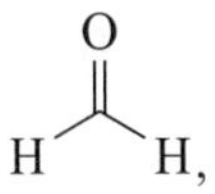

(c) Ethanol + H_2SO_4/140°C

(d) Ethanol + Na, then H(C=O)H, then H_3O^+

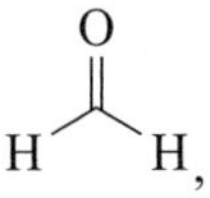

(e) Ethanol + H_2SO_4/180°C, then O (epoxide)

12.2 The principal product(s) formed when *1 mol* of methylmagnesium iodide reacts with 1 mol of O OH

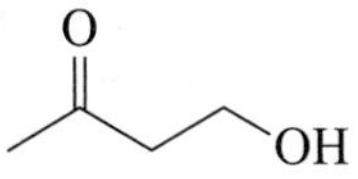

(a) CH_4 + O OMgI

(b) OMgI OH

(c) O CH_3 O

(d) OH CH_3 O

(e) None of the above

12.3 Supply the missing reagents.

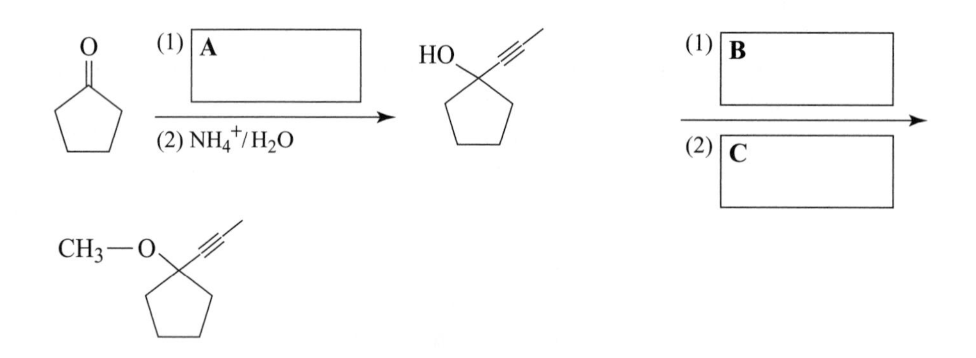

12.4 Supply the missing reagents and intermediates.

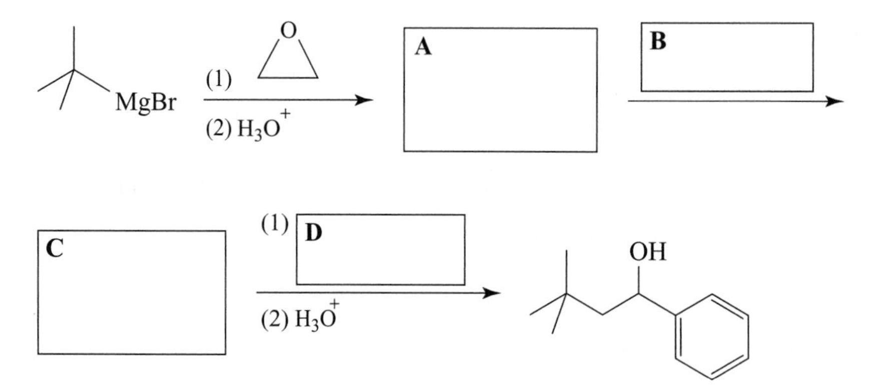

12.5 Supply the missing starting compound.

A

(1) CH_3MgBr

(2) NH_4^+/H_2O

OH

ANSWERS TO FIRST REVIEW PROBLEM SET

1. (a) :ÖH — H—A / A^- → $\overset{+}{O}H_2$ — $-H_2O$ / $+H_2O$ → 2° Carbocation

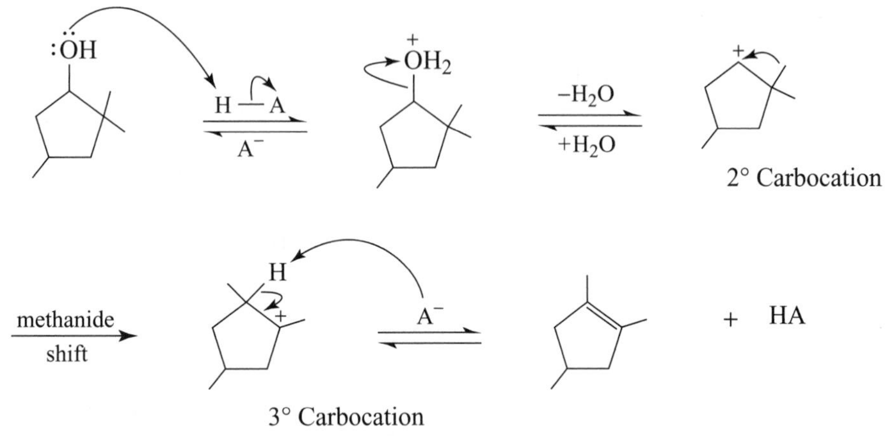

methanide shift → 3° Carbocation (H, +) — A^- → + HA

(b) + Br—Br ⟶ $\overset{+}{Br}$ + Br^-

then,

$\overset{+}{Br}$, :C̈l:⁻ ⟶ Br, Cl

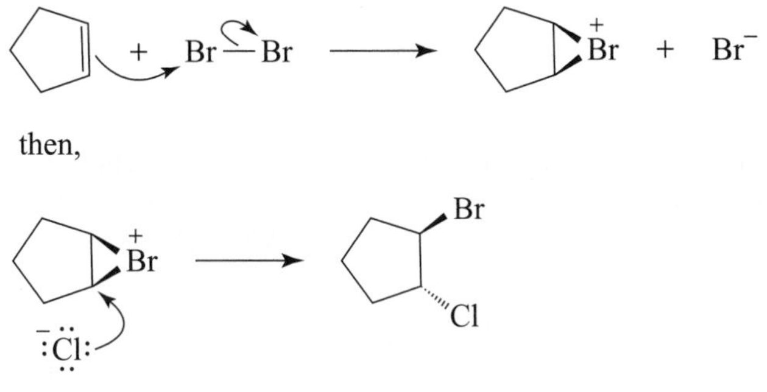

(c) The enantiomer of the product given would be formed in an equimolar amount via the following reaction:

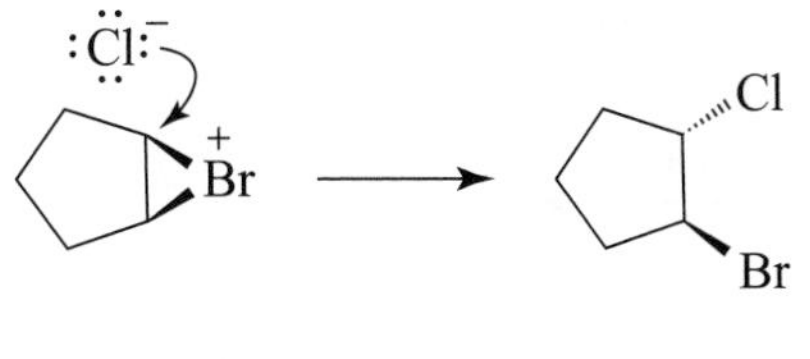

:C̈l:⁻, $\overset{+}{Br}$ ⟶ Cl, Br

The *trans*-1,2-dibromocyclopentane would be formed as a racemic form via the reaction of the bromonium ion with a bromide ion:

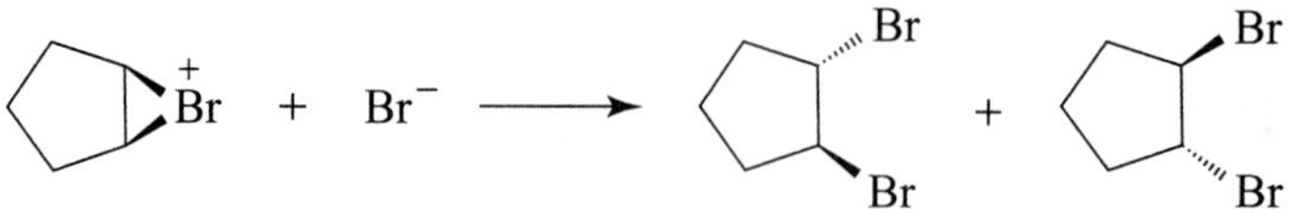

Racemic *trans*-1,2-dibromocyclopentane

And, *trans*-2-bromocyclopentanol (the bromohydrin) would be formed (as a racemic form) via the reaction of the bromonium ion with water.

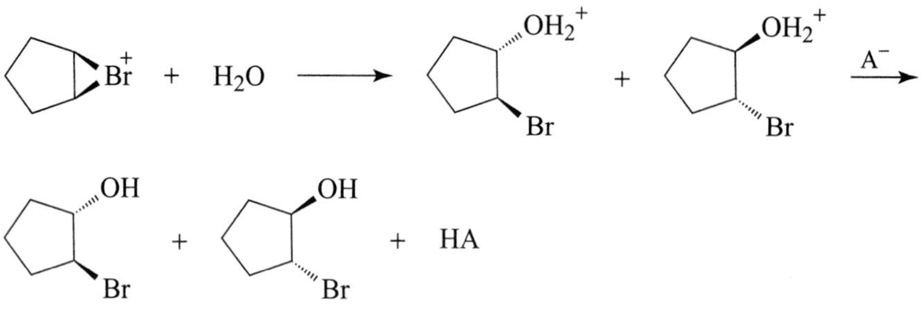

Racemic *trans*-2-bromocyclopentanol

(d)

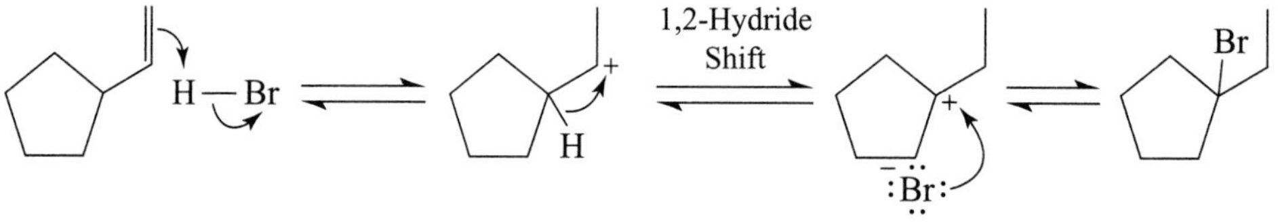

1,2-Hydride Shift

2. (a) $CHCl_3$ (b) The cis isomer (c) CH_3Cl

3. This indicates that the bonds in BF_3 are geometrically arranged so as to cancel each others' polarities in contrast to the case of NF_3. This, together with other evidence, indicates that BF_3 has trigonal planar structure and NF_3 has trigonal pyramidal structure.

4. (a) (1) Br_2, H_2O (2) NaOH (3) NaOMe, MeOH

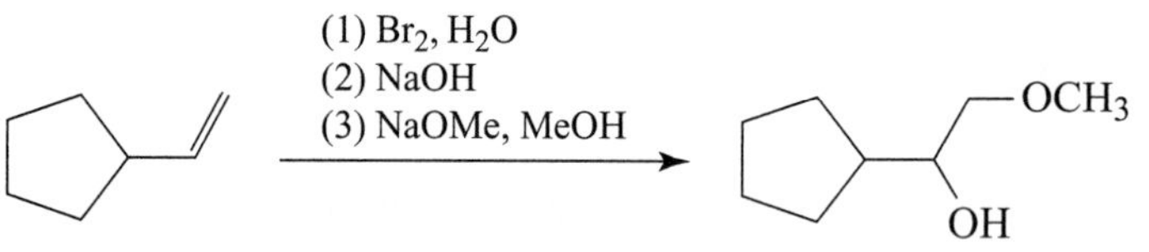

(b) (1) BH_3:THF (2) NaOH, H_2O_2 (3) NaH (4) Cl

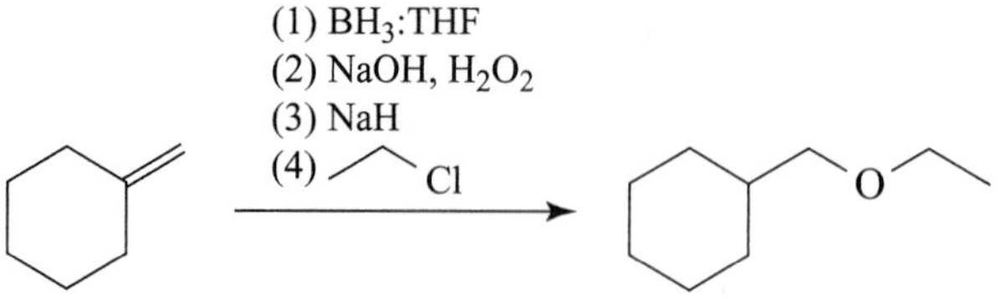

(c) (1) PBr_3 (2) $NaSCH_3$, $HSCH_3$

(d) (1) MCPBA
(2) CH_3CH_2OH, H_2SO_4 → OH, OEt

(e) OH (1) TsCl
(2) NaN_3 → N_3

(f) O concd. H_2SO_4, CH_3CH_2OH → O, OH

(g) OH $\xrightarrow{NaNH_2}$ ONa + NH_3

5. (a)

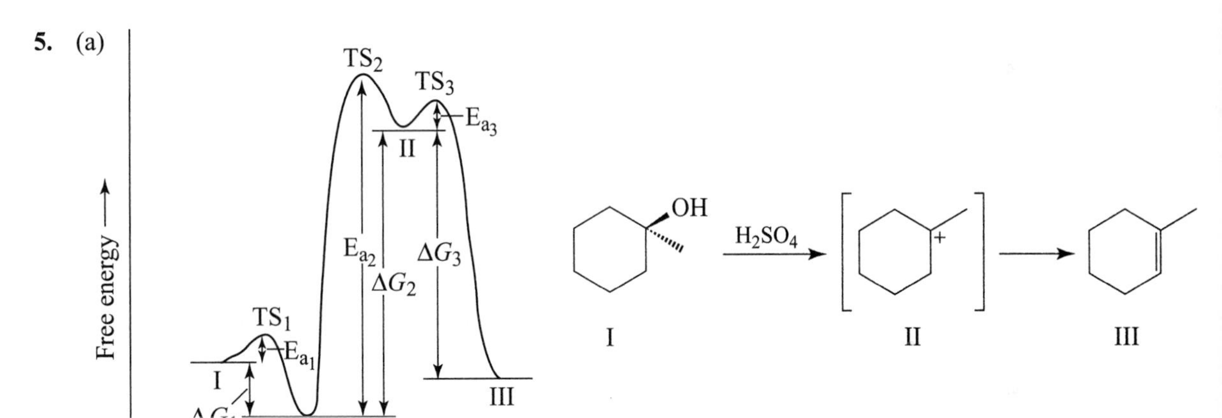

(b)

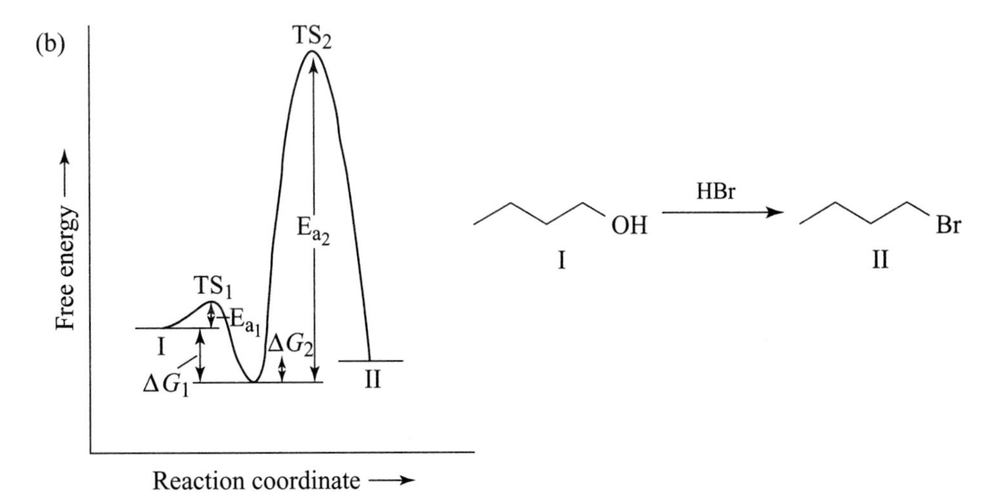

6.

A = 11 H

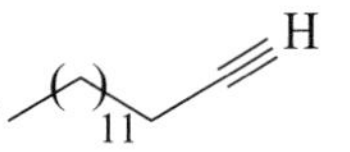

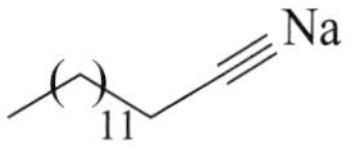

B = 11 Na

C = 11 6

Muscalure = 11 6

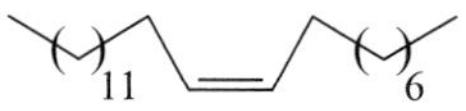

7.

C_6H_5

C_6H_5 C_6H_5 C_6H_5

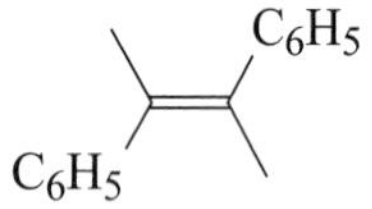

(*E*)-2,3-Diphenyl-2-butene 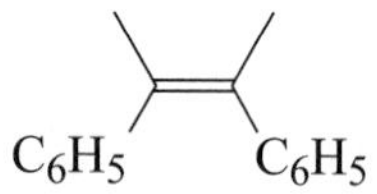 (*Z*)-2,3-Diphenyl-2-butene

Because catalytic hydrogenation is a syn addition, catalytic hydrogenation of the (*Z*) isomer would yield a meso compound.

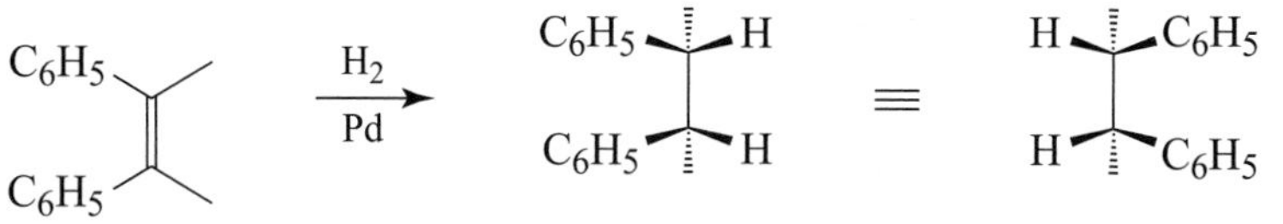

C_6H_5 C_6H_5 (*Z*) $\xrightarrow[\text{Pd}]{H_2}$ C_6H_5 H C_6H_5 H ≡ H C_6H_5 H C_6H_5

(by addition at one face) A meso compound

(by addition at the other face)

Syn addition of hydrogen to the (*E*) isomer would yield a racemic form:

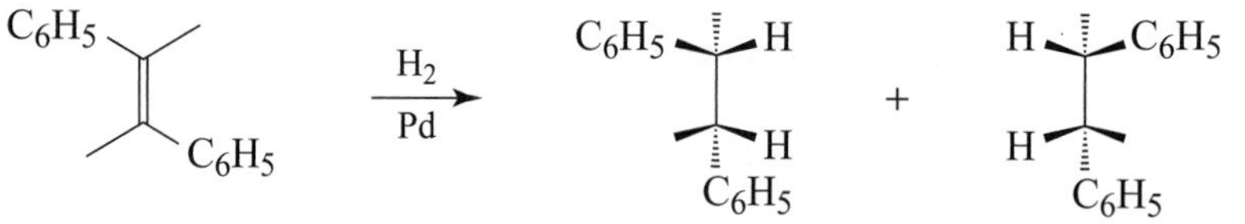

C_6H_5 C_6H_5 (*E*) $\xrightarrow[\text{Pd}]{H_2}$ C_6H_5 H H C_6H_5 + H C_6H_5 H C_6H_5

(by addition at one face)

(by addition at the other face)

Enantiomers - a racemic form

8. From the molecular formula of **A** and of its hydrogenation product **B**, we can conclude that **A** has two rings and a double bond. (**B** has two rings.)

From the product of strong oxidation with $KMnO_4$ and its stereochemistry (i.e., compound **C**), we can deduce the structure of **A**.

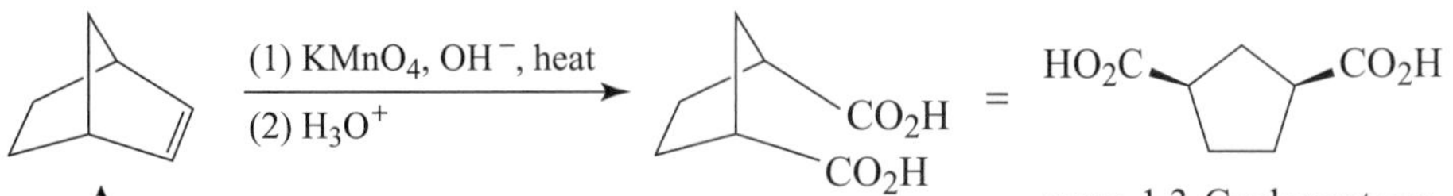

(1) $KMnO_4$, OH^-, heat
(2) H_3O^+

A

CO_2H CO_2H = HO_2C CO_2H

meso-1,3-Cyclopentane-dicarboxylic acid

Compound **B** is bicyclo[2.2.1]heptane and **C** is a glycol.

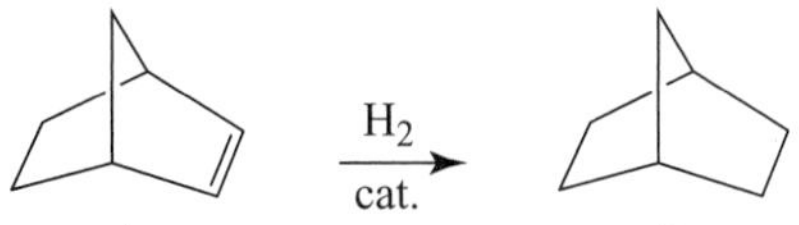

H_2 / cat.

A **B**

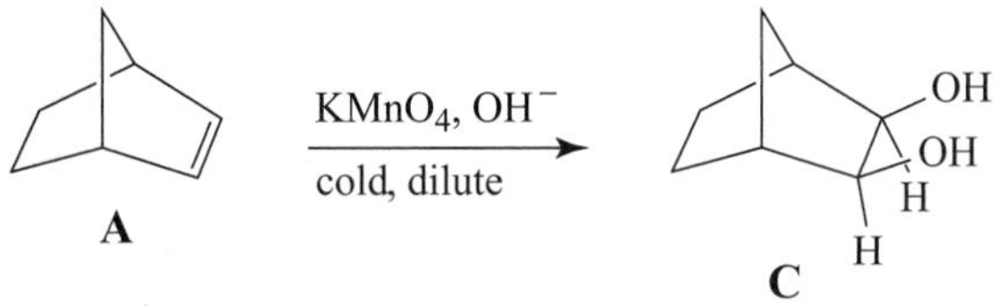

$KMnO_4$, OH^- / cold, dilute

A **C** OH OH H H

Notice that **C** is also a meso compound.

9. (a) (1) BH_3:THF (2) NaOH, H_2O_2 → OH

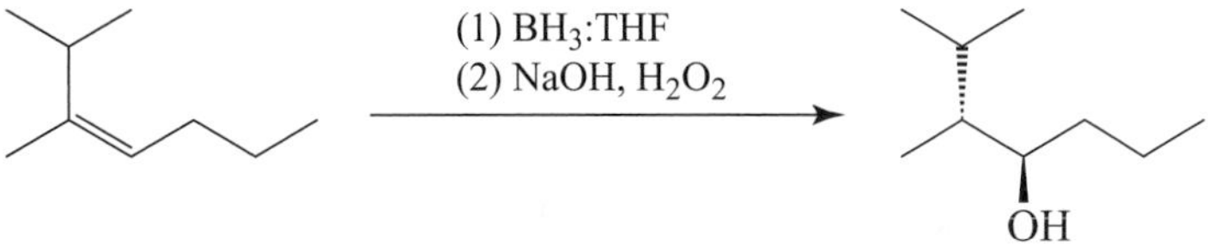

(b) (1) $Hg(OAc)_2$, H_2O, THF (2) $NaBH_4$, NaOH → OH

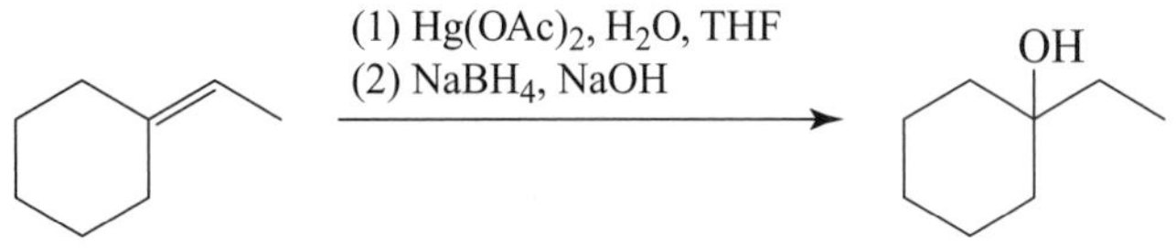

(c) MCPBA → O

(d) concd. H_2SO_4 → O

OH

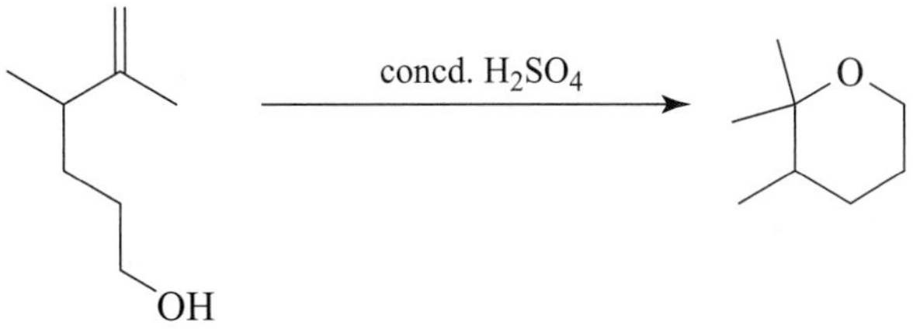

(e) (1) Br_2, H_2O (2) NaOH → O

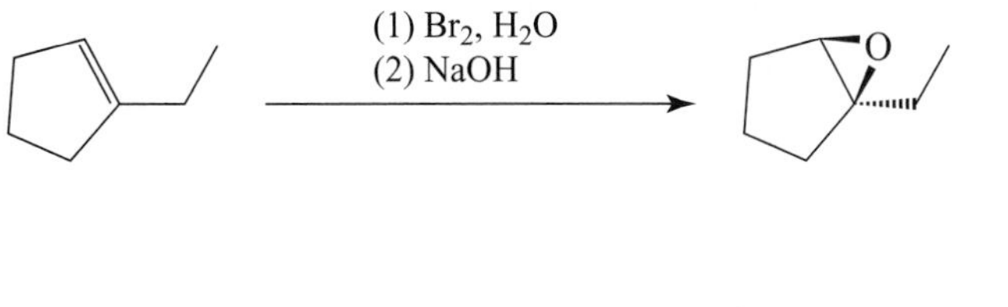

(f) 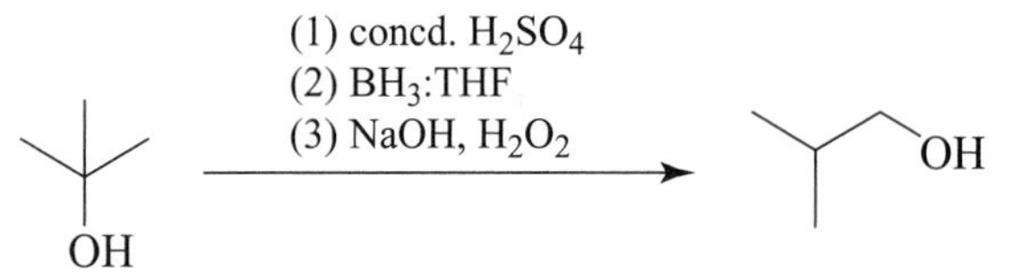 (1) concd. H_2SO_4
(2) BH_3:THF
(3) NaOH, H_2O_2

OH → OH

10. (a) 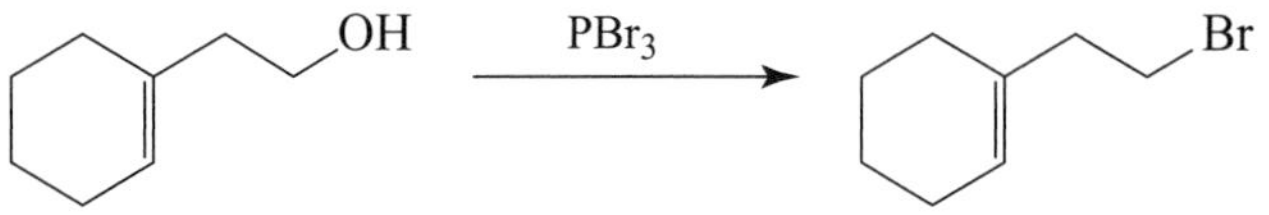 OH —PBr_3→ Br

(b) 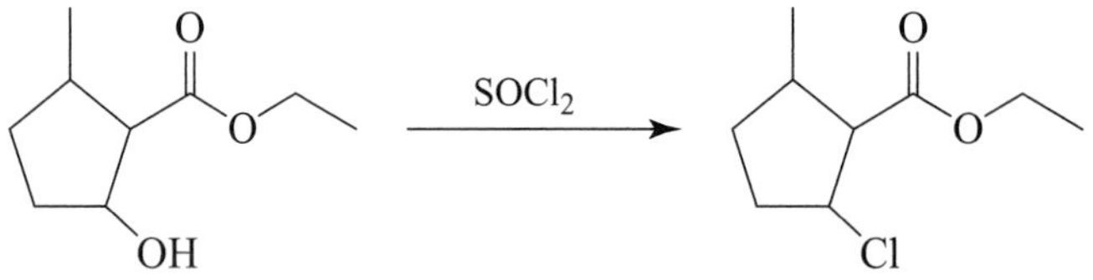 O, O, OH —$SOCl_2$→ O, O, Cl

(c) 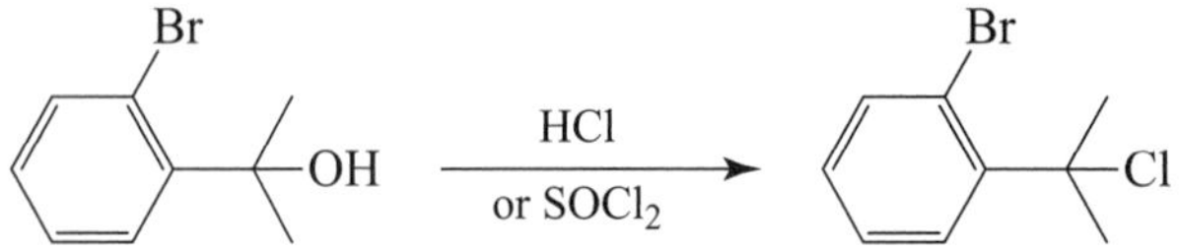 Br, OH —HCl or $SOCl_2$→ Br, Cl

(d) 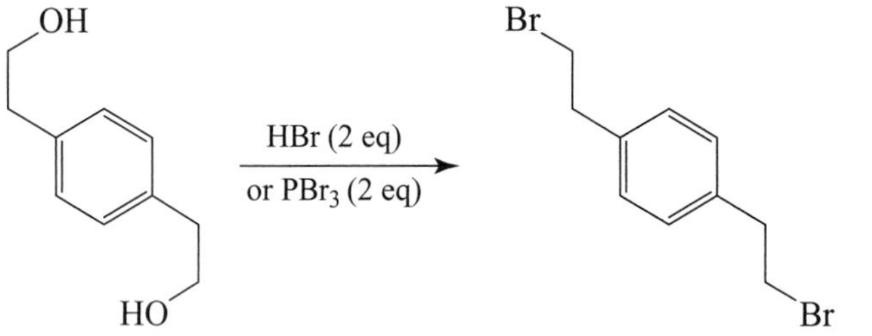 OH, HO —HBr (2 eq) or PBr_3 (2 eq)→ Br, Br

(e) 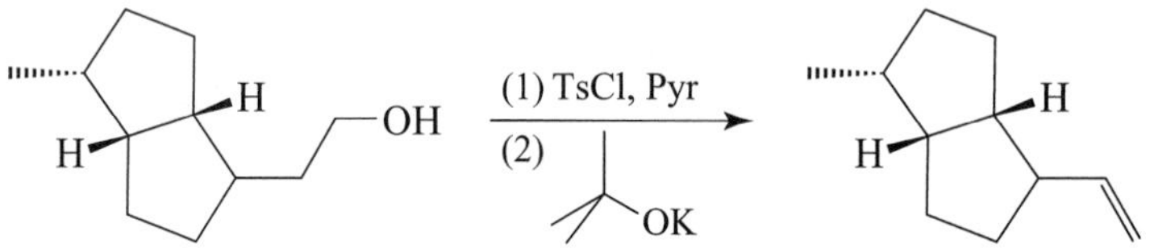 H, H, OH —(1) TsCl, Pyr (2) OK→ H, H

(f) 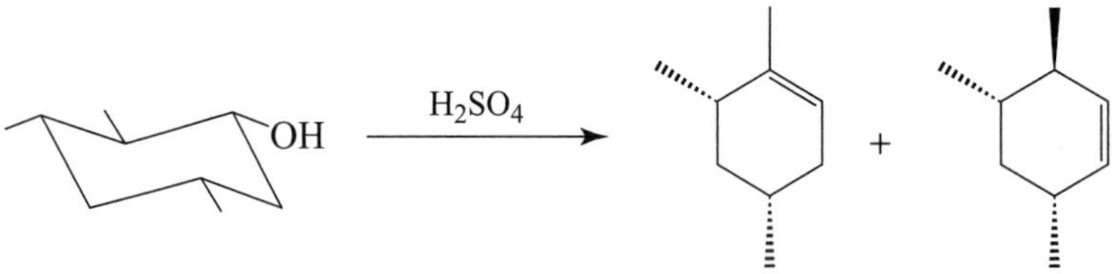 OH —H_2SO_4→ +

11. (a) 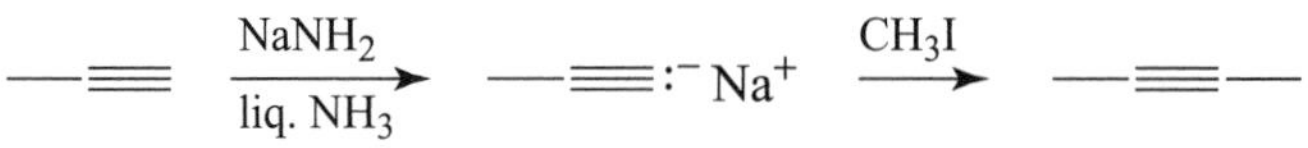 —$NaNH_2$, liq. NH_3→ $:^{-}\ Na^{+}$ —CH_3I→

(b) 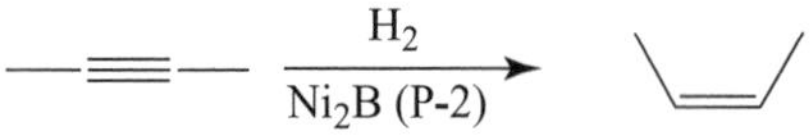 [from (a)] —H_2, Ni_2B (P-2)→

(c) 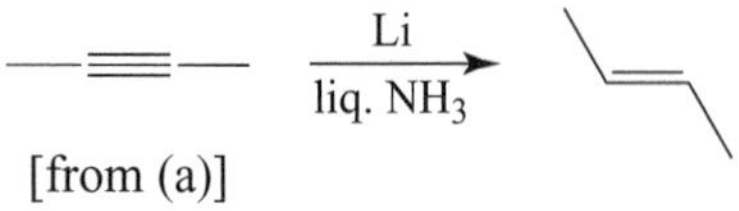 [from (a)] —Li, liq. NH_3→

(d) 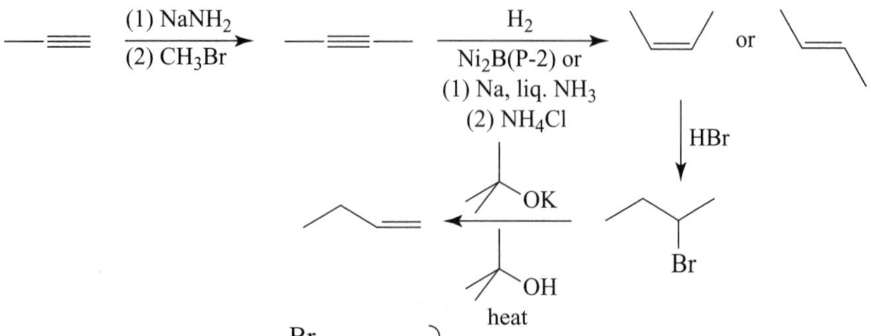 (1) $NaNH_2$ (2) CH_3Br; H_2, Ni_2B(P-2) or (1) Na, liq. NH_3 (2) NH_4Cl; or; HBr; OK; OH; heat; Br

(e) 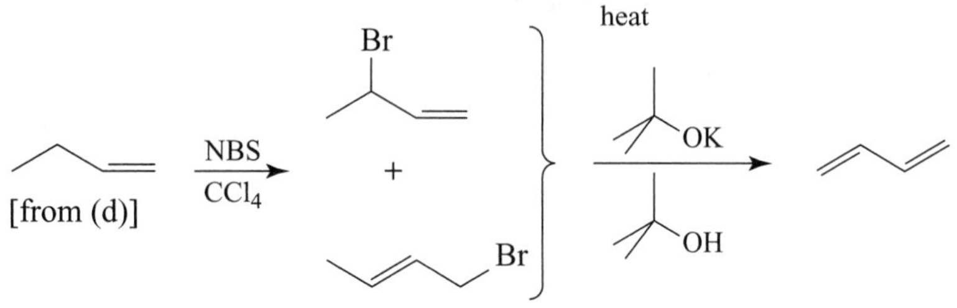 [from (d)] NBS, CCl_4; Br; +; Br; OK; OH

(f)  [from (d)] HBr, ROOR; Br

(g) 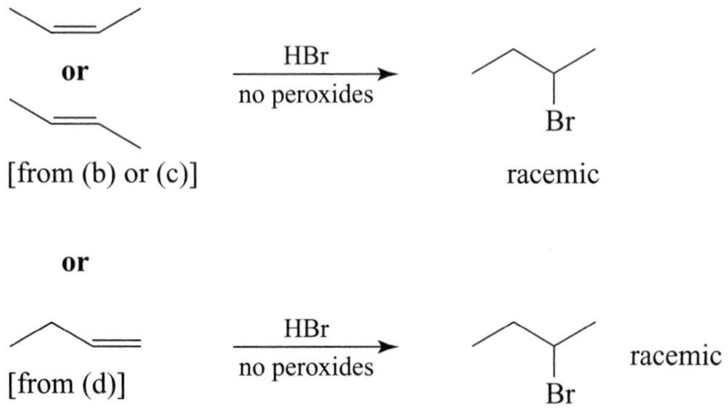 or; [from (b) or (c)]; HBr, no peroxides; Br; racemic

or

[from (d)]; HBr, no peroxides; Br; racemic

(h) 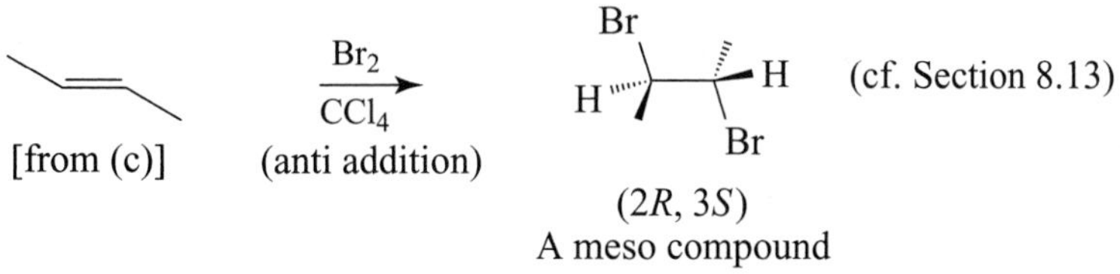 [from (c)]; Br_2, CCl_4 (anti addition); Br, H, H, Br (cf. Section 8.13)

(2*R*, 3*S*)
A meso compound

(i) 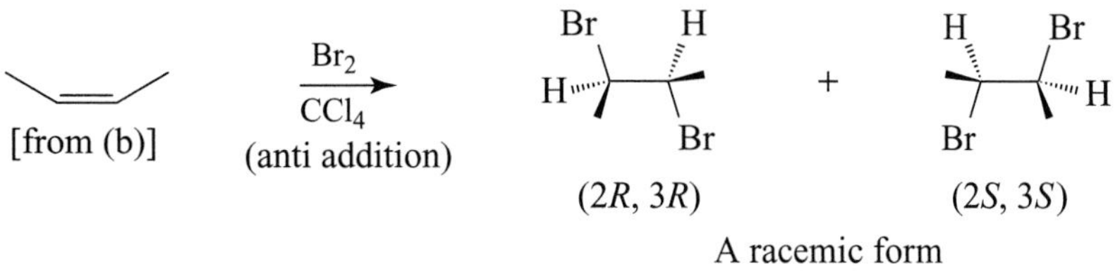 [from (b)]; Br_2, CCl_4 (anti addition); Br, H, H, Br + H, Br, H, Br

(2*R*, 3*R*) (2*S*, 3*S*)

A racemic form

(j) 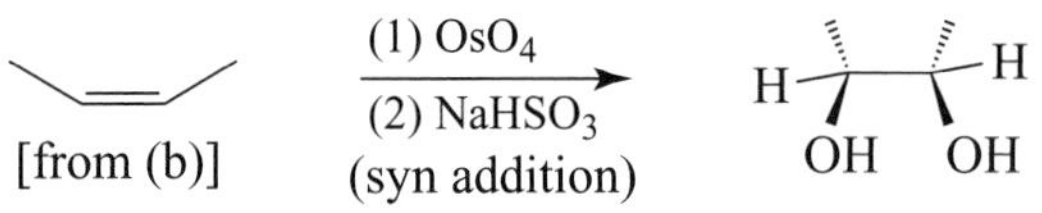 [from (b)] — (1) OsO_4 (2) $NaHSO_3$ (syn addition) → H, H, OH, OH (cf. Section 8.16)

or

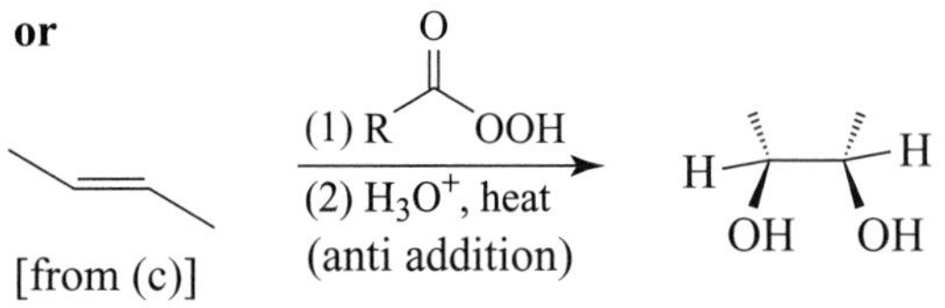

[from (c)] — (1) RC(=O)OOH (2) H_3O^+, heat (anti addition) → H, H, OH, OH (cf. Section 11.15)

(k) 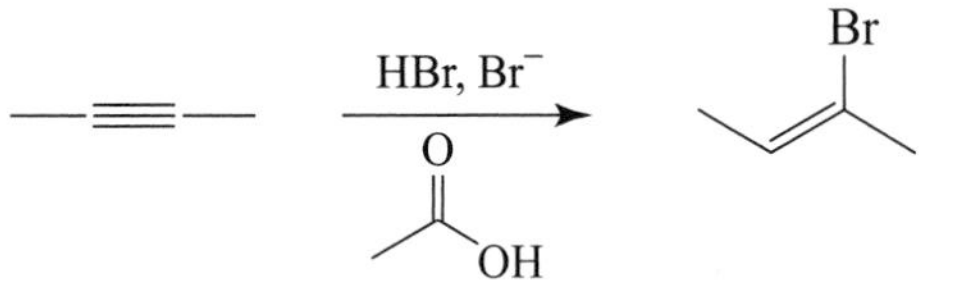 — HBr, Br^-, CH_3COOH → Br (cf. Section 8.19)

12. 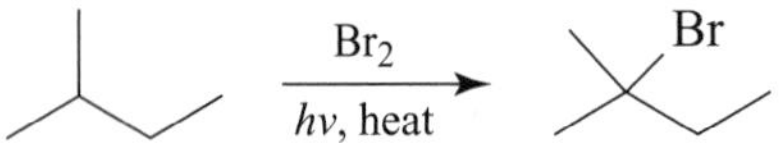 — Br_2, $h\nu$, heat → Br (cf. Section 10.6)

(a) 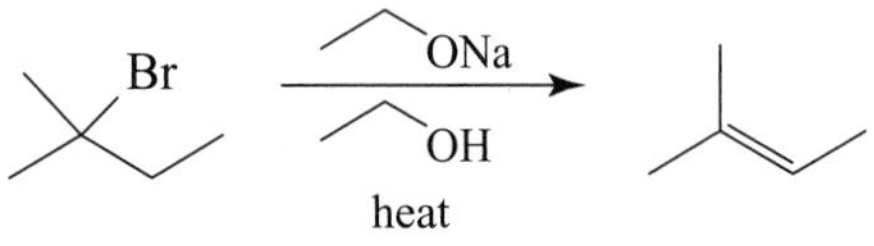 Br — ONa, OH, heat →

(b) 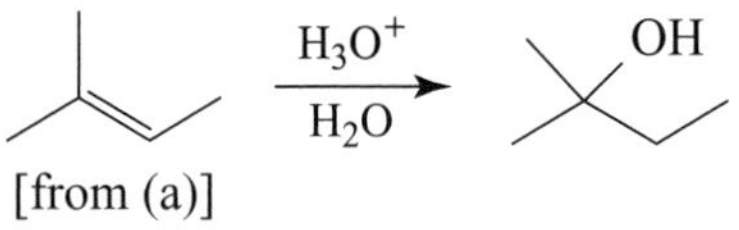 [from (a)] — H_3O^+, H_2O → OH

(c) 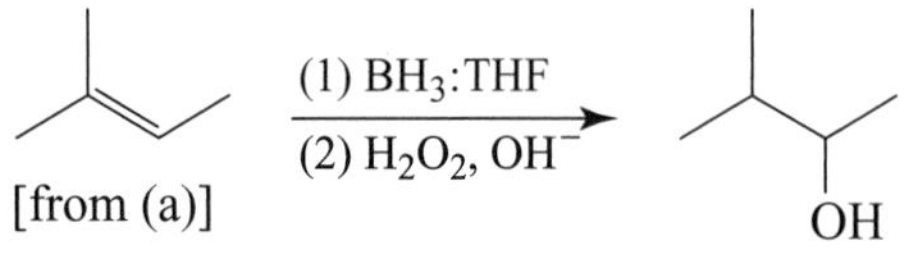 [from (a)] — (1) BH_3:THF (2) H_2O_2, OH^- → OH

(d) 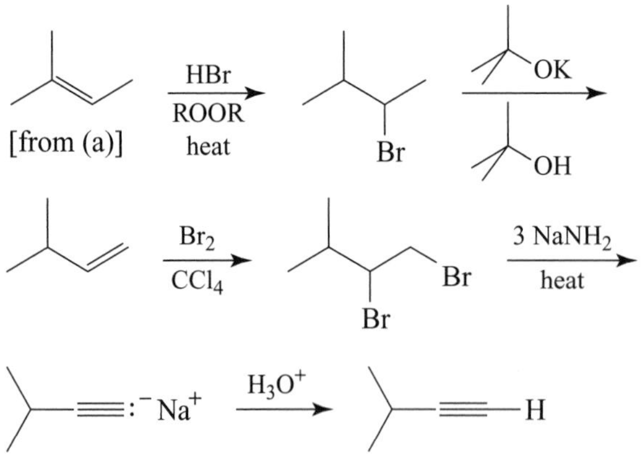 [from (a)] — HBr, ROOR, heat → Br — OK, OH →

— Br_2, CCl_4 → Br, Br — 3 $NaNH_2$, heat →

$:^-Na^+$ — H_3O^+ → H

(e) 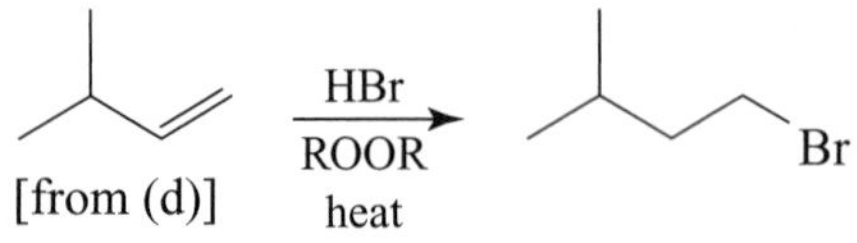 [from (d)] — HBr, ROOR, heat → Br

(f) 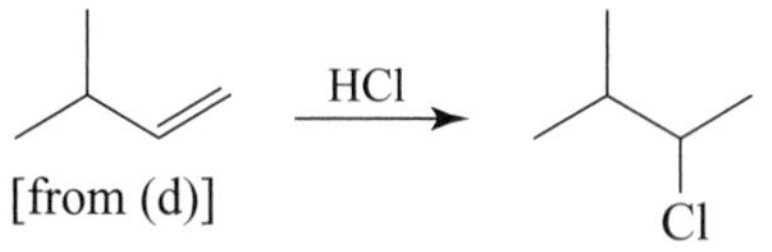 [from (d)] $\xrightarrow{HCl}$

(g) 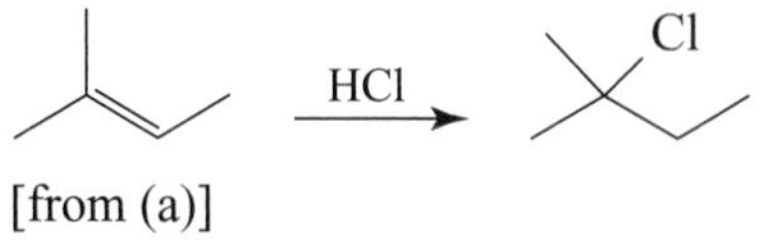 [from (a)] $\xrightarrow{HCl}$ Cl

(h) 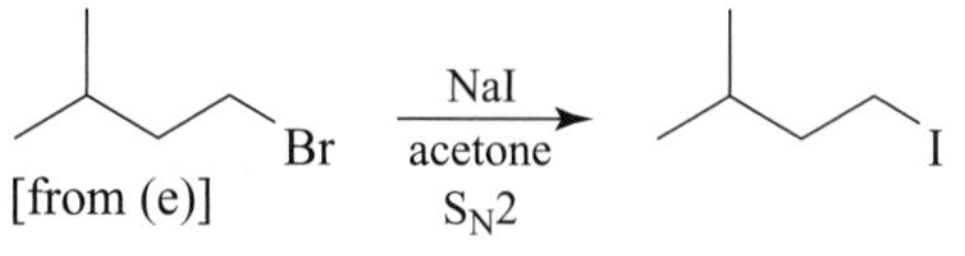 Br [from (e)] $\xrightarrow[\text{acetone}]{\text{NaI}}$ S_N2 I

(i) [from (a)] $\xrightarrow[(2)\ Me_2S]{(1)\ O_3}$ O + O H

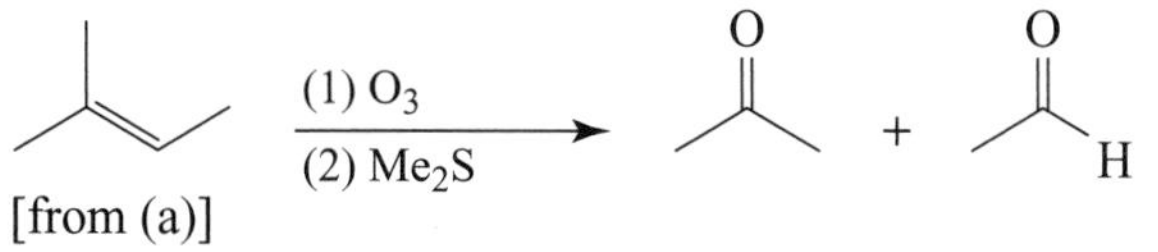

(j) 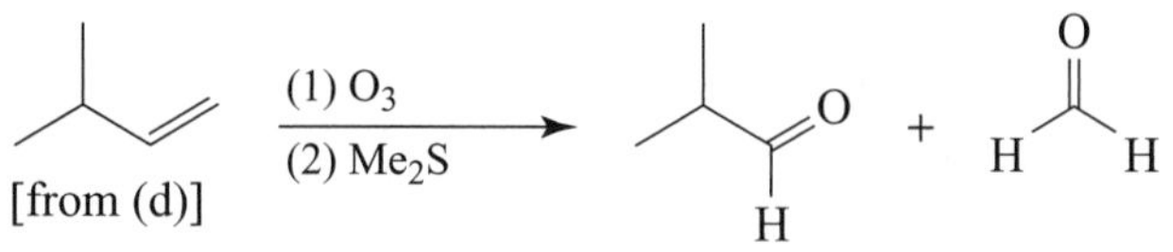 [from (d)] $\xrightarrow[(2)\ Me_2S]{(1)\ O_3}$ O H + O H H

13. 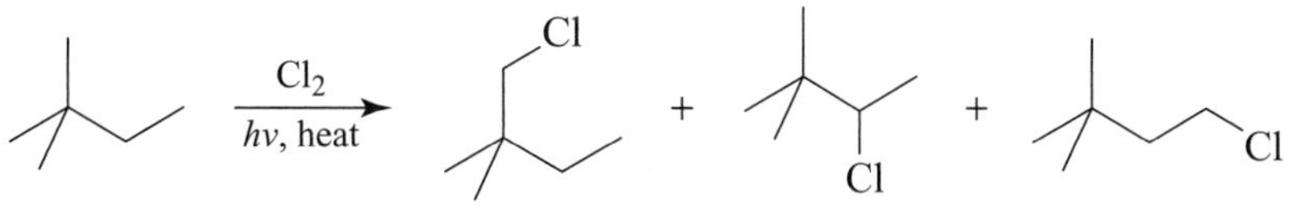 A $\xrightarrow[hv,\ \text{heat}]{Cl_2}$ B + C (racemic) + D

B cannot undergo dehydrohalogenation because it has no β hydrogen; however, **C** and **D** can, as shown next.

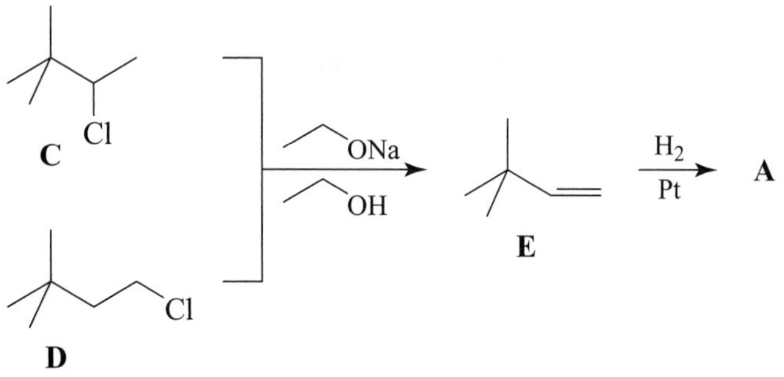

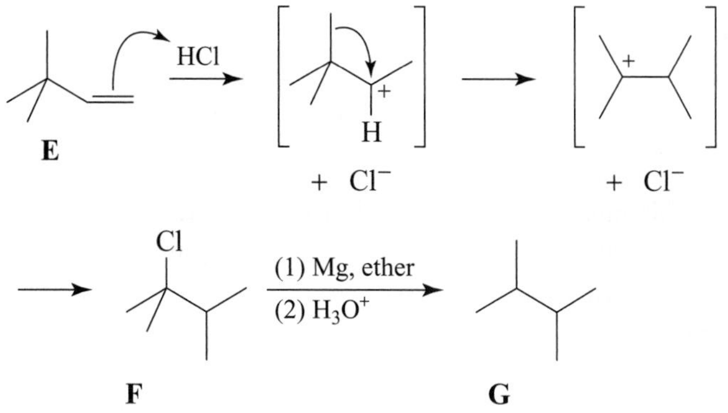

C, **D** $\xrightarrow[\text{EtOH}]{\text{EtONa}}$ **E** $\xrightarrow[Pt]{H_2}$ **A**

E $\xrightarrow{HCl}$ [carbocation] + Cl^- $\longrightarrow$ [carbocation] + Cl^-

$\longrightarrow$ **F** $\xrightarrow[(2)\ H_3O^+]{(1)\ \text{Mg, ether}}$ **G**

14. A, No IR absorption in 2200–2300 cm^{-1} region. $\xrightarrow{H_2, Pt}$

H_2 | Ni_2B (P-2)

B $\xrightarrow[\text{(2) } NaHSO_3]{\text{(1) } OsO_4}$ (syn hydroxylation)

HO OH H H

C (a meso compound)

15. The eliminations are anti eliminations, requiring an anti coplanar arrangement of the bromine atoms.

H Br Br H :I: $\xrightarrow{KI}$ OH H H + IBr

meso-2,3-Dibromobutane *trans*-2-Butene

H Br H Br :I: $\xrightarrow{KI}$ OH H H + IBr

(2*S*,3*S*)-2,3-Dibromobutane *cis*-2-Butene

Br H Br H :I: $\xrightarrow{KI}$ OH H H + IBr

(2*R*,3*R*)-2,3-Dibromobutane *cis*-2-Butene

16. The eliminations are anti eliminations, requiring an anti coplanar arrangement of the —H and —Br.

Br Br C_6H_5 H H C_6H_5 EtÖ: ⟶ Br H C_6H_5 C_6H_5

meso-1,2-Dibromo-1,2-diphenylethane (*E*)-1-Bromo-1,2-diphenylethene

Br Br C_6H_5 C_6H_5 H H EtÖ: ⟶ Br C_6H_5 C_6H_5 H

(2*R*,3*R*)-1,2-Dibromo-1,2-diphenylethane (*Z*)-1-Bromo-1,2-diphenylethene

(2*S*,3*S*)-1,2-Dibromo-1,2-diphenylethane will also give (*Z*)-1-bromo-1,2-diphenylethene in an anti elmination.

17. In all the following structures, notice that the large *tert*-butyl group is equatorial.

(a)

Br

H

Br

+ enantiomer
as a racemic form

(bromine addition is anti; cf. Section 8.13)

(b)

OH

OH

H

+ enantiomer
as a racemic form

(syn hydroxylation; cf. Section 8.16)

(c)

OH

H

OH

+ enantiomer
as a racemic form

(anti hydroxylation; cf. Section 11.15)

(d)

H

OH

H

+ enantiomer
as a racemic form

(syn and anti Markovnikov addition of —H and —OH; cf. Section 8.9)

(e)

OH

(Markovnikov addition of —H and —OH; cf. Section 8.6)

(f)

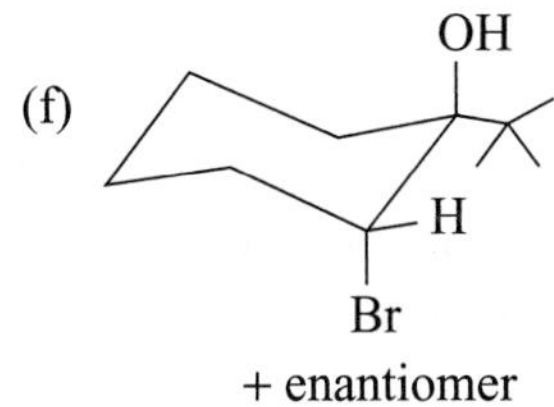

+ enantiomer
as a racemic form

(anti addition of —Br and —OH, with —Br and —OH placement resulting from the more stable partial carbocation in the intermediate bromonium ion; cf. Section 8.14)

(g)

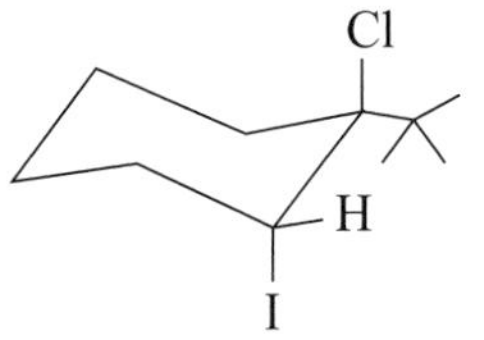

Cl

H

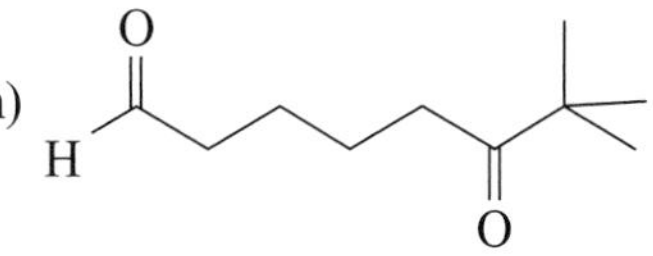

I

+ enantiomer
as a racemic form

(anti addition of —I and —Cl, following Markovnikov's rule; cf. Section 8.13)

(h)

O

H

O

(i)

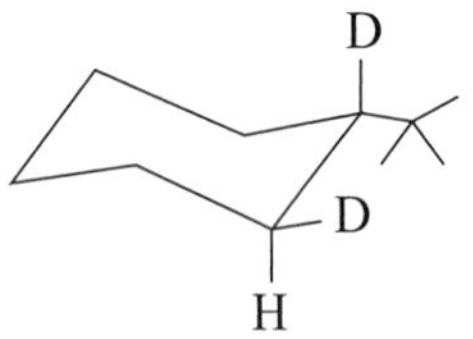

D

D

H

+ enantiomer
as a racemic form

(syn addition of deuterium; cf. Section 7.15A)

(j)

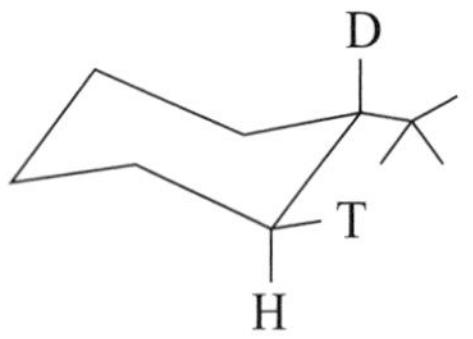

D

T

H

+ enantiomer
as a racemic form

(syn, anti Markovnikov addition of —D and B, with B being replaced by —T where it stands; cf. Section 8.11)

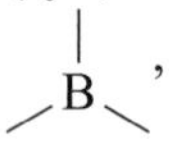

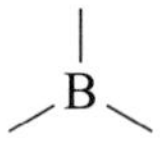

18. **A** = 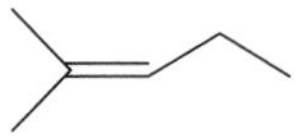 **B** = $(\ )_2$BH **C** = OH

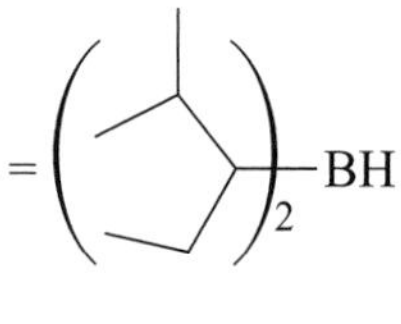

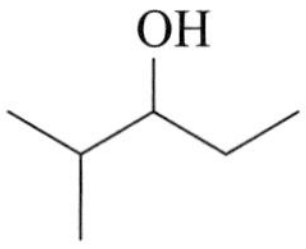

19. (a) The following products are diastereomers. They would have different boiling points and would be in separate fractions. Each fraction would be optically active.

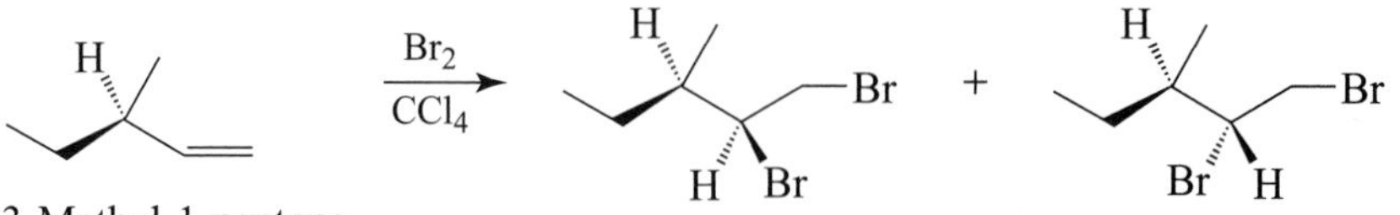

H

(*R*)-3-Methyl-1-pentene

$\xrightarrow[CCl_4]{Br_2}$

H Br H Br
(optically active)

+

H Br Br H
(optically active)

Diastereomers

(b) Only one product is formed. It is achiral, and, therefore, it would not be optically active.

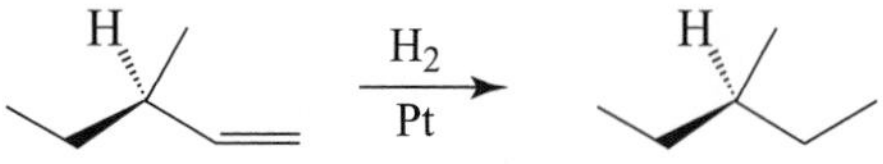

H $\xrightarrow[Pt]{H_2}$ H

(optically inactive)

(c) Two diastereomeric products are formed. Two fractions would be obtained. Each fraction would be optically active.

H

(1) OsO_4
(2) $NaHSO_3$

H OH + H OH

H OH
(optically active)

HO H
(optically active)

Diastereomers

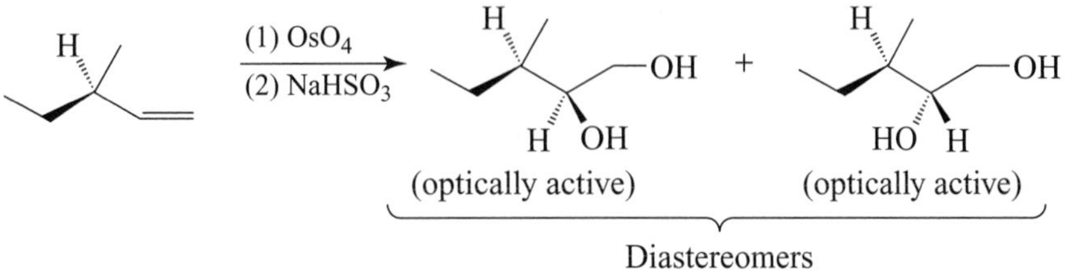

(d) One optically active compound is produced.

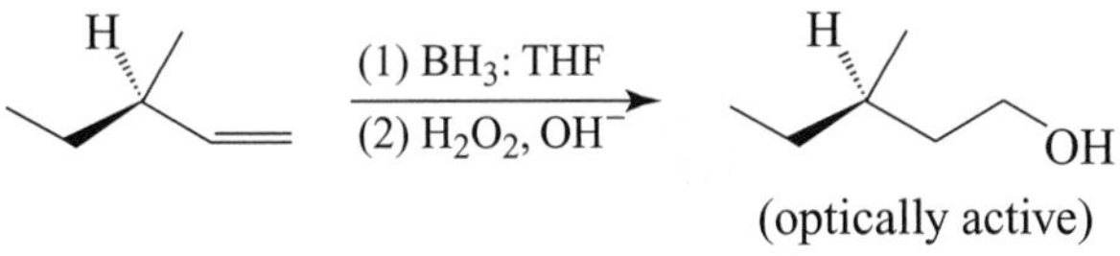

H

(1) BH_3: THF
(2) H_2O_2, OH^-

H OH
(optically active)

(e) Two diastereomeric products are formed. Two fractions would be obtained. Each fraction would be optically active.

H

(1) $(CH_3CO_2)_2$Hg, THF–H_2O
(2) $NaBH_4$, OH^-

H + H

H OH
(optically active)

HO H
(optically active)

Diastereomers

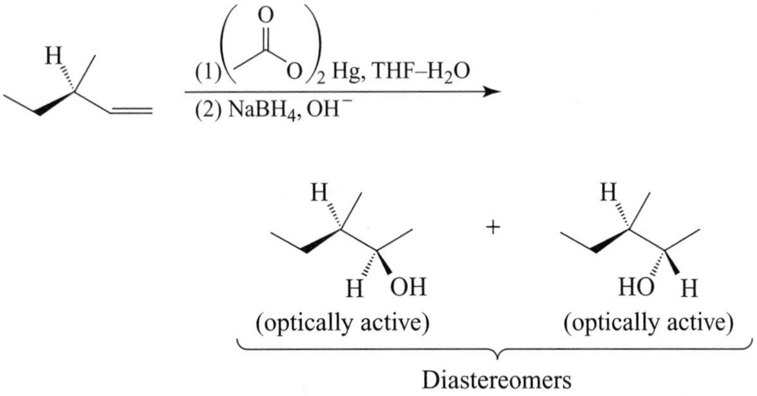

(f) Two diastereomeric products are formed. Two fractions would be obtained. Each fraction would be optically active.

H

(1) MCPBA
(2) H_3O^+, H_2O

H OH + H OH

H OH

HO H

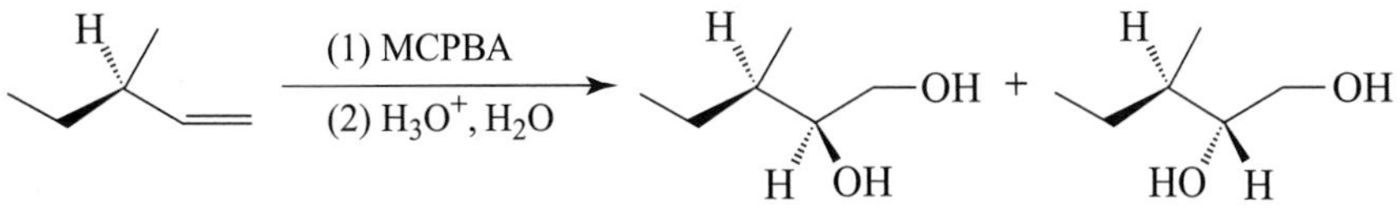

20.

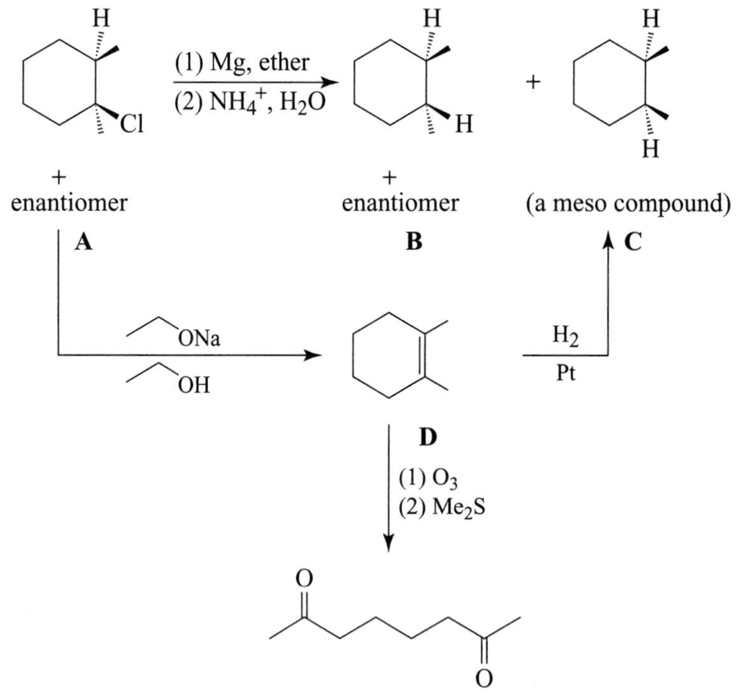

H, Cl; + enantiomer **A** → (1) Mg, ether (2) NH_4^+, H_2O → H, H; + enantiomer **B** + H, H (a meso compound) **C**

A → ONa / OH → **D** → H_2 / Pt → **C**

D → (1) O_3 (2) Me_2S → O, O

21. (a) O → (1) CH_3MgBr (2) H_3O^+ → HO

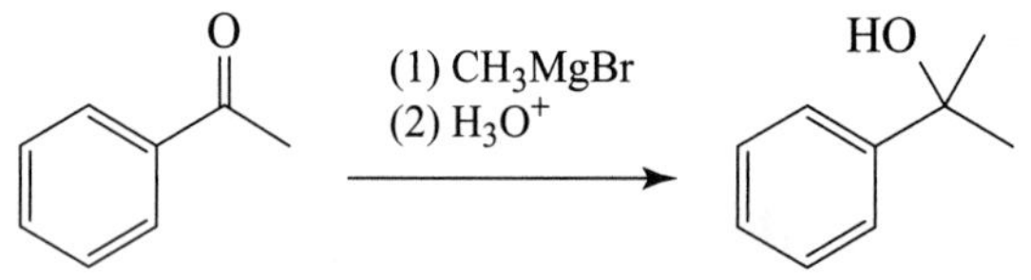

(b) O → (1) Li (2) H_3O^+ → OH

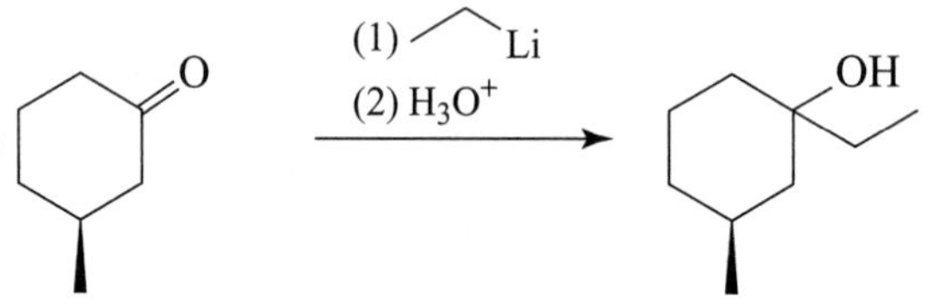

(c) O, O, CH_3 → (1) CH_3MgBr (excess), ether (2) H_3O^+ → OH + CH_3OH

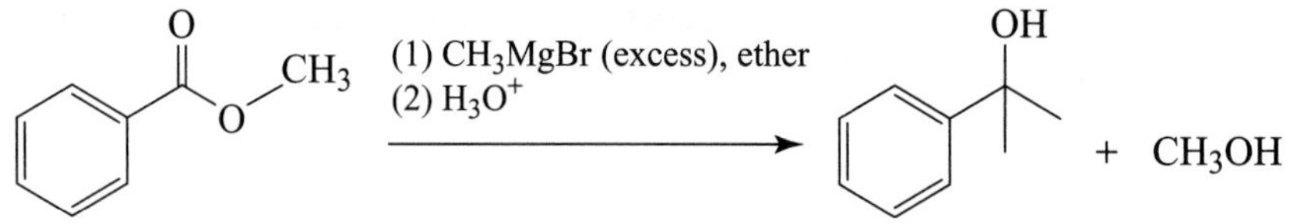

22.

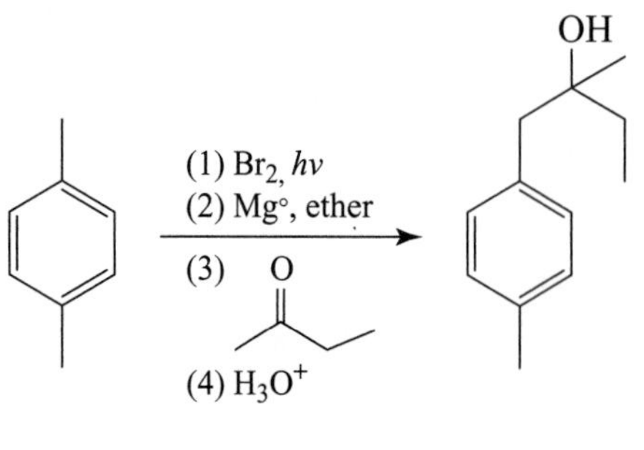

→ (1) Br_2, *hv* (2) Mg°, ether (3) O (4) H_3O^+ → OH

23. (a) OH → $SOCl_2$, pyridine → Cl

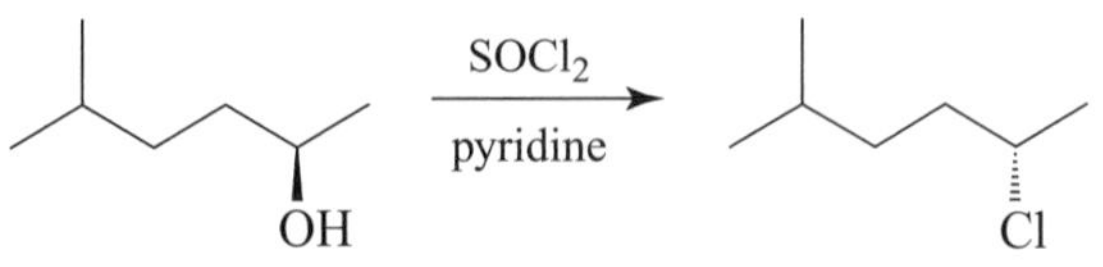

(b) (1) $LiAlH_4$ (excess) (2) H_3O^+ → OH, OH, HO

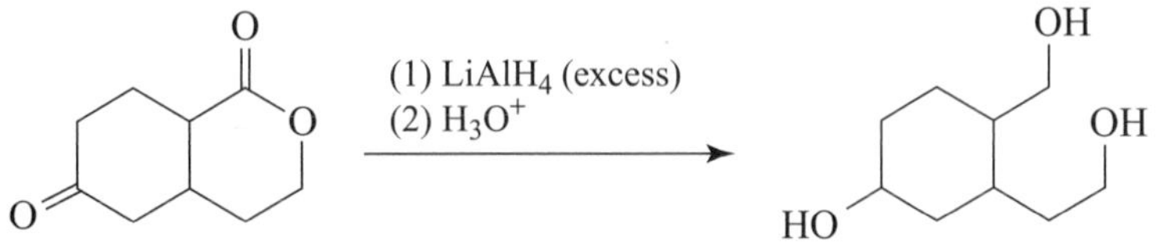

(c) O, O, CH_3 (1) $LiAlH_4$ (2) H_3O^+ → OH

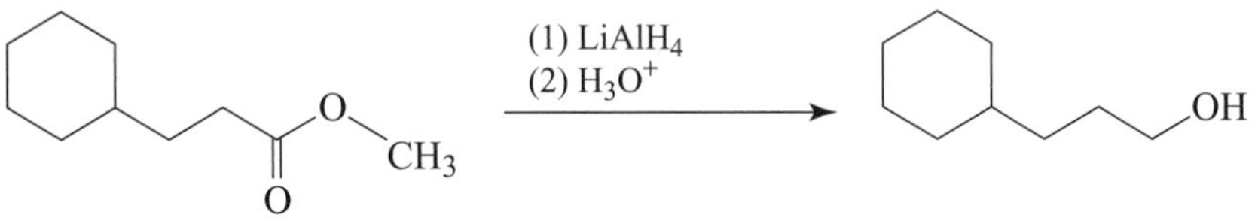

(d) (1) OsO_4 (2) $NaHSO_3$ → OH, OH

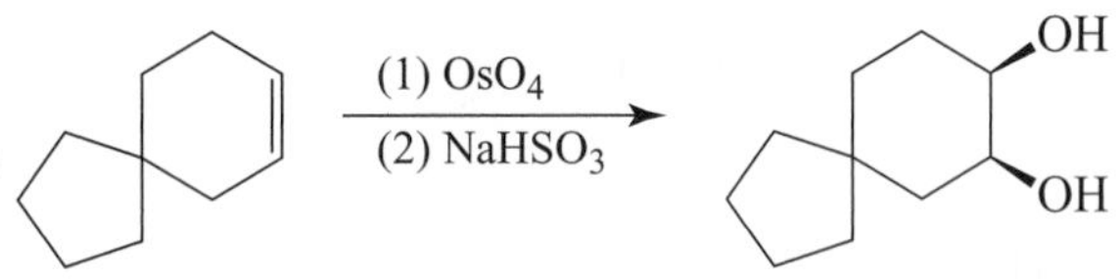

(e) Br, OH NaH → O

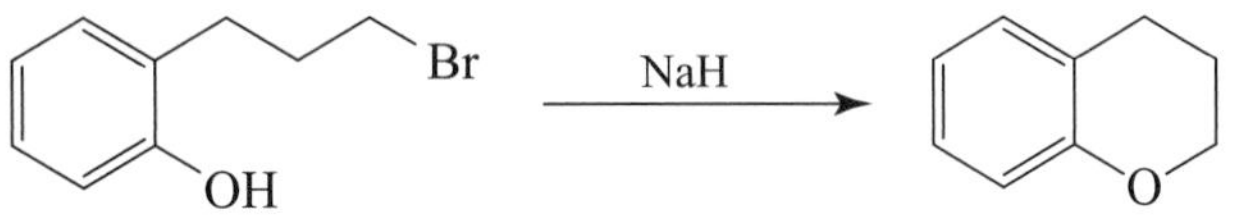

(f) OH H_2CrO_4 → O, OH

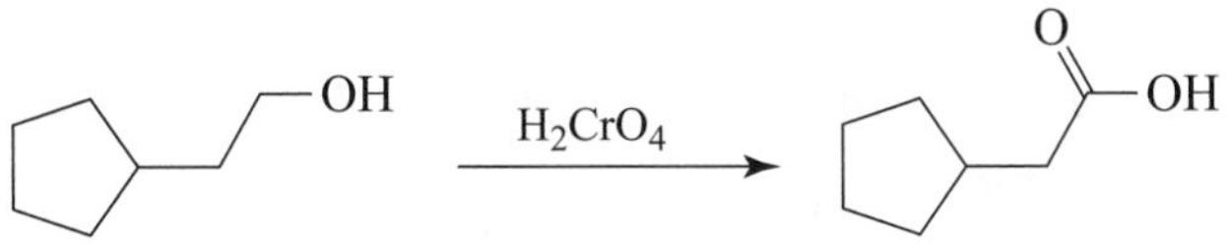

(g) O, HO, O (1) MgBr (1 equiv.) (2) H_3O^+ → O, HO, O +

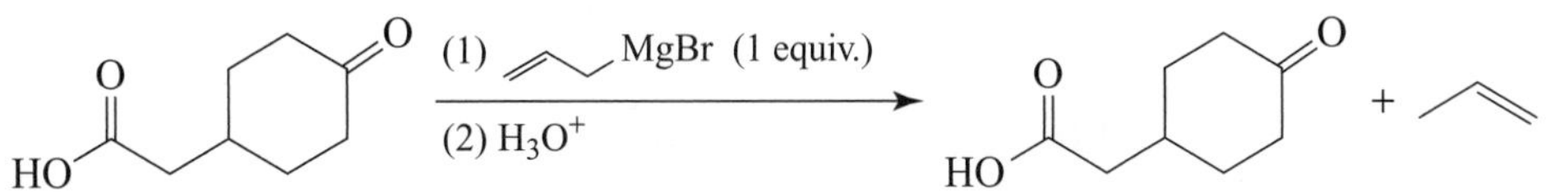

(h) O, O, O, CH_3 (1) $NaBH_4$ (2) H_3O^+ → HO, O, O, CH_3

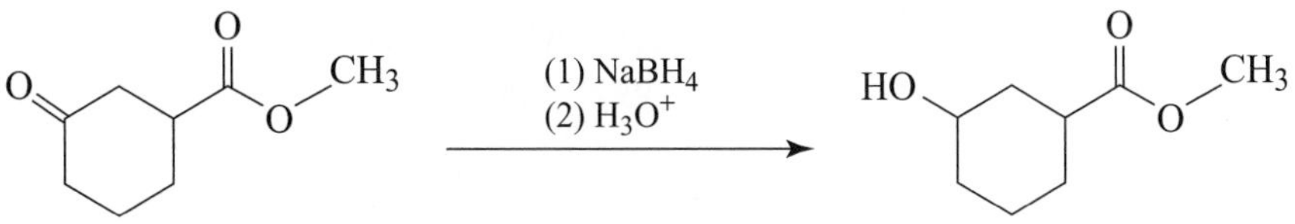

24.

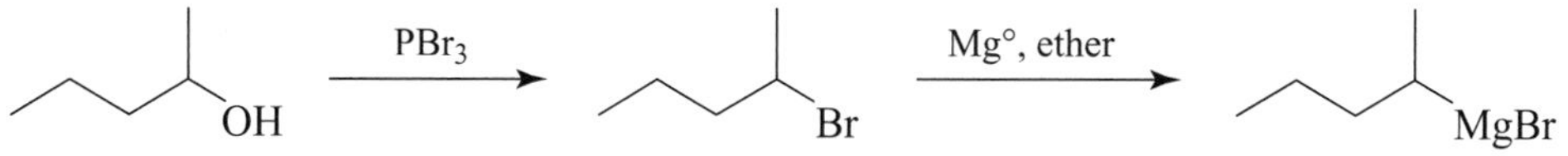

OH — PBr_3 → Br — Mg°, ether → MgBr

A

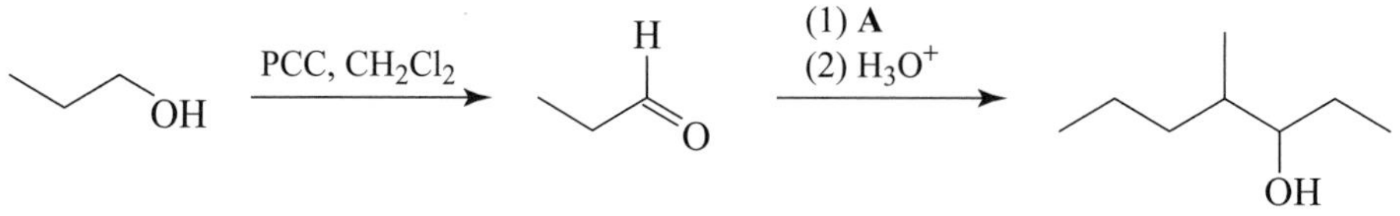

OH — PCC, CH_2Cl_2 → H, O — (1) **A** (2) H_3O^+ → OH

25.

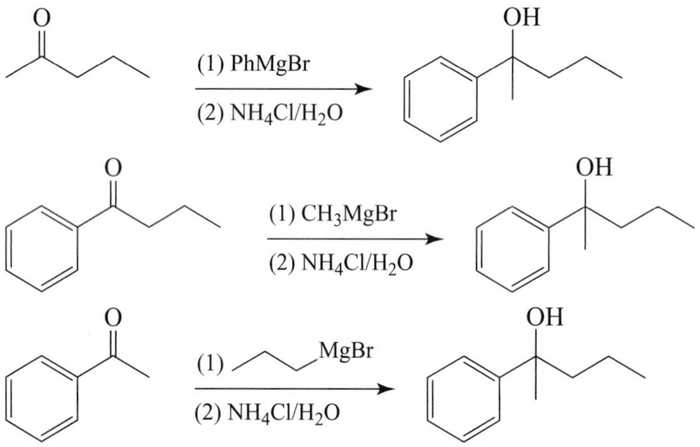

O — (1) PhMgBr (2) NH_4Cl/H_2O → OH

O — (1) CH_3MgBr (2) NH_4Cl/H_2O → OH

O — (1) MgBr (2) NH_4Cl/H_2O → OH

26.

O O O O O

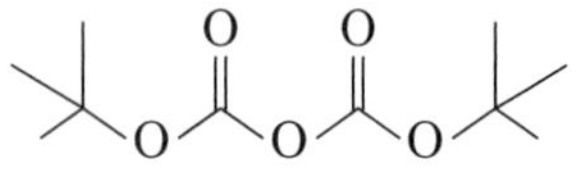

Y

^{1}H NMR

Singlet at δ 1.4 — All 18 protons equivalent

Values suggests —C(H)—O

^{13}C < NMR

δ 87 — C—O

δ 151 — C=O

IR

1750–1800 cm^{-1} — C=O

2 carbonyls indicated by split peak

27.

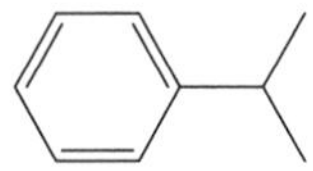

$m/z\ 120 = M^{+\cdot}$

$105 = M^{+\cdot} - 15(CH_3\cdot) = C_6H_5\overset{+}{C}H(CH_3)$

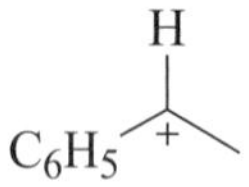

$77 = M^{+\cdot} - 43(i-Pr\cdot) = C_6H_5^+$

δ7.2–7.6	5 ring protons
2.95	CH of isopropyl group
1.29	equivalent CH_3s of isopropyl group

28. $C_5H_{10}O$ has IHD = 2
IR absorption indicates C=O
^{13}C NMR spectrum for **X** is consistent with structure

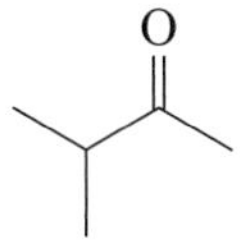

O

29. (a)

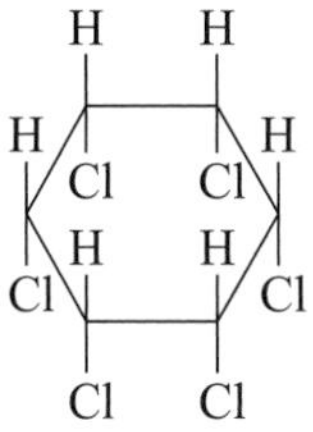

1 meso 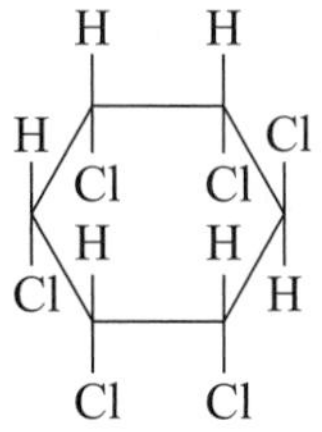 **2** meso 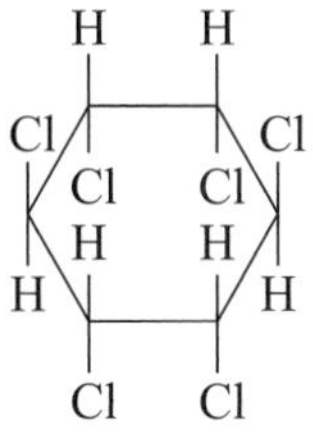 **3** meso 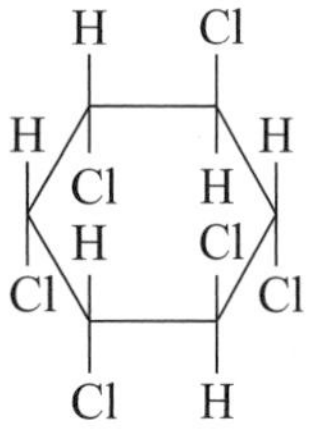 **4** meso

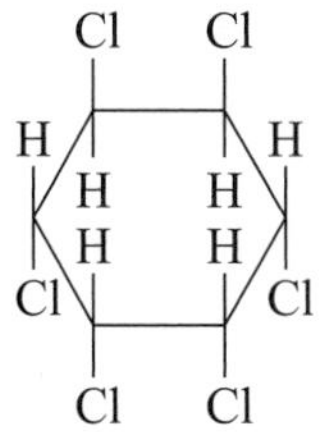

5 meso 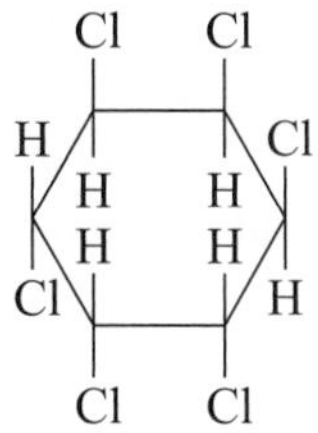 **6** meso 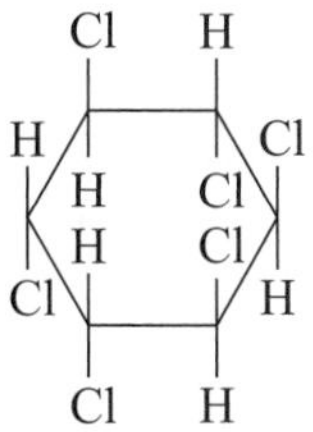 **7** 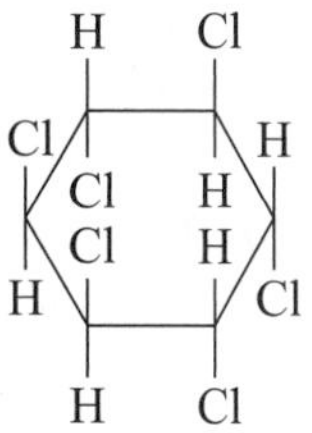 **8**

Enantiomers

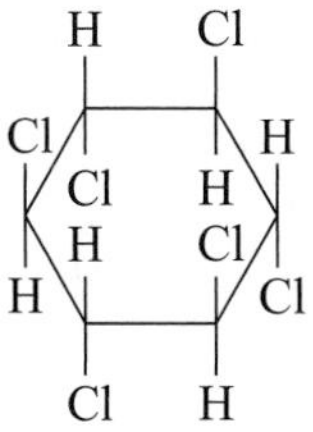

9 meso

(b) Isomer **9** is slow to react in an E2 reaction because in its more stable conformation (see following structure) all the chlorine atoms are equatorial and an anti coplanar transition state cannot be achieved. All other isomers **1–8** can have a —Cl axial and thus achieve an anti coplanar transition state.

9

30. (a)

$\xrightarrow{F_2}$ **1** + **2**

Enantiomers
(obtained in one fraction as an optically inactive racemic form)

+ **3** + **4** + **5** + **6**

3 (achiral and, therefore, optically inactive)

4, **5**: Enantiomers (obtained in one fraction as an optically inactive racemic form)

6 (achiral and, therefore, optically inactive)

(b) Four fractions. The enantiomeric pairs would not be separated by fractional distillation because enantiomers have the same boiling points.

(c) All of the fractions would be optically inactive.

(d) The fraction containing **1** and **2** and the fraction containing **4** and **5.**

31. (a)

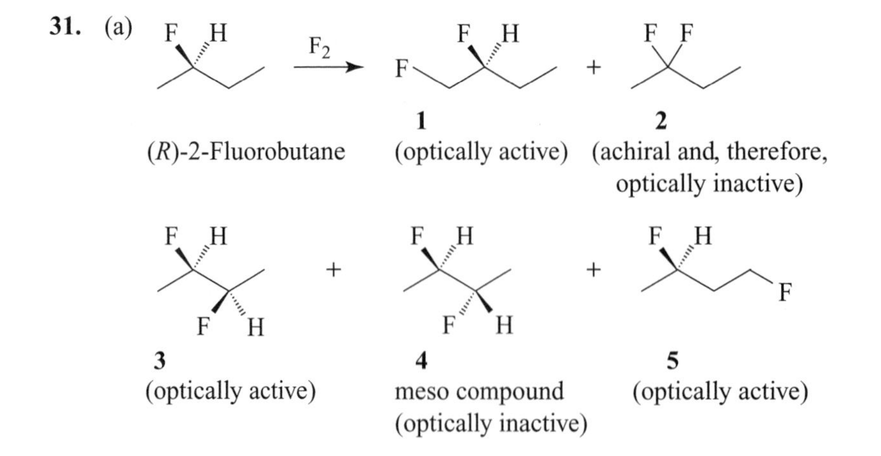

(b) Five. Compounds **3** and **4** are diastereomers. All others are constitutional isomers of each other.

(c) See above.

32.

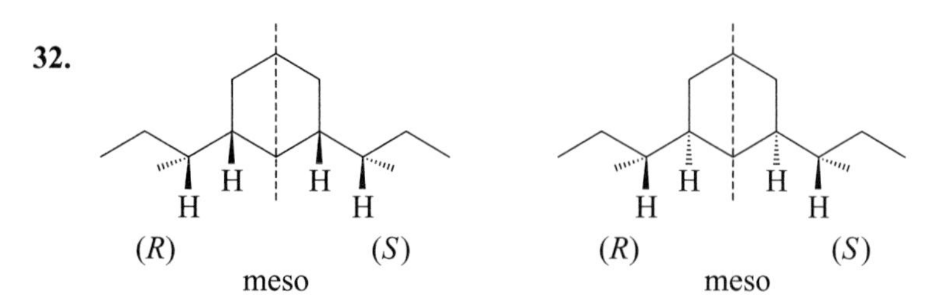

Each of the two structures just given has a plane of symmetry (indicated by the dashed line), and, therefore, each is a meso compound. The two structures are not superposable one on the other; therefore, they represent molecules of different compounds and are diastereomers.

33. Only a proton or deuteron anti to the bromine can be eliminated; that is, the two groups undergoing elimination (H and Br or D and Br) must lie in an anti coplanar arrangement. The two conformations of *erythro*-2-bromobutane-3-*d* in which a proton or deuteron is anti coplanar to the bromine are **I** and **II.**

Br D H EtO⁻ —HBr→ D

I

Br H D EtO⁻ —DBr→

II

Conformation **I** can undergo loss of HBr to yield *cis*-2-*d*-2-butene. Conformation **II** can undergo loss of DBr to yield *trans*-2-butene.

To a minor extent, a proton of the methyl group can be eliminated with the bromine.

Br H H H D EtO⁻ —HBr→ D

13 CONJUGATED UNSATURATED SYSTEMS

SOLUTIONS TO PROBLEMS

13.1 The following two allylic radicals are possible, differing only because of the isotopic label. Together they allow for four constitutional isomers with respect to the ^{13}C label. (In the absence of the isotopic label, only one constitutional isomer (as a racemic mixture) would be possible.)

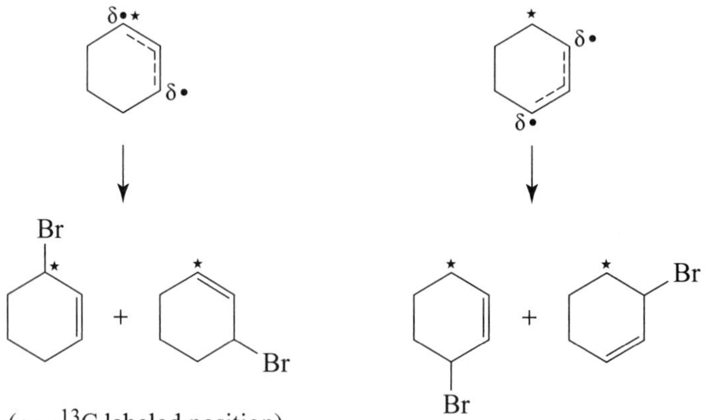

δ•⋆ δ• ⋆ δ• δ•

Br ⋆ + ⋆ Br ⋆ + ⋆ Br Br

(⋆ = ^{13}C labeled position)

13.2 (a) 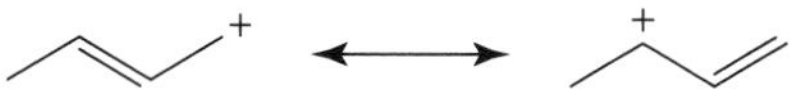 + ⟷ +

(b) 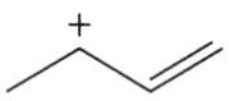 + because it represents a 2° carbocation.

(c) 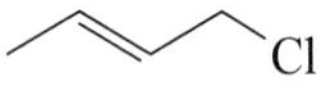 Cl and Cl (racemic)

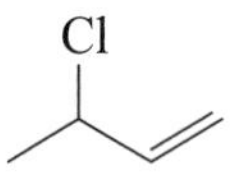

13.3 (a)  ⟷

(b) 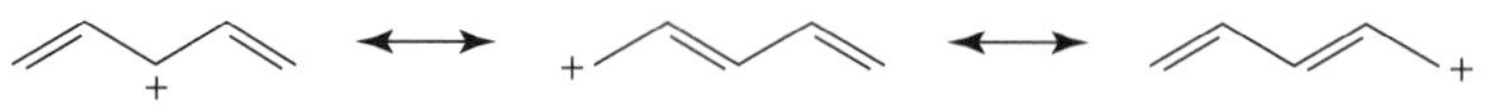 + ⟷ + ⟷ +

(c) 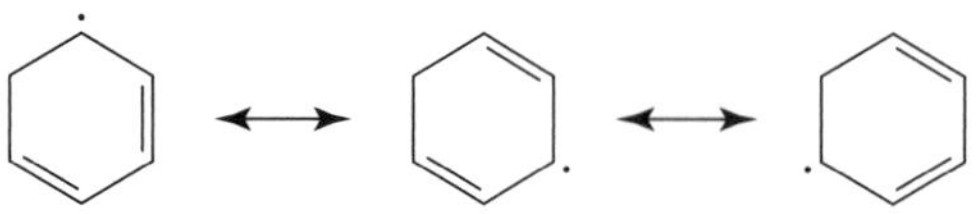 ⟷ ⟷

(d) 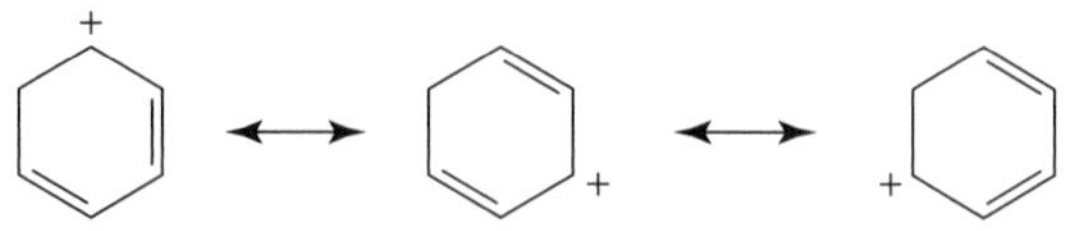 + ⟷ + ⟷ +

(e) 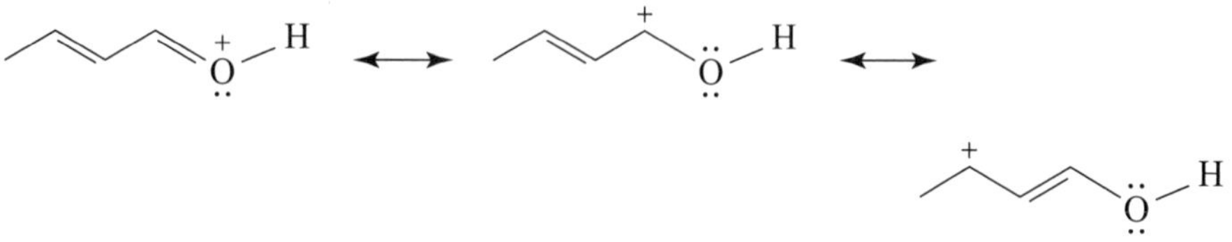 $\overset{+}{O}$—H ⟷ + $\ddot{O}$—H ⟷ + $\ddot{O}$—H

(f)  $\ddot{Br}$: ⟷ $\overset{-}{:}$ $\overset{+}{Br}$:

(g) 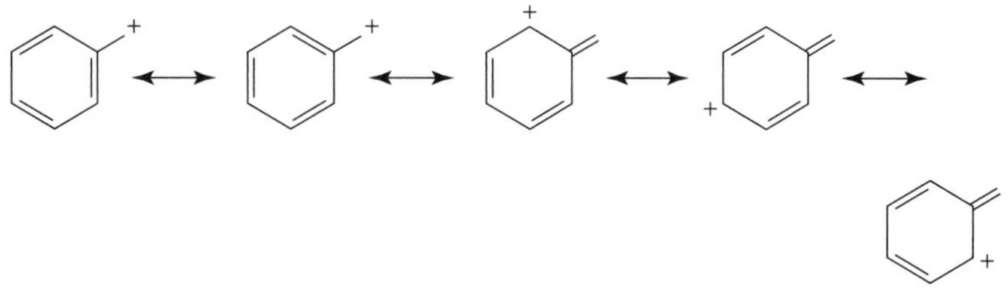 + ⟷ + ⟷ + ⟷ + ⟷ +

(h)  $\cdot\ddot{O}\cdot$ $:^{-}$ ⟷ $:\ddot{O}:^{-}$

(i)  $\overset{+}{\ddot{S}}$ ⟷ $\overset{+}{S}$

(j) 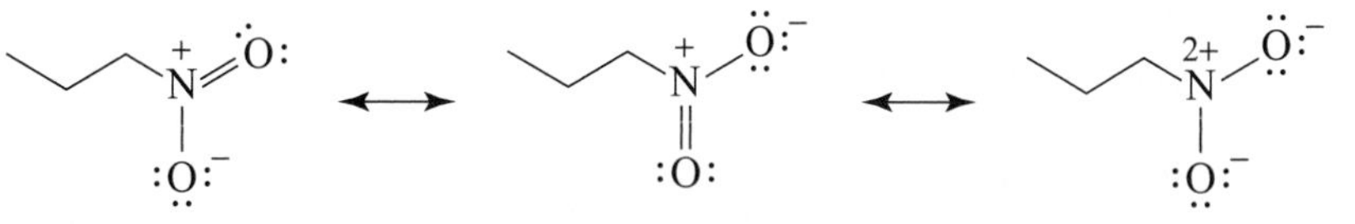 $\overset{+}{N}$=$\ddot{O}$: $:\ddot{O}:^{-}$ ⟷ $\overset{+}{N}$—$\ddot{O}:^{-}$ $:\ddot{O}:$ ⟷ $\overset{2+}{N}$—$\ddot{O}:^{-}$ $:\ddot{O}:^{-}$

(minor)

13.4 (a) 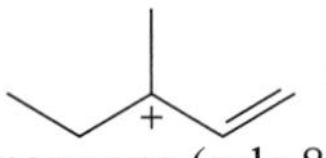 + because the positive charge is on a tertiary carbon atom rather than a primary one (rule 8).

(b) 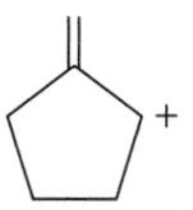 + because the positive charge is on a secondary carbon atom rather than a primary one (rule 8).

(c) 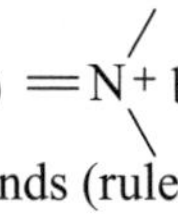 $=\!N^{+}$ because all atoms have a complete octet (rule 8b), and there are more covalent bonds (rule 8a).

(d) 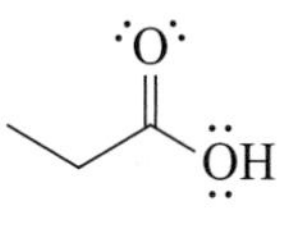 $\cdot\ddot{O}\cdot$ $\ddot{O}H$ because it has no charge separation (rule 8c).

(e)  because the radical is on a secondary carbon atom rather than a primary one (rule 8).

(f) :NH$_2$—≡N: because it has no charge separation (rule 8c).

13.5 In resonance structures, the positions of the nuclei must remain the same for all structures (rule 2). The keto and enol forms shown differ not only in the positions of their electrons, but also in the position of one of the hydrogen atoms. In the enol form, it is attached to an oxygen atom; in the keto form, it has been moved so that it is attached to a carbon atom.

13.6 (a) (3*Z*)-Penta-1,3-diene, (2*E*,4*E*)-2,4-hexadiene, (2*Z*,4*E*)-hexa-2,4-diene, and 1,3-cyclohexadiene are conjugated dienes.

(b) 1,4-Cyclohexadiene and 1,4-pentadiene are isolated dienes.

(c) Pent-1-en-4-yne (1-penten-4-yne) is an isolated enyne.

13.7 The formula, C_6H_8, tells us that **A** and **B** have six hydrogen atoms less than an alkane. This unsaturation may be due to three double bonds, one triple bond and one double bond, or combinations of two double bonds and a ring, or one triple bond and a ring. Since both **A** and **B** react with 2 mol of H_2 to yield cyclohexane, they are either cyclohexyne or cyclohexadienes. The absorption maximum of 256 nm for **A** tells us that it is conjugated. Compound **B,** with no absorption maximum beyond 200 nm, prossesses isolated double bonds. We can rule out cyclohexyne because of ring strain caused by the requirement of linearity of the —C≡C— system. Therefore, **A** is 1,3-cyclohexadiene; **B** is 1,4-cyclohexadiene. **A** has three signals in its ^{13}C NMR spectrum. With its higher symmetry, **B** shows only two ^{13}C NMR signals.

13.8 All three compounds have an unbranched five-carbon chain, because the product of hydrogenation is unbranched pentane. The formula, C_5H_6, suggests that they have one double bond and one triple bond. Compounds **D,** **E**, and **F** must differ, therefore, in the way the multiple bonds are distributed in the chain. Compounds **E** and **F** have a terminal —C≡CH [IR absorption at ~3300 cm^{-1}]. The UV absorption maximum near 230 nm for **D** and **E** suggests that in these compounds, the multiple bonds are conjugated. Absence of UV absorption beyond 200 nm indicates that the unsaturation sites are isolated in **F**. The structures are

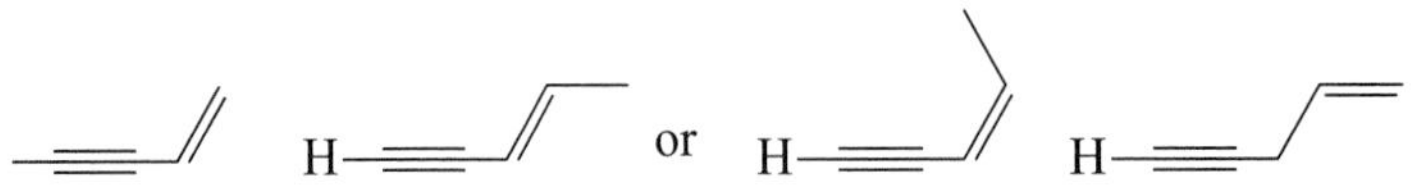

H— or H— H—

D **E** **F**

13.9 (a)

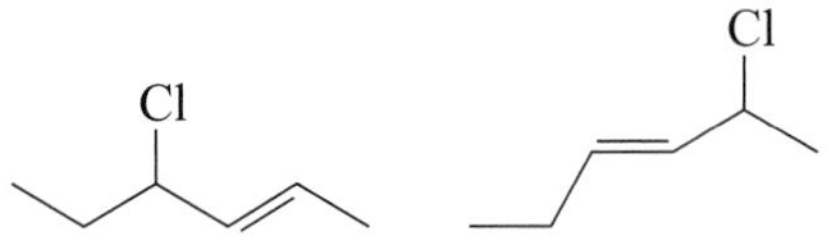

Cl Cl

\+ +

stereoisomers stereoisomers

(b) 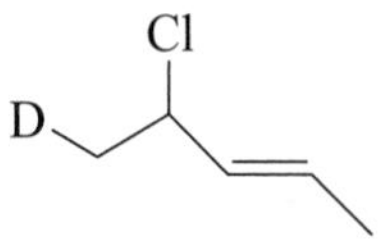 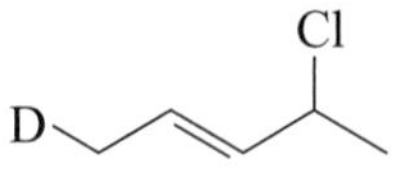 D Cl ... + stereoisomers; D Cl ... + stereoisomers; D Cl ... + stereoisomers; Cl D ... + stereoisomers

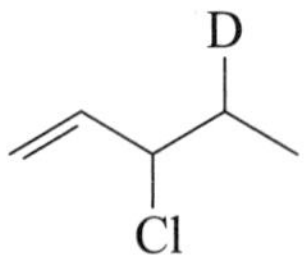

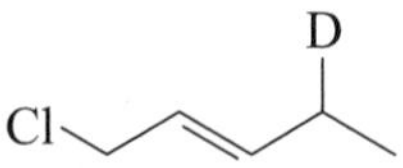

13.10 Addition of the proton gives the resonance hybrid.

(a)  I ⟷ II

The inductive effect of the methyl group in **I** stabilizes the positive charge on the adjacent carbon. Such stabilization of the positive charge does not occur in **II.** Because **I** contributes more heavily to the resonance hybrid than does **II,** C2 bears a greater positive charge and reacts faster with the bromide ion.

(b) In the 1,4-addition product, the double bond is more highly substituted than in the 1,2-addition product; hence it is the more stable alkene.

13.11 (a) 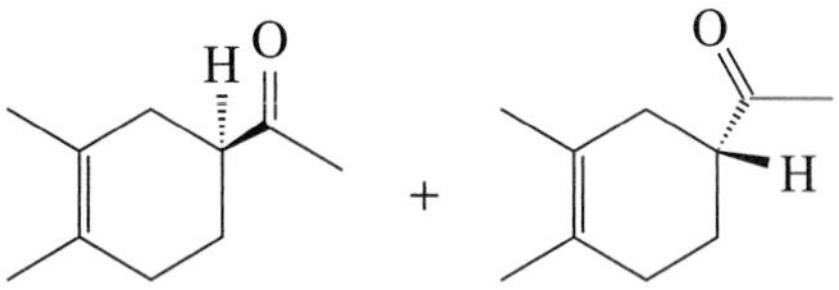 +

(b)

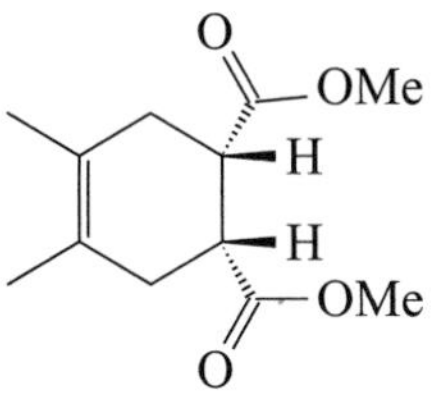

(c) 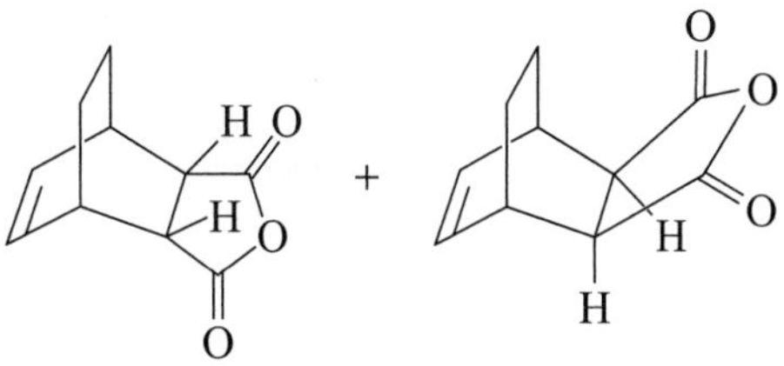 (major product) + (minor product)

(d) 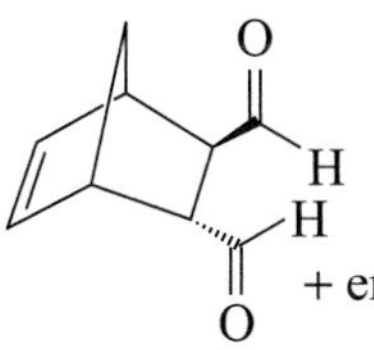 + enantiomer

(e)

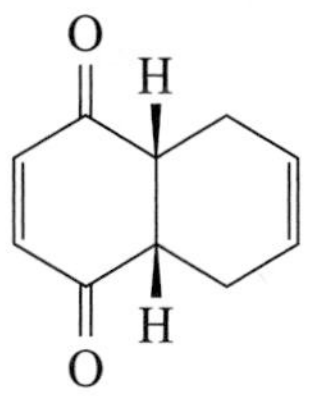

13.12 (a) Use the trans diester because the stereochemistry is retained in the adduct.

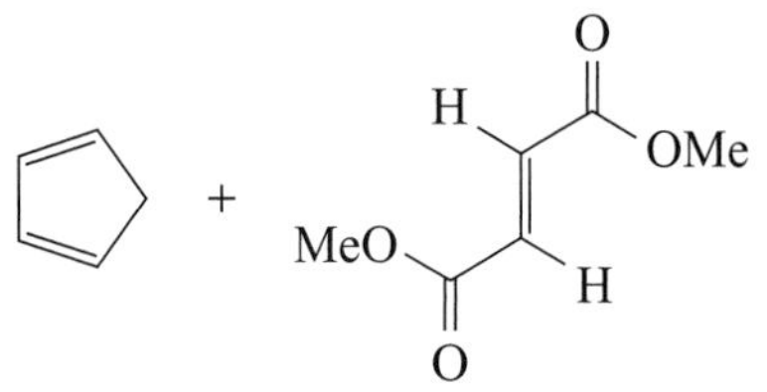

(b) Here, the cis relationship of the acetyl groups requires the use of the cis dienophile.

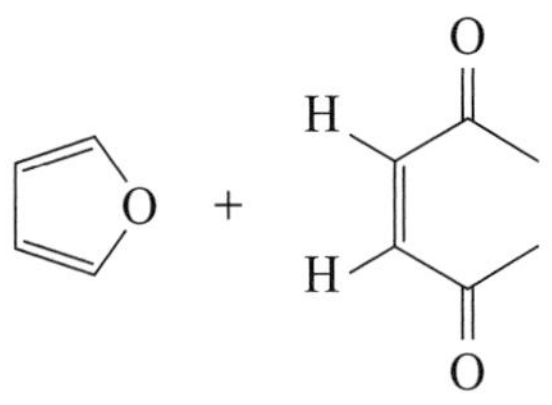

13.13

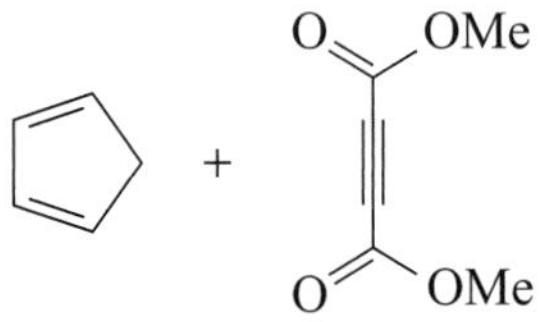

13.14

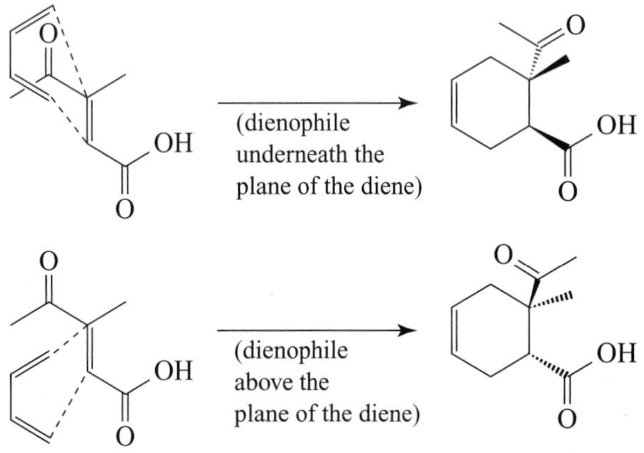

(dienophile underneath the plane of the diene)

(dienophile above the plane of the diene)

(Or, in each case, the other face of the dienophile could present itself to the diene, resulting in the respective enantiomer.)

Problems

Conjugated Systems

13.15 (a) 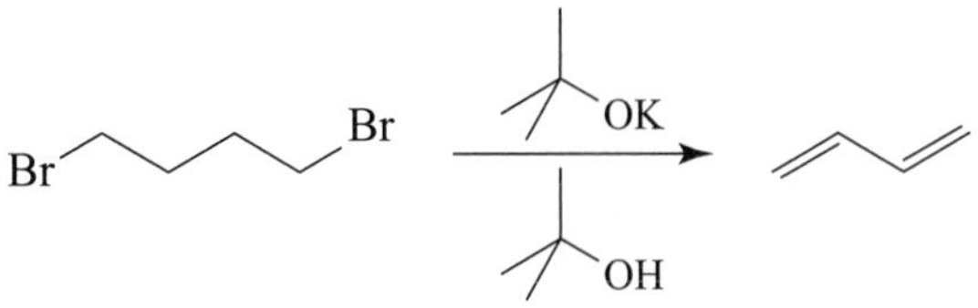 Br Br OK OH

(b) 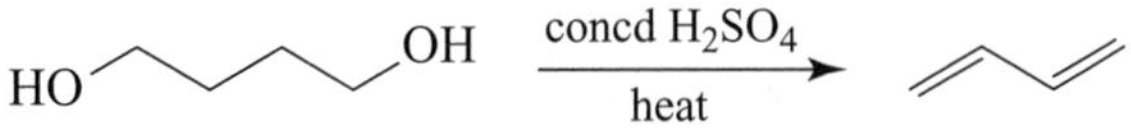 HO OH concd H_2SO_4 heat

(c) 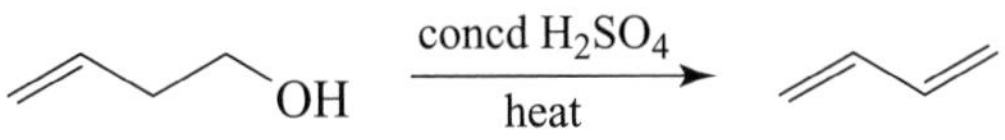 OH concd H_2SO_4 heat

(d) 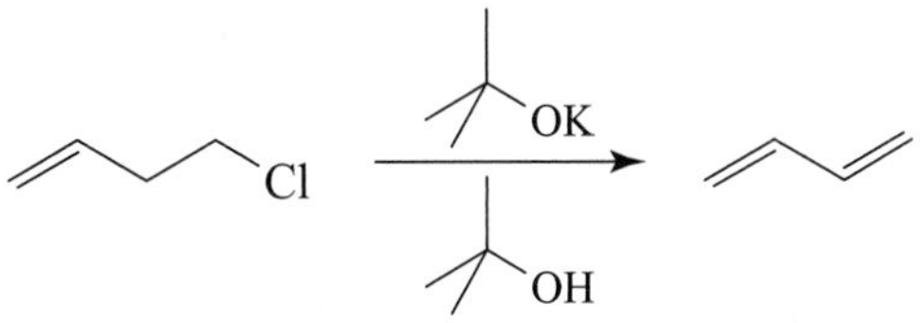 Cl OK OH

(e) 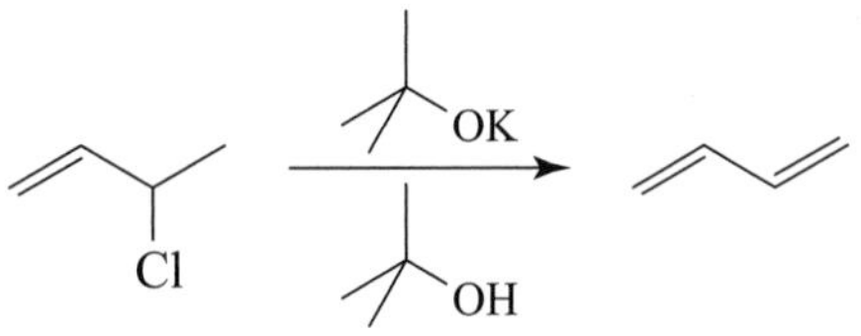 Cl OK OH

(f) 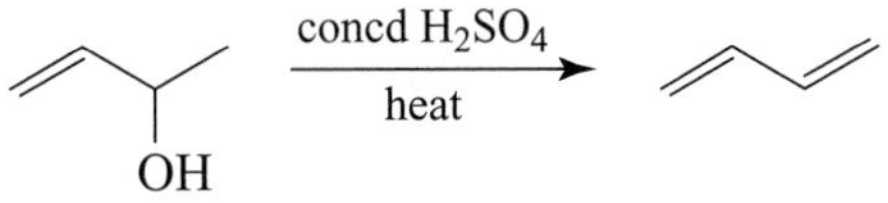 OH concd H_2SO_4 heat

(g) 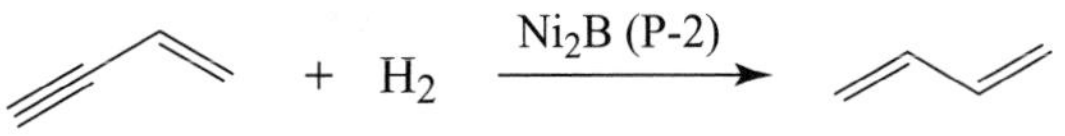 + H_2 Ni_2B (P-2)

13.16

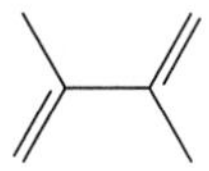

13.17 (a) 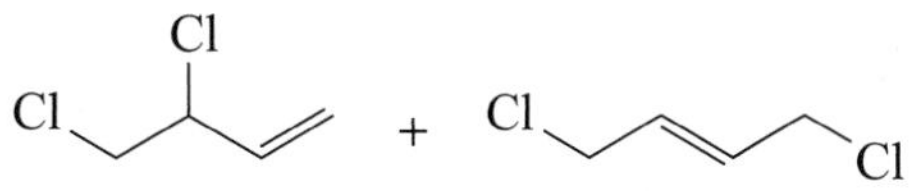 Cl Cl (racemic) + Cl Cl $(E) + (Z)$

(b) 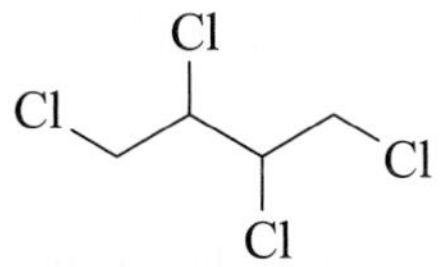 Cl Cl Cl Cl (3 stereoisomers)

(c) 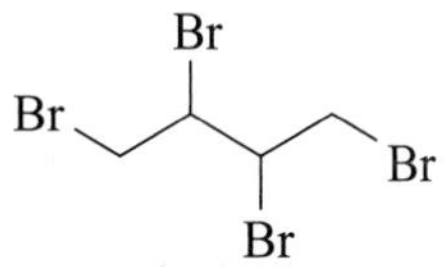 Br Br Br Br (3 stereoisomers)

(d)

(e) 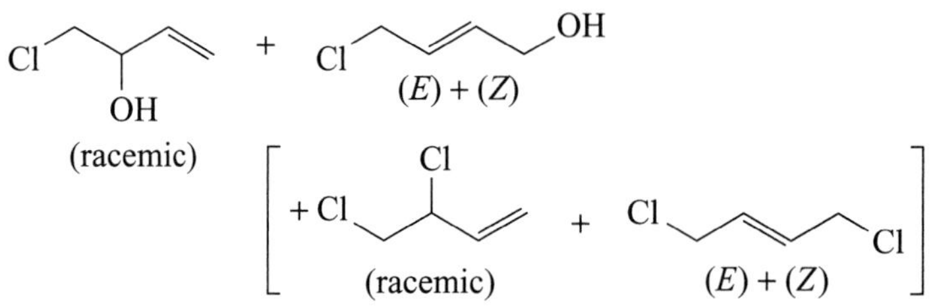 Cl OH (racemic) + Cl OH $(E) + (Z)$ [+ Cl Cl (racemic) + Cl Cl $(E) + (Z)$]

(f) 4 CO_2 (*Note:* $KMnO_4$ oxidizes $HO_2C—CO_2H$ to 2 CO_2.)

(g) 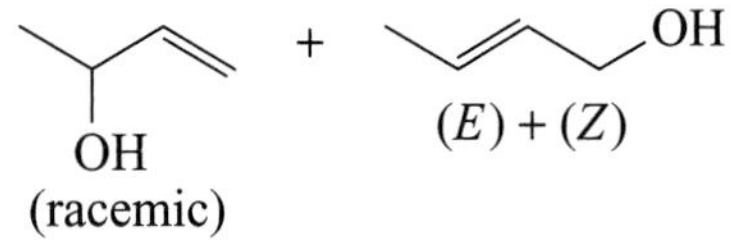 OH (racemic) + OH $(E) + (Z)$

13.18 (a) 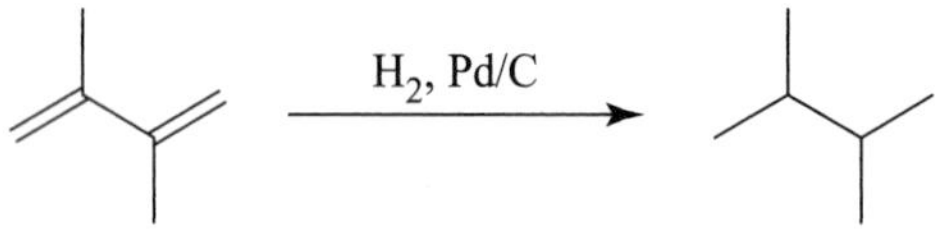 $\xrightarrow{H_2, \text{ Pd/C}}$

(b) 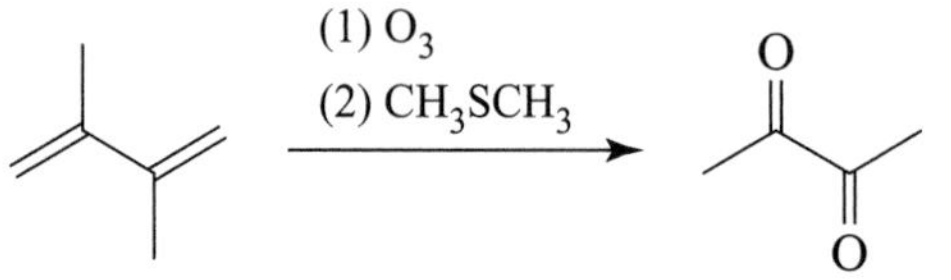 $\xrightarrow{(1)\ O_3;\ (2)\ CH_3SCH_3}$ O O

(c) $\xrightarrow{Br_2,\ h\nu}$ Br

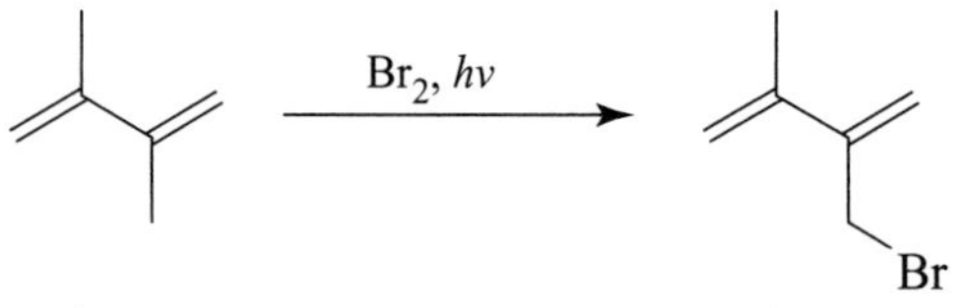

(d) 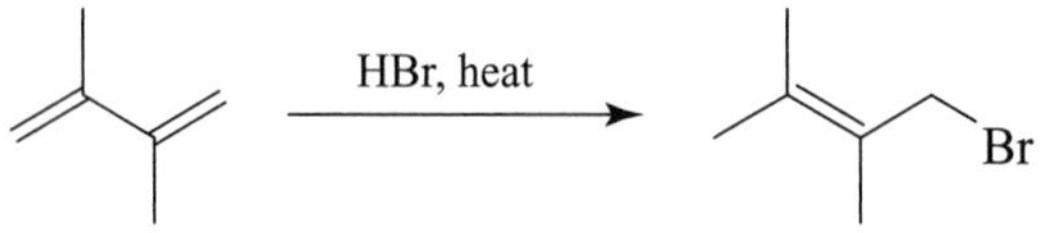 $\xrightarrow{\text{HBr, heat}}$ Br

13.19 (a) 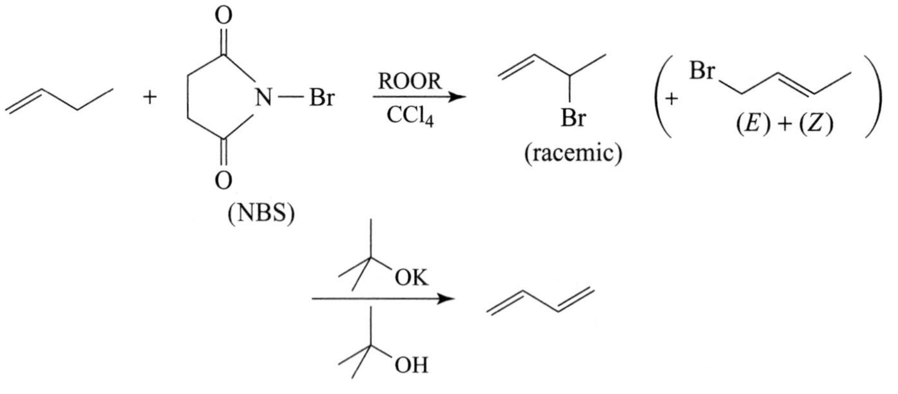 + N—Br (NBS) O O $\xrightarrow[CCl_4]{\text{ROOR}}$ Br (racemic) (+ Br $(E) + (Z)$)

$\xrightarrow[\text{OH}]{\text{OK}}$

Note: In the second step, both allylic halides undergo elimination of HBr to yield 1,3-butadiene; therefore, separating the mixture produced in the first step in unnecessary. The $BrCH_2CH{=}CHCH_3$ undergoes a 1,4 elimination (the opposite of a 1,4 addition).

(b) + NBS $\xrightarrow[CCl_4]{ROOR}$ Br (racemic)

(+ Br (E) + (Z)) $\xrightarrow[\text{t-BuOH}]{\text{t-BuOK}}$ OK OH

Here again both products undergo elimination of HBr to yield 1,3-pentadiene.

(c) OH $\xrightarrow[heat]{concd\ H_2SO_4}$ + [as in (a)] (+)

Br Br (E) + (Z) $\xleftarrow[heat]{Br_2}$

(d) + NBS $\xrightarrow[CCl_4]{ROOR}$ Br (E) + (Z) + Br (racemic)

(e) + Br_2 $\xrightarrow[heat]{light}$ Br $\xrightarrow[\text{t-BuOH}]{\text{t-BuOK}}$ OK OH $\xrightarrow[CCl_4]{NBS, ROOR}$ Br (racemic)

(excess)

(f) Br $\xrightarrow[\text{t-BuOH}]{\text{t-BuOK}}$ OK OH (same as)

13.20 $R{-}\ddot{\underset{\cdot\cdot}{O}}{-}\ddot{\underset{\cdot\cdot}{O}}{-}R \xrightarrow[\text{or light}]{\text{heat}} 2\,R{-}\ddot{\underset{\cdot\cdot}{O}}\cdot$

$R{-}\ddot{\underset{\cdot\cdot}{O}}\cdot + H{-}\ddot{\underset{\cdot\cdot}{Br}}: \longrightarrow R{-}\ddot{\underset{\cdot\cdot}{O}}{-}H + \cdot\ddot{\underset{\cdot\cdot}{Br}}:$

+ $\cdot\ddot{\underset{\cdot\cdot}{Br}}:$ $\longrightarrow$ [Br $\longleftrightarrow$ · Br]

$\xrightarrow{HBr}$ H Br + H Br [+ (Z) isomer] + $\cdot\ddot{\underset{\cdot\cdot}{Br}}:$

13.21 (a)

		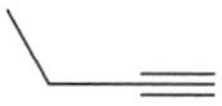
^{13}C NMR	2 peaks	4 peaks
UV-Vis	217 nm (s) [conj. system]	185 nm (w)
IR	~1600 cm^{-1} (s) [conj. system]	~3300 cm^{-1} (s) [≡C—H] 2100-2260 cm^{-1} (w) [C≡C]

(b)

	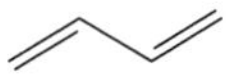	
UV-Vis	217 nm (s) [conj. system]	transparent in UV-Vis
^{1}H NMR	$\delta \sim$ 5.0 [=CH_2] $\delta \sim$ 6.5 [—CH=]	$\delta \sim$1.0 (t) [CH_3] $\delta \sim$1.4 (q) [CH_2]
^{13}C NMR	$\delta \sim$ 117 [=CH_2] $\delta \sim$ 137 [=CH—]	$\delta \sim$13 [CH_3] $\delta \sim$25 [CH_2]

(c)

		 OH
IR	2800–3300 cm^{-1} (s, sharp) [C—H]	3200–3550 cm^{-1} (s, broad) [O—H] ~900 cm^{-1} (s) [=CH—H] ~1000 cm^{-1} (s) [=C—H]
^{13}C NMR	2 peaks	4 peaks
MS	*m/z* 58 ($M^{+\cdot}$)	*m/z* 72 ($M^{+\cdot}$), 54 ($M^{+\cdot}$ − 18)

(d)

	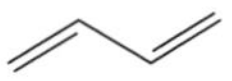	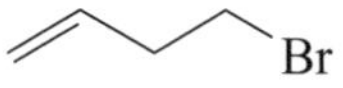 Br
MS	*m/z* 54 ($M^{+\cdot}$)	*m/z* 134 ($M^{+\cdot}$), 136 ($M^{+\cdot}$+2)
^{13}C NMR	2 peaks	4 peaks
UV-Vis	217 nm (s) [conj. system]	<200 nm

(e)

	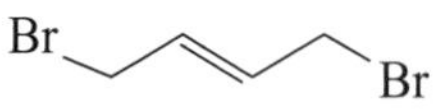 Br, Br	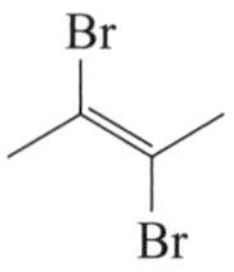 Br, Br
^{1}H NMR	2 signals	1 signal
IR	960–980 cm^{-1} [trans C—H]	no alkene C—H

s = strong; w = weak; q = quartet; t = triplet

13.22

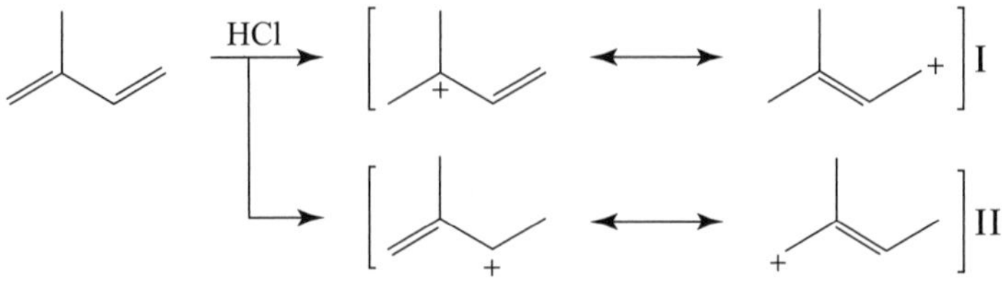

HCl

+ I

+ II

The resonance hybrid, **I,** has the positive charge, in part, on the tertiary carbon atom; in **II,** the positive charge is on primary and secondary carbon atoms only. Therefore, hybrid **I** is more stable and will be the intermediate carbocation. A 1,4 addition to **I** gives

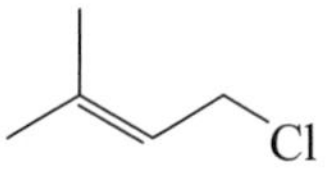

Cl

13.23 The products are Br (racemic) and 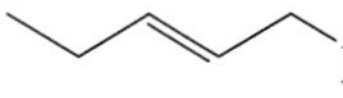 Br [also (*Z*)]. They are formed from an allylic radical in the following way:

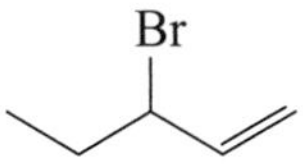

$Br_2 \longrightarrow 2\,Br\cdot$ (from NBS)

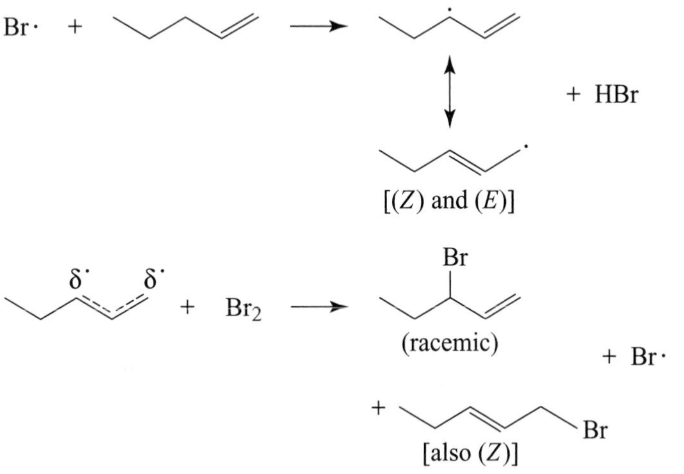

$Br\cdot$ + ⟶ + HBr

[(*Z*) and (*E*)]

δ· δ· + Br_2 ⟶ Br (racemic) + Br·

+ Br [also (*Z*)]

13.24 (a) Because a highly resonance-stabilized radical is formed:

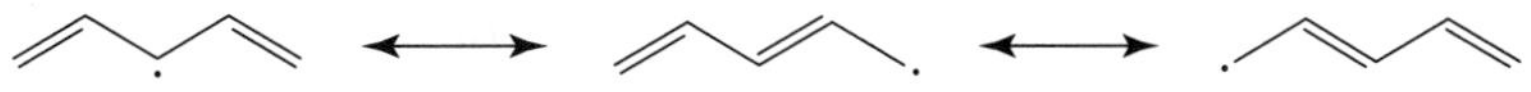

(b) Because the carbanion is more stable:

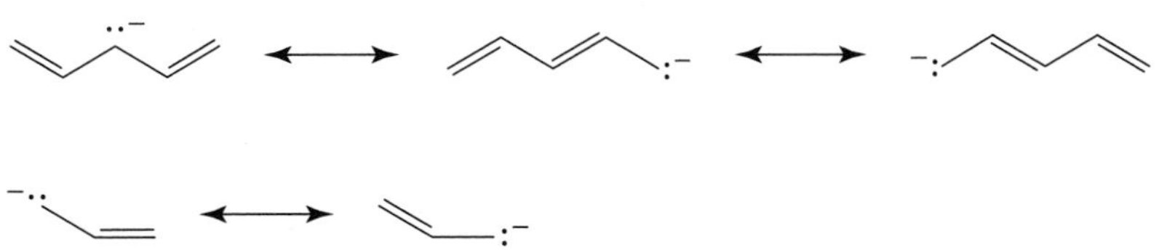

That is, for the carbanion derived from the diene we can write more resonance structures of nearly equal energies.

13.25

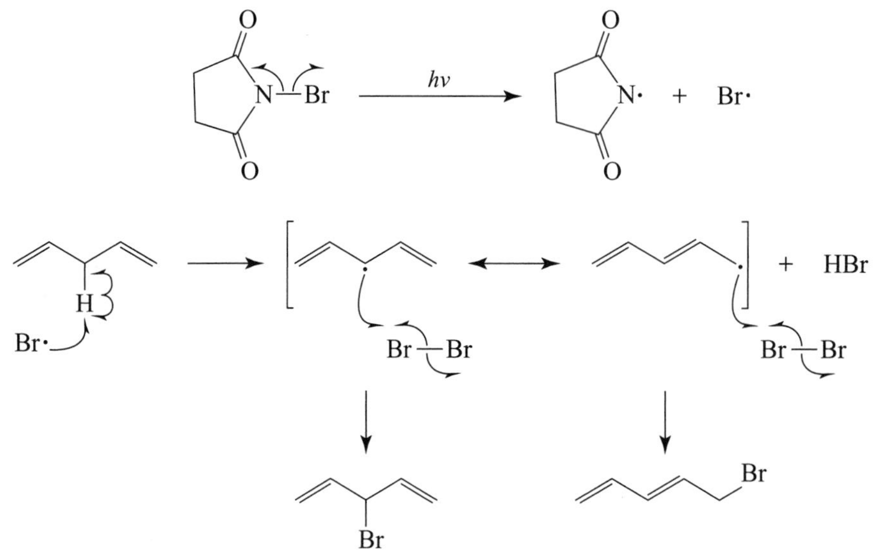

N—Br →($h\nu$) N· + Br·

Br· + H → [· ↔ ·] + HBr

Br—Br, Br—Br

Br, Br

Where did the Br_2 come from?

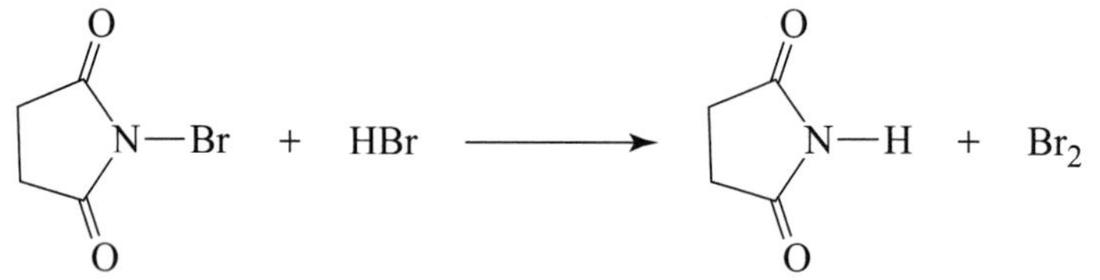

N—Br + HBr → N—H + Br_2

13.26

H—Br

A

Br⁻ B Br⁻

Br, Br

Kinetic product
(1,2 Addition)

C

Thermodynamic product
(1,4 Addition)

D

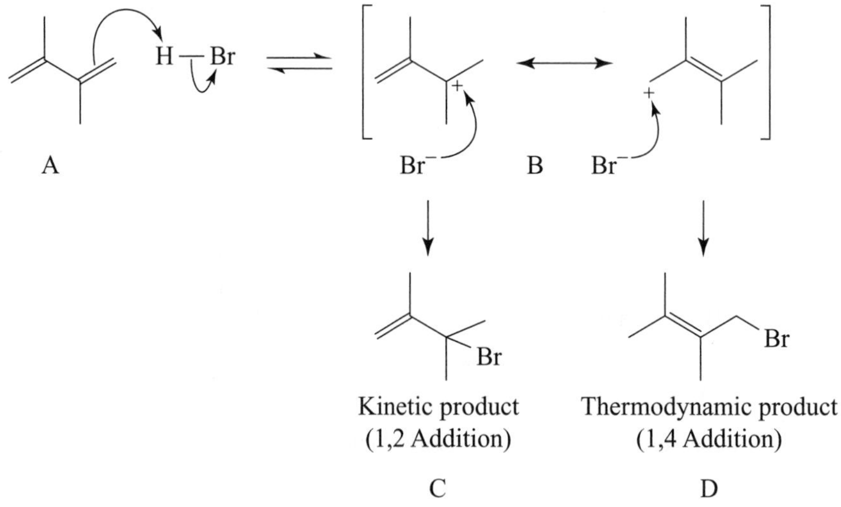

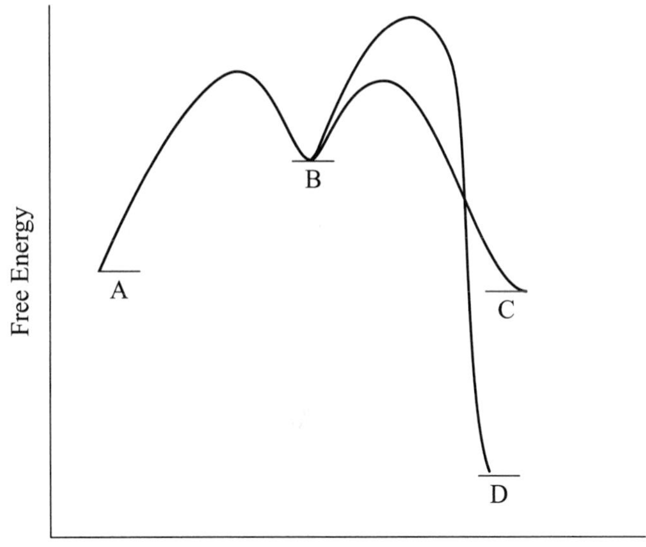

13.27 (a) HBr, −15 °C → Br

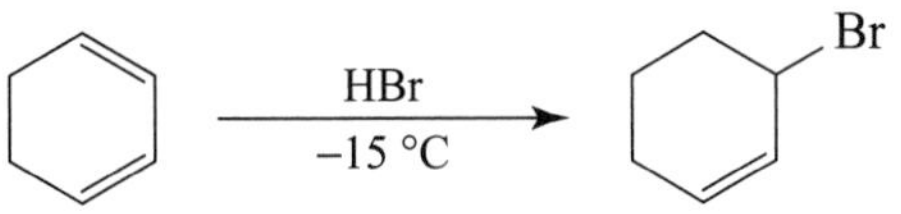

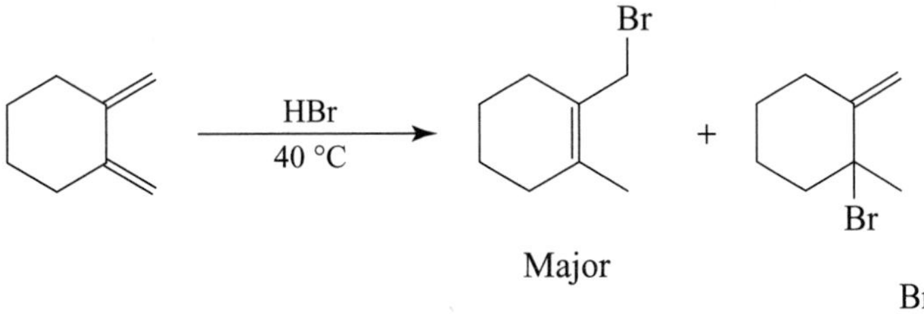

(b) HBr, 40 °C → Br (Major) + Br

(c) NBS, *hv*, heat → Br (Major) + Br

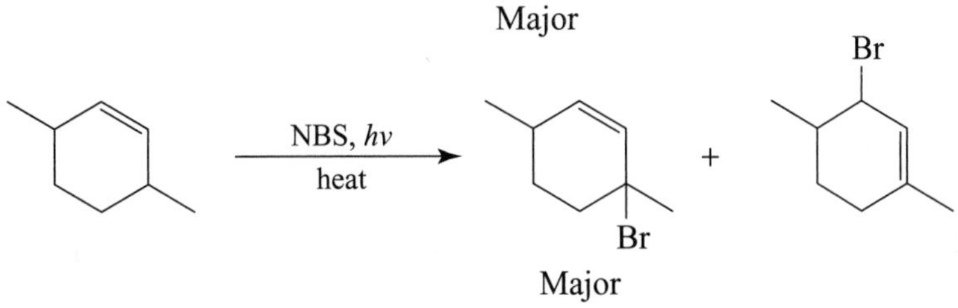

13.28 Protonation of the alcohol and loss of water leads to an allylic cation that can react with a chloride ion at either C1 or C3.

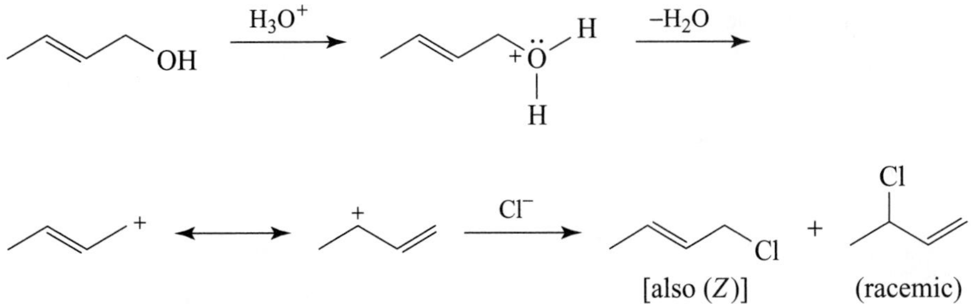

OH $\xrightarrow{H_3O^+}$ $\overset{+}{\ddot{O}}$(H)H $\xrightarrow{-H_2O}$

$^+$ ⟷ $^+$ $\xrightarrow{Cl^-}$ Cl [also (*Z*)] + Cl (racemic)

13.29 (1) + Cl_2 ⟶

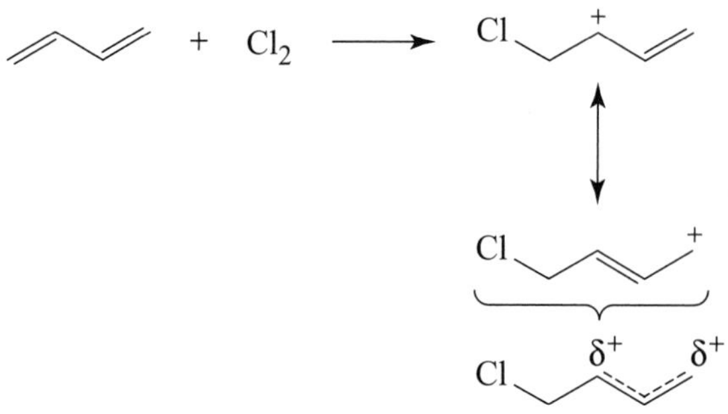

Cl +

Cl +

Cl δ+ δ+

(2) Cl δ+ δ+ $\xrightarrow[HA]{CH_3OH}$ Cl OMe (racemic) + Cl OMe [also (*Z*)]

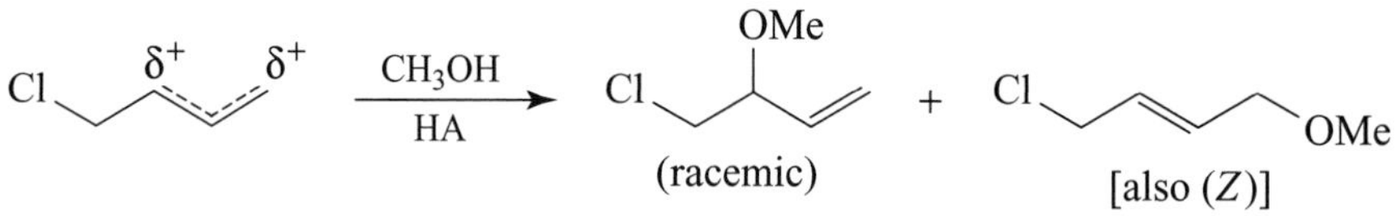

13.30 (a) The same carbocation (a resonance hybrid) is produced in the dissociation step:

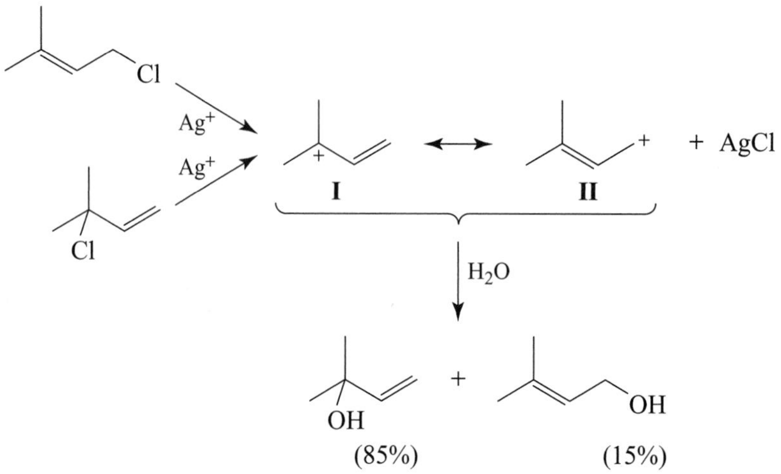

Cl Ag+ Ag+ Cl I II + AgCl

H_2O

OH (85%) + OH (15%)

(b) Structure **I** contributes more than **II** to the resonance hybrid of the carbocation (rule 8). Therefore, the hybrid carbocation has a larger partial positive charge on the tertiary carbon atom than on the primary carbon atom. Reaction of the carbocation with water will therefore occur more frequently at the tertiary carbon atom.

13.31 A six-membered ring cannot accommodate a triple bond because of the strain that would be introduced.

[1,2-Dibromocyclohexane $\xrightarrow{2\ RO^-Na^+}$ cyclohexyne (✗) — Too highly strained; instead → 1,3-cyclohexadiene]

13.32 (a) Propyne. (b) Base (: B^-) removes a proton, leaving the anion whose resonance structures are shown:

$$CH_2{=}C{=}CH_2 + :B^- \rightleftharpoons H{:}B + \underset{\mathbf{I}}{H_2C{=}C{=}\ddot{C}H^-} \longleftrightarrow \underset{\mathbf{II}}{H_2\ddot{C}^-{-}C{\equiv}C{-}H}$$

Reaction with H:B may then occur at the CH_2 carbanion (**II**). The overall reaction is

$$CH_2{=}C{=}CH_2 + :B^- \rightleftharpoons [CH_2{=}C{=}\ddot{C}H^- \longleftrightarrow \ddot{C}H_2^-{-}C{\equiv}CH] + H{:}B$$

$$\rightleftharpoons CH_3{-}C{\equiv}CH + :B^-$$

13.33 The product formed when 1-bromobutane undergoes elimination is 1-butene, a simple monosubstituted alkene. When 4-bromo-1-butene undergoes elimination, the product is 1,3-butadiene, a conjugated diene, and therefore, a more stable product. The transition states leading to the products reflect the relative stabilities of the products. Since the transition state leading to 1,3-butadiene has the lower free energy of activation of the two, the elimination reaction of 4-bromo-1-butene will occur more rapidly.

Diels-Alder Reactions

13.34

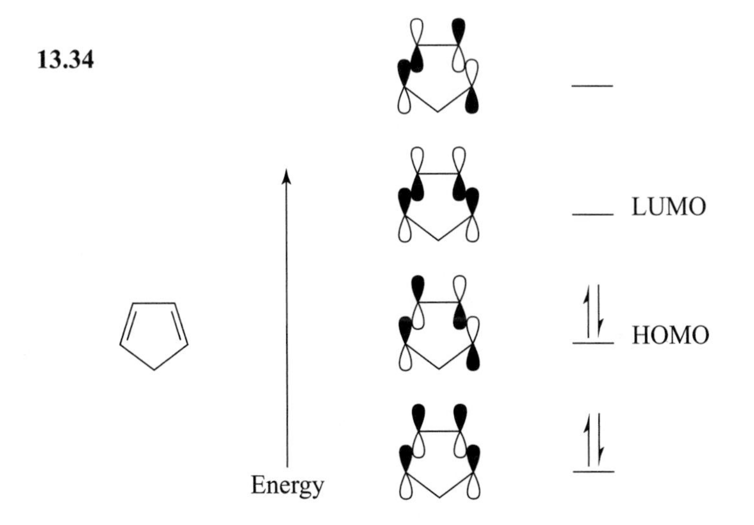

13.35 The diene portion of the molecule is locked into an s-trans conformation. It cannot, therefore, achieve the s-cis conformation necessary for a Diels-Alder reaction.

13.36 s-cis conformation is disfavored (steric strain)

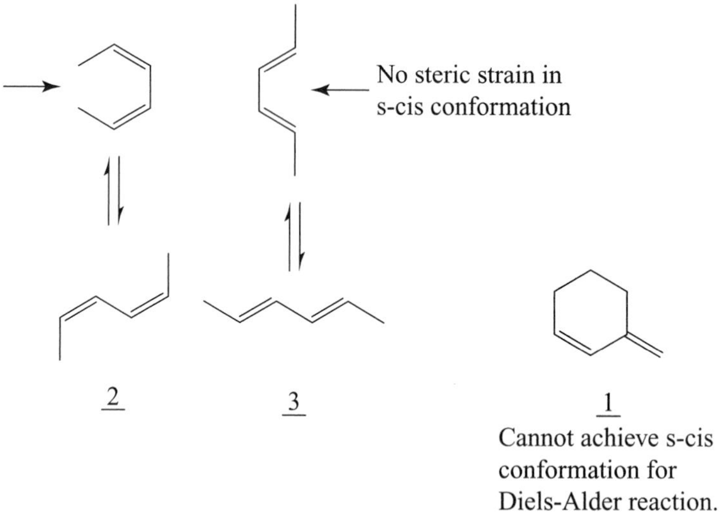

No steric strain in s-cis conformation

4
"Locked" s-cis conformation

2

3

1
Cannot achieve s-cis conformation for Diels-Alder reaction.

13.37 (a) OMe + enantiomer (b) OMe + enantiomer

(c) CN + enantiomer (d) CN + enantiomer

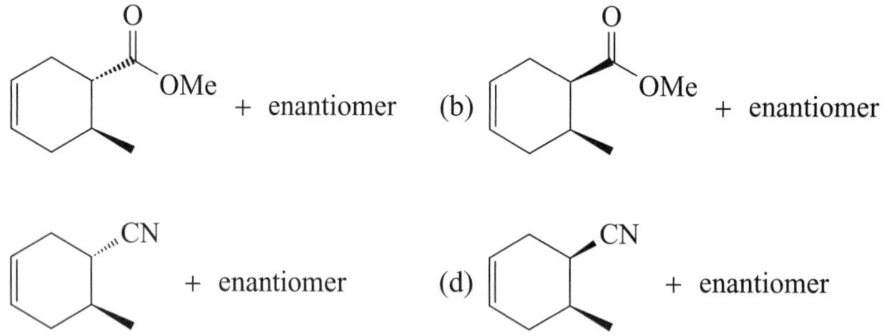

13.38

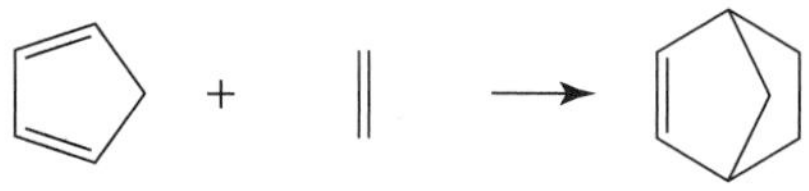

13.39 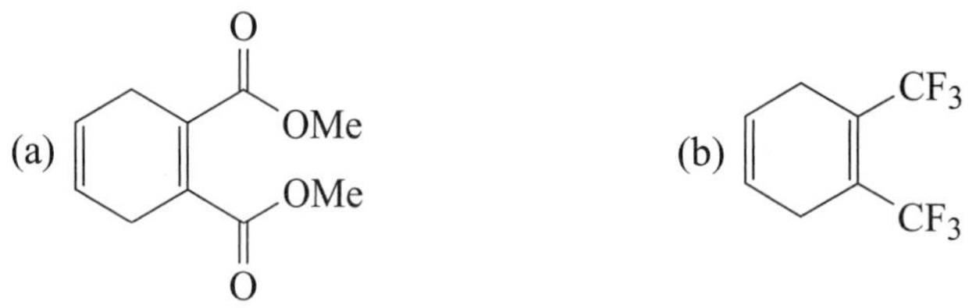 (a) OMe OMe (b) CF_3 CF_3

13.40 (a) + O Me $\xrightarrow{\Delta}$ MeO O Major + MeO O

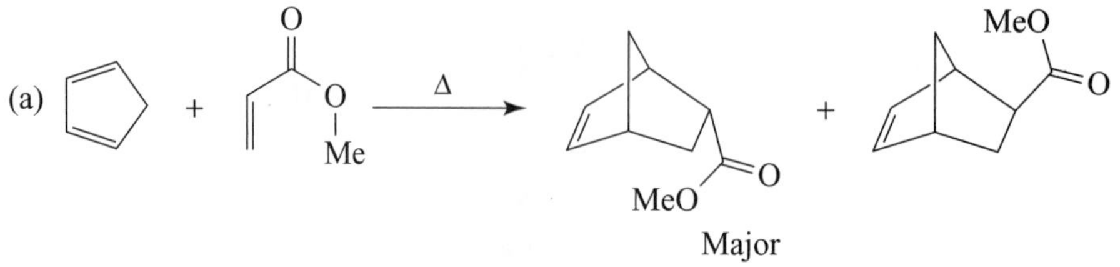

(b) + CN $\xrightarrow{\Delta}$ CN

(c) + O $\xrightarrow{\Delta}$ O

(d) O + O O Me O O Me $\xrightarrow{\Delta}$ O CO_2Me CO_2Me Major + O CO_2Me CO_2Me

(e) O O + 2 $\xrightarrow{\Delta}$ O O

(f) + O H H O (1) Δ (2) $NaBH_4$ (3) H_2O → OH OH Major + OH OH

(g) 2 O O—Me Me—O O Δ → CO_2Me CO_2Me MeO_2C CO_2Me

(h) + CN CN $\xrightarrow{\Delta}$ CN CN Major + CN CN

13.41 (a) 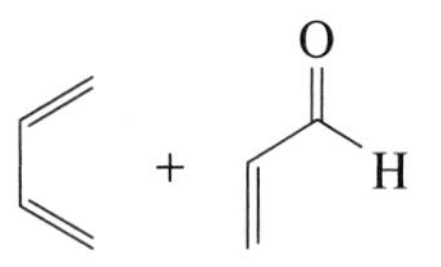 (b) 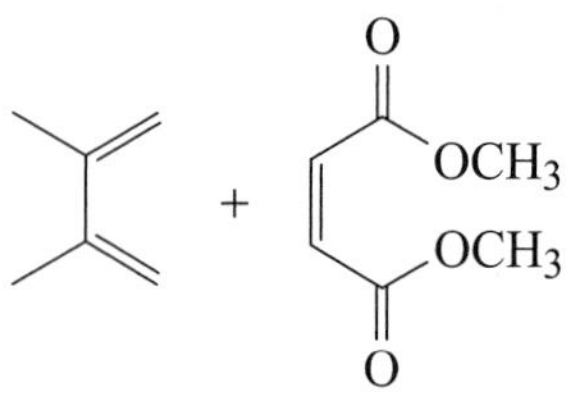 (c) 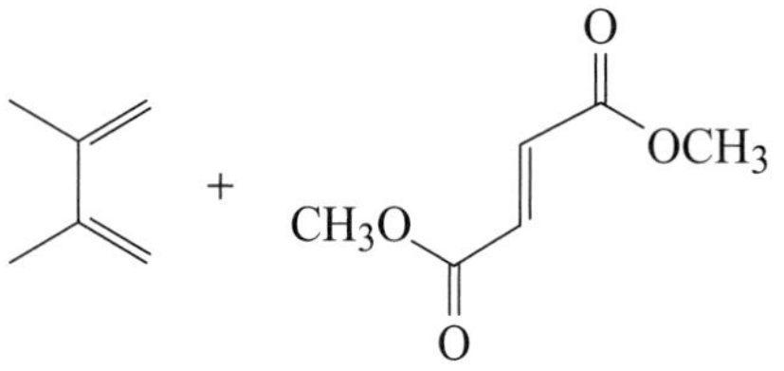 (d) 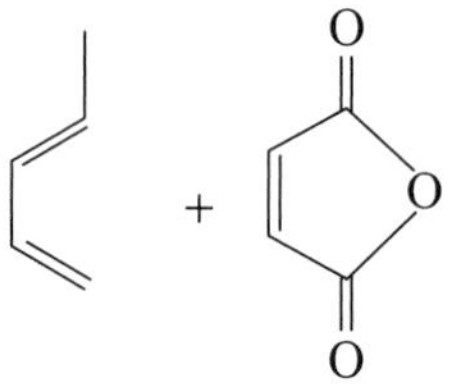 (e) 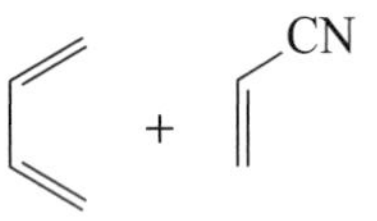 (f) 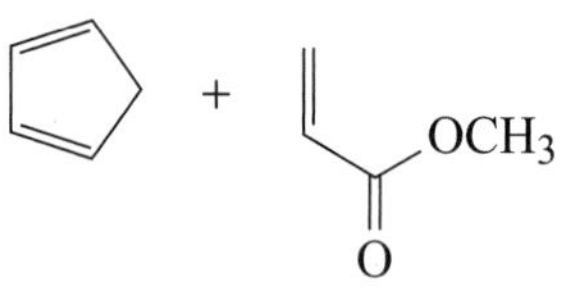 (g)

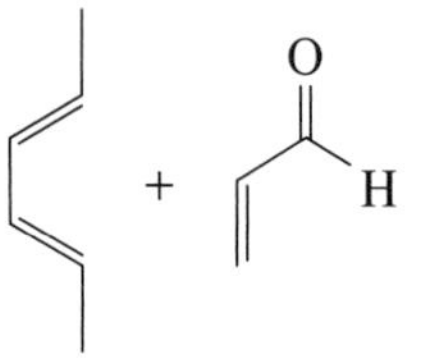

13.42 The endo adduct is less stable than the exo, but is produced at a faster rate at 25°C. At 90°C the Diels-Alder reaction becomes reversible; an equilibrium is established, and the more stable exo adduct predominates.

13.43

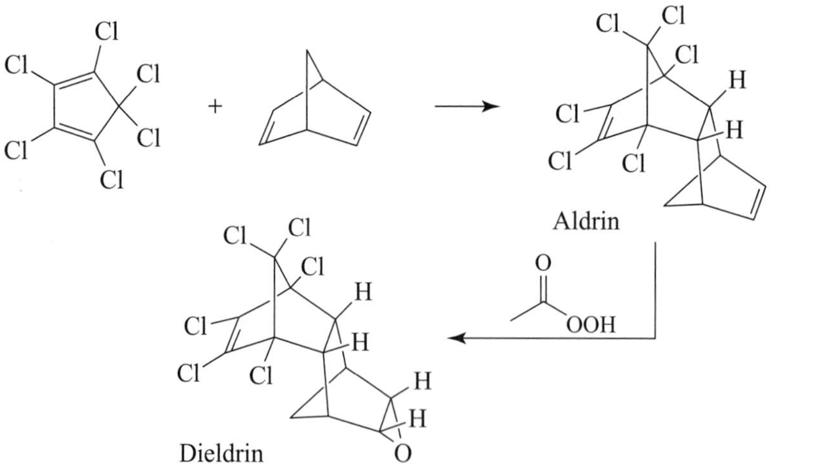

Aldrin

Dieldrin

13.44 (a)  + H–C≡C–H ⟶ Norbornadiene

(b) 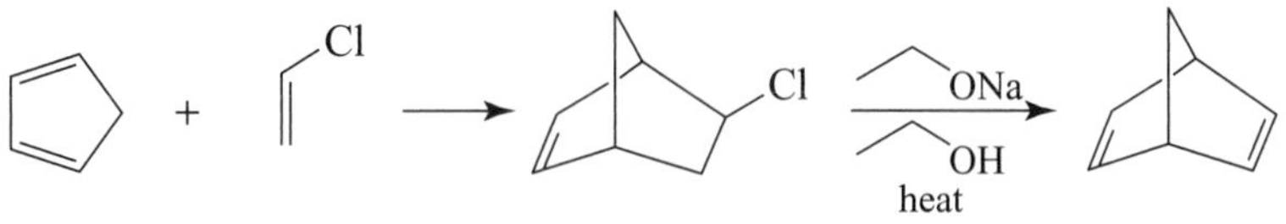 + ⟶ Cl; ONa / OH, heat ⟶

13.45 Cl + ⟶ Cl_2 (dark) ⟶ Chlordan

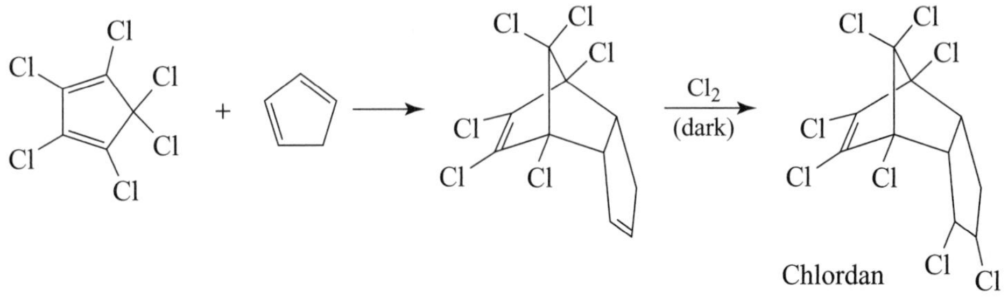

Note: The other double bond is less reactive because of the presence of the two chlorine substituents.

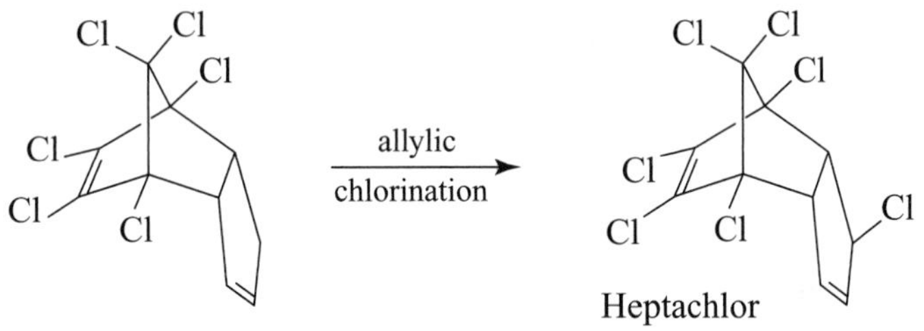

allylic chlorination ⟶ Heptachlor

13.46 ⟶ Isodrin

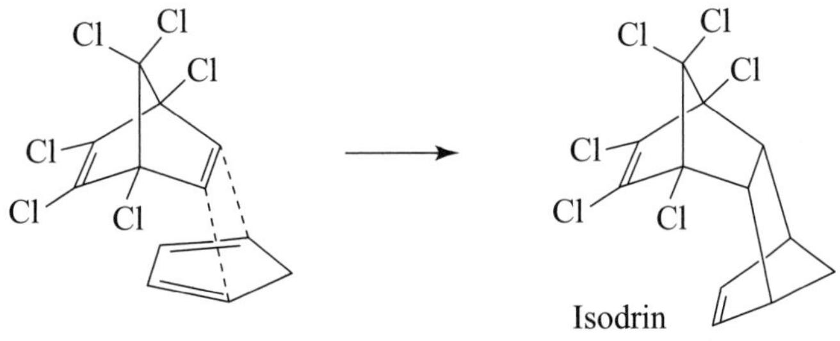

13.47 (a) 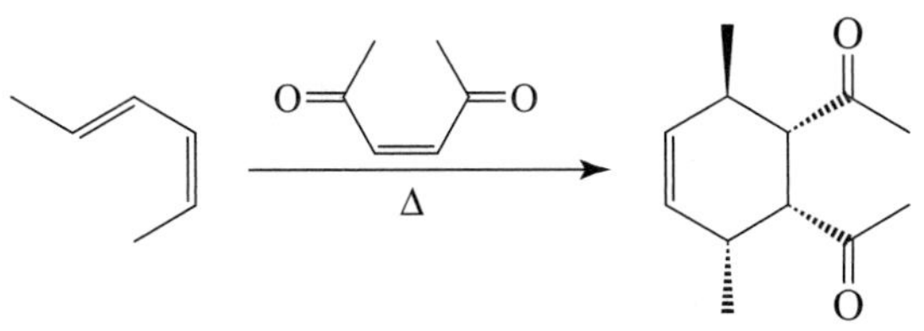 ⟶ Δ

(b) 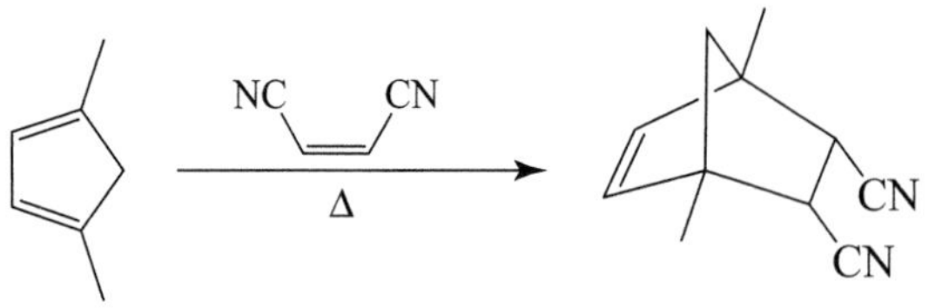 NC CN Δ CN CN

(c) 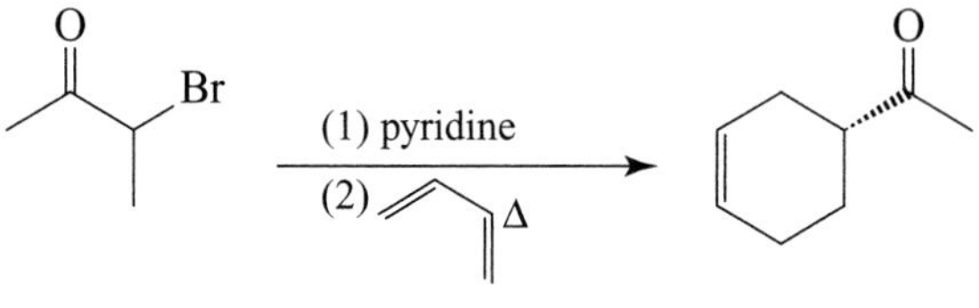 O Br (1) pyridine (2) Δ O

(d) 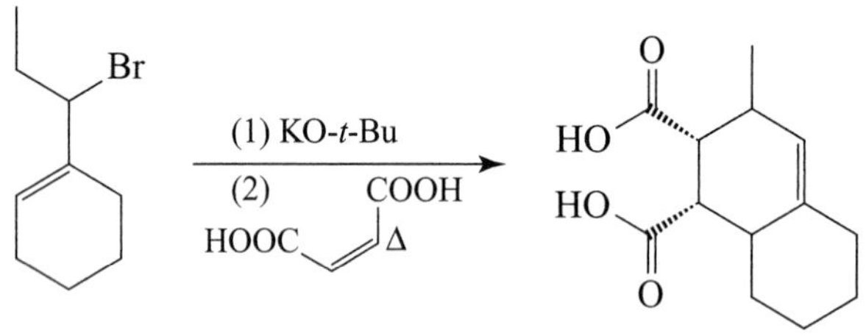 Br (1) KO-*t*-Bu (2) COOH HOOC Δ O HO HO O

(e) 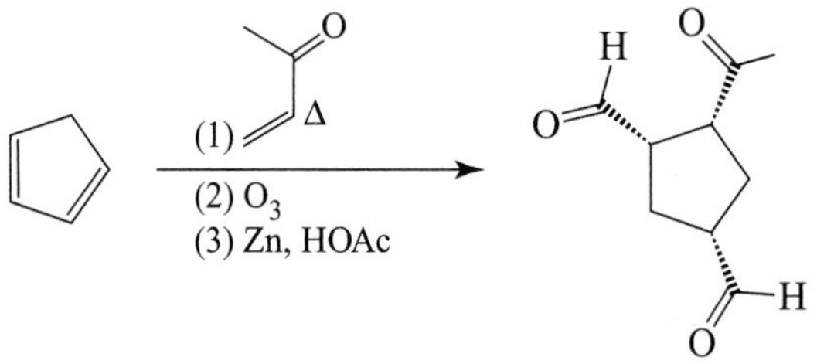 O (1) Δ (2) O_3 (3) Zn, HOAc H O O O H O

Challenge Problems

13.48 Diene OCH_3 $^{+}OCH_3$ Dienophile O H O^{-} H

$^{+}OCH_3$ O^{-} H + → CH_3O O H + OCH_3 H O

Good polarity interactions lead to major product.

Major Minor

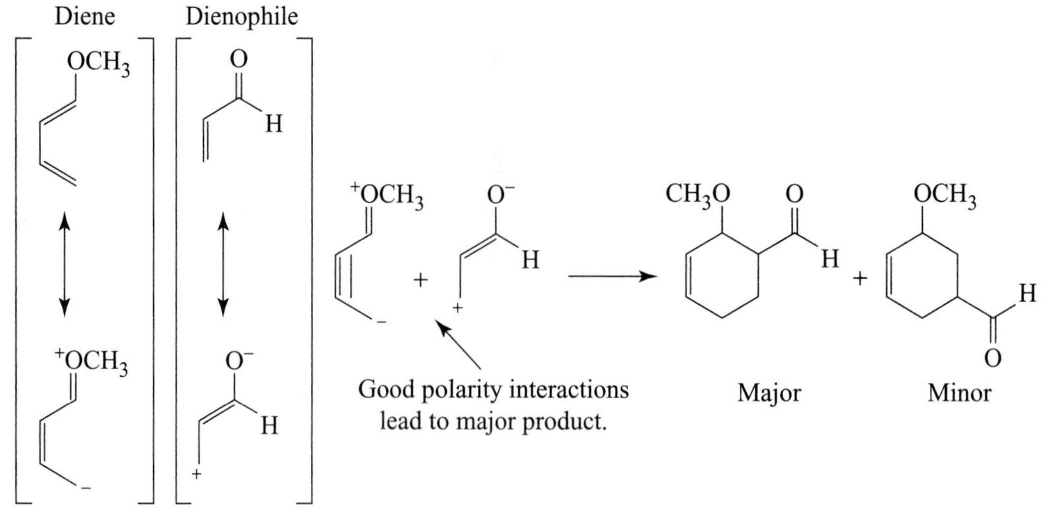

13.49 The first crystalline solid is the Diels-Alder adduct below, mp 125°C,

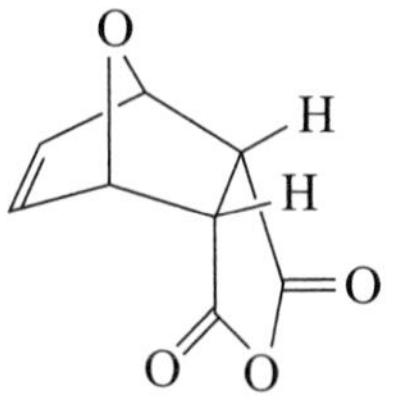

On melting, this adduct undergoes a reverse Diels-Alder reaction, yielding furan (which vaporizes) and maleic anhydride, mp 56°C,

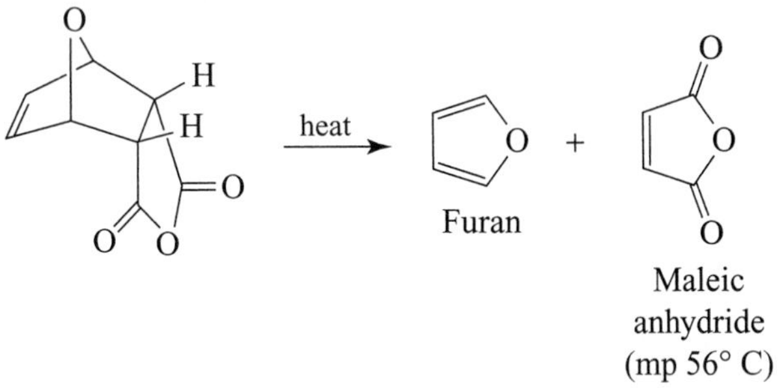

heat

Furan

Maleic anhydride (mp 56° C)

13.50 MeO_2C … $OSi(\textit{tert}\text{-Bu})Ph_2$

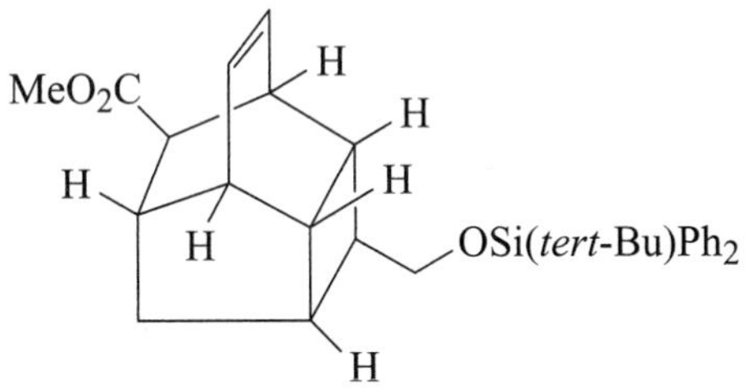

13.51

δ+ δ+ δ+

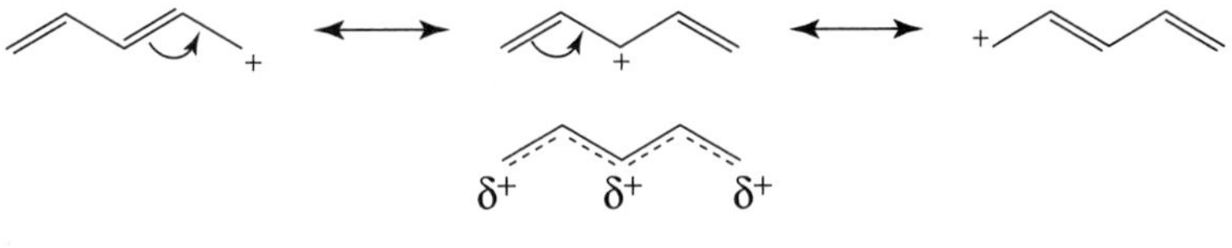

The map of electrostatic potential for the pentadienyl carbocation shows greater electron density or less positive charge (less blue color) in the vicinity of the C1—C2 and C4—C5 bonds, suggesting that the most important contributing resonance structure is the one with the most positive charge near C3, a secondary carbon. The other contributing resonance structures have a positive charge at primary carbon atoms.

QUIZ

13.1 Give the 1,4-addition product of the following reaction:

+ HCl ⟶ ?

(a) 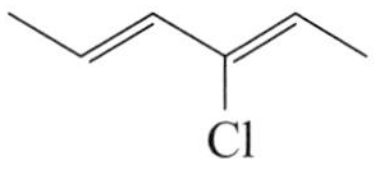 Cl

(b) 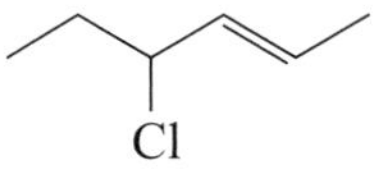 Cl

(c) 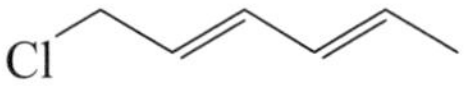 Cl

(d) 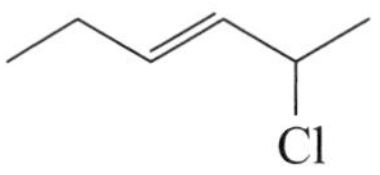 Cl

(e) 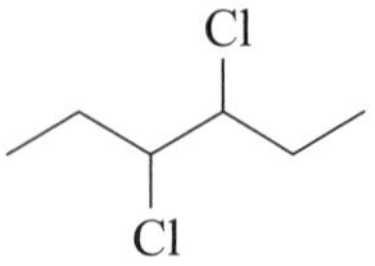 Cl Cl

13.2 Which diene and dienophile could be used to synthesize the following compound?

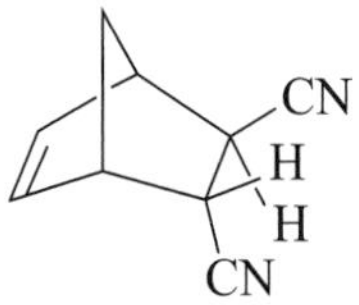

CN H H CN

(a) 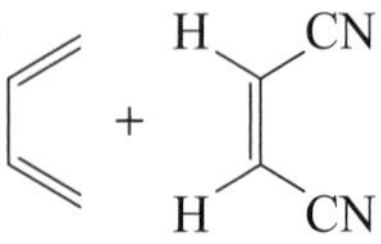 + H CN H CN

(b) 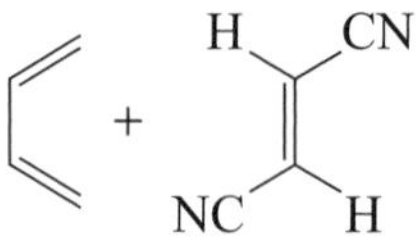 + H CN NC H

(c) 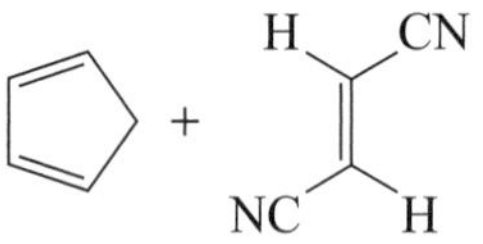 + H CN NC H

(d) 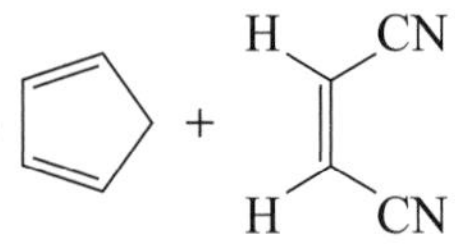 + H CN H CN

(e) 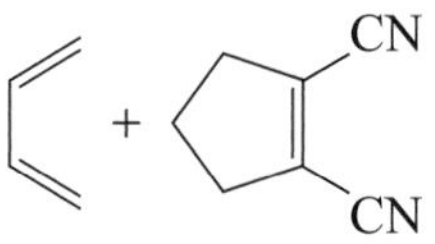 + CN CN

13.3 Which reagent(s) could be used to carry out the following reaction?

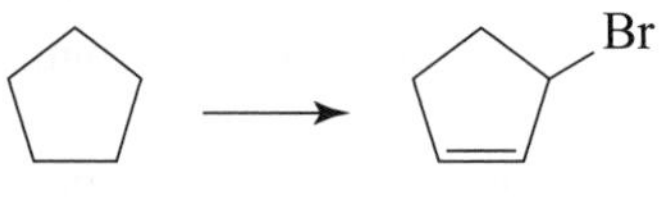

⟶ Br

(a) NBS/CCl_4 (NBS = O NBr O)

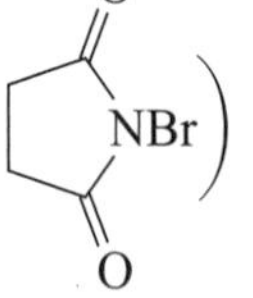

(b) NBS/CCl_4, then $Br_2/h\nu$

(c) $Br_2/h\nu$, then t-BuOK / t-BuOH, then NBS/CCl_4

(d) t-BuOK / t-BuOH, then NBS/CCl_4

13.4 Which of the following structures does not contribute to the hybrid for the carbocation formed when 4-chloro-2-pentene ionizes in an S_N1 reaction?

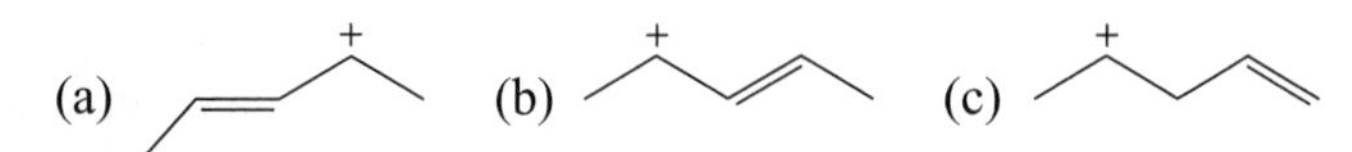

(d) All of these contribute to the resonance hybrid.

13.5 Which of the following resonance structures accounts, at least in part, for the lack of S_N2 reactivity of vinyl chloride?

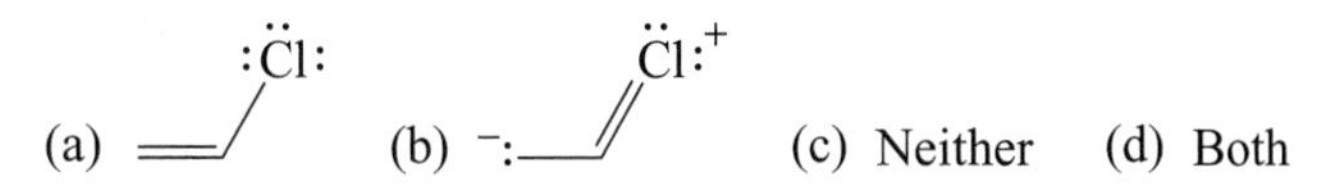

SUMMARY OF REACTIONS BY TYPE CHAPTERS 1–13

I. SUBSTITUTION REACTIONS

Type	Stereochemical Result	Favoring Conditions
S_N2 Chapter 6 See Section 6.14 for a summary of nucleophiles mentioned in early chapters	inversion	1°, 2°, benzylic (1° or 2°), or allylic (1° or 2°) leaving group (e.g., halide, tosylate, mesylate); strong nucleophile; polar aprotic solvent
S_N1 Chapter 6	racemization (via carbocation)	3°, benzylic, or allylic leaving group
Radical substitution Chapter 10 Section 13.2	racemization (via radical)	3°, benzylic, or allylic hydrogen; peroxide, heat/light

II. ELIMINATION REACTIONS

Type	Stereochemical/Regiochemical Result	Favoring Conditions
E2 (dehydrohalogenation) Section 7.6	elimination to form the most substituted alkene (Zaitsev elimination) with small bases formation of the less substituted alkene with use of a bulky base (e.g., *tert*-BuOK/ *tert*-BuOH)	strong base (e.g., NaOEt/EtOH, KOH/EtOH, *tert*-BuOK/*tert*-BuOH); 2° or 3° leaving group (e.g., halide, tosylate, mesylate, etc.); heat
E1 (dehydrohalogenation) Sections 6.17 and 7.8A	formation of most substituted alkene; may occur with carbocation rearrangement	3° leaving group; weak base; heat
Dehydration Section 7.7	formation of most substituted alkene; may occur with carbocation rearrangement	catalytic acid (H^+, e.g., concd H_2SO_4, H_3PO_4); heat

III. MECHANISTIC SUMMARY OF ALKENE AND ALKYNE ADDITION REACTIONS

Reactant	Stereochemical Result		Regiochemical Result	
	syn	**anti**	**Markovnikov**	**anti-Markovnikov**
Alkenes	H_2/Pt, Pd or Ni (Sections 4.17A, 7.13, 7.14)	*X_2/CCl_4 (Section 8.12)	*H_2O, H^+ (hydration) (Section 8.5)	(i) BH_3:THF (ii) OH^-, H_2O_2 (Section 8.7)
	(i) OsO_4, (ii) $NaHSO_3$ (Section 8.16)	* X_2,H_2O (Section 8.14)	* (i) $Hg(OAc)_2$, H_2O, THF (ii) $NaBH_4$, OH^- (Section 8.6)	
	RCO_3H (e.g., MCPBA) (Section 11.13)	(i) RCO_3H, (ii) H_3O^+ or OH^- (Section 11.14)	* HX (no peroxides) (Section 8.2)	HBr (w/peroxides) (Section 10.9)
	* (i) BH_3:THF (ii) H_2O_2, OH^- (Sections 8.7 and 8.8)	* X_2/ROH (Section 8.14)	* additon of other alkenes	
		* X_2, Nucleophile (e.g., RO^-, RMgX) (Section 8.14)		
	:C< Addition of carbenes (Section 8.15)			

* Shares mechanistic themes with other reactions denoted by the same symbol.

Alkynes	H_2, Pd on $BaCO_3$ w/quinoline (Lindlar's cat.) (Section 7.15A)	(i) Li, $EtNH_2$, −78°C (ii) NH_4^+ Cl^- (Section 7.15B)	HX (one or two molar equivalents) (Section 8.19)
	H_2, Ni_2B (P-2) (Section 7.15A)	X_2 (2 equiv.)/CCl_4 (Section 8.18)	
	H_2 (2 equiv.), Pt or Ni (complete hydrogenation) (Sections 4.16, 7.15)		

IV. ALKENE AND ALKYNE CLEAVAGE WITH OXIDATION

Conditions	Reactant		Product(s)
Alkenes			
(i) $KMnO_4/OH^-$, heat; (ii) H_3O^+	tetrasubstituted alkene	⟶	two ketones
	trisubstituted alkene	⟶	one ketone, one carboxylic acid
(Section 8.17A)	monosubstituted alkene	⟶	one carboxylic acid and CO_2
(i) O_3, (ii) Me_2S	tetrasubstituted alkene	⟶	two ketones
	trisubstituted alkene	⟶	one ketone, one aldehyde
(Section 8.17B)	monosubstituted alkene	⟶	one aldehyde and formaldehyde
Alkynes			
(i) $KMnO_4/OH^-$, heat; (ii) H_3O^+	terminal alkyne	⟶	a carboxylic acid and formic acid
	internal alkyne	⟶	two carboxylic acids
(Section 8.20)			
(i) O_3, (ii) HOAc	terminal alkyne	⟶	a carboxylic acid and formic acid
	internal alkyne	⟶	two carboxylic acids
(Section 8.20)			

V. CARBON—CARBON BOND-FORMING REACTIONS

(a) Alkylation of alkynide anions (with 1° alkyl halides, epoxides, and aldehydes or ketones) (Sections 7.12,8.21 and 12.8D)

(b) Grignard reaction (with aldehydes and ketones, or epoxides) (Sections 12.7B, C, and 12.8)

(c) Carbocation addition to alkenes (e.g., polymerization) (Special Topic A)

(d) Diels-Alder reaction (Section 13.11)

(e) Addition of a carbene to an alkene (Section 8.15)

VI. REDUCTIONS/OXIDATIONS (NOT INCLUDING ALKENES/ALKYNES)

(a) $2\ R{-}X\ (w/Zn/H_3O^+) \longrightarrow 2\ R{-}H + ZnX_2$ (Section 5.15)

(b) $R{-}X \xrightarrow[\text{ether}]{\text{Mg}} R{-}MgX \xrightarrow{H_3O^+} R{-}H$ (Section 12.7A)

(c) Lithium aluminum hydride ($LiAlH_4$) and sodium borohydride ($NaBH_4$) reduction of carbonyl compounds (Section 12.3)

(d) Oxidation of alcohols with chromic acid, pyridinium chlorochromate (PCC), or hot potassium permanganate (Section 12.4)

VII. MISCELLANEOUS

(a) R—CO—R′ + PCl_5 ⟶ R—CCl_2—R′ ⟶ alkynes (Section 7.10)

(b) R—COOH + $SOCl_2$ or PCl_5 ⟶ R—COCl⟶ acyl chlorides for Friedel-Crafts reactions, esters, amides, etc. (Sections 15.7 and 17.5)

(c) Terminal alkynes + $NaNH_2$ in liq. NH_3 ⟶ alkynide anions (Sections 8.21, 12.8D)

(d) ROH + TsCl, MsCl, or TfCl (with pyridine) ⟶ R—OTs, R—OMs, or R—OTf (Section 11.10)

(e) R—OH + Na (or NaH) ⟶ R—O^-Na^+ + H_2 (Section 6.15B, 11.11B)

VIII. CHEMICAL TESTS

(a) Alkenes/Alkynes: Br_2/CCl_4 (Section 8.12)

(b) Rings/Unsaturation/etc.: Index of Hydrogen Deficiency (Section 4.17)

(c) Position of Unsaturation: $KMnO_4$ (Section 8.17A); O_3 (Section 8.17B)

METHODS FOR FUNCTIONAL GROUP PREPARATION

CHAPTERS 1–13

I. ALKYL HALIDES

(a) HX addition to alkenes [Markovnikov (Section 8.2) and HBr anti-Markovnikov (Section 10.9)]

(b) X_2 addition to alkenes (Section 8.12)

(c) Radical halogenation

  (i) X_2/light/heat for alkanes (Sections 10.3, 10.4, 10.5, 10.6)

  (ii) *N*-Bromosuccinimide (NBS)/heat/light for allylic and benzylic substitution (Section 13.2B)

(d) $R—OH + SOCl_2$ (with pyridine) $\longrightarrow$ $R—Cl + SO_2$ + pyridinium hydrochloride (Section 11.9)

(e) $3\ R—OH + PBr_3 \longrightarrow 3\ R—Br + P(OH)_3$ (Section 11.9)

(f) $R—OH + HX \longrightarrow R—X + H_2O$ (Sections 11.7, 11.8)

II. ALCOHOLS

(a) Markovnikov addition of H_2O with catalytic H^+ (Section 8.5)

(b) Markovnikov addition of H_2O via (i) $Hg(OAc)_2$, H_2O, THF; (ii) $NaBH_4$, OH^- (Section 8.6)

(c) Anti-Markovnikov addition of H_2O via (i) BH_3:THF; (ii) OH^-, H_2O_2 (Section 8.7)

(d) OsO_4 addition to alkenes (syn) (Section 8.16)

(e) (i) RCO_3H (peroxycarboxylic acids); (ii) H_3O^+ (Section 11.15)

(f) Cleavage of ethers with HX (Section 11.12A)

(g) Opening of epoxides by a nucleophile (Sections 11.14, 11.15)

(h) Lithium aluminum hydride or sodium borohydride reduction of carbonyl compounds (Section 12.3)

(i) Grignard reaction with aldehydes, ketones, and epoxides (Sections 12.7B, C, 12.8)

(j) Cleavage of silyl ethers (Section 11.11E)

III. ALKENES

(a) E2 Dehydrohalogenation (preferred over E1 for alkene synthesis) (Section 7.6)

(b) Dehydration of alcohols (Section 7.7)

(c) Hydrogenation of alkynes: (Z) by catalytic hydrogenation, (E) by dissolving metal reduction (Sections 7.15A, B)

(d) Diels-Alder reaction (forms one new double bond with ring formation) (Section 13.11)

IV. ALKYNES

(a) Alkylation of alkynide anions (Sections 7.12 and 8.21)

(b) Double dehydrohalogenation of vicinal or geminal dihalides (Section 7.10)

V. CARBON—CARBON BONDS

(a) Alkylation of alkynide anions (Section 7.12 and 8.21)

(b) Organometallic addition to carbonyl compounds and epoxides (e.g., Grignard or RLi reactions) (Sections 12.7B, C, 12.8)

(c) Diels-Alder reaction (Section 13.11)

(d) Addition of alkenes to other alkenes (e.g., polymerization) (Section 10.10 and Special Topic A)

(e) Addition of a carbene to an alkene (Section 8.15)

VI. ALDEHYDES

(a) (i) O_3; (ii) Me_2S with appropriate alkenes (Section 8.17B)

(b) Pyridinium chlorochromate (PCC) oxidation of 1° alcohols (Section 12.4A)

VII. KETONES

(a) (i) O_3; (ii) Me_2S with appropriate alkenes (Section 8.17B)

(b) $KMnO_4/OH^-$ cleavage of appropriate alkenes (Section 8.17A)

(c) H_2CrO_4 oxidation of 2° alcohols (Section 12.4C)

VIII. CARBOXYLIC ACIDS

(a) (i) $KMnO_4/OH^-$; (ii) H_3O^+ with 1° alcohols (Section 12.4B)

(b) (i) O_3; (ii) HOAc with alkynes (Section 8.20)

(c) (i) $KMnO_4/OH^-$, heat; (ii) H_3O^+ with alkynes and alkenes (Sections 8.20 and 8.17A)

(d) H_2CrO_4 oxidation of 1° alcohols (Section 12.4D, E)

IX. ETHERS (INCLUDING EPOXIDES)

(a) $RO^- + R'X \longrightarrow ROR' + X^-$ (Williamson synthesis) (Section 11.11B)

(b) 2 ROH (+cat. H_2SO_4, heat) $\longrightarrow$ ROR + HOH (for symmetrical ethers only) (Section 11.11A)

(c) Alkene + RCO_3H (a peroxycarboxylic acid, e.g., MCPBA) $\longrightarrow$ an epoxide (Section 11.13)

(d) An epoxide + RO^- $\longrightarrow$ an α-hydroxy ether (also acid catalyzed) (Section 11.14)

(e) ROH + $ClSiR'_3$ $\longrightarrow$ $ROSiR'_3$ (silyl ether protecting groups, e.g., TBDMS ethers) (Section 11.11E)

14 AROMATIC COMPOUNDS

SOLUTIONS TO PROBLEMS

14.1 (a) 4-Bromobenzoic acid (or *p*-bromobenzoic acid)

(b) 2-Benzyl-1.3-cyclohexadiene

(c) 2-Chloro-2-phenylpentane

(d) Phenyl propyl ether

14.2 Compounds (a) and (b) would yield only one monosubstitution product.

14.3 Resonance structures are defined as being structures that differ *only* in the positions of the electrons. In the two 1,3,5-cyclohexatrienes shown, the carbon atoms are in different positions; therefore, they cannot be resonance structures.

14.4 The cyclopentadienyl cation would be a diradical. We would not expect it to be aromatic.

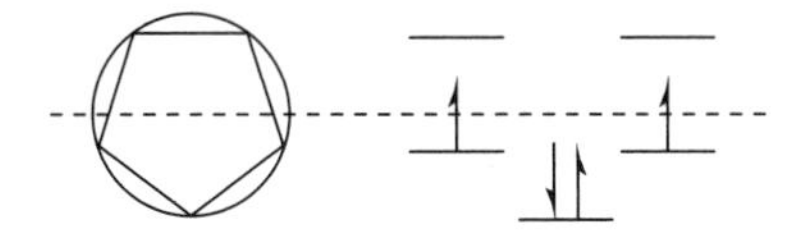

14.5 (a) No, the cycloheptatrienyl anion (below) would be a diradical.

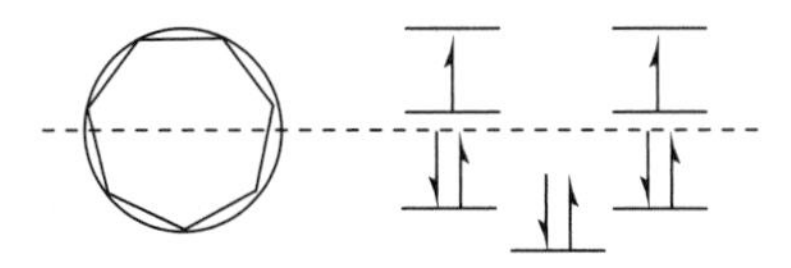

(b) The cycloheptatrienyl cation (below) would be aromatic because it would have a closed bonding shell of delocalized π electrons.

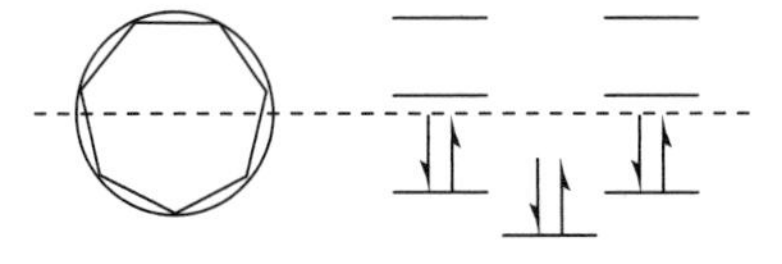

14.6 If the 1,3,5-cycloheptatrienyl anion *were* aromatic, we would expect it to be unusually stable. This would mean that 1,3,5-cycloheptatriene should be unusually acidic. The fact that 1,3,5-cycloheptatriene is not unusually acidic (it is less acidic than 1,3,5-heptatriene) confirms the prediction made in the previous problem, that the 1,3,5-cycloheptatrienyl anion should not be aromatic.

14.7 (a) (b) $\xrightarrow[-HBr]{heat}$ $+ \; Br^-$

Tropylium bromide

These results suggest that the bonding in tropylium bromide is ionic; that is, it consists of a positive tropylium ion and a negative bromide ion.

14.8 The fact that the cyclopentadienyl cation is antiaromatic means that the following hypothetical transformation would occur with an increase in π-electron energy.

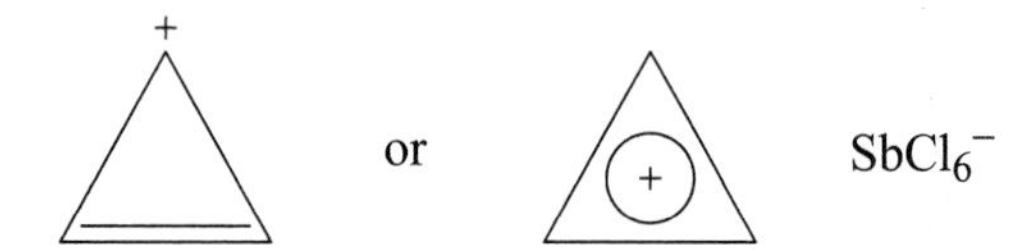

14.9 (a) The cyclopropenyl cation (below).

or $SbCl_6^-$

(b) Only one ^{13}C NMR signal is predicted for this ion.

14.10 (a) 3 (b) 4 (c) 7 (d) 5

14.11 The upfield signal arises from the six methyl protons of *trans*-15,16-dimethyldihydropyrene, which by virtue of their location are strongly shielded by the magnetic field created by the aromatic ring current (see Figure 14.8).

14.12 Major contributors to the hybrid must be ones that involve separated charges. Contributors like the following would have separated charges, and would have aromatic five- and seven-membered rings.

⟷ ⟷ ⟷ etc.

14.13 (a) (b)

14.14 Because of their symmetries, *p*-dibromobenzene would give two ^{13}C signals, *o*-dibromobenzene would give three, and *m*-dibromobenzene would give four.

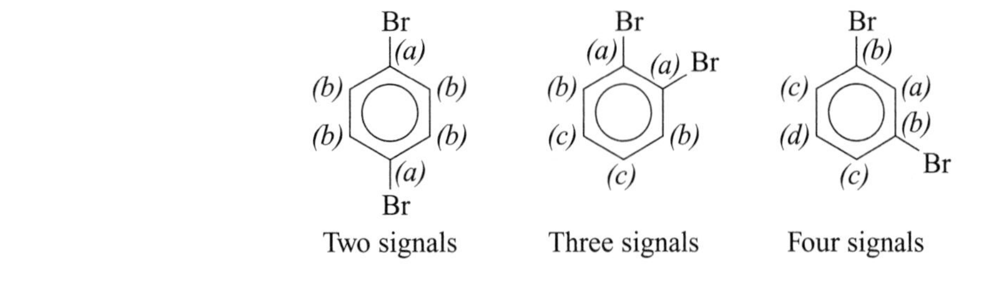

14.15

Br
Br
Br
Br

A **B** **C** **D**

A. Strong absorption at 740 cm^{-1} is characteristic of ortho substitution.

B. A very strong absorption peak at 800 cm^{-1} is characteristic of para substitution.

C. Strong absorption peaks at 680 and 760 cm^{-1} are characteristic of a meta substitution.

D. Very strong absorption peaks at 693 and 765 cm^{-1} are characteristic of a monosubstituted benzene ring.

Problems

Nomenclature

14.16 (a) O OH, NO_2 (b) Br (c) Br, Br

(d) NO_2, NO_2 (e) OH, O_2N, NO_2 (f) O OH, NO_2

(g) 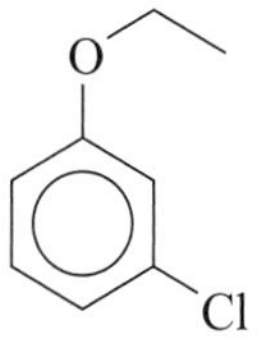 O; Cl

(h) 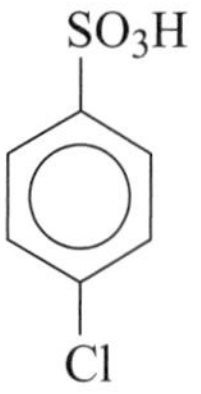 SO_3H; Cl

(i) 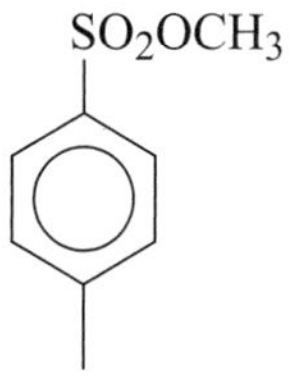 SO_2OCH_3

(j) 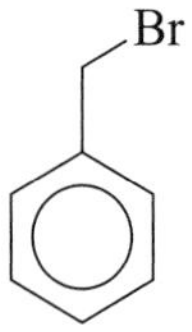 Br

(k) 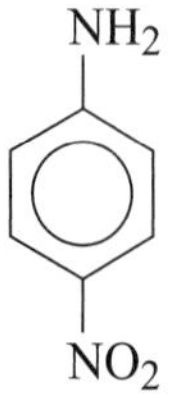 NH_2; NO_2

(l)

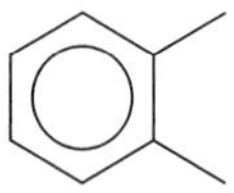

(m)

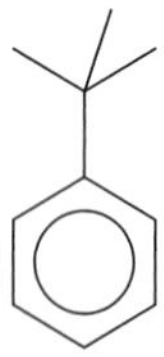

(n)  OH

(o) 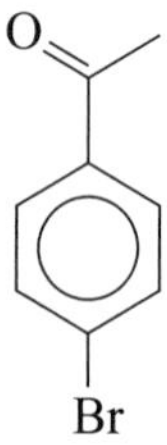 O; Br

(p) 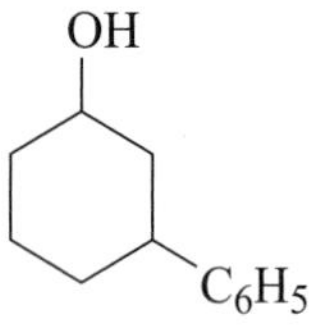 OH; C_6H_5

(q) 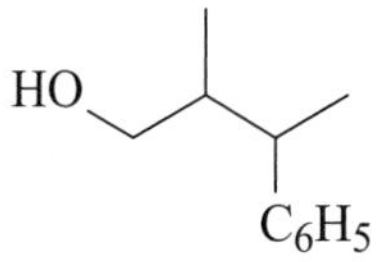 HO; C_6H_5

(r) 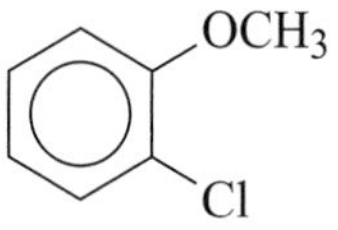 OCH_3; Cl

14.17 (a) Br; Br; Br — 1,2,3-Tribromo-benzene

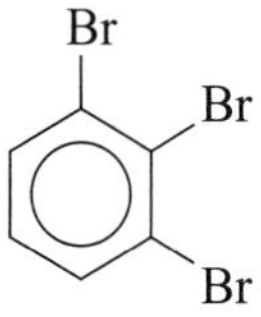

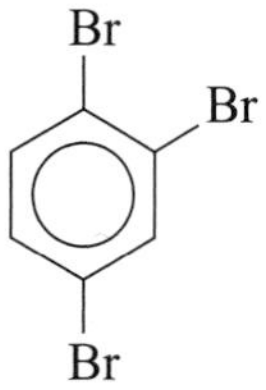

Br; Br; Br — 1,2,4-Tribromo-benzene

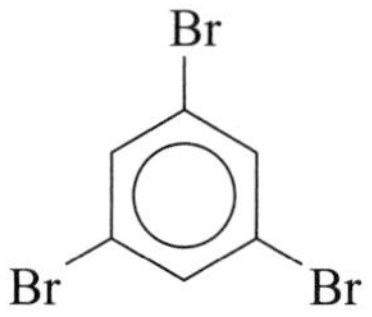

Br; Br; Br — 1,3,5-Tribromo-benzene

(b)

2,3-Dichloro-
phenol

2,4-Dichloro-
phenol

2,5-Dichloro-
phenol

2,6-Dichloro-
phenol

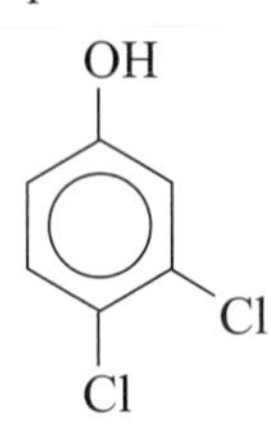

3,4-Dichloro-
phenol

3,5-Dichloro-
phenol

(c)

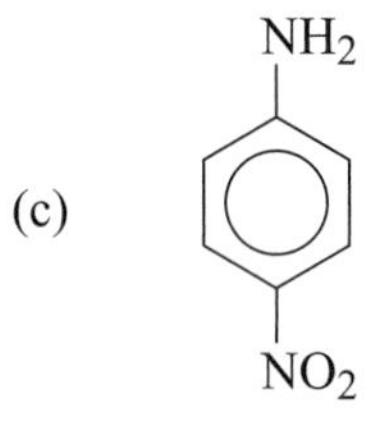

4-Nitroaniline
(*p*-nitroaniline)

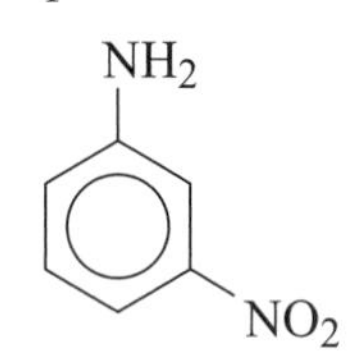

3-Nitroaniline
(*m*-nitroaniline)

2-Nitroaniline
(*o*-nitroaniline)

(d)

4-Methylbenzene-
sulfonic acid
(*p*-toluenesulfonic
acid)

3-Methylbenzene-
sulfonic acid
(*m*-toluenesulfonic
acid)

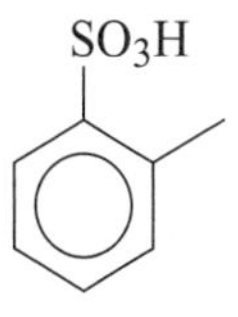

2-Methylbenzene-
sulfonic acid
(*o*-toluenesulfonic
acid)

(e)

Butylbenzene

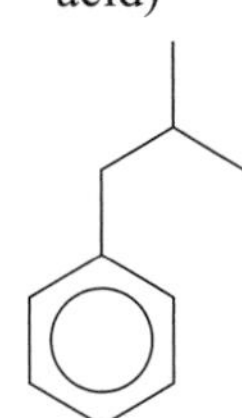

Isobutylbenzene

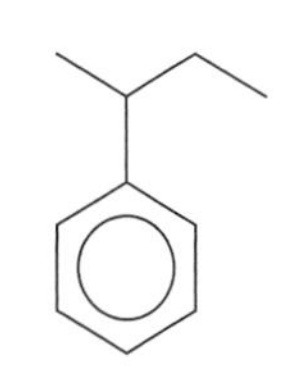

sec-Butylbenzene

tert-Butylbenzene

Aromaticity

14.18 (a) 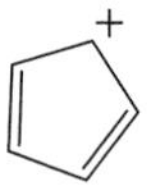 Antiaromatic

(b) 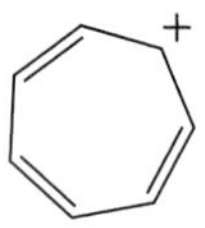 Aromatic

(c) 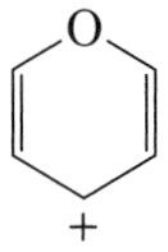 Aromatic

(d)  Antiaromatic

(e) 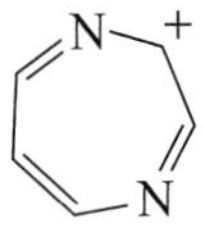 Aromatic

(f) 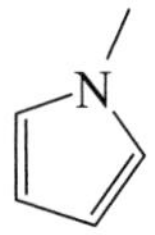 Aromatic

(g) 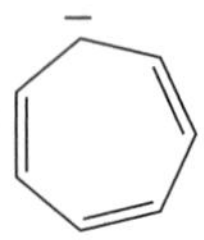 Antiaromatic

(h) 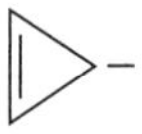 Antiaromatic

(i)  Aromatic

(j)  Nonaromatic

(k)  Aromatic

(l)  Aromatic

14.19 (a) 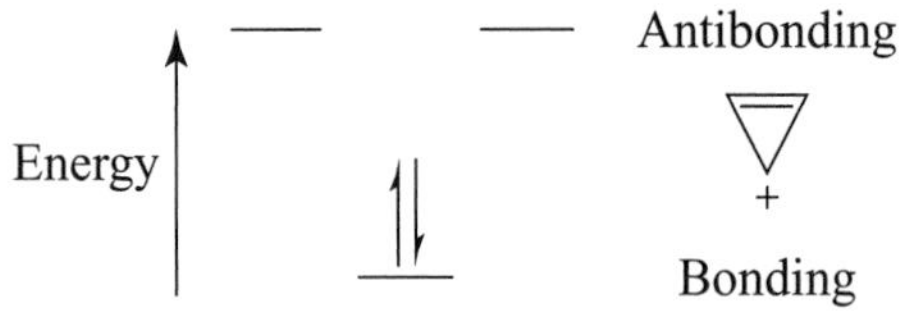 Energy

Antibonding

Bonding

(b) 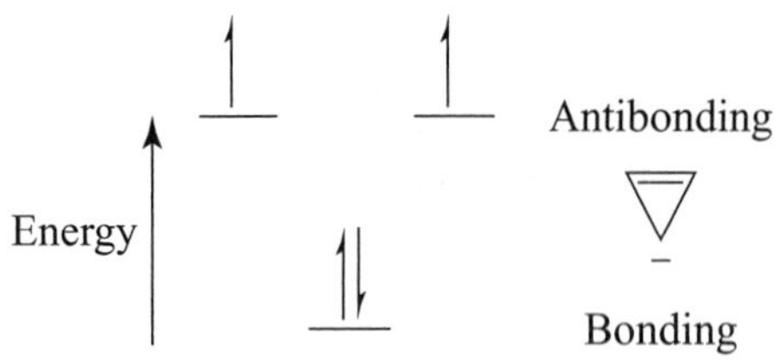 Energy

Antibonding

Bonding

14.20 (a) N, N, CH_3 — HCl (1 equiv.) $\longrightarrow$ H, ^+N, Cl^-, N, CH_3

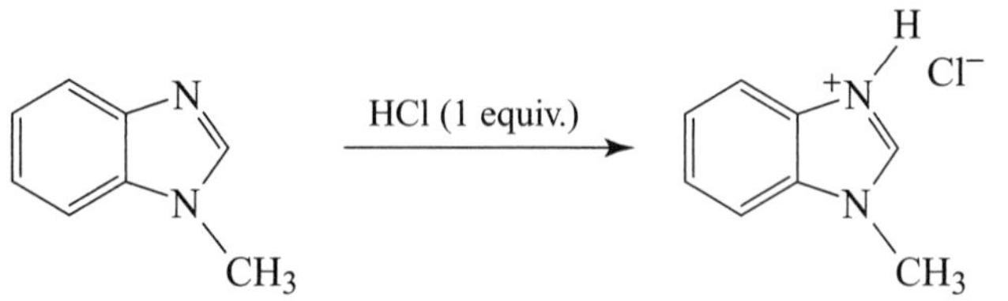

(b) H, N, N — HCl (1 equiv.) $\longrightarrow$ H, N, Cl^-, N^+–H

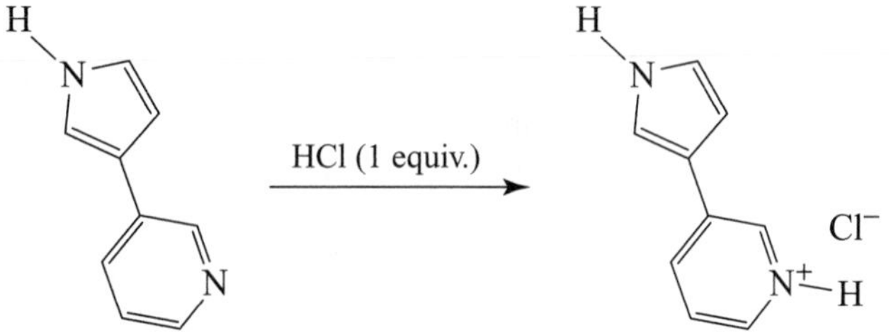

14.21

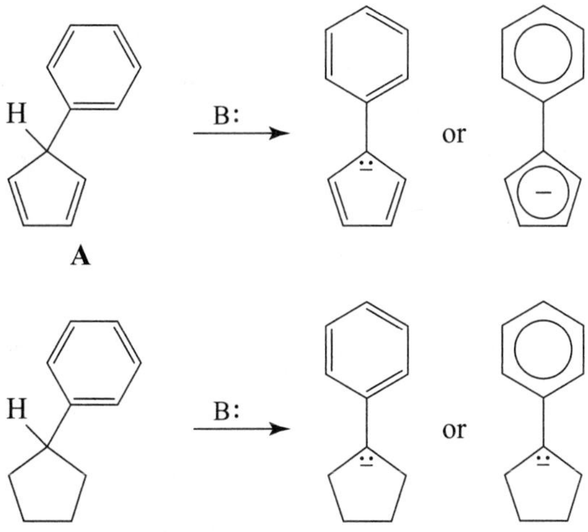

H — B: $\longrightarrow$ or

A

H — B: $\longrightarrow$ or

B

The conjugate base of **A** is a substituted cyclopentadienyl anion and both rings are aromatic. In **B**, the five-membered ring does not contribute to anion stability. The additional stabilization provided by the cyclopentyldienyl moiety is the reason for the greater acidity of **A**.

14.22

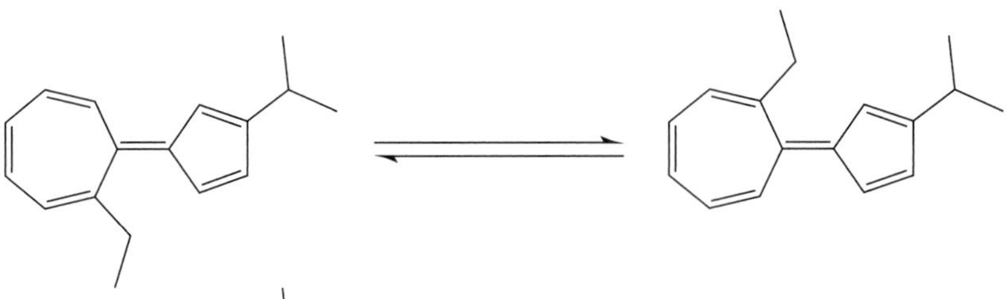

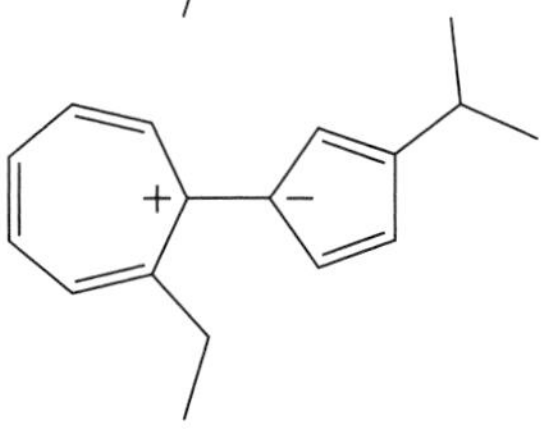

The major resonance form leaves the electrons in the cyclopentadiene ring to make both rings aromatic. Therefore, the central bond has more single bond character and rotation around that bond is easily achieved.

14.23 Hückel's rule should apply to both pentalene and heptalene. Pentalene's antiaromaticity can be attributed to its having 8 π electrons. Heptalene's lack of aromaticity can be attributed to its having 12 π electrons. Neither 8 nor 12 is a Hückel number.

14.24 (a) The extra two electrons go into the two partly filled (nonbonding) molecular orbitals (Fig. 14.7), causing them to become filled. The dianion, therefore, is not a diradical. Moreover, the cyclooctatetraene dianion has 10 π electrons (a Hückel number), and this apparently gives it the stability of an aromatic compound. (The highest occupied molecular orbitals may become slightly lower in energy and become bonding molecular orbitals.) The stability gained by becoming aromatic is apparently large enough to overcome the extra strain involved in having the ring of the dianion become planar.

(b) The strong base (butyllithium) removes two protons from the compound on the left. This acid-base reaction leads to the formation of the 10 π electron pentalene dianion, an aromatic dianion.

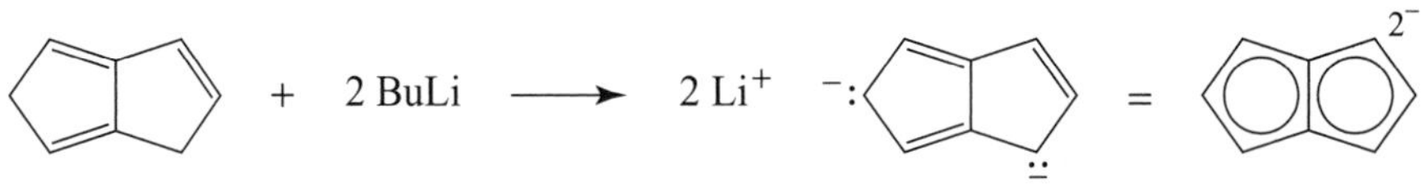

$$+ \ 2\,\text{BuLi} \longrightarrow 2\,\text{Li}^+ \quad = \quad 2^-$$

Pentalene dianion

14.25 The bridging $-CH_2-$ group causes the 10 π electron ring system (below) to become planar. This allows the ring to become aromatic.

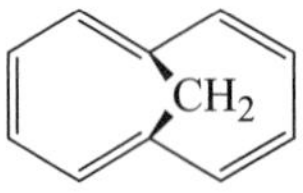

CH_2

14.26 (a) Resonance contributions that involve the carbonyl group of **I** resemble the *aromatic* cycloheptatrienyl cation and thus stabilize **I.** Similar contributors to the hybrid of **II** resemble the *antiaromatic* cyclopentadienyl cation (see Problem 14.8) and thus destabilize **II.**

(a) I ⟷ IA

Contributors like **IA** are exceptionally stable because they resemble an aromatic compound. They therefore make large stabilizing contributions to the hybrid.

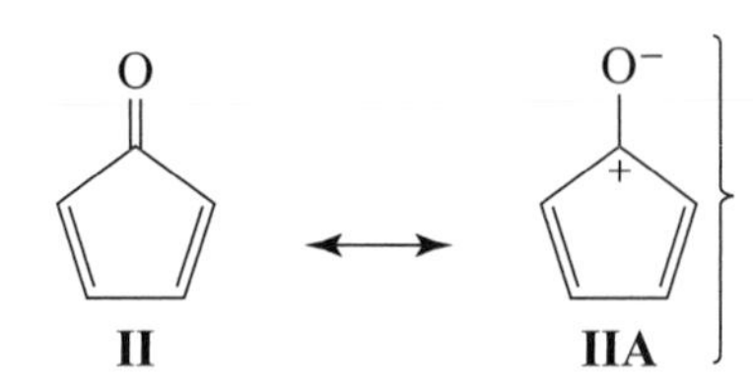

Contributors like **IIA** are exceptionally unstable because they resemble an antiaromatic compound. Any contribution they make to the hybrid is destabilizing.

(b)

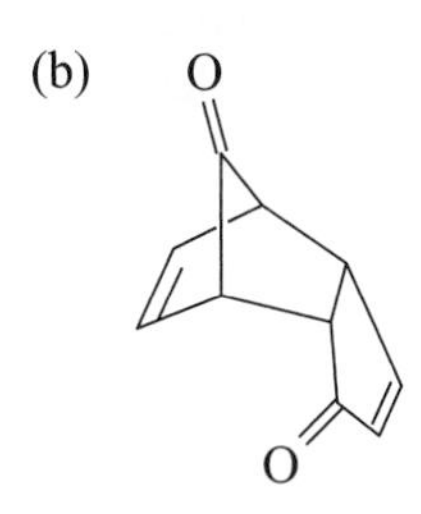

14.27 Ionization of 5-chloro-1,3-cyclopentadiene would produce a cyclopentadienyl cation, and the cyclopentadienyl cation (see Problem 14.8) would be highly unstable because it would be antiaromatic.

$$\text{C}_5\text{H}_5\text{—Cl} \xrightarrow{S_N1} \text{C}_5\text{H}_5^{+} + \text{Cl}^{-}$$

Antiaromatic ion
(highly unstable)

14.28 (a) The cyclononatetraenyl anion with 10 π electrons obeys Hückel's rule.

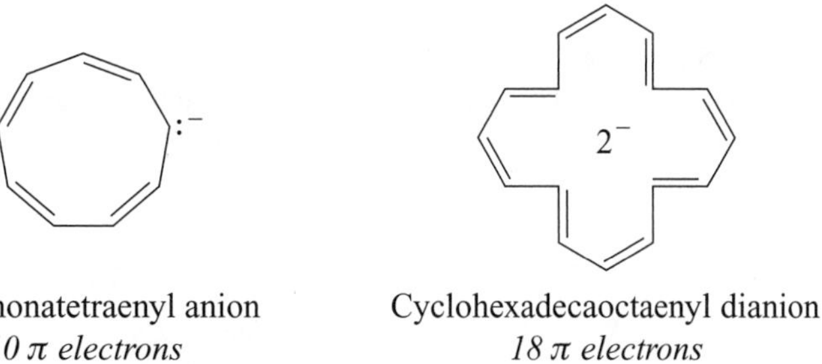

Cyclononatetraenyl anion	Cyclohexadecaoctaenyl dianion
10 π electrons	*18 π electrons*
Aromatic	**Aromatic**

(b) By adding 2 π electrons, [16] annulene becomes an 18 π electron system and therefore obeys Hückel's rule.

14.29 As noted in Problem 13.42, furan can serve as the diene component of Diels-Alder reactions, readily losing all aromatic character in the process. Benzene, on the other hand, is so unreactive in a Diels-Alder reaction that it can be used as a nonreactive solvent for Diels-Alder reactions.

Spectroscopy and Structure Elucidation

14.30 (a)

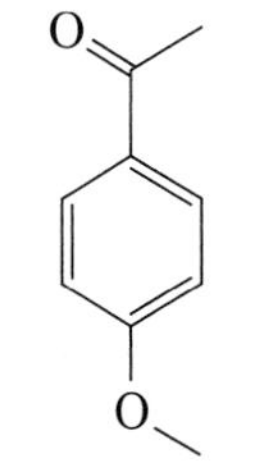

4 total signals
aromatic region, 2 doublets

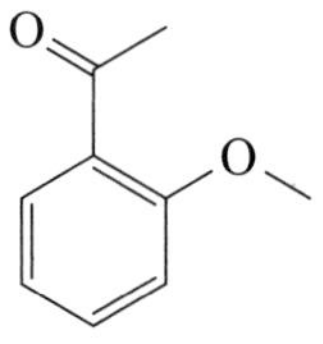

6 total signals
aromatic region, 2 doublets
and 2 doublets of doublets

(b)

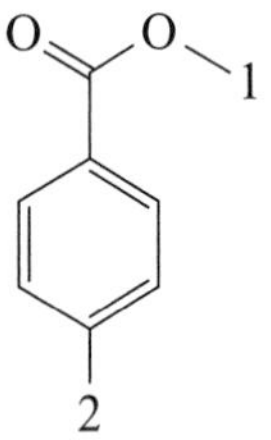

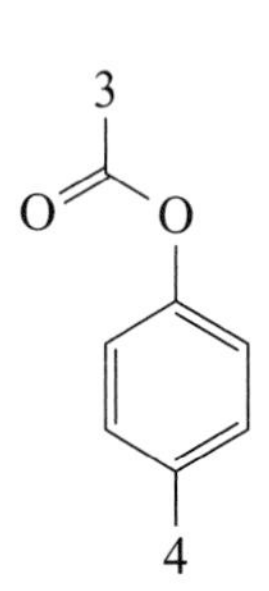

Signal 1 (~4 ppm, singlet)
Signal 2 (~2 ppm, singlet)

Signal 3 (~2 ppm, singlet)
Signal 4 (~2 ppm, singlet)

(c)

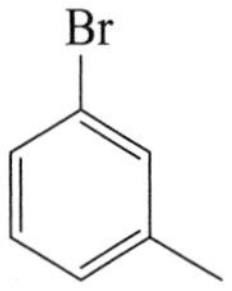

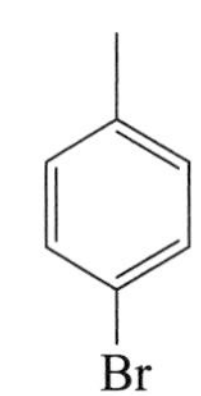

5 total signals
aromatic region: 1 singlet,
2 doublets, and 1 doublet
of doublets

3 total signals
aromatic region: 2 doublets

14.31 **A** C_6H_5 (*c*) – $CH(CH_3)_2$: CH_3 (*a*), H (*b*), CH_3 (*a*)

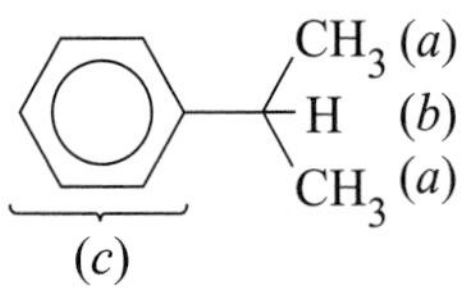

(*a*) doublet δ 1.25
(*b*) septet δ 2.9
(*c*) multiplet δ 7.3

B C_6H_5 (*d*) – $CH(CH_3)(NH_2)$: CH_3 (*a*), H (*b*), NH_2 (*c*)

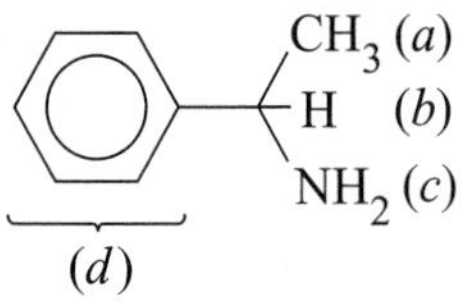

(*a*) doublet δ 1.35
(*b*) quartet δ 4.1
(*c*) singlet δ 1.7
(*d*) multiplet δ 7.3

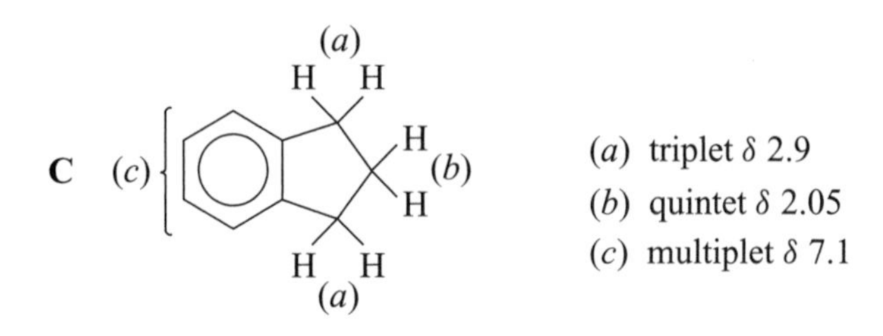

(*a*) triplet δ 2.9
(*b*) quintet δ 2.05
(*c*) multiplet δ 7.1

14.32 A ^{1}H NMR signal this far upfield indicates that cyclooctatetraene is a cyclic polyene and is not aromatic; its π electrons are not fully delocalized.

14.33 Compound F is *p*-isopropyltoluene. ^{1}H NMR assignments are shown in the following spectrum.

Strong IR absorption at $\sim$ 800 cm^{-1} indicates para substitution.

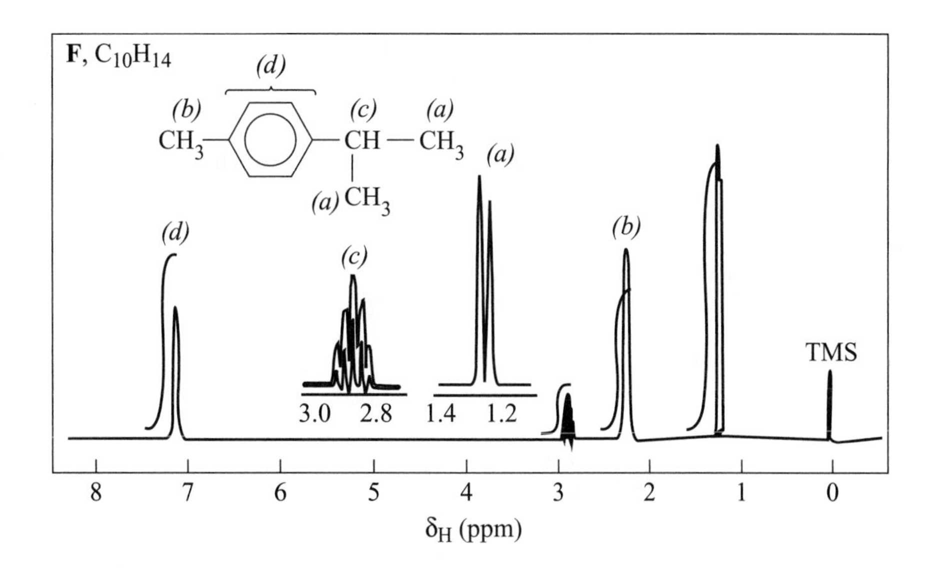

We can make the following ^{1}H NMR assignments:

(a) doublet δ 1.25
(b) singlet δ 2.3
(c) septet δ 2.85
(d) multiplet δ 7.1

14.34 Compound **L** is allylbenzene,

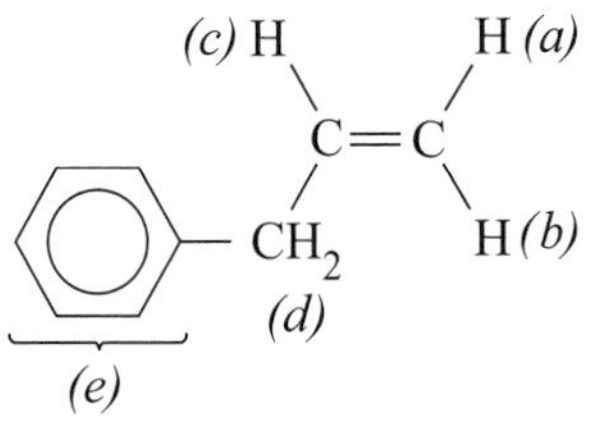

(c) H, H (a), C=C, H (b), CH_2 (d), (e)

(d) Doublet, δ 3.1 (2H)
(a) or *(b)* Multiplet, δ 4.8
(a) or *(b)* Multiplet, δ 5.1
(c) Multiplet, δ 5.8
(e) Multiplet, δ 7.1 (5H)

The following IR assignments can be made.

3035 cm^{-1}, C—H stretching of benzene ring
3020 cm^{-1}, C—H stretching of $—CH{=}CH_2$ group
2925 cm^{-1} and 2853 cm^{-1}, C—H stretching of $—CH_2—$ group
1640 cm^{-1}, C=C stretching
990 cm^{-1} and 915 cm^{-1}, C—H bendings of $—CH{=}CH_2$ group
740 cm^{-1} and 695 cm^{-1}, C—H bendings of $—C_6H_5$ group

The UV absorbance maximum at 255 nm is indicative of a benzene ring that is not conjugated with a double bond.

14.35 Compound **M** is *m*-ethyltoluene. We can make the following assignments in the spectrum.

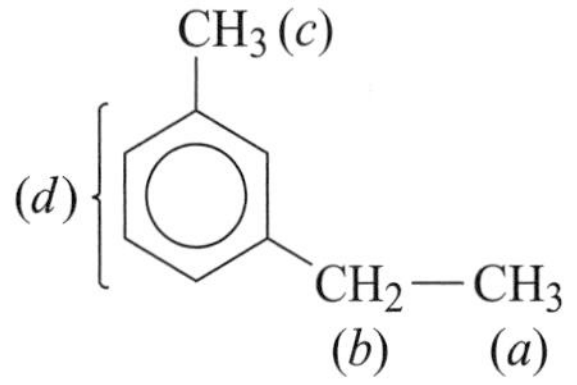

CH_3 (c), (d), $CH_2—CH_3$, (b) (a)

(*a*) triplet δ 1.4
(*b*) quartet δ 2.6
(*c*) singlet δ 2.4
(*d*) multiplet δ 7.05

Meta substitution is indicated by the very strong peaks at 690 and 780 cm^{-1} in the IR spectrum.

14.36 Compound **N** is $C_6H_5CH{=}CHOCH_3$. The absence of absorption peaks due to O—H or C=O stretching in the IR spectrum of **N** suggests that the oxygen atom is present as part of an ether linkage. The (5H) ^1H NMR multiplet between δ 7.1–7.6 strongly suggests the presence of a monosubstituted benzene ring; this is confirmed by the strong peaks at ~690 and ~770 cm^{-1} in the IR spectrum.

We can make the following assignments in the ^1H NMR spectrum:

$$\underset{(a)}{C_6H_5}-\underset{(b)}{CH}=\underset{(c)}{CH}-\underset{(d)}{OCH_3}$$

(*a*) Multiplet δ 7.1–7.6
(*b*) Doublet δ 6.1
(*c*) Doublet δ 5.2
(*d*) Singlet δ 3.7

14.37 Compound **X** is *m*-xylene. The upfield signal at δ 2.3 arises from the two equivalent methyl groups. The downfield signals at δ 6.9 and 7.1 arise from the protons of the benzene ring. Meta substitution is indicated by the strong IR peak at 680 cm^{-1} and very strong IR peak at 760 cm^{-1}.

14.38 The broad IR peak at 3400 cm^{-1} indicates a hydroxy group, and the two bands at 720 and 770 cm^{-1} suggest a monosubstituted benzene ring. The presence of these groups is also indicated by the peaks at δ 4.4 and δ 7.2 in the ^1H NMR spectrum. The ^1H NMR spectrum also shows a triplet at δ 0.85 indicating a $-CH_3$ group coupled with an adjacent $-CH_2-$ group. There is a complex multiplet at δ 1.7 and there is also a triplet at δ 4.5 (1H). Putting these pieces together in the only way possible gives us the following structure for **Y.**

OH

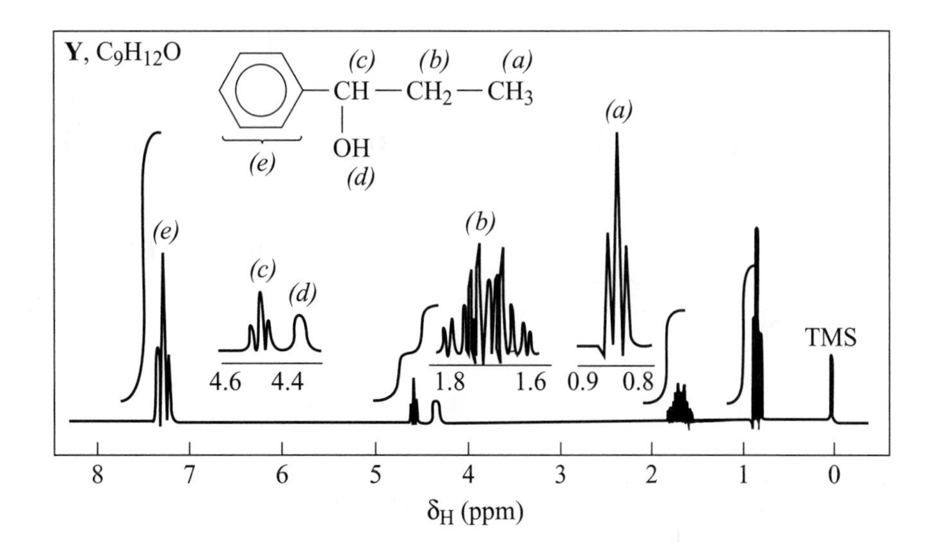

14.39 (a) Four unsplit signals.

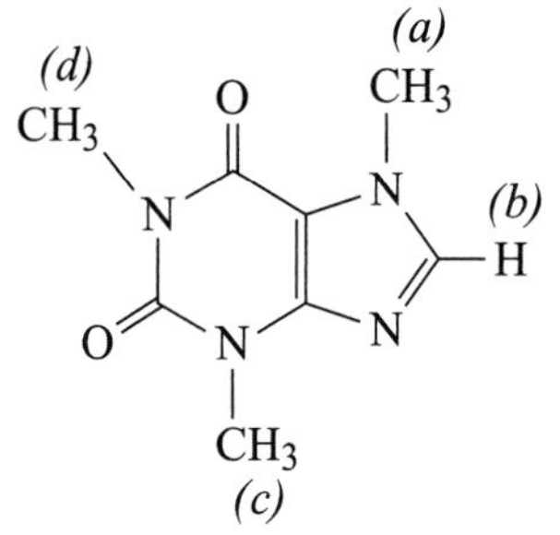

(b) Absorptions arising from: $=C-H$, $-CH_3$, and $C=O$ groups.

Challenge Problems

14.40 The vinylic protons of *p*-chlorostyrene should give a spectrum approximately like the following:

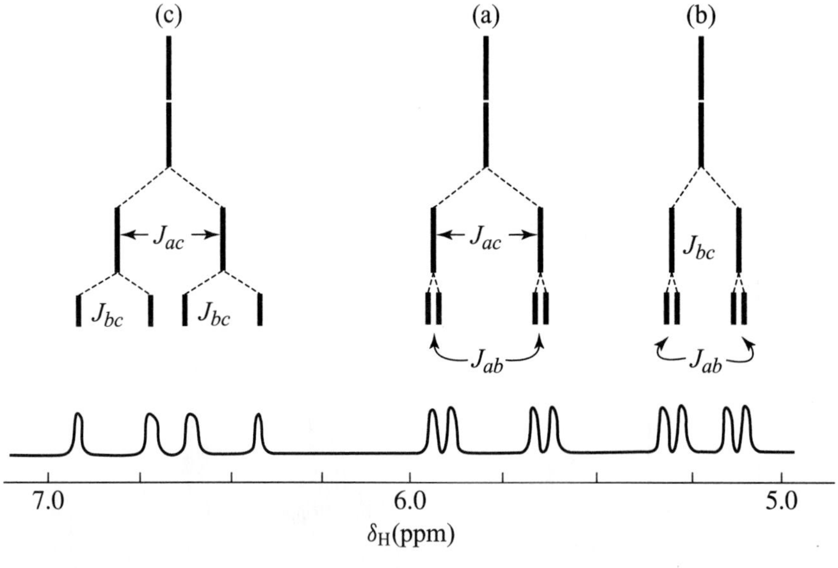

14.41

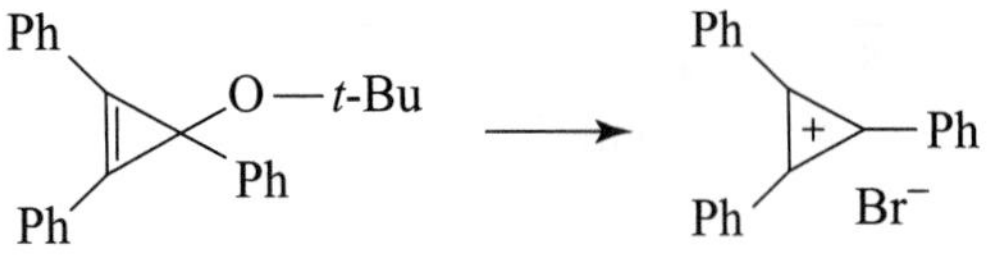

A **B**

14.42

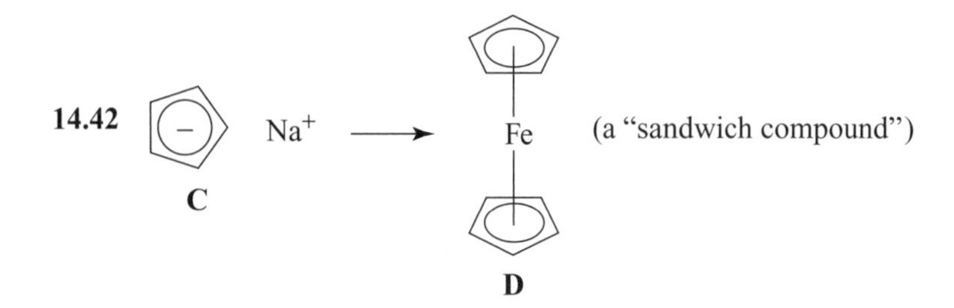

14.43 Ph—≡—≡—Ph

E

14.44

Third unoccupied π MO (five nodal planes)

Second unoccupied π MO (four nodal planes)

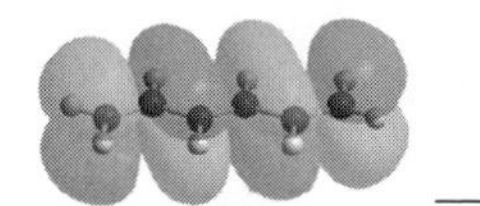

First unoccupied π MO (three nodal planes)

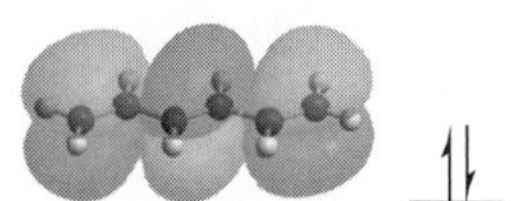

Highest energy occupied π MO (two nodal planes)

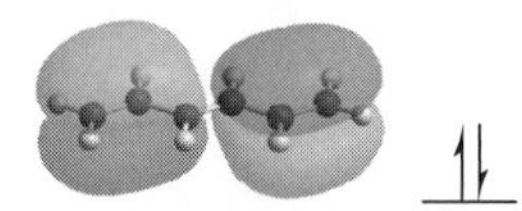

Second highest energy occupied π MO (one nodal plane)

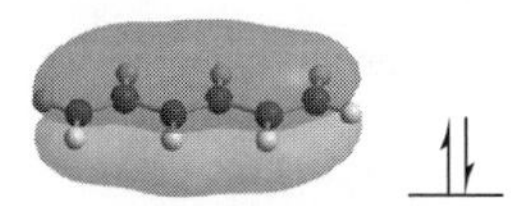

Lowest energy occupied π MO (no nodal planes)

QUIZ

14.1 Which of the following reactions of benzene is inconsistent with the assertion that benzene is aromatic?

(a) $Br_2/CCl_4/25°C \longrightarrow$ no reaction

(b) $H_2/Pt/25°C \longrightarrow$ no reaction

(c) $Br_2/FeBr_3 \longrightarrow C_6H_5Br + HBr$

(d) $KMnO_4/H_2O/25°C \longrightarrow$ no reaction

(e) None of the above

14.2 Which is the correct name of the compound shown?

NO_2

Cl

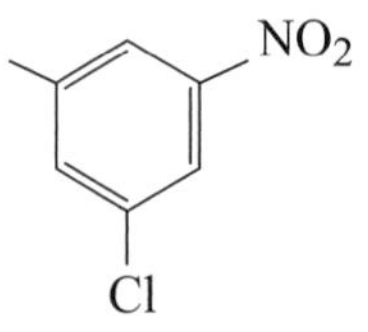

(a) 3-Chloro-5-nitrotoluene

(b) *m*-Chloro-*m*-nitrotoluene

(c) l-Chloro-3-nitro-5-toluene

(d) *m*-Chloromethylnitrobenzene

(e) More than one of these

14.3 Which is the correct name of the compound shown?

OH

F

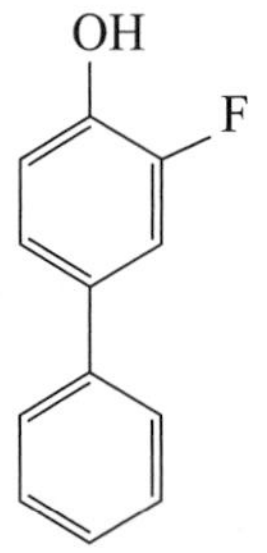

(a) 2-Fluoro-1-hydroxyphenylbenzene

(b) 2-Fluoro-4-phenylphenol

(c) *m*-Fluoro-*p*-hydroxybiphenyl

(d) *o*-Fluoro-*p*-phenylphenol

(e) More than one of these

14.4 Which of the following molecules or ions is not aromatic according to Hückel's rule?

(a) 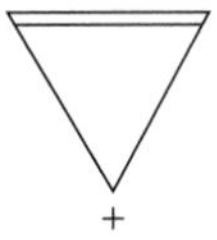 (b) (c) 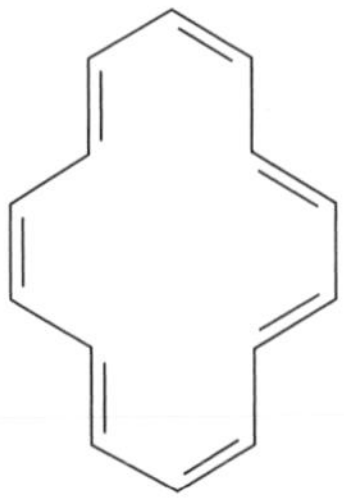 (d)

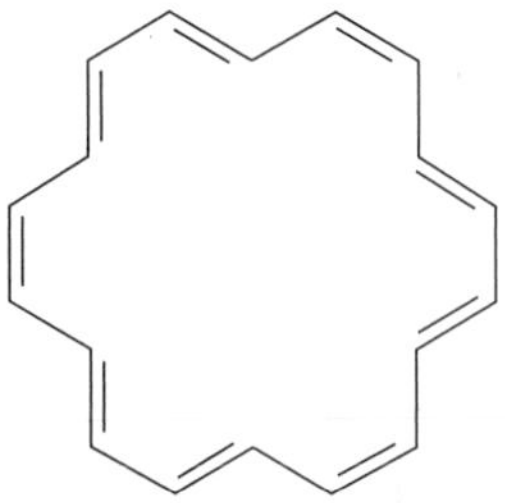

(e) All are aromatic.

14.5 Give the structure of a compound with the formula C_7H_7Cl that is capable of undergoing both S_N1 and S_N2 reactions.

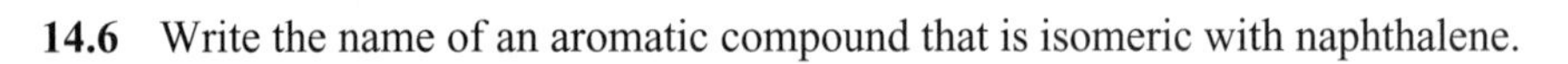

14.6 Write the name of an aromatic compound that is isomeric with naphthalene.

15 REACTIONS OF AROMATIC COMPOUNDS

SOLUTIONS TO PROBLEMS

15.1

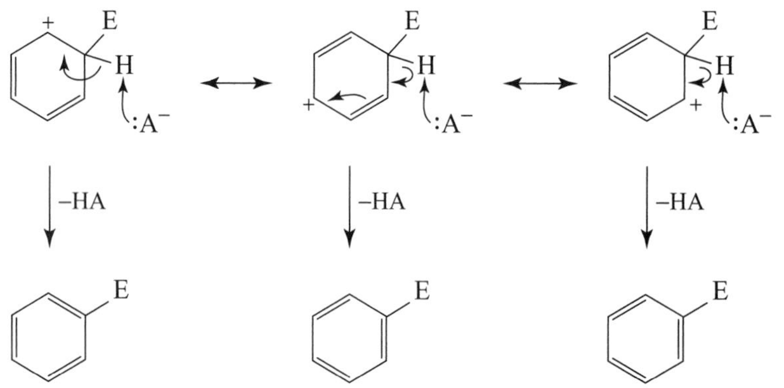

(Resonance structures of the arenium ion, each losing H^+ to $:A^-$ ($-HA$) to give the substituted benzene C_6H_5E.)

15.2 The rate is dependent on the concentration of NO_2^+ ion formed from protonated nitric acid.

$$H{-}\overset{+}{O}(H){-}NO_2 + HA \longrightarrow NO_2^+ + H_3O^+ + A^-$$

(where HA = HNO_3 or $HOSO_3H$)

Because H_2SO_4 ($HOSO_3H$) is a stronger acid, a mixture of it and HNO_3 will contain a higher concentration of protonated nitric acid than will nitric acid alone.

That is, the reaction,

$$H{-}O{-}NO_2 + HOSO_3H \rightleftharpoons H{-}\overset{+}{O}(H){-}NO_2 + HSO_4^-$$

Protonated nitric acid

produces more protonated nitric acid than the reaction,

$$H{-}O{-}NO_2 + H{-}O{-}NO_2 \rightleftharpoons H{-}\overset{+}{O}(H){-}NO_2 + NO_3^-$$

15.3 *Step 1*

+ H—F ⇌ + F^-

Step 2

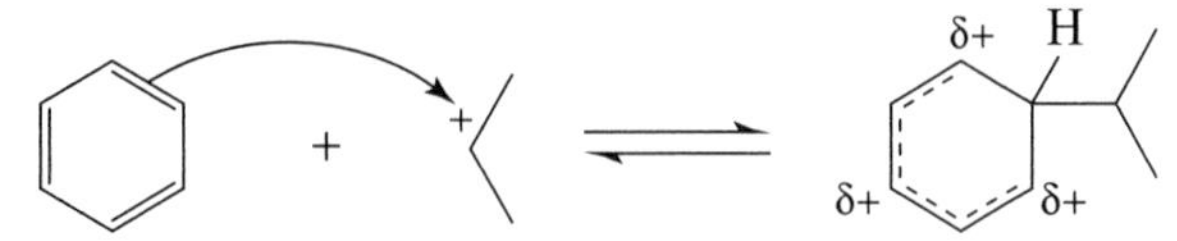

+ ⇌ δ+ H δ+ δ+

Step 3

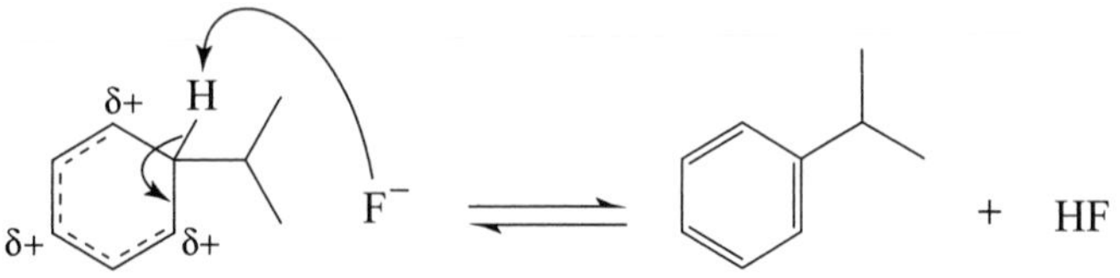

δ+ H δ+ δ+ F^- ⇌ + HF

15.4

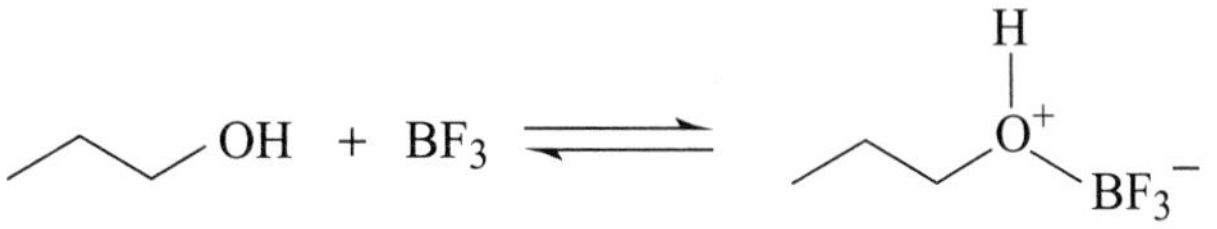

OH + BF_3 ⇌ H O^+ BF_3^-

Rearrangement to the secondary carbocation occurs as the leaving group departs.

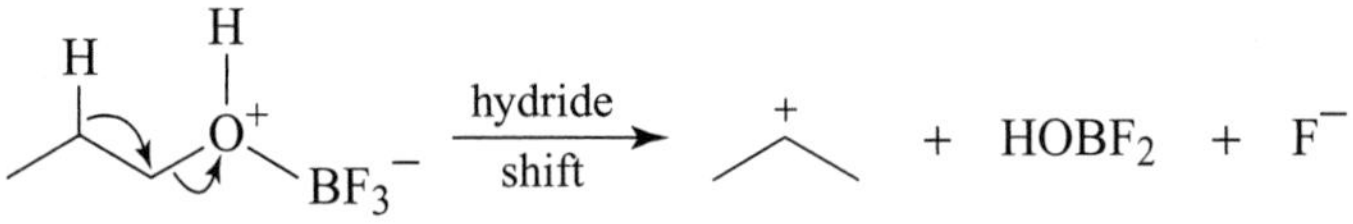

H H O^+ BF_3^- —hydride shift→ + + $HOBF_2$ + F^-

Both carbocations are then attacked by the ring.

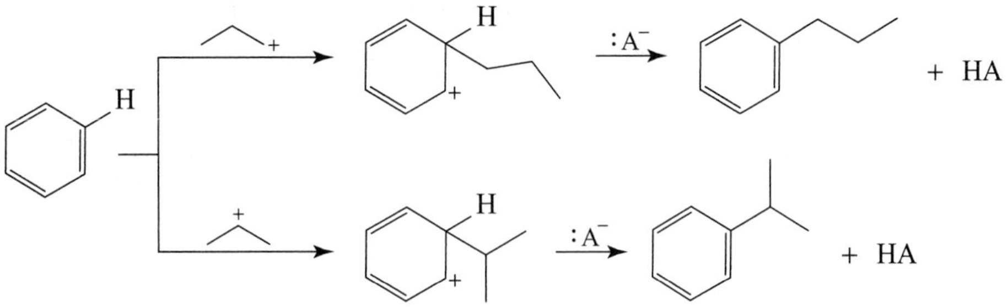

H + H :A^- + HA

+ H :A^- + HA

15.5 (a) —O, Cl, $AlCl_3$→ O —Zn(Hg), HCl reflux→

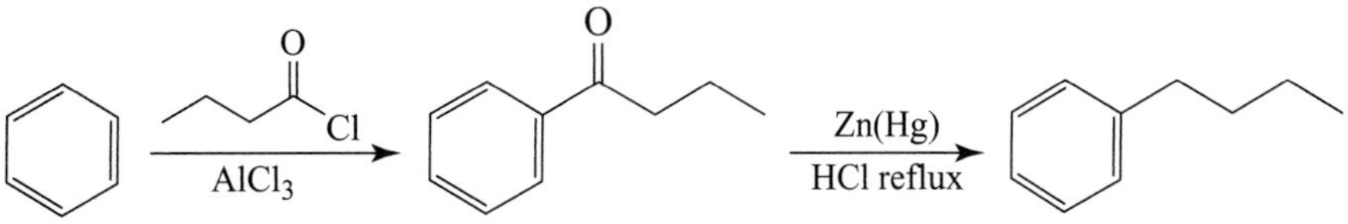

(b) —O, Cl, $AlCl_3$→ O —Zn(Hg), HCl reflux→

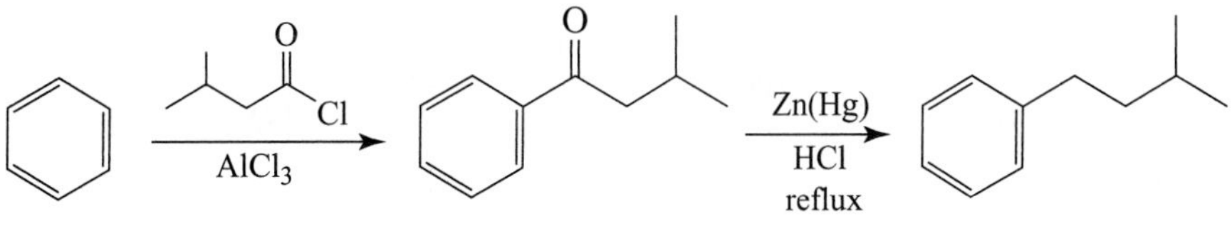

(c) [Benzene] $\xrightarrow[AlCl_3]{C_6H_5COCl}$ [Benzophenone]

(d) [Benzene] + [Phthalic anhydride] $\xrightarrow{AlCl_3}$ [2-Benzoylbenzoic acid] $\xrightarrow[AlCl_3]{Zn(Hg)}$

[2-Benzylbenzoic acid] $\xrightarrow{SOCl_2}$ [2-Benzylbenzoyl chloride] $\xrightarrow{AlCl_3}$

[Anthrone] $\xrightarrow[HCl]{Zn(Hg)}$ [9,10-Dihydroanthracene]

15.6 If the methyl group had no directive effect on the incoming electrophile, we would expect to obtain the products in purely statistical amounts. Since there are two ortho hydrogen atoms, two meta hydrogen atoms, and one para hydrogen, we would expect to get 40% ortho (2/5), 40% meta (2/5), and 20% para (1/5). Thus, we would expect that only 60% of the mixture of mononitrotoluenes would have the nitro group in the ortho or para position. And we would expect to obtain 40% of *m*-nitrotoluene. In actuality, we get 96% of combined *o*- and *p*-nitrotoluenes and only 4% *m*-nitrotoluene. This shows the ortho-para directive effect of the methyl group.

15.7 (a) [Toluene, CH_3] $\xrightarrow[H_2SO_4]{SO_3}$ [o-Toluenesulfonic acid, CH_3, SO_3H] minor + [p-Toluenesulfonic acid, CH_3, SO_3H] major

(b) [Benzoic acid, O, OH] $\xrightarrow[H_2SO_4]{HNO_3}$ [m-Nitrobenzoic acid, O, OH, NO_2]

(c) [Nitrobenzene, NO_2] $\xrightarrow[FeBr_3]{Br_2}$ [m-Bromonitrobenzene, NO_2, Br]

(d) 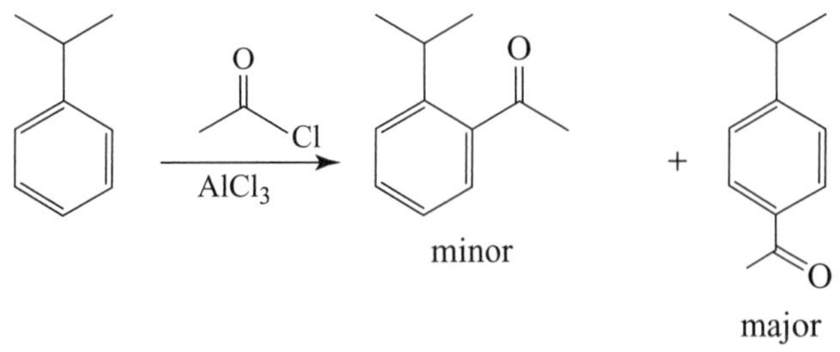 O Cl

AlCl$_3$

O

minor

+

O

major

15.8 As the following structures show, attack at the ortho and para positions of phenol leads to arenium ions that are more stable (than the one resulting from meta attack) because they are hybrids of four resonance structures, one of which is relatively stable. Only three resonance structures are possible for the meta arenium ion, and none is relatively stable.

Ortho attack

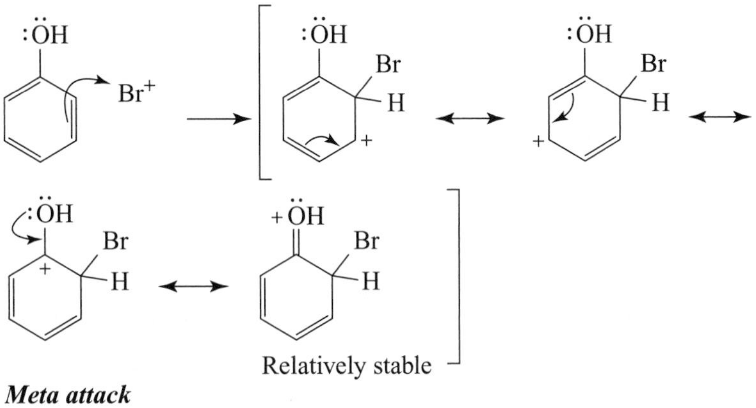

:ÖH Br$^+$ ⟶ [:ÖH Br H + ⟷ :ÖH Br H + ⟷

:ÖH Br H + ⟷ +ÖH Br H]

Relatively stable

Meta attack

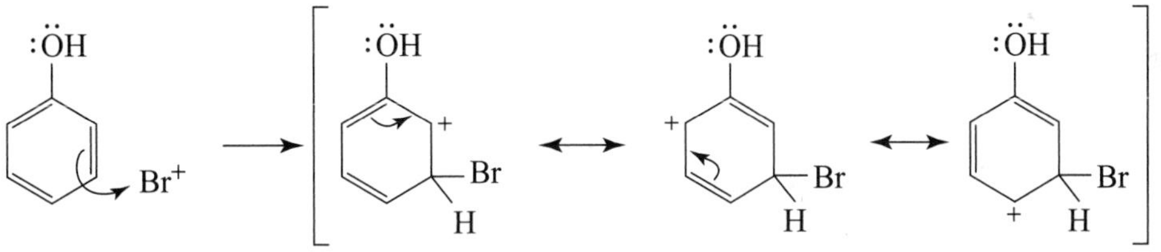

:ÖH Br$^+$ ⟶ [:ÖH + Br H ⟷ :ÖH + Br H ⟷ :ÖH Br + H]

Para attack

:ÖH Br$^+$ ⟶ [:ÖH + H Br ⟷ :ÖH + H Br ⟷ +ÖH H Br ⟷ :ÖH + H Br]

Relatively stable

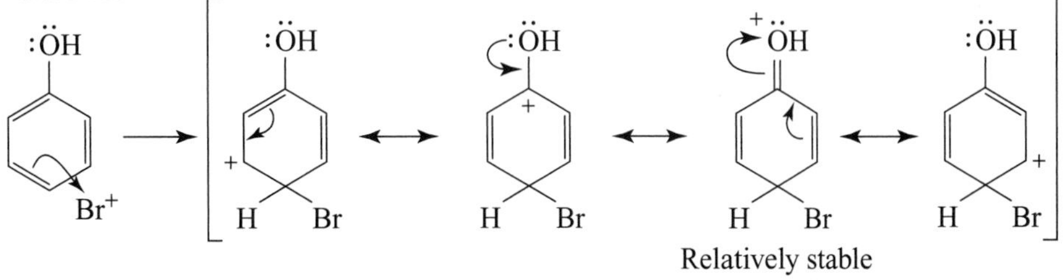

15.9 (a) The atom (an oxygen atom) attached to the benzene ring has an unshared electron pair that it can donate to the arenium ions formed from ortho and para attack, stabilizing them. (The arenium ions are analogous to the previous answer with a —$COCH_3$ group replacing the —H of the phenolic hydroxyl).

(b) Structures such as the following compete with the benzene ring for the oxygen electrons, making them less available to the benzene ring.

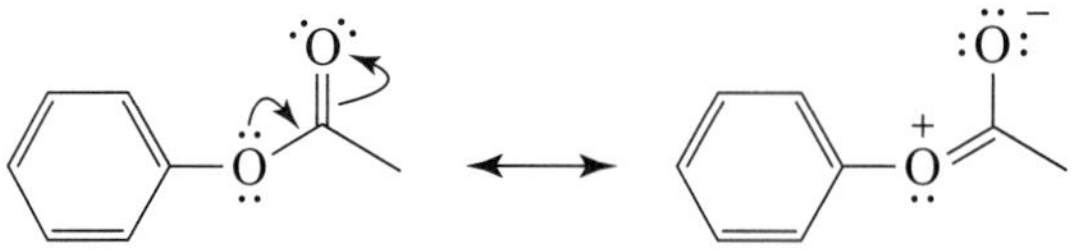

This effect makes the benzene ring of phenyl acetate less electron rich and, therefore, less reactive.

(c) Because the acetamido group has an unshared electron pair on the nitrogen atom that it can donate to the benzene ring, it is an ortho-para director.

(d) Structures such as the following compete with the benzene ring for the nitrogen electrons, making them less available to the benzene ring.

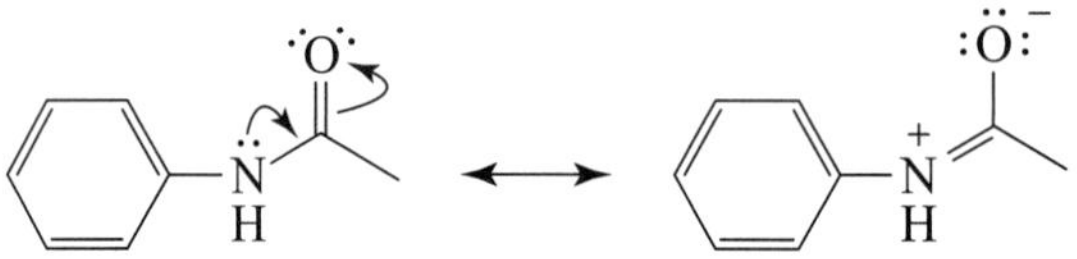

15.10 The electron-withdrawing inductive effect of the chlorine of chloroethene makes its double bond less electron rich than that of ethene. This causes the rate of reaction of chloroethene with an electrophile (i.e., a proton) to be slower than the corresponding reaction of ethene.

When chloroethene adds a proton, the orientation is governed by a resonance effect. In theory, two carbocations can form:

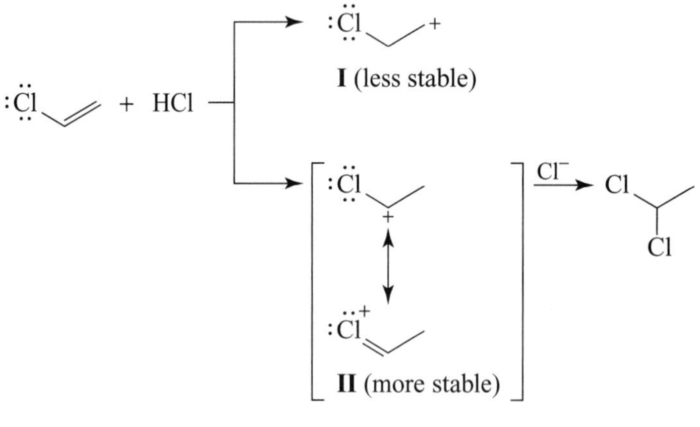

I (less stable)

II (more stable)

Carbocation **II** is more stable than **I** because of the resonance contribution of the extra structure just shown in which the chlorine atom donates an electron pair (see Section 15.11D).

15.11 ***Ortho attack***

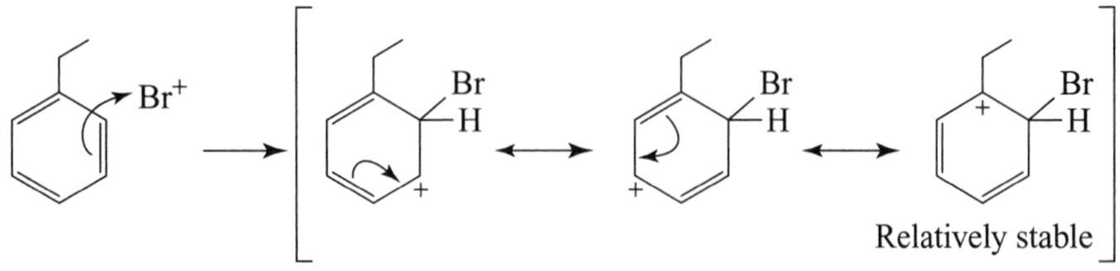

Br^+

Br H | Br H | Br H

Relatively stable

Para attack

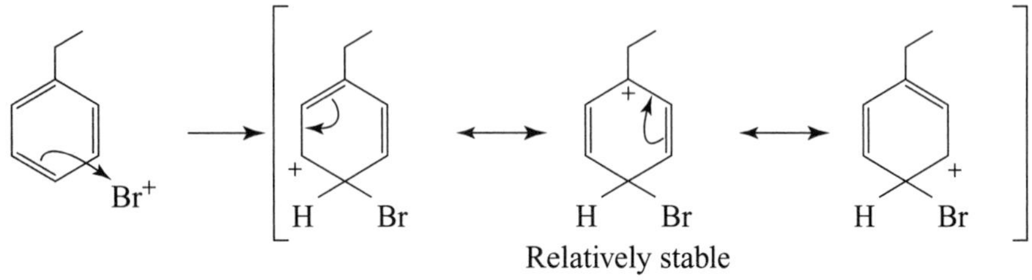

Br^+

H Br | H Br | H Br

Relatively stable

15.12 The phenyl group, as the following resonance structures show, can act as an electron-releasing group and can stabilize the arenium ions formed from ortho and para attack.

Ortho attack

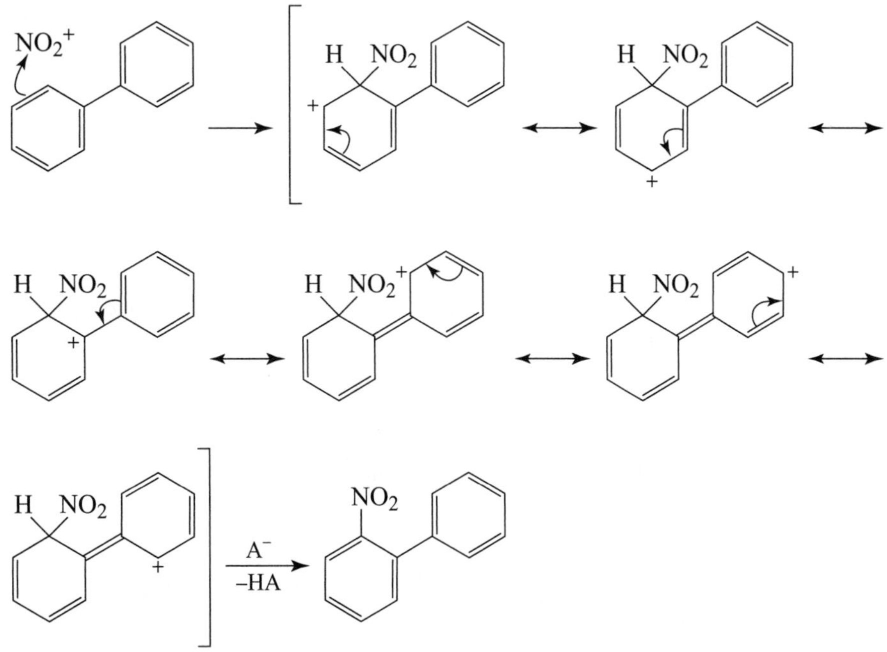

NO_2^+

H NO_2 | H NO_2

H NO_2 | H NO_2^+ | H NO_2

H NO_2 $\xrightarrow[-HA]{A^-}$ NO_2

In the case of the arenium ions above and following, the unsubstituted ring can also be shown in the alternative Kekule structure.

Para attack

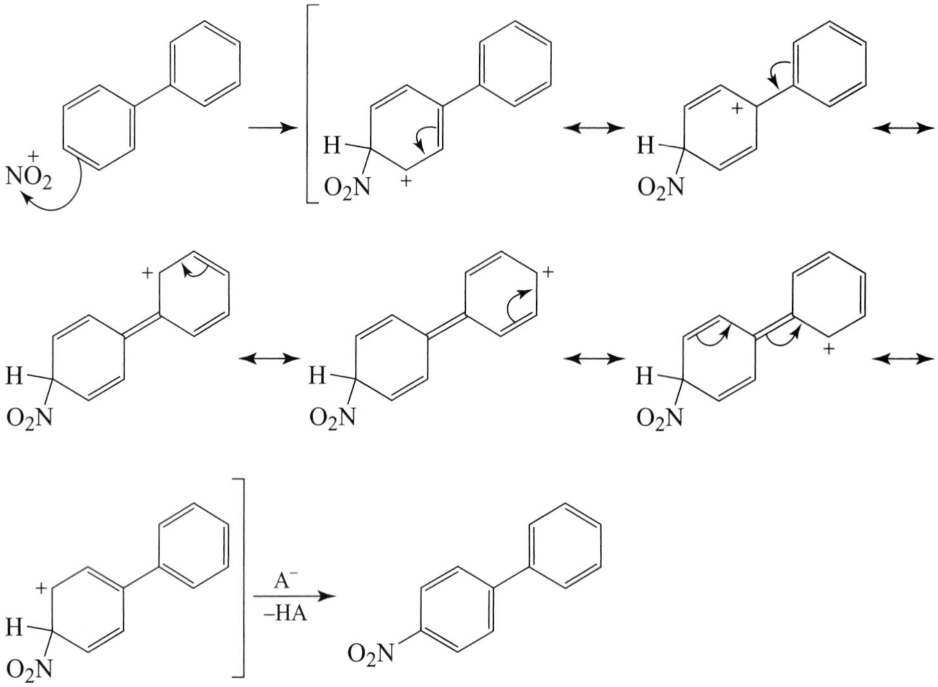

$\overset{+}{NO_2}$ → [H, O_2N, + ↔ H, O_2N, + ↔ + H, O_2N ↔ H, O_2N, + ↔ H, O_2N, + ↔ + H, O_2N ↔ + H, O_2N] $\xrightarrow[-HA]{A^-}$ O_2N

In the case of both ortho and para substitution, the phenyl group functions as an activating group, resulting in a faster reaction then in the case of benzene.

15.13 **I** leads to 1-chloro-1-phenylpropane

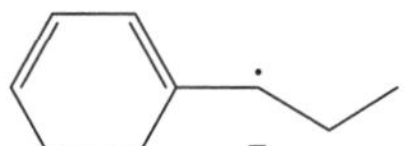

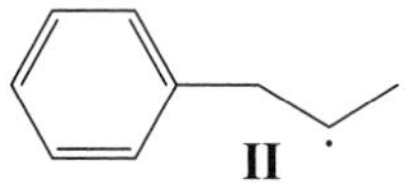

II leads to 2-chloro-1-phenylpropane

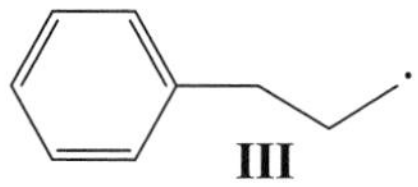

III leads to 1-chloro-3-phenylpropane

The major product is 1-chloro-1-phenylpropane because **I** is the most stable radical. It is a benzylic radical and therefore is stabilized by resonance.

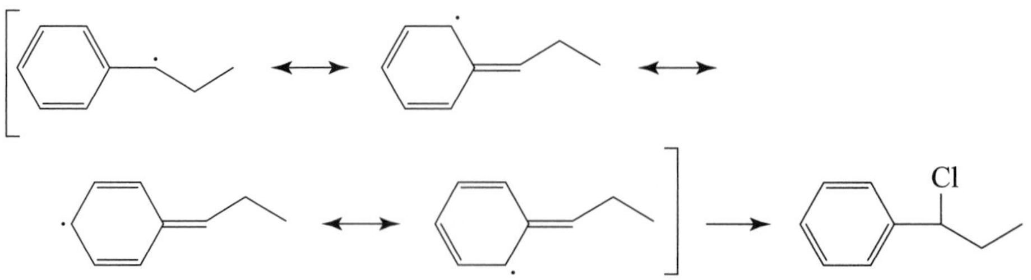

[↔ ↔ ↔] → Cl

15.14 (a) *Retrosynthetic Analysis*

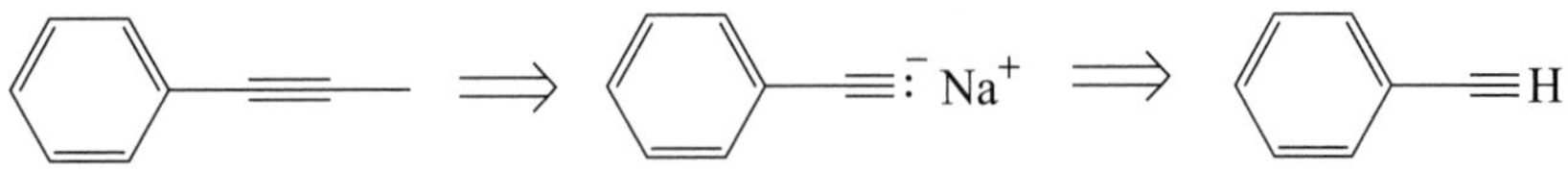

Synthesis

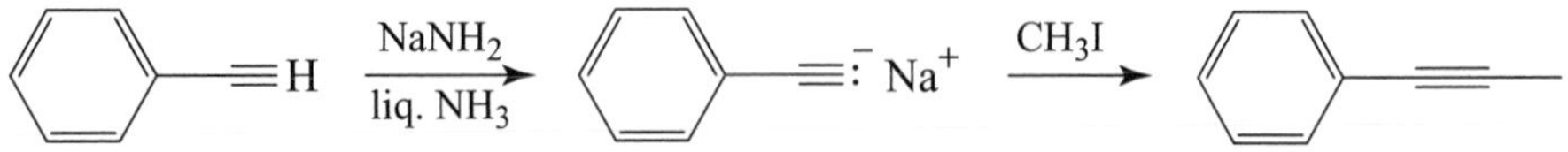

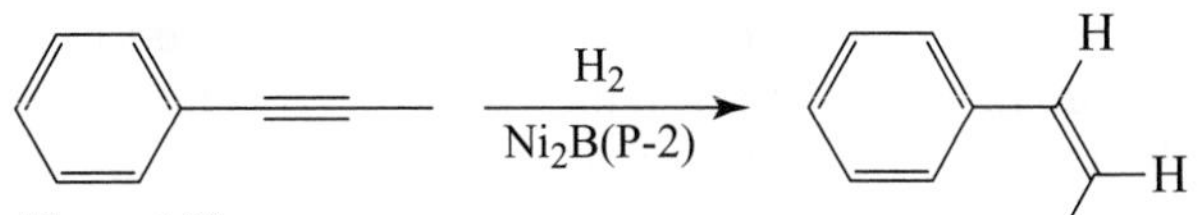

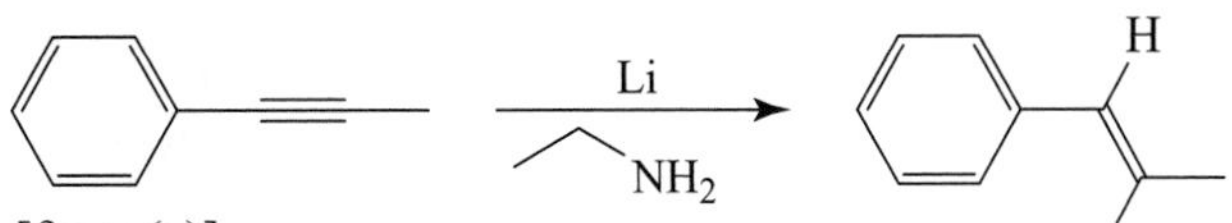

$\xrightarrow[\text{liq. NH}_3]{\text{NaNH}_2}$ $:^- Na^+$ $\xrightarrow{CH_3I}$

(b) *Retrosynthetic Analysis*

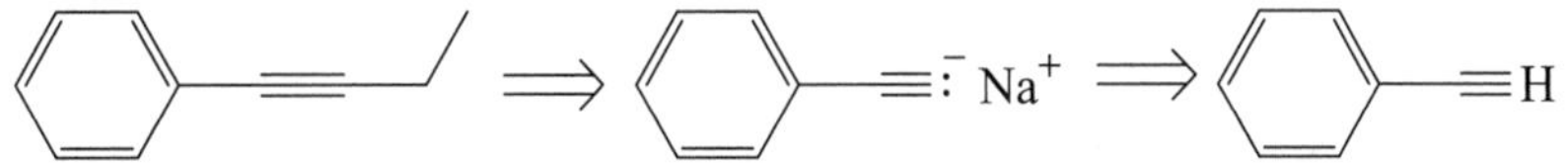

$:^- Na^+$ ⟹ ≡H

Synthesis

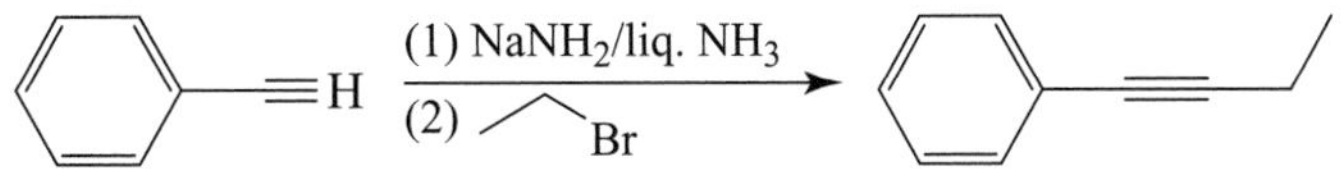

≡H (1) $NaNH_2$/liq. NH_3 (2) Br

(c) *Retrosynthetic Analysis*

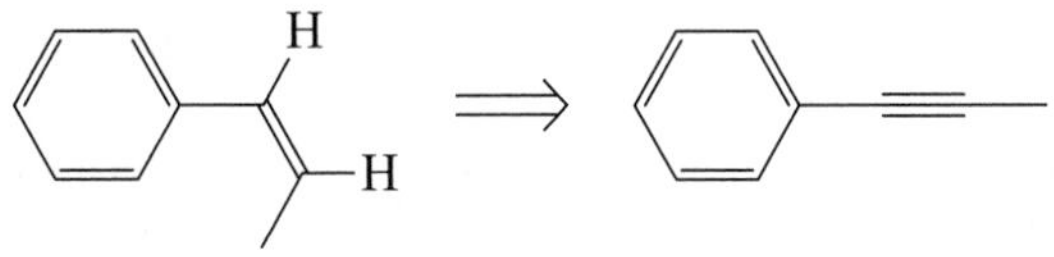

H H ⟹

Synthesis

$\xrightarrow[\text{Ni}_2\text{B(P-2)}]{\text{H}_2}$ H H

[from (a)]

(d) *Retrosynthetic Analysis*

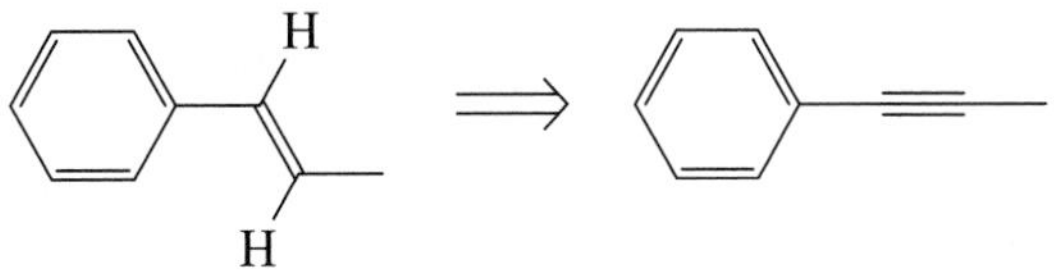

H H ⟹

Synthesis

Li, NH_2 H H

[from (a)]

15.15 The addition of hydrogen bromide to 1-phenylpropene proceeds through a benzylic radical in the presence of peroxides, and through a benzylic cation in their absence (cf., a and b as follows).

(a) Hydrogen bromide addition in the presence of peroxides.

Chain Initiation

Step 1 R—O—O—R ⟶ 2 R—O·

Step 2 RO· + H—Br ⟶ R—O—H + Br·

Step 3 Br· + C_6H_5CH=CHCH$_3$ ⟶ C_6H_5ĊH—CH(Br)CH$_3$

A benzylic radical

Chain Propagation

Step 4 C_6H_5ĊH—CH(Br)CH$_3$ + H—Br ⟶ C_6H_5CH$_2$—CH(Br)CH$_3$ + Br·

2-Bromo-1-phenylpropane

The mechanism for the addition of hydrogen bromide to 1-phenylpropene in the presence of peroxides is a chain mechanism analogous to the one we discussed when we described anti-Markovnikov addition in Section 10.9. The step that determines the orientation of the reaction is the first chain-propagating step. Bromine attacks the second carbon atom of the chain because by doing so the reaction produces a more stable benzylic radical. Had the bromine atom attacked the double bond in the opposite way, a less stable secondary radical would have been formed.

C_6H_5CH=CHCH$_3$ + Br· ⟶✕ C_6H_5CH(Br)—ĊHCH$_3$

A secondary radical

(b) Hydrogen bromide addition in the absence of peroxides.

C_6H_5CH=CHCH$_3$ + HBr ⟶ C_6H_5C$^+$H—CH$_2$CH$_3$ + Br$^-$

A benzylic cation

↓

C_6H_5CH(Br)CH$_2$CH$_3$

1-Bromo-1-phenylpropane

In the absence of peroxides, hydrogen bromide adds through an ionic mechanism. The step that determines the orientation in the ionic mechanism is the first, where the proton attacks the double bond to give the more stable benzylic cation. Had the proton attacked the double bond in the opposite way, a less stable secondary cation would have been formed.

C_6H_5 + HBr ⟶✕ C_6H_5 + Br$^-$
A secondary cation

15.16 (a) because the more stable carbocation intermediate is the benzylic carbocation, which then reacts with a chloride ion.

Cl

+

(b) because the more stable intermediate is a mercurinium ion in which a partial positive charge resides on the benzylic carbon, which then reacts with H_2O.

OH

15.17 (a) The first method would fail because introducing the chlorine substituent first would introduce an ortho-para directing group. Consequently, the subsequent Friedel-Crafts reaction would not then take place at the desired meta position.

The second method would fail for essentially the same reasons. Introducing the ethyl group first would introduce an ortho-para director, and subsequent ring chlorination would not take place at the desired meta position.

(b) If we introduce an acetyl group first, which we later convert to an ethyl group, we install a meta director. This allows us to put the chlorine atom in the desired position. Conversion of the acetyl group to an ethyl group is then carried out using the Clemmensen reduction.

O, Cl / $AlCl_3$ → O → Cl_2 / $FeCl_3$ → O, Cl → Zn(Hg) / HCl → Cl

15.18 (a) OH, NO_2, CF_3 (b) CN, O_2N, SO_3H (c) OCH_3, NO_2, NO_2 + OCH_3, O_2N, NO_2

15.19 (a) In concentrated base and ethanol (a relatively nonpolar solvent), the S_N2 reaction is favored. Thus, the rate depends on the concentration of both the alkyl halide and  ONa. Since no carbocation is formed, the only product is

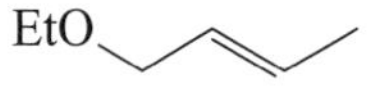

EtO

(b) When the concentration of 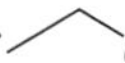 O⁻ ion is small or zero, the reaction occurs through the S_N1 mechanism. The carbocation that is produced in the first step of the S_N1 mechanism is a resonance hybrid.

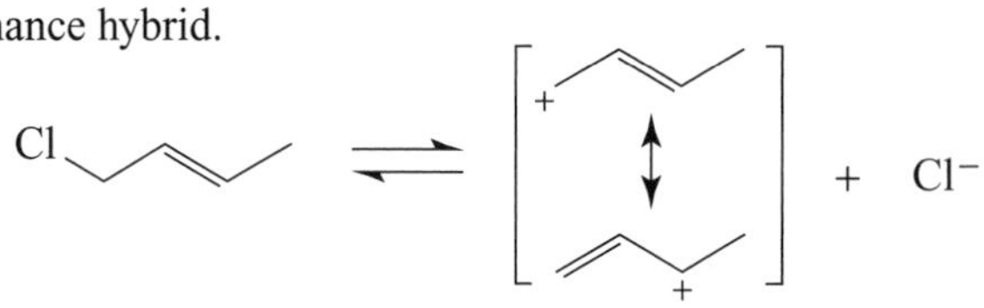

Cl ⇌ [+ ↕ +] + Cl^-

This ion reacts with the nucleophile (O⁻ or 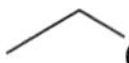 OH) to produce two isomeric ethers

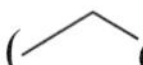

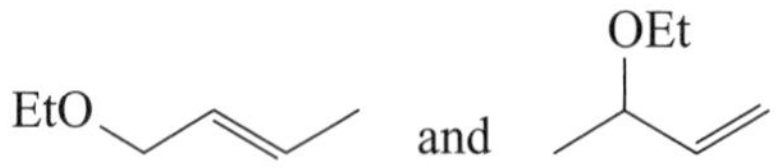

EtO and OEt

15.20 (a) The carbocation that is produced in the S_N1 reaction is exceptionally stable because one resonance contributor is not only allylic but also tertiary.

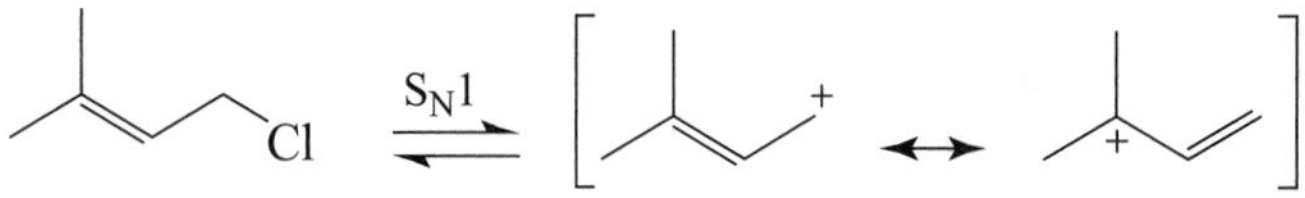

Cl $\underset{}{\overset{S_N1}{\rightleftharpoons}}$ [+ ⟷ +]

A 3° allylic carbocation

(b) OH + OH

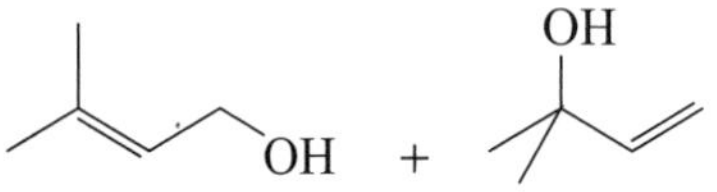

15.21 Compounds that undergo reactions by an S_N1 path must be capable of forming relatively stable carbocations. Primary halides of the type $ROCH_2X$ form carbocations that are stabilized by resonance:

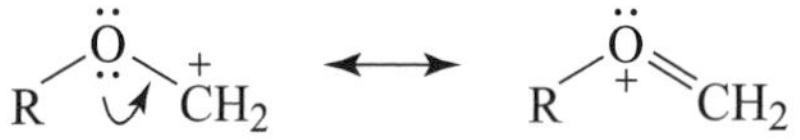

$R-\ddot{O}-\overset{+}{C}H_2 \longleftrightarrow R-\overset{+}{\ddot{O}}=CH_2$

15.22 The relative rates are in the order of the relative stabilities of the carbocations:

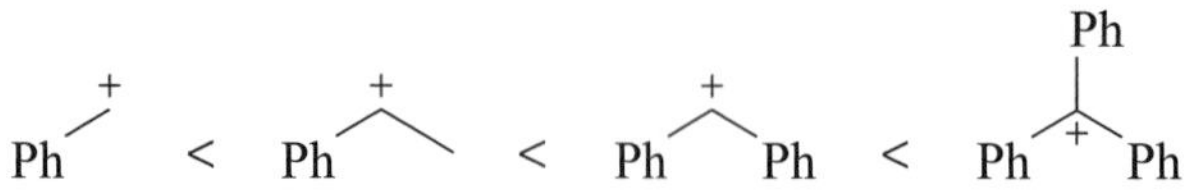

$Ph\overset{+}{C}H_2 < Ph\overset{+}{C}HCH_3 < Ph\overset{+}{C}HPh < Ph_3\overset{+}{C}$

The solvolysis reaction involves a carbocation intermediate.

15.23

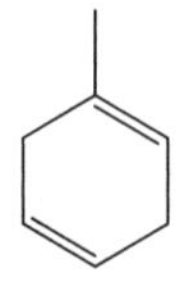

Mechanisms

15.24 (a) 1. Generate the electrophile

$HO-N^{+}(O^{-})=O + H-O-SO_2-O-H \rightleftharpoons H-O^{+}(H)-N^{+}(O^{-})=O \rightleftharpoons O=N^{+}=O + H_2O$

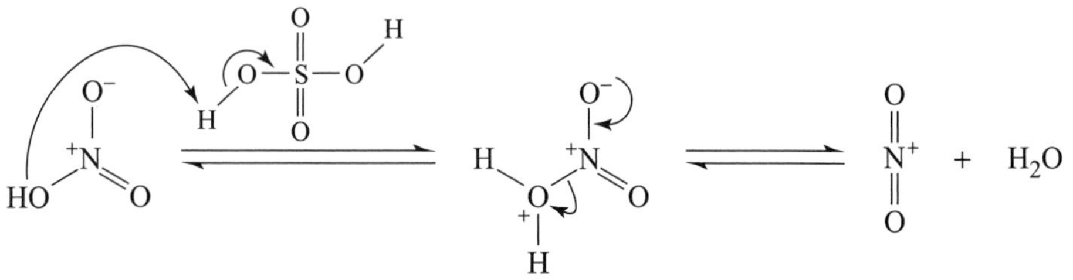

2. Attack the electrophile with aromatic ring to form the sigma complex.

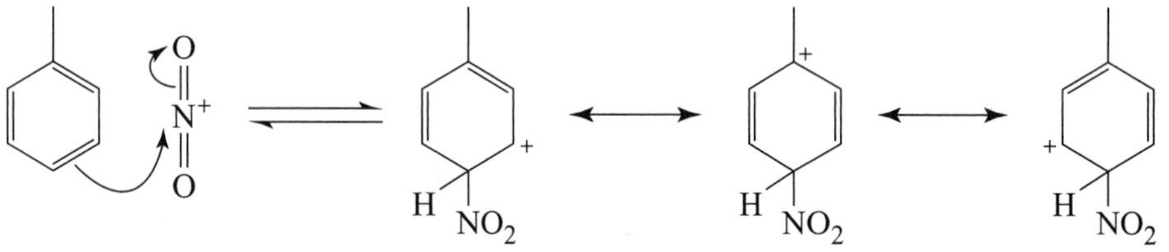

3. Elimination to regain aromaticity

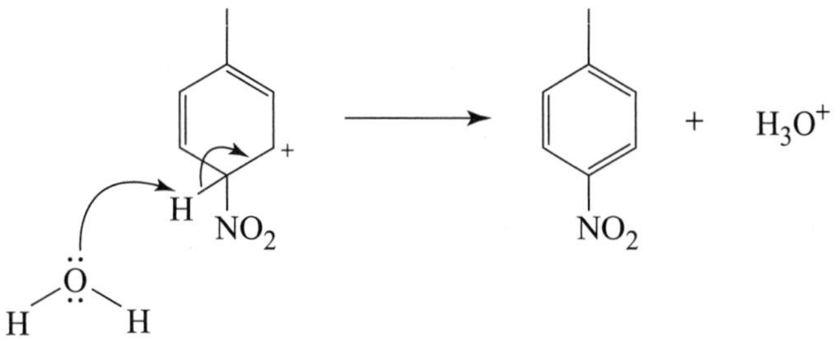

$\longrightarrow$ + H_3O^+

(b) 1. Generate the electrophile

$$\text{Br—Br} + \text{FeBr}_3 \longrightarrow \text{Br}^+ + \text{FeBr}_4^-$$

2. The aromatic ring attacks the electrophile to generate a resonance-stabilized sigma complex.

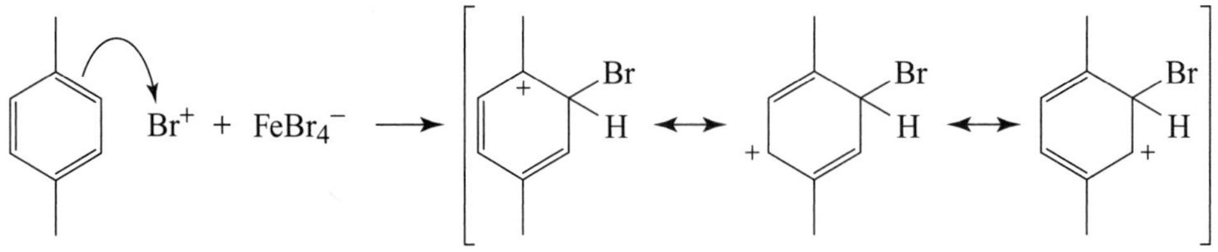

Br^+ + $FeBr_4^-$ ⟶ [Br H + ⟷ + Br H ⟷ Br H +]

3. Elimination to regain aromaticity

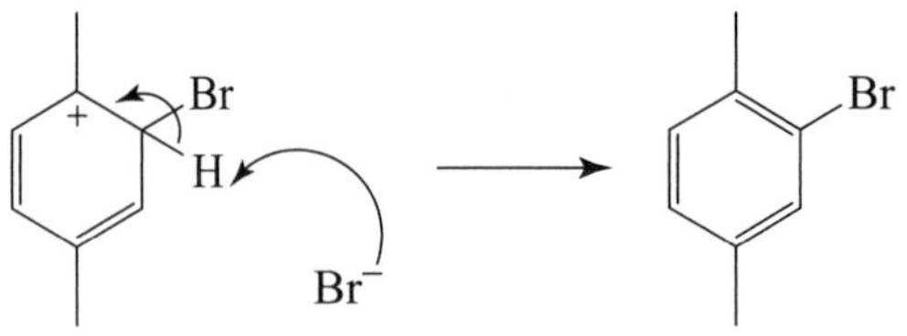

+ Br H Br^- ⟶ Br

(c) 1. Generate the electrophile

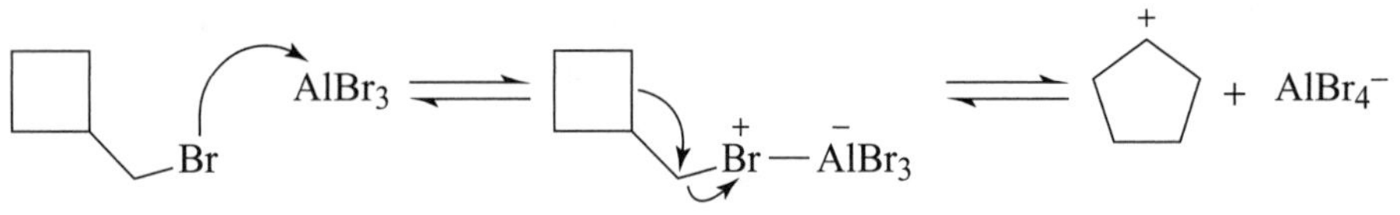

Br $AlBr_3$ ⇌ $\overset{+}{Br}$—$\overset{-}{Al}Br_3$ ⇌ + + $AlBr_4^-$

2. Attack the electrophile with aromatic ring to form the sigma complex.

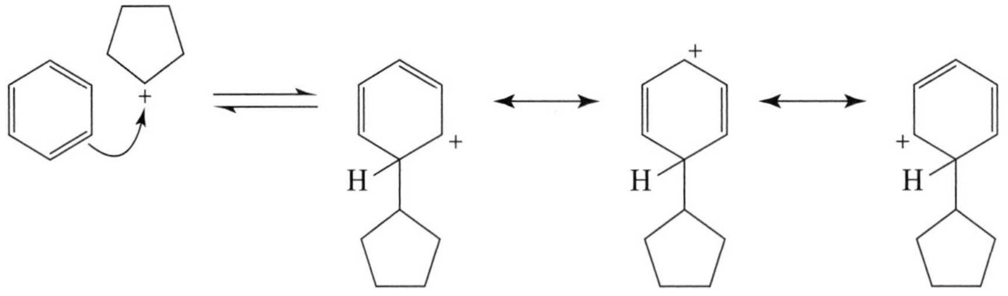

+ ⇌ H + ⟷ + H ⟷ + H

3. Elimination to regain aromaticity

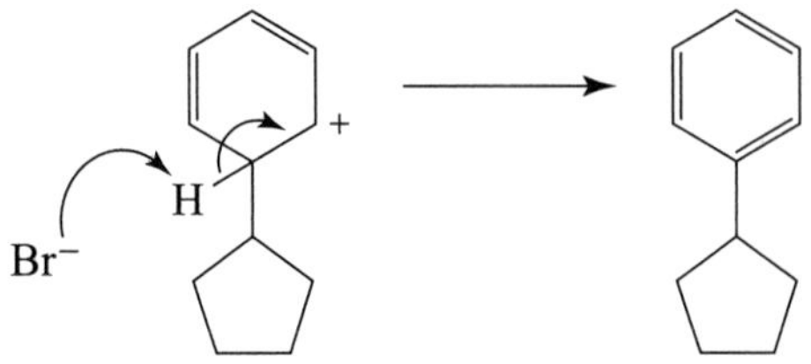

H

Br⁻

15.25

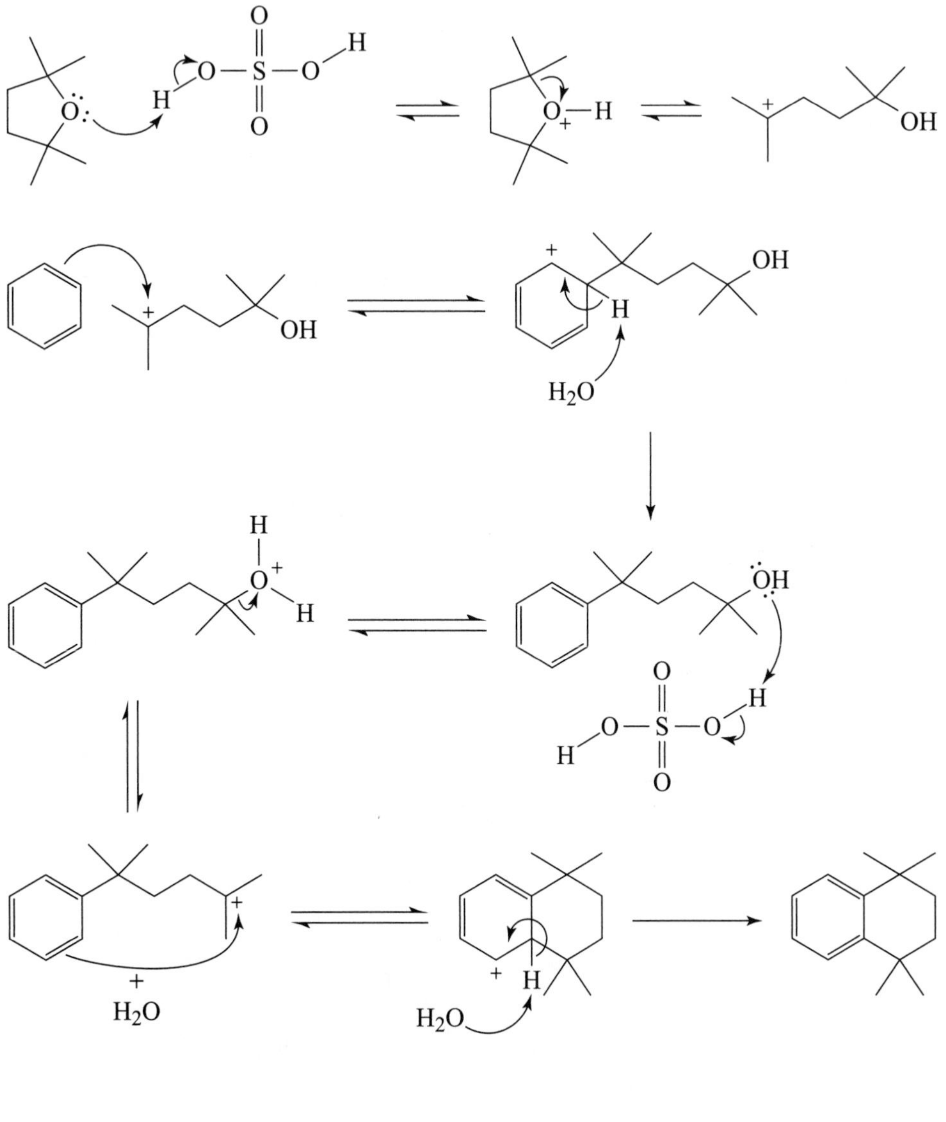

O

H

O—S—O

H

O

O—H

OH

OH

OH

H

H_2O

H

O

H

OH

O

H

H—O—S—O

O

+

H_2O

H

H_2O

15.26 (a) Electrophilic aromatic substitution will take place as follows:

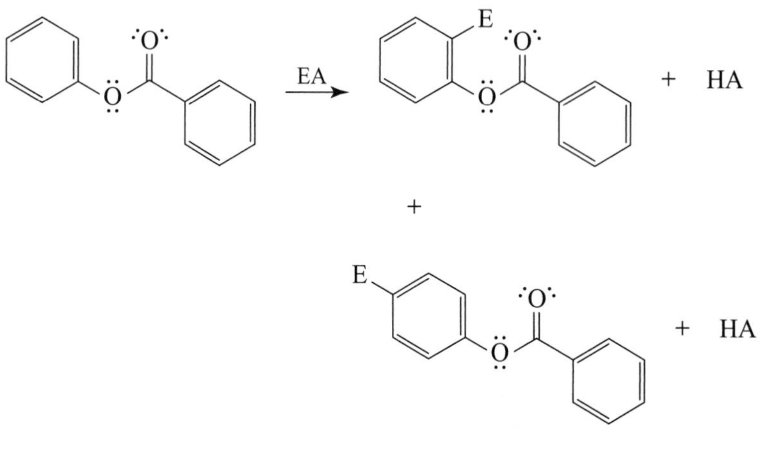

EA

+ HA

+

E

+ HA

(b) The ring directly attached to the oxygen atom is activated toward electrophilic attack because the oxygen atom can donate an unshared electron pair to it and stabilize the intermediate arenium ion when attack occurs at the ortho or para position.

15.27

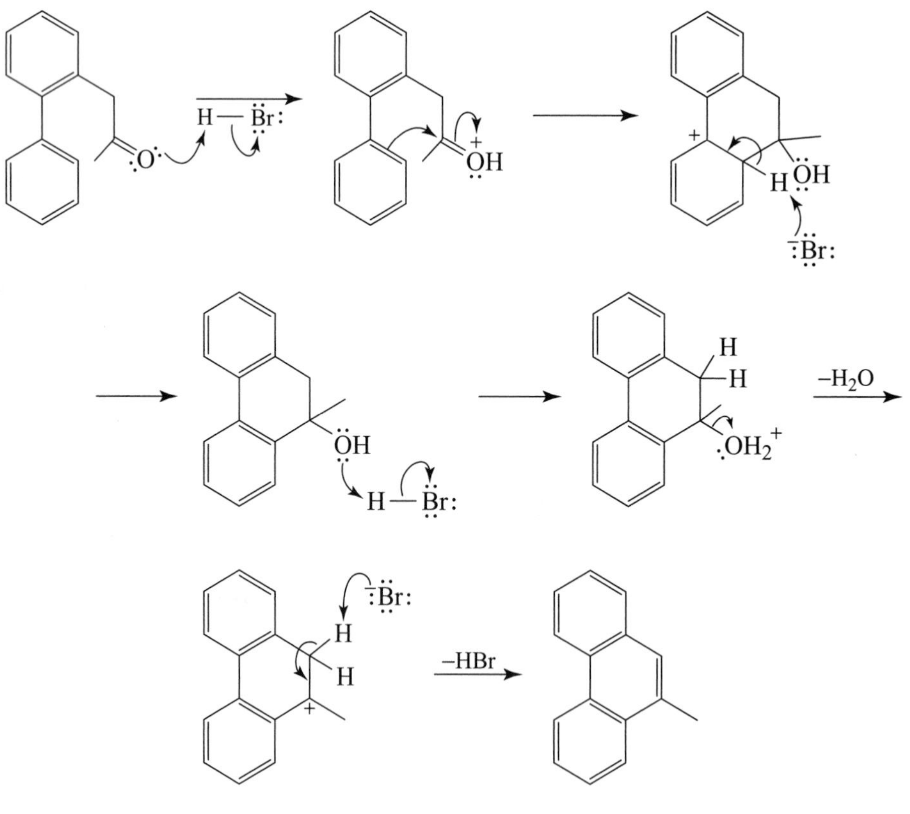

H—Br:

OH

H OH

:Br:

OH

H—Br:

H

H

OH_2^+

$-H_2O$

:Br:

H

H

–HBr

15.28 (a)

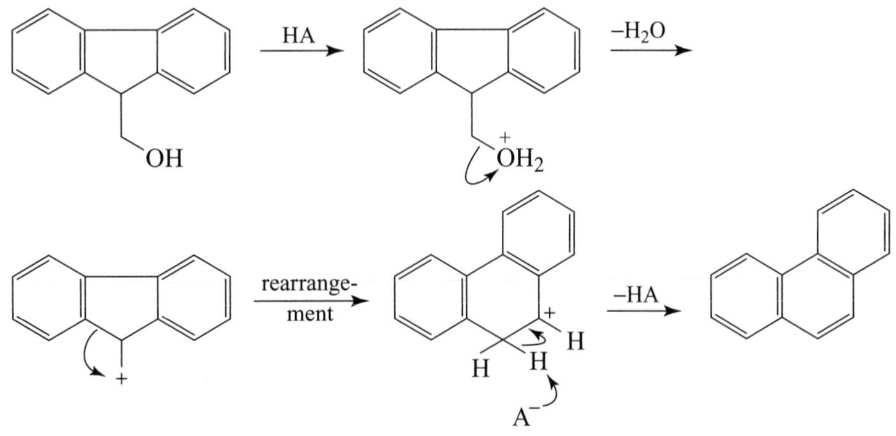

HA

$-H_2O$

OH

$\overset{+}{O}H_2$

rearrange-
ment

$-HA$

H H H

A^-

(b)

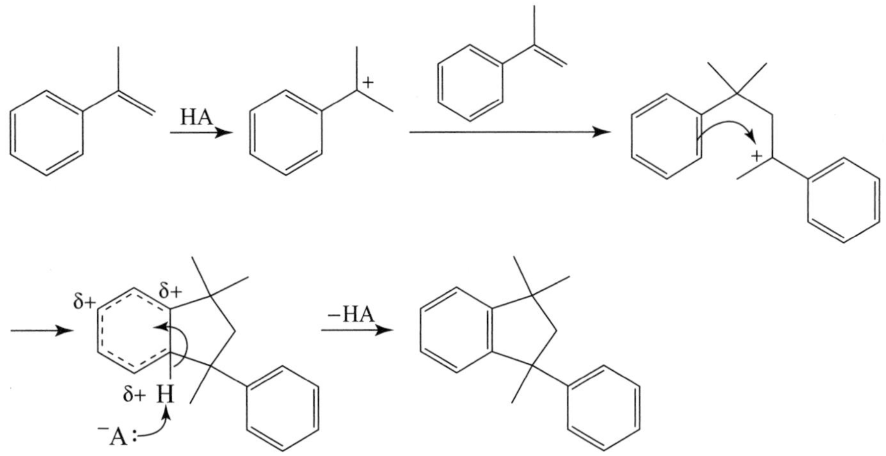

HA

+

δ+ δ+

$-HA$

δ+ H

$^-A:$

15.29 (a) C_6H_5 + H—X ⟶ C_6H_5 (+) ↕ C_6H_5 (+)

C_6H_5 δ^+ δ^+

C_6H_5 δ^+ δ^+ $\xrightarrow{X^-}$ C_6H_5 X

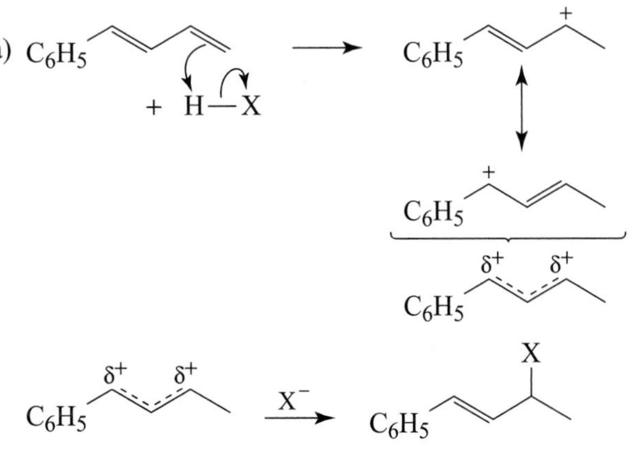

(b) 1,2 Addition.

(c) Yes. The carbocation given in (a) is a hybrid of *secondary allylic* and *benzylic* contributors and is therefore more stable than any other possibility; for example,

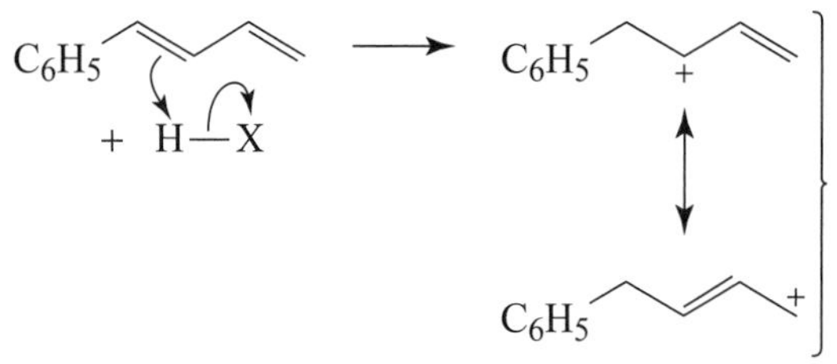

C_6H_5 + H—X ⟶ C_6H_5 (+) ⟷ C_6H_5 (+)

A hybrid of allylic contributors only

(d) Since the reaction produces only *the more stable isomer*—that is, the one in which the double bond is conjugated with the benzene ring—the reaction is likely to be under equilibrium control:

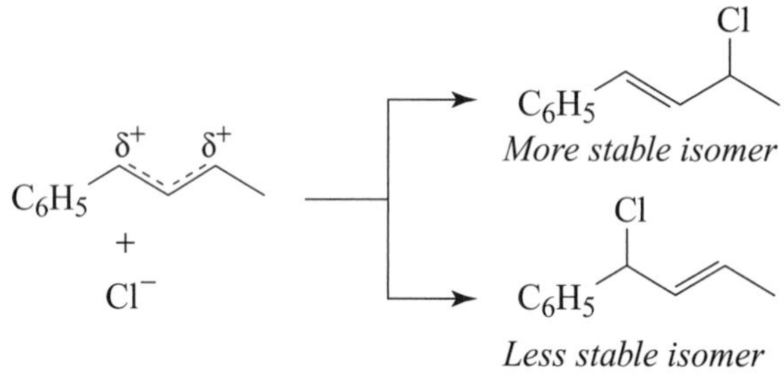

C_6H_5 (δ^+ δ^+) + Cl^- ⟶

Cl, C_6H_5 — *More stable isomer* — Actual product

Cl, C_6H_5 — *Less stable isomer* — Not formed

Reactions and Synthesis

15.30 (a) 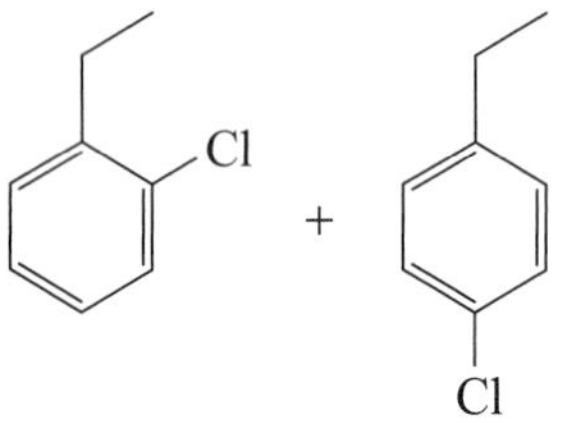 Cl + Cl (b) 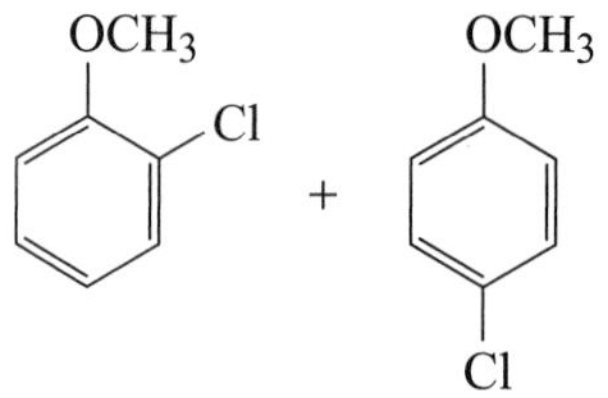 OCH_3, Cl + OCH_3, Cl

(c) 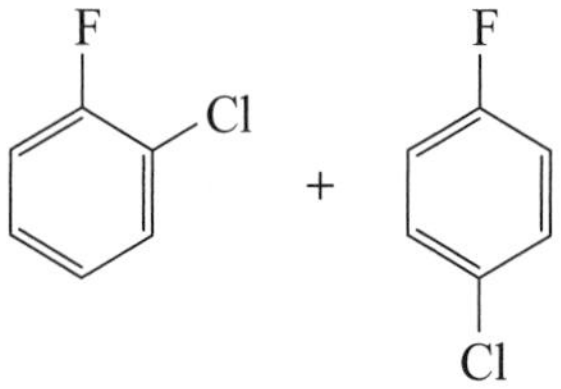 F, Cl + F, Cl (d) O, OH, Cl

(e) NO_2, Cl (f) Cl, Cl + Cl, Cl

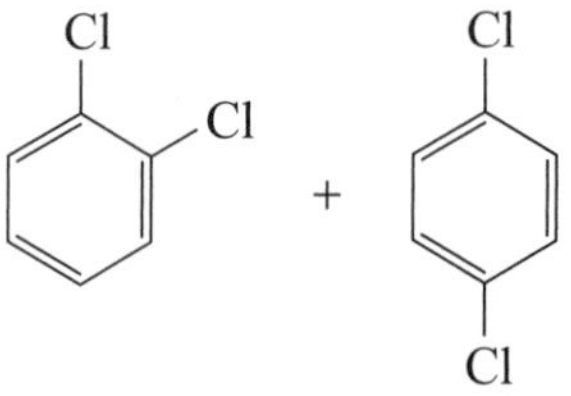

(g) 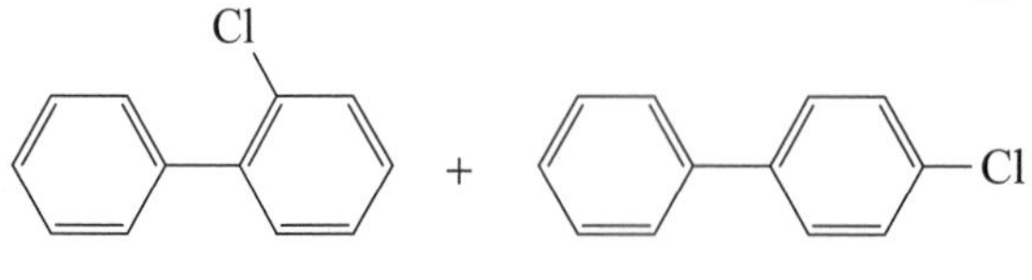 Cl + Cl

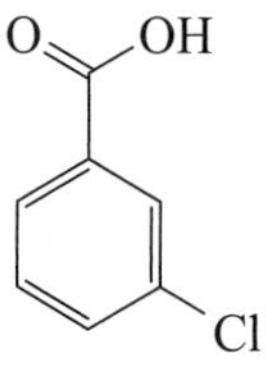

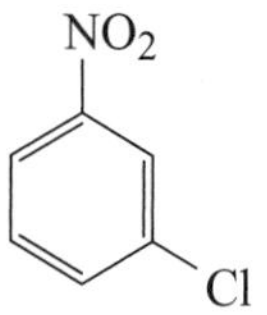

(h) 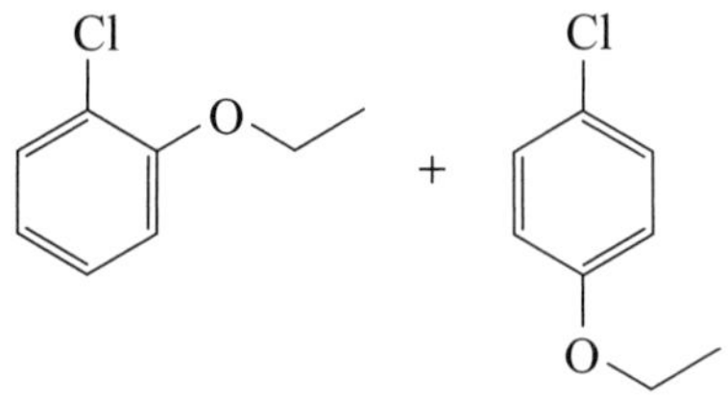 Cl, O + Cl, O

15.31 (a) 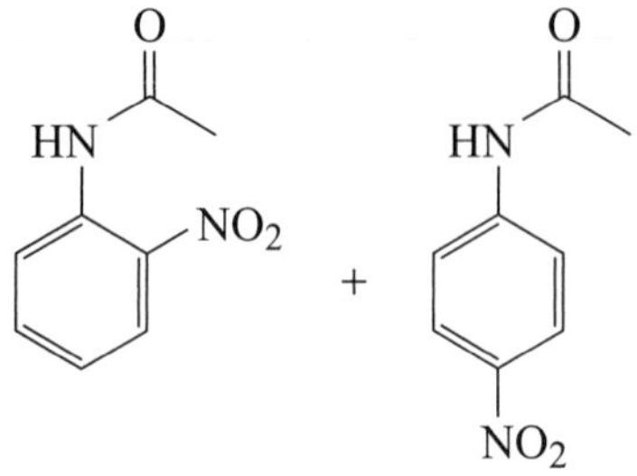 O, HN, NO_2 + O, HN, NO_2 (mainly) (b) 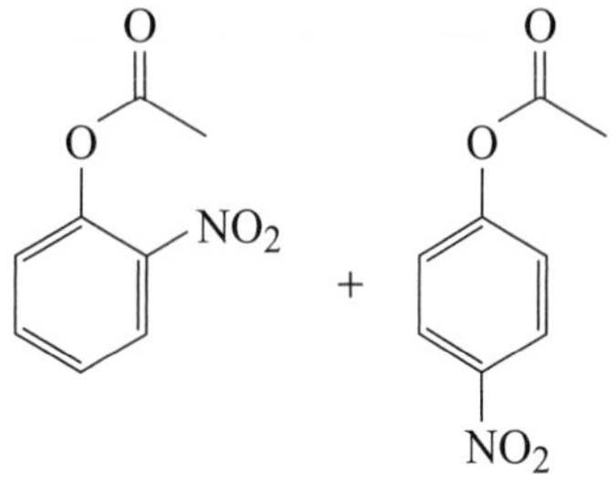 O, O, NO_2 + O, O, NO_2

(c) 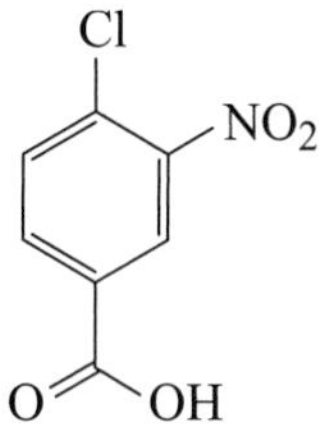 Cl, NO_2, O, OH (d) 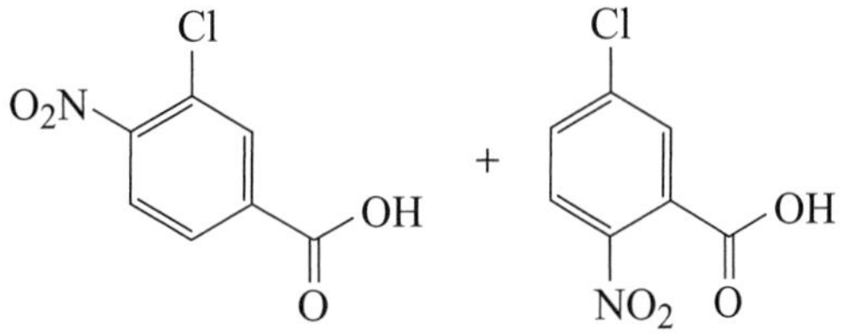 O_2N, Cl, OH, O + Cl, OH, NO_2, O

(e) 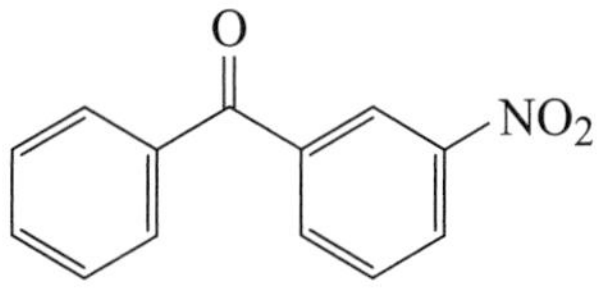 O, NO_2

15.32 (a) 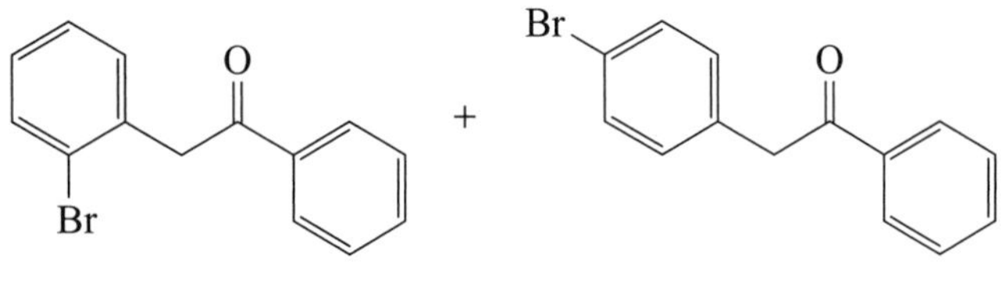 O, Br + Br, O

(b) 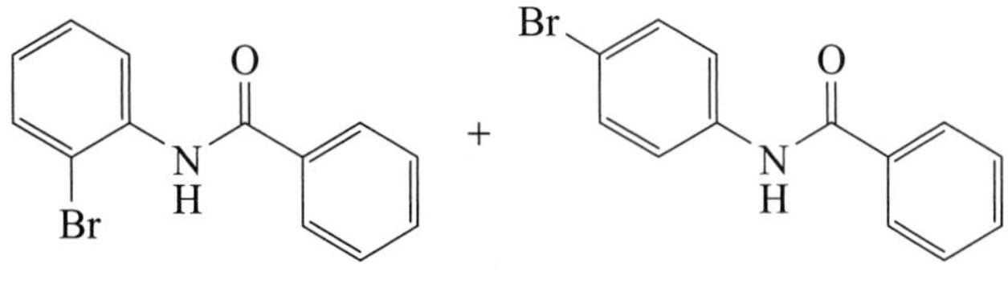 O, N, H, Br + Br, O, N, H

(c) 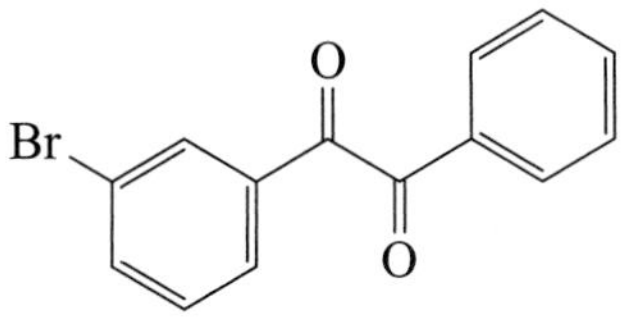 Br, O, O

15.33 (a) 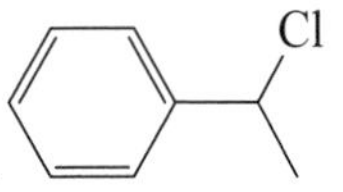 Cl (b) 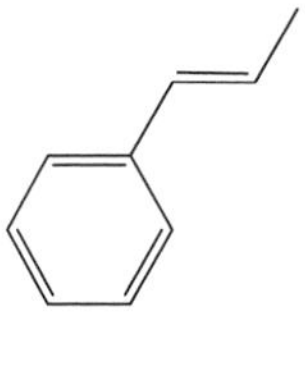 (c)

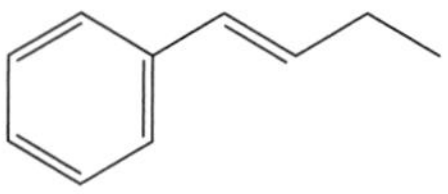

(d) 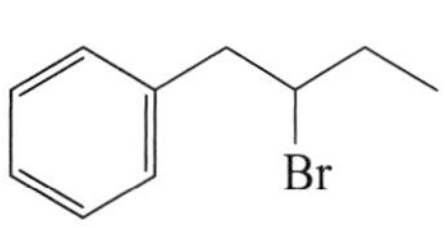 Br (e) 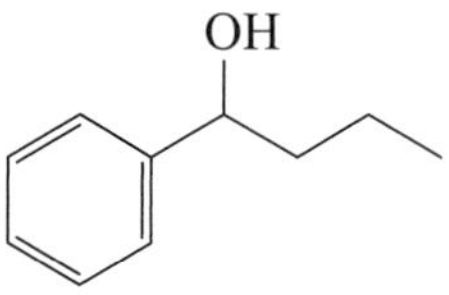 OH (f)

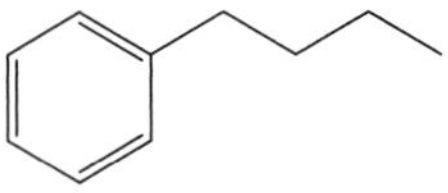

(g) 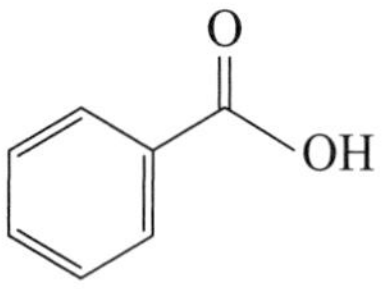 O OH

15.34 (a) 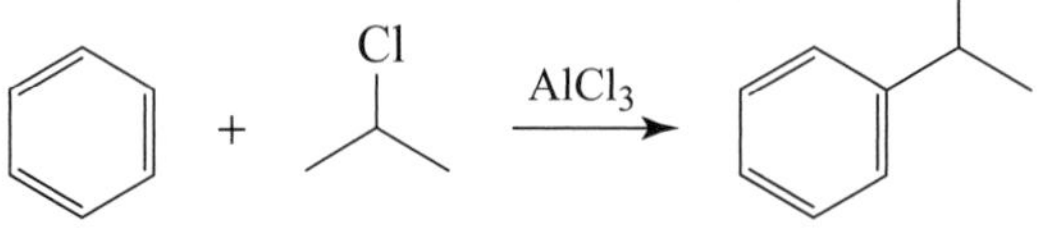 + Cl $\xrightarrow{AlCl_3}$

(b) 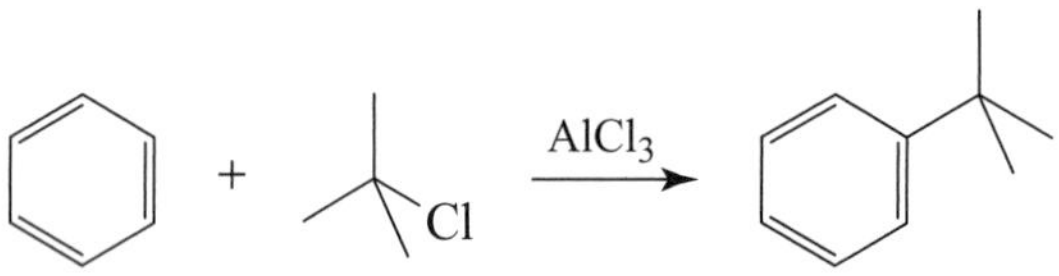 + Cl $\xrightarrow{AlCl_3}$

(c) 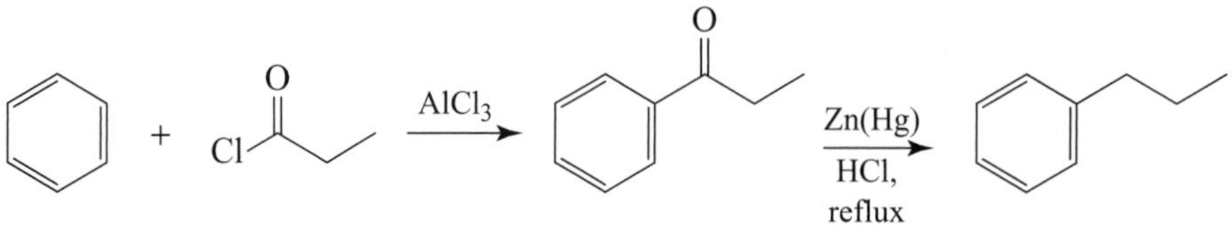 + Cl O $\xrightarrow{AlCl_3}$ O $\xrightarrow[\text{HCl, reflux}]{Zn(Hg)}$

(*Note:* The use of Cl in a Friedel-Crafts synthesis gives mainly the rearranged product, isopropylbenzene.)

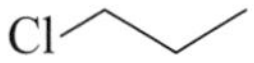

(d) 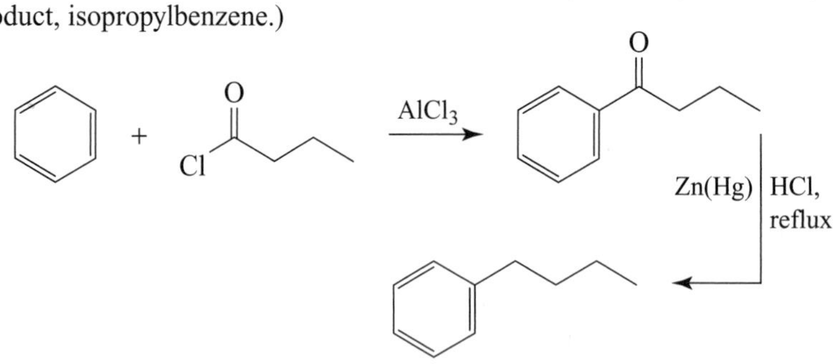 + O Cl $\xrightarrow{AlCl_3}$ O — Zn(Hg) HCl, reflux →

(e) 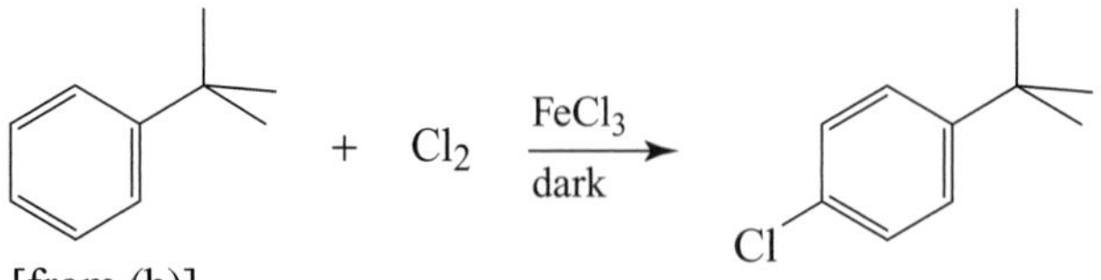 [from (b)] + Cl_2 $\xrightarrow[\text{dark}]{FeCl_3}$ Cl

(f) benzene + chlorocyclopentane $\xrightarrow{AlCl_3}$ phenylcyclopentane $\xrightarrow[hv]{Cl_2}$ 1-chloro-1-phenylcyclopentane $\xrightarrow[\text{heat}]{\text{EtONa, EtOH}}$ 1-phenylcyclopentene

Cl, $AlCl_3$, Cl_2, hv, ONa, OH, heat

(g) 1-phenylcyclopentene [from (f)] $\xrightarrow[\text{(2) } H_2O_2, OH^- \text{ (syn addition)}]{\text{(1) } BH_3{:}THF}$ product (H, OH, H) + enantiomer

(h) benzene + HNO_3 $\xrightarrow{H_2SO_4}$ nitrobenzene $\xrightarrow[H_2SO_4]{HNO_3}$ m-dinitrobenzene (NO_2, NO_2)

(i) nitrobenzene (NO_2) + Br_2 $\xrightarrow{FeBr_3}$ m-bromonitrobenzene (NO_2, Br)

(j) benzene + Br_2 $\xrightarrow{FeBr_3}$ bromobenzene (Br) $\xrightarrow[H_2SO_4]{HNO_3}$ p-bromonitrobenzene (NO_2, Br) + ortho isomer

(k) benzene + Cl_2 $\xrightarrow{FeCl_3}$ chlorobenzene (Cl) $\xrightarrow[H_2SO_4]{SO_3}$ p-chlorobenzenesulfonic acid (Cl, SO_3H) (+ o-isomer (Cl, SO_3H)) (separate)

ortho isomer $\xrightarrow{H_2O/H_2SO_4\text{, heat}}$ chlorobenzene

(l) Chlorobenzene [from (k)] $\xrightarrow[H_2SO_4]{SO_3}$ p-chlorobenzenesulfonic acid (Cl, SO_3H) (+ ortho isomer) $\xrightarrow[H_2SO_4]{HNO_3}$ 4-chloro-3-nitrobenzenesulfonic acid (Cl, NO_2, SO_3H) $\xrightarrow[heat]{H_2O/H_2SO_4}$ o-chloronitrobenzene (Cl, NO_2)

(m) Nitrobenzene [from (h)] $\xrightarrow[H_2SO_4]{SO_3}$ m-nitrobenzenesulfonic acid (NO_2, SO_3H)

15.35 (a) $C_6H_5CH{=}CH_2 \xrightarrow[CCl_4]{Cl_2} C_6H_5CHClCH_2Cl$

(b) $C_6H_5CH{=}CH_2 \xrightarrow[\text{Ni, pressure}]{H_2} C_6H_5CH_2CH_3$

(c) $C_6H_5CH{=}CH_2 \xrightarrow[\text{2) NaHSO}_3/H_2O]{\text{1) OsO}_4\text{, pyridine}} C_6H_5CH(OH)CH_2OH$

(d) $C_6H_5CH{=}CH_2 \xrightarrow[(2)\ H_3O^+]{(1)\ KMnO_4,\ OH^-,\ heat} C_6H_5COOH$

(e) $C_6H_5CH{=}CH_2 \xrightarrow[H_2SO_4]{H_2O} C_6H_5CH(OH)CH_3$

(f) $C_6H_5CH{=}CH_2 \xrightarrow[\text{no peroxides}]{HBr} C_6H_5CHBrCH_3$

(g) $C_6H_5CH{=}CH_2 \xrightarrow[(2)\ H_2O_2,\ OH^-]{(1)\ BH_3:THF} C_6H_5CH_2CH_2OH$

(h) $C_6H_5CH{=}CH_2 \xrightarrow[(2)\ CH_3COOD]{(1)\ BH_3:THF} C_6H_5CH_2CH_2D$

(i) $C_6H_5CH{=}CH_2 \xrightarrow[\text{peroxides}]{HBr} C_6H_5CH_2CH_2Br$

(j) 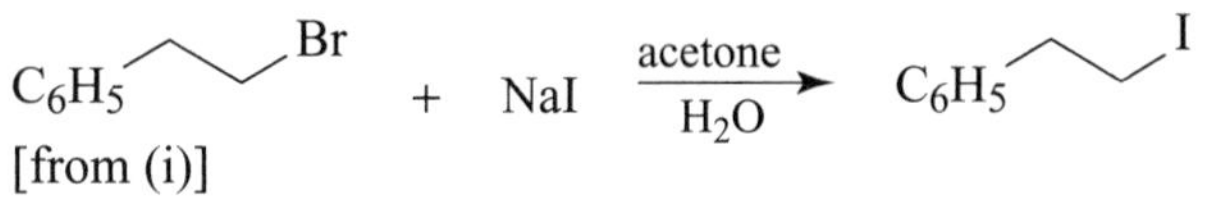 C_6H_5 Br [from (i)] + NaI $\xrightarrow[H_2O]{\text{acetone}}$ C_6H_5 I

(k) 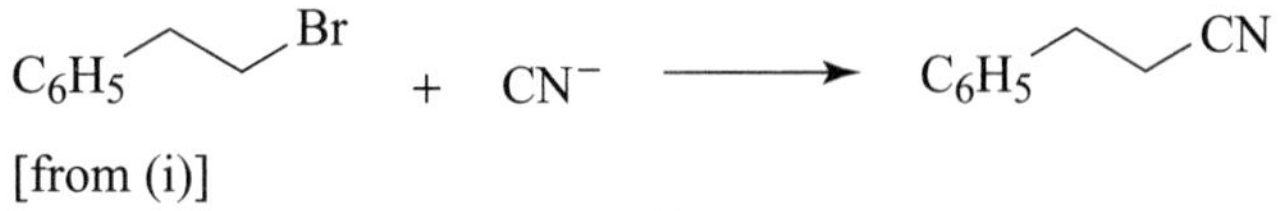 C_6H_5 Br [from (i)] + CN^- $\longrightarrow$ C_6H_5 CN

(l) 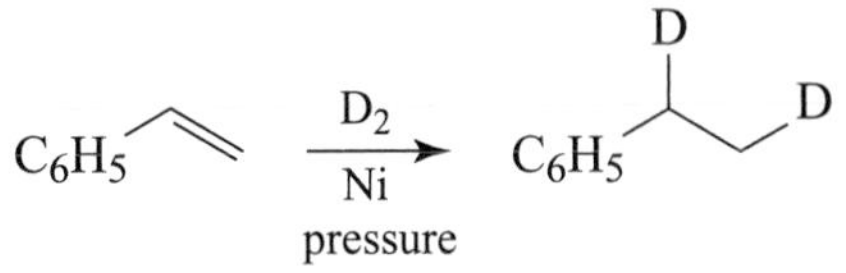 C_6H_5 $\xrightarrow[\text{Ni}]{D_2}$ pressure C_6H_5 D D

(m) 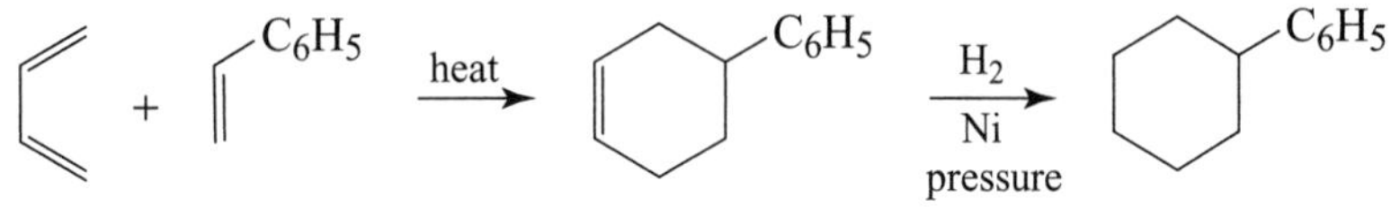 + C_6H_5 $\xrightarrow{\text{heat}}$ C_6H_5 $\xrightarrow[\text{Ni}]{H_2}$ pressure C_6H_5

(n) 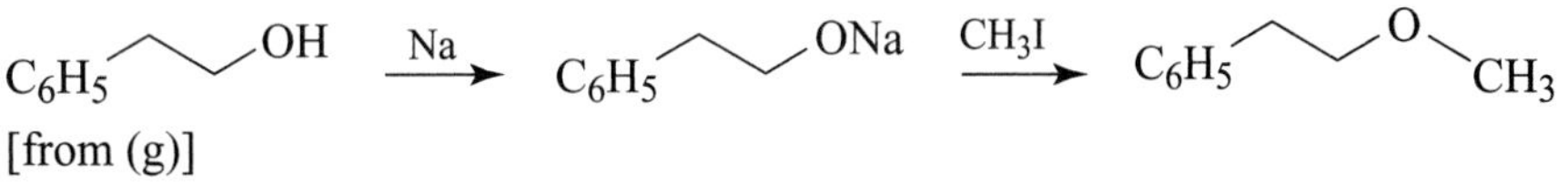 C_6H_5 OH [from (g)] $\xrightarrow{\text{Na}}$ C_6H_5 ONa $\xrightarrow{CH_3I}$ C_6H_5 O CH_3

15.36 (a) 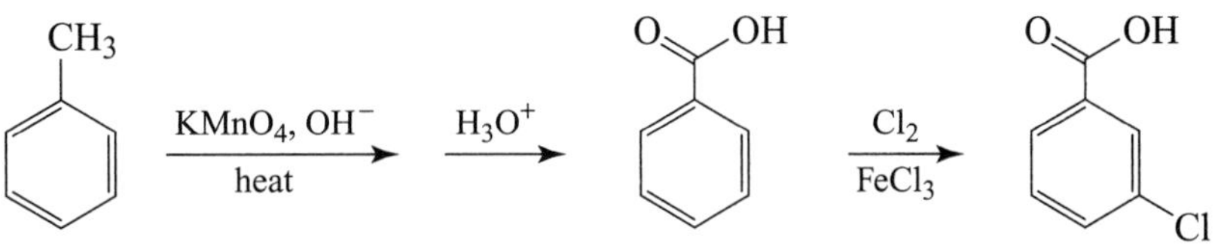 CH_3 $\xrightarrow[\text{heat}]{KMnO_4, OH^-}$ $\xrightarrow{H_3O^+}$ O OH $\xrightarrow[FeCl_3]{Cl_2}$ O OH Cl

(b) 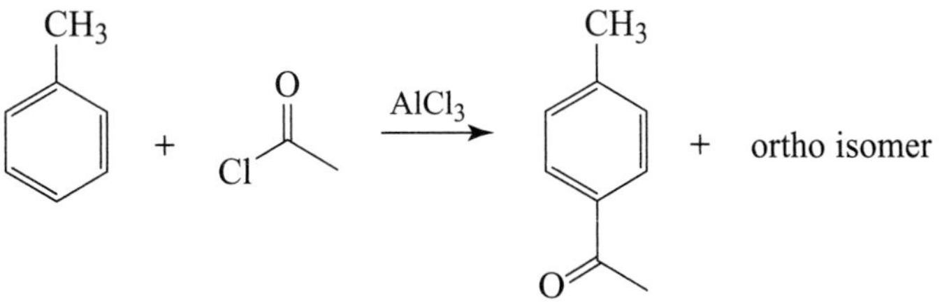 CH_3 + O Cl $\xrightarrow{AlCl_3}$ CH_3 O + ortho isomer

(c) 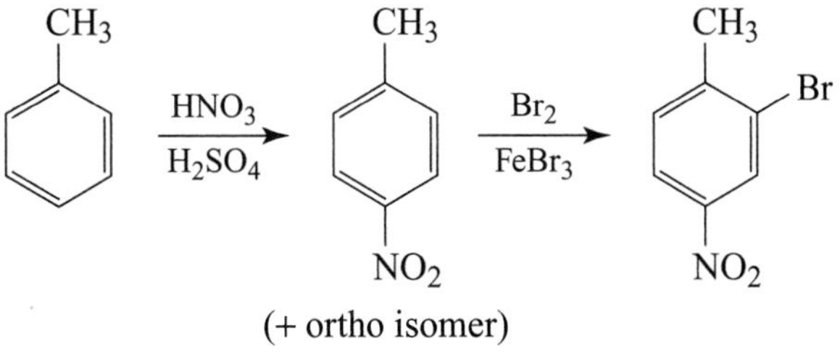 CH_3 $\xrightarrow[H_2SO_4]{HNO_3}$ CH_3 NO_2 (+ ortho isomer) $\xrightarrow[FeBr_3]{Br_2}$ CH_3 Br NO_2

(d) 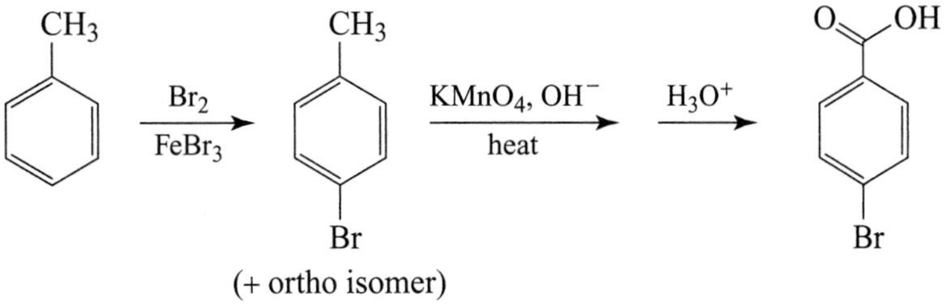 CH_3 $\xrightarrow[FeBr_3]{Br_2}$ CH_3 Br (+ ortho isomer) $\xrightarrow[\text{heat}]{KMnO_4, OH^-}$ $\xrightarrow{H_3O^+}$ O OH Br

(e) CH_3 → Cl_2 (excess), light → CCl_3 → Cl_2, $FeCl_3$ → CCl_3, Cl

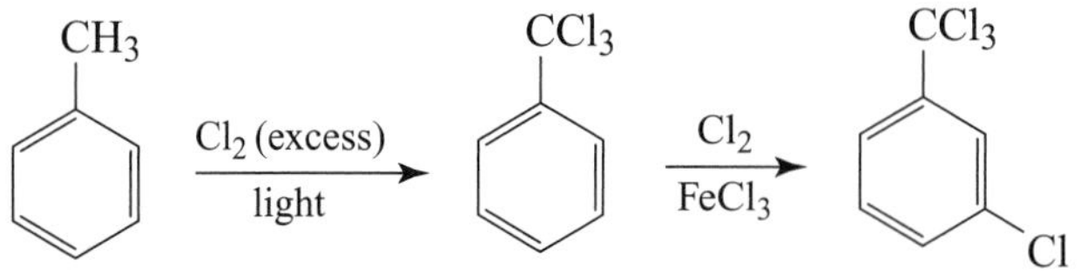

(f) CH_3 + Cl → $AlCl_3$ → CH_3 + ortho isomer

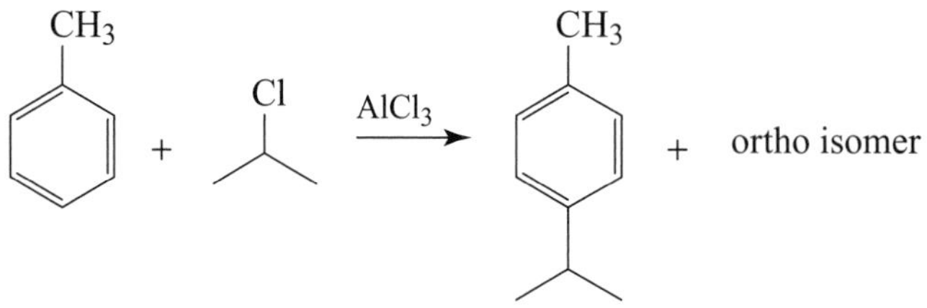

(g) CH_3 + Cl → $AlCl_3$ → H_3C— + ortho isomer

(h) CH_3 → HNO_3 (excess), H_2SO_4 → CH_3, O_2N, NO_2, NO_2

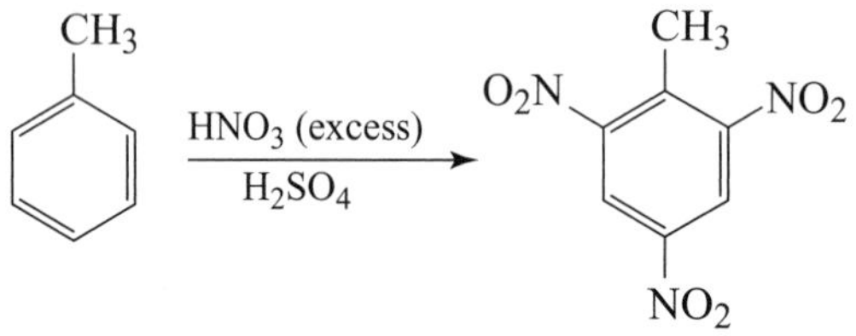

(i) CH_3 → SO_3, H_2SO_4 → CH_3, SO_3H (+ ortho isomer) → HNO_3, H_2SO_4 → CH_3, NO_2, SO_3H → H_2O/H_2SO_4, heat → CH_3, NO_2 → Cl_2, $FeCl_3$ → CH_3, NO_2, Cl → (1) $KMnO_4$, OH^-, heat (2) H_3O^+ → O, OH, NO_2, Cl

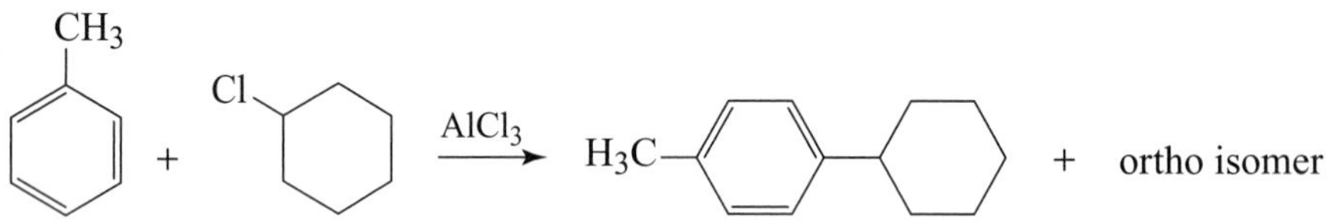

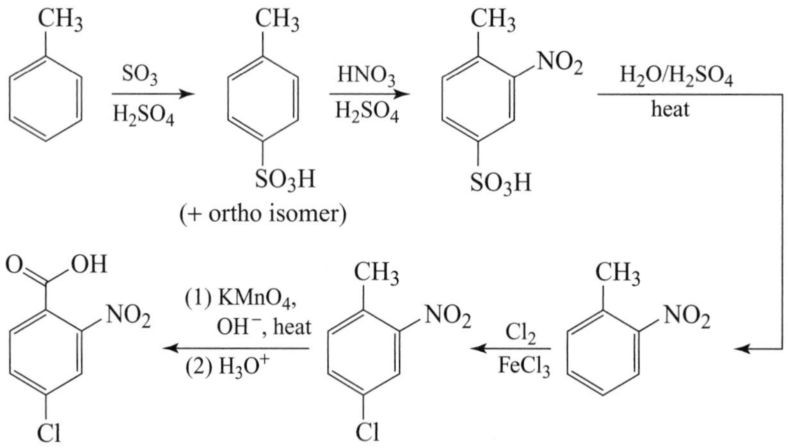

(j) CH_3 + O, Cl → $AlCl_3$ → CH_3, O (+ ortho isomer) → Zn(Hg), HCl → CH_3

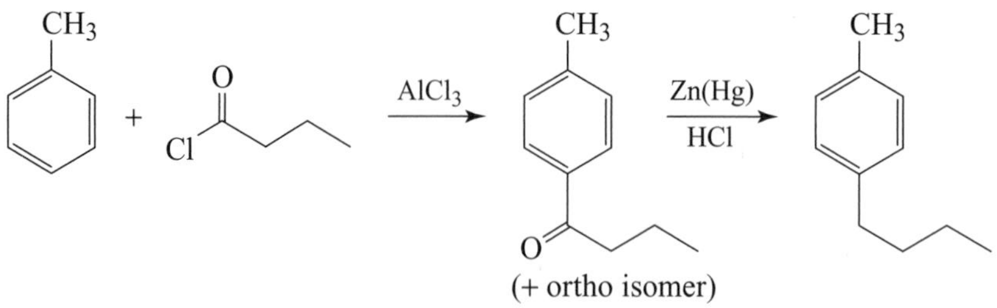

15.37 (a) Aniline (excess) $\xrightarrow{CH_3COCl}$ acetanilide $\xrightarrow[FeBr_3]{Br_2}$ 4-bromoacetanilide (+ ortho isomer) $\xrightarrow[H_2O \; (2)\; OH^-]{(1)\; H_3O^+}$ 4-bromoaniline

(b) Acetanilide [from (a)] $\xrightarrow[H_2SO_4]{SO_3}$ 2-acetamidobenzenesulfonic acid (minor product) + 4-acetamidobenzenesulfonic acid (major product) $\xrightarrow[FeBr_3]{Br_2}$ 3-bromo-4-acetamidobenzenesulfonic acid $\xrightarrow[heat \; (2)\; OH^-]{(1)\; H_2O/H_2SO_4}$ 2-bromoaniline

(c) 2-Bromoaniline (excess) [from (b)] $\xrightarrow{CH_3COCl}$ 2-bromoacetanilide $\xrightarrow[H_2SO_4]{HNO_3}$ 2-bromo-4-nitroacetanilide $\xrightarrow[H_2O \; (2)\; OH^-]{(1)\; H_3O^+}$ 2-bromo-4-nitroaniline

(d) 4-Acetamidobenzenesulfonic acid [from (b)] $\xrightarrow[H_2SO_4]{HNO_3}$ 4-acetamido-3-nitrobenzenesulfonic acid $\xrightarrow[FeBr_3]{Br_2}$ 4-bromo-2-nitroacetanilide $\xrightarrow[H_2O,\ heat \; (2)\; OH^-]{(1)\; H_3O^+}$ 4-bromo-2-nitroaniline

(e) $\xrightarrow[H_2O]{Br_2}$

NH$_2$ Br Br Br

15.38 (a) Step (2) will fail because a Friedel-Crafts reaction will not take place on a ring that bears an —NO$_2$ group (or any meta director).

$\xrightarrow[H_2SO_4]{HNO_3}$ NO$_2$ $\xrightarrow[AlCl_3]{}$ O Cl no Friedel-Crafts reaction

(b) The last step will first brominate the double bond.

$\xrightarrow[\text{CCl}_4,\ \text{light}]{\text{NBS}}$ Br $\xrightarrow[\text{OH, heat}]{\text{ONa}}$ $\xrightarrow[FeBr_3]{Br_2}$

Br Br + Br Br Br + Br Br Br

15.39 This problem serves as another illustration of the use of a sulfonic acid group as a blocking group in a synthetic sequence. Here we are able to bring about nitration between two meta substituents.

OH OH $\xrightarrow[60\text{–}65°C]{\text{concd } H_2SO_4}$ HO$_3$S OH OH SO$_3$H $\xrightarrow[\text{concd } H_2SO_4]{\text{concd } HNO_3}$

G

HO$_3$S OH NO$_2$ OH SO$_3$H $\xrightarrow[\text{heat}]{H_3O^+,\ H_2O}$ OH NO$_2$ OH

H **I**

15.40

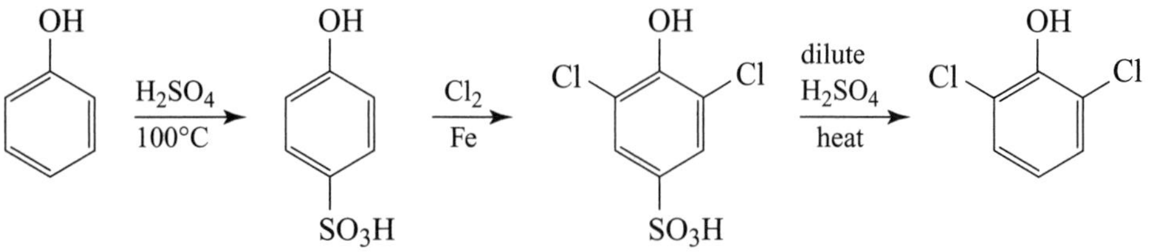

OH

H_2SO_4 / 100°C

OH, SO_3H

Cl_2 / Fe

OH, Cl, Cl, SO_3H

dilute H_2SO_4 / heat

OH, Cl, Cl

15.41

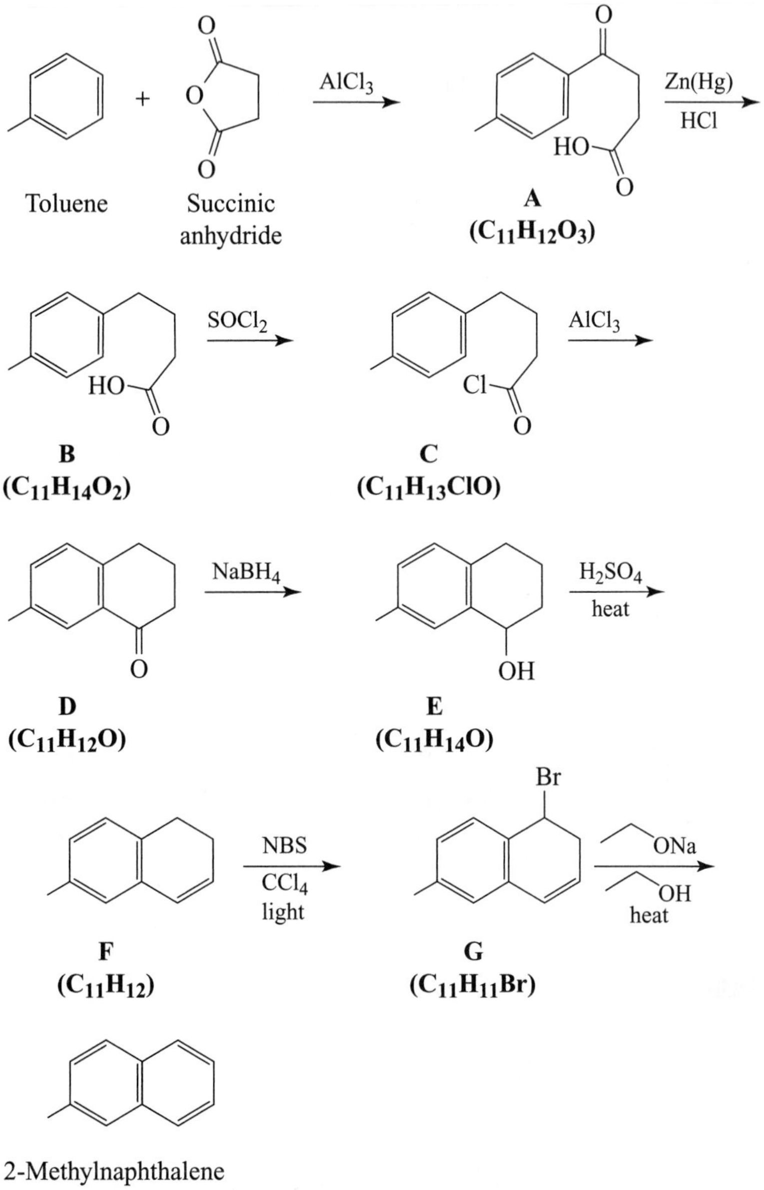

Toluene + Succinic anhydride (O, O, O)

$AlCl_3$

O, HO, O

A
($C_{11}H_{12}O_3$)

Zn(Hg) / HCl

HO, O

B
($C_{11}H_{14}O_2$)

$SOCl_2$

Cl, O

C
($C_{11}H_{13}ClO$)

$AlCl_3$

O

D
($C_{11}H_{12}O$)

$NaBH_4$

OH

E
($C_{11}H_{14}O$)

H_2SO_4 / heat

F
($C_{11}H_{12}$)

NBS / CCl_4 / light

Br

G
($C_{11}H_{11}Br$)

ONa / OH / heat

2-Methylnaphthalene

15.42 (a) $\xrightarrow[\text{HCl}]{\text{Zn(Hg)}}$

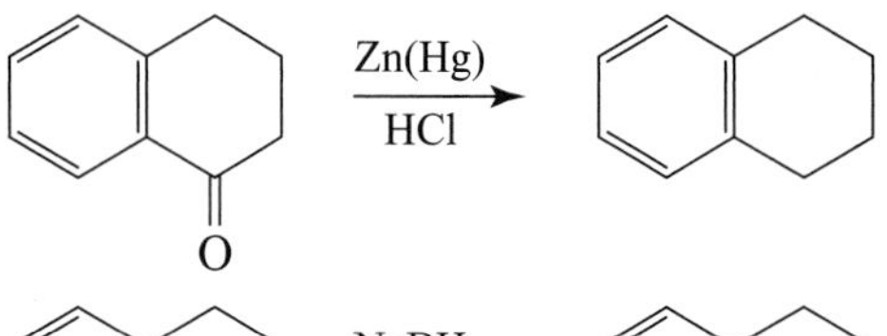

(b) $\xrightarrow[\text{OH}^-]{\text{NaBH}_4}$ OH

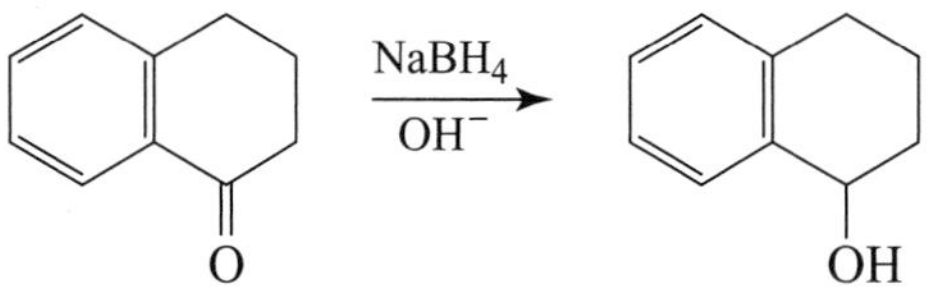

(c) $\xrightarrow[\text{(2) NH}_4^+]{\text{(1) CH}_3\text{MgI}}$ OH

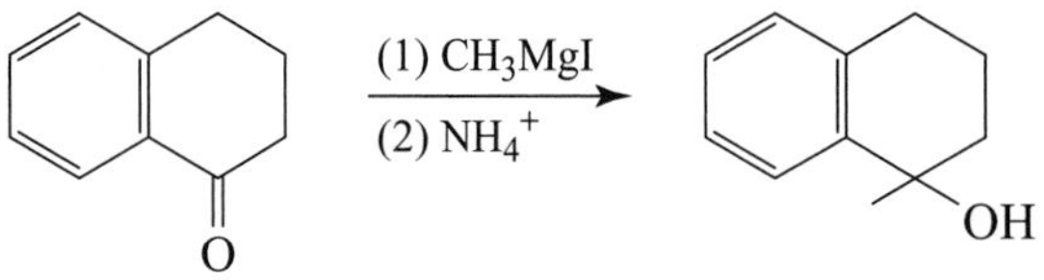

(d) $\xrightarrow[\text{(2) NH}_4^+]{\text{(1) C}_6\text{H}_5\text{MgBr}}$ C_6H_5 OH $\xrightarrow[\text{heat}]{\text{HA}}$ C_6H_5

$\xrightarrow[\text{pressure}]{\text{H}_2\text{/Ni}}$ C_6H_5

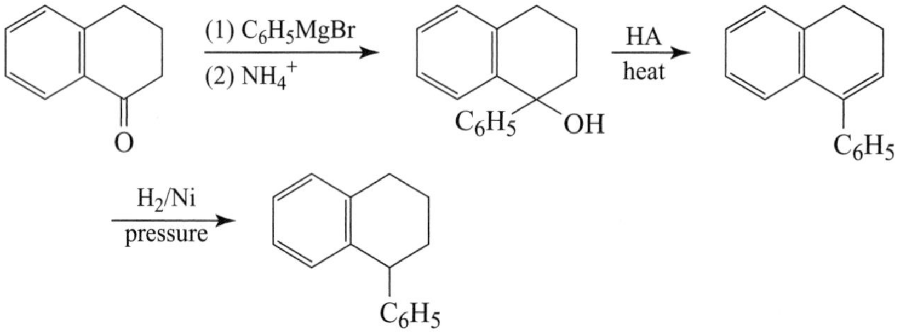

15.43 (a) 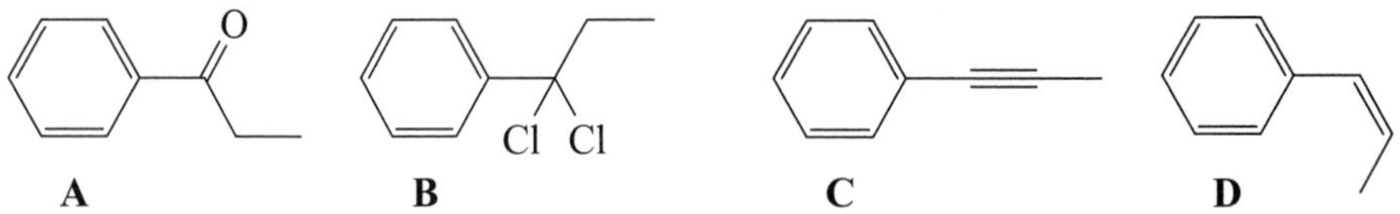 **A** (O) **B** (Cl Cl) **C** **D**

(b) **E** (c) Br, H, H, CH_3, C_6H_5, Br **F** (d) Br, CH_3, H, H, C_6H_5, Br **G**

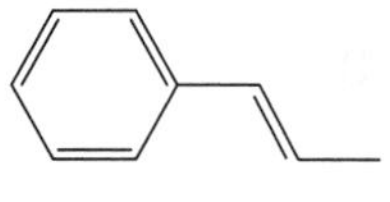

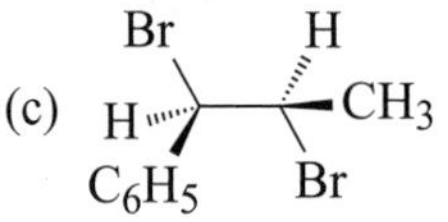

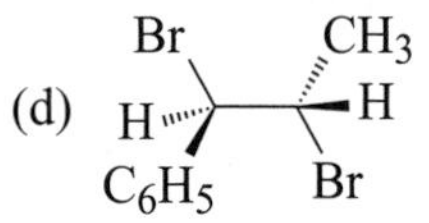

General Problems

15.44 (a) C_6H_5 Br $\xrightarrow[\text{DMF}]{\text{NaCN}}$ C_6H_5 CN
(−NaBr)

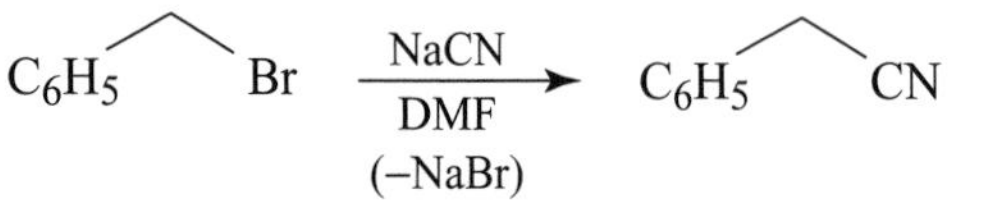

(b) C_6H_5 Br $\xrightarrow[\text{CH}_3\text{OH}]{\text{CH}_3\text{ONa}}$ C_6H_5 OCH_3
(−NaBr)

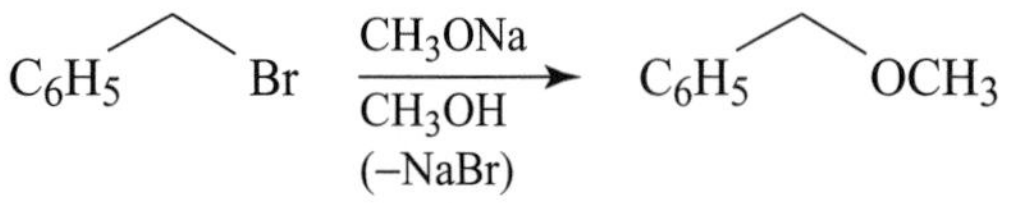

(c) $C_6H_5CH_2Br \xrightarrow[\text{(−NaBr)}]{CH_3CO_2Na,\ CH_3CO_2H} C_6H_5CH_2OC(O)CH_3$

(d) $C_6H_5CH_2CH_2Br \xrightarrow[\text{acetone (−NaBr)}]{NaI} C_6H_5CH_2CH_2I$

(e) $CH_2{=}CHCH_2Br \xrightarrow[\text{acetone (−NaBr)}]{NaN_3} CH_2{=}CHCH_2N_3$

(f) $CH_2{=}CHCH_2Br \xrightarrow[\text{(−NaBr)}]{(CH_3)_2CHCH_2ONa,\ (CH_3)_2CHCH_2OH} CH_2{=}CHCH_2OCH_2CH(CH_3)_2$

15.45 **A** = (cyclohexa-1,3-diene) **B** = (bromocyclohexadiene)

15.46 p-xylene (CH_3, CH_3) $\xrightarrow[H_2SO_4]{HNO_3}$ 2-nitro-p-xylene (CH_3, NO_2, CH_3) (only possible mononitro product)

15.47 (a) Large ortho substituents prevent the two rings from becoming coplanar and prevent rotation about the single bond that connects them. If the correct substitution patterns are present, the molecule as a whole will be chiral. Thus, enantiomeric forms are possible even though the molecules do not have a chirality center. The compound with 2-NO_2, 6-CO_2H, 2′-NO_2, 6′-CO_2H is an example.

(two enantiomeric biphenyl structures bearing O_2N, HO, O, NO_2, OH groups) and (mirror image)

The molecules are atropisomers (see Section 5.18). Furthermore, they are nonsuperposable mirror images and, thus, are enantiomers.

(b) Yes

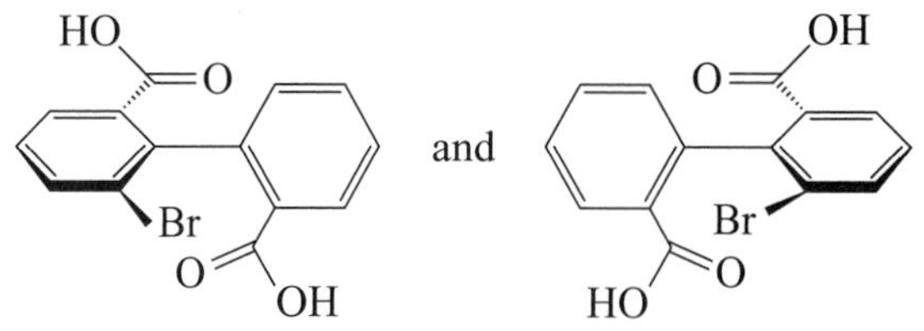

These molecules are enantiomeric atropisomers (Section 5.18).

(c) This molecule has a plane of symmetry; hence, its mirror image forms are equivalent.

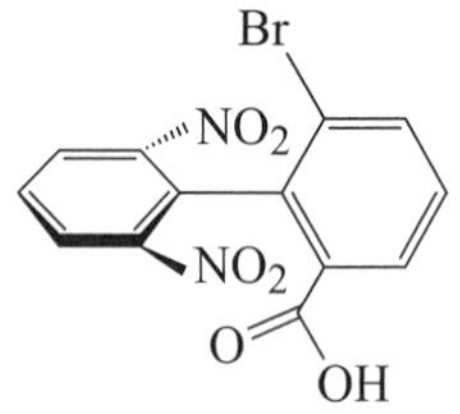

The plane of the page is a plane of symmetry.

15.48 $+ \; AlCl_3 \longrightarrow \; + \; AlCl_4^-$

$\xrightarrow{AlCl_4^-}$

$\xrightarrow{:A^-} \; + \; HA \; + \; Cl^-$

$C_8H_{13}OCl$

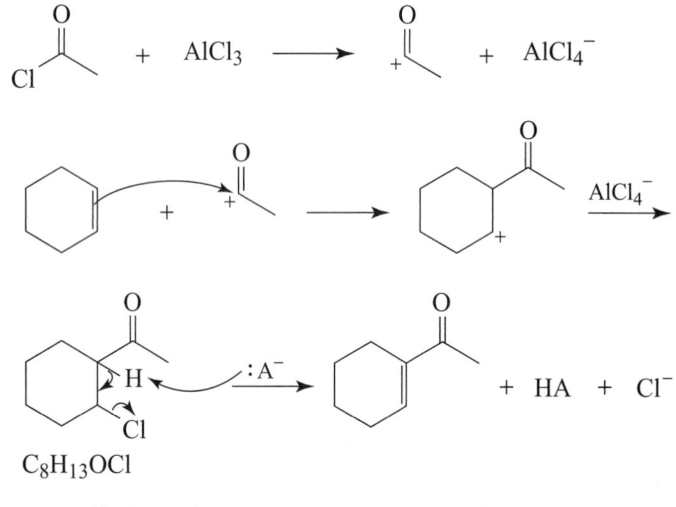

15.49 (a and b) The *tert*-butyl group is easily introduced by any of the variations of the Friedel-Crafts alkylation reaction, and, because of the stability of the *tert*-butyl cation, it is easily removed under acidic conditions.

(c) In contrast to the $-SO_3H$ group often used as a blocking group, $-C(CH_3)_3$ activates the ring to further electrophilic substitution.

15.50 At the lower temperature, the reaction is kinetically controlled, and the usual o/p directive effects of the $-CH_3$ group are observed. At higher temperatures, the reaction is thermodynamically controlled. At reaction times long enough for equilibrium to be reached, the most stable isomer, *m*-toluenesulfonic acid, is the principal product.

15.51 The evidence indicates that the mechanistic step in which the C—H bond is broken is *not* rate determining. (In the case cited, it makes no difference kinetically if a C—H or C—D bond is broken in electrophilic aromatic substitution.) This evidence is consistent with the two-step mechanism given in Section 15.2. The step in which the aromatic compound reacts with the electrophile (NO_2^+) is the slow rate-determining step. Proton (or deuteron) loss from the arenium ion to return to an aromatic system is a rapid step and has no effect on the overall rate.

15.52

$(C_6H_5)_3C{-}OH \underset{:A^-}{\overset{HA}{\rightleftharpoons}} (C_6H_5)_3C{-}\overset{+}{O}H_2 \underset{+H_2O}{\overset{-H_2O}{\rightleftharpoons}} (C_6H_5)_3C^+$ (Very stable carbocation) $\overset{+\ CH_3CH_2OH}{\underset{-\ CH_3CH_2OH}{\rightleftharpoons}}$

$(C_6H_5)_3C{-}\overset{+}{O}(H){-}CH_2CH_3 \underset{HA}{\overset{:A^-}{\rightleftharpoons}} (C_6H_5)_3C{-}O{-}CH_2CH_3$

15.53 (a) $CH_3CH_2CH{=}CHCH_2Br$ would be the most reactive in an S_N2 reaction because it is a 1° allylic halide. There would, therefore, be less steric hindrance to the attacking nucleophile.

(b) $CH_2{=}CH{-}C(CH_3)_2Br$ would be the most reactive in an S_N1 reaction because it is a 3° allylic halide. The carbocation formed in the rate-determining step, being both 3° and allylic, would be the most stable.

Challenge Problems

15.54 Resonance structures for adding the E^+ to the 2 position

[three resonance structures of the 2-substituted furan arenium ion: + at C3 ⟷ + at C5 ⟷ O^+]

Resonance structures for adding the E^+ to the 3 position

[two resonance structures of the 3-substituted furan arenium ion: + at C2 ⟷ O^+]

There are more resonance structures for substitution at the 2 position of furan. The pathway for EAS should be lower in energy for substitution at the 2 position over the 3 position.

15.55 The final product is *o*-nitroaniline. (The reactions are given in Section 15.14A.) The presence of six signals in the ^{13}C NMR spectrum confirms that the substitution in the final product is *ortho* and not *para.* A final product with para substitution (i.e., *p*-nitroaniline) would have given only four signals in the ^{13}C NMR spectrum.

15.56 HO–CH₂–CH(OH)–CH₂–OH (Glycerol). After sodium borohydride reduction of the aldehyde, ozonolysis oxidatively degrades the aromatic ring, leaving only the polyhydroxy side chain. Water is an alternative to HOAc sometimes used to work up ozonolysis reactions.

15.57 **D** (Threonic acid)

Ozonolysis oxidatively degrades the aromatic rings, leaving only the carboxyl carbon as a remnant of the alkyl-substituted benzene ring. Water is an alternative to HOAc used to work up the ozonolysis reaction.

QUIZ

15.1 Which of the following compounds would be most reactive toward ring bromination?

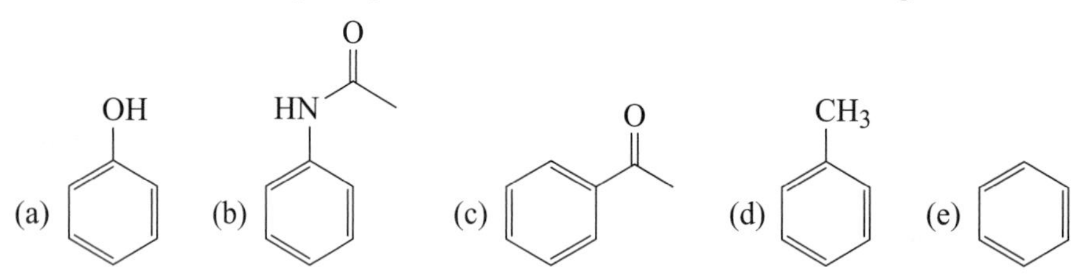

15.2 Which of the following is *not* a meta-directing substituent when present on a benzene ring?

(a) $—C_6H_5$ (b) $—NO_2$ (c) $—N(CH_3)_3^+$ (d) $—C{\equiv}N$ (e) $—CO_2H$

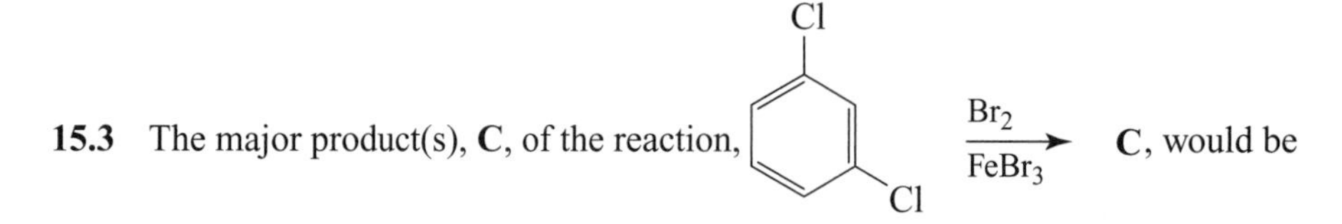

15.3 The major product(s), **C**, of the reaction, $\xrightarrow[FeBr_3]{Br_2}$ **C**, would be

(a) Cl, Br, Cl (b) Cl, Cl, Br (c) Cl, Br, Cl

(d) Equal amounts of (a) and (b) (e) Equal amounts of (a) and (c)

15.4 Complete the following syntheses.

(a) CH_3 (toluene) —**A** [ ]→ CH_3, SO_3H (p-toluenesulfonic acid) —HNO_3 / H_2SO_4→ **B** [ ]

—**C** [ ]→ CH_3, NO_2 (o-nitrotoluene) —(1) $KMnO_4$, OH^-, heat; (2) H_3O^+→ **D** [ ]

(b) benzoic acid (O, OH) —**A** [ ]→ benzoyl chloride (O, Cl) —**B** [ ]→

benzophenone (O) —**C** [ ]→ diphenylmethane

—**D** [ ]→ 4-bromodiphenylmethane (Br)

16 ALDEHYDES AND KETONES I. NUCLEOPHILIC ADDITION TO THE CARBONYL GROUP

SOLUTIONS TO PROBLEMS

16.1 (a)

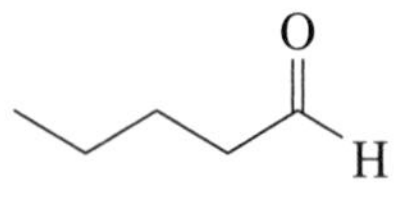

Pentanal 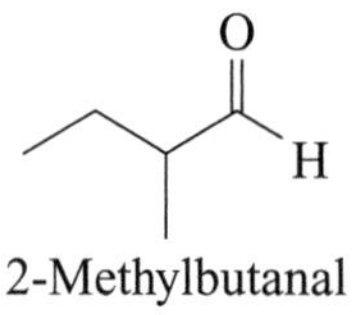　　2-Methylbutanal 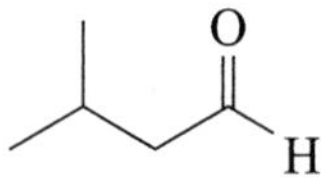　　3-Methylbutanal

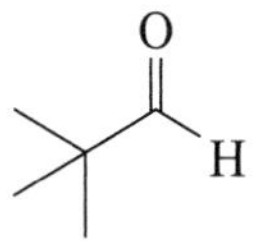

2,2-Dimethylpropanal 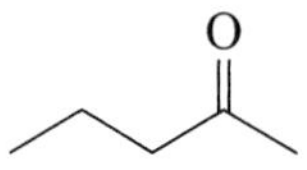　　2-Pentanone 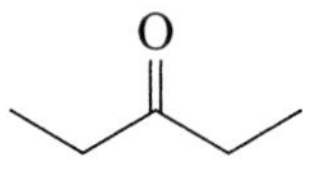　　3-Pentanone

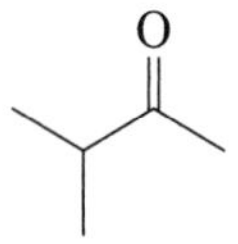

3-Methyl-2-butanone

(b)

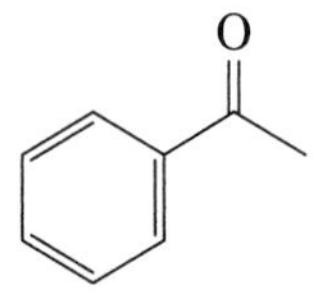

Acetophenone (methyl phenyl ketone)　　Phenylethanal (phenylacetaldehyde)　　2-Methylbenzaldehyde (*o*-tolualdehyde)

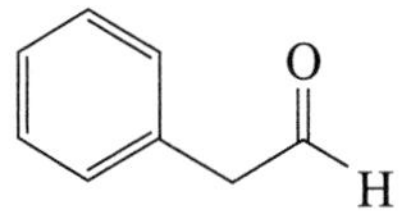

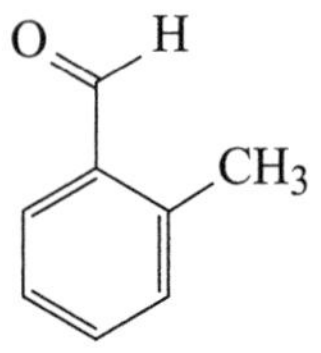

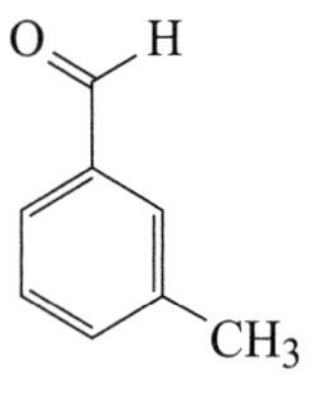

3-Methylbenzaldehyde (*m*-tolualdehyde)　　4-Methylbenzaldehyde (*p*-tolualdehyde)

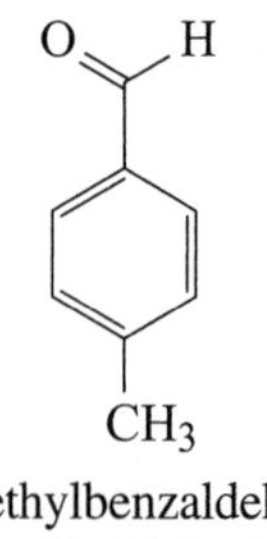

16.2 (a) 1-Pentanol, because its molecules form hydrogen bonds to each other.

(b) 2-Pentanol, because its molecules form hydrogen bonds to each other.

(c) Pentanal, because its molecules are more polar.

(d) 2-Phenylethanol, because its molecules form hydrogen bonds to each other.

(e) Benzyl alcohol because its molecules form hydrogen bonds to each other.

16.3 (a) OH $\xrightarrow[CH_2Cl_2]{PCC}$ O, H

(b) O, OH $\xrightarrow{SOCl_2}$ O, Cl $\xrightarrow[\text{(2) } H_2O]{\text{(1) } LiAlH(O\text{-}t\text{-Bu})_3,\ Et_2O,\ -78°\ C}$ O, H

16.4 (a) $\xrightarrow[Fe]{Br_2}$ Br $\xrightarrow[Et_2O]{Mg}$ MgBr $\xrightarrow[\text{(2) } H_3O^+]{\text{(1) } HCHO}$ OH $\xrightarrow[CH_2Cl_2]{PCC}$ O, H

(b) CH_3 $\xrightarrow[\text{(2) } H_3O^+]{\text{(1) } KMnO_4,\ OH^-,\ \text{heat}}$ O, OH $\xrightarrow{SOCl_2}$ O, Cl $\xrightarrow[\text{(2) } H_2O]{\text{(1) } LiAlH(O\text{-}t\text{-Bu})_3,\ Et_2O,\ -78°\ C}$ O, H

(c) OH $\xrightarrow[\text{acetone}]{H_2CrO_4}$ O

(d) $\xrightarrow[AlCl_3]{CH_3COCl}$ O

(e) 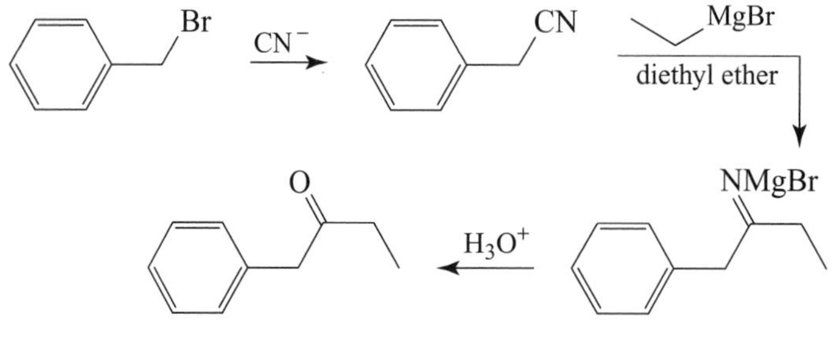 Br $\xrightarrow{CN^-}$ CN; MgBr, diethyl ether → NMgBr; H_3O^+ → O

(f) 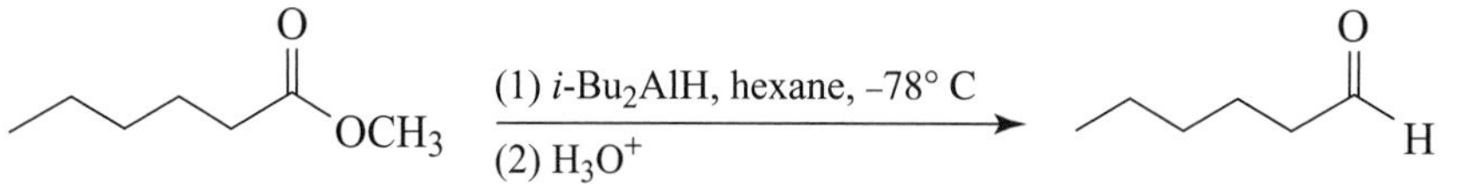 O, OCH_3 $\xrightarrow[(2)\ H_3O^+]{(1)\ i\text{-}Bu_2AlH,\ \text{hexane},\ -78°\ C}$ O, H

16.5 (a) The nucleophile is the negatively charged carbon of the Grignard reagent *acting as a carbanion.*

(b) The magnesium portion of the Grignard reagent acts as a Lewis acid and accepts an electron pair of the carbonyl oxygen. This acid-base interaction makes the carbonyl carbon even more positive and, therefore, even more susceptible to nucleophilic attack.

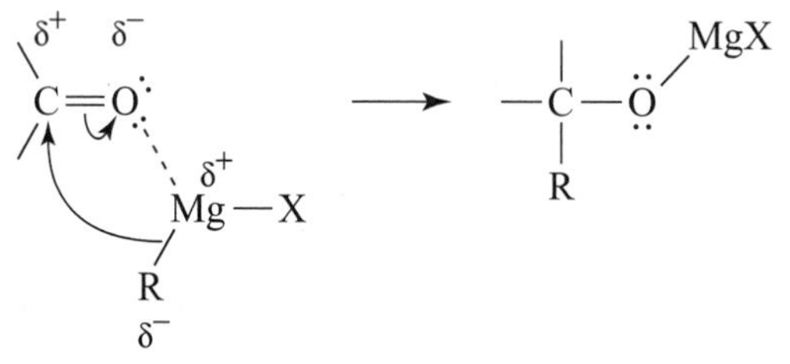

δ^+ δ^- C=O: ··· δ^+ Mg—X, R δ^- ⟶ —C—O—MgX, R

(c) The product that forms initially (above) is a magnesium alkoxide salt.

(d) On addition of water, the organic product that forms is an alcohol.

16.6 The nucleophile is a hydride ion.

16.7 $H_2C{=}O + H_2O \rightleftharpoons H_2C(O^-)(O^+H_2) \rightleftharpoons H_2C(OH)_2$

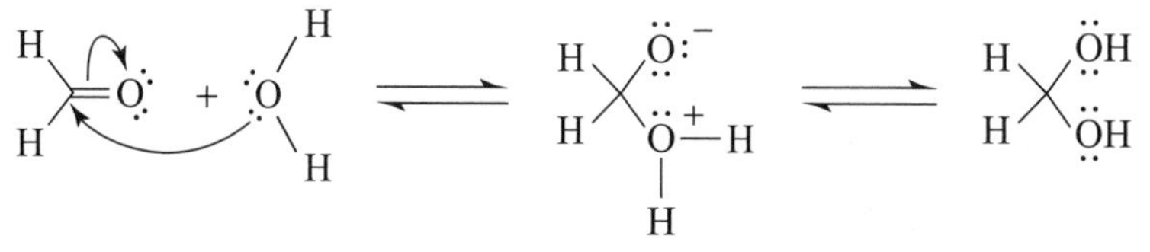

16.8 *Acid-Catalyzed Reaction*

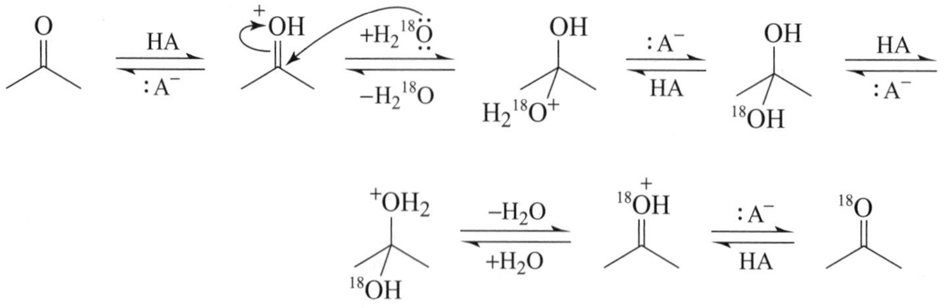

O $\underset{:A^-}{\overset{HA}{\rightleftharpoons}}$ ^+OH $\underset{-H_2{}^{18}O}{\overset{+H_2{}^{18}\ddot{O}}{\rightleftharpoons}}$ OH, $H_2{}^{18}O^+$ $\underset{HA}{\overset{:A^-}{\rightleftharpoons}}$ OH, ^{18}OH $\underset{:A^-}{\overset{HA}{\rightleftharpoons}}$

$^+OH_2$, ^{18}OH $\underset{+H_2O}{\overset{-H_2O}{\rightleftharpoons}}$ $^{18}\overset{+}{O}H$ $\underset{HA}{\overset{:A^-}{\rightleftharpoons}}$ ^{18}O

Base-Catalyzed Reaction

$$OH^- + H_2{}^{18}O \rightleftharpoons H_2O + {}^{18}OH^-$$

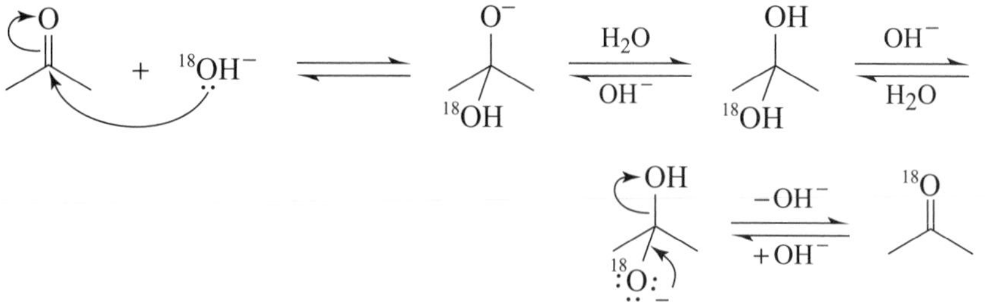

O + ^{18}ÖH$^-$ ⇌ O$^-$ ^{18}OH ⇌ (H_2O / OH^-) OH ^{18}OH ⇌ (OH^- / H_2O)

OH ^{18}Ö:$^-$ ⇌ ($-OH^-$ / $+OH^-$) ^{18}O

16.9

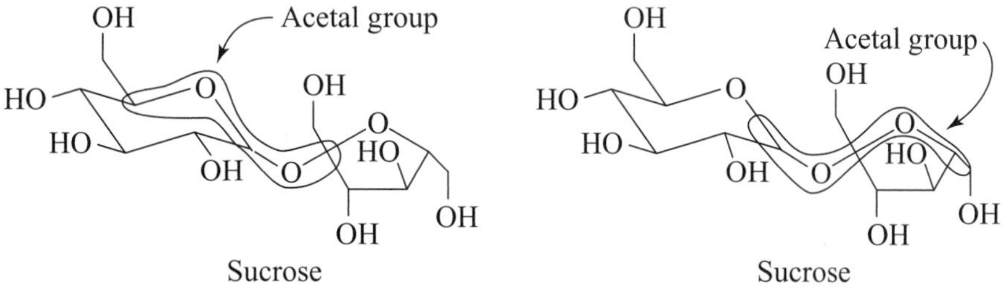

OH, Acetal group, HO, O, OH, HO, OH, O, O, HO, OH, OH

Sucrose

OH, Acetal group, OH, HO, O, HO, OH, O, O, HO, OH, OH

Sucrose

16.10

H :O: ⇌ (HA / :A$^-$) H $\overset{+}{Ö}H$ ⇌ ($+HÖ—CH_3$ / $-HÖ—CH_3$) H $\overset{+}{Ö}—CH_3$ H ÖH

⇌ (:A$^-$ / HA) $Ö—CH_3$ H ÖH

(hemiacetal)

⇌ (HA / :A$^-$) $:Ö—CH_3$ H $\overset{+}{Ö}H_2$ ⇌ ($-H_2O$ / $+H_2O$) $\overset{+}{O}—CH_3$ H

$\overset{+}{O}—CH_3$ H ⇌ ($+HÖ—CH_3$ / $-HÖ—CH_3$) H $\overset{+}{Ö}—CH_3$ H $Ö—CH_3$ ⇌ (:A$^-$ / HA)

$Ö—CH_3$ H $Ö—CH_3$

(acetal)

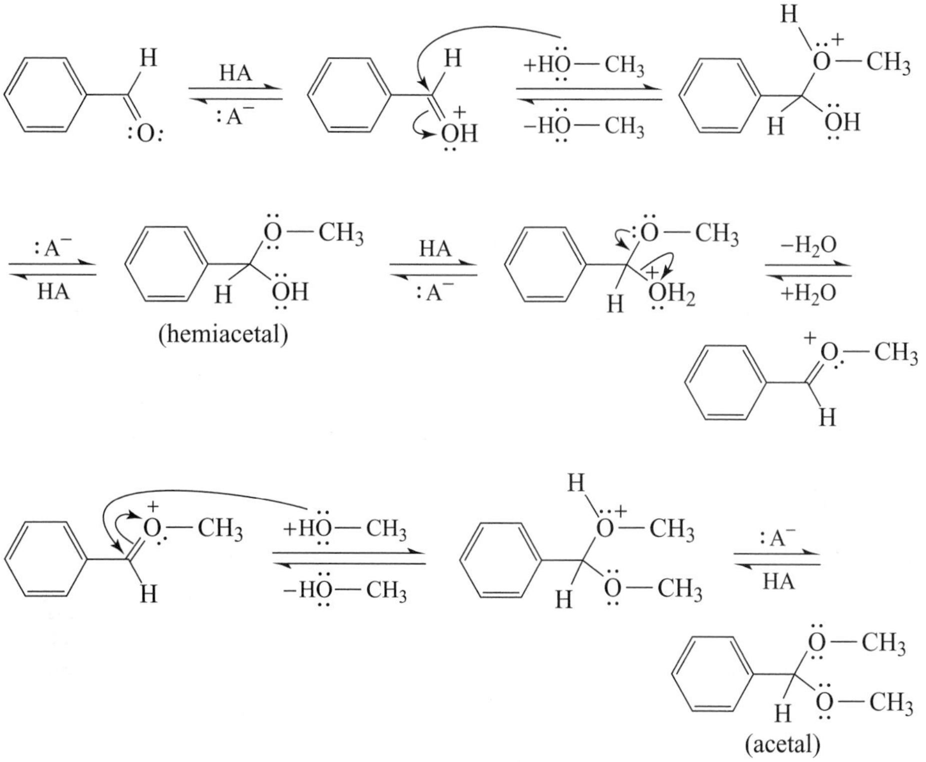

16.11

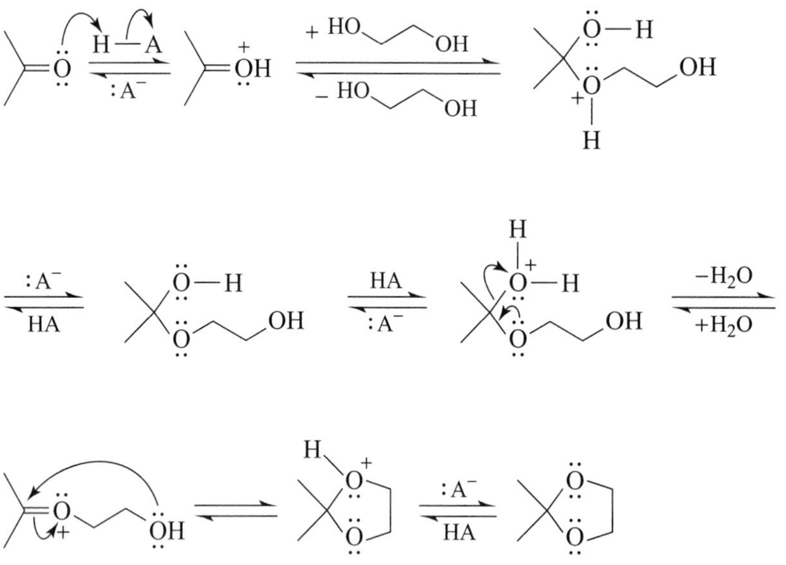

Ö H—A :A⁻ ÖH + HO OH − HO OH Ö—H Ö OH + H

:A⁻ HA Ö—H Ö OH HA :A⁻ H + Ö—H Ö OH $-H_2O$ $+H_2O$

Ö ÖH H + Ö Ö :A⁻ HA Ö Ö

16.12 HO OH

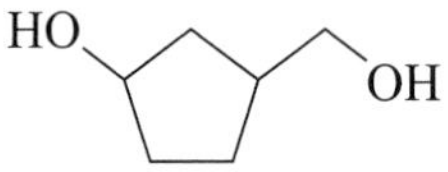

16.13 (a)

O O OEt A HO OH HA O O O OEt

2 CH_3MgI O O OMgI NH_4^+ H_2O OH O C

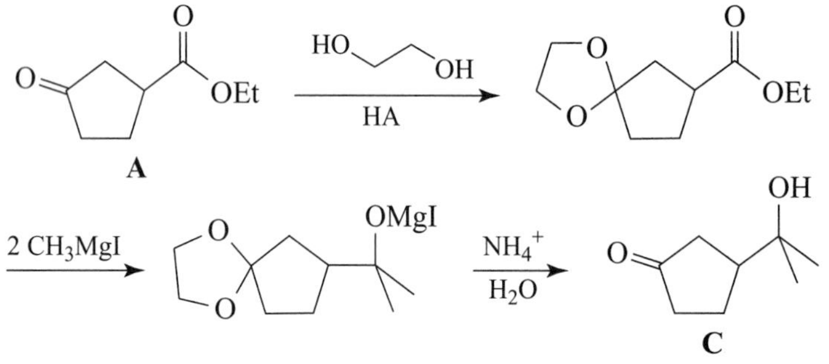

(b) Addition would take place at the ketone group as well as at the ester group. The product (after hydrolysis) would be

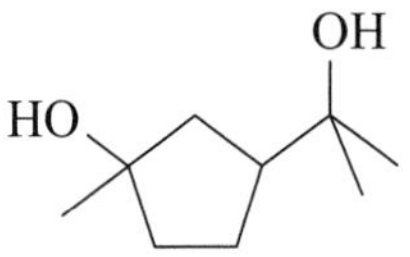

OH HO

16.14 (a)

H—A Ö Ö + HÖR Ö Ö—R + H :A⁻ Ö Ö—R

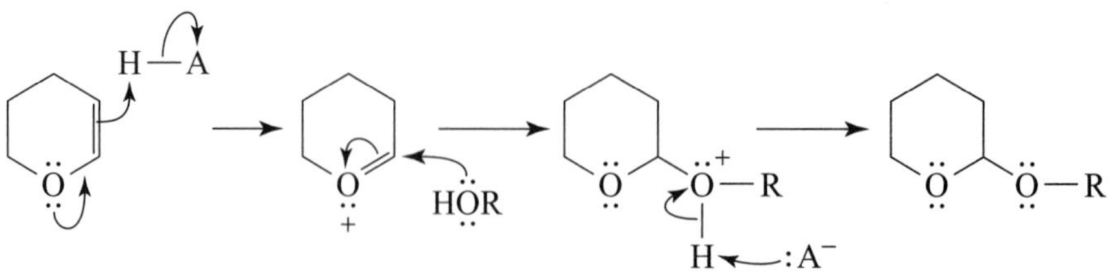

(b) Tetrahydropyranyl ethers are acetals; thus, they are stable in aqueous base and hydrolyze readily in aqueous acid.

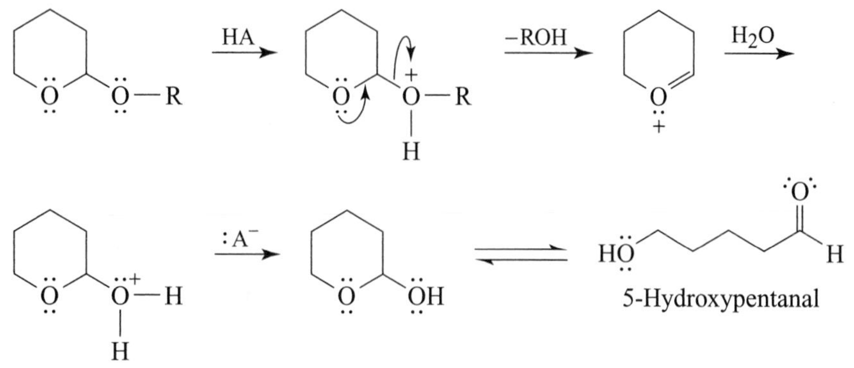

HA → −ROH → H_2O →

:A⁻ → ⇌ 5-Hydroxypentanal

(c) 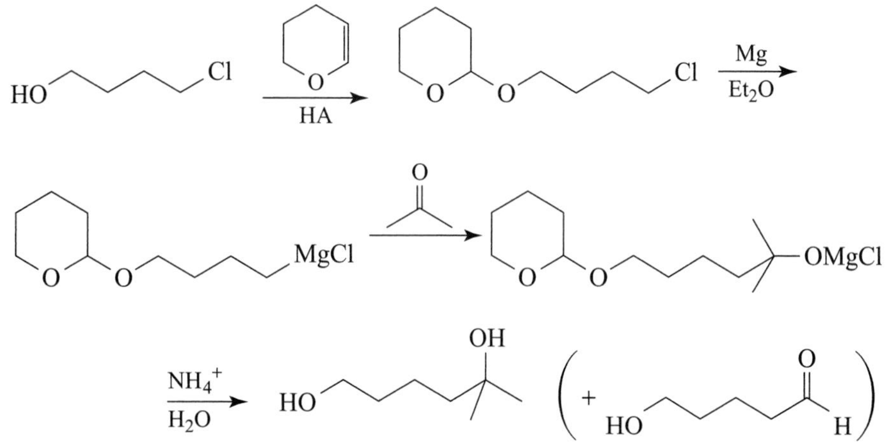 HO⁀⁀Cl → (HA) → Mg, Et_2O →

MgCl → OMgCl

NH_4^+, H_2O →

16.15 (a) 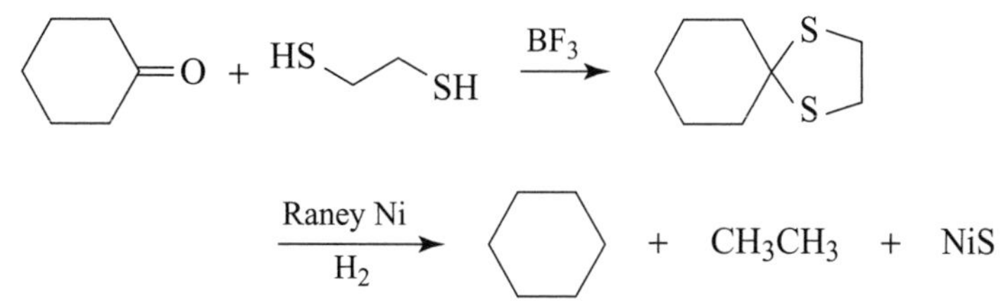 cyclohexanone + HS⁀SH → (BF_3) → dithiolane

Raney Ni, H_2 → cyclohexane + CH_3CH_3 + NiS

(b) 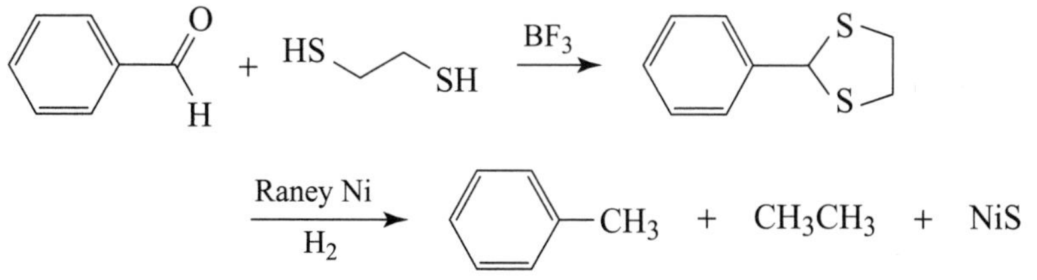 benzaldehyde + HS⁀SH → (BF_3) → dithiolane

Raney Ni, H_2 → C_6H_5–CH_3 + CH_3CH_3 + NiS

16.16 (a) 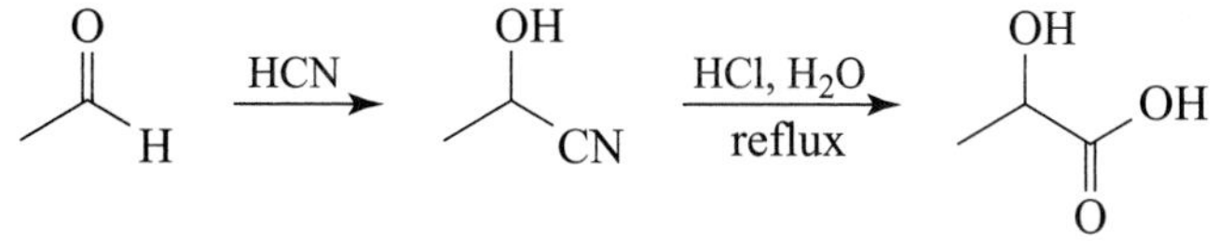 acetaldehyde → (HCN) → cyanohydrin → (HCl, H_2O, reflux) →

Lactic acid

(b) A racemic form

16.17 (a) $CH_3I \xrightarrow[(2)\ RLi]{(1)\ (C_6H_5)_3P:} H_2\ddot{C}^{-}-\overset{+}{P}(C_6H_5)_3 \longrightarrow$

(b) Br $\xrightarrow[(2)\ RLi]{(1)\ (C_6H_5)_3P:}$ $\overset{+}{P}(C_6H_5)_3 \longrightarrow$

(c) $H_2\ddot{C}^{-}-\overset{+}{P}(C_6H_5)_3 \longrightarrow$
[from part (a)]

(d) $H_2\ddot{C}^{-}-\overset{+}{P}(C_6H_5)_3 \longrightarrow$
[from part (a)]

(e) Br $\xrightarrow[(2)\ RLi]{(1)\ (C_6H_5)_3P:}$ $\overset{+}{P}(C_6H_5)_3 \longrightarrow$

(f) Br $\xrightarrow[(2)\ RLi]{(1)\ (C_6H_5)_3P:}$ $\overset{+}{P}(C_6H_5)_3 \longrightarrow$

(g) Br $\xrightarrow[(2)\ RLi]{(1)\ (C_6H_5)_3P:}$ $\overset{+}{P}(C_6H_5)_3 \longrightarrow$

16.18

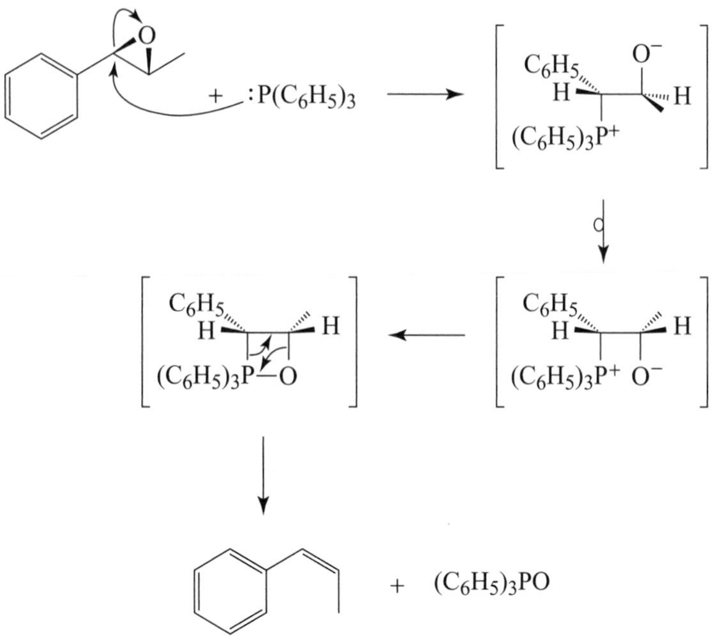

+ $:P(C_6H_5)_3$ ⟶ [C_6H_5, H, O^-, H, $(C_6H_5)_3P^+$]

↓

[C_6H_5, H, H, $(C_6H_5)_3P$—O] ⟵ [C_6H_5, H, H, $(C_6H_5)_3P^+$ O^-]

↓

+ $(C_6H_5)_3PO$

Problems

16.19 (a) 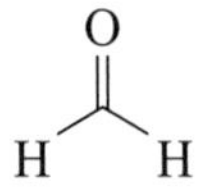 Methanal

(b) 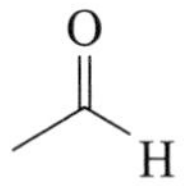 Ethanal

(c) 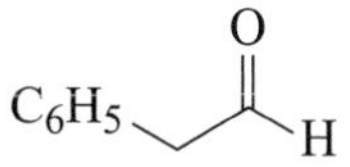 C_6H_5 ... H — Phenylethanal

(d) 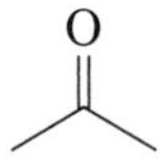 Propanone

(e) 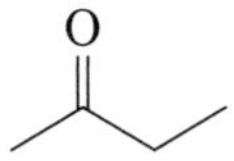 Butanone

(f) 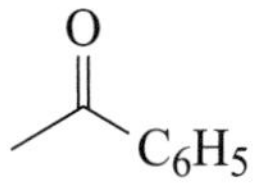 C_6H_5 — 1-Phenylethanone or methyl phenyl ketone

(g) 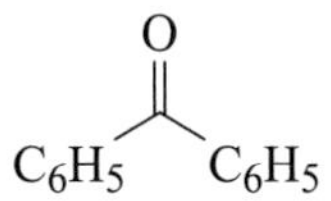 C_6H_5 C_6H_5 — Diphenylmethanone or diphenyl ketone

(h) 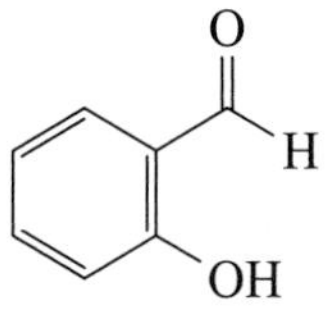 OH — 2-Hydroxybenzaldehyde or *o*-hydroxybenzaldehyde

(i) 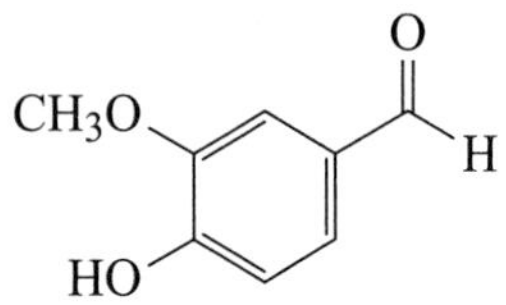 CH_3O, HO, O, H — 4-Hydroxy-3-methoxybenzaldehyde

(j) 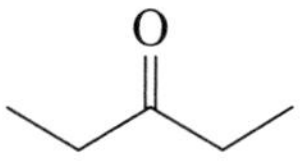 O — 3-Pentanone

(k) 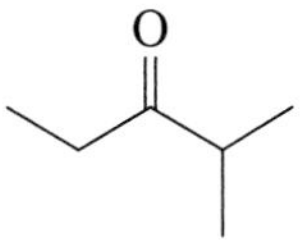 O — 2-Methyl-3-pentanone

(l) 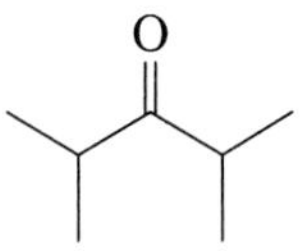 O — 2,4-Dimethyl-3-pentanone

(m) 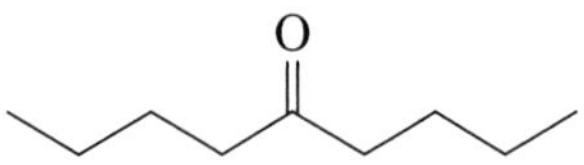 O — 5-Nonanone

(n) 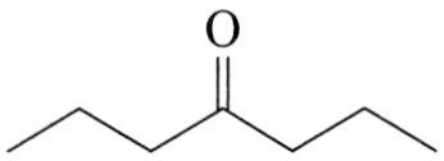 O — 4-Heptanone

(o) 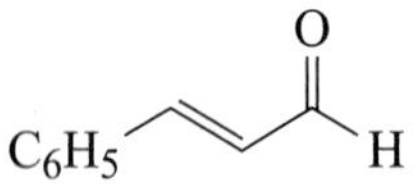 C_6H_5, O, H — (E)-3-Phenylpropenal

16.20

(a)  OH

(b) 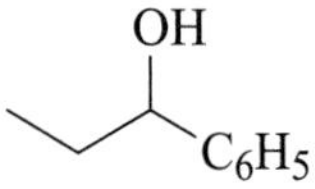 OH, C_6H_5

(c) OH

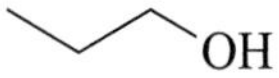

(d) 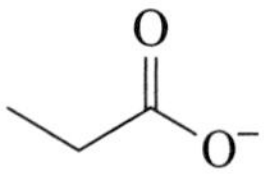 O, O^-

(e)

(f) 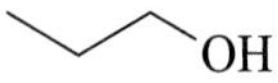 OH

(g) 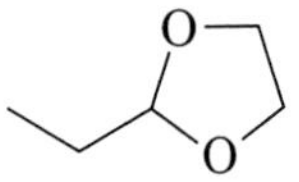 O, O

(h)

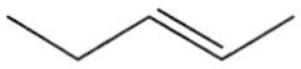

(i) 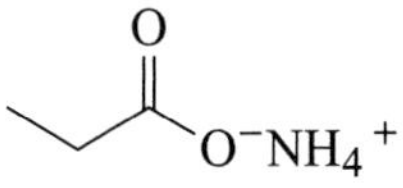 O, $O^-NH_4^+$ + Ag ↓

(j) 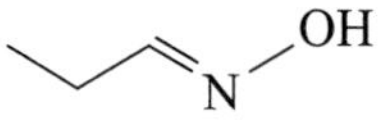 N, OH

(k) 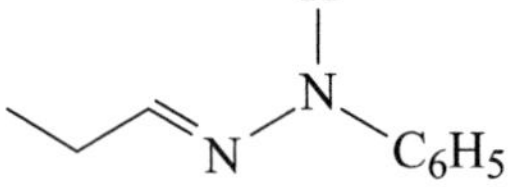 N, H, N, C_6H_5

(l) 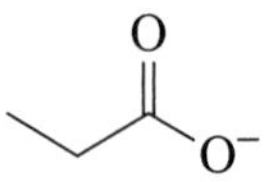 O, O^-

(m) 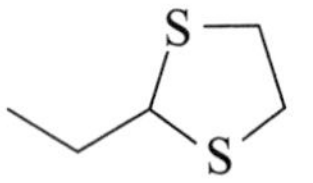 S, S

(n)  + CH_3CH_3 + NiS

16.21 (a) 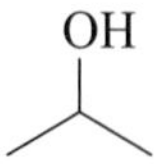 OH

(b) 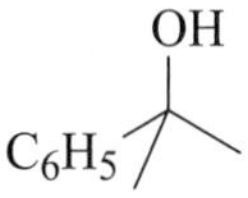 OH C_6H_5

(c) OH

(d) No reaction

(e)

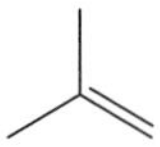

(f) 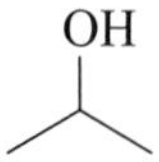 OH

(g) 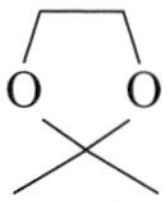 O O

(h)

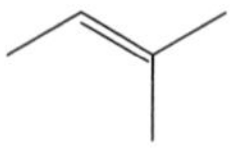

(i) No reaction

(j) 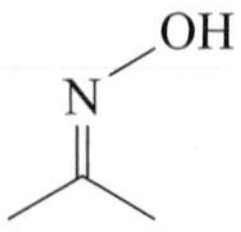 OH N

(k) 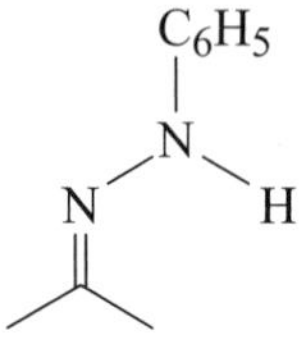 C_6H_5 N N H

(l) No reaction

(m) 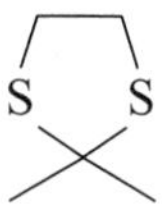 S S

(n)  + CH_3CH_3 + NiS

16.22 (a) 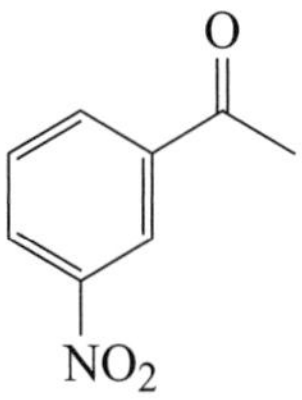 O NO_2

(b) 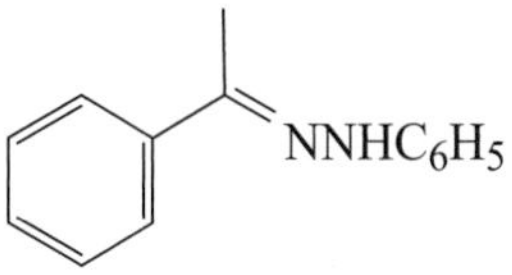 $NNHC_6H_5$

(c)

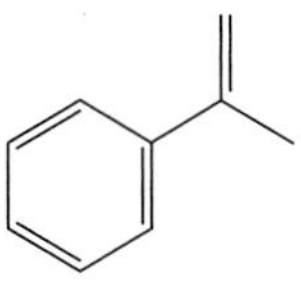

(d) 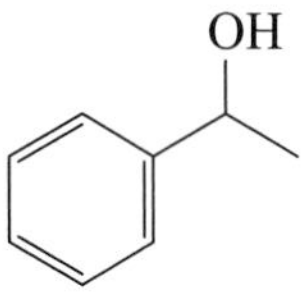 OH

(e) 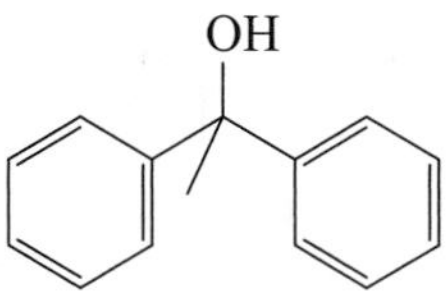 OH

16.23 (a) 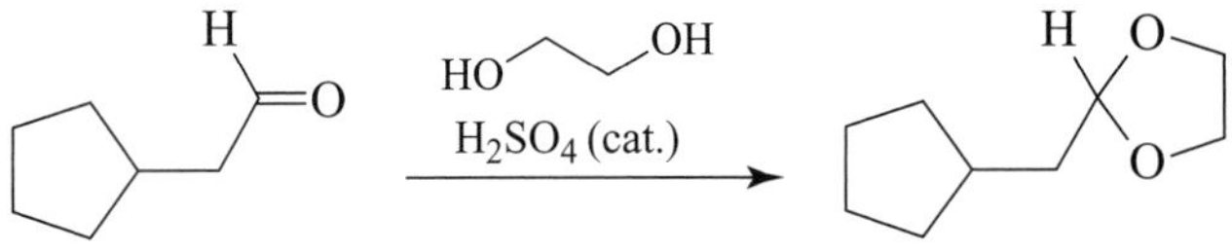 HO⌒OH, H_2SO_4 (cat.)

(b) 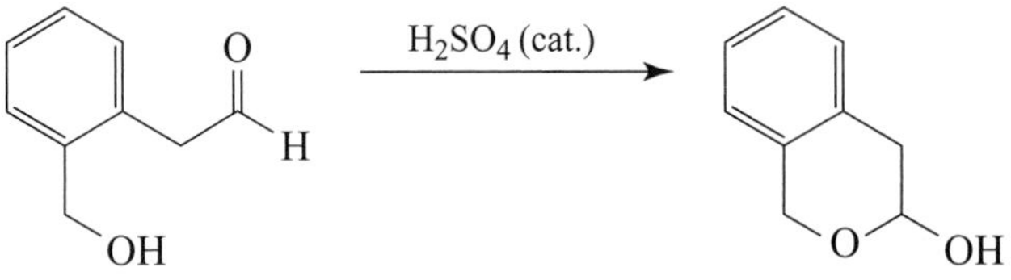 H_2SO_4 (cat.)

(c) 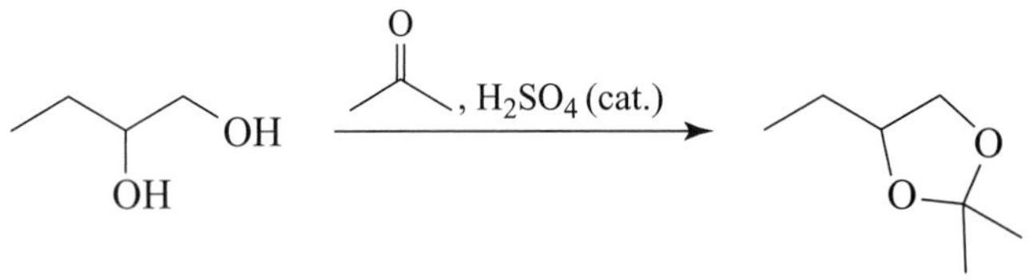 H_2SO_4 (cat.)

(d) 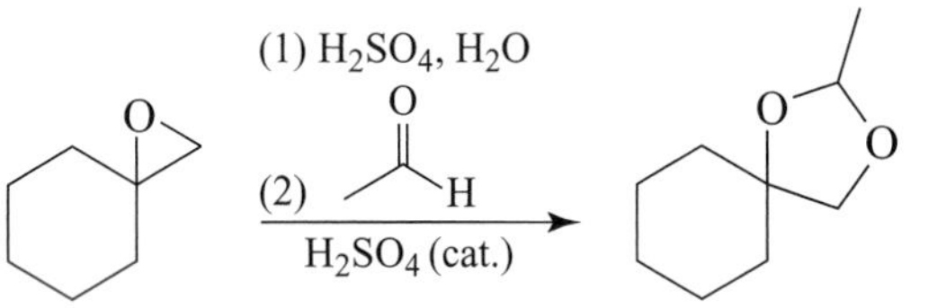 (1) H_2SO_4, H_2O (2) H_2SO_4 (cat.)

(e) 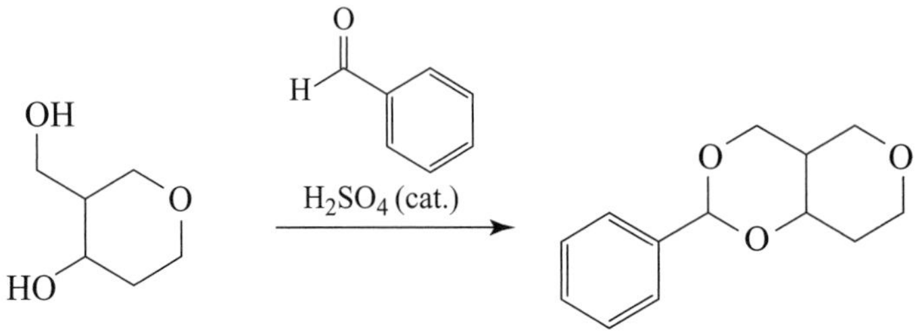 H_2SO_4 (cat.)

(f) 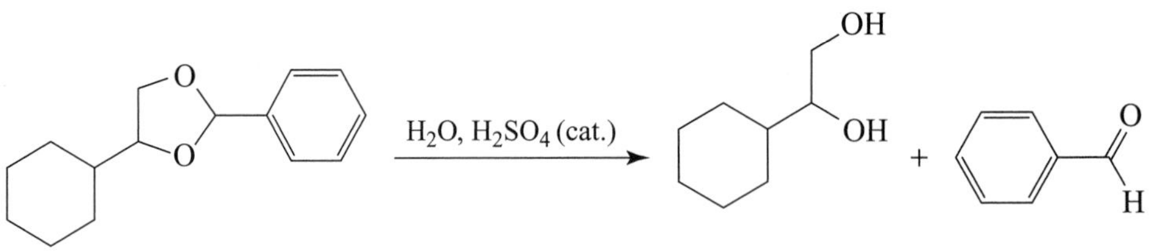 H_2O, H_2SO_4 (cat.) +

(g) 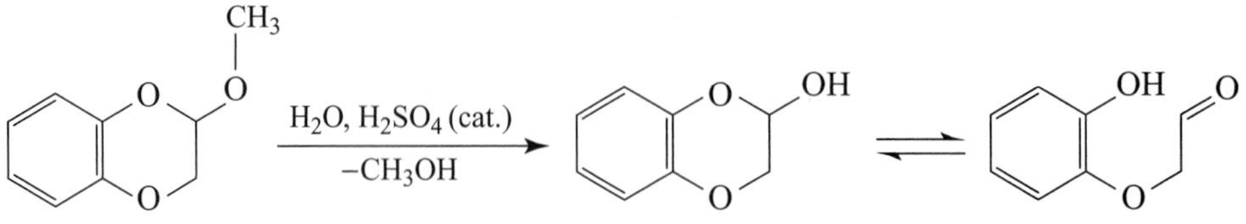 CH_3 H_2O, H_2SO_4 (cat.) $-CH_3OH$

16.24 (a) 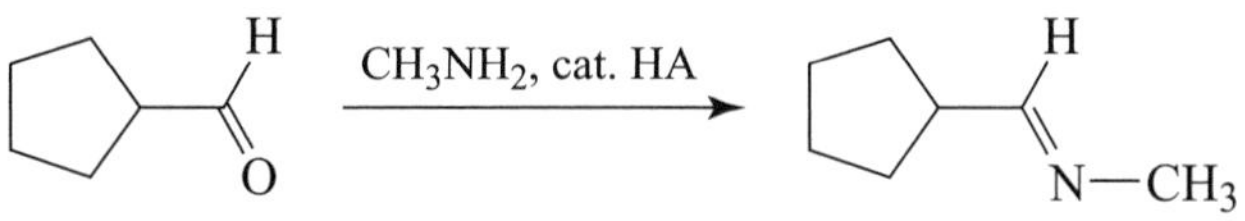 H, O $\xrightarrow{CH_3NH_2\text{, cat. HA}}$ H, N—CH_3

(b) 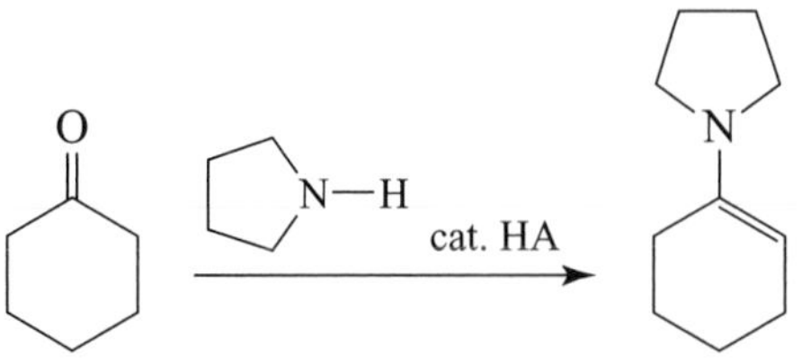 O, N—H, cat. HA, N

(c) 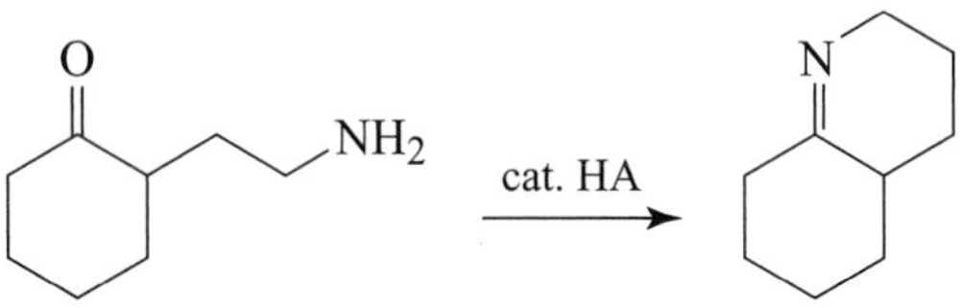 O, NH_2, cat. HA, N

(d) 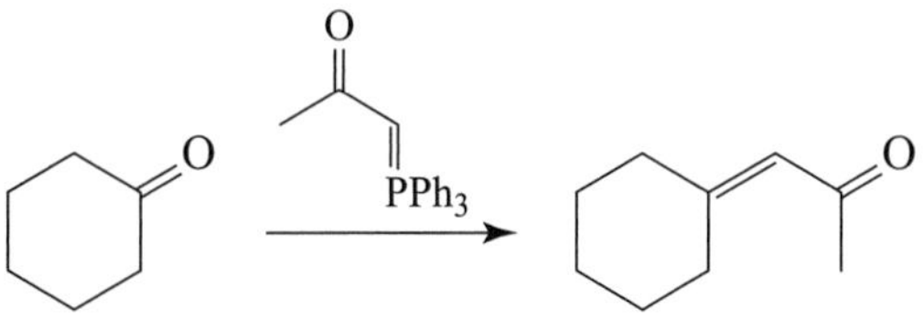 O, O, PPh_3, O

(e) 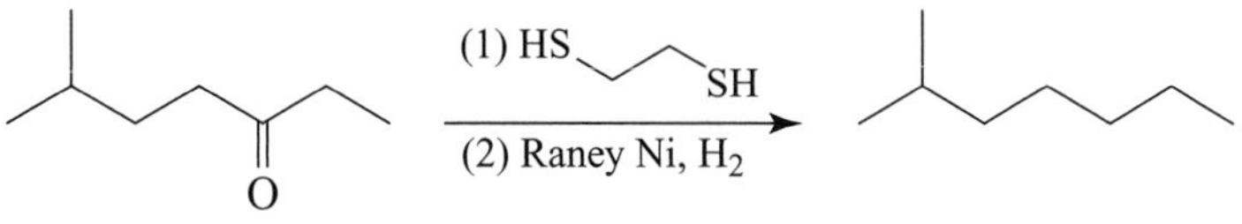 O, (1) HS, SH, (2) Raney Ni, H_2

(f) 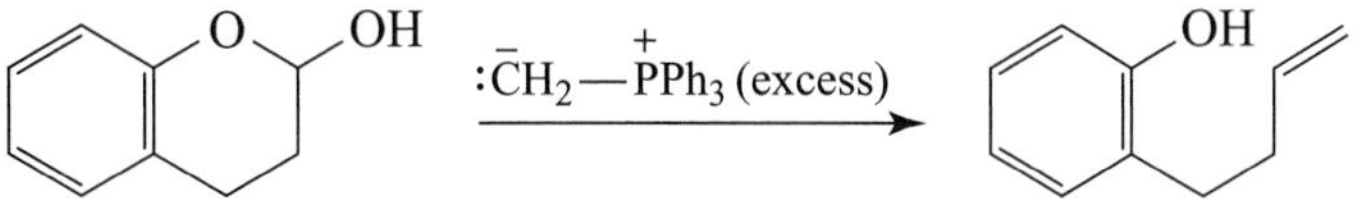 O, OH, $:\bar{C}H_2—\overset{+}{P}Ph_3$ (excess), OH

(g) 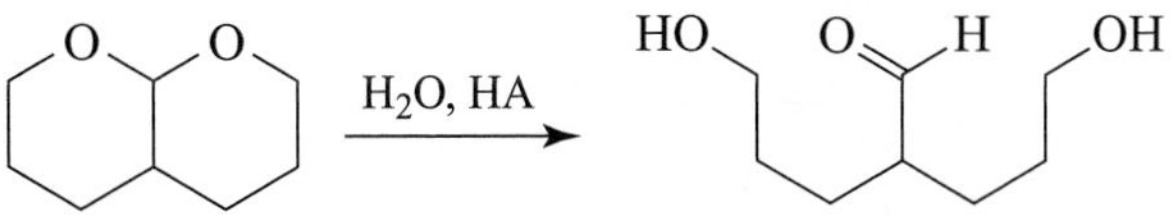 O, O, H_2O, HA, HO, O, H, OH

16.25 (a) 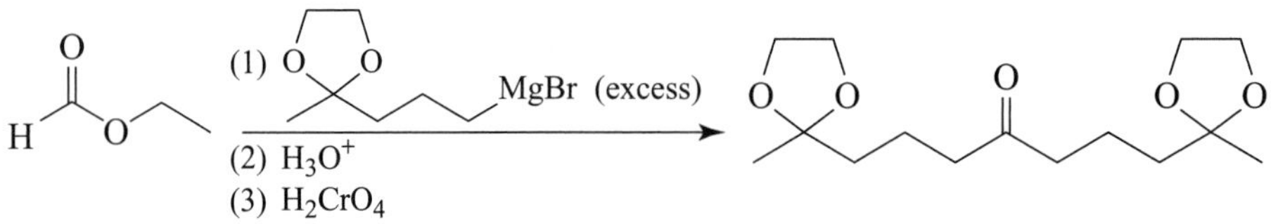 O H O (1) O O MgBr (excess) (2) H_3O^+ (3) H_2CrO_4 → O O O O O

(b) 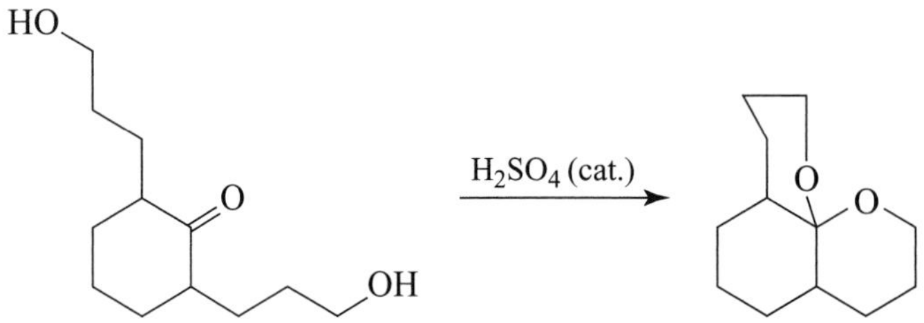 HO O OH H_2SO_4 (cat.) → O O

(c) 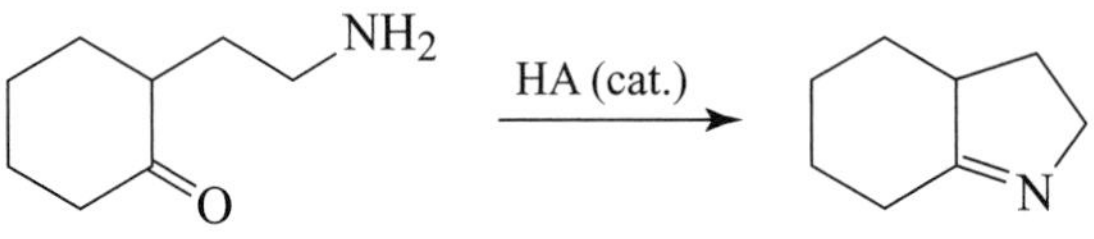 NH_2 O HA (cat.) → N

(d) 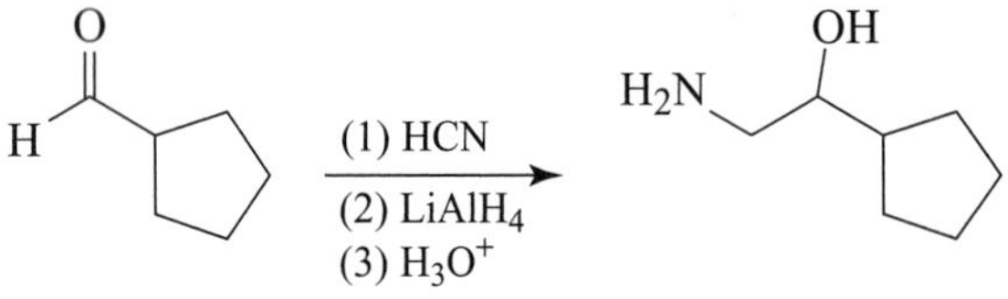 O H (1) HCN (2) $LiAlH_4$ (3) H_3O^+ → OH H_2N

16.26 (a) 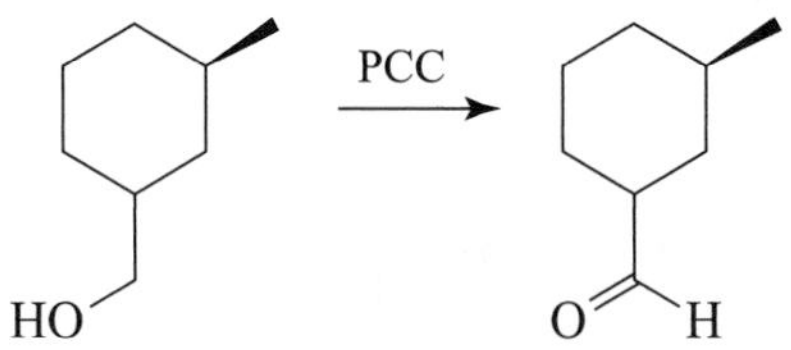 HO PCC → O H

(b) 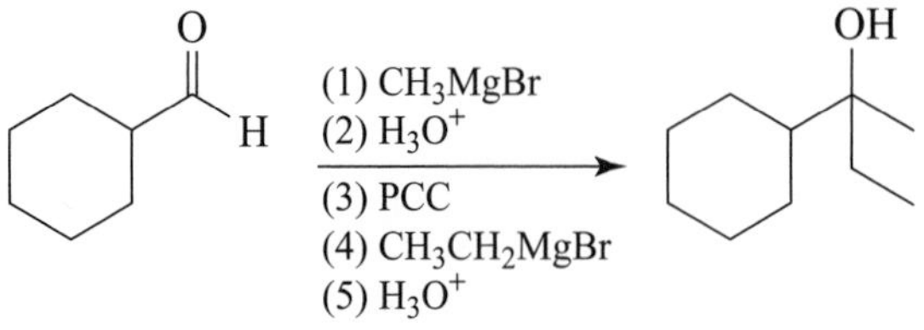 O H (1) CH_3MgBr (2) H_3O^+ (3) PCC (4) CH_3CH_2MgBr (5) H_3O^+ → OH

(c) 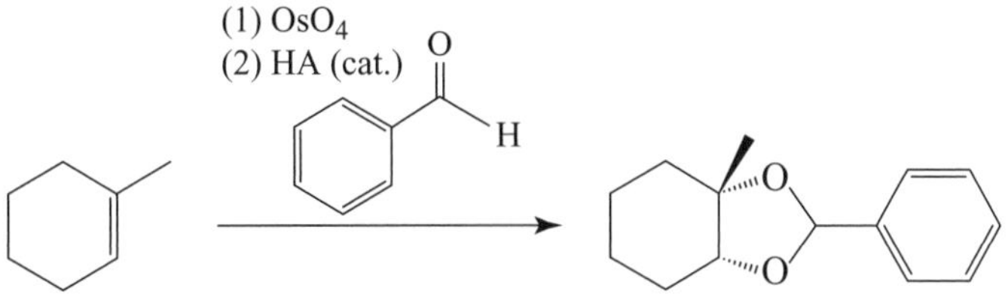 (1) OsO_4
(2) HA (cat.)

(d) 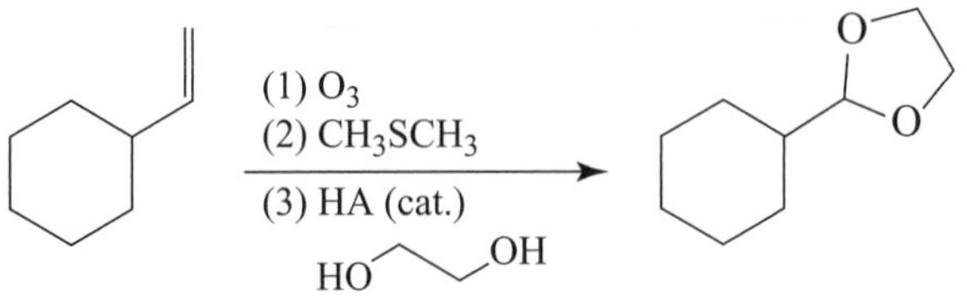 (1) O_3
(2) CH_3SCH_3
(3) HA (cat.)

HO OH

(e) 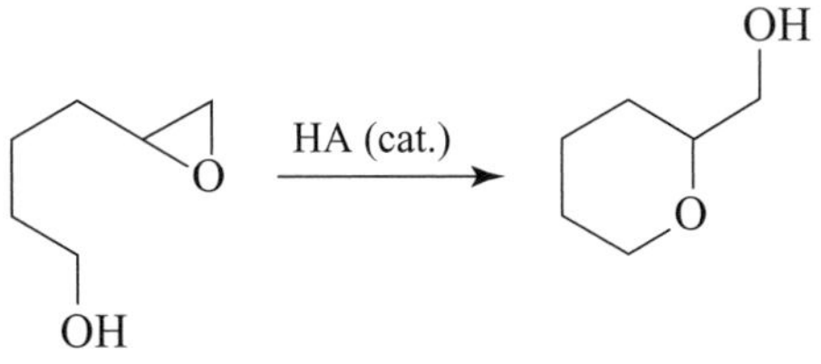 HA (cat.)

OH

O

OH

(f) 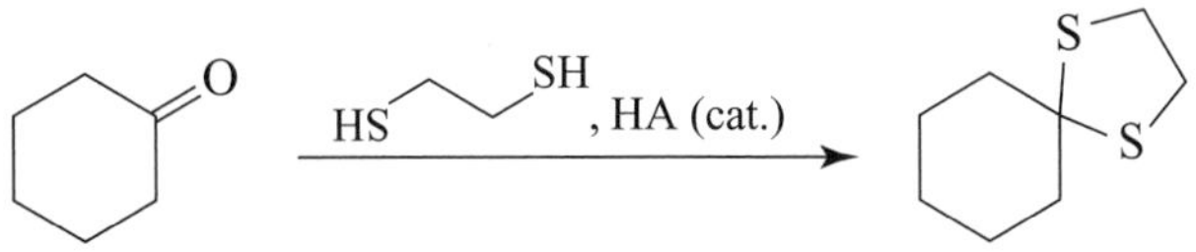 HS SH , HA (cat.)

16.27 (a)

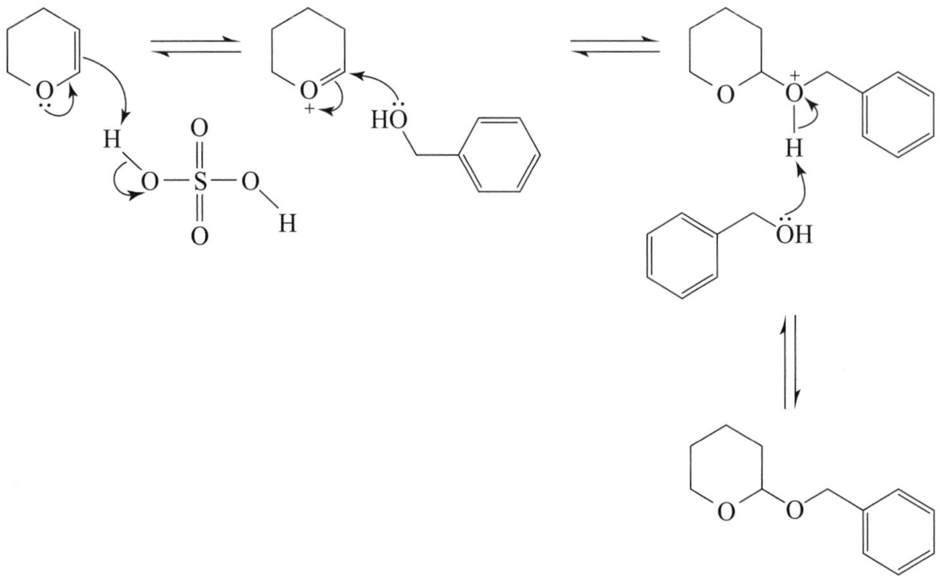

(b) $CH_3\overset{+}{O}H_2$ … OH ⇌ … $\ddot{O}H$ ⇌ $CH_3\ddot{O}H$ …

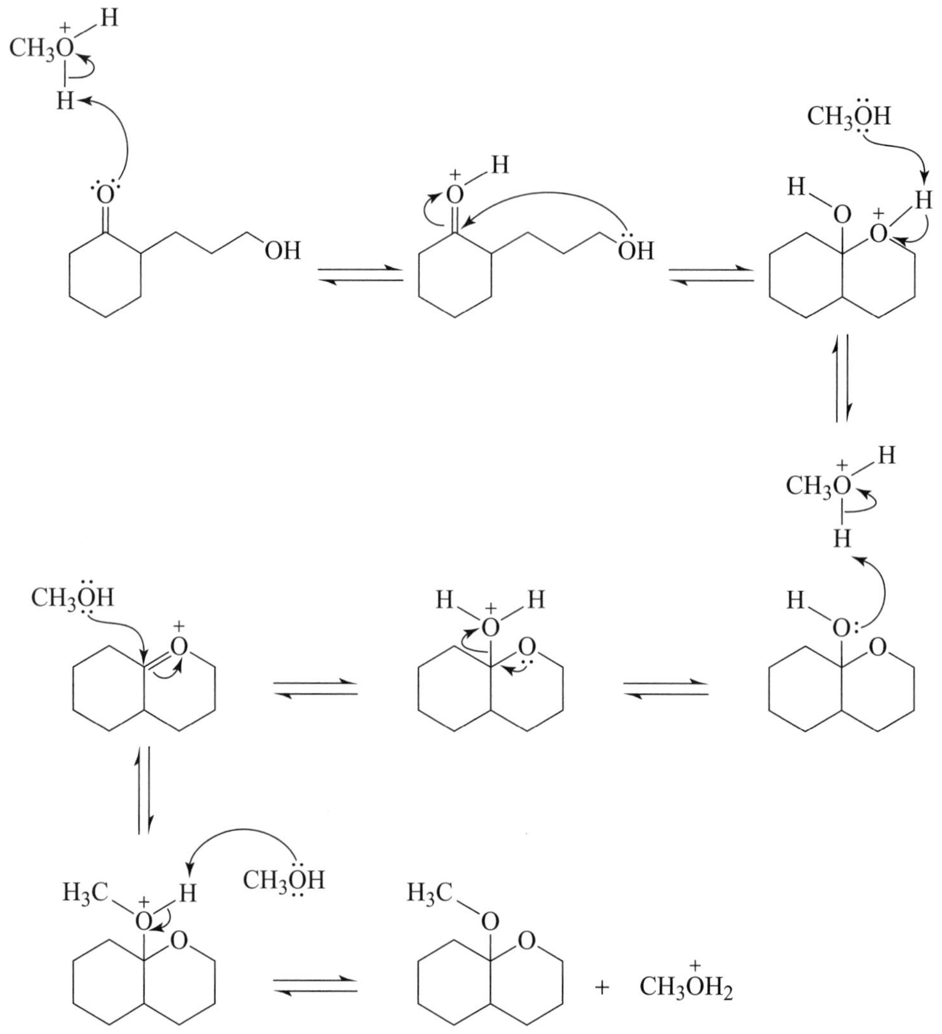

$CH_3\overset{+}{O}H_2$

$CH_3\ddot{O}H$ ⇌ ⇌

$CH_3\ddot{O}H$ ⇌ + $CH_3\overset{+}{O}H_2$

(c)

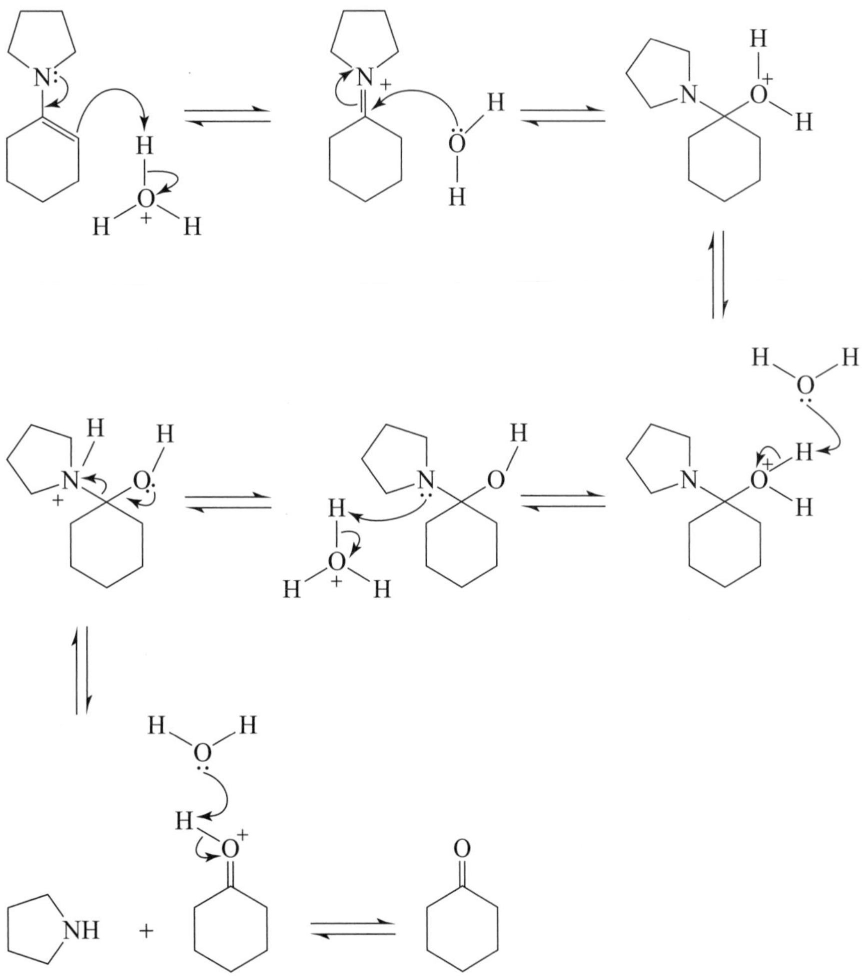

(d)

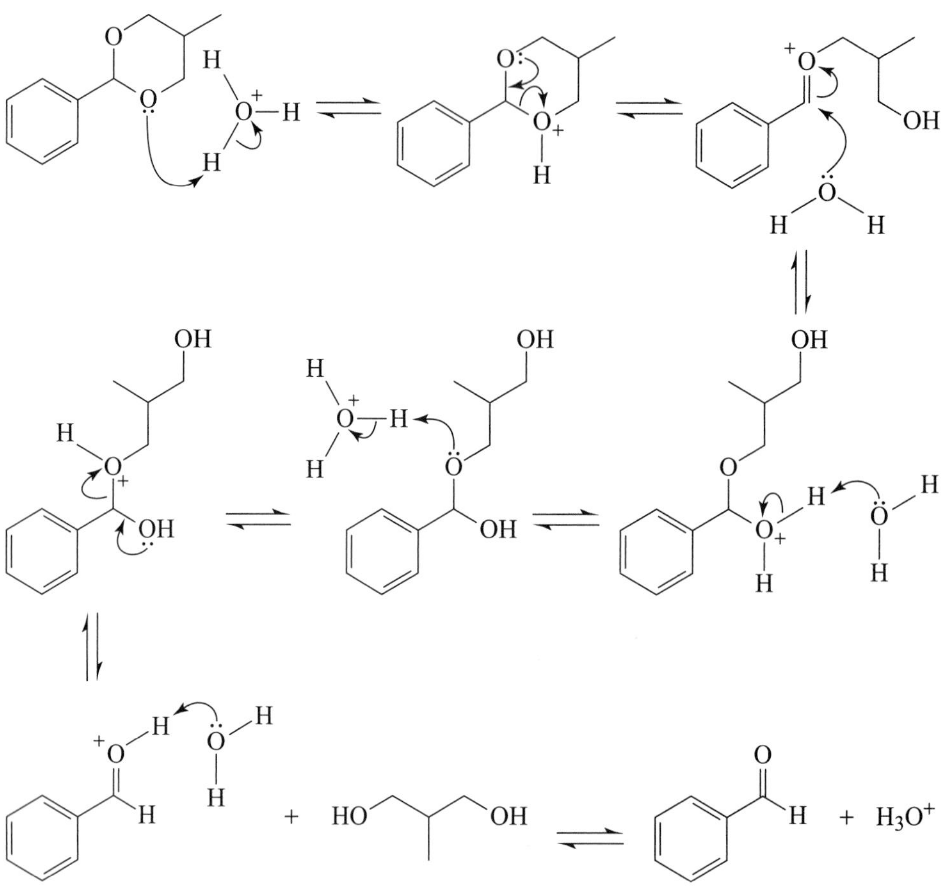

16.28

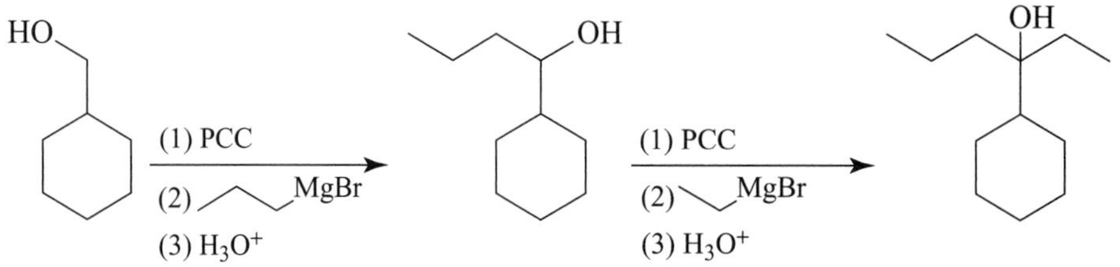

(1) PCC
(2) MgBr
(3) H_3O^+

(1) PCC
(2) MgBr
(3) H_3O^+

Synthesis

16.29 (a) benzene + butanoyl chloride $\xrightarrow{AlCl_3}$ butyrophenone

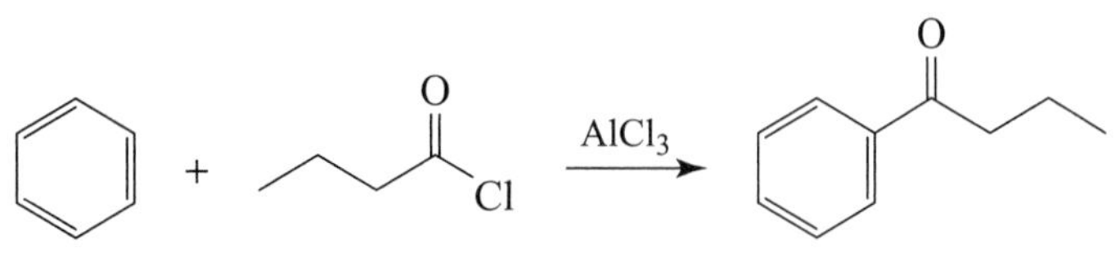

or

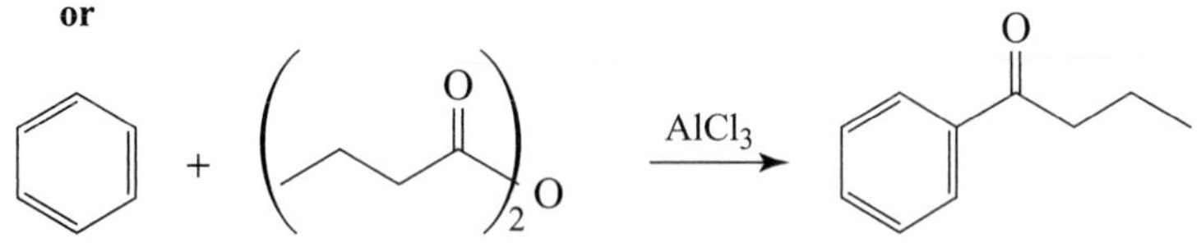

benzene + butanoic anhydride $\xrightarrow{AlCl_3}$ butyrophenone

(b) butyrophenone $\xrightarrow[HCl]{Zn(Hg)}$ butylbenzene

butyrophenone $\xrightarrow[HA]{HS\text{–}CH_2CH_2\text{–}SH}$ dithiolane $\xrightarrow[H_2]{\text{Raney Ni}}$ butylbenzene

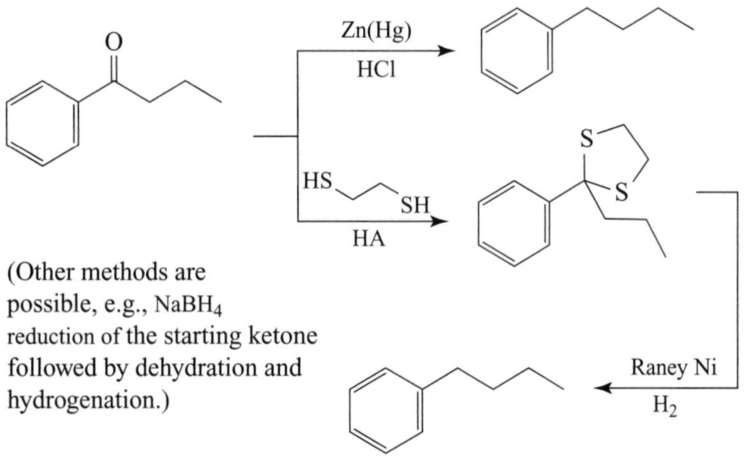

(Other methods are possible, e.g., $NaBH_4$ reduction of the starting ketone followed by dehydration and hydrogenation.)

16.30 (a) benzaldehyde $\xrightarrow{NaBH_4}$ benzyl alcohol (OH)

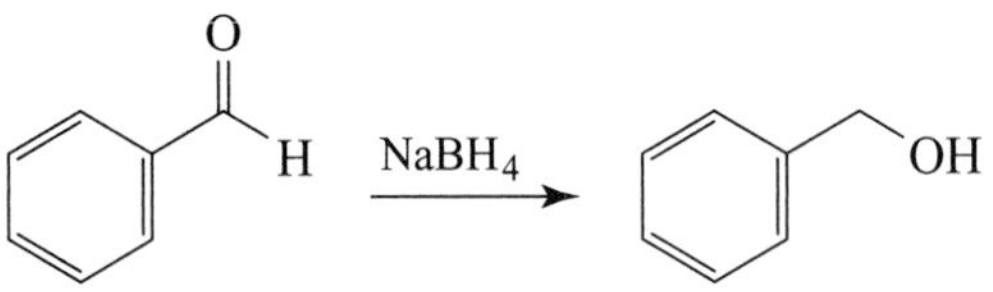

(b) benzaldehyde $\xrightarrow[NH_3/H_2O]{Ag(NH_3)_2^+}$ $\xrightarrow{H_3O^+}$ benzoic acid (OH)

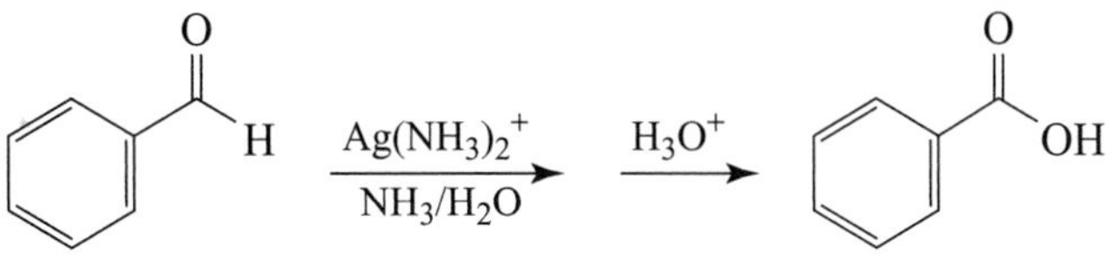

(c) benzoic acid [from (b)] $\xrightarrow{SOCl_2}$ benzoyl chloride (Cl)

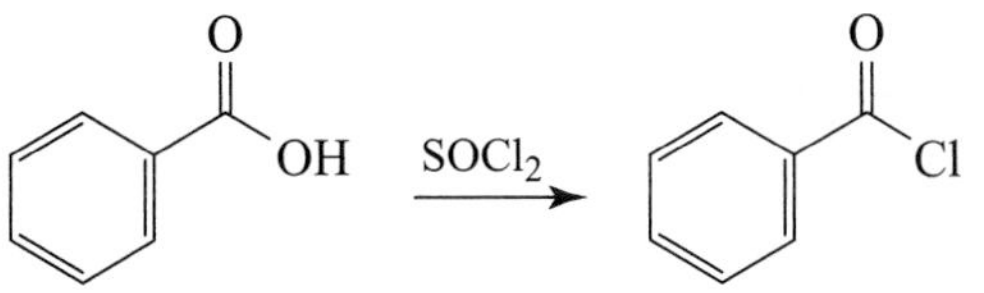

(d) 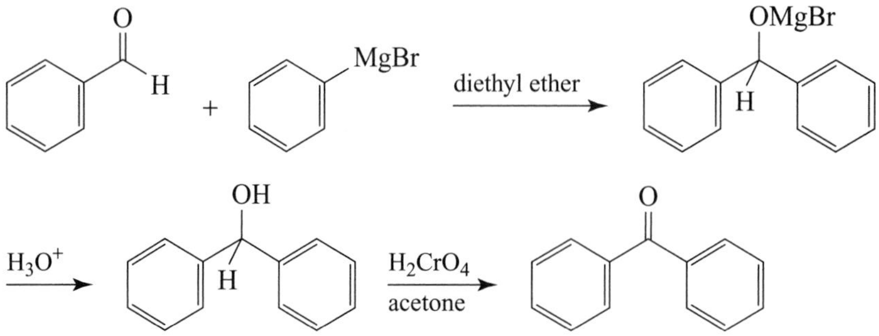 benzaldehyde + phenylmagnesium bromide (MgBr) —diethyl ether→ OMgBr, H adduct

$\xrightarrow{H_3O^+}$ OH, H alcohol $\xrightarrow[\text{acetone}]{H_2CrO_4}$ ketone (O)

or

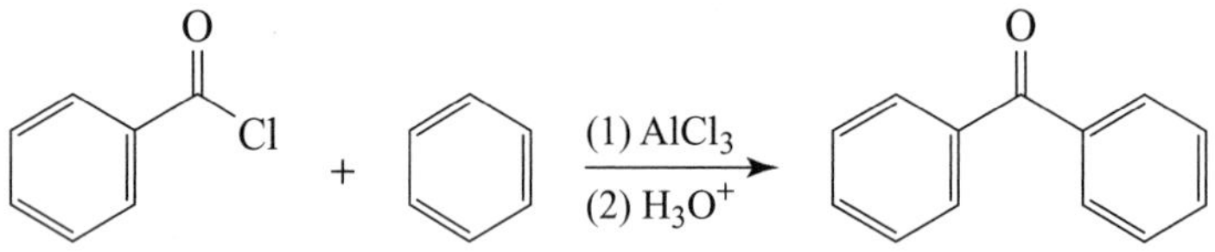

Cl (O) acid chloride + benzene $\xrightarrow[(2)\ H_3O^+]{(1)\ AlCl_3}$ ketone (O)

[from (c)]

(e) 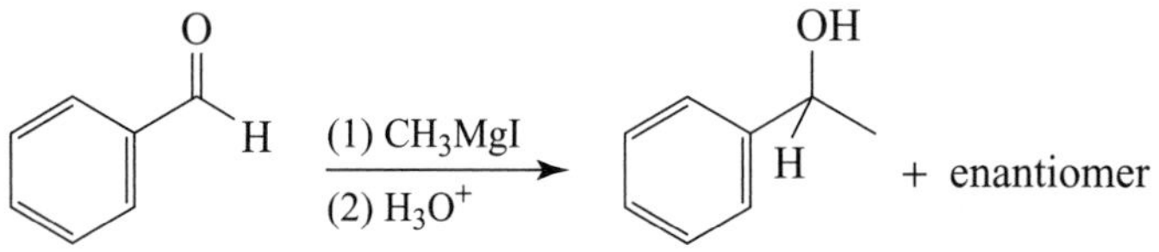 benzaldehyde (O, H) $\xrightarrow[(2)\ H_3O^+]{(1)\ CH_3MgI}$ OH, H + enantiomer

(f) 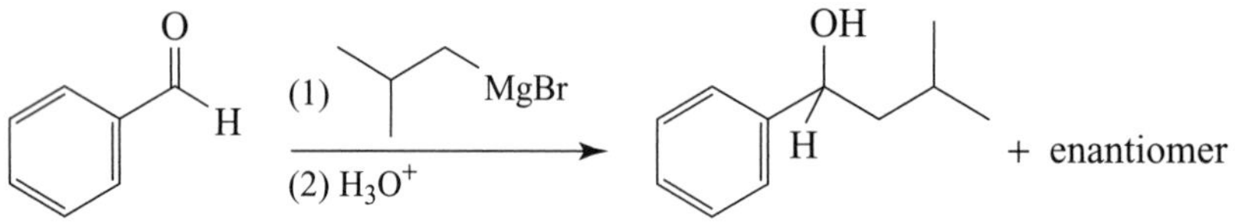 benzaldehyde (O, H) $\xrightarrow[(2)\ H_3O^+]{(1)\ \text{MgBr}}$ OH, H + enantiomer

(g) 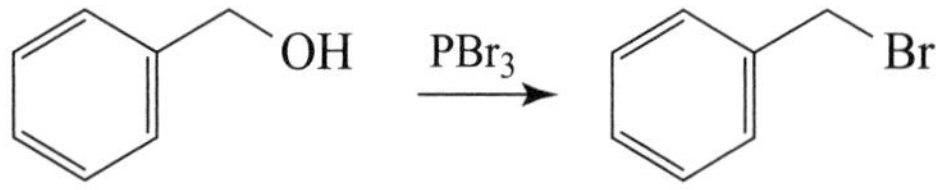 OH $\xrightarrow{PBr_3}$ Br

[from (a)]

(h) 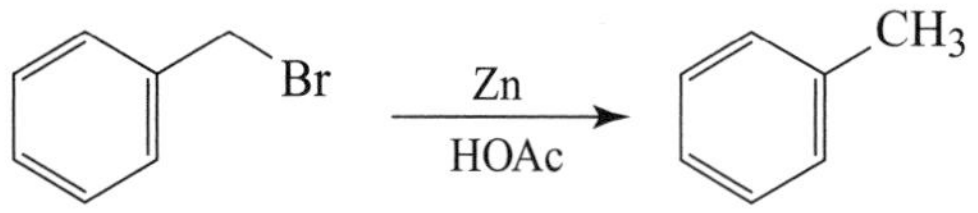 Br $\xrightarrow[\text{HOAc}]{\text{Zn}}$ CH_3

[from (g)]

or benzaldehyde (O, H) $\xrightarrow[BF_3]{\text{HS}\quad\text{SH}}$ S, S dithiolane $\xrightarrow[(H_2)]{\text{Raney Ni}}$ CH_3

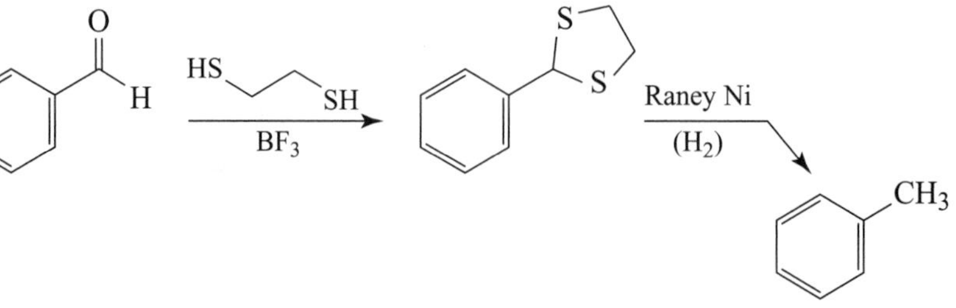

or Clemmensen reduction of benzaldehyde

(i) Benzaldehyde $\xrightarrow[HA]{CH_3OH}$ $C_6H_5CH(OCH_3)_2$

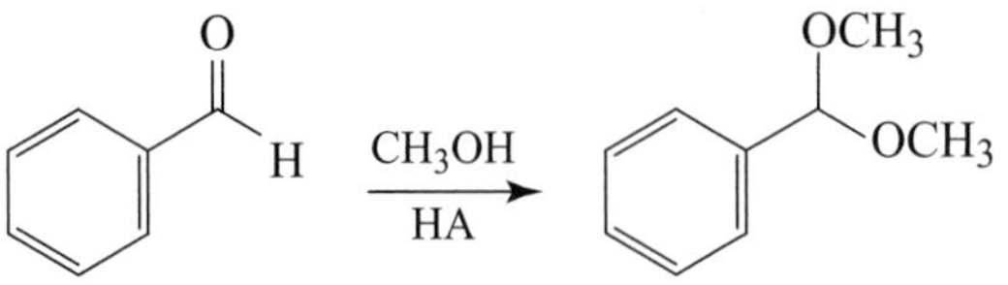

(j) Benzaldehyde $\xrightarrow[H_3{}^{18}O^+]{H_2{}^{18}O}$ $C_6H_5CH{=}^{18}O$ (See Problem 16.8 for the mechanism)

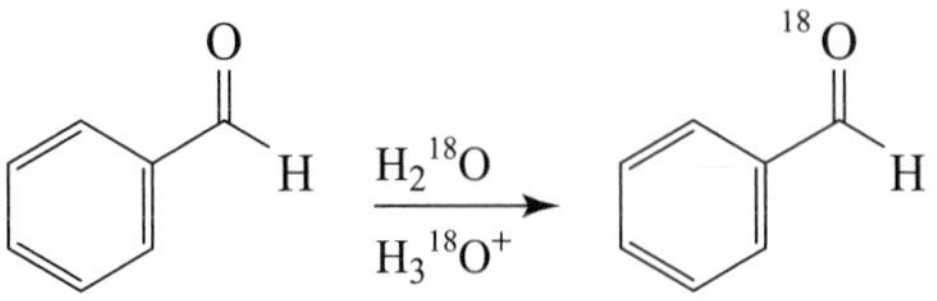

(k) Benzaldehyde $\xrightarrow[(2)\ H_3O^+]{(1)\ NaBD_4}$ $C_6H_5CHD(OH)$ + enantiomer

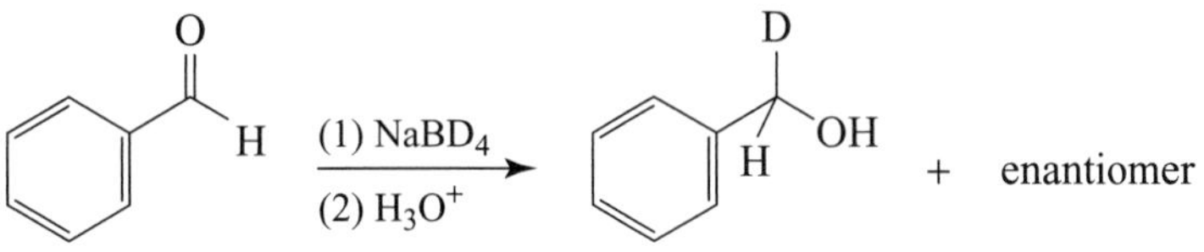

(l) Benzaldehyde $\xrightarrow{HCN}$ $C_6H_5CH(OH)CN$ + enantiomer

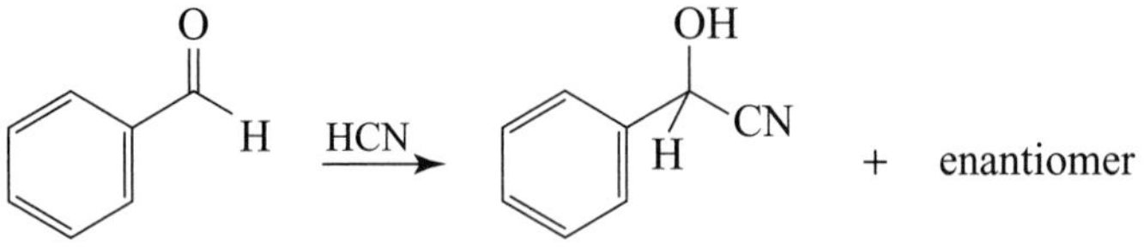

(a cyanohydrin)

(m) Benzaldehyde $\xrightarrow[HA]{NH_2OH}$ $C_6H_5CH{=}N{-}OH$ + stereoisomer

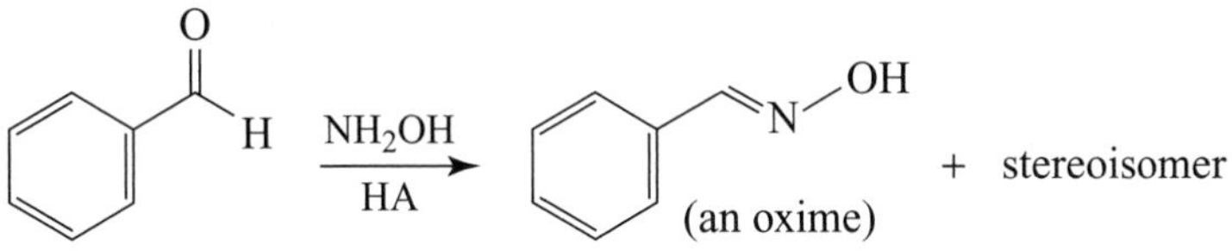

(an oxime)

(n) Benzaldehyde + $H_2N{-}NHC_6H_5$ $\xrightarrow[CH_3COOH]{H_3O^+}$ $C_6H_5CH{=}N{-}NHC_6H_5$ + stereoisomer

(a phenylhydrazone)

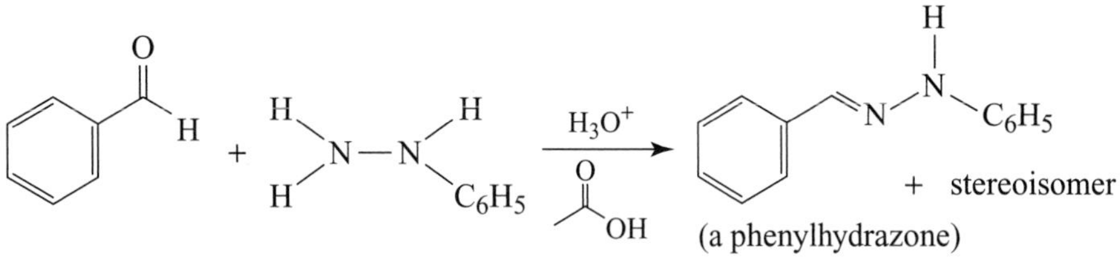

(o) Benzaldehyde + $(C_6H_5)_3\overset{+}{P}{-}\ddot{C}H^{-}{-}CH{=}CH_2$ $\longrightarrow$ $C_6H_5CH{=}CH{-}CH{=}CH_2$

(a Wittig reagent)

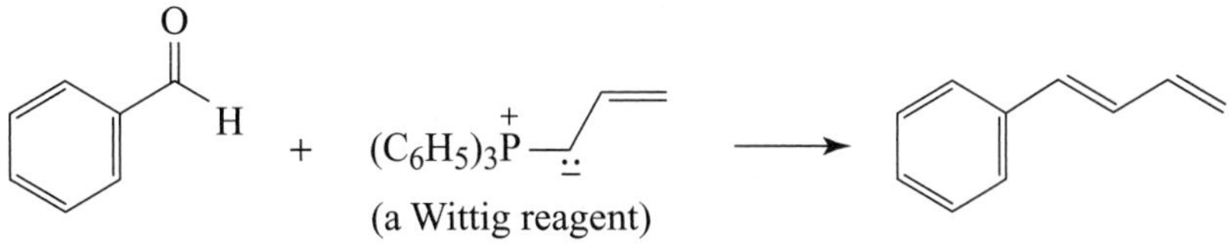

16.31 (a) Benzene + CH_3CH_2COCl $\xrightarrow{AlCl_3}$ $C_6H_5COCH_2CH_3$

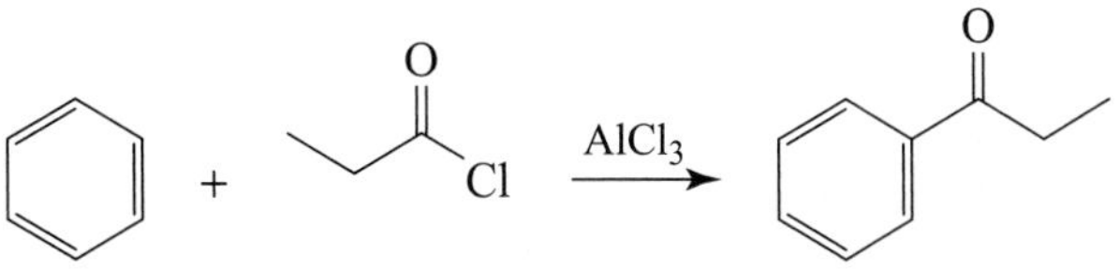

(b) + Li (2) H_3O^+ O

(c) O H + MgBr (2) H_3O^+ OH H_2CrO_4 acetone O

16.32 (a) OH PCC CH_2Cl_2 O H

(b) O OH $SOCl_2$ O Cl LiAlH(O-t-Bu)$_3$ Et_2O, −78° C O H

(c) (1) $KMnO_4$, OH^- (2) H_3O^+ O OH [as in (b)] O H

(d) (1) $KMnO_4$, OH^- (2) H_3O^+ O OH [as in (b)] O H

(e) O OCH_3 (1) ( )$_2$AlH, −78°C (2) H_2O O H

(f) N (1) ( )$_2$AlH, −78°C (2) H_2O O H

16.33 [cyclohexanol] $\xrightarrow[\text{acetone}]{H_2CrO_4}$ **A** $\xrightarrow[(2)\ NH_4^+/H_2O]{(1)\ CH_3MgI}$ **B** (OH) $\xrightarrow[\text{heat}]{HA}$ **C** $\xrightarrow[(2)\ Me_2S]{(1)\ O_3}$ **D** $\xrightarrow[(2)\ HA]{(1)\ Ag_2O,\ OH^-}$ **E**

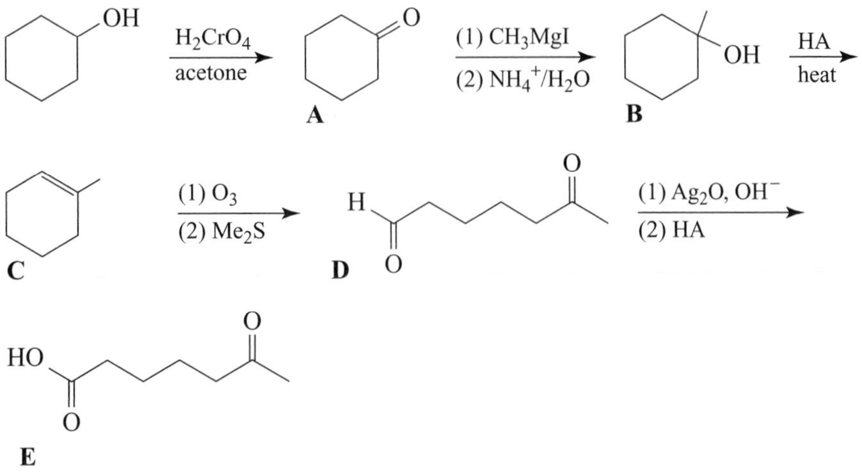

16.34 $\xrightarrow[\text{heat}]{H_3O^+}$ (OH, OH) + $H_2C{=}O$

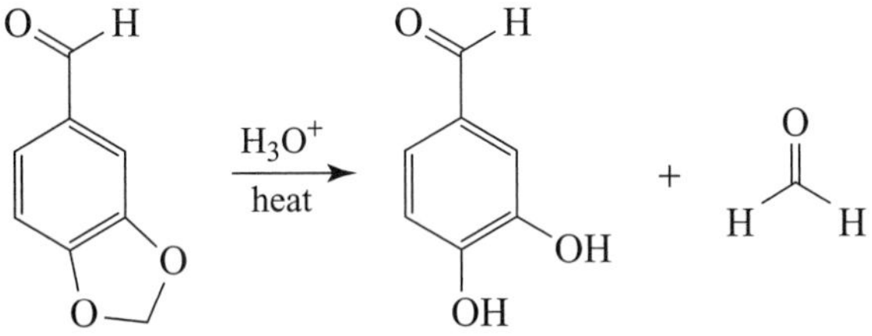

The compound $C_7H_6O_3$ is 3,4-dihydroxybenzaldehyde. The reaction involves hydrolysis of the acetal of formaldehyde.

16.35 (a) Br $\xrightarrow[Et_2O]{Mg}$ MgBr $\xrightarrow[(2)\ H_3O^+]{(1)\ CH_3CHO}$ OH racemic

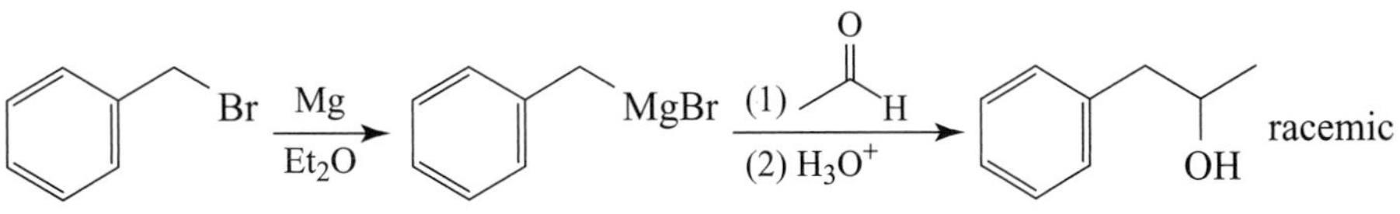

(b) MgBr [from (a)] $\xrightarrow[(2)\ H_3O^+]{(1)\ \text{ethylene oxide}}$ OH $\xrightarrow[CH_2Cl_2]{PCC}$ H

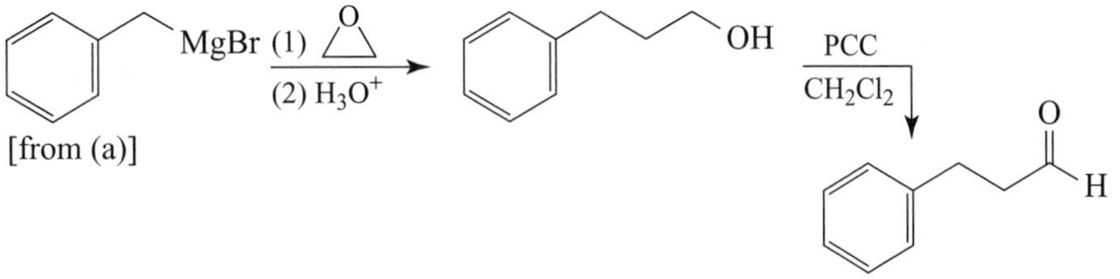

(c) Br $\xrightarrow[(2)\ C_6H_5Li]{(1)\ (C_6H_5)_3P}$ $\overset{+}{P}(C_6H_5)_3$ (a Wittig reagent) $\xrightarrow{C_6H_5CH{=}CHCHO}$

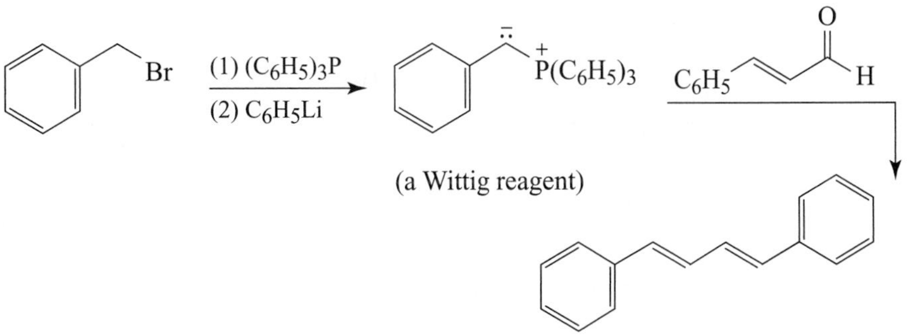

(d) 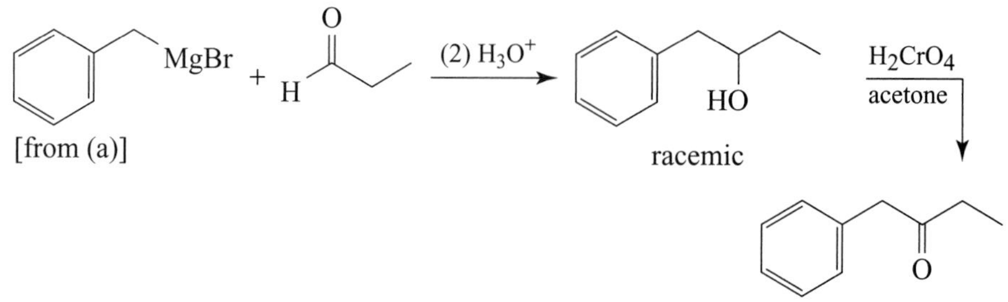 MgBr [from (a)] + H (2) H_3O^+ HO racemic H_2CrO_4 acetone

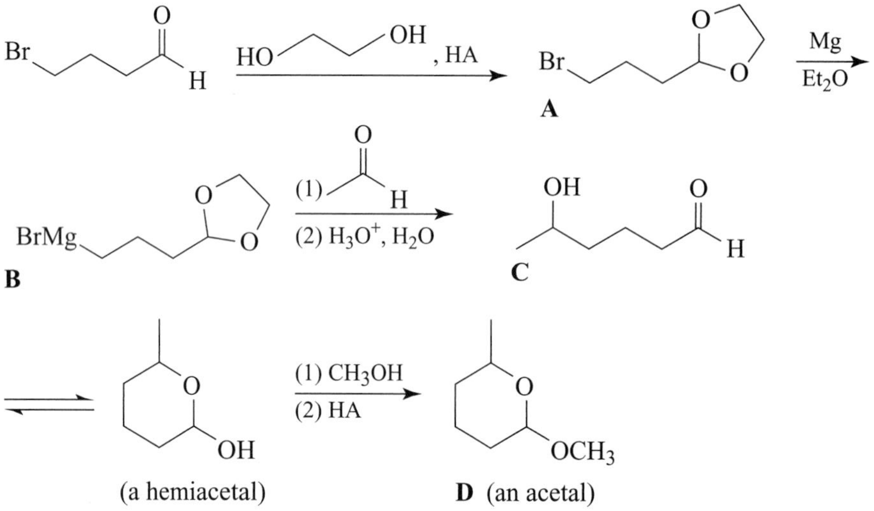

16.36 Br H HO OH , HA Br A O O Mg Et_2O

BrMg O O B (1) H (2) H_3O^+, H_2O OH C H

O OH (a hemiacetal) (1) CH_3OH (2) HA O OCH_3 **D** (an acetal)

16.37 (7*S*) (3*S*) O (7*R*) (3*S*) O

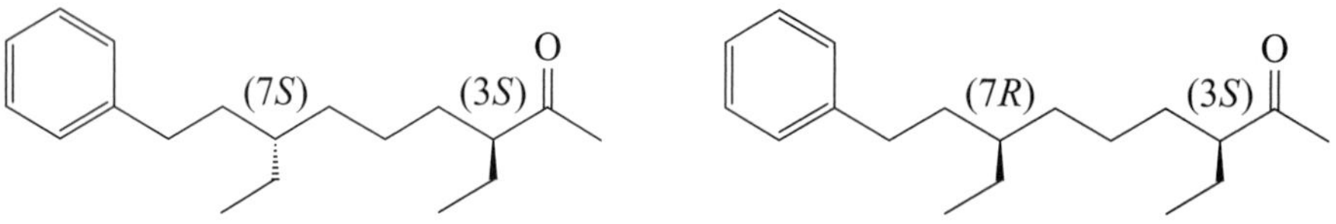

3,7-Diethyl-9-phenylnonan-2-one (dianeackerone) stereoisomers

16.38 OH PCC CH_2Cl_2 O H **A** CH_3OH HA OCH_3 OCH_3 **B**

$KMnO_4$, OH^- cold, dilute HO OCH_3 OCH_3 OH **C** H_3O^+ H_2O HO O H OH Glyceraldehyde

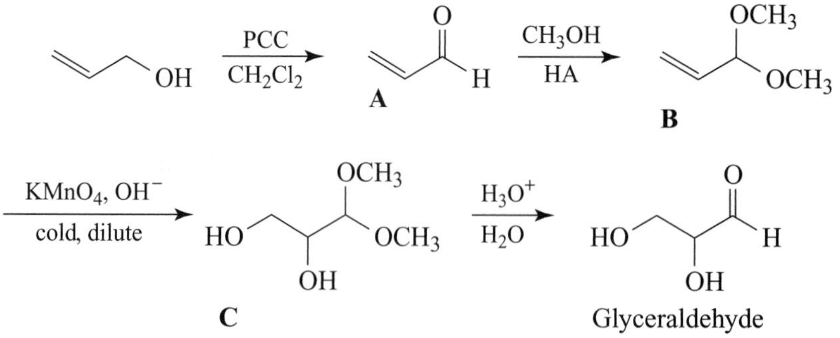

The product would be racemic since no chiral reagents were used.

16.39

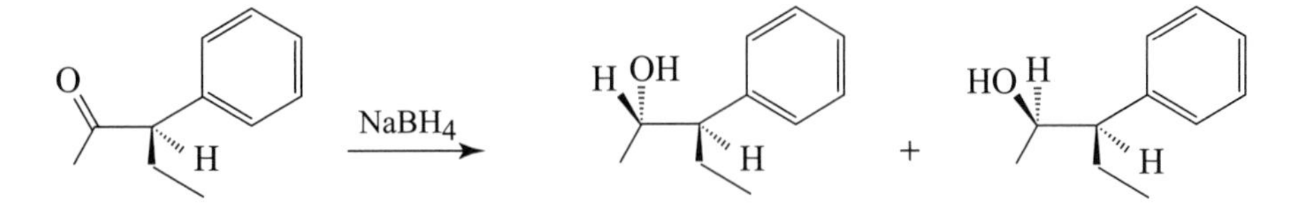

O, H $\xrightarrow{NaBH_4}$ H OH, H + HO H, H

(*R*)-3-Phenyl-2-pentanone

(*S*) (*R*) (*R*) (*R*)

Diastereomers

16.40 Br–$(CH_2)_7$–Br $\xrightarrow[(2)\ 2\ RLi]{(1)\ 2\ (C_6H_5)_3P}$ $(C_6H_5)_3\overset{+}{P}$–$\ddot{C}H$–$(CH_2)_7$–$\ddot{C}H$–$\overset{+}{P}(C_6H_5)_3$

A

2 (O, 11) $\longrightarrow$ (11, 7, 11) **B** $\xrightarrow[Pt,\ pressure]{H_2}$ (11, 9, 11) **C**

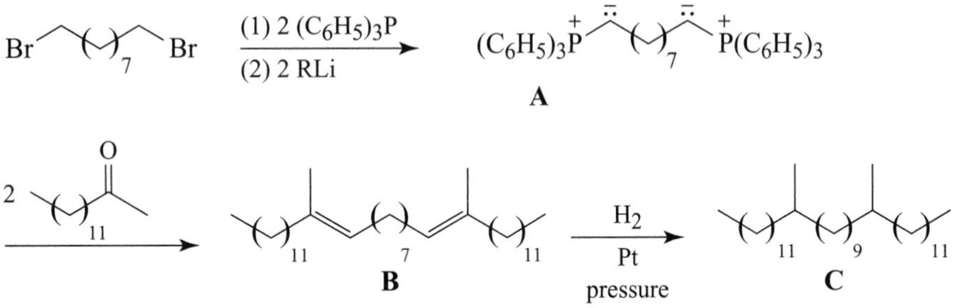

16.41 (a) ***Retrosynthetic Analysis***

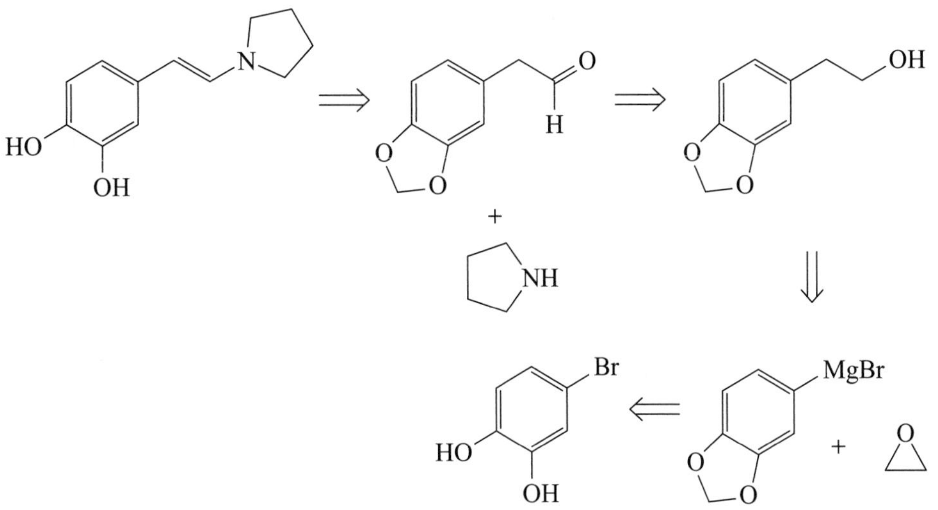

HO, OH, N ⟹ O, O, O, H + NH ⟹ O, O, OH

⇓

HO, OH, Br ⟸ O, O, MgBr + O

Synthesis

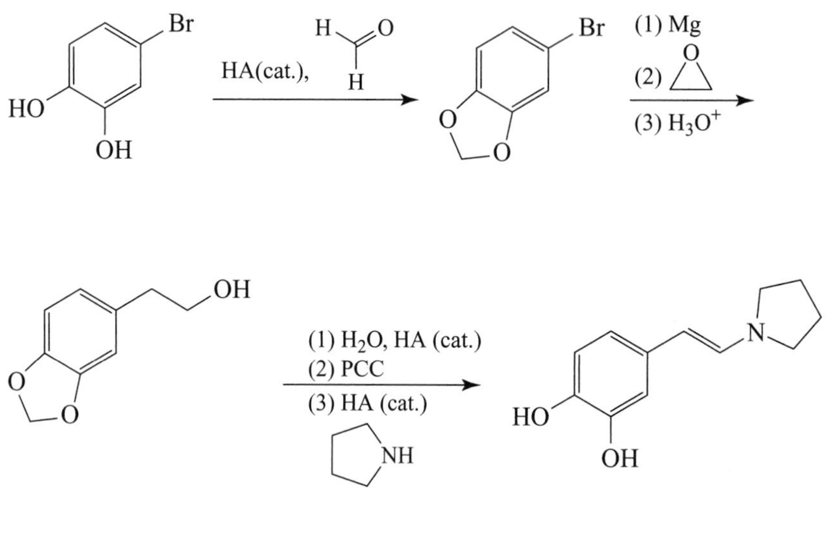

Br HO OH → HA(cat.), H(C=O)H → Br O O → (1) Mg (2) O (epoxide) (3) H_3O^+ →

OH O O → (1) H_2O, HA (cat.) (2) PCC (3) HA (cat.) NH → N HO OH

(b) ***Retrosynthetic Analysis***

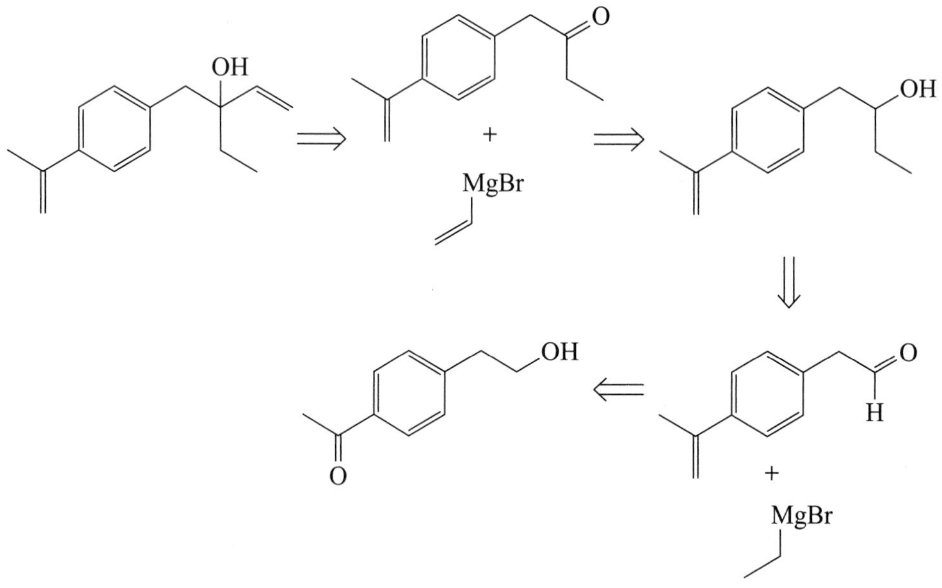

OH ⇒ O + MgBr ⇒ OH

⇓

OH O ⇐ O H + MgBr

Synthesis

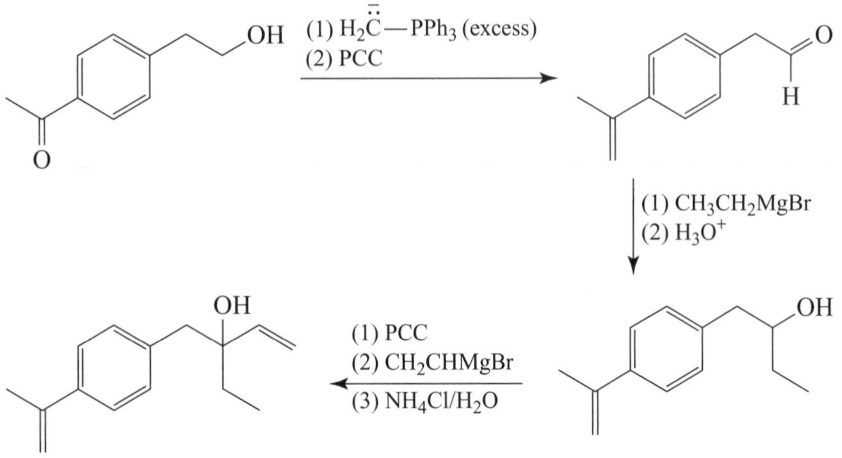

(1) $H_2\ddot{C}^{-}$—PPh_3 (excess)
(2) PCC

(1) CH_3CH_2MgBr
(2) H_3O^+

(1) PCC
(2) $CH_2CHMgBr$
(3) NH_4Cl/H_2O

16.42

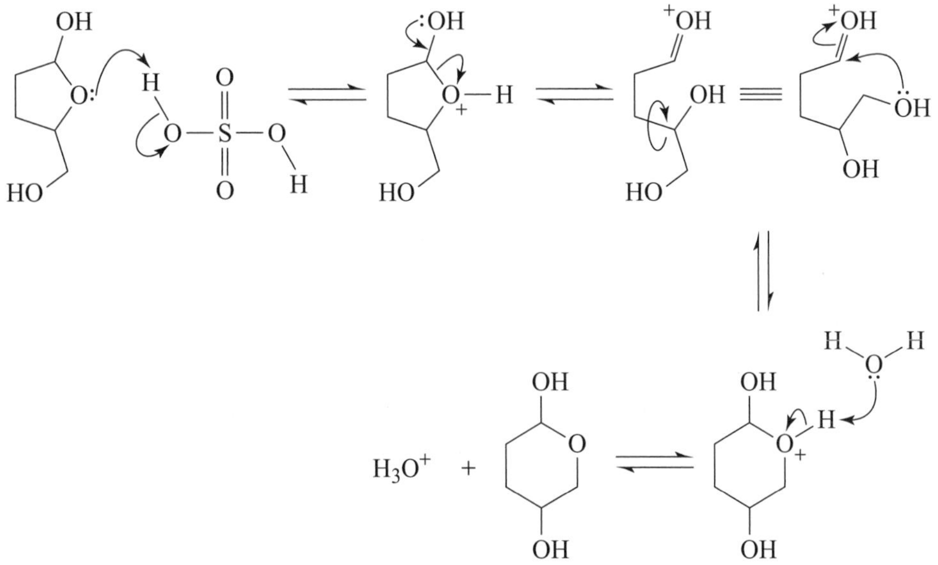

H_3O^+ +

16.43 The two nitrogen atoms of semicarbazide that are adjacent to the C=O group bear partial positive charges because of resonance contributions made by the second and third structures below.

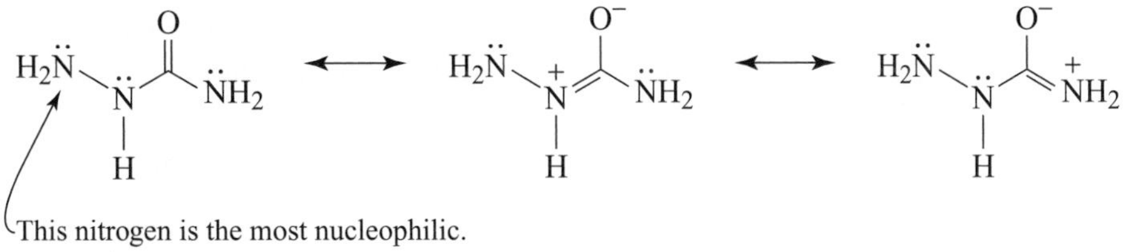

This nitrogen is the most nucleophilic.

16.44 Hydrolysis of the acetal linkage of multistriatin produces the ketodiol below.

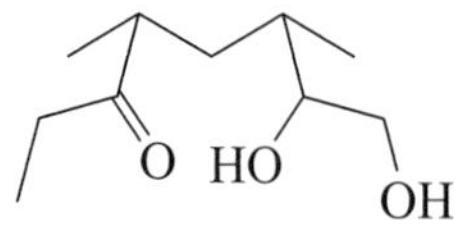

O HO OH

16.45

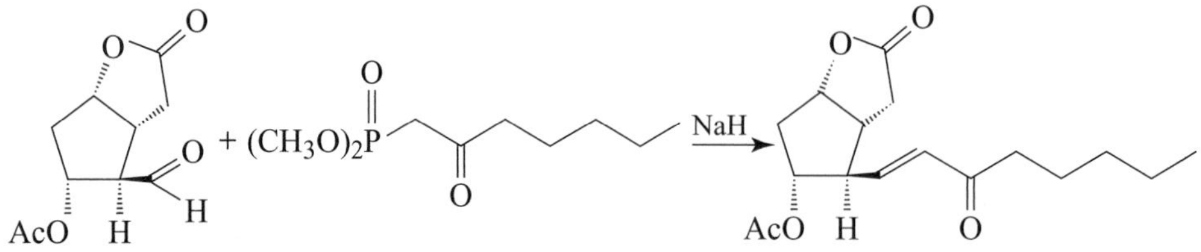

O O O + $(CH_3O)_2P$ O O AcO H H NaH O O AcO H O

16.46 **Compound W** is

singlet δ 3.4

multiplet δ 7.3 CH_2 C=O CH_2 IR peak near 1715 cm^{-1}

(1) $KMnO_4$, OH^-, heat

(2) H_3O^+

COH COH

Phthalic acid

Compound X is

multiplet δ 7.5 C CH_2 CH_2 triplet δ 2.5

triplet δ 3.1

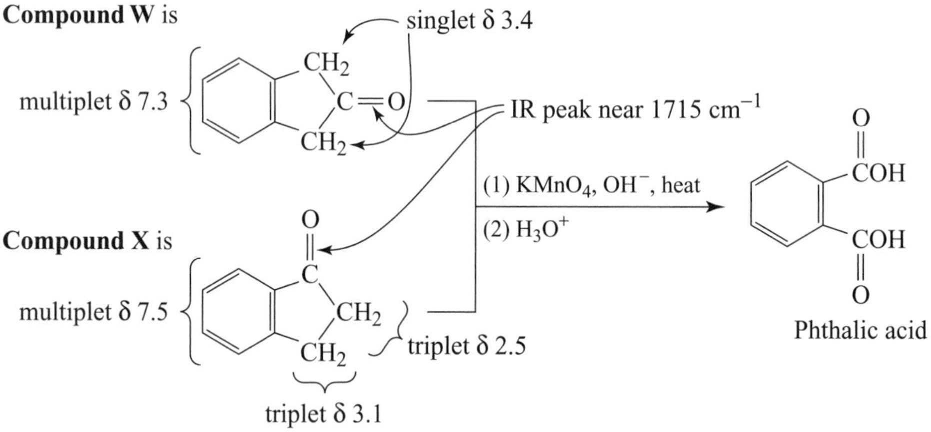

16.47 Each ^{1}H NMR spectrum (Figs. 16.4 and 16.5) has a five-hydrogen peak near *δ* 7.2, suggesting the **Y** and **Z** each has a C_6H_5— group. The IR spectrum of each compound shows a strong peak near 1710 cm^{-1}. This absorption indicates that each compound has a C=O group not adjacent to the phenyl group. We have, therefore, the following pieces,

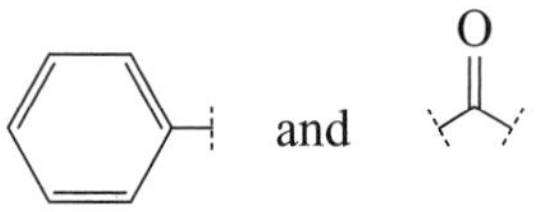

and

If we subtract the atoms of these pieces from the molecular formula,

$$\begin{array}{ll} & C_{10}H_{12}O \\ & \underline{C_7H_5O \quad (C_6H_5 + C{=}O)} \\ \text{we are left with} & C_3H_7 \end{array}$$

In the ^{1}H NMR spectrum of **Y**, we see an ethyl group [triplet, *δ* 1.0 (3H) and quartet, *δ* 2.45 (2H)] and an unsplit —CH_2— group signal [singlet, *δ* 3.7 (2H)]. This means that **Y** must be

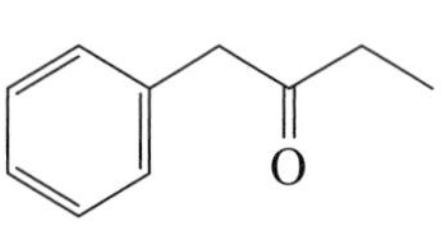

O

1-Phenyl-2-butanone

In the ^{1}H NMR spectrum of **Z**, we see an unsplit $-CH_3$ group signal [singlet, δ 2.1 (3H)] and two triplets at δ 2.7 and 2.9. This means **Z** must be

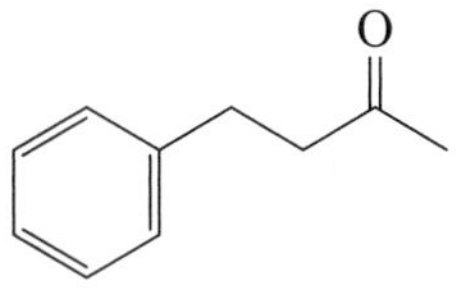

O

4-Phenyl-2-butanone

16.48 That compound **A** forms a phenylhydrazone, gives a negative Tollens' test, and gives an IR band near 1710 cm^{-1} indicates that **A** is a ketone. The ^{13}C spectrum of **A** contains only four signals indicating that **A** has a high degree of symmetry. The information from the DEPT ^{13}C NMR spectra enables us to conclude that **A** is diisobutyl ketone:

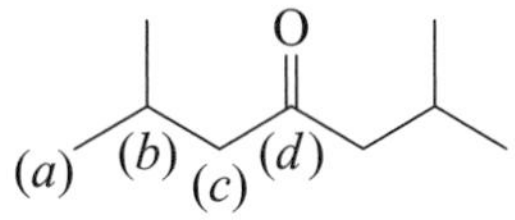

O

(a) (b) (c) (d)

Assignments:

(a) δ 22.6
(b) δ 24.4
(c) δ 52.3
(d) δ 210.0

16.49 That the ^{13}C spectrum of **B** contains only three signals indicates that **B** has a highly symmetrical structure. The information from DEPT spectra indicates the presence of equivalent methyl groups (CH_3 at δ 19), equivalent $-\overset{|}{\underset{|}{C}}-$ groups (at δ 71), and equivalent $>C{=}O$ groups (at δ 216). These features allow only one possible structure for **B**:

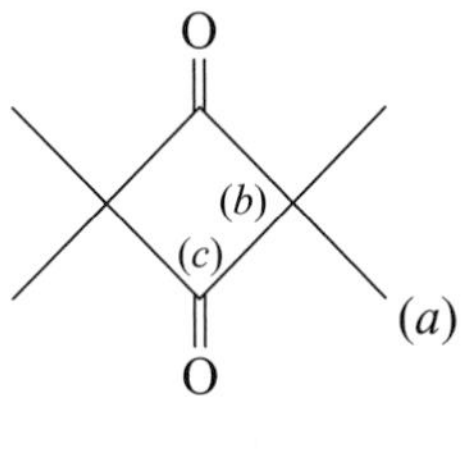

O

(b)
(c)
(a)

O

Assignments:

(a) δ 19
(b) δ 71
(c) δ 216

Challenge Problems

16.50 (a) O—H stretch at about 3300 cm^{-1}; C=O stretch at about 1710 cm^{-1}

(b) 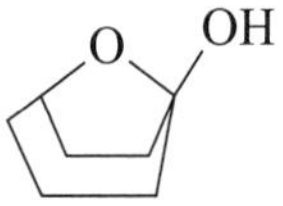 O, OH (Intramolecular hemiacetal from C)

16.51 H_3CO, H_3CO, OH

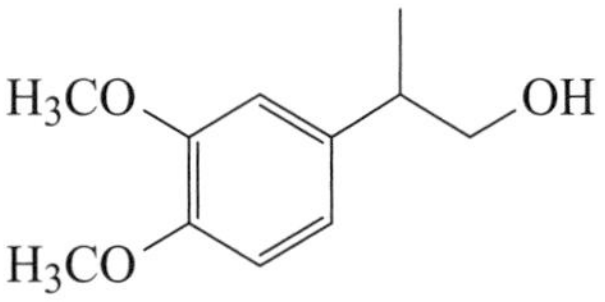

E

QUIZ

16.1 Which Wittig reagent could be used to synthesize 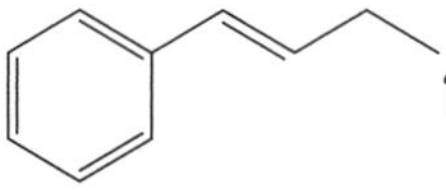 ? (Assume any other needed reagents are available.)

(a) 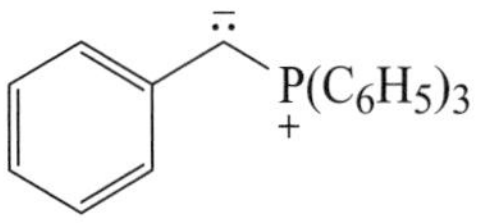 $P(C_6H_5)_3$ +

(b) $P(C_6H_5)_3$ +

(c) $P(C_6H_5)_3$ +

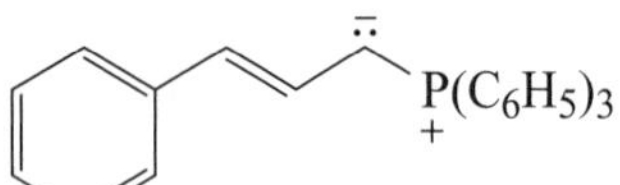

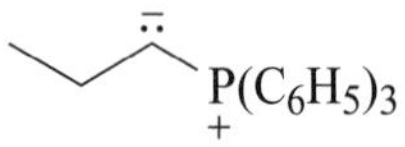

(d) More than one of these

(e) None of these

16.2 Which compound is an acetal?

(a) 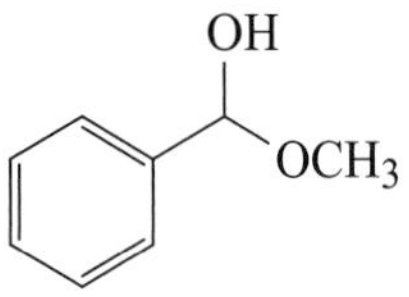 OH, OCH_3

(b) 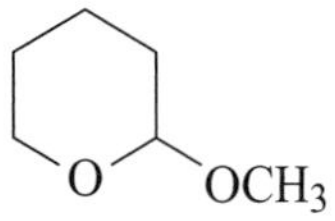 O, OCH_3

(c) 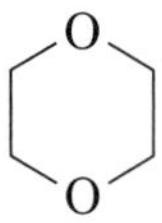 O, O

(d) More than one of these

(e) None of these

16.3 Which reaction sequence could be used to convert [structure: terminal alkyne, subscript 4, H] to [structure: methyl ketone, subscript 4]?

(a) O_3, then Me_2S, then $AlCl_3$, then acetic acid (CH_3COOH)

(b) (1) H_2, P-2 cat.; (2) $Hg(OAc)_2$/THF-H_2O; (3) $NaBH_4$, HO^-; (4) PCC

(c) HCl, then acetic acid (CH_3COOH)

(d) O_3, then Me_2S, then H_2SO_4, $HgSO_4$, H_2O, heat

(e) acetic acid (CH_3COOH), then H_2O_2, OH^-/H_2O

16.4 Complete the following syntheses. If more than one step is required for a transformation, list them as (1), (2), (3), and so on.

(a) C_6H_5–CH_3 $\xrightarrow[hv]{NBS}$ **A** [ ] $\xrightarrow[THF]{\textbf{B}\ [\ ]}$ $C_6H_5CH_2C{\equiv}N$ $\xrightarrow{\textbf{C}\ [\ ]}$ $C_6H_5CH_2CHO$

(b) Cl–C_6H_4–CH_2OH $\xrightarrow{\textbf{A}\ [\ ]}$ Cl–C_6H_4–CHO $\xrightarrow{\textbf{B}\ [\ ]}$ Cl–C_6H_4–(1,3-dioxolan-2-yl) $\xrightarrow{\textbf{C}\ [\ ]}$ Cl–C_6H_4–CHO

(c) Br —A→ [ ] B [ ] —C_6H_5Li→

$P(C_6H_5)_3$ + —C→ 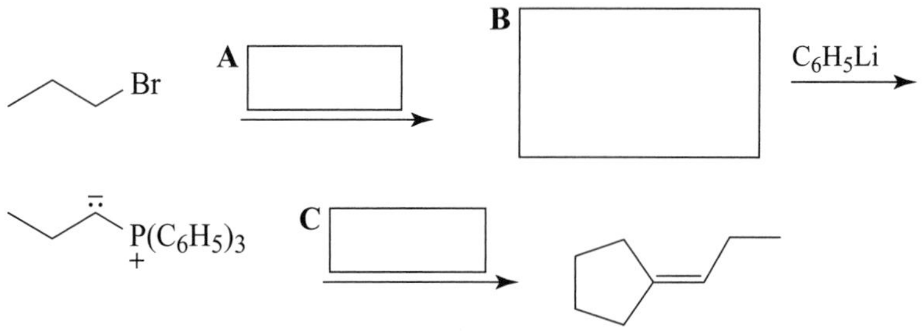 [ ]

(d) 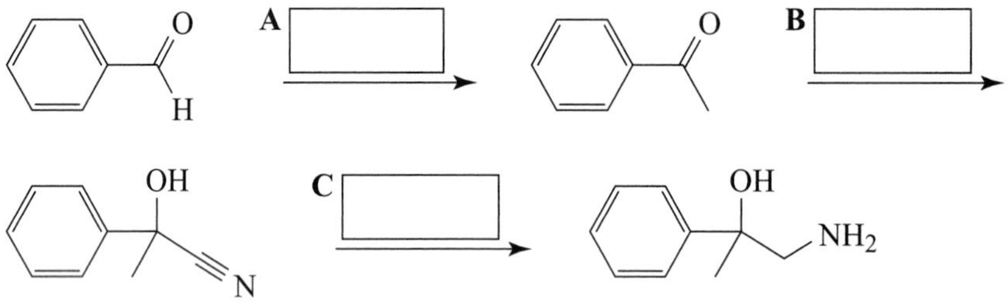 —A→ [ ] —B→ [ ]

OH —C→ [ ] OH NH_2

N

(racemic) (racemic)

16.5 An industrial synthesis of benzaldehyde makes use of toluene and molecular chlorine as starting materials to produce $C_6H_5CHCl_2$ (Cl, Cl). This compound is then converted to benzaldehyde. Suggest what steps are involved in the process.

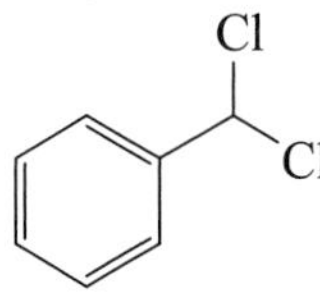

16.6 In the case of aldehydes and unsymmetrical ketones, two isomeric oximes are possible. What is the origin of this isomerism?

17 CARBOXYLIC ACIDS AND THEIR DERIVATIVES: NUCLEOPHILIC ADDITION-ELIMINATION AT THE ACYL CARBON

SOLUTIONS TO PROBLEMS

17.1 (a) 2-Methylbutanoic acid

(b) (*Z*)-3-Pentenoic acid or (*Z*)-pent-3-enoic acid

(c) Sodium 4-bromobutanoate

(d) 5-Phenylpentanoic acid

(e) (*E*)-3-Ethyl-3-pentenoic acid or (*E*)-3-Ethylpent-3-enoic acid

17.2 Acetic acid, in the absence of solvating molecules, exists as a dimer owing to the formation of two intermolecular hydrogen bonds:

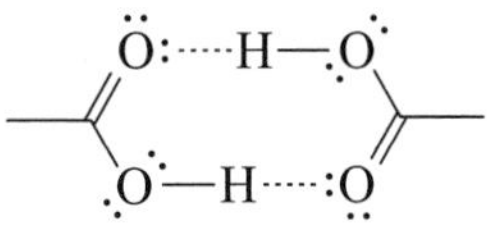

At temperatures much above the boiling point, the dimer dissociates into the individual molecules.

17.3 (a) 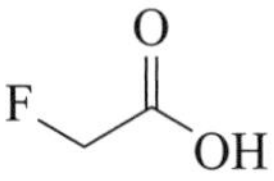 (F— is more electronegative than H—)

(b) 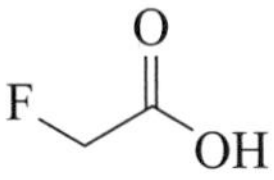 (F— is more electronegative than Cl—)

(c) 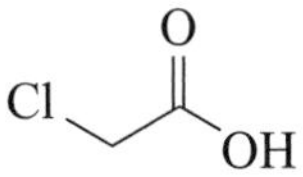 (Cl— is more electronegative than Br—)

(d) 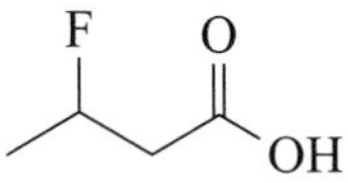 (F— is closer to $-CO_2H$)

(e) 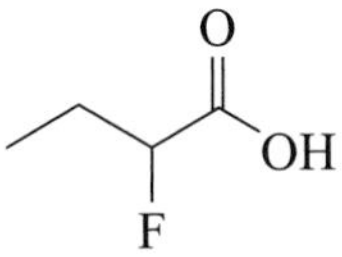 (F— is closer to $-CO_2H$)

(f) $Me_3\overset{+}{N}$–C₆H₄–COOH [$Me_3\overset{+}{N}$— is more electronegative than H—]

(g) CF_3–C₆H₄–COOH (CF_3— is more electronegative than CH_3—)

17.4 (a) $CH_3CH_2C(=O)OCH_3$ (b) O_2N–C₆H₄–C(=O)OCH₂CH₃

(c) $CH_3OC(=O)CH_2C(=O)OCH_3$ (d) C₆H₅C(=O)N(CH_3)$_2$

(e) CH₃CH₂CH₂CH₂CN (f) C₆H₄(C(=O)OCH₃)₂

(g) CH(C(=O)OCH₂CH₂CH₃)=CH(C(=O)OCH₂CH₂CH₃) (h) HC(=O)N(CH_3)$_2$

(i) $CH_3CH(Br)C(=O)Br$ (j) CH₂(C(=O)OCH₂CH₃)CH₂(C(=O)OCH₂CH₃)

17.5 (a) C₆H₅CH₂CH₃ $\xrightarrow[\text{(2) } H_3O^+]{\text{(1) } KMnO_4,\ OH^-,\ \text{heat}}$ C₆H₅COOH + CO_2

(b) Bromobenzene $\xrightarrow[Et_2O]{Mg}$ PhMgBr $\xrightarrow{CO_2}$ PhCO–OMgBr $\xrightarrow{H_3O^+}$ PhCO–OH

(c) Acetophenone $\xrightarrow[(2)\ H_3O^+]{(1)\ Cl_2/NaOH}$ PhCO–OH + $CHCl_3$

(d) Styrene $\xrightarrow[(2)\ H_3O^+]{(1)\ KMnO_4,\ OH^-,\ heat}$ PhCO–OH

(e) PhCH$_2$OH $\xrightarrow[(2)\ H_3O^+]{(1)\ KMnO_4,\ OH^-,\ heat}$ PhCO–OH

(f) PhCHO $\xrightarrow[(2)\ H_3O^+]{(1)\ KMnO_4,\ OH^-}$ PhCO–OH

17.6 These syntheses are easy to see if we work backward using a retrosynthetic analysis before writing the synthesis.

(a) ***Retrosynthetic analysis***

PhCH$_2$CO–OH $\Rightarrow$ PhCH$_2$MgBr + O=C=O $\Rightarrow$ PhCH$_2$Br

Synthesis

PhCH$_2$Br $\xrightarrow[Et_2O]{Mg}$ PhCH$_2$MgBr $\xrightarrow[(2)\ H_3O^+]{(1)\ CO_2}$ PhCH$_2$CO–OH

(b) ***Retrosynthetic analysis***

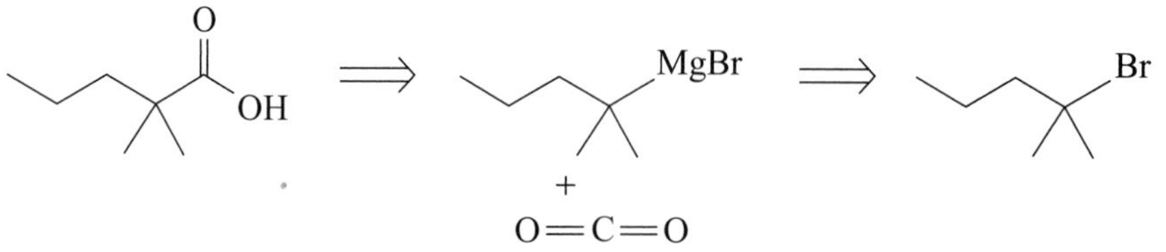

O, OH ⟹ MgBr + O=C=O ⟹ Br

Synthesis

Br $\xrightarrow[Et_2O]{Mg}$ MgBr + O=C=O $\xrightarrow[(2)\ H_3O^+]{(1)\ CO_2}$ O, OH

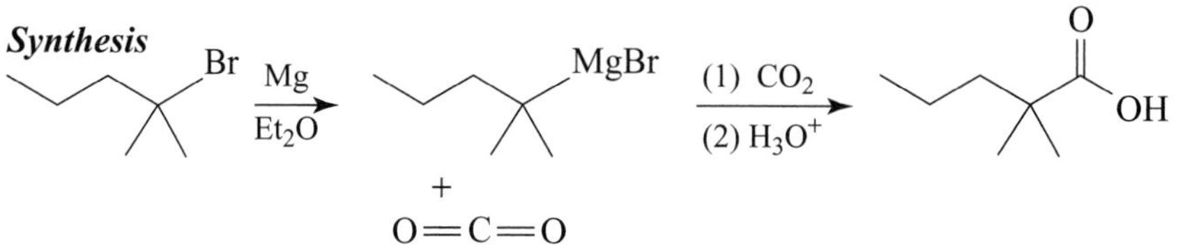

(c) ***Retrosynthetic analysis***

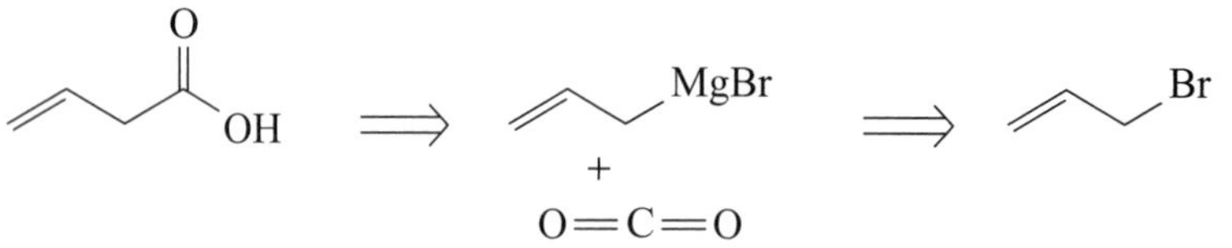

O, OH ⟹ MgBr + O=C=O ⟹ Br

Synthesis

Br $\xrightarrow[Et_2O]{Mg}$ MgBr + O=C=O $\xrightarrow[(2)\ H_3O^+]{(1)\ CO_2}$ O, OH

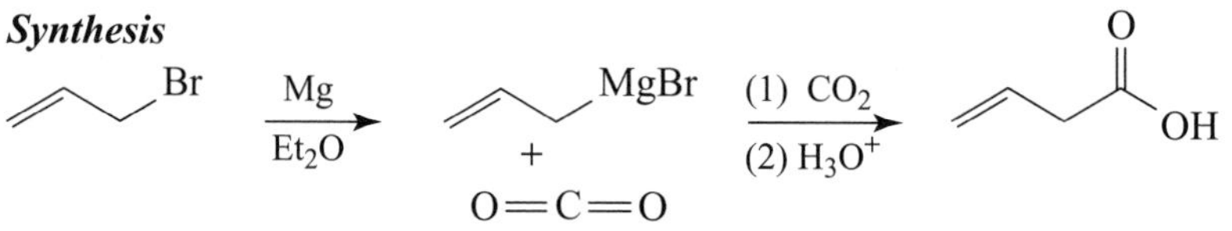

(d) ***Retrosynthetic analysis***

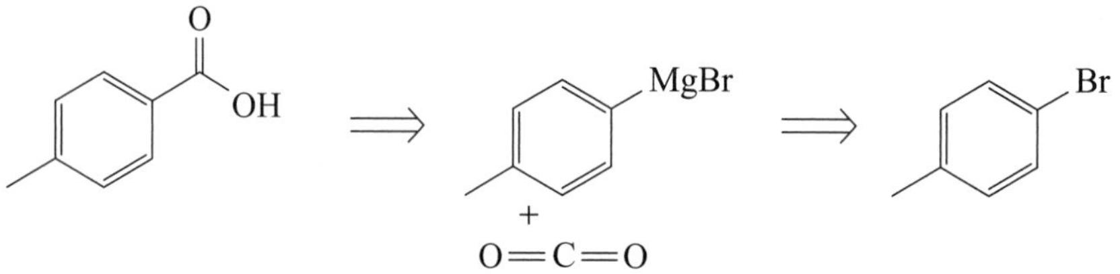

O, OH ⟹ MgBr + O=C=O ⟹ Br

Synthesis

Br $\xrightarrow[Et_2O]{Mg}$ MgBr + O=C=O $\xrightarrow[(2)\ H_3O^+]{(1)\ CO_2}$ O, OH

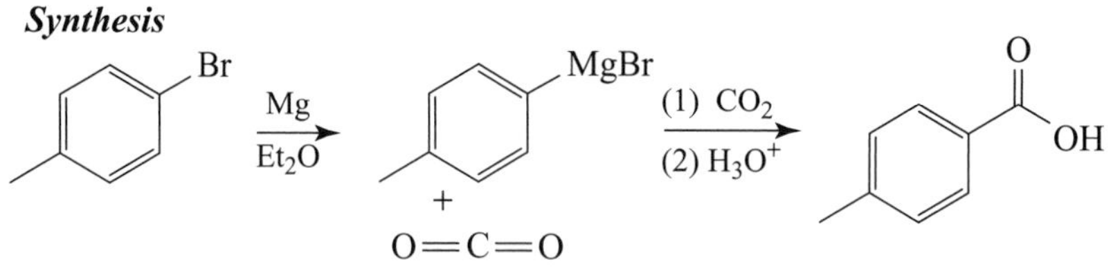

(e) ***Retrosynthetic analysis***

O, OH ⟹ MgBr + O=C=O ⟹ Br

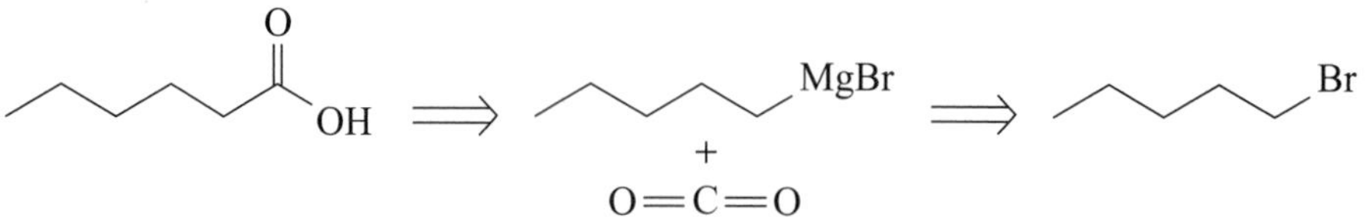

Synthesis

Br $\xrightarrow[Et_2O]{Mg}$ MgBr $\xrightarrow[(2)\ H_3O^+]{(1)\ CO_2}$ O, OH

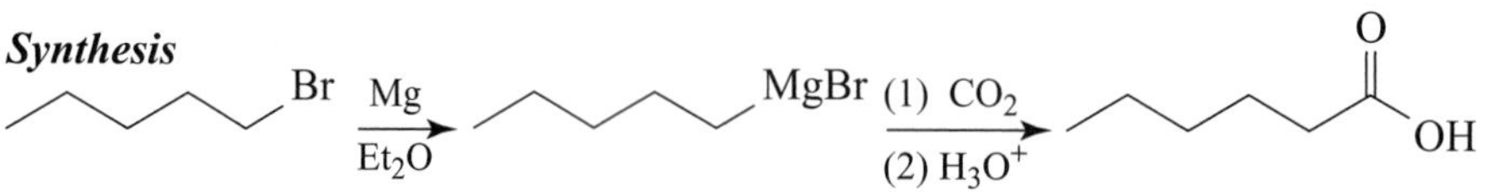

17.7 (a)

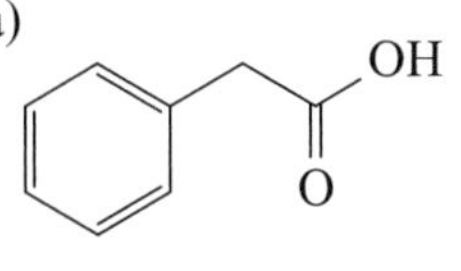

OH

O

Retrosynthetic analysis

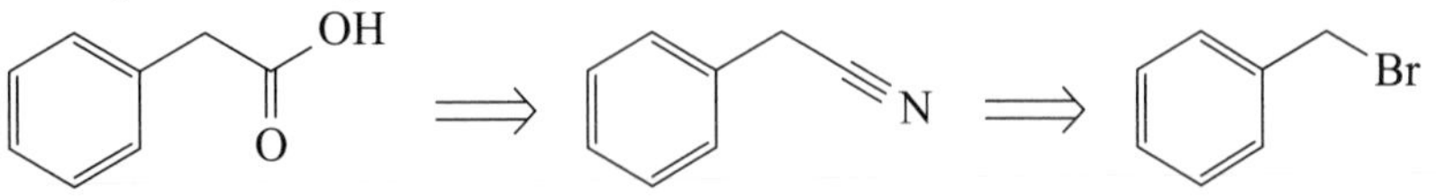

OH

O ⟹ N ⟹ Br

Synthesis

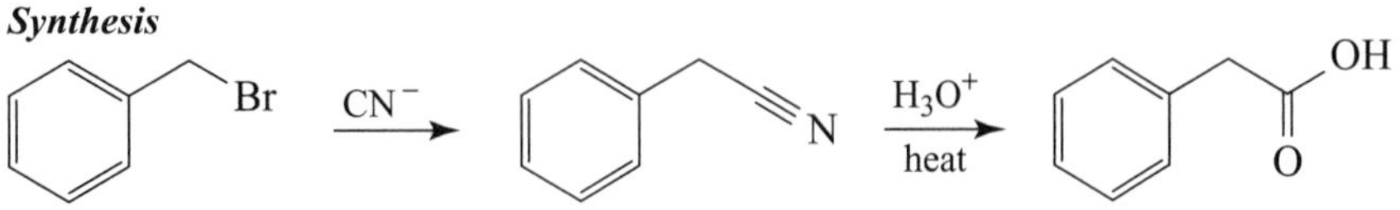

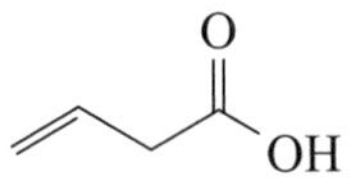

Br $\xrightarrow{CN^-}$ N $\xrightarrow[\text{heat}]{H_3O^+}$ OH O

O

OH

Retrosynthetic analysis

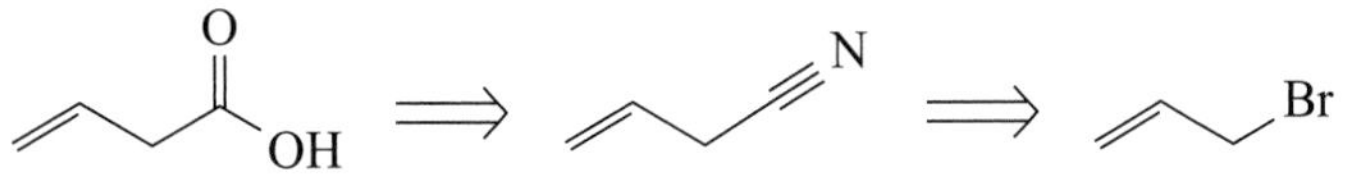

O

OH ⟹ N ⟹ Br

Synthesis

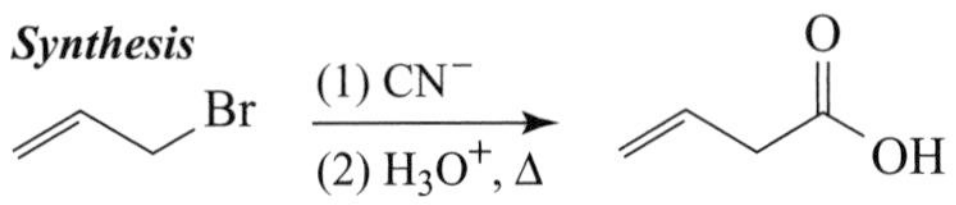

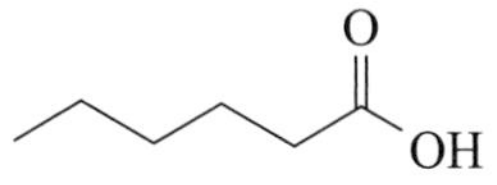

Br $\xrightarrow[(2)\ H_3O^+,\ \Delta]{(1)\ CN^-}$ O OH

O

OH

Retrosynthetic analysis

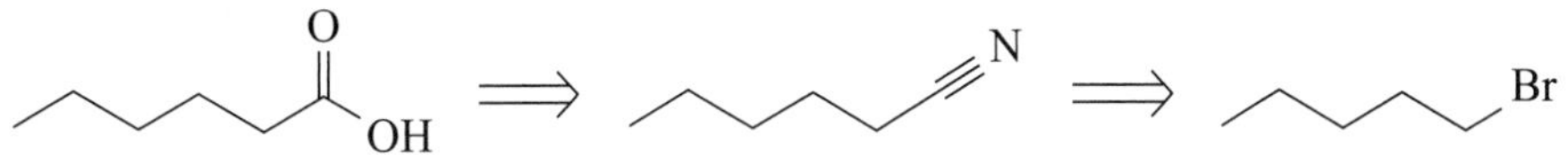

O

OH ⟹ N ⟹ Br

Synthesis

Br $\xrightarrow[(2)\ H_3O^+,\ \text{heat}]{(1)\ CN^-}$ O OH

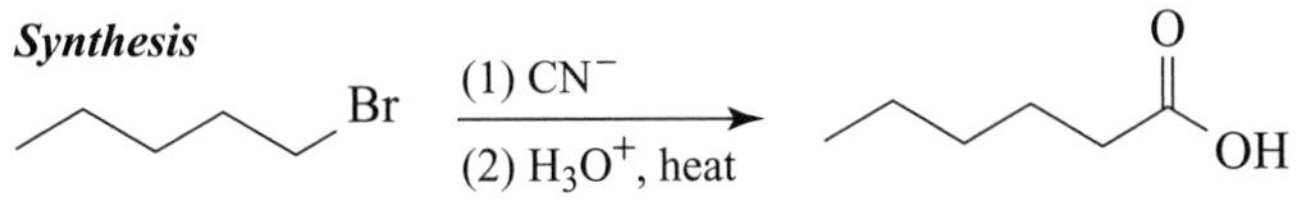

(b) A nitrile synthesis. Preparation of a Grignard reagent from HO Br would not be possible because of the presence of the acidic hydroxyl group.

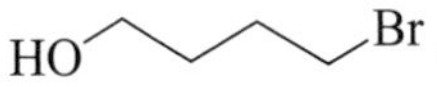

17.8 Since maleic acid is a cis dicarboxylic acid, dehydration occurs readily:

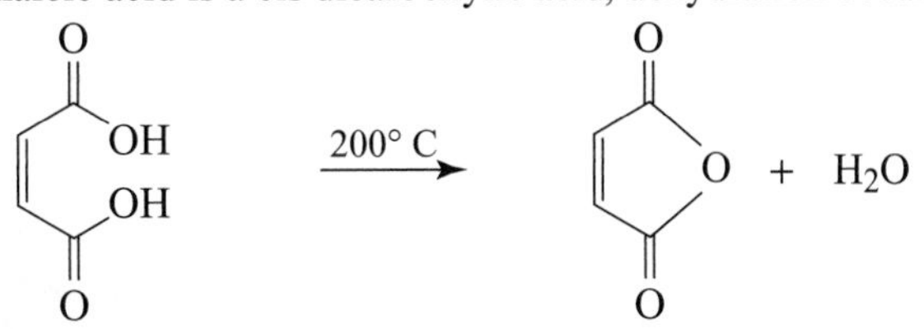

O OH OH O $\xrightarrow{200°\ C}$ O O O + H_2O

Maleic acid Maleic anhydride

Being a trans dicarboxylic acid, fumaric acid must undergo isomerization to maleic acid first. This isomerization requires a higher temperature.

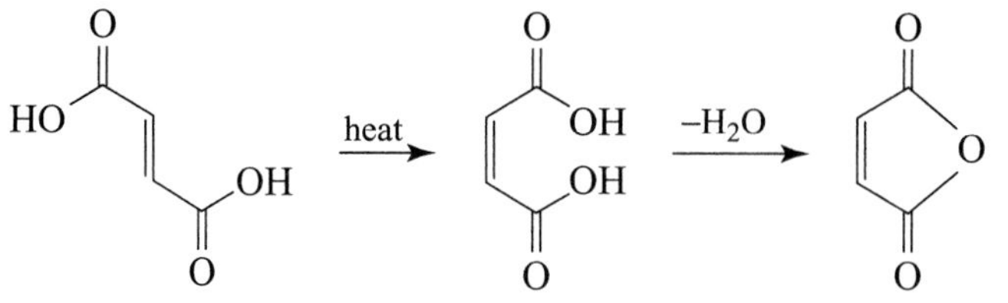

Fumaric acid $\xrightarrow{\text{heat}}$ maleic acid $\xrightarrow{-H_2O}$ maleic anhydride

17.9 The labeled oxygen atom should appear in the carboxyl group of the acid. (Follow the reverse steps of the mechanism in Section 17.7A of the text using $H_2{}^{18}O$.)

17.10

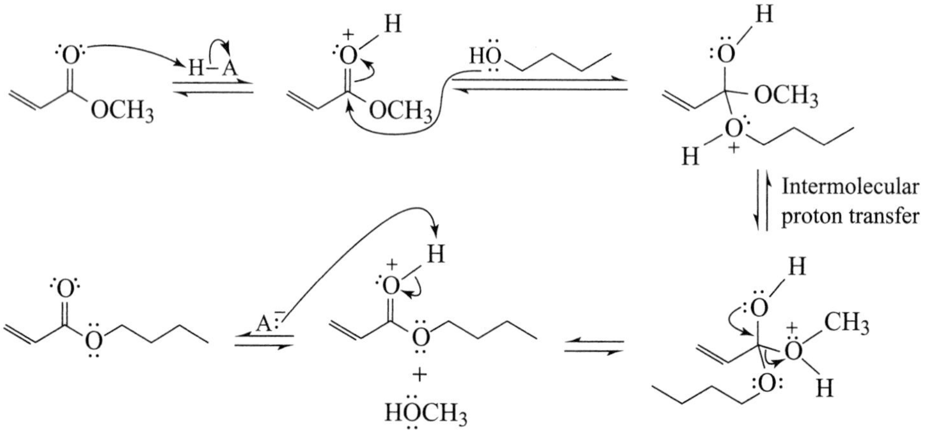

H–A; OCH_3; HO; Intermolecular proton transfer; CH_3; A:⁻; + $HOCH_3$

17.11 (a) (1)

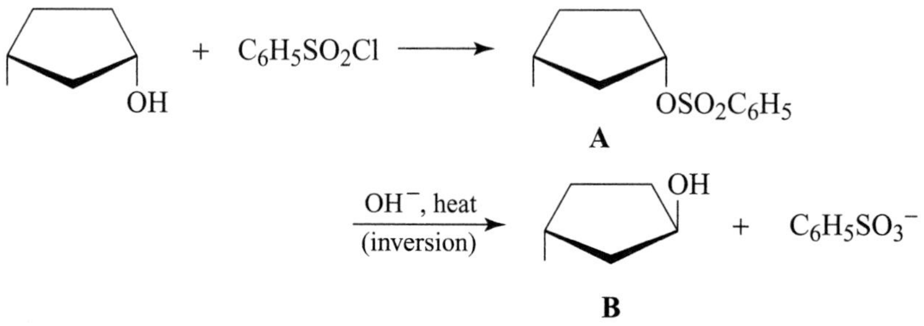

+ $C_6H_5SO_2Cl$ ⟶ $OSO_2C_6H_5$ **A**

$\xrightarrow[\text{(inversion)}]{OH^-,\ \text{heat}}$ **B** + $C_6H_5SO_3^-$

(2)

+ C_6H_5COCl ⟶ **C**

$\xrightarrow[\text{(retention)}]{OH^-,\ \text{heat}}$ **D** + $C_6H_5CO_2^-$

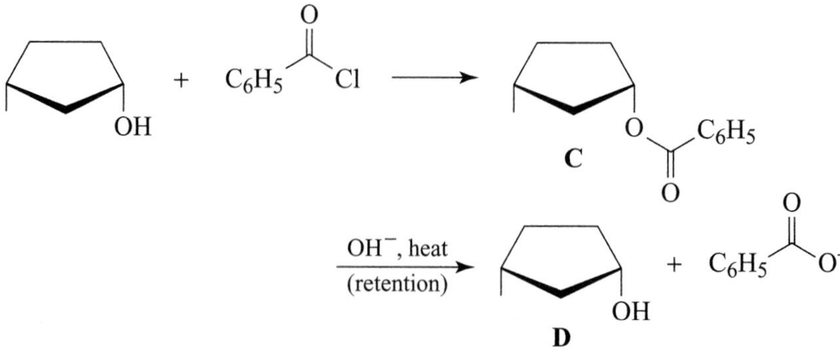

(3) [secondary alkyl bromide] $\xrightarrow{}$ + $CH_3CO_2^-Na^+$ $\xrightarrow{\text{(inversion)}}$ **E** + NaBr

$\xrightarrow[\text{(retention)}]{OH^-,\ \text{heat}}$ **F** + $CH_3CO_2^-Na^+$

(4) [secondary alkyl bromide] $\xrightarrow[\text{(inversion)}]{OH^-,\ \text{heat}}$ **F** + Br^-

(b) Method (3) should give a higher yield of **F** than method (4). Since the hydroxide ion is a strong base and since the alkyl halide is secondary, method (4) is likely to be accompanied by considerable elimination. Method (3), on the other hand, employs a weaker base, acetate ion, in the S_N2 step and is less likely to be complicated by elimination. Hydrolysis of the ester **E** that results should also proceed in high yield.

17.12 (a) Steric hindrance presented by the di-ortho methyl groups of methyl mesitoate prevents formation of the tetrahedral intermediate that must accompany attack at the acyl carbon.

(b) Carry out hydrolysis with labeled $^{18}OH^-$ in labeled $H_2{}^{18}O$. The label should appear in the methanol.

17.13 (a) $C_6H_5CON(CH_2CH_3)_2$

$\xrightarrow[H_2O]{OH^-}$ $C_6H_5CO_2^-$ + $HN(CH_2CH_3)_2$

$\xrightarrow[H_2O]{H_3O^+}$ $C_6H_5CO_2H$ + $H_2N^+(CH_2CH_3)_2$

(b) [lactam]

$\xrightarrow[H_2O]{OH^-}$ $^-O_2C(CH_2)_4NH_2$

$\xrightarrow[H_2O]{H_3O^+}$ $HO_2C(CH_2)_4NH_3^+$

(c) [Hydrolysis of the dipeptide with OH^-/H_2O gives the two carboxylate anions bearing NH_2 and C_6H_5 groups; with H_3O^+/H_2O gives the two carboxylic acids bearing $\overset{+}{N}H_3$ and C_6H_5 groups.]

17.14 (a) [Reaction scheme: carboxylic acid $\xrightarrow{SOCl_2}$ acid chloride $\xrightarrow{NH_3}$ amide $\xrightarrow[\text{heat}]{P_4O_{10}}$ nitrile (CN)]

(b) An elimination reaction would take place because CN^- is a strong base, and the substrate is a 3° alkyl halide.

$$NC:^- + \text{(CH}_3)_3\text{CBr} \longrightarrow HCN + (CH_3)_2C{=}CH_2 + Br^-$$

17.15 (a) $C_6H_5CH_2OH + O{=}C{=}N{-}C_6H_5 \longrightarrow C_6H_5CH_2{-}O{-}C(=O){-}NH{-}C_6H_5$

(b) $Cl{-}C(=O){-}Cl + 4\ CH_3NH_2 \longrightarrow CH_3NH{-}C(=O){-}NHCH_3 + 2\ CH_3\overset{+}{N}H_3 + 2\ Cl^-$

(c) $C_6H_5CH_2{-}O{-}C(=O){-}Cl + H_3\overset{+}{N}CH_2CO_2^- \xrightarrow{OH^-} C_6H_5CH_2{-}O{-}C(=O){-}NH{-}CH_2CO_2^- + Cl^-$

(d) $C_6H_5CH_2-O-C(=O)-NH-CH_2CO_2^-$ $\xrightarrow[Pd]{H_2}$ $H_3\overset{+}{N}CH_2CO_2^-$ + CO_2 + $C_6H_5CH_3$

(e) $C_6H_5CH_2-O-C(=O)-NH-CH_2CO_2^-$ $\xrightarrow[CH_3CO_2H]{HBr}$ $H_3\overset{+}{N}CH_2CO_2H$ + CO_2 + $C_6H_5CH_2Br$

(f) $H_2N-C(=O)-NH_2$ $\xrightarrow[heat]{OH^-, H_2O}$ $2\ NH_3 + CO_3^{2-}$

17.16 (a) By decarboxylation of a β-keto acid:

$CH_3CH_2CH_2CH_2C(=O)CH_2CO_2H$ $\xrightarrow{100–150°\ C}$ $CH_3CH_2CH_2CH_2C(=O)CH_3$ + CO_2

(b) By decarboxylation of a substituted malonic acid:

$HO_2C-C(CH_3)(CH_2CH_3)-CO_2H$ $\xrightarrow{100–150°\ C}$ $CH_3CH_2CH(CH_3)CO_2H$ + CO_2

(c) By decarboxylation of a β-keto acid:

2-oxocyclohexanecarboxylic acid $\xrightarrow{100–150°\ C}$ cyclohexanone + CO_2

(d) By decarboxylation of a substituted malonic acid:

$HO_2C-CH(CH_2CH_2CH_3)-CO_2H$ $\xrightarrow{100–150°\ C}$ $CH_3CH_2CH_2CH_2CO_2H$ + CO_2

17.17 (a) The oxygen-oxygen bond of the diacyl peroxide has a low homolytic bond dissociation energy ($DH^\circ \sim 139$ KJ mol^{-1}). This allows the following reaction to occur at a moderate temperature.

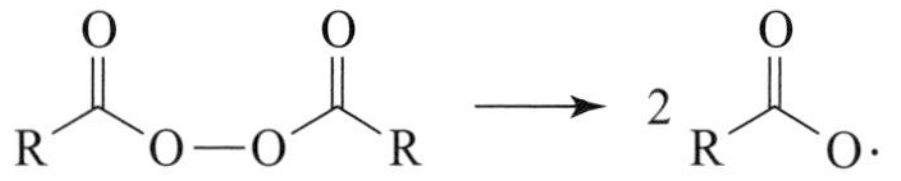

$$RC(=O)O{-}OC(=O)R \longrightarrow 2\ RC(=O)O\cdot \qquad \Delta H^\circ \simeq 139 \text{ kJ mol}^{-1}$$

(b) By decarboxylation of the carboxyl radical produced in part (a).

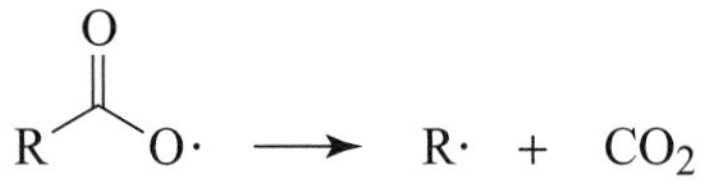

$$RC(=O)O\cdot \longrightarrow R\cdot + CO_2$$

(c) ***Chain Initiation***

Step 1 $RC(=O)O{-}OC(=O)R \xrightarrow{\text{heat}} 2\ RC(=O)O\cdot$

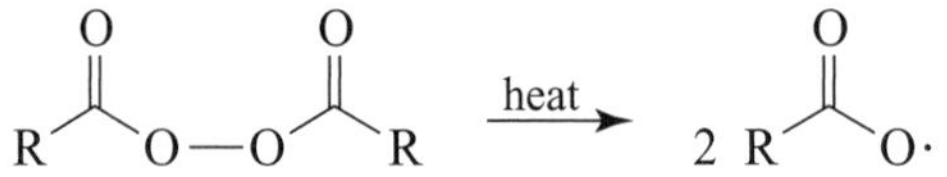

Step 2 $RC(=O)O\cdot \longrightarrow R\cdot + CO_2$

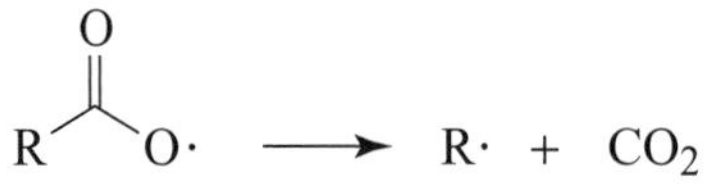

Chain Propagation

Step 3 $R\cdot + CH_2{=}CH_2 \longrightarrow RCH_2CH_2\cdot$

Step 4 $RCH_2CH_2\cdot + CH_2{=}CH_2 \longrightarrow RCH_2CH_2CH_2CH_2\cdot$

Steps 3, 4, 3, 4, and so on.

Problems

Structure and Nomenclature

17.18 (a) 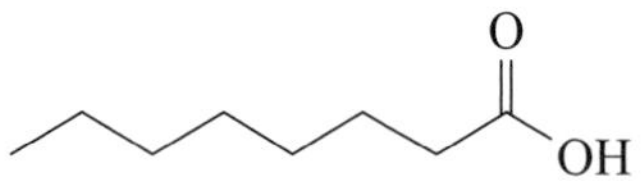 $CH_3(CH_2)_7COOH$ (b) 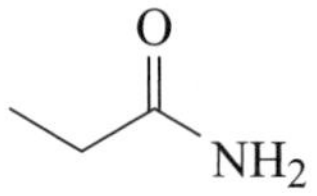 $CH_3CH_2C(=O)NH_2$

(c) 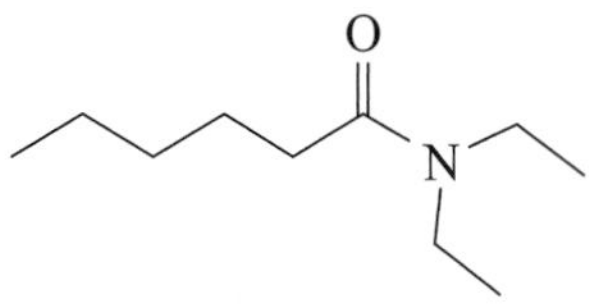 $CH_3(CH_2)_4C(=O)N(CH_2CH_3)_2$ (d) 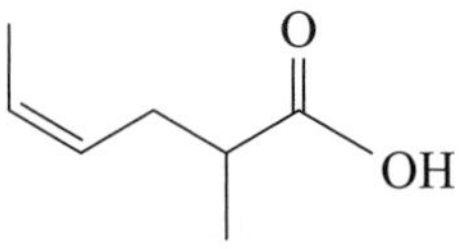 $CH_3CH{=}CHCH_2CH(CH_3)COOH$

[the (*Z*) isomer]

(e) 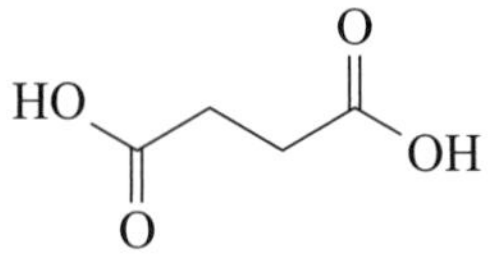 HO, O, O, OH

(f) 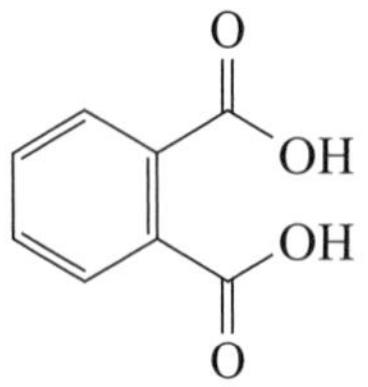 O, OH, OH, O

(g) 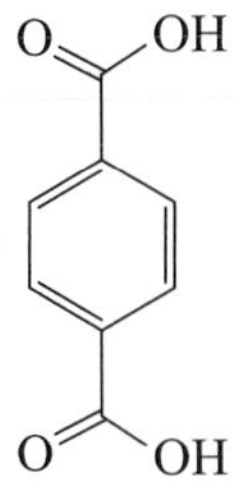 O, OH, O, OH

(h) 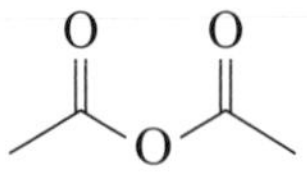 O, O, O

(i) 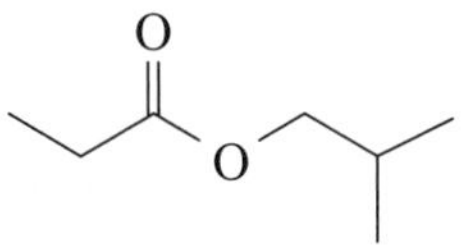 O, O

(j) 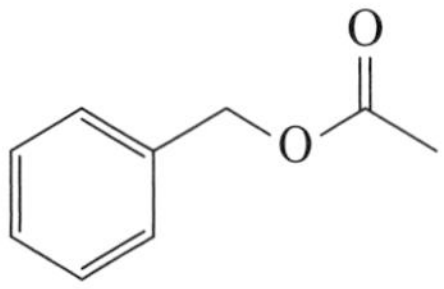 O, O

(k) 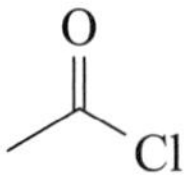 O, Cl

(l) 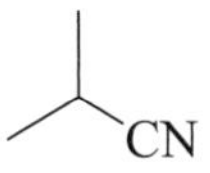 CN

(m) 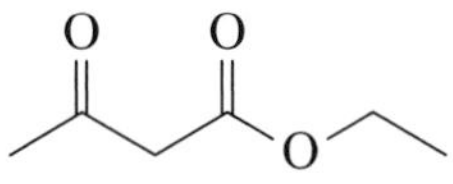 O, O, O

(n) 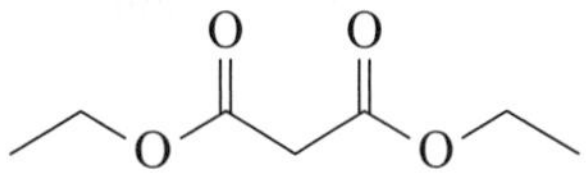 O, O, O, O

17.19 (a) Benzoic acid

(b) Benzoyl chloride

(c) Acetamide or ethanamide

(d) Acetic anhydride or ethanoic anhydride

(e) Benzyl benzoate

(f) Phenyl propanoate or phenyl propionate

(g) Isopropyl acetate or 1-methylethyl ethanoate

(h) Acetonitrile or ethanenitrile

17.20 Alkyl groups are electron releasing; they help disperse the positive charge of an alkylammonium salt and thereby help to stabilize it.

$$R\ddot{N}H_2 + H_3O^+ \longrightarrow R \rightarrow NH_3^+ + H_2O$$

Stabilized by electron-releasing alkyl group

Consequently, alkylamines are somewhat stronger bases than ammonia.

Amides, on the other hand, have acyl groups, RC(=O)–, attached to nitrogen, and acyl groups are electron withdrawing. They are especially electron withdrawing because of resonance contributions of the kind shown here,

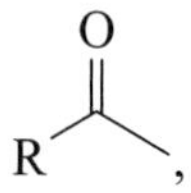

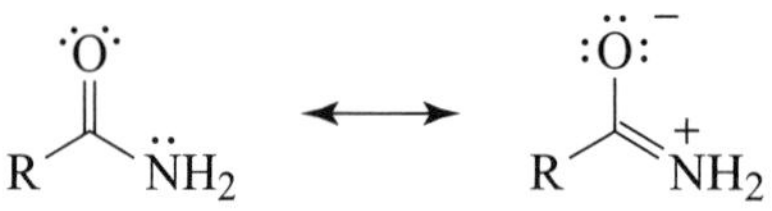

$$RC(=\ddot{O})\ddot{N}H_2 \longleftrightarrow RC(-\ddot{O}:^-)=\overset{+}{N}H_2$$

This kind of resonance also *stabilizes* the amide. The tendency of the acyl group to be electron withdrawing, however, *destabilizes* the conjugate acid of an amide, and reactions such as the following do not take place to an appreciable extent.

$$RC(=\ddot{O})\ddot{N}H_2 + H_3O^+ \rightleftharpoons RC(=O)\overset{+}{N}H_3 + H_2O$$

Stabilized by resonance — *Destabilized by electron-withdrawing acyl group*

17.21 (a) The conjugate base of an amide is stabilized by resonance.

$$RC(=\ddot{O})\ddot{N}H_2 + :B^- \rightleftharpoons RC(=\ddot{O})\ddot{N}H^- + BH$$

$$\updownarrow$$

$$RC(-\ddot{O}:^-)=\ddot{N}H$$

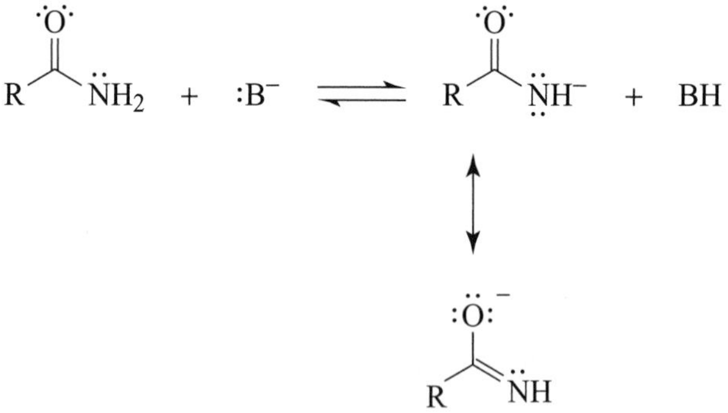

This structure is especially stable because the negative charge is on oxygen.

There is no resonance stabilization for the conjugate base of an amine, RNH^-.

(b) The conjugate base of an imide is stabilized by additional resonance structures,

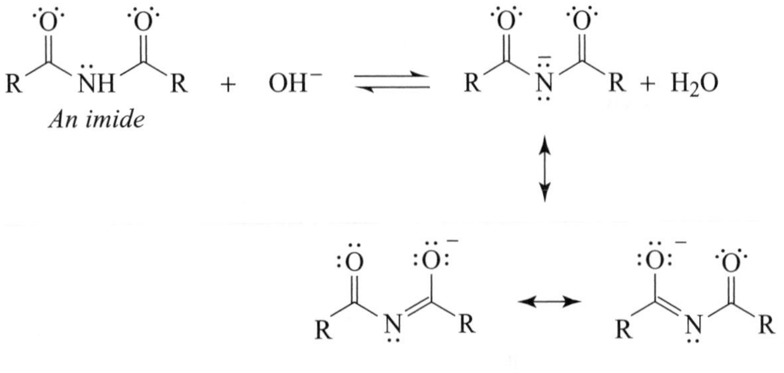

R–C(=O)–NH–C(=O)–R + OH^- ⇌ R–C(=O)–N^-–C(=O)–R + H_2O

An imide

↕

R–C(=O)–N=C(O^-)–R ⟷ R–C(O^-)=N–C(=O)–R

Functional Group Transformations

17.22 (a) 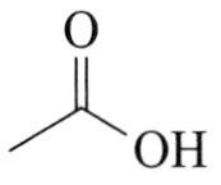 CH_3COOH + HCl

(b) 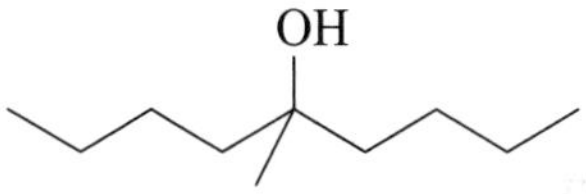 [tertiary alcohol structure with OH]

(c) 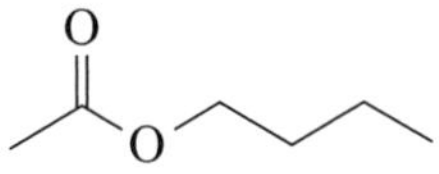 $CH_3COOCH_2CH_2CH_2CH_3$

(d) 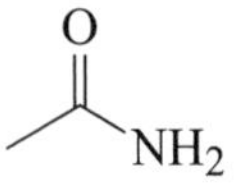 CH_3CONH_2

(e) 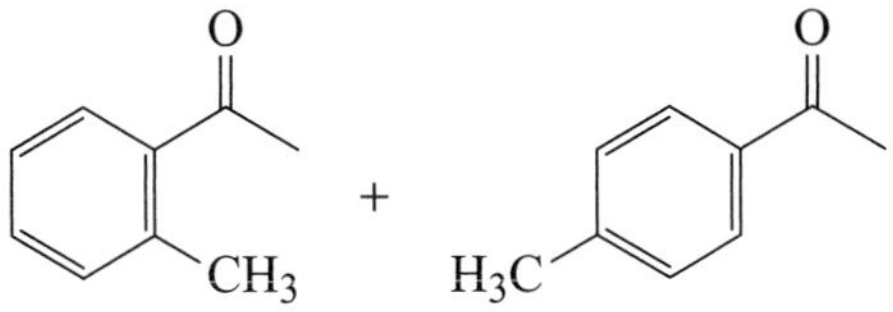 [ortho-methylacetophenone, with CH_3] + [para-methylacetophenone, with H_3C]

(f) CH_3CHO

(g) CH_3COONa

(h) $CH_3CONHCH_3$ + $CH_3NH_3^+$ Cl^-

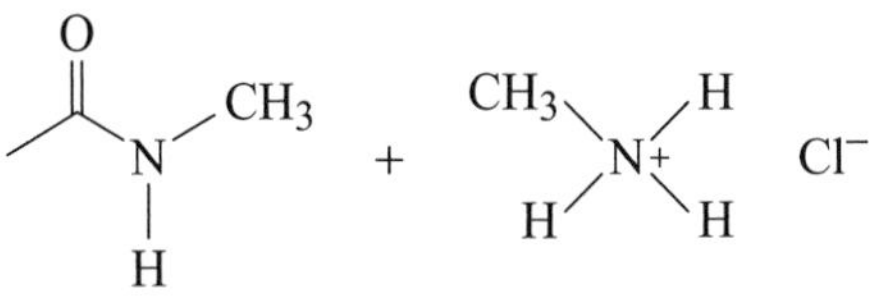

(i) 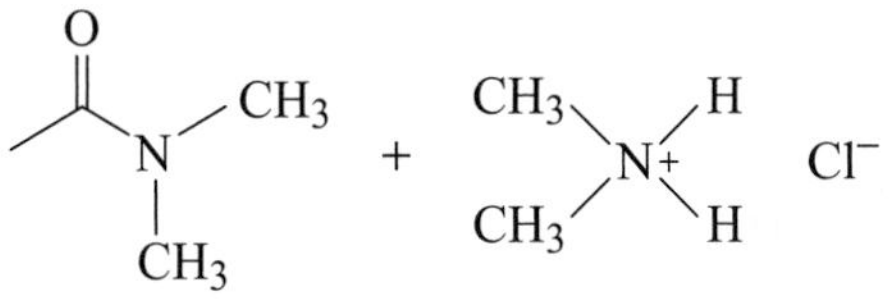 $CH_3CON(CH_3)_2$ + $(CH_3)_2NH_2^+$ Cl^-

(j) 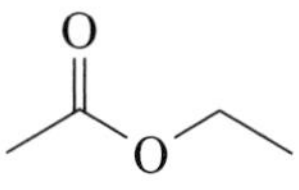 $CH_3COOCH_2CH_3$

(k) 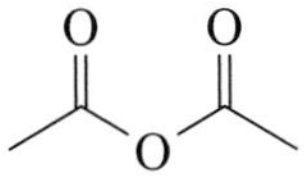 $CH_3COOCOCH_3$

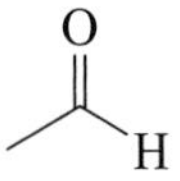

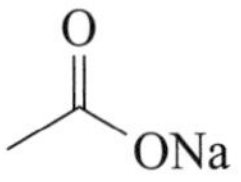

(l) 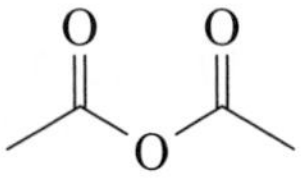 $CH_3COOCOCH_3$

17.23 (a) $CH_3CONH_2 + CH_3CO_2^- \ NH_4^+$

(b) $2\ CH_3CO_2H$

(c) $CH_3CO_2CH_2CH_2CH_3 + CH_3CO_2H$

(d) $C_6H_5COCH_3 + CH_3CO_2H$

(e) $CH_3CONHCH_2CH_3 + CH_3CO_2^- \ CH_3CH_2NH_3^+$

(f) $CH_3CON(CH_2CH_3)_2 + CH_3CO_2^-(CH_3CH_2)_2NH_2^+$

17.24 (a) 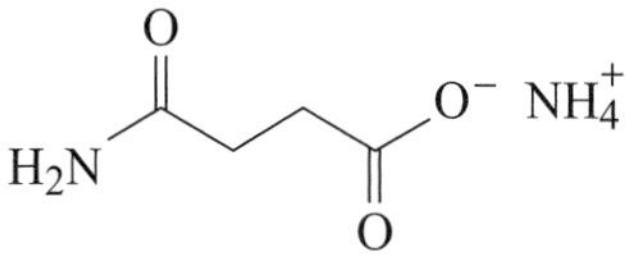 H_2N O O O^- NH_4^+

(b) 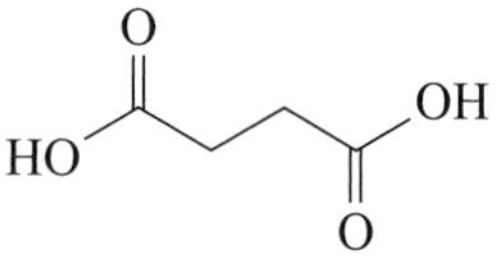 HO O O OH

(c) 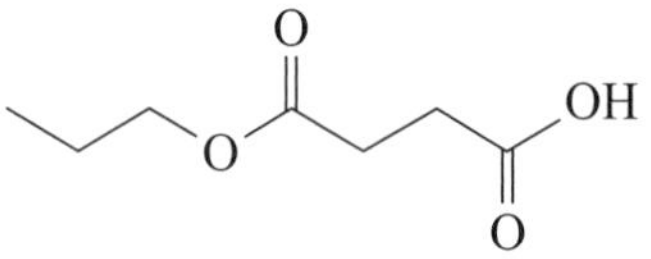 O O O OH

(d) 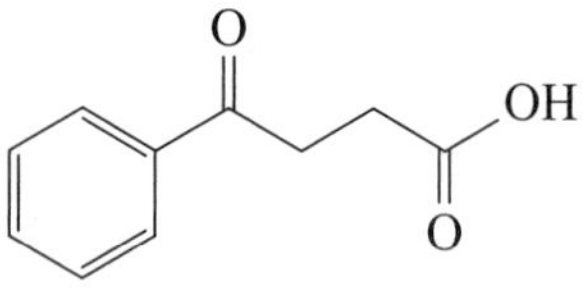 O O OH

(e) 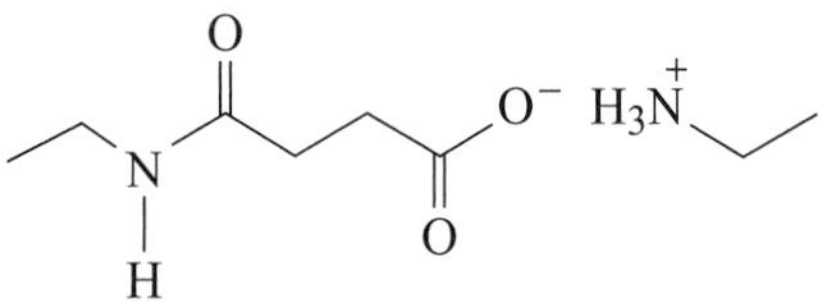 N H O O O^- $H_3\overset{+}{N}$

(f) 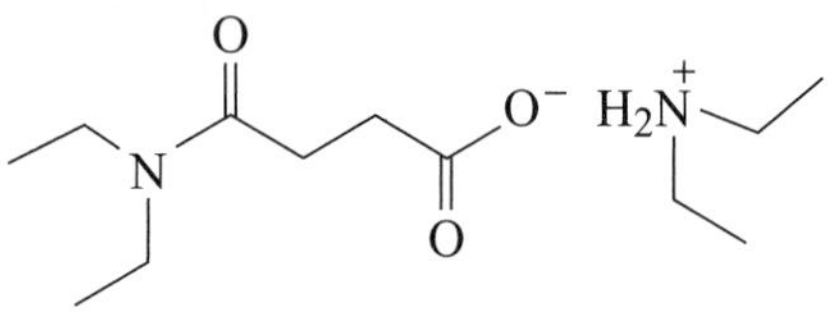 N O O O^- $H_2\overset{+}{N}$

17.25 (a) 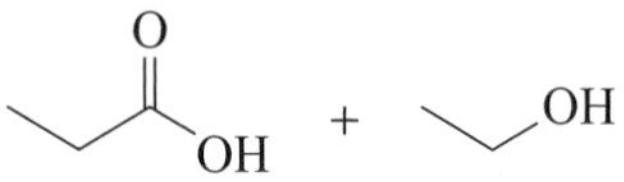 O OH + OH

(b) 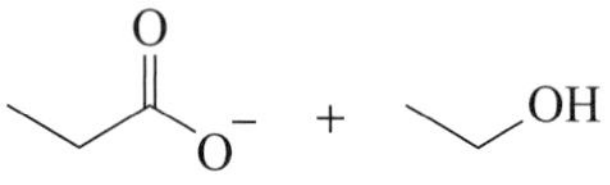 O O^- + OH

(c) 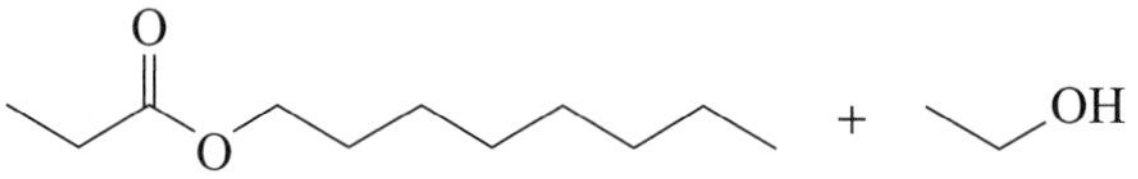 O O + OH

(d) 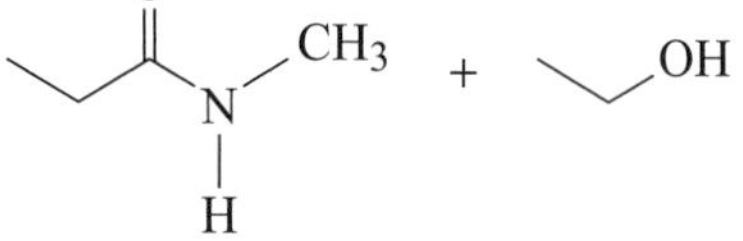 O N H CH_3 + OH

(e) 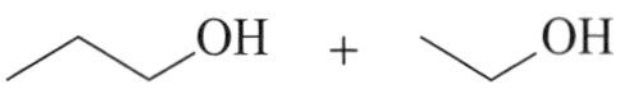 OH + OH

(f) 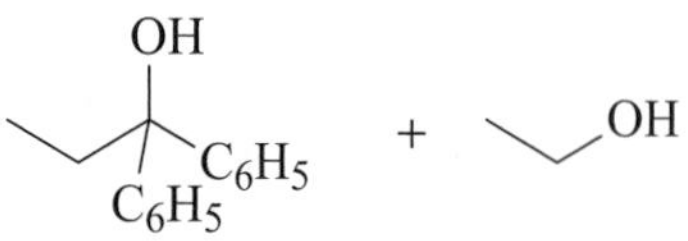 OH C_6H_5 C_6H_5 + OH

17.26 (a) 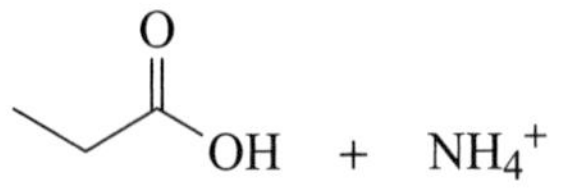 OH + NH_4^+

(b) 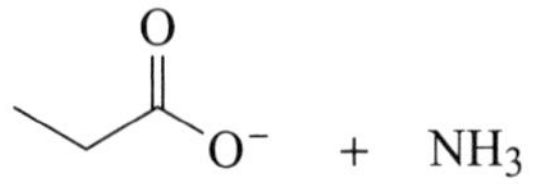 O^- + NH_3

(c) 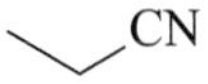 CN

17.27 (a) 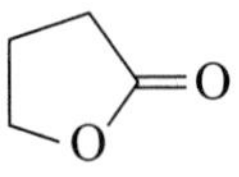 O

(b) 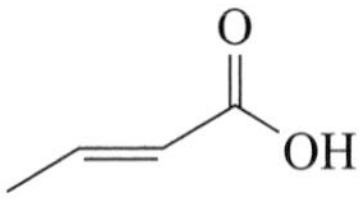 OH
[(*Z*) + (*E*)]

(c)

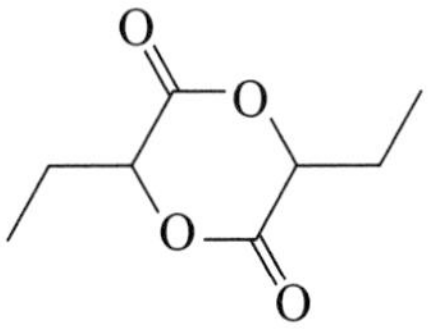

(d)

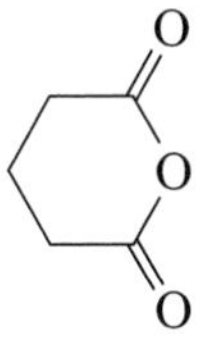

(e) 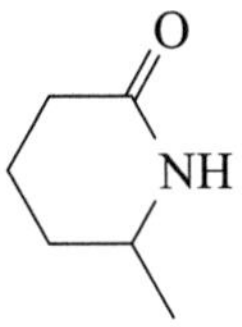 NH

(f) 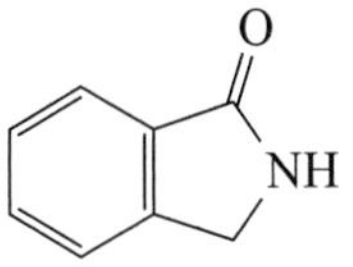 NH

General Problems

17.28 (a) 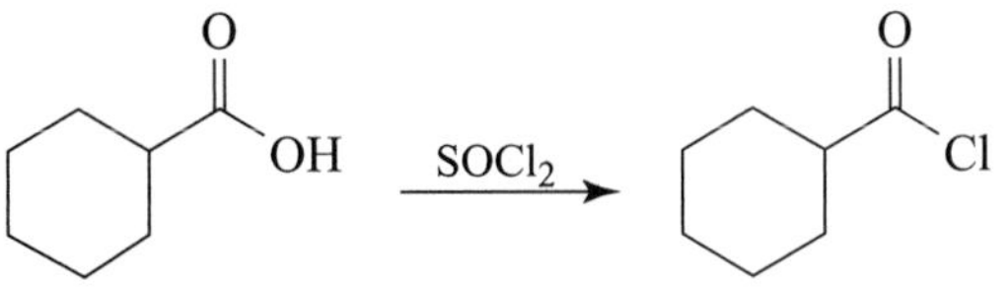 OH $\xrightarrow{SOCl_2}$ Cl

(b) 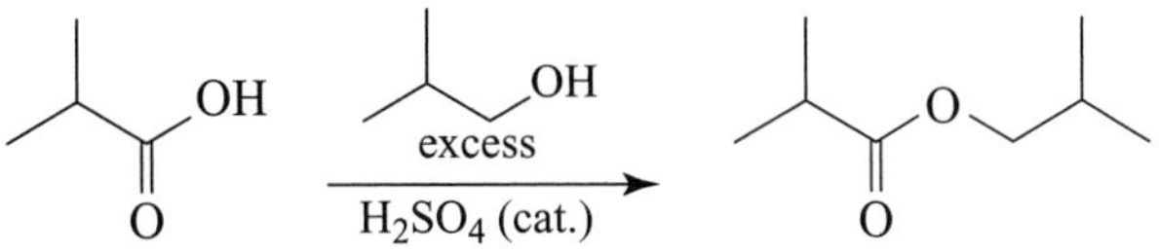 OH $\xrightarrow[H_2SO_4 \text{ (cat.)}]{\text{OH, excess}}$ O

(c) 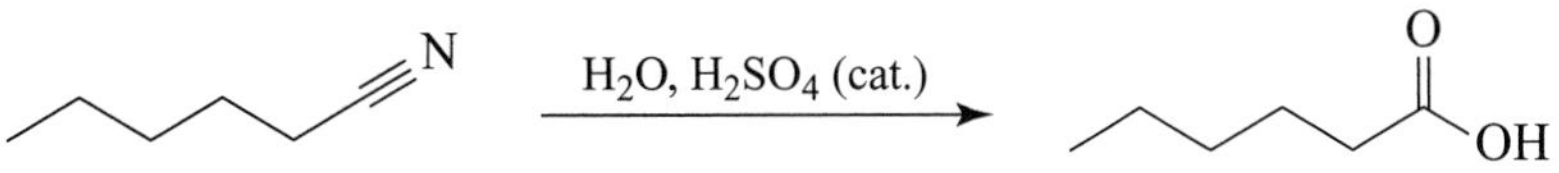 N $\xrightarrow{H_2O,\ H_2SO_4 \text{ (cat.)}}$ OH

(d) 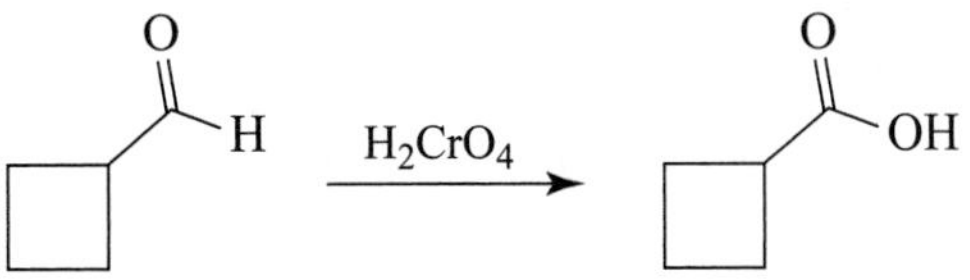 H $\xrightarrow{H_2CrO_4}$ OH

(e) 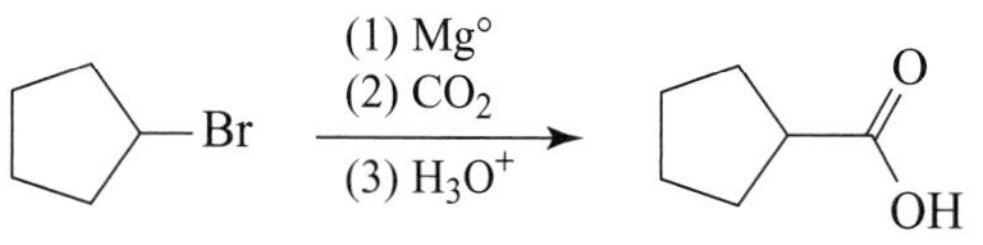 Br — (1) Mg°, (2) CO_2, (3) H_3O^+ → O, OH

(f) 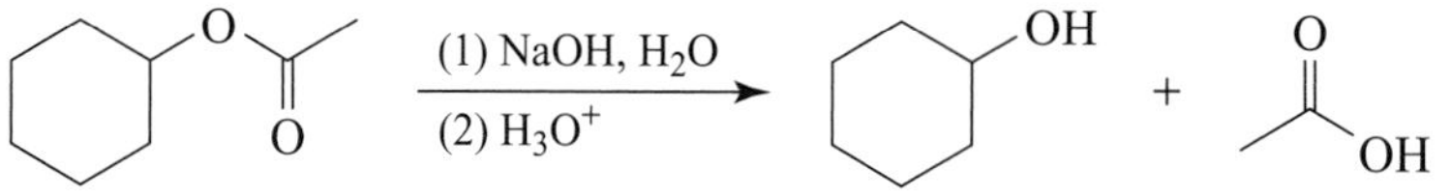 O, O — (1) NaOH, H_2O, (2) H_3O^+ → OH + O, OH

(g) 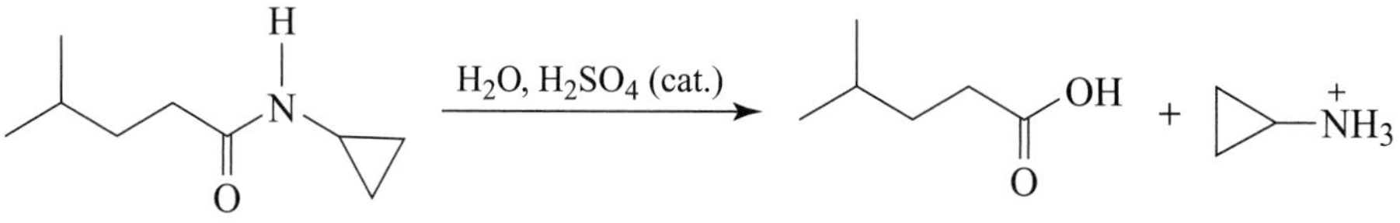 H, N, O — H_2O, H_2SO_4 (cat.) → OH, O + $\overset{+}{N}H_3$

(h) OH — (1) H_2CrO_4, (2) $SOCl_2$ → O, Cl

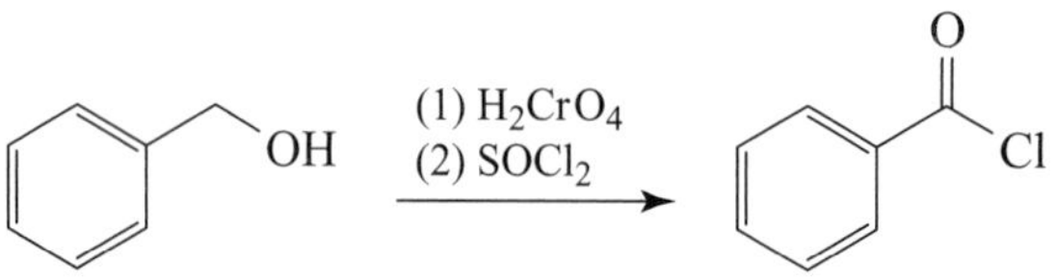

17.29 (a) 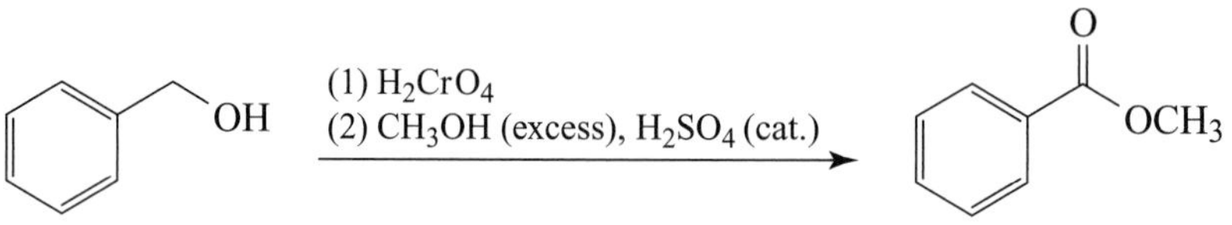 OH — (1) H_2CrO_4, (2) CH_3OH (excess), H_2SO_4 (cat.) → O, OCH_3

(b) 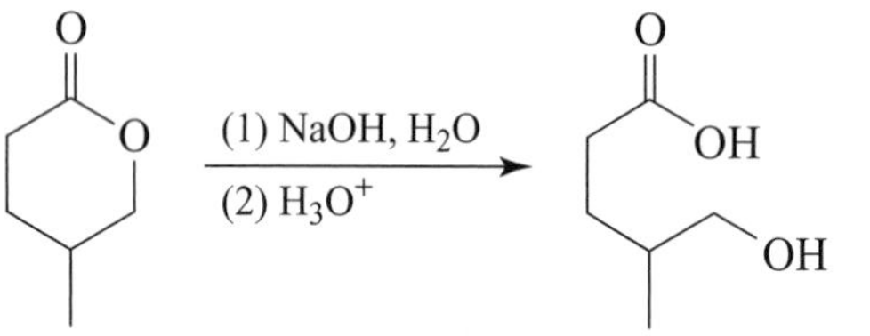 O, O — (1) NaOH, H_2O, (2) H_3O^+ → O, OH, OH

(c) 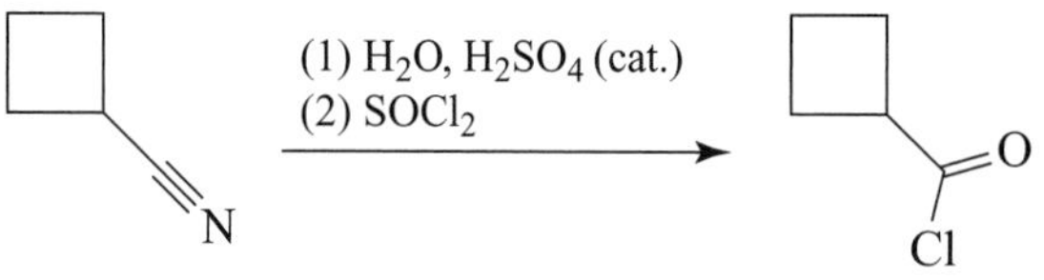 N — (1) H_2O, H_2SO_4 (cat.), (2) $SOCl_2$ → O, Cl

(d) 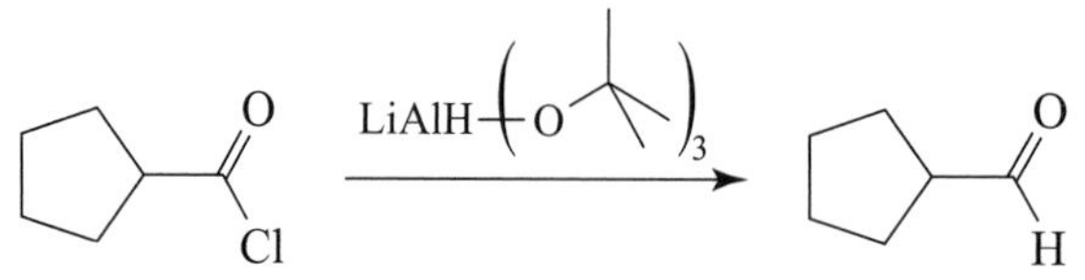 O, Cl — $LiAlH(O\text{-}t\text{-}Bu)_3$ → O, H

(e) 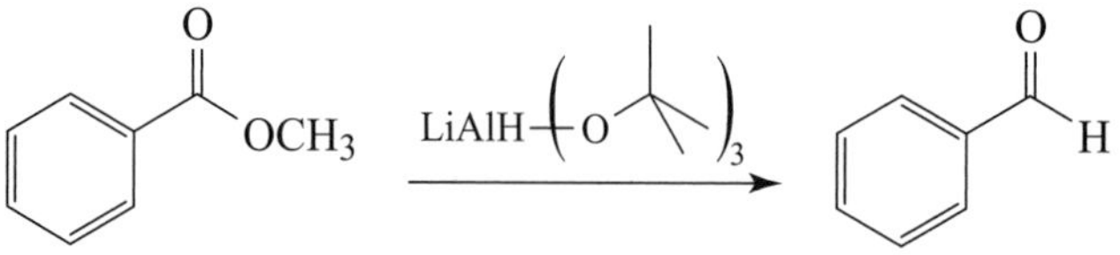 O, OCH_3 — $LiAlH(O\text{-}t\text{-}Bu)_3$ → O, H

(f) 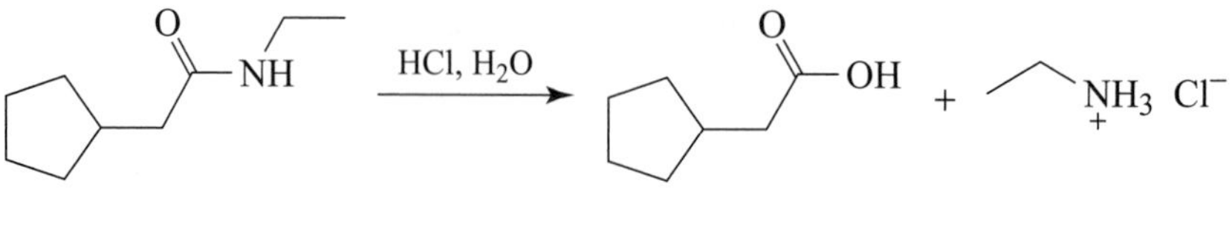 O, NH — HCl, H_2O → O, OH + $\overset{+}{N}H_3$ Cl^-

17.30 (a) O Cl → (SH) → O S

(b) O Cl → NH_2, N → O HN

(c) O O O → OH → O HO O O

(d) HO O OH → H_2SO_4 (cat.) → O O

(e) → O O O, $AlCl_3$ → O + O OH

17.31 (a) Br CN → (1) KCN (2) H_2O, H_2SO_4 (cat.) → O OH OH O

(b) N → (1) DIBAL-H (2) H_3O^+ → H O

(c) N → (1) CH_3MgBr (2) H_3O^+ → O

(d) 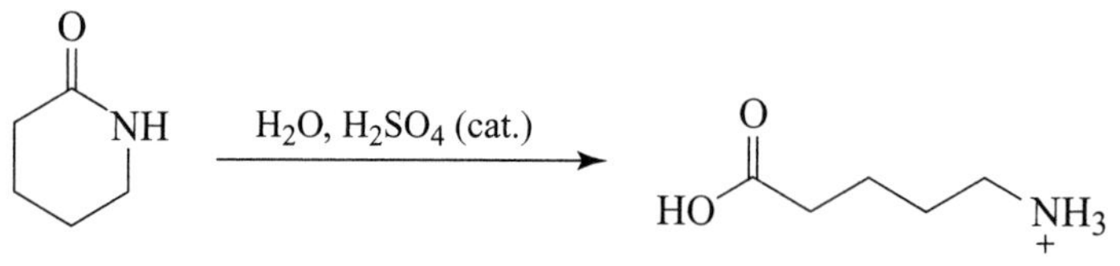 H_2O, H_2SO_4 (cat.)

(e) 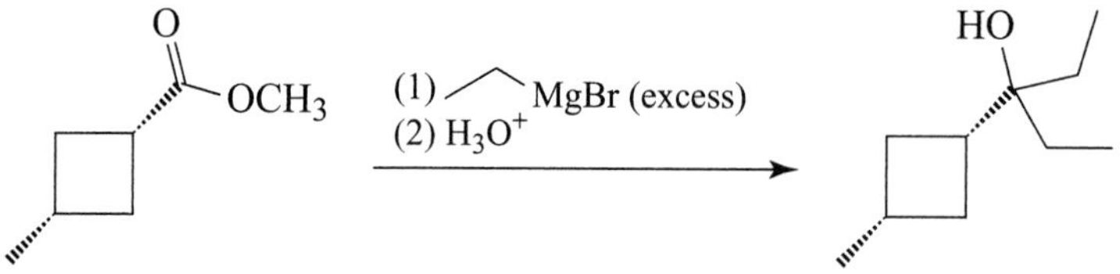 (1) MgBr (excess)
(2) H_3O^+

Mechanisms

17.32 See the mechanisms in Section 17.8F, where R = CH_2CH_3 for propanamide.

17.33 (a)

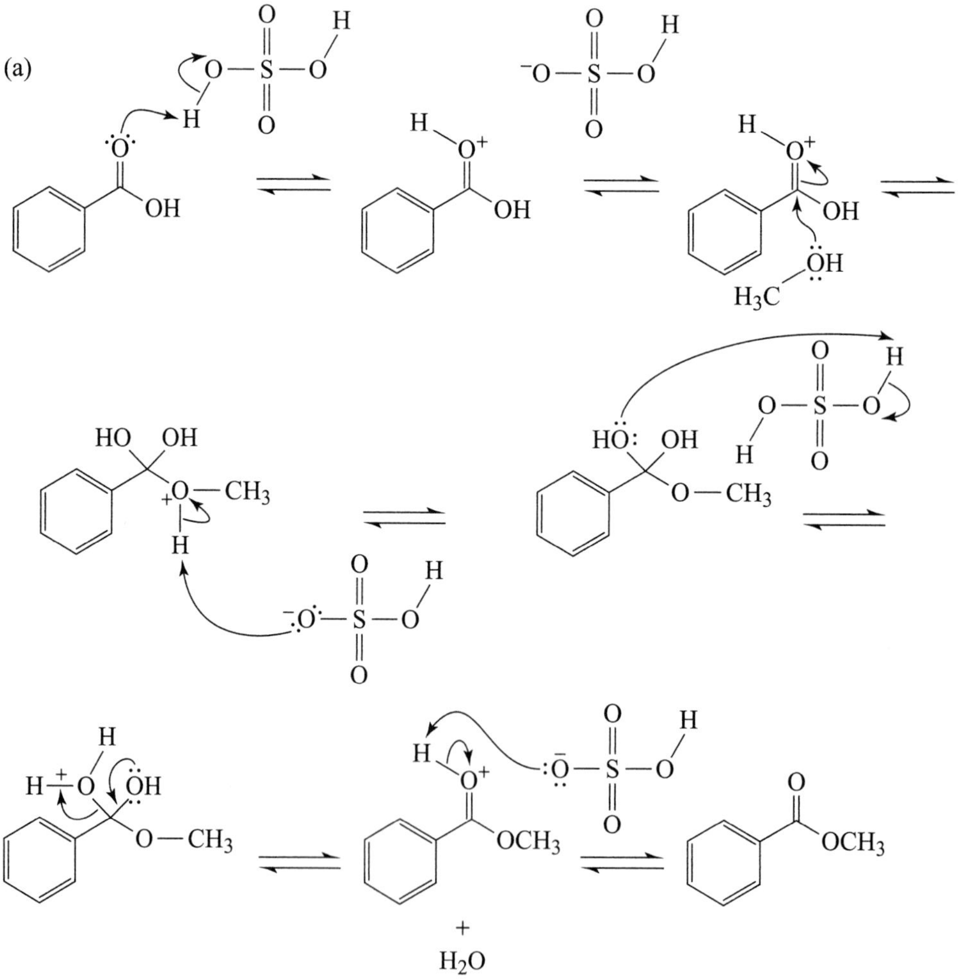

(b)

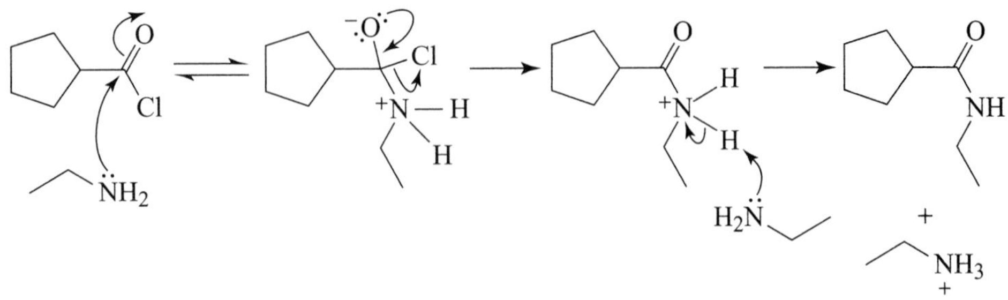

(c)

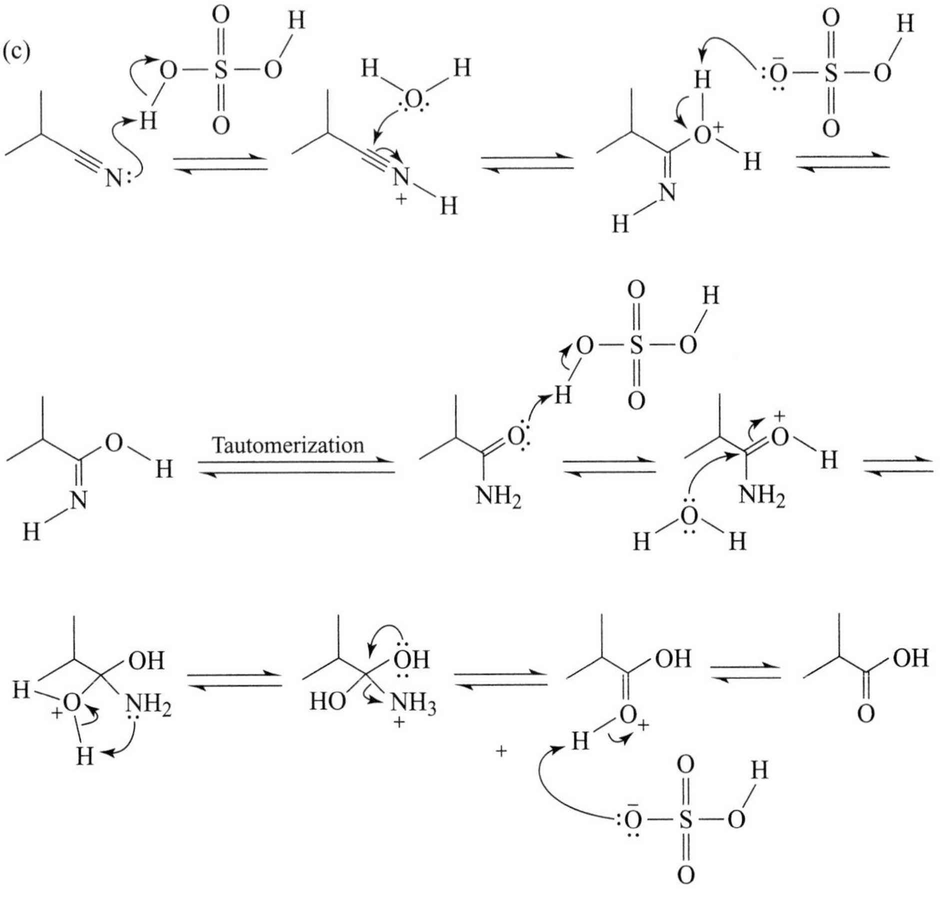

Tautomerization

17.34 *cis*-4-Hydroxycyclohexanecarboxylic acid can assume a boat conformation that permits lactone formation.

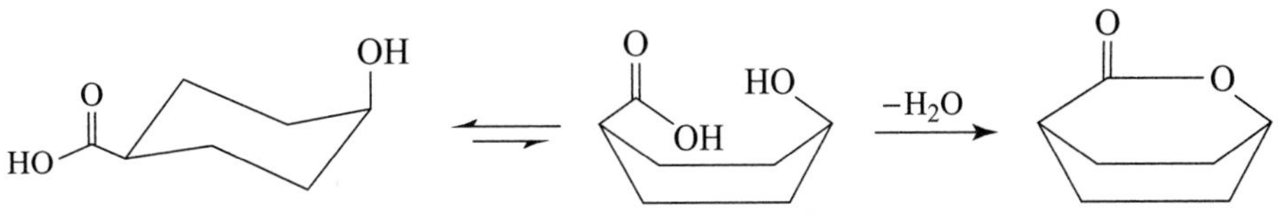

$-H_2O$

Neither of the chair conformations nor the boat form of *trans*-4-hydroxycyclohexanecarboxylic acid places the —OH group and the —CO_2H group close enough together to permit lactonization.

Synthesis

17.35 (a)

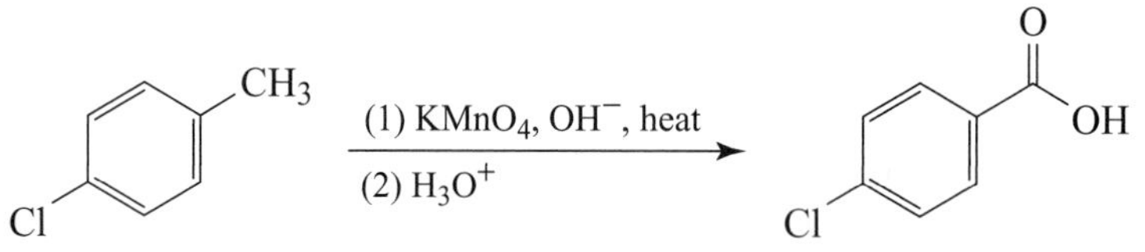

CH$_3$; Cl — (1) KMnO$_4$, OH$^-$, heat; (2) H$_3$O$^+$ → O, OH ; Cl

(b)

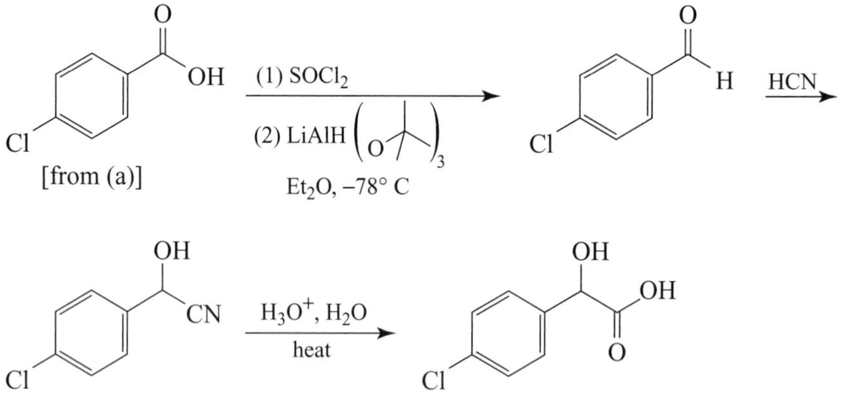

O, OH ; Cl ; [from (a)] — (1) SOCl$_2$; (2) LiAlH(O-t-Bu)$_3$, Et$_2$O, −78° C → O, H ; Cl — HCN →

OH, CN ; Cl — H$_3$O$^+$, H$_2$O, heat → OH, OH, O ; Cl

(c)

CH$_3$; Cl — NBS, *hv* or ROOR, CCl$_4$ → Br ; Cl — NaCN →

CN ; Cl — H$_3$O$^+$, H$_2$O, heat → O, OH ; Cl

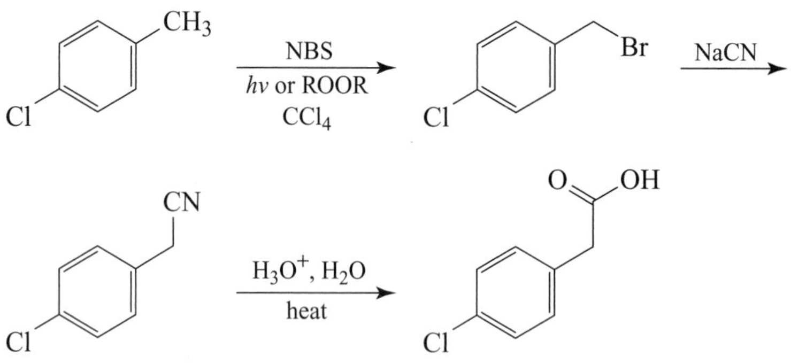

(d)

O, OH ; Cl ; [from (c)] — (1) SOCl$_2$; (2) LiAlH(O-t-Bu)$_3$, Et$_2$O, −78° C → O, H ; Cl — HCN → HO, CN ; Cl

H$_3$O$^+$, H$_2$O, heat → [HO, O, OH ; Cl] → O, OH ; Cl ; [(*E*) + (*Z*)]

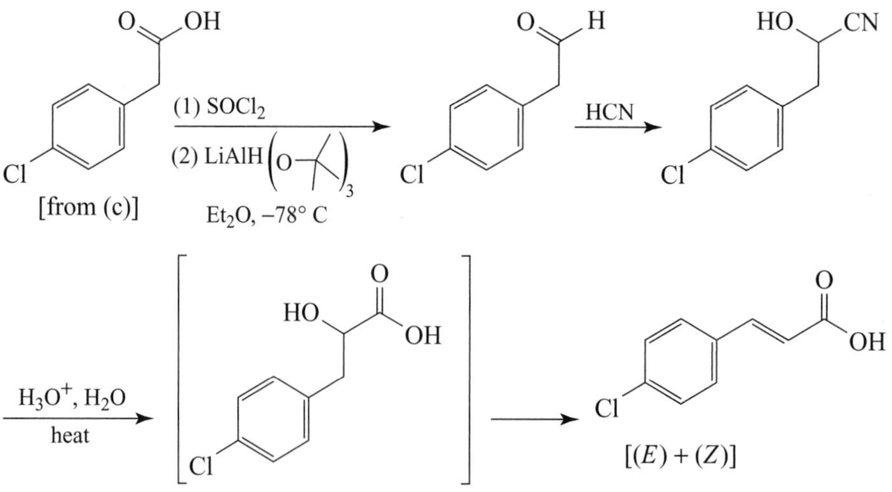

17.36 (a) 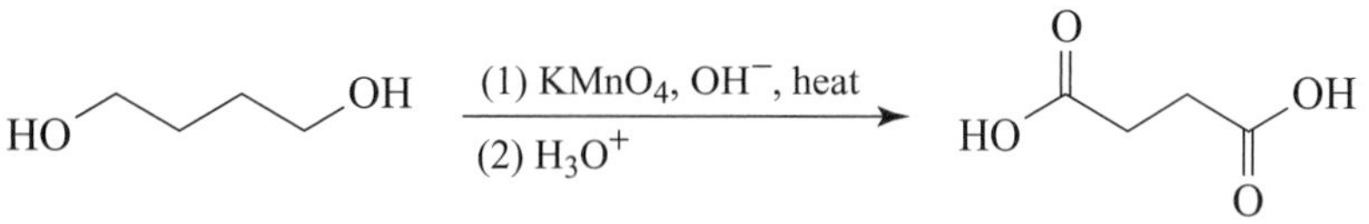 HO～～OH $\xrightarrow[(2)\ H_3O^+]{(1)\ KMnO_4,\ OH^-,\ heat}$ HO–C(=O)–CH₂CH₂–C(=O)–OH

(b) 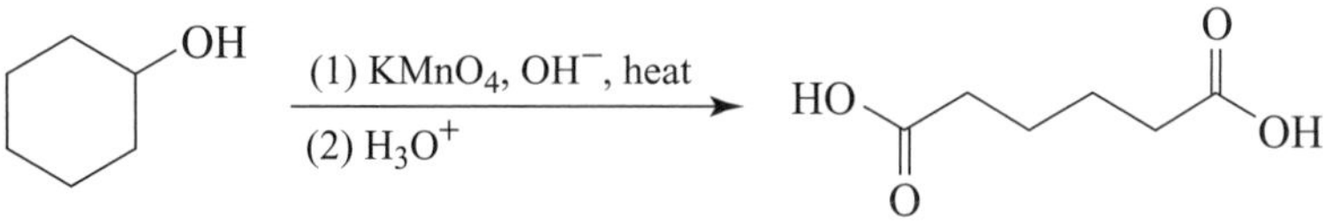 cyclohexanol (OH) $\xrightarrow[(2)\ H_3O^+]{(1)\ KMnO_4,\ OH^-,\ heat}$ HO–C(=O)–(CH₂)₄–C(=O)–OH

17.37 (a) 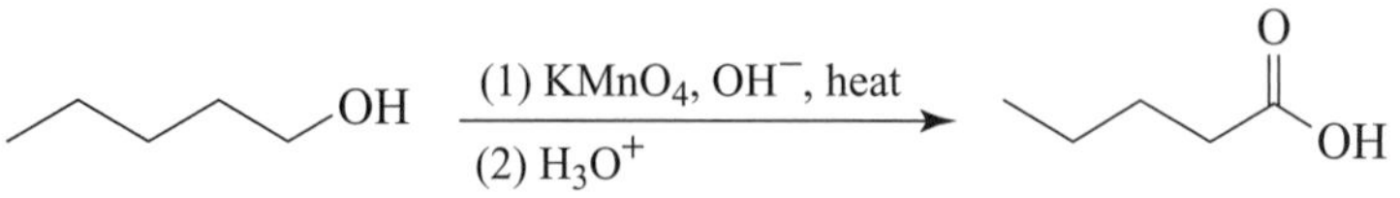 ～～OH $\xrightarrow[(2)\ H_3O^+]{(1)\ KMnO_4,\ OH^-,\ heat}$ ～～C(=O)OH

(b) ～～Br $\xrightarrow[(2)\ CO_2]{(1)\ Mg,\ Et_2O}$ ～～C(=O)OMgBr $\xrightarrow{H_3O^+}$ ～～C(=O)OH

～～Br $\xrightarrow{CN^-}$ ～～CN $\xrightarrow[heat]{H_3O^+,\ H_2O}$ ～～C(=O)OH

(c) ～～CH=CH～～ [(E) or (Z)] $\xrightarrow[(2)\ H_3O^+]{(1)\ KMnO_4,\ OH^-,\ heat}$ 2 ～～C(=O)OH

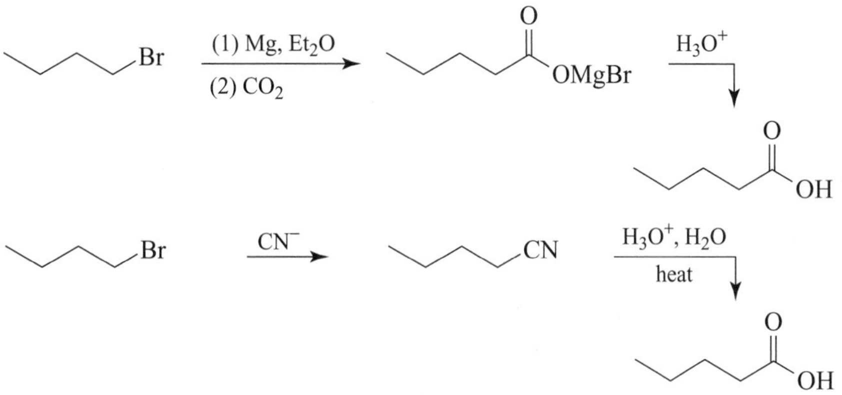

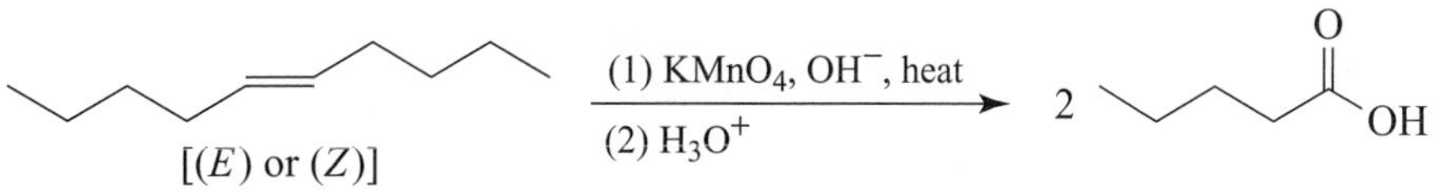

(d) 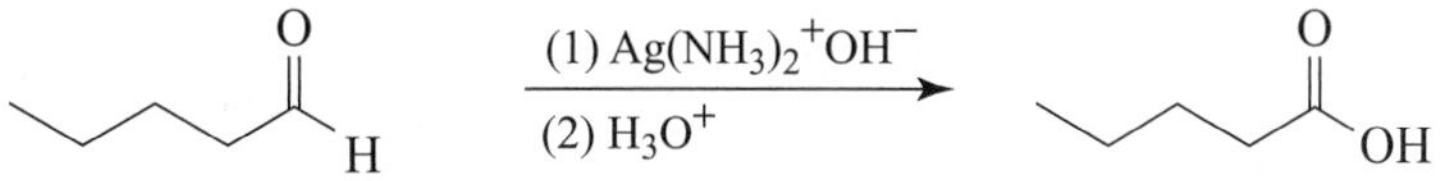 ～～C(=O)H $\xrightarrow[(2)\ H_3O^+]{(1)\ Ag(NH_3)_2{}^+OH^-}$ ～～C(=O)OH

17.38 (3-CH₃)C₆H₄–C(=O)OH + $SOCl_2$ ⟶ (3-CH₃)C₆H₄–C(=O)Cl $\xrightarrow{(\diagup\!\diagdown)_2NH}$ (3-CH₃)C₆H₄–C(=O)N(CH₂CH₃)₂

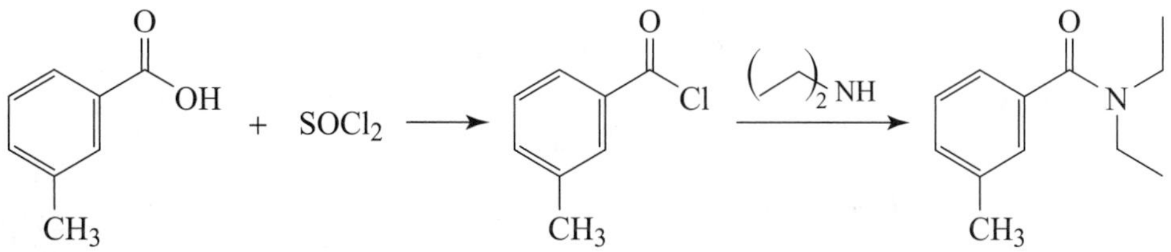

17.39 OH, O; $SOCl_2$ → O, Cl; $AlCl_3$ → O

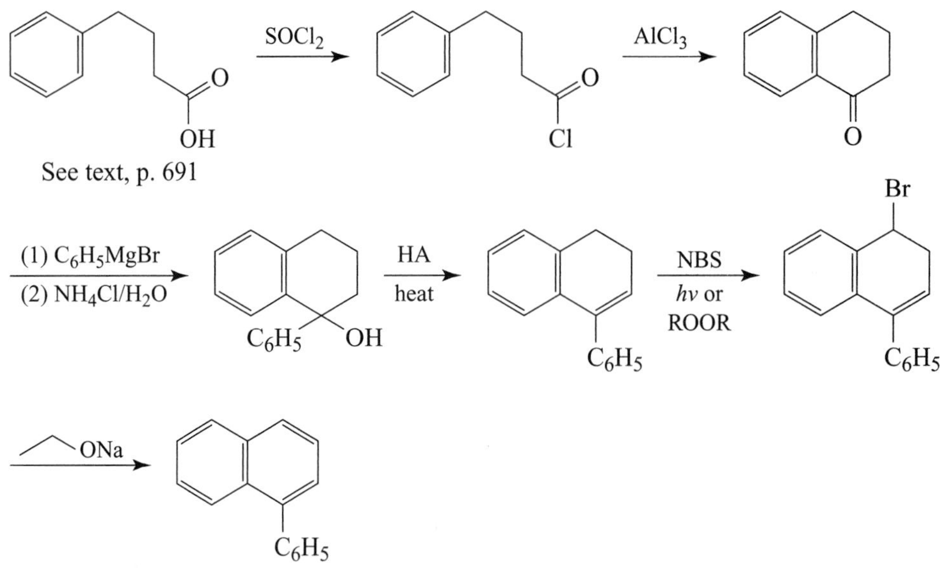

See text, p. 691

(1) C_6H_5MgBr (2) NH_4Cl/H_2O → C_6H_5 OH; HA heat → C_6H_5; NBS $h\nu$ or ROOR → Br, C_6H_5

ONa → C_6H_5

17.40 (a) + H, CO_2H, C=C, H, CO_2H → H, CO_2H, CO_2H, H

(b) + H, CO_2H, C=C, HO_2C, H → H, CO_2H, CO_2H, H

(c) + H, CO_2H, C=C, H, CO_2H → CO_2H, CO_2H

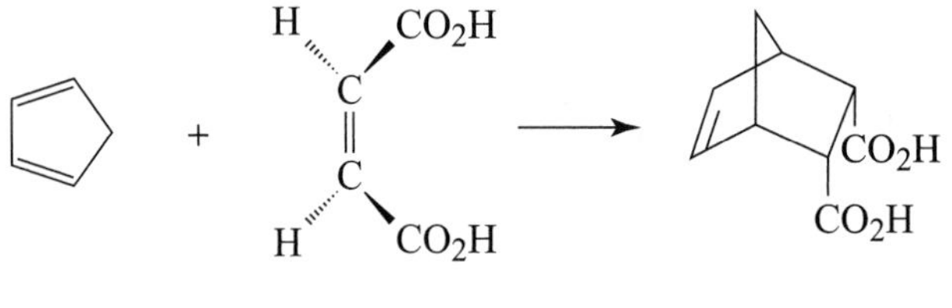

17.41 (a) OH (*R*)-(−)-2-butanol; TsCl, pyridine (retention) → OTs **A**; CN^- (inversion) → CN **B**

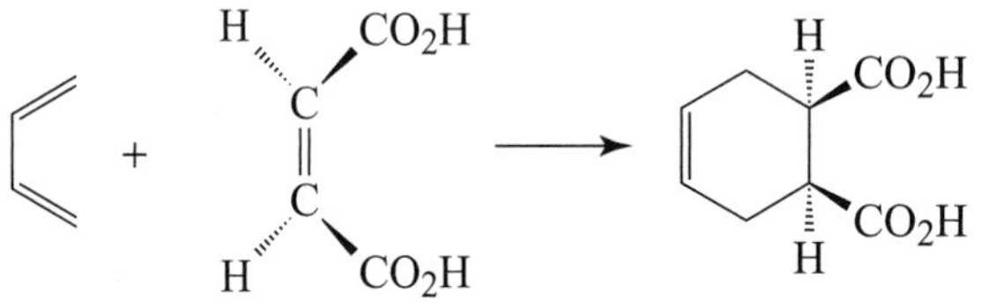

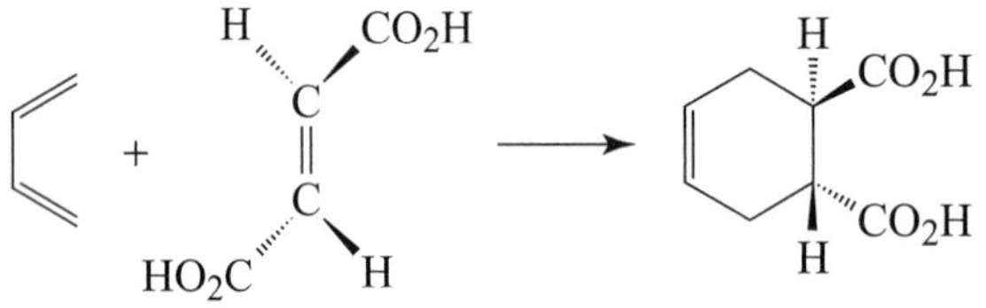

H_2SO_4, H_2O (retention) → CO_2H (+) **C**; (1) $LiAlH_4$ (2) H_2O (retention) → (−) **D** OH

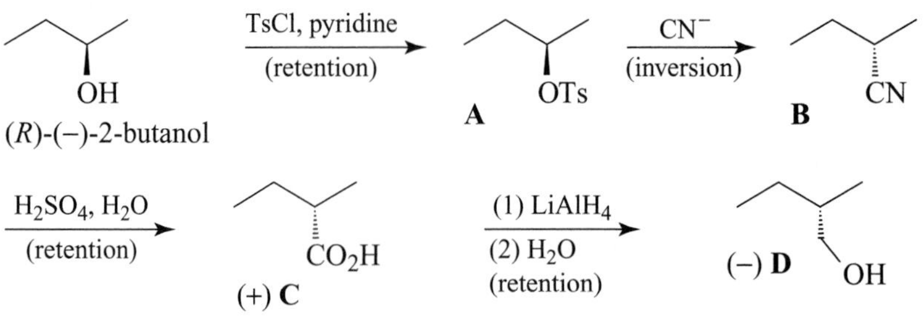

(b) (*R*)-(−)-2-butanol (OH) → PBr_3, pyridine (inversion) → **E** (Br) → CN^- (inversion) → **F** (CN)

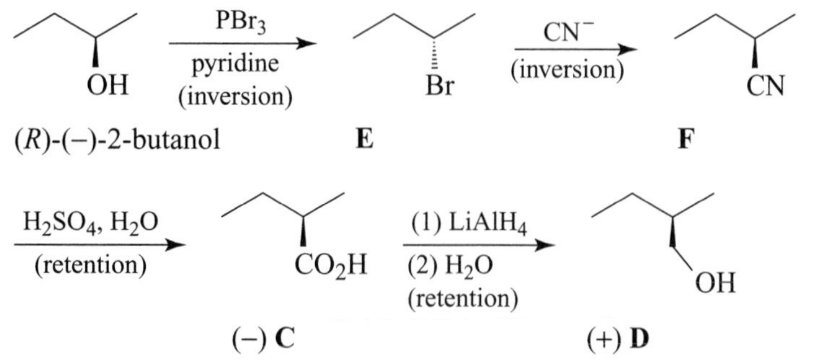

→ H_2SO_4, H_2O (retention) → (−) **C** (CO_2H) → (1) $LiAlH_4$ (2) H_2O (retention) → (+) **D** (OH)

(c) **A** (OTs) → $CH_3CO_2^-$ (inversion) → **G** → OH^- (retention) → (+) **H** (OH) + $CH_3CO_2^-$

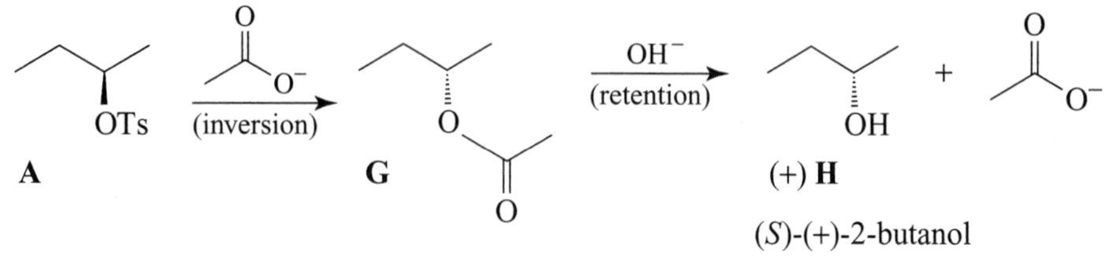

(*S*)-(+)-2-butanol

(d) (−) **D** (OH) → PBr_3 (retention) → **J** (Br) → Mg, diethyl ether (retention) →

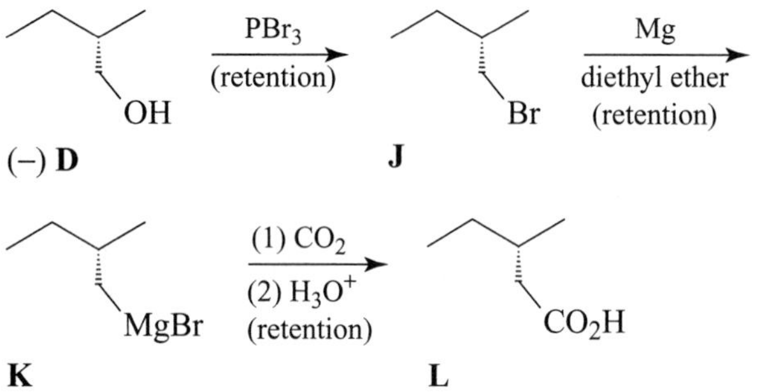

K (MgBr) → (1) CO_2 (2) H_3O^+ (retention) → **L** (CO_2H)

(e) (*R*)-(+)-Glyceraldehyde (HO, OH, O, H) → HCN → (HO, OH, CN) + (HO, OH, CN)

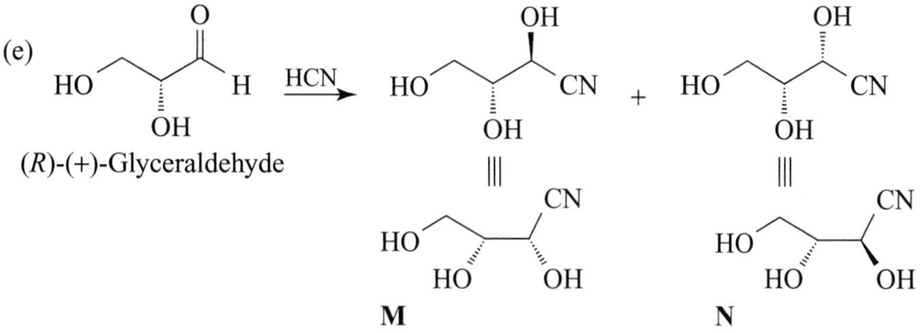

M (HO, HO, OH, CN) **N** (HO, HO, OH, CN)

(f) **M** → H_2SO_4, H_2O, heat → **P** (HO, HO, OH, O, OH) → [O], HNO_3 → *meso*-Tartaric acid (HO, HO, OH, O, O, OH)

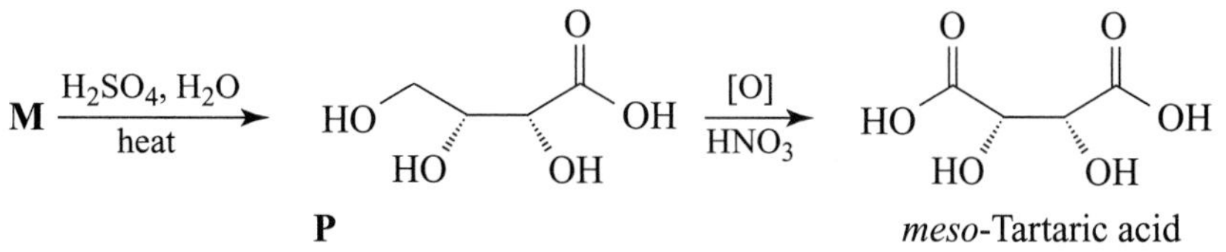

(g) N $\xrightarrow[\text{heat}]{H_2SO_4,\ H_2O}$ Q $\xrightarrow[HNO_3]{[O]}$ (−)-Tartaric acid

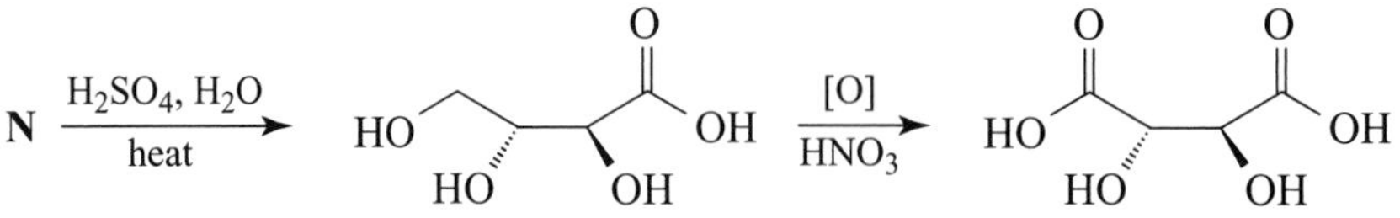

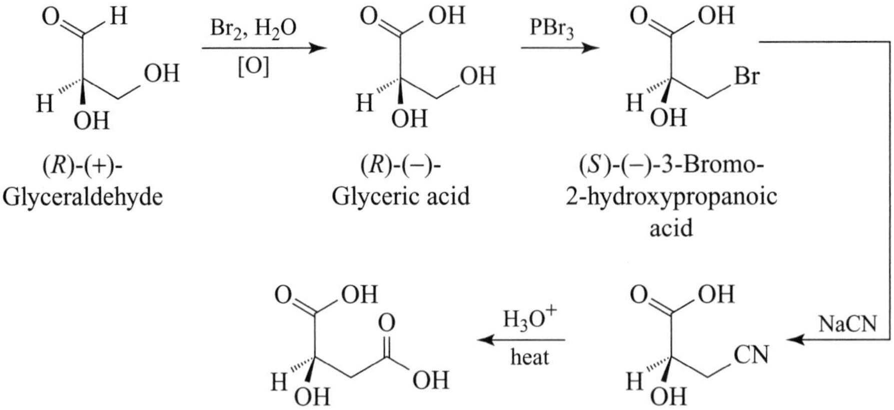

Q

(−)-Tartaric acid

17.42 (R)-(+)-Glyceraldehyde $\xrightarrow[[O]]{Br_2,\ H_2O}$ (R)-(−)-Glyceric acid $\xrightarrow{PBr_3}$ (S)-(−)-3-Bromo-2-hydroxypropanoic acid $\xrightarrow{NaCN}$ (R)-($C_4H_5NO_3$) $\xrightarrow[\text{heat}]{H_3O^+}$ (R)-(+)-Malic acid

17.43 (a) (R)-(+)-Glyceraldehyde $\xrightarrow{HCN}$ M + N [cf. Problem 17.41(e)]

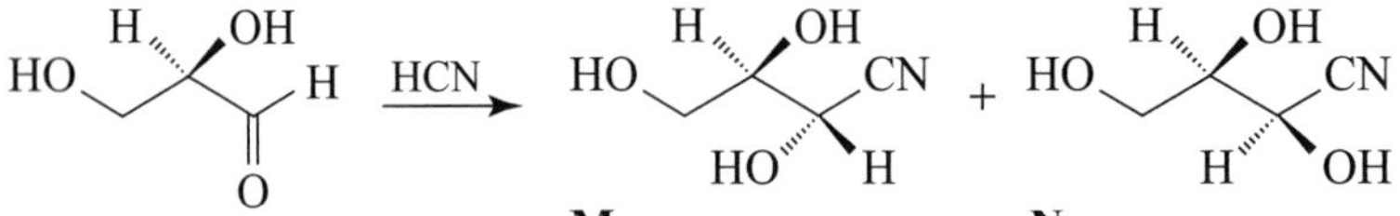

M

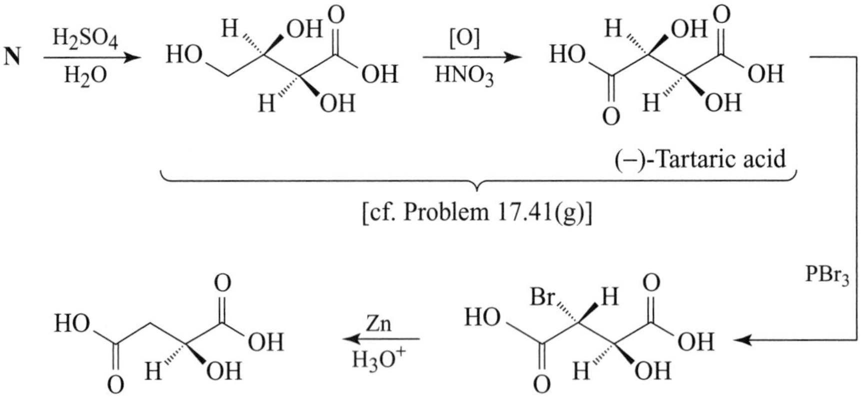

N

N $\xrightarrow[H_2O]{H_2SO_4}$ $\xrightarrow[HNO_3]{[O]}$ (−)-Tartaric acid

[cf. Problem 17.41(g)]

$\xrightarrow{PBr_3}$ $\xrightarrow[H_3O^+]{Zn}$ (S)-(−)-Malic acid

(b) Replacement of either alcoholic —OH by a reaction that proceeds with inversion produces the same stereoisomer.

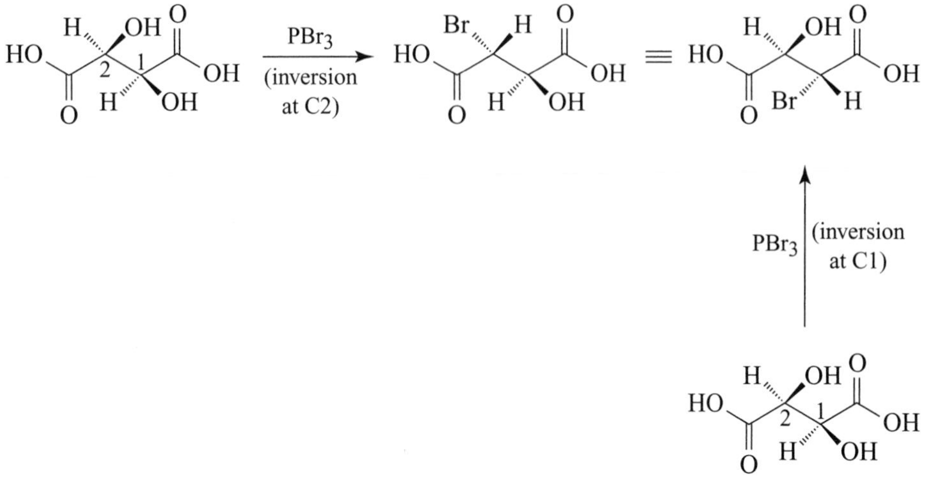

H OH O HO 2 1 OH O H OH

PBr_3 (inversion at C2)

Br H O HO OH O H OH ≡ H OH O HO OH O Br H

PBr_3 (inversion at C1)

H OH O HO 2 1 OH O H OH

(c) Two. The stereoisomer given in (b) and the one given next, below.

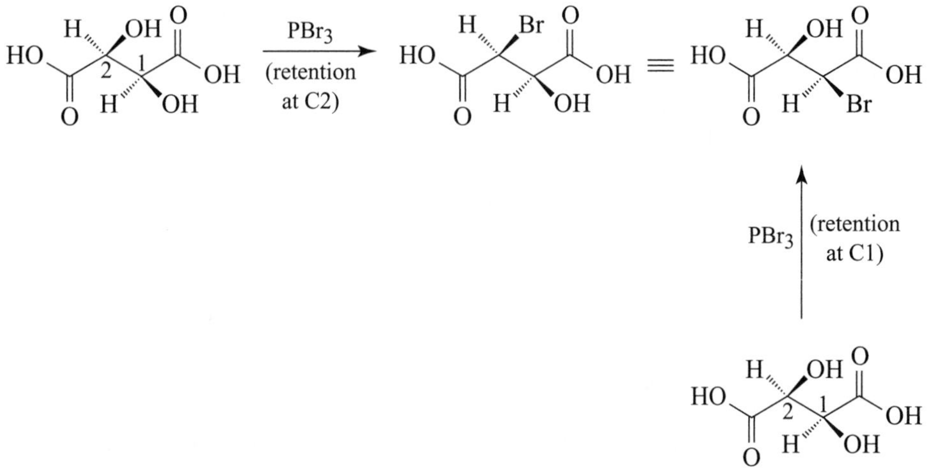

H OH O HO 2 1 OH O H OH

PBr_3 (retention at C2)

H Br O HO OH O H OH ≡ H OH O HO OH O H Br

PBr_3 (retention at C1)

H OH O HO 2 1 OH O H OH

(d) It would have made no difference because treating either isomer (or both together) with zinc and acid produces (−) malic acid.

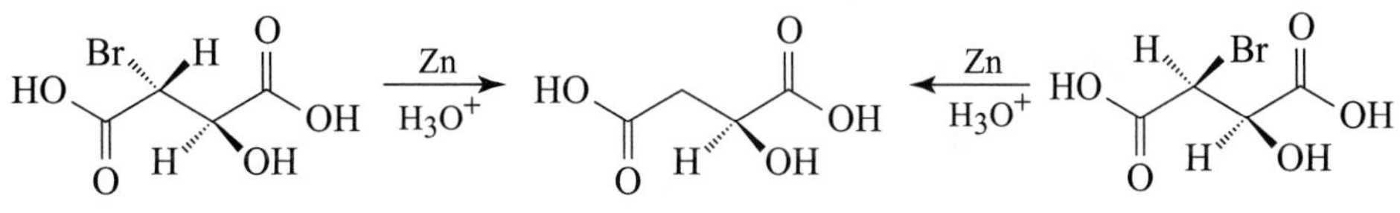

Br H O HO OH O H OH

Zn, H_3O^+

O HO OH O H OH

Zn, H_3O^+

H Br O HO OH O H OH

(−)-Malic acid

17.44 (a) $CH_3O_2C-C\equiv C-CO_2CH_3$. This is a Diels-Alder reaction.

(b) H_2, Pd. The disubstituted double bond is less hindered than the tetrasubstituted double bond and hence is more reactive.

(c) $CH_2=CH-CH=CH_2$. Another Diels-Alder reaction.

(d) $LiAlH_4$

(e) $CH_3-\overset{\overset{O}{\|}}{\underset{\underset{O}{\|}}{S}}-Cl$ and pyridine

(f) $CH_3CH_2S^-$

(g) OsO_4, then $NaHSO_3$

(h) Raney Ni

(i) Base. This is an aldol condensation.

(j) C_6H_5Li (or C_6H_5MgBr) followed by H_3O^+

(k) H_3O^+. This is an acid-catalyzed rearrangement of an allylic alcohol.

(l) $CH_3\overset{\overset{O}{\|}}{C}Cl$, pyridine,

(m) O_3, followed by oxidation

(n) Heat

Spectroscopy

17.45

$$\text{Phenacetin} \xrightarrow[\text{reflux}]{OH^-,\ H_2O} \text{Phenetidine } (-NH_2) + \text{acetate } (O^-)$$

Phenacetin

Phenetidine

An interpretation of the ^1H NMR spectral data for phenacetin is as follows:

(*a*) (*c*) O N H (*e*) (*d*) (*b*)

(*a*) triplet δ 1.4

(*b*) singlet δ 2.1

(*c*) quartet δ 3.95

(*d*) doublet δ 6.8; doublet, δ 7.4

(*e*) broad singlet δ 9.0

17.46 (a)

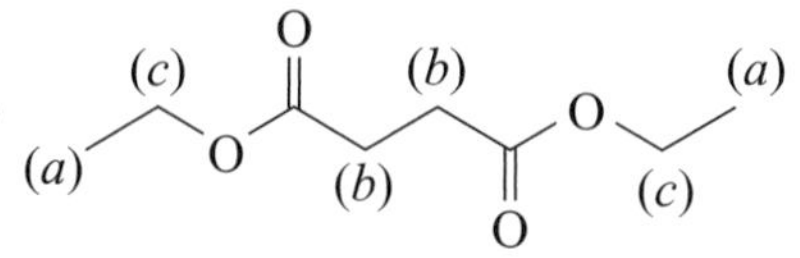

Interpretation:

(*a*) Triplet δ 1.2 (6H)
(*b*) Singlet δ 2.5 (4H)
(*c*) Quartet δ 4.1 (4H)

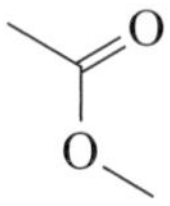

, 1740 cm^{-1} (ester)

(b)

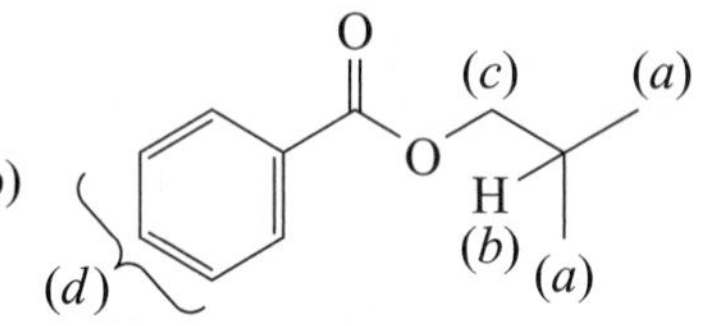

Interpretation:

(*a*) Doublet δ 1.0 (6H)
(*b*) Multiplet δ 2.1 (1H)
(*c*) Doublet δ 4.1 (2H)
(*d*) Multiplet δ 7.8 (5H)

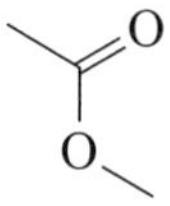

, 1720 cm^{-1} (ester)

(c)

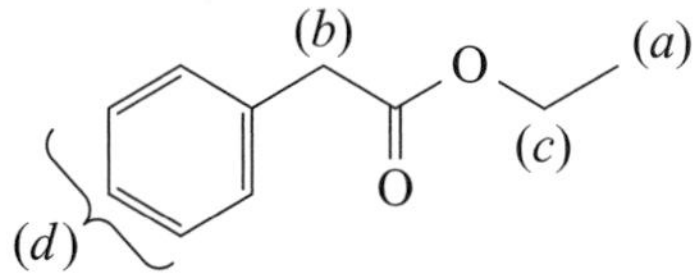

Interpretation:

(*a*) Triplet δ 1.2 (3H)
(*b*) Singlet δ 3.5 (2H)
(*c*) Quartet δ 4.1 (2H)
(*d*) Multiplet δ 7.3 (5H)

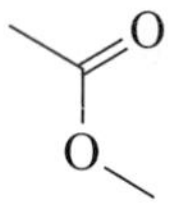

, 1740 cm^{-1} (ester)

(d)

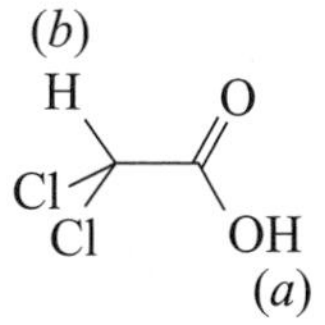

Interpretation:

(*a*) Singlet δ 6.0
(*b*) Singlet δ 11.70

—OH, 2500–2700 cm^{-1}

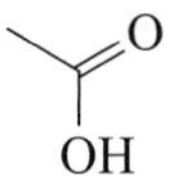

, 1705 cm^{-1} (acid)

(e) (*b*) H H (*b*) Cl O (*c*) O (*a*)

Interpretation:

(*a*) Triplet δ 1.3

(*b*) Singlet δ 4.0

(*c*) Quartet δ 4.2

O O , 1745 cm^{-1} (ester)

17.47 That compound **X** does not dissolve in aqueous sodium bicarbonate indicates that **X** is not a carboxylic acid. That **X** has an IR absorption peak at 1740 cm^{-1} indicates the presence of a carbonyl group, probably that of an ester (Figure 17.2). That the molecular formula of **X** ($C_7H_{12}O_4$) contains four oxygen atoms suggests that **X** is a diester.

The ^{13}C spectrum shows only four signals indicating a high degree of symmetry for **X**. The single signal at δ 166.7 is that of an ester carbonyl carbon, indicating that both ester groups of **X** are equivalent.

Putting these observations together with the information gathered from DEPT ^{13}C spectra and the molecular formula leads us to the conclusion that **X** is diethyl malonate. The assignments are

(a) (*c*) O (*d*) O (*b*) O (*d*) O (*c*) (*a*)

(*a*) δ 14.2

(*b*) δ 41.6

(*c*) δ 61.3

(*d*) δ 166.7

17.48 The very low hydrogen content of the molecular formula of **Y** ($C_8H_4O_3$) indicates that **Y** is highly unsaturated. That **Y** dissolves slowly in warm aqueous $NaHCO_3$ suggests that **Y** is a carboxylic acid anhydride that hydrolyzes and dissolves because it forms a carboxylate salt:

R O O R O (insoluble) $\xrightarrow[H_2O,\ heat]{NaHCO_3}$ R O O^-Na^+ + R O O^-Na^+ (soluble)

The infrared absorption peaks at 1779 and 1854 cm^{-1} are consistent with those of an aromatic carboxylic anhydride (Figure 17.2).

That only four signals appear in the ^{13}C spectrum of **Y** indicates a high degree of symmetry for **Y**. Three of the signals occur in the aromatic region (δ 120 - δ 140) and one signal is downfield (δ 162)

These signals and the information from the DEPT ^{13}C NMR spectra lead us to conclude that **Y** is phthalic anhydride. The assignments are

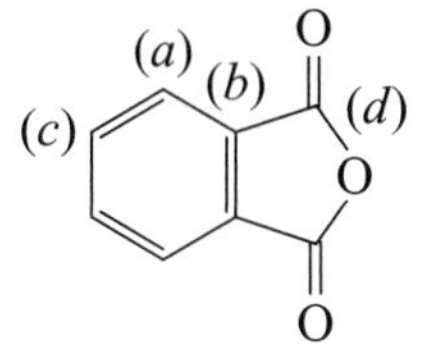

(a) δ 125
(b) δ 130
(c) δ 136
(d) δ 162

Y **Z** 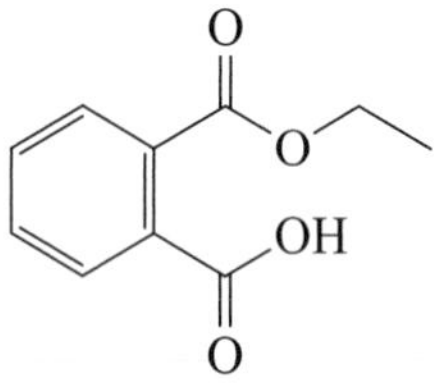 **AA**

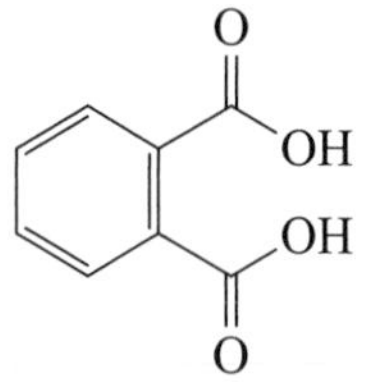

Z is phthalic acid and **AA** is ethyl hydrogen phthalate.

Challenge Problems

17.49 (a) Ethyl acetate (b) Acetic anhydride (c) *N*-Ethylacetamide.

17.50 In the first instance, nucleophilic attack by the amine occurs preferentially at the less hindered carbon of the formyl group. (Recall that aldehydes are more reactive than ketones toward nucleophiles for the same reason.) In the second case, $F_3CCO_2^-$ is a better leaving group than $H_3CCO_2^-$ since the former is the conjugate base of the stronger acid.

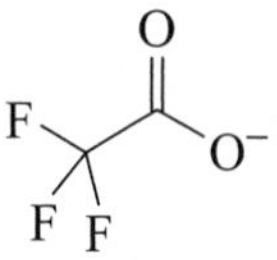

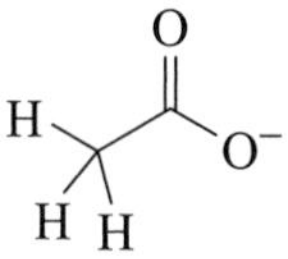

17.51 $\xrightarrow{COCl_2}$ $\xrightarrow{CH_3NH_2}$

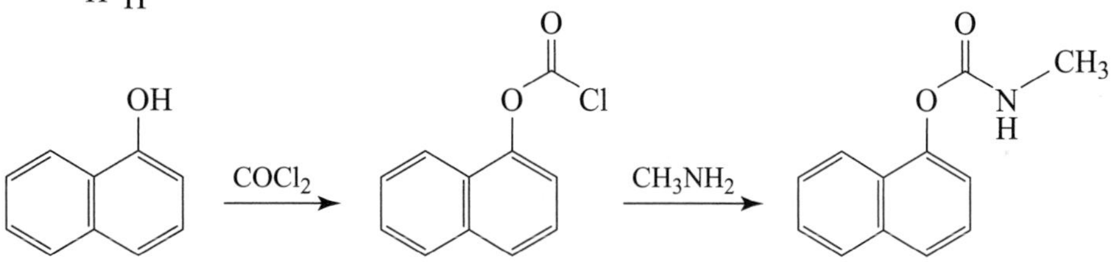

17.52 C_6H_6 + $\xrightarrow{AlCl_3}$ $\xrightarrow[\text{reduction}]{\text{Clemmensen}}$

$\xrightarrow[HCl,\ ZnCl_2]{HCHO}$ $\xrightarrow{KCN}$

$\xrightarrow[CH_3I]{NaNH_2}$ $\xrightarrow[\text{heat}]{H_3O^+}$

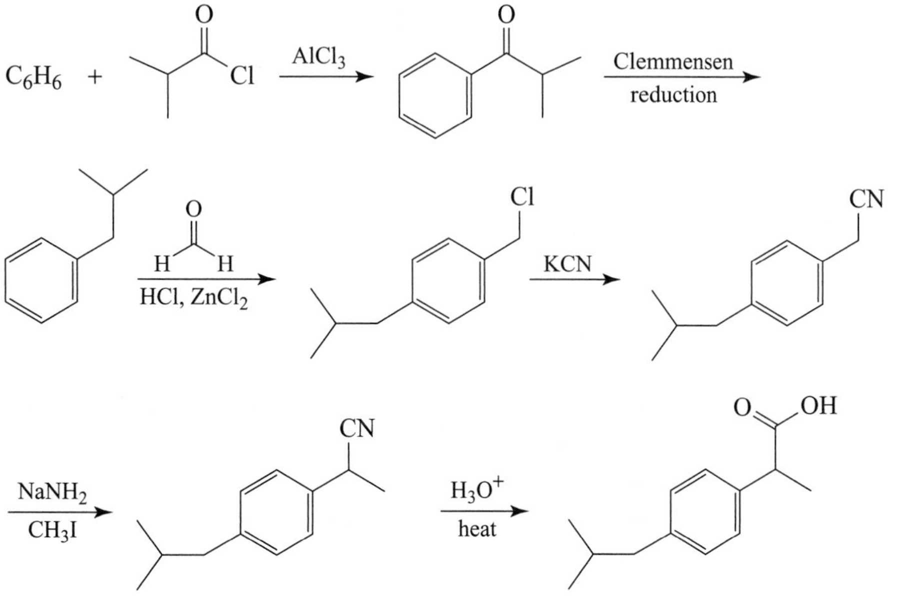

17.53 **A** = 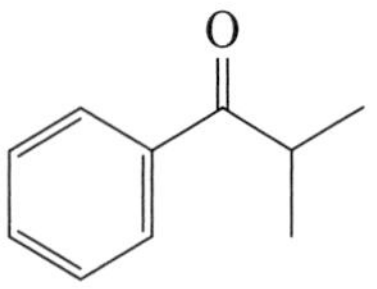   **C** =

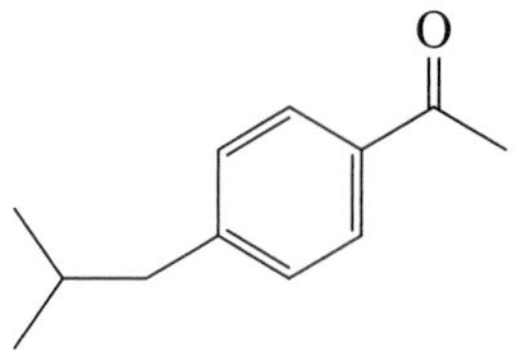

B = 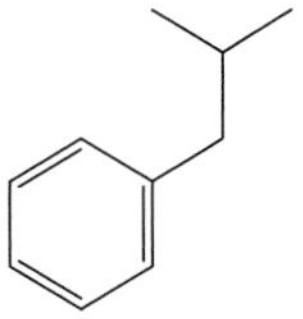   **D** =

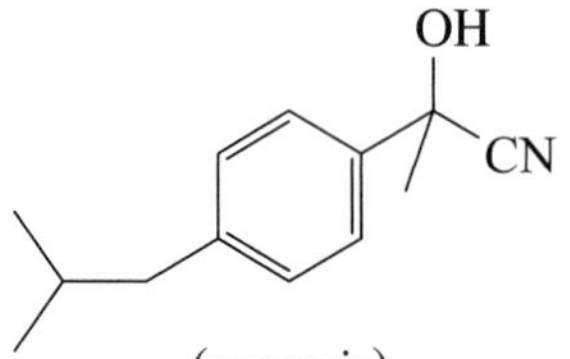

(*racemic*)

In the last step, HI/red P accomplishes both the reduction of —OH to —H and the hydrolysis of the nitrile function.

17.54 (a) The signal at δ 193.8 is consistent with the carbonyl carbon of an aldehyde and shows that the PCC reaction produced cinnamaldehyde.

(b) The signal at δ 164.5 is consistent with the carbonyl carbon of a carboxylic acid, and suggests that the oxidation with $K_2Cr_2O_7$ in sulfuric acid produced cinnamic acid.

QUIZ

17.1 Which of the following would be the strongest acid?

(a) Benzoic acid (b) 4-Nitrobenzoic acid (c) 4-Methylbenzoic acid

(d) 4-Methoxybenzoic acid (e) 4-Ethylbenzoic acid

17.2 Which of the following would yield (*S*)-2-butanol?

(a) (*R*)-2-Bromobutane + $CH_3CO_2^-\,Na^+$ ⟶ product $\xrightarrow[\text{heat}]{OH^-,\ H_2O}$

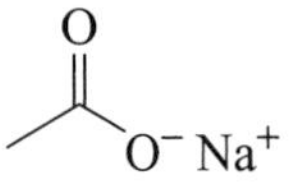

(b) (*R*)-2-Bromobutane $\xrightarrow[\text{heat}]{OH^-,\ H_2O}$

(c) (*S*)-2-Butyl acetate $\xrightarrow[\text{heat}]{OH^-,\ H_2O}$

(d) All of the above

(e) None of the above

17.3 Which reagent would serve as the basis for a simple chemical test to distinguish between hexanoic acid and hexanamide?

(a) Cold dilute NaOH (b) Cold dilute $NaHCO_3$

(c) Cold concd H_2SO_4 (d) More than one of these

(e) None of these

17.4 Give an acceptable name for:

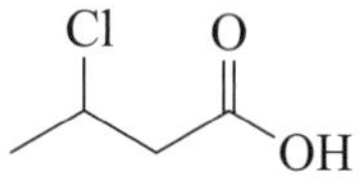

A: Cl, O, OH

B: O, OCH_3, NO_2

C: $NHCH_3$

A 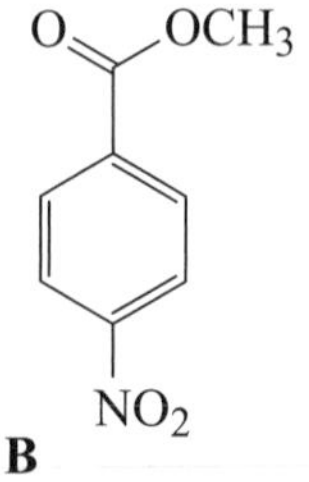 [ ] B 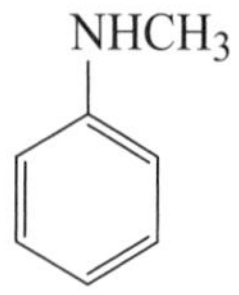 [ ] C [ ]

17.5 Complete the following synthesis.

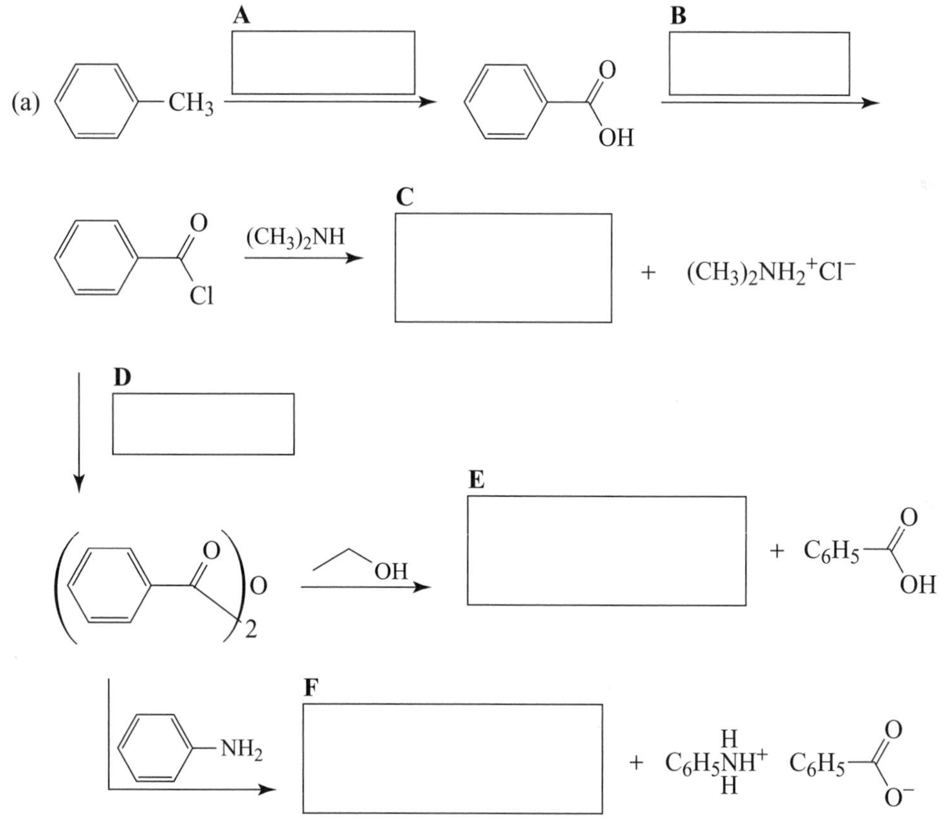

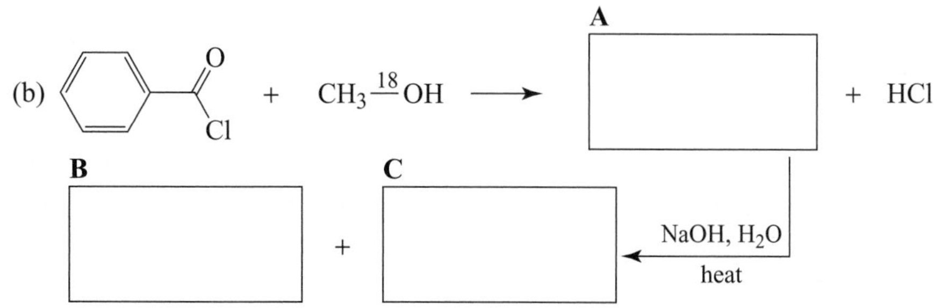

(a) CH_3 —A→ O, OH —B→

O, Cl —$(CH_3)_2NH$→ C [ ] + $(CH_3)_2NH_2^+Cl^-$

D [ ]

O, O, 2 —OH→ E [ ] + C_6H_5 O OH

NH_2 → F [ ] + $C_6H_5NH_3^+$ C_6H_5 O O^-

(b) O, Cl + CH_3—^{18}OH → A [ ] + HCl

NaOH, H_2O, heat → B [ ] + C [ ]

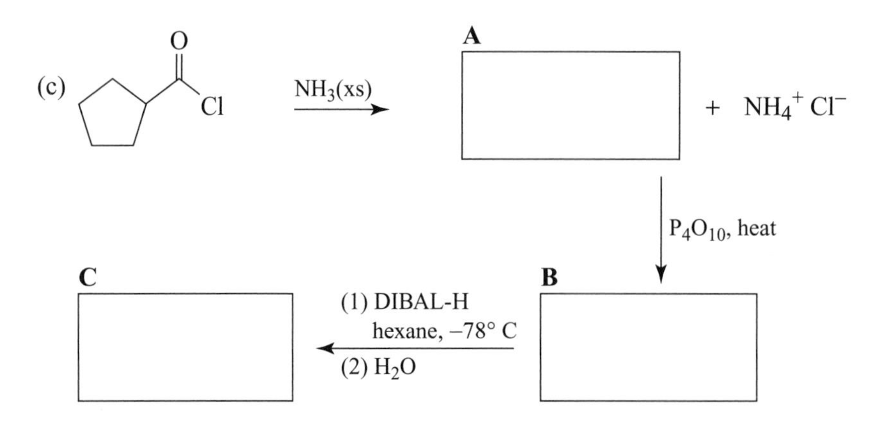

17.6 Which of these acids would undergo decarboxylation most readily?

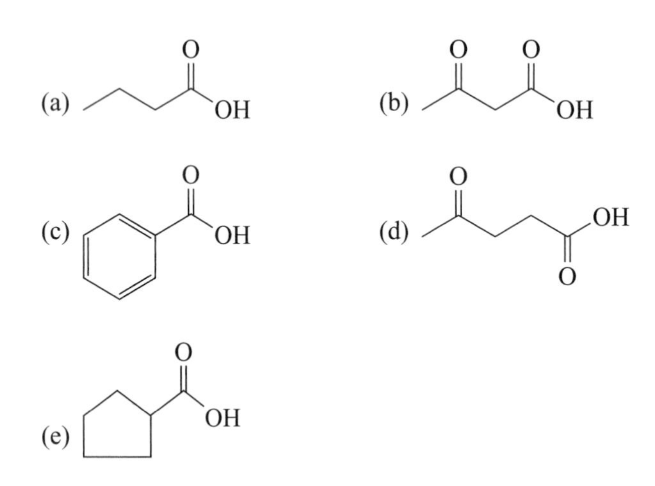

18 REACTIONS AT THE α CARBON OF CARBONYL COMPOUNDS: ENOLS AND ENOLATES

SOLUTIONS TO PROBLEMS

18.1

Cyclohexa-2,4-dien-1-one (keto form) ⇌ Phenol (enol form)

The enol form is aromatic, and it is therefore stabilized by the resonance energy of the benzene ring.

18.2 No. 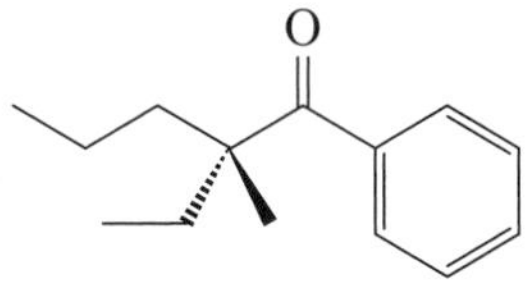 [structure] does not have a hydrogen attached to its α-carbon atom (which is a chirality center) and thus enol formation involving the chirality center is not possible. With [structure] the chirality center is a β carbon and thus enol formation does not affect it.

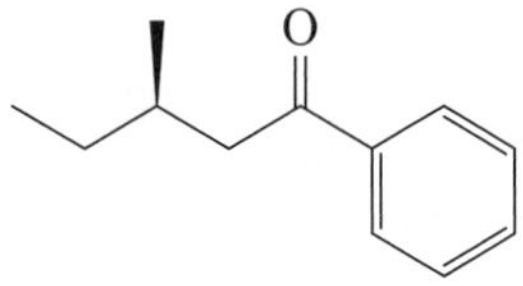

18.3 EtO^- EtO—H $+ EtO^-$

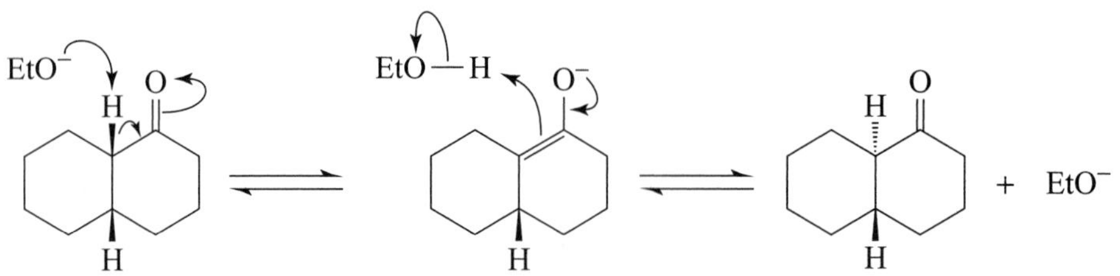

cis-Decalone

Protonation on the bottom face

trans-Decalone

A large group is axial in *cis*-decalone.

No bulky groups are axial in *trans*-decalone.

18.4 The reaction is said to be "base promoted" because base is consumed as the reaction takes place. A catalyst is, by definition, not consumed.

18.5 (a) The slow step in base-catalyzed racemization is the same as that in base-promoted halogenation—*the formation of an enolate anion.* (Formation of an enolate anion from 2-methyl-1-phenylbutan-1-one leads to racemization because the enolate anion is achiral. When it accepts a proton, it yields a racemic form.) The slow step in acid-catalyzed racemization is also the same as that in acid-catalyzed halogenation—*the formation of an enol.* (The enol, like the enolate anion, is achiral and tautomerizes to yield a racemic form of the ketone.)

(b) According to the mechanism given, the slow step for acid-catalyzed iodination (formation of the enol) is the same as that for acid-catalyzed bromination. Thus, we would expect both reactions to occur at the same rate.

(c) Again, the slow step for both reactions (formation of the enolate anion) is the same, and consequently, both reactions take place at the same rate.

18.6 (a)

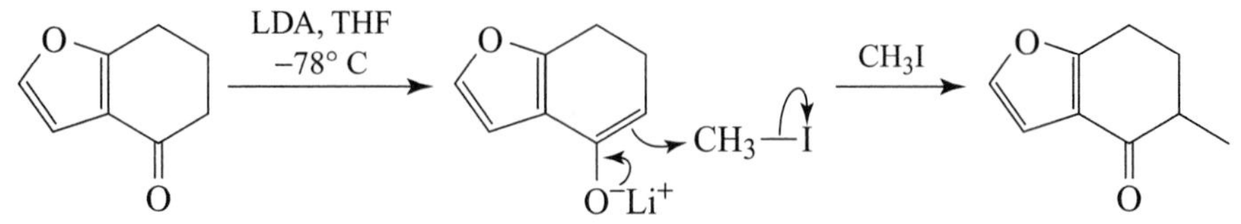

LDA, THF
−78° C

CH_3—I

CH_3I

O⁻Li⁺

(b)

OCH₃

2 eq. LDA,
THF

Br

O⁻Li⁺ O⁻Li⁺

OCH₃

$(CH_3)_3Si$

OCH₃

H_3O^+

$(CH_3)_3Si$

O⁻Li⁺

OCH₃

$(CH_3)_3Si$

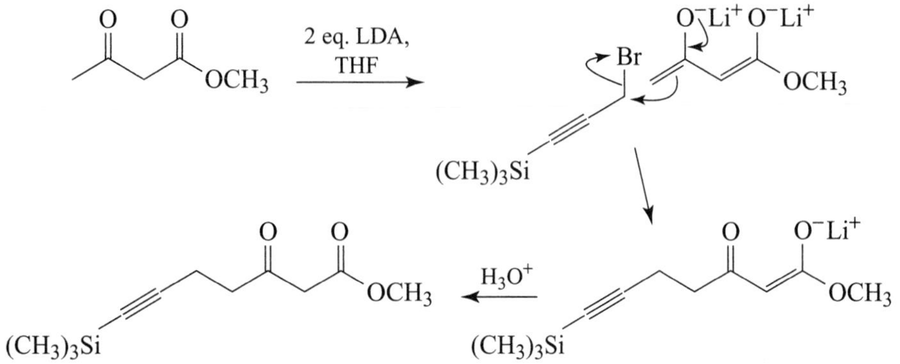

18.7 Again, working backward,

(a)

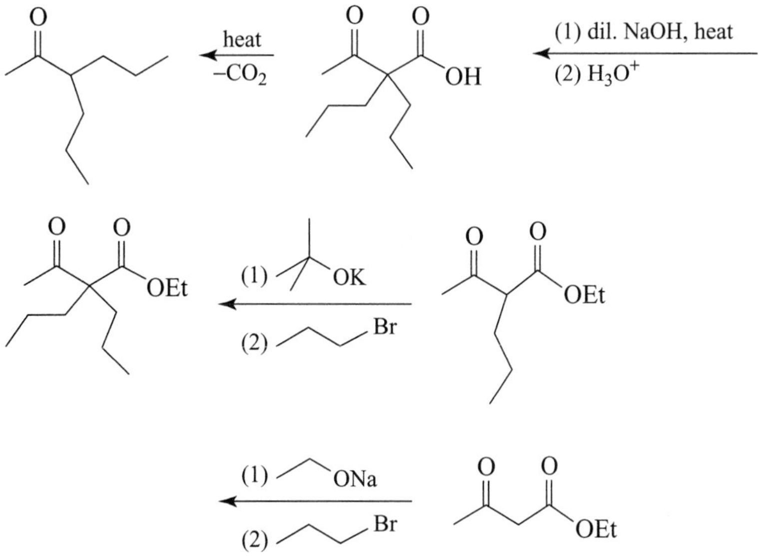

heat
$-CO_2$

OH

(1) dil. NaOH, heat
(2) H_3O^+

OEt

(1) OK
(2) Br

OEt

(1) ONa
(2) Br

OEt

(b)

C_6H_5

heat
$-CO_2$

OH

C_6H_5

(1) OH^-, heat
(2) H_3O^+

OEt

C_6H_5

(1) ONa
(2) C_6H_5 Br

OEt

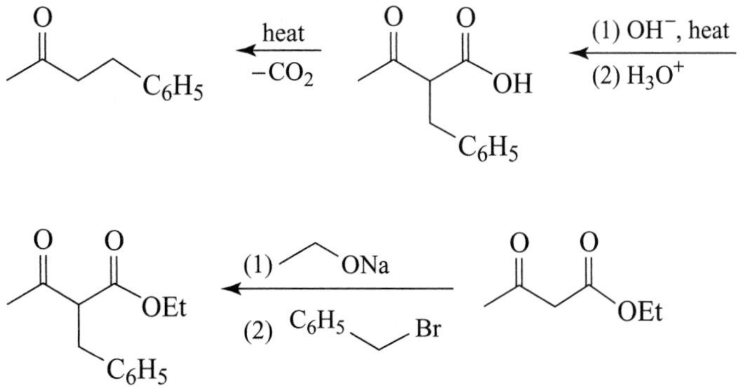

18.8 (a) Reactivity is the same as with any S_N2 reaction. With primary halides substitution is highly favored, with secondary halides elimination competes with substitution, and with tertiary halides elimination is the exclusive course of reaction.

(b) Acetoacetic ester and 2-methylpropene (isobutylene)

(c) Bromobenzene is unreactive to nucleophilic substitution.

18.9 The carboxyl group that is lost more readily is the one that is β to the keto group (cf. Section 17.10 of the text).

18.10 Working backward

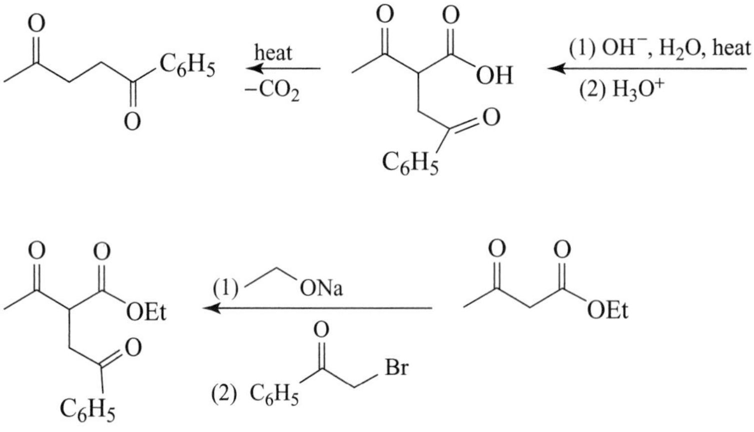

heat, $-CO_2$; (1) OH^-, H_2O, heat (2) H_3O^+

(1) ONa (2) C_6H_5 (O) Br

18.11 Working backward

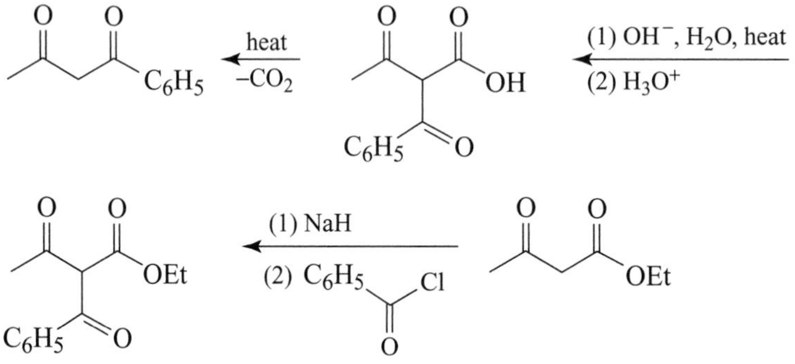

heat, $-CO_2$; (1) OH^-, H_2O, heat (2) H_3O^+

(1) NaH (2) C_6H_5 (O) Cl

18.12 Working backward,

(a) heat, $-CO_2$; (1) OH^-, H_2O, heat (2) H_3O^+

(1) ONa (2) Br

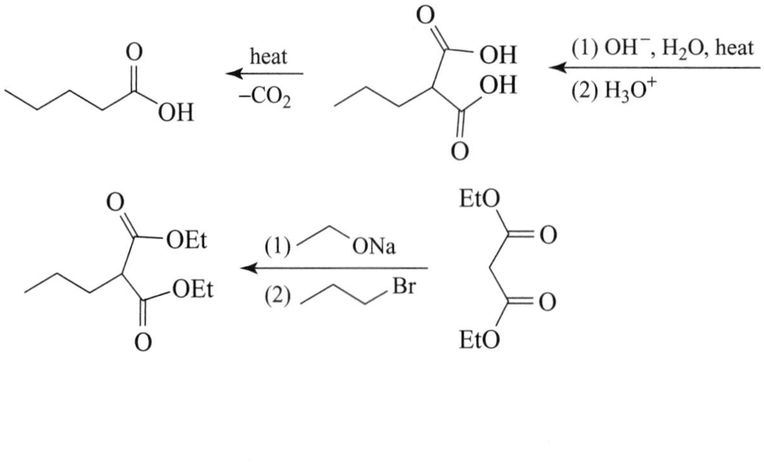

(b) OH ← heat, $-CO_2$ ← HO, O, O, HO ← (1) OH^-, H_2O, heat (2) H_3O^+

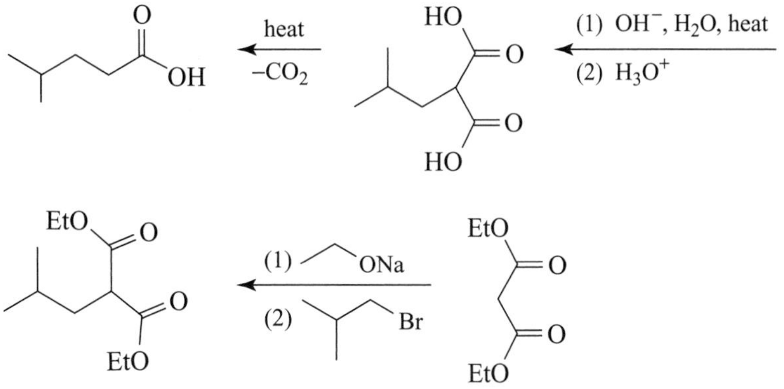

EtO, O, O, EtO ← MeI, OK ← EtO, O, O, EtO

← (1) ONa (2) Br ← EtO, O, O, EtO

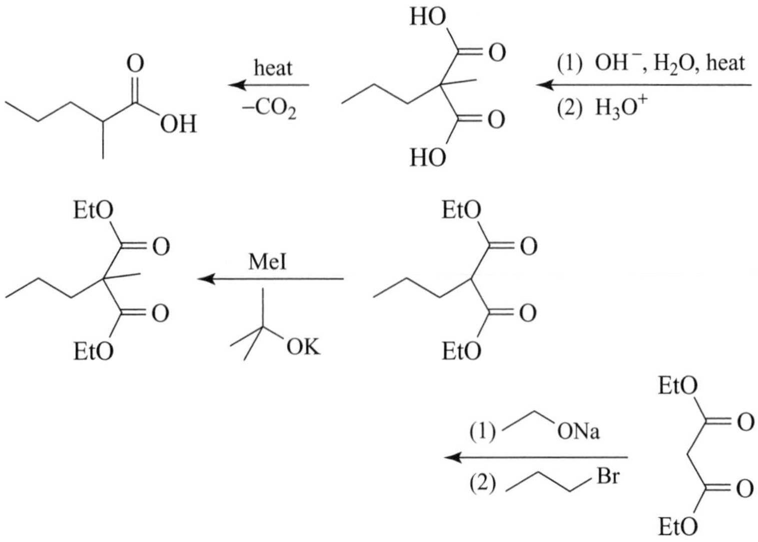

(c) OH ← heat, $-CO_2$ ← HO, O, O, HO ← (1) OH^-, H_2O, heat (2) H_3O^+

EtO, O, O, EtO ← (1) ONa (2) Br ← EtO, O, O, EtO

18.13 2 Br + CN, OEt, O $\xrightarrow[-2\text{ HBr}]{\text{base}}$ CN, OEt, O

$\xrightarrow[-CO_2,\ -EtOH]{OH^-,\ heat}$ CN $\xrightarrow[-NH_3\ (2)\ H_3O^+]{(1)\ OH^-,\ heat}$ O, OH

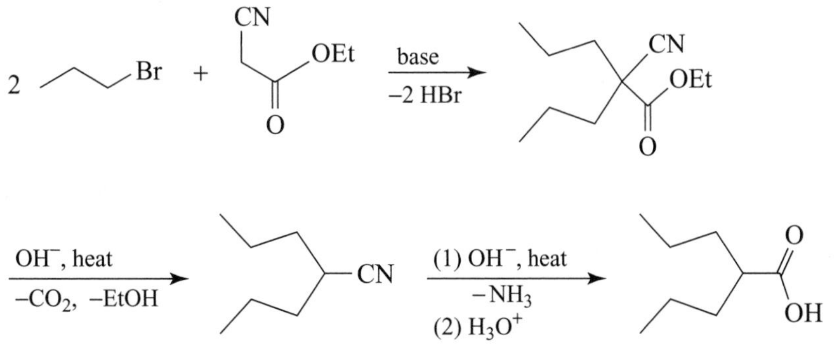

Valproic acid

18.14 These syntheses are easier to see if we work backward.

(a) H_2O; Cl⁻; Cl; (⟋⟍)$_3$N; H—N; HA, $-H_2O$

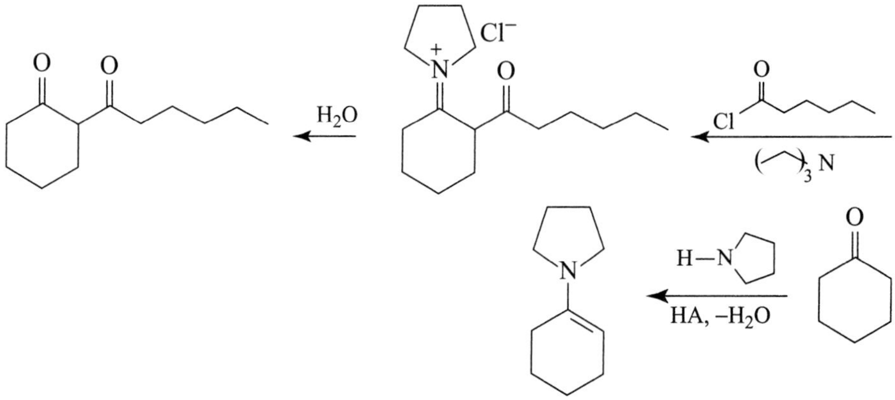

(b) H_2O; Br⁻; Br; H—N; HA, $-H_2O$

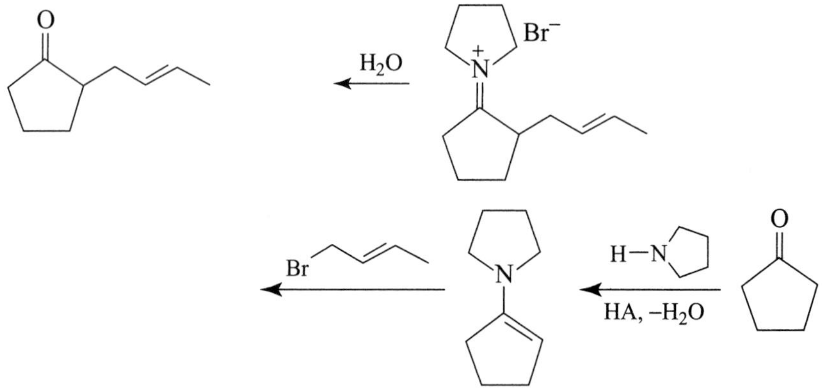

(c) (1) Br; (2) H_2O; H—N; HA, $-H_2O$

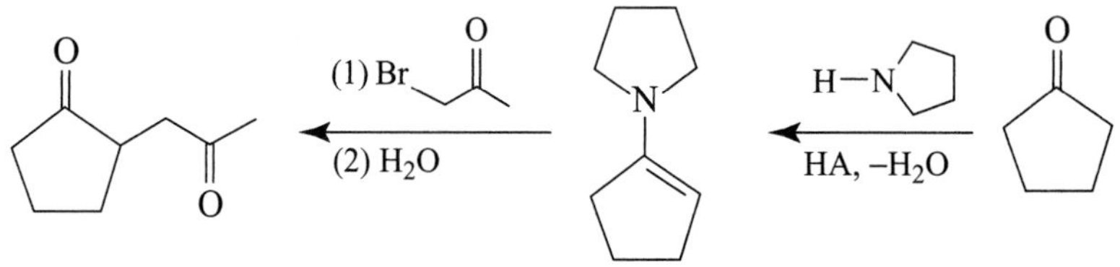

(d) OEt; (1) Br; OEt; (2) H_2O; H—N; HA, $-H_2O$

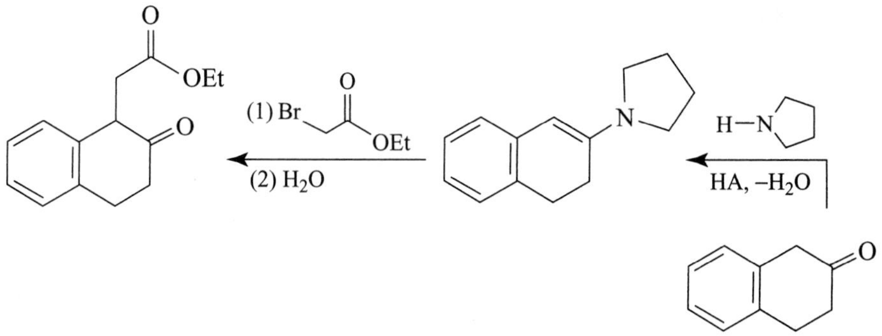

18.15 (a) **H_3C**–O–C(=O)–CH=CH_2 (1); H_3C–O–C(=O)–CH=CH–**CH_3** (2); **H**–O–C(=O)–CH=CH_2 (4); H_3C–C(=O)–C(**H**)(H)–C(=O)–CH_3 (3)

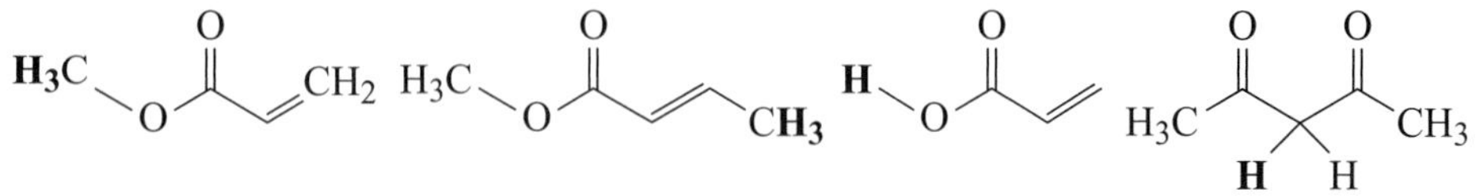

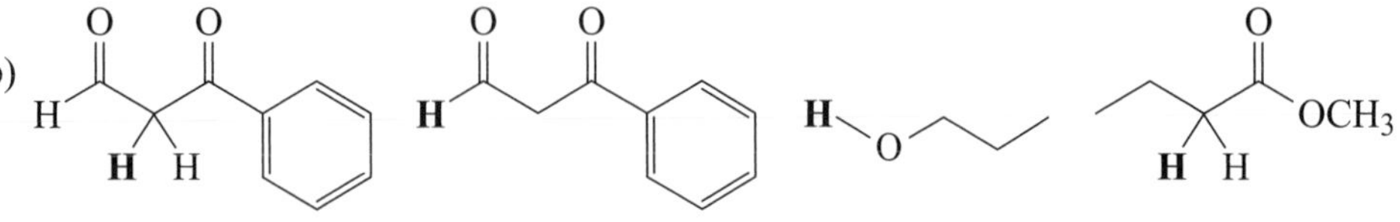

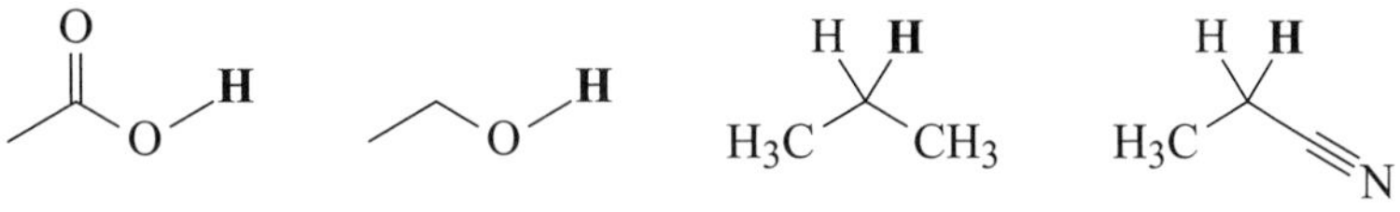

(b) H–C(=O)–C(**H**)(H)–C(=O)–Ph (4); **H**–C(=O)–CH_2–C(=O)–Ph (1); **H**–O–$CH_2CH_2CH_3$ (3); CH_3CH_2–C(**H**)(H)–C(=O)–OCH_3 (2)

(c) CH_3–C(=O)–O–**H** (4); CH_3CH_2–O–**H** (3); H_3C–C(H)(**H**)–CH_3 (1); H_3C–C(H)(**H**)–C≡N (2)

18.16 Abstraction of an α hydrogen at the ring junction yields an enolate anion that can then accept a proton to form either *trans*-1-decalone or *cis*-1-decalone. Since *trans*-1-decalone is more stable, it predominates at equilibrium.

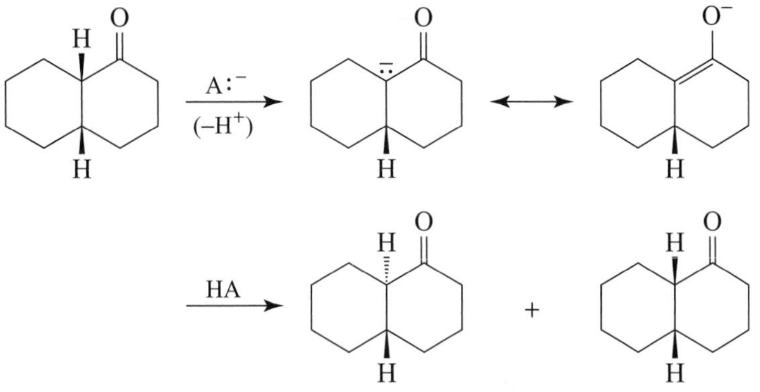

$$\xrightarrow[(-H^+)]{A:^-} \text{enolate} \longleftrightarrow \text{enolate} \xrightarrow{HA} trans\text{-1-Decalone} + cis\text{-1-Decalone}$$

(95%)
trans-1-Decalone
(more stable)

(5%)
cis-1-Decalone
(less stable)

18.17 In a polar solvent, such as water, the keto form is stabilized by solvation. When the interaction with the solvent becomes minimal, the enol form achieves stability by internal hydrogen bonding.

18.18 (a)

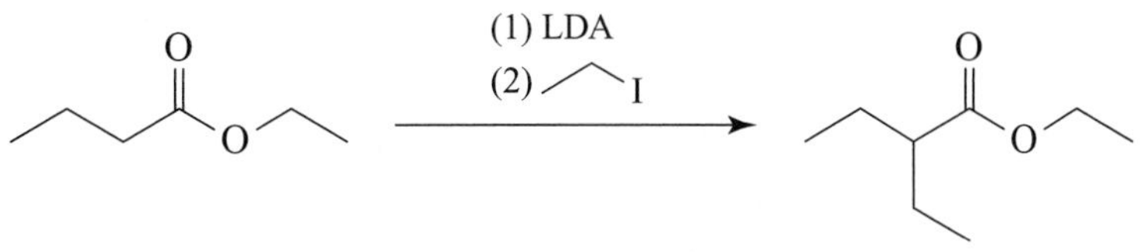

$$\xrightarrow[(2)\ \text{EtI}]{(1)\ \text{LDA}}$$

(b) 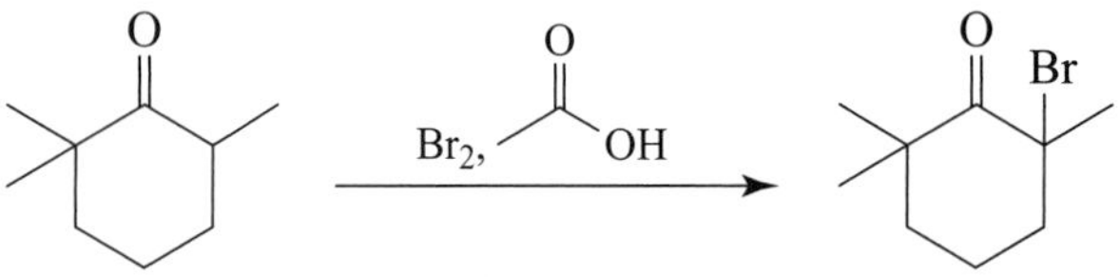 Br_2, CH$_3$COOH →

(c) 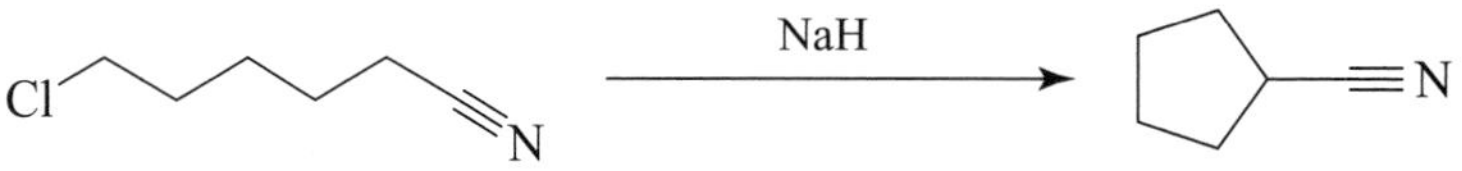 NaH →

(d) 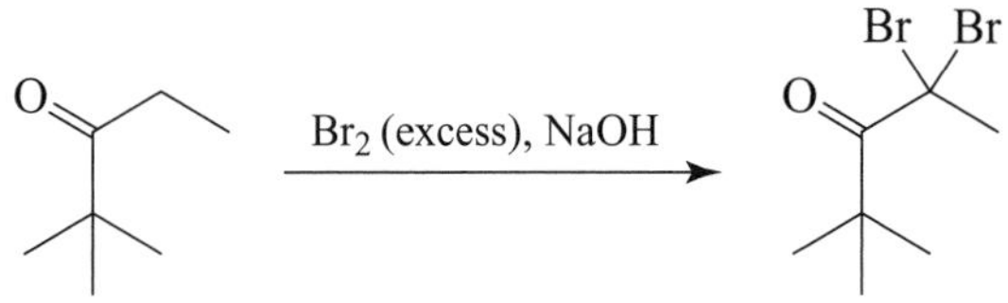 Br_2 (excess), NaOH →

(e) 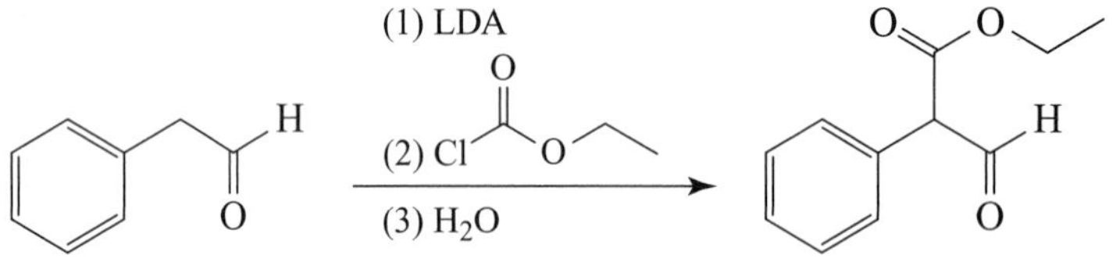 (1) LDA (2) ClCOOEt (3) H_2O →

(f) 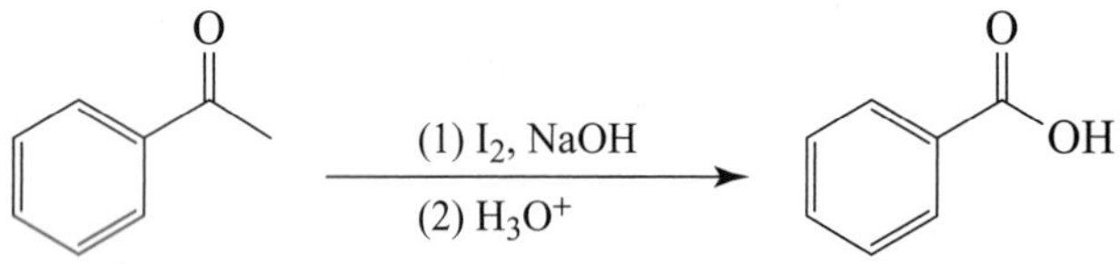 (1) I_2, NaOH (2) H_3O^+ →

18.19 (a)

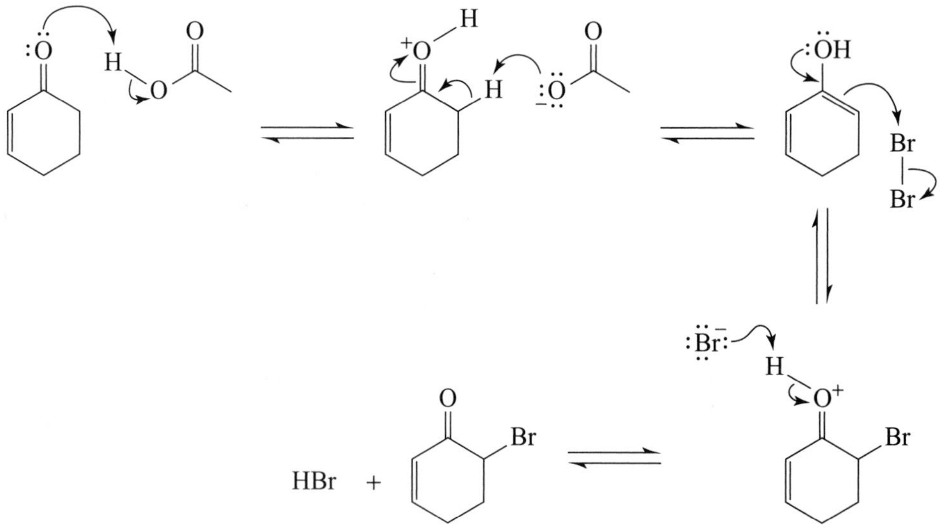

HBr +

(b)

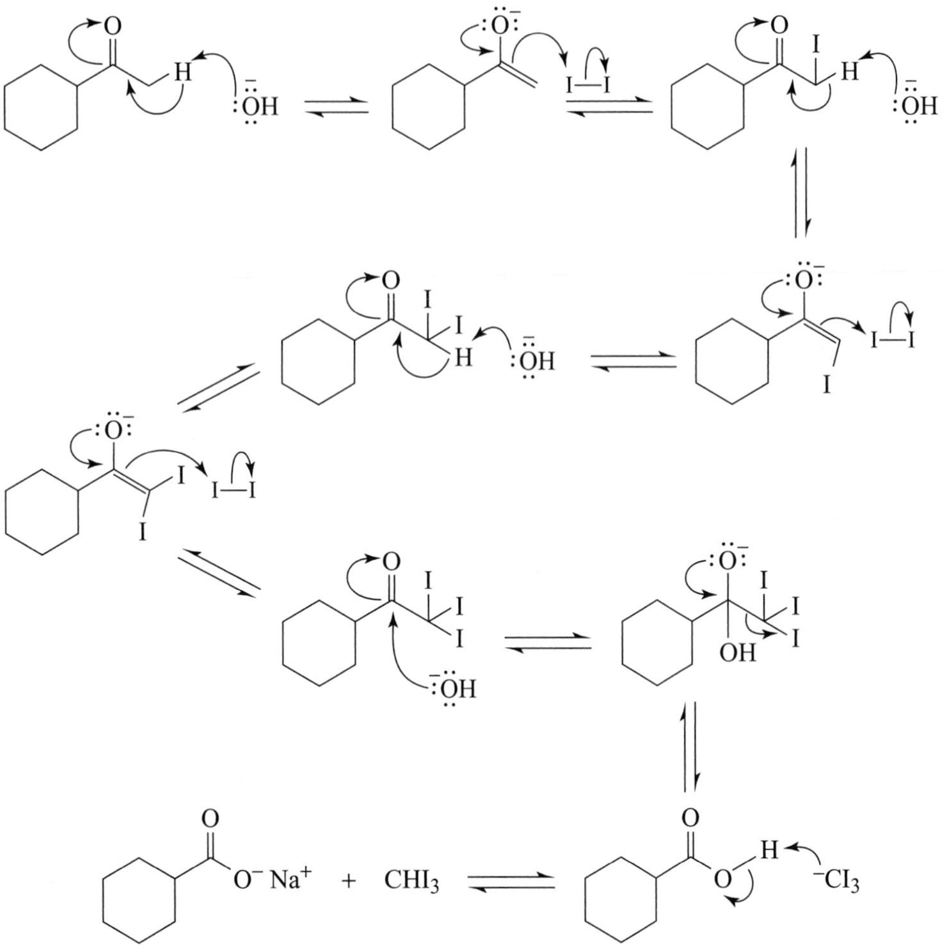

O H :ÖH ⇌ :Ö:⁻ I—I ⇌ O I H :ÖH

⇌

O I I H :ÖH ⇌ :Ö:⁻ I I—I

⇌

:Ö:⁻ I I I—I

⇌

O I I I ⁻:ÖH ⇌ :Ö:⁻ I I I OH

⇌

O $O^- Na^+$ + CHI_3 ⇌ O O H $^-CI_3$

(c)

H_3C O O H H H O+ H H_3C O+ O H H O H_3C O O+ H H O H H O

H_3C O O H H O+ H H O H_3C +O H O H O H O+ H H O H H O

+ CH_3OH

O H O

(d)

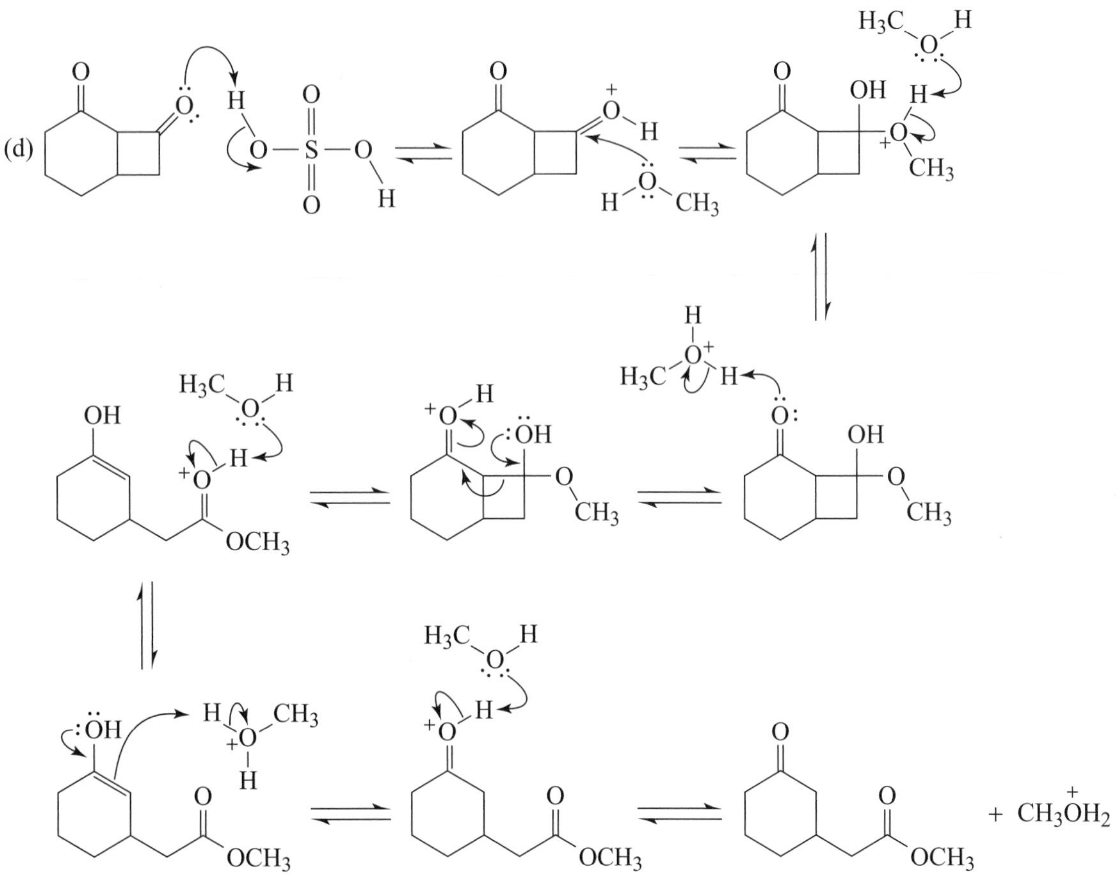

18.20 (a)

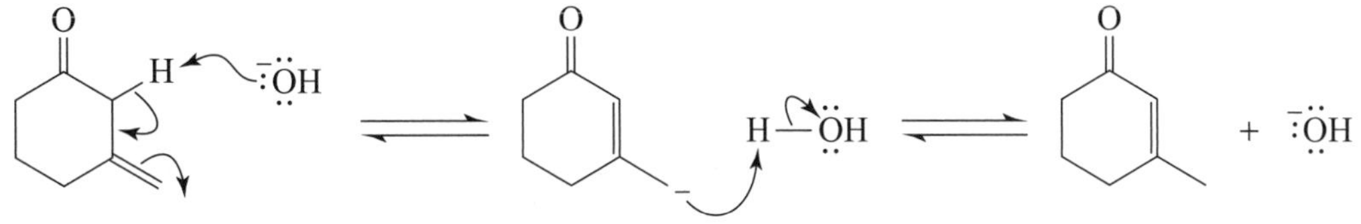

(b)

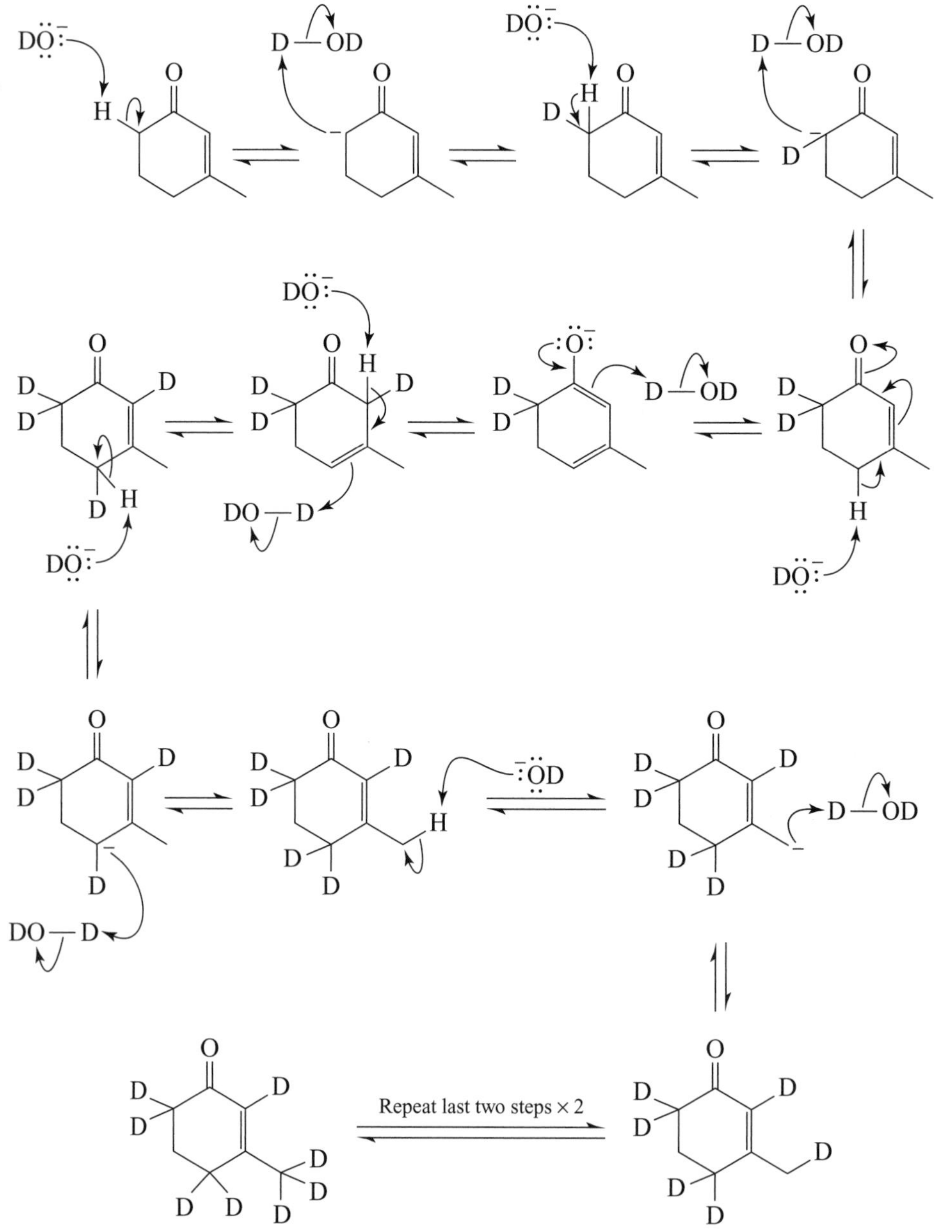

Repeat last two steps × 2

(c) H–OH, :ÖH⁻, HO–H

(d) N, Li +, H, I—CH_3, I—CH_3, +

SUMMARY OF ACETOACETIC ESTER AND MALONIC ESTER SYNTHESES

A. Acetoacetic Ester Synthesis

OEt —(1) ONa, (2) RX→ OEt (R, H) —(1) OK, (2) R'X→

OEt (R, R') —(1) OH^-, H_2O, (2) H_3O^+→ OH (R, R') —heat, $-CO_2$→ R', R

B. Malonic Ester Synthesis

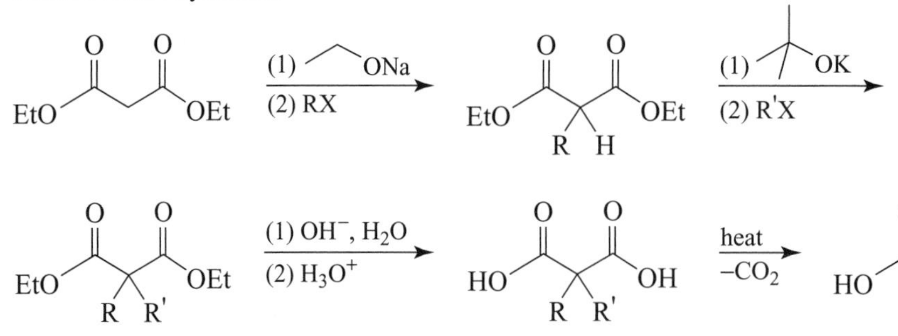

EtO(C=O)CH₂(C=O)OEt —(1) ONa (2) RX→ EtO(C=O)CH(R)(C=O)OEt —(1) OK (2) R'X→

EtO(C=O)C(R)(R')(C=O)OEt —(1) OH^-, H_2O (2) H_3O^+→ HO(C=O)C(R)(R')(C=O)OH —heat, $-CO_2$→ HO(C=O)CH(R)R'

Acetoacetic Ester and Malonic Ester Syntheses

18.21 (a) —(1) ONa (2) MeI→ —(1) OK (2) MeI→ OEt

—HO OH, HA→ OEt —(1) $LiAlH_4$ (2) H_2O→ OH —PBr_3→

Br —(1) Mg (2) H_3O^+→

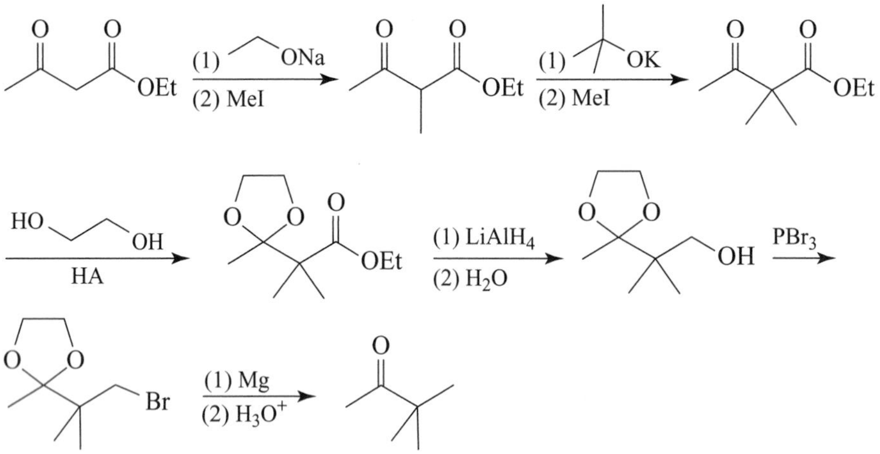

(b) OEt —(1) ONa (2) Br→ OEt —(1) OH^-, H_2O, heat (2) H_3O^+→

OH —heat, $-CO_2$→

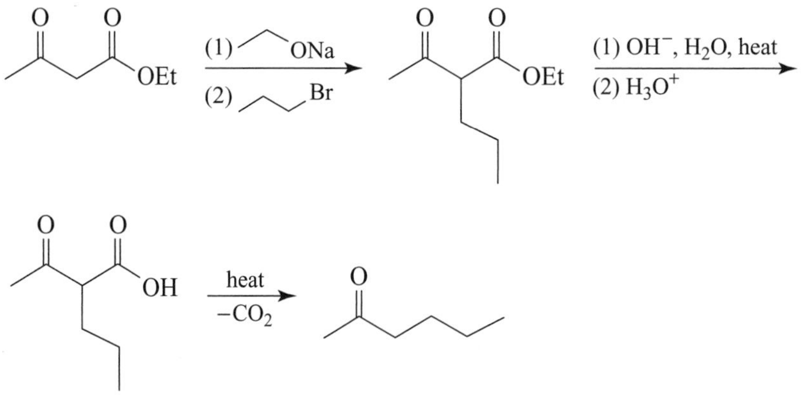

(c) OEt —(1) ONa (2) Br→ OEt —(1) OH^-, H_2O, heat (2) H_3O^+→

OH —heat, $-CO_2$→

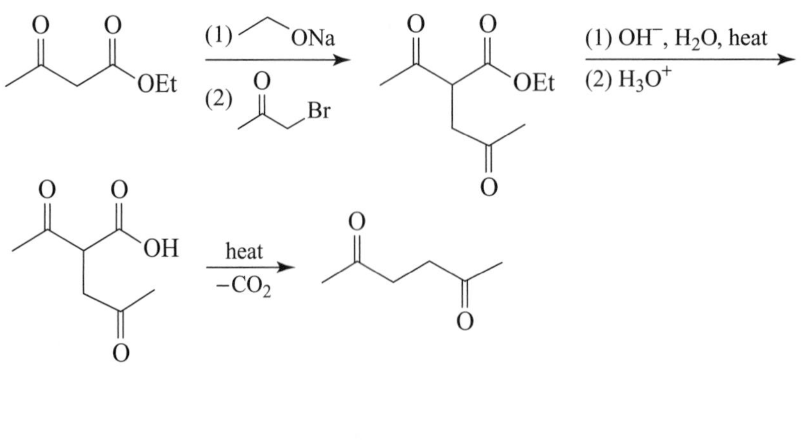

(d) 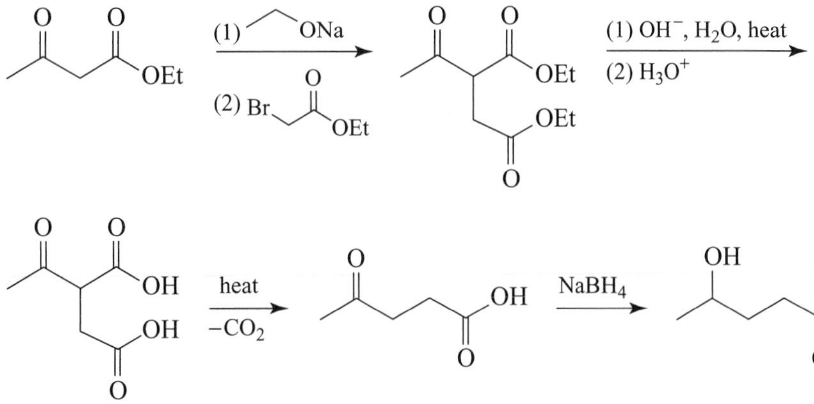 OEt (1) ONa (2) Br OEt → OEt OEt (1) OH^-, H_2O, heat (2) H_3O^+ →

OH OH heat $-CO_2$ → OH $NaBH_4$ → OH OH

(e) 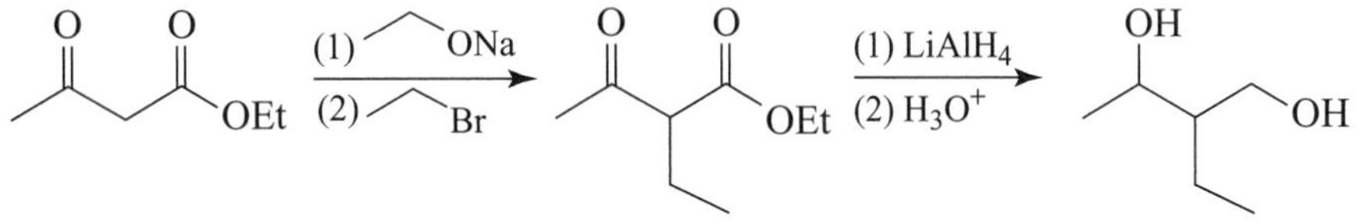 OEt (1) ONa (2) Br → OEt (1) $LiAlH_4$ (2) H_3O^+ → OH OH

(f) Compare Problem 18.11 → C_6H_5 $NaBH_4$ → OH OH C_6H_5

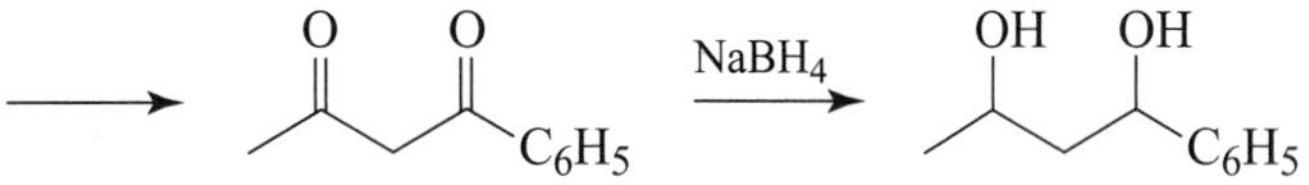

18.22 (a) 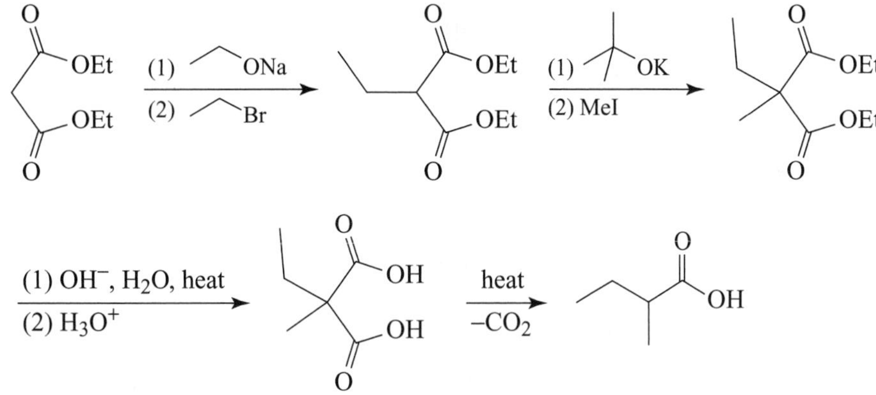 OEt OEt (1) ONa (2) Br → OEt OEt (1) OK (2) MeI → OEt OEt

(1) OH^-, H_2O, heat (2) H_3O^+ → OH OH heat $-CO_2$ → OH

(b) 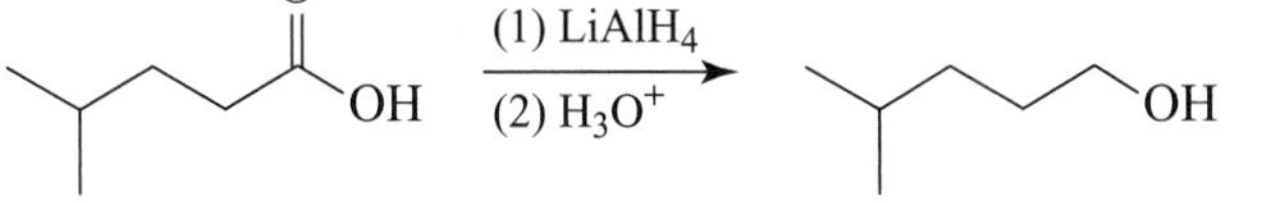 OH (1) $LiAlH_4$ (2) H_3O^+ → OH

[from Problem 18.12 (c)]

(c) 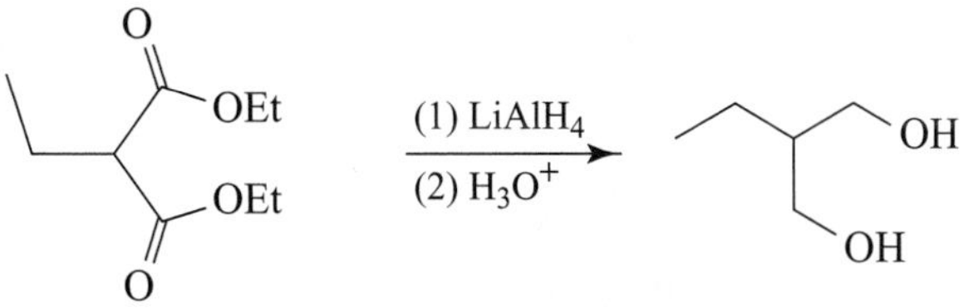 OEt OEt (1) $LiAlH_4$ (2) H_3O^+ → OH OH

[from (a) above]

(d) OEt, OEt —(1) ONa, (2) Br OEt→ EtO OEt OEt —HCl, heat→ HO OH OH

—heat, $-CO_2$→ HO OH —(1) $LiAlH_4$, (2) H_3O^+→ HO OH

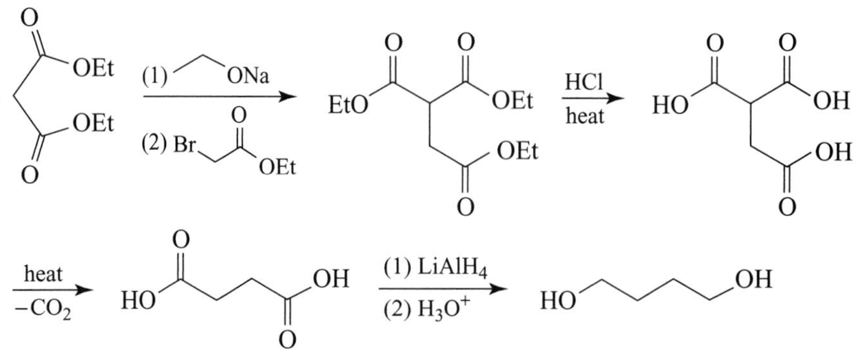

18.23 (a) —(1) $NaOCH_3$, CH_3OH (2) Br (3) HCl, H_2O, heat→

CH_3

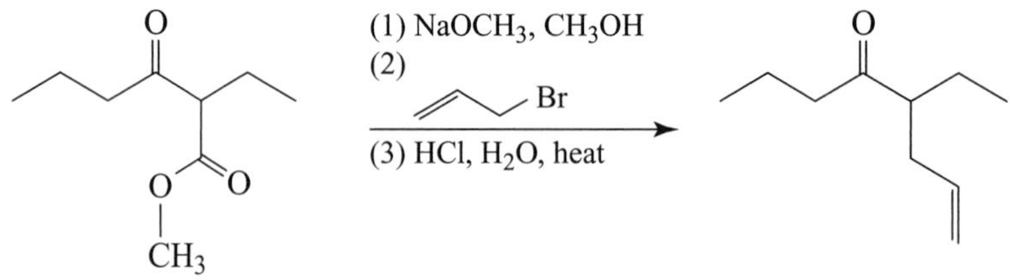

(b) OH, HO —heat→ HO

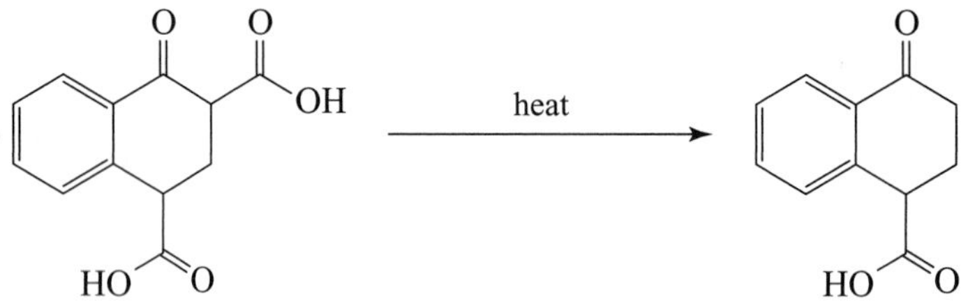

18.24 The following reaction took place:

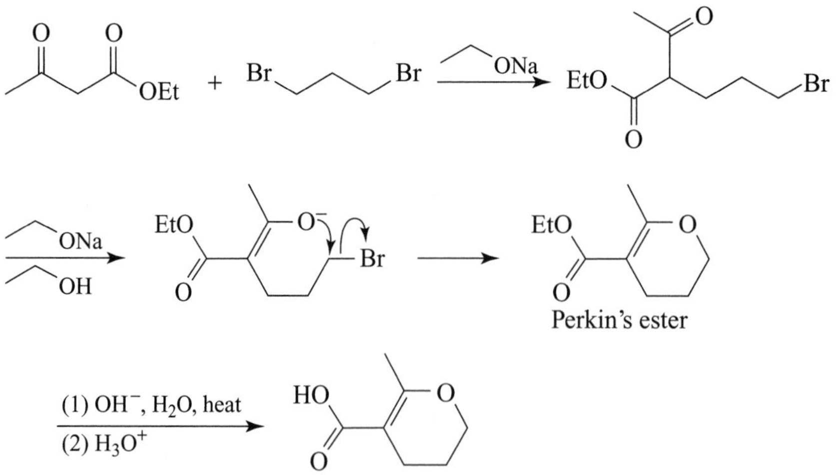

OEt + Br Br —ONa→ EtO Br

—ONa, OH→ EtO O^- Br → EtO O

Perkin's ester

—(1) OH^-, H_2O, heat (2) H_3O^+→ HO O

Perkin's acid

18.25 (a)

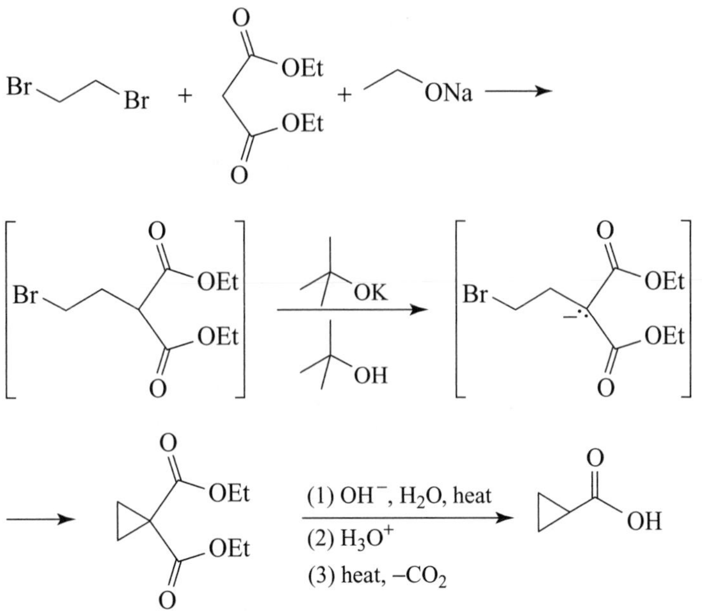
Br
Br
OEt
OEt
ONa
OK
OH
(1) OH⁻, H₂O, heat
(2) H₃O⁺
(3) heat, –CO₂
OH

(b)

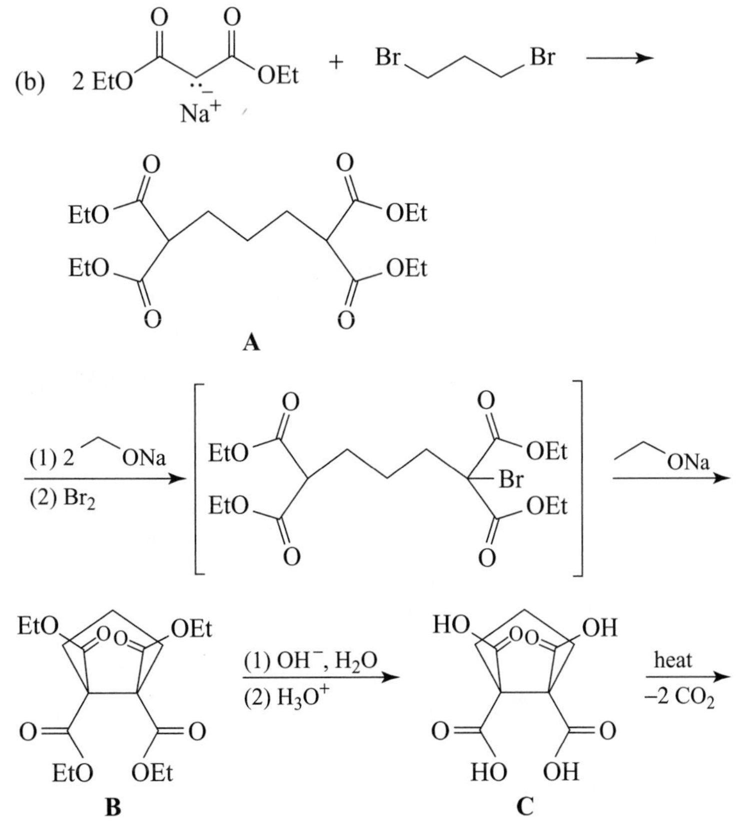
2 EtO
OEt
Na⁺
Br
Br
A
(1) 2 ONa
(2) Br₂
ONa
B
(1) OH⁻, H₂O
(2) H₃O⁺
C
heat
–2 CO₂

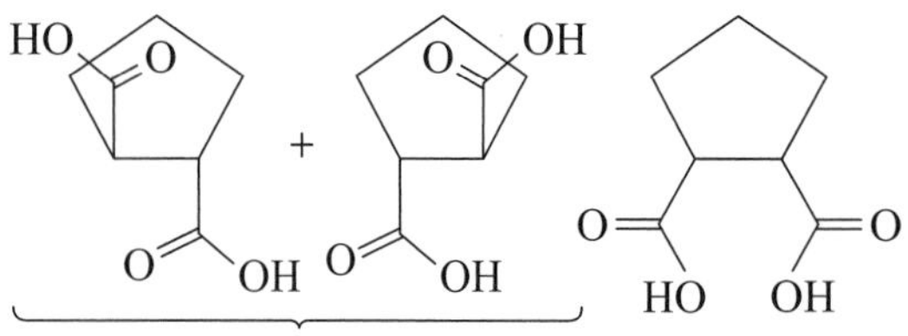

HO O + O OH

O OH O OH O O HO OH

D E

Racemic form Meso compound

(c) Br Br → (Na^{+} $^{-}$: malonate OEt OEt) → Br OEt OEt

(t-BuOK) → OEt OEt → (1) OH^{-}, H_2O (2) H_3O^{+} (3) heat, $-CO_2$ → OH

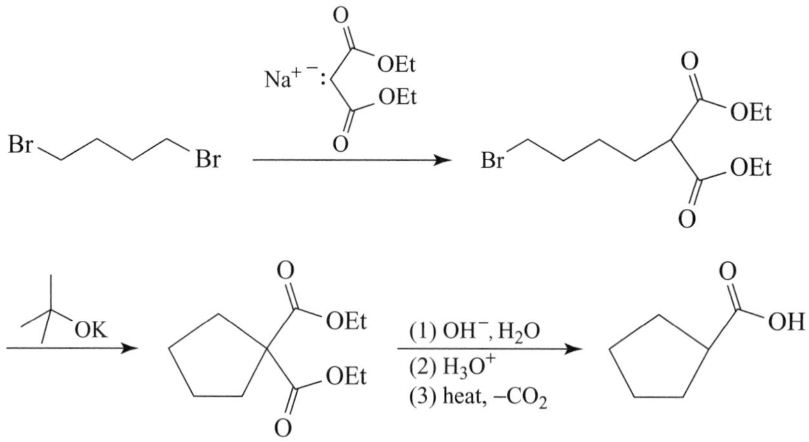

18.26 (a) (1) ONa (2) Cl Cl →

(1) NaOH (2) H_3O^{+} or H_3O^{+}

OH ← heat — HO OH

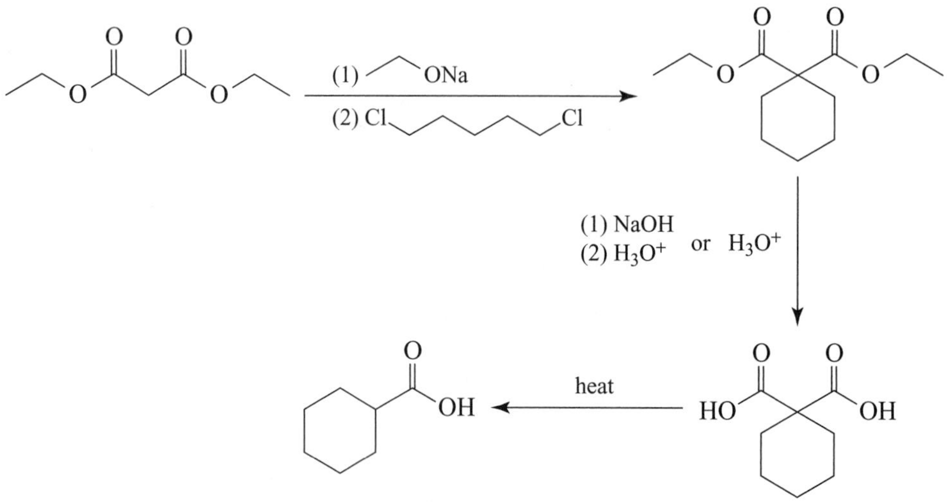

(b)

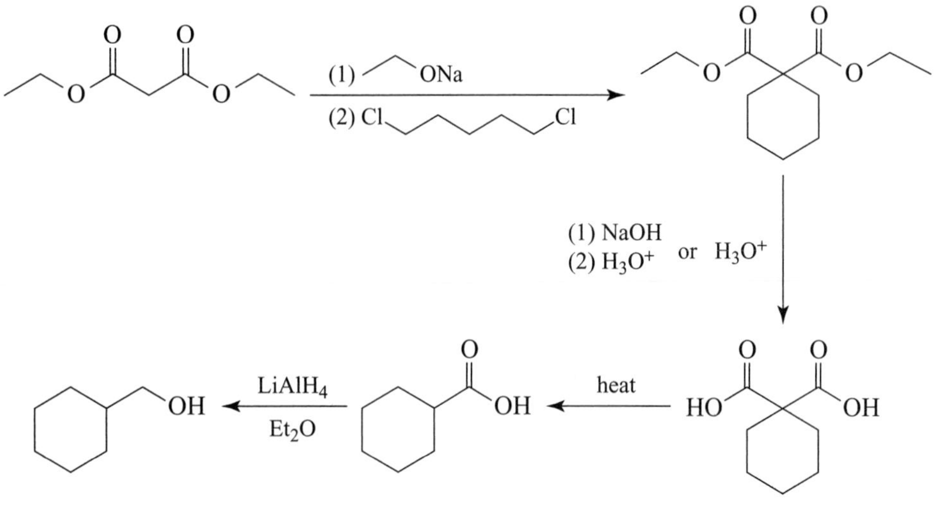

(1) ONa
(2) Cl Cl

(1) NaOH
(2) H_3O^+ or H_3O^+

heat

LiAlH$_4$
Et$_2$O

OH

HO OH

(c)

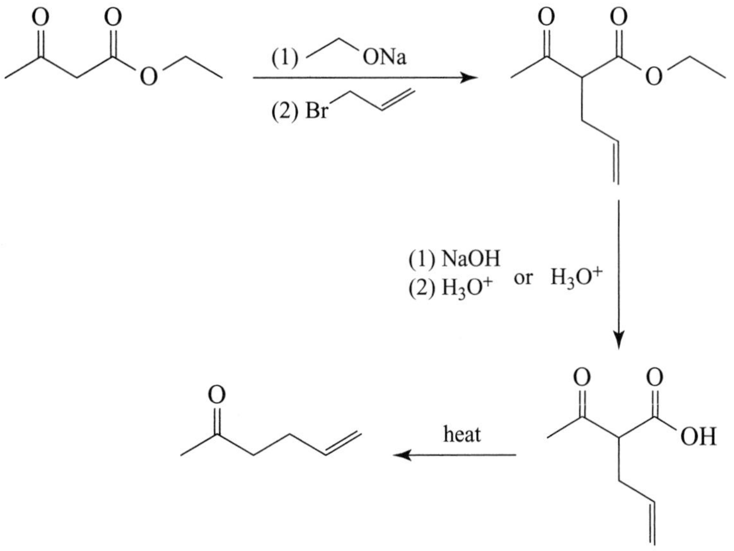

(1) ONa
(2) Br

(1) NaOH
(2) H_3O^+ or H_3O^+

heat

OH

(d) O O O O → (1) ONa (2) Cl Cl → O O O O

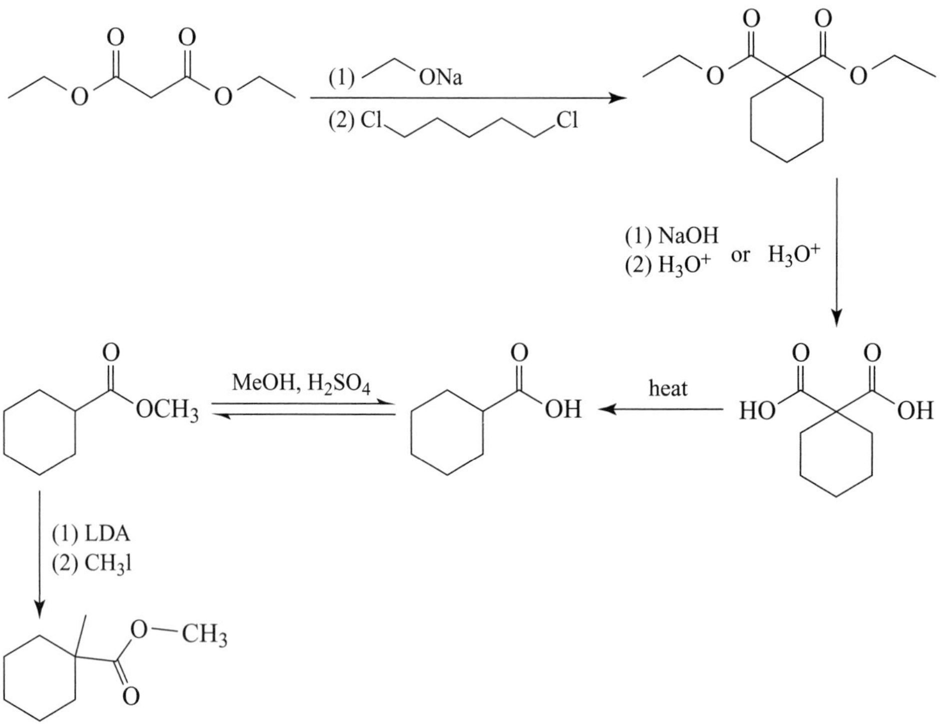

(1) NaOH
(2) H_3O^+ or H_3O^+

O OCH$_3$ ⇌ MeOH, H_2SO_4 O OH ← heat ← O O HO OH

(1) LDA
(2) CH_3I

O—CH$_3$ O

(e) O O O → (1) ONa (2) Br → O O O

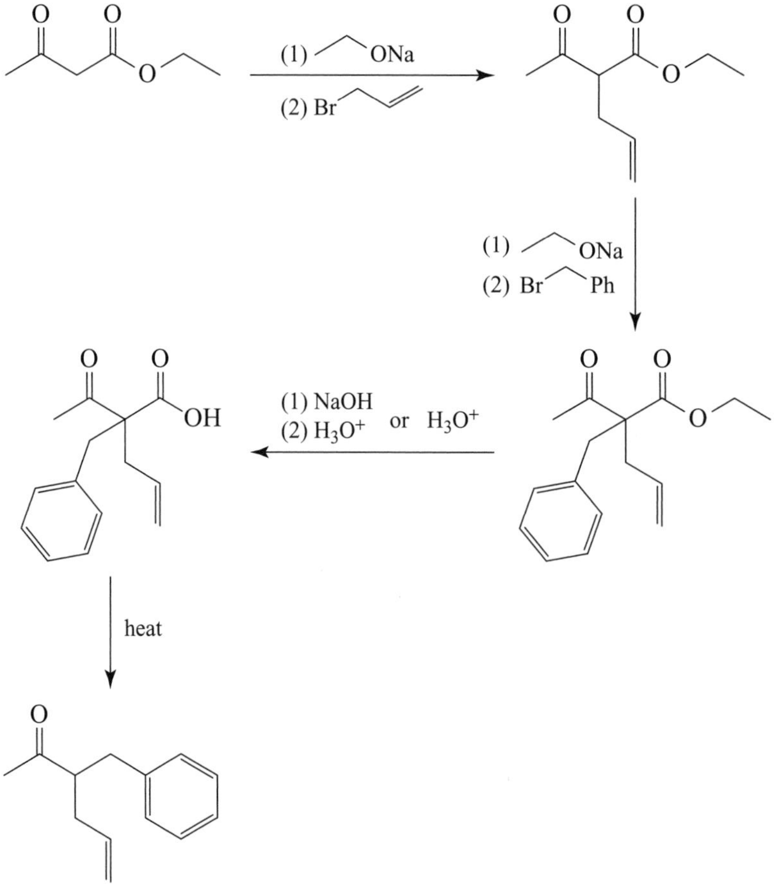

(1) ONa
(2) Br Ph

O O OH ← (1) NaOH (2) H_3O^+ or H_3O^+ ← O O O

heat

O

(f)

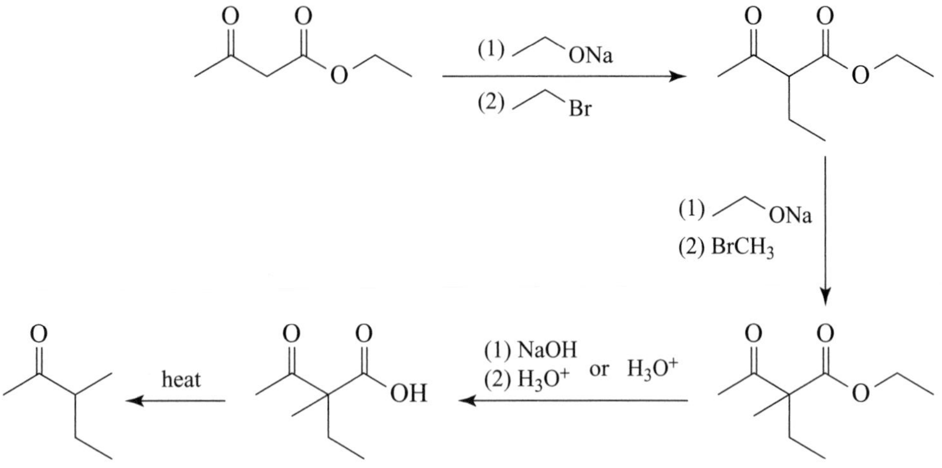

O O O

(1) ONa
(2) Br

O O O

(1) ONa
(2) $BrCH_3$

O

heat

O O OH

(1) NaOH
(2) H_3O^+ or H_3O^+

O O O

General Problems

18.27 (a)

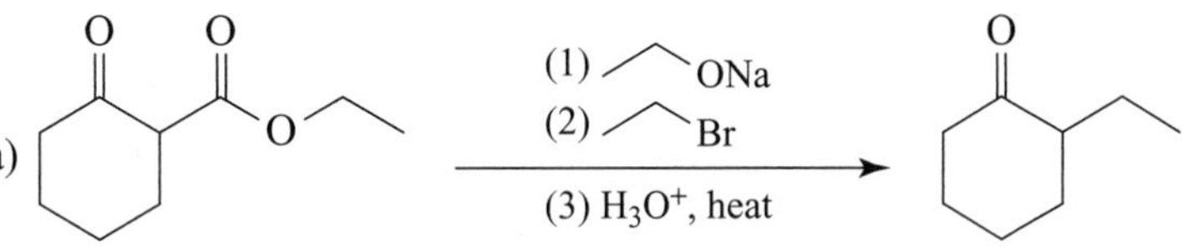

O O O

(1) ONa
(2) Br
(3) H_3O^+, heat

O

(b)

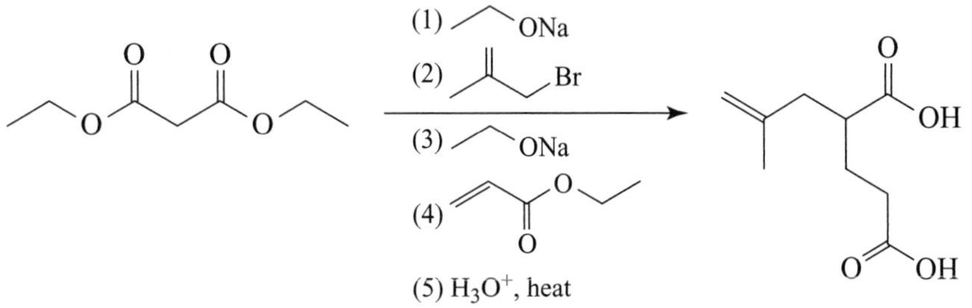

O O O O

(1) ONa
(2) Br
(3) ONa
(4) O O
(5) H_3O^+, heat

O OH

O OH

(c)

O O O

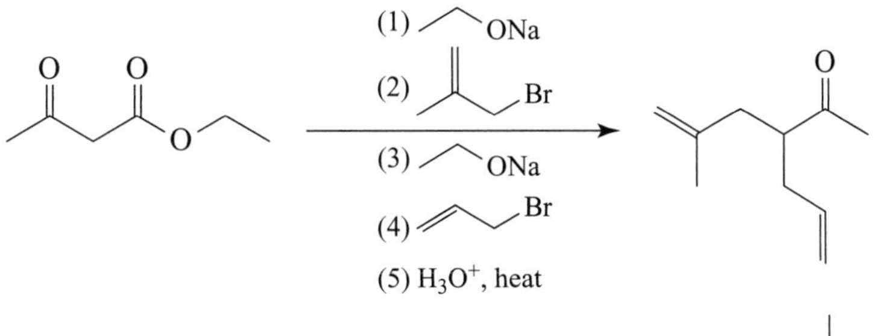

(1) ONa
(2) Br
(3) ONa
(4) Br
(5) H_3O^+, heat

O

(d)

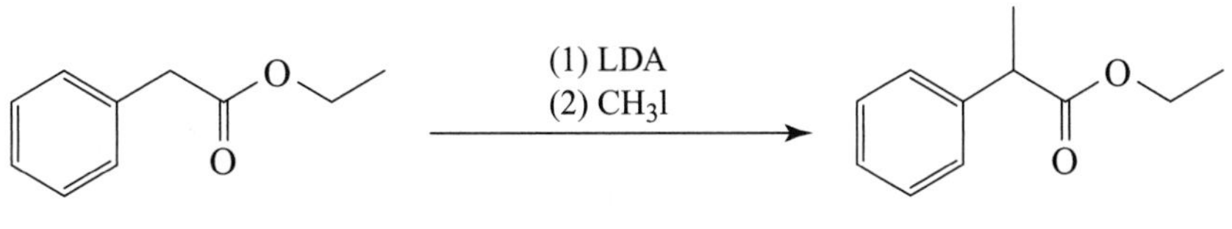

O O

(1) LDA
(2) CH_3I

O O

(e)

Br

Br

(1) O O O CH_3
$NaOCH_3$ (excess)
(2) H_3O^+, heat

O

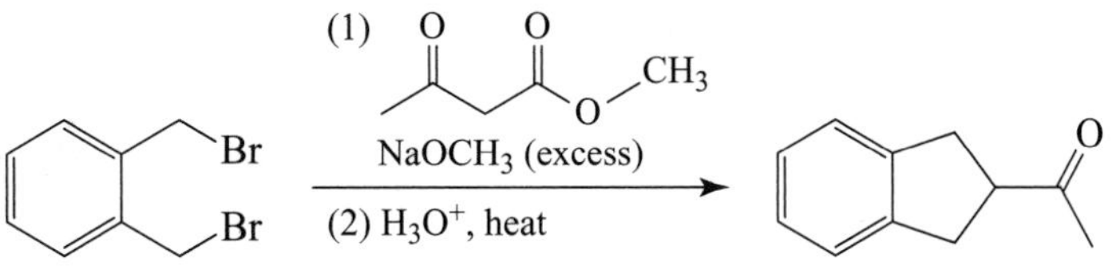

(f) 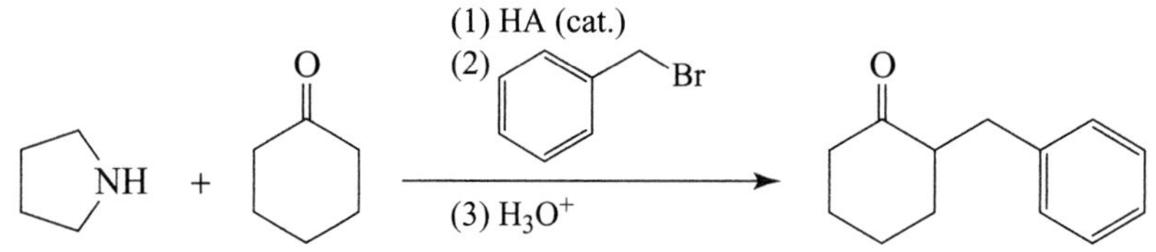 NH + O

(1) HA (cat.)
(2) Br
(3) H_3O^+

O

(g) 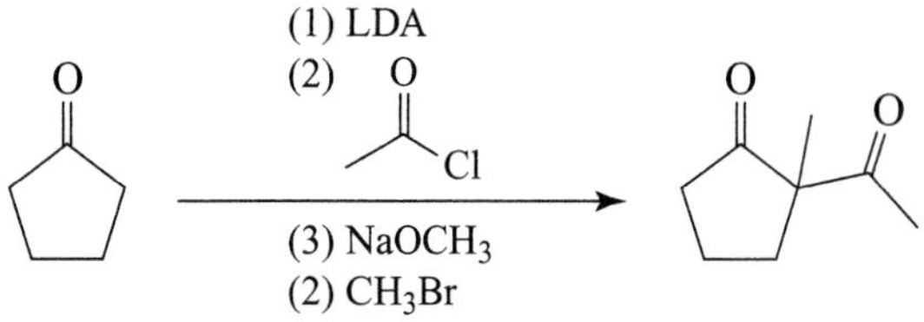 O

(1) LDA
(2) O Cl
(3) $NaOCH_3$
(2) CH_3Br

O O

18.28 + HBr → Br **F**

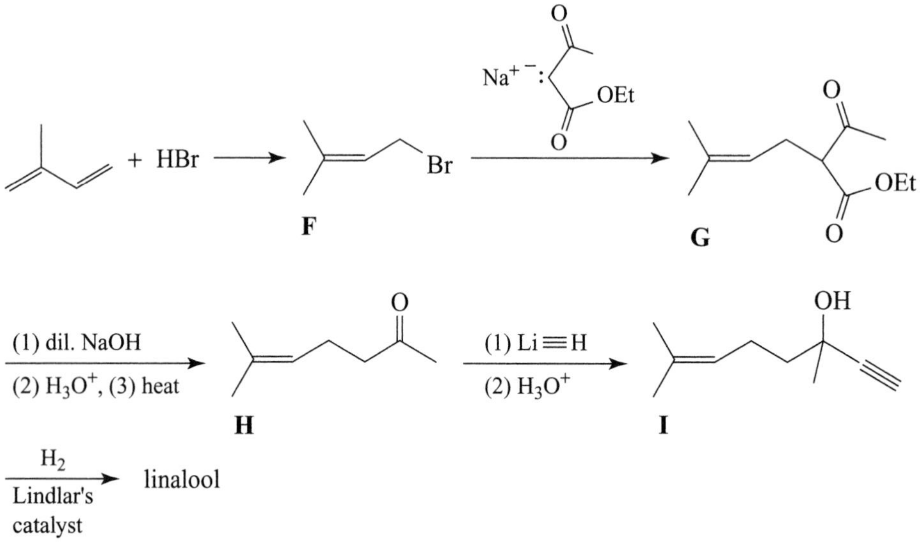

$Na^{+}\ ^{-}:$ O OEt O

→ O OEt O **G**

(1) dil. NaOH
(2) H_3O^+, (3) heat

O **H**

(1) Li≡H
(2) H_3O^+

OH **I**

H_2
Lindlar's catalyst

linalool

18.29 O EtO Na^+ : − EtO H O

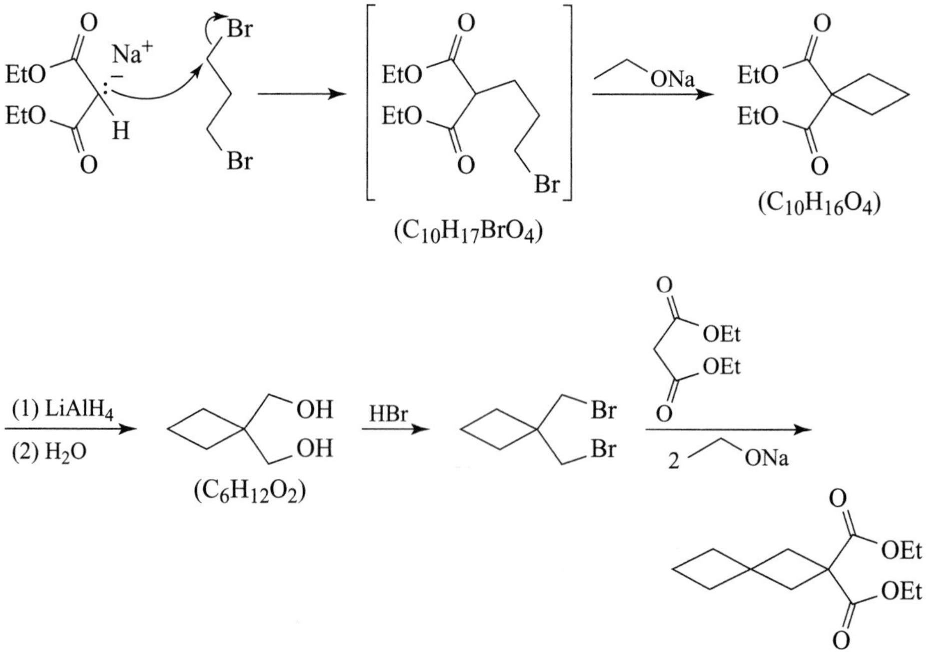

Br Br

[O EtO EtO O Br] ($C_{10}H_{17}BrO_4$)

ONa

O EtO EtO O ($C_{10}H_{16}O_4$)

(1) $LiAlH_4$
(2) H_2O

OH OH ($C_6H_{12}O_2$)

HBr

Br Br

O OEt OEt O
2 ONa

O OEt OEt O

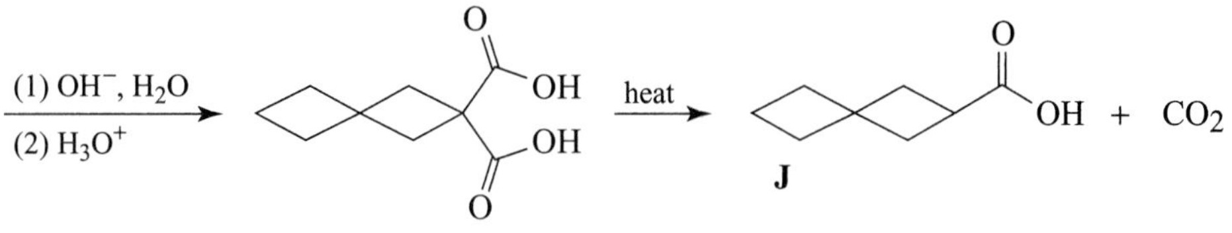

(1) OH^-, H_2O; (2) H_3O^+ → (diacid: O, OH, OH, O) ($C_9H_{12}O_4$) —heat→ **J** (O, OH) ($C_8H_{12}O_2$) + CO_2

18.30 (a) CH_3—O—CH_2—Br + $(C_6H_5)_3P$ —(2) RLi→ CH_3—O—$CH{=}P(C_6H_5)_3$

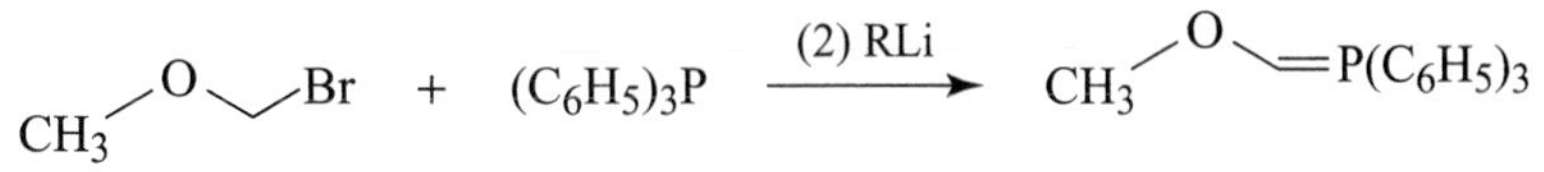

(b) Hydrolysis of the ether yields a hemiacetal that then goes on to form an aldehyde.

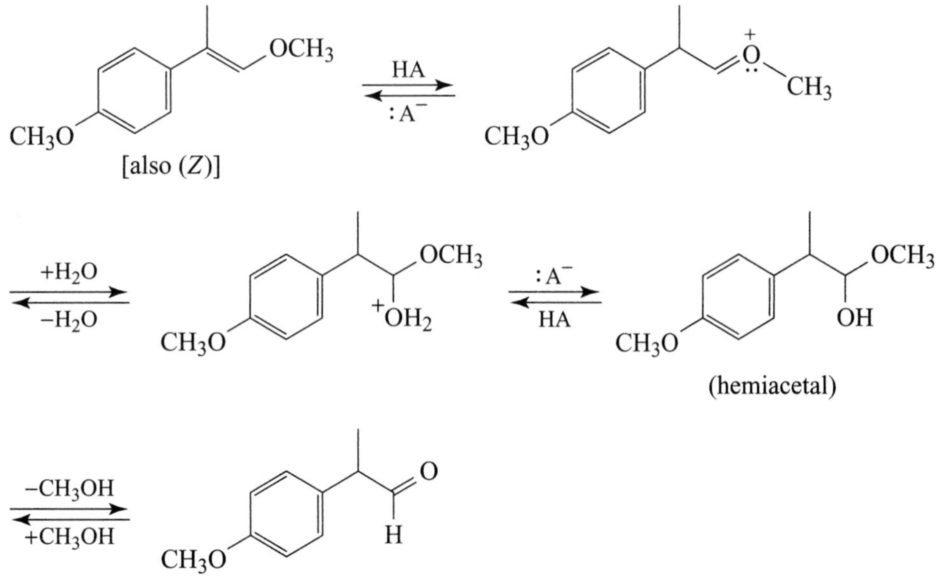

CH_3O … OCH_3 [also (Z)] $\underset{:A^-}{\overset{HA}{\rightleftharpoons}}$ CH_3O … $\overset{+}{O}$—CH_3

$\underset{-H_2O}{\overset{+H_2O}{\rightleftharpoons}}$ CH_3O … OCH_3, $^+OH_2$ $\underset{HA}{\overset{:A^-}{\rightleftharpoons}}$ CH_3O … OCH_3, OH (hemiacetal)

$\underset{+CH_3OH}{\overset{-CH_3OH}{\rightleftharpoons}}$ CH_3O … O, H

(c) (cyclohexanone, O) + CH_3—O—$CH{=}P(C_6H_5)_3$ ⟶ (cyclohexylidene) OCH_3 $\underset{:A^-}{\overset{HA}{\rightleftharpoons}}$

$\overset{+}{O}CH_3$ $\underset{-H_2O}{\overset{+H_2O}{\rightleftharpoons}}$ $^+OH_2$, OCH_3 $\underset{HA}{\overset{:A^-}{\rightleftharpoons}}$

OH, OCH_3 $\underset{+CH_3OH}{\overset{-CH_3OH}{\rightleftharpoons}}$ O, H

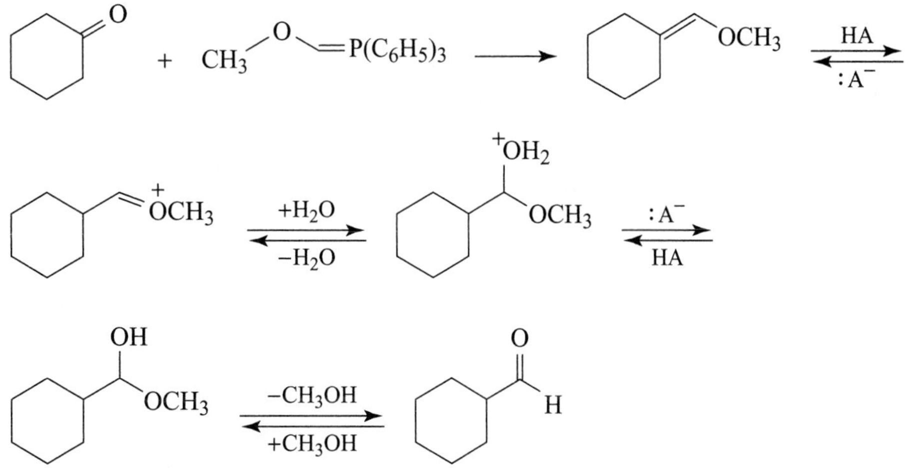

18.31 (a) The hydrogen atom that is added to the aldehyde carbon atom in the reduction must come from the other aldehyde rather than from the solvent. It must be transferred as a hydride ion and directly from molecule to molecule, since if it were ever a free species it would react immediately with the solvent. A possible mechanism is the following:

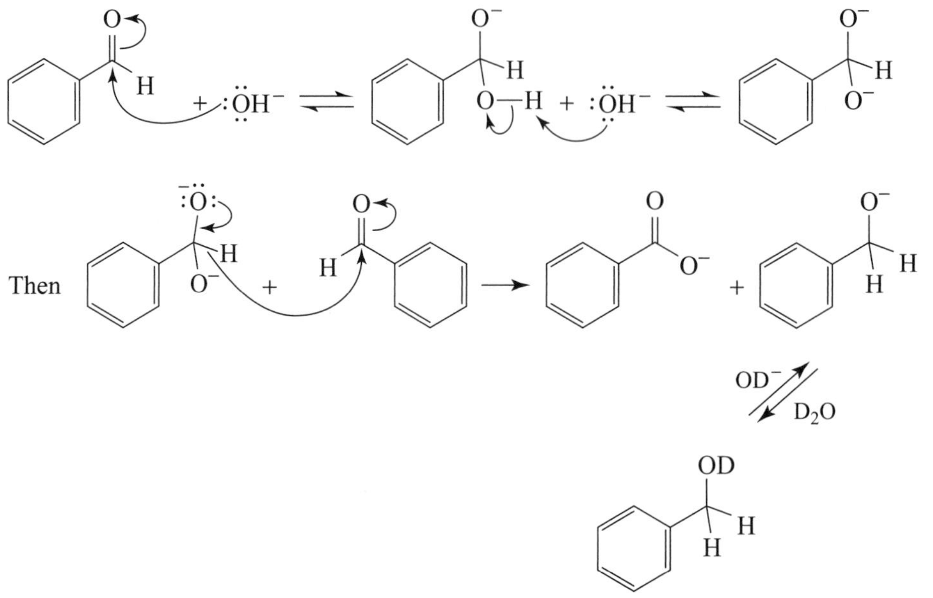

Then

(b) Although an aldol reaction occurs initially, the aldol reaction is reversible. The Cannizzaro reaction, though slower, is irreversible. Eventually, all the product is in the form of the alcohol and the carboxylate ion.

18.32

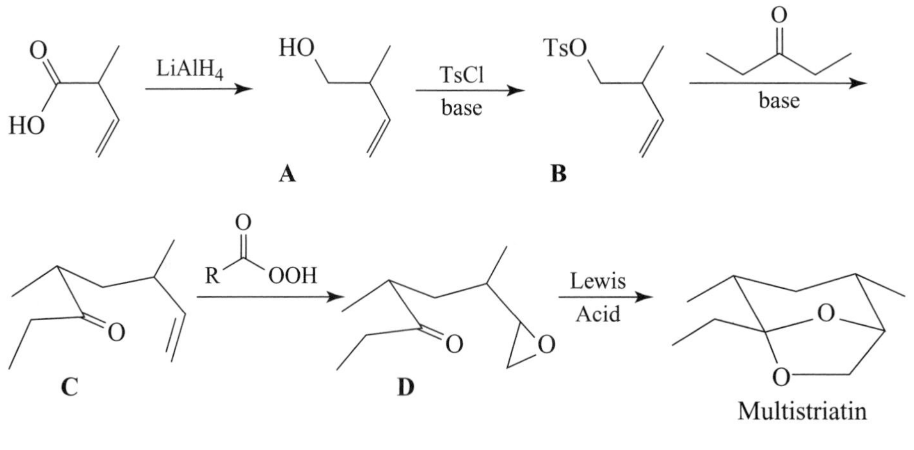

A **B** **C** **D** Multistriatin

Spectroscopy

18.33 (a) Compound **U** is ethyl phenyl ketone. (b) Compound **V** is benzyl methyl ketone:

O; 1690 cm^{-1}; δ 3.0, δ 1.2; δ 7.7; **U**

δ 2.0; 1705 cm^{-1}; O; δ 3.5; δ 7.1; **V**

18.34 **A** is O; OCH_3; OCH_3; δ 3.2; δ 2.1; δ 2.6; δ 4.7

O; OCH_3; OCH_3 — I_2, NaOH → CHI_3 ↓

$Ag(NH_3)_2{}^+OH^-$ → no reaction

H_3O^+ ↓

O; O; H — $Ag(NH_3)_2{}^+OH^-$ → Ag ↓ + O; O; O^-

Challenge Problem

18.35 TfO—Si— ; O: ; OH → —Si ; :NR_3 ; H ; O ; $\overset{+}{O}$; H ; H → Proton removal and hydride migration →

—Si ; Ö: ; O ; H ; H + $R_3\overset{+}{N}H$

QUIZ

18.1 What would be the major product of the following reaction?

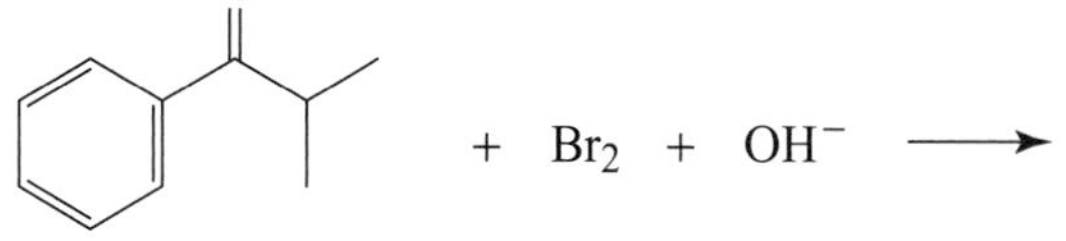

$+ Br_2 + OH^- \longrightarrow$

(a)

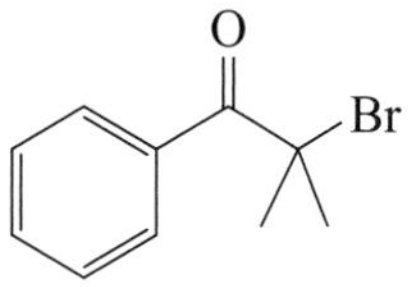

(b)

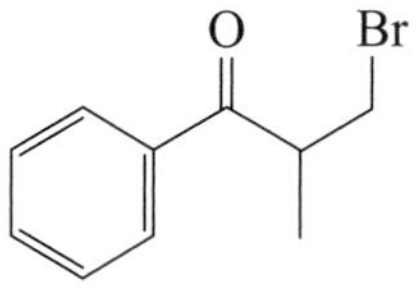

(c)

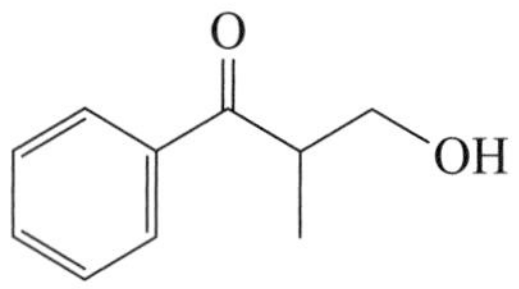

(d)

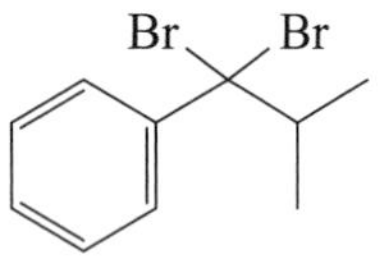

(e) None of these

18.2 Supply formulas for the missing reagents and intermediates in the following synthesis.

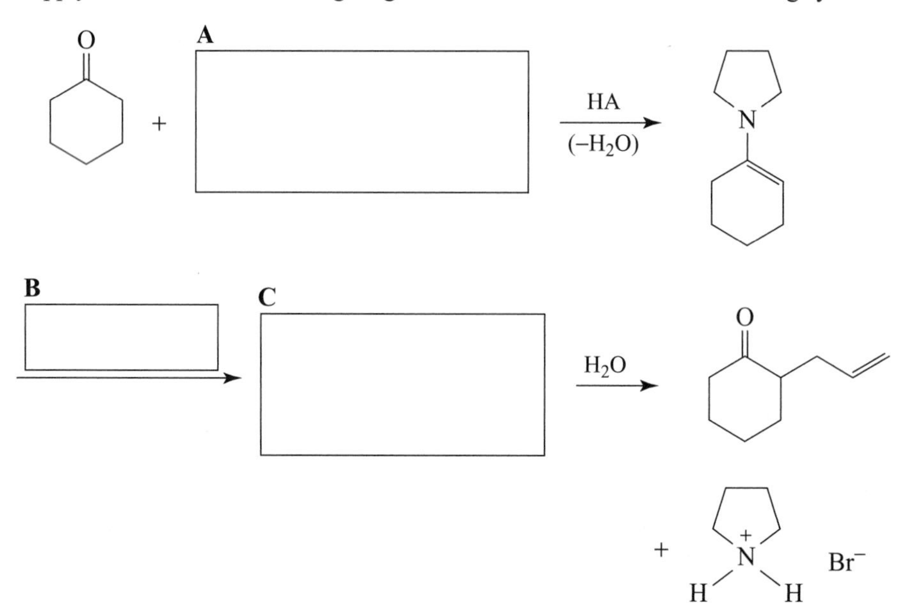

18.3 What alkyl halide would be used in the malonic ester synthesis of 4-methylpentanoic acid?

(a) 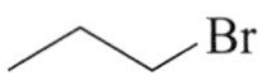 Br

(b) 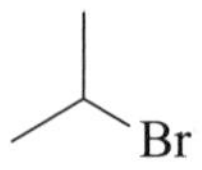 Br

(c) 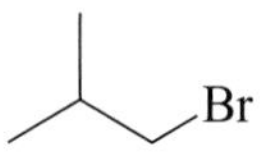 Br

(d) 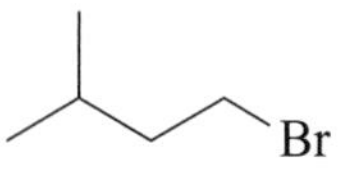 Br

(e) 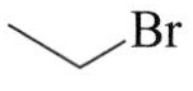 Br

18.4 Which is the preferred base for the alkylation of a monoalkylacetoacetic ester?

(a) NaOH (b) KOH

(c) 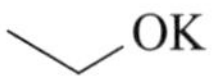 OK (d) 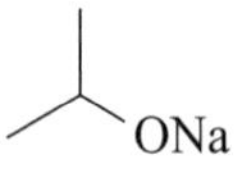 ONa

(e) 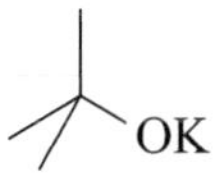 OK

18.5 To which of these can the haloform reaction be applied for the synthesis of a carboxylic acid?

(a) 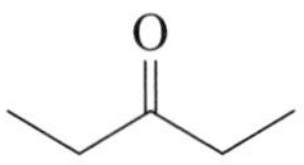 O (b) 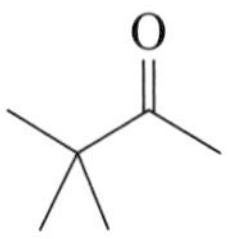 O

(c) 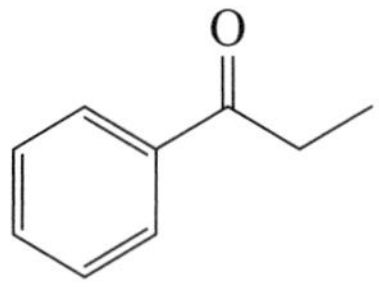 O (d) 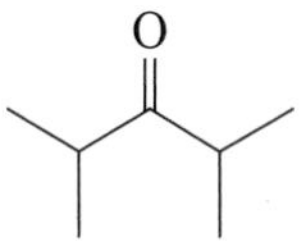 O

(e) 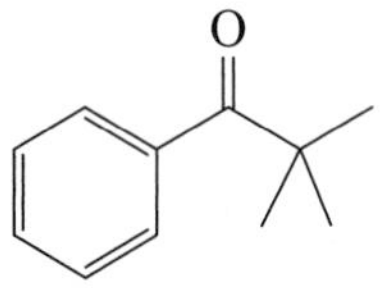 O

19 CONDENSATION AND CONJUGATE ADDITION REACTIONS OF CARBONYL COMPOUNDS: MORE CHEMISTRY OF ENOLATES

SOLUTIONS TO PROBLEMS

19.1 (a) *Step 1*

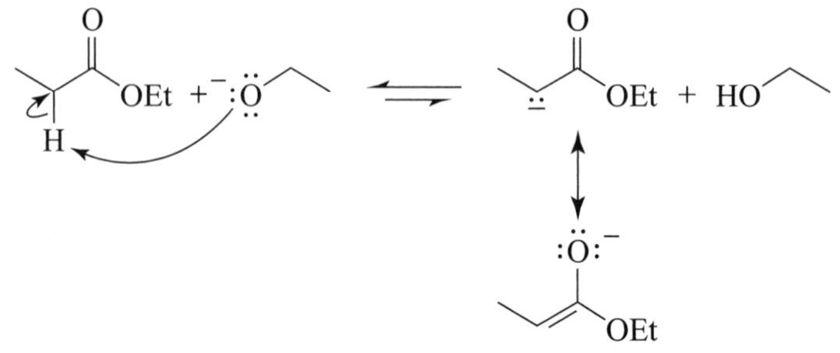

O; OEt + $^{-}$:Ö; H ⇌ O; OEt + HO; ↕ :Ö:$^{-}$; OEt

Step 2

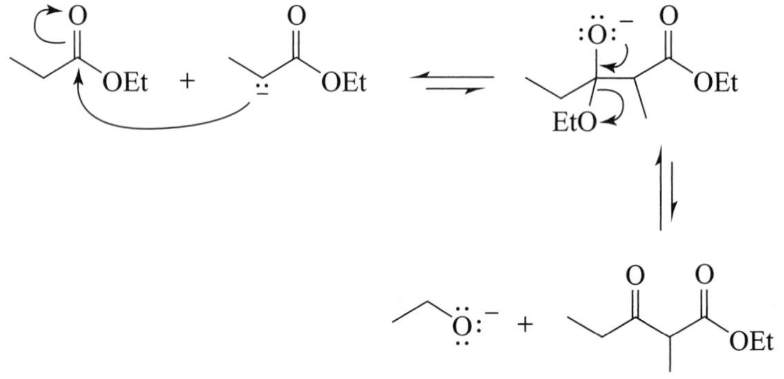

O; OEt + O; OEt ⇌ :Ö:$^{-}$; O; OEt; EtO ⇅ Ö:$^{-}$ + O O; OEt

Step 3

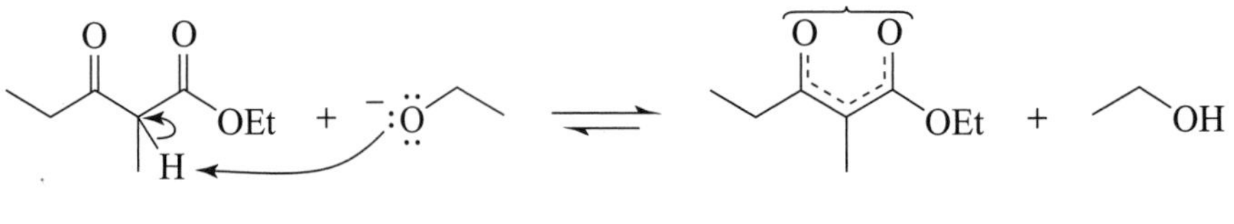

O O; OEt + $^{-}$:Ö; H ⇌ O O (−); OEt + OH

(b) O O; OEt + OH O; OEt

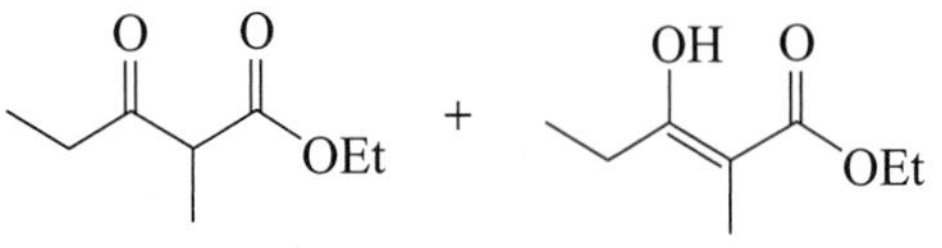

19.2 2 [ethyl butanoate] —(1) NaOEt, (2) H_3O^+→ [ethyl 2-ethyl-3-oxohexanoate] —(1) NaOH, H_2O, heat (2) H_3O^+→ [2-ethyl-3-oxohexanoic acid] —heat, $-CO_2$→ [4-heptanone]

19.3 (a) [ethyl 2-oxocyclohexanecarboxylate]

(b) To undergo a Dieckmann condensation, diethyl 1,5-pentanedioate would have to form a highly strained four-membered ring.

19.4 $C_6H_5CH_2CO_2Et$ + $CH_3CH_2O^-$ ⇌ $C_6H_5\ddot{C}H^-CO_2Et$ + CH_3CH_2OH

$C_6H_5\ddot{C}H^-CO_2Et$ + $EtOC(O)OEt$ ⇌ [tetrahedral intermediate: $(EtO)_2C(O^-)CH(C_6H_5)CO_2Et$] ⇌

$EtO_2CCH(C_6H_5)CO_2Et$ + $CH_3CH_2O^-$ ⇌ $EtO_2C\ddot{C}^-(C_6H_5)CO_2Et$ + CH_3CH_2OH

Resonance stabilized

—H_3O^+→ $EtO_2CCH(C_6H_5)CO_2Et$

19.5 (a) [ethyl propanoate] + [diethyl oxalate] —(1) NaOEt, (2) H_3O^+→ $EtO_2CC(O)CH(CH_3)CO_2Et$

(b) [ethyl acetate] + [ethyl formate, HCO_2Et] —(1) NaOEt, (2) H_3O^+→ $HC(O)CH_2CO_2Et$

19.6 (a) Cyclohexanone + HCOOEt —(1) EtONa, (2) H_3O^+→ 2-oxocyclohexanecarbaldehyde

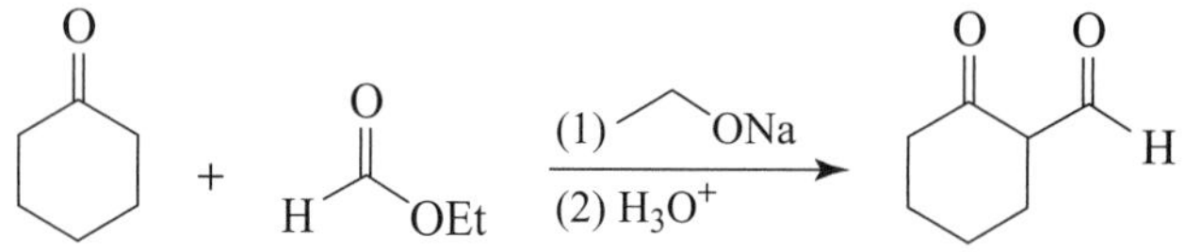

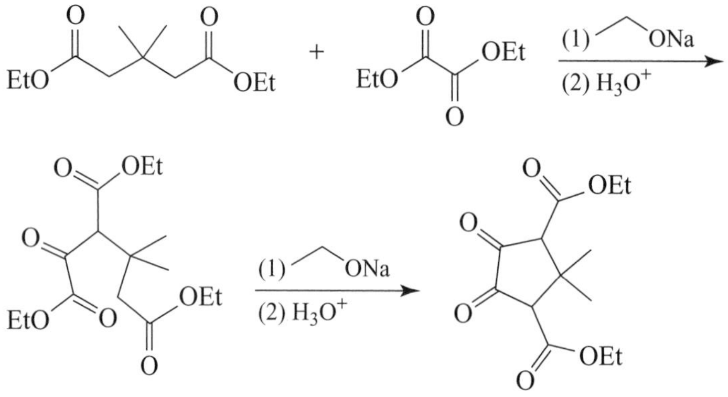

(b) —(1) EtONa, (2) H_3O^+→

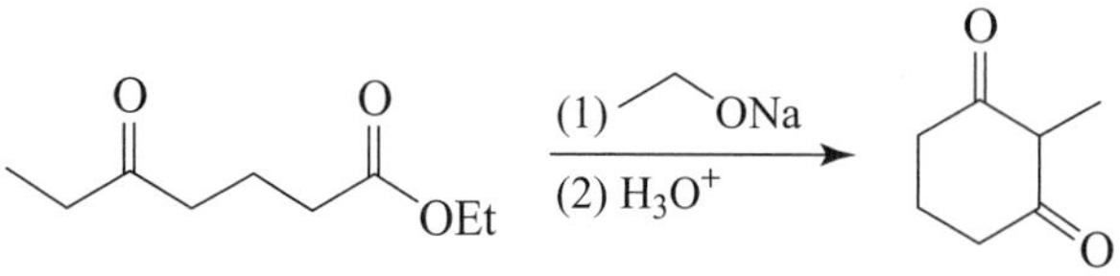

(c) EtO, OEt + EtO, OEt —(1) EtONa, (2) H_3O^+→

O, OEt, O, EtO, O, OEt, O —(1) EtONa, (2) H_3O^+→ O, OEt, O, O, O, OEt

19.7 + HCl ⇌ :ÖH⁺ + Cl^-

HO⁺ + :OH ⇌ OH :OH⁺

⇌ $\overset{+}{O}H_2$ O, H —$-H_2O$, $:A^-$→ + HA

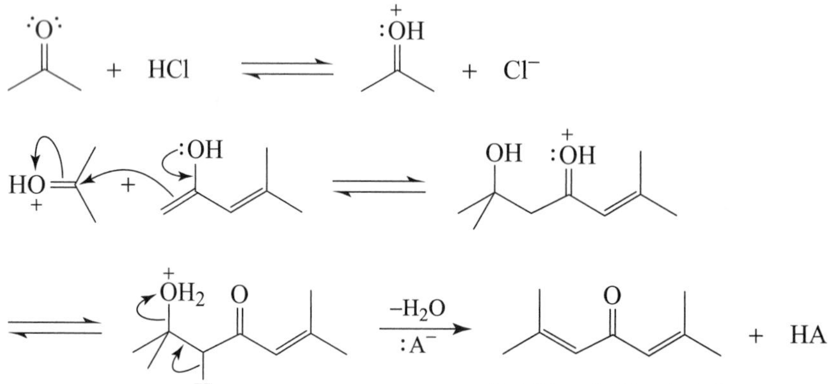

2,6-Dimethylhepta-2,5-dien-4-one

19.8 Drawing the molecules as they will appear in the final product helps to visualize the necessary steps:

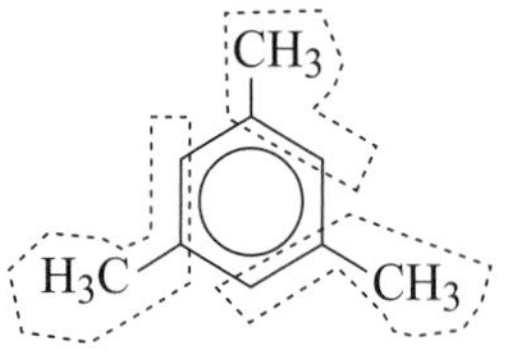

CH_3, H_3C, CH_3

Mesitylene

The two molecules that lead to mesitylence are shown as follows:

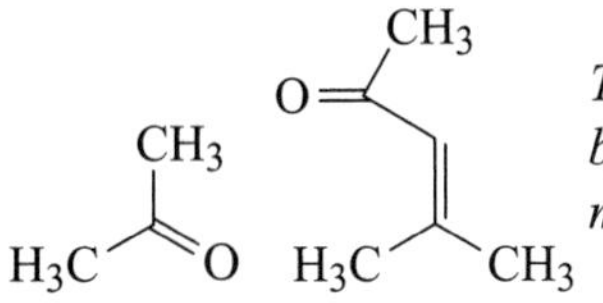

CH_3 O CH_3 H_3C O H_3C CH_3

This molecule (4-methylpent-3-en-2-one) is formed by an acid-catalyzed condensation between two molecules of acetone as shown in the text.

The mechanism is,

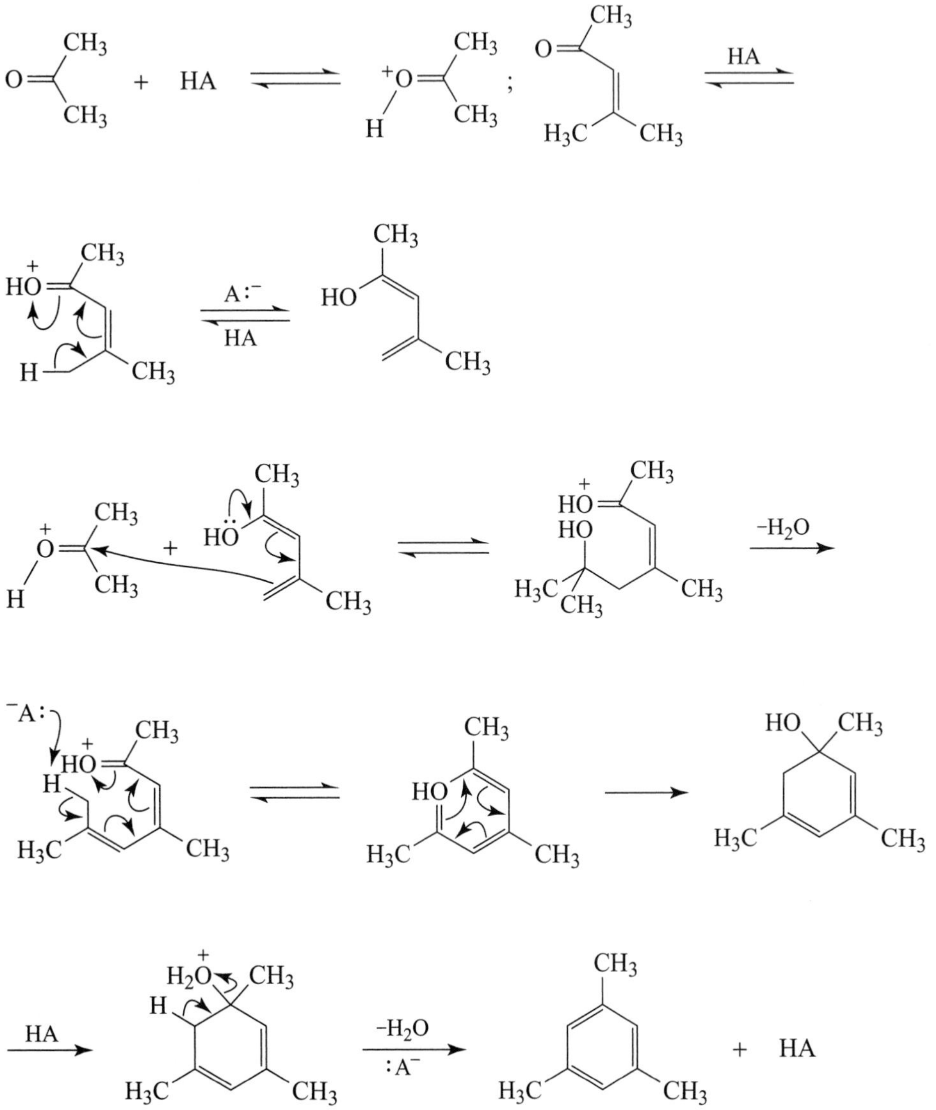

CH_3 O CH_3 + HA ⇌ ^+O H CH_3 CH_3 ; CH_3 O H_3C CH_3 $\overset{HA}{\rightleftharpoons}$

$\overset{+}{HO}$ CH_3 H CH_3 $\underset{HA}{\overset{A:^-}{\rightleftharpoons}}$ HO CH_3 CH_3

$\overset{+}{O}$ H CH_3 CH_3 + $\ddot{HO}$ CH_3 CH_3 ⇌ $\overset{+}{HO}$ CH_3 HO H_3C CH_3 CH_3 $\xrightarrow{-H_2O}$

$^-A:$ $\overset{+}{HO}$ H CH_3 H_3C CH_3 ⇌ CH_3 HO H_3C CH_3 ⟶ HO CH_3 H_3C CH_3

$\xrightarrow{HA}$ $H_2\overset{+}{O}$ CH_3 H H_3C CH_3 $\underset{:A^-}{\xrightarrow{-H_2O}}$ CH_3 H_3C CH_3 + HA

19.9 (a) α β H O + OH^- ⇌ H O + H_2O

H O + H O ⇌ H O– O

H O :O: + H—OH ⇌ H OH O + OH^-

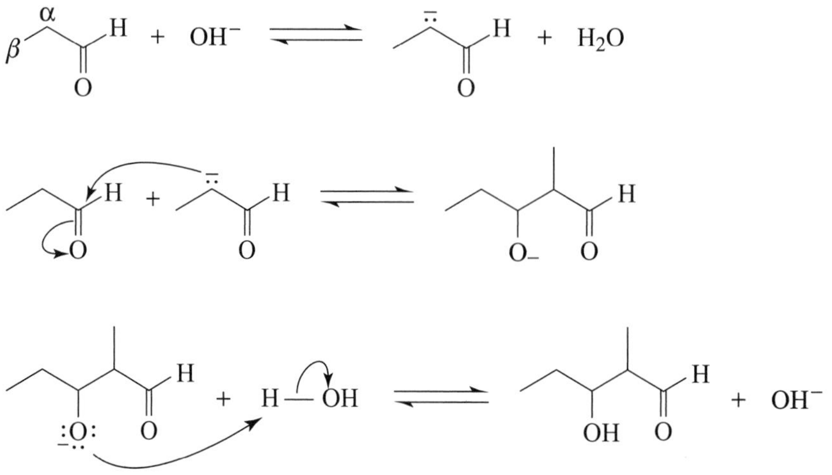

(b) To form O H OH, a hydroxide ion would have to remove a β proton in the first step. This does not happen because the anion that would be produced, H O, cannot be stabilized by resonance.

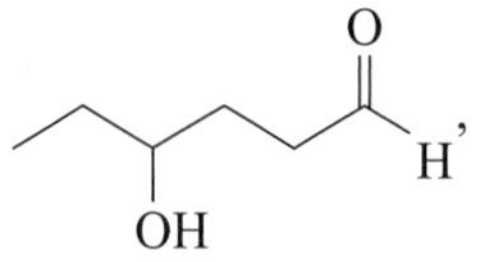

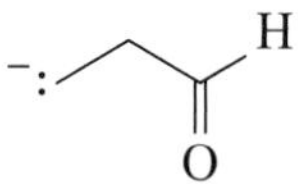

(c) O H

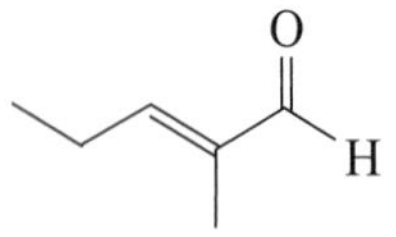

[both (E) and (Z)]

19.10 (a) 2 O H $\xrightarrow[H_2O]{OH^-}$ OH O H

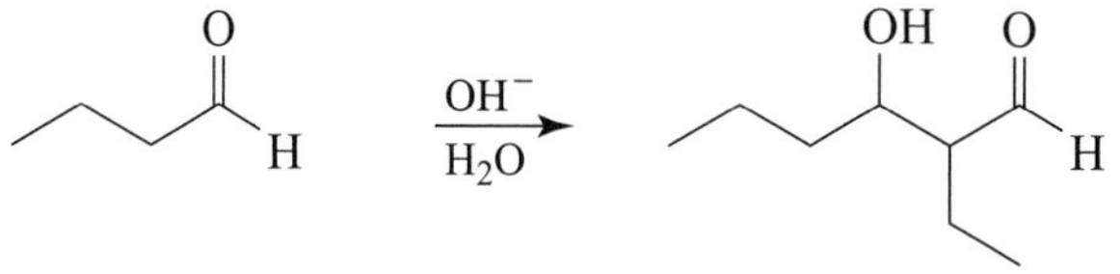

(b) Product of (a) $\xrightarrow[-H_2O]{HA}$ O H [(E) + (Z)] $\xrightarrow[(2)\ H_2O]{(1)\ LiAlH_4,\ Et_2O}$ OH [(E) + (Z)]

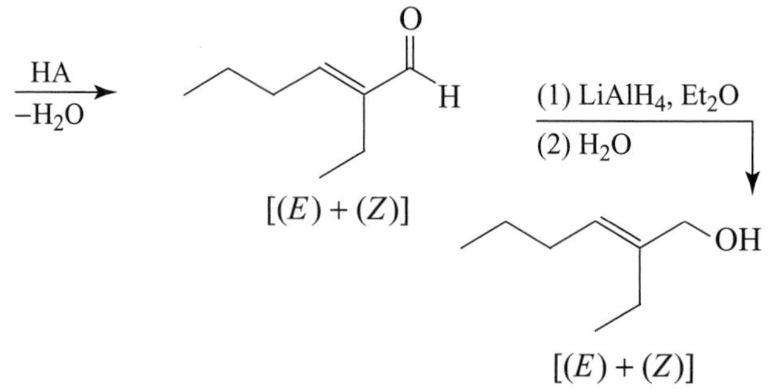

(c) Product of (b) $\xrightarrow[\text{Pt, pressure}]{H_2}$ OH

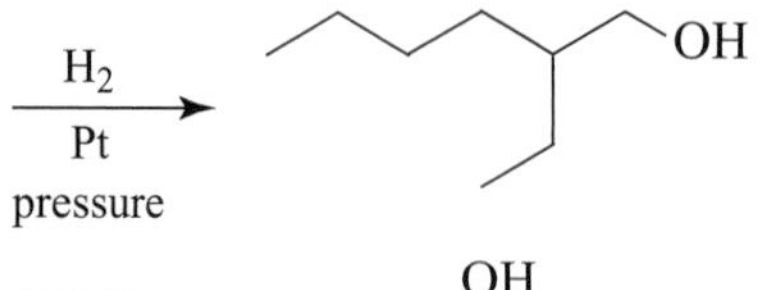

(d) Product of (a) $\xrightarrow{NaBH_4}$ OH OH

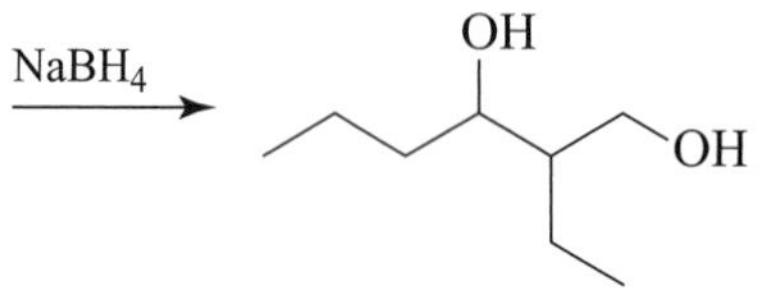

19.11

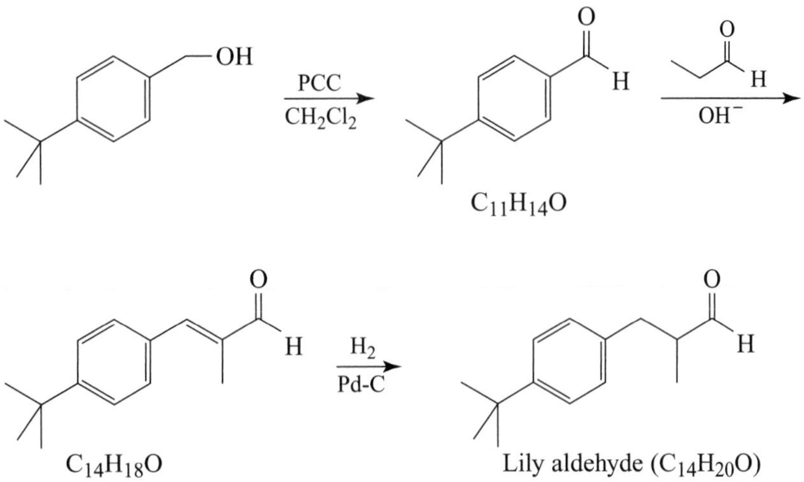

OH

PCC

CH_2Cl_2

O

H

O

H

OH^-

$C_{11}H_{14}O$

O

H

H_2

Pd-C

O

H

$C_{14}H_{18}O$

Lily aldehyde ($C_{14}H_{20}O$)

19.12 Three successive aldol additions occur.

First Aldol Addition

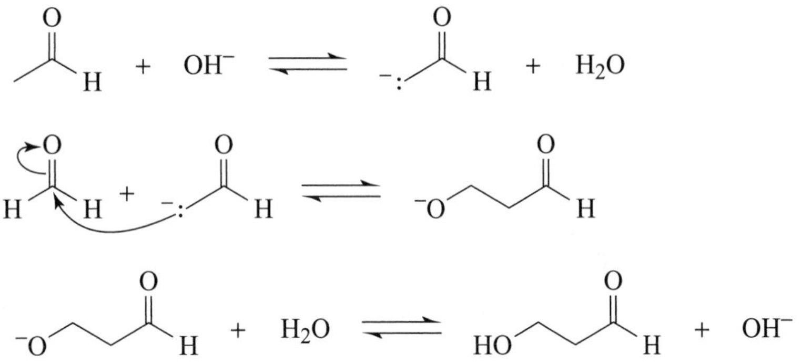

O

H + OH^- ⇌ O

$^-$: H + H_2O

O

H H + O

$^-$: H ⇌ O

$^-$O H

O

$^-$O H + H_2O ⇌ O

HO H + OH^-

Second Aldol Addition

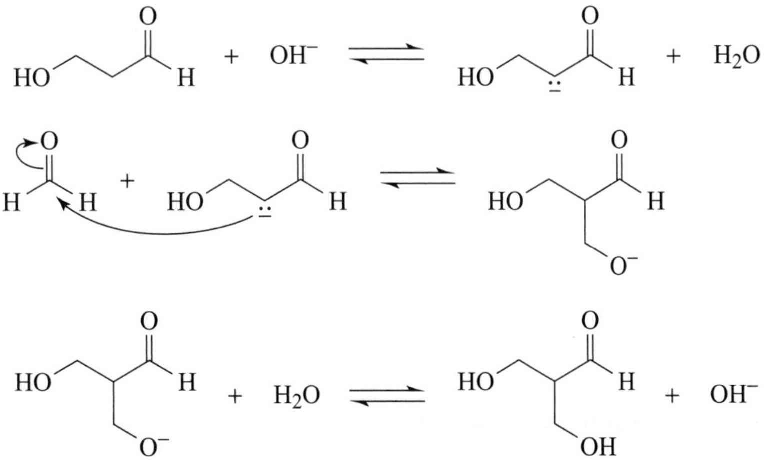

O

HO H + OH^- ⇌ O

HO : H + H_2O

O

H H + O

HO : H ⇌ O

HO H

O^-

O

HO H

O^- + H_2O ⇌ O

HO H

OH + OH^-

Third Aldol Addition

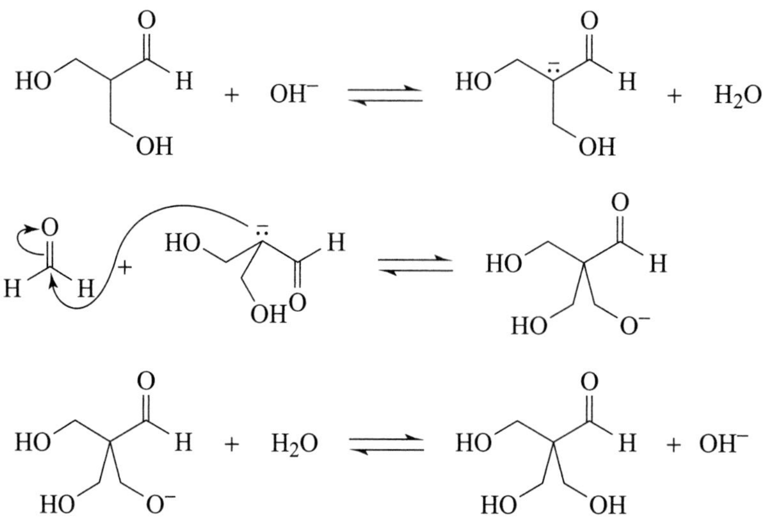

$+ OH^- \rightleftharpoons$ $+ H_2O$

$+ H_2O \rightleftharpoons$ $+ OH^-$

19.13 (a) 2 $CH_3COOH + BF_3 \rightleftharpoons CH_3COOBF_3^- + CH_3C(OH)=OH^+$

Pseudoionone

α-Ionone

β-Ionone

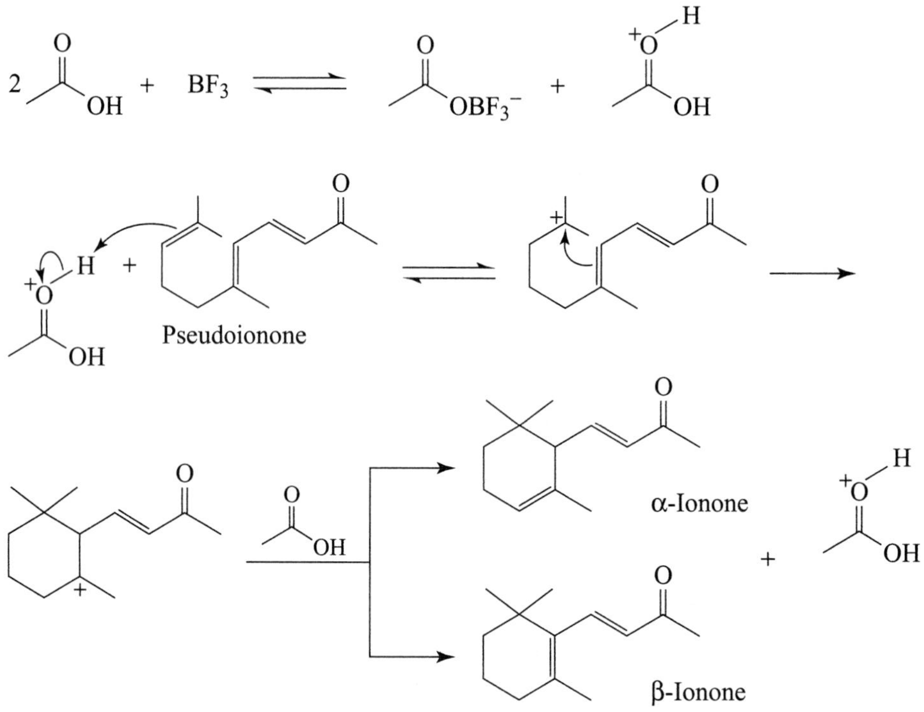

(b) In β-ionone both double bonds and the carbonyl group are conjugated; thus it is more stable.

(c) β-Ionone, because it is a fully conjugated unsaturated system.

19.14 (a) $^{-}$:—≡N: ⟷ =C=N̈:$^{-}$

(b) —≡N: ⇌ (O^{-} ethoxide) [$^{-}$:—≡N: ⟷ =C=N̈:$^{-}$] + OH

:Ö: H + $^{-}$:—≡N: ⟶ :Ö:$^{-}$ ≡N: ⟶ (OH)

OH ≡N: $\xrightarrow[-H_2O]{\text{heat}}$ ≡N: + (*Z*)

19.15 (a) O $\xrightarrow{\text{LDA}}$ $O^{-}Li^{+}$ Kinetic enolate $\xrightarrow[(2)\ H_2O]{(1)\ \text{O, H}}$ O OH

(b) O $\xrightarrow{\text{LDA}}$ $O^{-}Li^{+}$ Kinetic enolate $\xrightarrow[(2)\ H_2O]{(1)\ C_6H_5\text{—C(O)H}}$ O OH C_6H_5

(c) O $\xrightarrow{\text{LDA}}$ $O^{-}Li^{+}$ Kinetic enolate $\xrightarrow[(2)\ H_2O]{(1)\ \text{O, H}}$ O OH

19.16

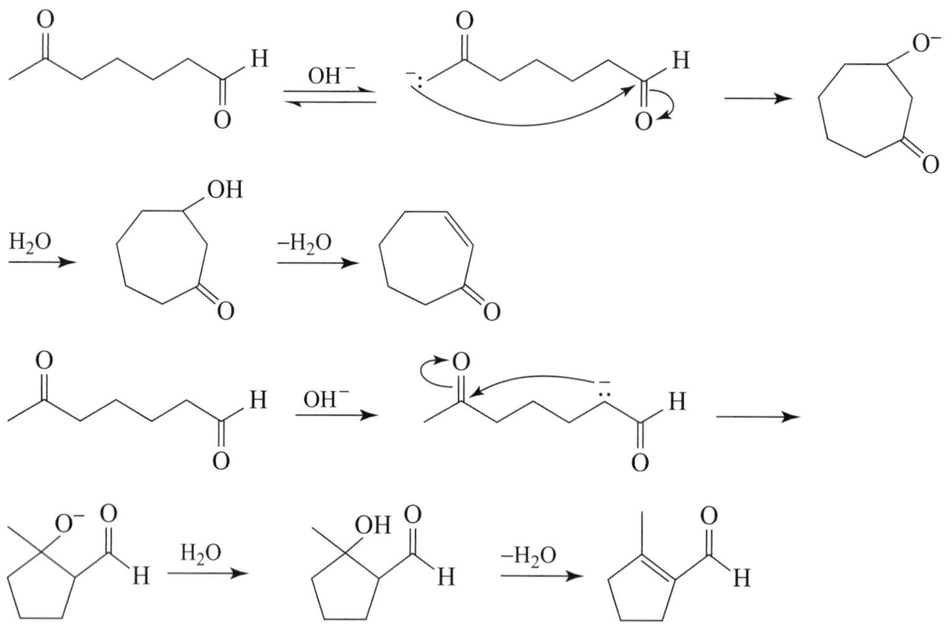

H_2O $-H_2O$

OH^- H_2O $-H_2O$

19.17 (a) 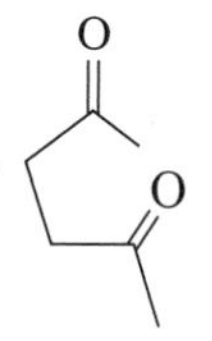 (b) 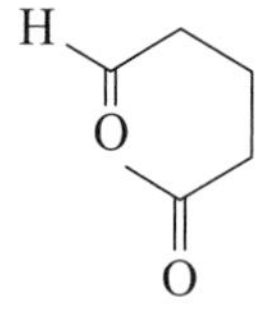 (c)

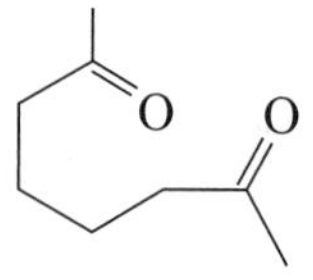

Notice that starting compounds are drawn so as to indicate which atoms are involved in the cyclization reaction.

19.18 It is necessary for conditions to favor the intramolecular reaction rather than the intermolecular one. One way to create these conditions is to use very dilute solutions when we carry out the reaction. When the concentration of the compound to be cyclized is very low (i.e., when we use what we call a "high dilution technique"), the probability is greater that one end of a molecule will react with the other end of that same molecule rather than with a different molecule.

19.19 (a)

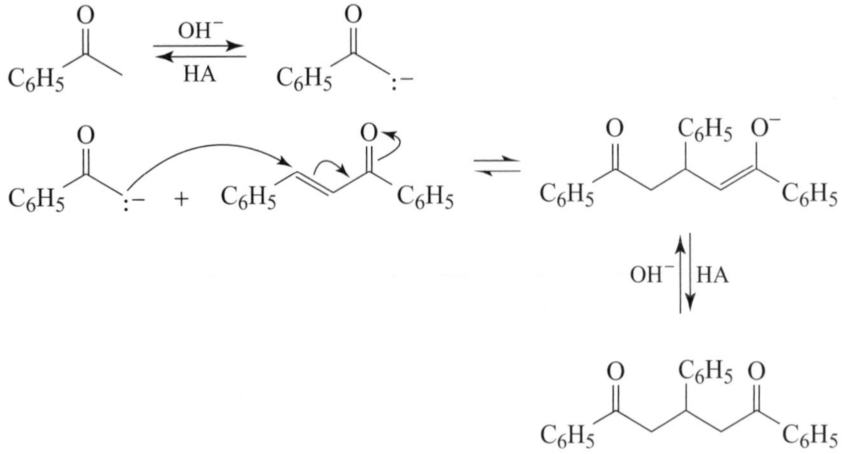

O, C_6H_5 $\xrightleftharpoons[HA]{OH^-}$ O, C_6H_5, :−

O, C_6H_5, :− + C_6H_5, O, C_6H_5 ⇌ O, C_6H_5, C_6H_5, O^-, C_6H_5

OH^- ⇅ HA

O, C_6H_5, C_6H_5, O, C_6H_5

(b)

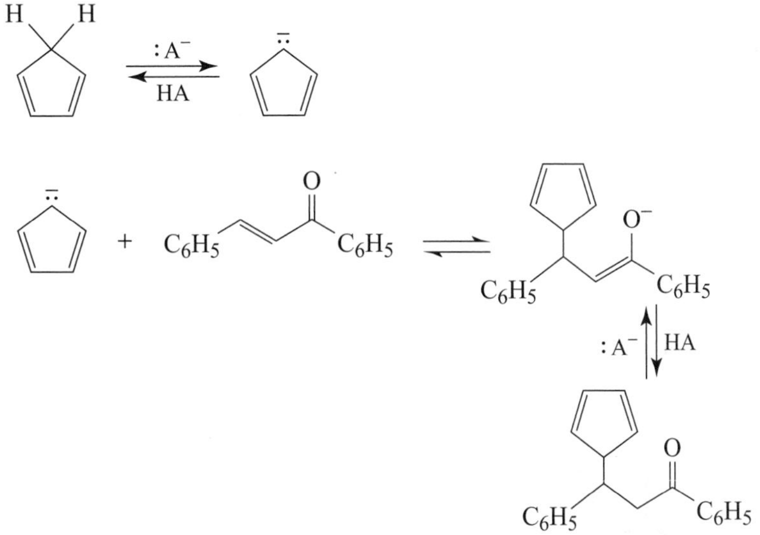

H, H $\xrightleftharpoons[HA]{:A^-}$

+ C_6H_5, O, C_6H_5 ⇌ O^-, C_6H_5, C_6H_5

$:A^-$ ⇅ HA

O, C_6H_5, C_6H_5

19.20 $H_2\ddot{N}-\ddot{N}H_2$ + O, H $\xrightarrow[\text{addition}]{\text{conjugate}}$ O, H, $\ddot{N}H_2$, N, H ⇌ O^-, N^+, H, H, N, H

⇌ OH, N, H, N, H $\xrightarrow{-H_2O}$ N, N, H

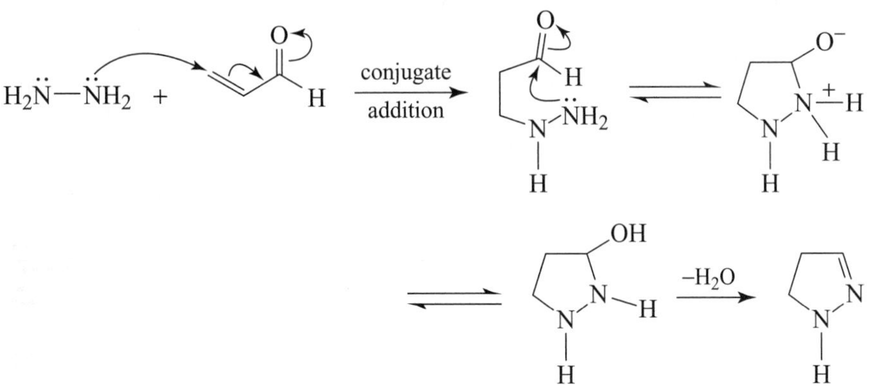

19.21 (a)

O, H, O + OH^- ⇌ O, :$^-$, O

O, :$^-$, O + O ⇌ O, O$^-$, O HA / :A^-

O, O, O :A^- / HA O, O, :, O ⇌ O, O, O^-

HA / :A^- O, O, OH $-H_2O$ O, O

(b) 2-Methylcyclohexane-1,3-dione is more acidic because its enolate ion is stabilized by an additional resonance structure.

O, H, O :A^- ⟶ [O, :, O ⟷ O^-, O ⟷ O, O^-]

O, H, H :A^- ⟶ [O, H, :$^-$ ⟷ O^-, H]

19.22 (a) $H_2C{=}O$ + Me_2NH $\underset{}{\overset{+HA}{\rightleftharpoons}}$ $CH_2{=}\overset{+}{N}Me_2$ + H_2O

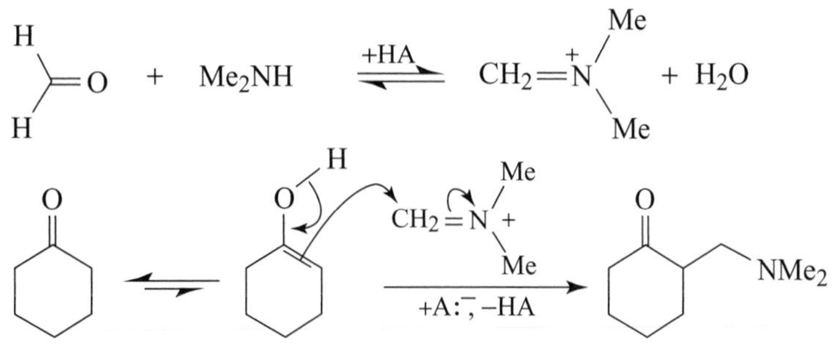

+A:⁻, –HA

NMe₂

(b) $H_2C{=}O$ + H—N (pyrrolidine) $\overset{+HA}{\rightleftharpoons}$ $CH_2{=}N^+$ + H_2O

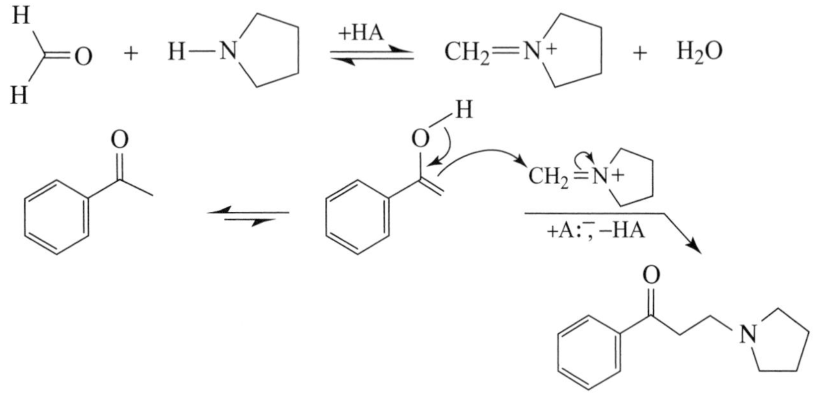

+A:⁻, –HA

(c) $H_2C{=}O$ + Me_2NH $\overset{+HA}{\rightleftharpoons}$ $CH_2{=}\overset{+}{N}Me_2$ + H_2O

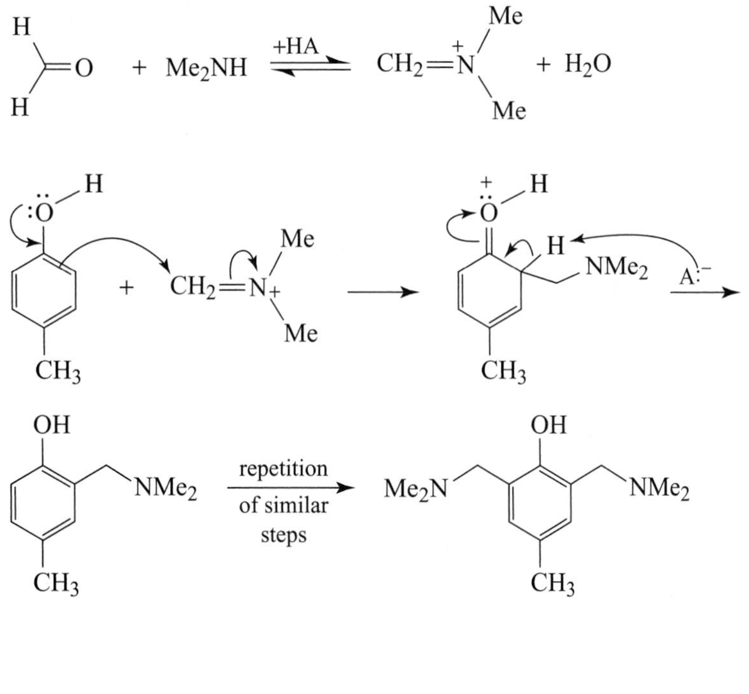

NMe₂ A:⁻

CH₃

OH NMe₂ repetition of similar steps Me₂N OH NMe₂

CH₃ CH₃

Problems

Claisen Condensation Reactions

19.23 (a) 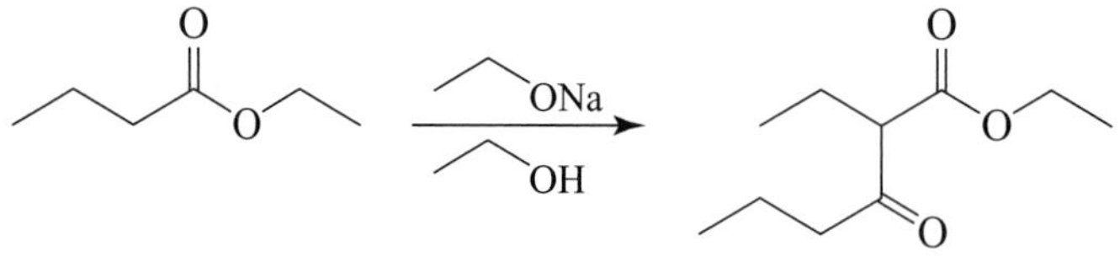 O ONa OH O O O O

(b) 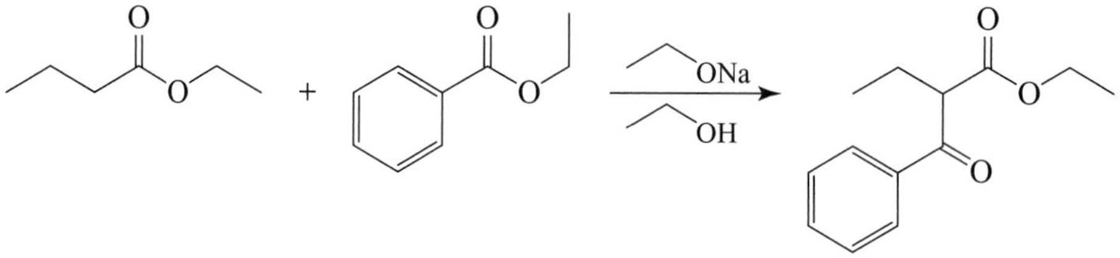 O O O O + ONa OH O O O

(c) 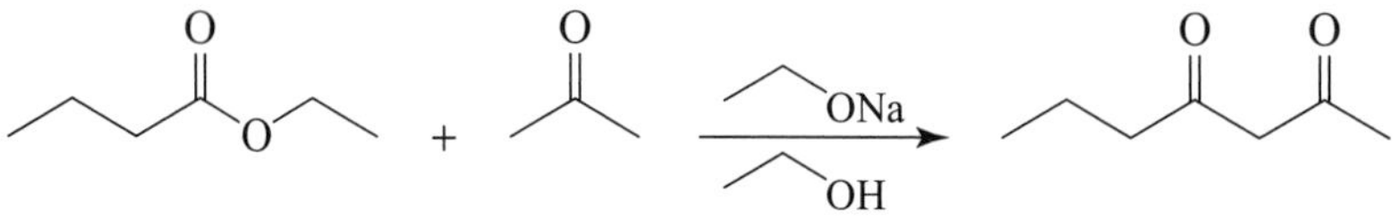 O O + ONa OH O O

(d) 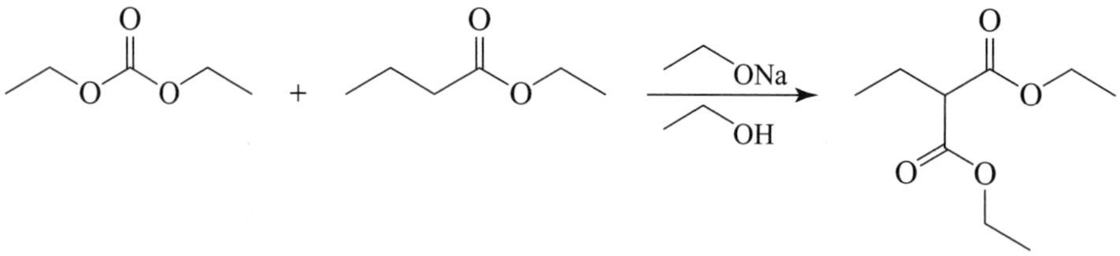 O O O O O + ONa OH O O O O O

(e) 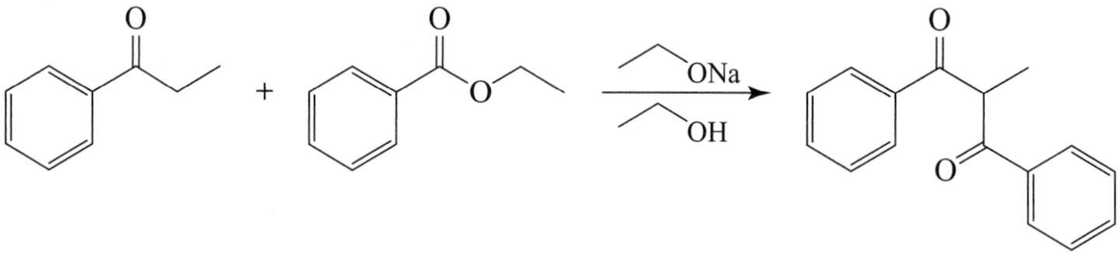 O O O + ONa OH O O

(f) 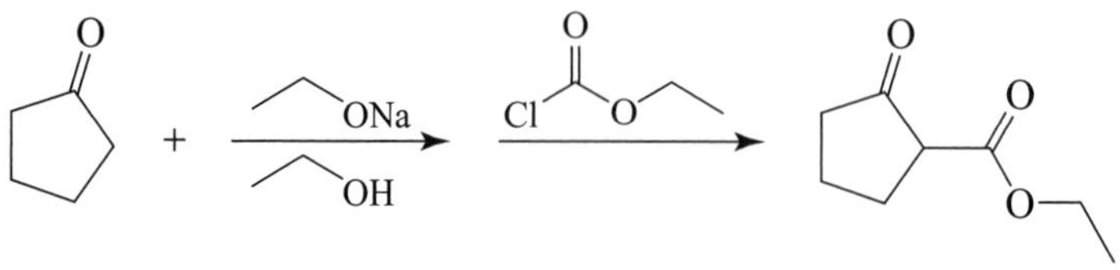 O + ONa OH Cl O O O O O O

19.24 Working backward

(a)

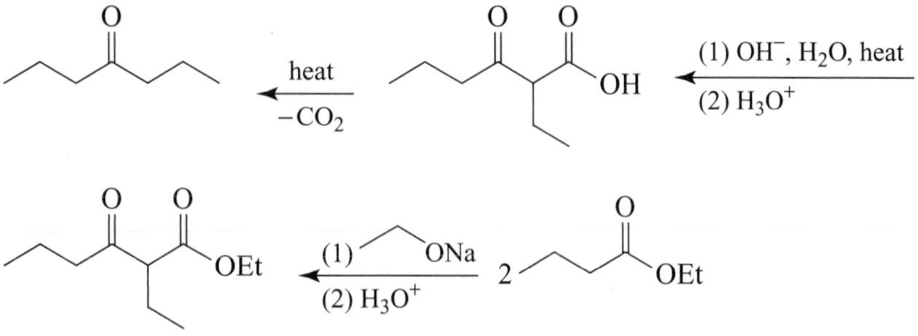

heat, $-CO_2$

(1) OH^-, H_2O, heat
(2) H_3O^+

OEt

(1) ONa
(2) H_3O^+

2 OEt

(b)

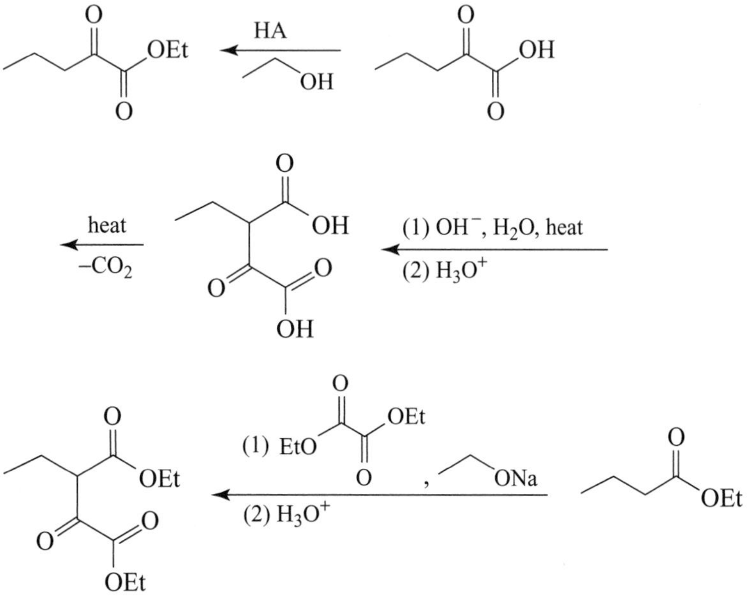

OEt

HA
OH

OH

heat
$-CO_2$

OH

(1) OH^-, H_2O, heat
(2) H_3O^+

OH

(1) EtO ... OEt , ONa
(2) H_3O^+

OEt

OEt

OEt

(c)

H

OEt

C_6H_5

(1) H OEt, ONa
(2) H_3O^+

C_6H_5 OEt

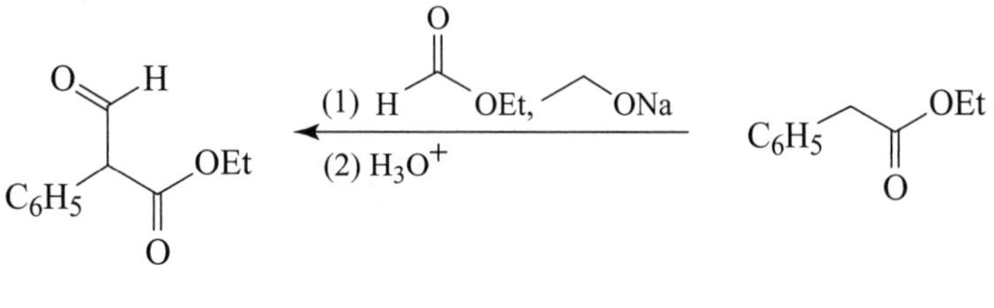

19.25 (a)

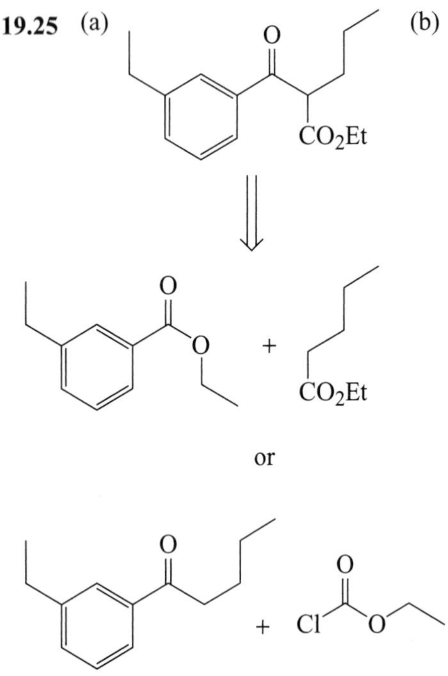
O
CO_2Et
O
O
+
CO_2Et
or
O
O
+ Cl O

(b)

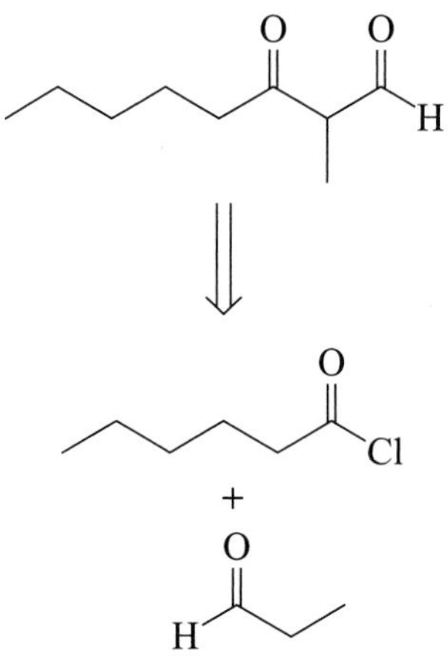
O
O
H
O
Cl
+
O
H

(c)

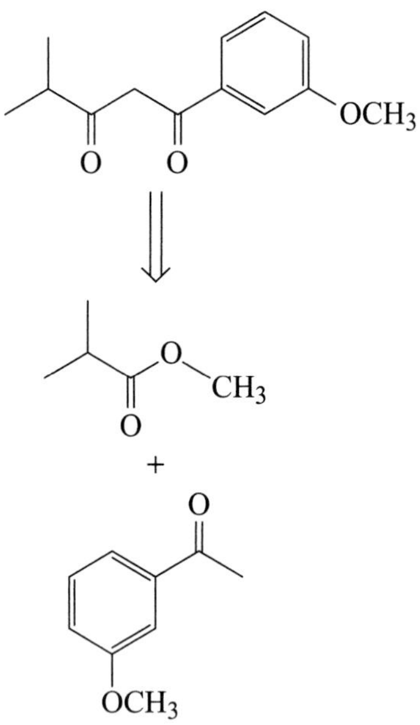
OCH_3
O
O
O
CH_3
O
+
O
OCH_3

19.26

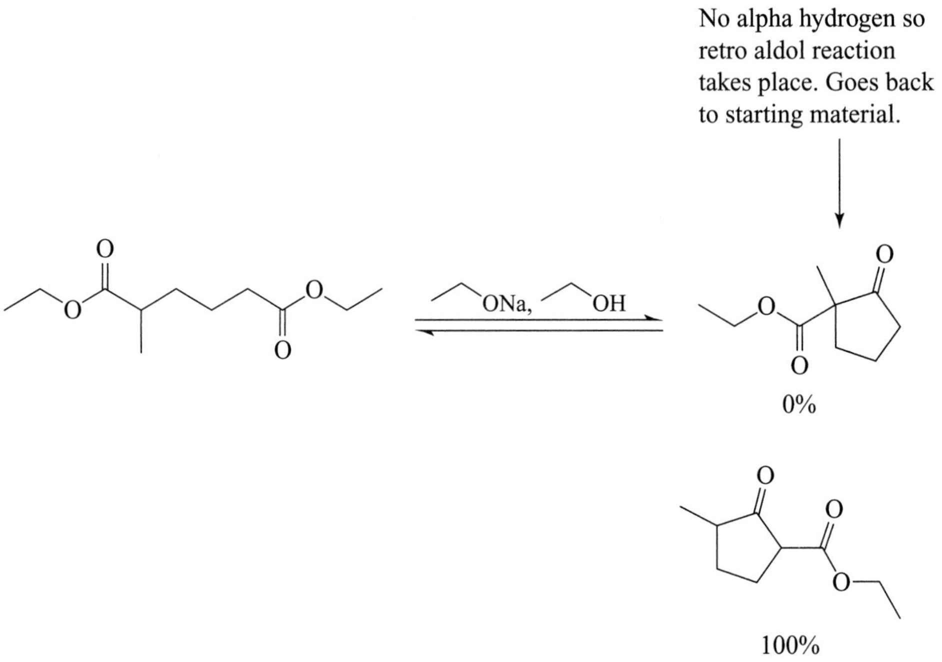
No alpha hydrogen so retro aldol reaction takes place. Goes back to starting material.
O
O
O
O
ONa, OH
O
O
O
0%
O
O
O
100%

The Claisen condensation is reversible. The driving force for this reaction is the deprotonation of the alpha hydrogen in the product between the two carbonyls. This deprotonation prevents the reaction from reversing, pulling the product out of the equilbrium.

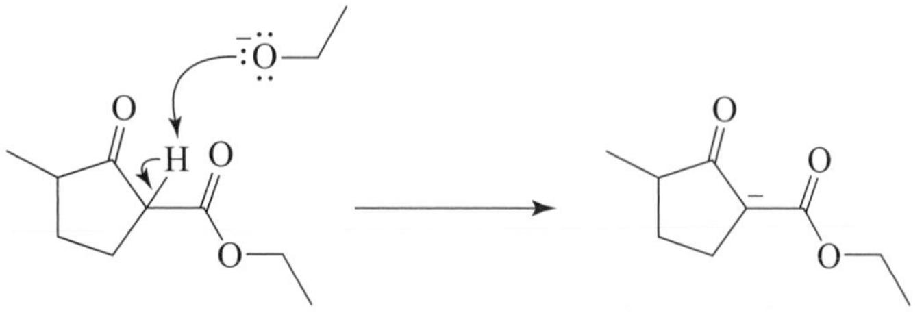

Molecule can not undergo retro Claisen in deprotonated form.

19.27 Intramolecular cyclization (which would give a product of formula $C_6H_8O_3$) is not favored because of ring strain. The formula of the product actually obtained suggests a 1:1 intermolecular reaction:

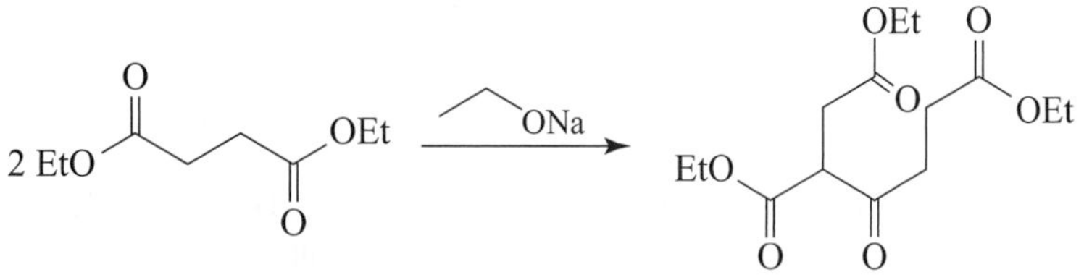

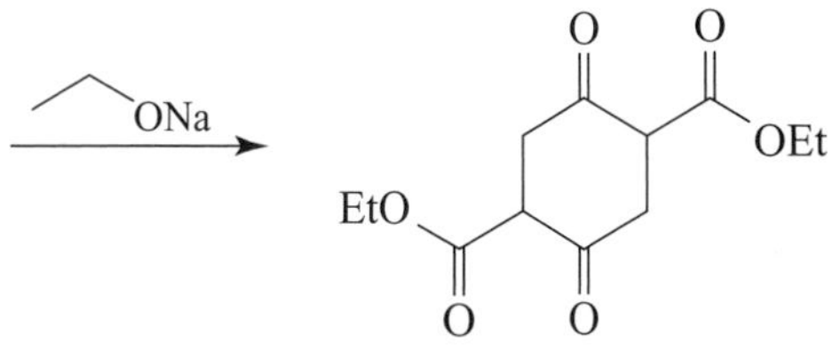

$C_{12}H_{16}O_6$

19.28 The synthesis actually involves two sequential Claisen condensations, with ethyl acetate serving as the source of the carbanionic species.

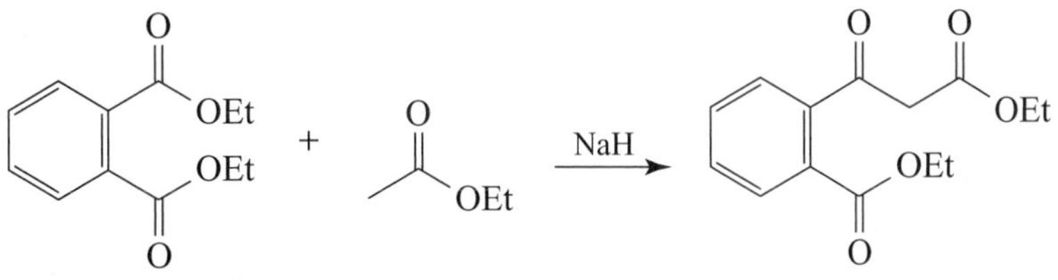

not isolated

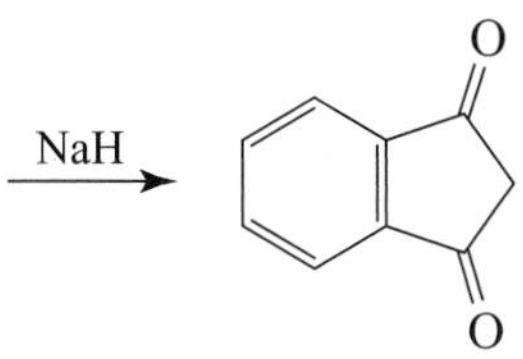

19.29

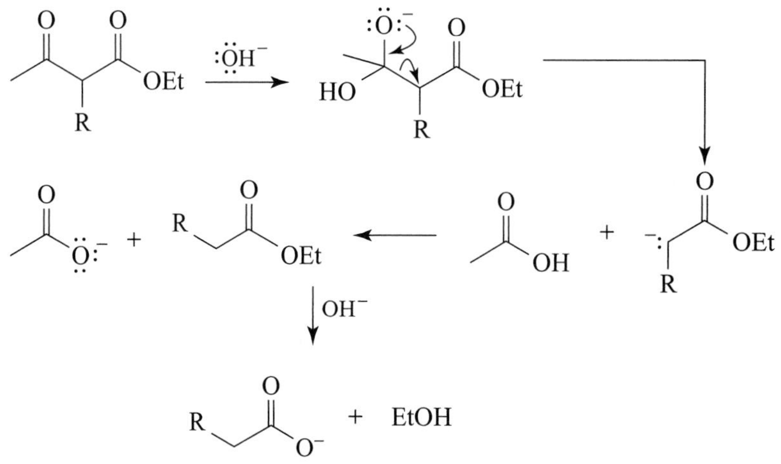

19.30 A gamma hydrogen is abstracted by base (as is an alpha hydrogen in the usual Claisen) to give a resonance-stabilized species:

Ethyl crotonate differs from ethyl acetate by –CH=CH–, a vinyl group. The transmission of the stabilizing effect of the $-COOC_2H_5$ group is an example of the **principle of vinylogy**.

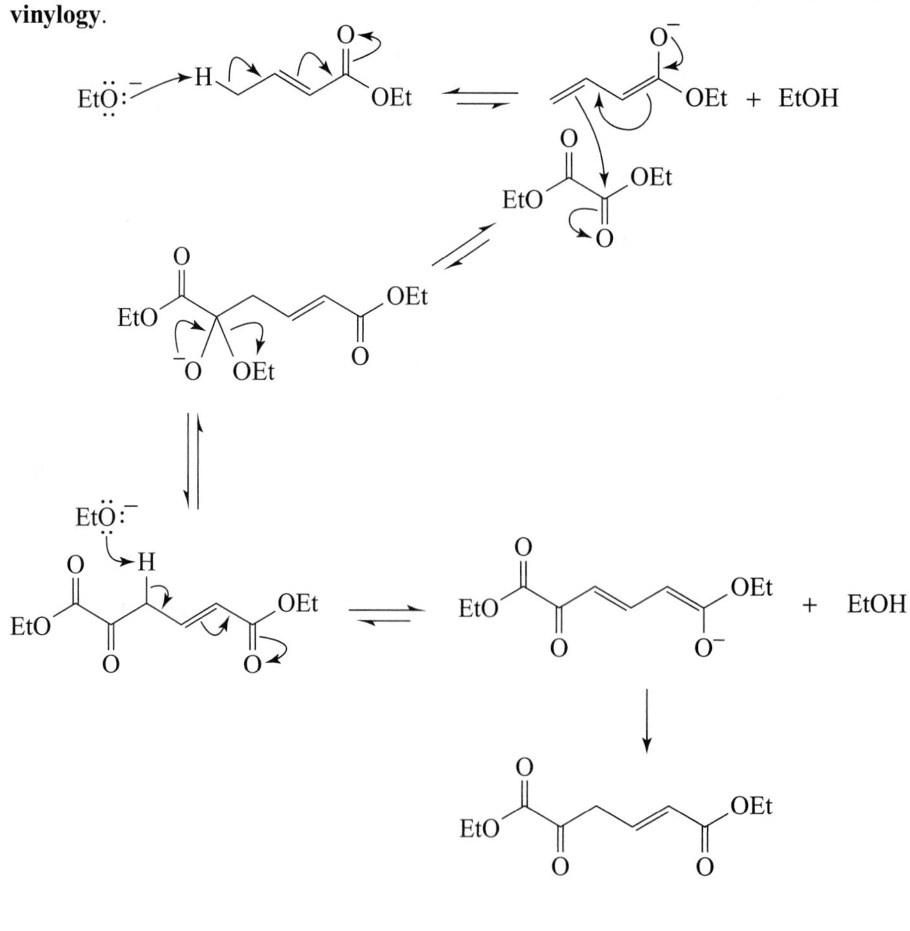

19.31

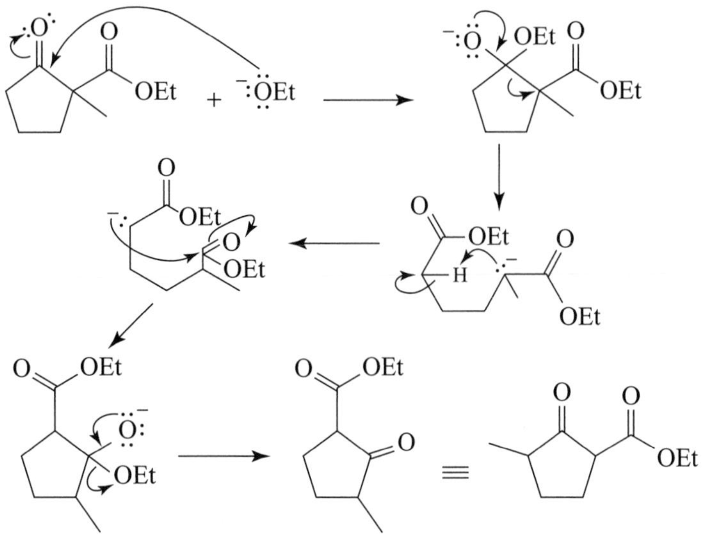

:O: O OEt + $^{-}$:ÖEt → $^{-}$:Ö OEt O OEt

↓

O OEt O H OEt ← O OEt O OEt

O OEt Ö: OEt → O OEt O ≡ O O OEt

The two-stage reaction involves a reverse Dieckmann condensation followed by a forward Dieckmann using a different enolate. The original enolate cannot lead back to the starting material since the crucial α-hydrogen is lacking.

19.32

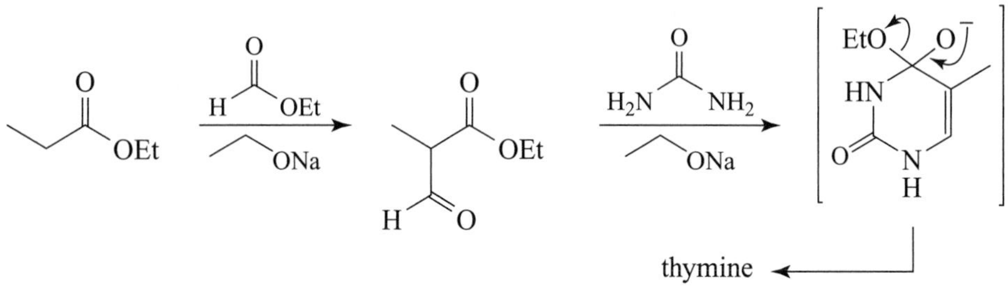

O OEt —(H, O, OEt / ONa)→ O OEt H O —(H_2N, O, NH_2 / ONa)→ [EtO O^{-} HN O N H]

thymine ←

<u>**Aldol Reactions**</u>

19.33 (a) 2 H O —(NaOH / H_2O)→ H OH O

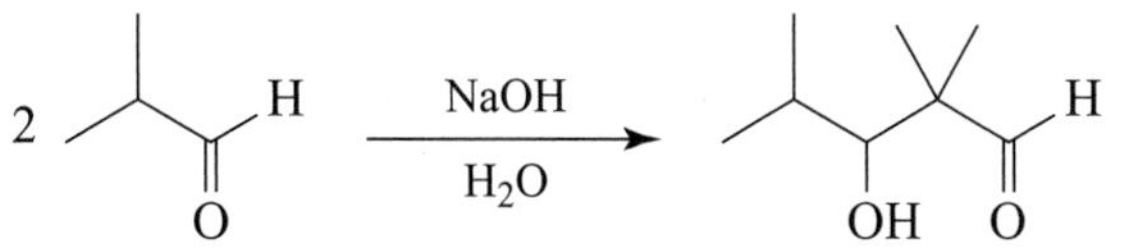

(b) H O + H H O —(NaOH / H_2O)→ H OH O

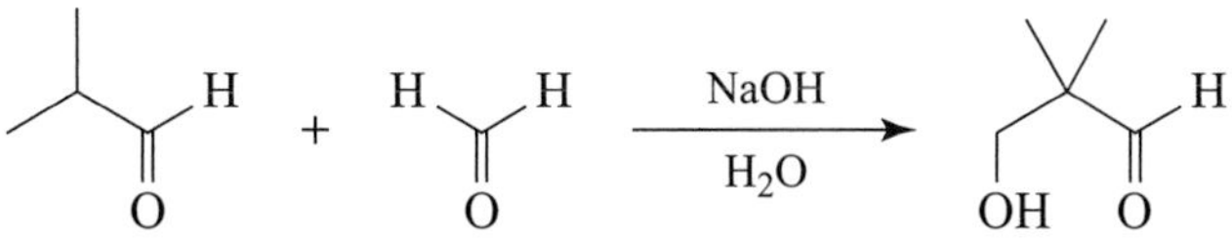

(c) O H + O H —(NaOH / H_2O)→ O H

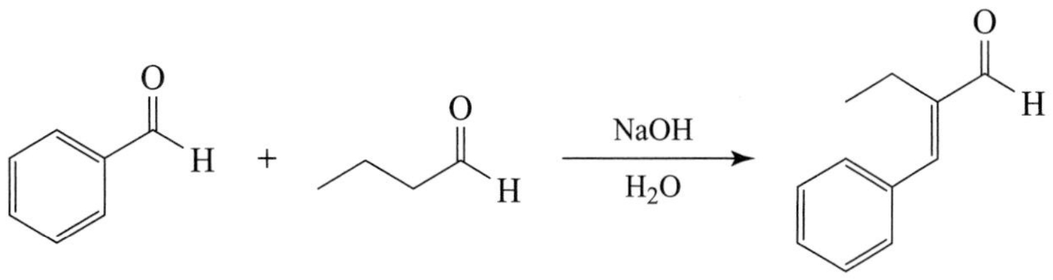

(d) 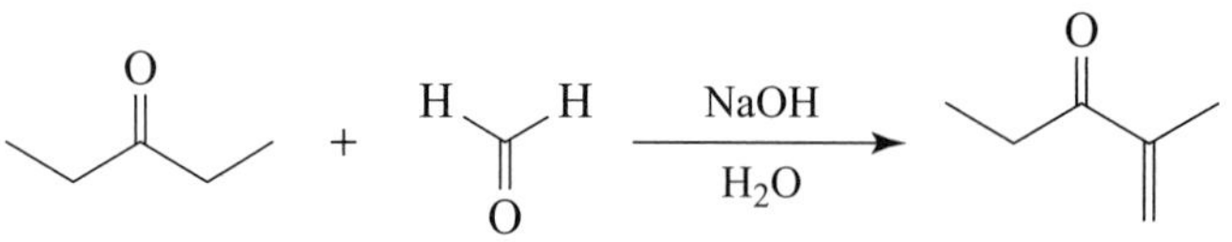 O + H H O NaOH H_2O → O

(e) 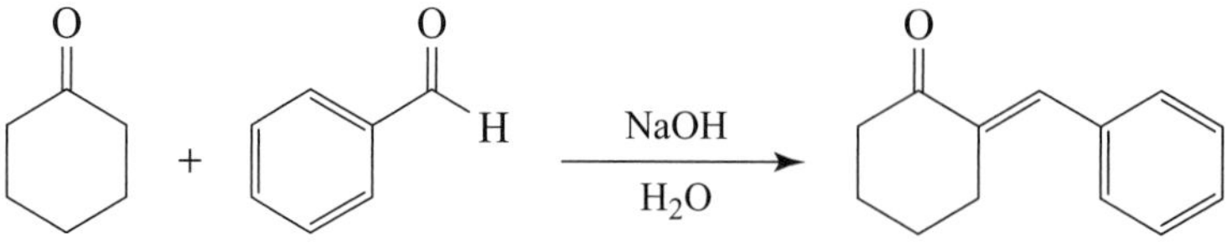 O + O H NaOH H_2O → O

19.34

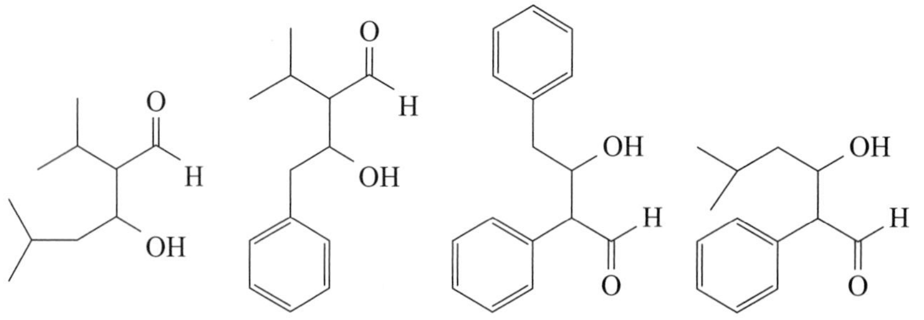

O H OH
O H OH
OH H O
OH H O

1 2 3 4

19.35 (a) 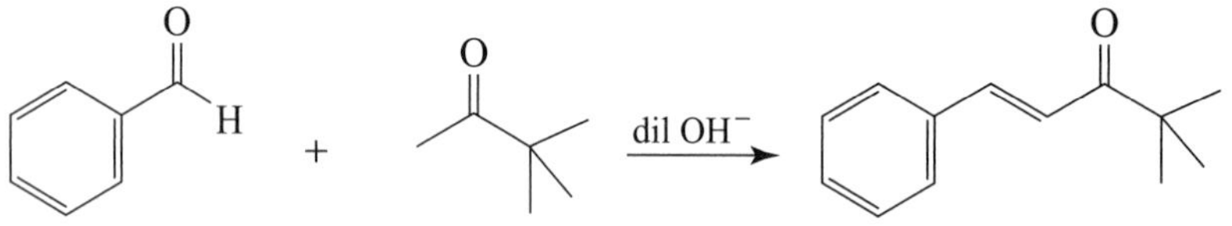 O H + O dil OH^- → O

(b) 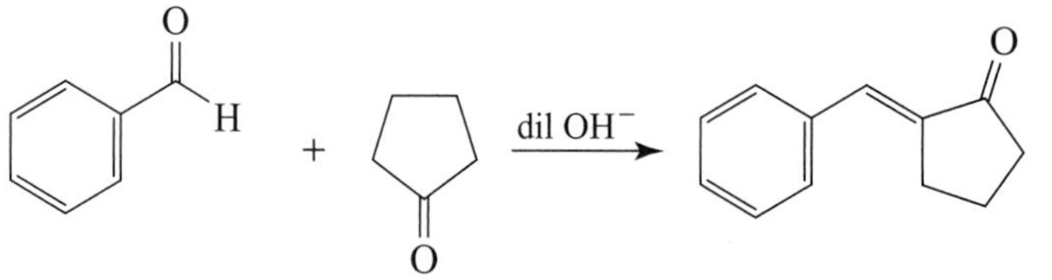 O H + O dil OH^- → O

(c) 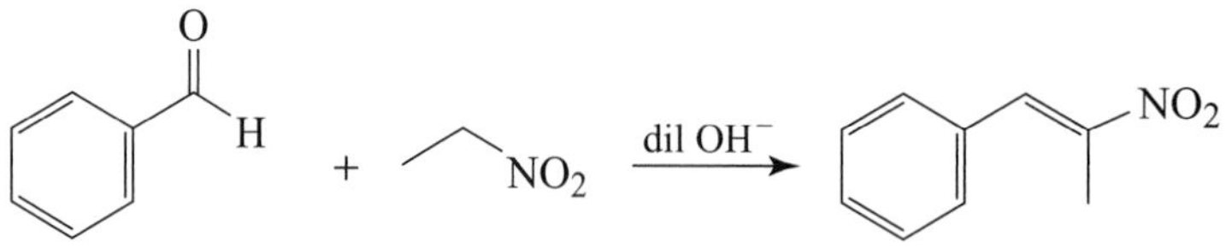 O H + NO_2 dil OH^- → NO_2

(d) 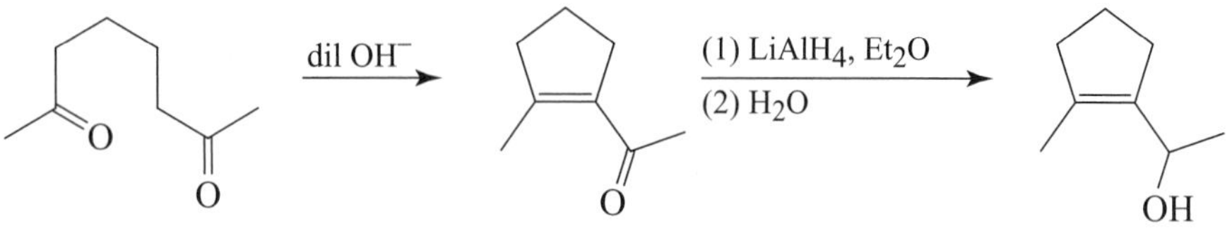 dil OH^- ; (1) $LiAlH_4$, Et_2O (2) H_2O
O O O OH

(e) 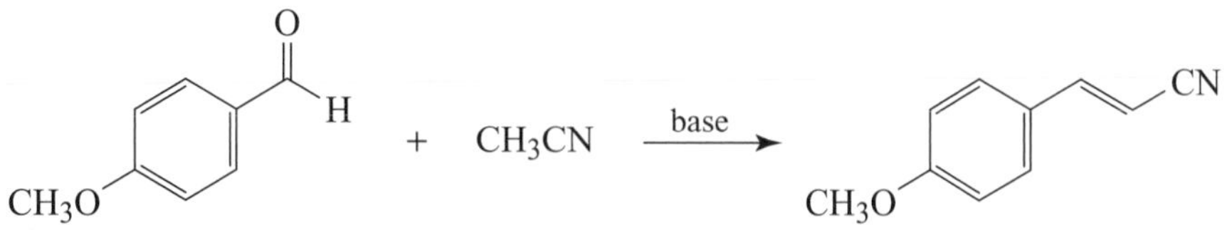 O H + CH_3CN base → CN
CH_3O CH_3O

(f) 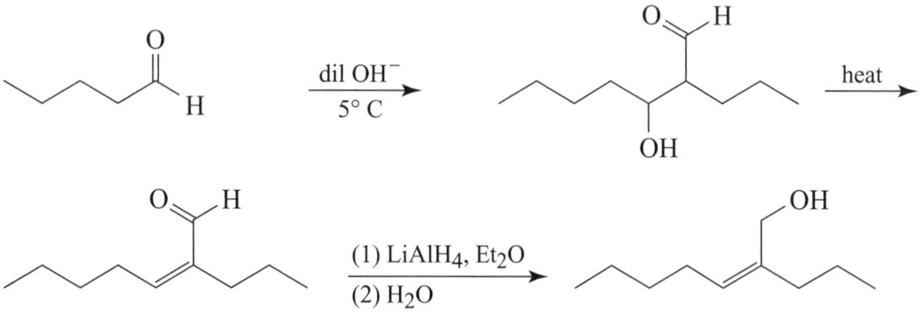 O H dil OH^- 5° C → O H OH heat →

O H (1) $LiAlH_4$, Et_2O (2) H_2O → OH

(g) 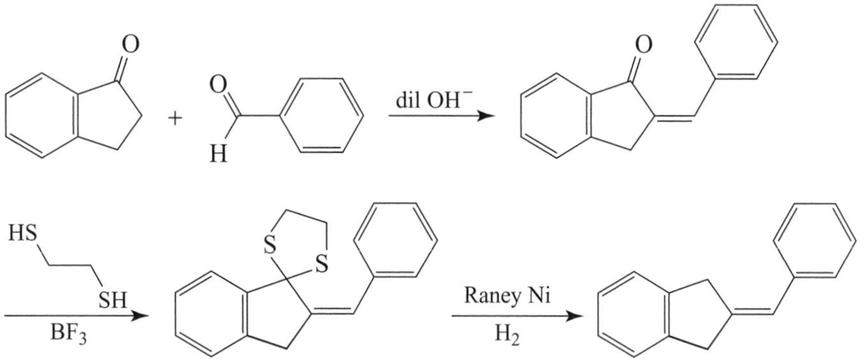 O + O H dil OH^- → O

HS SH BF_3 → S S Raney Ni H_2 →

19.36 (a) 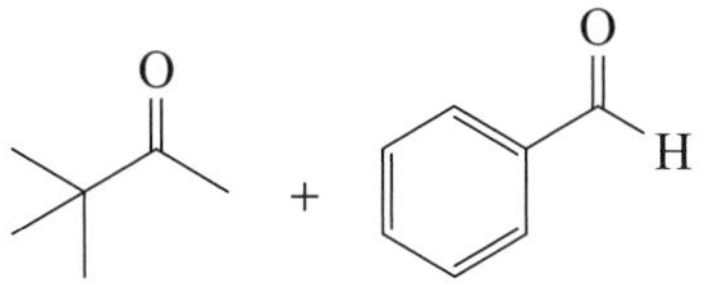 O + O H (b) 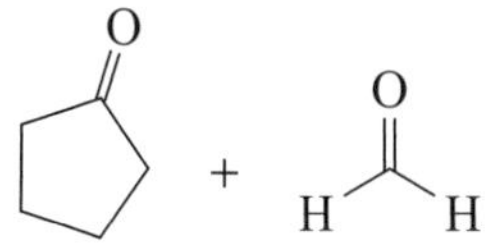 O + O H H

(c) 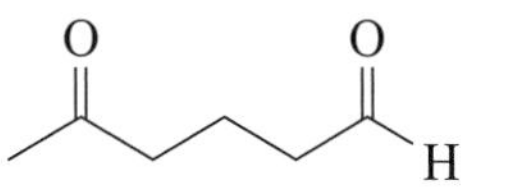 O O H (d) 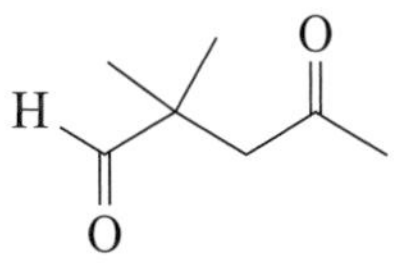 H O O

(e) 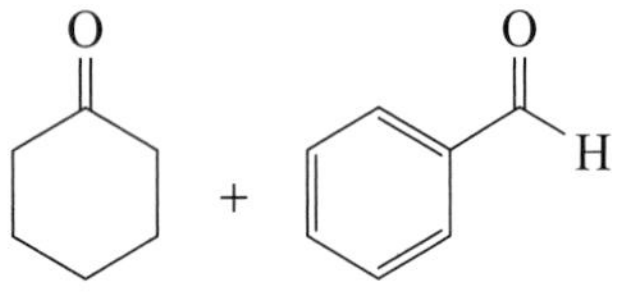 O + O H

(f) 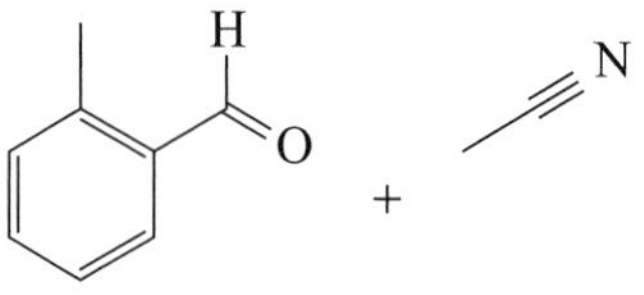 H O + N

(g) 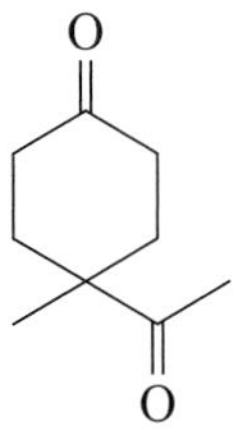 O O

(h) O O H

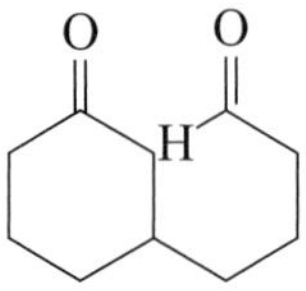

19.37 (a) 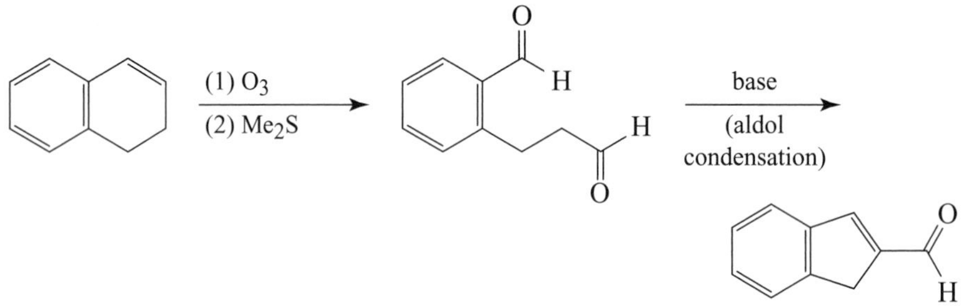 (1) O_3 (2) Me_2S O H H O base (aldol condensation) O H

(b) 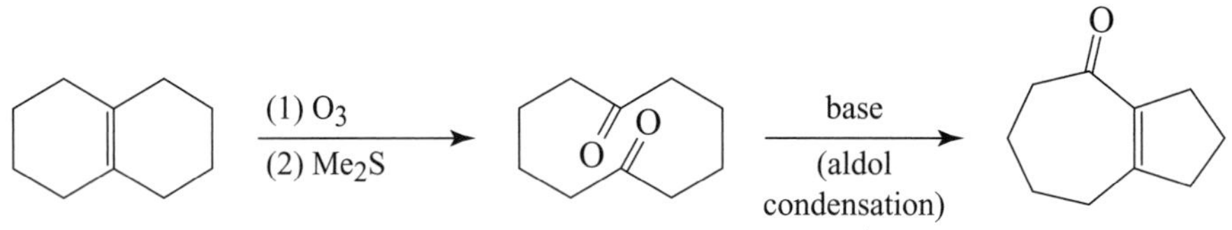 (1) O_3 (2) Me_2S O O base (aldol condensation) O

(c) 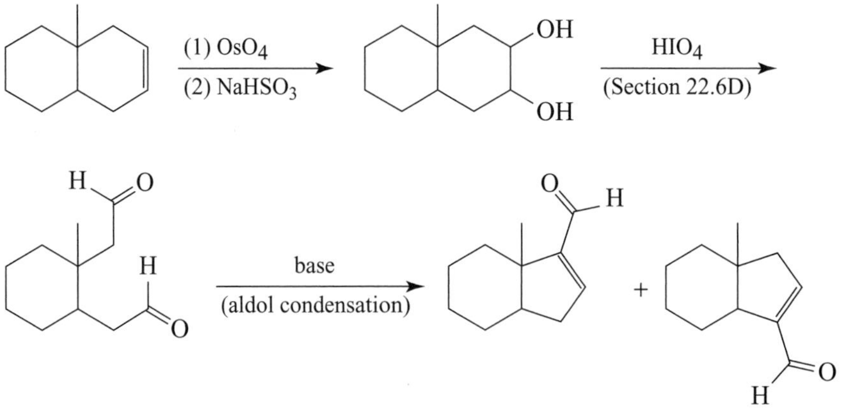 (1) OsO_4 (2) $NaHSO_3$ OH OH HIO_4 (Section 22.6D)

H O H O base (aldol condensation) O H + H O

(d)

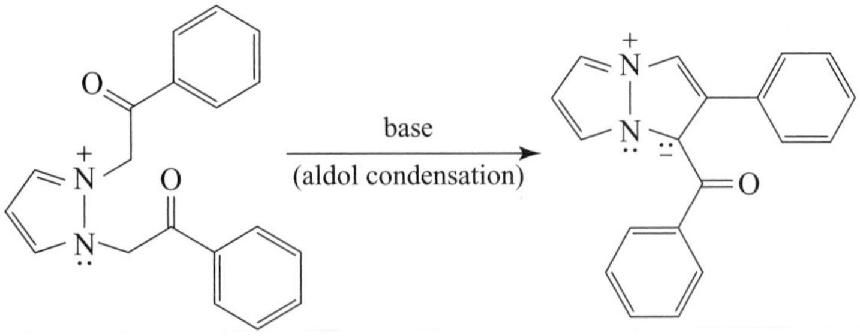

base

(aldol condensation)

19.38 (a) The conjugate base is a hybrid of the following structures (cf. Rev. Prob. 19.30):

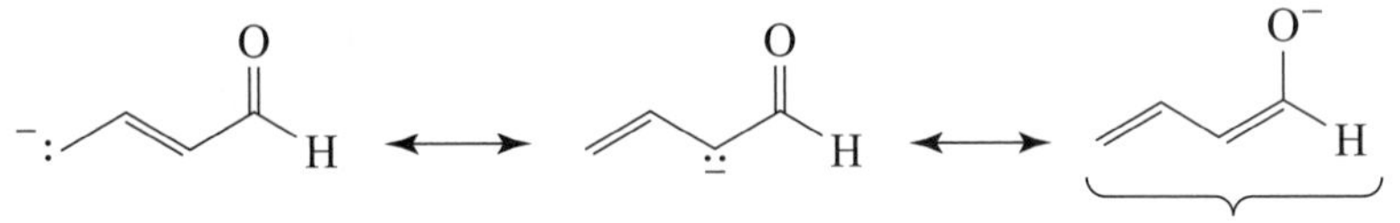

This structure is especially stable because the negative charge is on the oxygen atom.

(b)

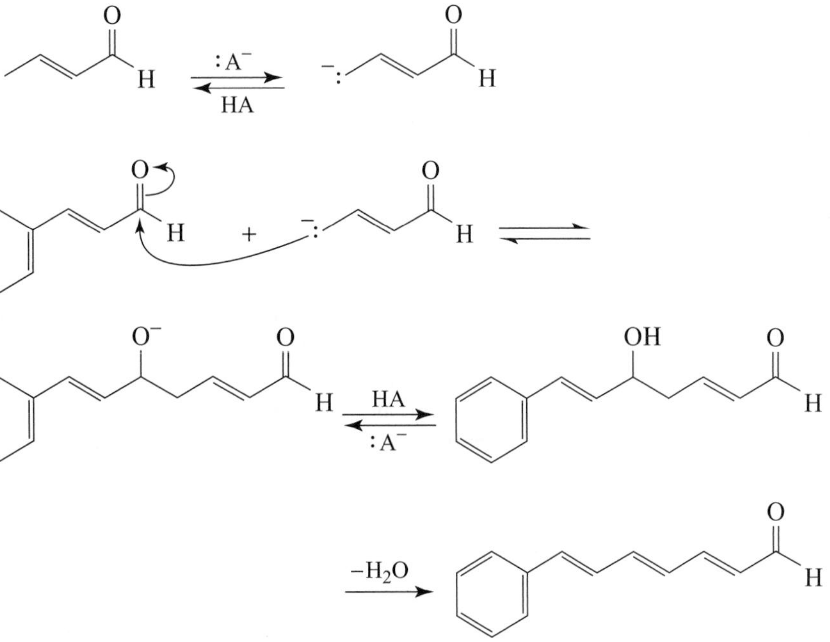

:A⁻

HA

HA

:A⁻

$-H_2O$

19.39

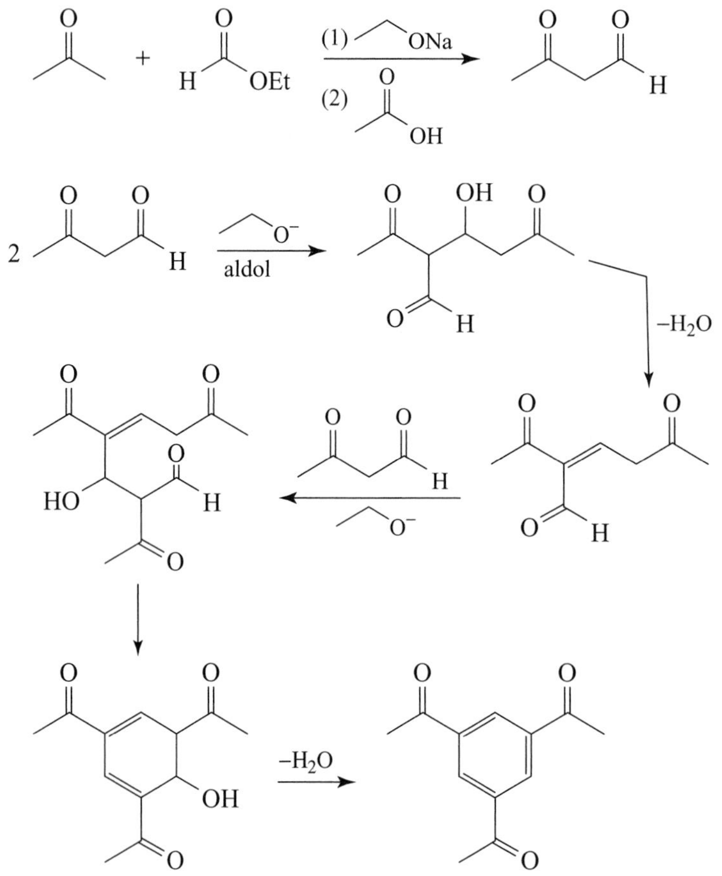

(1) ONa
(2) OH

aldol

$-H_2O$

HO H

OH

$-H_2O$

19.40 This difference in behavior indicates that, for acetaldehyde, the capture of a proton from the solvent (the reverse of the reaction by which the enolate anion is formed) occurs much more slowly than the attack by the enolate anion on another molecule.

When acetone is used, the equilibrium for the formation of the enolate anion is unfavorable, but more importantly, enolate attack on another acetone molecule is disfavored due to steric hindrance. Here proton capture (actually deuteron capture) competes very well with the aldol reaction.

Conjugate Addition Reactions

19.41 (a)

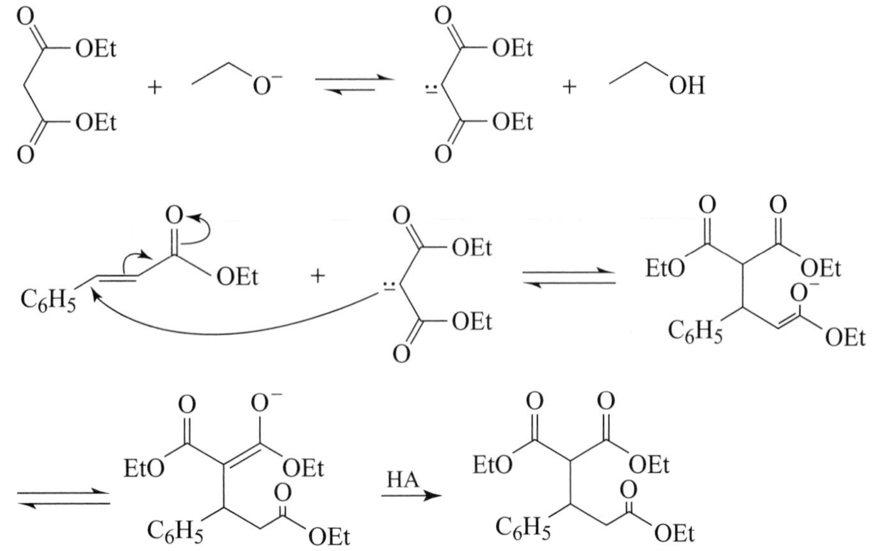

O OEt OEt O + O^- ⇌ O OEt OEt O + OH

C_6H_5 O OEt + O OEt OEt O ⇌ O O EtO OEt O^- C_6H_5 OEt

⇌ O O^- EtO OEt O C_6H_5 OEt $\xrightarrow{HA}$ O O EtO OEt O C_6H_5 OEt

(b) $CH_3\ddot{N}H_2$ + O OMe ⇌ H O^- $CH_3-\overset{+}{N}$ OMe H ⇌

CH_3 N H O OMe ⇌ (O OMe) O OMe CH_3N OMe O

$\xrightarrow{base}$ O OMe CH_3-N OMe O ⟶ O OMe CH_3-N OMe O^- ⟶

O OMe CH_3-N =O

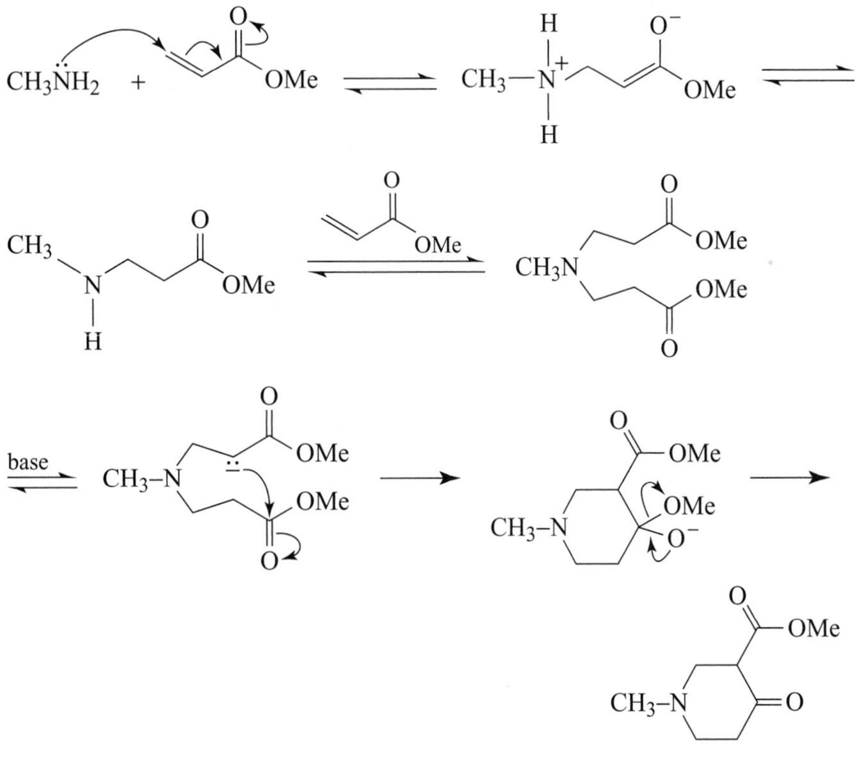

(c)

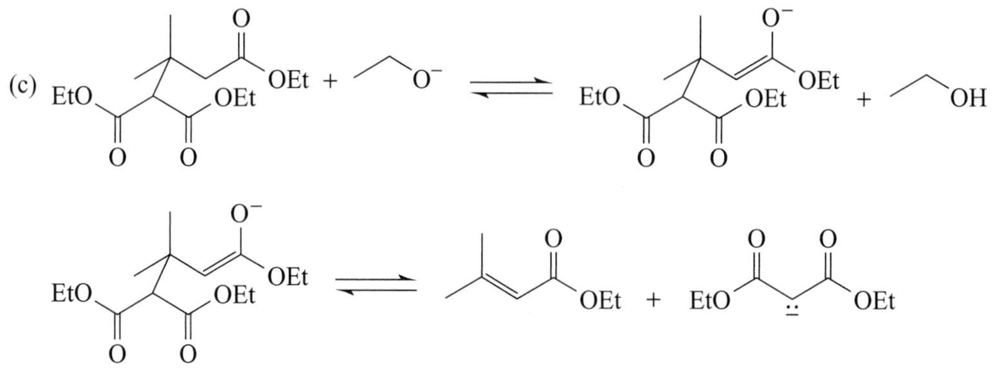

O
EtO OEt OEt + O⁻ ⇌ EtO OEt OEt + OH
O O O O
O⁻

O⁻
EtO OEt OEt ⇌ OEt + EtO OEt
O O O O O

The Michael reaction is reversible, and the reaction just given is an example of a reverse Michael reaction.

19.42 Two reactions take place. The first is a reaction called the Knoevenagel condensation, initiated by attack of the conjugate base of the dicarbonyl compound on the ketone,

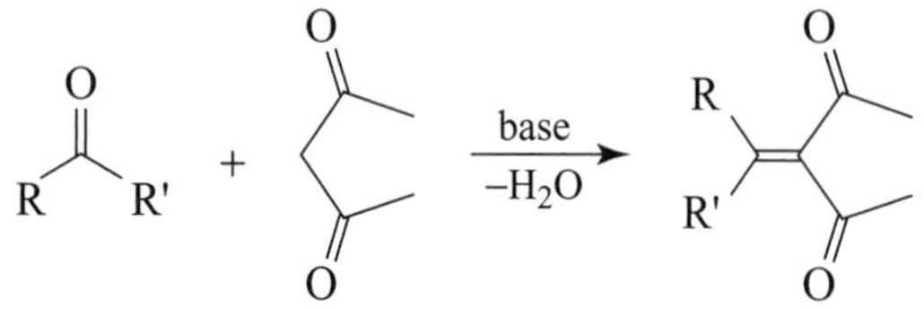

O O O
R R' + base, $-H_2O$ → R R'
O O

Then the α, β-unsaturated diketone reacts with a second mole of the active methylene compound in a Michael addition.

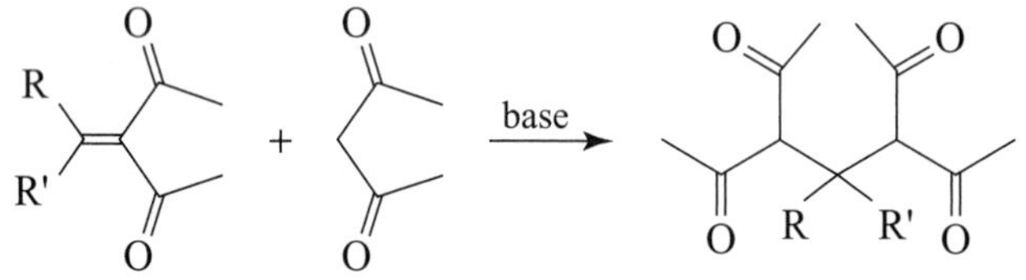

O O O O
R R' + base → R R'
O O O O

19.43

O O
C_6H_5 OH^- → C_6H_5 → O

O O O O
C_6H_5 → C_6H_5 $:\ddot{O}H^-$ → C_6H_5
HO—H :O: H O O

:O: OH $-H_2O$ C_6H_5
→ C_6H_5 → H—OH → C_6H_5 →
O O H $:\ddot{O}H^-$ O

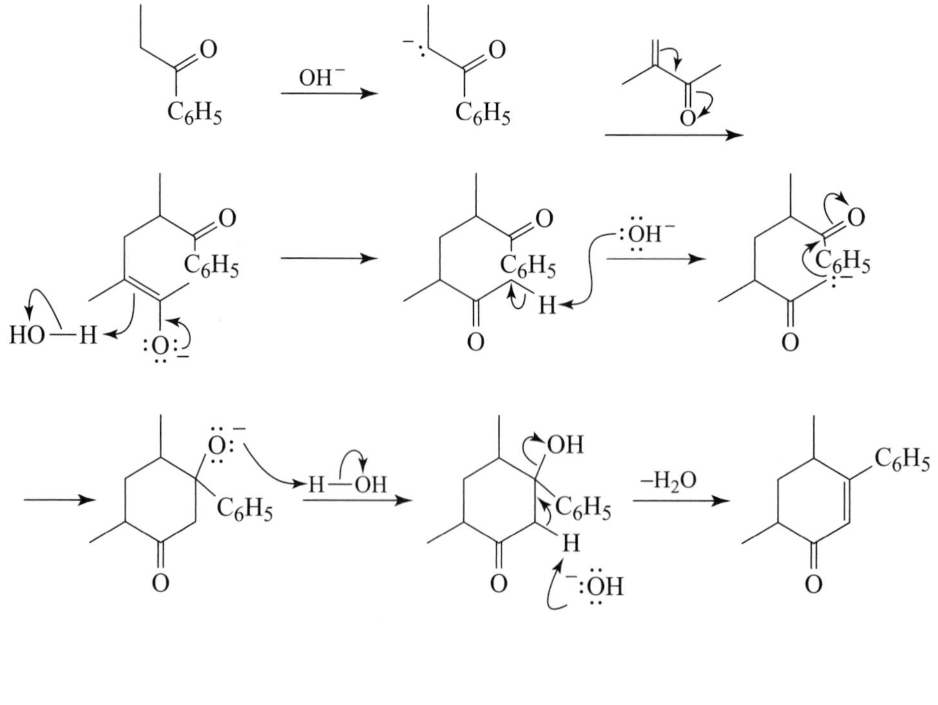

19.44 There are two stages to the reaction. The initial Michael addition is followed by the reaction of an enolate with an ester carbonyl to form a diketone (in equilibrium with an enol form).

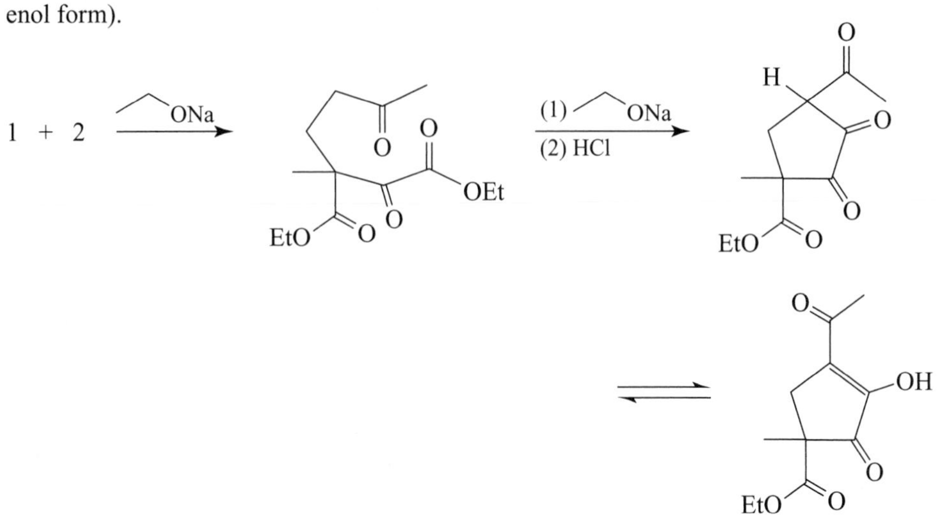

1 + 2 → (EtONa) → (1) EtONa (2) HCl → ⇌

Another (less likely) possibility is the formation of a seven-membered ring.

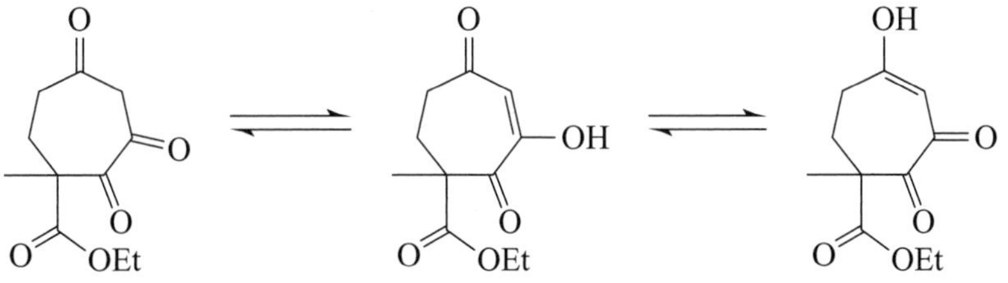

General Problems

19.45 (a) → (Br, NaOH) → H_3O^+ → NaOH, H_2O

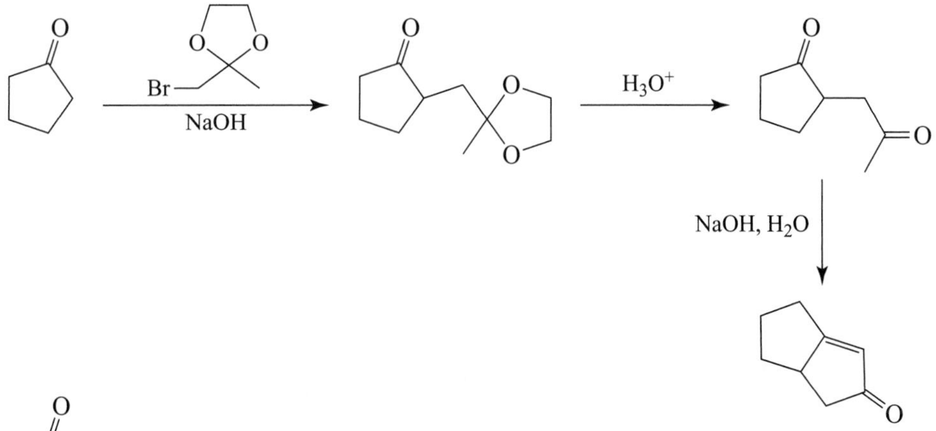

(b) + → NaOH, H_2O

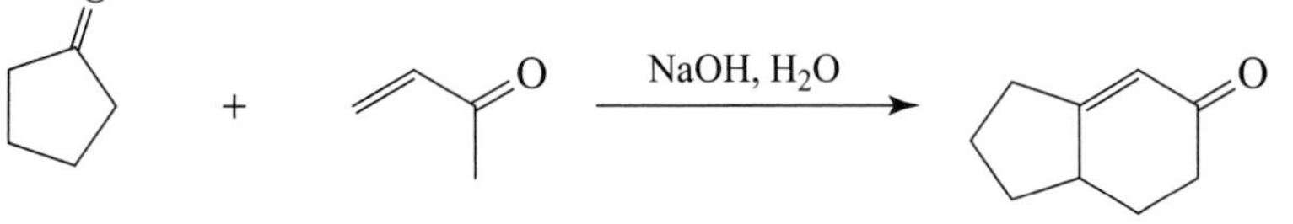

19.46

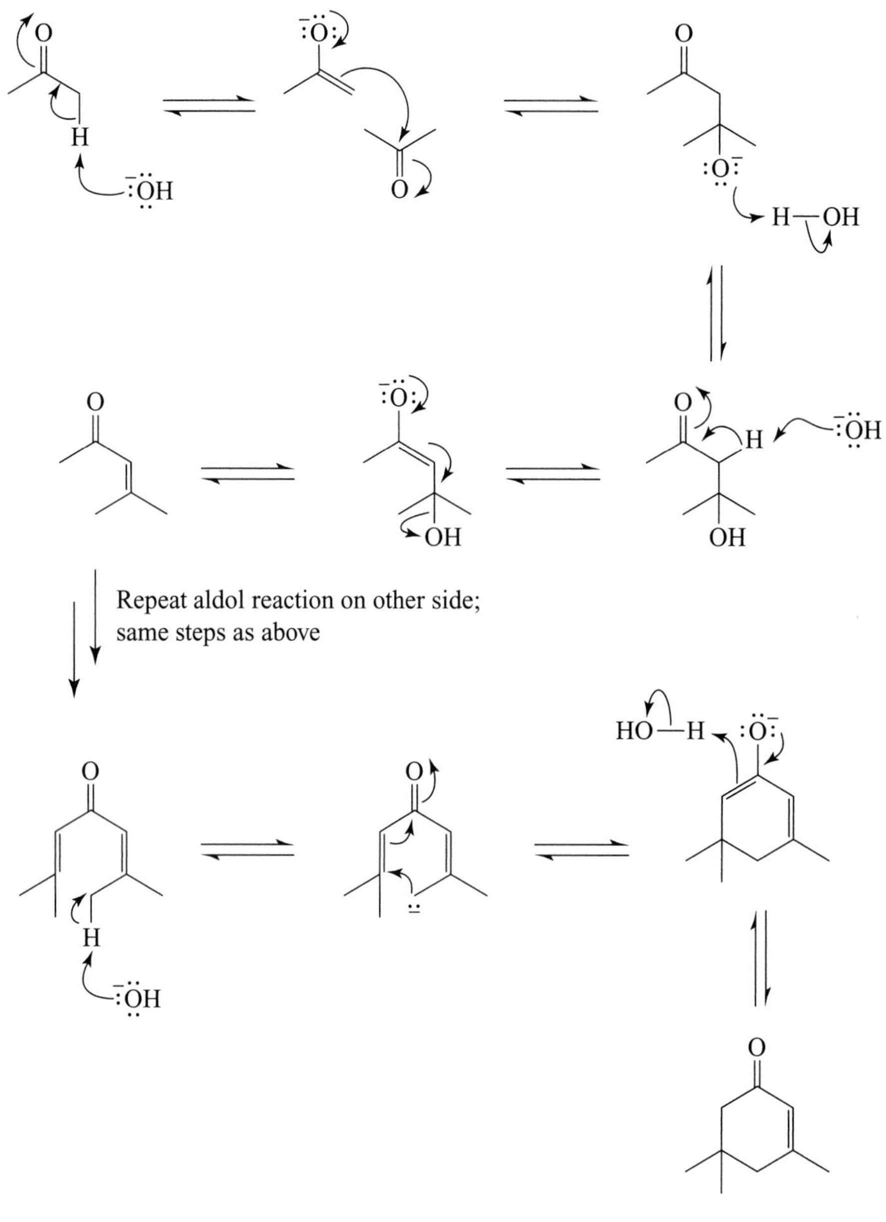

O H :ÖH O: O O :O: H—OH

O :O: OH O H :ÖH OH

Repeat aldol reaction on other side; same steps as above

O H :ÖH O HO—H :O: O

19.47 (a)

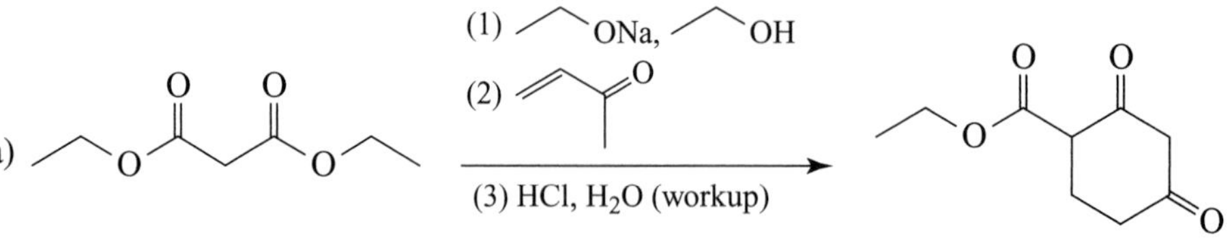

O O O O

(1) ONa, OH

(2) O

(3) HCl, H_2O (workup)

O O O O

(b) 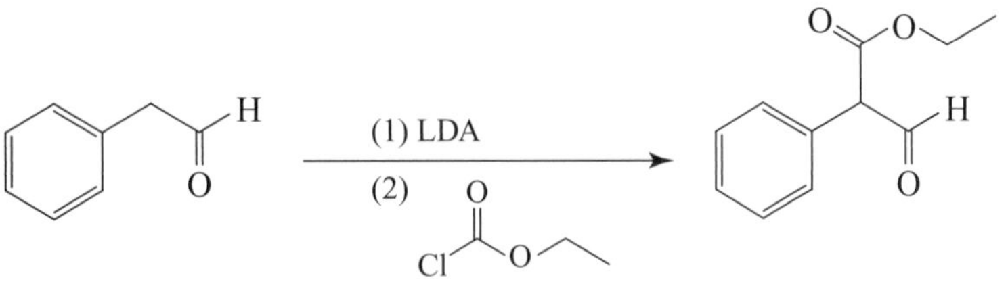 (1) LDA
(2) Cl–C(=O)–O–Et

(c) (1) EtONa, EtOH
(2) H_2O, HCl (workup)
(3) NaOH
(4) EtBr

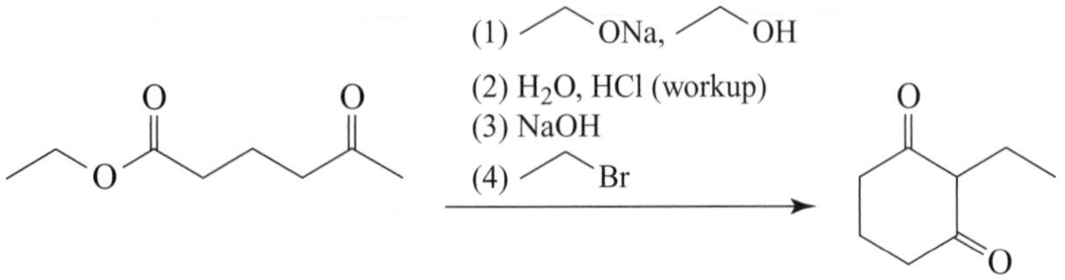

(d) 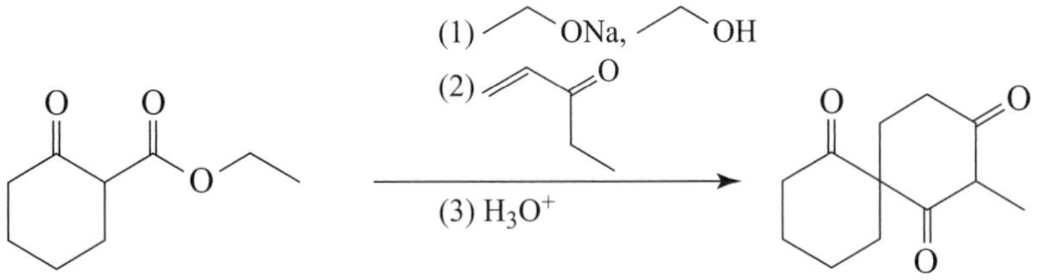 (1) EtONa, EtOH
(2) ethyl vinyl ketone
(3) H_3O^+

19.48 (a) 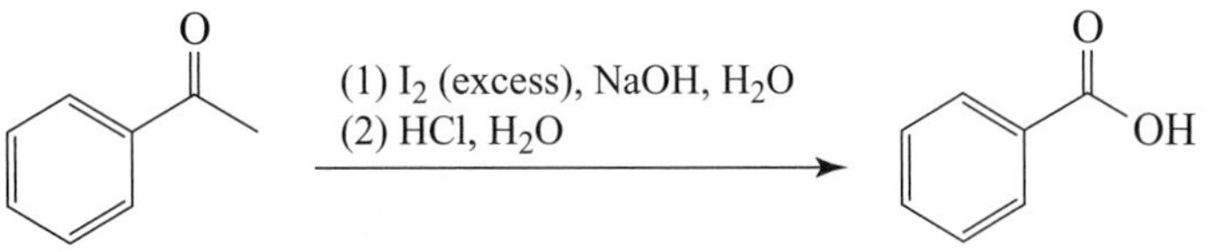 (1) I_2 (excess), NaOH, H_2O
(2) HCl, H_2O

(b) 2 [phthalaldehyde] + [cyclohexane-1,4-dione] KOH, H_2O, EtOH

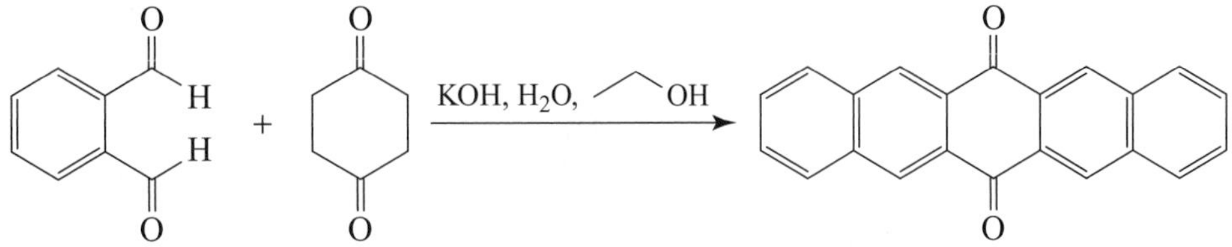

(c) 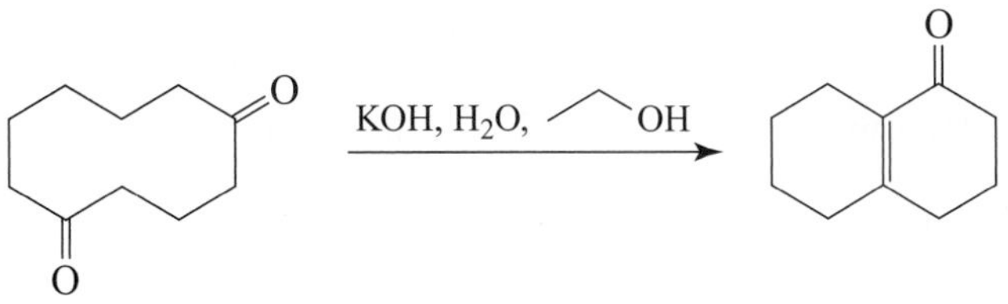 KOH, H_2O, EtOH

(d) 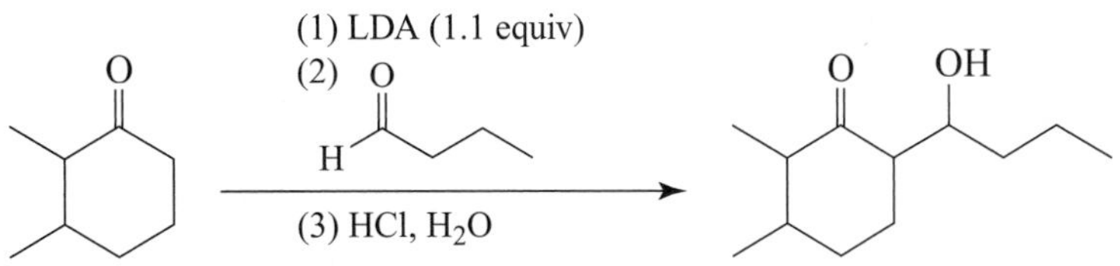 (1) LDA (1.1 equiv)
(2) butanal
(3) HCl, H_2O

19.49

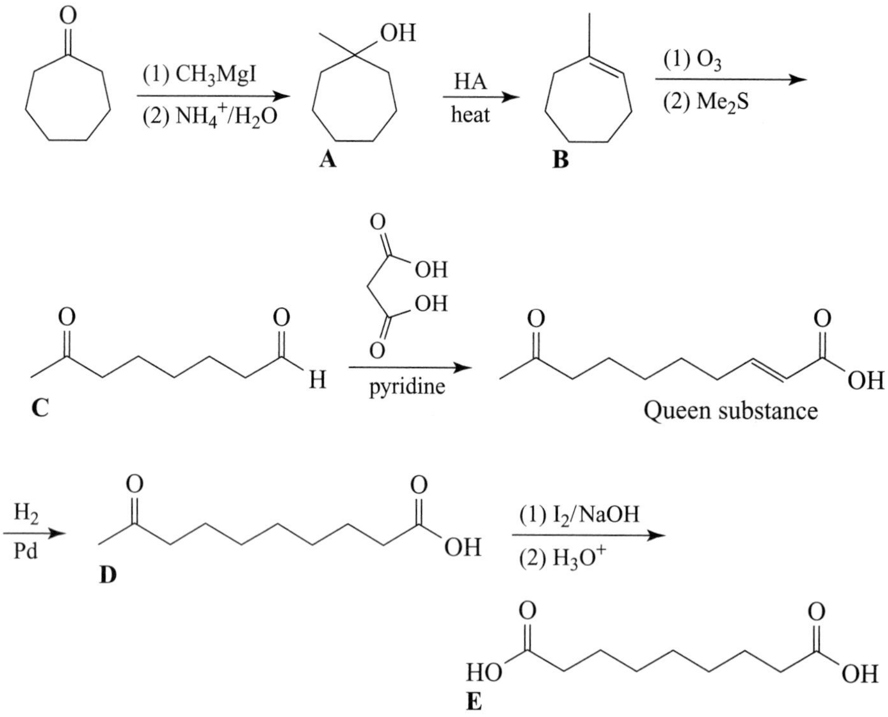

19.50 (a)

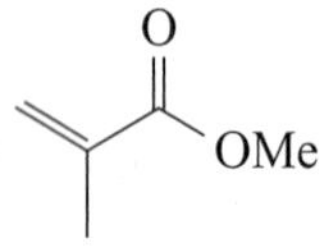

(b) $KMnO_4$, OH^-, then H_3O^+

(c) CH_3OH, HA

(d) CH_3ONa, then H_3O^+

(e) and (f)

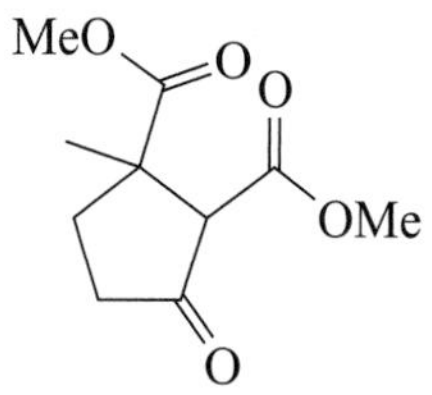

and

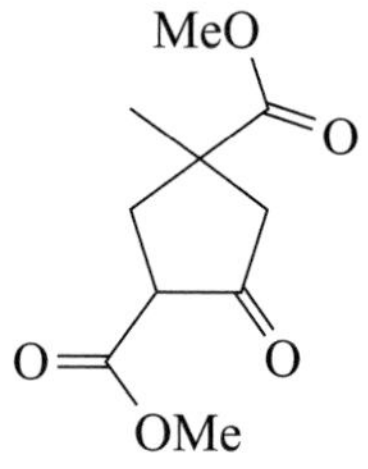

(g) OH^-, H_2O, then H_3O^+

(h) heat ($-CO_2$)

(i) CH_3OH, HA

(j)

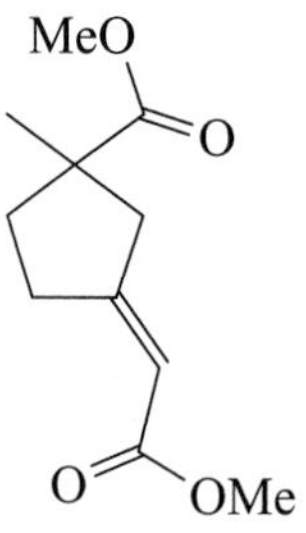

(k) H_2, Pt

(l) CH_3ONa, then H_3O^+

(m) 2 $NaNH_2$ + 2 CH_3I

19.51 C_6H_5 + H H + $HN(CH_3)_2$ Mannich reaction ($-H_2O$) → C_6H_5 ... CH_3 N CH_3

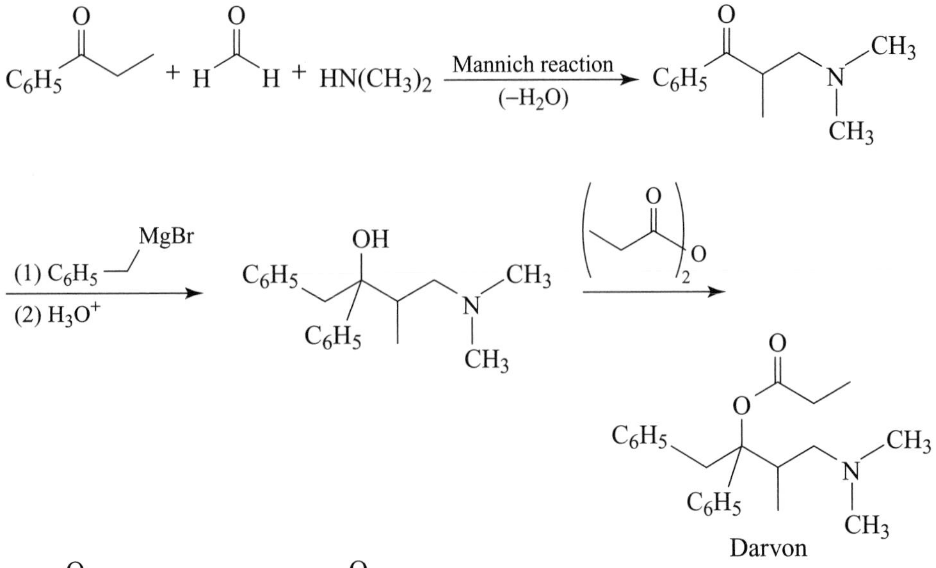

(1) C_6H_5 MgBr (2) H_3O^+ → C_6H_5 OH C_6H_5 N CH_3 CH_3 → Darvon

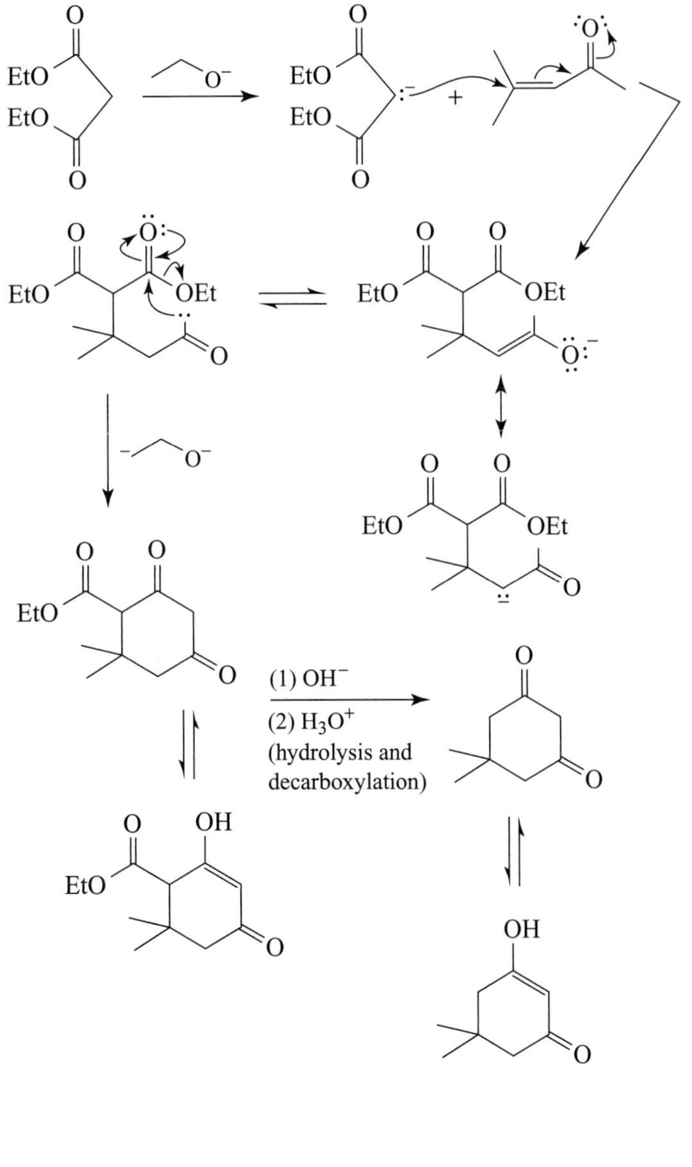

Darvon

19.52 EtO EtO → EtO EtO

(1) OH^-
(2) H_3O^+
(hydrolysis and decarboxylation)

19.53

H_2SO_4

(1) +A:⁻, –HA
(2) HA

A:⁻, –HA

19.54 (a)

(b) Decarboxylation of the epoxy acid gives an enolate anion which, on protonation, gives an aldehyde.

$-CO_2$

(c)

β-Ionone

(1) OH^-, H_2O
(2) H_3O^+

C_5H_5N, heat
$-CO_2$

19.55 (a) [Reaction scheme: benzaldehyde + propanoic anhydride, with $CH_3CH_2CO_2^-K^+$ → 2-methyl-3-phenylpropenoic acid + propanoic acid]

H + O O⁻ K⁺ OH + OH

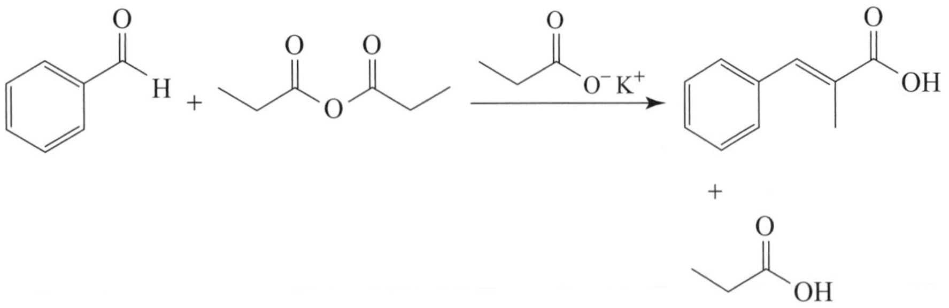

(b) [Reaction scheme: 4-chlorobenzaldehyde + acetic anhydride, with $CH_3CO_2^-K^+$ → 3-(4-chlorophenyl)propenoic acid + acetic acid]

Cl H + O O⁻ K⁺ Cl OH + OH

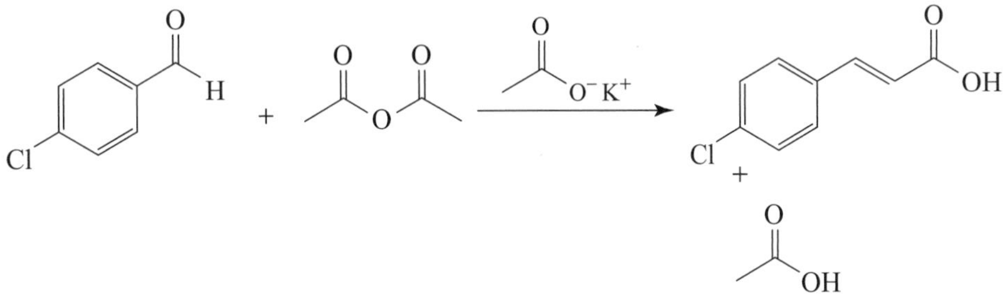

Spectroscopy

19.56 (a) In simple addition, the carbonyl peak (1665–1780-cm^{-1} region) does not appear in the product; in conjugate addition, it does.

(b) As the reaction takes place, the long-wavelength absorption arising from the conjugated system should disappear, One could follow the rate of the reaction by following the rate at which this absorption peak disappears.

19.57 [Structure: 1,5-diphenyl-1,4-pentadien-3-one]

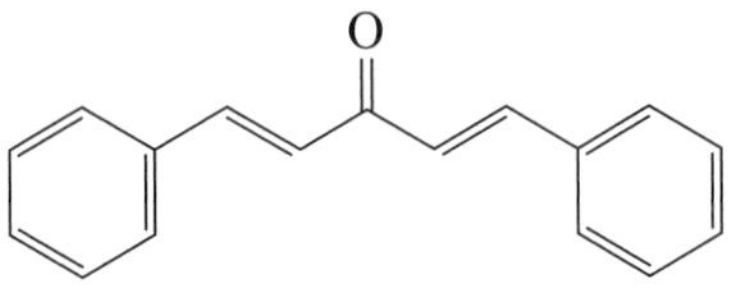

X

19.58 (a)

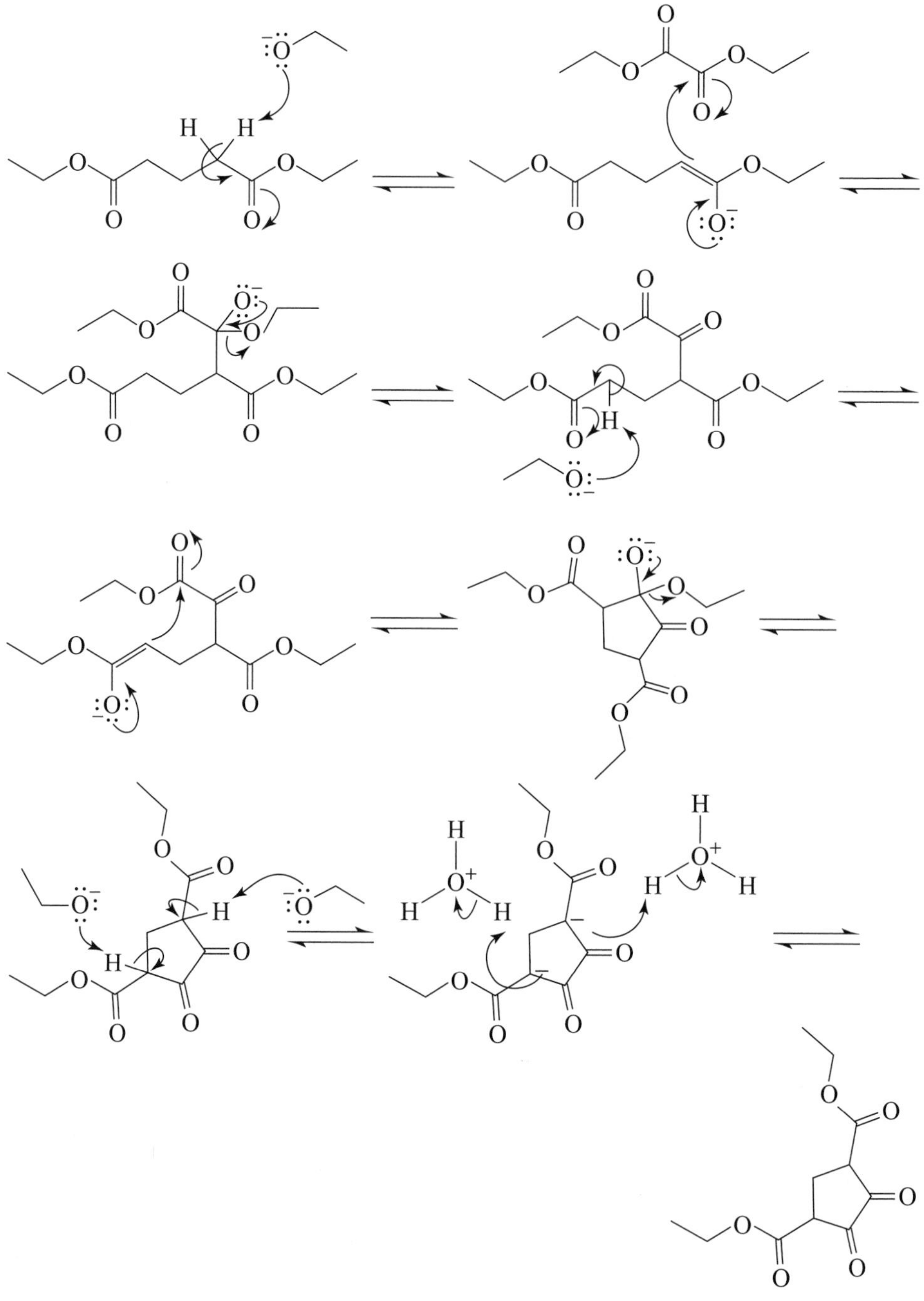

(b)

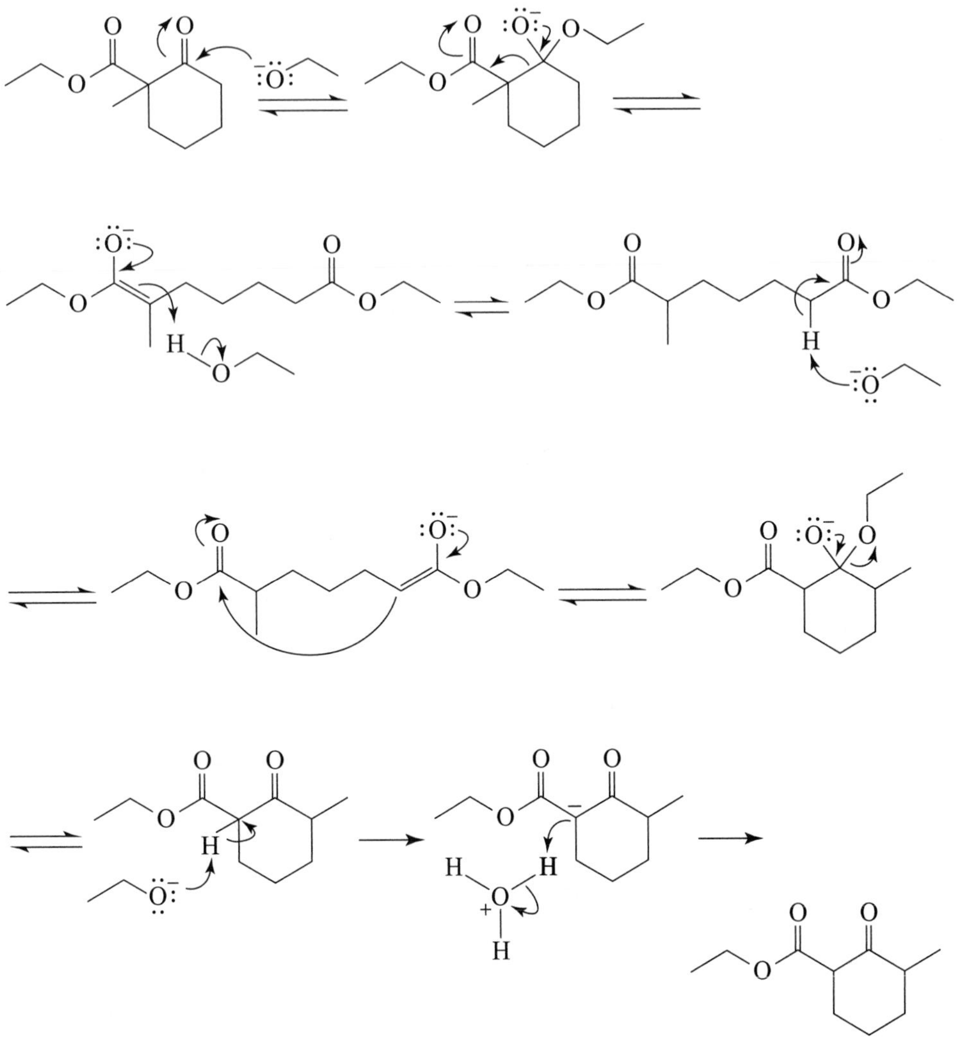

19.59 (a,b)

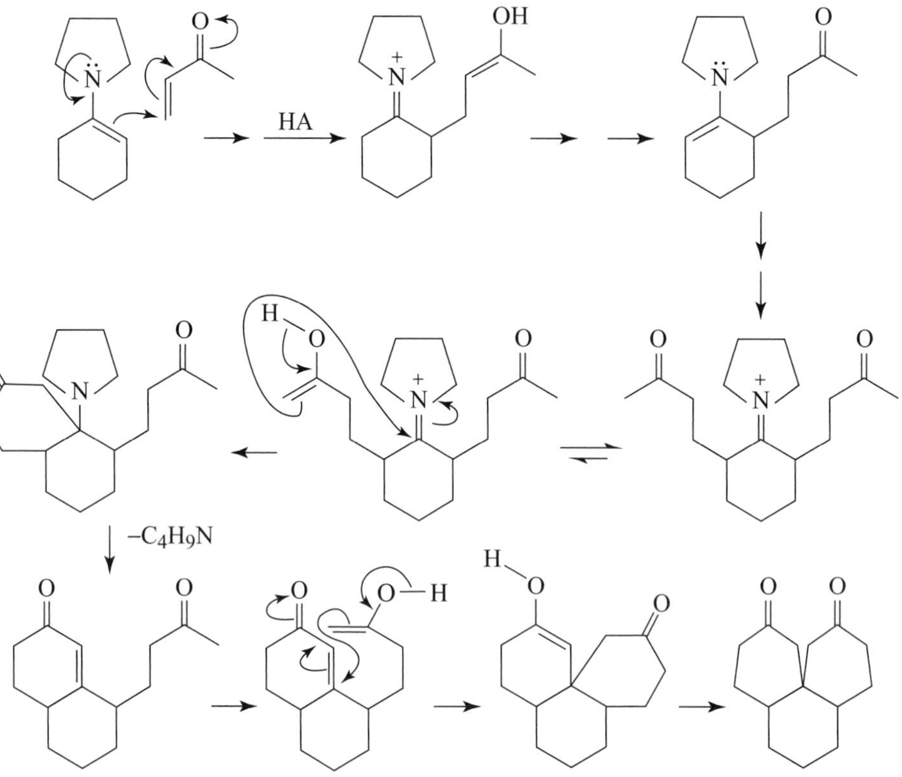

HA

$-C_4H_9N$

A

19.60

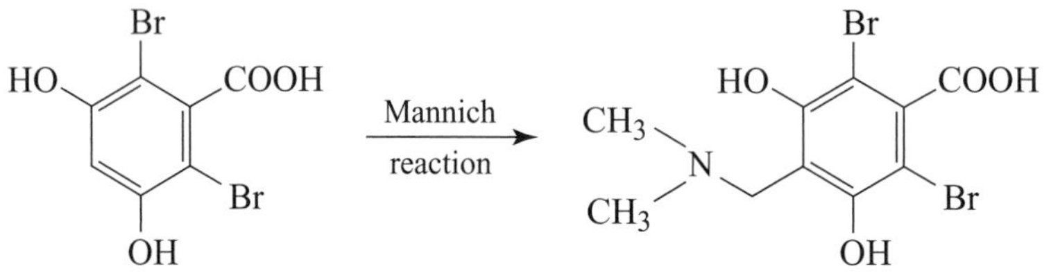

Mannich reaction

B

C

Hydrogenolysis

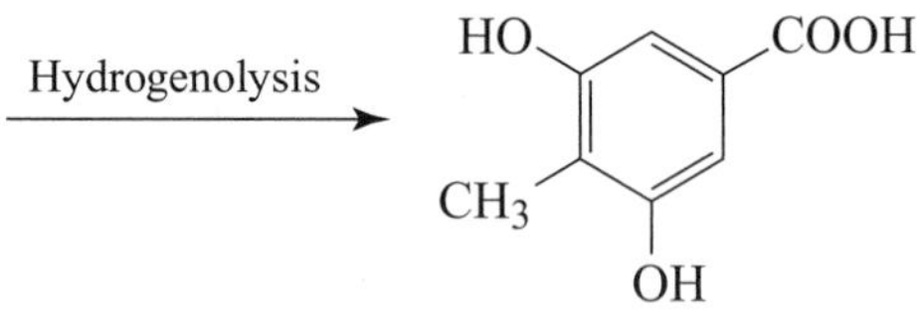

D

QUIZ

19.1 Which hydrogen atoms in the following ester are most acidic?

$$\overset{a}{CH_3}-\overset{b}{CH_2}-\overset{O}{\overset{\|}{C}}-\overset{c}{CH_2}-\overset{O}{\overset{\|}{C}}-O\overset{d}{CH_2}-\overset{e}{CH_3}$$

(a) *a* (b) *b* (c) *c* (d) *d* (e) *e*

19.2 What would be the product of the following reaction?

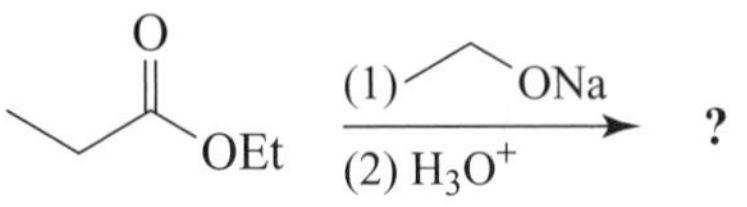

O

OEt (1) ONa (2) H_3O^+ ?

(a) 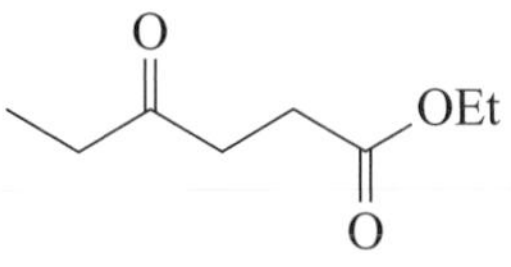 O OEt O

(b) 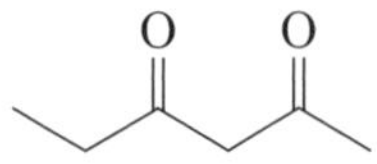 O O

(c) 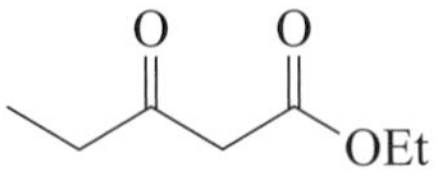 O O OEt

(d) 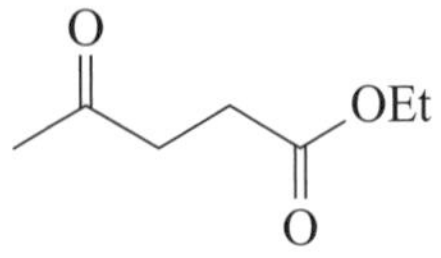 O OEt O

(e) 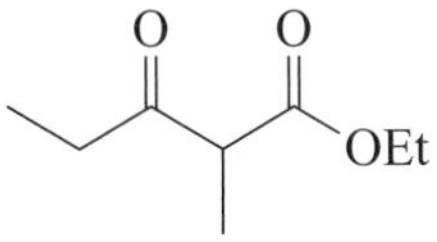 O O OEt

19.3 What starting materials could be used in a crossed Claisen condensation to prepare the following compound?

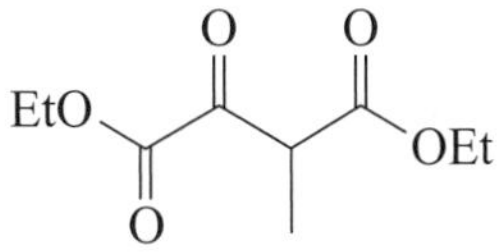

O O EtO OEt O

(a) 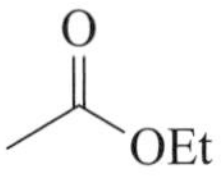 O OEt and O EtO O

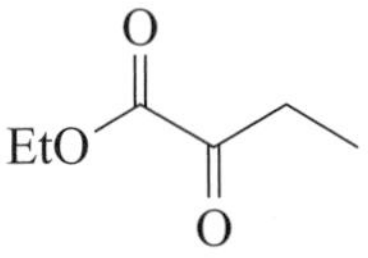

(b) 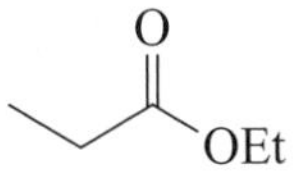 O OEt and O EtO OEt O

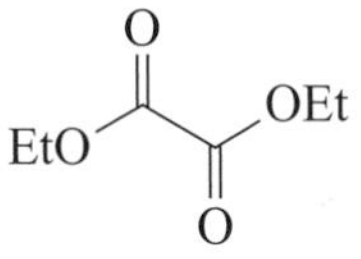

(c) 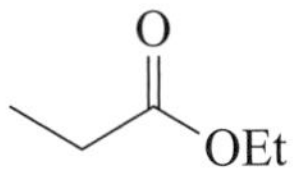 O OEt and O H OEt

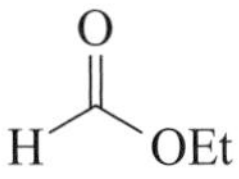

(d) 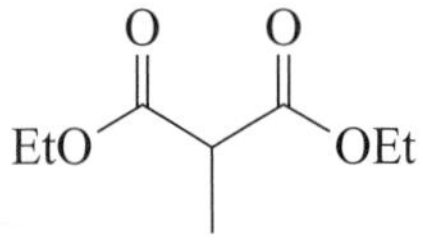 O O EtO OEt and O H OEt

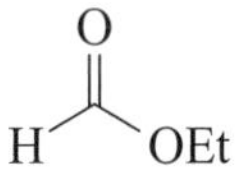

(e) More than one of the above

19.4 Supply the missing reagents, intermediates, and products.

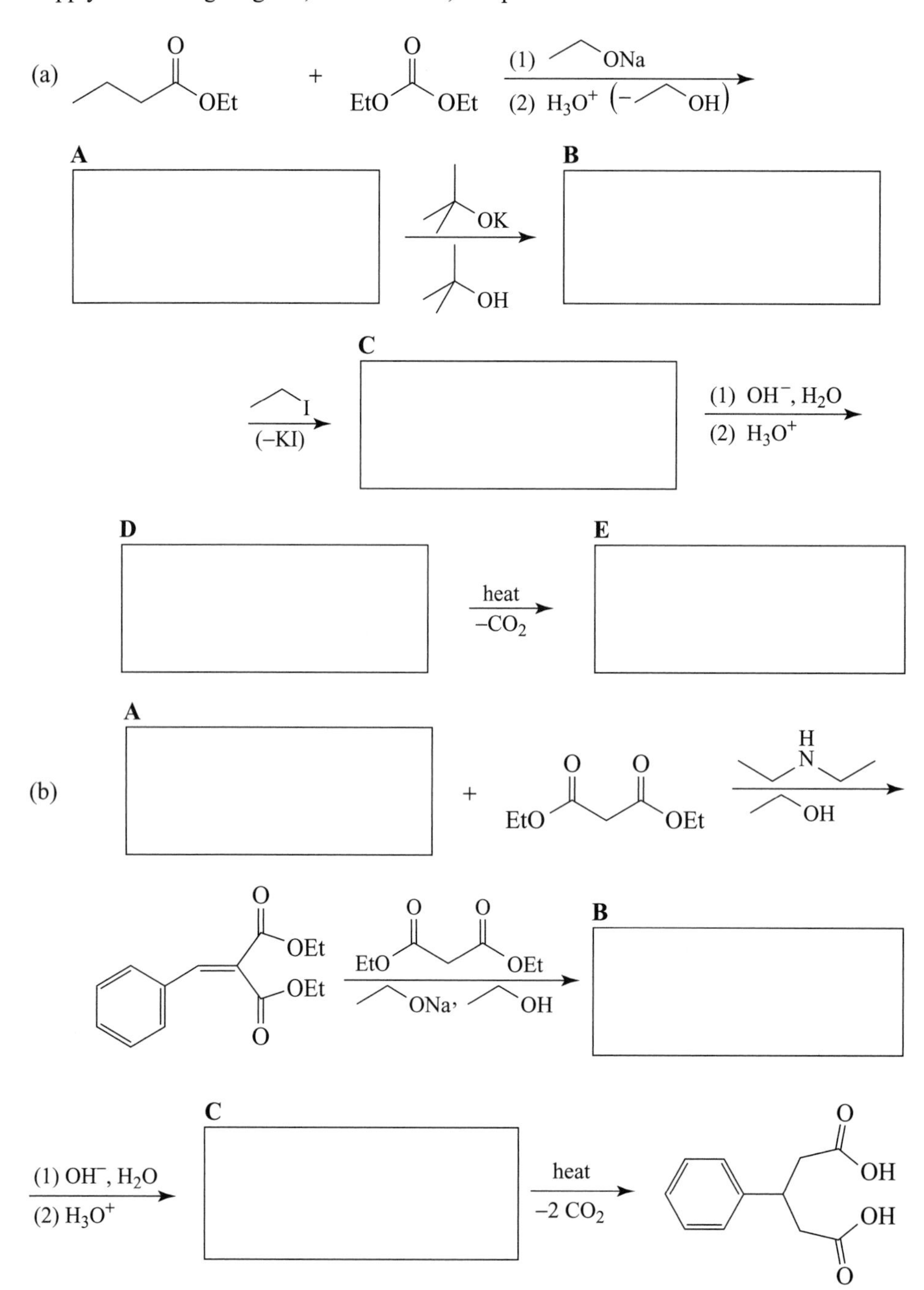

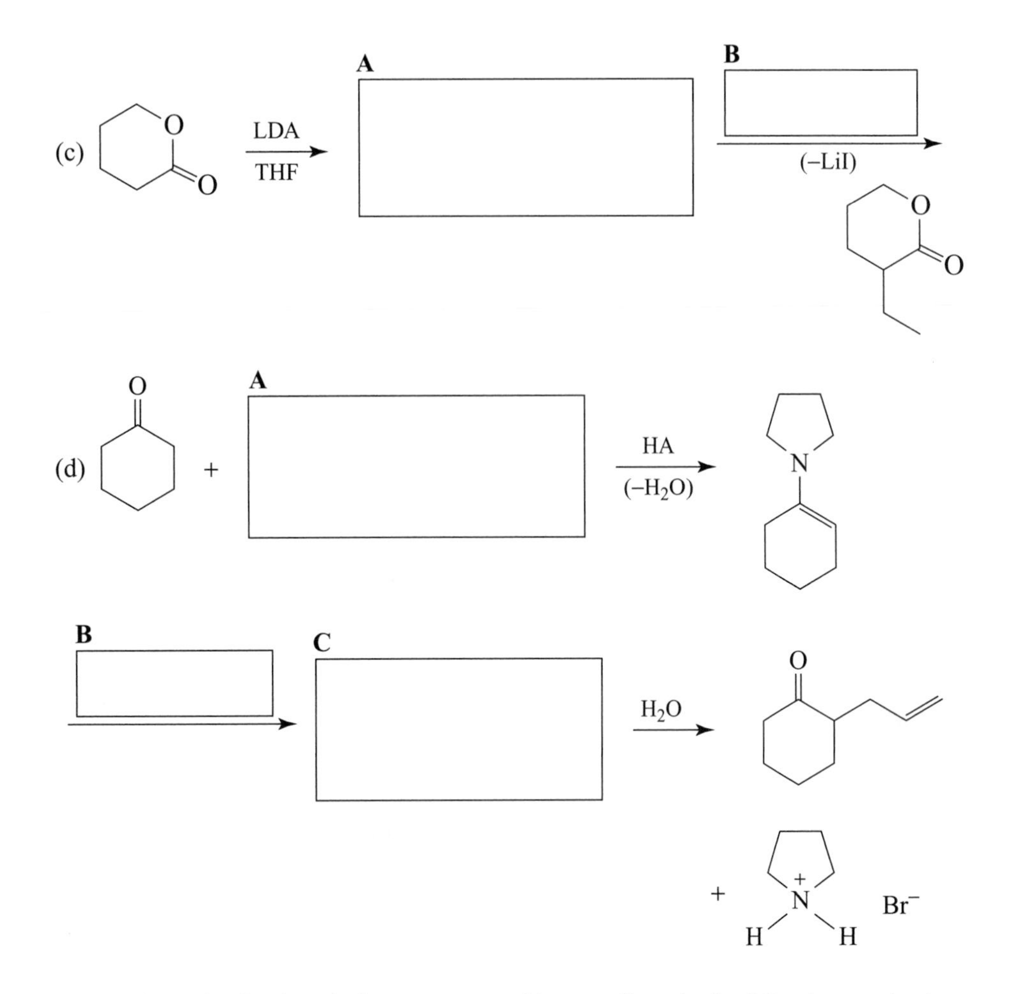

19.5 Supply formulas for the missing reagents and intermediates in the following synthesis.

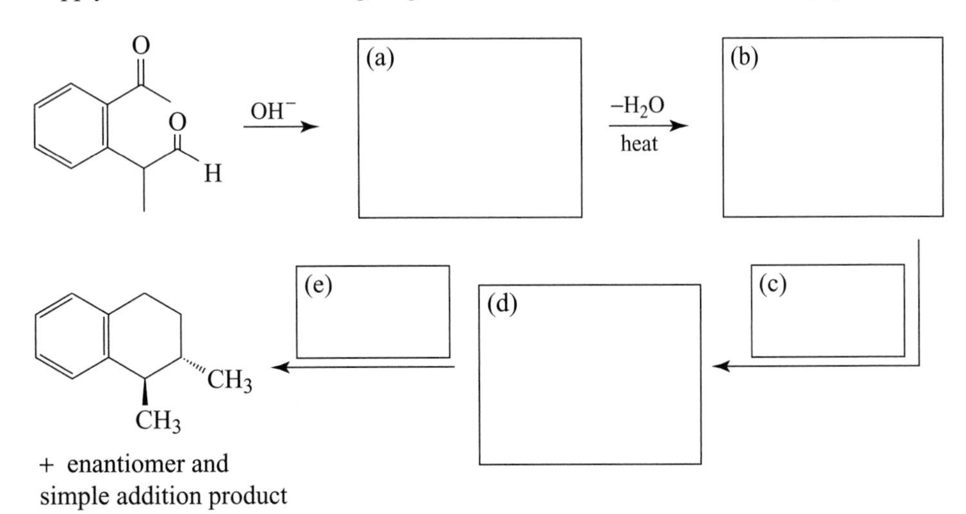

19.6 Which would be formed in the following reaction?

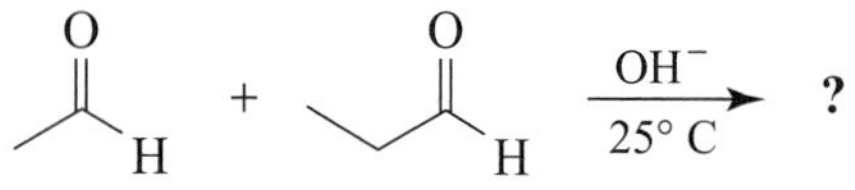

$CH_3CHO + CH_3CH_2CHO \xrightarrow[25° C]{OH^-}$?

(a) $CH_3CH(OH)CH_2CHO$ (b) $CH_3CH_2CH(OH)CH(CH_3)CHO$ (c) $CH_3CH(OH)CH(CH_3)CHO$

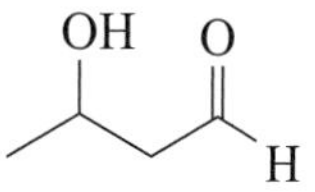

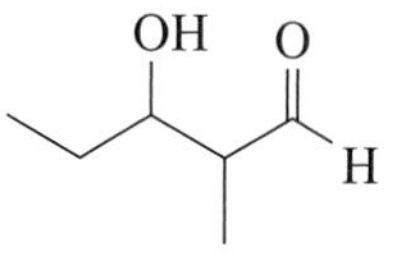

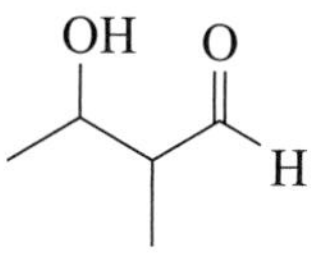

(d) 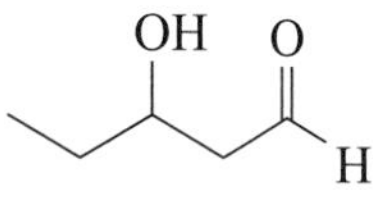 $CH_3CH_2CH(OH)CH_2CHO$ (e) All of these will be formed.

19.7 Supply formulas for the missing reagents and intermediates in the following synthesis.

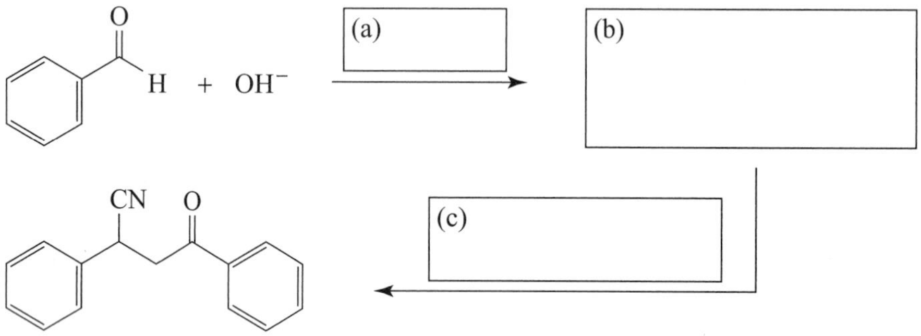

$C_6H_5CHO + OH^-$ → (a) → (b) → (c) → $C_6H_5CH(CN)CH_2COC_6H_5$

19.8 Supply formulas for the missing reagents and intermediates in the following synthesis.

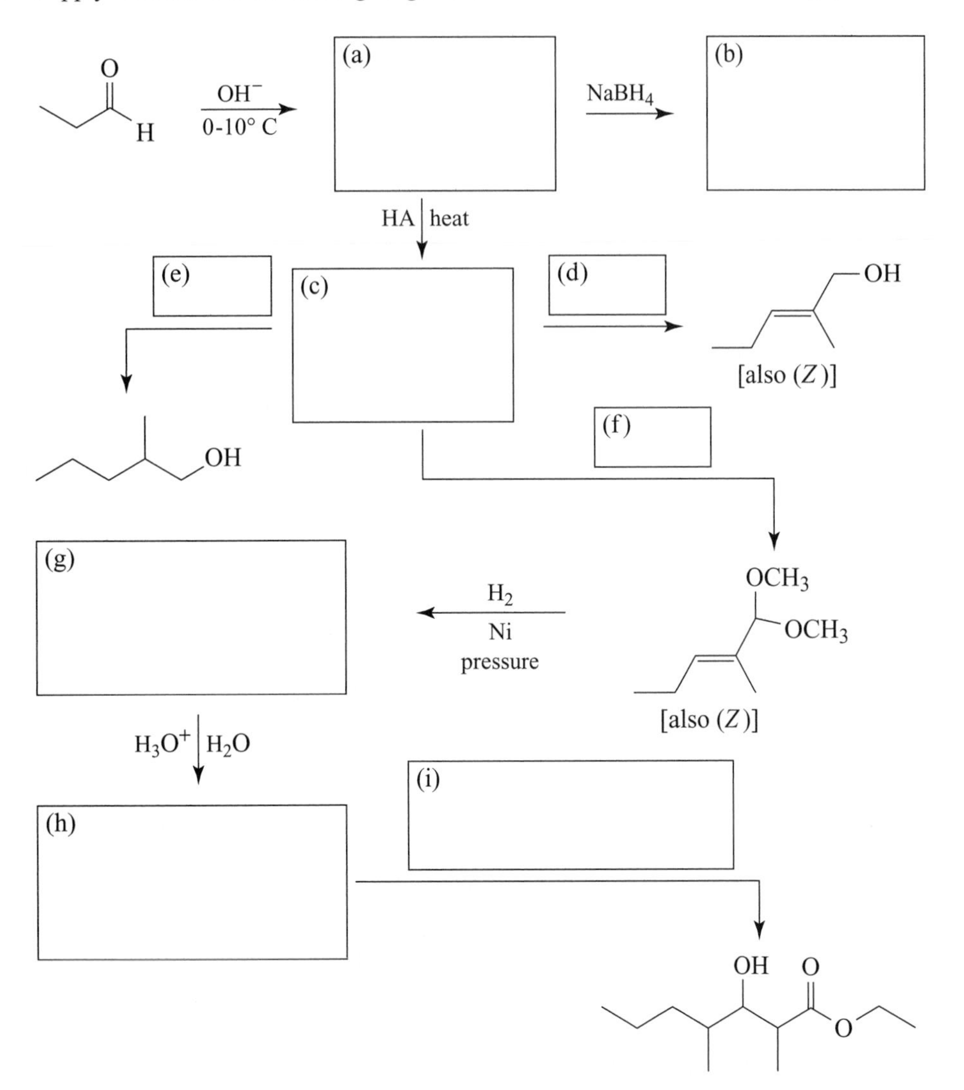

20 AMINES

PREPARATION AND REACTIONS OF AMINES

A. Preparation

1. Preparation via nucleophilic substitution reactions.

$$R\text{-}CH_2Br \xrightarrow{NaN_3,\ EtOH} RCH_2N_3 \xrightarrow[\text{or } LiAlH_4]{Na/EtOH} RCH_2NH_2$$

$$R\text{-}CH_2Br \xrightarrow{\text{potassium phthalimide (NK)}} \text{N-(RCH}_2\text{)phthalimide} \xrightarrow{NH_2NH_2} \text{phthalhydrazide} + RCH_2NH_2$$

$$R\text{-}CH_2Br \xrightarrow{NH_3} RCH_2NH_2 + (RCH_2)_2NH + (RCH_2)_3N \text{ (poor method)}$$

2. Preparation through reduction of nitro compounds.

$$C_6H_6 \xrightarrow[H_2SO_4]{HNO_3} C_6H_5NO_2 \xrightarrow[(2)\ OH^-]{H_2,\ \text{cat. or } (1)\ Fe,\ HCl} C_6H_5NH_2$$

3. Preparation via reductive amination.

$$RR'C{=}O \xrightarrow[\text{[H]}]{NH_3} RR'CH\text{-}NH_2$$

Aldehyde or ketone

$$RR'C{=}O \xrightarrow[\text{[H]}]{R''NH_2} RR'CH\text{-}NHR''$$

$$RR'C{=}O \xrightarrow[\text{[H]}]{R''R'''NH} RR'CH\text{-}NR''R'''$$

4. Preparation of amines through reduction of amides, oximes, and nitriles.

$$RCH_2Br \xrightarrow{CN^-} RCH_2C{\equiv}N \xrightarrow[Ni]{H_2} RCH_2CH_2NH_2$$

$$RR'C{=}O \xrightarrow{NH_2OH} RR'C{=}NOH \xrightarrow{Na/ethanol} RR'CH(NH_2)$$

$$R{-}NH_2 + R'COCl \longrightarrow RNHCOR' \xrightarrow[(2)\ H_2O]{(1)\ LiAlH_4} RNHCH_2R'$$

$$RNHR' + R''COCl \longrightarrow RN(R')COR'' \xrightarrow[(2)\ H_2O]{(1)\ LiAlH_4} RN(R')CH_2R''$$

5. Preparation through the Hofmann rearrangement of amides.

$$RCOOH \xrightarrow{SOCl_2} RCOCl \xrightarrow{NH_3} RCONH_2 \xrightarrow[(NaOBr)]{Br_2/NaOH} RNH_2 + CO_2^{2-}$$

B. Reaction of Amines

1. As a base or a nucleophile

As a base

$$\geq N{:} + H{-}\overset{+}{O}H_2 \rightleftharpoons \geq \overset{+}{N}{-}H + H_2O$$

As a nucleophile in alkylation

$$\geq N{:} + RCH_2{-}X \longrightarrow \geq \overset{+}{N}{-}CH_2R + X^-$$

As a nucleophile in acylation

$$\gt N{-}H{:} + RCOCl \xrightarrow{(-HCl)} \gt N{-}COR$$

2. With nitrous acid

$$R{-}NH_2 \xrightarrow[HX]{HONO} R{-}N_2^+X^- \xrightarrow{-N_2} R^+ \longrightarrow$$ alkenes, alcohols, and alkyl halides

1° aliphatic (unstable)

$$\underset{\textit{1° aromatic}}{Ar-NH_2} \xrightarrow[HX\ (0\text{–}5°C)]{HONO} Ar-N_2^+\,X^-$$

$$Ar-N_2^+X^- \xrightarrow{CuCl} ArCl + N_2$$

$$Ar-N_2^+X^- \xrightarrow{CuBr} ArBr + N_2$$

$$Ar-N_2^+X^- \xrightarrow{CuCN} ArCN + N_2$$

$$Ar-N_2^+X^- \xrightarrow{KI} ArI + N_2$$

$$Ar-N_2^+X^- \xrightarrow{HBF_4} ArN_2^+\,BF_4^- \xrightarrow{\Delta} \left\{ \begin{array}{l} ArF + \\ N_2 + \\ BF_3 \end{array} \right.$$

$$Ar-N_2^+X^- \xrightarrow[Cu^{2+},\ H_2O]{Cu_2O} ArOH + N_2$$

$$Ar-N_2^+X^- \xrightarrow{H_3PO_2} ArH + N_2$$

$$Ar-N_2^+X^- \xrightarrow{C_6H_5-OH} Ar-N{=}N-C_6H_4-OH$$

$$Ar-N_2^+X^- \xrightarrow{C_6H_5-NR_2} Ar-N{=}N-C_6H_4-NR_2$$

$$\underset{\textit{2° aliphatic}}{R_2NH} \xrightarrow{HONO} R_2N-N{=}O$$

$$\underset{\textit{2° aromatic}}{ArNHR} \xrightarrow{HONO} ArN(-N{=}O)-R$$

$$\underset{\textit{3° aliphatic}}{R_3N} \xrightarrow[HX]{NaNO_2} R_3NH^+\,X^- + R_3\overset{+}{N}-N{=}O\;X^-$$

$$\underset{\textit{3° aromatic}}{R_2N-C_6H_5} \xrightarrow{HONO} R_2N-C_6H_4-N{=}O$$

3. With sulfonyl chlorides

$$\underset{\textit{1° amine}}{R-NH_2} + ArSO_2Cl \xrightarrow[(-HCl)]{} RNHSO_2Ar \underset{H_3O^+}{\overset{OH^-}{\rightleftharpoons}} [RNSO_2Ar]^- + H_2O$$

$$\underset{\textit{2° amine}}{R_2NH} + ArSO_2Cl \xrightarrow[(-HCl)]{} R_2NSO_2Ar$$

4. The Hofmann elimination

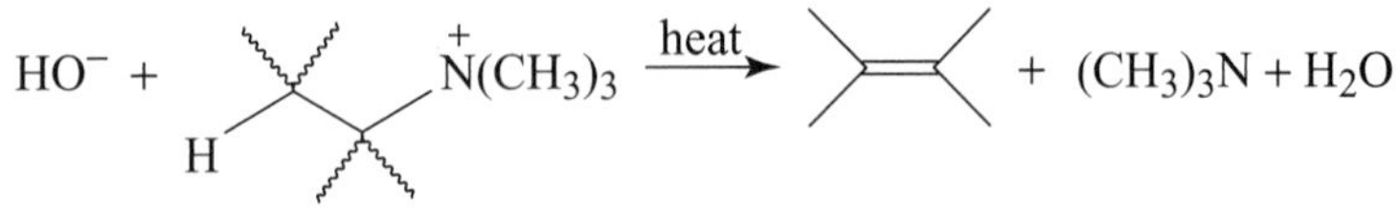

$$HO^- + \text{(H–C–C–}\overset{+}{N}(CH_3)_3) \xrightarrow{\text{heat}} \text{C=C} + (CH_3)_3N + H_2O$$

SOLUTIONS TO PROBLEMS

20.1 Dissolve both compounds in diethyl ether and extract with aqueous HCl. This procedure gives an ether layer that contains cyclohexane and an aqueous layer that contains hexylaminium chloride. Cyclohexane may then be recovered from the ether layer by distillation. Hexylamine may be recovered from the aqueous layer by adding aqueous NaOH (to convert hexylaminium chloride to the hexylamine) and then by ether extraction and distillation.

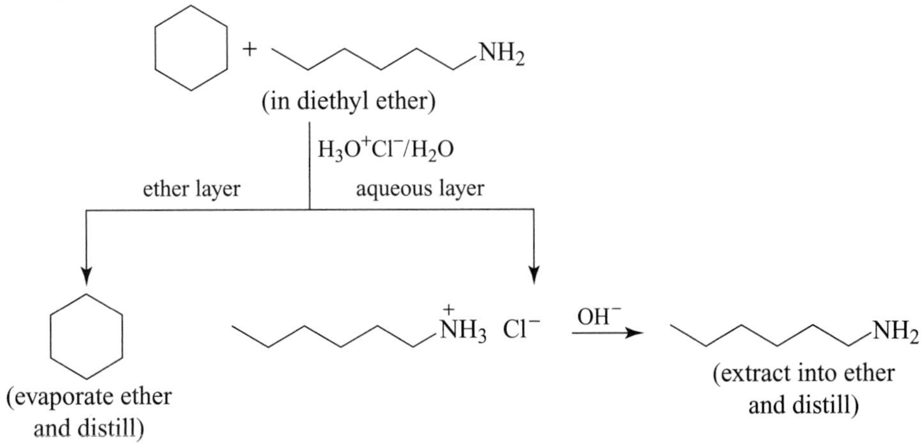

Cyclohexane + hexylamine (NH_2)
(in diethyl ether)
$H_3O^+Cl^-/H_2O$
ether layer → cyclohexane (evaporate ether and distill)
aqueous layer → hexyl-$\overset{+}{N}H_3$ Cl^- $\xrightarrow{OH^-}$ hexyl-NH_2 (extract into ether and distill)

20.2 We begin by dissolving the mixture in a water-immiscible organic solvent such as CH_2Cl_2 or diethyl ether. Then, extractions with aqueous acids and bases allow us to separate the components. [We separate 4-methylphenol (*p*-cresol) from benzoic acid by taking advantage of benzoic acid's solubility in the more weakly basic aqueous $NaHCO_3$, whereas *p*-cresol requires the more strongly basic, aqueous NaOH.]

(in CH_2Cl_2)

$NaHCO_3/H_2O$

aqueous layer

CH_2Cl_2 layer

O^-Na^+

H_3O^+

Separate and recrystallize

$NaOH, H_2O$

aqueous layer

CH_2Cl_2 layer

O^-Na^+

CH_3

H_3O^+

Extract into CH_2Cl_2 and distill

$H_3O^+Cl^-/H_2O$

aqueous layer

CH_2Cl_2 layer

$\overset{+}{N}H_3\ Cl^-$

OH^-

Extract into CH_2Cl_2 and distill

Isolate by distillation

20.3 (a) Neglecting Kekulé forms of the ring, we can write the following resonance structures for the phthalimide anion.

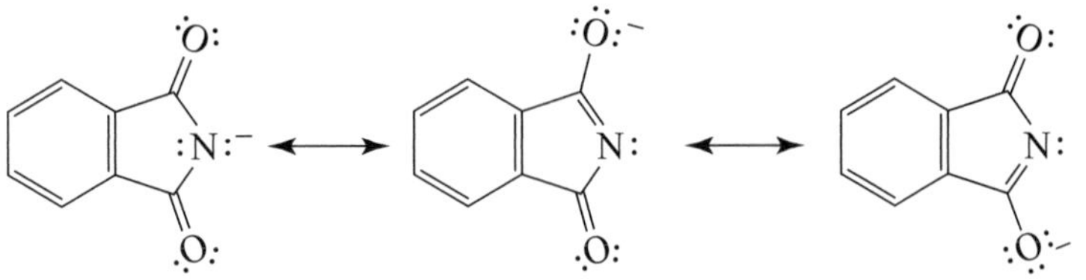

$\ddot{O}$: :N:$^-$ $\longleftrightarrow$:$\ddot{O}$:$^-$ N: $\longleftrightarrow$ $\ddot{O}$: N: :$\ddot{O}$:$^-$

(b) Phthalimide is more acidic than benzamide because its anion is stabilized by resonance to a greater extent than the anion of benzamide. (Benzamide has only one carbonyl group attached to the nitrogen atom, and thus fewer resonance contributors are possible.)

(c)

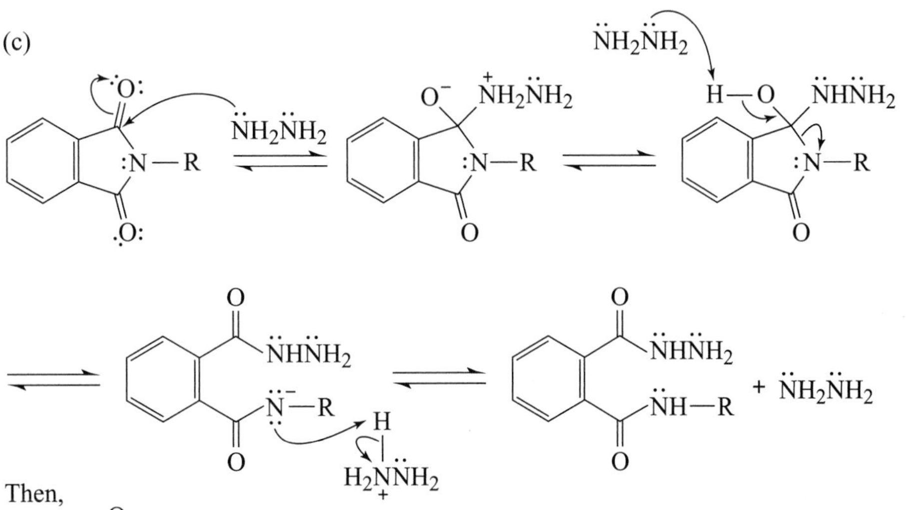

$\ddot{O}$: :N—R :$\ddot{O}$: + $\ddot{N}H_2\ddot{N}H_2$ $\rightleftharpoons$ O^- $\overset{+}{N}H_2\ddot{N}H_2$:N—R O $\rightleftharpoons$ ($\ddot{N}H_2\ddot{N}H_2$) H—O $\ddot{N}H\ddot{N}H_2$:N—R O

$\rightleftharpoons$ O $\ddot{N}H\ddot{N}H_2$ $\ddot{N}^-$—R O + H—$\overset{+}{N}H_2\ddot{N}H_2$ $\rightleftharpoons$ O $\ddot{N}H\ddot{N}H_2$ $\ddot{N}H$—R O + $\ddot{N}H_2\ddot{N}H_2$

Then,

O $\ddot{N}H\ddot{N}H_2$ $\ddot{N}H$—R O $\rightleftharpoons$ O NH NH_2^+ ^-O $\ddot{N}HR$ $\rightleftharpoons$

O N-H N-H ^-O $\overset{+}{N}H_2R$ $\rightleftharpoons$ O N—H N—H O + RNH_2

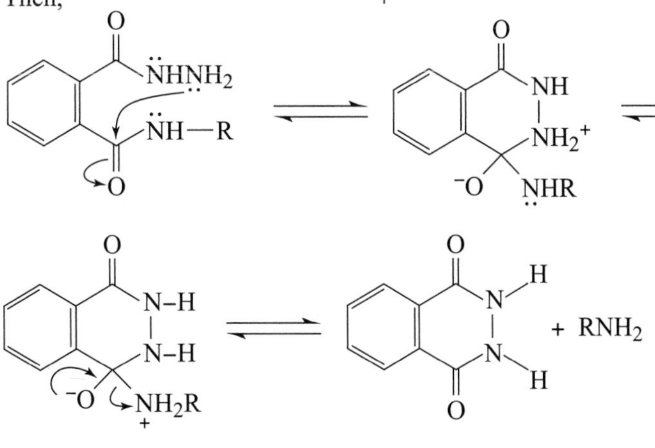

20.4 Phthalimide (:NH) $\xrightarrow{KOH}$:N:⁻ K^+ $\xrightarrow[(-KBr)]{C_6H_5CH_2Br}$ N-benzylphthalimide $\xrightarrow[\text{reflux}]{NH_2NH_2,\ CH_3CH_2OH}$ $C_6H_5CH_2NH_2$ + phthalhydrazide

Benzylamine

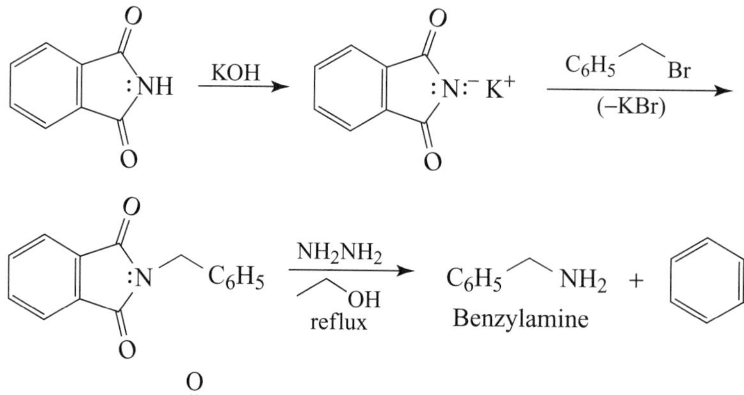

20.5 (a) Pentanal (H) + NH_3 $\xrightarrow[\text{Ni, pressure}]{H_2}$ pentylamine (NH_2)

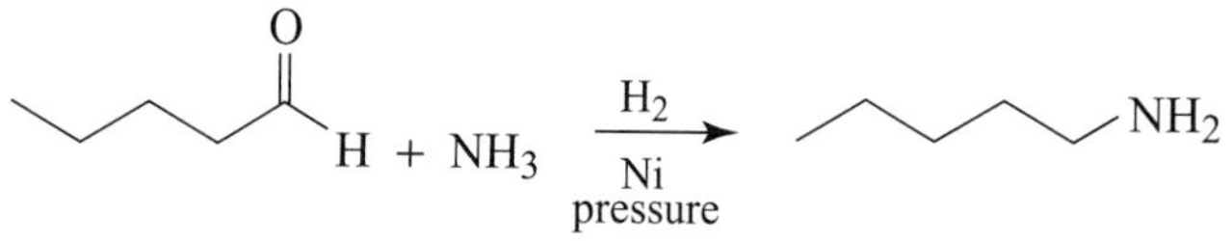

(b) Hexanal (H) + $C_6H_5NH_2$ $\xrightarrow[CH_3OH]{LiBH_3CN}$ N-hexylaniline (N–H, C_6H_5)

(c) C_6H_5CHO + $(CH_3)_2NH$ $\xrightarrow[CH_3OH]{LiBH_3CN}$ $C_6H_5CH_2N(CH_3)_2$

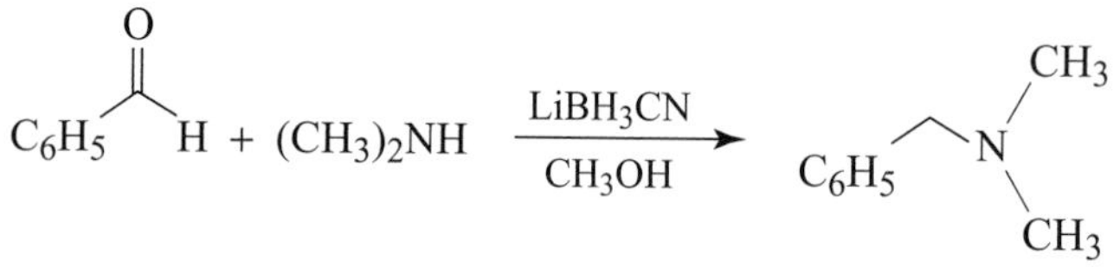

20.6 (a) C_6H_5COOH $\xrightarrow{SOCl_2}$ C_6H_5COCl $\xrightarrow{CH_3CH_2NH_2}$ $C_6H_5C(O)NHCH_2CH_3$ $\xrightarrow{LiAlH_4}$ $C_6H_5CH_2NHCH_2CH_3$

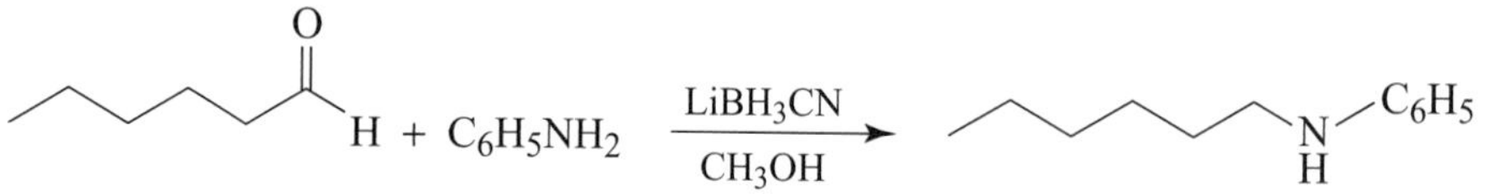

(b) 1-bromopentane (Br) $\xrightarrow{NaCN}$ hexanenitrile (C≡N) $\xrightarrow{LiAlH_4}$ hexylamine (NH_2)

(c) Propanoic acid (OH) $\xrightarrow{SOCl_2}$ propanoyl chloride (Cl) $\xrightarrow{\text{dipropylamine (N–H)}}$ N,N-dipropylpropanamide $\xrightarrow{LiAlH_4}$ tripropylamine (N)

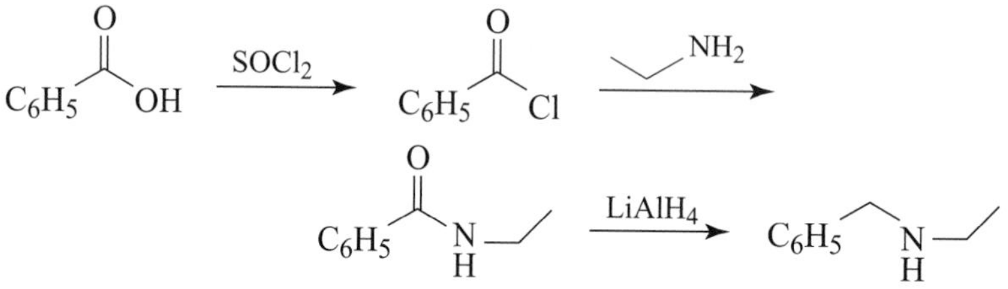

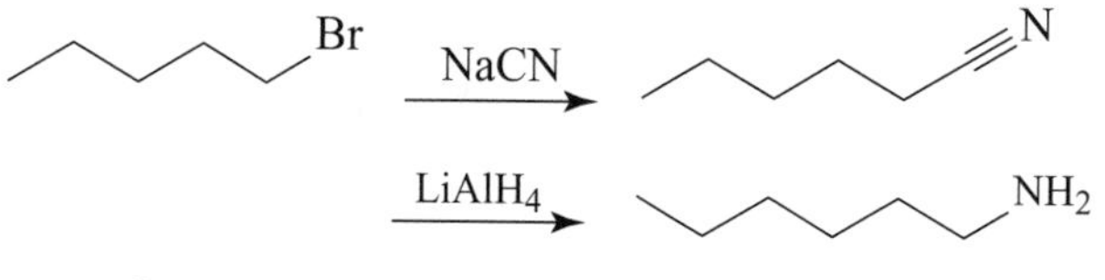

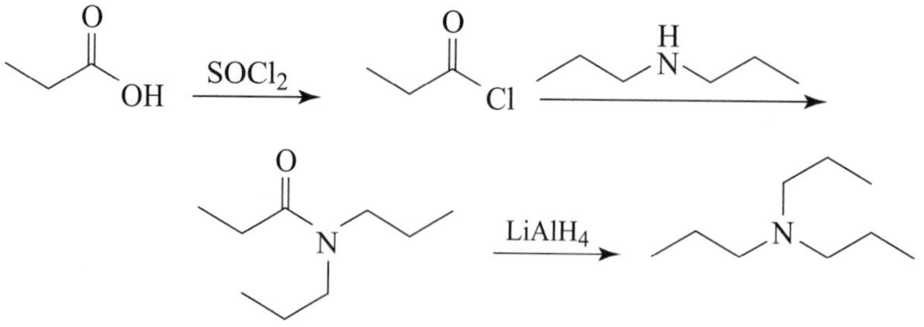

(d) Butanone $\xrightarrow{NH_2OH}$ oxime (HO–N=) $\xrightarrow{Na/CH_3CH_2OH}$ 2-butylamine (NH_2)

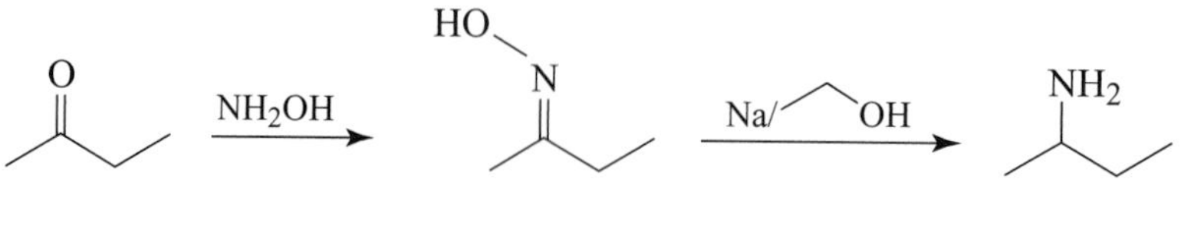

20.7 (a) CH_3O–C$_6$H$_5$ $\xrightarrow[H_2SO_4]{HNO_3}$ CH_3O–C$_6$H$_4$–NO_2 $\xrightarrow[HCl]{Fe}$ CH_3O–C$_6$H$_4$–NH_2

(b) CH_3O–C$_6$H$_5$ $\xrightarrow[AlCl_3]{CH_3COCl}$ CH_3O–C$_6$H$_4$–$COCH_3$ $\xrightarrow[Ni]{NH_3, H_2}$ CH_3O–C$_6$H$_4$–$CH(NH_2)CH_3$

(c) C$_6$H$_5$–CH_3 (excess) $\xrightarrow[hv]{Cl_2}$ C$_6$H$_5$–CH_2Cl $\xrightarrow{(CH_3)_3N}$ C$_6$H$_5$–$CH_2N^+(CH_3)_3$ Cl^-

(d) O_2N–C$_6$H$_4$–CH_3 $\xrightarrow[(2)\ H_3O^+]{(1)\ KMnO_4,\ OH^-}$ O_2N–C$_6$H$_4$–$COOH$ $\xrightarrow{SOCl_2}$

O_2N–C$_6$H$_4$–$COCl$ $\xrightarrow{NH_3}$ O_2N–C$_6$H$_4$–$CONH_2$ $\xrightarrow[OH^-]{Br_2}$ O_2N–C$_6$H$_4$–NH_2

(e) C$_6$H$_5$–CH_3 + NBS $\xrightarrow[hv]{ROOR}$ C$_6$H$_5$–CH_2Br $\xrightarrow{KCN}$

C$_6$H$_5$–$CH_2C{\equiv}N$ $\xrightarrow[(2)\ H_2O]{(1)\ LiAlH_4,\ Et_2O}$ C$_6$H$_5$–$CH_2CH_2NH_2$

20.8 An amine acting as a base.

$CH_3CH_2\ddot{N}H_2 + H{-}\ddot{O}^+{-}H$ (with H) $\rightleftharpoons$ $CH_3CH_2\overset{+}{N}H_3 + H{-}\ddot{O}{-}H$

An amine acting as a nucleophile in an alkylation reaction.

$(CH_3CH_2)_3\ddot{N} + CH_3{-}I \longrightarrow (CH_3CH_2)_3\overset{+}{N}CH_3\ \ I^-$

An amine acting as a nucleophile in an acylation reaction.

$$(CH_3)_2\ddot{N}H + CH_3C(=\ddot{O}:)Cl \longrightarrow (CH_3)_2NC(=O)CH_3 + (CH_3)_2\overset{+}{N}H_2\ Cl^-$$

(excess)

An amino group acting as an activating group and as an ortho-para director in electrophilic aromatic substitution.

$$C_6H_5NH_2 \xrightarrow[\text{room temp.}]{Br_2,\ H_2O} \text{2,4,6-}Br_3C_6H_2NH_2$$

20.9 (a, b) $^{-}O—N{=}O + H_3O^+ \rightleftharpoons HO—N{=}O + H_2O$

$$HO—N{=}O + H_3O^+ \rightleftharpoons H\overset{+}{O}(H)—N{=}O + H_2O$$

$$H\overset{+}{O}(H)—N{=}O \rightleftharpoons H_2O + \overset{+}{N}{=}O$$

$$C_6H_5\ddot{N}(CH_3)_2 + \overset{+}{N}{=}O \longrightarrow (CH_3)_2\overset{+}{N}{=}C_6H_5(H)(NO) \xrightarrow{:A^-} p\text{-}ON\text{-}C_6H_4\ddot{N}(CH_3)_2 + HA$$

(c) The $\overset{+}{N}O$ ion is a weak electrophile. For it to react with an aromatic ring, the ring must have a powerful activating group such as $-OH$ or $-NR_2$.

20.10 (a)

$$C_6H_6 \xrightarrow[H_2SO_4]{HNO_3} C_6H_5NO_2 \xrightarrow[Fe]{Cl_2} m\text{-}ClC_6H_4NO_2 \xrightarrow[(2)\ OH^-]{(1)\ Fe,\ HCl,\ heat} m\text{-}ClC_6H_4NH_2$$

(b)

$$C_6H_5NO_2 \xrightarrow[FeBr_3]{Br_2} m\text{-}BrC_6H_4NO_2 \xrightarrow[(2)\ OH^-]{(1)\ Fe,\ HCl,\ heat} m\text{-}BrC_6H_4NH_2$$

[from part (a)]

(c) NO_2-benzene [as in (a)] $\xrightarrow[(2)\ OH^-]{(1)\ Fe,\ HCl,\ heat}$ aniline (NH_2) $\xrightarrow{(CH_3CO)_2O}$ acetanilide (H–N–C(=O)CH$_3$) $\xrightarrow[H_2SO_4]{concd}$

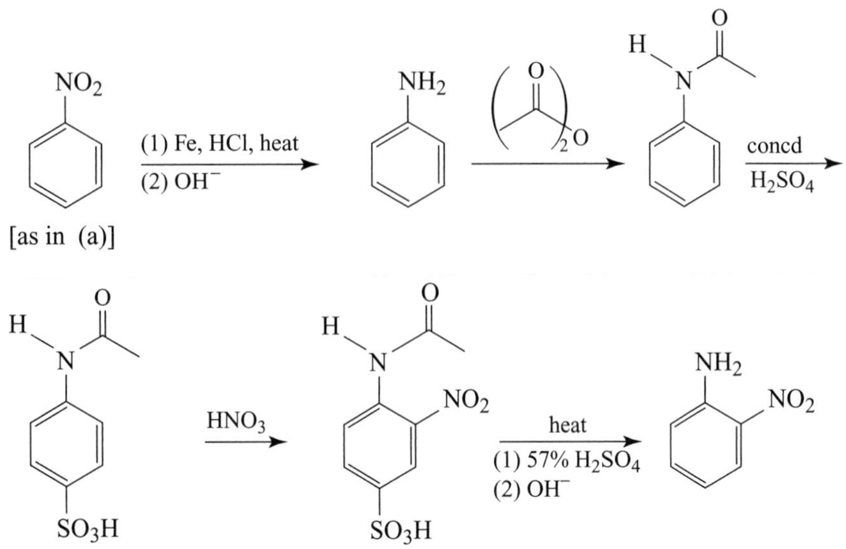

acetanilide with SO_3H para $\xrightarrow{HNO_3}$ NO_2 ortho to NHCOCH$_3$, SO_3H para $\xrightarrow[(1)\ 57\%\ H_2SO_4\ (2)\ OH^-]{heat}$ o-nitroaniline (NH_2, NO_2)

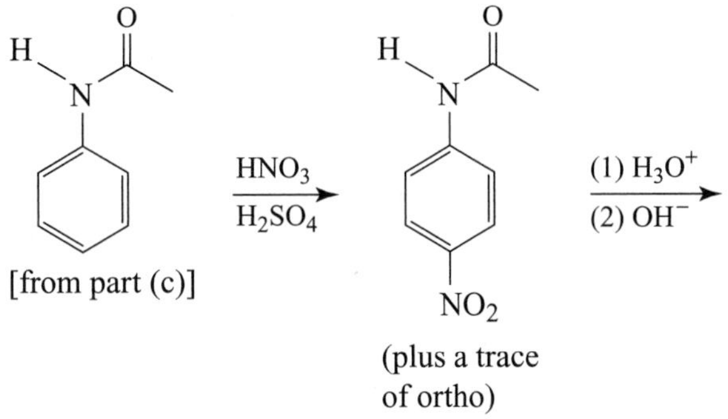

(d) acetanilide [from part (c)] $\xrightarrow[H_2SO_4]{HNO_3}$ p-nitroacetanilide (NO_2) (plus a trace of ortho) $\xrightarrow[(2)\ OH^-]{(1)\ H_3O^+}$ p-nitroaniline (NH_2, NO_2)

20.11 (a) Toluene $\xrightarrow[H_2SO_4]{HNO_3}$ *p*-Nitrotoluene (+ *o*-nitrotoluene) $\xrightarrow[(2)\ OH^-]{(1)\ Fe,\ HCl}$ p-toluidine (CH_3, NH_2) $\xrightarrow{(CH_3CO)_2O}$

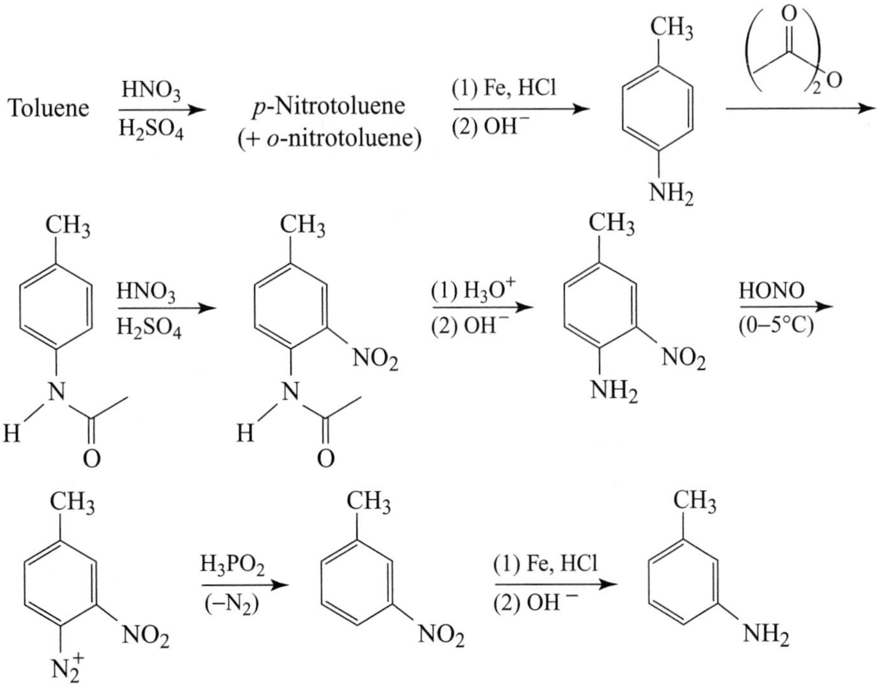

p-methylacetanilide (CH_3; N(H)C(=O)CH$_3$) $\xrightarrow[H_2SO_4]{HNO_3}$ (CH_3; NO_2 ortho to NHCOCH$_3$) $\xrightarrow[(2)\ OH^-]{(1)\ H_3O^+}$ (CH_3; NH_2; NO_2) $\xrightarrow[(0–5°C)]{HONO}$

(CH_3; N_2^+; NO_2) $\xrightarrow[(-N_2)]{H_3PO_2}$ m-nitrotoluene (CH_3; NO_2) $\xrightarrow[(2)\ OH^-]{(1)\ Fe,\ HCl}$ m-toluidine (CH_3; NH_2)

CH_3, NH_2 $\xrightarrow[(0\text{–}5°C)]{HONO}$ CH_3, N_2^+

(b) CuCl → CH_3, Cl

(c) CuBr → CH_3, Br

(d) KI → CH_3, I

(e) CuCN → CH_3, N

H_3O^+, H_2O ↓

(f) CH_3, OH, O

20.12 NH_2, NO_2 $\xrightarrow[FeBr_3]{Br_2}$ NH_2, Br, Br, NO_2 $\xrightarrow[(0\text{–}5°C)]{H_2SO_4/NaNO_2}$ N_2^+, Br, Br, NO_2 $\xrightarrow{CuBr}$

Br, Br, Br, NO_2 $\xrightarrow[Pt]{H_2}$ Br, Br, Br, NH_2 $\xrightarrow[(0\text{–}5°C)]{H_2SO_4/NaNO_2}$ Br, Br, Br, N_2^+

$\xrightarrow[-N_2]{H_3PO_2}$ Br, Br, Br

20.13 $HO_3S-C_6H_4-NH_2 \xrightarrow[(0\text{–}5°C)]{H_2SO_4/NaNO_2} HO_3S-C_6H_4-N_2^+$

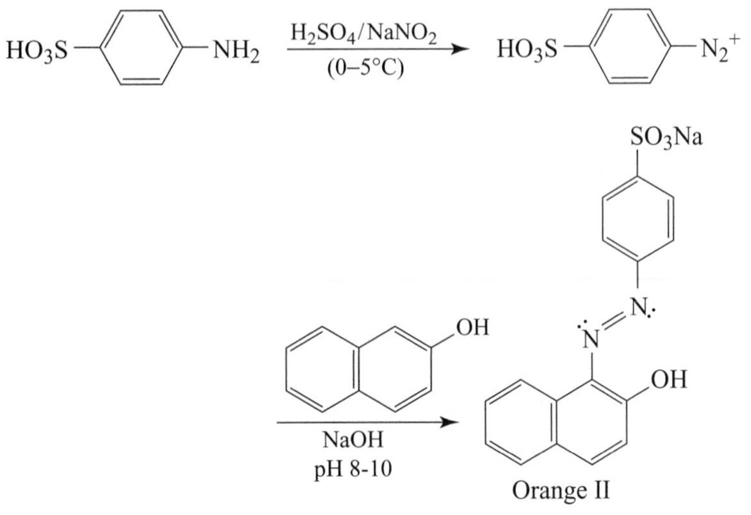

2-naphthol (OH), NaOH, pH 8-10 → Orange II (SO_3Na, N=N, OH)

Orange II

20.14 benzene $\xrightarrow[H_2SO_4]{HNO_3}$ NO_2-benzene $\xrightarrow[(2)\ OH^-]{(1)\ Fe,\ HCl}$ NH_2-benzene $\xrightarrow[(0\text{–}5°C)]{H_2SO_4/NaNO_2}$ N_2^+-benzene

$N(CH_3)_2$-benzene, pH 5–7 → N=N, $N(CH_3)_2$

Butter yellow

20.15 NH_2 (O-ethyl) $\xrightarrow[(0\text{–}5°C)]{H_2SO_4/NaNO_2}$ N_2^+ (O-ethyl); phenol (OH), OH^- →

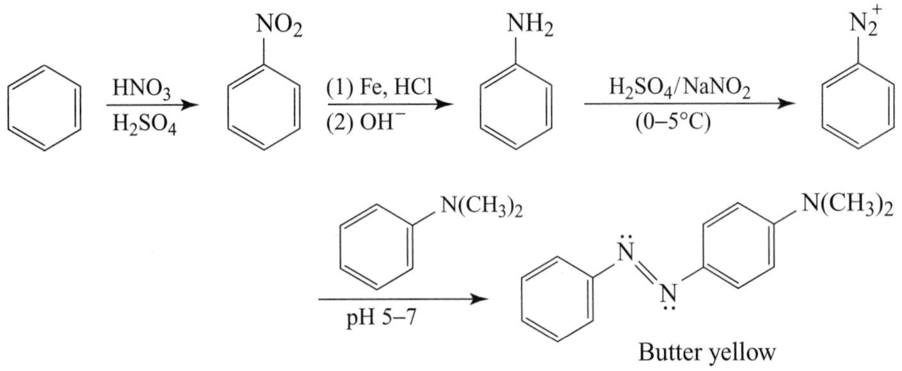

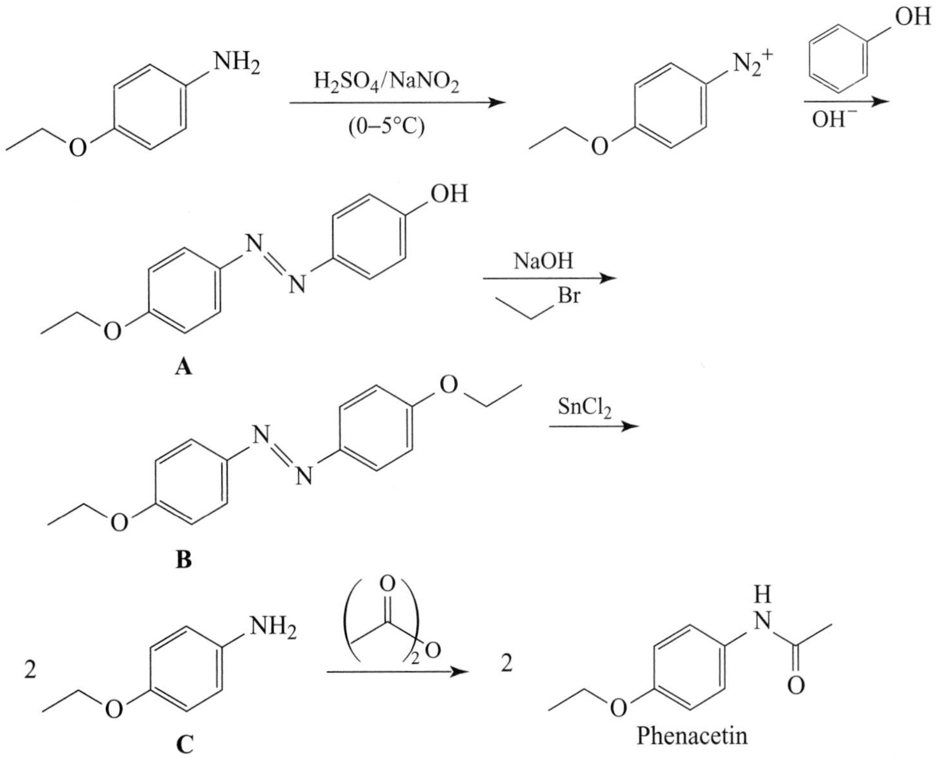

A (O-ethyl, N=N, OH) $\xrightarrow{NaOH,\ Br}$

A

B (O-ethyl, N=N, O-ethyl) $\xrightarrow{SnCl_2}$

B

2 C (NH_2, O-ethyl) $\xrightarrow{(CH_3CO)_2O}$ 2 Phenacetin (O-ethyl, N-H, C=O)

C

Phenacetin

20.16 (1) That **A** reacts with benzenesulfonyl chloride in aqueous KOH to give a clear solution, which on acidification yields a precipitate, shows that **A** is a primary amine.

(2) That diazotization of **A** followed by treatment with 2-naphthol gives an intensely colored precipitate shows that **A** is a primary aromatic amine; that is, **A** is a substituted aniline.

(3) Consideration of the molecular formula of **A** leads us to conclude that **A** is a methylaniline (i.e., a toluidine).

$$\frac{\begin{array}{r} C_7H_9N \\ -C_6H_6N \end{array}}{CH_3} = CH_3\text{–}C_6H_4\text{–}NH_2$$

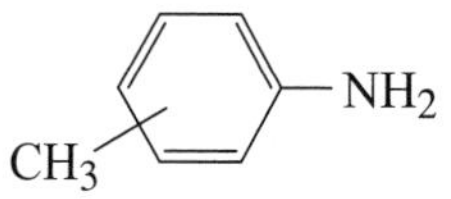

But is **A**, 2-methylaniline, 3-methylaniline, or 4-methylaniline?

(4) This question is answered by the IR data. A single absorption peak in the 680–840 cm^{-1} region at 815 cm^{-1} is indicative of a para-substituted benzene. Thus, **A** is 4-methylaniline (*p*-toluidine).

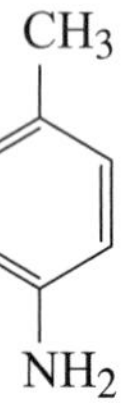

CH_3

NH_2 **A**

20.17 First convert the sulfonamide to its anion, then alkylate the anion with an alkyl halide, then remove the $-SO_2C_6H_5$ group by hydrolysis. For example,

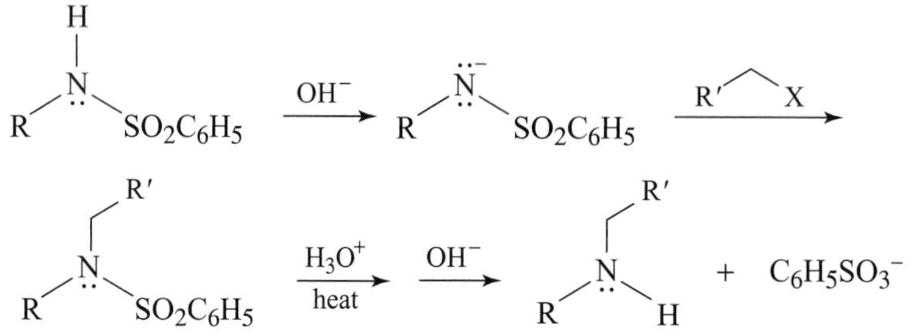

$$R\text{–}\ddot{N}H\text{–}SO_2C_6H_5 \xrightarrow{OH^-} R\text{–}\ddot{N}^{-}\text{–}SO_2C_6H_5 \xrightarrow{R'CH_2X}$$

$$R\text{–}\ddot{N}(CH_2R')\text{–}SO_2C_6H_5 \xrightarrow[heat]{H_3O^+} \xrightarrow{OH^-} R\text{–}\ddot{N}(CH_2R')\text{–}H + C_6H_5SO_3^-$$

20.18 (a)

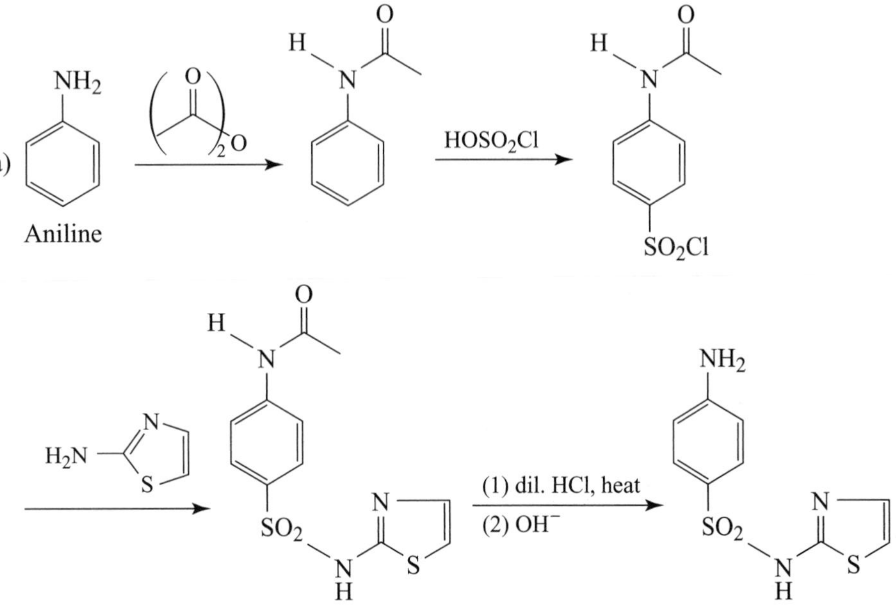

NH_2

Aniline

$\xrightarrow{((CH_3C(O))_2O}$ H–N–C(=O)–CH_3 (O)

$\xrightarrow{HOSO_2Cl}$ H–N–C(=O)–CH_3; SO_2Cl

$\xrightarrow{H_2N\text{–thiazole (N, S)}}$ H–N (O); SO_2–N(H)–thiazole (N, S)

$\xrightarrow[(2)\ OH^-]{(1)\ \text{dil. HCl, heat}}$ NH_2; SO_2–N(H)–thiazole (N, S)

Sulfathiazole

(b) $\xrightarrow{\text{succinic anhydride (O, O, O)}}$ H–N–C(=O)CH_2CH_2C(=O)OH; SO_2–N(H)–thiazole (N, S)

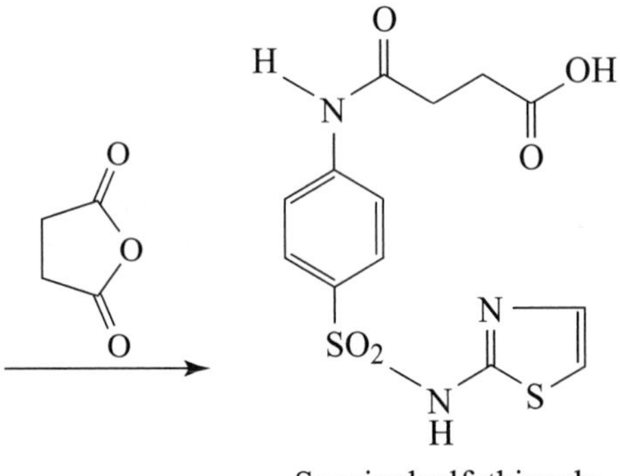

Succinylsulfathiazole

20.19 (a) $C_6H_5CH_2$–N(H)–CH_3

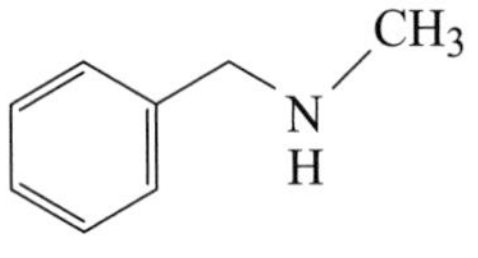

(b) $((CH_3)_2CH)_3N$

(c) C_6H_5–N(CH_3)–CH_2CH_3

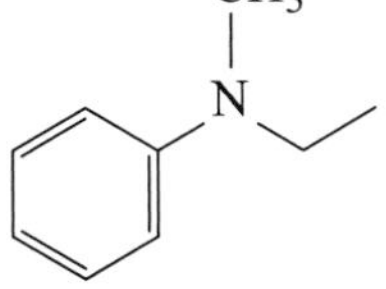

(d) CH_3, NH_2

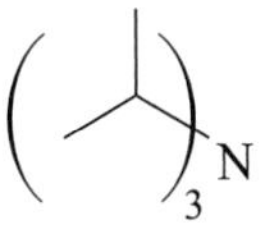

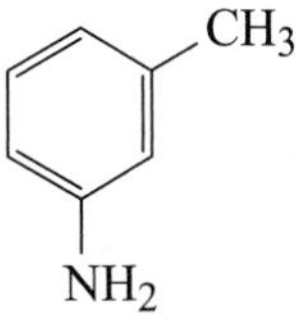

(e) N–H, CH_3

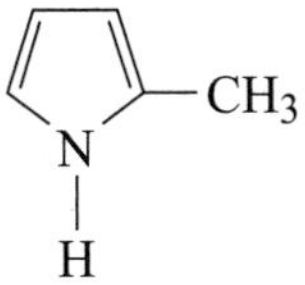

(f) N–CH_2CH_3

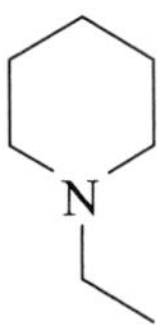

(g) 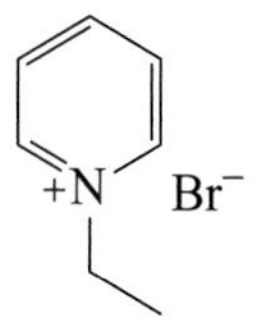 $+N$ Br^-

(h) 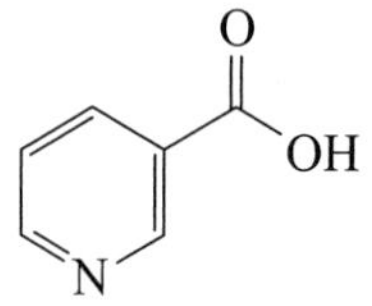 N, O, OH

(i) N, H

(j) H, N, O

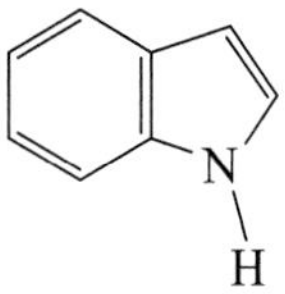

(k) $CH_3—N^+—H$ Cl^-, H, CH_3

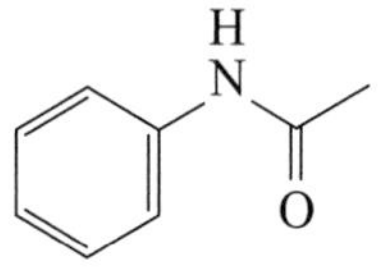

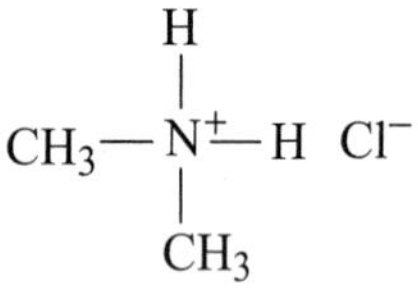

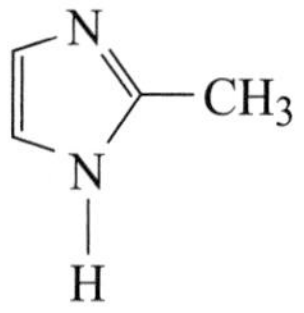

(l) N, N, H, CH_3

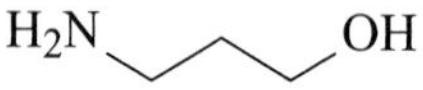

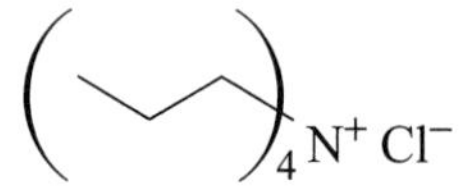

(m) H_2N OH

(n) $_4N^+$ Cl^-

(o) N, H

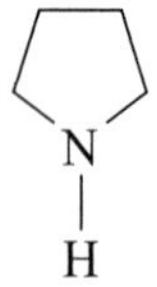

(p) 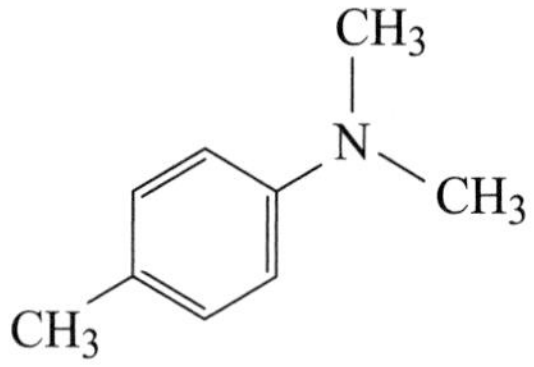 CH_3, N, CH_3, CH_3

(q) 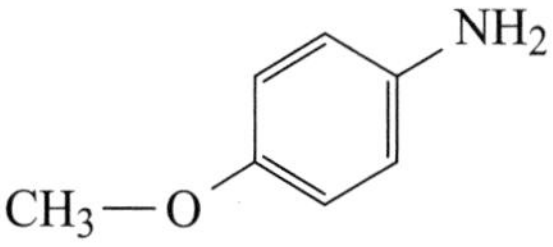 NH_2, $CH_3—O$

(r) $(CH_3)_4N^+\ OH^-$

(s) 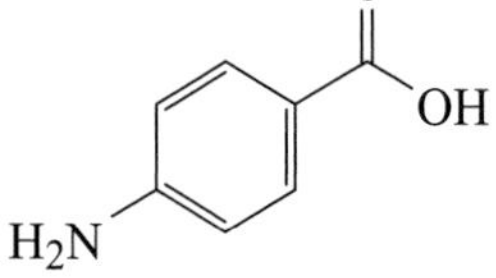 O, OH, H_2N

(t) 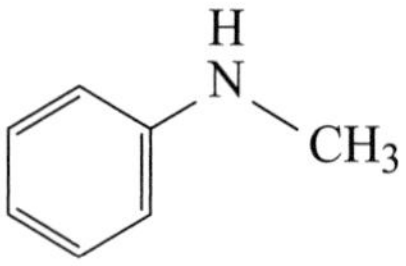 H, N, CH_3

20.20 (a) Propylamine (or 1-propanamine)
(b) *N*-Methylaniline
(c) Isopropyltrimethylammonium iodide
(d) 2-Methylaniline (*o*-toluidine)
(e) 2-Methoxyaniline (or *o*-methoxyaniline)
(f) Pyrazole
(g) 2-Aminopyrimidine
(h) Benzylaminium chloride
(i) *N*,*N*-Dipropylaniline
(j) Benzenesulfonamide
(k) Methylaminium acetate
(l) 3-Amino-1-propanol (or 3-aminopropan-1-ol)
(m) Purine
(n) *N*-Methylpyrrole

Amine Synthesis and Reactivity

20.21

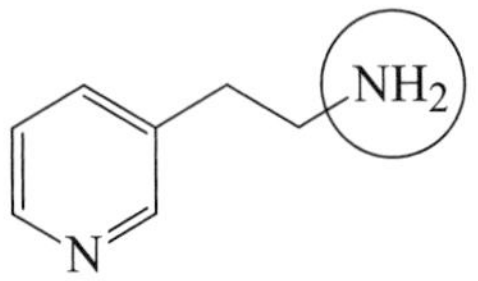

NH_2 H_2N N HN N

sp^3 hybridization stabilizes conjugate acid.

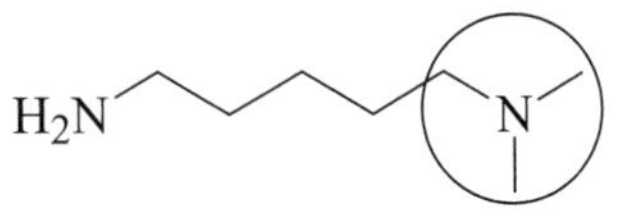

Alkyl groups are electron donating, stabilizing conjugate acid.

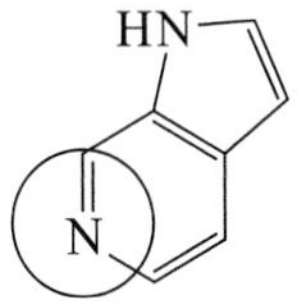

Lone pair is not involved in aromaticity.

20.22 (a) N + $LiAlH_4$ $\xrightarrow{(2)\ H_2O}$ NH_2

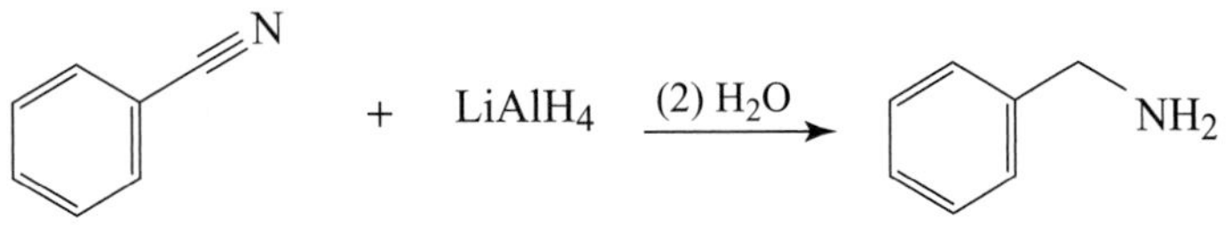

(b) O NH_2 + $LiAlH_4$ $\xrightarrow{(2)\ H_2O}$ NH_2

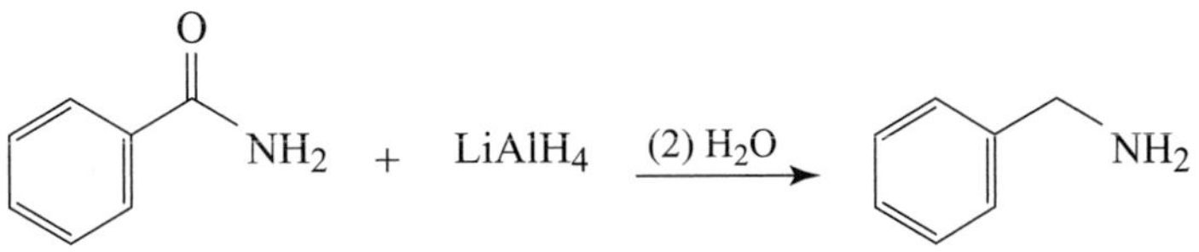

(c) Br + NH_3 (excess) $\longrightarrow$ NH_2

Br + O NK O $\longrightarrow$ O N O

$\xrightarrow{NH_2NH_2}$ NH_2 + O NH NH O

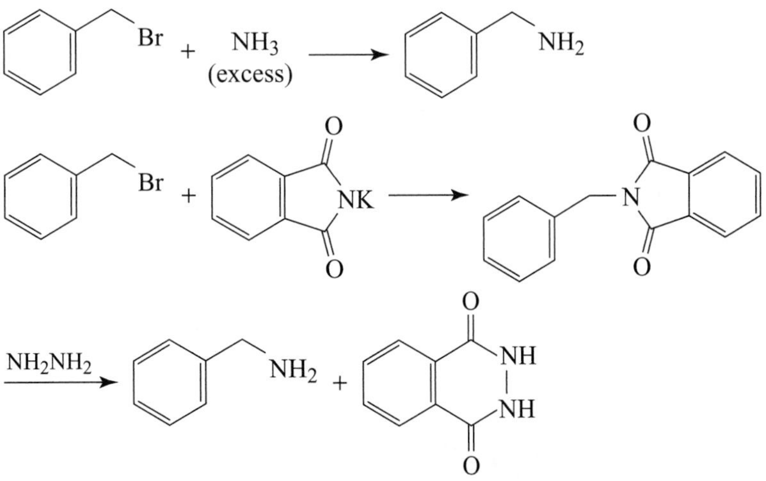

(d) $C_6H_5CH_2OTs + NH_3$ (excess) $\longrightarrow C_6H_5CH_2NH_2$

(e) $C_6H_5CHO + NH_3 \xrightarrow[\text{Ni, pressure}]{H_2} C_6H_5CH_2NH_2$

(f) $C_6H_5CH_2NO_2 + 3\ H_2 \xrightarrow[\text{pressure}]{\text{Pt}} C_6H_5CH_2NH_2$

(g) $C_6H_5CH_2CONH_2 \xrightarrow[OH^-]{Br_2} C_6H_5CH_2NH_2 + CO_3^{2-}$

20.23 (a) $C_6H_6 \xrightarrow[H_2SO_4]{HNO_3} C_6H_5NO_2 \xrightarrow[\text{(2) } OH^-]{\text{(1) Fe, HCl, heat}} C_6H_5NH_2$

(b) $C_6H_5Br \xrightarrow[Et_2O]{Mg} C_6H_5MgBr \xrightarrow[\text{(2) } H_3O^+]{\text{(1) } CO_2} C_6H_5COOH \xrightarrow{SOCl_2} C_6H_5COCl \xrightarrow[\text{(excess)}]{NH_3}$

$C_6H_5CONH_2 \xrightarrow[NaOH]{Br_2} C_6H_5NH_2$

Hofmann rearrangement

(c) $C_6H_5CONH_2 \xrightarrow[OH^-]{Br_2} C_6H_5NH_2$

20.24 (a) 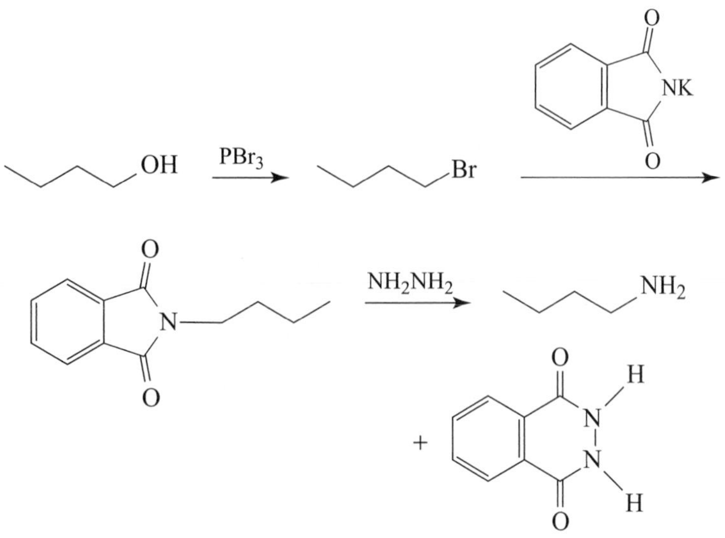 OH $\xrightarrow{PBr_3}$ Br $\xrightarrow{\text{potassium phthalimide (NK)}}$ N-butylphthalimide $\xrightarrow{NH_2NH_2}$ NH_2 + phthalhydrazide (N–H, N–H)

(b) 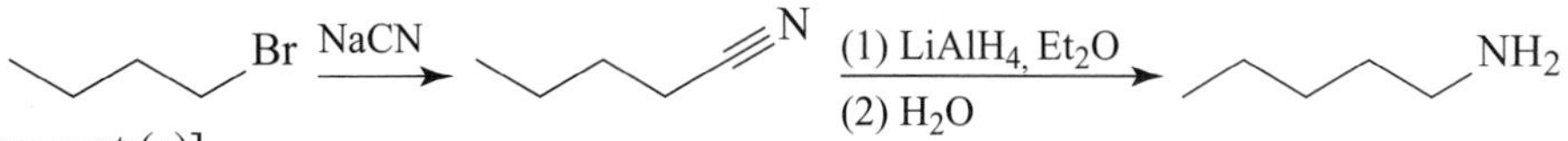 Br [from part (a)] $\xrightarrow{NaCN}$ ≡N $\xrightarrow[(2)\ H_2O]{(1)\ LiAlH_4,\ Et_2O}$ NH_2

(c) 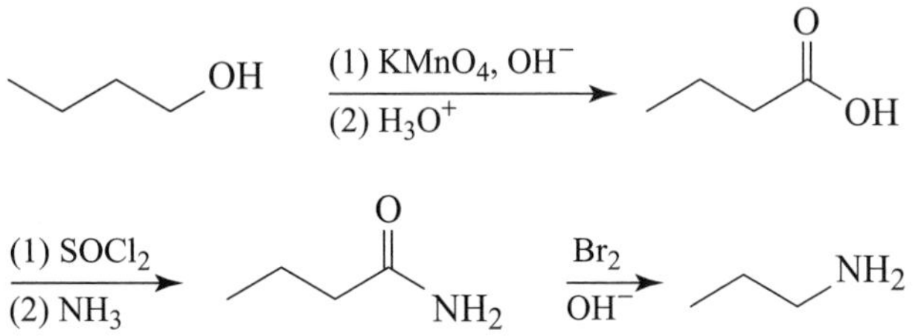 OH $\xrightarrow[(2)\ H_3O^+]{(1)\ KMnO_4,\ OH^-}$ (O) OH

$\xrightarrow[(2)\ NH_3]{(1)\ SOCl_2}$ (O) NH_2 $\xrightarrow[OH^-]{Br_2}$ NH_2

(d) 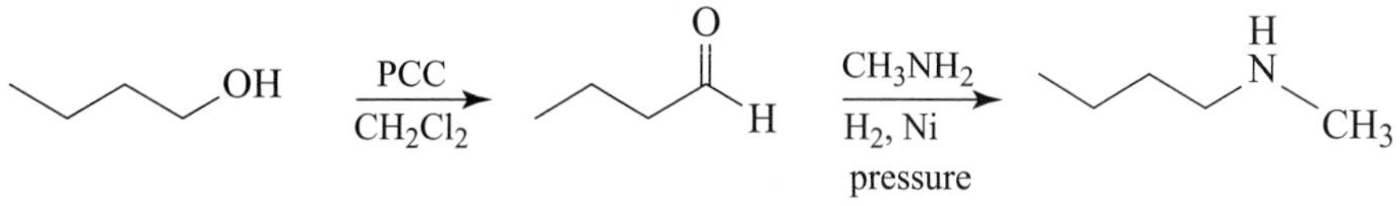 OH $\xrightarrow[CH_2Cl_2]{PCC}$ (O) H $\xrightarrow[H_2,\ Ni\ \text{pressure}]{CH_3NH_2}$ (H) N CH_3

20.25 (a) 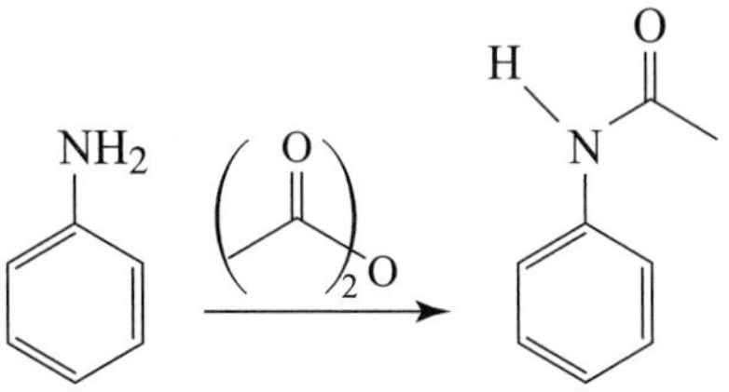 NH_2 (aniline) $\xrightarrow{(\text{acetyl})_2O}$ H–N(acetyl), phenyl (acetanilide)

(b) 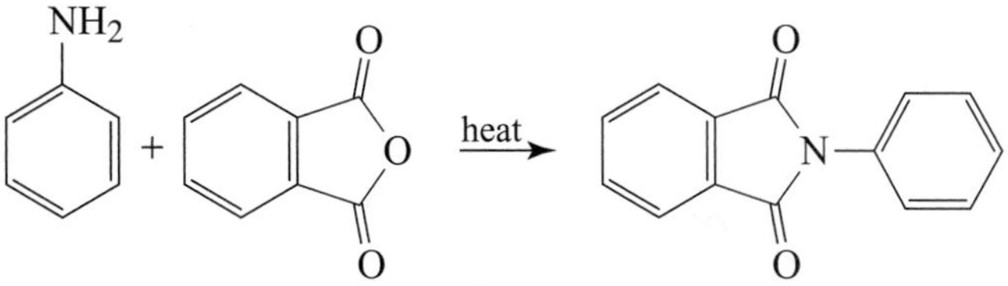 NH_2 (aniline) + phthalic anhydride (O, O, O) $\xrightarrow{heat}$ N-phenylphthalimide (O, N, O)

(c) [from part (a)] $\xrightarrow[H_2SO_4]{HNO_3}$ NO_2 (+ trace of ortho) $\xrightarrow[(2)\ OH^-]{(1)\ H_3O^+,\ H_2O}$ NH_2, NO_2

(d) [from part (a)] $\xrightarrow{HOSO_2Cl}$ SO_2Cl $\xrightarrow[(2)\ H_3O^+,\ heat;\ (3)\ OH^-]{(1)\ NH_3}$ NH_2, SO_2NH_2

(e) NH_2 $\xrightarrow[base]{2\ CH_3I}$ CH_3, CH_3, N

(f) NH_2 $\xrightarrow[(0–5°C)]{HONO}$ $N_2^+X^-$ $\xrightarrow{HBF_4}$ $N_2^+BF_4^-$ $\xrightarrow{heat}$ F

(g) $N_2^+X^-$ [from part (f)] $\xrightarrow{CuCl}$ Cl

(h) $N_2^+X^-$ [from part (f)] $\xrightarrow{CuBr}$ Br

(i) $N_2^+X^-$ [from part (f)] $\xrightarrow{KI}$ I

(j) $N_2^+X^-$ [from part (f)] $\xrightarrow{CuCN}$ N

(k) N [from part (j)] $\xrightarrow[(2)\ OH^-]{(1)\ H_3O^+,\ H_2O}$ O, OH

(l) $N_2^+X^-$ benzene [from part (f)] $\xrightarrow[Cu^{2+},\ H_2O]{Cu_2O}$ phenol (OH)

(m) $N_2^+X^-$ benzene [from part (f)] $\xrightarrow[H_2O]{H_3PO_2}$ benzene

(n) $N_2^+X^-$ [from part (f)] + OH [from part (l)] $\xrightarrow[pH\ 8\text{-}10]{OH^-}$ Ph–N=N–C$_6$H$_4$–OH

(o) $N_2^+X^-$ [from part (f)] + $N(CH_3)_2$ [from part (e)] $\xrightarrow[pH\ 5\text{-}7]{H_3O^+}$ Ph–N=N–C$_6$H$_4$–N(CH$_3$)(CH$_3$)

20.26 (a) 2-(2-aminoethyl)cyclohexanone (NH_2, O) $\xrightarrow{(1)\ cat.\ HA\quad (2)\ NaBH_3CN}$ octahydroindole (N–H)

(b) cyclohexylmethyl bromide (Br) $\xrightarrow{(1)\ NaCN\quad (2)\ LiAlH_4\quad (3)\ H_3O^+}$ 2-cyclohexylethylamine (NH_2)

(c) cyclohexylmethyl bromide (Br) $\xrightarrow{(1)\ NaN_3\quad (2)\ LiAlH_4\quad (3)\ H_3O^+}$ cyclohexylmethylamine (NH_2)

(d) cyclopentanone (O) $\xrightarrow{(1)\ NH_2OH,\ cat.\ HA\quad (2)\ NaBH_3CN}$ cyclopentylamine (NH_2)

(e) cyclohexylamine (NH_2) $\xrightarrow{(1)\ acetic\ anhydride\quad (2)\ LiAlH_4\quad (3)\ H_3O^+}$ N-ethylcyclohexylamine (HN)

20.27 (a) [propylamine] NH_2 $\xrightarrow[\text{NaNO}_2/\text{HCl, 0-5°C}]{\text{HONO}}$ [N_2^+] $\xrightarrow{-N_2}$

[primary carbocation, +] $\xrightarrow{\text{hydride shift}}$ [secondary carbocation, +]

From the primary cation: Cl^- → 1-chloropropane (Cl); H_2O → 1-propanol (OH); $-H^+$ → propene

From the secondary cation: $-H^+$ → propene; H_2O → 2-propanol (OH); Cl^- → 2-chloropropane (Cl)

(b) $($propyl$)_2$NH $\xrightarrow[\text{NaNO}_2/\text{HCl, 0–5°C}]{\text{HONO}}$ $($propyl$)_2$N—N=O

(c) N-propylaniline (N–H) $\xrightarrow[\text{NaNO}_2/\text{HCl, 0–5°C}]{\text{HONO}}$ N-nitroso-N-propylaniline (N–N=O)

(d) N,N-dipropylaniline $\xrightarrow[\text{NaNO}_2/\text{HCl, 0–5°C}]{\text{HONO}}$ O=N–(para)–C$_6$H$_4$–N(propyl)$_2$

(e) 4-propylaniline (NH_2) $\xrightarrow[\text{NaNO}_2/\text{HCl}]{\text{HONO, 0–5°C}}$ 4-propylbenzenediazonium $N_2^+\ Cl^-$

20.28 (a) [propyl]$\ddot{N}H_2$ + $C_6H_5SO_2Cl$ $\xrightarrow[H_2O]{KOH}$ [propyl]$\ddot{N}^-$(K^+)$SO_2C_6H_5$

Clear solution

$\xrightarrow{H_3O^+}$ [propyl]N(H)$SO_2C_6H_5$

Precipitate

(b) $($propyl$)_2$NH + $C_6H_5SO_2Cl$ $\xrightarrow[H_2O]{KOH}$ $($propyl$)_2$N—$SO_2C_6H_5$

Precipitate

$\xrightarrow{H_3O^+}$ No reaction (precipitate remains)

(c) N-propylaniline + $C_6H_5SO_2Cl$ $\xrightarrow[H_2O]{KOH}$ N-phenyl-N-propyl-$SO_2C_6H_5$

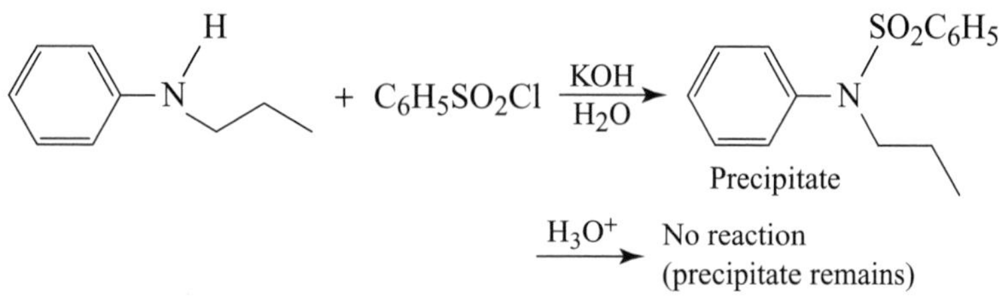

Precipitate

$\xrightarrow{H_3O^+}$ No reaction (precipitate remains)

(d) N,N-dipropylaniline + $C_6H_5SO_2Cl$ $\xrightarrow[H_2O]{KOH}$ No reaction (3° amine is insoluble)

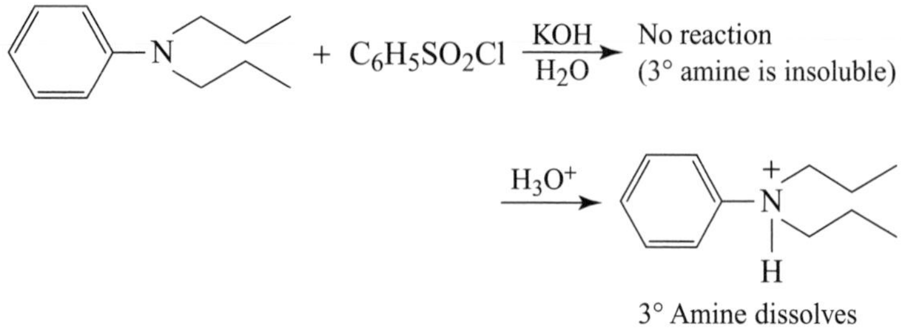

$\xrightarrow{H_3O^+}$ $C_6H_5\overset{+}{N}H(CH_2CH_2CH_3)_2$

3° Amine dissolves

(e) 4-propylaniline ($\ddot{N}H_2$) + $C_6H_5SO_2Cl$ $\xrightarrow[H_2O]{KOH}$ 4-propyl-C_6H_4-$\ddot{N}^-$ K^+ $SO_2C_6H_5$

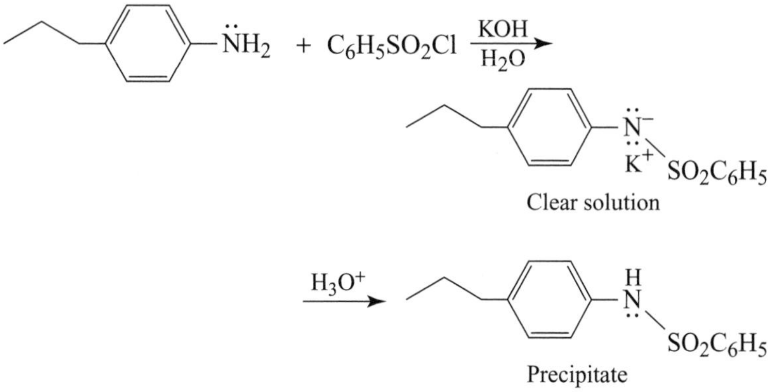

Clear solution

$\xrightarrow{H_3O^+}$ 4-propyl-C_6H_4-$\ddot{N}H$-$SO_2C_6H_5$

Precipitate

20.29 (a) piperidine N—H $\xrightarrow[NaNO_2/HCl]{HONO}$ piperidine N—N=O

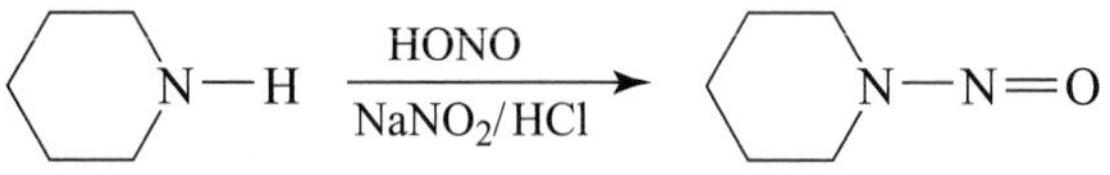

(b) piperidine N—H + $C_6H_5SO_2Cl$ $\xrightarrow[H_2O]{KOH}$ piperidine N—$SO_2C_6H_5$

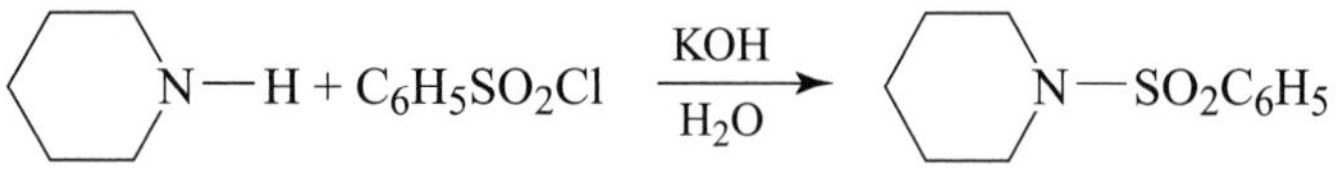

20.30 (a) 2 ⁄⁀NH_2 + C_6H_5—C(=O)—Cl ⟶ ⁄⁀N(H)—C(=O)—C_6H_5 + ⁄⁀$\overset{+}{N}H_3$ Cl^-

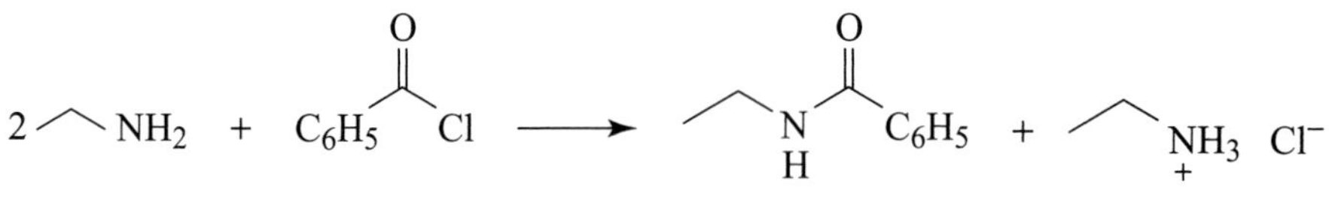

(b) 2 CH_3NH_2 + $\left(CH_3C(=O)\right)_2O$ ⟶ CH_3—N(H)—C(=O)CH_3 + $CH_3\overset{+}{N}H_3$ $CH_3C(=O)O^-$

(c) [succinic anhydride: ring with O, two C=O] + 2 CH_3NH_2 ⟶ O=C(NHCH$_3$)—CH_2CH_2—C(=O)O^- $H_3\overset{+}{N}CH_3$

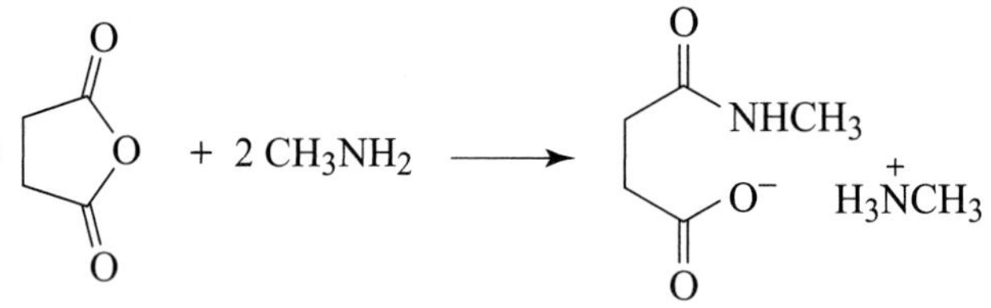

(d) [product of (c)] $\xrightarrow{\text{heat}}$ [succinimide ring, two C=O] N—CH_3 + H_2O + CH_3NH_2

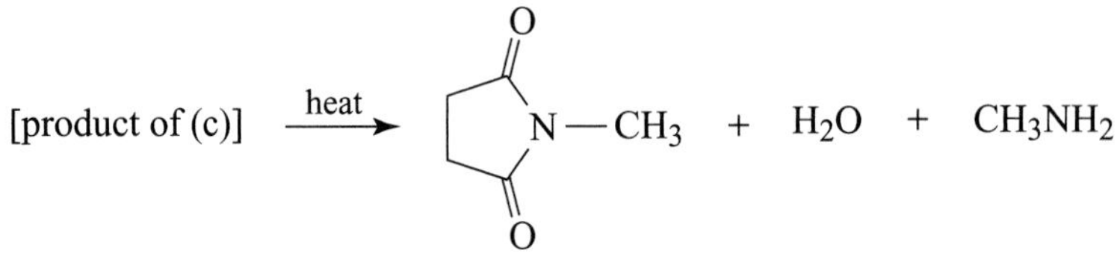

(e) [pyrrolidine] N—H + [phthalic anhydride: O, two C=O] ⟶ [benzene ring with C(=O)—N(pyrrolidine) and C(=O)OH]

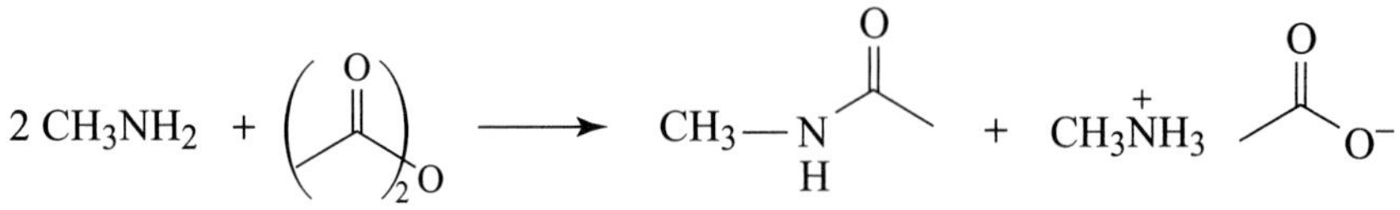

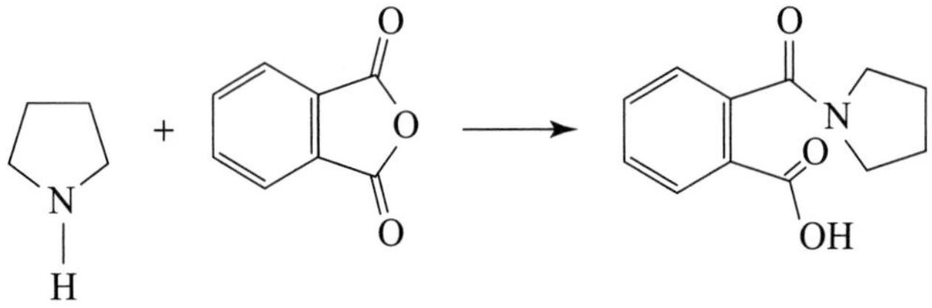

(f) [pyrrole] N—H + $\left(CH_3C(=O)\right)_2O$ ⟶ [pyrrole] N—C(=O)CH_3 + $CH_3C(=O)OH$

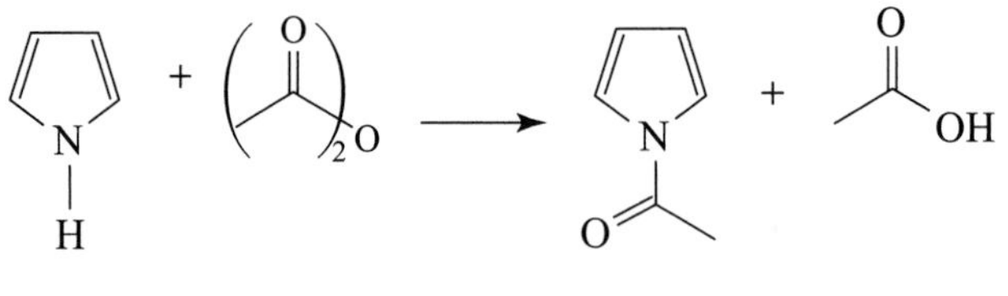

(g) 2 C_6H_5—NH_2 + $CH_3CH_2C(=O)Cl$ ⟶ C_6H_5—N(H)—C(=O)CH_2CH_3 +

C_6H_5—NH_3^+ Cl^-

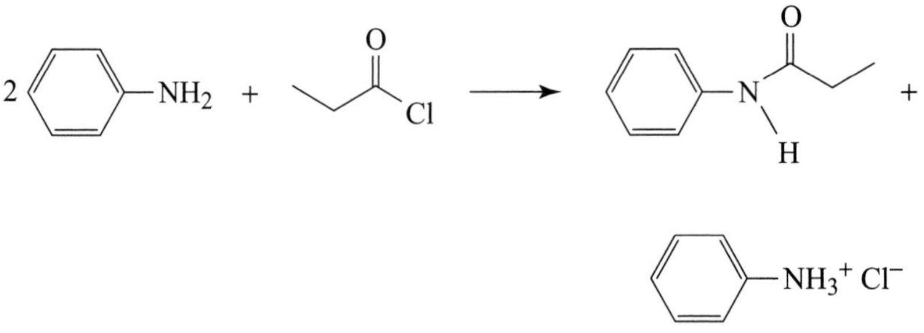

(h) $(CH_3CH_2)_4N^+ \; OH^-$ $\xrightarrow{\text{heat}}$ $H_2C=CH_2$ + $(CH_3CH_2)_3N$ + H_2O

(i) 4-methylaniline (CH_3, NH_2) + Br_2 (excess) $\xrightarrow{H_2O}$ 2,6-dibromo-4-methylaniline (CH_3, Br, Br, NH_2)

20.31 (a) toluene (CH_3) $\xrightarrow[H_2SO_4]{HNO_3}$ o-nitrotoluene (CH_3, NO_2) + p-nitrotoluene (CH_3, NO_2)

Separated

p-nitrotoluene (CH_3, NO_2) $\xrightarrow[(2)\ OH^-]{(1)\ Fe,\ HCl}$ p-toluidine (CH_3, NH_2) $\xrightarrow[0\text{–}5°C]{HONO}$ (CH_3, $N_2^+\ X^-$) $\xrightarrow{HBF_4}$ $\xrightarrow{\text{heat}}$ p-fluorotoluene (CH_3, F)

(b) o-nitrotoluene (CH_3, NO_2) [from part (a)] $\xrightarrow[(2)\ OH^-]{(1)\ Fe,\ HCl}$ o-toluidine (CH_3, NH_2) $\xrightarrow[0\text{–}5°C]{HONO}$

(CH_3, $N_2^+X^-$) $\xrightarrow{KI}$ o-iodotoluene (CH_3, I)

(c) p-nitrotoluene (CH_3, NO_2) [from part (a)] $\xrightarrow[(2)\ OH^-]{(1)\ Fe,\ HCl,\ heat}$ p-toluidine (CH_3, NH_2) $\xrightarrow[(2)\ Cu_2O,\ Cu^{2+},\ H_2O]{(1)\ H_2SO_4/NaNO_2,\ 0\text{–}5°C}$ p-cresol (CH_3, OH)

(d) [3-chloroaniline] $\xrightarrow[0\text{–}5°C]{HCl/NaNO_2}$ [3-chlorobenzenediazonium $N_2^+X^-$] $\xrightarrow{CuCl}$ [1,3-dichlorobenzene]

[from Problem 20.10(a)]

(e) [benzene] $\xrightarrow[H_2SO_4,\ heat]{fuming\ HNO_3}$ [1,3-dinitrobenzene] $\xrightarrow{Fe,\ HCl,\ heat}$ [1,3-diaminobenzene] $\xrightarrow[0\text{–}5°C]{HCl/NaNO_2}$ [bis-diazonium $N_2^+\ X^-$] $\xrightarrow{CuCN}$ [1,3-dicyanobenzene]

(f) [nitrobenzene] $\xrightarrow[FeBr_3]{Br_2}$ [3-bromonitrobenzene] $\xrightarrow[(2)\ OH^-]{(1)\ Fe,\ HCl,\ heat}$ [3-bromoaniline]

[from Problem 20.10(a)]

$\xrightarrow[(2)\ CuCN]{(1)\ H_2SO_4/NaNO_2,\ 0\text{–}5°C}$ [3-bromobenzonitrile]

(g) [4-nitroaniline] $\xrightarrow[FeBr_3]{Br_2}$ [2,6-dibromo-4-nitroaniline] $\xrightarrow[0\text{–}5°C]{H_2SO_4/NaNO_2}$

[from Problem 20.10(d)]

[2,6-dibromo-4-nitrobenzenediazonium N_2^+] $\xrightarrow{H_3PO_2}$ [1,3-dibromo-5-nitrobenzene]

(h) [1,3-dibromo-5-nitrobenzene] $\xrightarrow[(2)\ OH^-]{(1)\ Fe,\ HCl,\ heat}$ [3,5-dibromoaniline]

[from part (g)]

(i) NO_2, Br, Br, NH_2 [from part (g)] $\xrightarrow[\text{(2) CuBr}]{\text{(1) HBr/NaNO}_2\text{, 0–5°C}}$ NO_2, Br, Br, Br $\xrightarrow[\text{(2) OH}^-]{\text{(1) Fe, HCl, heat}}$

NH_2, Br, Br, Br $\xrightarrow[\text{(2) Cu}_2\text{O, Cu}^{2+}\text{, H}_2\text{O}]{\text{(1) H}_2\text{SO}_4\text{/NaNO}_2\text{, 0–5°C}}$ OH, Br, Br, Br

(j) NH_2, Br, Br, Br [from part (i)] $\xrightarrow[\text{(2) CuCN}]{\text{(1) H}_2\text{SO}_4\text{/NaNO}_2\text{, 0–5°C}}$ N, Br, Br, Br

(k) NO_2, Br, Br, NH_2 [from part (g)] $\xrightarrow[\text{(2) CuCN}]{\text{(1) H}_2\text{SO}_4\text{/NaNO}_2\text{, 0–5°C}}$ NO_2, Br, Br, N $\xrightarrow[\text{heat}]{H_3O^+}$

NO_2, Br, Br, O, OH $\xrightarrow[\text{Pt, pressure}]{H_2}$

NH_2, Br, Br, O, OH $\xrightarrow[\text{(2) H}_3\text{PO}_2]{\text{(1) H}_2\text{SO}_4\text{/NaNO}_2\text{, 0–5°C}}$ Br, Br, O, OH

(l)

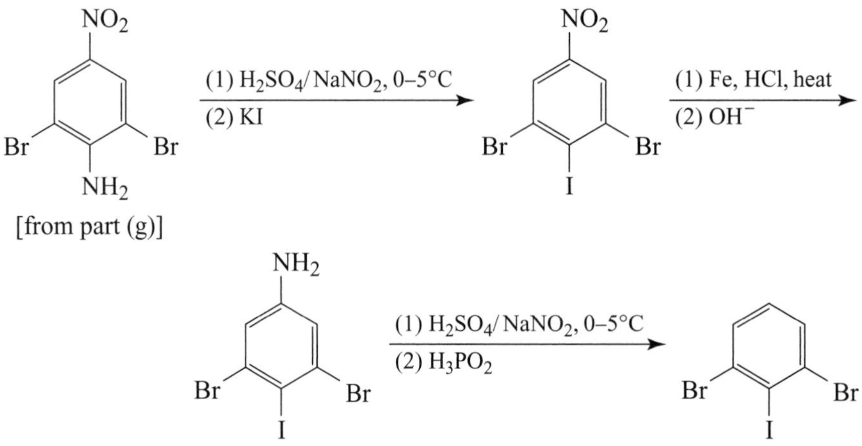

NO_2, Br, Br, NH_2

[from part (g)]

(1) H_2SO_4/ $NaNO_2$, 0–5°C
(2) KI

NO_2, Br, Br, I

(1) Fe, HCl, heat
(2) OH^-

NH_2, Br, Br, I

(1) H_2SO_4/ $NaNO_2$, 0–5°C
(2) H_3PO_2

Br, Br, I

(m)

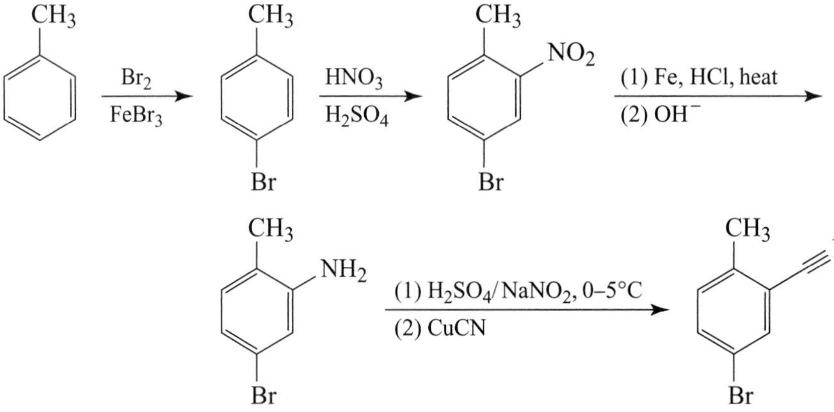

CH_3

Br_2
$FeBr_3$

CH_3, Br

HNO_3
H_2SO_4

CH_3, NO_2, Br

(1) Fe, HCl, heat
(2) OH^-

CH_3, NH_2, Br

(1) H_2SO_4/$NaNO_2$, 0–5°C
(2) CuCN

CH_3, N, Br

(n)

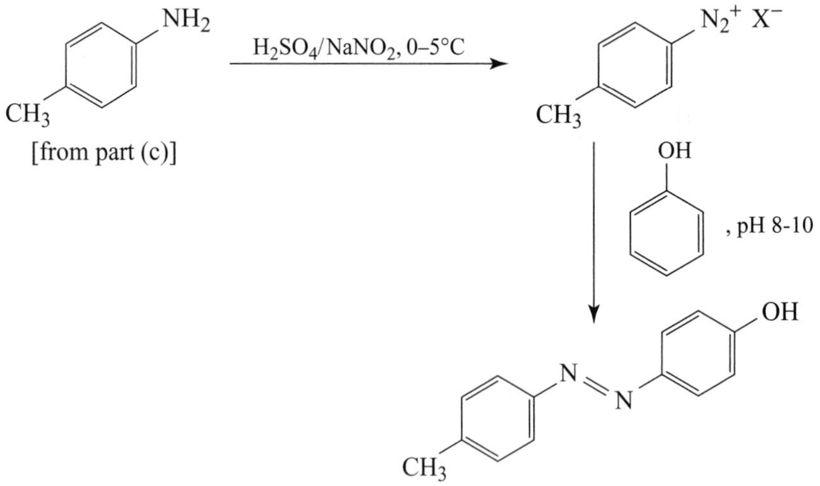

NH_2, CH_3

[from part (c)]

H_2SO_4/$NaNO_2$, 0–5°C

N_2^+ X^-, CH_3

OH, pH 8-10

OH, N, N, CH_3

(o) [4-methylbenzenediazonium] $N_2^+ X^-$ [from part (n)] + [4-methylphenol] [from part (c)] $\xrightarrow{\text{pH 8-10}}$ [azo product: HO, $N{=}N$, CH_3, CH_3]

20.32 (a) Benzylamine dissolves in dilute HCl at room temperature,

$$C_6H_5CH_2NH_2 + H_3O^+ + Cl^- \xrightarrow{25°\,C} C_6H_5CH_2\overset{+}{N}H_3\ Cl^-$$

benzamide does not dissolve:

$$C_6H_5CONH_2 + H_3O^+ + Cl^- \xrightarrow{25°C} \text{No reaction}$$

(b) Allylamine reacts with (and decolorizes) bromine in carbon tetrachloride instantly,

$$CH_2{=}CHCH_2NH_2 + Br_2 \xrightarrow{CCl_4} BrCH_2CHBrCH_2NH_2$$

propylamine does not:

$$CH_3CH_2CH_2NH_2 + Br_2 \xrightarrow{CCl_4} \text{No reaction if the mixture is not heated or irradiated}$$

(c) The Hinsberg test:

$$H_3C{-}C_6H_4{-}\ddot{N}H_2 + C_6H_5SO_2Cl \xrightarrow[H_2O]{KOH} H_3C{-}C_6H_4{-}\ddot{\underset{..}{N}}^{-}SO_2C_6H_5\ K^+$$

Soluble

$$\xrightarrow{H_3O^+} H_3C{-}C_6H_4{-}\ddot{N}HSO_2C_6H_5$$

Precipitate

$$C_6H_5{-}NHCH_3 + C_6H_5SO_2Cl \xrightarrow[H_2O]{KOH} C_6H_5{-}N(CH_3)SO_2C_6H_5$$

Precipitate

$$\xrightarrow{H_3O^+} \text{Precipitate remains}$$

(d) The Hinsberg test:

$$\text{C}_6\text{H}_{11}\ddot{\text{N}}\text{H}_2 + \text{C}_6\text{H}_5\text{SO}_2\text{Cl} \xrightarrow[\text{H}_2\text{O}]{\text{KOH}} \text{C}_6\text{H}_{11}\ddot{\text{N}}^{-}\text{SO}_2\text{C}_6\text{H}_5\ \text{K}^+ \xrightarrow{\text{H}_3\text{O}^+}$$

Soluble

$$\text{C}_6\text{H}_{11}\ddot{\text{N}}\text{HSO}_2\text{C}_6\text{H}_5$$

Precipitate

$$\text{C}_5\text{H}_{10}\text{N—H} + \text{C}_6\text{H}_5\text{SO}_2\text{Cl} \xrightarrow[\text{H}_2\text{O}]{\text{KOH}} \text{C}_5\text{H}_{10}\text{N—SO}_2\text{C}_6\text{H}_5 \xrightarrow{\text{H}_3\text{O}^+} \text{Precipitate remains}$$

Precipitate

(e) Pyridine dissolves in dilute HCl,

$$\text{C}_5\text{H}_5\text{N} + \text{H}_3\text{O}^+ + \text{Cl}^- \longrightarrow \text{C}_5\text{H}_5\overset{+}{\text{N}}\text{—H}\ \ \text{Cl}^-$$

benzene does not:

$$\text{C}_6\text{H}_6 + \text{H}_3\text{O}^+ + \text{Cl}^- \longrightarrow \text{No reaction}$$

(f) Aniline reacts with nitrous acid at 0–5°C to give a stable diazonium salt that couples with 2-naphthol, yielding an intensely colored azo compound.

$$\text{C}_6\text{H}_5\text{NH}_2 \xrightarrow[0\text{–}5^\circ\text{C}]{\text{H}_2\text{SO}_4/\text{NaNO}_2} \text{C}_6\text{H}_5\overset{+}{\text{N}_2} \xrightarrow{\text{2-naphthol}} \text{C}_6\text{H}_5\text{N=N-(2-hydroxy-1-naphthyl)}$$

Cyclohexylamine reacts with nitrous acid at 0–5°C to yield a highly unstable diazonium salt—one that decomposes so rapidly that the addition of 2-naphthol gives no azo compound.

$$\text{C}_6\text{H}_{11}\text{NH}_2 \xrightarrow[0\text{–}5^\circ\text{C}]{\text{H}_2\text{SO}_4/\text{NaNO}_2} \left[\text{C}_6\text{H}_{11}\overset{+}{\text{N}_2}\right] \xrightarrow{-\text{N}_2} \left[\text{C}_6\text{H}_{11}^{+}\right] \longrightarrow$$

$$\text{alkenes, alcohols, and alkyl halides} \xrightarrow{\text{2-naphthol}} \text{No reaction}$$

(g) The Hinsberg test:

$$(\text{Pr})_3N + C_6H_5SO_2Cl \xrightarrow[H_2O]{KOH} \text{No reaction} \xrightarrow{H_3O^+} (\text{Pr})_3\overset{+}{N}H$$

Soluble

$$(\text{Pr})_2NH + C_6H_5SO_2Cl \xrightarrow[H_2O]{KOH} (\text{Pr})_2NSO_2C_6H_5 \xrightarrow{H_3O^+} \text{Precipitate remains}$$

Precipitate

(h) Tripropylaminium chloride reacts with aqueous NaOH to give a water insoluble tertiary amine.

$$(\text{Pr})_3\overset{+}{N}H\ Cl^- \xrightarrow[H_2O]{NaOH} (\text{Pr})_3N$$

Water soluble — Water insoluble

Tetrapropylammonium chloride does not react with aqueous NaOH (at room temperature), and the tetrapropylammonium ion remains in solution.

$$(\text{Pr})_4\overset{+}{N}\ Cl^- \xrightarrow[H_2O]{NaOH} (\text{Pr})_4\overset{+}{N}\ [Cl^- \text{ or } OH^-]$$

Water soluble — Water soluble

(i) Tetrapropylammonium chloride dissolves in water to give a neutral solution. Tetrapropylammonium hydroxide dissolves in water to give a strongly basic solution.

20.33 Follow the procedure outlined in the answer to Problem 20.2. Toluene will show the same solubility behavior as benzene.

Mechanisms

20.34

$$\xrightarrow[Pd]{H_2} \left[\ \ddot{N}H_2\ \right] \xrightarrow[\text{(several steps)}]{-H_2O}$$

$$\left[\ N\ \right] \xrightarrow[Pd]{H_2} \text{H–N}$$

20.35 (a)

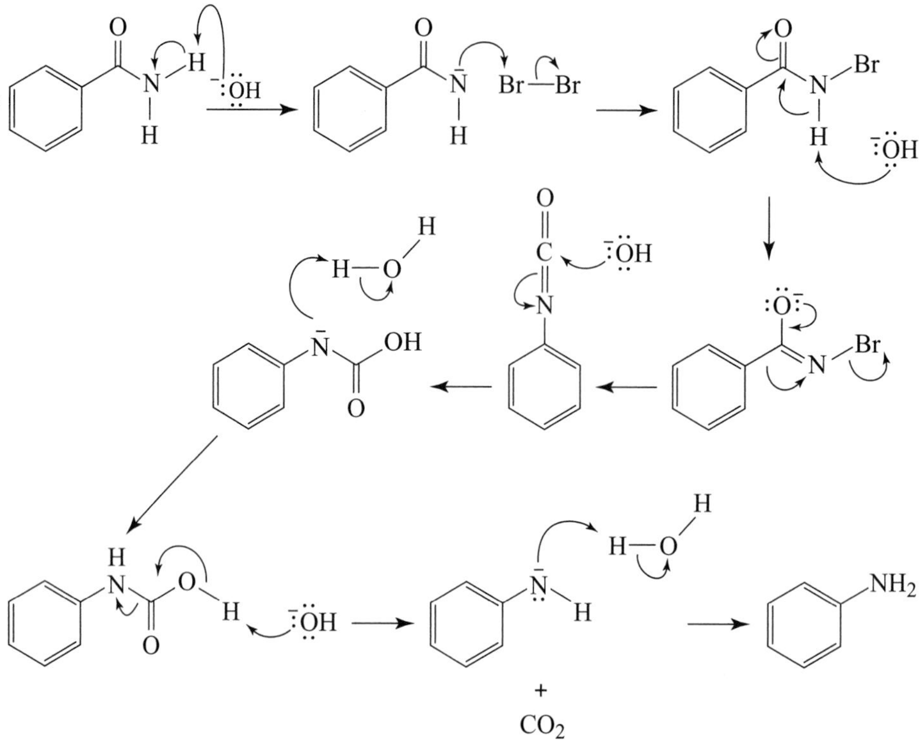

O H N H $^{-}$:ÖH O $\bar{N}$ H Br—Br O N Br H :ÖH

O C N :ÖH :Ö: N Br H H—O $\bar{N}$ OH O

H N O H O :ÖH $\bar{N}$ H H H—O NH_2

+ CO_2

(b)

O $\ddot{N}H_2$ O P Cl Cl Cl :Ö: N—H H O Cl—P Cl O $\overset{+}{N}$—H H :Cl:$^{-}$

N $\overset{+}{N}$ H :Cl:$^{-}$

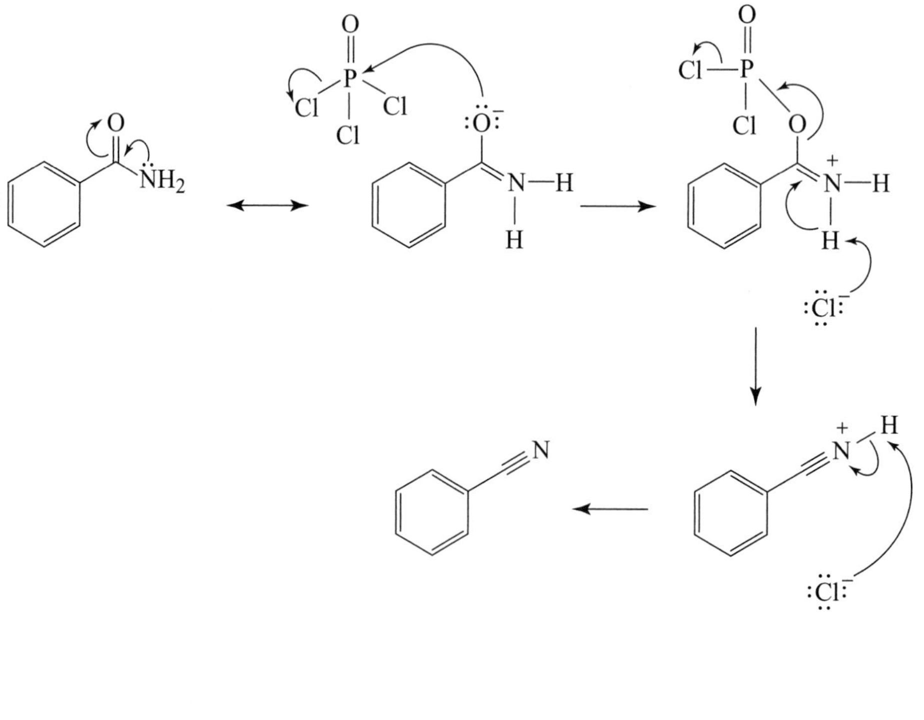

20.36 Carry out the Hofmann reaction using a mixture of ^{15}N labeled benzamide, $C_6H_5CO{*}NH_2$, and *p*-chlorobenzamide. If the process is intramolecular, only labeled aniline, $C_6H_5{*}NH_2$, and *p*-chloroaniline, will be produced.

If the process is one in which the migrating moiety truly separates from the remainder of the molecule, then, in addition to the two products mentioned above, there should be produced both the unlabeled aniline and labeled *p*-chloroaniline, *p*-$ClC_6H_4{*}NH_2$.

Note: When this experiment is actually carried out, analysis of the reaction mixture by mass spectrometry shows that the process is intramolecular.

General Synthesis

20.37 (succinic anhydride) + NH_3 $\xrightarrow{(2)\ H_3O^+}$ H_2N–C(=O)–CH_2CH_2–C(=O)–OH

$\xrightarrow[-CO_3^{2-}]{Br_2,\ OH^-}$ H_2N–CH_2CH_2–C(=O)–O^- $\xrightarrow{H_3O^+}$ $H_3\overset{+}{N}$–CH_2CH_2–C(=O)–O^-

20.38 (a) HO–CH_2CH_2–$(CH_2)_8$–CH_2CH_2–OH $\xrightarrow{2\ PBr_3}$ Br–CH_2CH_2–$(CH_2)_8$–CH_2CH_2–Br $\xrightarrow{2\ (CH_3)_3N}$

$(CH_3)_3\overset{+}{N}$–CH_2CH_2–$(CH_2)_8$–CH_2CH_2–$\overset{+}{N}(CH_3)_3$ 2 Br^-

(b) HO–C(=O)–CH_2CH_2–C(=O)–OH + 2 Br–CH_2CH_2–OH $\xrightarrow[(-2\ H_2O)]{HA}$

Br–CH_2CH_2–O–C(=O)–CH_2CH_2–C(=O)–O–CH_2CH_2–Br $\xrightarrow{2\ (CH_3)_3N}$

$(CH_3)_3\overset{+}{N}$–CH_2CH_2–O–C(=O)–CH_2CH_2–C(=O)–O–CH_2CH_2–$\overset{+}{N}(CH_3)_3$ 2 Br^-

20.39 H_2N ... NH_2 ... NH_2 ... OH + Br ... O ... Br ... Br $\xrightarrow{-H_2O}$ H_2N ... $\ddot{N}H_2$ ... Br ... Br ... Br ... OH

$\xrightarrow[-(HBr)]{}$ H ... H_2N ... N ... Br ... Br ... OH $\xrightarrow[-HBr]{HCO_3^-}$

H_2N ... Br + $H_2\ddot{N}$ ... O ... N H ... OH ... O ... OH ... O ... OH $\xrightarrow{-HBr}$ folic acid

20.40 + CH_3I ⟶ I^- H_3C CH_3 **R** $\xrightarrow[H_2O]{Ag_2O}$ OH^- H_3C CH_3 **S** $\xrightarrow[-H_2O]{heat}$

CH_3 N CH_3 **T** $\xrightarrow{CH_3I}$ $(CH_3)_3\overset{+}{N}$ I^- **U** $\xrightarrow[H_2O]{Ag_2O}$ $\overset{+}{N}(CH_3)_3$ OH^- **V**

$\xrightarrow{heat}$ **W** + H_2O + $(CH_3)_3N$

20.41 $(CH_3)_2NH$ + (ethylene oxide) ⟶ CH_3 N CH_3 OH $\xrightarrow[\text{base (–HCl)}]{CH_3COCl}$

CH_3 N CH_3 O O $\xrightarrow{CH_3I}$ I^- CH_3—$\overset{+}{N}$(CH_3)(CH_3) O O

Acetylcholine iodide

20.42 (reaction scheme: ethylene oxide $\xrightarrow{NH_3}$ HO–CH₂CH₂–NH₂ $\xrightarrow{\text{ethylene oxide}}$ HO–CH₂CH₂–NH–CH₂CH₂–OH)

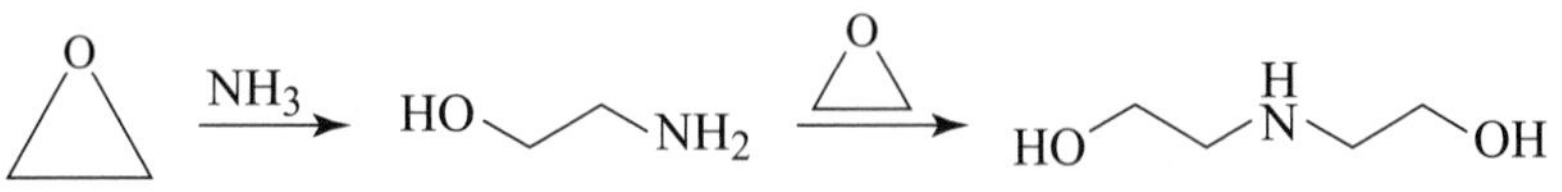

HO NH$_2$ HO N H OH

20.43 (reaction scheme: benzene, propanoyl chloride / $AlCl_3$ → propiophenone; acetic acid, Br_2 → 2-bromo-1-phenylpropan-1-one; diethylamine → diethylpropion)

Cl AlCl$_3$ OH Br$_2$ Br

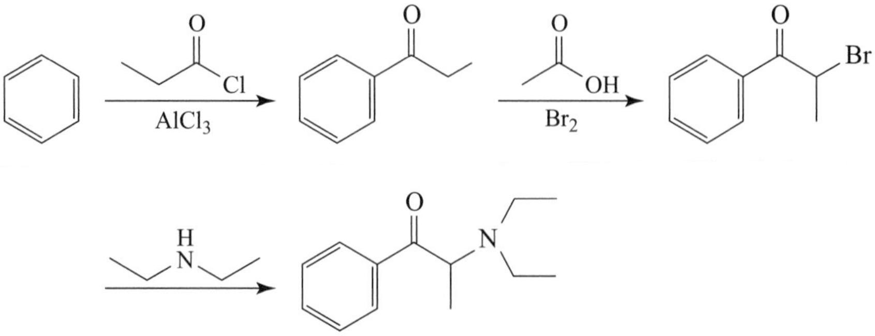

Diethylpropion

20.44 Cl OH $\xrightarrow{SOCl_2}$ Cl Cl (aniline, NH$_2$) → Cl N H (2-naphthol, OH / base) → O N H

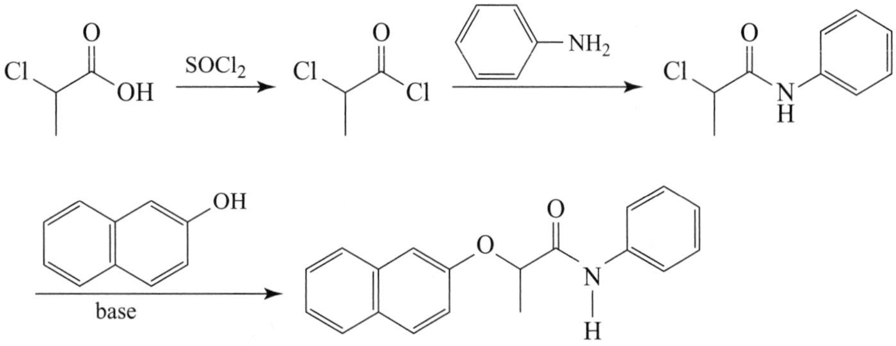

Spectroscopy

20.45 The results of the Hinsberg test indicate that compound **W** is a tertiary amine. The ^{1}H NMR provides evidence for the following:

(1) Two different C_6H_5— groups (one absorbing at δ 7.2 and one at δ 6.7).
(2) A CH_3CH_2— group (the quartet at δ 3.5 and the triplet at δ 1.2).
(3) An unsplit —CH_2— group (the singlet at δ 4.5).

There is only one reasonable way to pull all of this together.

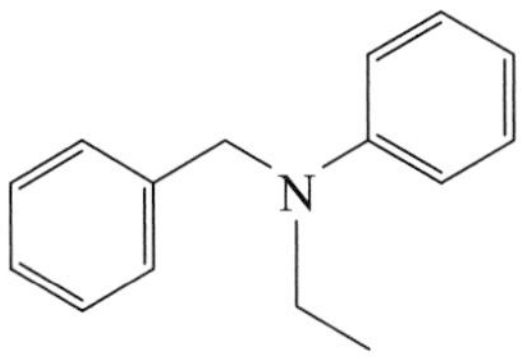

N

Thus **W** is *N*-benzyl-*N*-ethylaniline.

20.46 Compound **X** is benzyl bromide, $C_6H_5CH_2Br$. This is the only structure consistent with the ^1H NMR and IR data. (The monosubstituted benzene ring is strongly indicated by the (5H), δ 7.3 ^1H NMR absorption and is confirmed by the peaks at 690 and 770 cm^{-1} in the IR spectrum.)

Compound **Y**, therefore, must be phenylacetonitrile, $C_6H_5CH_2CN$, and **Z** must be 2-phenylethylamine, $C_6H_5CH_2CH_2NH_2$.

Br NaCN → N LiAlH$_4$ → NH$_2$

X C_7H_7Br **Y** C_8H_7N **Z** $C_8H_{11}N$

Interpretations of the IR and ^1H NMR spectra of **Z** are as follows.

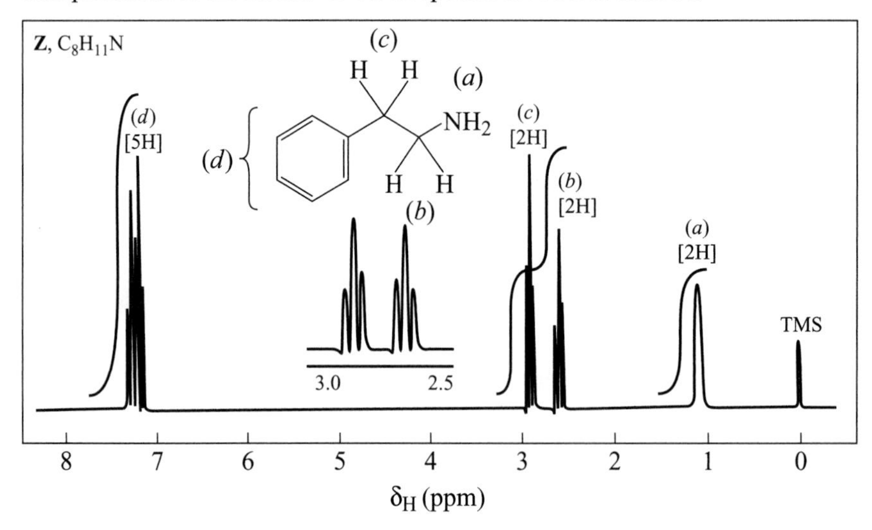

(c) H H (a) NH$_2$ (d) H H (b)

(*a*) singlet δ 1.0
(*b*) triplet δ 2.7
(*c*) triplet δ 2.9
(*d*) multiplet δ 7.25

20.47 That **A** contains nitrogen and is soluble in dilute HCl suggests that **A** is an amine. The two IR absorption bands in the 3300–3500-cm^{-1} region suggest that **A** is a primary amine. The ^{13}C spectrum shows only two signals in the upfield aliphatic region. There are four signals downfield in the aromatic region. The information from the DEPT spectra suggests an ethyl group *or two equivalent ethyl groups.* Assuming the latter, and assuming that **A** is a primary amine, we can conclude from the molecular formula that **A** is 2,6-diethylaniline. The assignments are

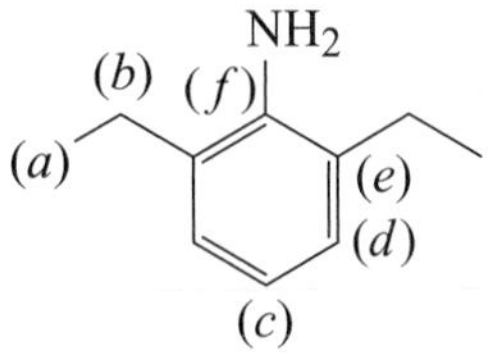

(*a*) δ 12.9
(*b*) δ 24.2
(*c*) δ 118.1
(*d*) δ 125.9
(*e*) δ 127.4
(*f*) δ 141.5

(An equally plausible answer would be that **A** is 3,5-diethylaniline.)

20.48 That **B** dissolves in dilute HCl suggests that **B** is an amine. That the IR spectrum of **B** lacks bands in the 3300–3500-cm^{-1} region suggests that **B** is a tertiary amine. The upfield signals in the ^{13}C spectrum and the DEPT information suggest two equivalent ethyl groups (as was also true of **A** in the preceding problem). The DEPT information for the downfield peaks (in the aromatic region) is consistent with a monosubstituted benzene ring. Putting all of these observations together with the molecular formula leads us to conclude that **B** is *N*,*N*-diethylaniline. The assignments are

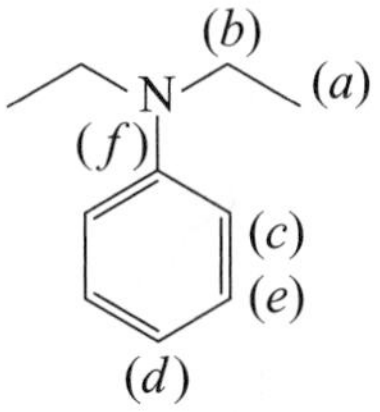

(*a*) δ 12.5
(*b*) δ 44.2
(*c*) δ 112.0
(*d*) δ 115.5
(*e*) δ 128.1
(*f*) δ 147.8

20.49 That **C** gives a positive Tollens' test indicates the presence of an aldehyde group; the solubility of **C** in aqueous HCl suggests that **C** is also an amine. The absence of bands in the 3300–3500-cm^{-1} region of the IR spectrum of **C** suggests that **C** is a tertiary amine. The signal at δ 189.7 in the ^{13}C spectrum can be assigned to the aldehyde group. The signal at δ 39.7 is the only one in the aliphatic region and is consistent with a methyl group or with two equivalent methyl groups. The remaining signals are in the aromatic region. If we assume that **C** has a benzene ring containing a $-\overset{\text{O}}{\overset{\|}{\text{C}}}\text{H}$ group and a $-N(CH_3)_2$ group, then the aromatic signals and their DEPT spectra are consistent with **C** being *p*-(*N*,*N*-dimethylamino)benzaldehyde. The assignments are

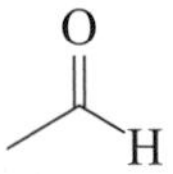

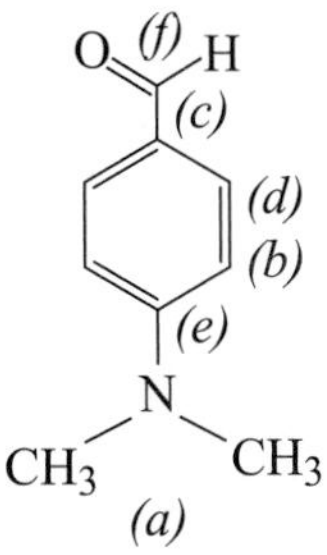

(*a*) δ 39.7
(*b*) δ 110.8
(*c*) δ 124.9
(*d*) δ 131.6
(*e*) δ 154.1
(*f*) δ 189.7

Challenge Problems

20.50 (a) $C_6H_5\ddot{N}(CH_3)H$

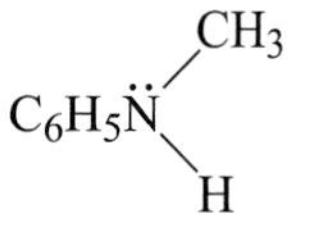

I

(b) I $\xrightarrow{-e^-}$ (during electron impact mass spectrometry) $C_6H_5-\overset{+\cdot}{N}(CH_2-H)H$ *m/z* 107

$\downarrow$ –H·

$C_6H_5(H)\overset{+}{N}=CH_2$ *m/z* 106

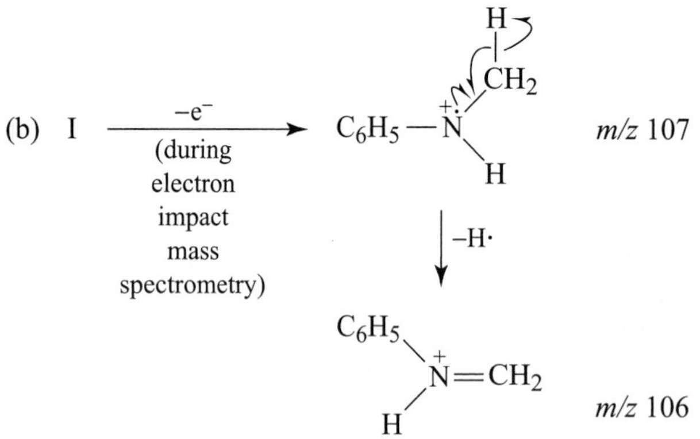

20.51 (a,b)

H H N N :O:

A

CH_3 $O=S=O$ Cl

+

H H N N $+O$ O S O CH_3 +

– N^+ H

H N $=\ddot{N}$ O S O CH_3 +

N=C=N

B Diphenylcarbodiimide

+

CH_3 O S $\ddot{O}:^-$ O

– N_+ H

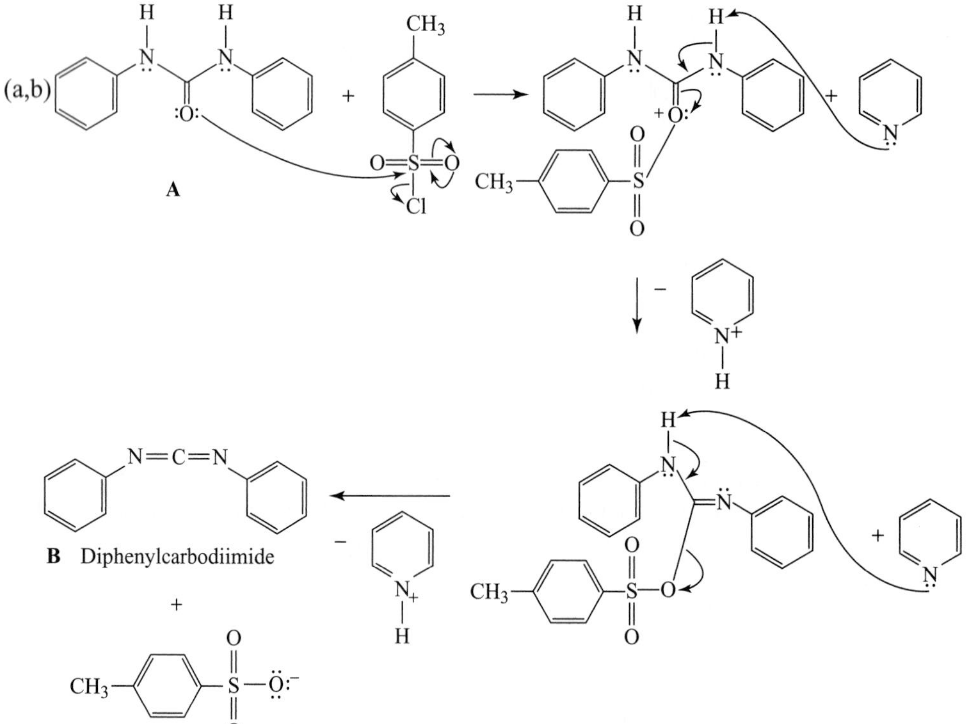

20.52

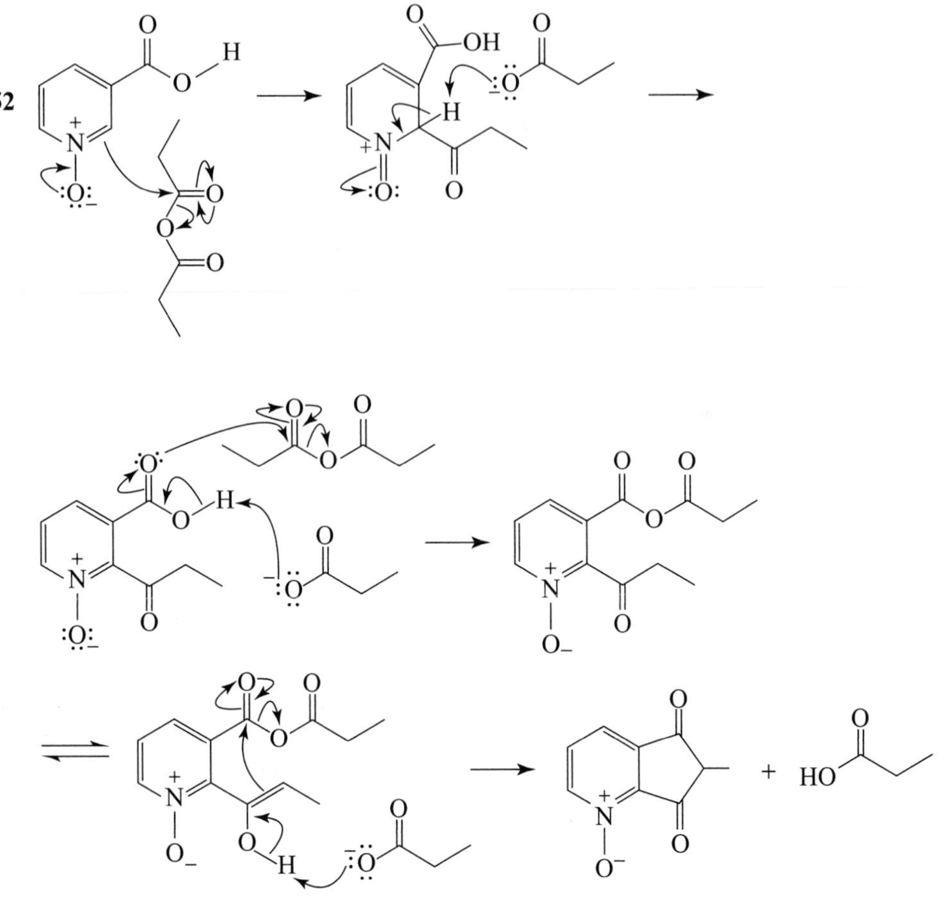

20.53 An abbreviated version of one of several possible routes:

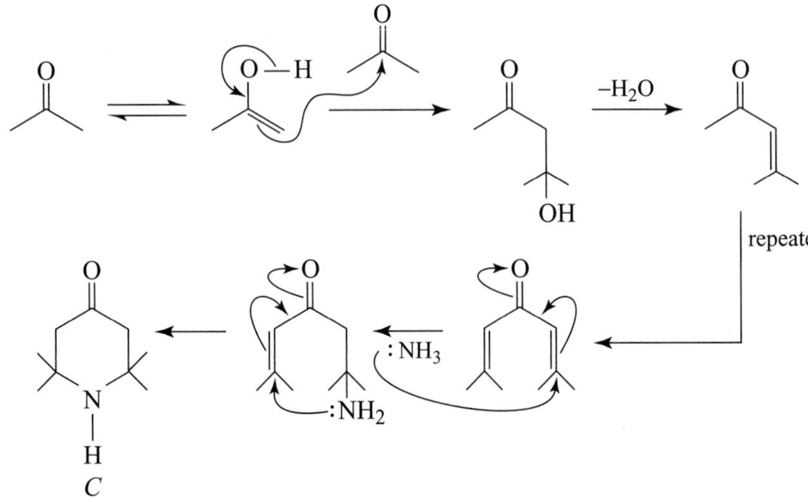

$-H_2O$

repeated sequence

$:NH_3$

$:NH_2$

C

"triacetoneamine"

20.54

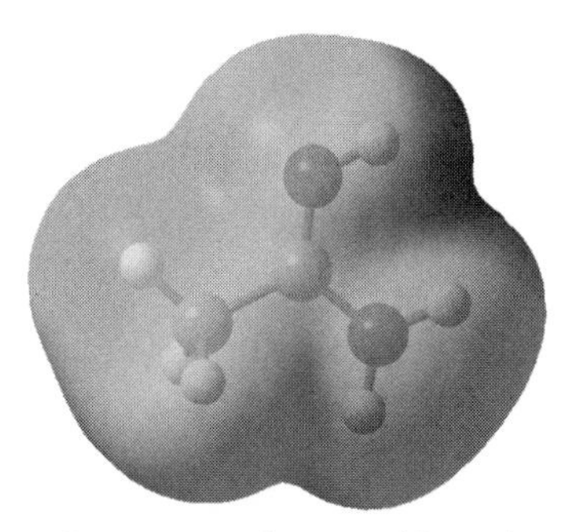

O-protonated acetamide cation

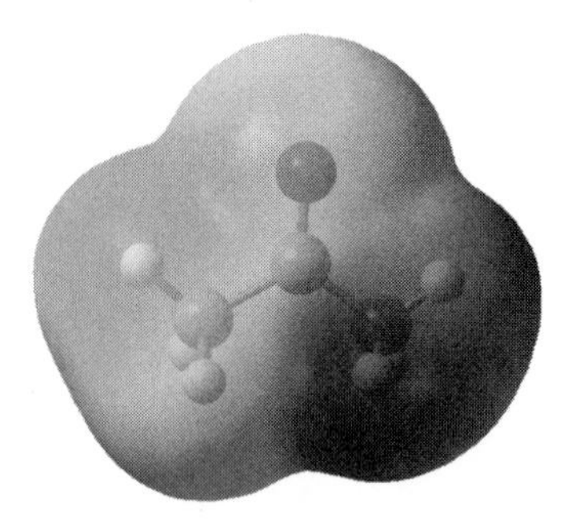

N-protonated acetamide cation

The *N*-protonated acetamide cation shows significant localization of positive charge near its NH_3^+ group, as indicated by the strong blue color at that location in its electrostatic potential map. The *N*-protonated acetamide model also shows noticeable negative charge (or less positive charge) near its oxygen atom, as indicated by yellow in that region. These observations suggest a lack of delocalization of the positive charge.

On the other hand, the *O*-protonated acetamide cation has some blue as well as green mapped to both its NH_2 group and O—H (protonated carbonyl) group, suggesting delocalization of the positive charge between these two locations.

An attempt to draw resonance structures for these two protonated forms shows that two resonance contributors are possible for *O*-protonated acetamide, whereas only one proper Lewis structure is possible for *N*-protonated acetamide. The resonance contributors for *O*-protonated acetamide distribute the formal positive charge over both the oxygen and nitrogen atoms.

H, $^+$O:, $\ddot{N}H_2$ ⟷ H, ·O:, $\overset{+}{N}H_2$ versus   :O:, $\overset{+}{N}H_3$

O-protonated acetamide cation   *N*-protonated acetamide cation

QUIZ

20.1 Which of the following would be soluble in dilute aqueous HCl?

(a) NH_2

(b) NH_2

(c) 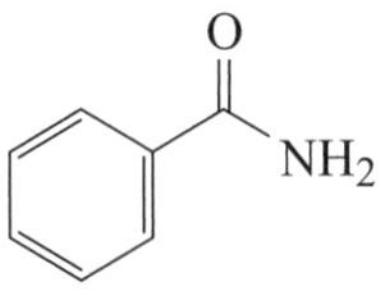 [structure: benzamide, $C_6H_5C(=O)NH_2$]

(d) Two of the above

(e) All of the above

20.2 Which would yield propylamine?

(a) 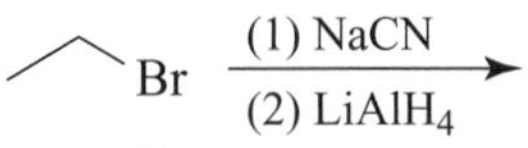 $CH_3CH_2Br \xrightarrow[(2)\ LiAlH_4]{(1)\ NaCN}$

(b) 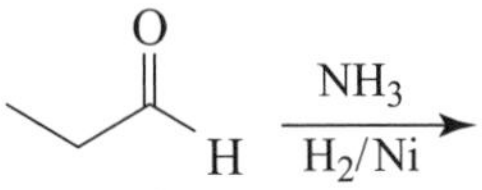 $CH_3CH_2CHO \xrightarrow[H_2/Ni]{NH_3}$

(c) 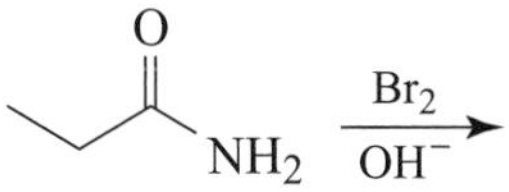 $CH_3CH_2C(=O)NH_2 \xrightarrow[OH^-]{Br_2}$

(d) Two of the above

(e) All of the above

20.3 Select the reagent from the list below that could be the basis of a simple chemical test that would distinguish between each of the following:

(a) 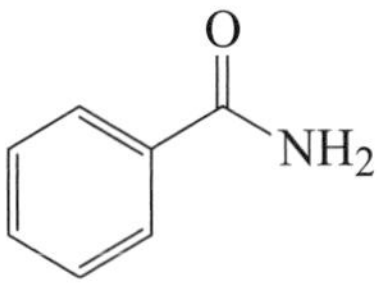 $C_6H_5C(=O)NH_2$ and $C_6H_5CH_2NH_2$

(b) $C_6H_5CH_2NH_2$ and $C_6H_5NHCH_3$

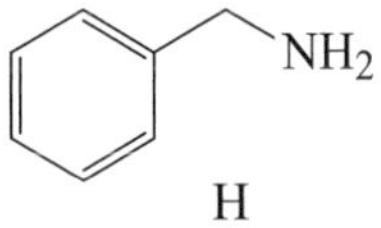

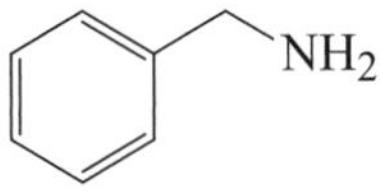

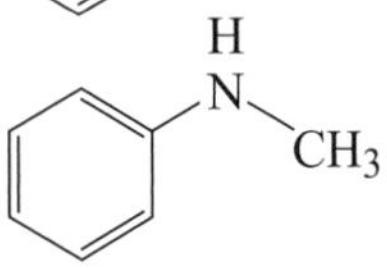

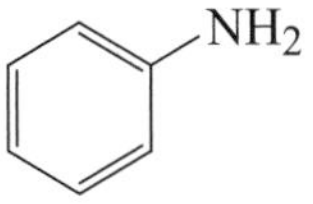

(c) $C_6H_5NH_2$ and $C_6H_{11}NH_2$ (cyclohexylamine)

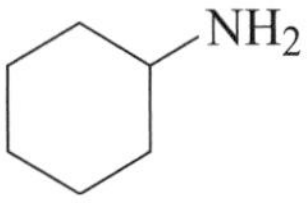

1. Cold dilute $NaHCO_3$
2. Cold dilute HCl
3. $NaNO_2$, HCl, 5°C, then 2-naphthol
4. $C_6H_5SO_2Cl$, OH^-, then HCl
5. Cold dilute NaOH

20.4 Complete the following syntheses:

(a) Toluene —A→ B (+ isomer) —(1) Fe, HCl; (2) OH^-→ p-toluidine (H_2N, CH_3) —C→ p-methylbenzenediazonium chloride ($\overset{+}{N_2}$, Cl^-, CH_3) —D→ p-tolunitrile (N≡C, CH_3) —E→ 4-methylbenzylamine (H_2N, CH_3)

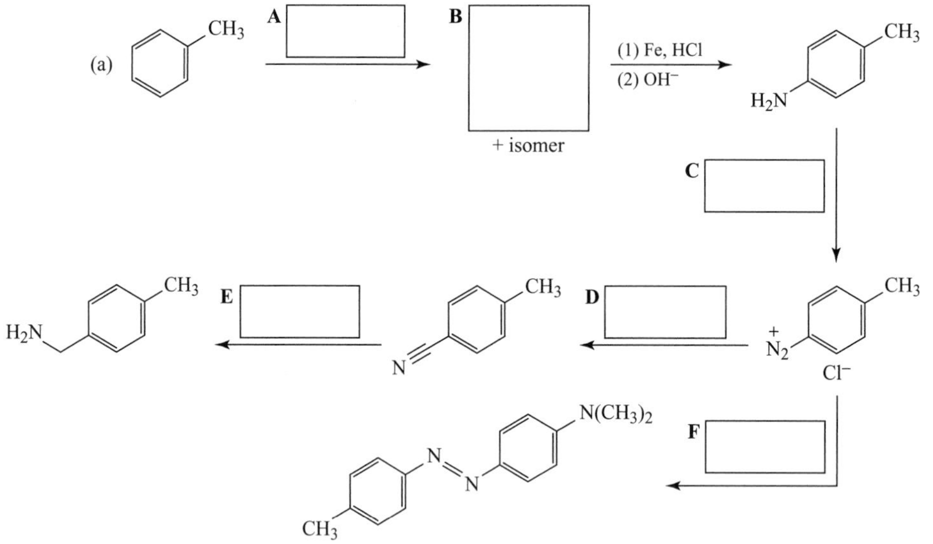

p-methylbenzenediazonium chloride —F→ azo compound (N=N, CH_3, $N(CH_3)_2$)

(b) Benzoyl chloride (O, Cl) —A→ B —heat→ phenyl isocyanate (N=C=O) —H_2O→ C (+ CO_2) —Br_2 (xs), H_2O→ D —(1) HONO, 0–5°C; (2) E→ 1,3,5-tribromobenzene (Br, Br, Br)

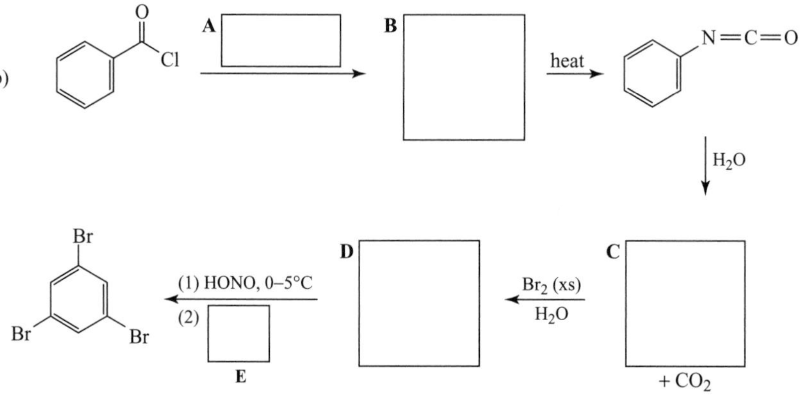

20.5 Select the stronger base from each pair (in aqueous solution):

(a) (1) p-toluidine (NH_2, CH_3) or (2) benzylamine (NH_2)

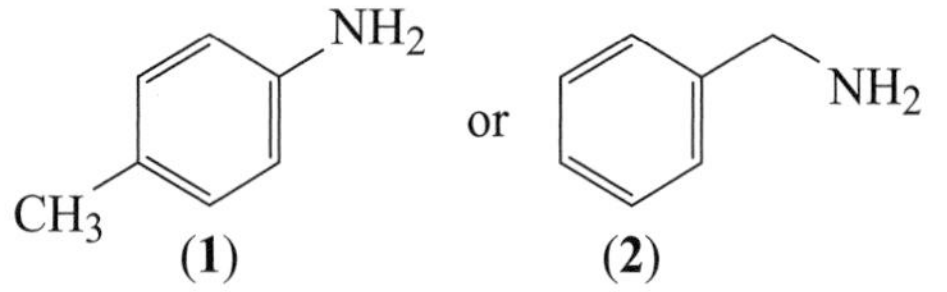

(b) (1) benzamide (O, NH_2) or (2) benzylamine (NH_2)

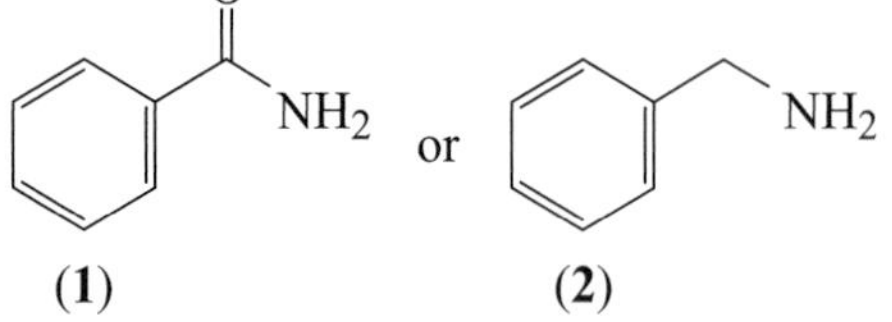

(c) (1) p-toluidine (NH_2, CH_3) or (2) p-nitroaniline (NH_2, O_2N)

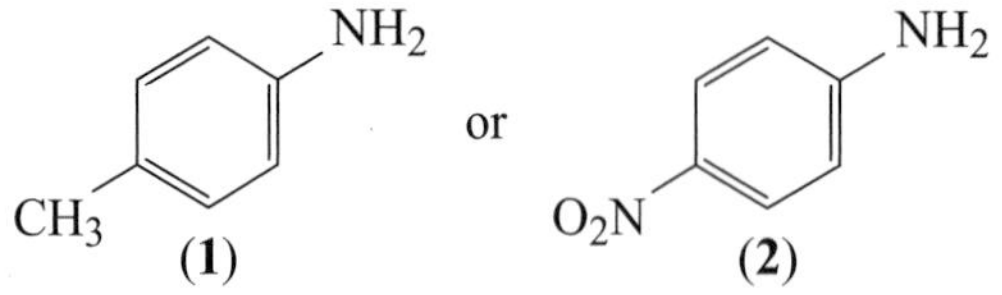

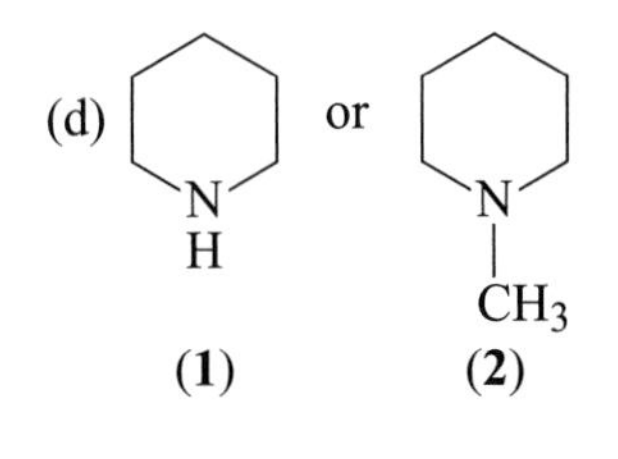

(e) H N or H N CH_3

(1) (2)

(f) N or N CH_3

(1) (2)

21 PHENOLS AND ARYL HALIDES: NUCLEOPHILIC AROMATIC SUBSTITUTION

SOLUTIONS TO PROBLEMS

21.1 The electron-releasing group (i.e., $-CH_3$) changes the charge distribution in the molecule so as to make the hydroxyl oxygen less positive, causing the proton to be held more strongly; it also destabilizes the phenoxide anion by intensifying its negative charge. These effects make the substituted phenol less acidic than phenol itself.

OH, CH_3 + H_2O ⇌ O^-, CH_3 δ^+ + H_3O^+

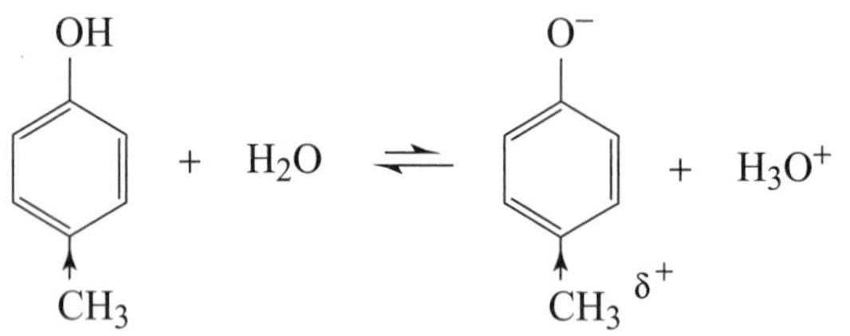

Electron-releasing $-CH_3$ destabilizes the anion more than the acid. pK_a is larger than for phenol.

21.2 An electron-withdrawing group such as chlorine changes the charge distribution in the molecule so as to make the hydroxyl oxygen more positive, causing the proton to be held less strongly; it also can stabilize the phenoxide ion by dispersing its negative charge. These effects make the substituted phenol more acidic than phenol itself.

OH, Cl + H_2O ⇌ O^-, Cl δ^- + H_3O^+

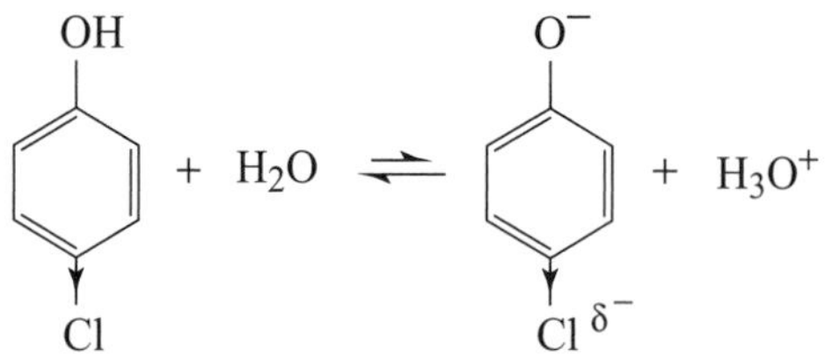

Electron-withdrawing chlorine stabilizes the anion by dispersing the negative charge. pK_a is smaller than for phenol.

Nitro groups are very powerful electron-withdrawing groups by their inductive and resonance effects. Resonance structures (**B–D**) below place a positive charge on the hydroxyl oxygen. This effect makes the hydroxyl oxygen dramatically more positive, causing the proton to be held much less strongly. These contributions explain why 2,4,6-trinitrophenol (picric acid) is so exceptionally acidic.

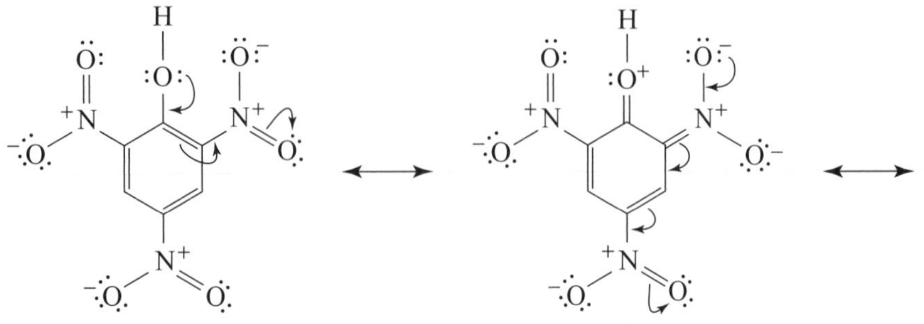

A

B

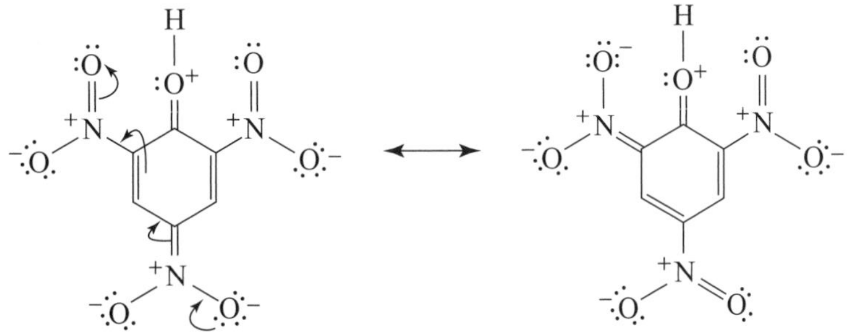

C

D

21.3 Dissolve the mixture in a solvent such as CH_2Cl_2 (one that is immiscible with water). Using a separatory funnel, extract this solution with an aqueous solution of sodium bicarbonate. This extraction will remove the benzoic acid from the CH_2Cl_2 solution and transfer it (as sodium benzoate) to the aqueous bicarbonate solution. Acidification of this aqueous extract will cause benzoic acid to precipitate; it can then be separated by filtration and purified by recrystallization.

The CH_2Cl_2 solution can now be extracted with an aqueous solution of sodium hydroxide. This will remove the 4-methylphenol (as its sodium salt). Acidification of the aqueous extract will cause the formation of 4-methylphenol as a water-insoluble layer. The 4-methylphenol can then be extracted into ether, the ether removed, and the 4-methylphenol purified by distillation.

The CH_2Cl_2 solution will now contain only toluene (and CH_2Cl_2). These can be separated easily by fractional distillation.

21.4 (a) 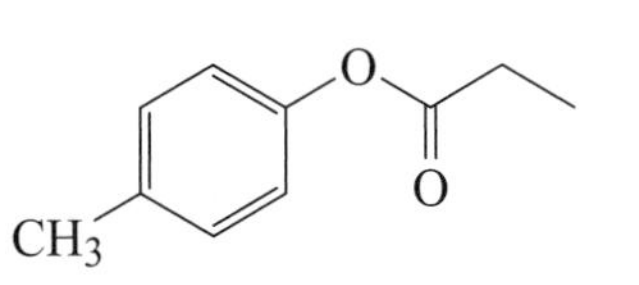 (b)

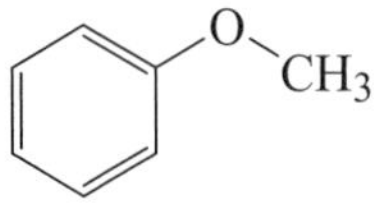

(c) 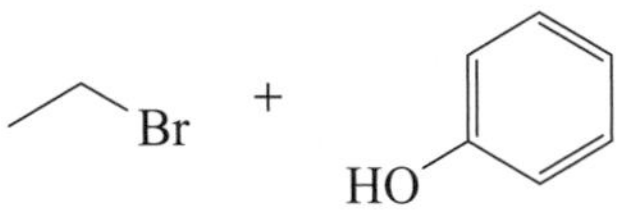 Br + HO (d)

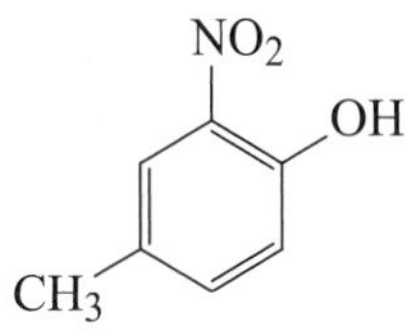

21.5 If the mechanism involved dissociation into an allyl cation and a phenoxide ion, then recombination would lead to two products: one in which the labeled carbon atom is bonded to the ring and one in which an unlabeled carbon atom is bonded to the ring.

dissociation

$H_2\overset{\delta+}{C}$ $\overset{14\ \delta+}{CH_2}$

(recombination)

OH OH

$^{14}CH_2$ $^{14}CH_2$

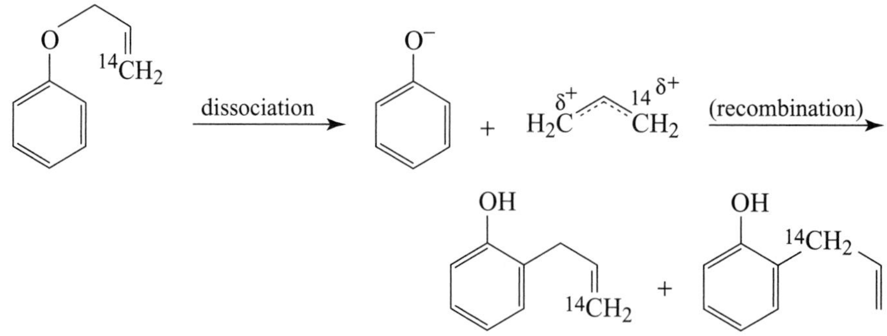

The fact that all of the product has the labeled carbon atom bonded to the ring eliminates this mechanism from consideration.

21.6

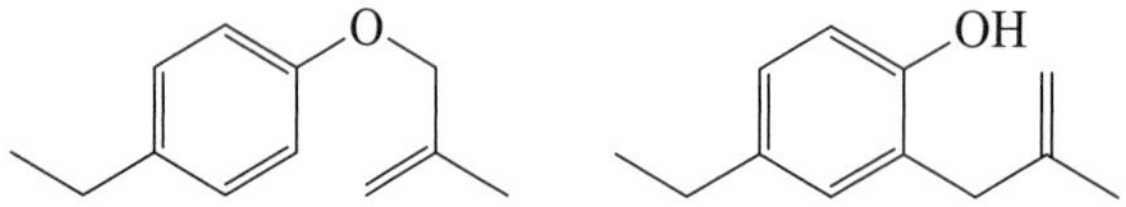

A **B**

21.7 (a) (b)

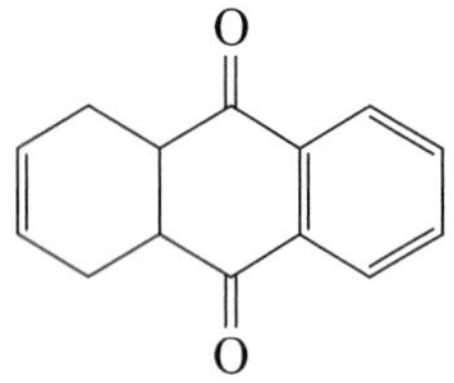

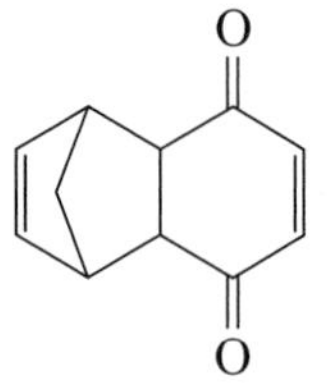

21.8

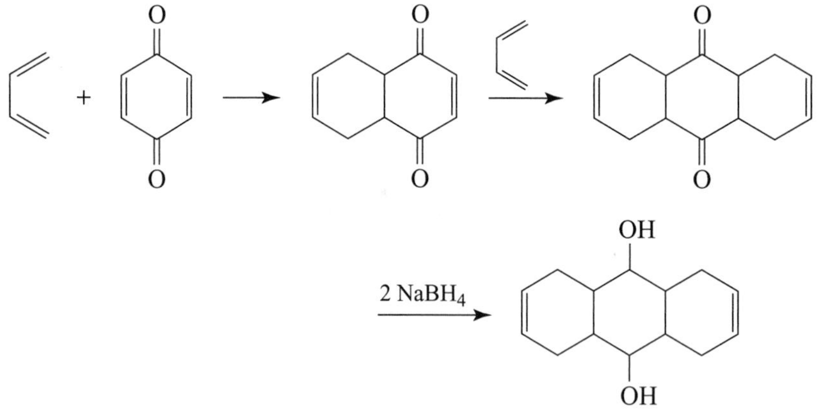

2 $NaBH_4$

OH

OH

21.9 (a) 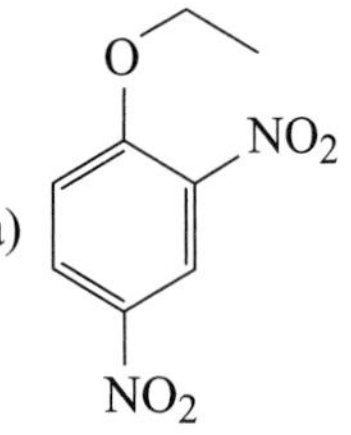 (b) 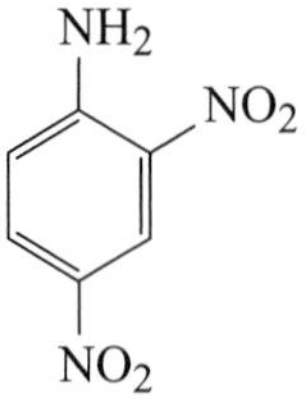 (c) 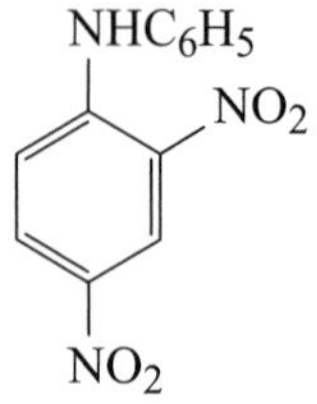 (d)

NO_2 NH_2 NHC_6H_5 NO_2

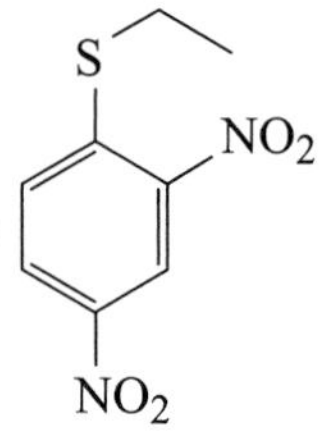

21.10 That *o*-chlorotoluene leads to the formation of two products (*o*-cresol and *m*-cresol) when submitted to the conditions used in the Dow process suggests that an elimination-addition mechanism takes place.

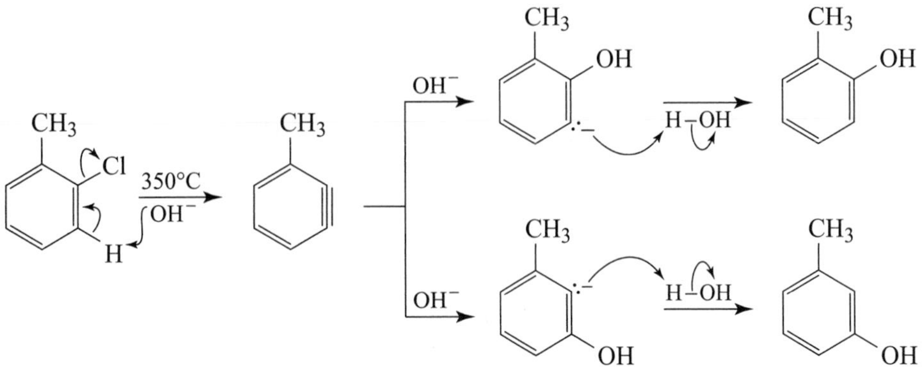

Inasmuch as chlorobenzene and *o*-chlorotoluene should have similar reactivities under these conditions, it is reasonable to assume that chlorobenzene also reacts by an elimination-addition mechanism in the Dow process.

21.11 2-Bromo-1,3-dimethylbenzene, because it has no α-hydrogen, cannot undergo an elimination. Its lack of reactivity toward sodium amide in liquid ammonia suggests that those compounds (e.g., bromobenzene) that do react do so by a mechanism that begins with an elimination.

21.12 (a) One molar equivalent of $NaNH_2$ converts acetoacetic ester to its anion,

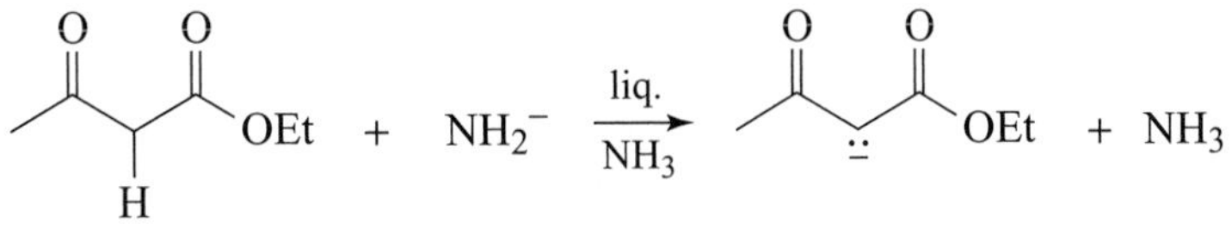

and one molar equivalent of $NaNH_2$ converts bromobenzene to benzyne (cf. Section 21.11B):

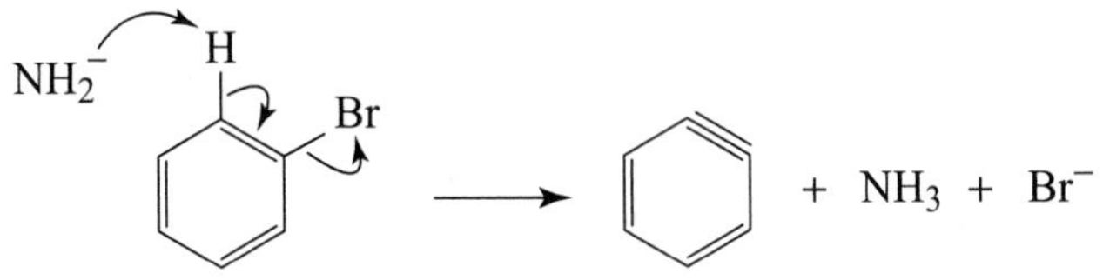

Then the anion of acetoacetic ester adds to the benzyne as it forms in the mixture.

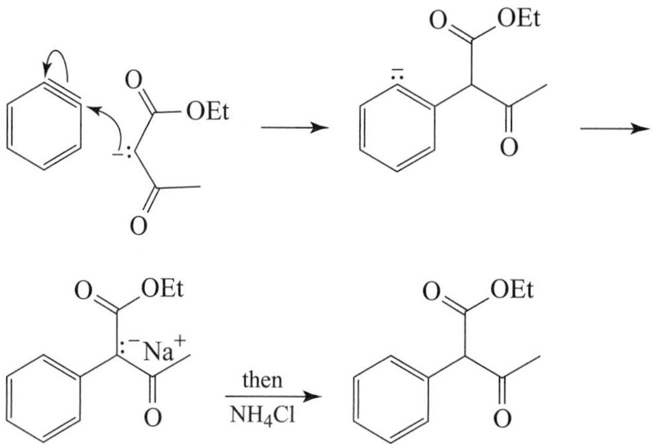

O OEt O OEt

then NH_4Cl

O O O

This is the end product of the addition.

(b) 1-phenyl-2-propanone, as follows:

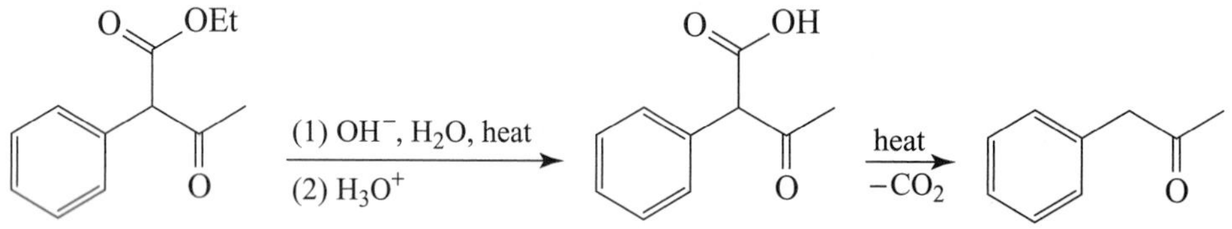

O OEt O OH

(1) OH^-, H_2O, heat

(2) H_3O^+

heat $-CO_2$

O O O

(c) By treating bromobenzene with diethyl malonate and two molar equivalents of $NaNH_2$ to form diethyl phenylmalonate.

Br + OEt OEt

2 $NaNH_2$

liq. NH_3

O OEt OEt O

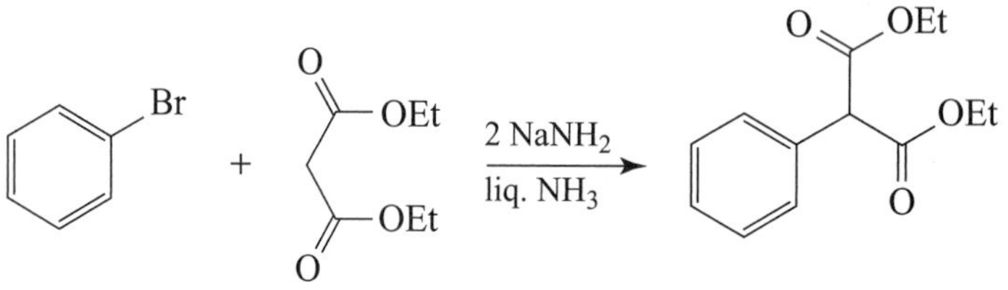

[The mechanism for this reaction is analogous to that given in part (a).]

Then hydrolysis and decarboxylation will convert diethyl phenylmalonate to 2-phenylethanoic acid.

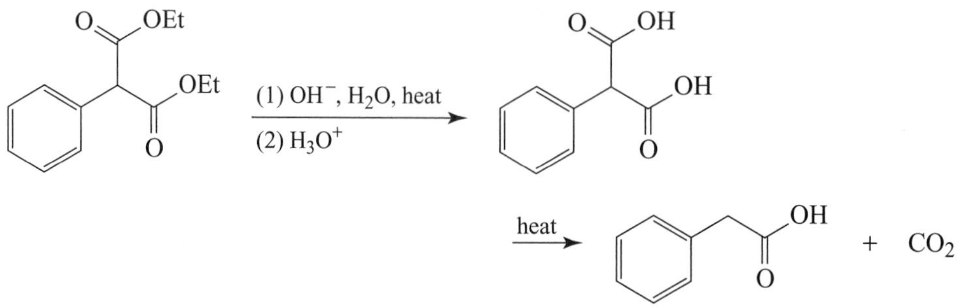

O OEt OEt O

(1) OH^-, H_2O, heat

(2) H_3O^+

O OH OH O

heat

OH O + CO_2

Problems

Physical Properties

21.13 4-methoxyphenol (2), 4-nitrocyclohexanol (1), 4-(trifluoromethyl)phenol (3), 4-(trifluoromethyl)benzoic acid (5), benzoic acid (4)

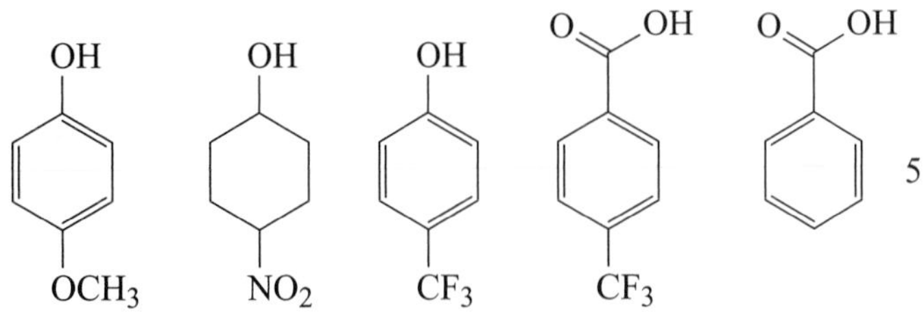

5 = most acidic

21.14 (a) 4-Fluorophenol because a fluorine substituent is more electron withdrawing than a methyl group.

(b) 4-Nitrophenol because a nitro group is more electron withdrawing than a methyl group.

(c) 4-Nitrophenol because the nitro group can exert an electron-withdrawing effect through a resonance effect.

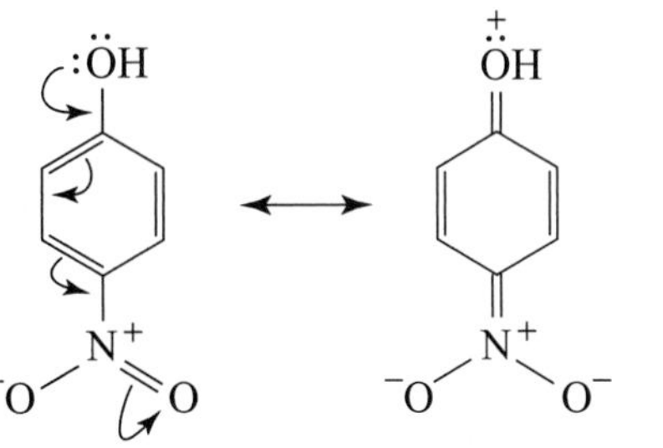

In 3-nitrophenol this resonance effect is not possible.

(d) 4-Methylphenol because it is a phenol, not an alcohol.

(e) 4-Fluorophenol because fluorine is more electronegative than bromine.

21.15 (a) 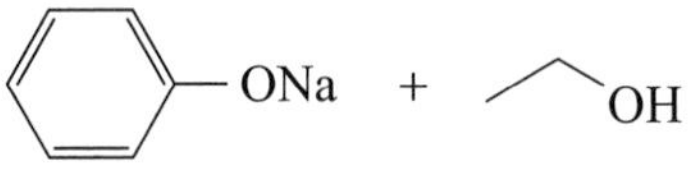 C_6H_5—ONa + ethanol (OH)   (b) C_6H_5—ONa + H_2O

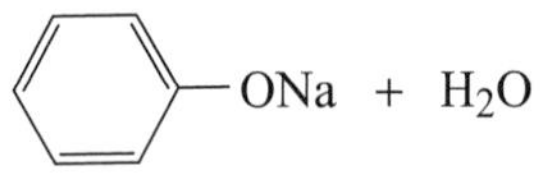

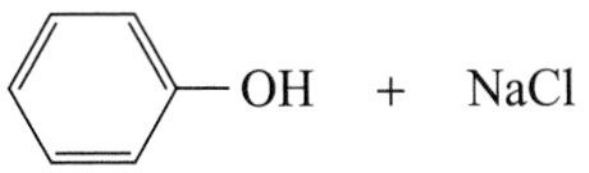

(c) C_6H_5—OH + NaCl   (d) 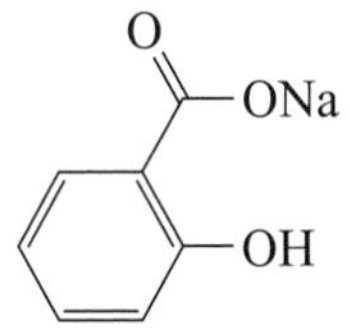 2-hydroxybenzoate (C(=O)ONa, OH)

21.16 (a) 4-Chlorophenol will dissolve in aqueous NaOH; 4-chloro-1-methylbenzene will not.

(b) 4-Methylbenzoic acid will dissolve in aqueous $NaHCO_3$; 4-methylphenol will not.

(c) 2,4,6-Trinitrophenol, because it is so exceptionally acidic ($pK_a = 0.38$), will dissolve in aqueous $NaHCO_3$; 4-methylphenol ($pK_a = 10.17$) will not.

(d) 4-Ethylphenol will dissolve in aqueous NaOH; ethyl phenyl ether will not.

General Reactions

21.17 (a) 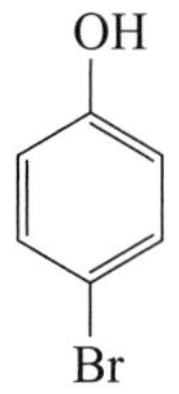 OH, Br (major)

(b) 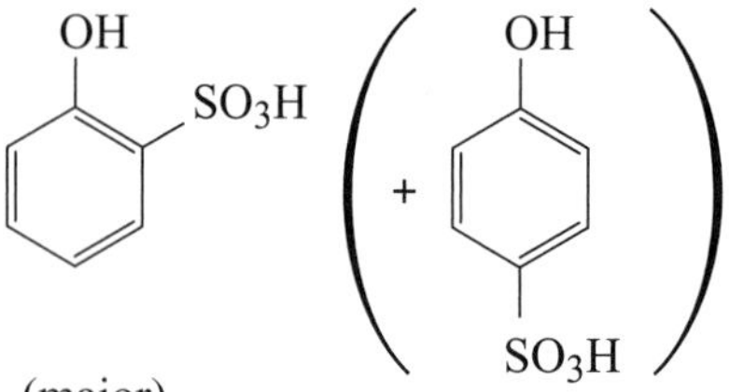 OH, SO_3H (major) (+ OH, SO_3H)

(c) 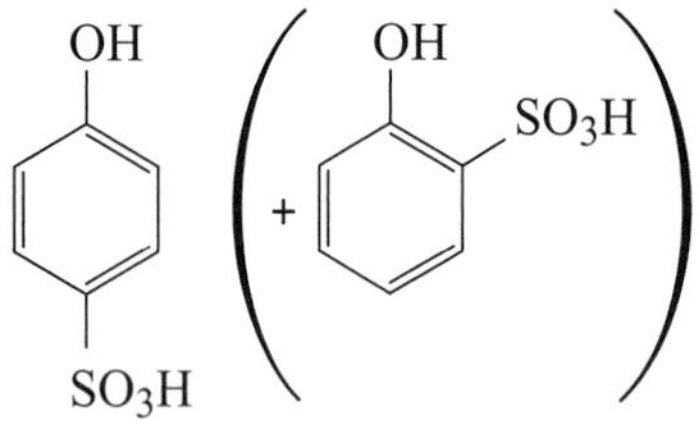 OH, SO_3H (major) (+ OH, SO_3H)

(d) 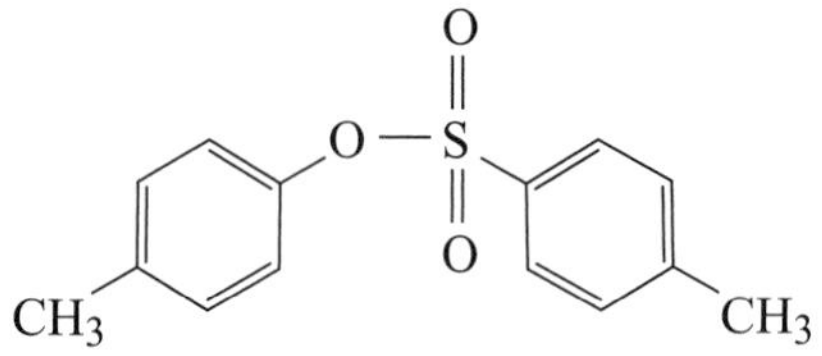 O, O—S, O, CH_3, CH_3

(e) 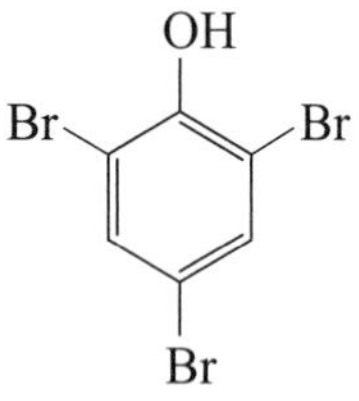 OH, Br, Br, Br

(f) 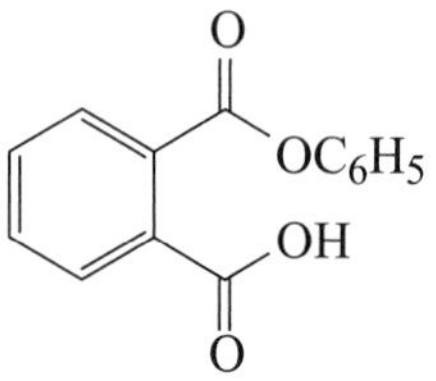 O, OC_6H_5, OH, O

(g) 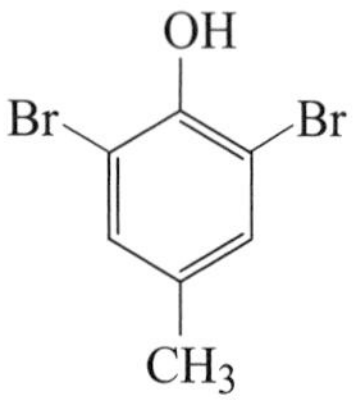 OH, Br, Br, CH_3

(h) 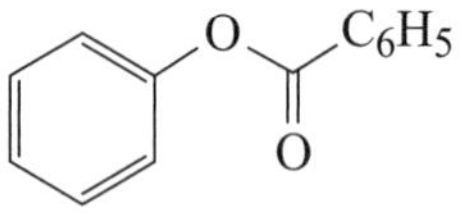 O, C_6H_5, O

(i) Same as (h) + O, C_6H_5, O^-

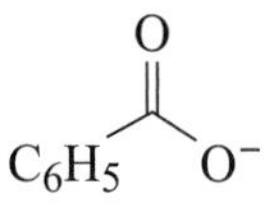

(j) 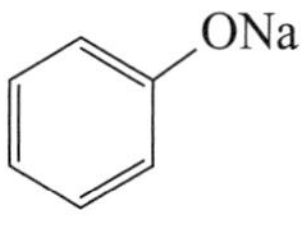 ONa

(k) 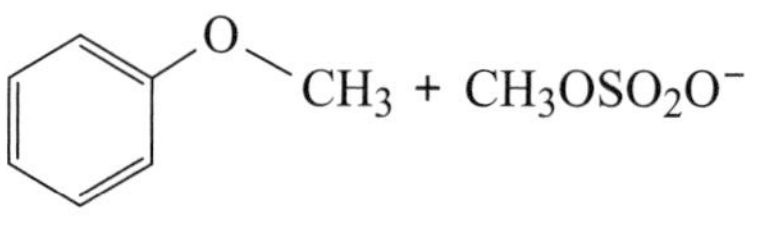 O, CH_3 + $CH_3OSO_2O^-$

(l) $C_6H_5OCH_3$

(m) 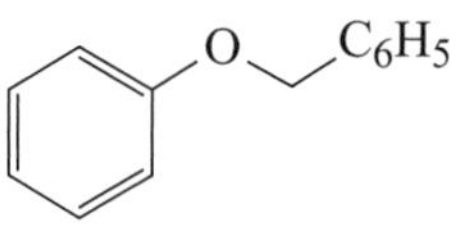 O, C_6H_5

21.18 Predict the product of the following reactions.

(a) 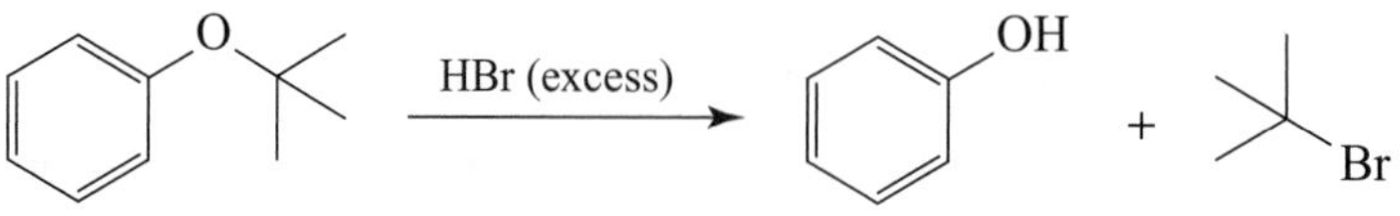 O —HBr (excess)→ OH + Br

(b) OH (1) NaH (2) Br O

(c) OH O CH$_3$ O O O O O + O OH O CH$_3$

(d) OH H_3C Br_2 (excess) H_2O Br OH H_3C Br

(e) OH OH O O O Excess O O O O O + O OH

(f) H_3C OH (1) NaH, Br (2) heat H_3C OH

(g) OH HNO_3, H_2SO_4 NO_2 OH

(h) 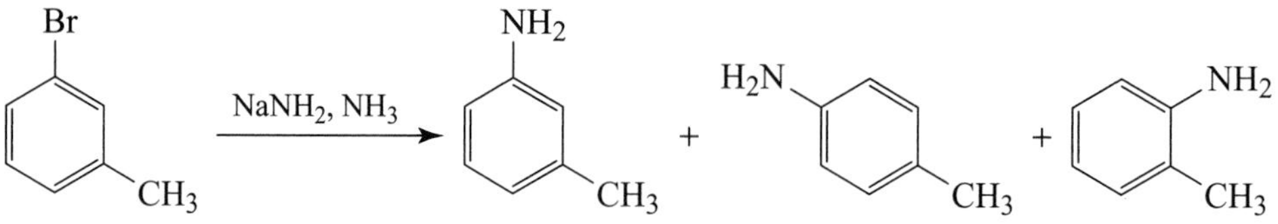 Br, CH_3 — $NaNH_2$, NH_2 → NH_2, CH_3 + H_2N, CH_3 + NH_2, CH_3

(i) 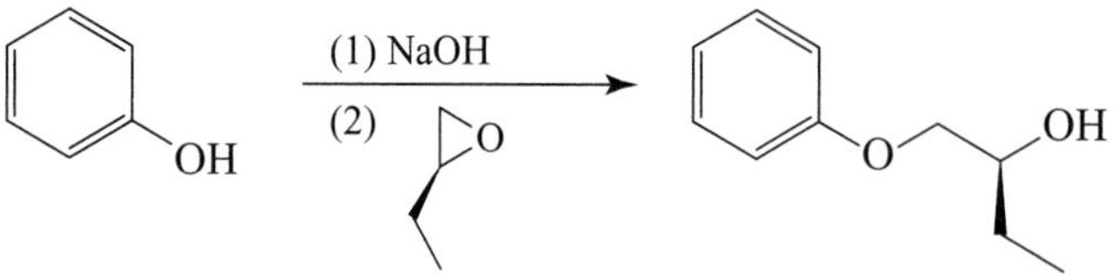 OH — (1) NaOH (2) O → O, OH

Mechanisms and Synthesis

21.19 $\ddot{O}H$, CH_3 + Cl, O (Epichlorohydrin) → O, OH, Cl, CH_3

OH^- → O, O, CH_3 — $H_2\ddot{N}$ → O, OH, H, N, CH_3 Toliprolol

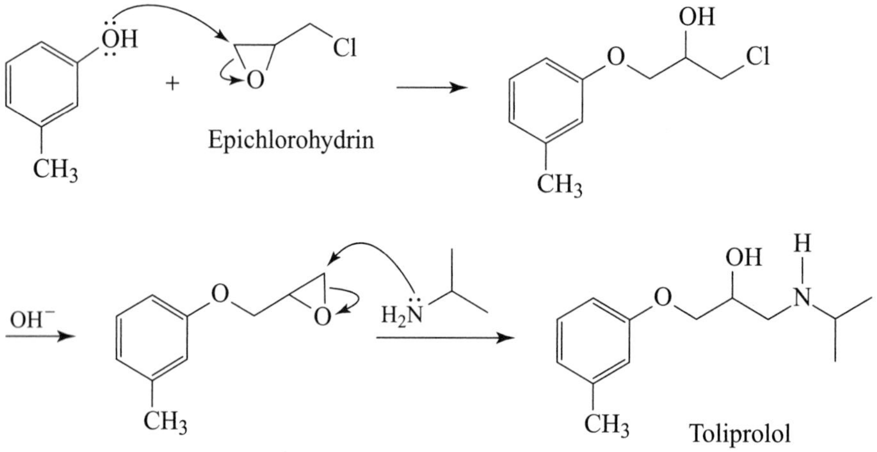

21.20 One can draw a resonance structure for the 2,6-di-*tert*-butylphenoxide ion which shows the carbon para to the oxygen as a nucleophilic center; that is, the species is an ambident nucleophile.

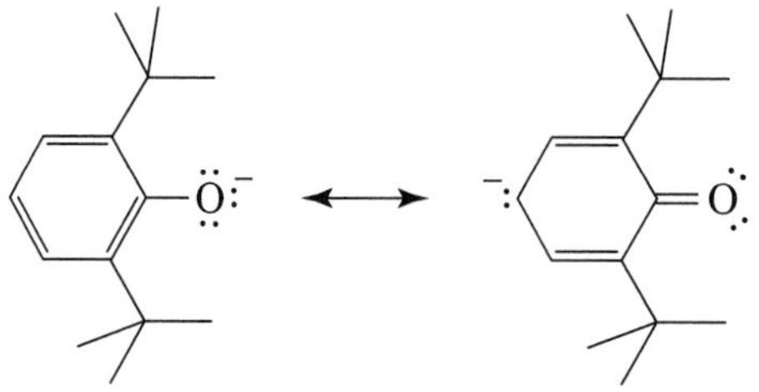

Given the steric hindrance about the oxygen, the nucleophilic character of the para carbon is dominant and an S_NAr reaction occurs at that position to produce this biphenyl derivative:

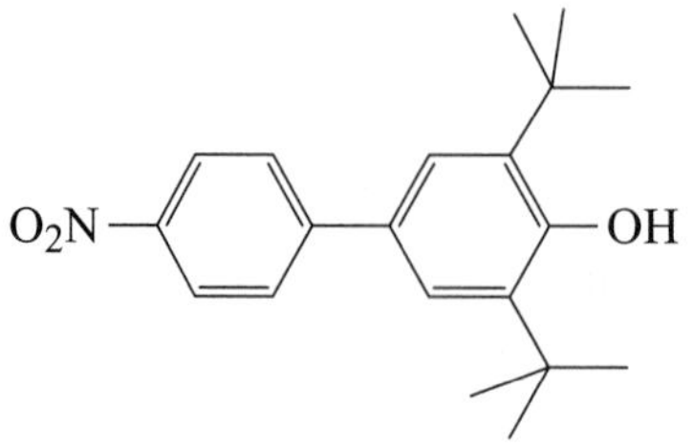

O_2N — OH

21.21 Both *o*- and *m*-toluidine are formed in the following way:

CH_3 H   Cl $^-:NH_2$ 350° C CH_3

NH_2^- CH_3 NH_2 H–NH_2 CH_3 NH_2

NH_2^- CH_3 H–NH_2 NH_2 CH_3 NH_2

Both *m*- and *p*-toluidine are formed from another benzyne-type intermediate.

CH_3 Cl   H $^-:NH_2$ 350° C CH_3

NH_2^- CH_3 NH_2 H–NH_2 CH_3 NH_2

NH_2^- CH_3 NH_2 H – NH_2 CH_3 NH_2

21.22 OH H_2SO_4 / 25° C   OH SO_3H 2 Cl_2 Cl   OH SO_3H Cl H_3O^+ / H_2O (steam)

Cl   OH   Cl   NaOH / Cl–CH₂–C(=O)–OH   Cl $:\ddot{O}:^-$ Na^+ Cl   +   Cl–CH₂–C(=O)–ONa ⟶

Cl   O–CH₂–C(=O)–ONa   Cl H_3O^+ Cl   O–CH₂–C(=O)–OH   Cl

2,4–D

21.23

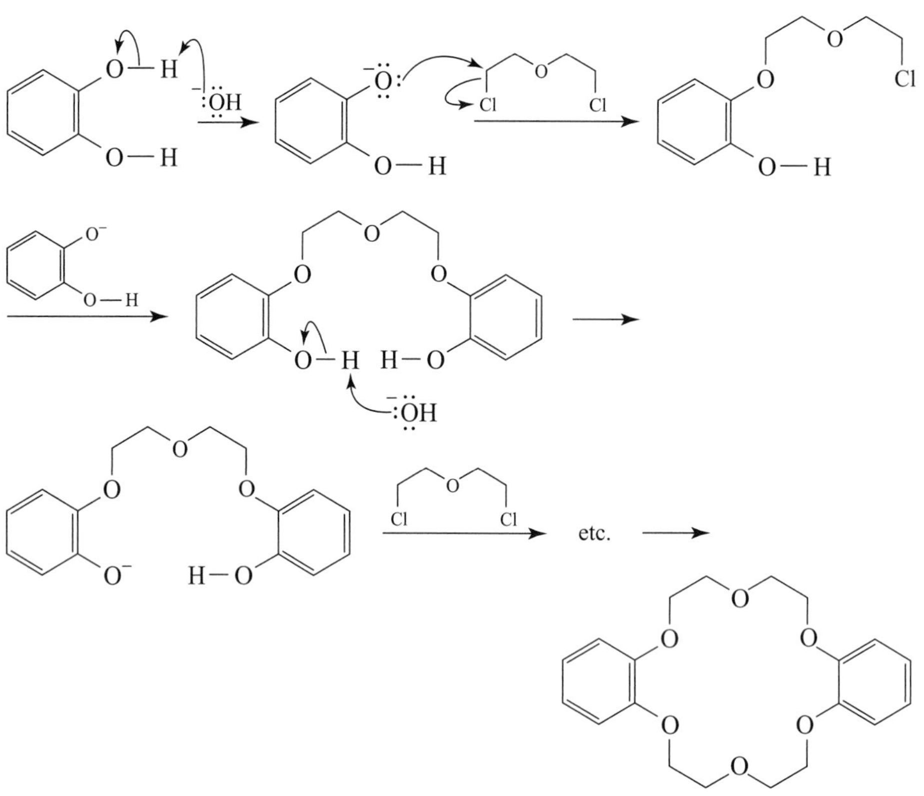

Dibenzo-18-crown-6

21.24

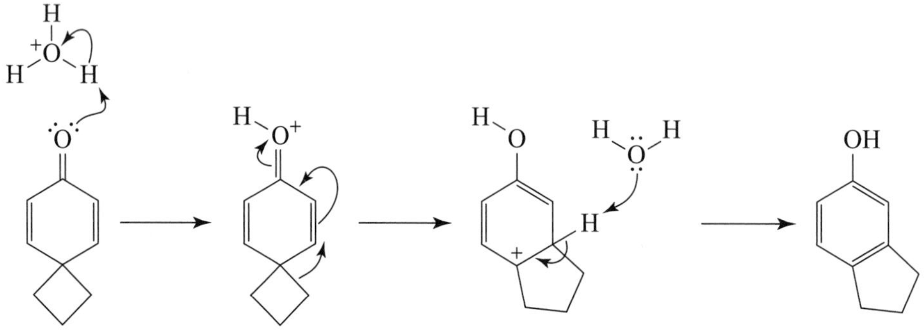

21.25

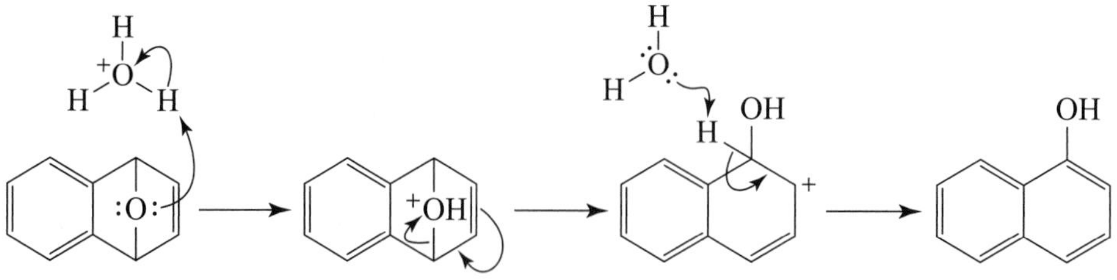

21.26 (a) [4-methoxyphenol] + [isobutylene] $\xrightarrow{HA}$ [2-tert-butyl-4-methoxyphenol] + [3-tert-butyl-4-methoxyphenol]

OH, OCH_3, OH, OCH_3, OH, OCH_3

BHA

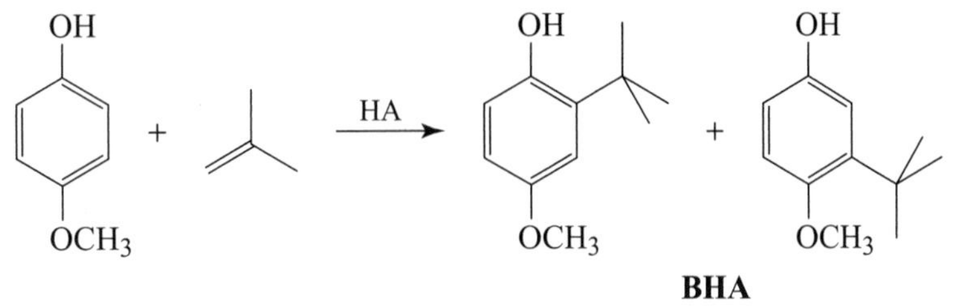

(b) [4-methylphenol] + 2 [isobutylene] $\xrightarrow{HA}$ [2,6-di-tert-butyl-4-methylphenol]

OH, CH_3, OH, CH_3

BHT

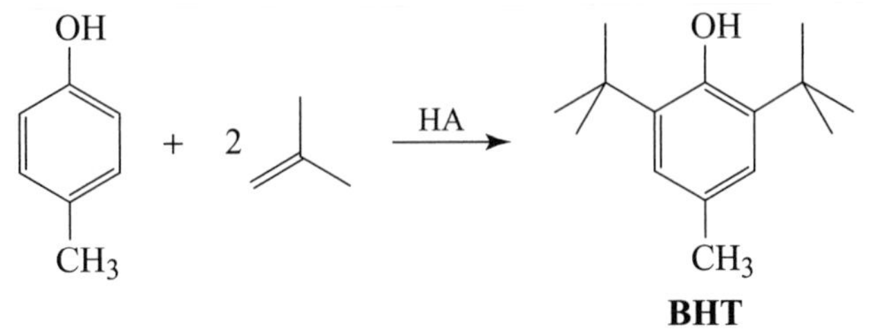

Notice that both reactions are Friedel-Crafts alkylations.

21.27

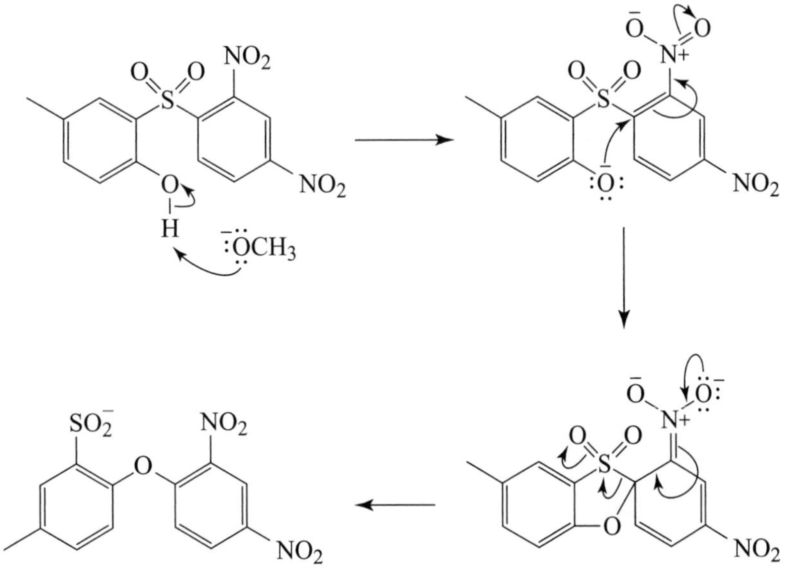

NO_2, NO_2, $:\ddot{O}CH_3$, SO_2^-, NO_2, NO_2

21.28 The phenoxide ion has nucleophilic character both at the oxygen and at the carbon para to it; it is ambident.

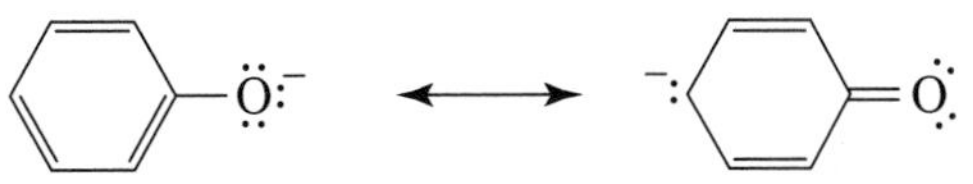

The benzyne produced by the elimination phase of the reaction can then undergo addition by attack by either nucleophilic center: reaction at oxygen producing **1** (diphenyl ether), and reaction at carbon producing **2** (4-hydroxybiphenyl).

21.29 (a) 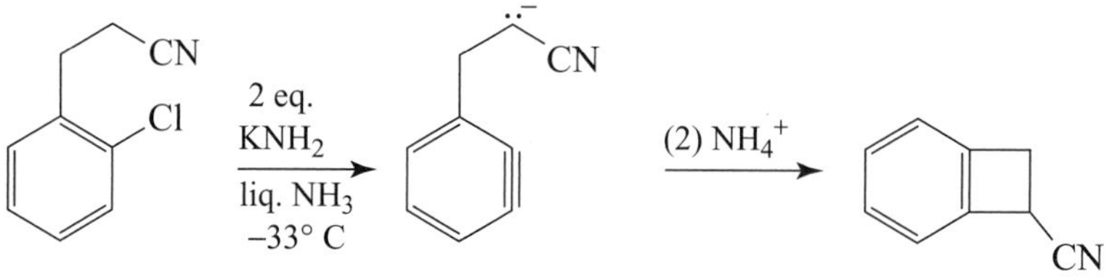

(b) 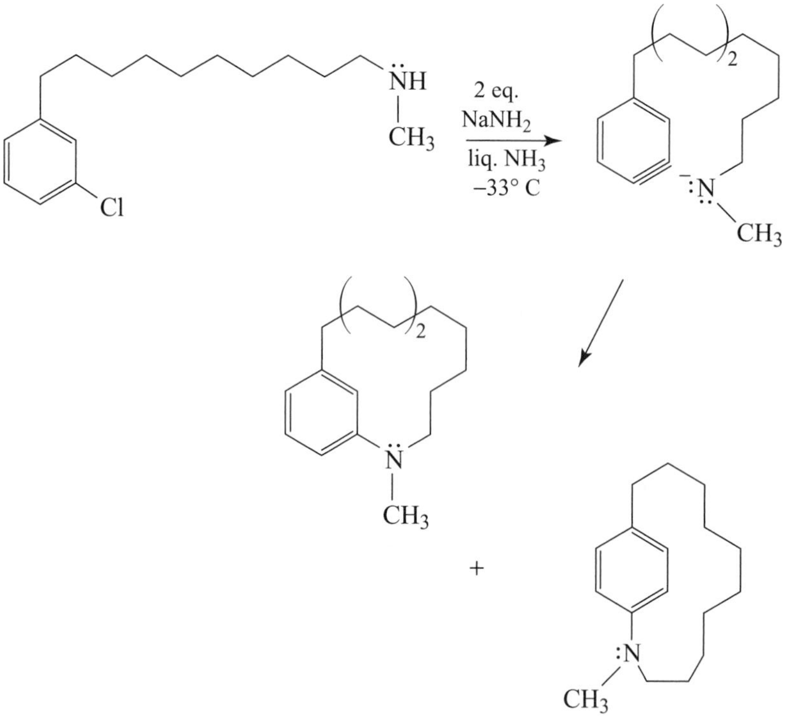

21.30 The proximity of the two —OH groups results in the two naphthalene nuclei being non-coplanar. As a result, the two enantiomeric forms are nonequivalent and can be separated by a resolution technique.

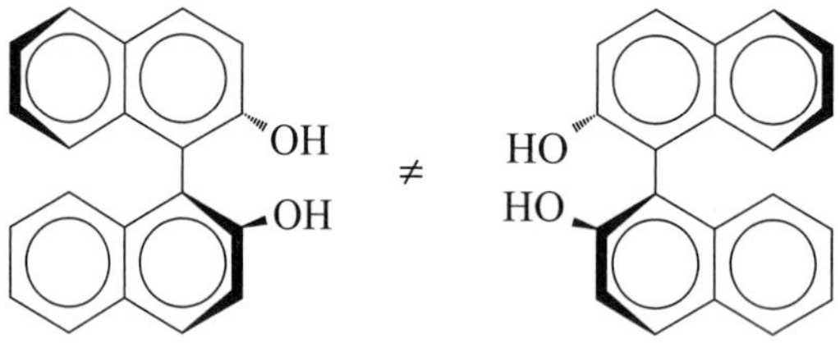

21.31 (a) 

(b) Because the phenolic radical is highly stabilized by resonance (see the following structures), it is relatively unreactive.

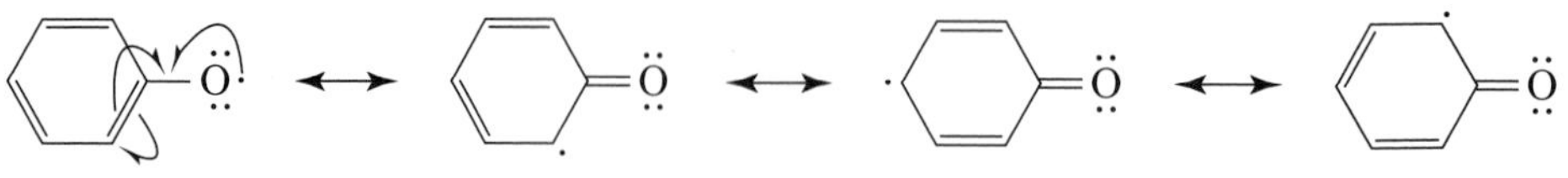

Spectroscopy

21.32 **X** is a phenol because it dissolves in aqueous NaOH but not in aqueous $NaHCO_3$. It gives a dibromo derivative and must therefore be substituted in the ortho or para position. The broad IR peak at 3250 cm^{-1} also suggests a phenol. The peak at 830 cm^{-1} indicates para substitution. The ^1H NMR singlet at δ 1.3 (9H) suggests nine equivalent methyl hydrogen atoms, which must be a *tert*-butyl group. The structure of **X** is

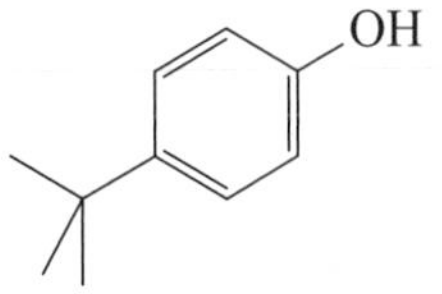

OH

21.33 The broad IR peak at 3200–3600 cm^{-1} suggests a hydroxyl group. The two ^1H NMR peaks at δ 1.67 and δ 1.74 are not a doublet because their separation is not equal to other splittings; therefore, these peaks are singlets. Reaction with Br_2/CCl_4 suggests an alkene. If we put these bits of information together, we conclude that **Z** is 3-methyl-2-buten-1-ol (3-methylbut-2-en-1-ol).

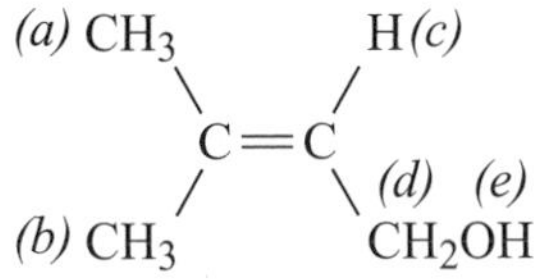

(*a*) CH_3, (*b*) CH_3, C=C, H(*c*), (*d*) (*e*) CH_2OH

(*a*) and (*b*) singlets δ 1.67 and δ 1.74
(*c*) multiplet δ 5.40
(*d*) doublet δ 4.10
(*e*) singlet δ 2.3

Challenge Problems

21.34 Initially, rearrangement to the ortho position does occur, but subsequent aromatization cannot take place. A second rearrangement occurs to the para position (with the labeled carbon again being at the end of the allyl group); now tautomerization can result in reestablishing an aromatic system.

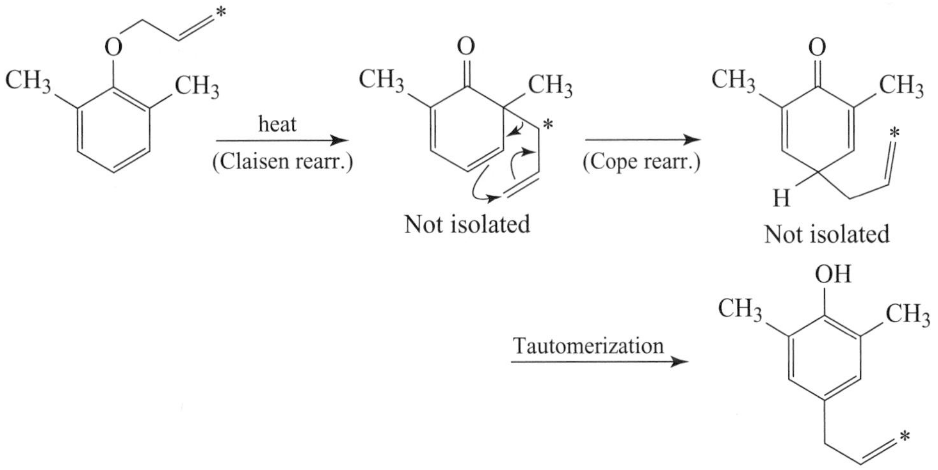

heat (Claisen rearr.) → Not isolated → (Cope rearr.) → Not isolated → Tautomerization →

21.35 There are alternative representations for **I**, namely:

A ⟷ **B**

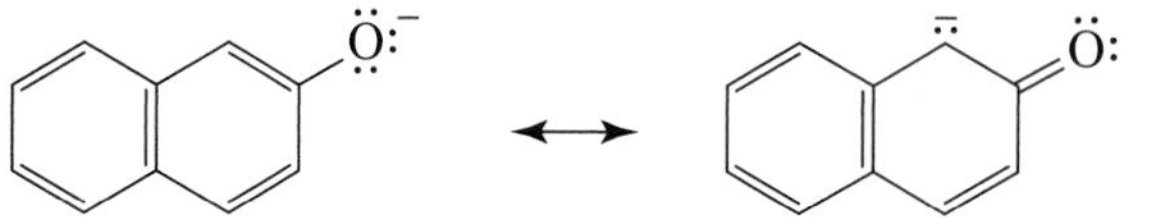

Thus, **I** is an ambident nucleophile. *O*-alkylation via the nucleophile represented by **A** is anticipated since the negative charge resides primarily on the more electronegative atom. Use of DMF as solvent provides for a nucleophile unencumbered with solvent.

When protic solvents are used, they strongly hydrogen bond to the oxygen atom and so reduce the nucleophilicity of the oxygen; *C*-alkylation then is favored.

21.36 In those cases where the S_NAr mechanism holds, the first step is rate-determining; the C—Nu bond forms, but there is no C—F bond breaking. The strongly electronegative fluorine facilitates this step by helping to disperse the negative charge on the carbanionic intermediate. The other halogens, being less electronegative, are less effective in this.

In S_N1 reactions the considerable strength of the C—F bond results in very slow dissociation of the RF compound. In S_N2 reactions, since F^- is the most basic halide ion it is the poorest leaving group.

21.37 Examination of the resonance structures that can be drawn for the S_NAr intermediate in the first case reveals that the five-membered ring has cyclopentadienyl anion character. This is a stabilizing feature.

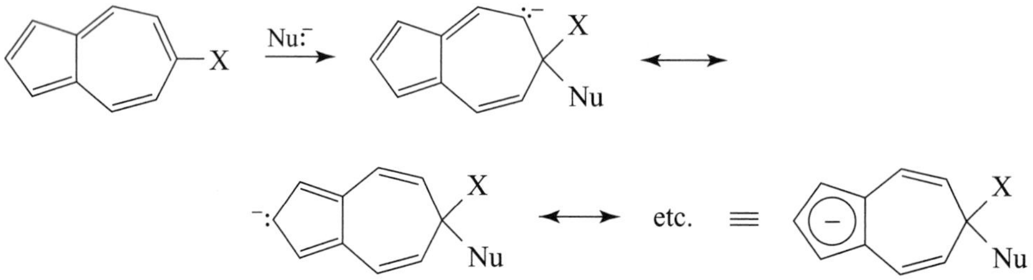

X —Nu:⁻→ X, Nu ⟷ X, Nu ⟷ etc. ≡ X, Nu

In the latter case, electron-delocalization into the other (seven-membered) ring results in eight electrons in that ring, not a factor which makes for stabilization.

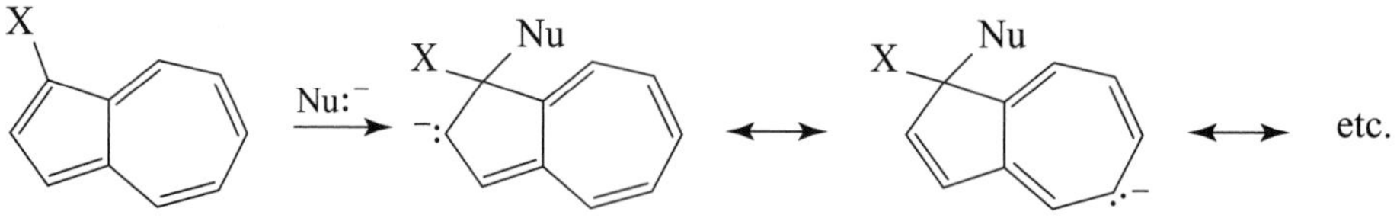

X —Nu:⁻→ X, Nu ⟷ X, Nu ⟷ etc.

21.38 NH_2^- ; Tautomeric shift

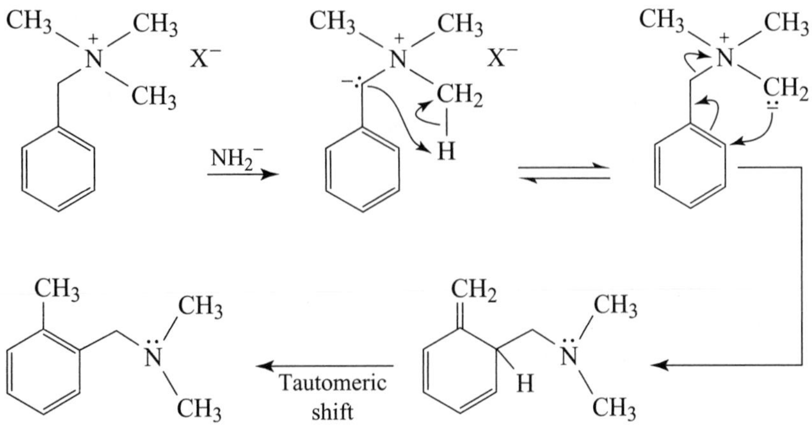

21.39 excess Cl_2, Fe ; $H_2C{=}O + HA$; 1. OH^- 2. H_3O^+ ; 2,4,5-TCP, HA ; Hexachlorophene

2,4,5-Trichlorophenol (2,4,5-TCP)

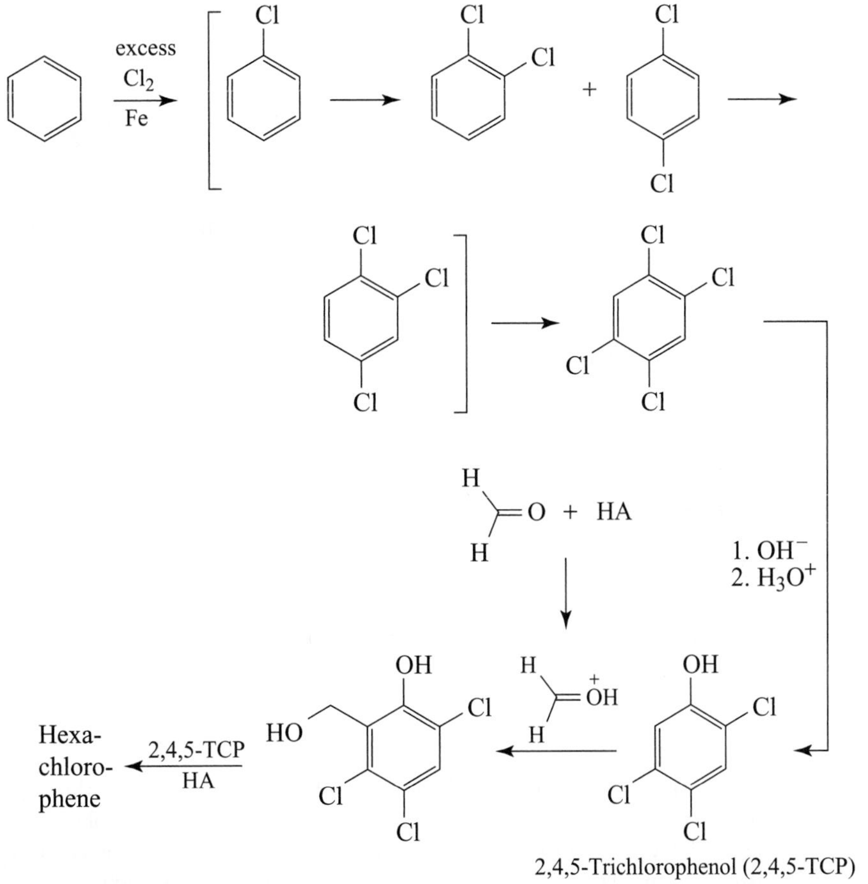

21.40 (a) Carry out the reaction in the presence of another aromatic compound, ideally another phenol. If the reaction is intermolecular, acylation of the other phenol would be expected. If that acylation does not occur, the reaction apparently is intramolecular. (In principle, this

approach should provide the answer to the question. However, the situation is more complex since other research supports an intermolecular process.)

(b) The temperature effects suggest that the para product is kinetically favored with the higher temperatures leading to the thermodynamically favored (more stable) ortho product. The ortho product is stabilized by hydrogen bonding or complex formation with $AlCl_3$.

21.41

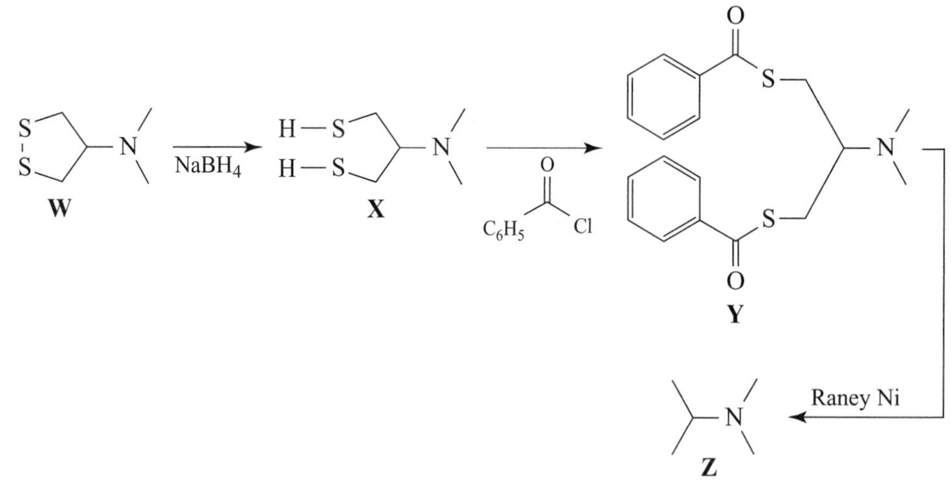

W —($NaBH_4$)→ X —(C_6H_5COCl)→ Y —(Raney Ni)→ Z

21.42 (a, b)

BH_4^-; H_3O^+ (or BH_3); $-H_2O$ (or $-HOBH_3^-$); H(or BH_3^-)

A

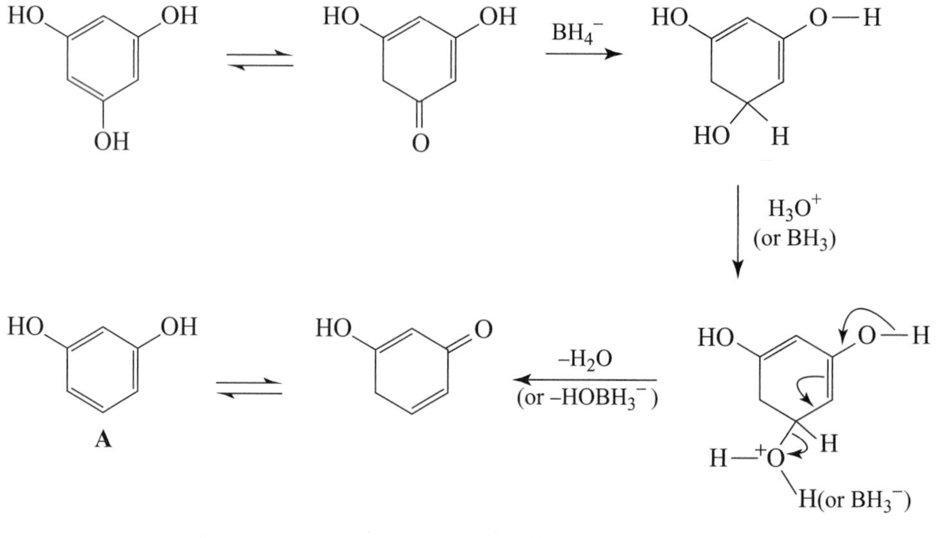

21.43 (a) HOMO-1 best represents the region where electrons of the additional π bond would be found because it has a large lobe extending toward the periphery of the carbon ring, signifying the high probability of finding the additional π electrons in this region. HOMO-2 also has a small lobe inside the ring, corresponding to the second lobe expected for a typical π bond. Overall, the molecular orbital bears less resemblance to that of a simple alkene π bond, however, because of the complexity of the molecule.

(b) Only a vacant orbital can accept an electron pair from a Lewis base, hence it is the LUMO+1 (next to the lowest *unoccupied* molecular orbital) that is involved in the nucleophilic addition step.

(c) The HOMO, HOMO-2 and HOMO-3 are the orbitals associated with the six electrons of the aromatic π system. These orbitals approximate the shape of and have the same symmetry as the three bonding aromatic π molecular orbitals of benzene.

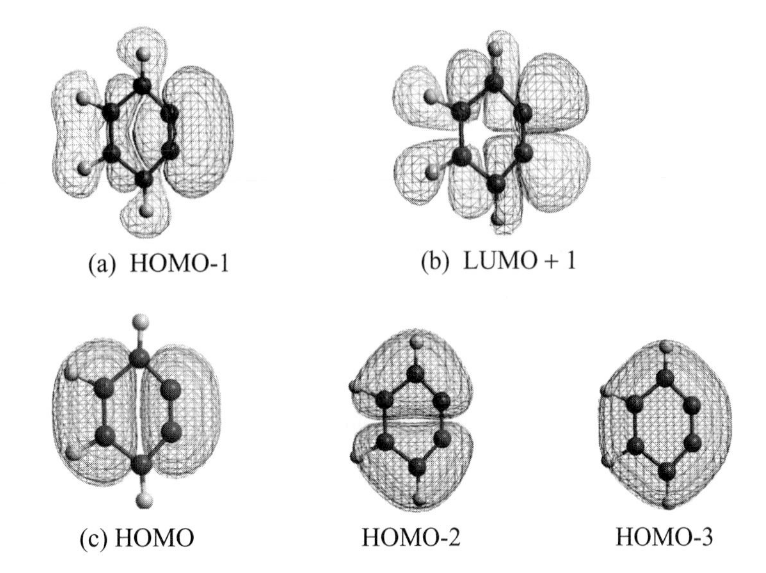

QUIZ

21.1 Which of the following would be the strongest acid?

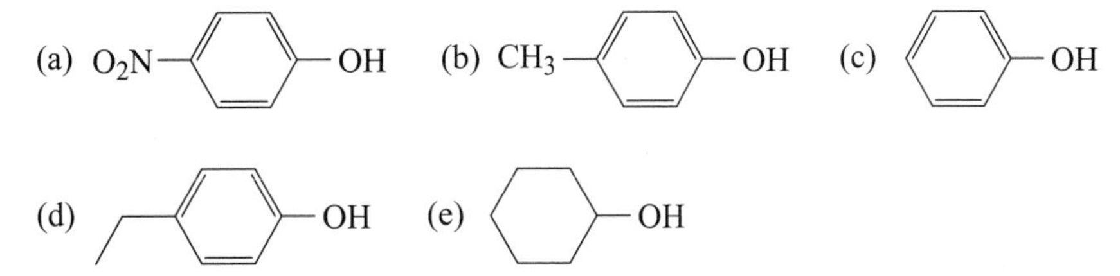

21.2 What products would you expect from the following reaction?

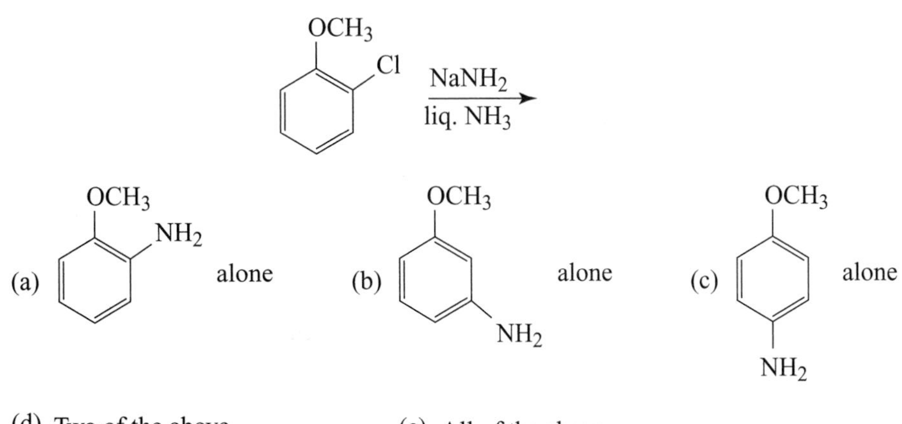

(d) Two of the above

(e) All of the above

21.3 Which of the reagents listed here would serve as the basis for a simple chemical test to distinguish between

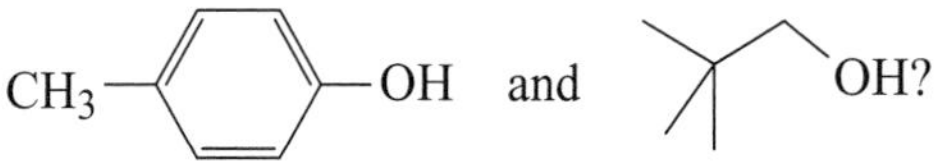

CH_3–C$_6$H$_4$–OH and (CH$_3$)$_3$C–CH$_2$–OH?

(a) $Ag(NH_3)_2OH$

(b) $NaOH/H_2O$

(c) Dilute HCl

(d) Cold concd H_2SO_4

(e) None of the above

21.4 Indicate the correct product, if any, of the following reaction.

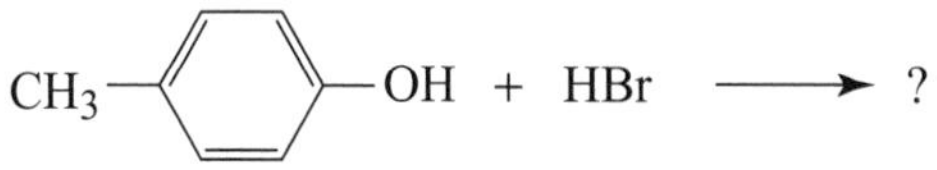

CH_3–C$_6$H$_4$–OH + HBr ⟶ ?

(a) 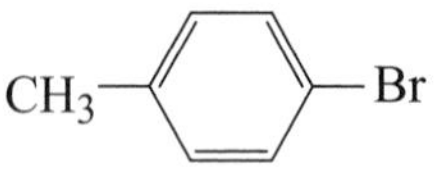 CH_3–C$_6$H$_4$–Br (b) 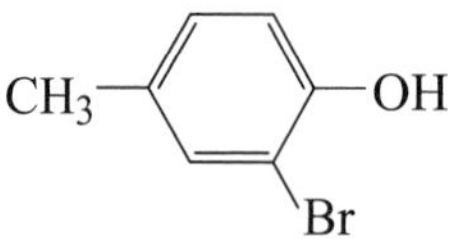 CH_3–C$_6$H$_3$(Br)–OH (c) CH_3–C$_6$H$_3$(Br)–OH

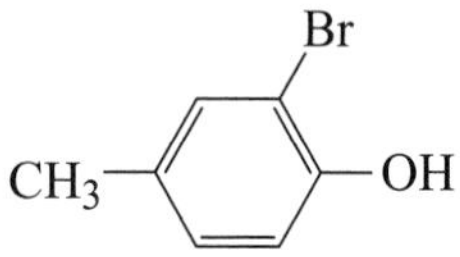

(d) 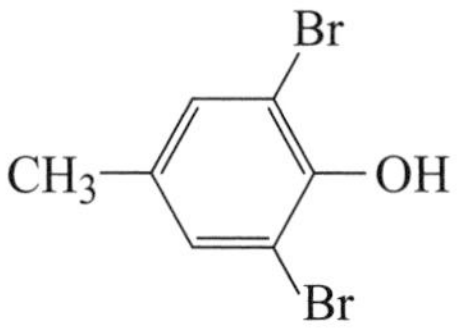 CH_3–C$_6$H$_2$(Br)$_2$–OH (e) There is no reaction.

21.5 Complete the following synthesis:

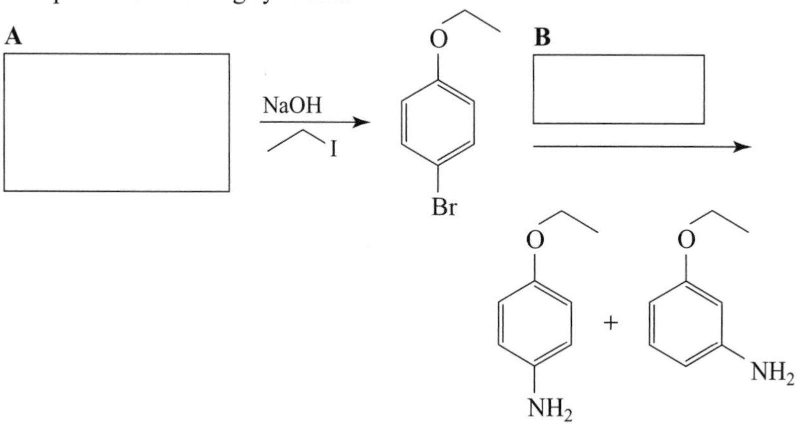

A [ ] $\xrightarrow[\text{CH}_3\text{CH}_2\text{I}]{\text{NaOH}}$ 4-bromo-1-ethoxybenzene (O, Br) **B** [ ] ⟶ 4-ethoxyaniline (O, NH_2) + 3-ethoxyaniline (O, NH_2)

21.6 Give the products:

CH_3O–C₆H₄–Br $\xrightarrow[\text{heat}]{\text{concd HBr}}$ [ ] + [ ]

21.7 Select the stronger acid.

(a) C_6H_5COOH **(1)** or CH_3–C₆H₄–OH **(2)**

(b) O_2N–C₆H₄–OH **(1)** or CH_3–C₆H₄–OH **(2)**

(c) Cl–C₆H₄–OH **(1)** or (2-chlorophenol: Cl ortho to OH) **(2)**

The solutions in this section correspond with problems that can be found only in the WileyPLUS course for this text. WileyPLUS is an online teaching and learning solution. For more information about WileyPLUS, please visit http://www.wiley.com/college/wileyplus

ANSWERS TO SECOND REVIEW PROBLEM SET

These problems review concepts from Chapters 13–21.

1. Increasing acidity →

(a) 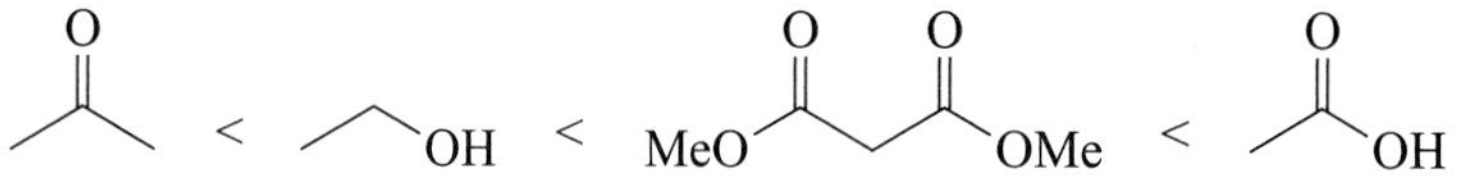

(b) 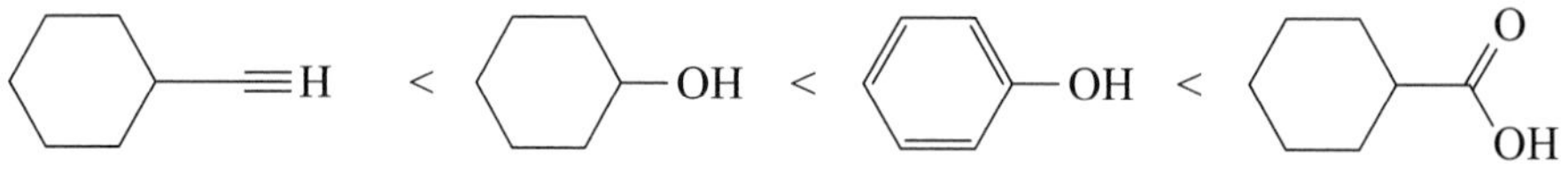

(c) 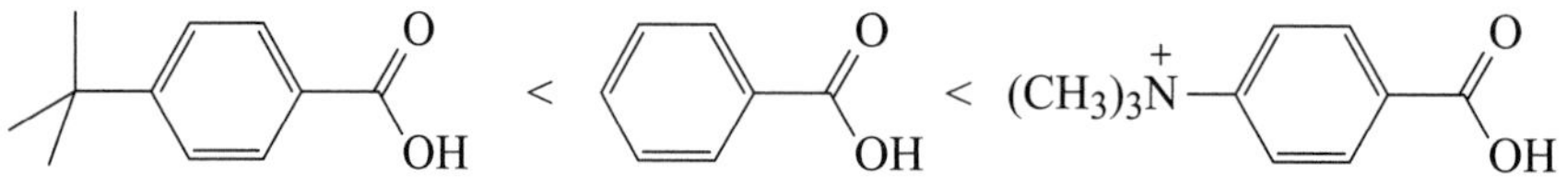

(d) 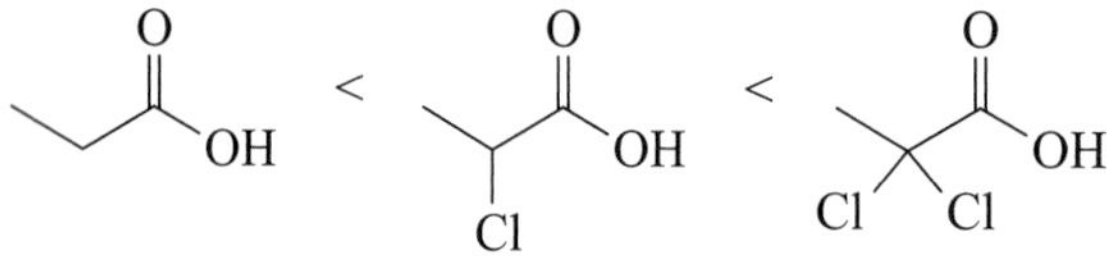

(e) 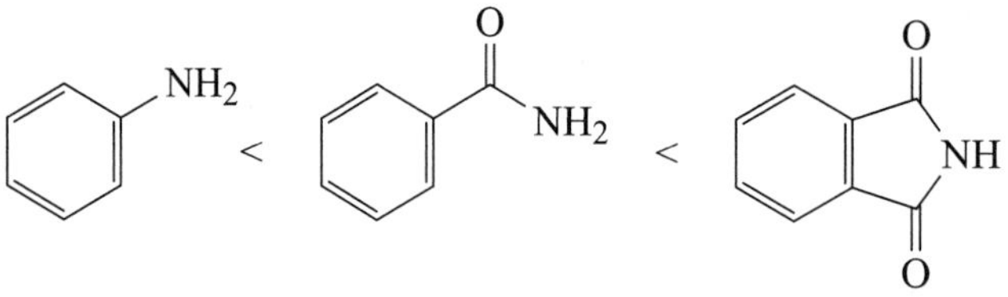

2. Increasing basicity →

(a) 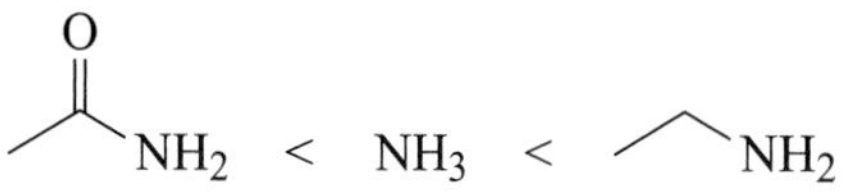

(b) 

(c) 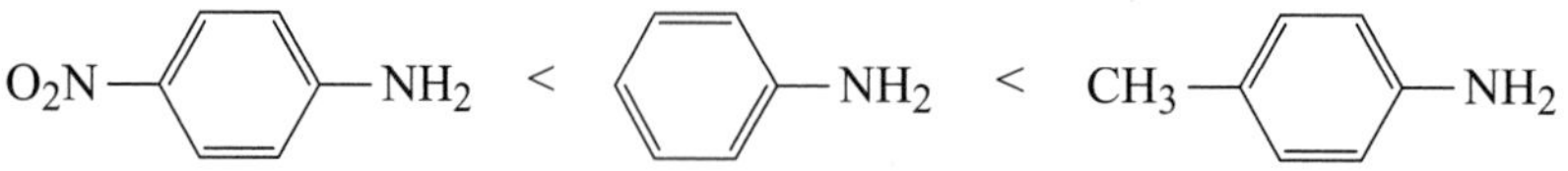

(d) 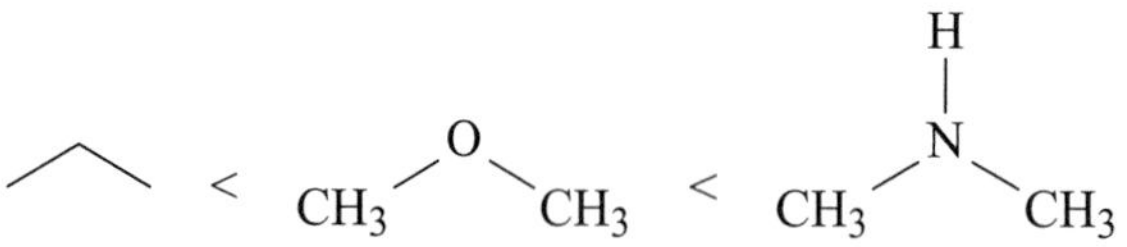

3.

(a) 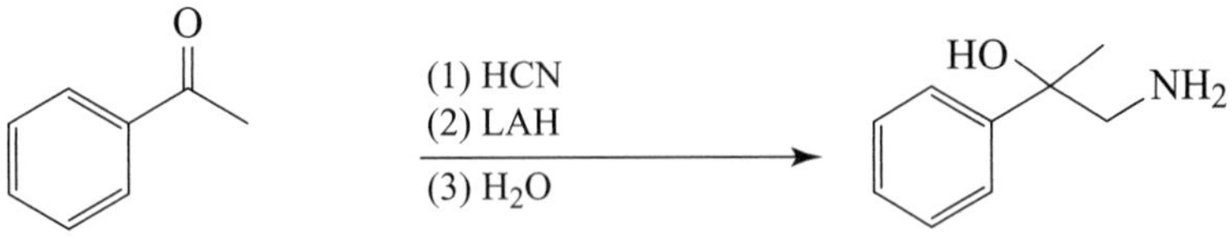 (1) HCN, (2) LAH, (3) H_2O →

(b) 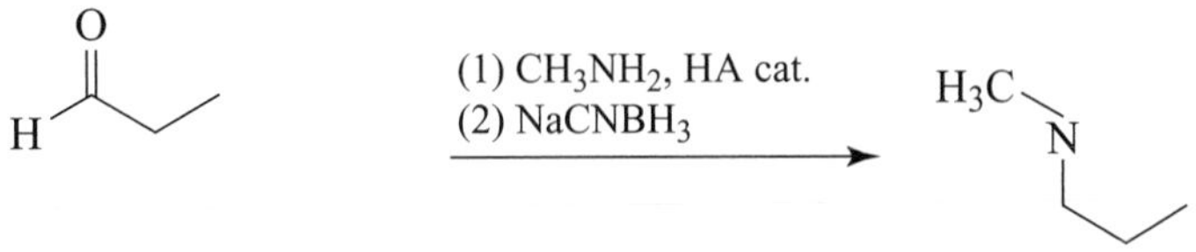 (1) CH_3NH_2, HA cat. (2) $NaCNBH_3$ →

H_3C N

(c) 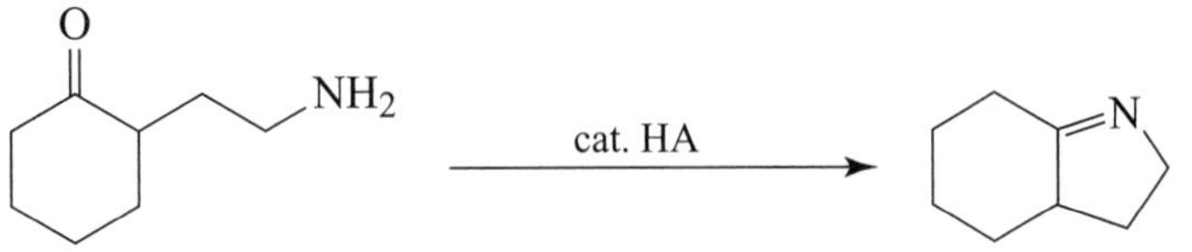 NH_2 cat. HA → N

(d) 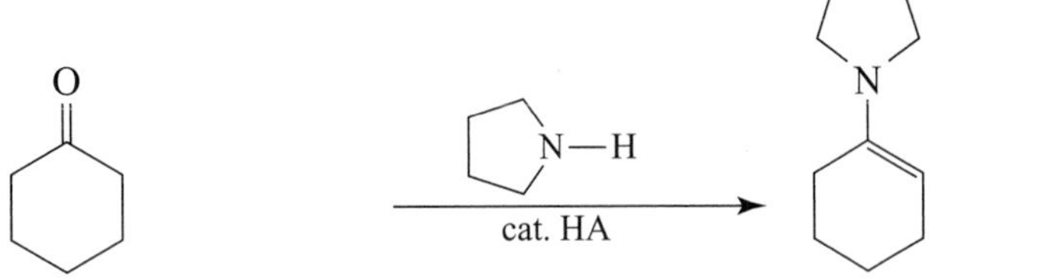 N—H cat. HA → N

4.

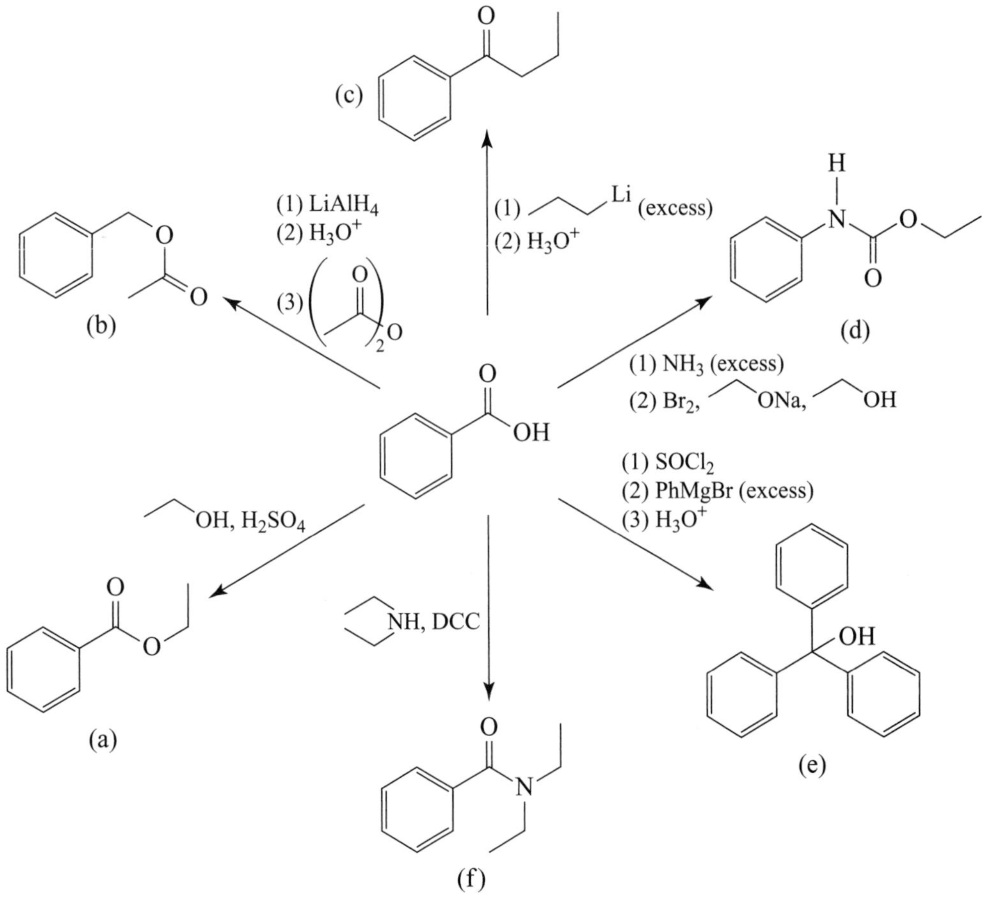

(c) O

(1) Li (excess) (2) H_3O^+

(1) $LiAlH_4$ (2) H_3O^+ (3) O O 2 O

(b) O O

H N O O O

(d)

O OH

(1) NH_3 (excess) (2) Br_2, ONa, OH

(1) $SOCl_2$ (2) PhMgBr (excess) (3) H_3O^+

OH, H_2SO_4

O O

(a)

NH, DCC

OH

(e)

O N

(f)

5. (a)

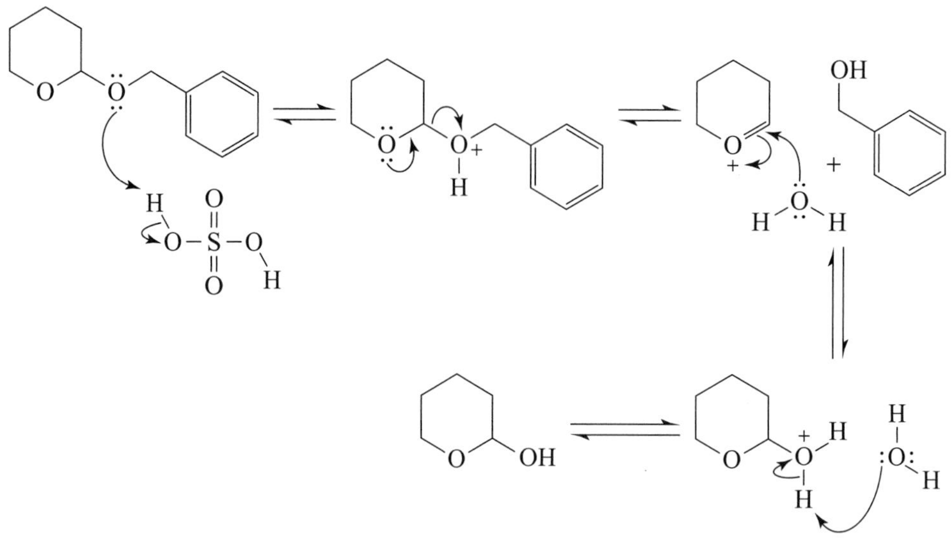

(b)

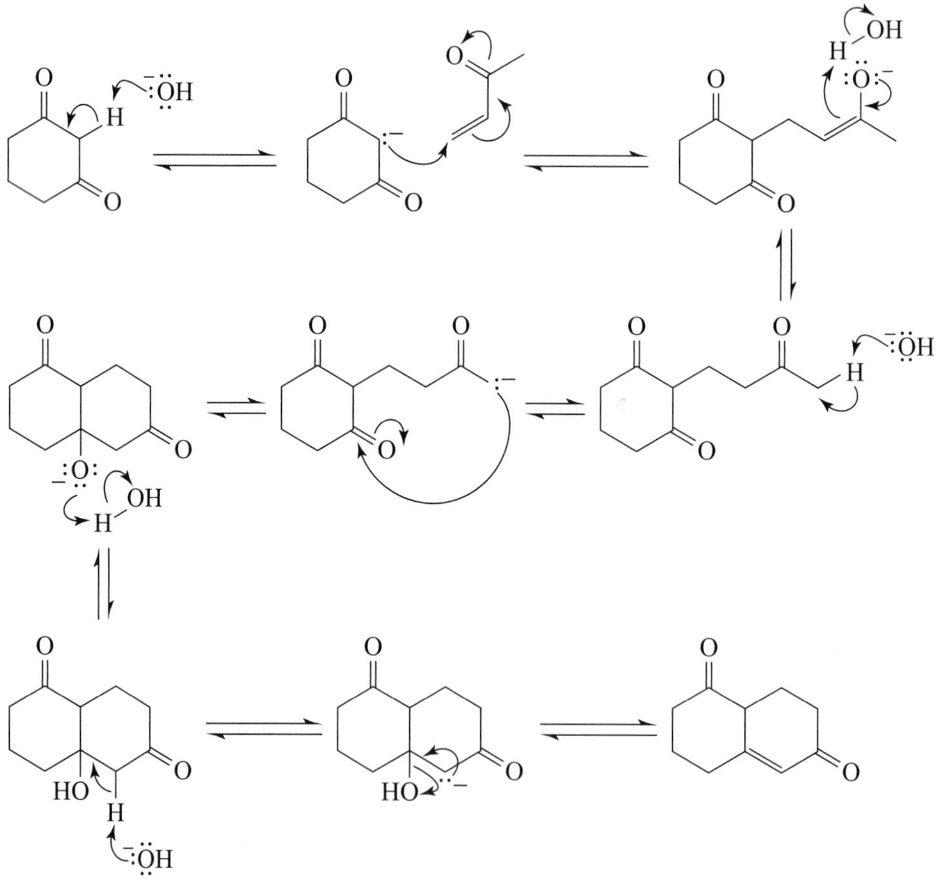

6. (a) OH $\xrightarrow[\text{(or HBr)}]{PBr_3}$ Br

(b) Br [from part (a)] → (potassium phthalimide, NK) → N-alkylphthalimide

$\xrightarrow[\text{heat}]{H_2NNH_2}$ NH_2 + phthalhydrazide (H–N, N–H)

(c) Br [from part (a)] $\xrightarrow{NaCN}$ CN $\xrightarrow[Et_2O]{LiAlH_4}$ NH_2

(d) OH $\xrightarrow[(2)\ H_3O^+]{(1)\ KMnO_4,\ OH^-,\ \text{heat}}$ OH (carboxylic acid, O)

(e) CN [from part (c)] $\xrightarrow[\text{heat}]{H_3O^+,\ H_2O}$ OH (carboxylic acid, O) + NH_4^+

(f) OH (O) [from part (d)] $\xrightarrow{SOCl_2}$ Cl (O)

(g) Cl (O) [from part (f)] $\xrightarrow{NH_3}$ NH_2 (O)

(h) Cl (O) [from part (f)] $\xrightarrow[\text{base}]{\text{OH (butanol)}}$ ester (O, O)

(i) NH_2 (O) [from part (g)] $\xrightarrow[(2)\ H_3O^+]{(1)\ Br_2,\ OH^-}$ NH_2 + CO_3^{2-}

(j) [from part (f)] $\xrightarrow{\text{benzene, } AlCl_3}$ $\xrightarrow[\text{HCl}]{\text{Zn(Hg)}}$

(k) [from part (f)] $\xrightarrow{\text{ONa salt}}$ $(\ )_2O$

(l) Br [from part (a)] $\xrightarrow{Na^+ \ ^-\!:CH(CO_2Et)_2}$ OEt, OEt $\xrightarrow[(2)\ H_3O^+]{(1)\ OH^-,\ H_2O,\ heat}$

OH, OH $\xrightarrow[-CO_2]{\text{heat}}$ OH

7. (a) CH_3 $\xrightarrow[FeBr_3]{Br_2}$ CH_3, Br (separate from ortho isomer) $\xrightarrow[Et_2O]{Mg}$ CH_3, MgBr

$\xrightarrow[(2)\ H_3O^+]{(1)\ \text{ethylene oxide}}$ CH_3, OH $\xrightarrow[CH_2Cl_2]{PCC}$ CH_3, H, O

$\xrightarrow[\text{(aldol condensation)}]{OH^-}$ CH_3, O, H, CH_3 [(*E*) or (*Z*)]

(b)

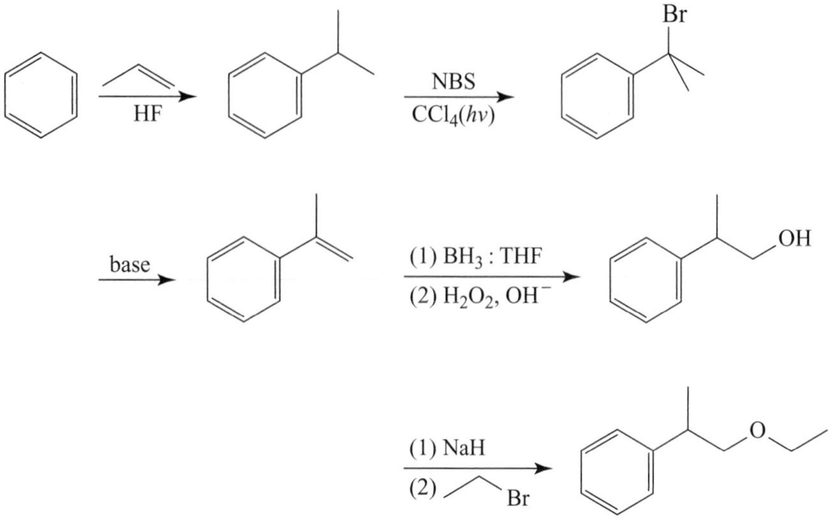

HF

NBS

$CCl_4(h\nu)$

Br

base

(1) BH_3 : THF

(2) H_2O_2, OH^-

OH

(1) NaH

(2) Br

O

(c)

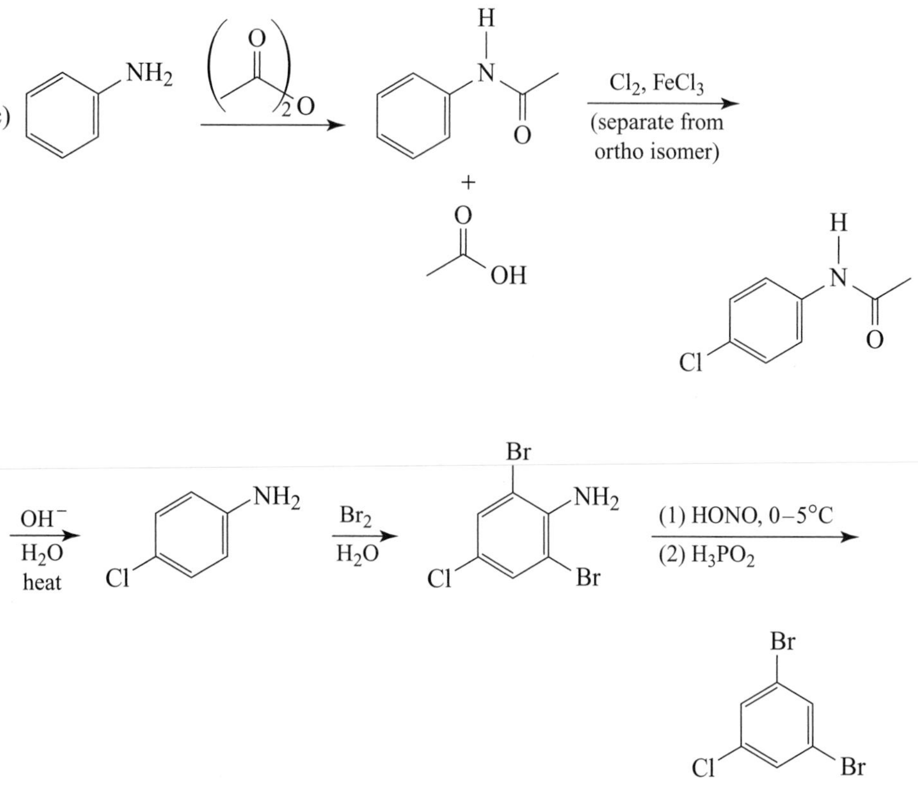

NH_2

O

2 O

H

N

O

+

O

OH

Cl_2, $FeCl_3$

(separate from ortho isomer)

H

N

O

Cl

OH^-

H_2O

heat

Cl

NH_2

Br_2

H_2O

Br

NH_2

Cl

Br

(1) HONO, 0–5°C

(2) H_3PO_2

Br

Cl

Br

(d) CH_3 — HNO_3, H_2SO_4 (separate from ortho isomer) → O_2N … CH_3 — (1) $KMnO_4$, OH^-, heat (2) H_3O^+ →

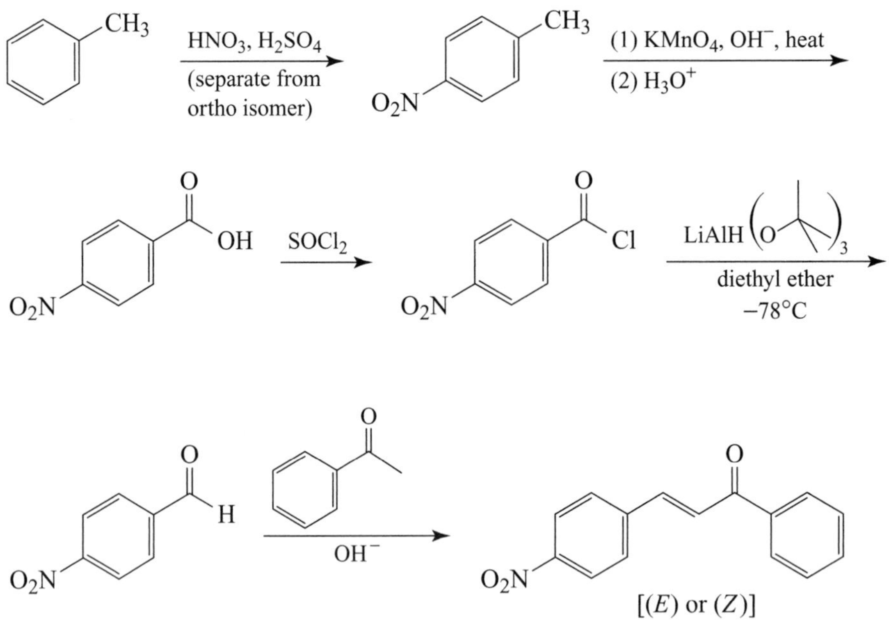

O ‖ OH, O_2N — $SOCl_2$ → O ‖ Cl, O_2N — LiAlH(O-t-Bu)$_3$, diethyl ether, −78°C →

O ‖ H, O_2N — (acetophenone) O, OH^- → O, O_2N [(*E*) or (*Z*)]

(e) CH_3 — NBS, CCl_4 (*hv*) → Br — Na^+ $^-$:≡H →

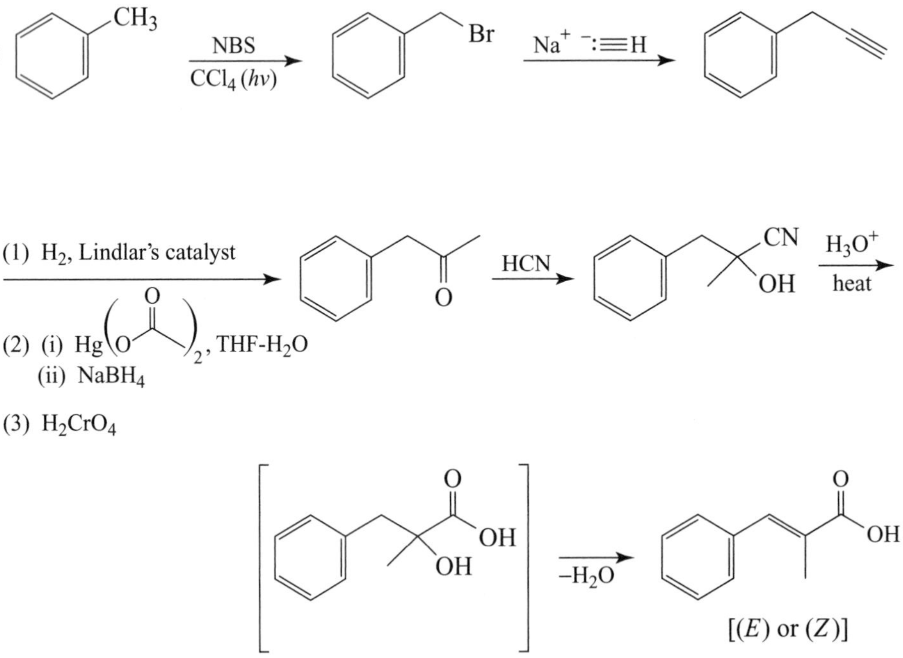

(1) H_2, Lindlar's catalyst
(2) (i) Hg(OAc)$_2$, THF-H_2O
(ii) $NaBH_4$
(3) H_2CrO_4
→ O — HCN → CN, OH — H_3O^+, heat →

[O, OH, OH] — $-H_2O$ → O, OH [(*E*) or (*Z*)]

8.

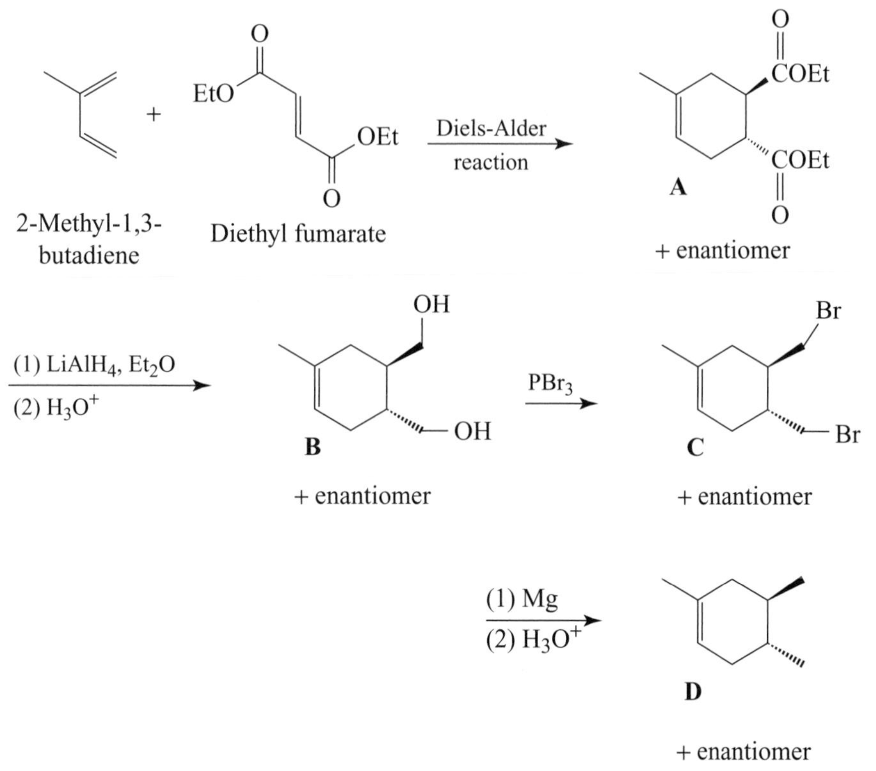

2-Methyl-1,3-butadiene + Diethyl fumarate $\xrightarrow{\text{Diels-Alder reaction}}$ **A** + enantiomer

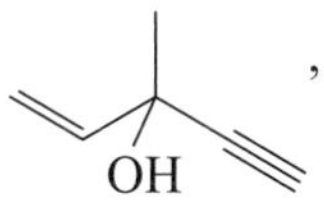

$\xrightarrow[\text{(2) } H_3O^+]{\text{(1) } LiAlH_4,\ Et_2O}$ **B** + enantiomer $\xrightarrow{PBr_3}$ **C** + enantiomer

$\xrightarrow[\text{(2) } H_3O^+]{\text{(1) Mg}}$ **D** + enantiomer

9. (a) **A** is [structure: OH], **C** is [structure: BrMgO ... MgBr]

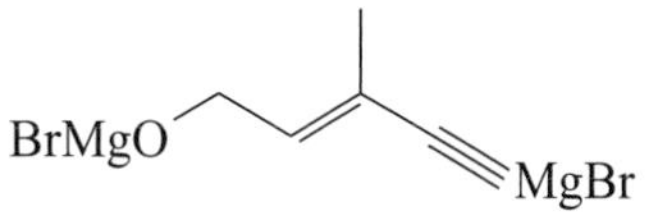

(b) **A** is an allylic alcohol and thus forms a carbocation readily. **B** is a conjugated enyne and is therefore more stable than **A**.

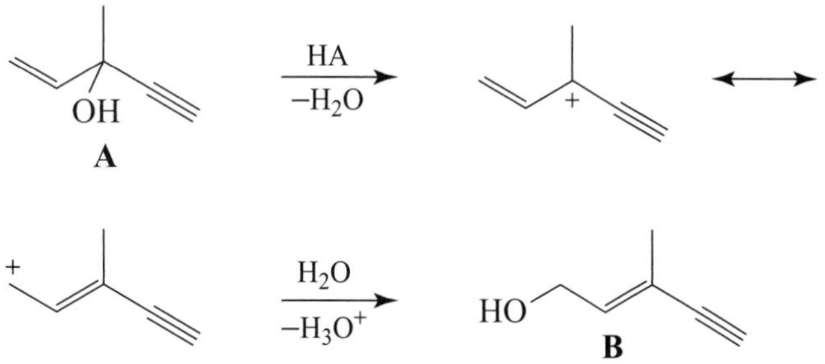

A $\xrightarrow[-H_2O]{HA}$ [carbocation] $\longleftrightarrow$ [carbocation] $\xrightarrow[-H_3O^+]{H_2O}$ **B** (HO)

10.

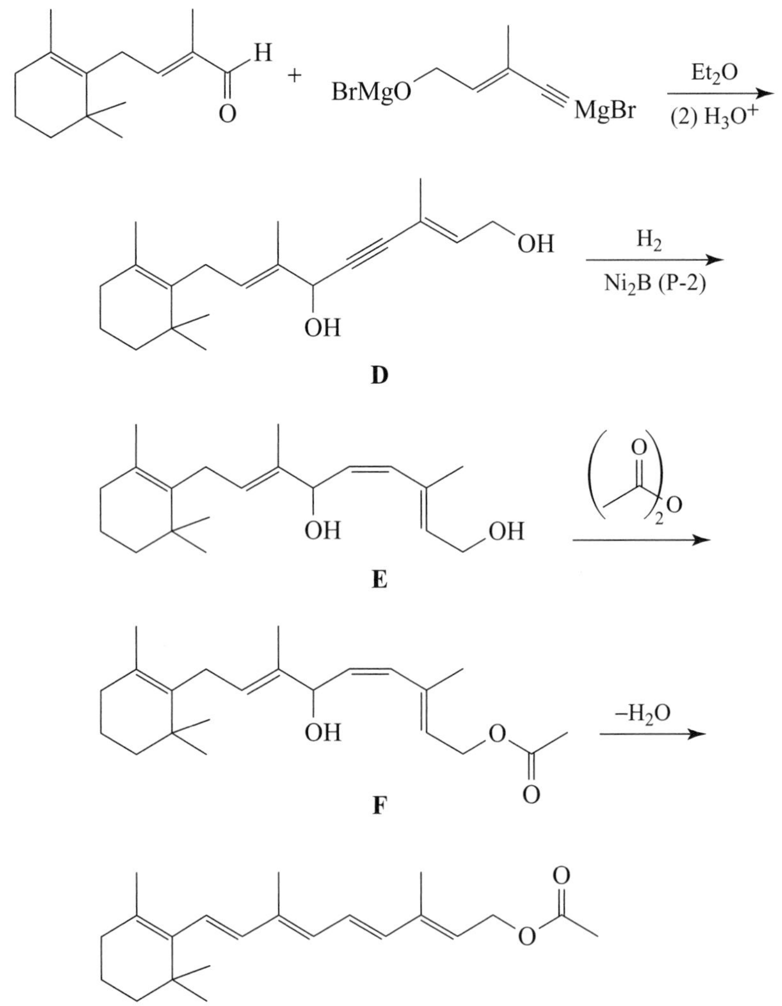

H

+ BrMgO MgBr

Et$_2$O

(2) H_3O^+

O

OH

H_2

Ni_2B (P-2)

OH

D

O

O

2

OH

OH

E

$-H_2O$

OH

O

O

F

O

O

Vitamin A acetate

11.

O

+ HCl

+OH

OH

+

OH

+

H

HO

OH

:A$^-$

$-$HA

HO

OH

HA

H

O

H

+

OH

$-H_2O$

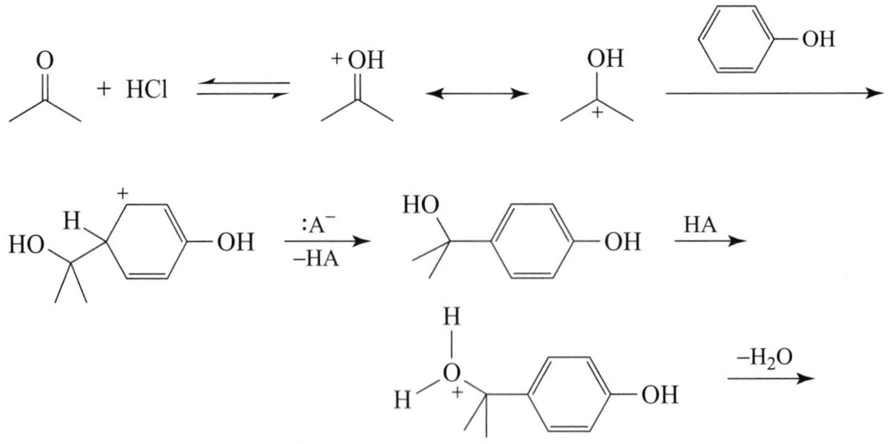

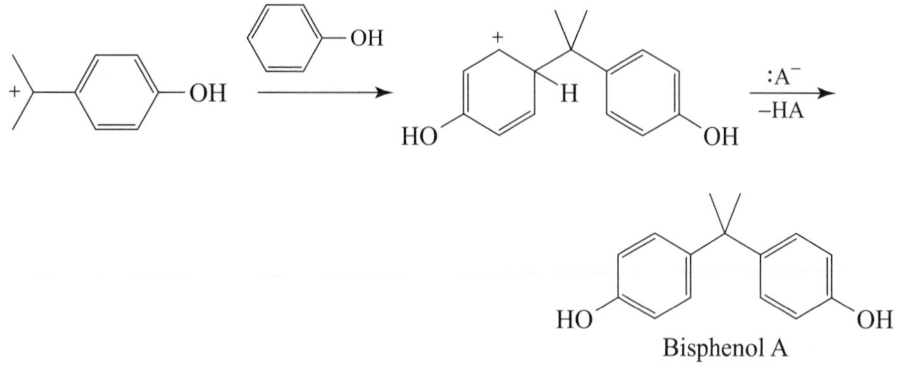

OH · OH · + · H · HO · OH · $:A^-$ · $-HA$

HO · OH

Bisphenol A

12. CH_3 · O_2N · (1) $KMnO_4$, OH^-, heat · (2) H_3O^+ · O · OH · O_2N · **A** · $SOCl_2$

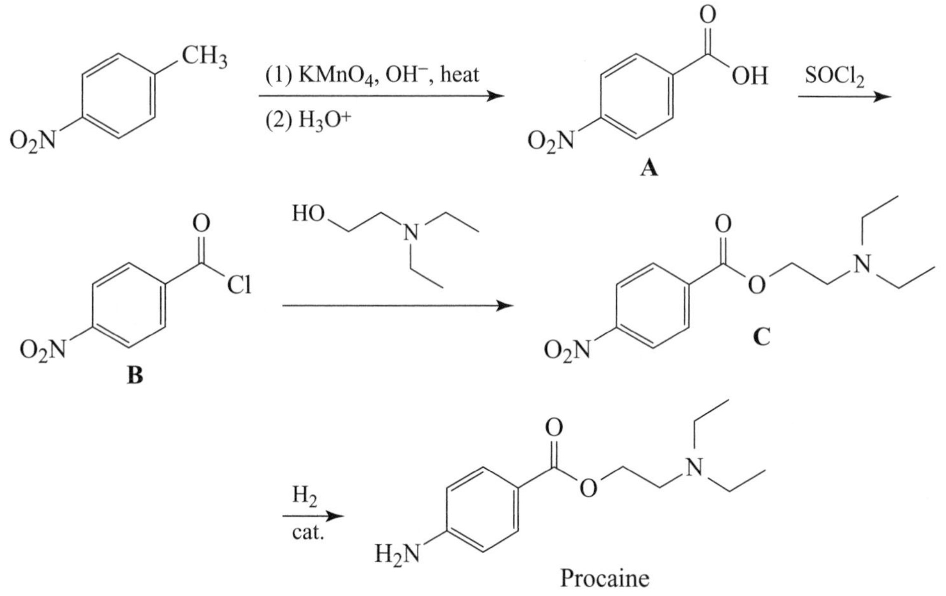

O · Cl · O_2N · **B** · HO · N · O · O · N · O_2N · **C**

H_2 · cat. · O · O · N · H_2N

Procaine

13. O · (1) H≡:⁻ Na^+ · (2) NH_4Cl/H_2O · OH · **A** · O · Cl · Cl · O · Cl · O · **B**

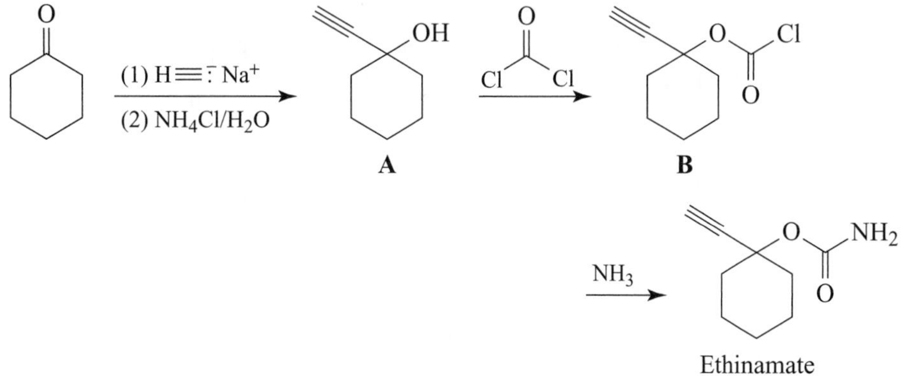

NH_3 · O · NH_2 · O

Ethinamate

14. (a) C_6H_5CHO $\xrightarrow[(2)\ H_3O^+]{(1)\ C_6H_5MgBr}$ $(C_6H_5)_2CHOH$ (**A**) $\xrightarrow{PBr_3}$ $(C_6H_5)_2CHBr$ (**B**)

$\xrightarrow[-HBr]{(CH_3)_2NCH_2CH_2OH}$ $(C_6H_5)_2CH{-}O{-}CH_2CH_2N(CH_3)_2$

Diphenhydramine

(b) The last step probably takes place by an S_N1 mechanism. Diphenylmethyl bromide, **B**, ionizes readily because it forms the resonance-stabilized benzylic carbocation.

$(C_6H_5)_2CH^+$

15. (a) For this synthesis we need to prepare the benzylic halide, p-$BrC_6H_4CH(Br)C_6H_5$, and then allow it to react with $(CH_3)_2NCH_2CH_2OH$ as in Problem 14.

This benzylic halide can be made as follows:

p-BrC_6H_4CHO $\xrightarrow[(2)\ H_3O^+]{(1)\ C_6H_5MgBr}$ p-$BrC_6H_4CH(OH)C_6H_5$ $\xrightarrow{PBr_3}$ p-$BrC_6H_4CH(Br)C_6H_5$

(b) For this synthesis we can prepare the requisite benzylic halide in two ways:

o-$CH_3C_6H_4CHO$ $\xrightarrow[(2)\ H_3O^+]{(1)\ C_6H_5MgBr}$ o-$CH_3C_6H_4CH(OH)C_6H_5$ $\xrightarrow{PBr_3}$ o-$CH_3C_6H_4CH(Br)C_6H_5$

or

C_6H_5CHO $\xrightarrow[(2)\ H_3O^+]{(1)\ o\text{-}CH_3C_6H_4MgBr}$ o-$CH_3C_6H_4CH(OH)C_6H_5$ $\xrightarrow{PBr_3}$ o-$CH_3C_6H_4CH(Br)C_6H_5$

We shall then allow the benzylic halide to react with $(CH_3)_2NCH_2CH_2OH$ as in Problem 14.

16.

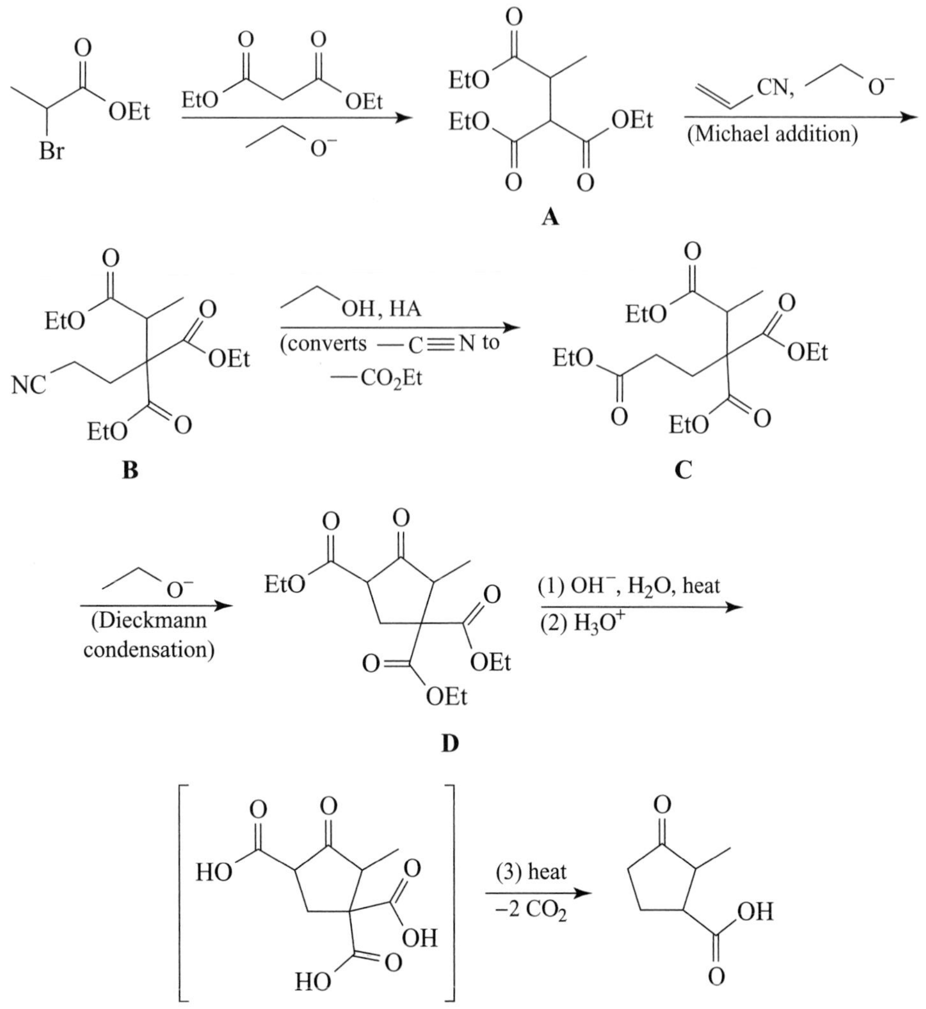

O, OEt, Br

EtO, O, O, OEt; O^-

O, EtO, EtO, OEt, O, O

A

CN, O^-

(Michael addition)

O, EtO, O, OEt, NC, EtO, O

B

OH, HA

(converts —C≡N to —CO_2Et

O, EtO, O, EtO, OEt, O, EtO, O

C

O^-

(Dieckmann condensation)

O, O, EtO, O, O, OEt, OEt

D

(1) OH^-, H_2O, heat

(2) H_3O^+

[O, O, HO, O, OH, HO, O]

(3) heat

$-2\ CO_2$

O, OH, O

17.

O

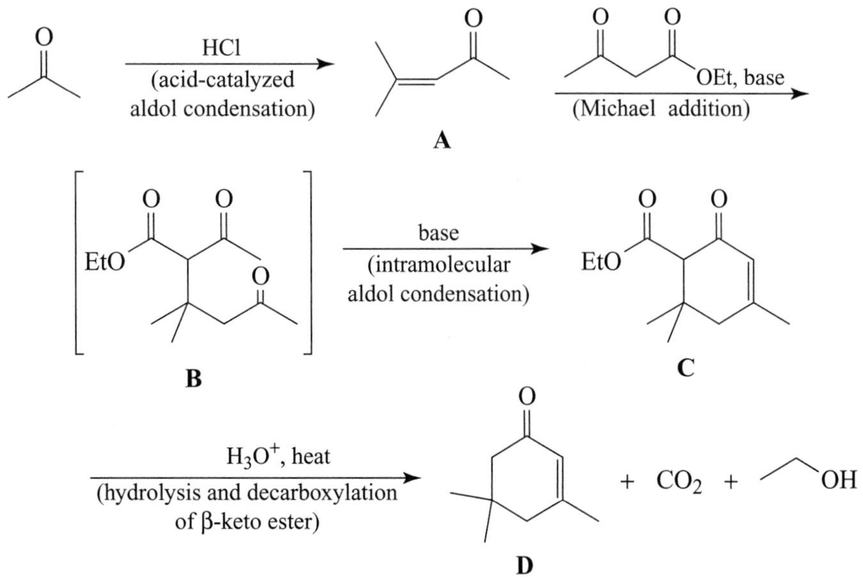

HCl

(acid-catalyzed aldol condensation)

O

A

O, O, OEt, base

(Michael addition)

[O, O, EtO, O]

B

base

(intramolecular aldol condensation)

O, O, EtO

C

H_3O^+, heat

(hydrolysis and decarboxylation of β-keto ester)

O

D

+ CO_2 + OH

18.

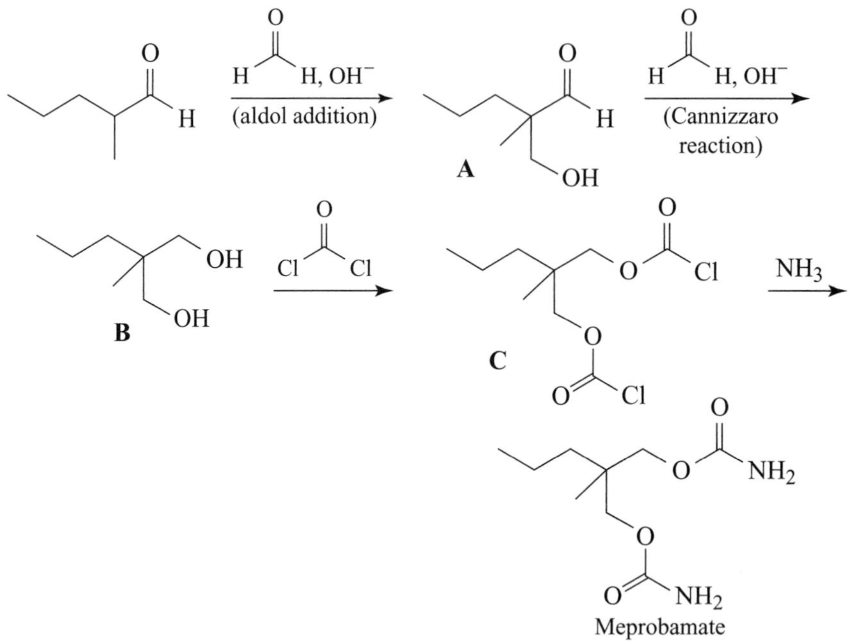

O H

O H H, OH^- (aldol addition)

O H OH A

O H H, OH^- (Cannizzaro reaction)

OH OH B

O Cl Cl

O O Cl O O Cl C

NH_3

O O NH_2 O O NH_2

Meprobamate

19.

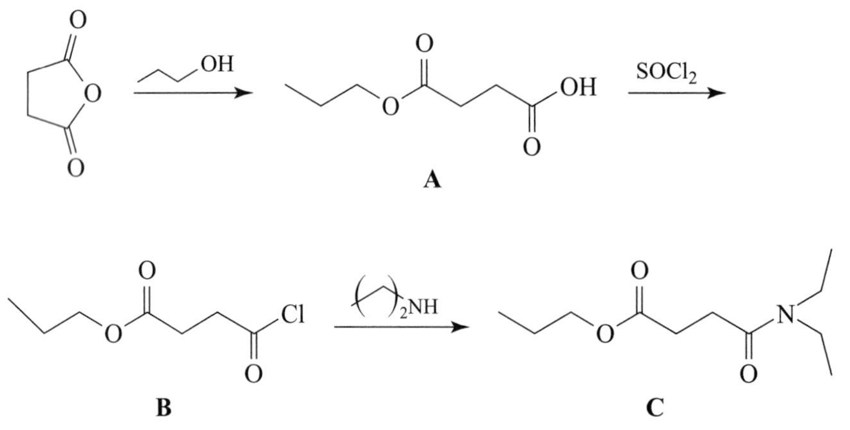

O O O

OH

O O OH O A

$SOCl_2$

O O Cl O B

$(\diagup\!\!\diagdown)_2NH$

O O N O C

20.

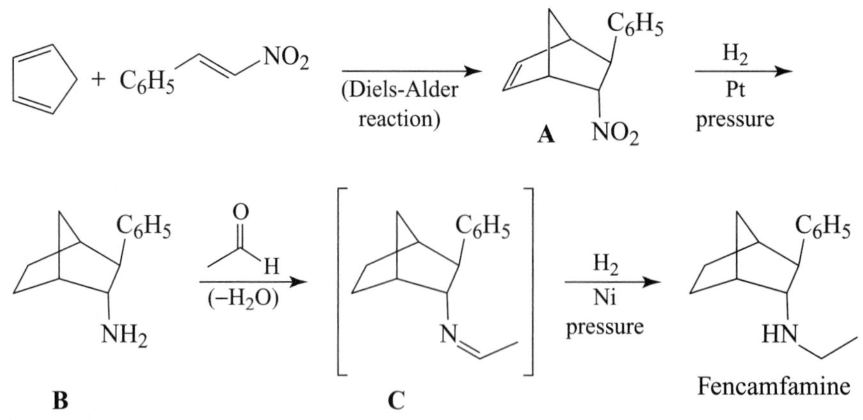

+ C_6H_5 NO_2

(Diels-Alder reaction)

C_6H_5 NO_2 A

H_2 Pt pressure

C_6H_5 NH_2 B

O H ($-H_2O$)

[C_6H_5 N C]

H_2 Ni pressure

C_6H_5 HN

Fencamfamine

21. (1) OsO_4 (2) $NaHSO_3$ → **A** (OH, OH) → CrO_3, acetic acid (O, OH) →

B OH ← Infrared band in 3200–3550 cm^{-1} region
O ← Infrared band in 1650–1730 cm^{-1} region

Notice that the second step involves the oxidation of a secondary alcohol in the presence of a tertiary alcohol. This selectivity is possible because tertiary alcohols do not undergo oxidation readily (Sections 12.4D, E).

22. Working backward, we notice that methyl *trans*-4-isopropylcyclohexanecarboxylate has both large groups equatorial and is, therefore, more stable than the corresponding cis isomer. This stability of the trans isomer means that, if we were to synthesize the cis isomer or a mixture of both the cis and trans isomers, we could obtain the desired trans isomer by a base-catalyzed isomerization (epimerization):

(more stable trans isomer) H, H, OCH_3, O ← ^{-}OH ← (cis isomer or mixture of cis and trans isomers) OCH_3, O

We could synthesize a mixture of the desired isomers from phenol in the following way:

OCH_3, O ← CH_3OH, HA ← OH, O

← (1) CO_2 (2) H_3O^+ ← MgBr ← Mg, Et_2O

Br ← PBr_3 ← OH ← H_2, cat., pressure

OH ← HF (propene), (Friedel-Crafts alkylation) ← OH

23. The molecular formula indicates that the compound is saturated and is an alcohol or an ether. The IR data establish the compound as an alcohol. That **Y** gives a green opaque solution when treated with CrO_3 in aqueous H_2SO_4 indicates that **Y** is a primary or secondary alcohol. That **Y** gives a negative iodoform test indicates that **Y** does not contain the grouping —CH(OH)CH$_3$. The ^{13}C spectrum of **Y** contains only four signals indicating that some of the carbons in **Y** are equivalent. The information from DEPT spectra helps us conclude that **Y** is 2-ethyl-1-butanol.

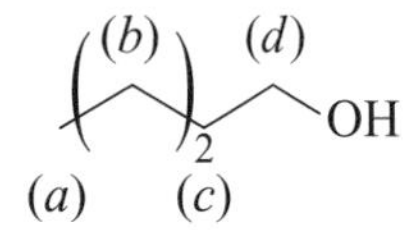

(*a*) δ 11.1
(*b*) δ 23.0
(*c*) δ 43.6
(*d*) δ 64.6

Notice that the most downfield signal is a CH_2 group. This indicates that this carbon atom bears the —OH group and that **Y** is a primary alcohol. The most upfield signals indicate the presence of the ethyl groups.

24. That **Z** decolorizes bromine in CCl_4 indicates that **Z** is an alkene. We are told that **Z** is the more stable isomer of a pair of stereoisomers. This fact suggests that **Z** is a trans alkene. That the ^{13}C spectrum contains only three signals, even though **Z** contains eight carbon atoms, indicates that **Z** is highly symmetric. The information from DEPT spectra indicates that the upfield signals of the alkyl groups arise from equivalent isopropyl groups. We conclude, therefore, that **Z** is *trans*-2,5-dimethyl-3-hexene.

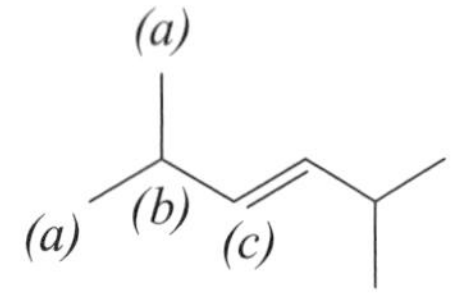

(*a*) δ 22.8
(*b*) δ 31.0
(*c*) δ 134.5

25. (a, b)

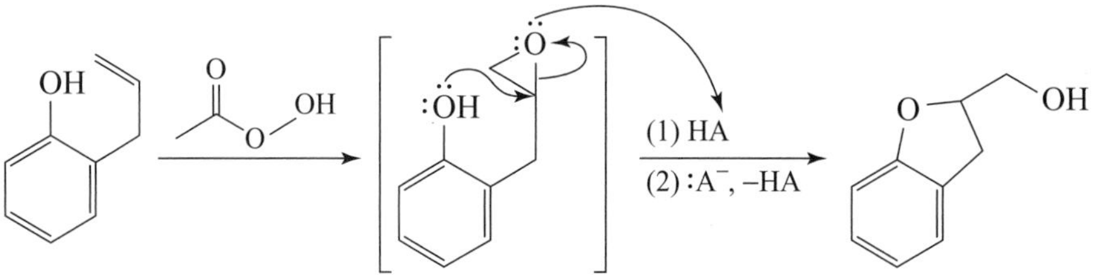

(1) HA
(2) :A$^-$, –HA

B

26. Compounds and reagents **A—M,** regarding the synthesis of dianeackerone:

(a)

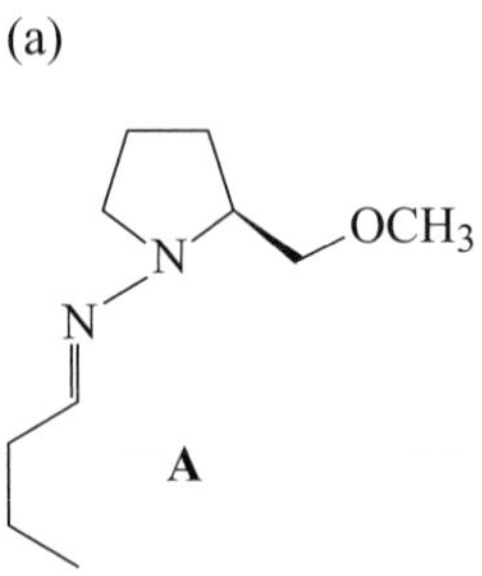

OCH3 N N **A**

OCH3 N N **B**

H O **C**

OH **D**

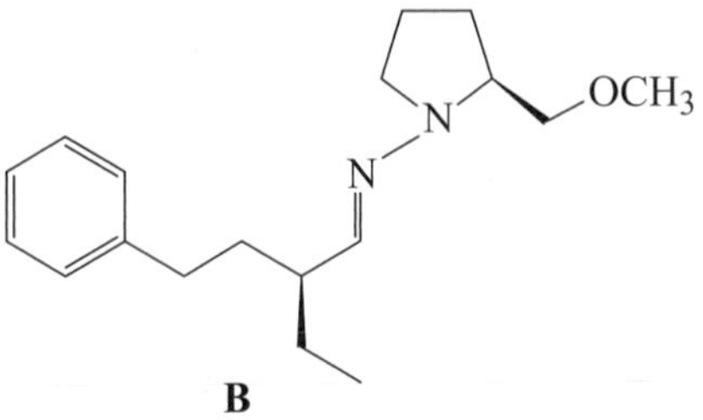

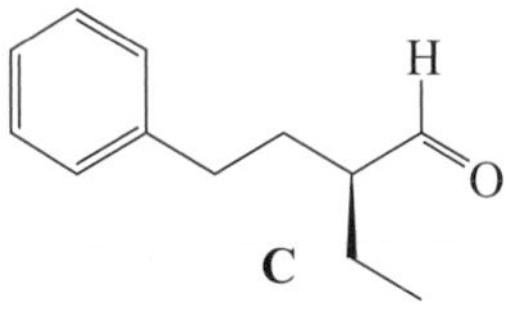

O Br O **E**

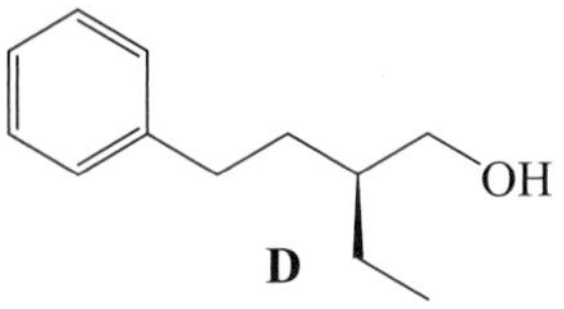

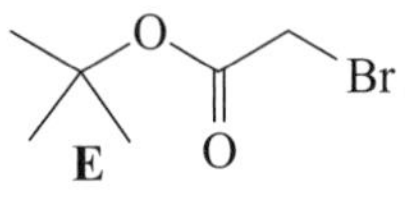

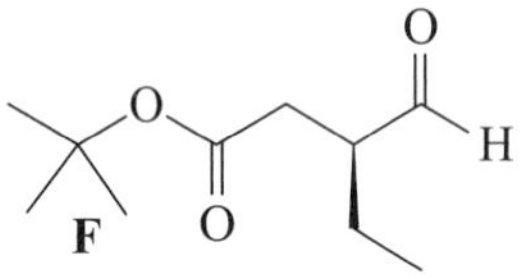

O O H O **F**

G CH_3I

H H_2CrO_4

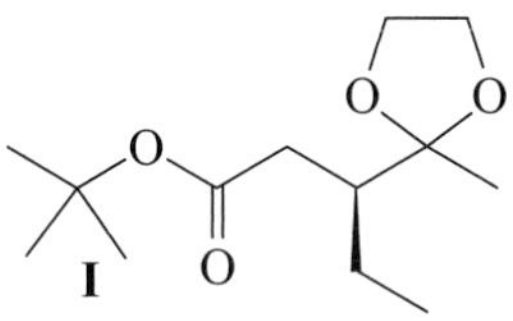

O O O O **I**

J LAH

K PCC

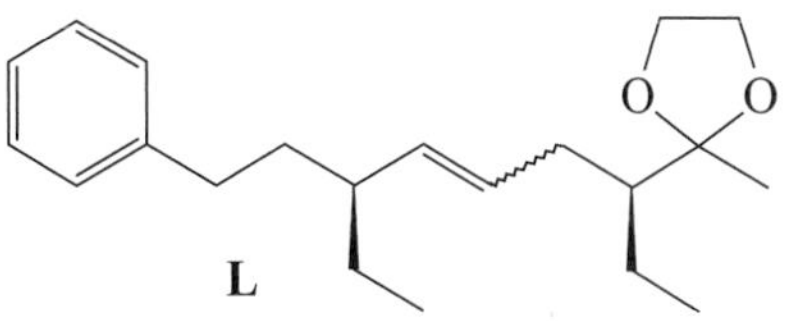

O O **L**

M H_2, Pd

(b)

Br

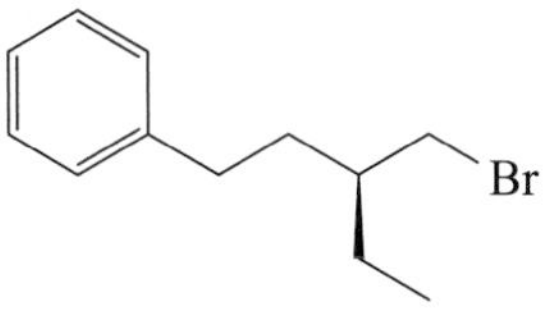

(c) (3*S*,7*R*)-Dianeackerone

22 CARBOHYDRATES

SUMMARY OF SOME REACTIONS OF MONOSACCHARIDES

Open-chain form of aldose: $O=C(H)-(CHOH)_n-CH_2OH$

- $\xrightarrow[H_2O]{Br_2}$ $CO_2H-(CHOH)_n-CH_2OH$ Aldonic acid
- $\xrightarrow{HNO_3}$ $CO_2H-(CHOH)_n-CO_2H$ Aldaric acid
- $\xrightarrow{C_6H_5NHNH_2}$ $CH=NNHC_6H_5-C=NNHC_6H_5-(CHOH)_{n-1}-CH_2OH$ Osazone
- $\xrightarrow{HCN}$ (Kiliani-Fischer synthesis) $CN-(CHOH)_{n+1}-CH_2OH$ Cyanohydrin $\xrightarrow{\text{several steps}}$ $CHO-(CHOH)_{n+1}-CH_2OH$ Aldose with one more carbon
- $\xrightarrow{NaBH_4}$ $CH_2OH-(CHOH)_n-CH_2OH$ Alditol
- $\xrightarrow{(1)\ Br_2/H_2O,\ (2)\ H_2O_2/Fe^{3+}}$ (Ruff degradation) $CHO-(CHOH)_{n-1}-CH_2OH$ Aldose with one less carbon atom

Cyclic form of D-glucose (CH_2OH, O, HO, HO, OH, CHOH) $\xrightarrow[HA]{CH_3OH}$ Methyl glucoside (CH_2OH, O, HO, HO, OH, $CHOCH_3$) $\xrightarrow[OH^-]{(CH_3)_2SO_4}$

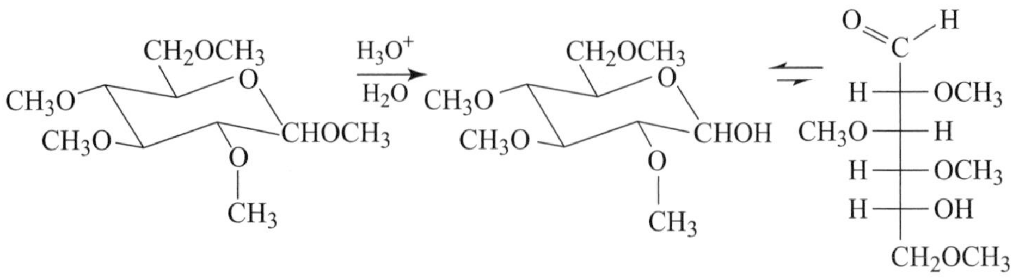

$$\xrightarrow[H_2O]{H_3O^+}$$

CHOHCH$_3$ → CHOH ⇌ CHO (open chain: H–OCH$_3$, CH$_3$O–H, H–OCH$_3$, H–OH, CH$_2$OCH$_3$)

SOLUTIONS TO PROBLEMS

22.1 (a) Two, CHO–*CHOH–*CHOH–CH_2OH (b) Two, CH_2OH–C=O–*CHOH–*CHOH–CH_2OH

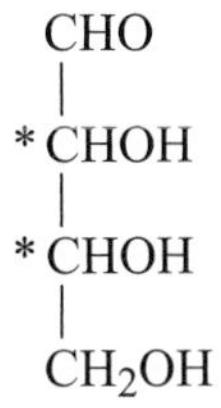

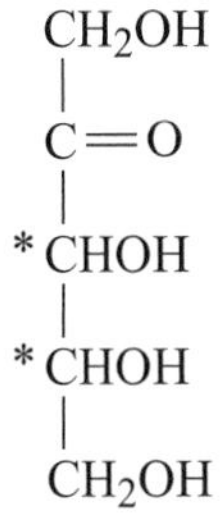

(c) There would be four stereoisomers (two sets of enantiomers) for each general structure: $2^2 = 4$.

22.2

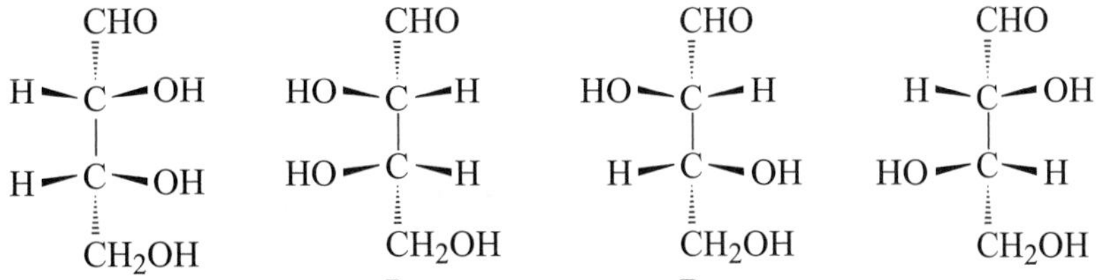

CHO; H–C–OH; H–C–OH; CH_2OH	CHO; HO–C–H; HO–C–H; CH_2OH	CHO; HO–C–H; H–C–OH; CH_2OH	CHO; H–C–OH; HO–C–H; CH_2OH
D	**L**	**D**	**L**

CH_2OH; C=O; H–C–OH; H–C–OH; CH_2OH	CH_2OH; C=O; HO–C–H; HO–C–H; CH_2OH	CH_2OH; C=O; HO–C–H; H–C–OH; CH_2OH	CH_2OH; C=O; H–C–OH; HO–C–H; CH_2OH
D	**L**	**D**	**L**

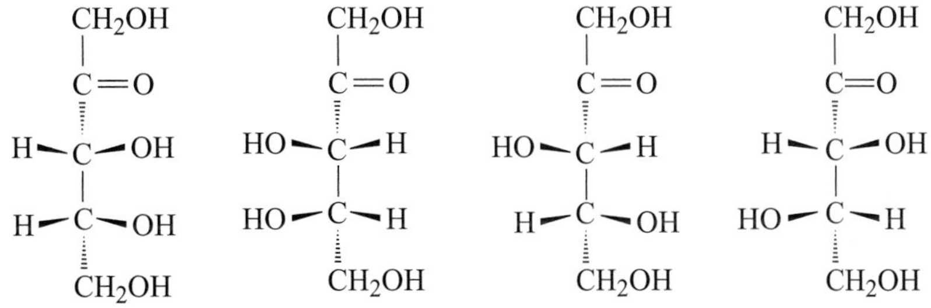

22.3 (a) HO OH O OH OH OH (b) OH O H OH H OH HO H CH_2OH

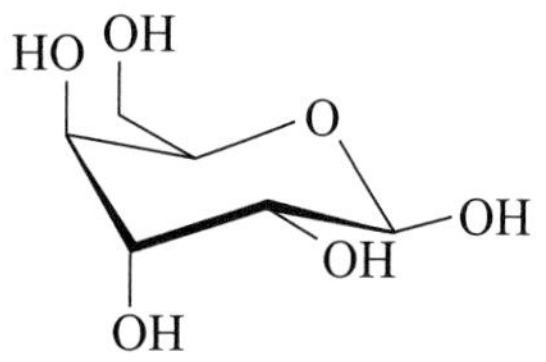

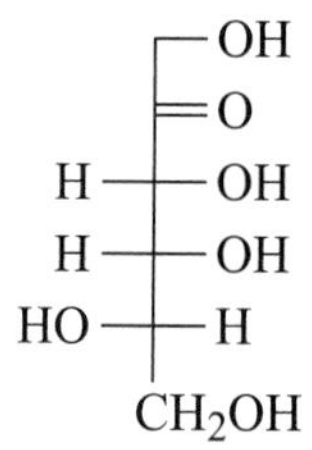

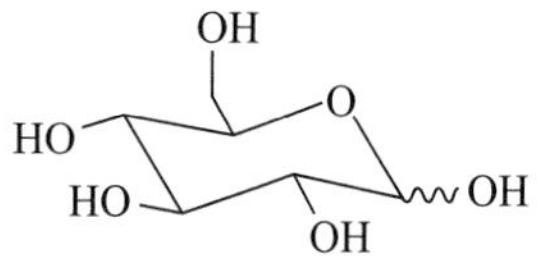

22.4 (a) OH O HO HO OH OH and OH HO

D-(+)-Glucose 2-Hydroxybenzyl alcohol

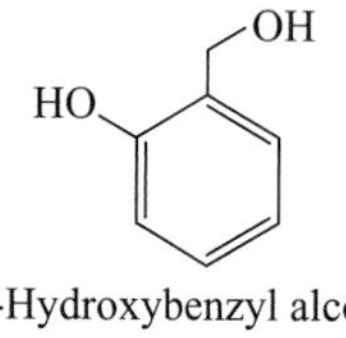

(b) OH O HO HO OH O OH H H—O—H HO HO OH O OH H O OH

Salicin

OH HO − OH O+ HO HO OH H O—H HO HO OH O $\overset{+}{O}H_2$ OH

A:⁻, –HA OH O HO HO OH OH

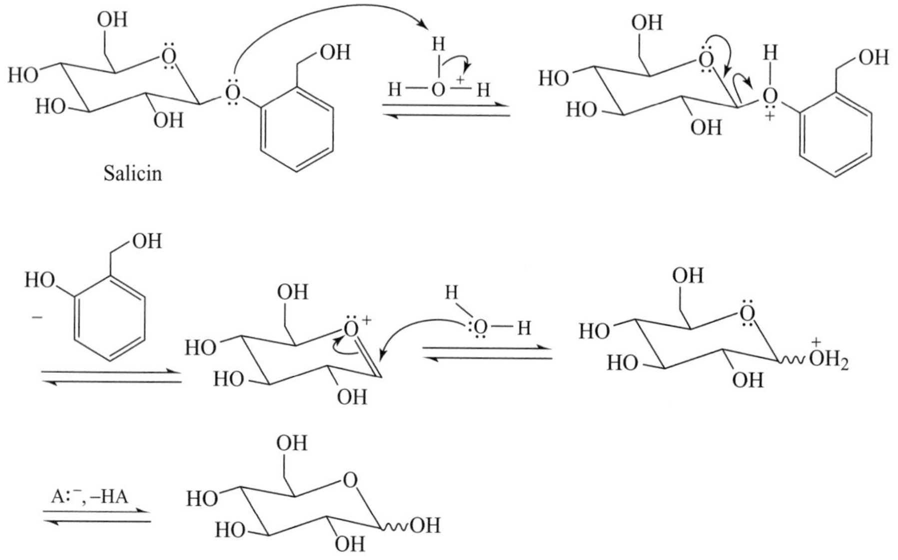

22.5 Dissolve D-glucose in ethanol and then bubble in gaseous HCl.

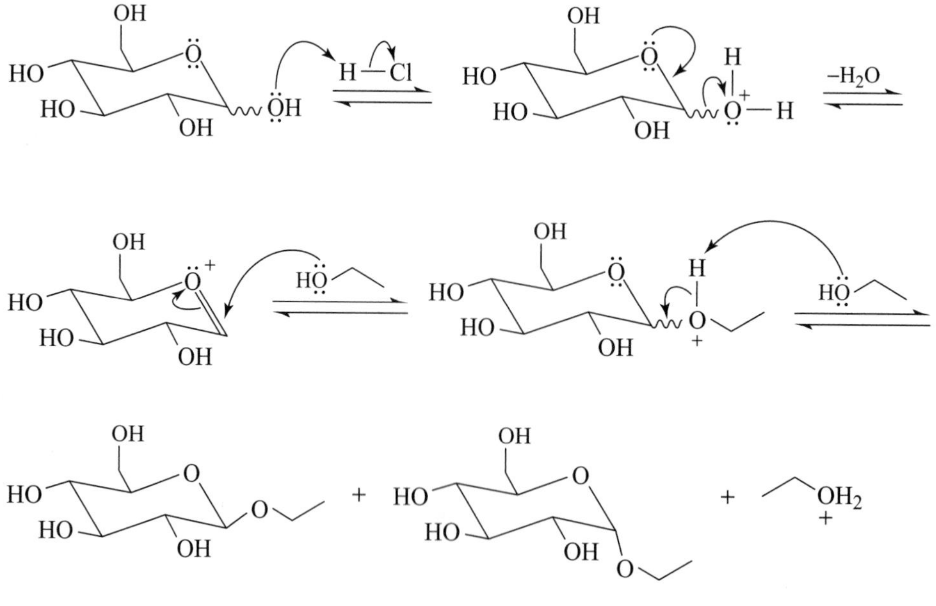

OH O HO HO OH OH H—Cl HO HO OH O H O—H OH $-H_2O$

OH O+ HO HO OH HO HO HO OH O H O OH HO

OH O HO HO OH O + HO HO OH O OH O + OH_2 +

22.6 α-D-Glucopyranose will give a positive test with Benedict's or Tollens' solution because it is a cyclic hemiacetal. Methyl α-D-glucopyranoside, because it is a cyclic acetal, will not.

22.7 (a) Yes (b) (c) Yes (d)

(e) No (f)

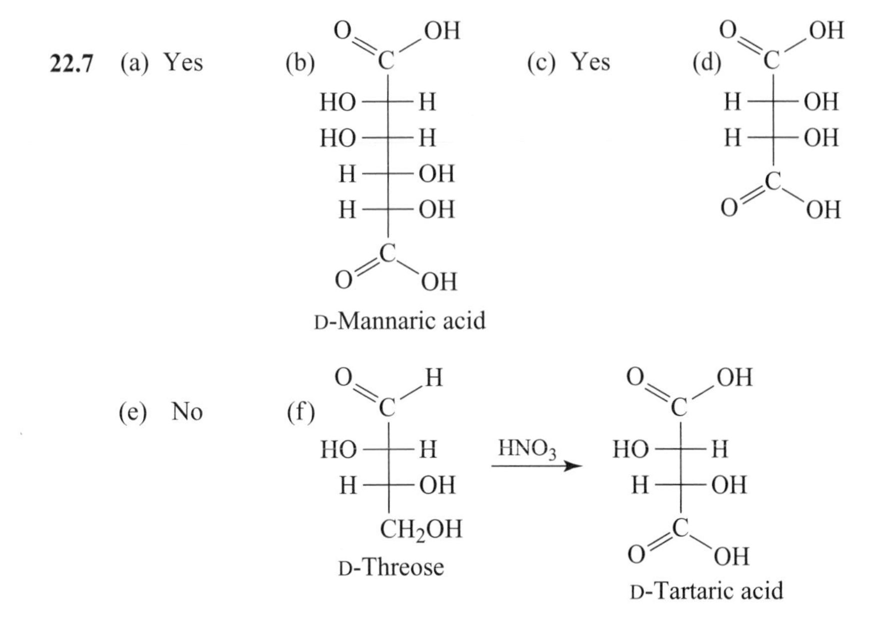

(g) The aldaric acid obtained from D-erythrose is *meso*-tartaric acid; the aldaric acid obtained from D-threose is D-tartaric acid.

22.8

or

and

or

22.9 One way of predicting the products from a periodate oxidation is to place an —OH group on each carbon atom at the point where C—C bond cleavage has occurred:

$$\begin{array}{c} | \\ -\mathrm{C}-\mathrm{OH} \\ | \\ -\mathrm{C}-\mathrm{OH} \\ | \end{array} \xrightarrow{IO_4^-} \begin{array}{c} | \\ -\mathrm{C}-\mathrm{OH} \\ | \\ \mathrm{OH} \end{array} + \begin{array}{c} \mathrm{OH} \\ | \\ -\mathrm{C}-\mathrm{OH} \\ | \end{array}$$

Then if we recall (Section 16.7A) that *gem*-diols are usually unstable and lose water to produce carbonyl compounds, we get the following results:

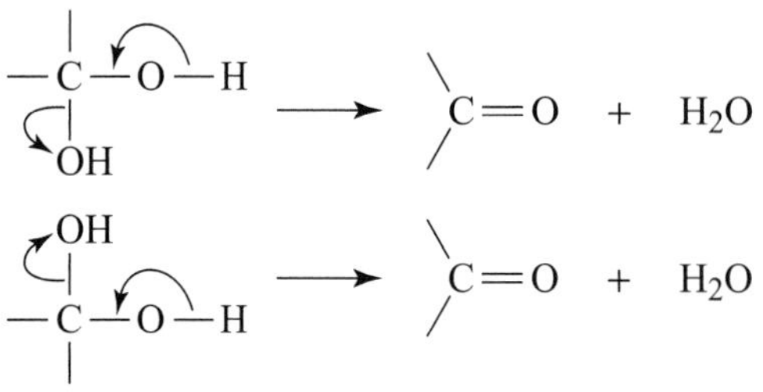

—C—O—H (with OH) ⟶ >C=O + H_2O

OH—C—O—H ⟶ >C=O + H_2O

Let us apply this procedure to several examples here while we remember that for every C—C bond that is broken 1 mol of HIO_4 is consumed.

(a) 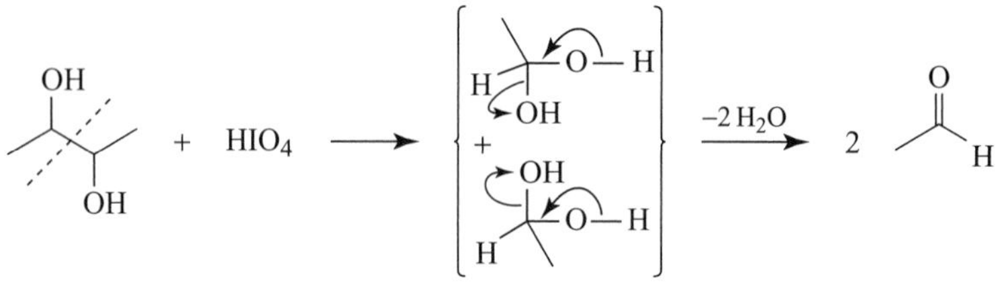 + HIO_4 ⟶ [ ] $\xrightarrow{-2\,H_2O}$ 2

(b) 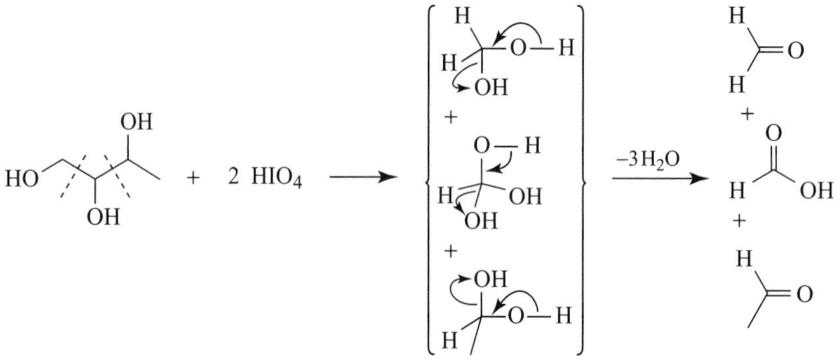 HO + 2 HIO_4 ⟶ [ ] $\xrightarrow{-3\,H_2O}$ + +

(c) 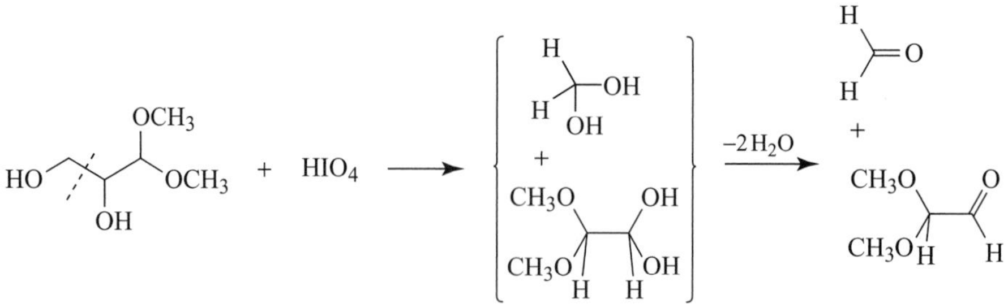 HO OCH_3 OCH_3 OH + HIO_4 ⟶ [+ CH_3O OH CH_3O OH H H] $\xrightarrow{-2\,H_2O}$ + CH_3O O CH_3O H H

(d) 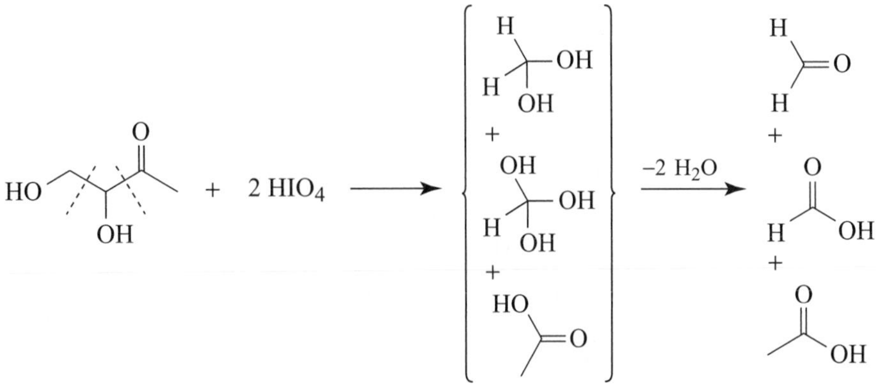 HO, OH, O + 2 HIO_4 ⟶ [H, OH, H, OH + OH, OH, H, OH + HO, O] $\xrightarrow{-2\ H_2O}$ H, O, H + O, H, OH + O, OH

(e) 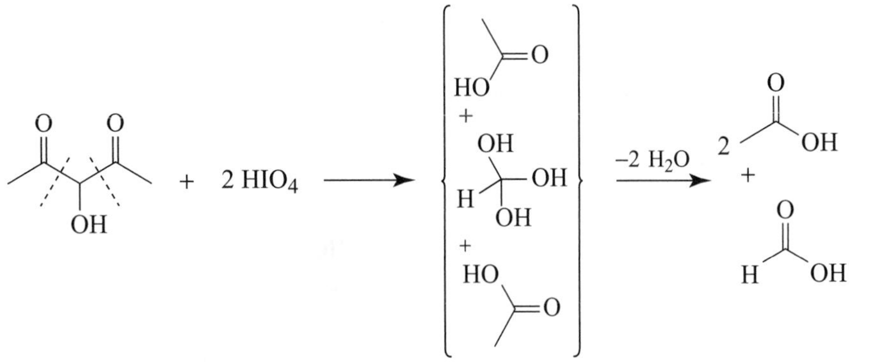 O, O, OH + 2 HIO_4 ⟶ [O, HO + OH, OH, H, OH + HO, O] $\xrightarrow{-2\ H_2O}$ 2 O, OH + O, H, OH

(f) 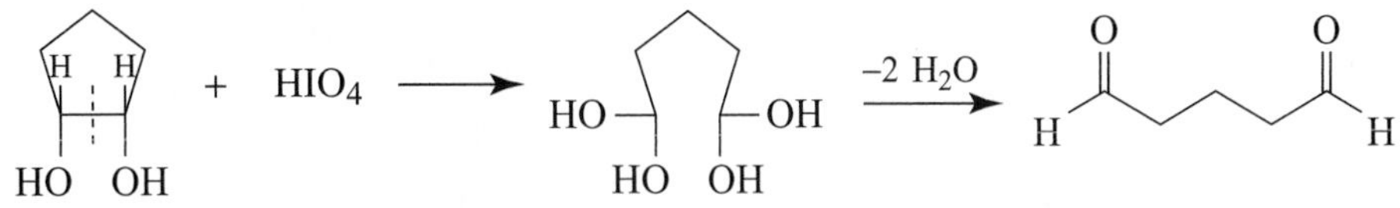 H, H, HO, OH + HIO_4 ⟶ HO, OH, HO, OH $\xrightarrow{-2\ H_2O}$ O, O, H, H

(g) 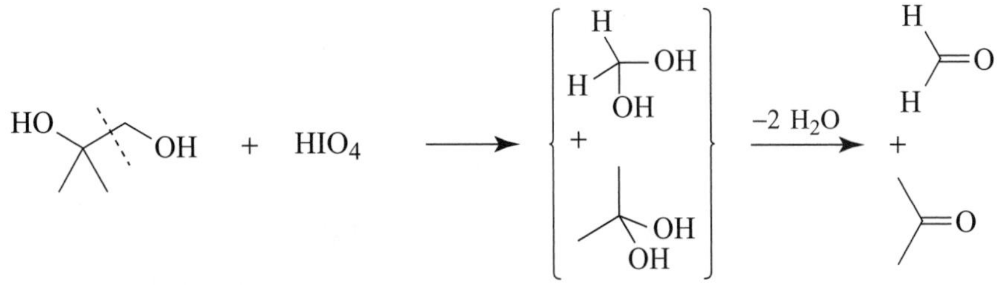 HO, OH + HIO_4 ⟶ [H, OH, H, OH + OH, OH] $\xrightarrow{-2\ H_2O}$ H, O, H + O

(h)

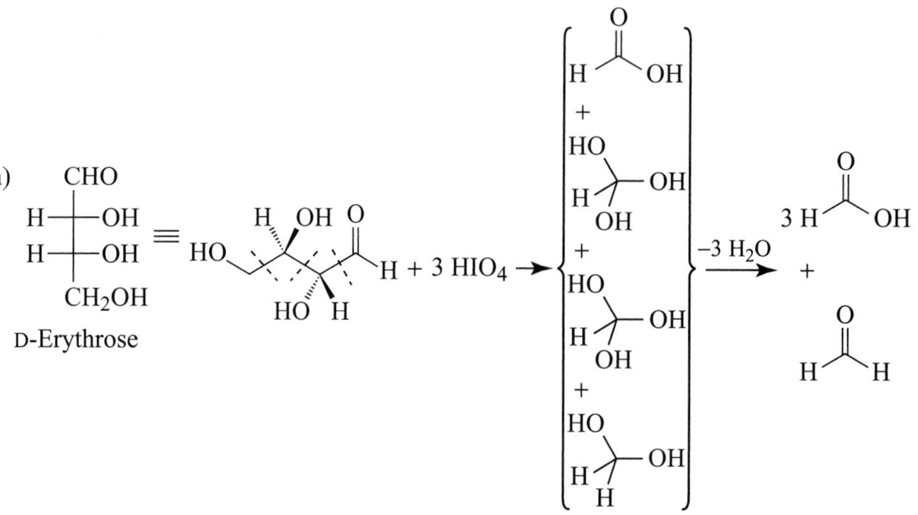

CHO
H—OH
H—OH
CH_2OH
D-Erythrose

≡ HO, H, OH, O, HO, H (zigzag structure) H + 3 HIO_4 →

[H–C(=O)–OH + HO–CH(OH)–OH + HO–CH(OH)–OH + HO–CH(H)–OH]

$\xrightarrow{-3\,H_2O}$ 3 H–C(=O)–OH + H–C(=O)–H

22.10 Oxidation of an aldohexose and a ketohexose would each require 5 mol of HIO_4 but would give different results.

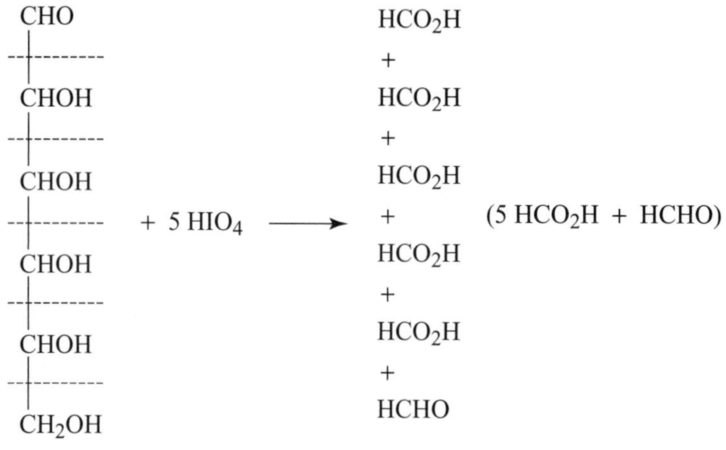

CHO
CHOH
CHOH
CHOH
CHOH
CH_2OH
Aldohexose

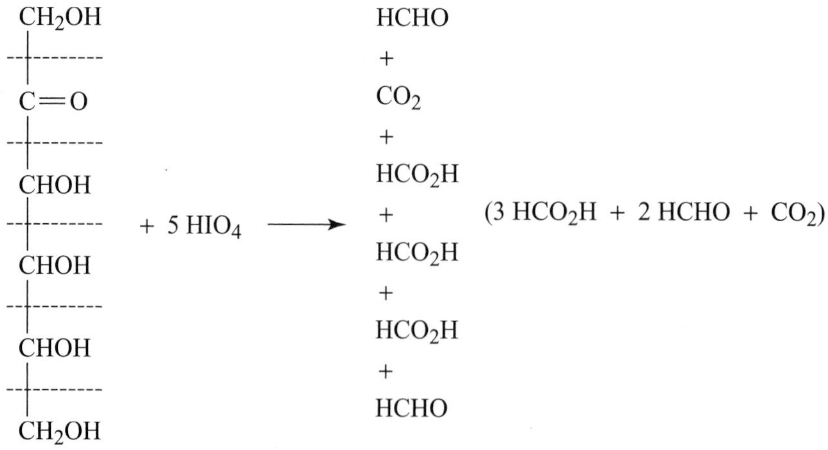

+ 5 HIO_4 ⟶ HCO_2H + HCO_2H + HCO_2H + HCO_2H + HCO_2H + HCHO (5 HCO_2H + HCHO)

CH_2OH
C=O
CHOH
CHOH
CHOH
CH_2OH
Ketohexose

+ 5 HIO_4 ⟶ HCHO + CO_2 + HCO_2H + HCO_2H + HCO_2H + HCHO (3 HCO_2H + 2 HCHO + CO_2)

22.11 (a) Yes, D-glucitol would be optically active; only those alditols whose molecules possess a plane of symmetry would be optically inactive.

(b)

H–C=O
H–C–OH
H–C–OH
H–C–OH
H–C–OH
CH₂OH

$\xrightarrow{NaBH_4}$

CH₂OH
H–C–OH
H–C–OH
- - - - - - Plane of symmetry
H–C–OH
H–C–OH
CH₂OH

Optically inactive

H–C=O
H–C–OH
HO–C–H
HO–C–H
H–C–OH
CH₂OH

$\xrightarrow{NaBH_4}$

CH₂OH
H–C–OH
HO–C–H
- - - - - - Plane of symmetry
HO–C–H
H–C–OH
CH₂OH

Optically inactive

22.12 (a)

CH₂OH
C=O
HO–C–H
H–C–OH
H–C–OH
CH₂OH

$\xrightarrow{C_6H_5NHNH_2}$

$CH{=}NNHC_6H_5$
$C{=}NNHC_6H_5$
HO–C–H
H–C–OH
H–C–OH
CH₂OH

(b) This experiment shows that D-glucose and D-fructose have the same configurations at C3, C4, and C5.

22.13 (a)

H–C=O
HO–C–H
HO–C–H
CH₂OH

L-(+)-Erythrose

H–C=O
H–C–OH
HO–C–H
CH₂OH

L-(+)-Threose

(b) L-Glyceraldehyde

H–C=O
HO–C–H
CH₂OH

22.14 (a)

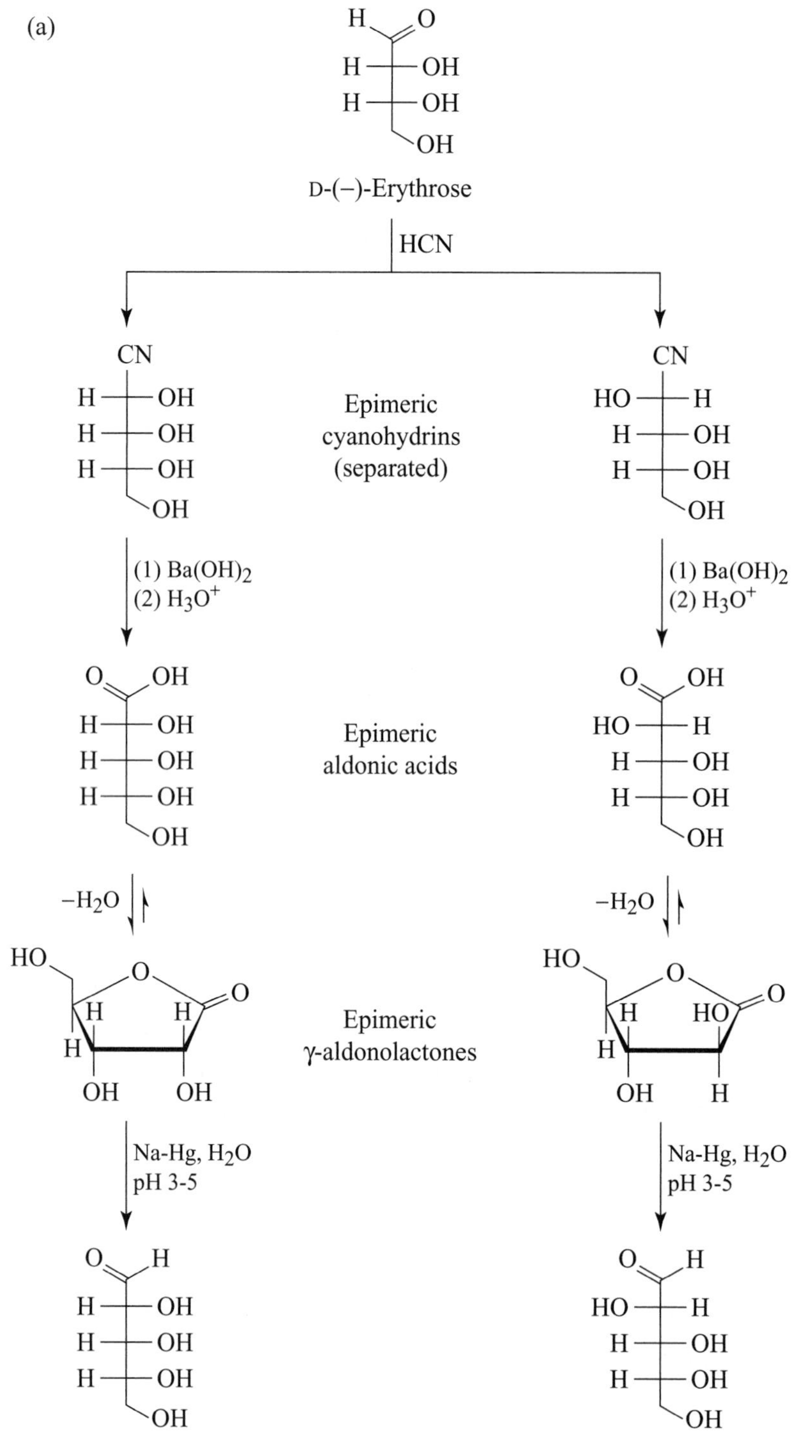

D-(−)-Erythrose

HCN

Epimeric cyanohydrins (separated)

(1) $Ba(OH)_2$
(2) H_3O^+

Epimeric aldonic acids

$-H_2O$

Epimeric γ-aldonolactones

Na-Hg, H_2O
pH 3-5

(b)

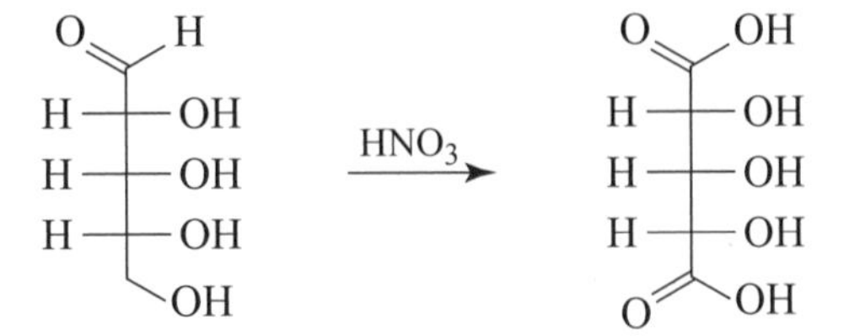

O H
H—OH
H—OH
H—OH
OH

D-(−)-Ribose

$\xrightarrow{HNO_3}$

O OH
H—OH
H—OH
H—OH
O OH

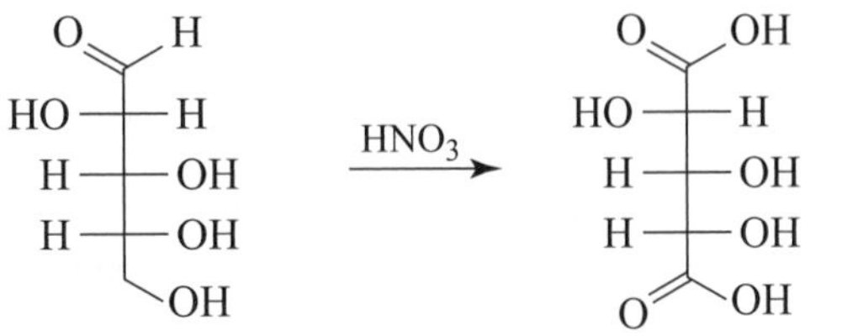

Optically inactive

O H
HO—H
H—OH
H—OH
OH

D-(−)-Arabinose

$\xrightarrow{HNO_3}$

O OH
HO—H
H—OH
H—OH
O OH

Optically active

22.15 A Kiliani-Fischer synthesis starting with D-(−)-threose would yield **I** and **II**.

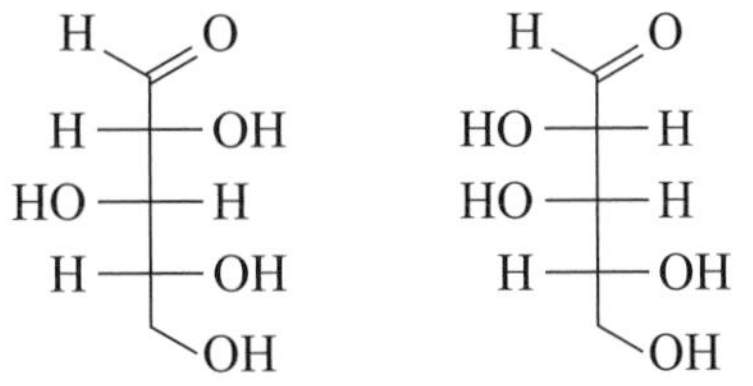

H O
H—OH
HO—H
H—OH
OH

I
D-(+)-Xylose

H O
HO—H
HO—H
H—OH
OH

II
D-(−)-Lyxose

I must be D-(+)-xylose because, when oxidized by nitric acid, it yields an optically inactive aldaric acid:

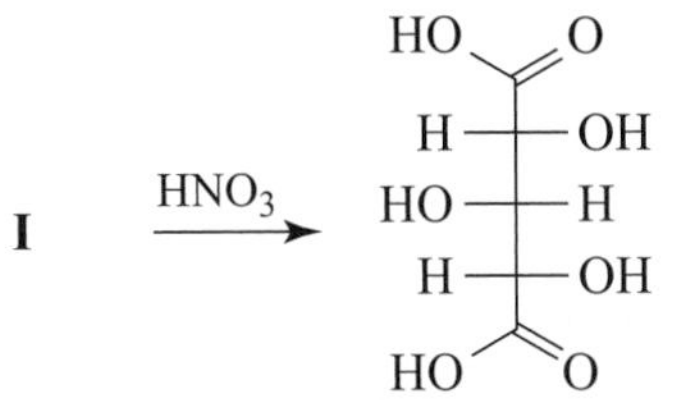

I $\xrightarrow{HNO_3}$

HO O
H—OH
HO—H
H—OH
HO O

Optically inactive

II must be D-(−)-lyxose because, when oxidized by nitric acid, it yields an optically active aldaric acid:

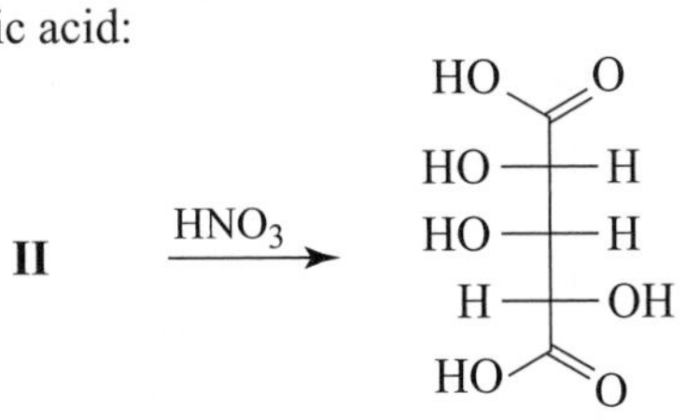

II $\xrightarrow{HNO_3}$

HO O
HO—H
HO—H
H—OH
HO O

Optically active

22.16

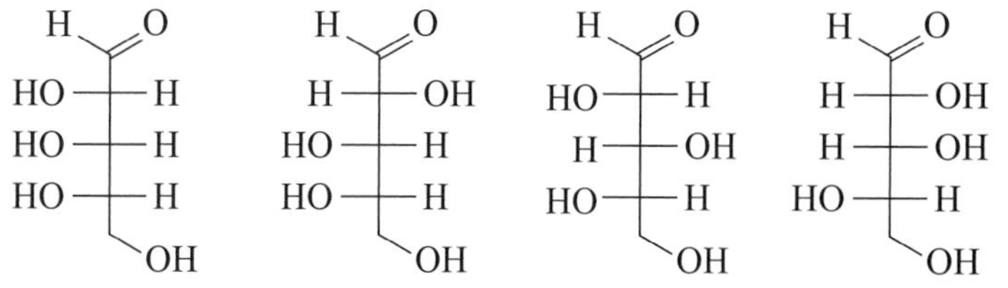

L-(+)-Ribose	L-(+)-Arabinose	L-(−)-Xylose	L-(+)-Lyxose
H O (CHO)	H O (CHO)	H O (CHO)	H O (CHO)
HO—H	H—OH	HO—H	H—OH
HO—H	HO—H	H—OH	H—OH
HO—H	HO—H	HO—H	HO—H
OH	OH	OH	OH

22.17 Since D-(+)-galactose yields an optically inactive aldaric acid, it must have either structure **III** or structure **IV**.

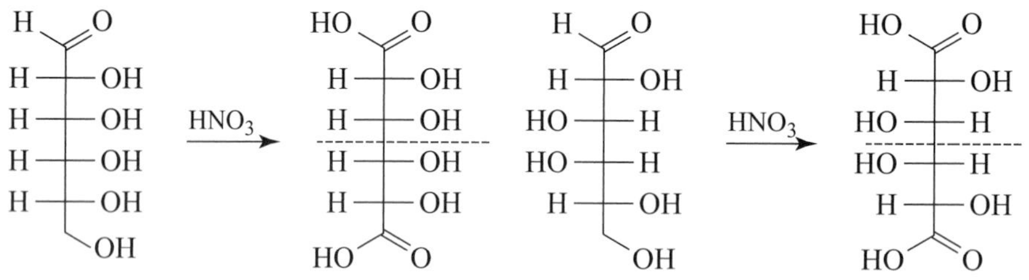

III	$\xrightarrow{HNO_3}$	Optically inactive	IV	$\xrightarrow{HNO_3}$	Optically inactive
H O		HO O	H O		HO O
H—OH		H—OH	H—OH		H—OH
H—OH		H—OH	HO—H		HO—H
H—OH		H—OH	HO—H		HO—H
H—OH		H—OH	H—OH		H—OH
OH		HO O	OH		HO O

A Ruff degradation beginning with **III** would yield D-(−)-ribose

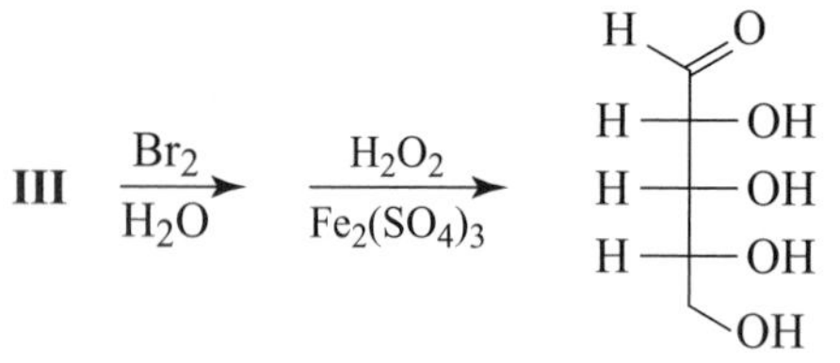

III $\xrightarrow[H_2O]{Br_2}$ $\xrightarrow[Fe_2(SO_4)_3]{H_2O_2}$ H O; H—OH; H—OH; H—OH; OH

D-(−)-Ribose

A Ruff degradation beginning with **IV** would yield D-(−)-lyxose: thus, D-(+)-galactose must have structure **IV**.

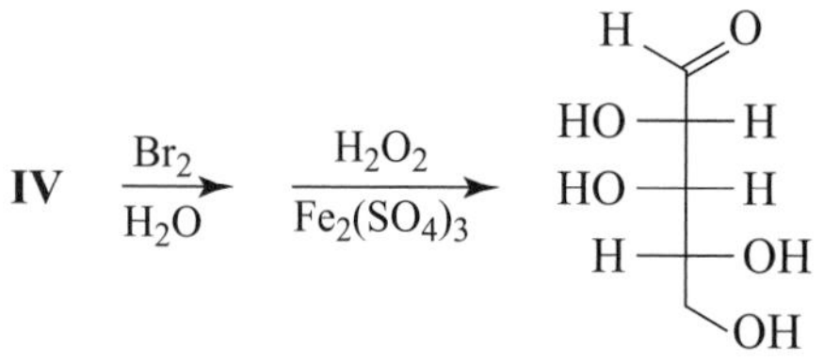

IV $\xrightarrow[H_2O]{Br_2}$ $\xrightarrow[Fe_2(SO_4)_3]{H_2O_2}$ H O; HO—H; HO—H; H—OH; OH

D-(−)-Lyxose

22.18 D-(+)-glucose, as shown here.

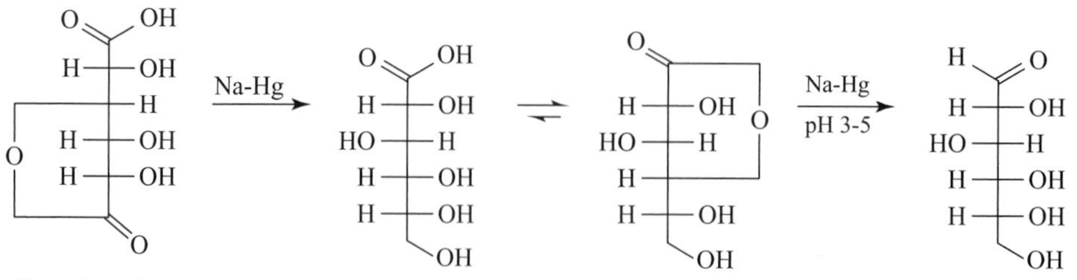

The other γ-lactone of D-glucaric acid	$\xrightarrow{Na\text{-}Hg}$		⇌		$\xrightarrow[pH\ 3\text{-}5]{Na\text{-}Hg}$	D-(+)-Glucose
O OH		O OH		O		H O
H—OH		H—OH		H—OH O		H—OH
O —H		HO—H		HO—H		HO—H
H—OH		H—OH		H—		H—OH
H—OH		H—OH		H—OH		H—OH
O		OH		OH		OH

22.19

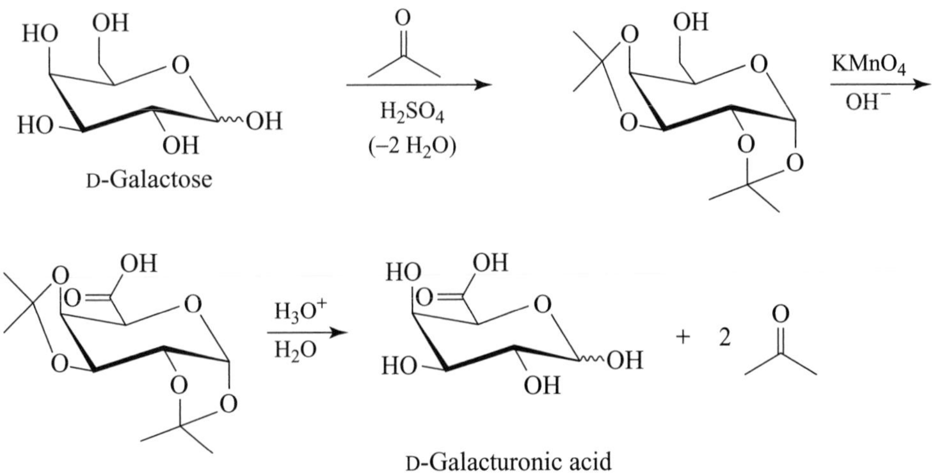

HO OH O OH HO OH D-Galactose → (acetone, H_2SO_4, $(-2\ H_2O)$) → OH O O O O O → $KMnO_4$, OH^- →

O OH O O O O O → H_3O^+, H_2O → HO OH O O HO OH OH + 2 acetone

D-Galacturonic acid

Problems

Carbohydrate Structure and Reactions

22.20 (a) H–C(=O)–CHOH–CHOH–CHOH–CH_2OH

(b) CH_2OH–C(=O)–CHOH–CHOH–CHOH–CH_2OH

(c) H–C(=O)–$(CHOH)_n$–C(HO)(H)–CH_2OH or CH_2OH–C(=O)–$(CHOH)_n$–C(HO)(H)–CH_2OH

(d) CHOR–$(CHOH)_n$–CH(–O– ring to CHOR)–CH_2OH

(e) HO–C(=O)–$(CHOH)_n$–CH_2OH

(f) HO–C(=O)–$(CHOH)_n$–C(=O)–OH

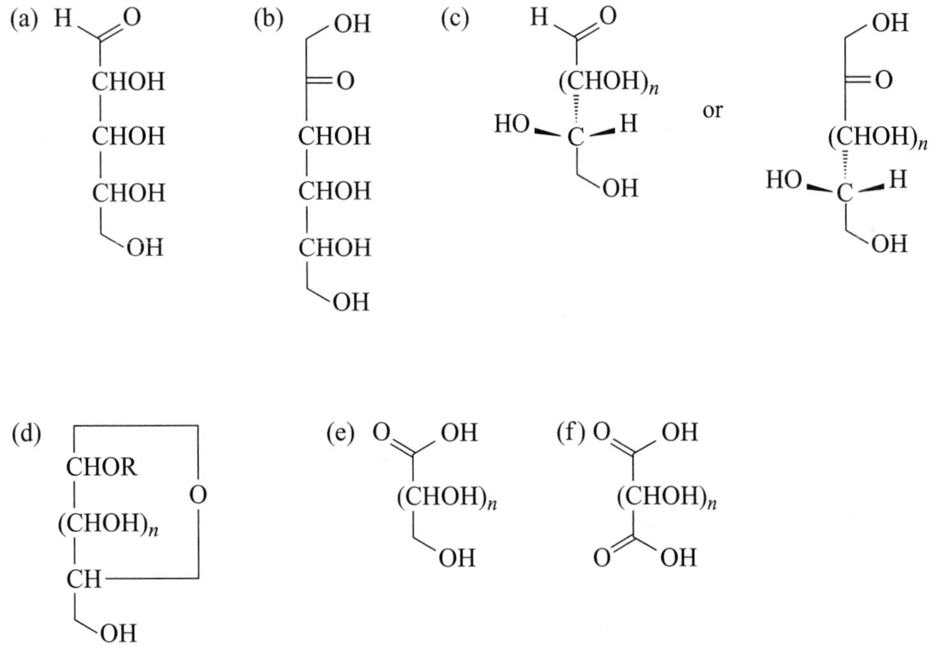

(g)

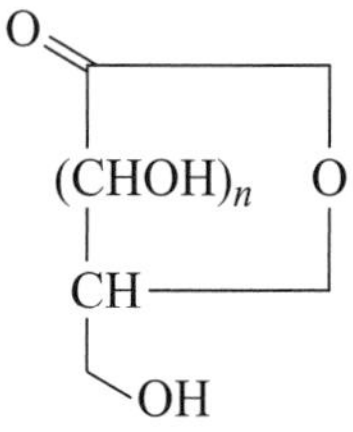

O
(CHOH)$_n$ O
CH
OH

(h)

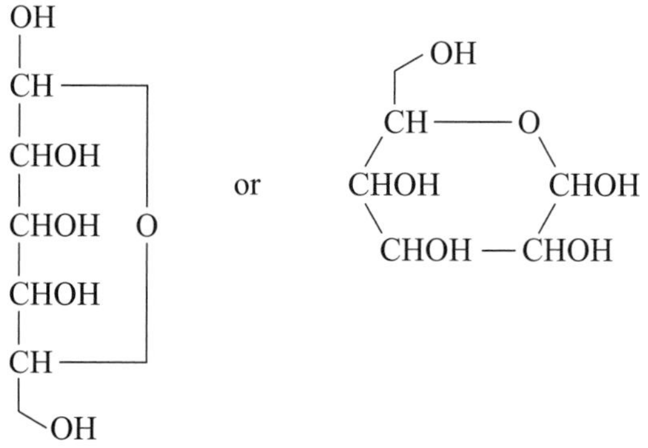

OH
CH
CHOH
CHOH O
CHOH
CH
OH

or

OH
CH — O
CHOH CHOH
CHOH — CHOH

(i)

OH
CH
CHOH
O
CHOH
CH
CHOH
OH

or

OH
CHOH
CH O CHOH
CHOH — CHOH

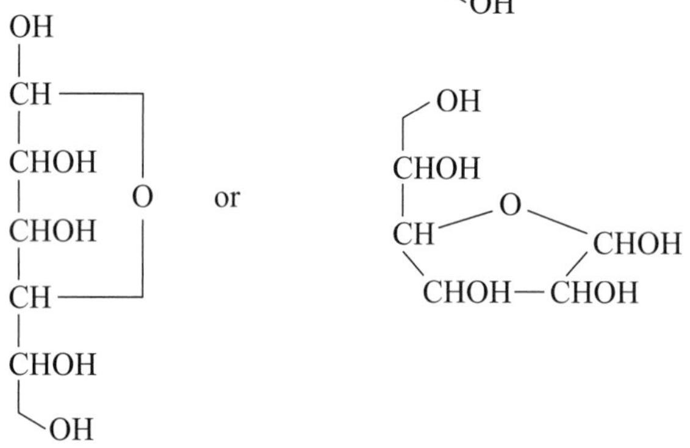

(j) Any sugar that has a free aldehyde or ketone group or one that exists as a cyclic hemiacetal. The following are examples:

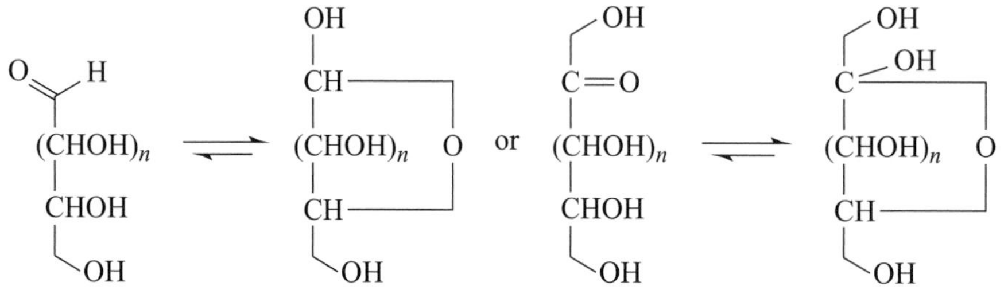

O H
(CHOH)$_n$
CHOH
OH

⇌

OH
CH
(CHOH)$_n$ O
CH
OH

or

OH
C=O
(CHOH)$_n$
CHOH
OH

⇌

OH
C OH
(CHOH)$_n$ O
CH
OH

(k)

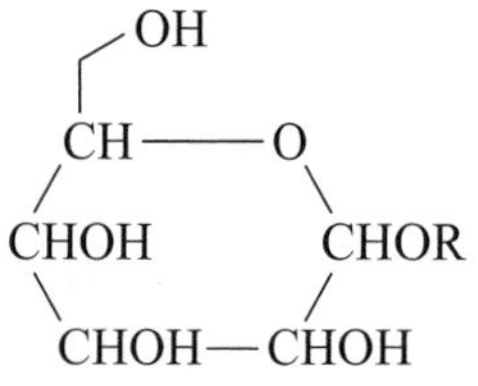

OH
CH — O
CHOH CHOR
CHOH — CHOH

(l)

OH
CHOH
CH O CHOR
CHOH — CHOH

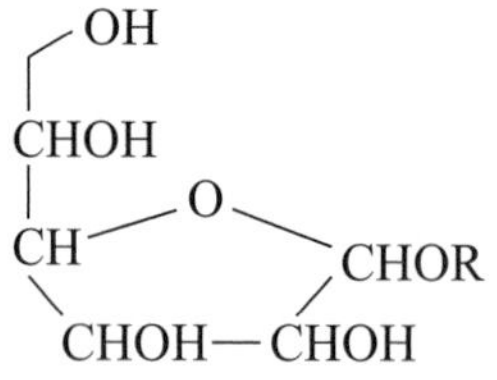

(m) Any two aldoses that differ only in configuration at C2. (See also Section 22.8 for a broader definition.) D-Erythrose and D-threose are examples.

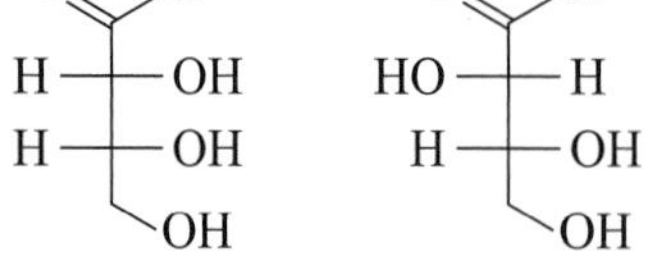

O H
H — OH
H — OH
OH

D-Erythrose

O H
HO — H
H — OH
OH

D-Threose

(n) Cyclic sugars that differ only in the configuration of C1. The following are examples:

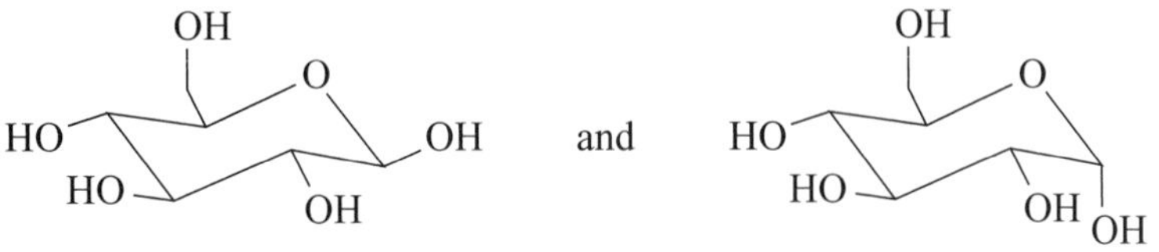

(o)

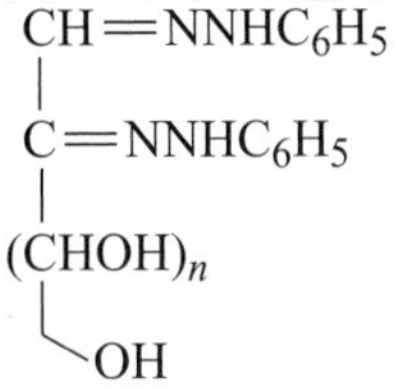

(p) Maltose is an example:

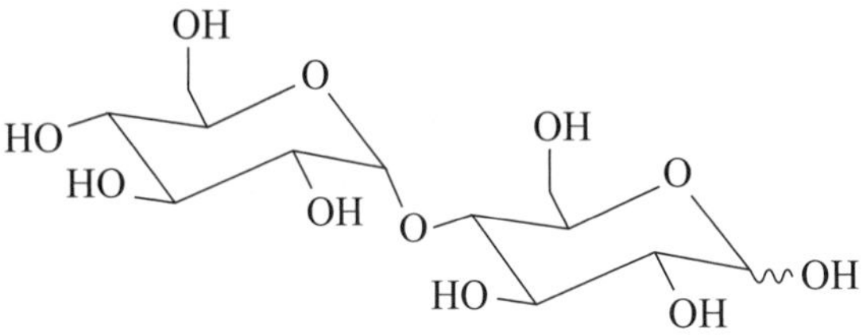

(q) Amylose is an example:

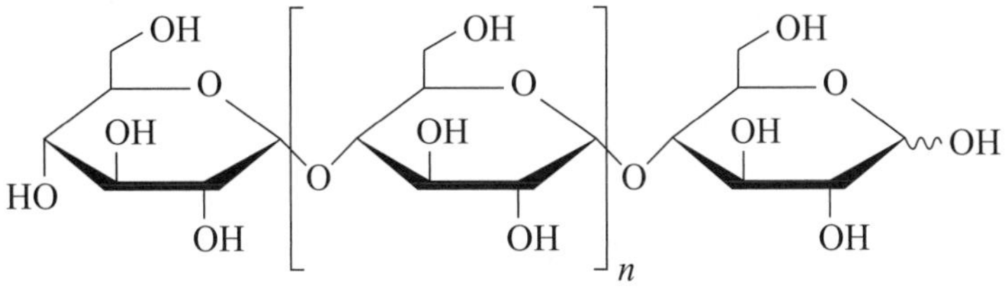

(r) Any sugar in which all potential carbonyl groups are present as acetals (i.e., as glycosides). Sucrose (Section 22.12A) is an example of a nonreducing disaccharide; the methyl D-glucopyranosides (Section 22.4) are examples of nonreducing monosaccharides.

22.21 (a)

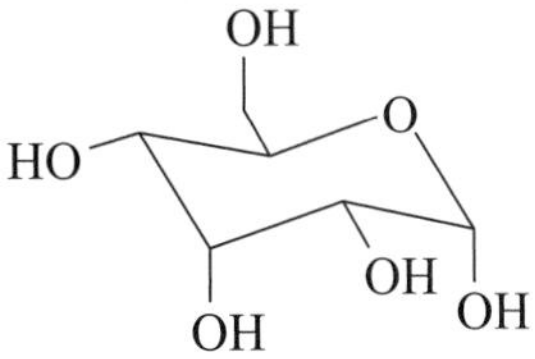

(b)

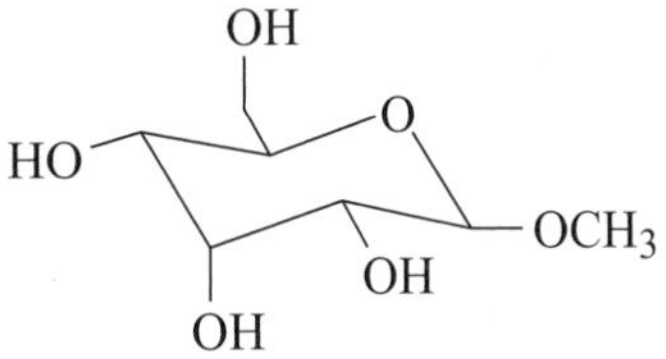

(c)

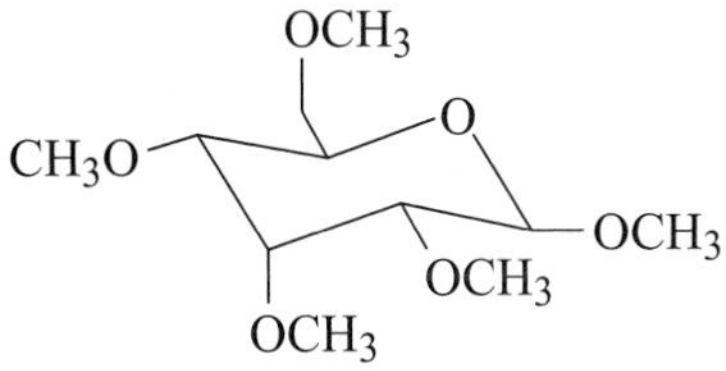

22.22

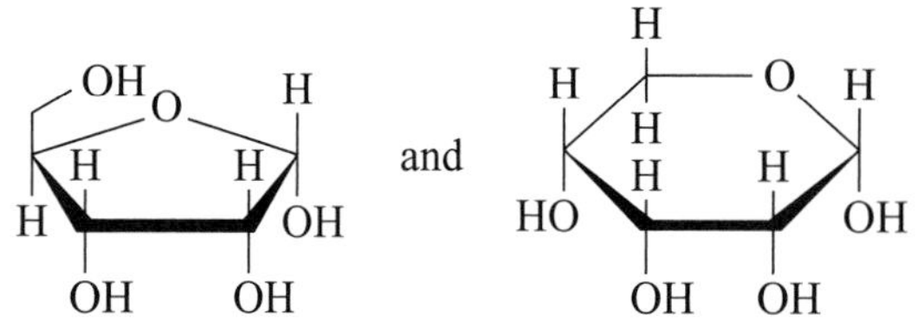

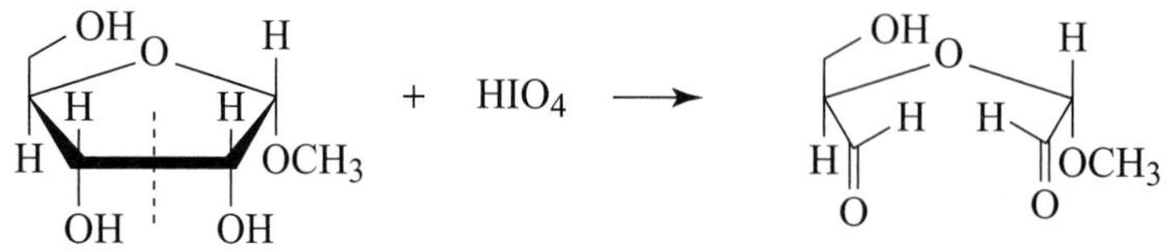

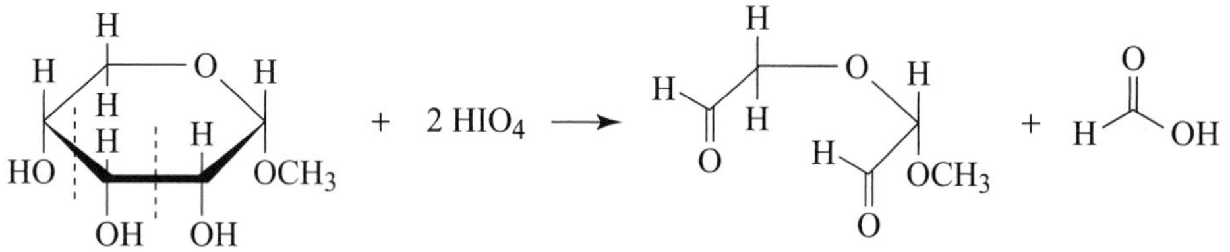

The above would apply in the same way to the β anomers. A methyl ribofuranoside would consume only 1 mol of HIO_4; a methyl ribopyranoside would consume 2 mol of HIO_4 and would also produce 1 mol of formic acid.

22.23 One anomer of D-mannose is dextrorotatory ($[\alpha]_D^{25} = +29.3$); the other is levorotatory ($[\alpha]_D^{25} = -17.0$).

22.24 The microorganism selectively oxidizes the $-$CHOH group of D-glucitol that corresponds to C5 of D-glucose.

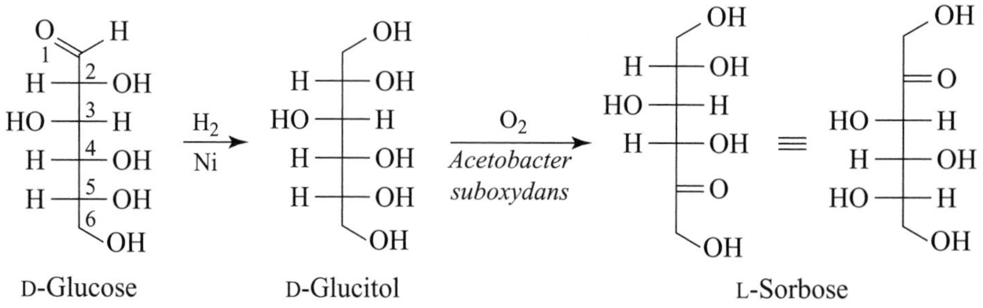

D-Glucose $\xrightarrow[\text{Ni}]{H_2}$ D-Glucitol $\xrightarrow[\textit{Acetobacter suboxydans}]{O_2}$ L-Sorbose

22.25 L-Gulose and L-idose would yield the same phenylosazone as L-sorbose.

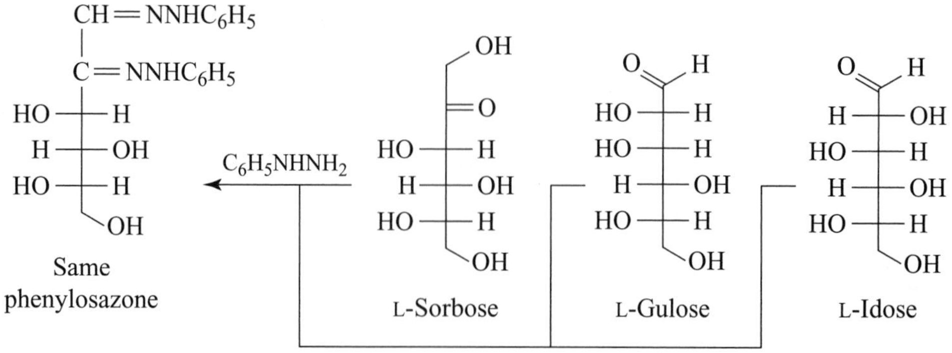

L-Sorbose, L-Gulose, L-Idose $\xrightarrow{C_6H_5NHNH_2}$ Same phenylosazone

22.26

D-Psicose $\xrightarrow{C_6H_5NHNH_2}$ phenylosazone $\xleftarrow{C_6H_5NHNH_2}$ D-Allose

D-Psicose: CH_2OH; C=O; H–C–OH; H–C–OH; H–C–OH; CH_2OH

Phenylosazone: $CH{=}NNHC_6H_5$; $C{=}NNHC_6H_5$; H–C–OH; H–C–OH; H–C–OH; CH_2OH

D-Allose: CHO; H–C–OH; H–C–OH; H–C–OH; H–C–OH; CH_2OH

D-Tagatose $\xrightarrow{C_6H_5NHNH_2}$ phenylosazone $\xleftarrow{C_6H_5NHNH_2}$ D-Galactose

D-Tagatose: CH_2OH; C=O; HO–C–H; HO–C–H; H–C–OH; CH_2OH

Phenylosazone: $CH{=}NNHC_6H_5$; $C{=}NNHC_6H_5$; HO–C–H; HO–C–H; H–C–OH; CH_2OH

D-Galactose: CHO; H–C–OH; HO–C–H; HO–C–H; H–C–OH; CH_2OH

22.27 **A** is D-altrose, **B** is D-talose, **C** is D-galactose.

D-Altrose (**A**) $\xrightarrow[Ni]{H_2}$ Same alditol $\xleftarrow[Ni]{H_2}$ D-Talose (**B**)

D-Altrose: CHO; HO–C–H; H–C–OH; H–C–OH; H–C–OH; CH_2OH

Same alditol: CH_2OH; HO–C–H; H–C–OH; H–C–OH; H–C–OH; CH_2OH ≡ CH_2OH; HO–C–H; HO–C–H; HO–C–H; H–C–OH; CH_2OH

D-Talose: CHO; HO–C–H; HO–C–H; HO–C–H; H–C–OH; CH_2OH

A $\xrightarrow{C_6H_5NHNH_2}$ $CH{=}NNHC_6H_5$; $C{=}NNHC_6H_5$; H–C–OH; H–C–OH; H–C–OH; CH_2OH

B $\xrightarrow{C_6H_5NHNH_2}$ $CH{=}NNHC_6H_5$; $C{=}NNHC_6H_5$; HO–C–H; HO–C–H; H–C–OH; CH_2OH

Different phenylosazones

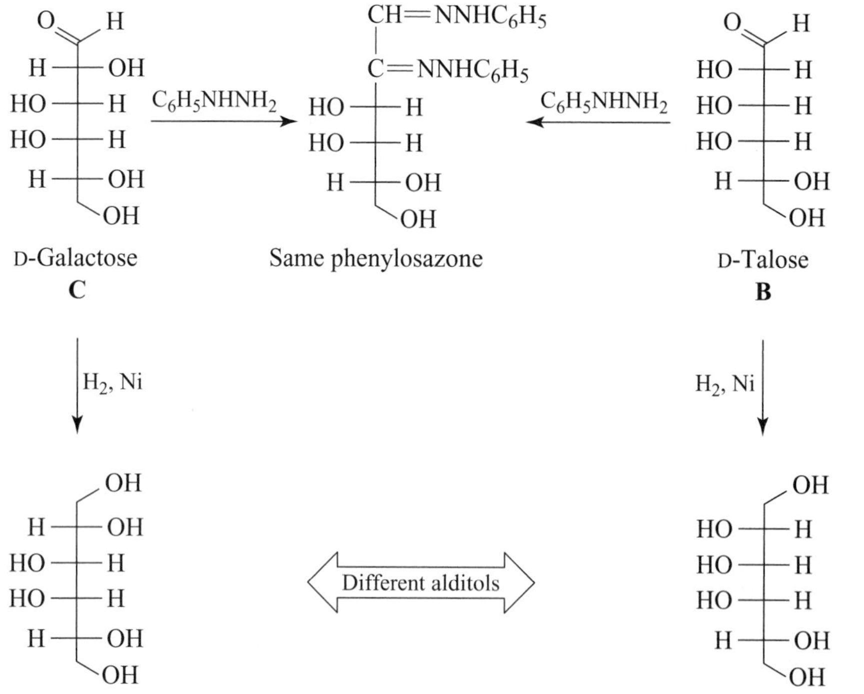

D-Galactose **C** —$C_6H_5NHNH_2$→ Same phenylosazone ←$C_6H_5NHNH_2$— D-Talose **B**

Same phenylosazone: CH=$NNHC_6H_5$, C=$NNHC_6H_5$, HO—H, HO—H, H—OH, OH

D-Galactose **C**: O H; H—OH; HO—H; HO—H; H—OH; OH

D-Talose **B**: O H; HO—H; HO—H; HO—H; H—OH; OH

D-Galactose **C** —H_2, Ni→ alditol: OH; H—OH; HO—H; HO—H; H—OH; OH

D-Talose **B** —H_2, Ni→ alditol: OH; HO—H; HO—H; HO—H; H—OH; OH

Different alditols

(*Note*: If we had designated D-talose as **A**, and D-altrose as **B**, then **C** is D-allose).

22.28

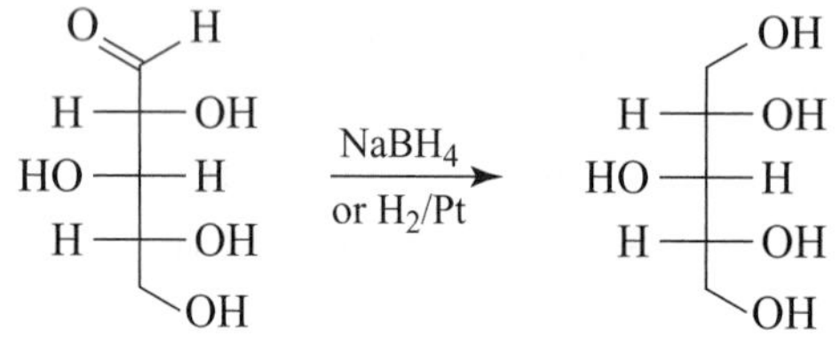

D-Xylose (O H; H—OH; HO—H; H—OH; OH) —$NaBH_4$ or H_2/Pt→ D-Xylitol (OH; H—OH; HO—H; H—OH; OH)

22.29

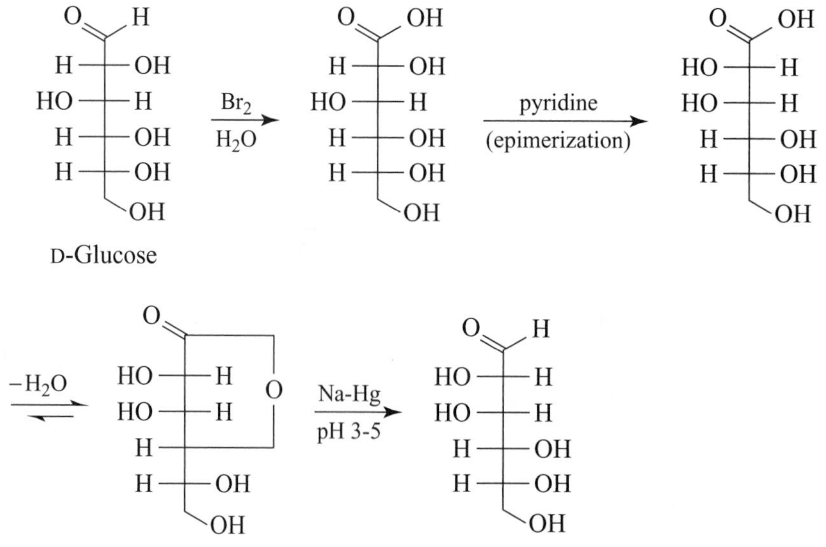

D-Glucose (O H; H—OH; HO—H; H—OH; H—OH; OH) —Br_2, H_2O→ (O OH; H—OH; HO—H; H—OH; H—OH; OH) —pyridine (epimerization)→ (O OH; HO—H; HO—H; H—OH; H—OH; OH)

$-H_2O$ ⇌ lactone (O; HO—H; HO—H; H—O (ring); H—OH; OH) —Na-Hg, pH 3-5→ D-Mannose (O H; HO—H; HO—H; H—OH; H—OH; OH)

22.30 The conformation of D-idopyranose with four equatorial —OH groups and an axial $—CH_2OH$ group is more stable than the one with four axial —OH groups and an equatorial $—CH_2OH$ group.

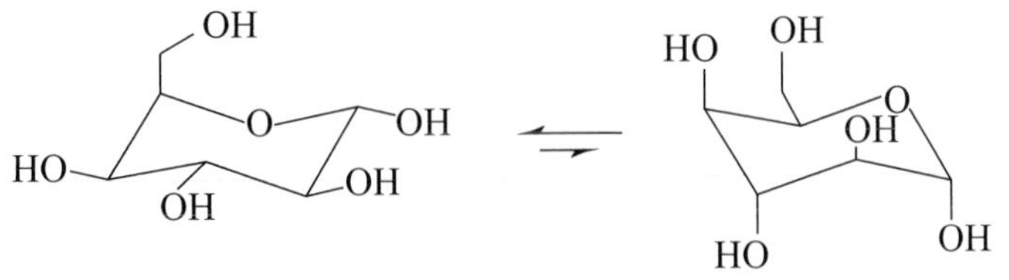

More stable

4 Equatorial —OH groups
1 Axial $—CH_2OH$ group

Less stable

4 Axial —OH groups
1 Equatorial $—CH_2OH$ group

Structure Elucidation

22.31 (a) The anhydro sugar is formed when the axial $—CH_2OH$ group reacts with C1 to form a cyclic acetal.

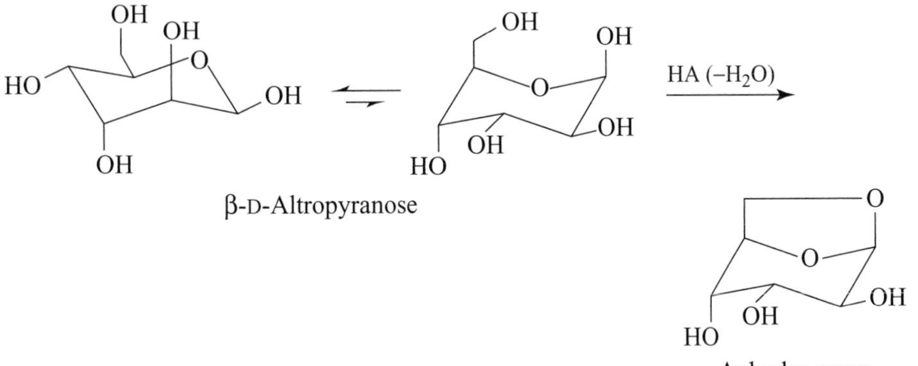

β-D-Altropyranose

HA $(-H_2O)$

Anhydro sugar

Because the anhydro sugar is an acetal (i.e., an internal glycoside), it is a nonreducing sugar.
Methylation followed by acid hydrolysis converts the anhydro sugar to 2,3,4-tri-*O*-methyl-D-altrose:

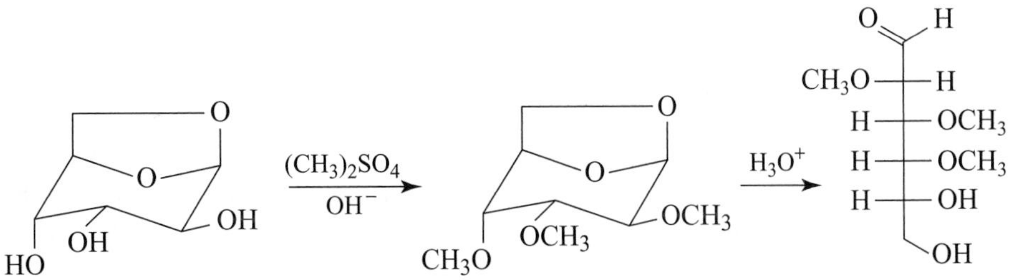

Anhydro β-D-altropyranose

$(CH_3)_2SO_4$, OH^-

H_3O^+

2,3,4-Tri-*O*-methyl-D-altrose

(b) Formation of an anhydro sugar requires that the monosaccharide adopt a chair conformation with the —CH_2OH group axial. With β-D-altropyranose this requires that two —OH groups be axial as well. With β-D-glucopyranose, however, it requires that all four —OH groups become axial and thus that the molecule adopt a very unstable conformation:

Highly unstable conformation

HA ($-H_2O$)

β-D-Glucopyranose

Anhydro-β-D-glucopyranose

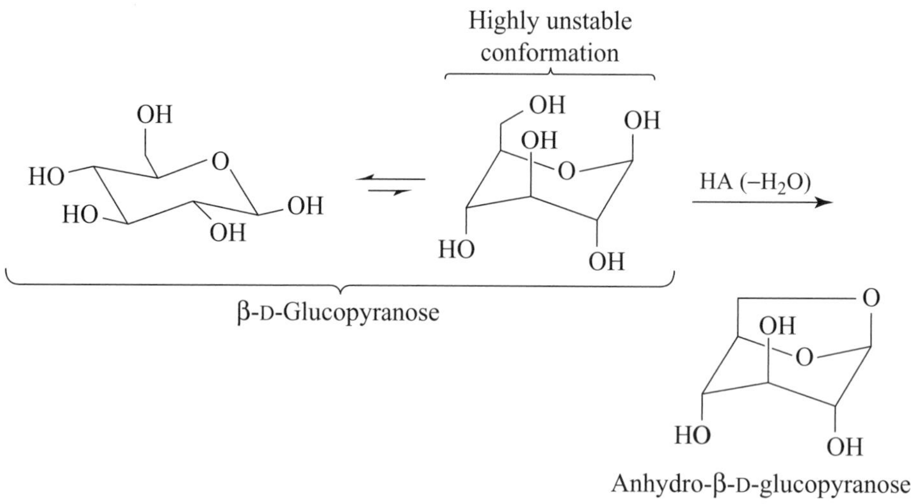

22.32 1. The molecular formula and the results of acid hydrolysis show that lactose is a disaccharide composed of D-glucose and D-galactose. The fact that lactose is hydrolyzed by a *β-galactosidase* indicates that galactose is present as a glycoside and that the glycosidic linkage is beta to the galactose ring.

2. That lactose is a reducing sugar, forms a phenylosazone, and undergoes mutarotation indicates that one ring (presumably that of D-glucose) is present as a hemiacetal and thus is capable of existing to a limited extent as an aldehyde.

3. This experiment confirms that the D-glucose unit is present as a cyclic hemiacetal and that the D-galactose unit is present as a cyclic glycoside.

4. That 2,3,4,6-tetra-*O*-methyl-D-galactose is obtained in this experiment indicates (by virtue of the free —OH at C5) that the galactose ring of lactose is present as a pyranoside. That the methylated gluconic acid obtained from this experiment has a free —OH group at C4 indicates that the C4 oxygen atom of the glucose unit is connected in a glycosidic linkage to the galactose unit.

Now only the size of the glucose ring remains in question, and the answer to this is provided by experiment 5.

5. That methylation of lactose and subsequent hydrolysis gives 2,3,6-tri-*O*-methyl-D-glucose—that it gives a methylated glucose derivative with a free —OH at C4 and C5 demonstrates that the glucose ring is present as a pyranose. (We know already that the oxygen at C4 is connected in a glycosidic linkage to the galactose unit; thus, a free —OH at C5 indicates that the C5 oxygen atom is a part of the hemiacetal group of the glucose unit and that the ring is six membered.)

22.33

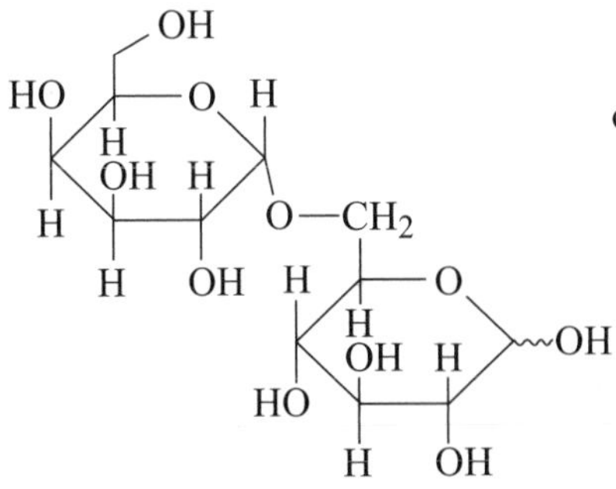

or

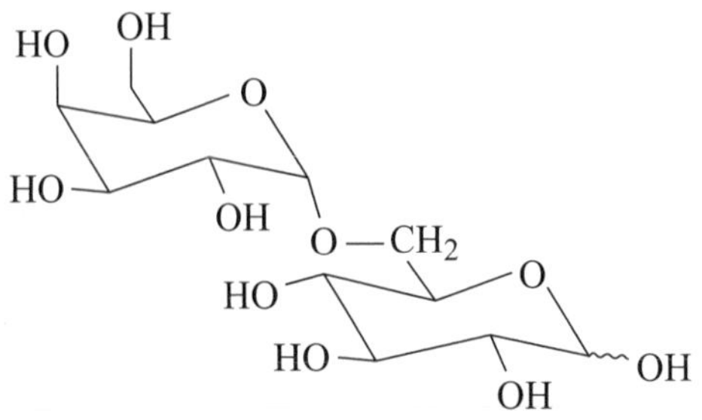

6-O-(α-D-Galactopyranosyl)-D-glucopyranose

We arrive at this conclusion from the data given:

1. That melibiose is a reducing sugar and that it undergoes mutarotation and forms a phenylosazone indicate that one monosaccharide is present as a cyclic hemiacetal.

2. That acid hydrolysis gives D-galactose and D-glucose indicates that melibiose is a disaccharide composed of one D-galactose unit and one D-glucose unit. That melibiose is hydrolyzed by an α-galactosidase suggests that melibiose is an α-D-galactosyl-D-glucose.

3. Oxidation of melibiose to melibionic acid and subsequent hydrolysis to give D-galactose and D-gluconic acid confirms that the glucose unit is present as a cyclic hemiacetal and that the galactose unit is present as a glycoside. (Had the reverse been true, this experiment would have yielded D-glucose and D-galactonic acid.)

Methylation and hydrolysis of melibionic acid produces 2,3,4,6-tetra-O-methyl-D-galactose and 2,3,4,5-tetra-O-methyl-D-gluconic acid. Formation of the first product—a galactose derivative with a free —OH at C5—demonstrates that the galactose ring is six membered; formation of the second product—a gluconic acid derivative with a free —OH at C6—demonstrates that the oxygen at C6 of the glucose unit is joined in a glycosidic linkage to the galactose unit.

4. That methylation and hydrolysis of melibiose gives a glucose derivative (2,3,4-tri-O-methyl-D-glucose) with free —OH groups at C5 and C6 shows that the glucose ring is also six membered. Melibiose is, therefore, 6-O-(α-D-galactopyranosyl-D-glucopyranose.

22.34 Trehalose has the following structure:

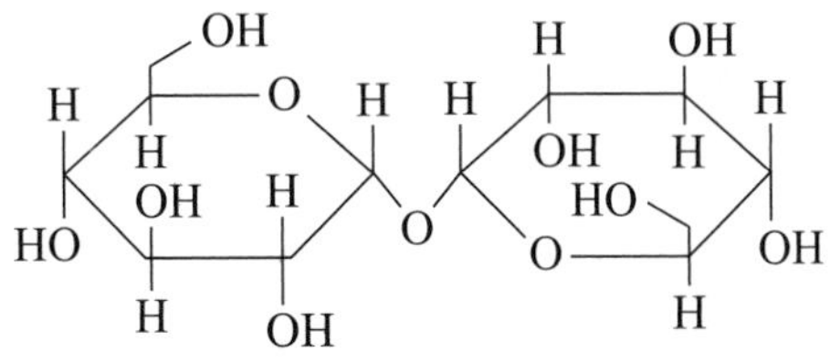

or

α-D-Glucopyranosyl-α-D-glucopyranoside

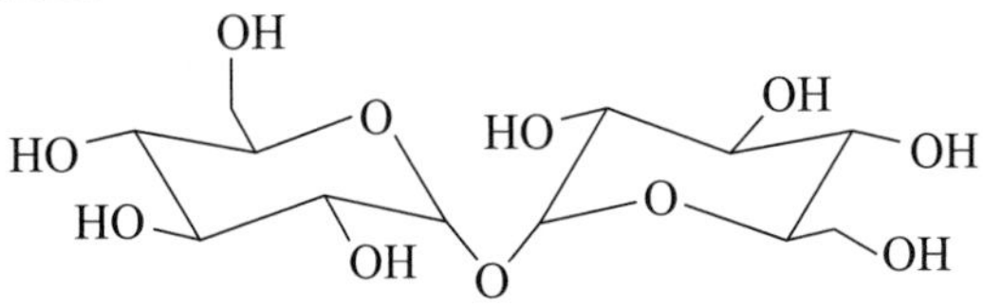

We arrive at this structure in the following way:

1. Acid hydrolysis shows that trehalose is a disaccharide consisting only of D-glucose units.

2. Hydrolysis by α-glucosidases and not by β-glucosidases shows that the glycosidic linkages are alpha.

3. That trehalose is a nonreducing sugar, that it does not form a phenylosazone, and that it does not react with bromine water indicate that no hemiacetal groups are present. This means that C1 of one glucose unit and C1 of the other must be joined in a glycosidic linkage. Fact 2 (just cited) indicates that this linkage is alpha to each ring.

4. That methylation of trehalose followed by hydrolysis yields only 2,3,4,6-tetra-*O*-methyl-D-glucose demonstrates that both rings are six membered.

22.35 (a) Tollens' reagent or Benedict's reagent will give a positive test with D-glucose but will give no reaction with D-glucitol.

(b) D-Glucaric acid will give an acidic aqueous solution that can be detected with blue litmus paper. D-Glucitol will give a neutral aqueous solution.

(c) D-Glucose will be oxidized by bromine water and the red brown color of bromine will disappear. D-Fructose will not be oxidized by bromine water since it does not contain an aldehyde group.

(d) Nitric acid oxidation will produce an *optically active* aldaric acid from D-glucose but an *optically inactive* aldaric acid will result from D-galactose.

(e) Maltose is a reducing sugar and will give a positive test with Tollens' or Benedict's solution. Sucrose is a nonreducing sugar and will not react.

(f) Maltose will give a positive Tollens' or Benedict's test; maltonic acid will not.

(g) 2,3,4,6-Tetra-*O*-methyl-β-D-glucopyranose will give a positive test with Tollens' or Benedict's solution; methyl β-D-glucopyranoside will not.

(h) Periodic acid will react with methyl α-D-ribofuranoside because it has hydroxyl groups on adjacent carbons. Methyl 2-deoxy-α-D-ribofuranoside will not react.

22.36 That the Schardinger dextrins are nonreducing shows that they have no free aldehyde or hemiacetal groups. This lack of reaction strongly suggests the presence of a *cyclic* structure. That methylation and subsequent hydrolysis yields only 2,3,6-tri-*O*-methyl-D-glucose indicates that the glycosidic linkages all involve C1 of one glucose unit and C4 of the next. That α-glucosidases cause hydrolysis of the glycosidic linkages indicates that they are α-glycosidic linkages. Thus, we are led to the following general structure.

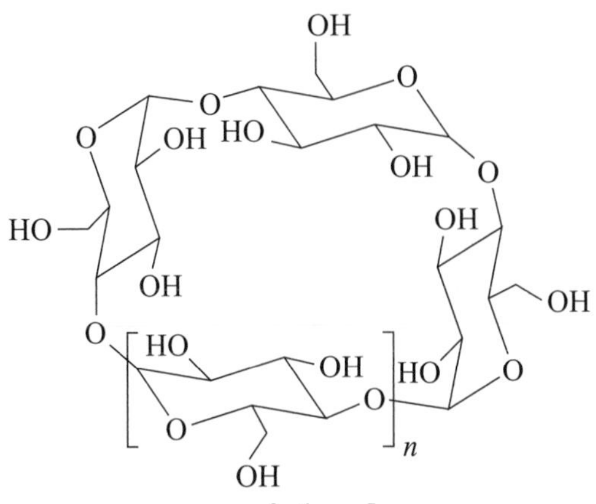

$n = 3, 4, \text{ or } 5$

Note: Schardinger dextrins are extremely interesting compounds. They are able to form complexes with a wide variety of compounds by incorporating these compounds in the cavity in the middle of the cyclic dextrin structure. Complex formation takes place, however, only when the cyclic dextrin and the guest molecule are the right size. Anthracene molecules, for example, will fit into the cavity of a cyclic dextrin with eight glucose units but will not fit into one with seven. For more information about these fascinating compounds, see Bergeron, R. J., "Cycloamyloses," *J. Chem. Educ.* **1977,** *54,* 204–207.

22.37 Isomaltose has the following structure:

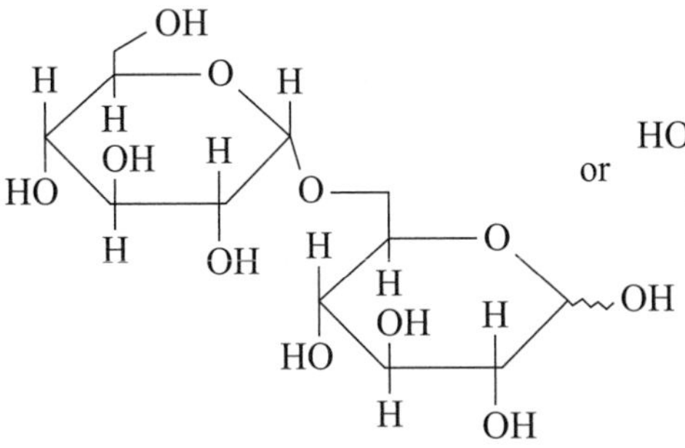

6-*O*-(α-D-Glucopyranosyl)-D-glucopyranose

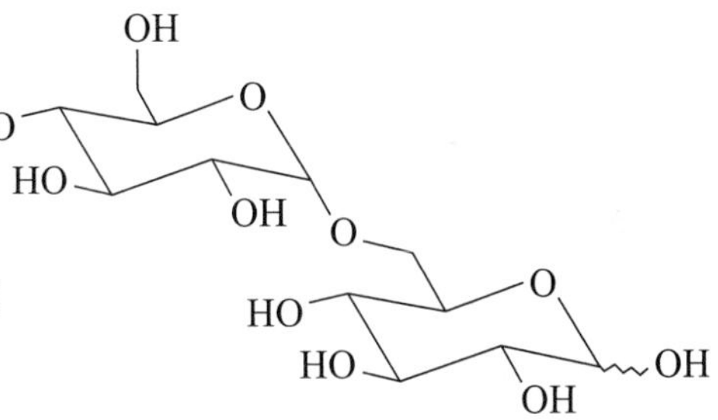

(1) The acid and enzymic hydrolysis experiments tell us that isomaltose has two glucose units linked by an α linkage.

(2) That isomaltose is a reducing sugar indicates that one glucose unit is present as a cyclic hemiacetal.

(3) Methylation of isomaltonic acid followed by hydrolysis gives us information about the size of the nonreducing pyranoside ring and about its point of attachment to the reducing ring. The formation of the first product (2,3,4,6-tetra-*O*-methyl-D-glucose)—a compound with an —OH at C5—tells us that the nonreducing ring is present as a pyranoside. The formation of 2,3,4,5-tetra-*O*-methyl-D-gluconic acid—a compound with an —OH at C6—shows that the nonreducing ring is linked to C6 of the reducing ring.

(4) Methylation of maltose itself tells the size of the reducing ring. That 2,3,4-tri-*O*-methyl-D-glucose is formed shows that the reducing ring is also six membered; we know this because of the free —OH at C5.

22.38 Stachyose has the following structure:

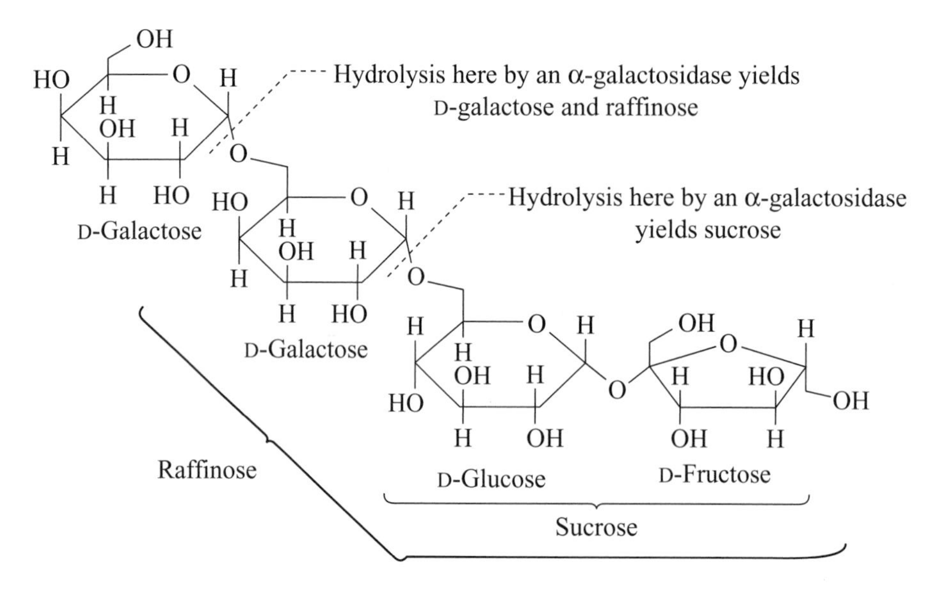

Raffinose has the following structure:

Hydrolysis here by an α-galactosidase yields D-galactose and sucrose

Hydrolysis here by an invertase yields melibiose and fructose

Melibiose

D-Galactose

D-Glucose

D-Fructose

Sucrose

The enzymic hydrolyses (as just indicated) give the basic structure of stachyose and raffinose. The only remaining question is the ring size of the first galactose unit of stachyose. That methylation of stachyose and subsequent hydrolysis yields 2,3,4,6-tetra-*O*-methyl-D-galactose establishes that this ring is a pyranoside.

22.39 Arbutin has the following structure:

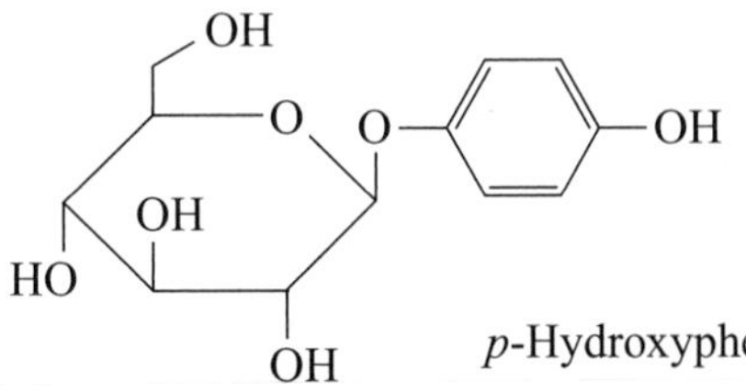

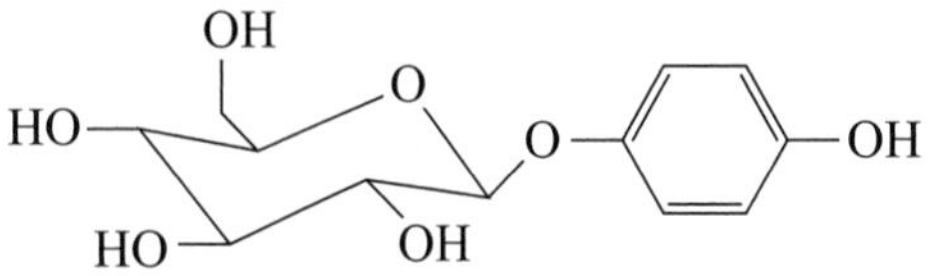

p-Hydroxyphenyl-β-D-glucopyranoside

Compounds **X**, **Y**, and **Z** are hydroquinone, *p*-methoxyphenol, and *p*-dimethoxybenzene, respectively.

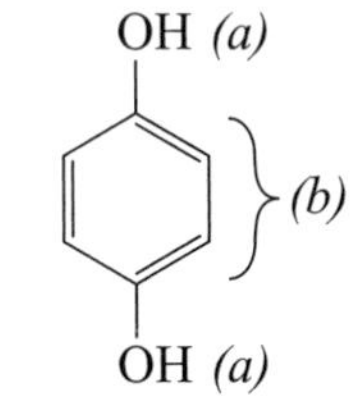

OH *(a)* / *(b)* / OH *(a)*

X

Hydroquinone

(*a*) Singlet δ 7.9 [2H]
(*b*) Singlet δ 6.8 [4H]

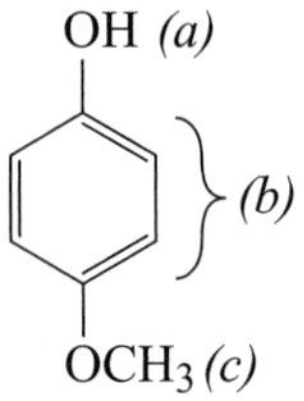

OH *(a)* / *(b)* / OCH_3 *(c)*

Y

p-Methoxyphenol

(*a*) Singlet δ 4.8 [1H]
(*b*) Multiplet δ 6.8 [4H]
(*c*) Singlet δ 3.9 [3H]

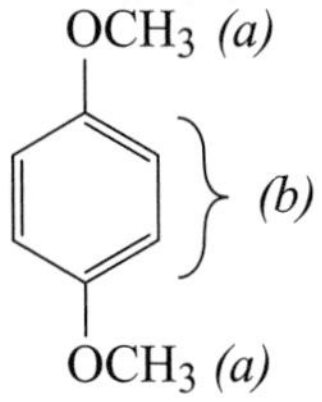

OCH_3 *(a)* / *(b)* / OCH_3 *(a)*

Z

p-Dimethoxybenzene

(*a*) Singlet δ 3.75 [6H]
(*b*) Singlet δ 6.8 [4H]

The reactions that take place are the following:

Arbutin $\xrightarrow{H_3O^+ \text{ or } \beta\text{-glucosidase}}$ D-Glucose + HO–C$_6$H$_4$–OH (X, Hydroquinone)

Arbutin $\xrightarrow[OH^-]{(CH_3)_2SO_4 \text{ (xs)}}$ (permethylated arbutin) $\xrightarrow{H_3O^+}$ 2,3,4,6-Tetra-*O*-methyl-D-glucose + HO–C$_6$H$_4$–OCH$_3$ (Y, *p*-Methoxyphenol)

p-Methoxyphenol $\xrightarrow[OH^-]{(CH_3)_2SO_4}$ CH$_3$O–C$_6$H$_4$–OCH$_3$ (Z, *p*-Dimethoxybenzene)

22.40 Aldotetrose **B** must be D-threose because the alditol derived from it (D-threitol) is optically active (the alditol from D-erythrose, the other possible D-aldotetrose, would be meso). Due to rotational symmetry, however, the alditol from **B** (D-threitol) would produce only two ^{13}C NMR signals. Compounds **A-F** are thus in the family of aldoses stemming from D-threose. Since reduction of aldopentose **A** produces an optically inactive alditol, **A** must be D-xylose. The two diastereomeric aldohexoses **C** and **D** produced from **A** by a Kiliani-Fischer synthesis must therefore be D-idose and D-gulose, respectively. **E** and **F** are the alditols derived from **C** and **D**, respectively. Alditol **E** would produce only three ^{13}C NMR signals due to rotational symmetry while **F** would produce six signals.

22.41 There are four closely spaced upfield alkyl signals in the ^{13}C NMR spectrum (δ 26.5, δ 25.6, δ 24.9, δ 24.2), corresponding to the four methyls of the two acetonide protecting groups. (The compound is, therefore, the 1,2,5,6-bis-acetonide of mannofuranose, below.)

22.42 The final product is the acetonide of glyceraldehyde (below); two molar equivalents are formed from each molar equivalent of the 1,2,5,6-bis-acetonide of mannitol.

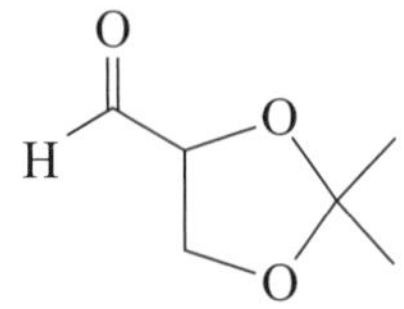

Challenge Problems

22.43 The β-anomer can hydrogen bond intramolecularly, as in:

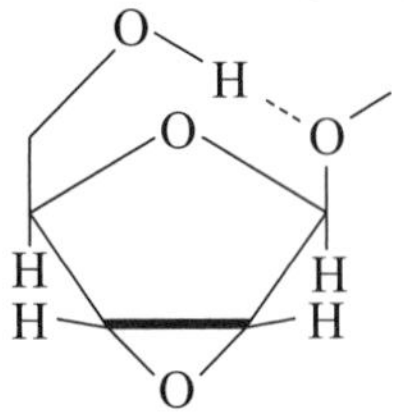

In contrast, the α-anomer can only hydrogen bond intermolecularly, leading to a higher boiling point.

22.44

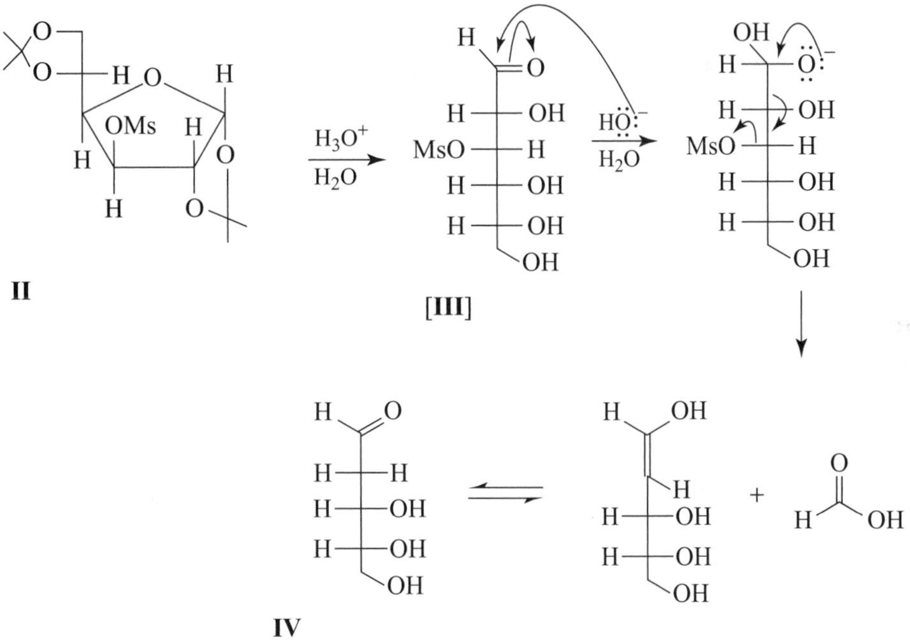

II

$\xrightarrow[H_2O]{H_3O^+}$

[III]

$\xrightarrow[H_2O]{HO^-}$

IV

22.45 (a) The proton at Cl. (b) Because of the single neighboring hydrogen (at C2). (c and d).

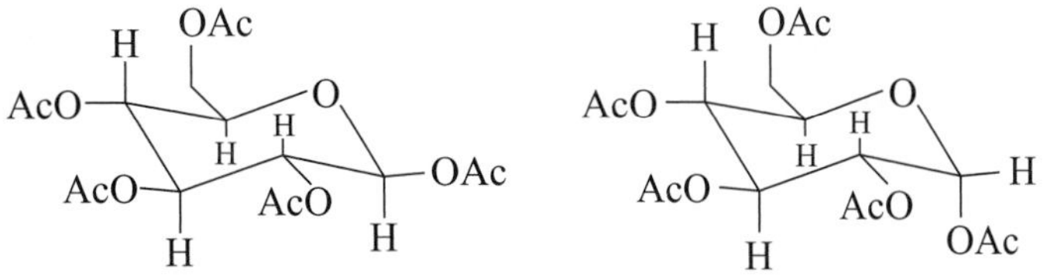

β-anomer
V

α-anomer
VI

QUIZ

22.1 Supply the appropriate structural formula or complete the partial formula for each of the following:

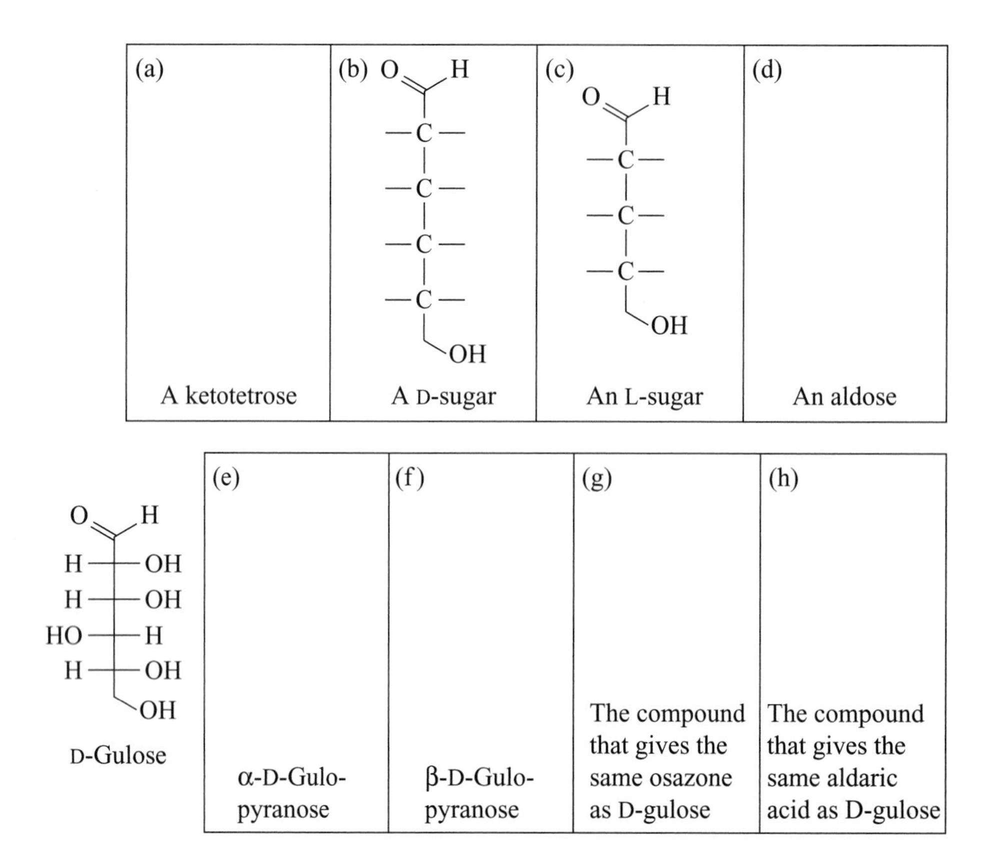

22.2 Which of the following monosaccharides yields an optically inactive alditol on $NaBH_4$ reduction?

	A	B	C	D
C1	CHO	CHO	CHO	CHO
C2	HO–C–H	HO–C–H	H–C–OH	HO–C–H
C3	HO–C–H	H–C–OH	HO–C–H	HO–C–H
C4	H–C–OH	HO–C–H	HO–C–H	HO–C–H
C5	H–C–OH	H–C–OH	H–C–OH	H–C–OH
C6	CH_2OH	CH_2OH	CH_2OH	CH_2OH

Answer: ☐

22.3 Give the structural formula of the monosaccharide that you could use as starting material in the Kiliani-Fischer synthesis of the following compound:

Kiliani-Fischer synthesis →

O H
H — OH
HO — H
H — OH
OH

+ epimer

22.4 The D-aldopentose, (a), is oxidized to an aldaric acid, (b), which is optically active. Compound (a) undergoes a Ruff degradation to form an aldotetrose, (c), which undergoes oxidation to an optically inactive aldaric acid, (d). Supply the reagents for these transformations and the structural formulas of (a), (b), (c), and (d).

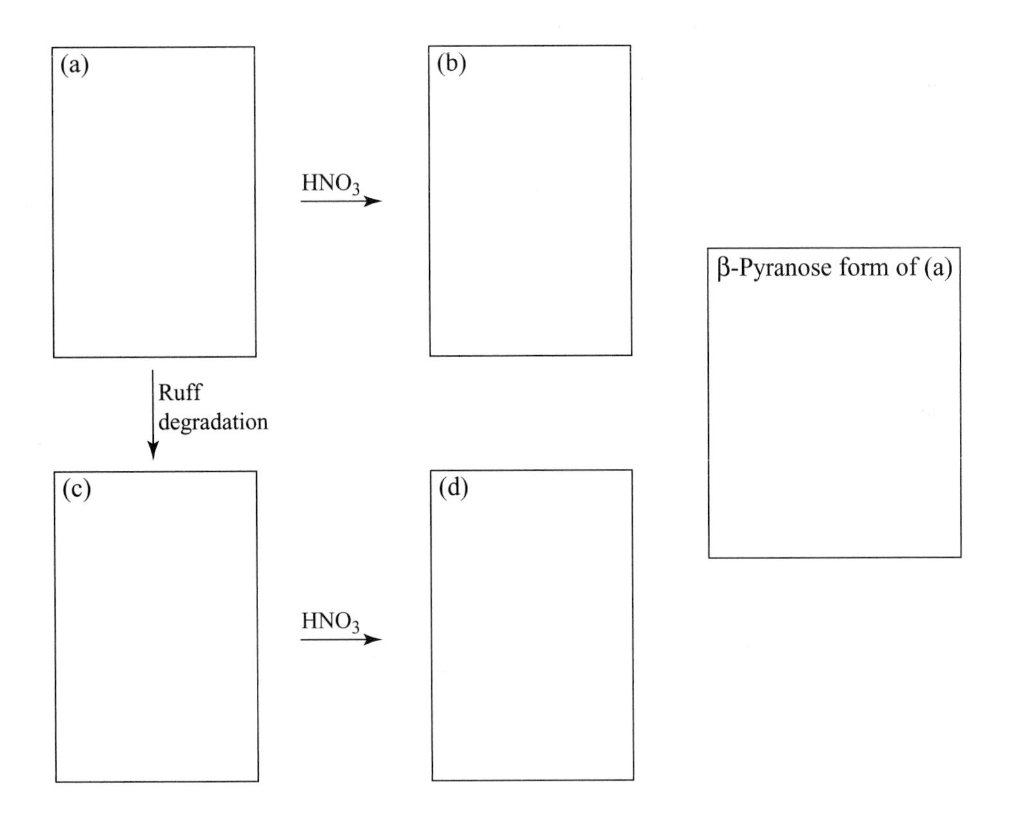

22.5 Give the structural formula of the β-pyranose form of (a) in the space just given.

22.6 Complete the following skeletal formulas and statements by filling in the blanks and circling the words that make the statements true.

The Haworth and conformational formulas of the β-cyclic hemiacetal

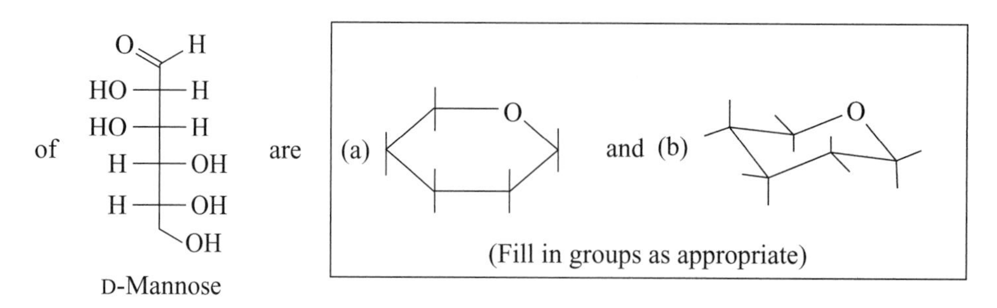

This cyclic hemiacetal is (c) reducing, nonreducing; on reaction with Br_2/H_2O it gives an optically (d) active, inactive (e) aldaric, aldonic acid. On reaction with dilute HNO_3 it gives an optically (f) active, inactive (g) aldaric, aldonic acid. Reaction of the cyclic hemiacetal with (h) ______ converts it into an optically (i) active, inactive alditol.

22.7 Outline chemical tests that would allow you to distinguish between:

(a) Glucose (CHO; H–OH; HO–H; H–OH; H–OH; CH_2OH) and galactose (CHO; H–OH; HO–H; HO–H; H–OH; CH_2OH)

(b) Glucose and fructose

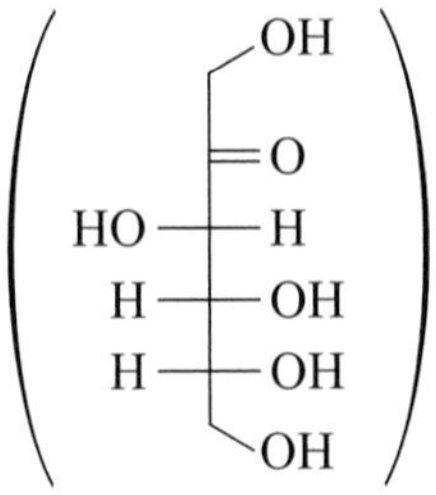

22.8 Hydrolysis of (+)-sucrose (ordinary table sugar) yields
(a) D-glucose
(b) D-mannose
(c) D-fructose
(d) D-galactose
(e) More than one of the above.

22.9 Select the reagent needed to perform the following transformation:

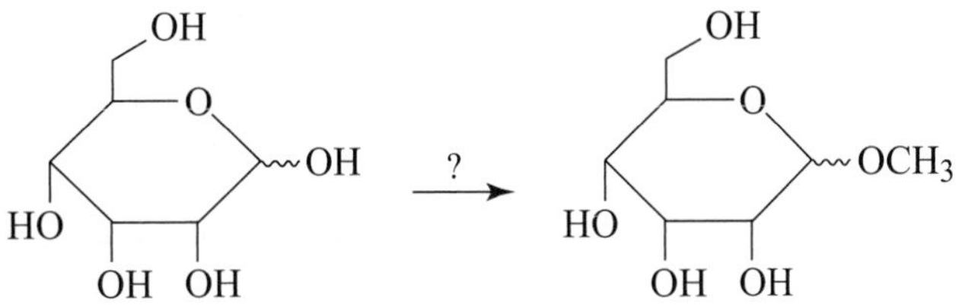

(a) CH_3OH, KOH (b) 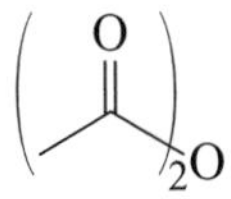 $(CH_3CO)_2O$ (c) $(CH_3)_2SO_4$, OH^-

(d) CH_3OH, HCl (e) CH_3OCH_3, HCl

23 LIPIDS

SOLUTIONS TO PROBLEMS

23.1 (a) There are two sets of enantiomers, giving a total of four stereoisomers:

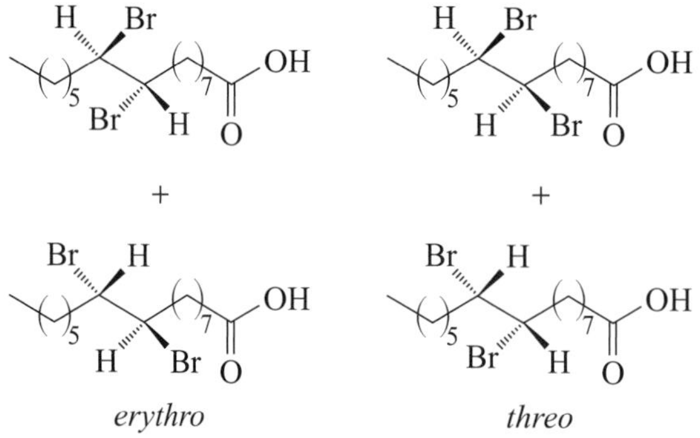

erythro *threo*

(b) + Br_2 ⟶

(a) (b)

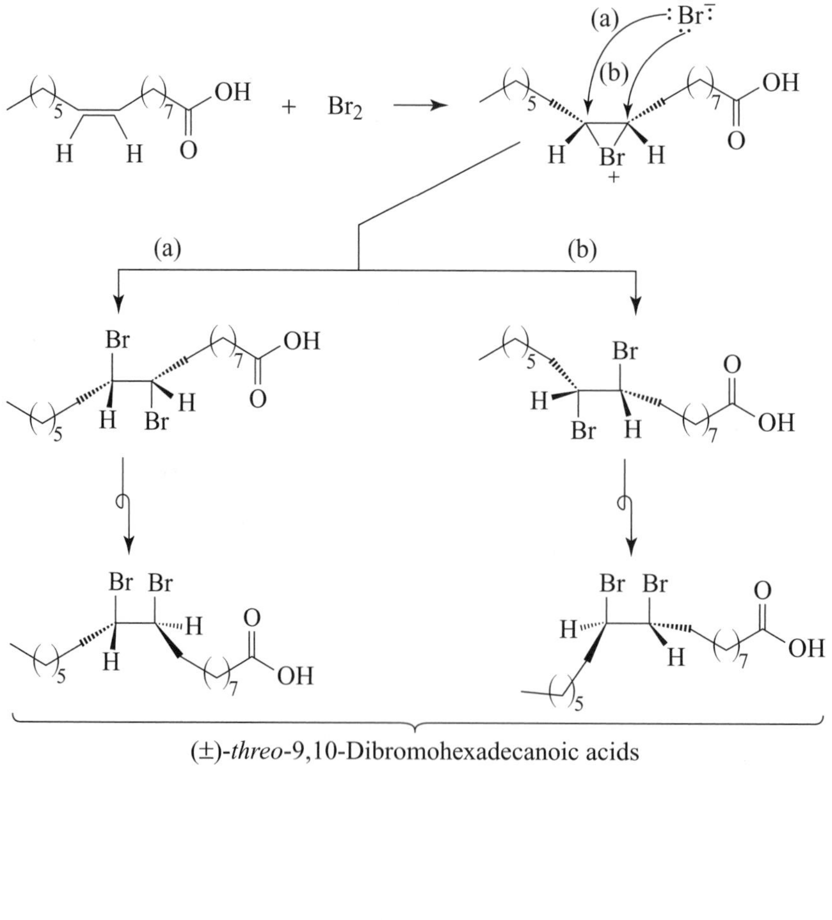

(±)-*threo*-9,10-Dibromohexadecanoic acids

Formation of a bromonium ion at the other face of palmitoleic acid gives a result such that the *threo* enantiomers are the only products formed (obtained as a racemic modification).

The designations *erythro* and *threo* come from the names of the sugars called *erythrose* and *threose* (Section 22.9A).

23.2 (a,b)

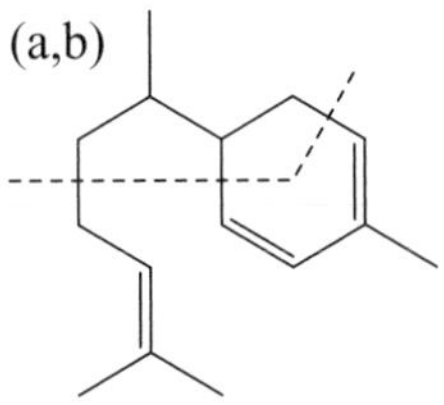

Zingiberene
(a sesquiterpene)

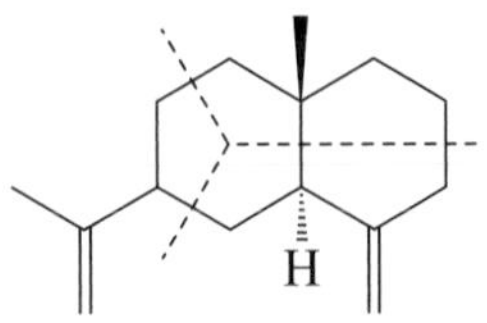

β-Selinene
(a sesquiterpene)

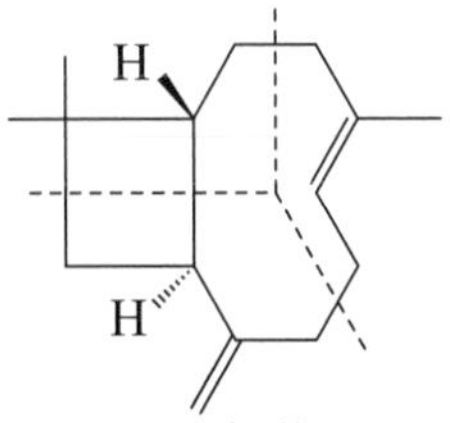

Caryophyllene
(a sesquiterpene)

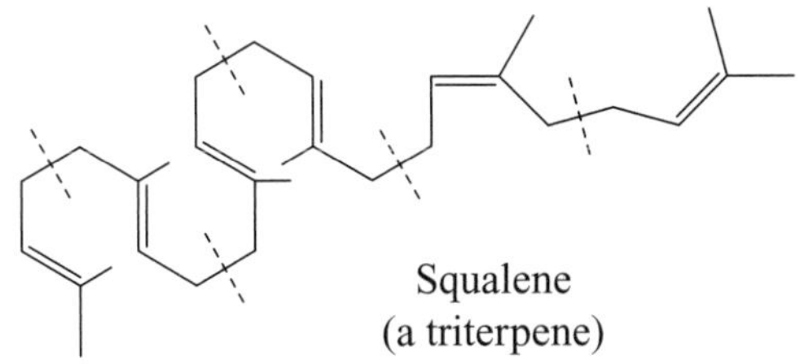

Squalene
(a triterpene)

23.3 (a)

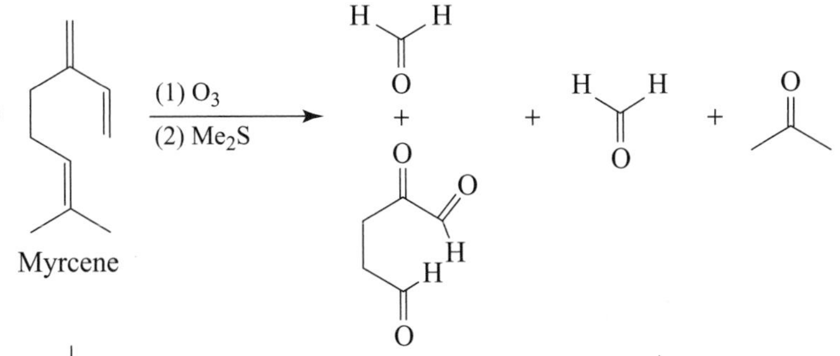

Myrcene

(1) O_3
(2) Me_2S

(b)

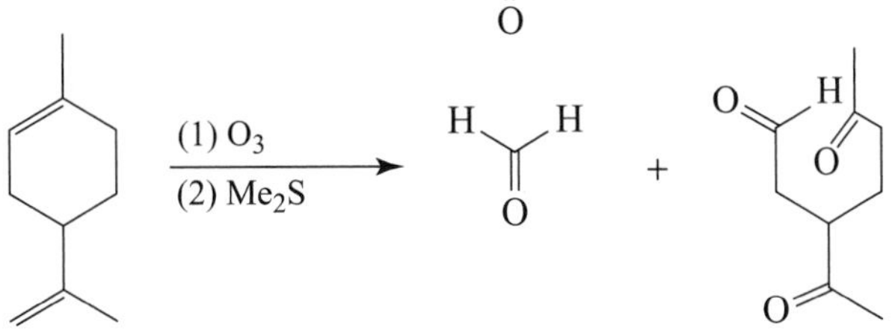

Limonene

(1) O_3
(2) Me_2S

(c) α-Farnesene
(see Section 23.3)

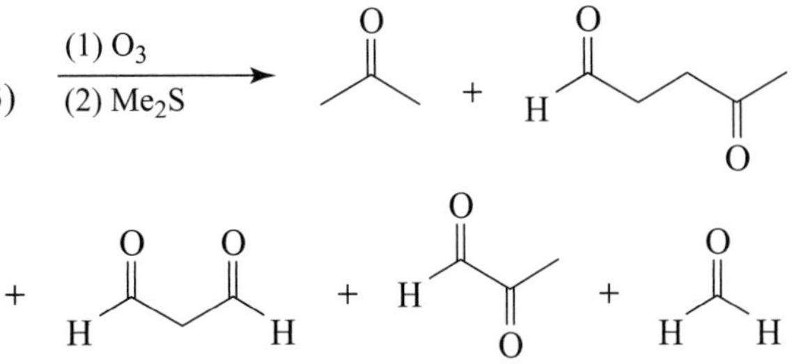

(1) O_3
(2) Me_2S

(d) Geraniol (see Section 23.3) $\xrightarrow[(2)\ Me_2S]{(1)\ O_3}$ O + H O O + H O OH

(e) Squalene (see Section 23.3) $\xrightarrow[(2)\ Me_2S]{(1)\ O_3}$ 2 O + H O H O + 4 O H O

23.4 (a) O + CO_2

(+ further oxidation products)

(b)

(c) H Cl H Cl

(+ rearranged products)

(d) H OH OH

23.5 Br_2 in CCl_4 or $KMnO_4$ in H_2O at room temperature. Either reagent would give a positive result with geraniol and a negative result with menthol.

23.6

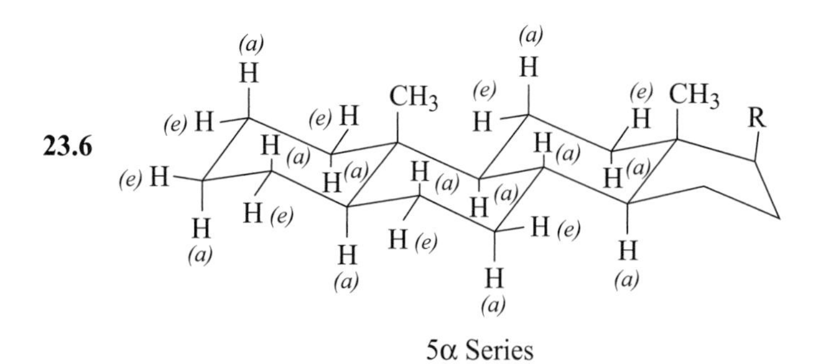

5α Series

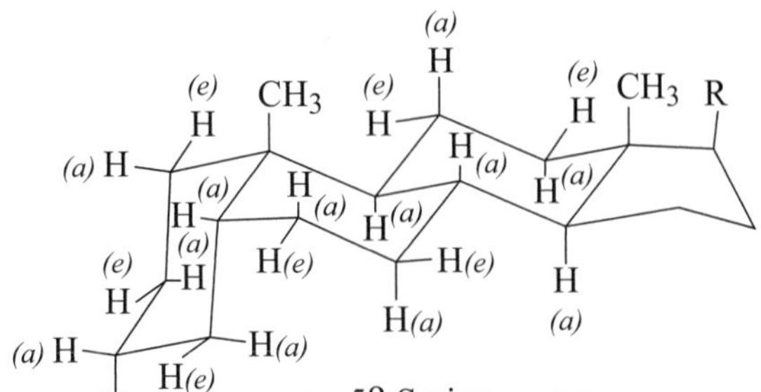

5β Series

23.7 (a)

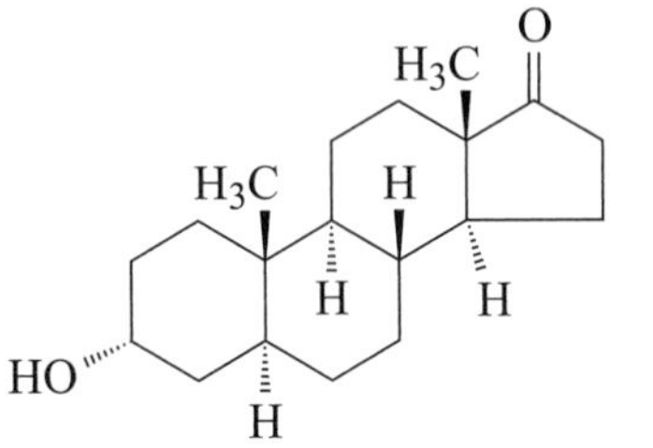

3α-Hydroxy-5α-androstan-17-one
(androsterone)

(b)

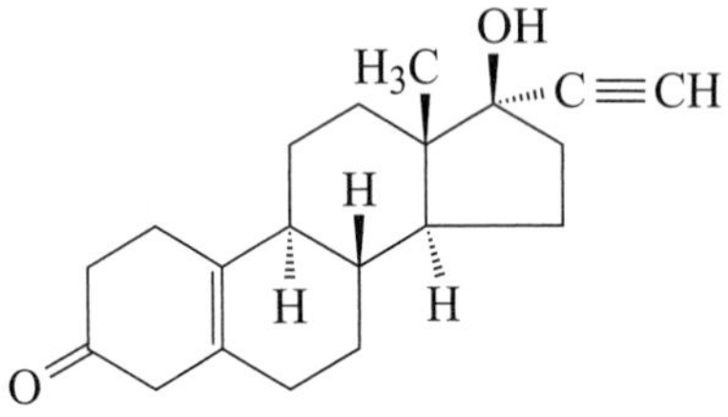

17α-Ethynyl-17β-hydroxy-5(10)-estren-3-one
(norethynodrel)

23.8

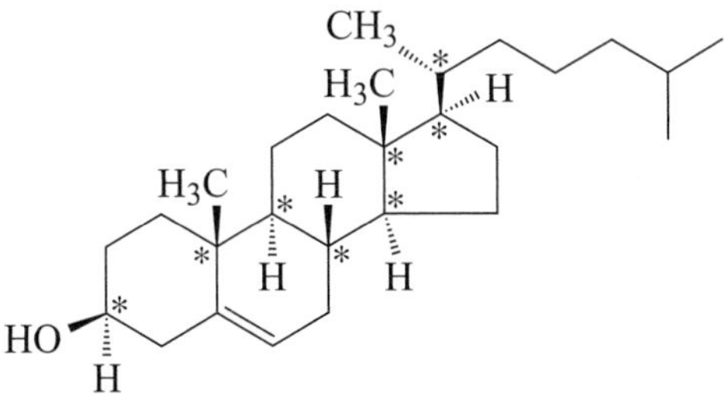

Absolute configuration of cholesterol
(5-cholesten-3β-ol)

23.9 Estrone and estradiol are *phenols* and thus are soluble in aqueous sodium hydroxide. Extraction with aqueous sodium hydroxide separates the estrogens from the androgens.

23.10 (a)

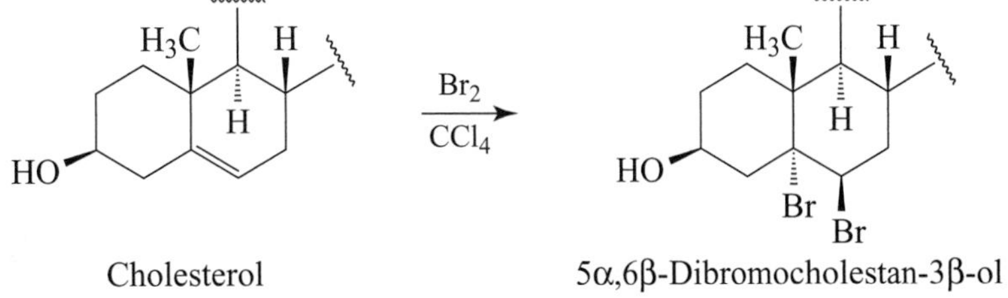

$$\text{Cholesterol} \xrightarrow[CCl_4]{Br_2} 5\alpha,6\beta\text{-Dibromocholestan-}3\beta\text{-ol}$$

(b)

H_3C H H HO O

$\xrightarrow[H_2O]{H_3O^+}$

H_3C H H HO OH OH

5α,6α-Epoxycholestan-3β-ol (prepared by epoxidation of cholesterol; cf. Section 23.4G)

Cholestan-3β,5α,6β-triol

(c)

H_3C H H HO H

$\xrightarrow[acetone]{H_2CrO_4}$

H_3C H H O H

5α–Cholestan-3β-ol (prepared by hydrogenation of cholesterol; cf. Section 23.4G)

5α–Cholestan-3-one

(d)

H_3C H H HO

Cholesterol

$\xrightarrow[(cf. Sect. 23.4G)]{BH_3:THF}$

H_3C H H HO H B

O OD

H_3C H H HO H D

6α-Deuterio-5α-cholestan-3β-ol

(e)

H_3C H H HO O

$\xrightarrow{HBr}$

H_3C H H HO OH Br

5α, 6α-Epoxycholestan-3β-ol (cf. (b) above)

6β-Bromocholestan-3β,5α-diol

23.11 (a) HOCH₂CH(OH)CH₂OH + RCOOH + R'COOH + H_3PO_4 + $HOCH_2CH_2\overset{+}{N}(CH_3)_3\ X^-$

(b) HOCH₂CH(OH)CH₂OH + RCOOH + R'COOH + H_3PO_4 + $HOCH_2CH_2NH_2$

(c) HOCH₂CH(OH)CH₂OH + $CH_3(CH_2)_nCH_2CHO$ + R'COOH + H_3PO_4 + $HOCH_2CH_2\overset{+}{N}(CH_3)_3\ X^-$

23.12 (a) $CH_3(CH_2)_{16}COOH + CH_3CH_2OH \overset{HA}{\rightleftharpoons} CH_3(CH_2)_{16}COOCH_2CH_3 + H_2O$

$CH_3(CH_2)_{16}COOH \xrightarrow{SOCl_2} CH_3(CH_2)_{16}COCl \xrightarrow{CH_3CH_2OH} CH_3(CH_2)_{16}COOCH_2CH_3 + HCl$

(b) $CH_3(CH_2)_{16}COCl$ (from a) $\xrightarrow{(CH_3)_3COH} CH_3(CH_2)_{16}COOC(CH_3)_3 + HCl$

(c) $CH_3(CH_2)_{16}COCl \xrightarrow[\text{(excess)}]{NH_3} CH_3(CH_2)_{16}CONH_2 + NH_4Cl$

(d) $CH_3(CH_2)_{16}COCl \xrightarrow[\text{(excess)}]{(CH_3)_2NH} CH_3(CH_2)_{16}CON(CH_3)_2 + (CH_3)_2\overset{+}{N}H_2\ Cl^-$

(e) $CH_3(CH_2)_{16}CONH_2$ (from c) $\xrightarrow[(2)\ H_2O]{(1)\ LiAlH_4,\ Et_2O} CH_3(CH_2)_{16}CH_2NH_2$

(f) $CH_3(CH_2)_{16}CONH_2 \xrightarrow[OH^-]{Br_2} CH_3(CH_2)_{15}CH_2NH_2$

(g) $CH_3(CH_2)_{16}COCl$ (from a) $\xrightarrow[Et_2O,\ -78°\ C]{LiAlH(OC(CH_3)_3)_3} CH_3(CH_2)_{16}CHO$

(h) (from a) $\xrightarrow[\text{Ni, pressure}]{H_2}$ OH; O Cl → O O

(i) OH $\xrightarrow[(2)\ H_2O]{(1)\ LiAlH_4,\ Et_2O}$ OH

O $\xrightarrow[\text{Ni, pressure}]{H_2}$ OH + OH

(j) H (from g) $\xrightarrow[(2)\ H_2O,\ H_3O^+]{(1)\ CH_3MgI,\ Et_2O}$ OH $\xrightarrow[\text{acetone}]{H_2CrO_4}$ O

(k) OH (from h) $\xrightarrow{PBr_3}$ Br

(l) Br (from k) $\xrightarrow[(2)\ H_3O^+,\ \text{heat}]{(1)\ NaCN}$ OH O

23.13 (a) H O $\xrightarrow[H_3O^+]{Br_2}$ Br H O $\xrightarrow[(2)\ H_3O^+]{(1) Ag_2O}$ Br OH O

(b) Br OH O (from a) $\xrightarrow[(2)\ H_3O^+]{(1)\ OH^-,\ \text{heat}}$ OH OH O

(c) Br OH O (from a) $\xrightarrow[(2)\ H_3O^+]{(1)\ NaCN}$ CN OH O

(d) Br OH O (from a) $\xrightarrow[(2)\ H_3O^+]{(1)\ NH_3\ (\text{excess})}$ NH_2 OH O or $^+NH_3$ O^- O

23.14 (a) 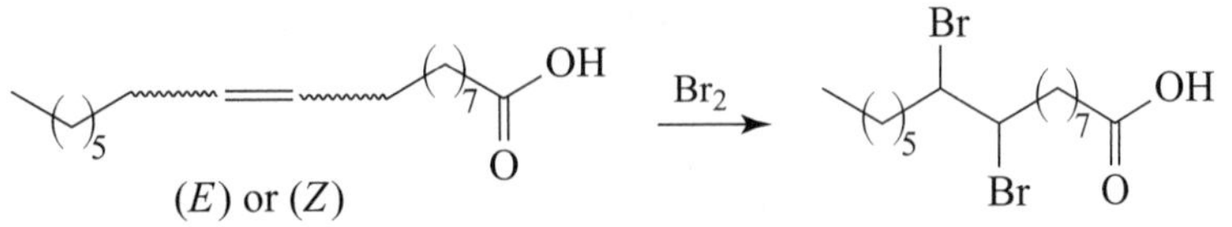 (*E*) or (*Z*) $\xrightarrow{Br_2}$

(b) 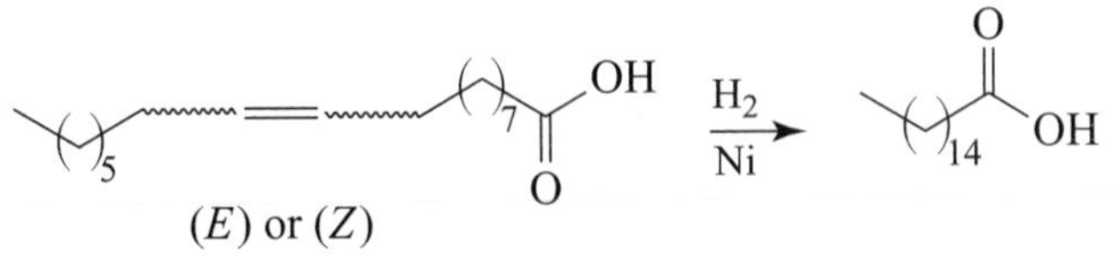 (*E*) or (*Z*) $\xrightarrow[Ni]{H_2}$

(c) 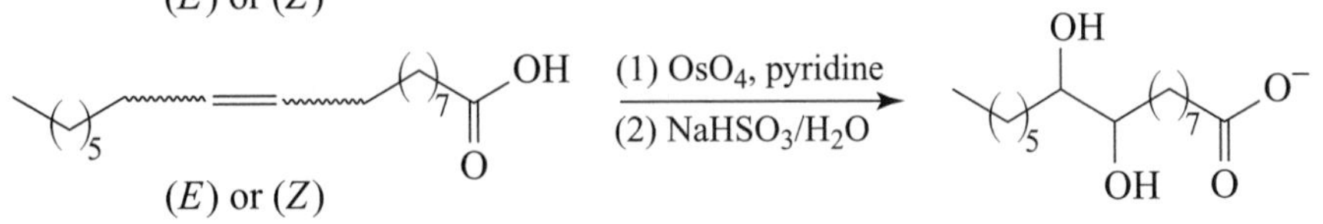 (*E*) or (*Z*) $\xrightarrow[(2)\ NaHSO_3/H_2O]{(1)\ OsO_4,\ pyridine}$

(d) 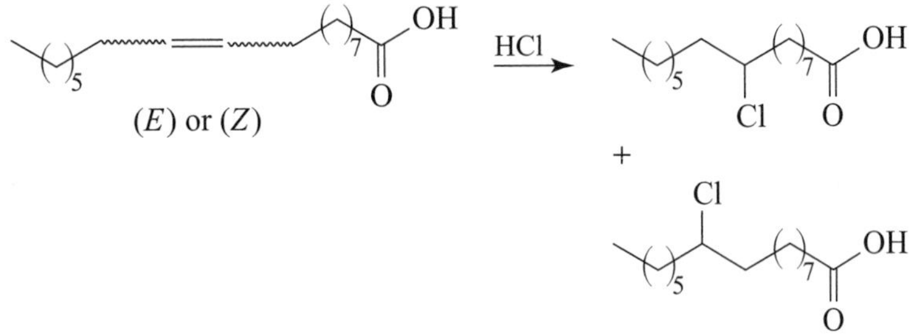 (*E*) or (*Z*) $\xrightarrow{HCl}$ +

23.15 Elaidic acid is *trans*-9-octadecenoic acid (*trans*-octadec-9-enoic acid):

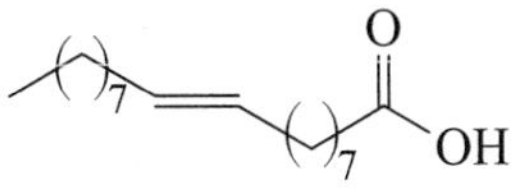

It is formed by the isomerization of oleic acid.

23.16 A reverse Diels-Alder reaction takes place.

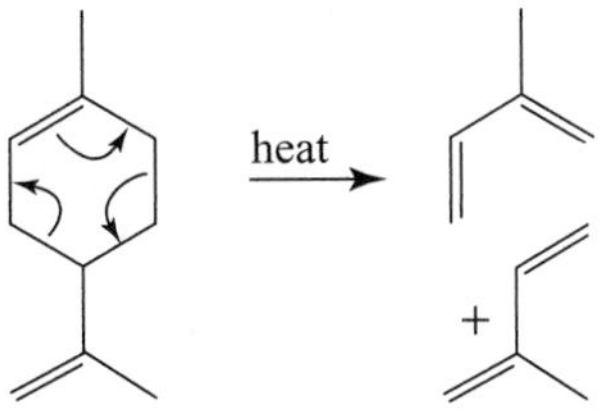

$\xrightarrow{heat}$ +

23.17 (a) 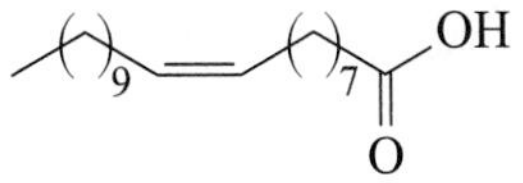 and

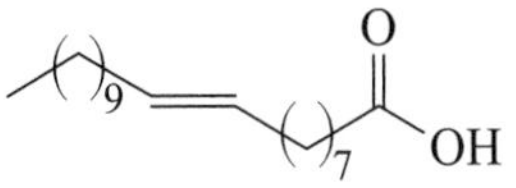

(b) Infrared spectroscopy

(c) A peak in the 675–730-cm^{-1} region would indicate that the double bond is cis; a peak in the 960–975-cm^{-1} region would indicate that it is trans.

23.18

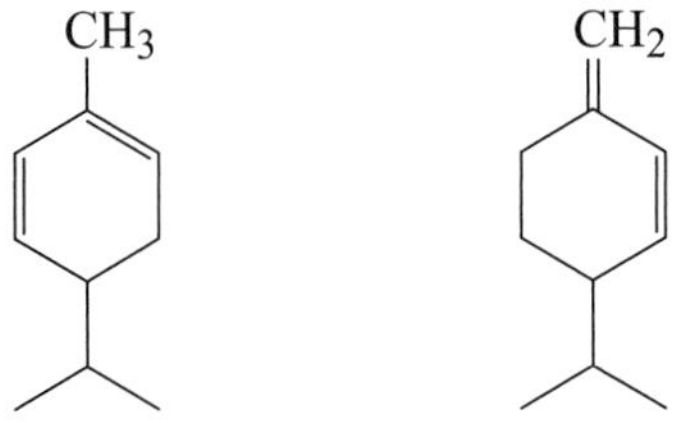

CH_3 CH_2

α-Phellandrene β-Phellandrene

Note: On permanganate oxidation, the $=CH_2$ group of β-phellandrene is converted to CO_2 and thus is not detected in the reaction.

Roadmap Syntheses

23.19

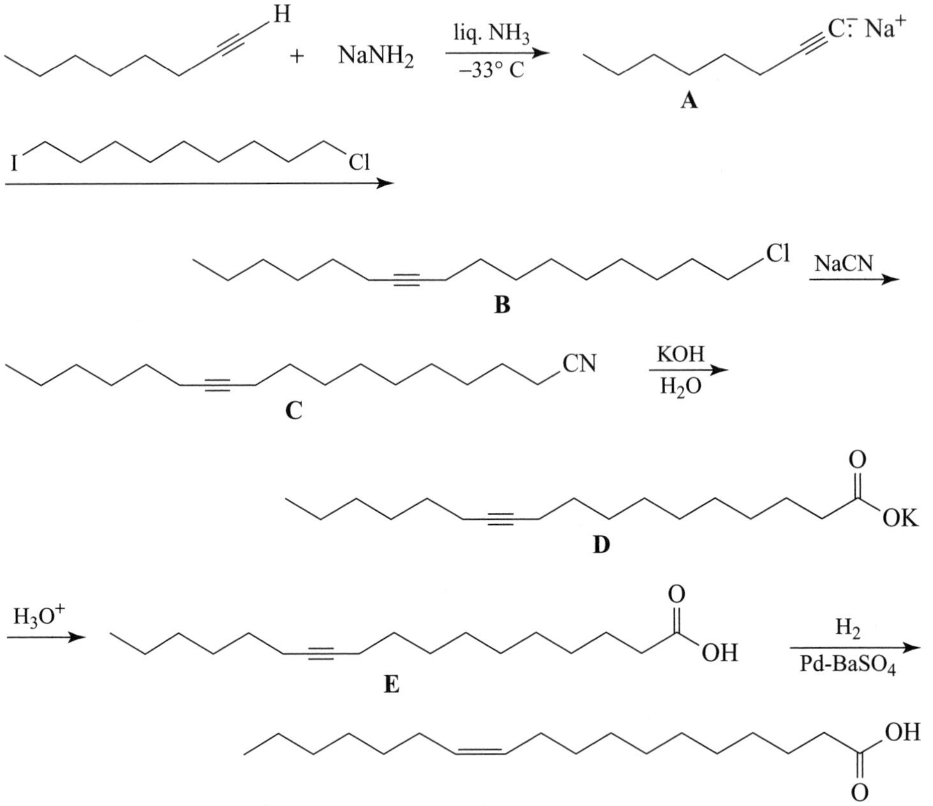

H + $NaNH_2$ $\xrightarrow[-33^\circ C]{\text{liq. } NH_3}$ $C\bar{:}$ Na^+

A

I Cl

Cl $\xrightarrow{NaCN}$

B

CN $\xrightarrow[H_2O]{KOH}$

C

O

OK

D

$\xrightarrow{H_3O^+}$ O

OH $\xrightarrow[Pd\text{-}BaSO_4]{H_2}$

E

OH

O

Vaccenic acid

23.20 F~~~~Br + $H-\equiv C:^- Na^+$ ⟶ F~~~~≡—H (**F**)

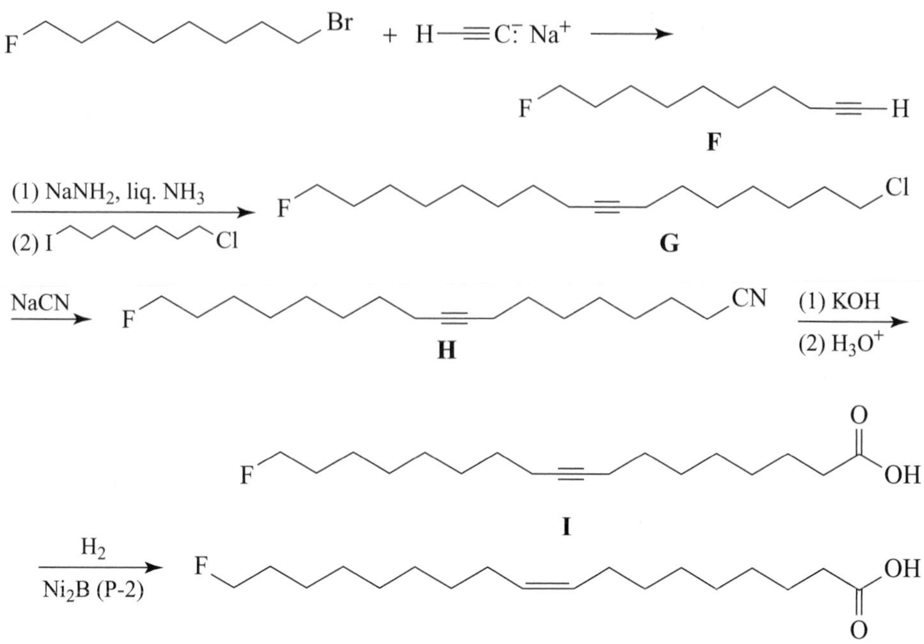

(1) $NaNH_2$, liq. NH_3; (2) I~~~~Cl ⟶ F~~~~≡~~~~Cl (**G**)

NaCN ⟶ F~~~~≡~~~~CN (**H**) —(1) KOH, (2) H_3O^+⟶

F~~~~≡~~~~COOH (**I**)

—H_2, Ni_2B (P-2)⟶ F~~~~CH=CH~~~~COOH

23.21 5α-Cholest-2-ene —$C_6H_5CO_3H$ (OOH)⟶ **A** —HBr⟶ **B**

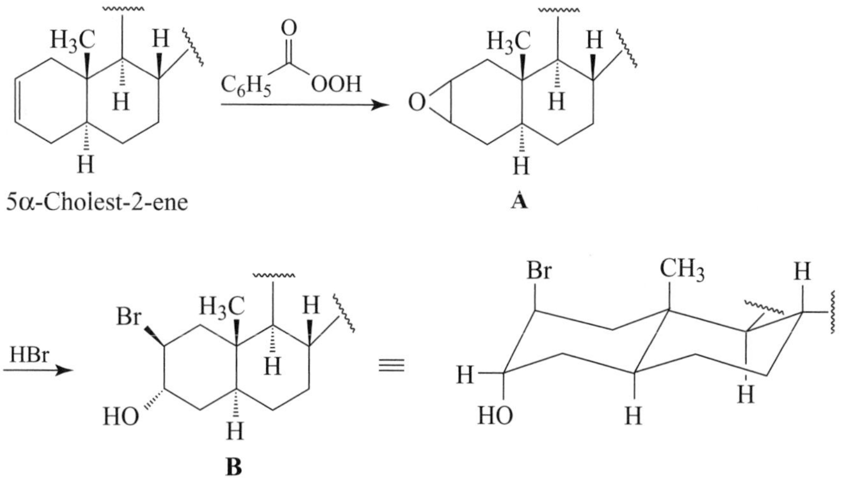

Here we find that epoxidation takes place at the less hindered α face (cf. Section 23.4G). Ring opening by HBr takes place in an anti fashion to give a product with diaxial substituents.

23.22 (a) 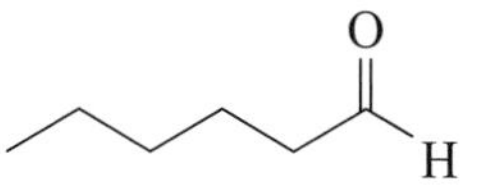 hexanal (CH₃CH₂CH₂CH₂CH₂CHO) (b) BuLi, Et_2O (c) 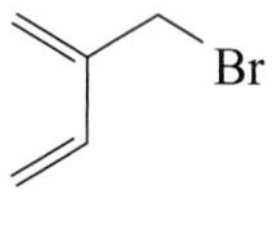 2-(bromomethyl)-1,3-butadiene

(d) 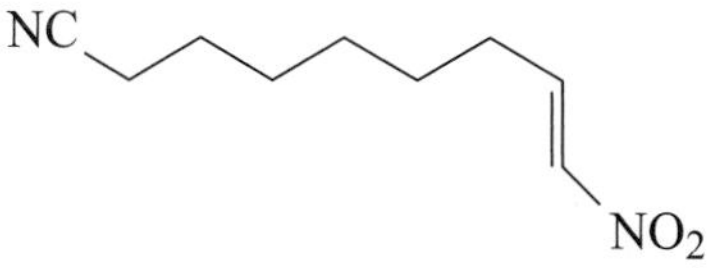 NC~~~~CH=CH–NO_2 (e) Michael addition using a basic catalyst.

23.23 First: an elimination takes place,

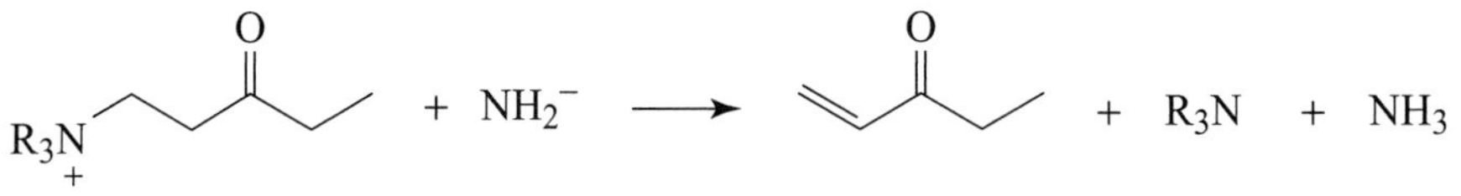

+ NH_2^- ⟶ + R_3N + NH_3

$R_3\overset{+}{N}$

Then a conjugate addition occurs, followed by an aldol addition:

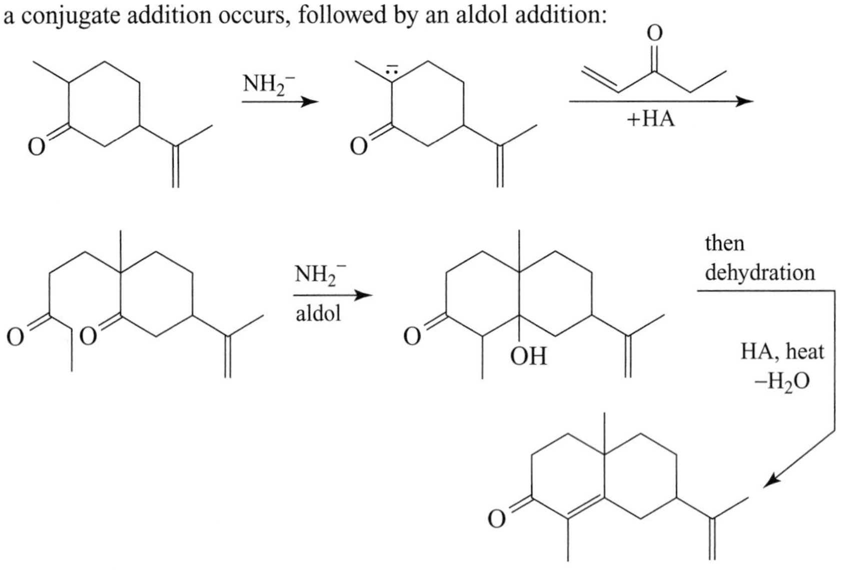

NH_2^-

+HA

NH_2^-
aldol

OH

then
dehydration

HA, heat
$-H_2O$

A cyperone

Challenge Problems

23.24

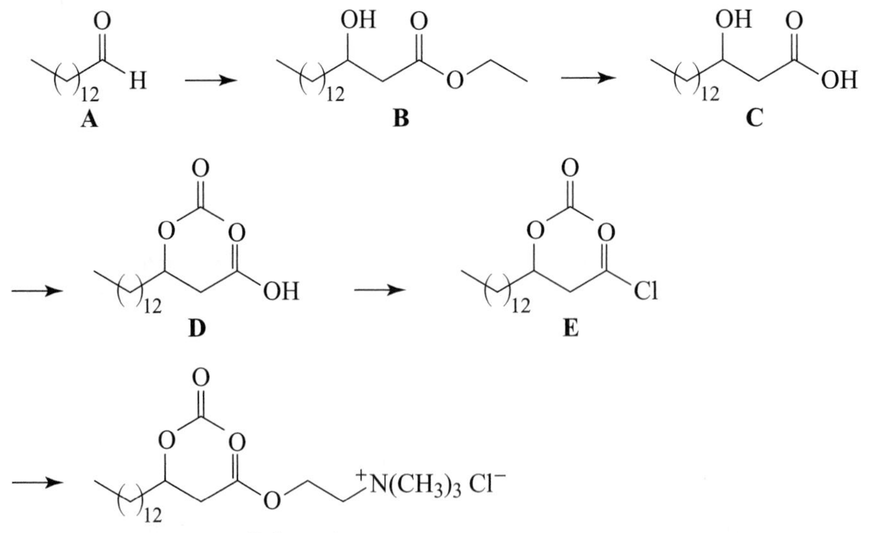

H

A **B** **C**

OH

D **E**

Cl

$^+N(CH_3)_3\ Cl^-$

Pahutoxin

23.25 (a, b) The reaction is an intramolecular transesterification.

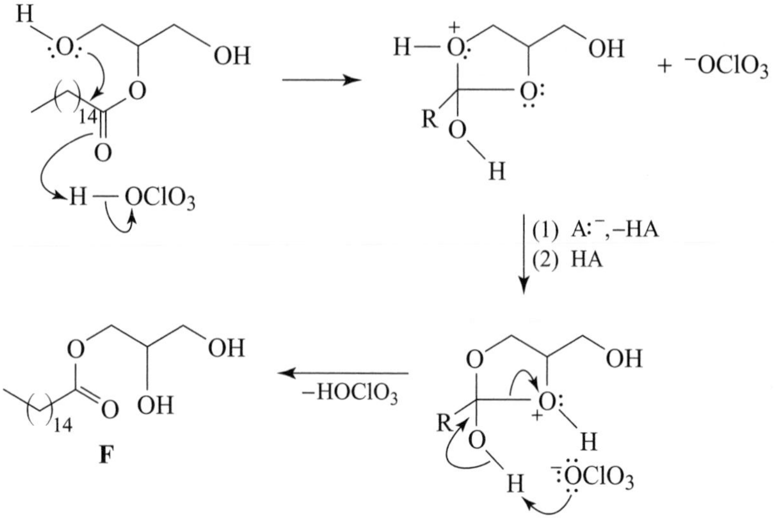

H
:O:
OH
O
14
O
H — $OClO_3$

H — $\overset{+}{O}$:
OH
+ $^{-}OClO_3$
O:
R
O
H

(1) $A:^{-}$, $-HA$
(2) HA

O
OH
O
R
O
+
O:
H
H
$^{-}$:ÖClO_3

$-HOClO_3$

O
OH
O OH
14
F

QUIZ

23.1 Write an appropriate formula in each box.

(a)

A naturally occurring fatty acid

(b)

A soap

(c)

A solid fat

(d)

An oil

(e)

A synthetic detergent

(f)

5α-Estran-17-one

23.2 Give a reagent that would distinguish between each of the following:

(a) Pregnane and 20-pregnanone

(b) Stearic acid and oleic acid

(c) 17α-Ethynyl-1,3,5(10)-estratriene-3,17β-diol (ethynylestradiol) and 1,3,5(10)-estratriene-3,17β-diol (estradiol)

23.3 What product would be obtained by catalytic hydrogenation of 4-androstene?

23.4 Supply the missing compounds:

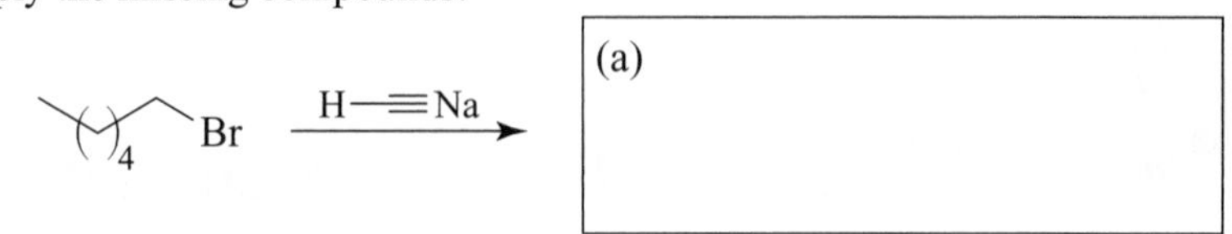

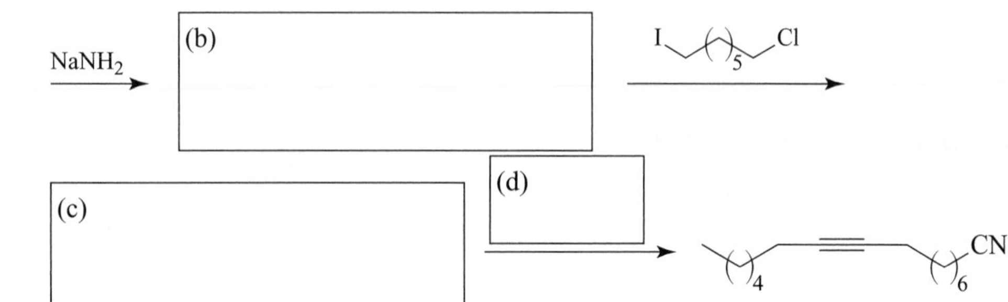

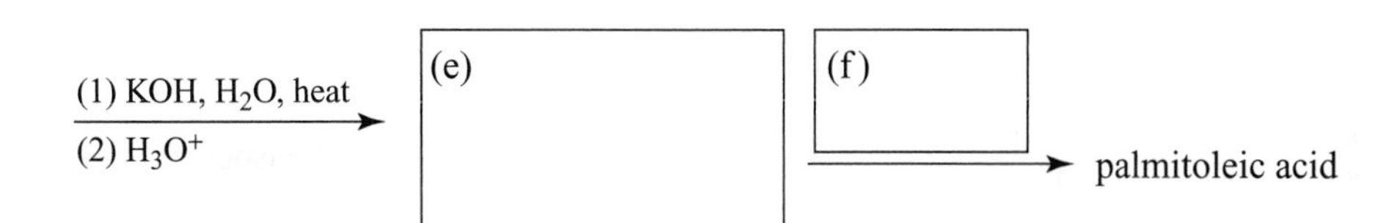

23.5 The following compound is a:

(a) Monoterpene (b) Sesquiterpene (c) Diterpene
(d) Triterpene (e) Tetraterpene

23.6 Mark off the isoprene units in the previous compound.

23.7 Which is a systematic name for the steroid shown here?
(a) 5α-Androstan-3α-ol
(b) 5β-Androstan-3β-ol
(c) 5α-Pregnan-3α-ol
(d) 5β-Pregnan-3β-ol
(e) 5α-Estran-3α-ol

CH_3 H H HO H

24 AMINO ACIDS AND PROTEINS

SOLUTIONS TO PROBLEMS

24.1 (a) HO–C(=O)–CH$_2$–CH$_2$–CH(NH_3^+)–C(=O)–OH

(b) $^-$O–C(=O)–CH$_2$–CH$_2$–CH(NH_2)–C(=O)–O$^-$

(c) HO–C(=O)–CH$_2$–CH$_2$–CH(NH_3^+)–C(=O)–O$^-$ predominates at the isoelectric point rather than

$^-$O–C(=O)–CH$_2$–CH$_2$–CH(NH_3^+)–C(=O)–OH because of the acid-strengthening inductive effect of the α-aminium group.

(d) Since glutamic acid is a dicarboxylic acid, acid must be added (i.e., the pH must be made lower) to suppress the ionization of the second carboxyl group and thus achieve the isoelectric point. Glutamine, with only one carboxyl group, is similar to glycine or phenylalanine and has its isoelectric point at a higher pH.

24.2 The conjugate acid is highly stabilized by resonance.

R–NH–C(=NH)–NH$_2$ $\xrightarrow{HA}$ R–NH–C(=NH_2^+)–NH$_2$ ⟷ R–NH–C(NH$_2$)=NH_2^+ ⟷ R–NH^+=C(NH$_2$)–NH$_2$

24.3 (a) O NK O + O O EtO OEt Br ⟶ O EtO O N O O EtO

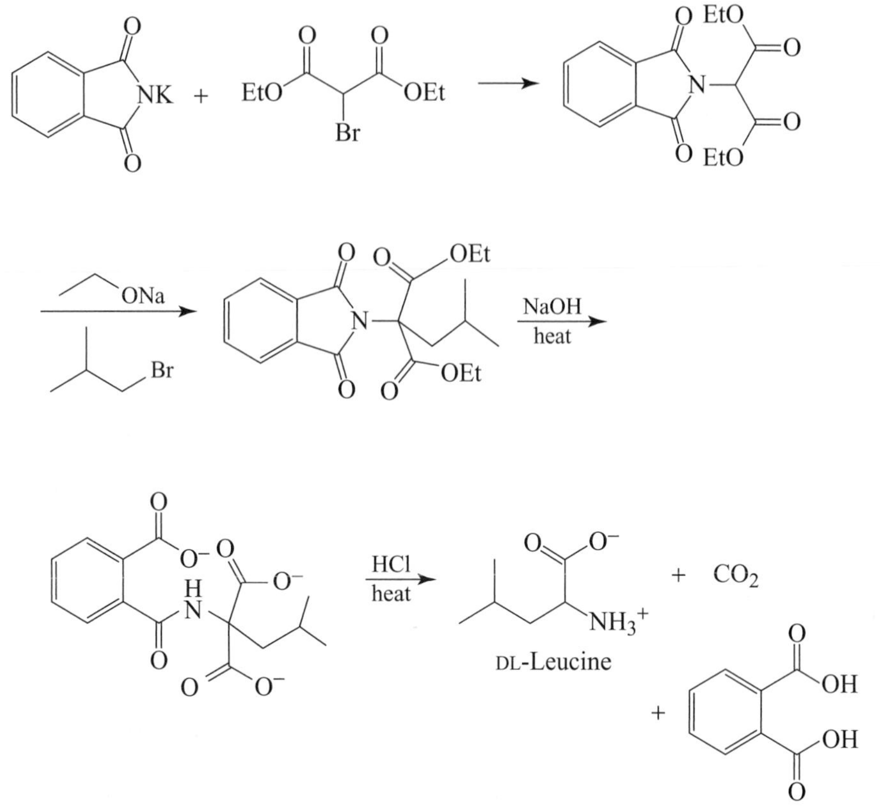

$\xrightarrow[\text{Br}]{\text{ONa}}$ O O OEt N O O OEt $\xrightarrow[\text{heat}]{\text{NaOH}}$

O O⁻ O H N O⁻ O O O⁻ $\xrightarrow[\text{heat}]{\text{HCl}}$ O O⁻ NH_3^+ + CO_2 + O OH OH O

DL-Leucine

(b) O O OEt N OEt O O $\xrightarrow[CH_3I]{\text{ONa}}$ O O OEt N O O OEt $\xrightarrow[\text{heat}]{\text{NaOH}}$

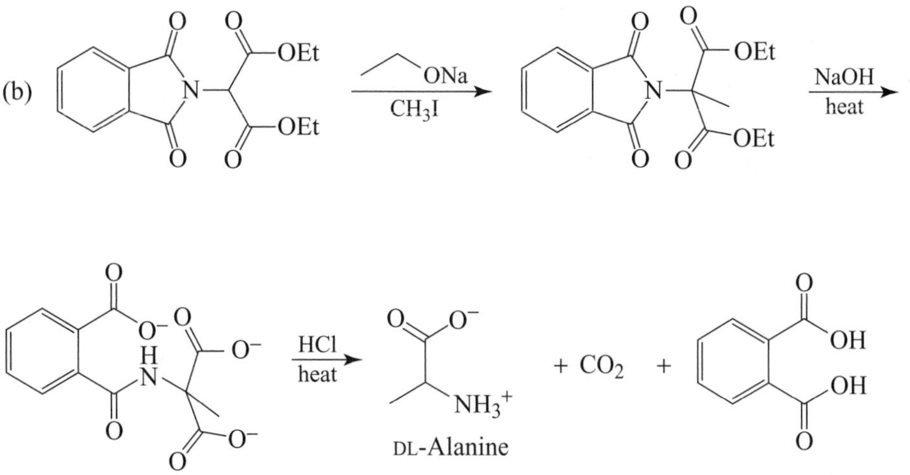

O O⁻ O H N O⁻ O O O⁻ $\xrightarrow[\text{heat}]{\text{HCl}}$ O O⁻ NH_3^+ + CO_2 + O OH OH O

DL-Alanine

(c) [Reaction scheme: diethyl phthalimidomalonate + NaOEt, then $C_6H_5CH_2Br$ → benzylated phthalimidomalonic ester; NaOH, heat → ring-opened dicarboxylate; HCl, heat → DL-Phenylalanine + CO_2 + phthalic acid]

ONa

C_6H_5 Br

OEt

C_6H_5

NaOH / heat

O^-

HCl / heat

NH_3^+

DL-Phenylalanine

+ CO_2

OH

OH

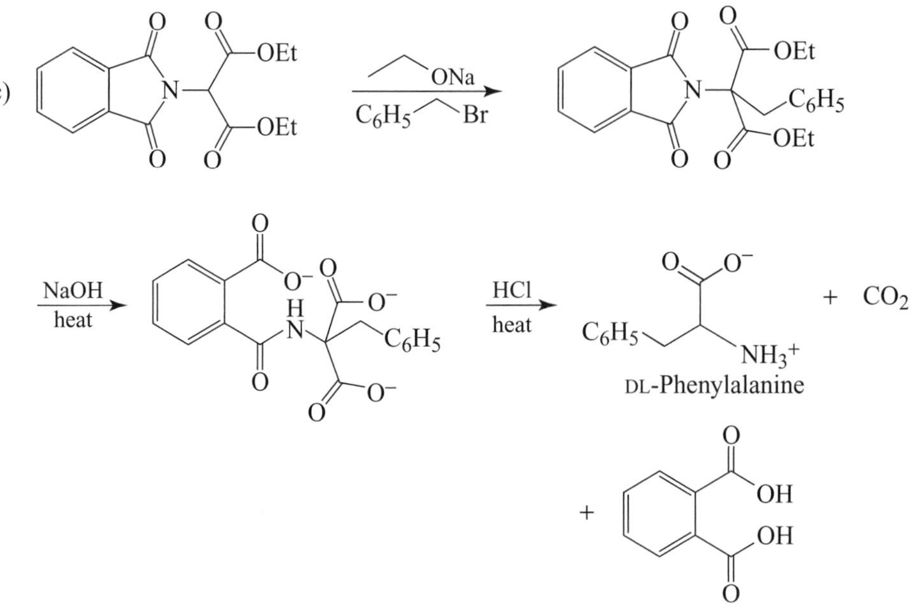

24.4 (a) Phenylacetaldehyde $\xrightarrow[HCN]{NH_3}$ [aminonitrile] $\xrightarrow{H_3O^+}$ DL-Phenylalanine

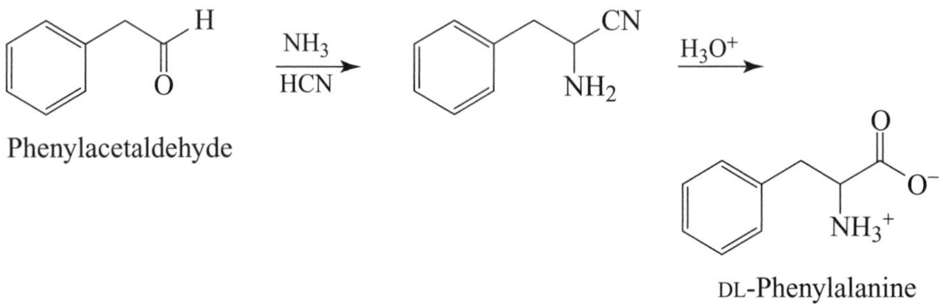

(b) CH_3SH + [acrolein] $\xrightarrow{base}$ CH_3S–CH_2CH_2CHO $\xrightarrow[HCN]{NH_3}$

CH_3S–$CH_2CH_2CH(CN)NH_2$ $\xrightarrow{H_3O^+}$ CH_3S–$CH_2CH_2CH(NH_3^+)CO_2^-$

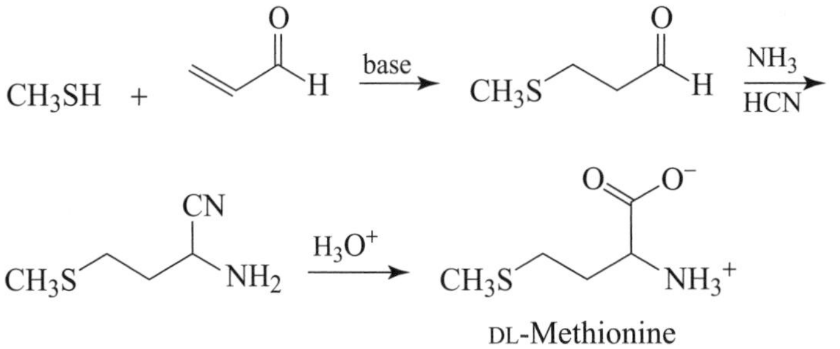

DL-Methionine

24.5 Because of the presence of an electron-withdrawing 2,4-dinitrophenyl group, the labeled amino acid is relatively nonbasic and is, therefore, insoluble in dilute aqueous acid. The other amino acids (those that are not labeled) dissolve in dilute aqueous acid.

24.6 (a) H_3N^+ ... VAG $\xrightarrow[HCO_3^-]{O_2N\text{-}C_6H_3(NO_2)\text{-}F}$

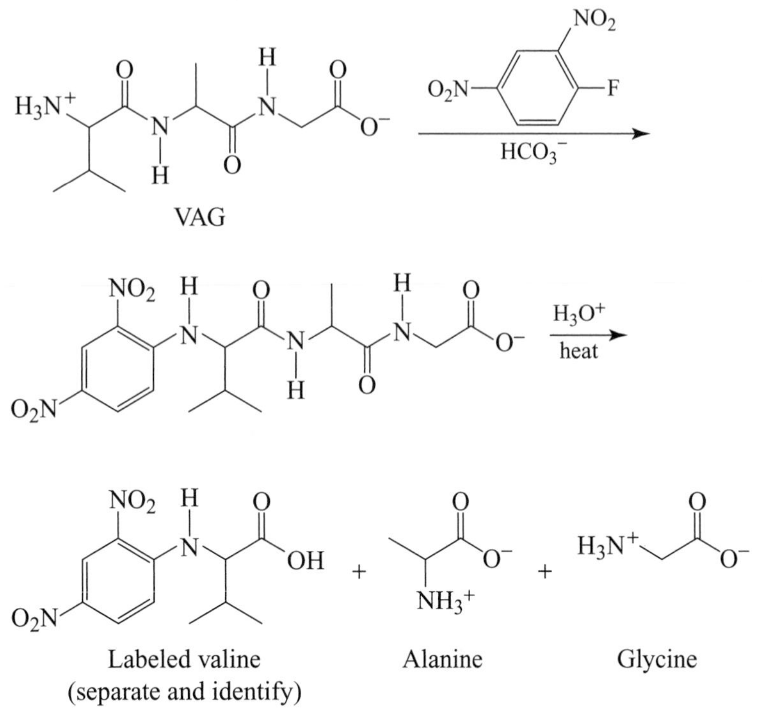

$\xrightarrow[\text{heat}]{H_3O^+}$

Labeled valine (separate and identify) + Alanine + Glycine

NO_2 H O OH O_2N O O^- NH_3^+ H_3N^+ O O^-

Labeled valine
(separate and identify)

Alanine

Glycine

(b)

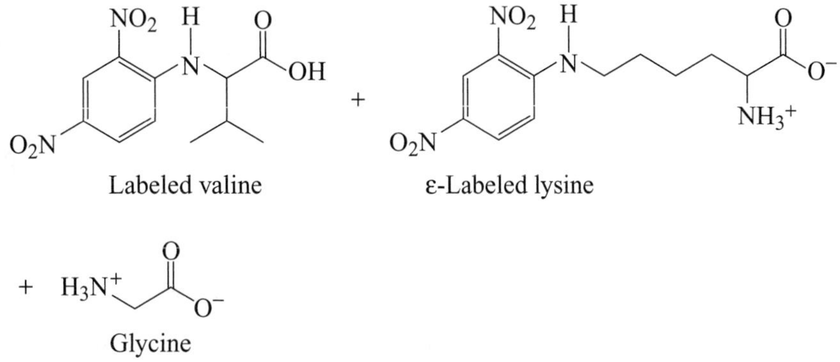

NO_2 H O OH O_2N + NO_2 H O O^- NH_3^+ O_2N

Labeled valine

ε-Labeled lysine

+ H_3N^+ O O^-

Glycine

24.7 C_6H_5–N=C=S + $H_2\ddot{N}$– ... MIR $\xrightarrow{OH^-}$

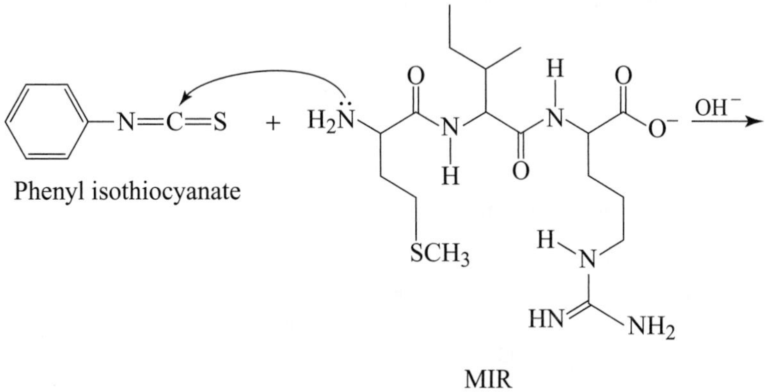

Phenyl isothiocyanate

SCH_3 H–N HN NH_2

MIR

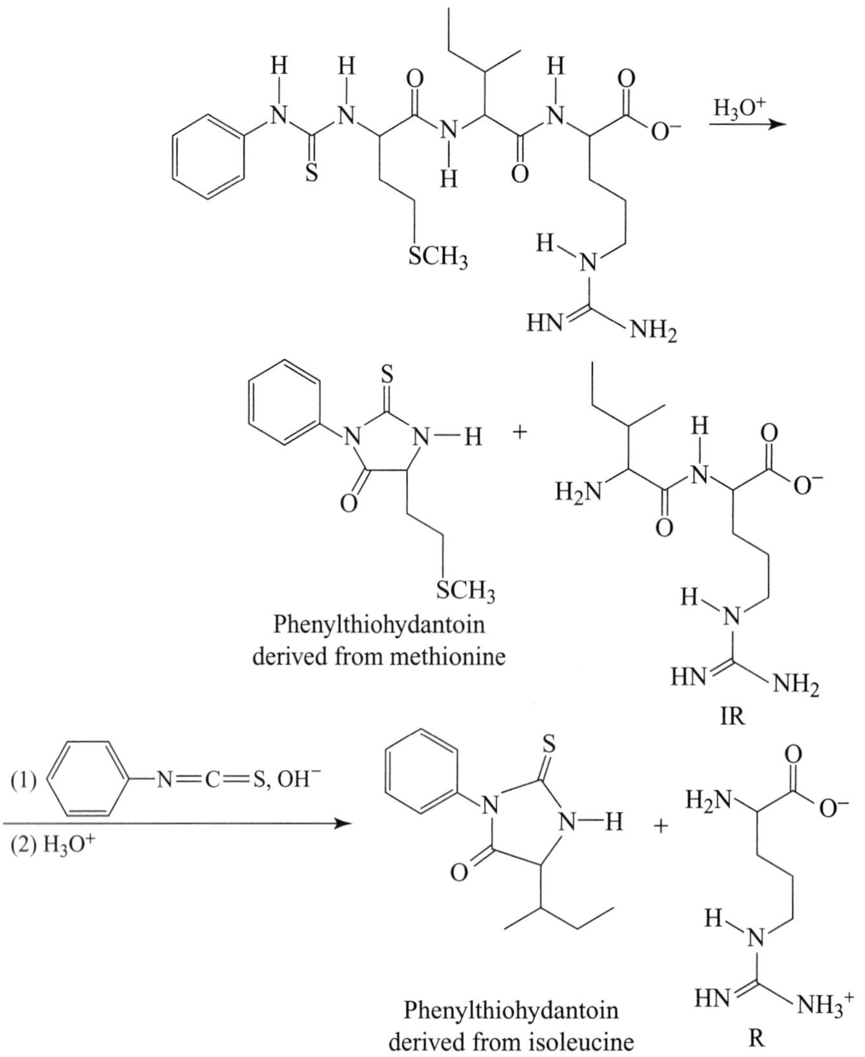

$\xrightarrow{H_3O^+}$

Phenylthiohydantoin derived from methionine + IR

$\xrightarrow[(2)\ H_3O^+]{(1)\ C_6H_5N{=}C{=}S,\ OH^-}$

Phenylthiohydantoin derived from isoleucine + R

24.8 (a) Two structures are possible with the sequence ECG. Glutamic acid may be linked to cysteine through its α-carboxyl group.

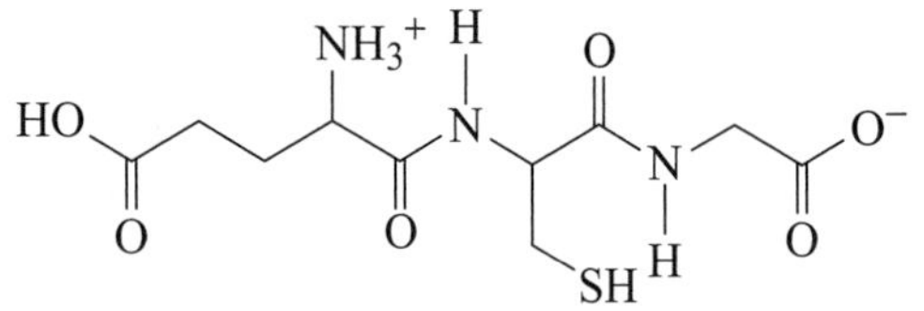

or through its γ-carboxyl group,

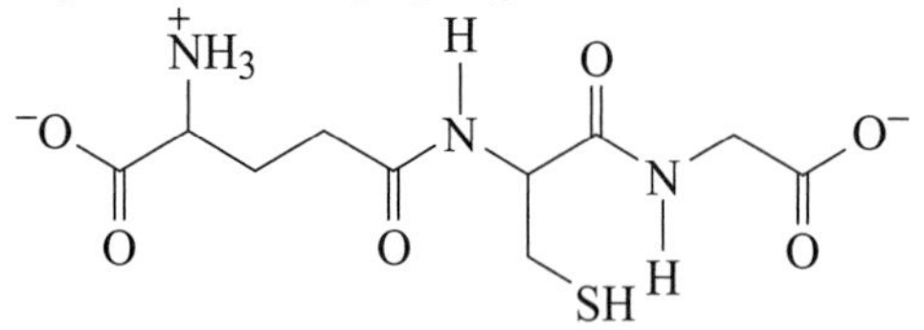

(b) This result shows that the second structure is correct, that in glutathione the γ-carboxyl group is linked to cysteine.

24.9 We look for points of overlap to determine the amino acid sequence in each case.

(a)
```
  S T
    T O
P S
-------
P S T O
```

(b)
```
  A C
    C R
      R V
L A
---------
L A C R V
```

24.10 Sodium in liquid ammonia brings about reductive cleavage of the disulfide linkage of oxytocin to two thiol groups; then air oxidizes the two thiol groups back to a disulfide linkage:

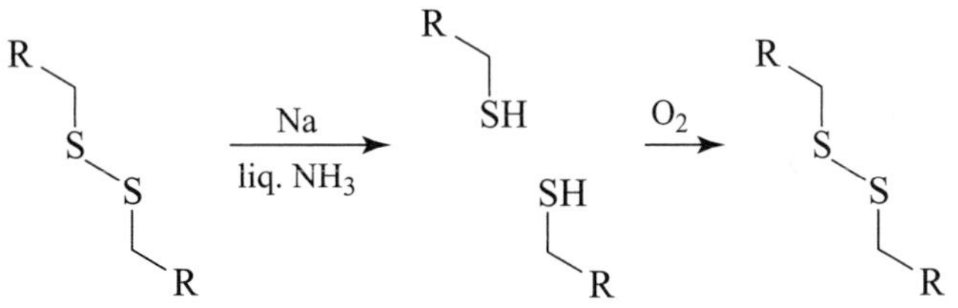

R–CH₂–S–S–CH₂–R $\xrightarrow[\text{liq. } NH_3]{Na}$ R–CH₂–SH + HS–CH₂–R $\xrightarrow{O_2}$ R–CH₂–S–S–CH₂–R

24.11 Removal of the Fmoc group initially involves an elimination reaction promoted by piperidine to form the carbamic acid derivative of the amino acid and 9-methylidenefluorene, which reacts further with piperidine by nucleophilic addition to form the byproduct. Spontaneous decarboxylation of the carbamic acid generates CO_2 and the free amino acid.

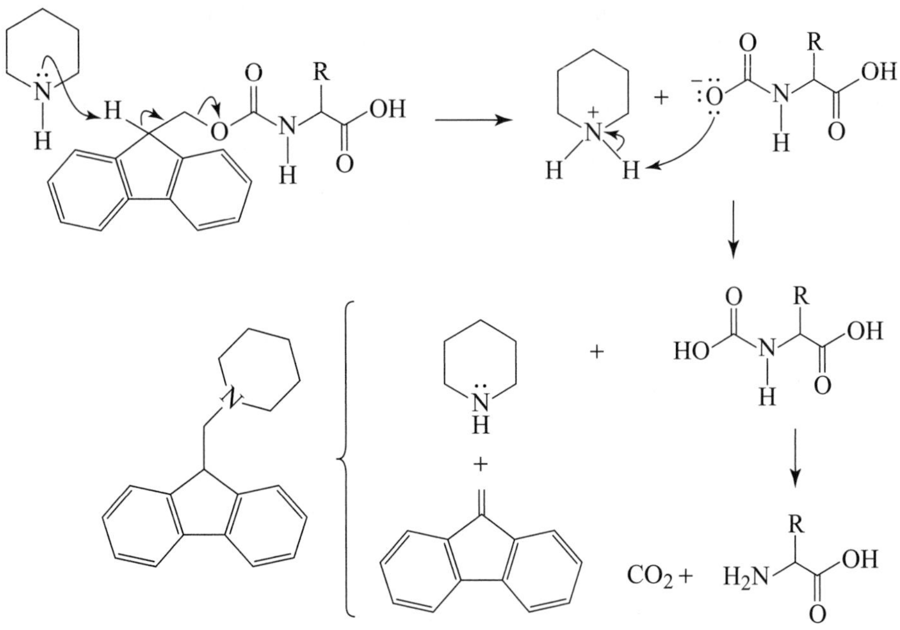

24.12 Glycine + Di-*tert*-butyl carbonate $\xrightarrow{OH^-}$

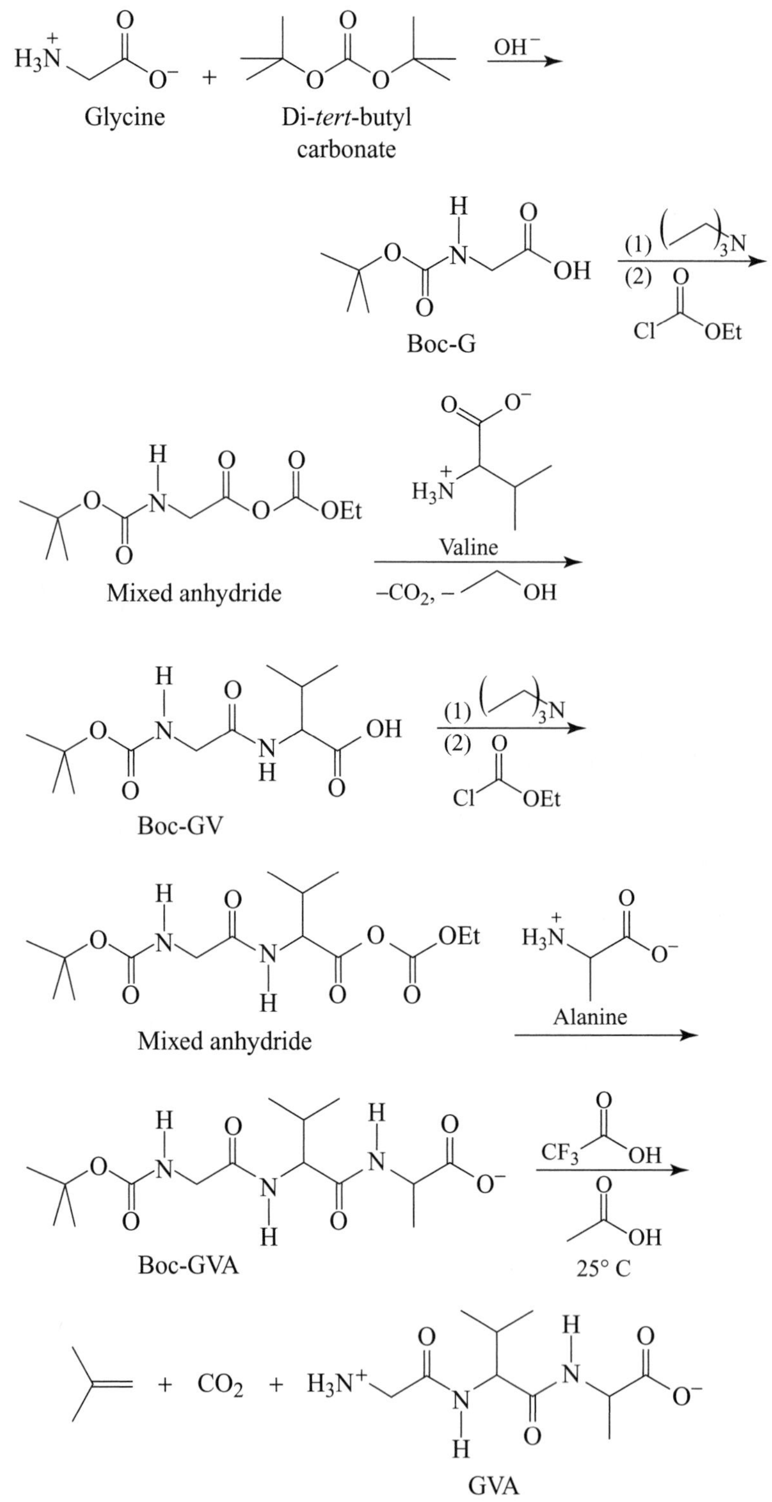

Boc-G $\xrightarrow[\text{(2) } Cl\text{–}C(=O)\text{–}OEt]{\text{(1) } (CH_3CH_2)_3N}$

Mixed anhydride $\xrightarrow[-CO_2,\ -CH_3CH_2OH]{\text{Valine}}$

Boc-GV $\xrightarrow[\text{(2) } Cl\text{–}C(=O)\text{–}OEt]{\text{(1) } (CH_3CH_2)_3N}$

Mixed anhydride $\xrightarrow{\text{Alanine}}$

Boc-GVA $\xrightarrow[CH_3COOH,\ 25°\ C]{CF_3COOH}$

Isobutylene + CO_2 + H_3N^+–GVA

GVA

24.13 (a)

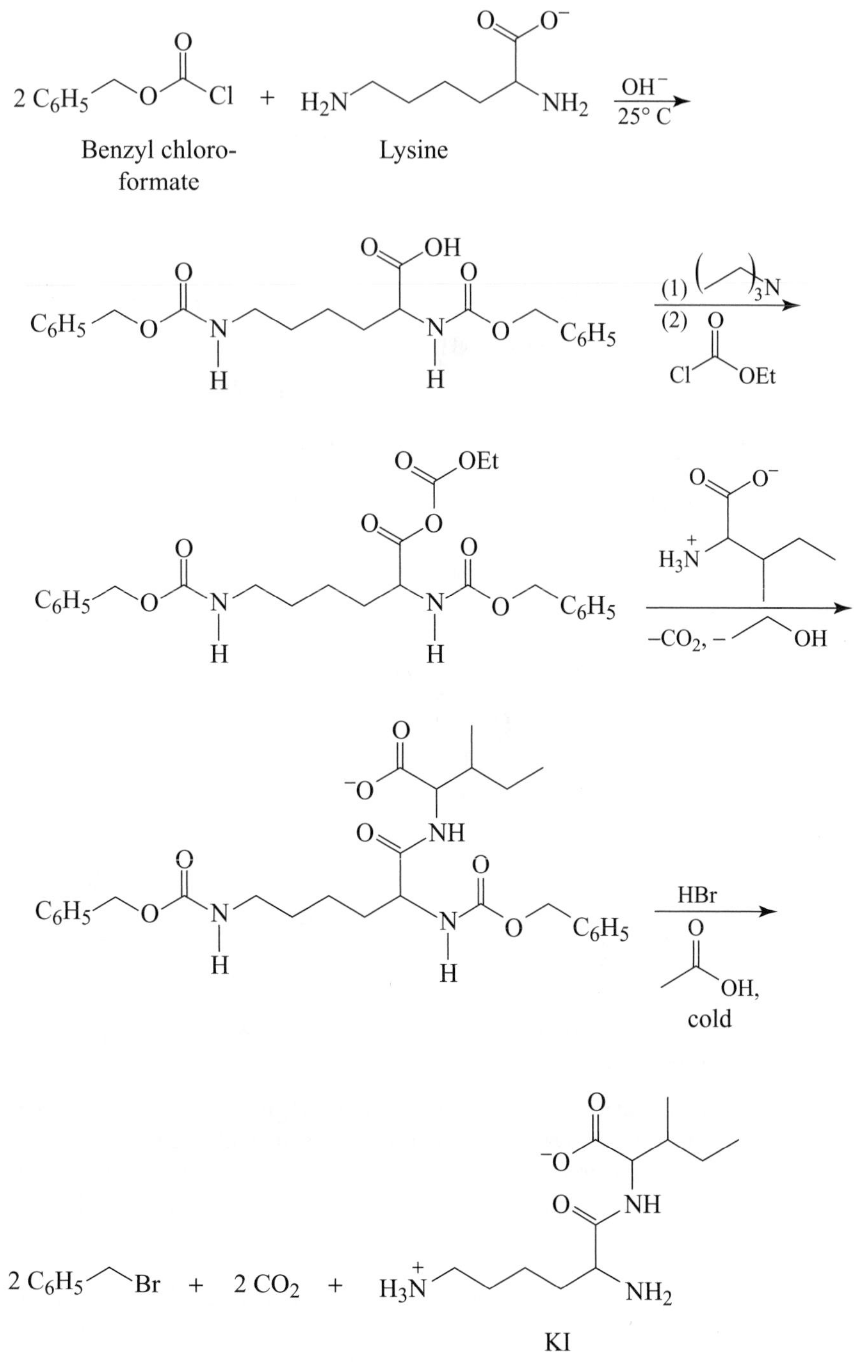

(b) $3\ C_6H_5CH_2OC(O)Cl$ + $H_2NC(=NH)NH(CH_2)_3CH(NH_2)CO_2^-$ $\xrightarrow[25°C]{OH^-}$

(Cbz-protected arginine: $C_6H_5CH_2OC(O)NHC(=NH)N[C(O)OCH_2C_6H_5](CH_2)_3CH[NHC(O)OCH_2C_6H_5]CO_2H$) $\xrightarrow[(2)\ ClC(O)OEt]{(1)\ (CH_3CH_2)_3N}$

(mixed anhydride, $-CO_2$–O–$C(O)OEt$) $\xrightarrow[-CO_2,\ -CH_3CH_2OH]{H_3\overset{+}{N}CH(CH_3)CO_2^-}$

(protected dipeptide, $-C(O)NHCH(CH_3)CO_2H$) $\xrightarrow[CH_3CO_2H,\ cold]{HBr}$

$3\ C_6H_5CH_2Br$ + $3\ CO_2$ + $H_3\overset{+}{N}C(=NH)NH(CH_2)_3CH(NH_2)C(O)NHCH(CH_3)CO_2^-$

RA

24.14 The weakness of the benzyl-oxygen bond allows these groups to be removed by catalytic hydrogenolysis.

24.15 Trifluoroacetic acid protonates the carbonyl group of the ester linkage joining the resin to the peptide. Heterolysis of the ester linkage then yields the relatively stable benzyl-type carbocation at the point of attachment. Acid hydrolysis of the amide linkages (peptide bonds) requires more stringent conditions because the fragments produced are not similarly stabilized.

Polymer–CH_2–O–C_6H_4–CH_2–O–C(=:O:)–peptide + H–O–C(O)–CF_3 $\longrightarrow$

Polymer–CH_2–O–C_6H_4–CH_2–O–C(=$\overset{+}{O}$H)–peptide $\longrightarrow$

Polymer–CH_2–O–C_6H_4–$\overset{+}{C}H_2$ + HO–C(=O)–peptide

24.16

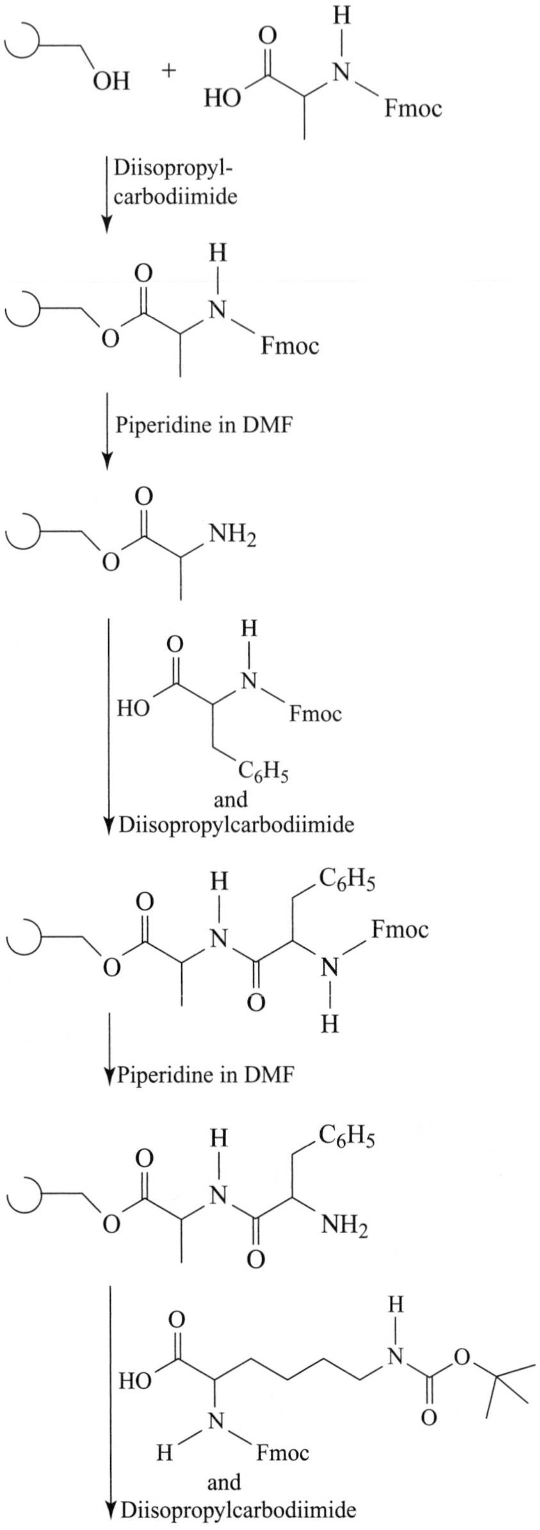

Resin–CH_2OH + HO–C(=O)–CH(CH_3)–NH–Fmoc — 1. Add Ala-Fmoc.

↓ Diisopropyl-carbodiimide

Resin–CH_2–O–C(=O)–CH(CH_3)–NH–Fmoc — 2. Purify by washing.

↓ Piperidine in DMF — 3. Remove protecting group.

Resin–CH_2–O–C(=O)–CH(CH_3)–NH_2 — 4. Purify by washing.

↓ HO–C(=O)–CH($CH_2C_6H_5$)–NH–Fmoc and Diisopropylcarbodiimide — 5. Add Phe-Fmoc.

Resin–CH_2–O–C(=O)–CH(CH_3)–NH–C(=O)–CH($CH_2C_6H_5$)–NH–Fmoc — 6. Purify by washing.

↓ Piperidine in DMF — 7. Remove protecting group.

Resin–CH_2–O–C(=O)–CH(CH_3)–NH–C(=O)–CH($CH_2C_6H_5$)–NH_2 — 8. Purify by washing.

↓ HO–C(=O)–CH(NH–Fmoc)–$(CH_2)_4$–NH–C(=O)–O–C(CH_3)$_3$ and Diisopropylcarbodiimide — 9. Add protected Lys. (Side-chain amino group protected by Boc.)

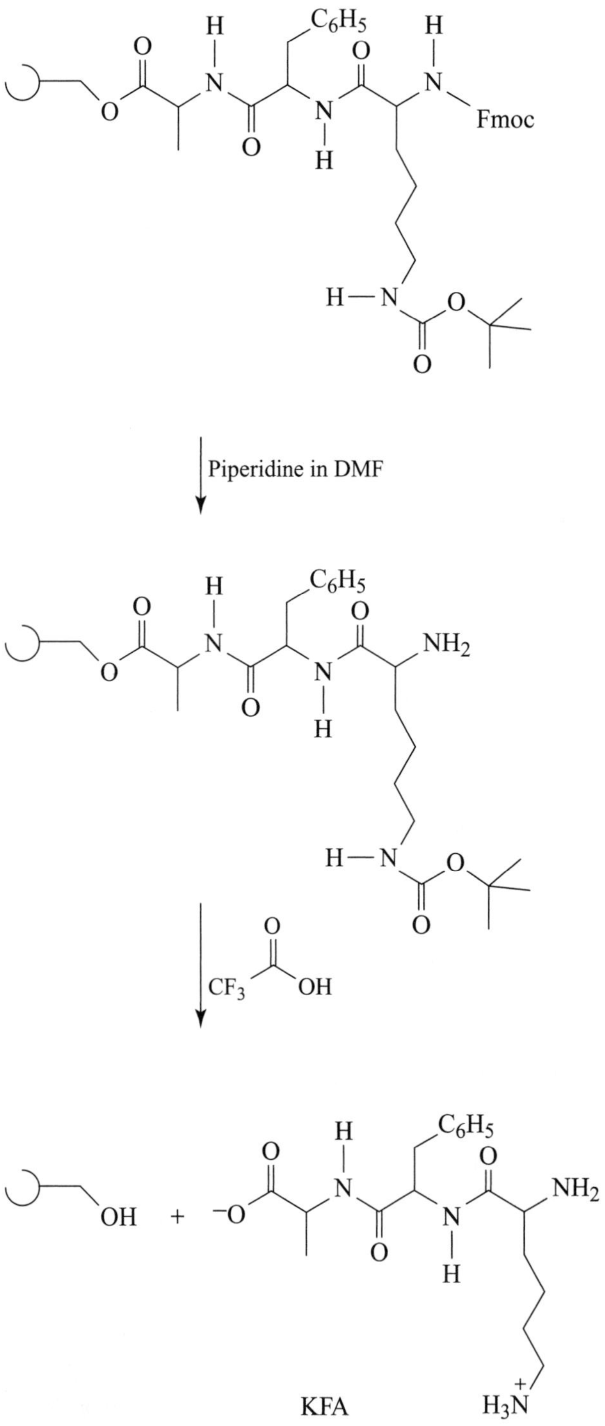

H, C_6H_5, H, O, O, N, N, O, Fmoc, O, N, H, H—N, O, O

10. Purify by washing.

Piperidine in DMF

11. Remove Fmoc protecting group.

H, C_6H_5, O, O, N, NH_2, O, N, O, H, H—N, O, O

12. Purify by washing.

CF_3, O, OH

13. Detach tripeptide and remove Lys side-chain Boc group.

OH + $^{-}$O, H, C_6H_5, O, O, N, NH_2, N, O, H, KFA, $H_3\overset{+}{N}$

14. Isolate product.

Problems

Structure and Reactivity

24.17 (a) Isoleucine, threonine, hydroxyproline, and cystine.

(b)
CO_2^-
$H_3\overset{+}{N}$—H
CH_3—H
CH_2
CH_3

and

CO_2^-
$H_3\overset{+}{N}$—H
H—CH_3
CH_2
CH_3

CO_2^-
$H_3\overset{+}{N}$—H
H—OH
CH_3

and

CO_2^-
$H_3\overset{+}{N}$—H
HO—H
CH_3

CO_2^-
$H_2\overset{+}{N}$—H
H OH

and

CO_2^-
$H_2\overset{+}{N}$—H
HO H

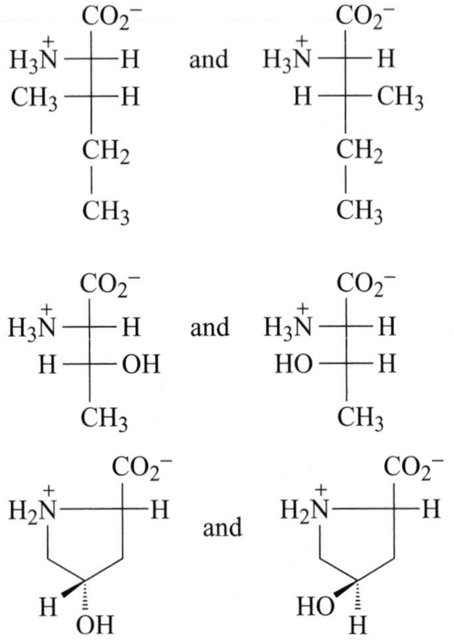

(With cystine, both chirality centers are α-carbon atoms; thus, according to the problem, both must have the L-configuration, and no isomers of this type can be written.)
(c) Diastereomers

24.18 (a) Br, HO, Br, O, O^-, NH_3^+ (b) O, OEt, NH_3^+

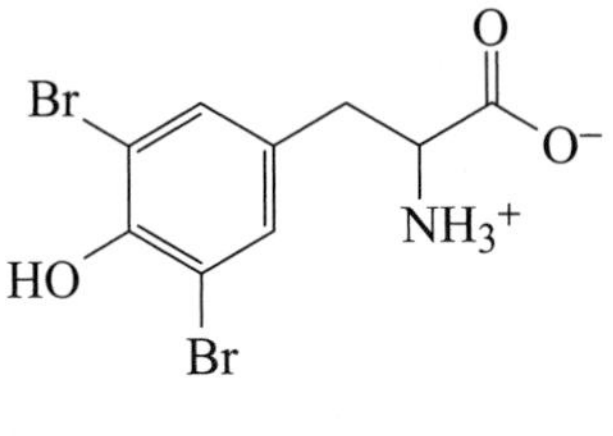

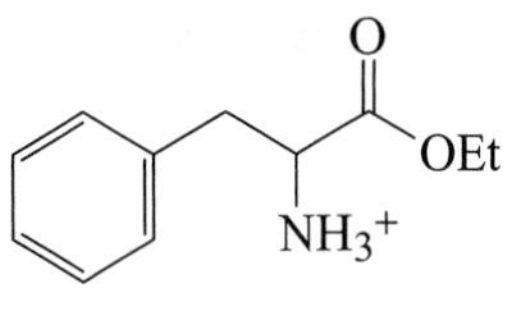

(c) C_6H_5, O, H, N, O, O^-

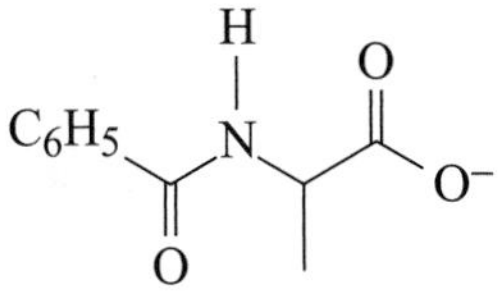

24.19 (a) (−)-Serine (CO_2^-, $H_3\overset{+}{N}$—C—H, CH_2OH) $\xrightarrow[CH_3OH]{HCl}$ **A** (CO_2CH_3, $H_3\overset{+}{N}$—C—H, CH_2OH, Cl^-) ($C_4H_{10}ClNO_3$) $\xrightarrow{PCl_5}$ **B** (CO_2CH_3, $H_3\overset{+}{N}$—C—H, CH_2Cl, Cl^-) ($C_4H_9Cl_2NO_2$)

$\xrightarrow[(2)\ OH^-]{(1)\ H_3O^+,\ H_2O,\ heat}$ **C** (CO_2^-, $H_3\overset{+}{N}$—C—H, CH_2Cl) ($C_3H_6ClNO_2$) $\xrightarrow[dil\ H_3O^+]{Na\text{-}Hg}$ L-(+)-Alanine (CO_2^-, $H_3\overset{+}{N}$—C—H, CH_3)

(b) **B** $\xrightarrow{OH^-}$ **D** (CO_2CH_3, H_2N—C—H, CH_2Cl) ($C_4H_8ClNO_2$) $\xrightarrow{NaSH}$ **E** (CO_2CH_3, H_2N—C—H, CH_2SH) ($C_4H_9NO_2S$)

$\xrightarrow[(2)\ OH^-]{(1)\ H_3O^+,\ H_2O,\ heat}$ L-(+)-Cysteine (CO_2^-, $H_3\overset{+}{N}$—C—H, CH_2SH)

(c) L-(−)-Asparagine (CO_2^-, $H_3\overset{+}{N}$—C—H, CH_2CNH_2 with C=O) $\xrightarrow[\text{Hofmann rearrangement}]{NaOBr,\ OH^-}$ **F** (CO_2^-, H_2N—C—H, CH_2NH_2) ($C_3H_7N_2O_2$) $\xleftarrow{NH_3}$ **C** (CO_2^-, $H_3\overset{+}{N}$—C—H, CH_2Cl) [from part (a)]

24.20 (a) Diethyl acetamidomalonate + acrylonitrile (CN) $\xrightarrow[\text{EtOH (95\% yield)}]{EtONa}$ **G** (diethyl 2-acetamido-2-(2-cyanoethyl)malonate: N–H, OEt, OEt, CN)

$\xrightarrow[\text{reflux 6 h (66\% yield)}]{concd\ HCl}$ DL-Glutamic acid ($HO_2CCH_2CH_2CH(\overset{+}{N}H_3)CO_2^-$) + CH_3CO_2H + 2 EtOH + NH_4^+ + CO_2

(b) G → (H_2, Ni, 68° C, 1000 psi, (90% yield)) → [intermediate] → (– EtOH) → H → (concd HCl, reflux 4 h, (97% yield)) → DL-Ornithine hydrochloride + acetic acid + EtOH + CO_2

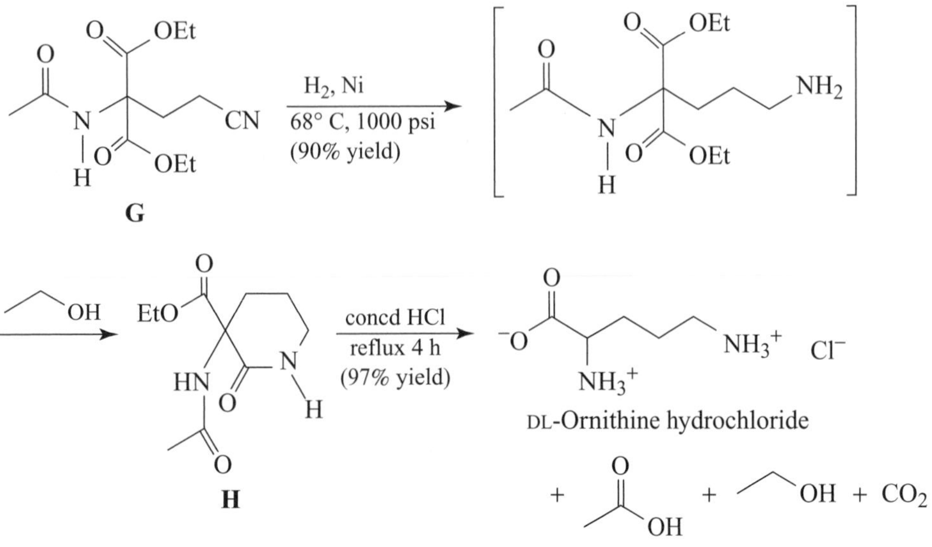

24.21 At pH 2–3 the γ-carboxyl groups of polyglutamic acid are uncharged. (They are present as $-CO_2H$ groups.) At pH 5 the γ-carboxyl groups ionize and become negatively charged. (They become γ-CO_2^- groups.) The repulsive forces between these negatively charged groups cause an unwinding of the α helix and the formation of the random coil.

<u>**Peptide Sequencing**</u>

24.22 We look for points of overlap:

```
        F S
      P G F
    P P     S P F
  R P           F R
  -----------------
  R P P G F S P F R
      Bradykinin
```

24.23 1. This experiment shows that valine is the N-terminal amino acid and that valine is attached to leucine. (Lysine labeled at the ϵ-amino group is to be expected if lysine is not the N-terminal amino acid and if it is linked in the polypeptide through its α-amino group.)

2. This experiment shows that alanine is the C-terminal amino acid and that it is linked to glutamic acid.

At this point, then, we have the following information about the structure of the heptapeptide.

V L (A, K, F) E A

(A, K, F): Sequence unknown

3. (a) This experiment shows that the dipeptide, **A**, is: L K
   (b) The carboxypeptidase reaction shows that the C-terminal amino acid of the tripeptide, **B**, is glutamic acid; the DNP labeling experiment shows that the N-terminal amino acid is phenylalanine. Thus, the tripeptide **B** is : F A E

Putting these pieces together in the only way possible, we arrive at the following amino acid sequence for the heptapeptide.

```
V L
  L K
      F A E
          E A
-------------
V L K F A E A
```

Challenge Problem

24.24 The observation that the ^1H NMR spectrum taken at room temperature shows two different signals for the methyl groups suggests that they are in different environments. This would be true if rotation about the carbon-nitrogen bond was not taking place.

```
δ 8.05  H           CH3  δ 2.95
         \         /
          C — N
         //        \
        O           CH3  δ 2.80
```

We assign the δ 2.80 signal to the methyl group that is on the same side as the electronegative oxygen atom.

The fact that the methyl signals appear as doublets (and that the formyl proton signal is a multiplet) indicates that long-range coupling is taking place between the methyl protons and the formyl proton.

That the two doublets are not simply the result of spin-spin coupling is indicated by the observation that the distance that separates one doublet from the other changes when the applied magnetic field strength is lowered. [*Remember!* The magnitude of a chemical shift is proportional to the strength of the applied magnetic field, while the magnitude of a coupling constant is not.]

That raising the temperature (to 111°C) causes the doublets to coalesce into a single signal indicates that at higher temperatures the molecules have enough energy to surmount the energy barrier of the carbon-nitrogen bond. Above 111°C, rotation is taking place so rapidly that the spectrometer is unable to discriminate between the two methyl groups.

QUIZ

24.1 Write the structural formula of the principal ionic species present in aqueous solutions at pH 2, 7, and 12 of isoleucine (2-amino-3-methylpentanoic acid).

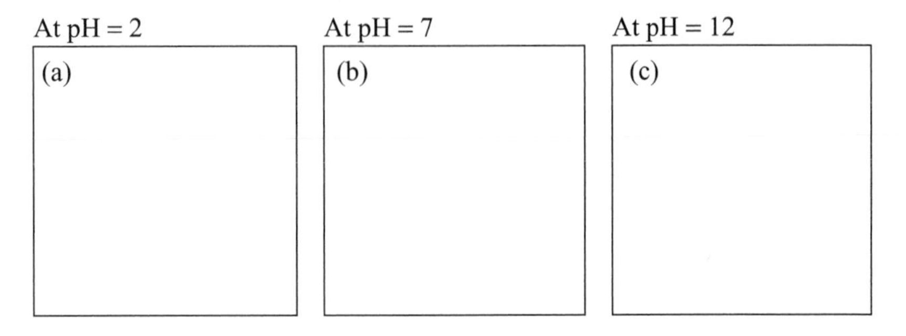

24.2 A hexapeptide gave the following products:

Hexapeptide $\xrightarrow[\text{HCO}_3^-]{\text{O}_2\text{N}-\text{C}_6\text{H}_3(\text{NO}_2)-\text{F}}$ $\xrightarrow{\text{H}_3\text{O}^+}$ O_2N–$C_6H_3(NO_2)$–N(proline ring)–COOH

(HN(proline ring)–COOH = Proline (P))

Hexapeptide $\xrightarrow{3N\ \text{HCl, } 100°\ \text{C}}$ 2 G, 1 L, 1 F, 1 P, 1 Y

Hexapeptide $\xrightarrow{1N\ \text{HCl, } 80°\ \text{C}}$ FGY + GFG + PLG + LGF

The structure of the hexapeptide (using abbreviations such as G, L etc.) is

25 NUCLEIC ACIDS AND PROTEIN SYNTHESIS

SOLUTIONS TO PROBLEMS

25.1 Adenine:

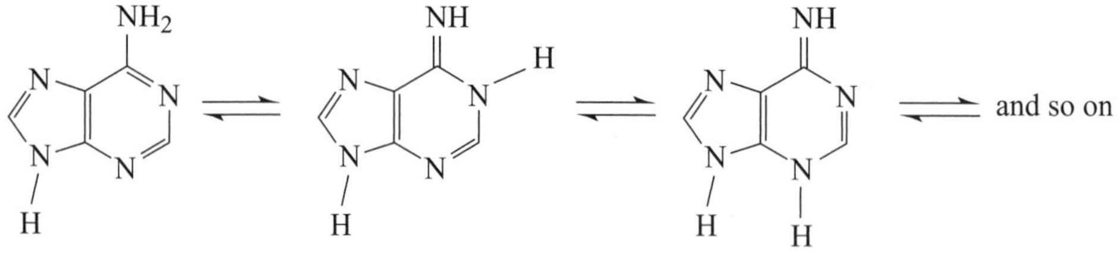

NH$_2$ N N N N H ⇌ NH H N N N N H ⇌ NH N N N N H H ⇌ and so on

Guanine:

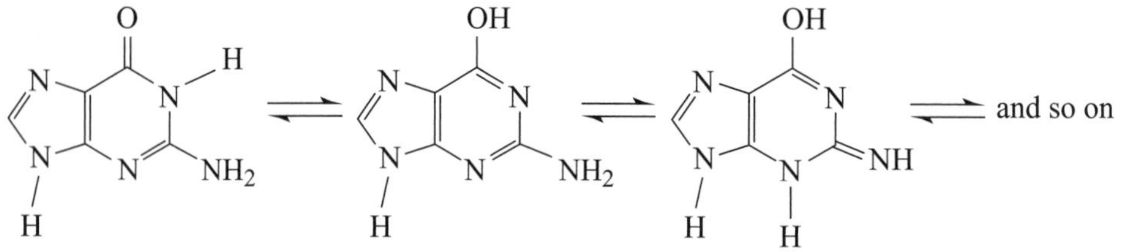

O H N N N N NH$_2$ H ⇌ OH N N N N NH$_2$ H ⇌ OH N N N N NH H H ⇌ and so on

Cytosine:

NH$_2$ N N O H ⇌ NH H N N O H ⇌ NH H N N OH ⇌ and so on

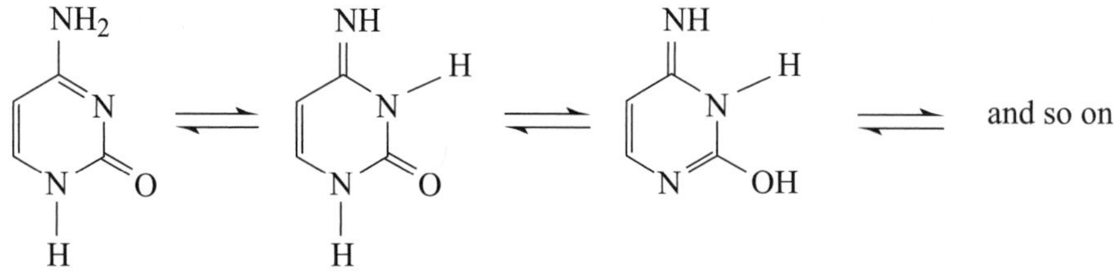

Thymine (R = CH_3) or Uracil (R = H):

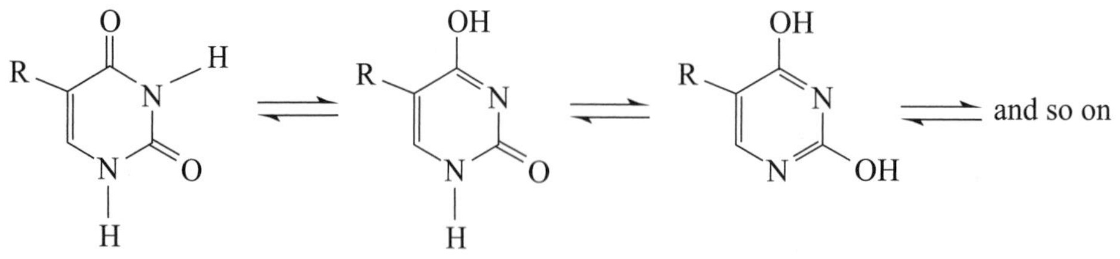

O R H N N O H ⇌ OH R N N O H ⇌ OH R N N OH ⇌ and so on

25.2 (a) The nucleosides have an *N*-glycosidic linkage that (like an *O*-glycosidic linkage) is rapidly hydrolyzed by aqueous acid but is one that is stable in aqueous base.

(b) HO HO H N O N + Ö +H₃O⁺ ⇌ −H₃O⁺ + H_2O ⇌ OH OH Nucleoside

HO + O H + N H OH Heterocyclic base

$-H_2O$ ⇅ $+H_2O$

HO $\overset{+}{OH_2}$ O + H_2O $\underset{+H_3O^+}{\overset{-H_3O^+}{\rightleftharpoons}}$ HO O OH ⇌ HO OH H =O OH OH Deoxyribose OH

25.3 C_6H_5 O O ṄH₂ O + O NH Michael addition C_6H_5 O O C_6H_5 EtO EtO O O O

O NH amide formation (−⁄OH) EtO HN: OEt O → H O N H EtO N O − ⁄OH → O N H N O

25.4 (a) The isopropylidene group is part of a cyclic acetal and is thus susceptible to hydrolysis by mild acid.

(b) It can be installed by treating the nucleoside with acetone and a trace of acid and by simultaneously removing the water that is produced.

HO base OH OH + O — cat. HA, $-H_2O$ → HO base O O

25.5 (a) Cordycepin is

HO Ad O H H H H H OH (3'-Deoxyadenosine)

(b) HO Ad O SEt H H H H OH — $SOCl_2$ → Cl S O O Ad O SEt H H H H O S=O Cl — $-SO_2$, $-Cl^-$ →

Cl S O O O Ad Et S^+ H H H H H O H H O H — $-SO_2$, $-HCl$ → HO Ad O H EtS H H OH H — H_2, Raney Ni → HO Ad O H H H H OH H

2'-Deoxyadenosine

25.6 (a) $3 \times 10^9 \text{ base pairs} \times \dfrac{34\text{Å}}{10 \text{ base pairs}} \times \dfrac{10^{-10}\text{m}}{\text{Å}} \cong 1 \text{ m}$

(b) $6 \times 10^{-12} \dfrac{\text{g}}{\text{ovum}} \times 6.5 \times 10^9 \text{ ova} = 4 \times 10^{-2}\text{g}$

25.7 (a)

N OH ---------- :O: CH_3 N N:------- H–N H N= N NH-------- :O: H H

Lactim form of guanine   Thymine

(b) Thymine would pair with adenine and thus adenine would be introduced into the complementary strand where guanine should occur.

25.8 (a) A diazonium salt and a heterocyclic analog of a phenol.

NH_2 (adenine nucleotide, R) $\xrightarrow{HONO}$ N_2^+ (diazonium salt, R) $\xrightarrow[H_2O]{-N_2}$ OH (R) $\rightleftharpoons$ Hypoxanthine nucleotide

(b) Hypoxanthine ··· Cytosine

Hypoxanthine Cytosine

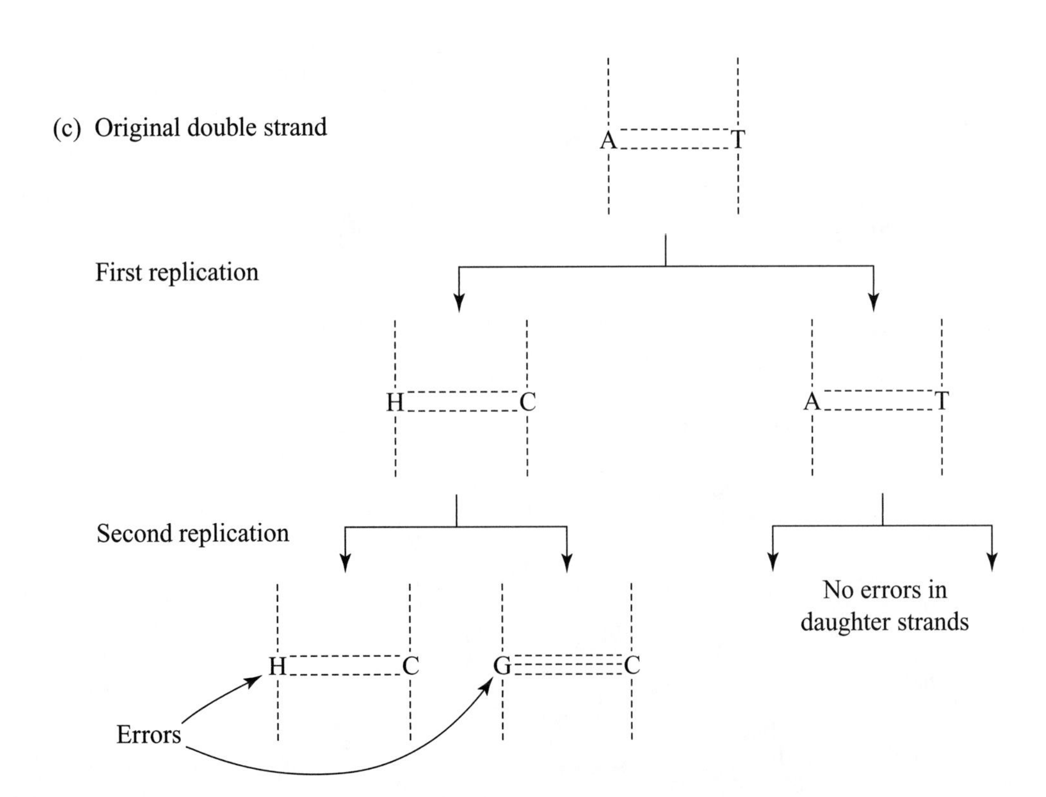

25.9

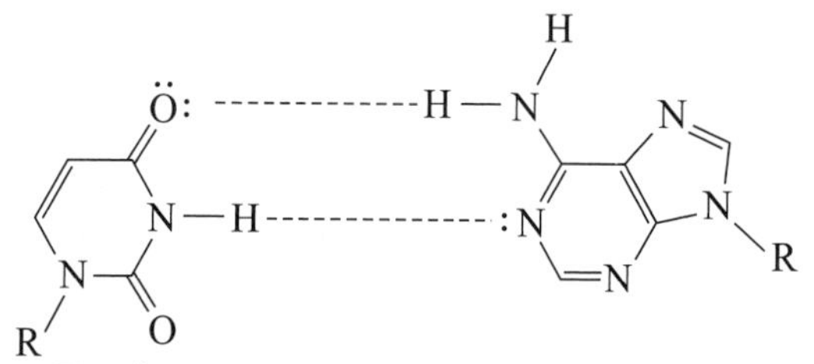

Uracil (in mRNA) — Adenine (in DNA)

25.10

(a)	ACC	CCC	AAA	AUG	UCG	*m*RNA
(b)	T	P	K	M	S	Amino acids
(c)	UGG	GGC	UUU	UAC	AGC	Anticodons

25.11

	R	I	C	Y	V	Amino acids
(a)	AGA	AUA	UGC	UGG	GUA	*m*RNA
(b)	TCT	TAT	ACG	ACC	CAT	DNA
(c)	UCU	UAU	ACG	ACC	CAU	Anticodons

Problems

Nucleic Acid Structure

25.12

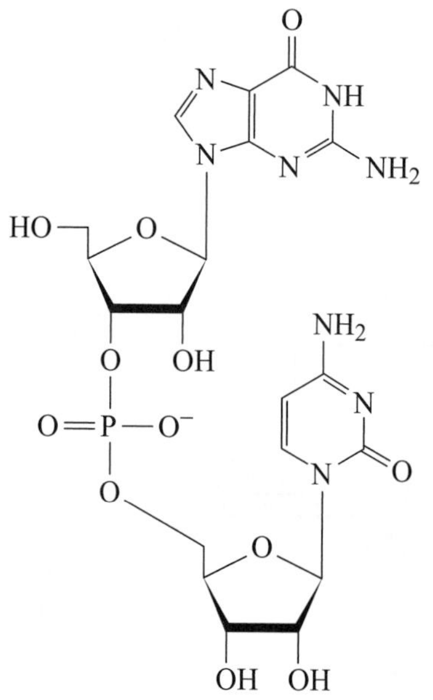

25.13

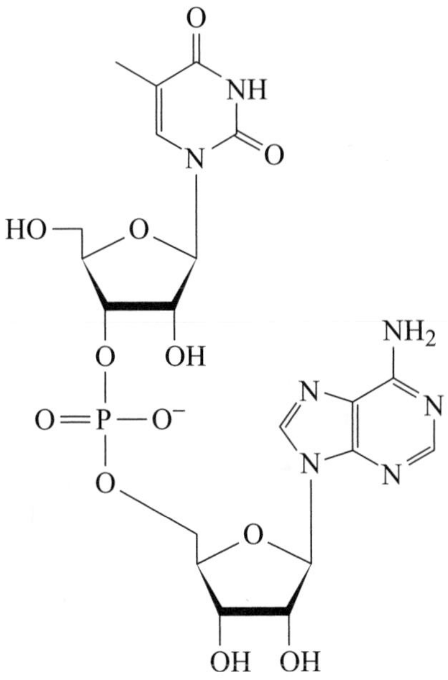

O NH N O HO O O OH NH$_2$ N N O=P—O$^-$ N N O O OH OH

Mechanisms

25.14 Ionization of the benzoyl group at the anomeric carbon (C1) is assisted by the carbonyl oxygen of the C2 benzoyl group, resulting in a stabilized cation blocked from attack on the α-face.

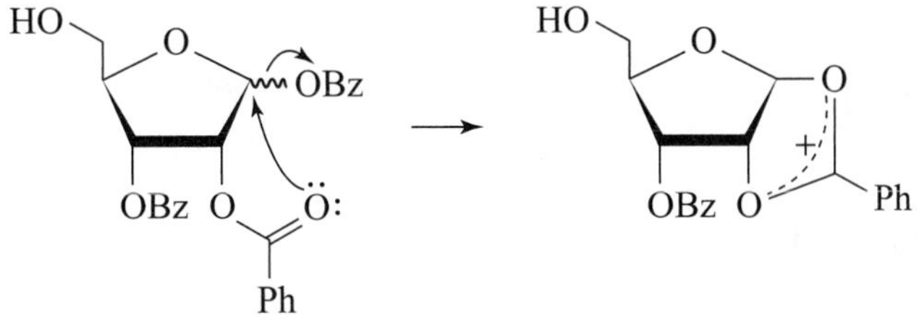

HO O OBz OBz O Ö: Ph → HO O O + OBz O Ph

25.15

Mitomycin A → Leuco-aziridinomitosene

Protonation and ring opening

Resonance-stabilized cation intermediate

Alkylation by N^2 of a deoxyguanosine in DNA

Deprotonation $:A^-$

A monoadduct with DNA

Tautomerization

Second alkylation by N^2 of another deoxyguanosine in DNA

1-Dihydromitosene A

A cross-linked adduct with DNA

25.16

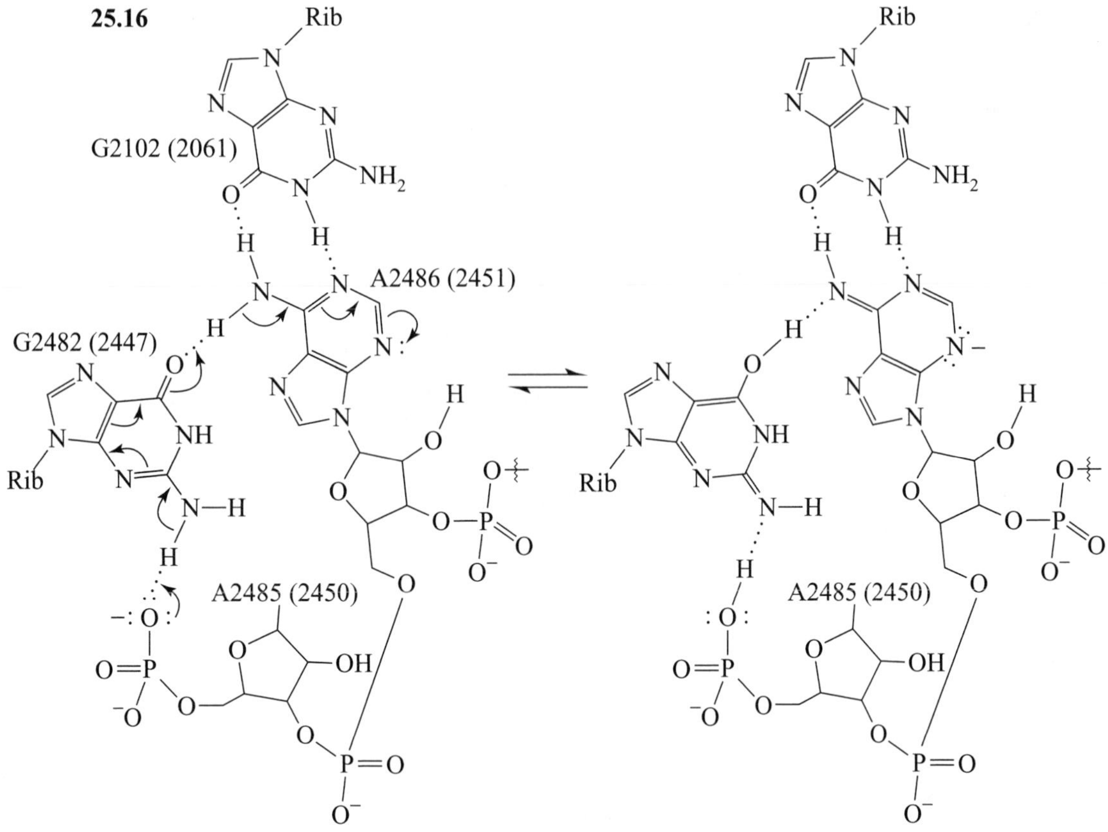

G2102 (2061)
A2486 (2451)
G2482 (2447)
A2485 (2450)
Rib

A2485 (2450)
Rib

A SPECIAL TOPIC
^{13}C NMR Spectroscopy

A.1 Analysis of the molecular formula $C_5H_{10}O$ indicates an IHD = 1. Structural possibilities are: a C=C, a C=O, or a ring.

The δ 211.0 signal must be due to a carbonyl group. The remaining C_4H_{10} is represented by only two signals in the alkyl group region, suggesting symmetry in the molecule and two unique carbons.

Two CH_3CH_2 groups must be present, with the CH_3 giving rise to the signal at δ ~10 and the CH_2 giving the signal at δ ~37.

Hence, the structure is

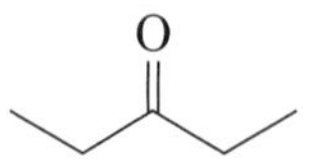

A.2 Qualifying structures are:

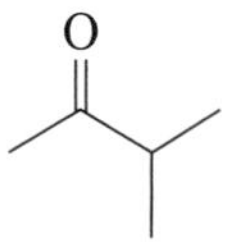

and

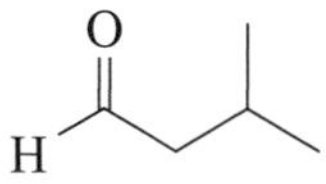

Each will give four ^{13}C NMR signals, one of which is due to the carbonyl group (δ 211.0).

A.3 (a)

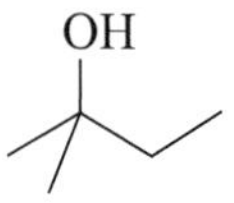

4 signals

(b)

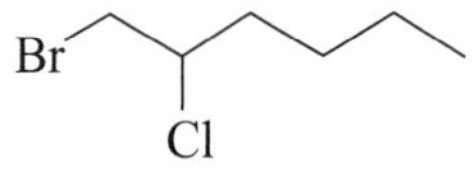

6 signals

(c)

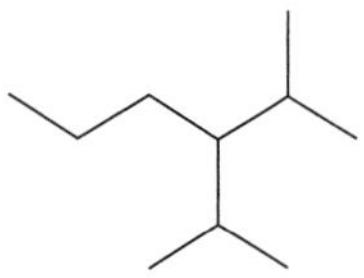

6 signals

A.4 (a) 5 signals
(b) 7 signals
(c) 8 signals
(d) 6 signals
(e) 4 signals

A.5 (a) 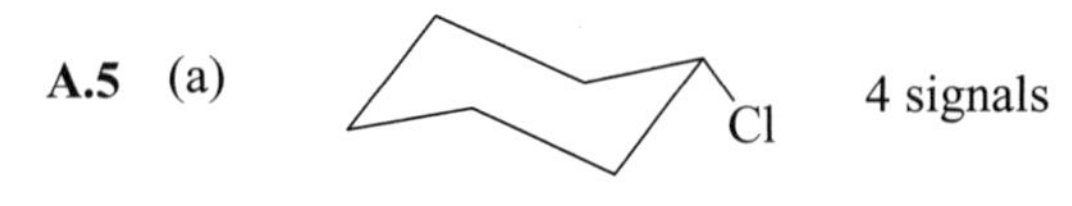

4 signals

(b) 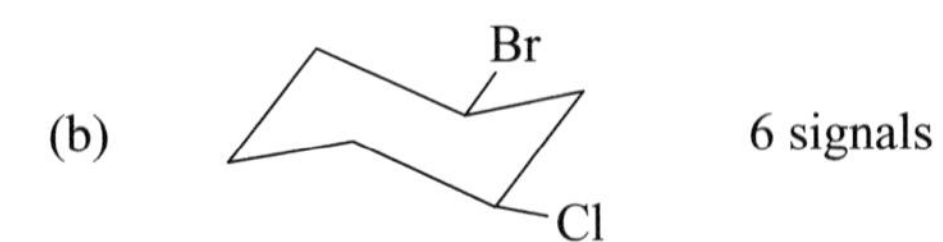

6 signals

(c) 

4 signals

B SPECIAL TOPIC
Chain-Growth Polymers

B.1 (a)

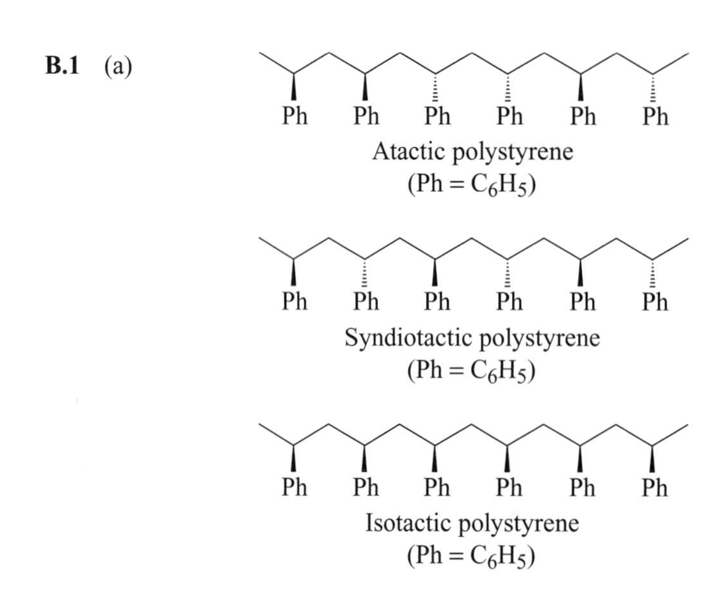

(b) The solution of isotactic polystyrene.

C

SPECIAL TOPIC
Step-Growth Polymers

SOLUTIONS TO PROBLEMS

C.1 (a) cyclohexane $\xrightarrow[\text{cat.}]{O_2}$ cyclohexanol (OH) + cyclohexanone (O) $\xrightarrow[\text{cat.}]{HNO_3}$ HO(O)C–(CH$_2$)$_4$–C(O)OH + N_2O

(b) HO(O)C–(CH$_2$)$_4$–C(O)OH $\xrightarrow{2\ NH_3}$ $^{+}NH_4$ ^{-}O(O)C–(CH$_2$)$_4$–C(O)O^{-} $^{+}NH_4$

$\xrightarrow[-2\ H_2O]{\text{heat}}$ H_2N(O)C–(CH$_2$)$_4$–C(O)NH_2 $\xrightarrow[\text{catalyst}]{350°C}$ NC–(CH$_2$)$_4$–CN $\xrightarrow[\text{catalyst}]{4\ H_2}$

H_2N–(CH$_2$)$_6$–NH_2

(c) 1,3-butadiene $\xrightarrow{Cl_2}$ Cl–CH$_2$CH=CHCH$_2$–Cl $\xrightarrow{2\ NaCN}$ NC–CH$_2$CH=CHCH$_2$–CN $\xrightarrow[Ni]{H_2}$

NC–(CH$_2$)$_4$–CN $\xrightarrow[\text{catalyst}]{4\ H_2}$ H_2N–(CH$_2$)$_6$–NH_2

(d) tetrahydrofuran (O) $\xrightarrow{2\ HCl}$ Cl–(CH$_2$)$_4$–Cl $\xrightarrow{2\ NaCN}$ NC–(CH$_2$)$_4$–CN

$\xrightarrow[\text{catalyst}]{4\ H_2}$ H_2N–(CH$_2$)$_6$–NH_2

C.2 (a) $HOCH_2CH_2OH + :B^- \rightleftharpoons {}^-OCH_2CH_2OH + HB$

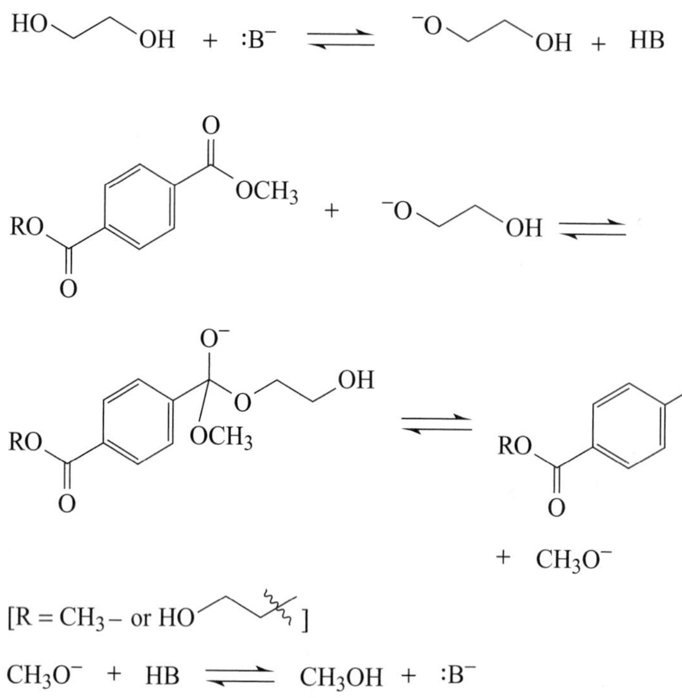

$ROC(=O)C_6H_4CO_2CH_3 + {}^-OCH_2CH_2OH \rightleftharpoons ROC(=O)C_6H_4C(O^-)(OCH_3)OCH_2CH_2OH \rightleftharpoons ROC(=O)C_6H_4CO_2CH_2CH_2OH + CH_3O^-$

[R = CH3– or HOCH2CH2–]

$CH_3O^- + HB \rightleftharpoons CH_3OH + :B^-$

(b) $ROC(=O)C_6H_4CO_2CH_3 \underset{:A^-}{\overset{HA}{\rightleftharpoons}} ROC(=O)C_6H_4C(=\overset{+}{O}H)OCH_3 \underset{-\,HOCH_2CH_2OH}{\overset{+\,HOCH_2CH_2OH}{\rightleftharpoons}}$

$ROC(=O)C_6H_4C(OH)(OCH_3)\overset{+}{O}(H)CH_2CH_2OH \rightleftharpoons ROC(=O)C_6H_4C(OH)(\overset{+}{O}HCH_3)OCH_2CH_2OH$

$\underset{+CH_3OH}{\overset{-CH_3OH}{\rightleftharpoons}} ROC(=O)C_6H_4C(=\overset{+}{O}H)OCH_2CH_2OH \underset{HA}{\overset{:A^-}{\rightleftharpoons}} ROC(=O)C_6H_4CO_2CH_2CH_2OH$

[R = CH3– or HOCH2CH2–]

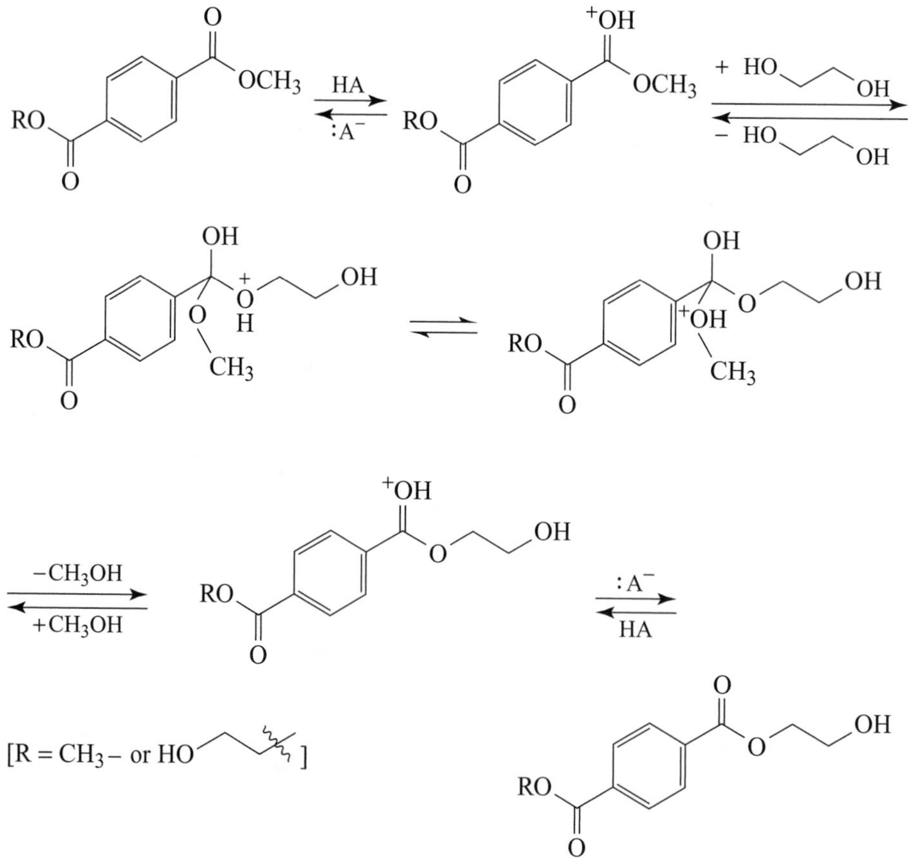

C.3 (a)

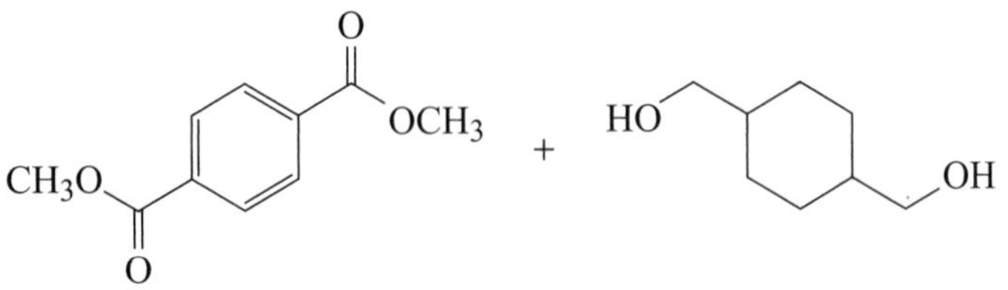

CH_3O ... OCH_3 + HO ... OH

(b) By high-pressure catalytic hydrogenation

C.4 etc. ... etc.

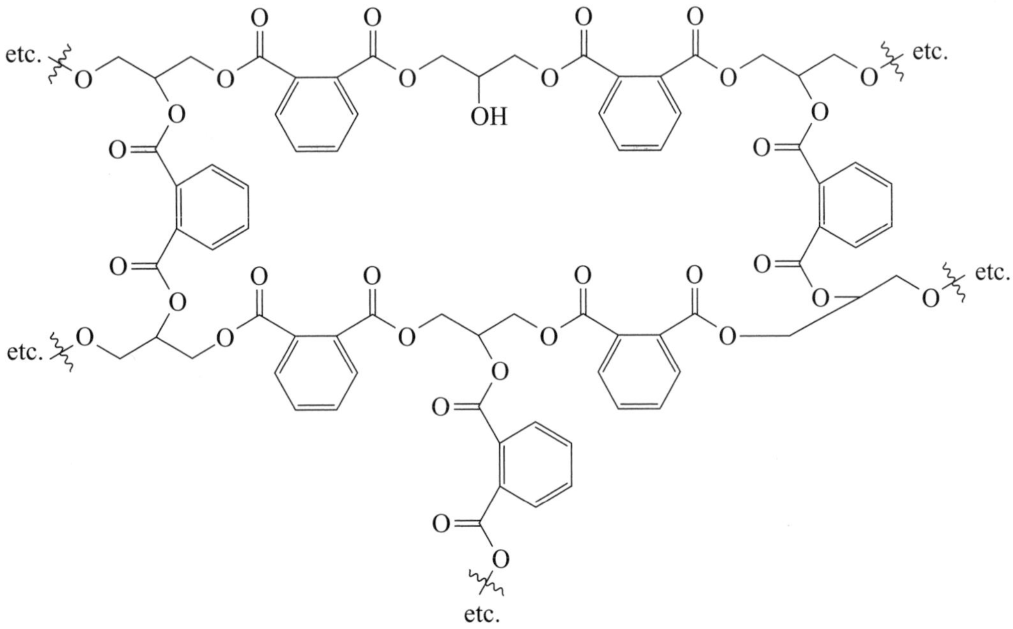

OH

etc. ... etc.

etc.

C.5 HO ... OH + Cl ... Cl $\xrightarrow[-HCl]{\text{pyridine}}$

Lexan

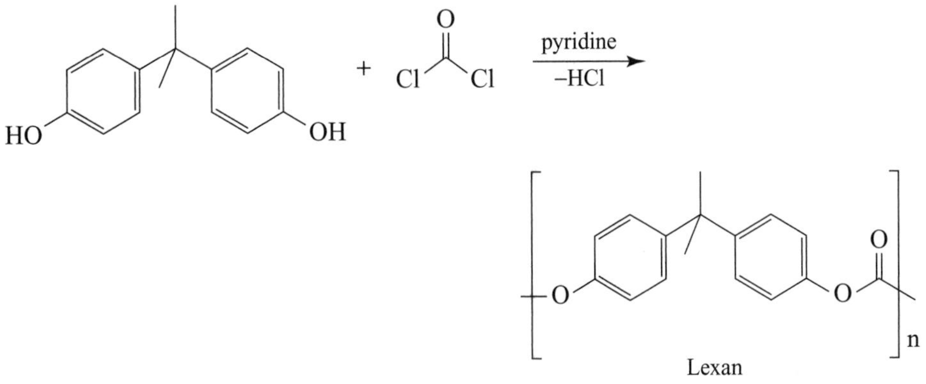

C.6 (a) The resin is probably formed in the following way. Base converts the bisphenol A to a double phenoxide ion that attacks a carbon atom of the epoxide ring of each epichlorohydrin:

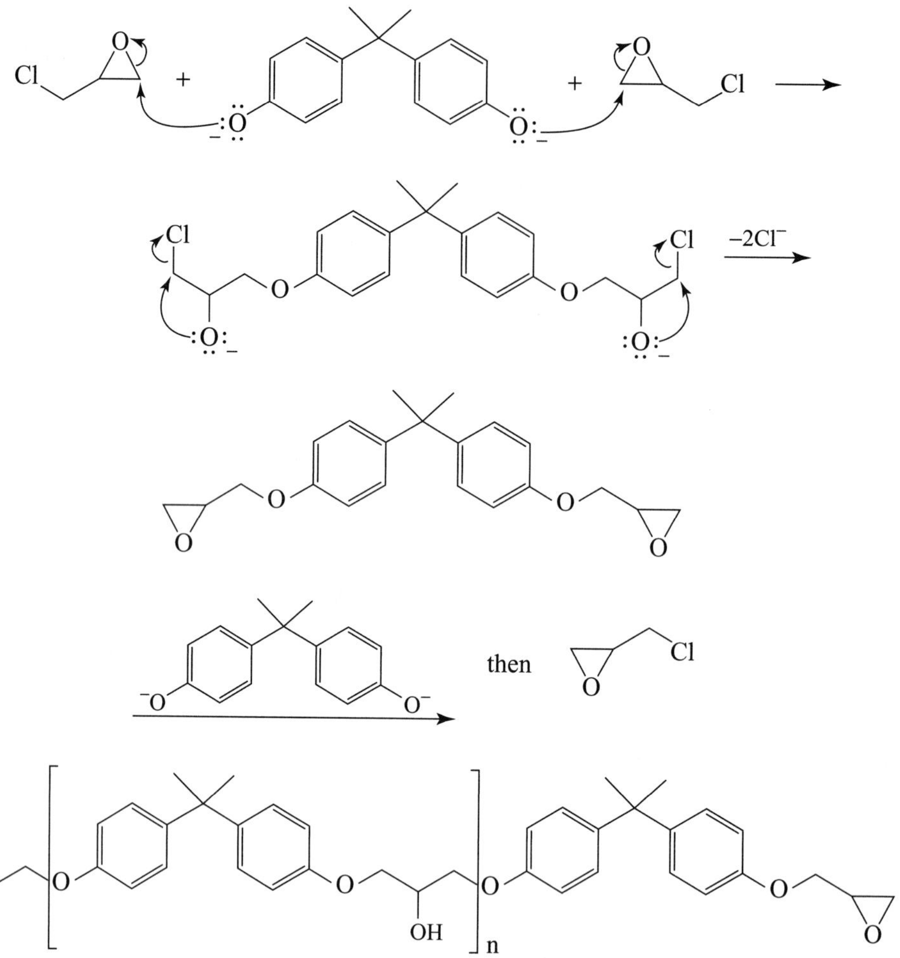

then

(b) The excess of epichlorohydrin limits the molecular weight and ensures that the resin has epoxy ends.

(c) Adding the hardener brings about cross linking by reacting at the terminal epoxide groups of the resin:

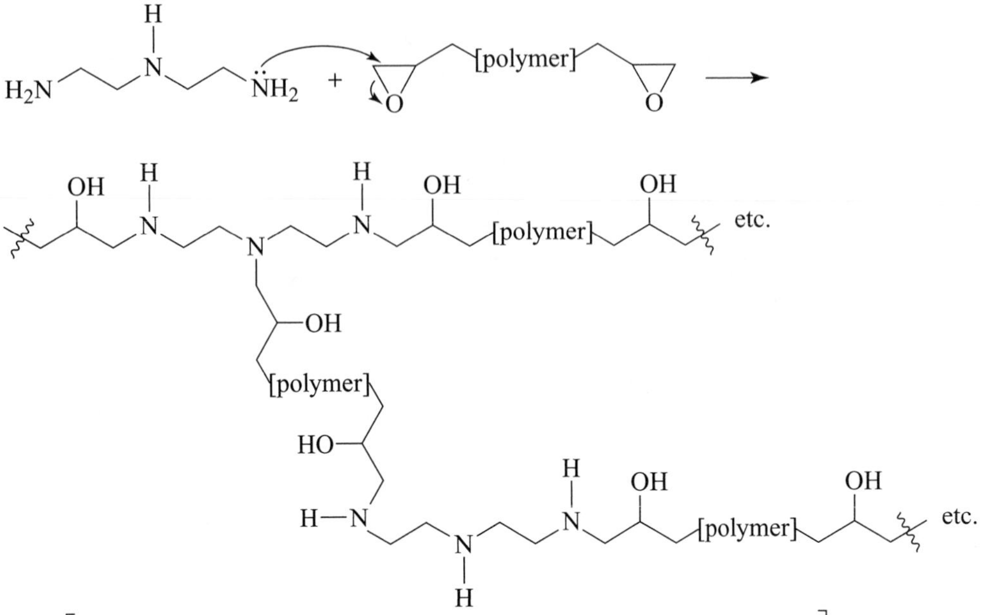

H_2N, N–H, $\ddot{N}H_2$ + [polymer] (epoxide) →

OH, H, N, N, H, N, OH, [polymer], OH, etc.
OH
[polymer]
HO
H—N, N–H, H, N, OH, [polymer], OH, etc.

C.7 (a) 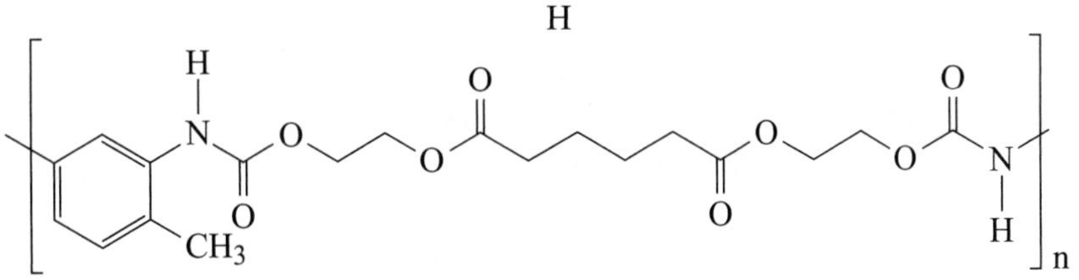 [H, N, O, O, O, O, O, O, O, O, N–H, CH_3]$_n$

(b) To ensure that the polyester chain has $-CH_2OH$ end groups.

C.8 Because the para position is occupied by a methyl group, cross-linking does not occur and the resulting polymer remains thermoplastic (See Section C.4.)

C.9

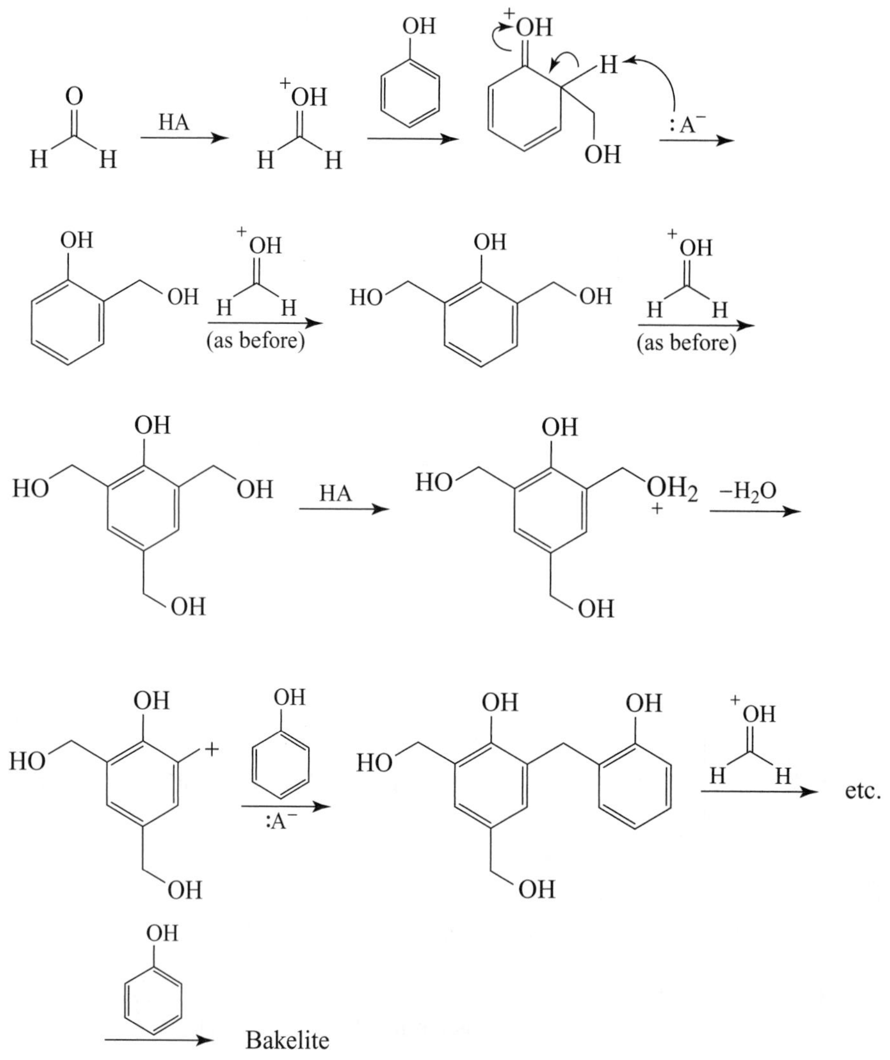

O, H, H $\xrightarrow{HA}$ $^{+}$OH, H, H $\xrightarrow{\text{phenol (OH)}}$ $^{+}$OH, H, OH $\xrightarrow{:A^-}$

OH, OH $\xrightarrow[\text{(as before)}]{^{+}\text{OH, H, H}}$ HO, OH, OH $\xrightarrow[\text{(as before)}]{^{+}\text{OH, H, H}}$

OH, HO, OH, OH $\xrightarrow{HA}$ OH, HO, $\overset{+}{O}H_2$, OH $\xrightarrow{-H_2O}$

OH, HO, +, OH $\xrightarrow[:A^-]{\text{phenol (OH)}}$ OH, OH, HO, OH $\xrightarrow{^{+}\text{OH, H, H}}$ etc.

$\xrightarrow{\text{phenol (OH)}}$ Bakelite

D SPECIAL TOPIC
Thiols, Sulfur Ylides, and Disulfides

SOLUTIONS TO PROBLEMS

D.1 (a) cyclohexanone (=O) + $CH_2{=}S(CH_3)_2$ ⟶ spiro epoxide (O) + CH_3SCH_3

(b) acetone (=O) + $CH_2{=}S(CH_3)_2$ ⟶ 2,2-dimethyloxirane (O) + CH_3SCH_3

D.2 (a) $C_6H_5CH_2\overset{+}{S}{=}C(NH_2)_2$ Br^- (b) $C_6H_5CH_2SH$

(c) $C_6H_5CH_2S{-}SCH_2C_6H_5$ (d) $C_6H_5CH_2S^-\ Na^+$

(e) $C_6H_5CH_2SCH_2C_6H_5$

D.3 $CH_2{=}CHCH_2Br$ + $S{=}C(NH_2)_2$ $\xrightarrow[\text{(2) } OH^-,\ H_2O]{\text{(1) ethanol (OH)}}$ $CH_2{=}CHCH_2SH$

$\xrightarrow{H_2O_2}$ $CH_2{=}CHCH_2S{-}SCH_2CH{=}CH_2$

D.4 $CH_2{=}CHCH_2OH$ $\xrightarrow{Br_2}$ $BrCH_2CH(Br)CH_2OH$ $\xrightarrow{NaSH}$ $HSCH_2CH(SH)CH_2OH$

D.5 (a) 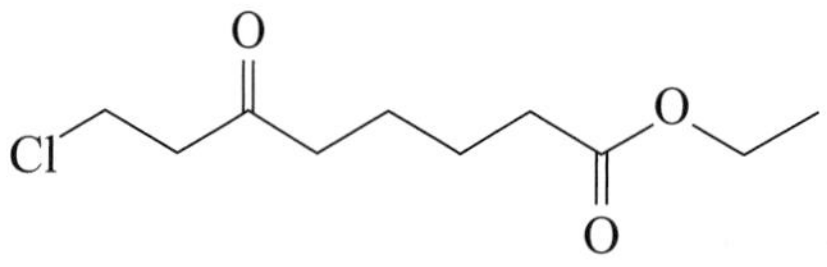 Cl–CH₂CH₂–C(=O)–CH₂CH₂CH₂CH₂–C(=O)–O–CH₂CH₃ (This step is the Friedel-Crafts alkylation of an alkene.)

(b) $SOCl_2$

(c) 2 $C_6H_5CH_2SH$ and KOH

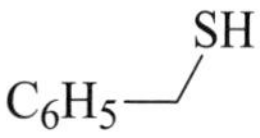

(d) H_3O^+

(e) 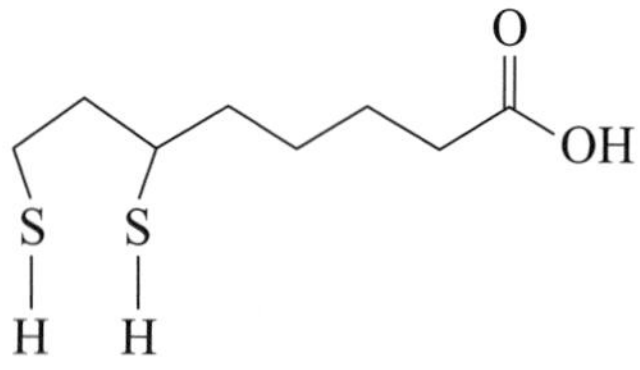 HS–CH₂CH₂–CH(SH)–CH₂CH₂CH₂CH₂–C(=O)OH

E SPECIAL TOPIC
Thiol Esters and Lipid Biosynthesis

SOLUTION TO PROBLEM

E.1

Farnesol $\xrightarrow{HA}$ $\xrightarrow{-H_2O}$ $\longleftrightarrow$ $\longrightarrow$ $\xrightarrow{:A^-}$ Bisabolene

SPECIAL TOPIC
Alkaloids

SOLUTIONS TO PROBLEMS

F.1 (a) The first step is similar to a crossed Claisen condensation (see Section 19.2B):

O OEt + O N CH_3 —ONa→ O N O N CH_3

(b) This step involves hydrolysis of an amide (lactam) and can be carried out with either acid or base. Here we use acid.

$\xrightarrow[H_2O]{H_3O^+}$ O H N CH_3 N HO O $\xrightarrow{\text{heat}}$ O H N CH_3 N

(c) This step is the decarboxylation of a β-keto acid; it requires only the application of heat and takes place during the acid hydrolysis of step (b).

(d) This is the reduction of a ketone to a secondary alcohol. A variety of reducing agents can be used, sodium borohydride, for example.

$\xrightarrow{NaBH_4}$ OH H N CH_3 N

(e) Here we convert the secondary alcohol to an alkyl bromide with hydrogen bromide; this reagent also gives a hydrobromide salt of the aliphatic amine.

$\xrightarrow[\text{heat}]{HBr}$ Br H N^+ H CH_3 Br^- N

(f) Treating the salt with base produces the secondary amine; it then acts as a nucleophile and attacks the carbon atom bearing the bromine. This reaction leads to the formation of a five-membered ring and (±) nicotine.

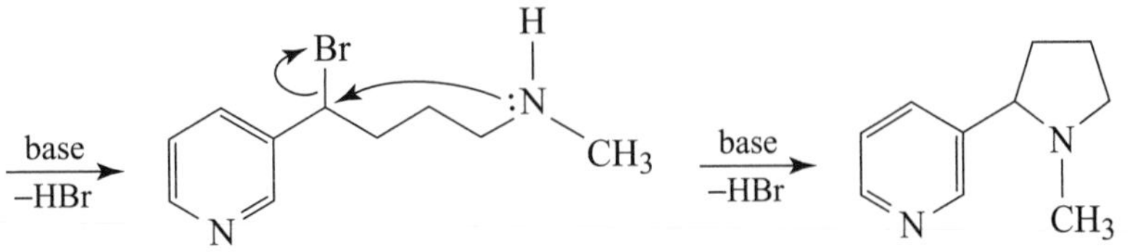

base, −HBr → (Br ... :N(H)CH$_3$ intermediate) → base, −HBr → (nicotine, N–CH$_3$)

F.2 (a) The chirality center adjacent to the ester carbonyl group is racemized by base (probably through the formation of an anion that can undergo inversion of configuration; cf. Section 18.3A).

(b) [structure: CH$_3$, N, O, H, O, OH, C$_6$H$_5$, H]

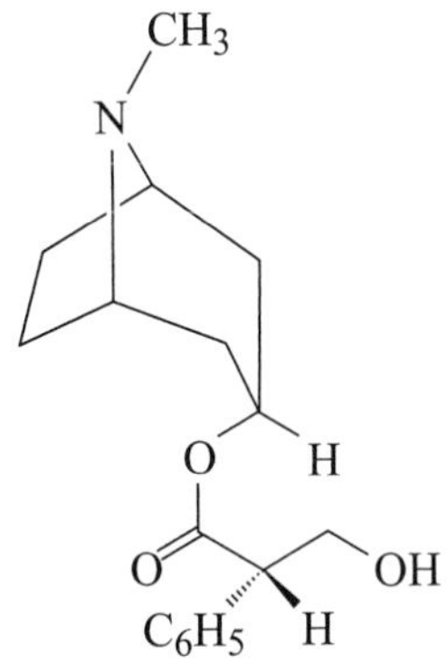

F.3 (a) [structures: CH$_3$, N, HO, H, O, OH, OH]

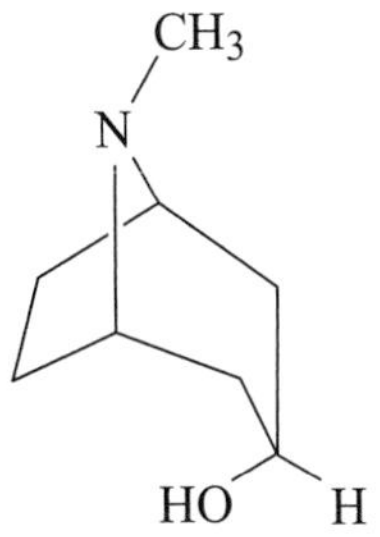

Tropine 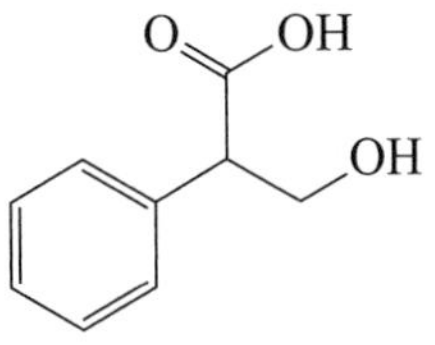 (±) Tropic acid

(b) Tropine is a meso compound; it has a plane of symmetry that passes through the >CHOH group, the >NCH$_3$ group, and between the two —CH$_2$– groups of the five-membered ring.

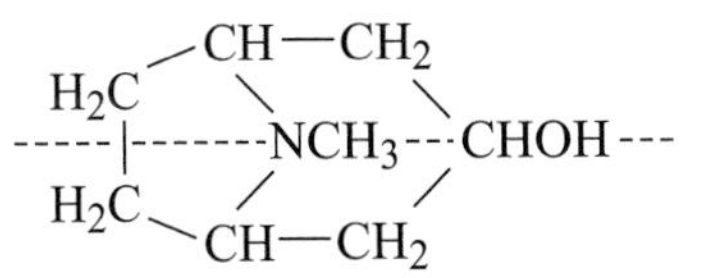

[structure: CH—CH$_2$, H$_2$C, NCH$_3$, CHOH, H$_2$C, CH—CH$_2$] plane of symmetry

(c)

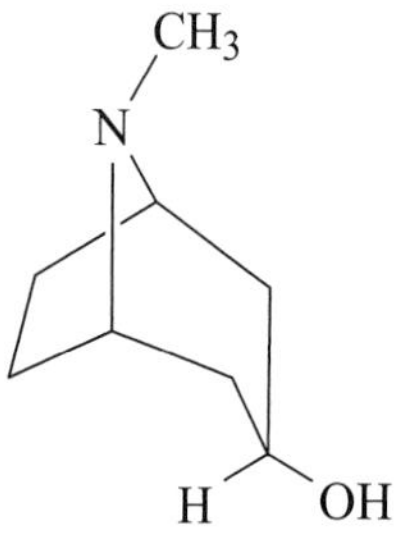

CH$_3$ N H OH

ψ-Tropine

F.4

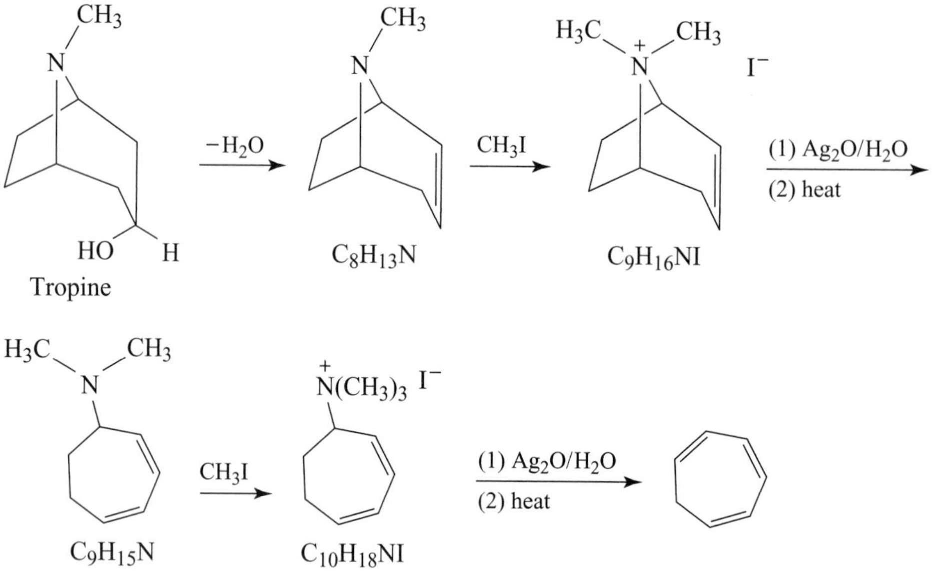

Tropine (HO, H) $\xrightarrow{-H_2O}$ $C_8H_{13}N$ $\xrightarrow{CH_3I}$ $C_9H_{16}NI$ (I^-) $\xrightarrow[(2)\ heat]{(1)\ Ag_2O/H_2O}$

$C_9H_{15}N$ $\xrightarrow{CH_3I}$ $C_{10}H_{18}NI$ ($\overset{+}{N}(CH_3)_3\ I^-$) $\xrightarrow[(2)\ heat]{(1)\ Ag_2O/H_2O}$

F.5 One possible sequence of steps is the following:

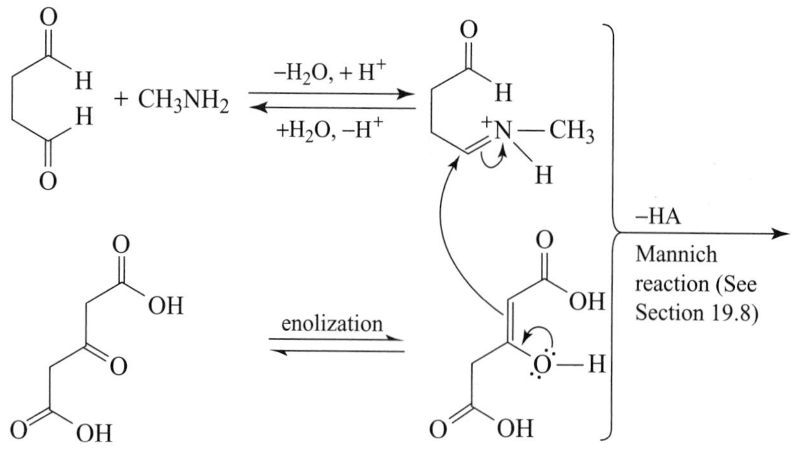

$+ CH_3NH_2$ $\underset{+H_2O,\ -H^+}{\overset{-H_2O,\ +H^+}{\rightleftharpoons}}$ (H, $^+N—CH_3$, H)

(O, OH, O, O, OH) $\underset{}{\overset{enolization}{\rightleftharpoons}}$ (O, OH, Ö—H, O, OH)

$\xrightarrow{-HA}$ Mannich reaction (See Section 19.8)

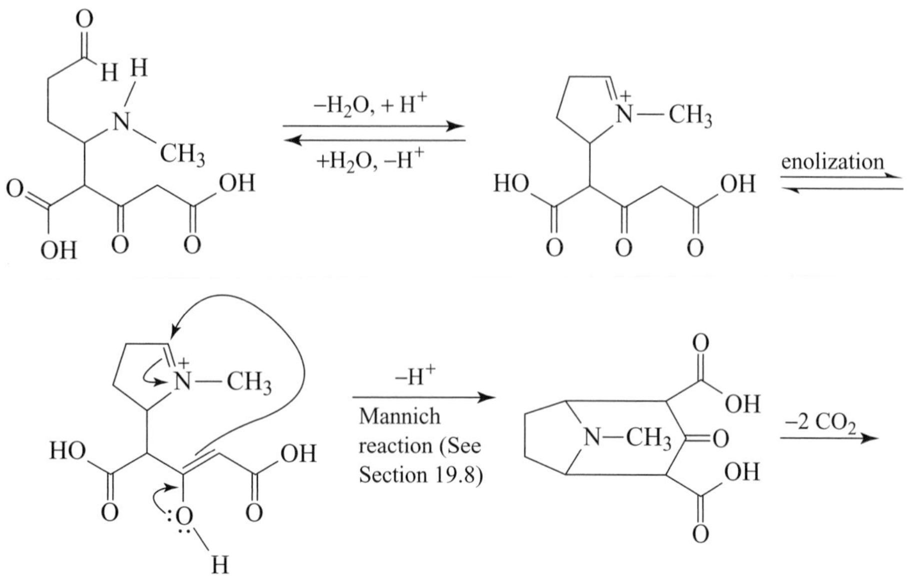

$-H_2O, + H^+$

$+H_2O, -H^+$

enolization

$-H^+$

Mannich reaction (See Section 19.8)

$-2\ CO_2$

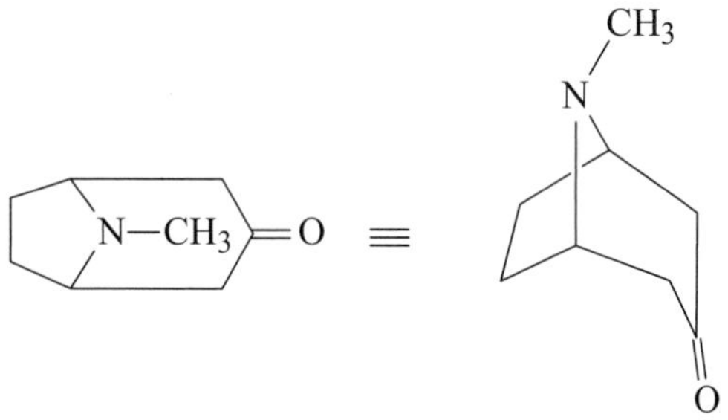

Tropinone

F.6

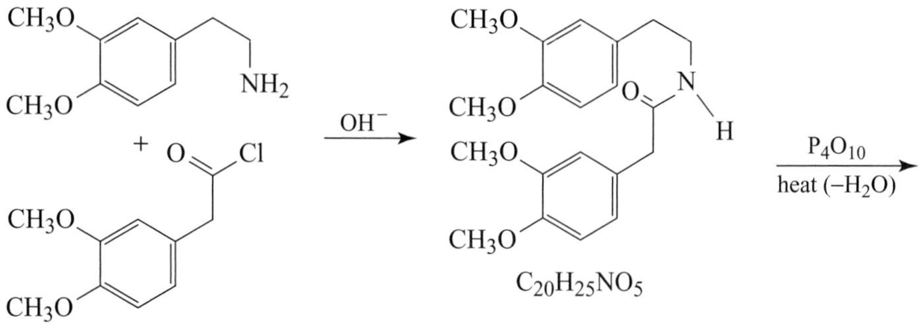

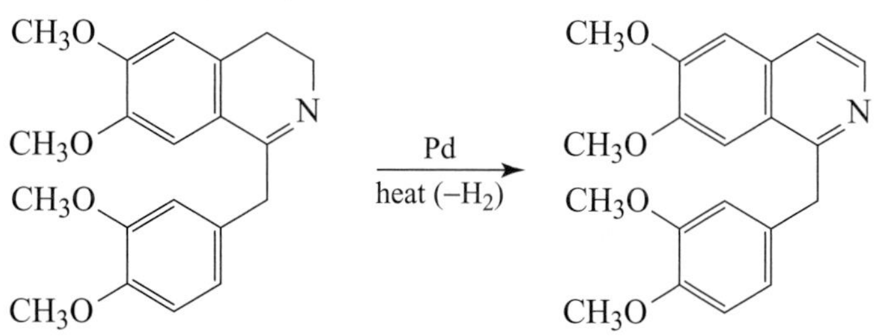

OH^-

$C_{20}H_{25}NO_5$

P_4O_{10}

heat ($-H_2O$)

Pd

heat ($-H_2$)

Dihydropapaverine

Papaverine

F.7 A Diels-Alder reaction was carried out using 1,3-butadiene as the diene component.

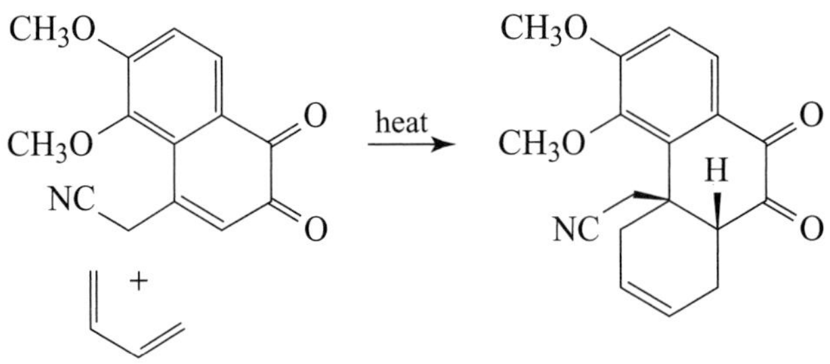

CH_3O CH_3O NC O O + heat CH_3O CH_3O H NC O O

F.8 Acetic anhydride acetylates both —OH groups.

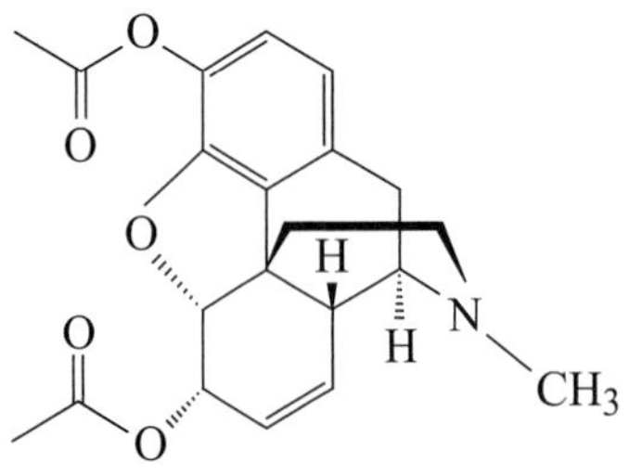

O O O O O H H N CH_3

Heroin

F.9 (a) A Mannich reaction (see Section 19.8).

(b)

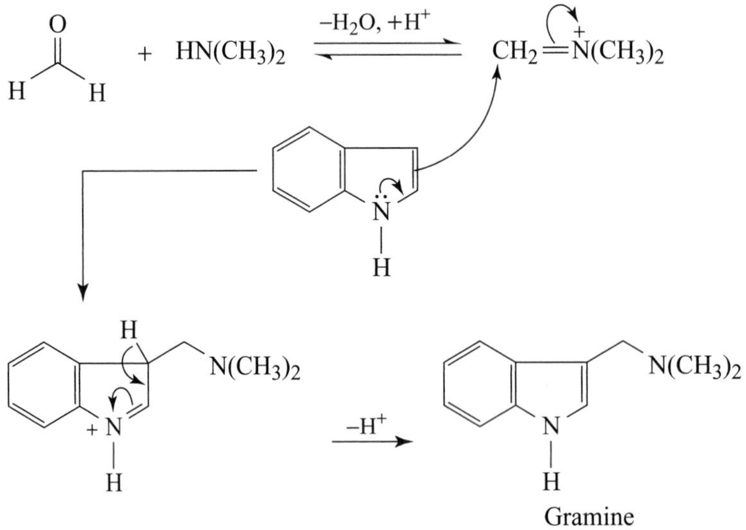

O H H + $HN(CH_3)_2$ $-H_2O, +H^+$ $CH_2{=}\overset{+}{N}(CH_3)_2$

N H

H N(CH$_3$)$_2$ +N H $-H^+$ N(CH$_3$)$_2$ N H

Gramine

SPECIAL TOPIC
Carbon-Carbon Bond-Forming and Other Reactions of Transition Metal Organometallic Compounds

G.1 (a) CO_2H CO_2CH_3 (b) H_3CO

G.2 (a) O X + X = Cl, Br or I

(b) X NC +

G.3

G.4 $B(OH)_2$ + O H Br

G.5 (a) O O (b) CF_3 *t*-Bu

(c) 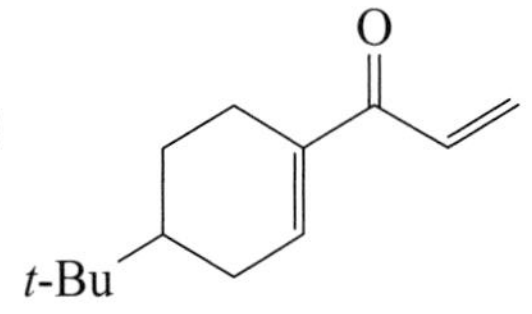 t-Bu

(d) 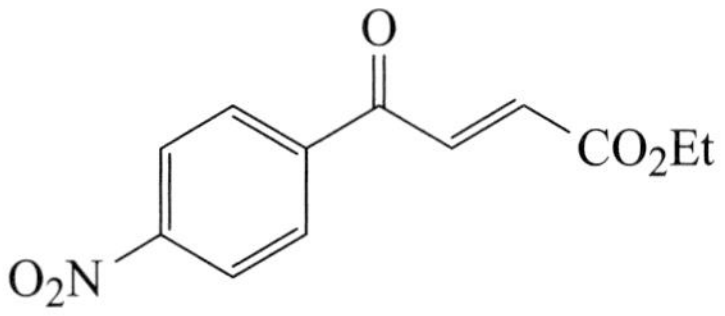 O_2N CO_2Et

G.6 (a) 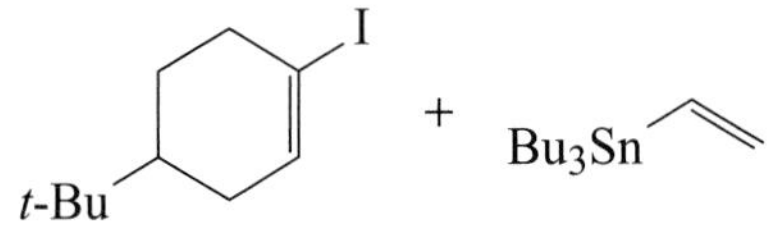 t-Bu I + Bu_3Sn

(b) 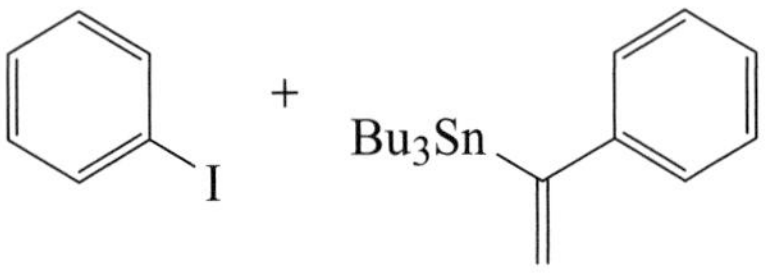 I + Bu_3Sn

G.7 (a) 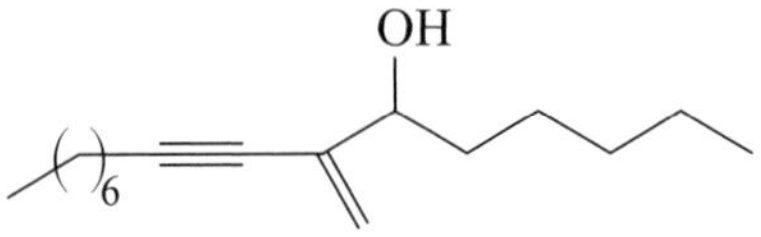 OH 6

(b) 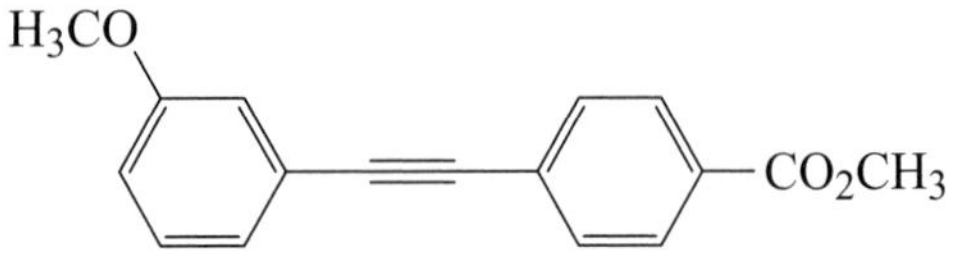 H_3CO CO_2CH_3

G.8 (a) 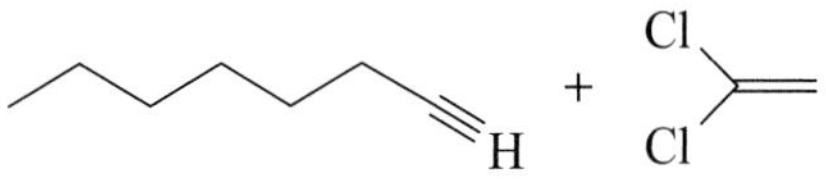 H + Cl Cl

(b) 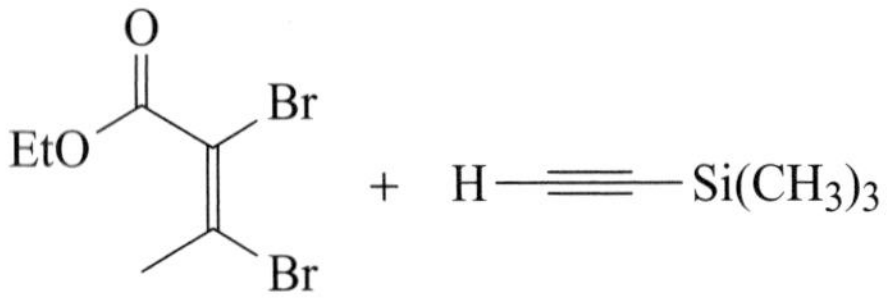 EtO O Br Br + H—≡—$Si(CH_3)_3$

G.9 (a) 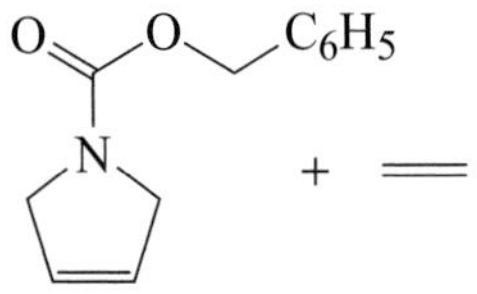 O O C_6H_5 N + =

(b) 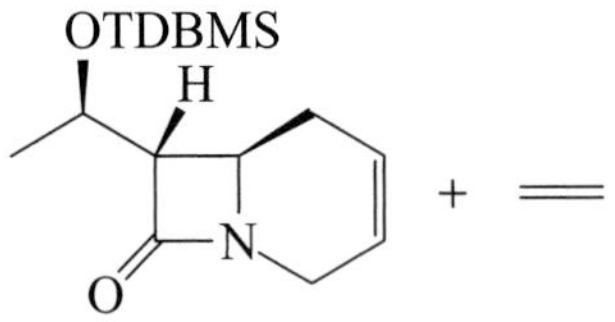 OTDBMS H N O + =

(c) O O + 2 =

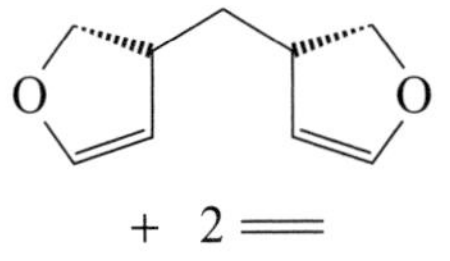

(d) 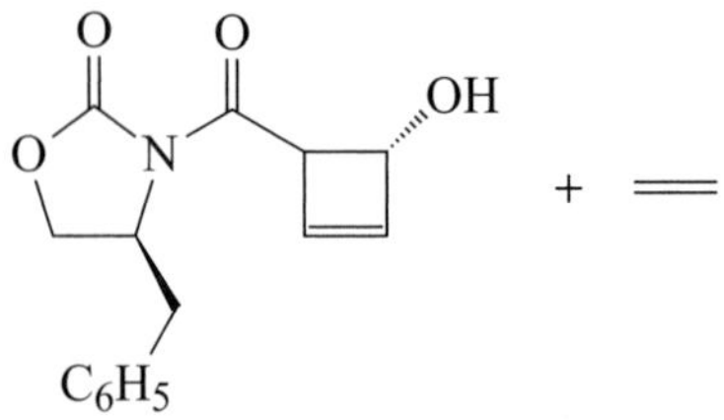 O O O N OH C_6H_5 + =

G.10 Br $\xrightarrow{\text{Li}}$ Li $\xrightarrow{\text{CuI}}$ $(\ )_2$CuLi

(a) $(\ )_2$CuLi + Br ⟶

(b) $(\ )_2$CuLi + I ⟶

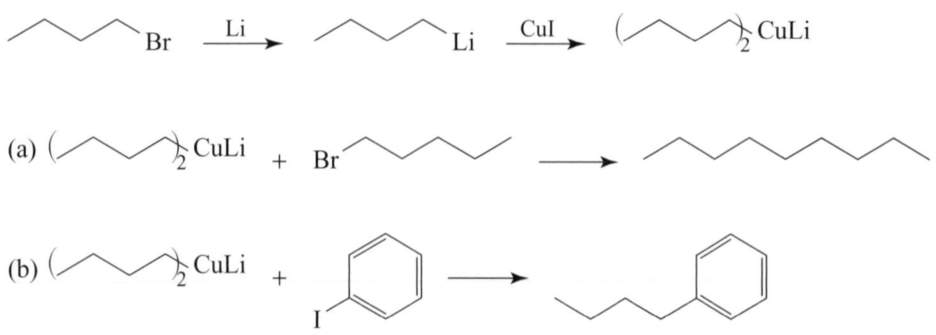

G.11 A syn addition of D_2 to the *trans* alkene would produce the following racemic form.

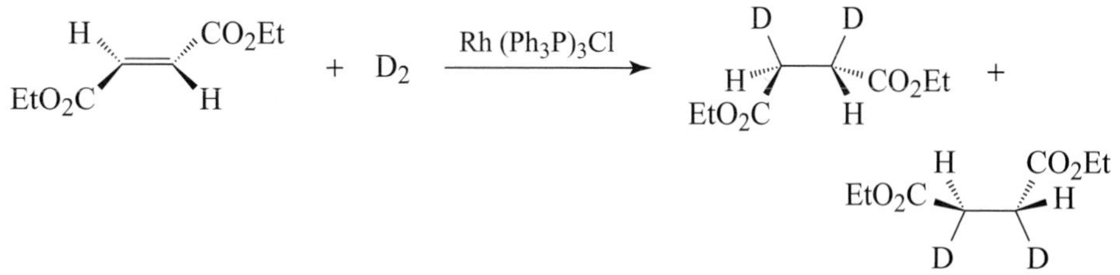

H, CO_2Et, EtO_2C, H + D_2 $\xrightarrow{\text{Rh (Ph}_3\text{P)}_3\text{Cl}}$ D, D, H, CO_2Et, EtO_2C, H + H, CO_2Et, EtO_2C, H, D, D

H SPECIAL TOPIC
Electrocyclic and Cycloaddition Reactions

SOLUTIONS TO PROBLEMS

H.1 According to the Woodward-Hoffmann rule for electrocyclic reactions of $4n$ π-electron systems (Section H.2A), the photochemical cyclization of *cis, trans*-2,4-hexadiene should proceed with *disrotatory motion*. Thus, it should yield *trans*-3,4-dimethylcyclobutene:

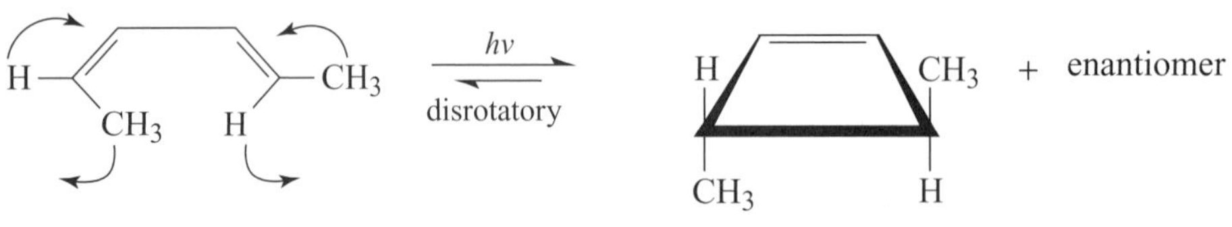

cis, trans-2,4-Hexadiene   *trans*-3,4-Dimethylcyclobutene

H.2 (a)

CH3 CH3 H3C CH3 (−) (+) (+) (−) H H $\xrightarrow{200°\,C}$ CH3 (+) (−) CH3 (+) (−) (+) (−) H3C H (−) H CH3 (+)

ψ_2 of a hexadiene
(Section H.2A)

(b) This is a thermal electrocyclic reaction of a $4n$ π-electron system; it should, *and does*, proceed with conrotatory motion.

H.3

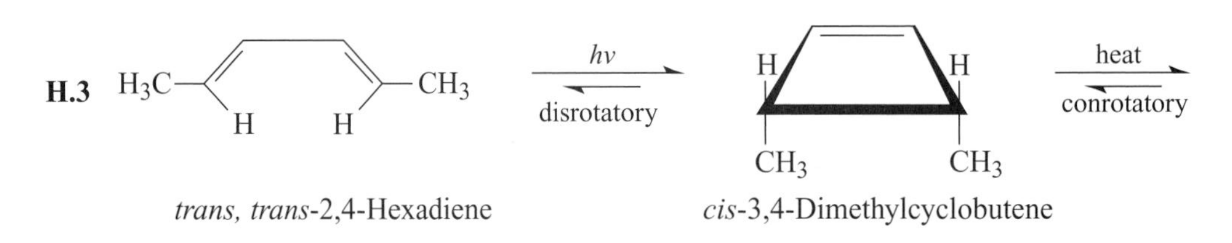

trans, trans-2,4-Hexadiene   *cis*-3,4-Dimethylcyclobutene

H CH3 CH3 H

cis, trans-2,4-Hexadiene

Here we find that two consecutive electrocyclic reactions (the first photochemical, the second thermal), provide a stereospecific synthesis of *cis, trans*-2,4-hexadiene from *trans, trans*-2,4-hexadiene.

H.4 (a) This is a photochemical electrocyclic reaction of an eight π-electron system—a $4n$ π-electron system where $n = 2$. It should, therefore, proceed with disrotatory motion.

CH_3 H H CH_3 — $h\nu$ (disrotatory) → CH_3 H H CH_3

cis-7,8-Dimethyl-1,3,5-cyclooctatriene

(b) This is a thermal electrocyclic reaction of the eight π-electron system. It should proceed with conrotatory motion.

H CH_3 H CH_3 — heat (conrotatory) — CH_3 H H CH_3

cis-7,8-Dimethyl-1,3,5-cyclooctatriene

H.5 (a) This is conrotatory motion, and since this is a 4n π-electron system (where $n = 1$) it should occur under the influence of heat.

H_3CO_2C CO_2CH_3 H H — heat (conrotatory) → H CO_2CH_3 CO_2CH_3 H

(b) This is conrotatory motion, and since this is also a 4n π-electron system (where $n = 2$) it should occur under the influence of heat.

CH_3 H H CH_3 — heat (conrotatory) — H CH_3 H_3C H

+ enantiomer

(c) This is disrotatory motion. This, too, is a 4n π-electron system (where $n = 1$); thus it should occur under the influence of light.

H H — $h\nu$ (disrotatory) → H H

H.6 (a) This is a $(4n + 2)$ π-electron system (where $n = 1$); a thermal reaction should take place with disrotatory motion:

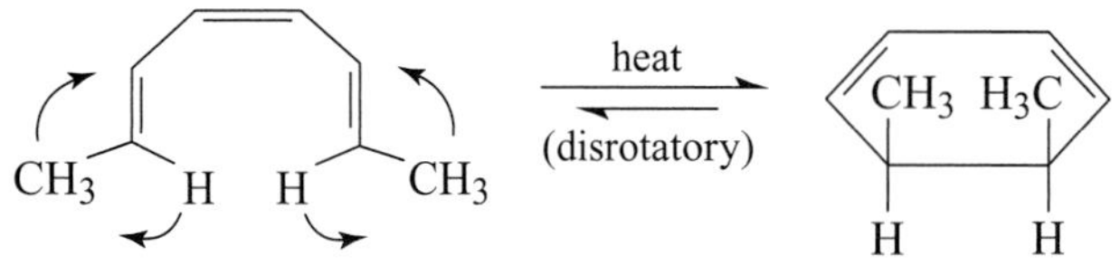

CH_3 H H CH_3 heat (disrotatory) CH_3 H_3C H H

(b) This is also a $(4n + 2)$ π-electron system; a photochemical reaction should take place with conrotatory motion.

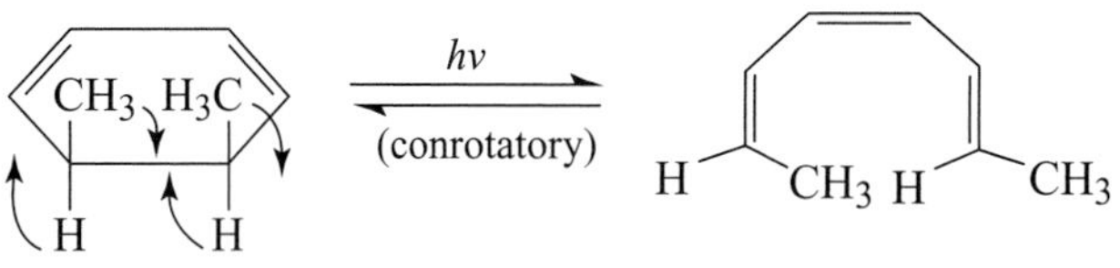

CH_3 H_3C H H $h\nu$ (conrotatory) H CH_3 H CH_3

H.7 Here we need a conrotatory ring opening of *trans*-5,6-dimethyl-1,3-cyclohexadiene (to produce *trans,cis,trans*-2,4,6-octatriene); then we need a disrotatory cyclization to produce *cis*-5,6-dimethyl-1,3-cyclohexadiene.

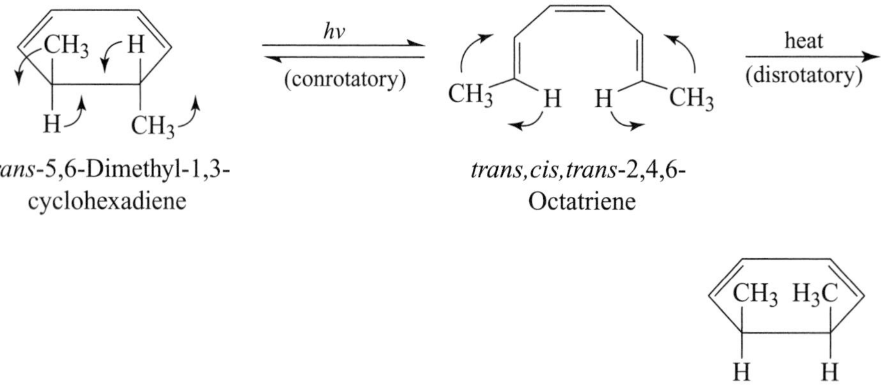

CH_3 H H CH_3 $h\nu$ (conrotatory) CH_3 H H CH_3 heat (disrotatory)

trans-5,6-Dimethyl-1,3-cyclohexadiene

trans,cis,trans-2,4,6-Octatriene

CH_3 H_3C H H

cis-5,6-Dimethyl-1,3-cyclohexadiene

Since both reactions involve $(4n + 2)$ π-electron systems, we apply light to accomplish the first step and heat to accomplish the second. It would also be possible to use heat to produce *trans,cis,cis*-2,4,6-octatriene and then use light to produce the desired product.

H.8 The first electrocyclic reaction is a thermal, conrotatory ring opening of a $4n$ π-electron system. The second electrocyclic reaction is a thermal, disrotatory ring closure of a $(4n + 2)$ π-electron system.

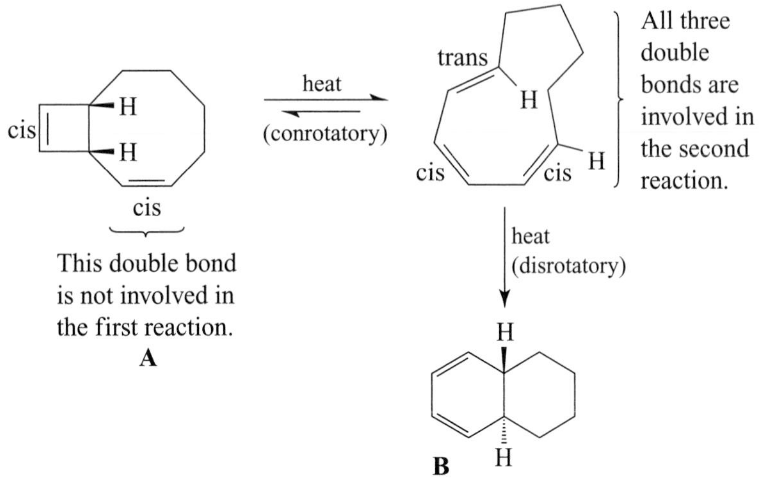

cis H H cis

This double bond is not involved in the first reaction.

A

heat

(conrotatory)

trans H H cis cis

All three double bonds are involved in the second reaction.

heat

(disrotatory)

H H

B

H.9 (a) There are two possible products that can result from a concerted cycloaddition. They are formed when *cis*-2-butene molecules come together in the following ways:

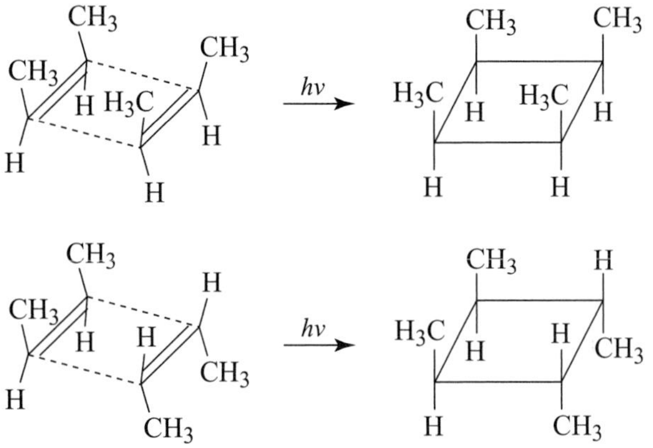

CH_3 CH_3 CH_3 H H_3C H H H $\xrightarrow{h\nu}$ CH_3 CH_3 H_3C H H_3C H H H

CH_3 H CH_3 H H CH_3 H CH_3 $\xrightarrow{h\nu}$ CH_3 H H_3C H H CH_3 H CH_3

(b) There are two possible products that can be obtained from *trans*-2-butene as well.

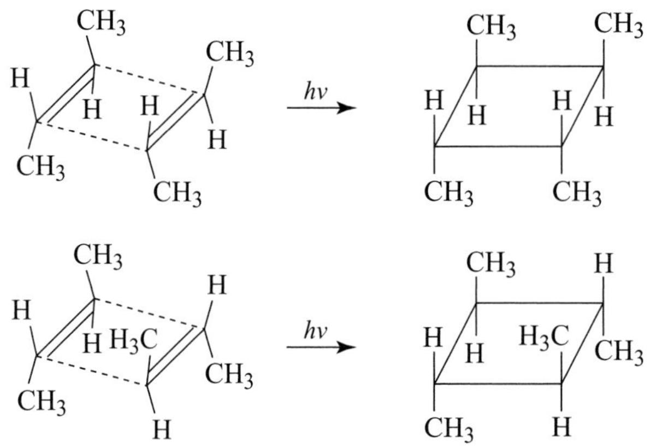

CH_3 CH_3 H H H H CH_3 CH_3 $\xrightarrow{h\nu}$ CH_3 CH_3 H H H H CH_3 CH_3

CH_3 H H H H_3C CH_3 CH_3 H $\xrightarrow{h\nu}$ CH_3 H H H H_3C CH_3 CH_3 H

H.10 This is an intramolecular [2 + 2] cycloaddition.

H.11 (a) CN CN H_3C CH_3 CN CN H H

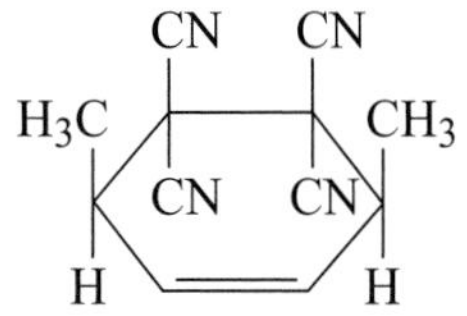

(b) CN CN H_3C H CN CN H CH_3

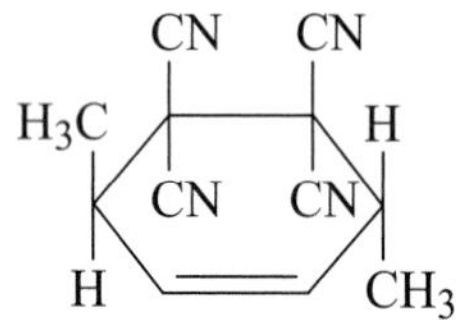

+
Enantiomer

A APPENDIX
Empirical and Molecular Formulas

In the early and middle 19th century methods for determining formulas for organic compounds were devised by J. J. Berzelius, J. B. A. Dumas, Justus Liebig and Stanislao Cannizzaro. Although the experimental procedures for these analyses have been refined, the basic methods for determining the elemental composition of an organic compound today are not substantially different from those used in the nineteenth century. A carefully weighed quantity of the compound to be analyzed is oxidized completely to carbon dioxide and water. The weights of carbon dioxide and water are carefully measured and used to find the percentages of carbon and hydrogen in the compound. The percentage of nitrogen is usually determined by measuring the volume of nitrogen (N_2) produced in a separate procedure.

Special techniques for determining the percentage composition of other elements typically found in organic compounds have also been developed, but the direct determination of the percentage of oxygen is difficult. However, if the percentage composition of all the other elements is known, then the percentage of oxygen can be determined by difference. The following examples will illustrate how these calculations can be carried out.

EXAMPLE A

A new organic compound is found to have the following elemental analysis.

Carbon	67.95%
Hydrogen	5.69
Nitrogen	26.20
Total:	99.84%

Since the total of these percentages is very close to 100% (within experimental error), we can assume that no other element is present. For the purpose of our calculation it is convenient to assume that we have a 100-g sample. If we did, it would contain the following:

67.95 g of carbon
5.69 g of hydrogen
26.20 g of nitrogen

In other words, we use percentages *by weight* to give us the ratios *by weight* of the elements in the substance. To write a formula for the substance, however, we need *ratios by moles.*

We now divide each of these weight-ratio numbers by the atomic weight of the particular element and obtain the number of moles of each element, respectively, in 100 g of the compound. This operation gives us the ratios *by moles* of the elements in the substance:

$$\text{C} \quad \frac{67.95 \text{ g}}{12.01 \text{ g mol}^{-1}} = 5.66 \text{ mol}$$

$$\text{H} \quad \frac{5.69 \text{ g}}{1.008 \text{ g mol}^{-1}} = 5.64 \text{ mol}$$

$$\text{N} \quad \frac{26.20 \text{ g}}{14.01 \text{ g mol}^{-1}} = 1.87 \text{ mol}$$

One possible formula for the compound, therefore, is $C_{5.66}H_{5.64}N_{1.87}$.

By convention, however, we use *whole* numbers in formulas. Therefore, we convert these fractional numbers of moles to whole numbers by dividing each by 1.87, the smallest number.

$$\text{C} \quad \frac{5.66}{1.87} = 3.03 \text{ which is} \sim 3$$

$$\text{H} \quad \frac{5.64}{1.87} = 3.02 \text{ which is} \sim 3$$

$$\text{N} \quad \frac{1.87}{1.87} = 1.00$$

Thus, within experimental error, the ratios by moles are 3 C to 3 H to 1 N, and C_3H_3N is the *empirical formula.* By empirical formula, we mean the formula in which the subscripts are the smallest integers that give the ratio of atoms in the compound. In contrast, a *molecular* formula discloses the complete composition of one molecule. The molecular formula of this particular compound could be C_3H_3N or some whole number multiple of C_3H_3N; that is, $C_6H_6N_2$, $C_9H_9N_3$, $C_{12}H_{12}N_4$, and so on. If, in a separate determination, we find that the molecular weight of the compound is 108 ± 3, we can be certain that the *molecular formula* of the compound is $C_6H_6N_2$.

FORMULA	MOLECULAR WEIGHT
C_3H_3N	53.06
$C_6H_6N_2$	106.13 (which is within the range 108 $\pm$3)
$C_9H_9N_3$	159.19
$C_{12}H_{12}N_4$	212.26

The most accurate method for determining molecular weights is by high-resolution mass spectrometry (Section 9.17C). A variety of other methods based on freezing point depression, boiling point elevation, osmotic pressure, and vapor density can also be used to determine molecular weights.

EXAMPLE B

Histidine, an amino acid isolated from protein, has the following elemental analysis:

Carbon	46.38%
Hydrogen	5.90
Nitrogen	27.01
Total:	79.29
Difference	20.71 (assumed to be oxygen)
	100.00%

Since no elements, other than carbon, hydrogen, and nitrogen, are found to be present in histidine, the difference is assumed to be oxygen. Again, we assume a 100-g sample and divide the weight of each element by its gram-atomic weight. This gives us the ratio of moles (A).

	(A)	(B)	(C)
C	$\frac{46.38}{12.01} = 3.86$	$\frac{3.86}{1.29} = 2.99$	$\times\ 2 = 5.98 \sim 6$ carbon atoms
H	$\frac{5.90}{1.008} = 5.85$	$\frac{5.85}{1.29} = 4.53$	$\times\ 2 = 9.06 \sim 9$ hydrogen atoms
N	$\frac{27.01}{14.01} = 1.93$	$\frac{1.93}{1.29} = 1.50$	$\times\ 2 = 3.00 = 3$ nitrogen atoms
O	$\frac{20.71}{16.00} = 1.29$	$\frac{1.29}{1.29} = 1.00$	$\times\ 2 = 2.00 = 2$ oxygen atoms

Dividing each of the moles (A) by the smallest of them does not give a set of numbers (B) that is close to a set of whole numbers. Multiplying each of the numbers in column (B) by 2 does, however, as seen in column (C). The empirical formula of histidine is, therefore, $C_6H_9N_3O_2$.

In a separate determination, the molecular weight of histidine was found to be 158 ± 5. The empirical formula weight of $C_6H_9N_3O_2$ (155.15) is within this range; thus, the molecular formula for histidine is the same as the empirical formula.

PROBLEMS

A.1 What is the empirical formula of each of the following compounds?

(a) Hydrazine, N_2H_4

(b) Benzene, C_6H_6

(c) Dioxane, $C_4H_8O_2$

(d) Nicotine, $C_{10}H_{14}N_2$

(e) Cyclodecane, $C_{10}H_{20}$

(f) Acetylene, C_2H_2

A.2 The empirical formulas and molecular weights of several compounds are given next. In each case, calculate the molecular formula for the compound.

EMPIRICAL FORMULA	MOLECULAR WEIGHT
(a) CH_2O	179 ± 5
(b) CHN	80 ± 5
(c) CCl_2	410 ± 10

A.3 The widely used antibiotic, penicillin G, gave the following elemental analysis: C, 57.45%; H, 5.40%; N, 8.45%; S, 9.61%. The molecular weight of penicillin G is 330 ± 10. Assume that no other elements except oxygen are present and calculate the empirical and molecular formulas for penicillin G.

ADDITIONAL PROBLEMS

A.4 Calculate the percentage composition of each of the following compounds.

(a) $C_6H_{12}O_6$

(b) $CH_3CH_2NO_2$

(c) $CH_3CH_2CBr_3$

A.5 An organometallic compound called *ferrocene* contains 30.02% iron. What is the minimum molecular weight of ferrocene?

A.6 A gaseous compound gave the following analysis: C, 40.04%; H, 6.69%. At standard temperature and pressure, 1.00 g of the gas occupied a volume of 746 mL. What is the molecular formula of the compound?

A.7 A gaseous hydrocarbon has a density of 1.251 g L^{-1} at standard temperature and pressure. When subjected to complete combustion, a 1.000-L sample of the hydrocarbon gave 3.926 g of carbon dioxide and 1.608 g of water. What is the molecular formula for the hydrocarbon?

A.8 Nicotinamide, a vitamin that prevents the occurrence of pellagra, gave the following analysis: C, 59.10%; H, 4.92%; N, 22.91%. The molecular weight of nicotinamide was shown in a separate determination to be 120 ± 5. What is the molecular formula for nicotinamide?

A.9 The antibiotic chloramphenicol gave the following analysis: C, 40.88%; H, 3.74%; Cl, 21.95%; N, 8.67%. The molecular weight was found to be 300 ± 30. What is the molecular formula for chloramphenicol?

SOLUTIONS TO PROBLEMS OF APPENDIX A

A.1 (a) NH_2 (b) CH (c) C_2H_4O (d) C_5H_7N (e) CH_2 (f) CH

A.2

	EMPIRICAL FORMULA	EMPIRICAL FORMULA WEIGHT	$\left(\frac{\text{MOLECULAR WEIGHT}}{\text{EMP. FORM. WT.}}\right)$	MOLECULAR FORMULA
(a)	CH_2O	30	$\frac{179}{30} \cong 6$	$C_6H_{12}O_6$
(b)	CHN	27	$\frac{80}{27} \cong 3$	$C_3H_3N_3$
(c)	CCl_2	83	$\frac{410}{83} \cong 5$	C_5Cl_{10}

A.3 If we assume that we have a 100-g sample, the amounts of the elements are

	WEIGHT	Moles (A)	(B)
C	57.45	$\frac{57.45}{12.01} = 4.78$	$\frac{4.78}{0.300} = 15.9 \cong 16$
H	5.40	$\frac{5.40}{1.008} = 5.36$	$\frac{5.36}{0.300} = 17.9 \cong 18$
N	8.45	$\frac{8.45}{14.01} = 0.603$	$\frac{0.603}{0.300} = 2.01 \cong 2$
S	9.61	$\frac{9.61}{32.06} = 0.300$	$\frac{0.300}{0.300} = 1.00 = 1$
O*	19.09	$\frac{19.09}{16.00} = 1.19$	$\frac{1.19}{0.300} = 3.97 \cong 4$
	100.00		

(* by difference from 100)

The empirical formula is thus $C_{16}H_{18}N_2SO_4$. The empirical formula weight (334.4) is within the range given for the molecular weight (330 ± 10). Thus, the molecular formula for penicillin G is the same as the empirical formula.

A.4 (a) To calculate the percentage composition from the molecular formula, first determine the weight of each element in 1 mol of the compound. For $C_6H_{12}O_6$,

$$C_6 = 6 \times 12.01 = 72.06 \qquad \frac{72.06}{180.2} = 0.400 = 40.0\%$$

$$H_{12} = 12 \times 1.008 = 12.10 \qquad \frac{12.10}{180.2} = 0.0671 = 6.7\%$$

$$O_6 = 6 \times 16.00 = 96.00 \qquad \frac{96.00}{180.2} = 0.533 = 53.3\%$$

$$\text{MW} \qquad 180.16$$

(MW = molecular weight)

Then determine the percentage of each element using the formula

$$\text{Percentage of A} = \frac{\text{Weight of A}}{\text{Molecular Weight}} \times 100$$

(b)

$$\begin{aligned}
C_2 &= 2 \times 12.01 = 24.02 & \frac{24.02}{75.07} &= 0.320 = 32.0\% \\
H_5 &= 5 \times 1.008 = 5.04 & \frac{5.04}{75.07} &= 0.067 = 6.7\% \\
N &= 1 \times 14.01 = 14.01 & \frac{14.01}{75.07} &= 0.187 = 18.7\% \\
O_2 &= 2 \times 16.00 = \underline{32.00} & \frac{32.00}{75.07} &= 0.426 = 42.6\% \\
\text{Total} &= 75.07
\end{aligned}$$

(c)

$$\begin{aligned}
C_3 &= 3 \times 12.01 = 36.03 & \frac{36.03}{280.77} &= 0.128 = 12.8\% \\
H_5 &= 5 \times 1.008 = 5.04 & \frac{5.04}{280.77} &= 0.018 = 1.8\% \\
Br_3 &= 3 \times 79.90 = \underline{239.70} & \frac{239.70}{280.77} &= 0.854 = 85.4\% \\
\text{Total} &= 280.77
\end{aligned}$$

A.5 If the compound contains iron, each molecule must contain at least one atom of iron, and 1 mol of the compound must contain at least 55.85 g of iron. Therefore,

$$\begin{aligned}
\text{MW of ferrocene} &= 55.85 \, \frac{\text{g of Fe}}{\text{mol}} \times \frac{1.000 \text{ g}}{0.3002 \text{ g of Fe}} \\
&= 186.0 \, \frac{\text{g}}{\text{mol}}
\end{aligned}$$

A.6 First, we must determine the empirical formula. Assuming that the difference between the percentages given and 100% is due to oxygen, we calculate:

$$\begin{aligned}
&\text{C} \quad 40.04 & \frac{40.04}{12.01} &= 3.33 & \frac{3.33}{3.33} &= 1 \\
&\text{H} \quad 6.69 & \frac{6.69}{1.008} &= 6.64 & \frac{6.64}{3.33} &\cong 2 \\
&\text{O} \quad \underline{53.27} & \frac{53.27}{16.00} &= 3.33 & \frac{3.33}{3.33} &= 1 \\
&\quad\quad 100.00
\end{aligned}$$

The empirical formula is thus CH_2O.

To determine the molecular formula, we must first determine the molecular weight. At standard temperature and pressure, the volume of 1 mol of an ideal gas is 22.4 L. Assuming ideal behavior,

$$\frac{1.00 \text{ g}}{0.746 \text{ L}} = \frac{\text{MW}}{22.4 \text{ L}} \quad \text{where MW = molecular weight}$$

$$\text{MW} = \frac{(1.00)(22.4)}{0.746} = 30.0 \text{ g}$$

The empirical formula weight (30.0) equals the molecular weight; thus, the molecular formula is the same as the empirical formula.

A.7 As in Problem A.6, the molecular weight is found by the equation

$$\frac{1.251 \text{ g}}{1.00 \text{ L}} = \frac{\text{MW}}{22.4 \text{ L}}$$

$$\text{MW} = (1.251)(22.4)$$

$$\text{MW} = 28.02$$

To determine the empirical formula, we must determine the amount of carbon in 3.926 g of carbon dioxide, and the amount of hydrogen in 1.608 g of water.

$$\text{C} \quad \left(3.926 \, \cancel{\text{g CO}_2}\right)\left(\frac{12.01 \text{g C}}{44.01 \, \cancel{\text{g CO}_2}}\right) = 1.071 \text{ g carbon}$$

$$\text{H} \quad \left(1.608 \, \cancel{\text{g H}_2\text{O}}\right)\left(\frac{2.016 \text{ g H}}{18.016 \, \cancel{\text{g H}_2\text{O}}}\right) = \frac{0.180 \text{ g hydrogen}}{1.251 \text{ g sample}}$$

The weight of C and H in a 1.251-g sample is 1.251 g. Therefore, there are no other elements present.

To determine the empirical formula, we proceed as in Problem A.6 except that the sample size is 1.251 g instead of 100 g.

$$\text{C} \quad \frac{1.071}{12.01} = 0.0892 \qquad \frac{0.0892}{0.0892} = 1$$

$$\text{H} \quad \frac{0.180}{1.008} = 0.179 \qquad \frac{0.179}{0.0892} = 2$$

The empirical formula is thus CH_2. The empirical formula weight (14) is one-half the molecular weight. Thus, the molecular formula is C_2H_4.

A.8 Use the procedure of Problem A.3.

$$\text{C} \quad 59.10 \quad \frac{59.10}{12.01} = 4.92 \qquad \frac{4.92}{0.817} = 6.02 \cong 6$$

$$\text{H} \quad 4.92 \quad \frac{4.92}{1.008} = 4.88 \qquad \frac{4.88}{0.817} = 5.97 \cong 6$$

N	22.91	$\frac{22.91}{14.01} = 1.64$	$\frac{1.64}{0.817} = 2$	
O	13.07	$\frac{13.07}{16.00} = 0.817$	$\frac{0.817}{0.817} = 1$	
	100.00			

The empirical formula is thus $C_6H_6N_2O$. The empirical formula weight is 122.13, which is equal to the molecular weight within experimental error. The molecular formula is thus the same as the empirical formula.

A.9

C	40.88	$\frac{40.88}{12.01} = 3.40$	$\frac{3.40}{0.619} = 5.5$	$5.5 \times 2 = 11$
H	3.74	$\frac{3.74}{1.008} = 3.71$	$\frac{3.71}{0.619} = 6$	$6 \times 2 = 12$
Cl	21.95	$\frac{21.95}{35.45} = 0.619$	$\frac{0.619}{0.619} = 1$	$1 \times 2 = 2$
N	8.67	$\frac{8.67}{14.01} = 0.619$	$\frac{0.619}{0.619} = 1$	$1 \times 2 = 2$
O	24.76	$\frac{24.76}{16.00} = 1.55$	$\frac{1.55}{0.619} = 2.5$	$2.5 \times 2 = 5$
	100.00			

The empirical formula is thus $C_{11}H_{12}Cl_2N_2O_5$. The empirical formula weight (323) is equal to the molecular weight; therefore, the molecular formula is the same as the empirical formula.

B APPENDIX Answers to Quizzes

EXERCISE 1

1.1 (d) **1.2** (d) **1.3** (e) **1.4** (d) **1.5** (c) **1.6** $CH_3-C(=\ddot{O}:)-\ddot{O}:^-$

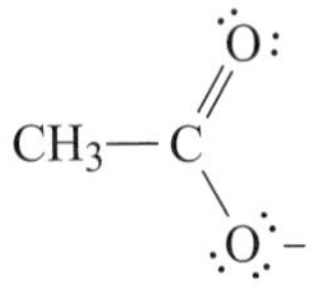

1.7 $CH_3CH(CH_3)CH_2CH_3$ and $CH_3-C(CH_3)_2-CH_3$

1.8 H, H, H, C, C, H, Cl, CH_3

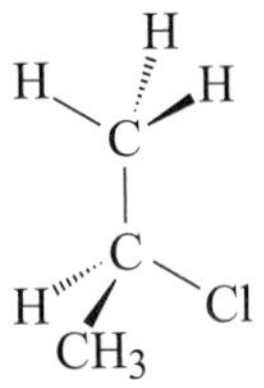

1.9 (a) sp^2 (b) sp^3 (c) 0 (d) trigonal planar (e) 0 D

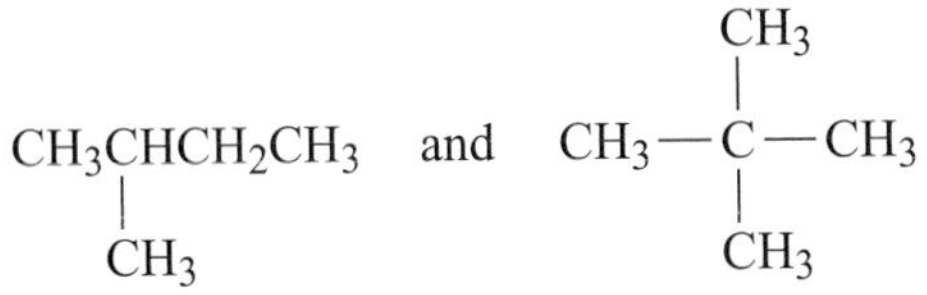

1.10 (a) + 1 (b) 0 (c) − 1

1.11 O, C, H, N, H, H

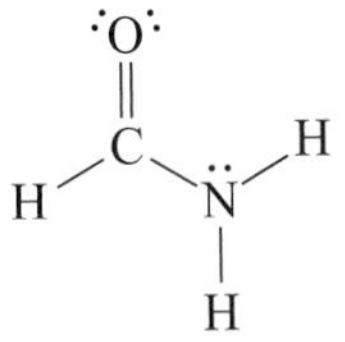

1.12 H_3C, Cl, C, H_3C, F

1.13 OH, OH, OH, OH

unstable

OCH_3, O, O

O, H, O, OH

unstable

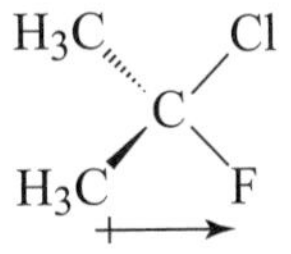

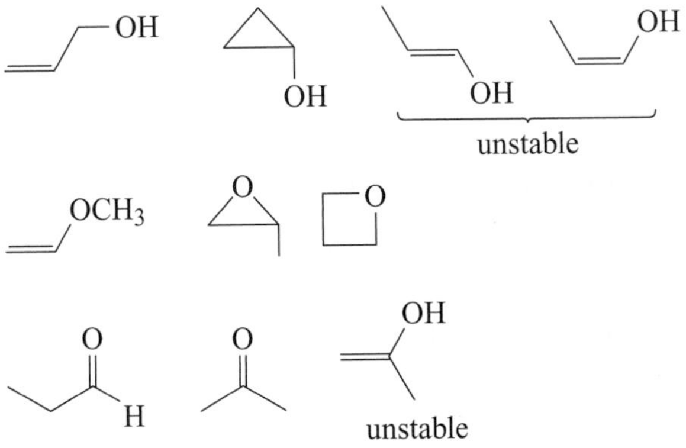

EXERCISE 2

2.1 (e) **2.2** (a) **2.3** (b)

2.4 (a) 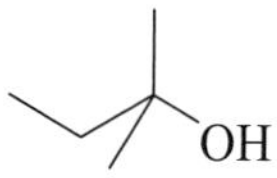 OH (b) 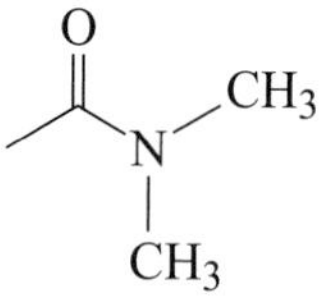 O, N, CH_3, CH_3 (c) 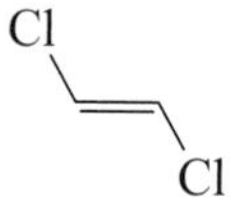 Cl, Cl

(d) 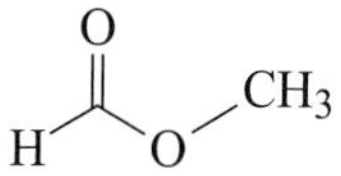 O, H, O, CH_3 (e) 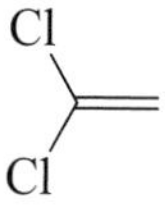 Cl, Cl (f) 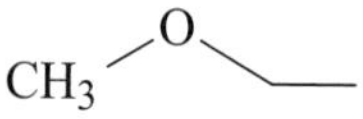 CH_3, O

(g) 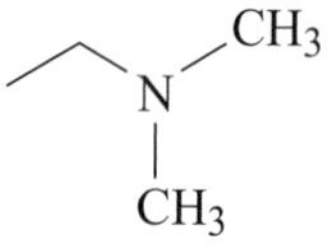 N, CH_3, CH_3

2.5 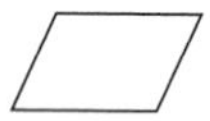 or

2.6 (a)  OH (b) 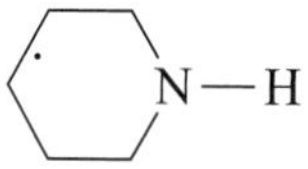 N—H (c) 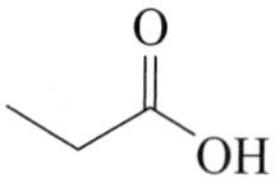 O, OH

(d) 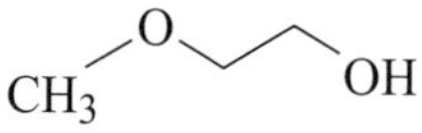 CH_3, O, OH (e) 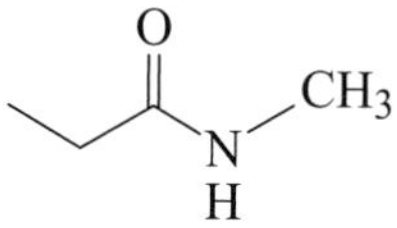 O, N, H, CH_3

2.7 (a) Isopropyl phenyl ether

(b) Ethylmethylphenylamine

(c) Isopropylamine

EXERCISE 3

3.1 (a) **3.2** (c) **3.3** (b) **3.4** (e) **3.5** (b)

3.6 (b)

3.7 $H_2SO_4 + NaF \longrightarrow NaHSO_4 + HF$

3.8 $(CH_3)_2NH$

3.9 (a) CH_3CH_2Li (b) D_2O

3.10 (a) CH_3Li (b) $CH_3CH(CH_3)CH_2OH$ (c) $CH_3CH(CH_3)CH_2OLi$

EXERCISE 4

4.1 (c) **4.2** (c) **4.3** (b) **4.4** (a) **4.5** (b)

4.6 (a) **4.7** (a)

4.8 (a) 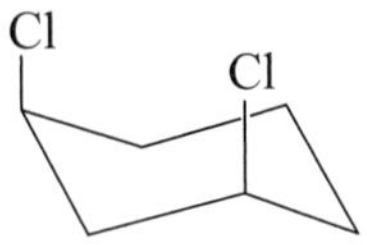 Cl, Cl (b) 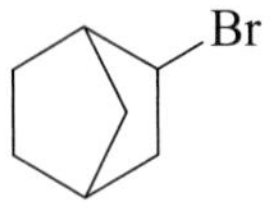 Br or Br

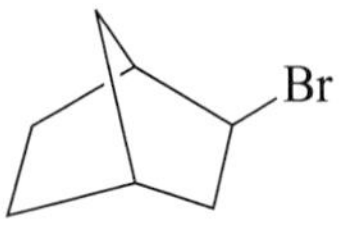

(c) 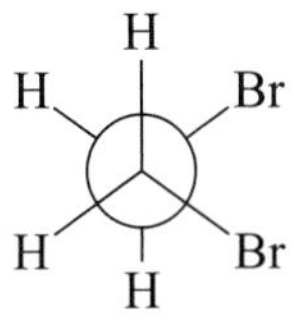 H, H, Br, H, Br, H

4.9 H_2, Pt, pressure or H_2, Ni, pressure

4.10 $CH_3-CH(CH_3)$ — CH_3

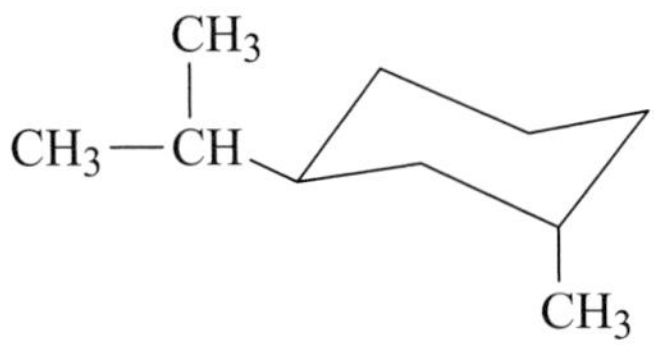

EXERCISE 5

5.1 (a) **5.2** (b) **5.3** (b) **5.4** (e) **5.5** (b)

5.6 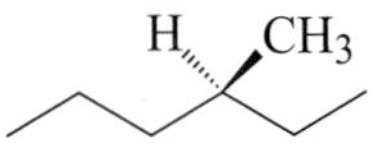 H, CH_3 and H, CH_3

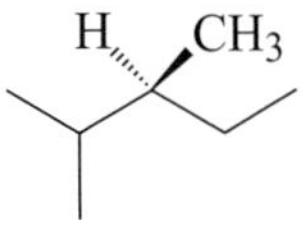

5.7 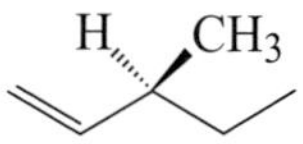 H, CH_3 **5.8** 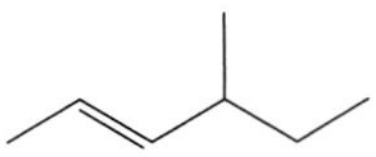 (*E* and *Z*)

5.9 (d) **5.10** (e)

EXERCISE 6

6.1 (b) **6.2** (b) **6.3** (a)

6.4 Br Br Br Br

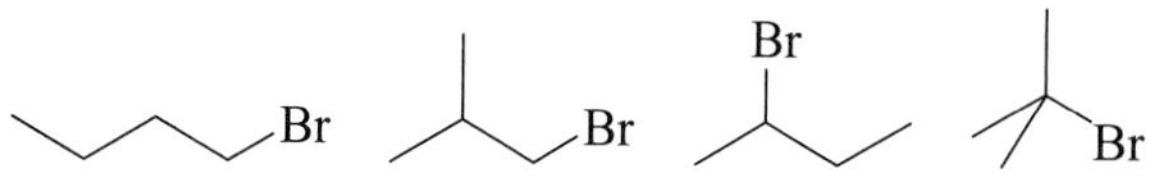

6.5 **A** = H Br **B** =

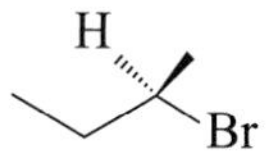

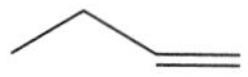

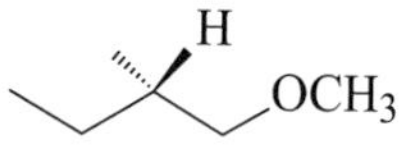

C = H OCH_3 **D** = CN

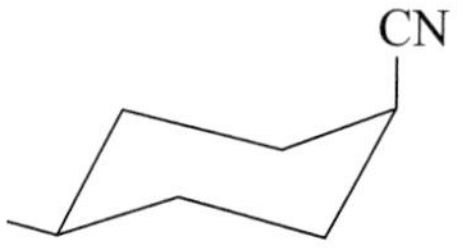

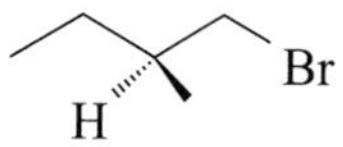

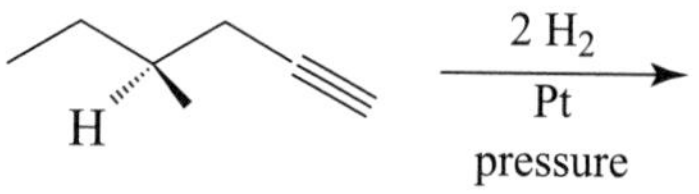

6.6 (b)

6.7 **A** = H Br **B** = H $\xrightarrow[\text{Pt, pressure}]{2\ H_2}$

C = H

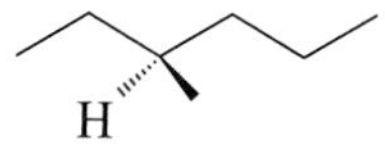

EXERCISE 7

7.1 (c) **7.2** (d) **7.3** (a)

7.4 (a) Li, $C_2H_5NH_2$, −78 °C, then NH_4Cl

(b) H_2/Ni_2B(P−2) or $H_2/Pd/CaCO_3$ (Lindlar's catalyst)

(c) H_2/Ni, pressure or H_2/Pt, pressure (using at least 2 molar equivalents of H_2)

(d) C_2H_5ONa/C_2H_5OH

(e) $(CH_3)_3COK/(CH_3)_3COH$

7.5 > > >

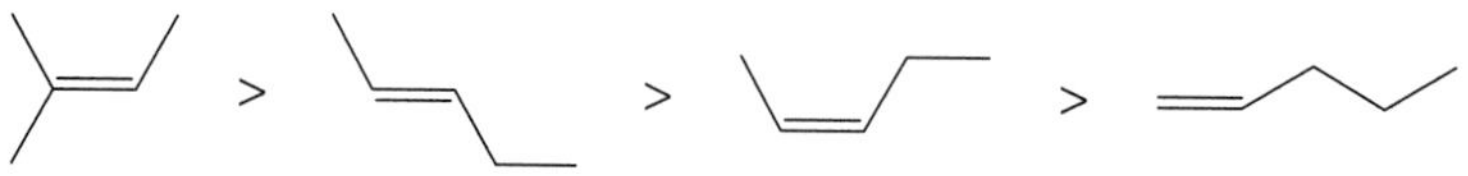

7.6 (a) 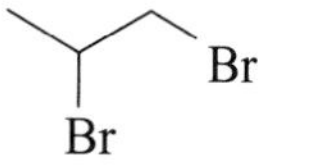 Br Br (b) 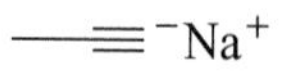 $\equiv^-Na^+$ (c) 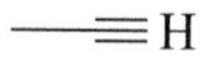 $\equiv$H

(d) 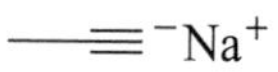 $\equiv^-Na^+$ (e) 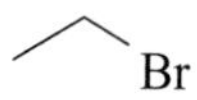 Br

EXERCISE 8

8.1 (e) **8.2** (c) **8.3** (e) **8.4** (a) **8.5** (d) **8.6** (c)

8.7 (c) **8.8** (b)

EXERCISE 9

9.1 (a) 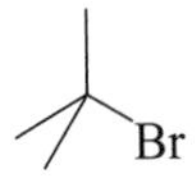 Br (b) 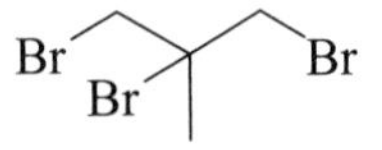 Br Br Br (c)

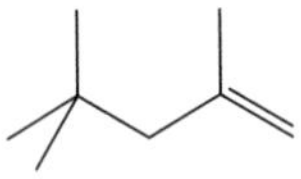

(d) 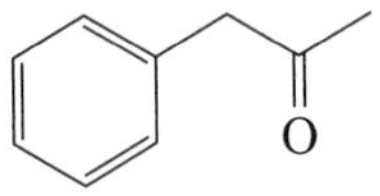 O (e) 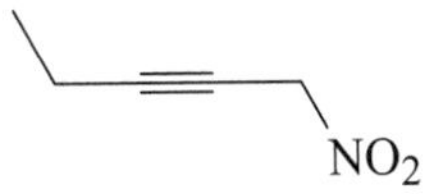 NO_2

9.2 (c) **9.3** (a) **9.4** (b) **9.5** (c) **9.6** (c,d,e) **9.7**

EXERCISE 10

10.1 (d) **10.2** (b) **10.3** (c) **10.4** (b) **10.5** (c)

10.6

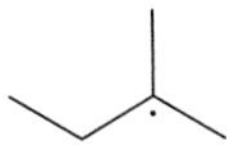

10.7 Six

10.8 (d)

EXERCISE 11

11.1 (d) **11.2** (a) **11.3** (e)

11.4 **A** = 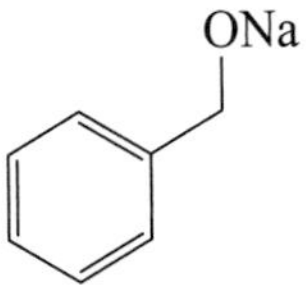 ONa **B** = 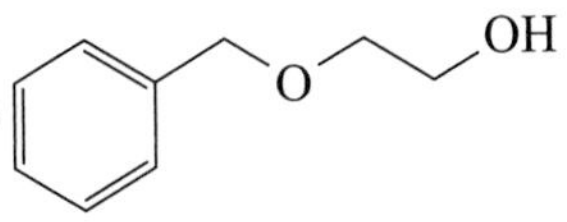 O OH

C = 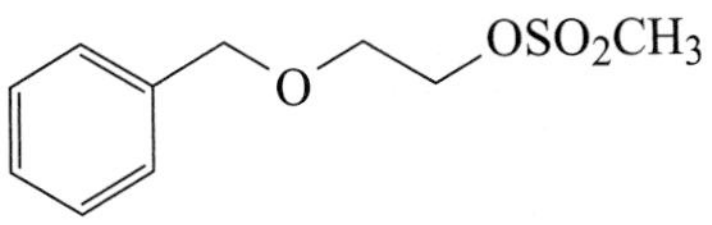 O OSO_2CH_3 **D** = O O

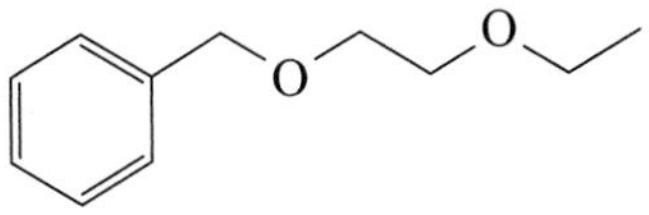

EXERCISE 12

12.1 (b) **12.2** (a)

12.3 **A** = —≡Li or —≡MgBr

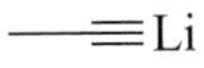

B = NaH

C = CH_3I

12.4 **A** = (3,3-dimethylbutan-1-ol, skeletal structure with OH)

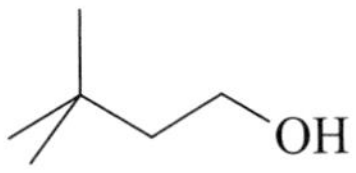

B = PCC/CH_2Cl_2

C = (3,3-dimethylbutanal, skeletal structure with O and H)

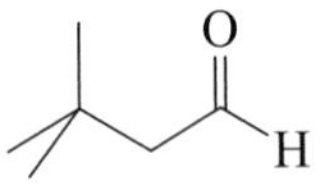

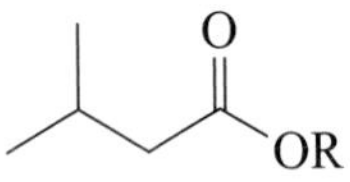

D = (phenyl–MgBr, benzene ring with MgBr)

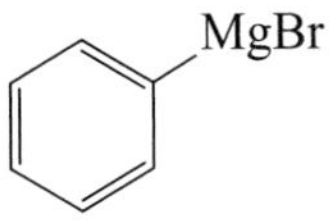

12.5 **A** = (isopropyl-CH₂–C(=O)–OR ester structure) or (4-methylpentan-2-one structure)

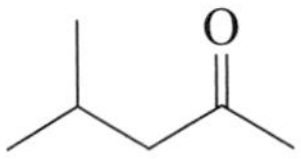

EXERCISE 13

13.1 (d) **13.2** (c) **13.3** (c) **13.4** (c) **13.5** (b)

EXERCISE 14

14.1 (e) **14.2** (a) **14.3** (b) **14.4** (b)

14.5 (benzene ring–CH₂–Cl structure) **14.6** Azulene

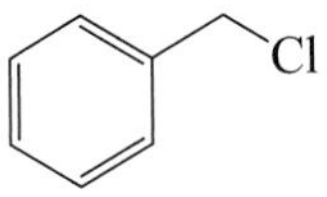

EXERCISE 15

15.1 (a) **15.2** (a) **15.3** (b)

15.4 (a) **A** = SO_3/H_2SO_4 **B** = [ring: CH_3, NO_2, SO_3H]

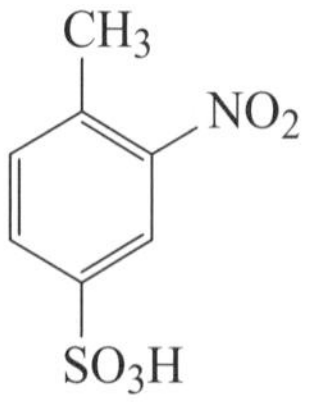

C = H_2O, H_2SO_4, heat **D** = [ring: O, OH, NO_2]

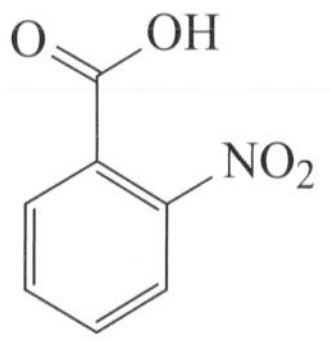

(b) **A** = $SOCl_2$ or PCl_5 **B** = [benzene] + $AlCl_3$

C = Zn(Hg), HCl, reflux **D** = $Br_2/FeBr_3$

EXERCISE 16

16.1 (d) **16.2** (b) **16.3** (b)

16.4 (a) **A** = [benzyl Br] **B** = NaCN

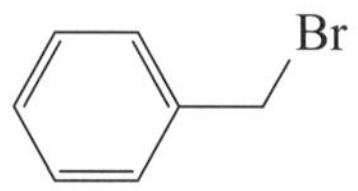

C (1) DIBAL-H, hexane, −78°C (2) H_2O

(b) **A** = PCC, CH_2Cl_2 **B** = HO⁄⁀OH, H_3O^+ **C** = H_2O, H_3O^+

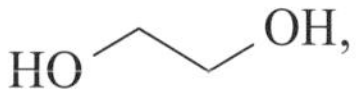

(c) **A** = $(C_6H_5)_3P$ **B** = $\overset{+}{P}(C_6H_5)_3$ Br^- **C** = O [cyclopentanone]

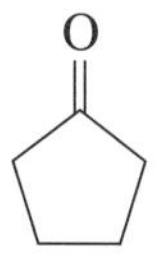

(d) **A** = (1) CH_3MgBr (2) H_2CrO_4 **B** = HCN **C** = (1) $LiAlH_4$, Et_2O (2) H_2O

16.5 CH_3 $\xrightarrow[\text{heat, } h\nu, \text{ or peroxide}]{Cl_2}$ Cl, H, Cl $\xrightarrow[H_2O]{OH^-}$ [OH, H, OH]

$\xrightarrow{-H_2O}$ O, H

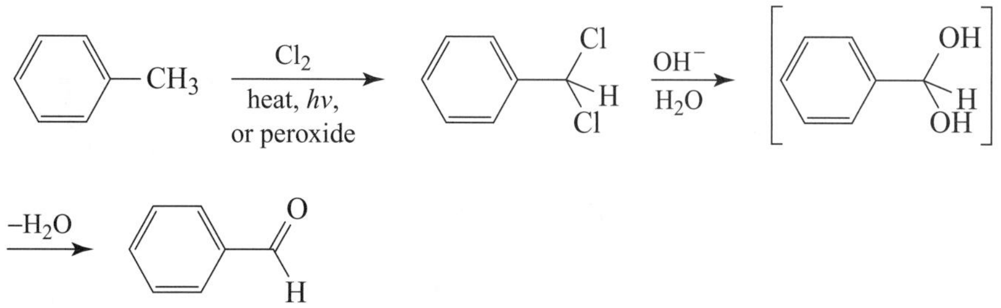

The *gem*-diol formed in the alkaline hydrolysis step readily loses water to form the aldehyde.

16.6 The general formula for an oxime is

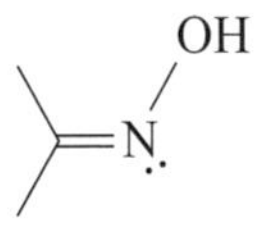

OH

═N:

Both carbon and nitrogen are sp^2 hybridized; the electron pair on nitrogen occupies one sp^2 orbital. Aldoximes and ketoximes can exist in either of these two stereoisomeric forms:

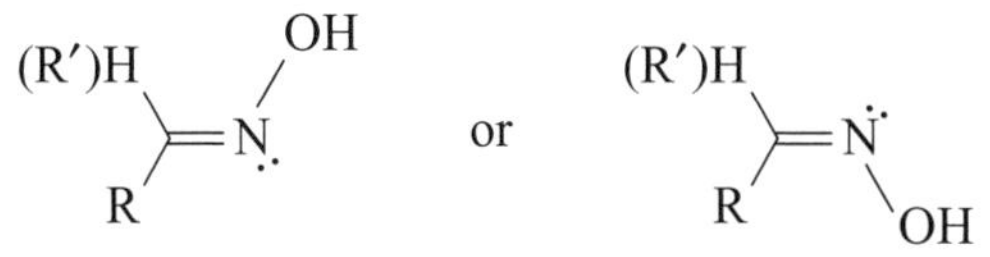

(R′)H, R, ═N:, OH or (R′)H, R, ═N:, OH

This type of stereoisomerism is also observed in the case of other compounds that possess the C═N group, for example, phenylhydrazones and imines.

EXERCISE 17

17.1 (b) **17.2** (d) **17.3** (d)

17.4 **A** = 3-Chlorobutanoic acid

B = Methyl 4-nitrobenzoate or methyl *p*-nitrobenzoate

C = *N*-Methylaniline

17.5 (a) **A** = (1) $KMnO_4$, OH^-, heat **B** = $SOCl_2$ or PCl_5

(2) H_3O^+

C = $C_6H_5C(=O)N(CH_3)_2$ **D** = $C_6H_5C(=O)O^-\ Na^+$

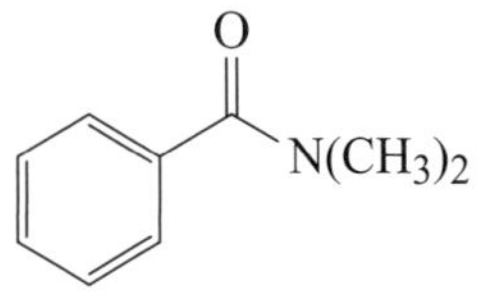

E = $C_6H_5C(=O)OCH_2CH_3$ **F** = $C_6H_5C(=O)NHC_6H_5$

(b) **A** = $C_6H_5C(=O)\,^{18}O\,CH_3$ **B** = $CH_3—^{18}OH$

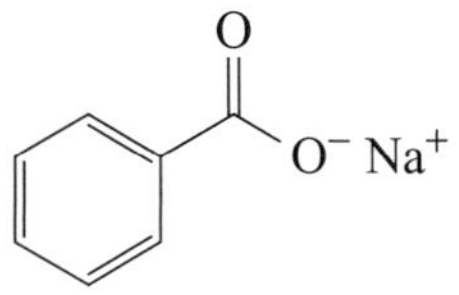

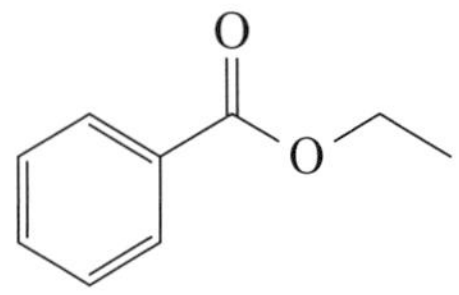

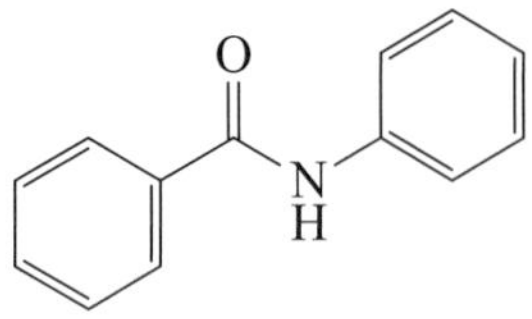

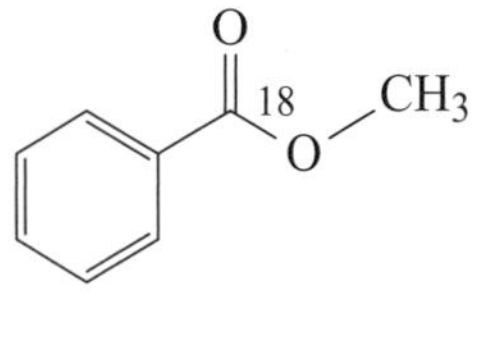

C = $C_6H_5C(=O)O^-\ Na^+$

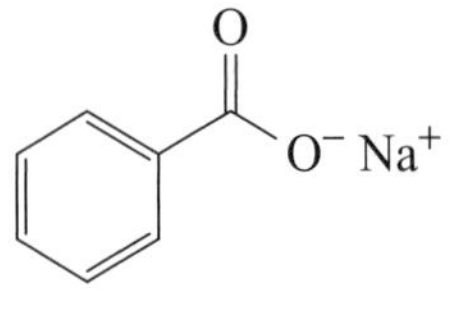

(c) **A** = [cyclopentanecarboxamide, NH_2] **B** = [cyclopentanecarbonitrile, N] **C** = [cyclopentanecarbaldehyde, O, H]

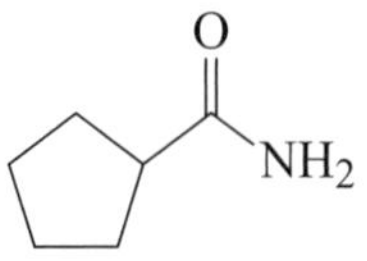

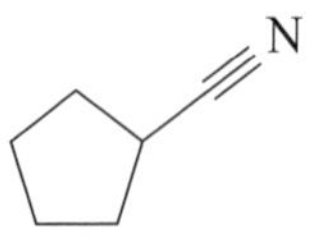

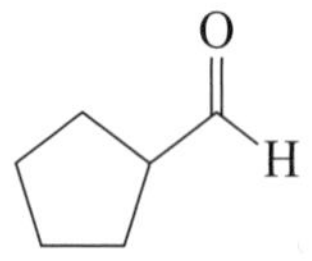

17.6 (b)

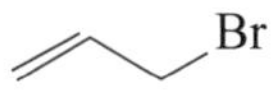

EXERCISE 18

18.1 (a)

18.2 **A** = [pyrrolidine, N–H] **B** = [allyl bromide, Br]

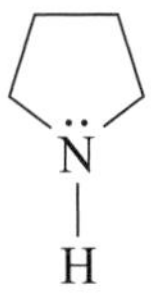

C = [iminium salt, N^+, Br^-]

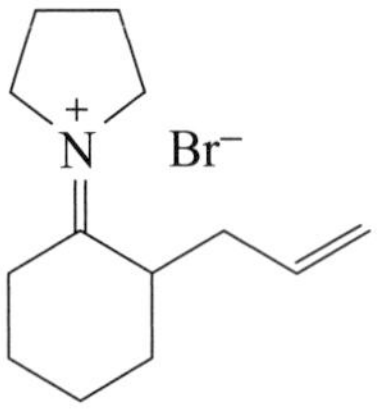

18.3 (c) **18.4** (e) **18.5** (b)

EXERCISE 19

19.1 (c) **19.2** (e) **19.3** (b)

19.4 (a) **A** = EtO, O, O, OEt **B** = EtO, O, O, OEt, K^+

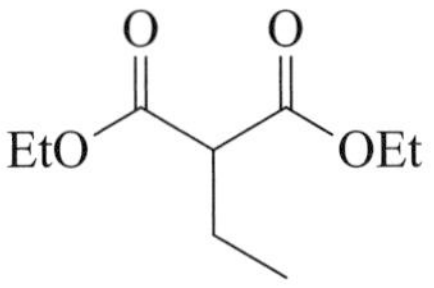

C = EtO, O, O, OEt **D** = HO, O, O, OH

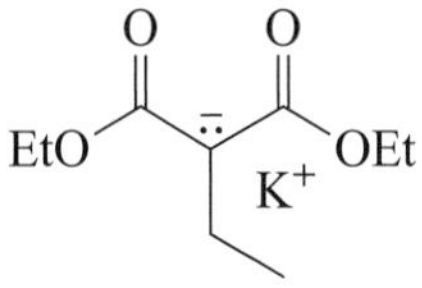

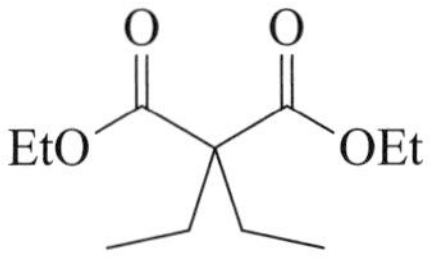

E = O, OH

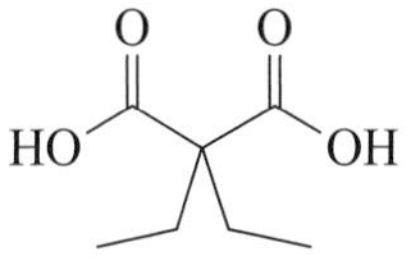

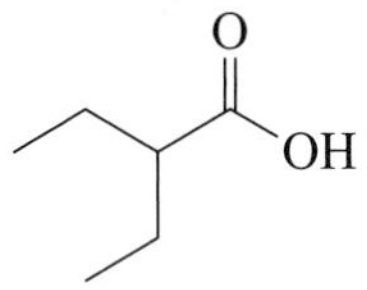

(b) **A** = O H

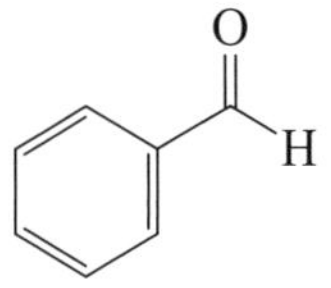

B = OEt OEt O O O OEt O OEt

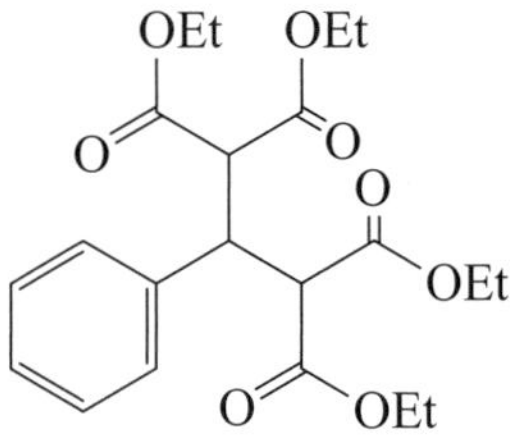

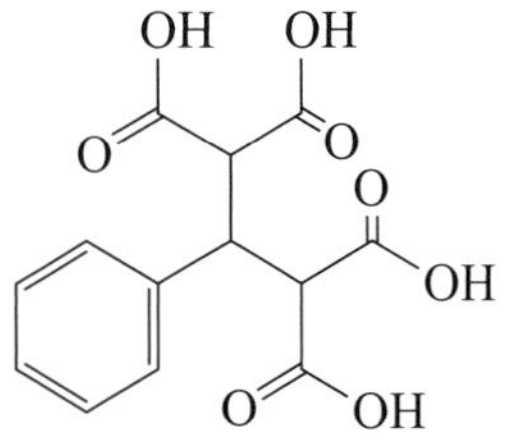

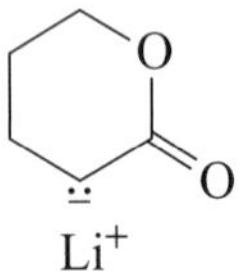

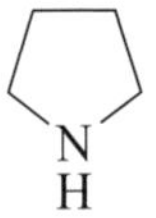

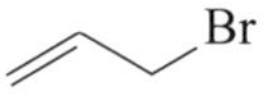

C = OH OH O O O OH O OH

(c) **A** = O O Li$^+$ **B** = I

(d) **A** = N H **B** = Br

C = N$^+$ Br$^-$

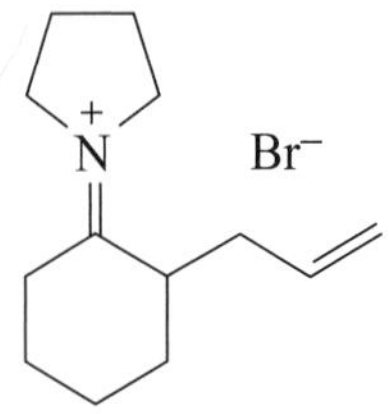

19.5 (a) 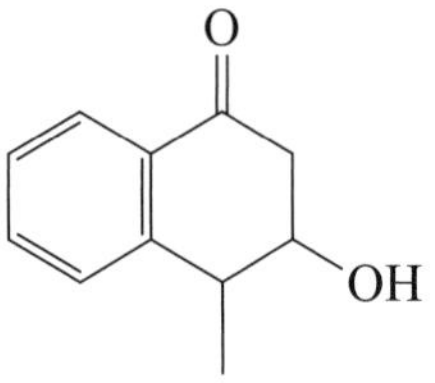 O OH (b) 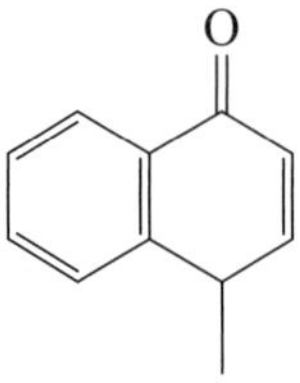 O (c) CH_3MgI, Et_2O

(d) 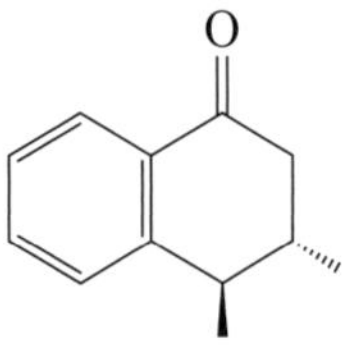 O

\+

simple addition

(e) Zn(Hg)/HCl

19.6 (e)

19.7 (a) 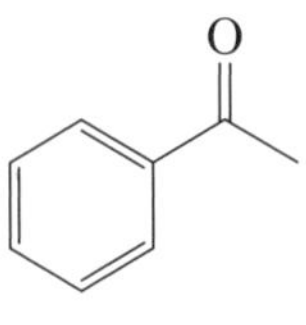 [structure] (b) 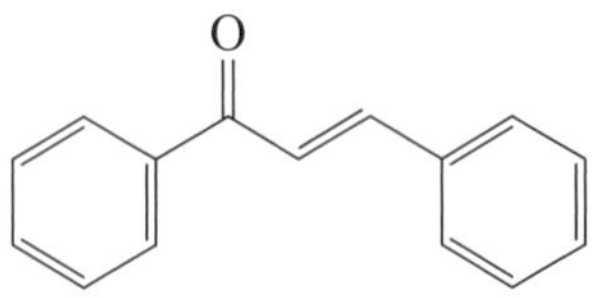 [structure] (c) HCN

19.8 (a) 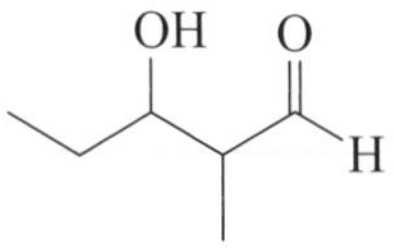 [structure] (b) 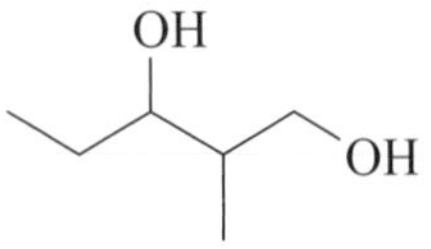 [structure] (c) 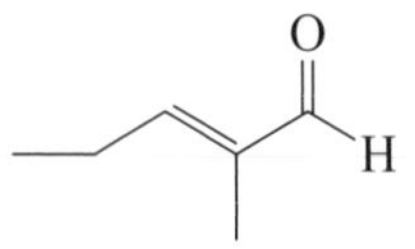 [structure]

(d) $LiAlH_4$, Et_2O (e) H_2, Ni, pressure (f) CH_3OH (excess), HA

(g) 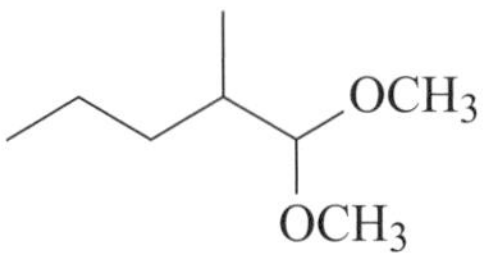 [structure with OCH_3, OCH_3] (h) 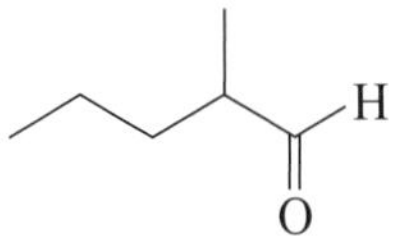 [structure]

(i) (1) 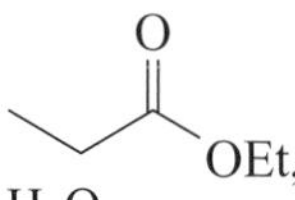 [structure] OEt, LDA, −78°C
(2) H_2O

EXERCISE 20

20.1 (d) **20.2** (e)

20.3 (a) (2) (b) (4) (c) (3)

20.4 (a) **A** = HNO_3/H_2SO_4 **B** = [structure: O_2N, CH_3] **C** = $NaNO_2$, HCl, 0–5°C

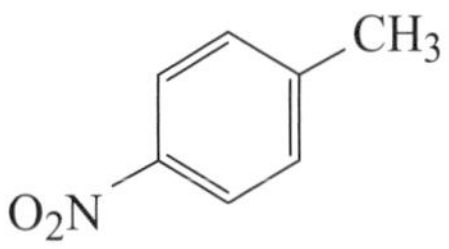

D = CuCN **E** = $LiAlH_4$, Et_2O **F** = [structure: $N(CH_3)_2$]

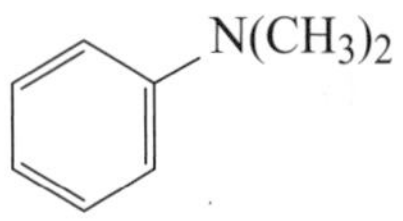

(b) **A** = NaN_3 **B** = [structure] **C** = [structure: NH_2]

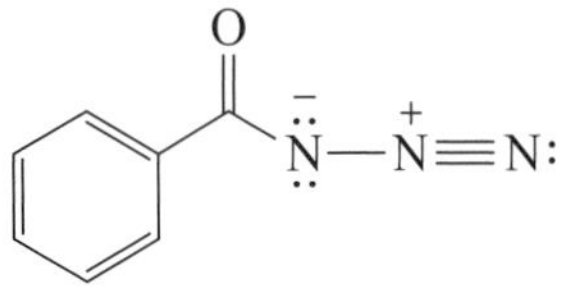

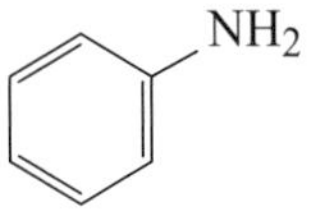

D = [structure: Br, NH_2, Br, Br] **E** = H_3PO_2

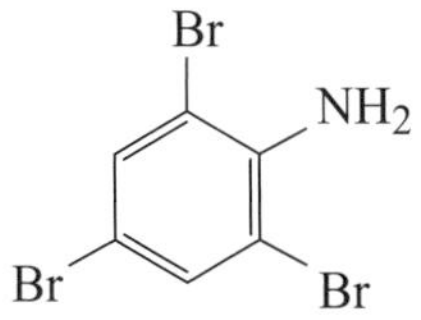

20.5 (a) (2) (b) (2) (c) (1) (d) (1) (e) (2) (f) (2)

EXERCISE 21

21.1 (a) **21.2** (b) **21.3** (b) **21.4** (e)

21.5 **A** = 4-bromophenol (benzene ring with OH and Br para) **B** = KNH_2, liq. NH_3, −33°C

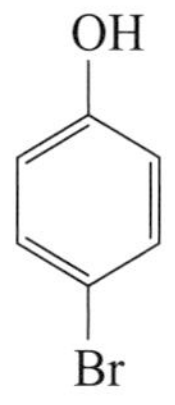

21.6 Br–C_6H_4–OH + CH_3Br

21.7 (a) (1) (b) (1) (c) (2)

EXERCISE 22

22.1 (a) 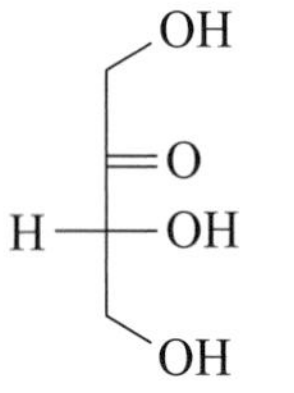 $HOCH_2$–C(=O)–CH(OH)–CH_2OH (H–C–OH)

(b) 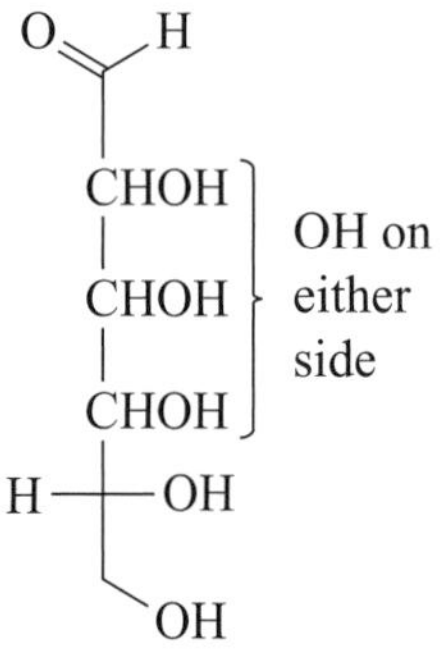 O=CH–CHOH–CHOH–CHOH–(H–C–OH)–CH_2OH; OH on either side

(c) O=CH–CHOH–CHOH–(HO–C–H)–CH_2OH; OH on either side

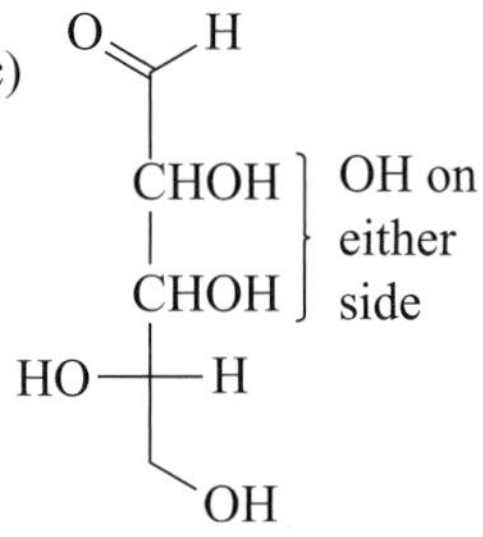

(d) 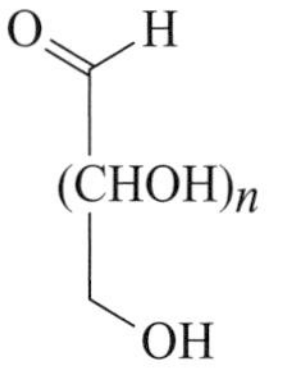 O=CH–$(CHOH)_n$–CH_2OH, n = 1,2,3...

(e) 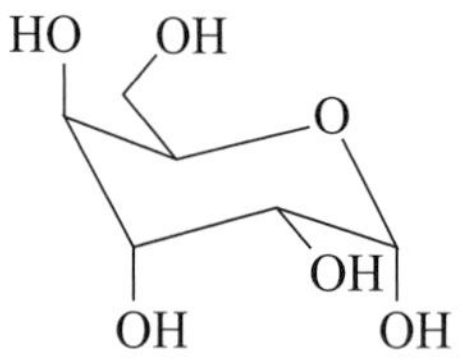 HO, OH, O, OH, OH, OH

(f) 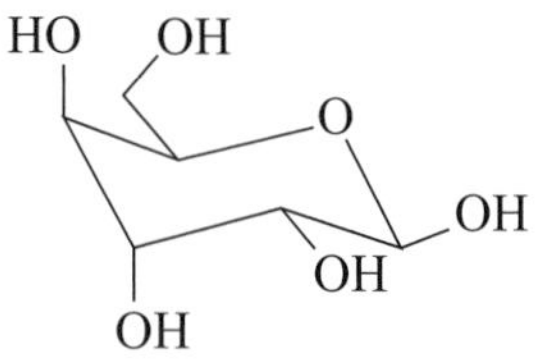 HO, OH, O, OH, OH, OH

(g) 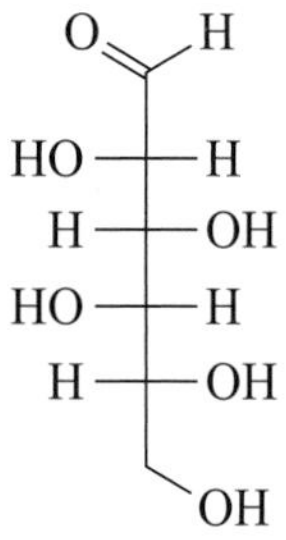 O=CH; HO–C–H; H–C–OH; HO–C–H; H–C–OH; CH_2OH

(h) O=CH; HO–C–H; H–C–OH; HO–C–H; HO–C–H; CH_2OH

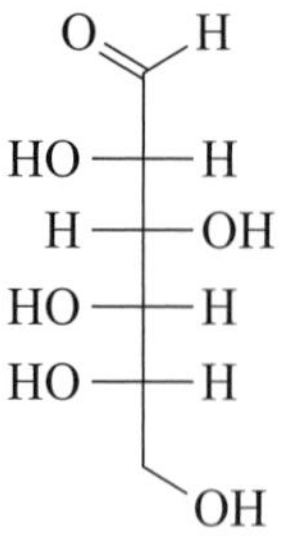

22.2 C

22.3 [structure: O=CH; HO–C–H; H–C–OH; CH₂OH]

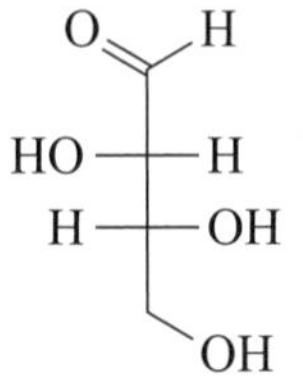

22.4 (a) 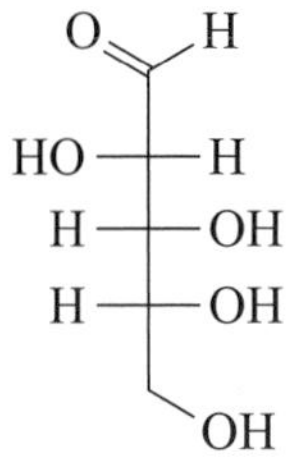 (b) 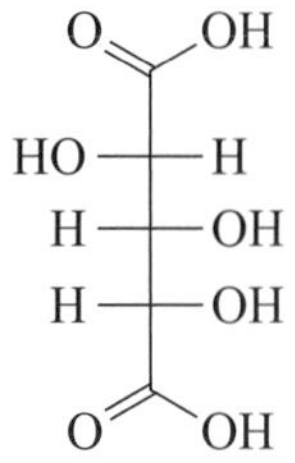 (c) 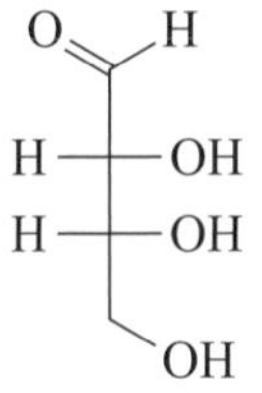 (d)

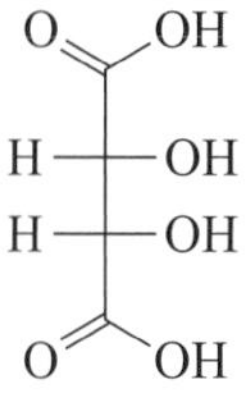

22.5

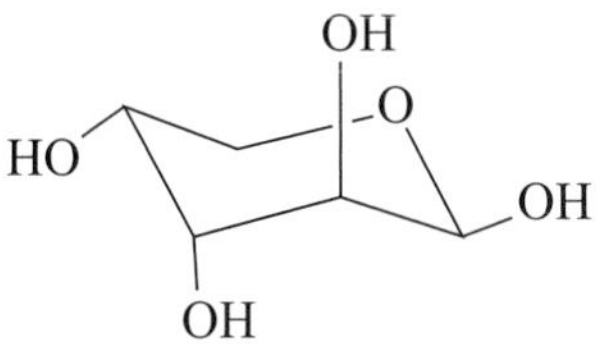

22.6 (a) 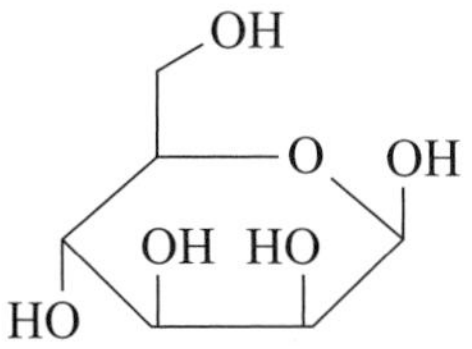 (b) 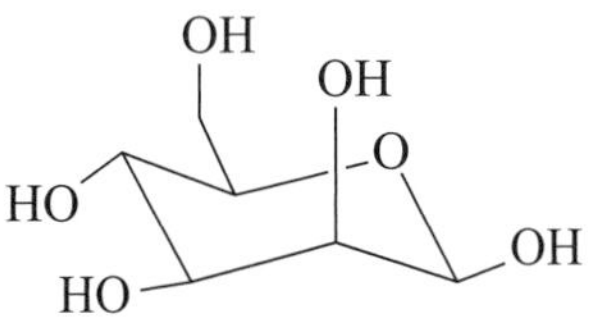 (c) Reducing

(d) Active (e) Aldonic (f) Active (g) Aldaric (h) $NaBH_4$

(i) Active

22.7 (a) Galactose $\xrightarrow{NaBH_4}$ optically *inactive* alditol; glucose → optically active alditol

Galactose $\xrightarrow[HNO_3]{dil}$ optically *inactive* aldaric acid; glucose → optically active aldaric acid

(b) HIO_4 oxidation ⟶ different products:

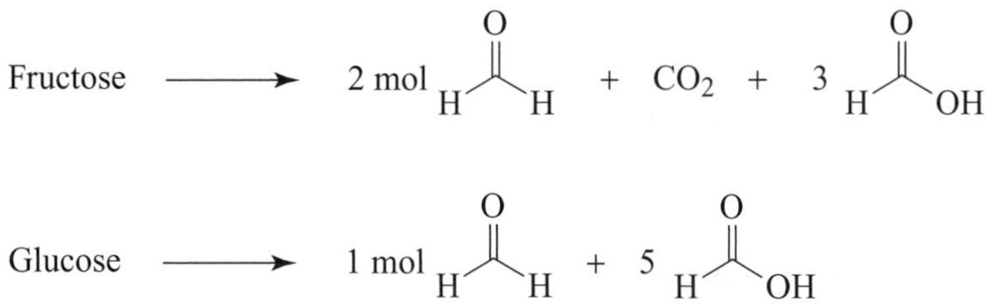

Fructose ⟶ 2 mol HCHO + CO_2 + 3 HCOOH

Glucose ⟶ 1 mol HCHO + 5 HCOOH

22.8 (e) **22.9** (d)

EXERCISE 23

23.1 (a) 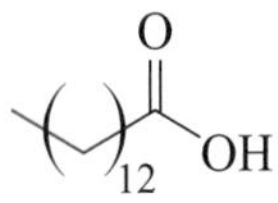 or C_{16} or C_{18}

(b) 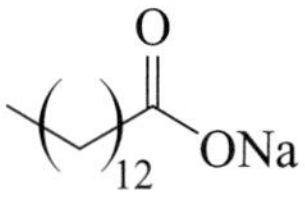

(c) 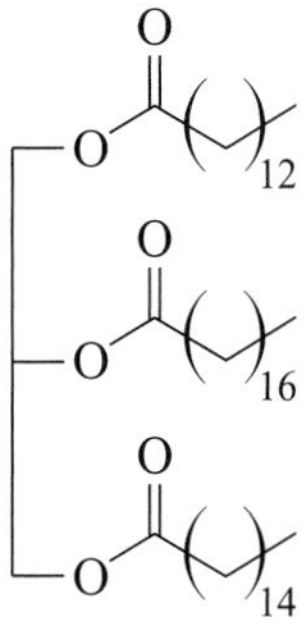

(d) 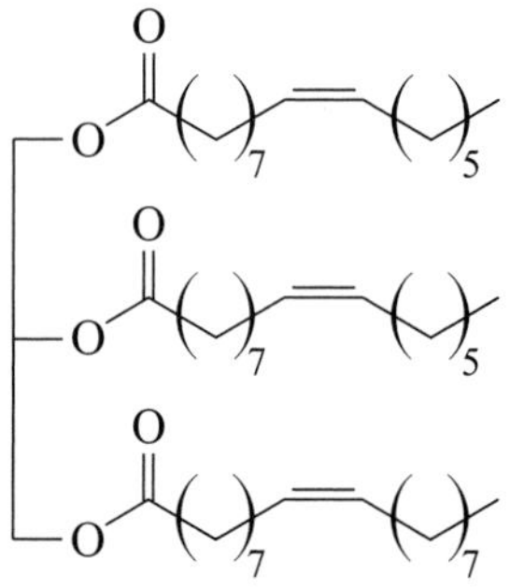

(e) 

(f) 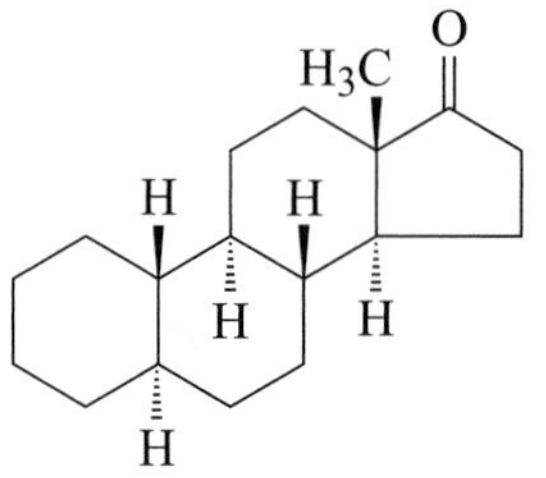

23.2 (a) I_2/OH^- (iodoform test) (b) Br_2/CCl_4 (c) Ethynylestradiol only shows IR absorption at $\sim$3300 cm^{-1}.

23.3 5α-Androstane

23.4 (a) 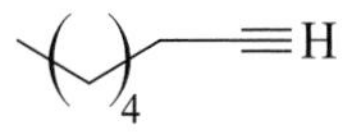

(b) 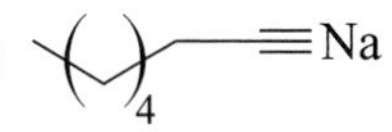

(c) 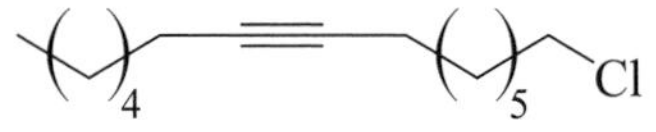

(d) KCN

(e) 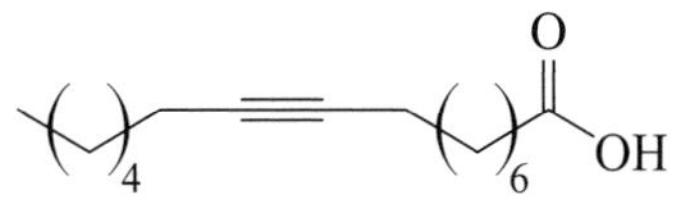

(f) H_2/Pd

23.5 (b) Sesquiterpene

23.6

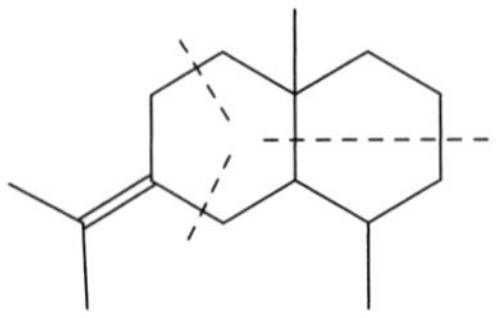

23.7 (e)

EXERCISE 24

24.1 (a) 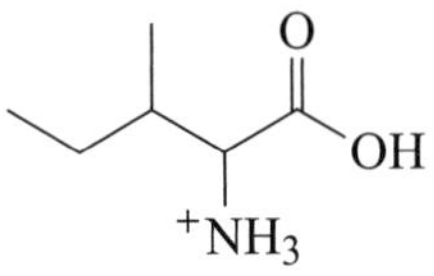 (b)

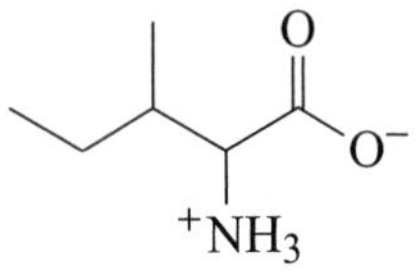

$^{+}NH_3$ OH O O^-

(c)

NH_2 O O^-

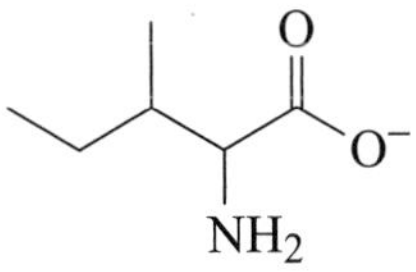

24.2 PLGFGY

C APPENDIX
Molecular Model Set Exercises

The exercises in this appendix are designed to help you gain an understanding of the three-dimensional nature of molecules. You are encouraged to perform these exercises with a model set as described.

These exercises should be performed as part of the study of the chapters shown below.

Chapter in Text	Accompanying Exercises
4	1, 3, 4, 5, 6, 8, 10, 11, 12, 14, 15, 16, 17, 18, 20, 21
5	2, 7, 9, 13, 24, 25, 26, 27
7	9, 19, 22, 28
22	29
24	30
13	31
14	23, 27

The following molecular model set exercises were originally developed by Ronald Starkey.

Refer to the instruction booklet that accompanies your model set for details of molecular model assembly.

Exercise 1 (Chapter 4)

Assemble a molecular model of methane, CH_4. Note that the hydrogen atoms describe the apexes of a regular tetrahedron with the carbon atom at the center of the tetrahedron. Demonstrate by attempted superposition that two models of methane are identical.

Replace any one hydrogen atom on each of the two methane models with a halogen to form two molecules of CH_3X. Are the two structures identical? Does it make a difference which of the four hydrogen atoms on a methane molecule you replace? How many different configurations of CH_3X are possible?

Repeat the same considerations for two disubstituted methanes with two identical substituents (CH_2X_2), and then with two different substituents (CH_2XY). Two colors of atom-centers could be used for the two different substituents.

Exercise 2 (Chapter 5)

Construct a model of a trisubstituted methane molecule (CHXYZ). Four different colored atom-centers are attached to a central tetrahedral carbon atom-center. Note that the carbon now has four different substituents. Compare this model with a second model of CHXYZ. Are the two structures identical (superposable)?

Interchange any two substituents on one of the carbon atoms. Are the two CHXYZ molecules identical now? Does the fact that interchange of any two substituents on the carbon interconverts the stereoisomers indicate that there are only two possible configurations of a tetrahedral carbon atom?

Compare the two models that were not identical. What is the relationship between them? Do they have a mirror-image relationship? That is, are they related as an object and its mirror image?

Exercise 3 (Chapter 4)

Make a model of ethane, CH_3CH_3. Does each of the carbon atoms retain a tetrahedral configuration? Can the carbon atoms be rotated with respect to each other without breaking the carbon-carbon bond?

Rotate about the carbon-carbon bond until the carbon-hydrogen bonds of one carbon atom are aligned with those of the other carbon atom. This is the eclipsed conformation. When the C—H bond of one carbon atom bisects the H—C—H angle of the other carbon atom the conformation is called staggered. Remember, conformations are arrangements of atoms in a molecule that can be interconverted by bond rotations.

In which of the two conformations of ethane you made are the hydrogen atoms of one carbon closer to those of the other carbon?

Exercise 4 (Chapter 4)

Prepare a second model of ethane. Replace one hydrogen, any one, on each ethane model with a substituent such as a halogen, to form two models of CH_3CH_2X. Are the structures identical? If not, can they be made identical by rotation about the C—C bond? With one of the models, demonstrate that there are three equivalent staggered conformations (see Exercise 3) of CH_3CH_2X. How many equivalent eclipsed conformations are possible?

Exercise 5 (Chapter 4)

Assemble a model of a 1,2-disubstituted ethane molecule, CH_2XCH_2X. Note how the orientation of and the distance between the X groups changes with rotation about the carbon-carbon bond. The arrangement in which the X substituents are at maximum separation is the *anti*-staggered conformation. The other staggered conformations are called *gauche*. How many *gauche* conformations are possible? Are they energetically equivalent? Are they identical?

Exercise 6 (Chapter 4)

Construct two models of butane,  . Note that the structures can be viewed as dimethyl-substituted ethanes. Show that rotations about the C2, C3 bond of butane produce eclipsed, *anti*-staggered, and *gauche*-staggered conformations. Measure the distance between C1 and C4 in the conformations just mentioned. [The scale of the Darling Framework Molecular Model Set, for example, is: 2.0 inches in a model corresponds to approximately 1.0 Å (0.1 nm) on a molecular scale.] In which eclipsed conformation are the C1 and C4 atoms closest to each other? How many eclipsed conformations are possible?

Exercise 7 (Chapter 5)

Using two models of butane, verify that the two hydrogen atoms on C2 are not stereochemically equivalent. Replacement of one hydrogen leads to a product that is not identical to that obtained by replacement of the other C2 hydrogen atom. Both replacement products have the same condensed formula, $CH_3CHXCH_2CH_3$. What is the relationship of the two products?

Exercise 8 (Chapter 4)

Make a model of hexane, . Extend the six-carbon chain as far as it will go. This puts C1 and C6 at maximum separation. Notice that this *straight-chain* structure maintains the tetrahedral bond angles at each carbon atom and therefore the carbon chain adopts a zigzag arrangement. Does this extended chain adopt staggered or eclipsed conformations of the hydrogen atoms? How could you describe the relationship of C1 and C4?

Exercise 9 (Chapters 5 and 7)

Prepare models of the four isomeric butenes, C_4H_8. Note that the restricted rotation about the double bond is responsible for the cis-trans stereoisomerism. Verify this by observing that breaking the π bond of *cis*-2-butene allows rotation and thus conversion to *trans*-2-butene. Is any of the four isomeric butenes chiral (nonsuperposable with its mirror image)? Indicate pairs of butene isomers that are structural (constitutional) isomers. Indicate pairs that are diastereomers. How does the distance between the C1 and C4 atoms in *trans*-2-butene compare with that of the *anti* conformation of butane? Compare the C1 to C4 distance in *cis*-2-butene with that in the conformation of butane in which the methyls are eclipsed.

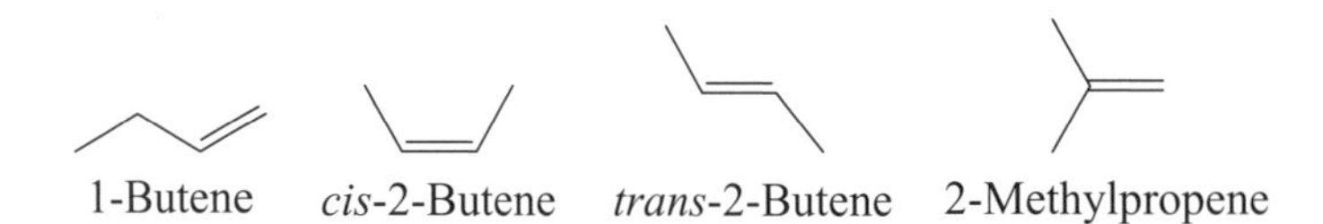

Exercise 10 (Chapter 4)

Make a model of cyclopropane. Take care not to break your models due to the angle strain of the carbon-carbon bonds of the cyclopropane ring. It should be apparent that the ring carbon atoms must be coplanar. What is the relationship of the hydrogen atoms on adjacent carbon atoms? Are they staggered, eclipsed, or skewed?

Exercise 11 (Chapter 4)

A model of cyclobutane can be assembled in a conformation that has the four carbon atoms coplanar. How many eclipsed hydrogen atoms are there in the conformation? Torsional strain (strain caused by repulsions between the aligned electron pairs of eclipsed bonds) can be relieved at the expense of increased angle strain by a slight folding of the ring. The deviation of one ring carbon from the plane of the other three carbon atoms is about 25°. This folding compresses the C—C—C bond angle to about 88°. Rotate the ring carbon bonds of the planar conformation to obtain the folded conformation. Are the hydrogen atoms on adjacent carbon atoms eclipsed or skewed? Considering both structural and stereoisomeric forms, how many dimethylcyclobutane structures are possible? Do deviations of the ring from planarity have to be considered when determining the number of possible dimethylcyclobutane structures?

Exercise 12 (Chapter 4)

Cyclopentane is a more flexible ring system than cyclobutane or cyclopropane. A model of cyclopentane in a conformation with all the ring carbon atoms coplanar exhibits minimal deviation of the C—C—C bond angles from the normal tetrahedral bond angle. How many eclipsed hydrogen interactions are there in this planar conformation? If one of the ring carbon atoms is pushed slightly above (or below) the plane of the other carbon atoms, a model of the envelope conformation is obtained. Does the envelope conformation relieve some of the torsional strain? How many eclipsed hydrogen interactions are there in the envelope conformation?

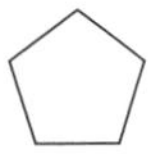

Cyclopentane

Exercise 13 (Chapter 5)

Make a model of 1,2-dimethylcyclopentane. How many stereoisomers are possible for this compound? Identify each of the possible structures as either cis or trans. Is it apparent that cis-trans isomerism is possible in this compound because of restricted rotation? Are any of the stereoisomers chiral? What are the relationships of the 1,2-dimethylcyclopentane stereoisomers?

Exercise 14 (Chapter 4)

Assemble the six-membered ring compound cyclohexane. Is the ring flat or puckered? Place the ring in a chair conformation and then in a boat conformation. Demonstrate that the chair and boat are indeed conformations of cyclohexane—that is, they may be interconverted by rotations about the carbon-carbon bonds of the ring.

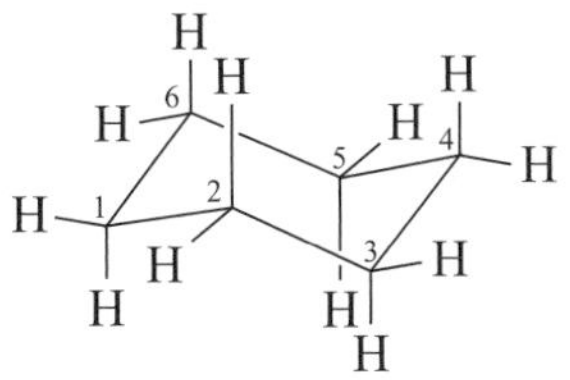

Chair form

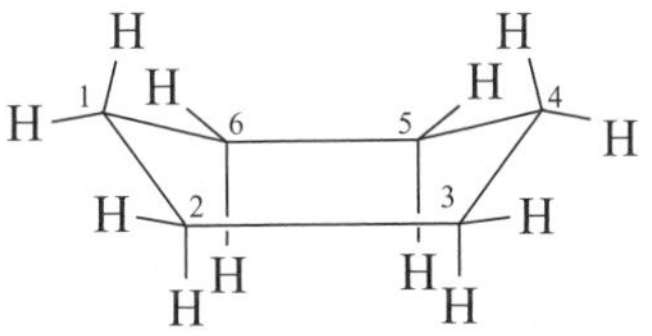

Boat form

Note that in the chair conformation carbon atoms 2, 3, 5, and 6 are in the same plane and carbon atoms 1 and 4 are below and above the plane, respectively. In the boat conformation, carbon atoms 1 and 4 are both above (they could also both be below) the plane described by carbon atoms 2, 3, 5, and 6. Is it apparent why the boat is sometimes associated with the flexible form? Are the hydrogen atoms in the chair conformation staggered or eclipsed? Are any hydrogen atoms eclipsed in the boat conformation? Do carbon atoms 1 and 4 have an *anti* or *gauche* relationship in the chair conformation? (*Hint*: Look down the C2, C3 bond).

A twist conformation of cyclohexane may be obtained by slightly twisting carbon atoms 2 and 5 of the boat conformation as shown.

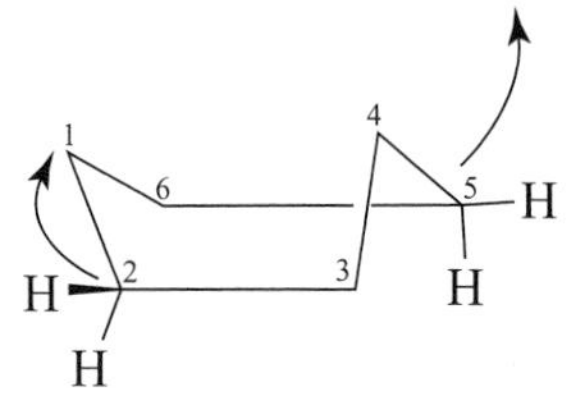

Boat form

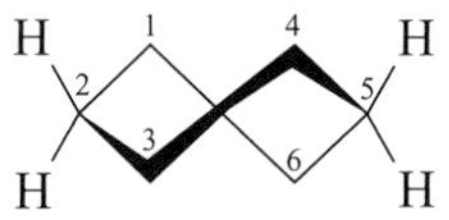

Twist form

Note that the C2, C3 and the C5, C6 sigma bonds no longer retain their parallel orientation in the twist conformation. If the ring system is twisted too far, another boat conformation results. Compare the nonbonded (van der Waals repulsion) interactions and the torsional strain present in the boat, twist, and chair conformations of cyclohexane. Is it apparent why the relative order of thermodynamic stabilities is: chair > twist > boat?

Exercise 15 (Chapter 4)

Construct a model of methylcyclohexane. How many chair conformations are possible? How does the orientation of the methyl group change in each chair conformation?

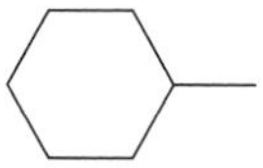

Identify carbon atoms in the chair conformation of methylcyclohexane that have intramolecular interactions corresponding to those found in the *gauche* and *anti* conformations of butane. Which of the chair conformations has the greatest number of *gauche* interactions? How many more? If we assume, as in the case for butane, that the *anti* interaction is 3.8 kJ mol^{-1} more favorable than *gauche*, then what is the relative stability of the two chair conformations of methylcyclohexane? *Hint*: Identify the relative number of *gauche* interactions in the two conformations.

Exercise 16 (Chapter 4)

Compare models of the chair conformations of monosubstituted cyclohexanes in which the substituent alkyl groups are methyl, ethyl, isopropyl, and *tert*-butyl.

Rationalize the relative stability of axial and equatorial conformations of the alkyl group given in the table for each compound. The chair conformation with the alkyl group equatorial is more stable by the amount shown.

ALKYL GROUP	ΔG° (kJ mol^{-1}) EQUATORIAL $\rightleftharpoons$ AXIAL
CH_3	7.3
CH_2CH_3	7.5
$CH(CH_3)_2$	9.2
$C(CH_3)_3$	21 (approximate)

Exercise 17 (Chapter 4)

Make a model of 1,2-dimethylcyclohexane. Answer the questions posed in Exercise 13 with regard to 1,2-dimethylcyclohexane.

Exercise 18 (Chapter 4)

Compare models of the neutral and charged molecules shown next. Identify the structures that are isoelectronic, that is, those that have the same electronic structure. How do those structures that are isoelectronic compare in their molecular geometry?

CH_3CH_3	CH_3NH_2	CH_3OH
$CH_3CH_2^-$	$CH_3NH_3^+$	$CH_3OH_2^+$
CH_3NH^-		

Exercise 19 (Chapter 7)

Prepare a model of cyclohexene. Note that chair and boat conformations are no longer possible, as carbon atoms 1, 2, 3, and 6 lie in a plane. Are cis and trans stereoisomers possible for the double bond? Attempt to assemble a model of *trans*-cyclohexene. Can it be done? Are cis and trans stereoisomers possible for 2,3-dimethylcyclohexene? For 3,4-dimethylcyclohexene?

Cyclohexene

Assemble a model of *trans*-cyclooctene. Observe the twisting of the π-bond system. Would you expect the cis stereoisomer to be more stable than *trans*-cyclooctene? Is *cis*-cyclooctene chiral? Is *trans*-cyclooctene chiral?

Exercise 20 (Chapter 4)

Construct models of *cis*-decalin (*cis*-bicyclo[4.4.0]decane) and *trans*-decalin. Observe how it is possible to convert one conformation of *cis*-decalin in which both rings are in chair conformations to another all-chair conformation. This interconversion is not possible in the case of the *trans*-decalin isomer. Suggest a reason for the difference in the behavior of the cis and trans isomers. *Hint*: What would happen to carbon atoms 7 and 10 of *trans*-decalin if the other ring (indicated by carbon atoms numbered 1–6) is converted to the alternative chair conformation. Is the situation the same for *cis*-decalin?

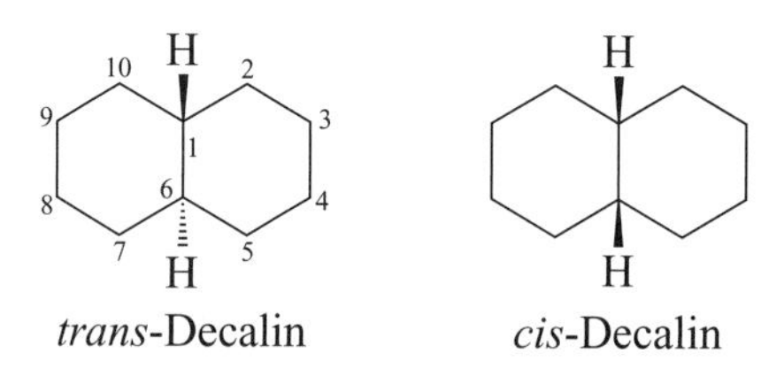

trans-Decalin *cis*-Decalin

Exercise 21 (Chapter 4)

Assemble a model of norbornane (bicyclo[2.2.1]heptane). Observe the two cyclopentane ring systems in the molecule. The structure may also be viewed as having a methylene (CH_2) bridge between carbon atoms 1 and 4 of cyclohexane. Describe the conformation of the cyclohexane ring system in norbornane. How many eclipsing interactions are present?

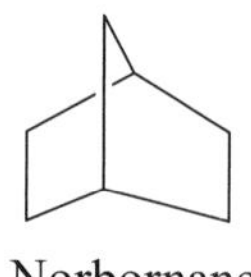
Norbornane

Using a model of twistane, identify the cyclohexane ring systems held in twist conformations. In adamantane, find the chair conformation cyclohexane systems. How many are present? Evaluate the torsional and angle strain in adamantane. Which of the three compounds in this exercise are chiral?

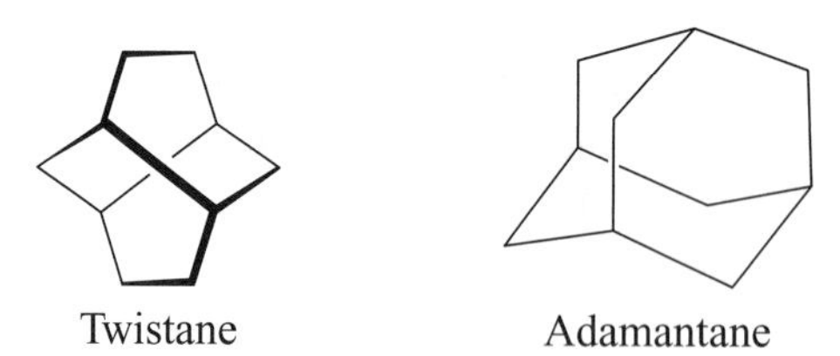
Twistane Adamantane

Exercise 22 (Chapter 7)

An hypothesis known as Bredt's Rule states that a double bond to a bridgehead of a small-ring bridged bicyclic compound is not possible. The basis of this rule can be seen if you attempt to make a model of bicyclo[2.2.1]hept-1-ene, **A.** One approach to the assembly of this model is to try to bridge the number 1 and number 4 carbon atoms of cyclohexene with a methylene (CH_2) unit. Compare this bridging with the ease of installing a CH_2 bridge between the 1 and 4 carbon atoms of cyclohexane to form a model of norbornane (see Exercise 21). Explain the differences in ease of assembly of these two models.

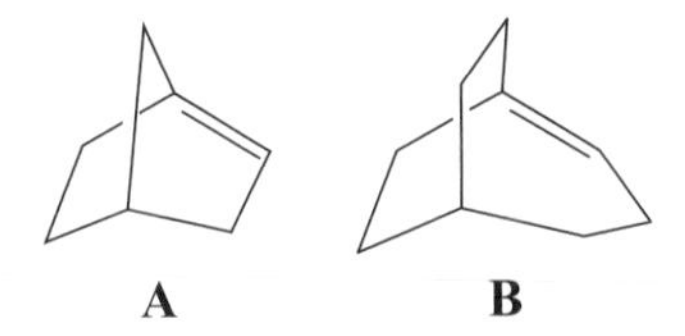

Bridgehead double bonds can be accommodated in larger ring-bridged bicyclic compounds such as bicyclo[3.2.2]non-1-ene, **B.** Although this compound has been prepared in the laboratory, it is an extremely reactive alkene.

Exercise 23 (Chapter 14)

Not all cyclic structures with alternating double and single bonds are aromatic. Cyclooctatetraene shows none of the aromatic characteristics of benzene. From examination of molecular models of cyclooctatetraene and benzene, explain why there is π-electron delocalization in benzene but not in cyclooctatetraene. *Hint*: Can the carbon atoms of the eight-membered ring readily adopt a planar arrangement?

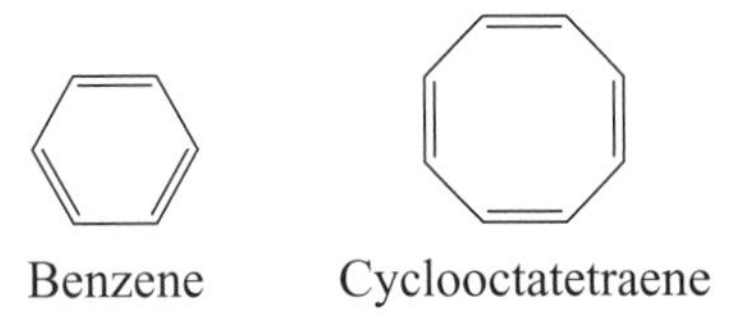

Note that benzene can be represented in several different ways with most molecular model sets. In this exercise, the Kekulé representation with alternating double and single bonds is appropriate. Alternative representations of benzene, such as a form depicting molecular orbital lobes, are shown in your model set instruction booklet.

Exercise 24 (Chapter 5)

Consider the $CH_3CHXCHYCH_3$ system. Assemble all possible stereoisomers of this structure. How many are there? Indicate the relationship among them. Are they all chiral?

Repeat the analysis with the $CH_3CHXCHXCH_3$ system.

Exercise 25 (Chapter 5)

The $CH_3CHXCHXCH_3$ molecule can exist as the stereoisomers shown here. In the eclipsed conformation (meso) shown on the left (**E**), the molecule has a plane of symmetry that bisects the C2, C3 bond. This is a more energetic conformation than any of the three staggered conformations, but it is the only conformation of this configurational stereoisomer that has a plane of symmetry. Can you consider a molecule achiral if only one conformation, and in this case not even the most stable conformation, has a plane of symmetry? Are any of the staggered conformations achiral (superposable on its mirror image)? Make a model of the staggered conformation shown here (**S**) and make another model that is the mirror image of it. Are these two structures different conformations of the same configurational stereoisomer (e.g., are they conformers that can be interconverted by bond rotations), or are they configurational stereoisomers? Based on your answer to the last question, suggest an explanation for the fact that the molecule is not optically active.

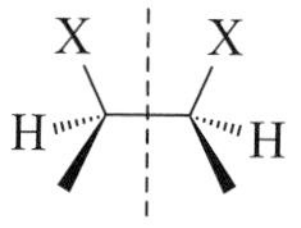

X X H H
E

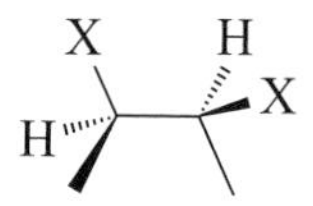

X H X H
S

Exercise 26 (Chapter 5)

Not all molecular chirality is a result of a tetrahedral chirality center, such as CHXYZ. Cumulated dienes (1,2-dienes, or allenes) are capable of generating molecular chirality. Identify, using models, which of the following cumulated dienes are chiral.

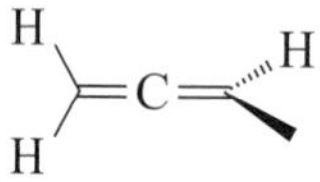

H H C H
A

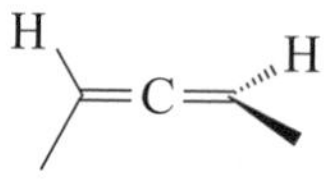

H C H
B

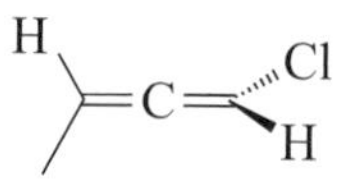

H C Cl H
C

Are the following compounds chiral? How are they structurally related to cumulated dienes?

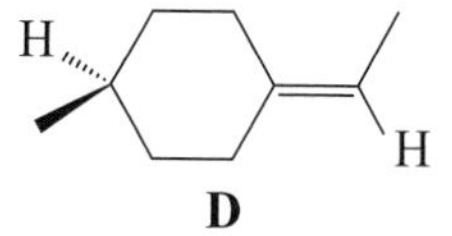

H H
D

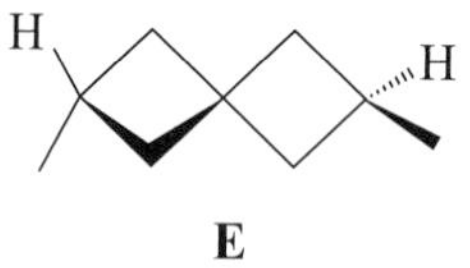

H H
E

Is the cumulated triene **F** chiral? Explain the presence or absence of molecular chirality. More than one stereoisomer is possible for triene **F**. What are the structures, and what is the relationship between those structures?

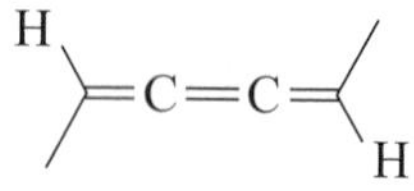

H C=C H
F

Exercise 27 (Chapters 5 and 14)

Substituted biphenyl systems can produce molecular chirality if the rotation about the bond connecting the two rings is restricted. Which of the three biphenyl compounds indicated here are chiral and would be expected to be optically active?

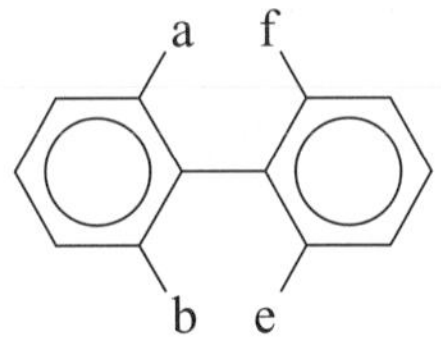

a f

b e

J. a = f = CH_3
b = e = $\overset{+}{N}(CH_3)_3$

K. a = b = CH_3
e = f = $\overset{+}{N}(CH_3)_3$

L. a = f = CH_3
b = e = H

Exercise 28 (Chapter 7)

Assemble a simple model of ethyne (acetylene). The linear geometry of the molecule should be readily apparent. Now, use appropriate pieces of your model set to depict the σ and both the π bonds of the triple bond system using *sp* hybrid carbon atoms and pieces that represent orbitals. Based on attempts to assemble cycloalkynes, predict the smallest cycloalkyne that is stable.

Exercise 29 (Chapter 22)

Construct a model of β-D-glucopyranose. Note that in one of the chair conformations all the hydroxyl groups and the CH_2OH group are in an equatorial orientation. Convert the structure of β-D-glucopyranose to α-D-glucopyranose, to β-D-mannopyranose, and to β-D-galactopyranose. Indicate the number of large ring substituents (OH or CH_2OH) that are axial in the more favorable chair conformation of each of these sugars. Is it reasonable that the β-anomer is more stable than the α-anomer of D-glucopyranose?

Make a model of β-L-glucopyranose. What is the relationship between the D and L configurations? Which is more stable?

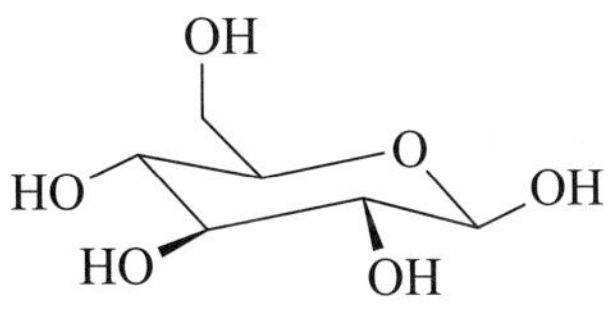

OH
HO O OH
HO OH

β-D-Glucopyranose

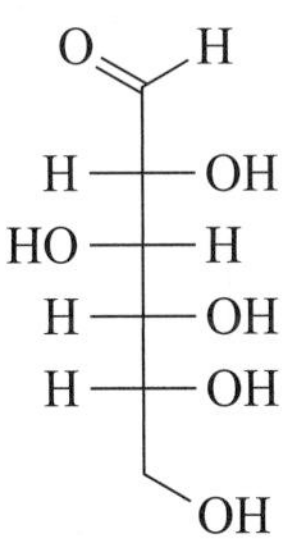

D-(+)-Glucose	D-(+)-Mannose	D-(+)-Galactose
O=CH	O=CH	O=CH
H—OH	HO—H	H—OH
HO—H	HO—H	HO—H
H—OH	H—OH	HO—H
H—OH	H—OH	H—OH
OH	OH	OH

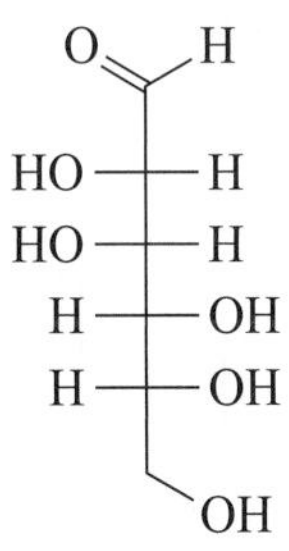

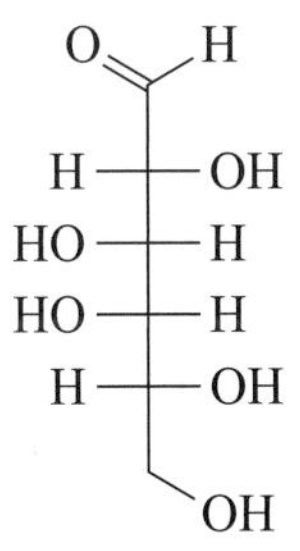

Exercise 30 (Chapter 24)

Assemble a model of tripeptide **A** shown here. Restricted rotation of the C—N bond in the amide linkage results from resonance contribution of the nitrogen nonbonding electron pair. Note the planarity of the six atoms associated with the amide portions of the molecule caused by this resonance contribution. Which bonds along the peptide chain are free to rotate? The amide linkage can either be cisoid or transoid. How does the length (from the N-terminal nitrogen atom to the C-terminal carbon atom) of the tripeptide chain that is transoid compare with one that is cisoid? Which is more "linear"? Convert a model of tripeptide **A** in the transoid arrangement to a model of tripeptide **B.** Which tripeptide has a longer chain?

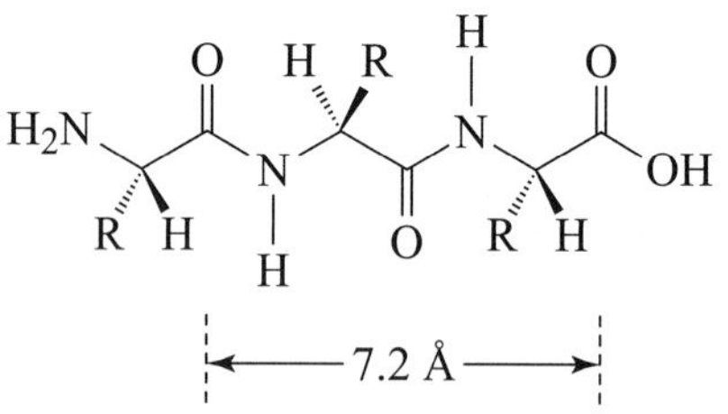

7.2 Å

Tripeptide **A**	$R = CH_3$	(L-Alanine)
Tripeptide **B**	$R = CH_2OH$	(L-Serine)

Exercise 31 (Chapter 13)

Make models of the π molecular orbitals for the following compounds. Use the phase representation of each contributing atomic orbital shown in your molecular model set instruction booklet. Compare each model with π molecular orbital diagrams shown in the textbook.

(a) π_1 and π_2 of ethene ($CH_2{=}CH_2$, ═)

(b) π_1 through π_4 of 1,3-butadiene ($CH_2{=}CH{-}CH{=}CH_2$, 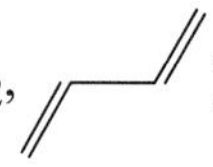)

(c) π_1, π_2, and π_3 of the allyl (propenyl) radical ($CH_2{=}CH{-}\dot{C}H_2$, )

EXERCISE 32

Use your model set to construct several of the interesting representative natural product structures shown here.

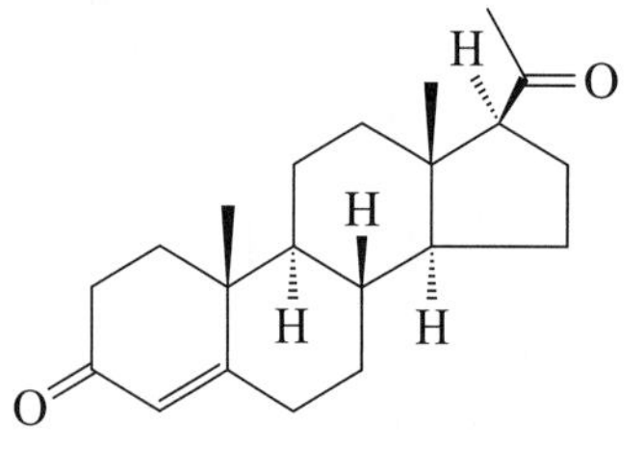

Progesterone 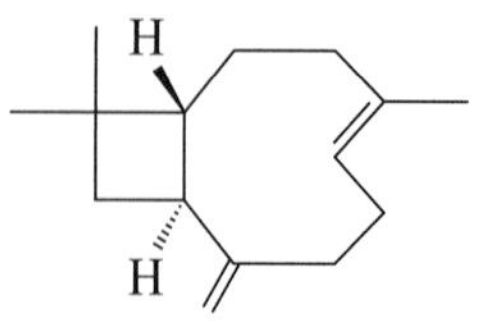　　Caryophyllene 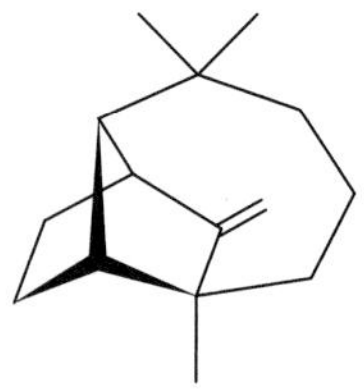　　Longifolene

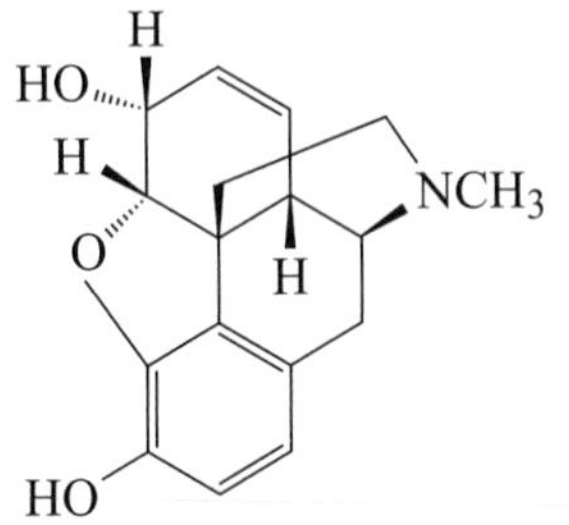

Morphine

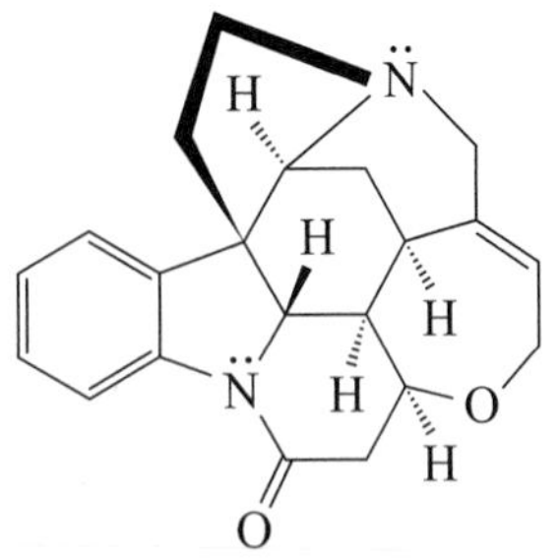

Strychnine

MOLECULAR MODEL SET EXERCISES — SOLUTIONS

Solution 1 Replacement of any hydrogen atom of methane leads to the same monosubstituted product CH_3X. Therefore, there is only one configuration of a monosubstituted methane. There is only one possible configuration for a disubstituted methane of either the CH_2X_2 or CH_2XY type.

Solution 2 Interchange of any two substituents converts the configuration of a tetrahedral chirality center to that of its enantiomer. There are only two possible configurations. If the models are not identical, they will have a mirror-image relationship.

Solution 3 The tetrahedral carbon atoms may be rotated without breaking the carbon-carbon bond. There is no change in the carbon-carbon bond orbital overlap during rotation. The eclipsed conformation places the hydrogen atoms closer together than they are in the staggered conformation.

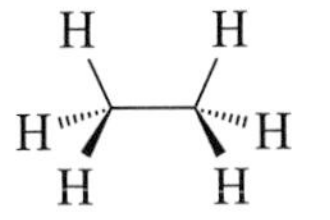

Eclipsed conformation

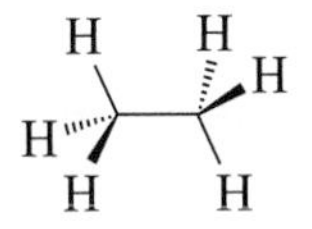

Staggered conformation

Solution 4 All monosubstituted ethanes (CH_3CH_2X) may be made into identical structures by rotations about the C—C bond. The following structures are three energetically equivalent staggered conformations.

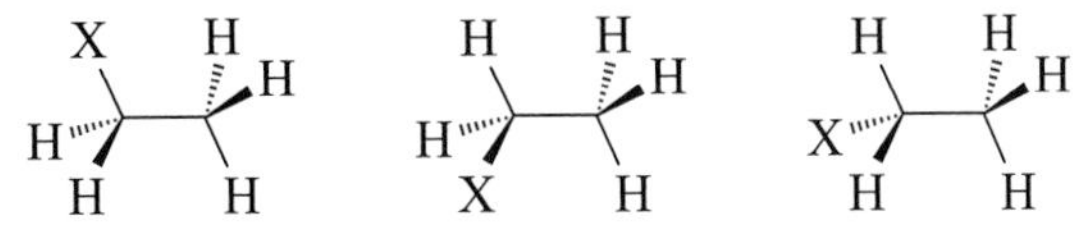

The three equivalent eclipsed conformations are

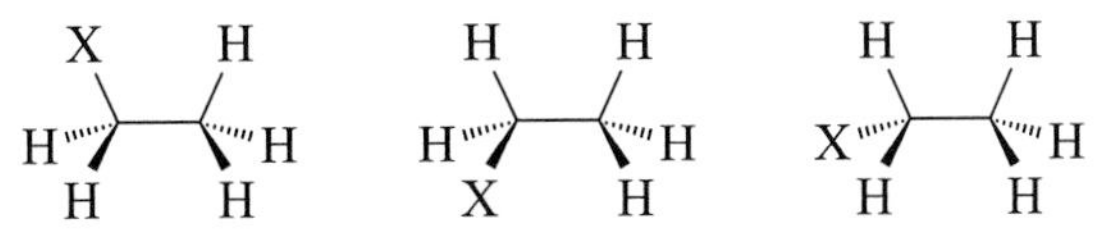

Solution 5 The two *gauche* conformations are energetically equivalent, but not identical (superposable) since they are conformational enantiomers. They bear a mirror-image relationship and are interconvertible by rotation about the carbon-carbon bond.

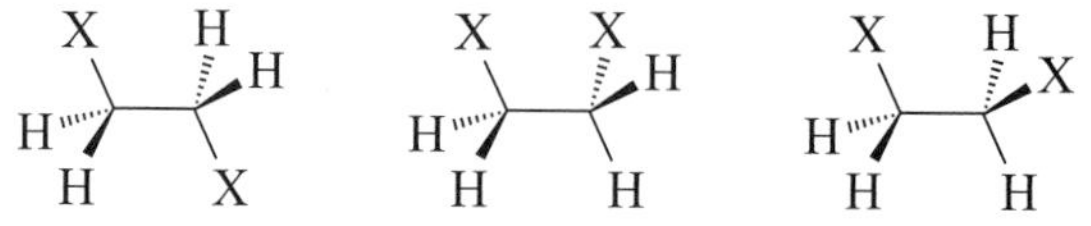

anti Conformation *gauche* Conformations

Solution 6 There are three eclipsed conformations. The methyl groups (C1 and C4) are closest together in the methyl-methyl eclipsed conformation. The carbon-carbon internuclear distances between C1 and C4 are shown in the following table. The number of conformations of each type and the molecular distances in angstroms (Å) are shown.

CONFORMATION	NUMBER	DISTANCES (Å)
Eclipsed (CH_3, CH_3)	1	2.5
Gauche	2	2.8
Eclipsed (H, CH_3)	2	3.3
Anti	1	3.7

Solution 7 The enantiomers formed from replacement of the C2 hydrogen atoms of butane are

Solution 8 The extended chain assumes a staggered arrangement. The relationship of C1 and C4 is *anti*.

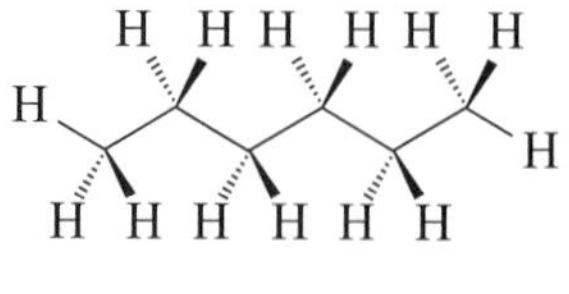

Solution 9 None of the isomeric butenes is chiral. They all have a plane of symmetry. All the isomeric butenes are related as constitutional (or structural) isomers except *cis*-2-butene and *trans*-2-butene, which are diastereomers.

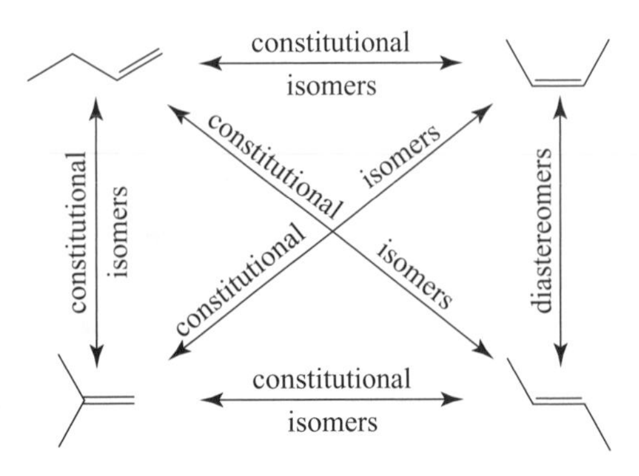

Molecular Model Set C1 to C4 Distances:

COMPOUND	DISTANCES (Å)
cis-2-Butene	2.0
trans-2-Butene	3.7
Butane (*gauche*)	2.8
Butane (*anti*)	3.7

Solution 10 The hydrogen atoms are all eclipsed in cyclopropane.

Solution 11 All the hydrogen atoms are eclipsed in the planar conformation of cyclobutane. The folded ring system has skewed hydrogen interactions. There are six possible isomers of dimethylcyclobutane. Since the ring is not held in one particular folded conformation, deviations of the ring planarity need not be considered in determining the number of possible dimethyl structures.

Solution 12 In the planar conformation of cyclopentane, all five pairs of methylene hydrogen atoms are eclipsed. That produces 10 eclipsed hydrogen interactions. Some torsional strain is relieved in the envelope conformation since there are only six eclipsed hydrogen interactions.

Solution 13 The three configurational stereoisomers of 1,2-dimethylcyclopentane are shown here. Both trans stereoisomers are chiral, while the cis configuration is an achiral meso compound.

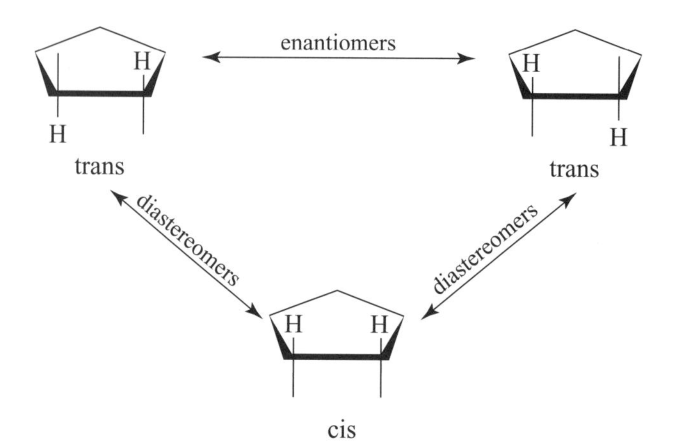

Solution 14 The puckered ring of the chair and the boat conformations can be interconverted by rotation about the carbon-carbon bonds. The chair is more rigid than the boat conformation. All hydrogen atoms in the chair conformation have a staggered arrangement. In the boat conformation, there are eclipsed relationships between the hydrogen atoms on C2 and C3, and also between those on C5 and C6. Carbon atoms that are 1,4 to each other in the chair conformation have a *gauche* relationship. An evaluation of the three conformations confirms the relative stability: chair > twist > boat. The boat conformation has considerable eclipsing strain and nonbonded (van der Waals repulsion) interactions, the twist conformation has slight eclipsing strain, and the chair conformation has a minimum of eclipsing and nonbonded interactions.

Solution 15 Interconversion of the two chair conformations of methylcyclohexane changes the methyl group from an axial to a less crowded equatorial orientation, or the methyl that is equatorial to the more crowded axial position.

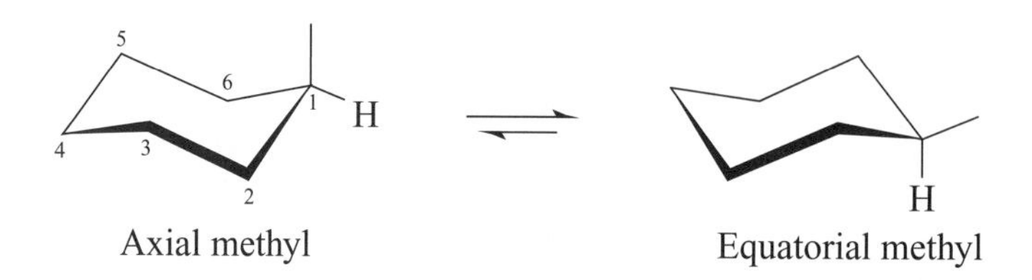

The conformation with the axial methyl group has two *gauche* (1,3-diaxial) interactions that are not present in the equatorial methyl conformation. These *gauche* interactions are axial methyl to C3 and axial methyl to C5. The methyl to C3 and methyl to C5 relationships with methyl groups in an equatorial orientation are anti.

Solution 16 The ΔG° value reflects the relative energies of the two chair conformations for each structure. The crowding of the alkyl group in an axial orientation becomes greater as the bulk of the group increases. The increased size of the substituent has little effect on the steric interactions of the conformation that has the alkyl group equatorial. The *gauche* (1,3-diaxial) interactions are responsible for the increased strain for the axial conformation. Since the ethyl and isopropyl groups can rotate to minimize the nonbonded interactions, their effective size is less than their actual size. The *tert*-butyl group cannot relieve the steric interactions by rotation and thus there is a considerably greater difference in potential energy between the axial and equatorial conformations.

Solution 17 All four stereoisomers of 1,2-dimethylcyclohexane are chiral. The *cis*-1,2-dimethylcyclohexane conformations have equal energy and are readily interconverted, as shown here.

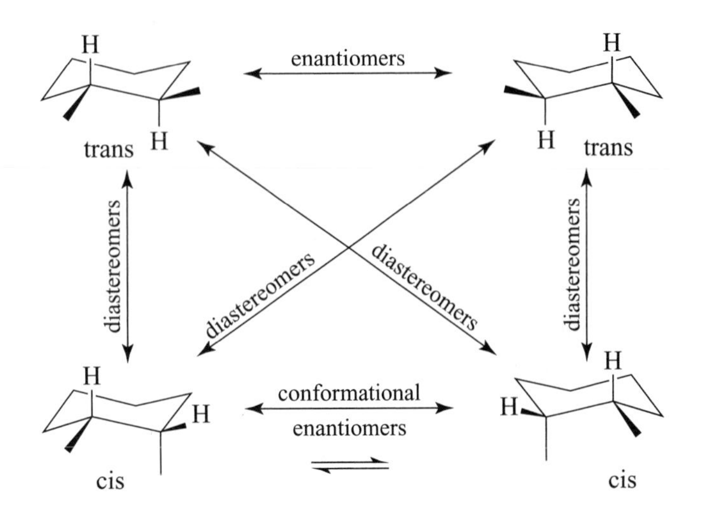

Solution 18 The structures that are isoelectronic have the same geometry. Isoelectronic structures are

CH_3CH_3 and $CH_3NH_3^+$

CH_3NH_2 $CH_3CH_2^-$ and $CH_3OH_2^+$

Structure CH_3NH^- would be isoelectronic to CH_3OH.

Solution 19 Cis-trans stereoisomers are possible only for 3,4-dimethylcyclohexene. The ring size and geometry of the double bond prohibit a trans configuration of the double bond. Two configurational isomers (they are enantiomers) are possible for 2,3-dimethylcyclohexene.

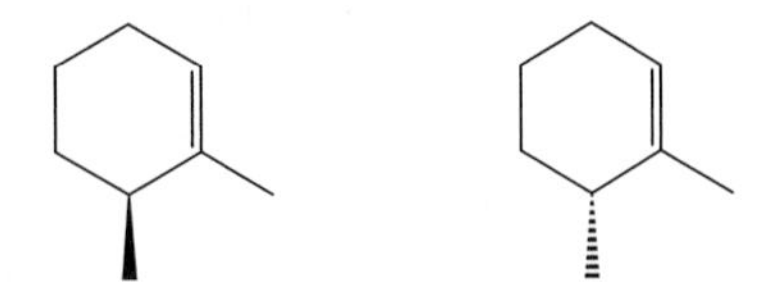

cis-Cyclooctene is more stable because it has less strain than the *trans*-cyclooctene structure. The relative stability of cycloalkene stereoisomers in rings larger than cyclodecene generally favors trans. The *trans*-cyclooctene structure is chiral.

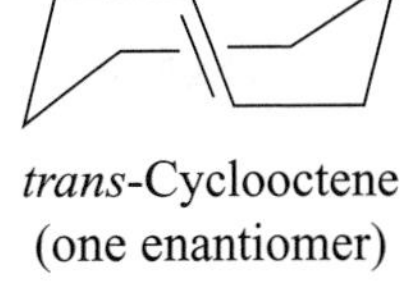

trans-Cyclooctene
(one enantiomer)

Solution 20 The ring fusion in *trans*-decalin is equatorial, equatorial. That is, one ring is attached to the other as 1,2-diequatorial substituents would be. Interconversion of the chair conformations of one ring (carbon atoms 1 through 6) in *trans*-decalin would require the other ring to adopt a 1,2-diaxial orientation. Carbon atoms 7 and 10 would both become axial substituents to the other ring. The four carbon atoms of the *substituent* ring (carbon atoms 7 through 10) cannot bridge the diaxial distance. In *cis*-decalin both conformations have an axial, equatorial ring fusion. Four carbon atoms can easily bridge the axial, equatorial distance.

Solution 21 The cyclohexane ring in norbornane is held in a boat conformation, and therefore has four hydrogen eclipsing interactions. All the six-membered ring systems in twistane are in twist conformations. All four of the six-membered ring systems in adamantane are chair conformations.

Solution 22 Bridging the 1 and 4 carbon atoms of cyclohexane is relatively easy since in the boat conformation the flagpole hydrogen atoms (on C1 and C4) are fairly close and their C—H bonds are directed toward one another. With cyclohexene, the geometry of the double bond and its inability to rotate freely make it impossible to bridge the C1, C4 distance with a single methylene group. Note, however, that a cyclohexene ring can accommodate a methylene bridge between C3 and C6. This bridged bicyclic system (bicyclo[2.2.1]hept-2-ene) does not have a bridgehead double bond.

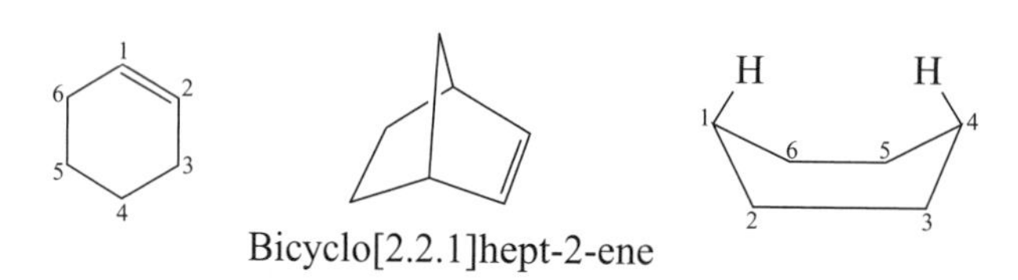

Bicyclo[2.2.1]hept-2-ene

Solution 23 The 120° geometry of the double bond is ideal for incorporation into a planar six-membered ring, as the internal angle of a regular hexagon is 120°. Cyclooctatetraene cannot adopt a planar ring system without considerable angle strain. The eight-membered ring adopts a "tub" conformation that minimizes angle strain and does not allow significant *p*-orbital overlap other than that of the four double bonds in the system. Thus, cyclooctatetraene has four isolated double bonds and is not a delocalized π-electron system.

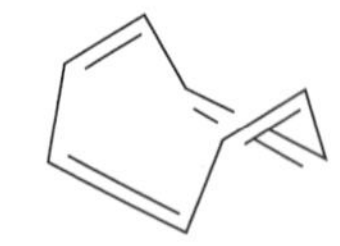

Cyclooctatetraene (tub conformation)

Solution 24 In the $CH_3CHXCHYCH_3$ system, there are four stereoisomers, all of which are chiral.

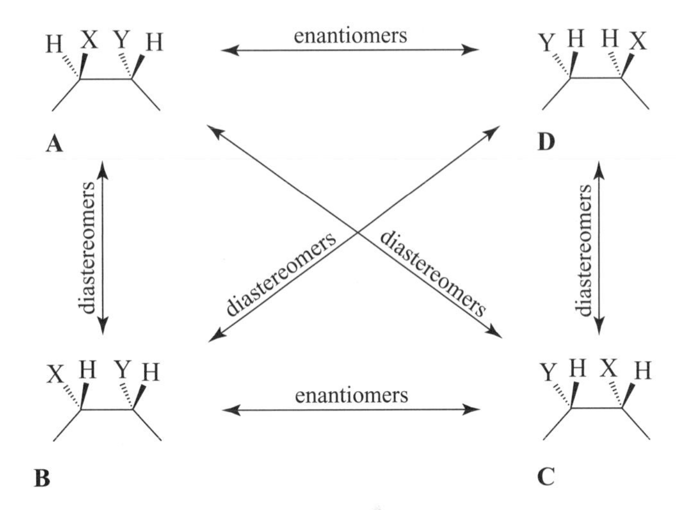

In the $CH_3CHXCHXCH_3$ system, there are three stereoisomers, two of which are chiral. The third stereoisomer (**E**) (shown on page 698) is an achiral meso structure.

Solution 25 If at least one conformation of a molecule in which free rotation is possible has a plane of symmetry, the molecule is achiral. For a molecule with the configurations specified, there are two achiral conformations: the eclipsed conformation **E** shown in the exercise and staggered conformation **F.**

A model of **F** is identical with its mirror image. It is achiral, although it does not have a plane of symmetry, due to the presence of a *center* of symmetry that is located between C2 and C3. A center of symmetry, like a plane of symmetry, is a reflection symmetry element. A center of symmetry involves reflection through a point; a plane of symmetry requires reflection about a plane. A model of the mirror image of **S** (structure **T**) is not identical to **S**, but is a conformational enantiomer of **S**. They can be made identical by rotation about the C2, C3 bond. Since **S** and **T** are conformational enantiomers, the two will be present in equal amounts in a solution of this configurational stereoisomer. Both conformation **S** and conformation **T** are chiral and therefore should rotate the plane of plane polarized light.

Since they are enantiomeric, the rotations of light will be equal in magnitude but *opposite* in direction. The net result is a racemic form of conformational enantiomers, and thus optically inactive. A similar argument can be made for any other chiral conformation and this configuration of $CH_3CHXCHXCH_3$.

Chemical interchange of two groups at either chirality center in meso compound **E** leads to a pair of enantiomers (**G** and **H**).

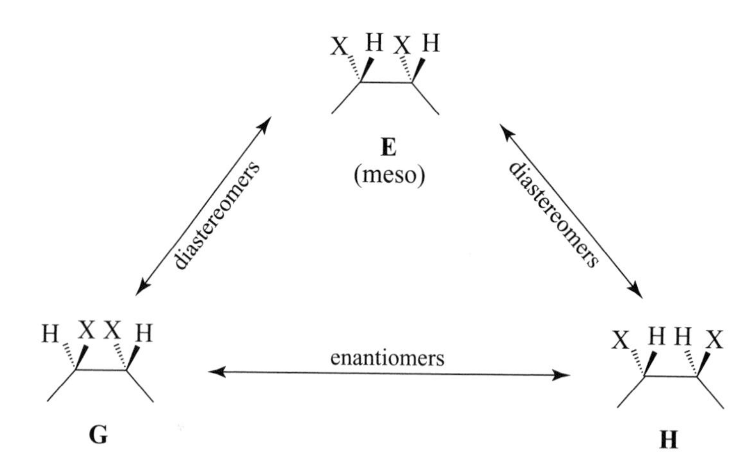

Solution 26 Structures **B** and **C** are chiral. Structure **A** has a plane of symmetry and is therefore achiral. Compounds **D** and **E** are both chiral. The relative orientation of the terminal groups in **D** and **E** is perpendicular, as is the case in the cumulated dienes.

Cumulated triene **F** is achiral. It has a plane of symmetry passing through all six carbon atoms. Structure **F** has a trans configuration. The cis diastereomer is the only other possible stereoisomer.

Solution 27 Structure **J** can be isolated as a chiral stereoisomer because of the large steric barrier to rotation about the bond connecting the rings. Biphenyl **K** has a plane of symmetry and is therefore achiral. The symmetry plane of **K** is shown here. Any chiral conformation of **L** can easily be converted to its enantiomer by rotation. It is only when a $\neq$ b and f $\neq$ e and rotation is restricted by bulky groups that chiral (optically active) stereoisomers can be isolated.

CH_3 $\overset{+}{N}(CH_3)_3$

CH_3 $\underset{+}{N}(CH_3)_3$

A plane of symmetry

Solution 28 A representation of the molecular orbitals in ethyne is given in Section 1.14 of the text. The smallest stable cycloalkyne is the nine-membered ring cyclononyne.

Solution 29 As shown here, the alternative chair conformation of β-D-glucopyranose has all large substituents in an axial orientation. The structures α-D-glucopyranose, β-D-mannopyranose, and β-D-galactopyranose all have one large axial substitutent in the most favorable conformation. β-L-Glucopyranose is the enantiomer (mirror image) of β-D-glucopyranose. Enantiomers are of equal thermodynamic stability.

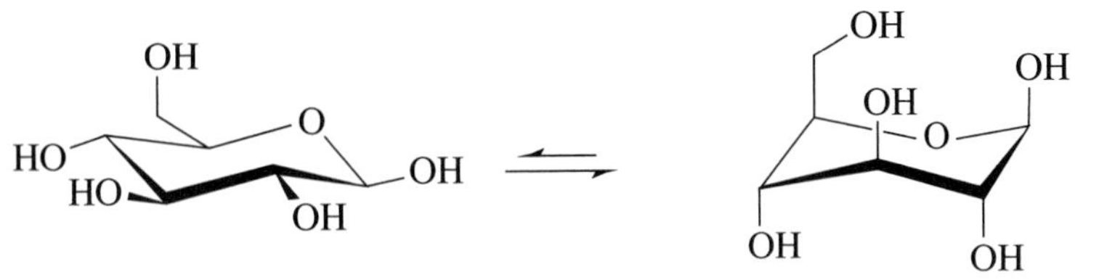

β-D-Glucopyranose

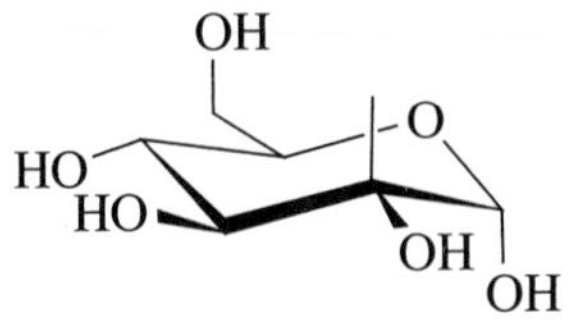

α-D-Glucopyranose

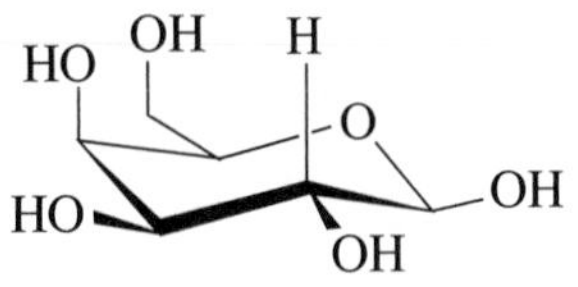

β-D-Galactopyranose

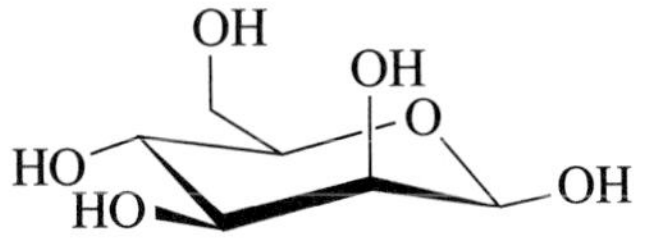

β-D-Mannopyranose

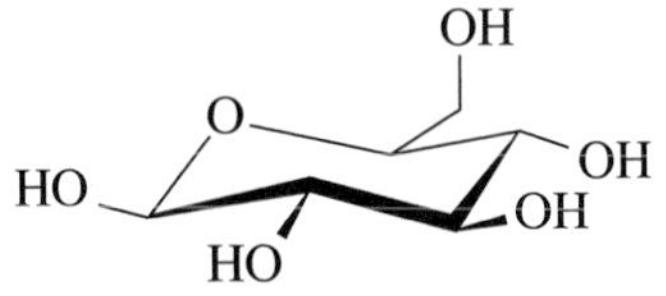

β-L-Glucopyranose

Solution 30 The peptide chain bonds not free to rotate are those indicated by the bold lines in the structure shown here. The transoid arrangement produces a more linear tripeptide chain. The length of the tripeptide chain does not change if you change the substituent **R** groups.

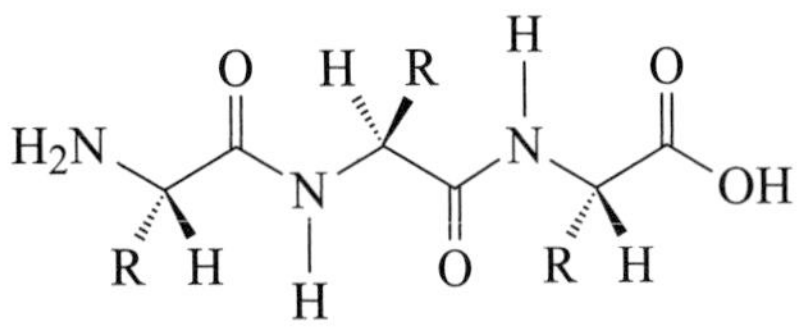

Solution 31 The models of the π molecular orbitals for ethene are shown in the Orbital Symmetry section of the Darling Framework Molecular Model Set instruction booklet. A representation of these orbitals can be found in the text in Section 1.13.

The π molecular orbitals for 1,3-butadiene are shown in the text in Section 13.7C. A model of the π molecular orbitals of 1,3-butadiene is also shown in the Orbital Symmetry section of the Darling Framework Molecular Model instruction booklet. The phases of the contributing atomic orbitals to the molecular orbitals of the allyl radical can be found in the text in Section 13.3A. The π molecular orbital of the allyl radical has a node at C2.